Clubroot Disease of Crucifers

Govind Singh Saharan • Naresh K. Mehta •
Prabhu Dayal Meena

Clubroot Disease of Crucifers

Biology, Ecology and Disease Management

Springer

Govind Singh Saharan
Department of Plant Pathology
CCS Haryana Agricultural University
Hisar, Haryana, India

Naresh K. Mehta
Department of Plant Pathology
CCS Haryana Agricultural University
Hisar, Haryana, India

Prabhu Dayal Meena
Crop Protection Unit
ICAR-Directorate of Rapeseed-Mustard Research
Bharatpur, Rajasthan, India

ISBN 978-981-16-2132-1 ISBN 978-981-16-2133-8 (eBook)
https://doi.org/10.1007/978-981-16-2133-8

This Springer imprint is published by the registered company Springer Nature Singapore Pte Ltd.
The registered company address is: 152 Beach Road, #21-01/04 Gateway East, Singapore 189721, Singapore

Foreword

Trust for Advancement of Agricultural Sciences
Avenue-II, Indian Agricultural Research Institute, New Delhi - 110 012
Phone: 011-65437870 Telefax: 011-25843243
E-mail: taasiari@gmail.com Website: www.taas.in
taas
Progress Through Science

Dr. R.S. Paroda
Founder Chairman

FOREWORD

The clubroot is a serious disease of oil yielding *Brassica* crops, including vegetables. It is a potential threat to cruciferous crops covering 2.69 per cent of total cropped area of the world. It is widely distributed all over the world causing both quantitative and qualitative yield losses ranging from 10-100 per cent with world average of 10-15 per cent. In India, it is restricted to a few states. It is widely distributed in Australia, Canada, China, Germany, Japan, Korea, Poland, UK, and USA affecting trade adversely at the international level.

The book entitled *"Clubroot disease of crucifers: Biology, Ecology and Disease Management"* authored by Drs. G. S Saharan, Former Professor and Head, Department of Plant Pathology, Naresh K. Mehta, Former Assoc. Dean, College of Agriculture, CCS Haryana Agricultural University, Hisar; and Prabhu Dayal Meena, Principal Scientist, DRMR, Bharatpur, is the sixth book on diseases of crucifers being published in series by Springer Nature after the books on *Sclerotinia, Albugo, Alternaria, Hyaloperonospora*, and *Erysiphe*. The book has been written after analysis of world literature with encyclopedic treatment for better comprehension by the readers. The information has been arranged in 15 chapters which have been elucidated with photographs, graphs, tables and figures. Detailed information on major aspects like pathogen, its taxonomy, infection process, pathogenesis, fine structures of the pathogen, biochemistry of host pathogen interaction, pathogenic variability, its distribution, yield losses, disease cycle, epidemiology and forecasting, resistance, its sources, various techniques for germplasm screening, biotechnological approaches, and disease management practices, has been provided in each chapter. The chapter on priority areas of clubroot research will motivate the researchers to undertake further research On this disease, I am sure this book will serve as a reference document for the Brassicalogists for further investigations concerning this important host-parasite system.

I congratulate the authors for bringing out this timely publication. It will be useful for the students, researchers, extension workers and all others concerned with the future growth and development of cruciferous crops.

Dr. R. S. Paroda
Chairman, TAAS &
Former, Secretary (DARE) & DG (ICAR)
Ministry of Agriculture and Farmers Welfare,
Government of India, New Delhi

Preface

The book "*Clubroot Disease of Crucifers: Biology, Ecology, and Disease Management*" is presenting a comprehensive information available in literature on fundamental and applied knowledge of *Plasmodiophora brassicae* Woronin. infecting cruciferous crops and weeds. Clubroot of crucifers which was noticed for the first time in the year 1736 has now spread over more than 88 countries of the world with an average annual loss of crops from 10–15% at a global level. It is considered as a disease of cultivation since once introduced in a field, its inoculum piles up year by year in the form of very hard, resilient to harsh environmental conditions, the resting spores of *P. brassicae* which spreads in the field through cultivation operations. This disease is very unique since the pathogen can survive in the soil in the rhizosphere of non-host plants in addition to its main cruciferous species, cultivated or wild hosts. We have complied the information generated so far in the form of published reports, research articles, popular scientific articles, books, bulletins, reviews, etc., in this book which are arranged in 16 chapters, with appropriate headings and sub-headings with the sections of references to consult original publications. It is a sixth book on the diseases of cruciferous crops series after *Sclerotinia, Albugo, Alternaria, Hyaloperonospora,* and *Erysiphe* published by Springer Nature. The chapter wise sections include the information, viz., the disease, its geographical distribution, symptoms, host range, yield losses, and disease assessment scales, while detailed description on pathogen includes taxonomic position, phylogeny, variability, sporulation, perpetuation, and resting spore germination; host–parasite interactions in the form of seed infection, disease cycle, process of infection, and pathogenesis, epidemiology, forecasting, fine structures, biochemical changes, host defense mechanism, genetical and molecular mechanisms of host–parasite relationships; genomics of clubroot pathogen and pathogenesis, and management practices including cultural, chemical, biological control practices, and deployment of host resistance and integrated approaches. A chapter on standardized, reproducible techniques has been included for the researchers of cruciferous crops. The last section deals with gaps in our understanding, knowledge about management of the disease, and offers suggestions for future research priorities. The subject matter has been vividly illustrated with photographs, graphs, figures, histogram, tables, and colored plates, which makes it stimulating, effective, and easy to comprehend by the readers. The headings and sub-headings of each chapter have been arranged in numbered series to make the subject matter contiguous.

We do hope and believe that this book will be immensely useful to the researchers, teachers, extension specialists, students, industrialists, farmers, and all others who are interested to grow healthy and profitable cruciferous crops all over the world. Any suggestions by the readers are always a source of inspiration for the authors. Any shortcomings, lacunae, flaws in the book are responsibility of ours, and suggestions are welcome.

Hisar, India Govind Singh Saharan
Hisar, India Naresh K. Mehta
Bharatpur, India Prabhu Dayal Meena

Acknowledgements

The authors are highly grateful to the following persons/scientists/publishers/societies/journals/institutes/websites, and all others whose valuable materials such as photographs (macroscopic, microscopic, electron micrographs, scanning electron micrographs), drawings, figures, histograms, graphs, tables, and flow charts have been used through reproduction in the present document. The address of the author (s)/source(s) from where material was adopted can be obtained from the reference which has been cited in the reference section of the book. We have duly acknowledged all the authors of their excellent and valuable publications/articles in this manuscript by citing their names against each table/figure/plate/dendrogram, etc. We sought the necessary permission to reproduce their material in our document and majority of them have granted the permission to reuse them. The authors are sincerely thankful to all the scientists/publishers/journals/institutes/societies and websites whose materials have been reproduced in one or the other form in the present manuscript but forget to acknowledge their name(s) inadvertently. Their prompt help in providing permission, information, photographs, figures, and consents for reproduction in the present manuscript has been duly acknowledged. There might be some errors, mistakes, and shortcomings in this manuscript. We shall greatly appreciate the healthy criticism and suggestions for the improvement of this publication in future.

Contents

About the Authors

Govind Singh Saharan Ex. Professor and Head, Department of Plant Pathology, CCS HAU, Hisar. Dr. Saharan has conducted research in diverse fields of Plant Pathology and had published 250 articles in National and International Journals. He has been the editor/author of several books, monographs, and Crop Production Compendium (CABI). He is on the panel of Experts of SAU, ICAR, IARI, CSIR, UGC, and Department of Biotechnology. Dr. Saharan has been a visiting Professor at the University of Alberta, Edmonton, Canada; Agriculture and Agri-Food Canada, Saskatoon, Canada; and Rothamsted Research, IACR, Harpenden, UK.

He has been the President (NZ) of the Indian Phytopathological Society, Editor-in-Chief, Journal of Mycology and Plant Pathology, Journal of Oilseed Brassica, and the President of the Indian Society of Mycology and Plant Pathology. He has been awarded with Y. L. Nene, Outstanding Plant Pathology Teacher by ISMPP, Udaipur, and Life Time Achievement Award by the Society for Rapeseed-Mustard Research, Bharatpur, India.

Naresh K. Mehta Former Associate Dean, Professor and Consultant Faculty, Department of Plant Pathology, CCS HAU, Hisar. He has been teaching Plant Pathology courses to UG and PG students and conducted research in diverse fields of Plant Pathology especially on rapeseed-mustard. He has guided several M.Sc. and Ph.D. students. He has published about 200 research articles and has been the editor/author of several books, book chapters, review articles, and teaching manuals. Dr. Mehta has been the President of INSOPP, Ludhiana and an Editor-in-Chief of ISMPP and served as the member of Editorial board of various phytopathological societies in India. Dr. Mehta has been awarded Y. L. Nene, Outstanding Plant Pathology Teacher by ISMPP, Udaipur, and Ms. Manju Utereja Memorial Gold Medal, HAU. He is on the panel of Experts of SAU, ICAR, UGC, and a member of various National and International committees. Dr. Mehta has been a visiting scientist to the University of Alberta, Edmonton, Canada.

Prabhu Dayal Meena Principal Scientist (Plant Pathology) at the ICAR-Directorate of Rapeseed-Mustard Research, Bharatpur, Rajasthan, India. He has conducted research on different aspects of rapeseed-mustard diseases including host resistance, management, epidemiology, development of screening techniques and biology of crucifer's pathogen. He has been associated for the development and registration of 7 trait specific germplasm and 6 varieties of rapeseed-mustard. He has published 150 research papers, reviews, and book chapters in reputed journals, and ten books including five by Springer.

Dr. Meena has been honored with the Fellow of Indian Society of Mycology and Plant Pathology, Plant Protection Association of India, Indian Society of Oilseed Research, Society for Rapeseed-Mustard Research, Indian Phytopathological Society, and awarded Dr. P.R. Kumar Outstanding *Brassica* Scientist Award, Society for Rapeseed-Mustard Research, Best Scientist Award 2012–13 of ICAR-Directorate of Rapeseed-Mustard Research. He is the founder Secretary of the Society for Rapeseed-Mustard Research and Managing editor of the Journal of Oilseed Brassica. He visited UK during 2007 under Indo-UK Collaborative Research on Oilseed Brassica crops. He has been guided 12 M.Sc. and one Ph.D. students.

Abbreviations

2-DE	Two-dimensional electrophoresis
ABA	Abscisic acid/abscisic acid
ACC	Aminocyclopropane carboxylic acid
ADK	Adenosine kinase
AFLP	Amplified Fragment Length Polymorphism
ANOVA	Analysis of variance
AUDPC	Area under Disease Progression Curve
AUX	Auxin
B	Boron
BCA	Biocontrol agent
Bj	*Brassica juncea*
Bn	*Brassica napus*
Bo	*Brassica oleracea*
Br	*Brassica rapa*
bp	Base pairs
BR	Brassinolide
BR	Brassinosteroids
BSGs	*Brassica* specific genes
$Ca\ (OH)_2$	Calcium hydroxide
CaO	Calcium oxide
$CaCO_2$	Calcium carbonate
$CaNO_3$	Calcium nitrate
CABI	Centre for Agriculture and Bioscience International
CAPP	Canadian Association of Petroleum Producers
CAZymes	Carbohydrate-active Enzymes
CC	Coiled-coil
CCD	Canadian clubroot differential
CEC	Cation exchange capacity
cDNA	Complementary deoxyribonucleic acid
CHS	Chalcone Synthase
CK	Cytokinins
CMI	Commonwealth Mycological Institute
CR	Clubroot resistant

CSGs	Cruciferae-Specific Genes
CTAB	Cetyltrimethyl Ammonium Bromide
CTOS	Chitin Oligosaccharides
cvs.	Cultivars
CYPs	Cyclophilins
DAB	Diamino-benzidine
DAI/dai	Days after Inoculation/days after inoculation
DAPs	Differentially Accumulated Proteins
DAS	Days after Seeding
DEGs	Differentially Expressed Genes
DH	Doubled Haploid
DI	Disease Index/disease indices/Disease Intensity
DNA	Deoxyribonucleic acid
dpi	Days Post-Infection
DS	Disease Severity
e.g.	For example
EBC	Ethyl Benzimidazol-2-ylcarbamate
ECD	European Clubroot Differential
ECGs	Evolutionary-Conserved Genes
EDTA	Ethylene Diamine Tetraacetic Acid
ELISA	Enzyme-Linked Immunosorbent Assays
EPPO	European and Mediterranean Plant Protection Organization
ESTs	Expressed Sequence Tags
ET/Et	Ethylene
et al.	And others
ETI	Effector Triggered Immunity
f. sp.	Forma specials
FBA	Fructose Bisphosphate Aldolase
FKBPs	FK506-Binding Proteins
FSC	Forest Stewardship Council
FYM	Farmyard manure
g	Gram
GBA	Ground burnt agricultural lime
GBS	Genotyping-by-Sequencing
GC-MS	Gas chromatography–mass spectrometry
GDR	German Democratic Republic
GIS	Geographic Information System
GO	Gene ontology
GPCR	G-protein-coupled receptors
GPS	Global Positioning System
GST	Glutathione s-transferase
GUS	β-glucuronidase
GWAS	Genome-Wide Association Study
h	Hours
ha	Hectare

hpi	Hours Post-Inoculation
HR	Hypersensitive Response
$H_2 SO_4$	Sulfuric acid
IAA	Indole-3-Acetic Acid
IAN	Indole-3-Acetonitrile
IAOx	Indole-3-Acetaldoxime
ICBN	International Code of Botanical Nomenclature
ICP	Inductively Coupled Plasma
ICS	Isochorismate Synthase
ICWG	International Clubroot Working Group
ID	Index of Disease
IG/IGL	Indole glucosinolates
IGS	Intergenic spaces
IP	Intron Polymorphic
IPM	Integrated Pest Management
ISHS	International Society for Horticultural Science
ITS	Internal Transcribed Spacer
JA	Jasmonic Acid
KOH	Potassium hydroxide
kg	Kilogram
LAMP	Loop mediated isothermal DNA amplification
LB	Luria–Bertani
LSGs	Lineage-Specific Genes
LSU	Large subunit
M	Million
MAMPs	Microbe-Associated Molecular Patterns
MAS	Marker-assisted selection
Mb	Million bases
MBC	Benzimidazol-2-ylcarbamate
MeSA	Methyl Salicylate
MgO	Magnesium oxide
min	Minute
miRNA	Micro Ribonucleic acid
MPSS	Massively Parallel Signature Sequencing
MS	Murashige and Skoog
NBS	Nucleotide-Binding Site
NBS-LRR	Nucleotide-Binding Site and Leucine-Rich Repeat
NBSs	Nucleotide-Binding Sites
NCBI	National Center for Biotechnology Information
NGS	Next Generation Sequencing
NILs	Near-Isogenic Lines
NIT	Nitrilase
NPA	N-1-Naphthylphthalamic Acid
NVZs	Nitrate Vulnerable Zones
ORFs	Opening Reading Frames

OSR	Oilseed Rape
P	Phosphorus
PAMP's	Pathogen-associated Molecular Pattern
PAR	Photosynthetically Active Radiation
PCNB	Pentachloronitrobenzene
Pb	*Plasmodiophora brassicae*
PBP	Bacterial Periplasmic Binding
PCR	Polymerase Chain Reaction
PDA	Potato Dextrose Agar
PFGE	Pulsed-Field Gel Electrophoresis
PMA	Propidium monoazide
ppm	*parts per million*
PR	Pathogenesis Related
PRRs	Pattern Recognition Receptors
PTI	Pathogen-associated Molecular Patterns-Triggered Immunity
qPCR	Quantitative Polymerase Chain Reaction
QQS	Qua-Quine starch
QTL	Quantitative Trait Loci
R	Resistance
RACE	Rapid Amplification of cDNA Ends
RAPD	Random Amplification of Polymorphic DNA/ Random Amplified Polymorphic DNA
rDNA	Recombinant Deoxyribonucleic acid
RFLP	Restriction Fragment Length Polymorphism
RNA	Ribonucleic acid
rRNA	Recombinant Ribonucleic acid
RISC	RNA Induced Silencing Complex
RLPs	Receptor-like Protein
ROS	Reactive Oxygen Species
RS	Resting Spores
RT-PCR	Real-time PCR
S	Susceptible
SA	Salicylic Acid
SAH	S-adenosylhomocysteine
SAR	Systemic Acquired Resistance
SBP	Solute-Binding Protein
SCAR	Sequence Characterized Amplified Regions
SCF	Skp1-Cullin1-F-Box Protein
SEM	Scanning Electron Microscopy
SNPs	Single-nucleotide Polymorphisms
spp.	Species
SSGs	Small, Spheroids, Resistant-Type Galls/Small Spheroid Galls
SSH	Suppression Subtractive Hybridization
SSI	Single-spore Isolates
SSR	Simple Sequence Repeats

SSU	Small Subunit
sid2	Salicylic acid induction-deficient2
SZ	Secondary Zoospores
TAG	Triacylglycerol
TAM	Tryptomine
TEA	Tetraethylammonium
TEM	Transmission Electron Microscopy
TIR-NB-LRR	Toll-Interlinked-1 Receptor/Nucleotide Binding Set/ Leucine-Rich Repeat
TPS	Trehalose phosphate synthase/trehalose-6-phosphate synthase
TRGs	Taxonomically Restricted Genes
UK	United Kingdom
USA	United States of America
USDA	United States Department of Agriculture
USSR	Union of Soviet Socialist Republics
UTRs	Un-translated Regions
WCFA	Whole-Cell Fatty Acid
WPBS	Welsh Plant Breeding Station

Symbols

^{o}C	Degree *Celsius*
<	Less than
>	More than
%	Percent
@	At a rate of
£	Pound
m^2	Square Meter
$	Dollar
Σ	Sigma means "sum up"
≥	Is greater than or equal to
≤	Is less than or equal to
>	Is greater than
≈	Is approximately equal to
μm	Micrometer
kb	Kilobytes
Mb	Megabyte
ml	Milliliter
mg	Milligram
g	Gram
kg	Kilogram
ml^{-1}	Per milliliter
dpi	Days post-inoculation

cm	Centimeter
mm	Millimeter
α	Alpha
β	Beta
±	Plus or minus
Δ	Delta
g^{-1}	Per gram
μg	Microgram
w/v	Weight by Volume
v/v	Volume by Volume
cm	Centimeter
w/w	Weight by Weight
a.i.	Active ingredient
L	Liter
t	Tone
N	Nitrogen
P	Potassium
S	Sulfur
Ca	Calcium
Mg	Magnesium
B	Boron

1 Clubroot Perspective

Abstract

Clubroot of crucifers was recorded as a disease of turnip in 1736. Several causes were suspected till Woronin in 1875 established its causal organism which he named as *Plasmodiophora brassicae.* The characteristics distinguishing symptoms are visible only after uprooting the infected plants in the form of clubroot. The disease is distributed on all continents of the world on cruciferous economic vegetables and oil yielding crops causing heavy quantitative and qualitative yield losses. The pathogen is obligate, intracellular, non-axenic, microscopic, single-celled, soil borne with dormant, well-protected resting spores, and has two stages of infections as Primaryroot hair and Secondarycortical cells infection. The pathogen, for the majority of its life cycle, exists in a highly protected environment, and has immediate access to all the nutrition that is required for its growth and reproduction. Being protected in this manner is a big challenge to control this pathogen. The pathogens survival source is resting spores which are very robust and well protected by five spore walls composed of fungal chitin, and carbohydrates which render the spores resilient to degradation by extracellular enzymes produced by predatory soil organisms. *P. brassicae* genome is small in size (24.2–25.5 Mb), and has high gene density. It contains genes potentially involved in host hormone metabolism and metabolic pathways. The *P. brassicae* is a pathogen of cultivation, and there has been continuous co-evolution of cruciferous hosts with *P. brassicae* pathogen interactions. The characteristics symptoms of the disease, the root galls ultimately function as strong metabolic sinks in infected plants drawing nutrients and other resources away from the shoots and leaves contributing to above-ground symptoms development. Several genes governing quantitative and qualitative resistance to *P. brassicae* have been identified and characterized in different *Brassica* species. However, only a few genes have been postulated or demonstrated to be related to the pathogenicity of *P. brassicae*. The isolates of

G. S. Saharan et al., *Clubroot Disease of Crucifers*,
https://doi.org/10.1007/978-981-16-2133-8_1

the pathogen display great variations in the form of races or pathotypes which show tendencies for overcoming resistance sources from *Brassica* species. Biologically *P. brassicae* is an intrinsically interesting pathogen, which has a novel life cycle well adapted to the hosts which provide food and energy for use by the pathogen for its future generations. Several traditional and modern molecular techniques have been developed and are being used to detect and quantify pathogen from soil, water, and host tissues with fair amount of precision and sensitivity. Disease management through cultural, chemical, biological, nutrition, soil regulation along with integrated approach has been advocated but still there are number of challenges before researchers to get satisfactory results. The interactions of physical, chemical, biological, and interactions between *P. brassicae* and *Brassica* hosts along with non-hosts in the soil environment are very complex and challenging to determine conduciveness and suppressiveness conditions for the disease and pathogen.

Keywords

Importance · Host crops · Disease · Pathogen · Resting spores · Environmental factors · Biotic interactions · Host pathogen interactions · *P. brassicae* genome · Historical research development · Host metabolism and disease development · Challenging aspects in disease management · Pathogen diagnosis and quantification

1.1 Introduction

Clubroot disease and the causal organism *Plasmodiophora brassicae* Woronin offer abundant challenges to agriculturists and biological scientists. This pathogen is well fitted for the environments which it inhabits. *Plasmodiophora brassicae* exists in soil as microscopic well-protected resting spores, and then grows actively, and reproduces while shielded inside the roots of host plants. The pathogen is active outside the host for only short periods. Consequently, scientific studies are made challenging by the biological context of the host, and pathogen, and the technology required to investigate and understand that relationship. Controlling clubroot disease is a challenge for farmers, crop consultants, and plant pathology practitioners because of the limited options for its management strategies. Full symptom expression happens solely in members of the Brassicaceae family. The characteristic distinguishing symptoms are visible only after uprooting the infected plants in the form of clubroot. Currently, only a few genes expressing strong resistance to *P. brassicae* are known and readily available. Agrochemical control is similarly limited by difficulties in molecule formulation which combines efficacy with environmental acceptability. Manipulation of husbandry encouraging improvements in soil structure, texture, nutrient composition, and moisture content can reduce populations of *P. brassicae*. Integrating such strategies with rotation and crop management will reduce but not eliminate this disease. There are indications that forms of biological competition may be mobilized as additions to integrated control strategies.

1.2 Host Crops and Their Importance

Cruciferae is one of the most diverse plant families, containing over 3660 species including economically important edible, and industrial oilseeds, and vegetable crops as well (Kagale et al. 2014). The genus *Brassica* is one of 51 genera in the Cruciferae family, and it is comprised of several valuable crops that are used for human nutrition, and that provide important insights into genome evolution research (Rakow 2004; Wang et al. 2014). Cultivated *Brassica* species include six *Brassica* crops described by the US's triangle (Nagaharu 1935). These crops include diploid *Brassica rapa* (A genome), *B. nigra* (B genome), and *B. carinata* (C genome), as well as allotetraploid *B. juncea* (AB genome), *Brassica napus* (AC genome), and *B. carinata* (BC genome) (Cheng et al. 2013). *Brassica* species display extreme morphological characteristics resulting from artificial selection during domestication, and breeding, such as leafy heads, enlarged organs, and extensive axillary branching. Leaf heading traits are agronomic characteristics with great economic and breeding values that directly influence crop yield, and its marketability, which often attracts the attention of breeders. Several studies have focused on determining how several factors such as light intensity, temperature, carbohydrates to nitrogen ratios, and auxin concentrations affect the leafy head formation (Wang et al. 2014; Zhang et al. 2016).

Crucifers and the Brassicas groups of crops after cereals are the most economically important and nutritionally essential crops worldwide. Members of the family Brassicaceae provide mankind with fresh, and in some cases processed food derived from the leaves, flowers, stems, and roots. Increasingly, it is recognized that consumption of these foods has health benefits which reduce the risks from cancer, heart diseases, and strokes, the ailments of affluence. Vegetable oils derived from *Brassica* seed constitute the world's major source of unsaturated fats carrying similar benefits for human health and welfare. *Brassica* oilseeds also provide sources of saturated fats used for oils and industrial lubricants. Other Brassicas provide animal fodder and forage, food condiments, and flowering amenity ornamentals (Dixon 2006). The biggest biological service for mankind may yet come from a tiny rockcress ornamental and weed *Arabidopsis thaliana*. This has become the workhorse of choice for plant molecular biology being used in laboratories worldwide as a basic model vehicle for unraveling and understanding genetic control of cell, and organism form, structure, inheritance, evolution, and ancestry in all biological entities. The importance of *A. thaliana* is illustrated by its being the first plant where the entire genome was sequenced. That places it on par with the human genome (Anon 2000; Adams 2000).

1.3 The Disease and Pathogen

The greatest challenges to plant pathologists and agriculturists are the pathogen *Plasmodiophora brassicae*, which is the causal agent of clubroot disease of crucifer's world over. This is a member of the order Plasmodiophorales, all of which are obligate parasites (Tommerup and Ingram 1971). The challenge for plant pathologists is its nature as a non-axenic, microscopic, single-celled, soil-borne pathogen which exists as a dormant, well-protected resting spore which after germination releases a mobile biflagellate Primaryzoospore. Existence in the soil moisture films is very short term, and following invasion of *Brassica* root hairs, the pathogen enters a Primaryphase of multiplication leading on to Secondarystages which are accompanied by massive host cell hypertrophy that too is induced by disrupted growth regulator production in the cortical cells. On completion of multiplication cycles, resultant resting spores are liberated back into the soil in vast quantities. Plant pathologists are challenged by this secretive lifestyle of *P. brassicae*, where even the occurrence and position of meiosis is still open to speculation (Dixon 1981, 2014).

1.4 The Resting Spores

The resting spores of *P. brassicae* are produced in profusion. Soil samples containing tens of millions of resting spores per gram are not unusual in heavily infested land. These spores are very robust, and have evolved an intricate protection mechanism of five spore walls composed of fungal chitin and carbohydrates (Moxham and Buczacki 1983). This combination renders the spores resilient to degradation by extracellular enzymes produced by predatory soil organisms. The life cycle for the majority, the component stages—zoospores, sporangia, and sporangiospore develop and grow protected within the host body. Only for a short period, *P. brassicae* is exposed to unfavorable environmental conditions, either biotic or abiotic in nature, when it exists as minute, naked, free-swimming primary zoospores (Dixon 2006, 2009a). It is known that some of the requirements for favorable edaphic environments include: acidic pH values (<pH 6.8), minimal calcium contents, nitrogen in an ammonical (N-NH_4) state, poor soil structure, and impeded drainage that the soil becomes waterlogged, and temperatures above 15 °C (Dixon 2002). The stages of life cycle of *P. brassicae* exists in a highly protected environment. *P. brassicae* has immediate access to all the nutrition that is required for growth and reproduction. The presence of *P. brassicae* leads to the disruption of the host growth regulator metabolism and subsequent formation of distorted structures of host tissue which provide this pathogen with an ideal habitat (Dixon 2009a). Being protected in this manner is probably one reason why chemical control of this pathogen proves so difficult to achieve. Potential agrochemicals need a systemic mode of action, being translocated through the host roots, and being capable of moving into the cortical cells. Alternatively, the molecule must be capable of foliar uptake, and then transfer downwards into the root systems. That is a mode

of transfer which is unusual. The alternative of this disease is developing chemical or biological means of control which disrupt either the resting spore or the Primaryzoospore. Dormant resting spores have a substantial armory of protection, making their destruction very difficult (Dixon and Tilston 2010). The resting spore germination appears to be triggered by the presence of host root exudates, evidence with regard to whether *Brassica* root exudates are specifically required is not clear. It is likely that an organism which is as highly evolved as *P. brassicae* which would possess a defense against germination being triggered by non-specific exudates. But this is not proven, and it is thought that some plants which are not members of the Brassicaceae, such as poppy (*Papaver* spp.), strawberry (*Fragaria* spp.), and ryegrass (*Lolium* spp.), can provide host environments in which the first phases of the life cycle may take place (Kowalski and Bochow 1996; Dixon 2009b). It indicates that exudates from these plants may encourage resting spore germination. It is clear that from the period when germination commences to that where there is entry into a host root hair or epidermal cell is the most vulnerable time for *P. brassicae*. In this period, the pathogen is limited in the distance that it may travel by the energy available in the naked cell, which provides locomotion achieved through the swimming action of the two flagella which move it towards the root hairs (Kageyama and Asanoo 2009). The resting spore must also possess a reserve of energy sufficient for the changes which take place within the Primaryzoospore when it reaches the surface of the root hair, and to provide the structures which enable *P. brassicae* cytoplasm to be injected forcefully into the root hair or epidermal cell. In all of these processes, and activities, the Primaryzoospore exists solely on the energy stores laid down when the resting spore formed within the clubbed roots of its host. This energy is derived originally via the photosynthesis which took place in the host where the resting spores developed. *Plasmodiophora brassicae* relies, therefore, on storing and transporting energy produced by one host generation for invasion of the next. That means, there are considerable inflexible energy limitations placed on the distance that Primaryzoospores may move. Considerations of the availability of energy in this host–pathogen system have received little attention (Ludwig-Müller et al. 2009; Neuhauser et al. 2011; Dixon 2014).

1.5 The Environmental Factors of the Soil

There are number of environmental factors which limit primary zoospore movement. There is a requirement for continuous moisture films existing on and between the soil particles in which the Primaryzoospore can move by swimming movements produced by the two flagellae. Dry and drying soils reduce the capacity of *P. brassicae* for movement. Wet and waterlogged soils favor the movement of this pathogen. But total water-logging of the soil environment is likely to deter movement since the zoospore must also respire during motion in order to convert the potential energy stored as carbohydrates, and lipids into kinetic energy. Respiration will demand a supply of oxygen from the soil atmosphere, unless of course *P. brassicae* has some as yet undiscovered unusual anaerobic metabolic pathways. Being the nature of this organism, suggesting such a trait would not be entirely fanciful. It is also possible that germinated zoospores are moved physically and involuntarily through the soil profile by water movements, and in that way are transferred onto root hairs. This pathway is quite likely to be a route by which *P. brassicae* is washed downwards in the soil profile, reaching the root hairs of well-established, deeply rooted maturing crops. The soil environment must also be chemically favorable towards *P. brassicae*. The most obvious component of the chemical environment appears to be an acidic pH (<6.5) for mineral soils. High concentrations of hydrogen ions favor *P. brassicae* whereas large concentrations of calcium are apparently deleterious. Boron has a similar property in relation to the growth and reproduction of this pathogen (Dixon 2010) and appears to have even greater impact on the metabolism of *P. brassicae*. The calcium effect is particularly intriguing since a closely related organism, *Spongospora subterranea* var. *subterranea*, the causal organism of powdery scab disease of potato, is encouraged by an environment with large amounts of calcium, and is inhibited by high concentrations of hydrogen ions. One of the challenges which *P. brassicae* poses for agriculture is the correct and sustainable use of forms of calcium as feasible control system. Success or failure of infection by *P. brassicae* also relies on the presence or absence of other chemical elements (Dixon 1996; Dixon and Page 1998).

1.6 Soil Temperature and Other Biotic Interactions

Heat is the third element of the soil environment which affects the success of primary zoospore movement, and infection into root hairs and epidermal cells. Temperatures in excess of 20 °C have been cited as necessary in order to initiate the movement of *P. brassicae,* and subsequent infection processes. In all probability, the field temperature required in practice is considerably less since that figure comes from laboratory experiments. Clubroot symptoms are known to occur at lower temperatures, and these must be preceded by movement and infection. Heat is essential for the enzyme reactions which liberate energy within the zoospore cell. Within reason (<35 °C) enzyme reaction rates increase as temperature rises. Consequently, rising temperatures accelerate zoospore motion and infection. It is considered that temperatures required for symptom development and expression are less than those required for zoospore motion and penetration. There might be scientific justification for that hypothesis since zoospore motion and penetration rely on activities in a vulnerable, single-celled pathogen with a very thin wall. Declines or increases in temperature would be rapidly translated into none or enhanced activity by the pathogen. Symptom initiation, and expression, on the other hand, results from metabolic activity in large, complex, multicellular angiosperm hosts, and might be less influenced by changing temperatures. The soil environment in which *P. brassicae* and its hosts reside is biologically complex. Studies indicate considerable interaction between *P. brassicae* and other microbes (Einhorn et al. 1991). Results suggest that predatory microbes are able to disable *P. brassicae,* and some of their secondary products can inhibit the metabolic activities of the pathogen. The bacterium *Bacillus subtilis* appears to engage with *P. brassicae* in a vigorously antagonistic manner (Feng et al. 2012). There are also suggestions that aromatic plants such as Mint (*Mentha piperita*), Basil (*Ocimum basilicum*), Bardana (*Arctium minus*), Chive (*Allium fistulosum*), and Parsley (*Petroselinum crispum*) may mitigate the severity of clubroot disease (Hasse et al. 2007).

1.7 Evolution of Host–Pathogen Interaction

Relationships between *P. brassicae* and crucifers host genotypes result in either susceptibility or resistant reactions, and in some instances the expression of partial resistance. These interactions between the pathogen and different host genotypes appear to be complex (Crute et al. 1980; Dixon 1980; Gustafsson and Falt 1986). Evolution of resistance happens geographically where the pathogen and host come into co-existence. Original ancestral wild *Brassicas* came from north-eastern Africa, possibly not too far from the regions where early humans developed, and then spread out across the Middle East, Europe, and Asia. Ultimately, some ancestral *Brassica* types which were plants inhabiting arid or possibly semi-desert areas became adapted for life on maritime coasts in the Mediterranean basin. Others moved considerable distances, and laid the genetic foundations for Asian *Brassicas* (mainly forms of *B. rapa*). Still others moved towards the Atlantic seaboard of Western Europe, the Iberian Peninsula, and towards what is now Scandinavia. In some of these areas, the natural mutability of the Brassicaceae expressed itself by the development of amphidiploid, fertile new sub-species, and species such as *B. napus*. Additionally in these areas, an enormous range of cultivated forms developed through evolution in domestication. The result is the vast multiplicity of phenotypes found currently within this family. The family Brassicaceae is the most flexible and mutable of all angiosperms by virtue of both natural and artificial hybridization. Consider the huge range of *Occidental Brassicas* within the species *Brassica oleracea* which contains all of the *European Brassica* vegetables, and then compare that with the equally huge range of *Oriental Brassicas* within the species *B. rapa* (*B. campestris*) providing the Asian *Brassica* vegetables. Almost every individual crop form found in Oriental types is matched by an Occidental form (Toxopeus et al. 1984; Dixon 2006).

At some stage, a soil-borne microbe which was an ancestor of *P. brassicae* came into contact with some of the evolving *Brassicas*. Where and under what circumstances this encounter took place has yet to be uncovered? The encounter resulted in a single-celled vulnerable pathogen taking up residence in the lumen of root hairs, and exploiting the food sources available within this environment. Ultimately, that pathogen was able to complete a life cycle by further multiplication in the cortical cells of the host. Where is it likely that this encounter was most successful with wild *Brassicas* or with early ancestral cultivated forms? The scant evidence available suggests that *P. brassicae* is a pathogen of cultivation. Infrequently, it is found on native plants although, it is fairly commonly present on weeds in cultivated fields. Additionally, surveys of wild *Brassica* species sampled around the Mediterranean Basin failed to detect the presence of *P. brassicae*. More likely, the pathogen encountered relatively well-fed and irrigated cultivated forms of *Brassica,* and exploited their succulence and the readily available energy sources. Whether that encounter took place once in a single location or several times, and in different places is a fascinating question for the researchers (Dixon 2014).

1.8 The *Plasmodiophora brassicae* Genome

Plasmodiophora brassicae genomes are available, including one reference genome from a European isolate, and additional re-sequencing of five Canadian and one Chinese isolates. The pathotype classification of these isolates is difficult to compare because they have been characterized using different differential hosts set. In these, the main characteristics of the *P. brassicae* genome were described, such as its small size (24.2–25.5 Mb) and high gene density. Thirteen chitin synthases were described, suggesting an important role for this gene family in resting spore formation (Bi et al. 2016, 2019; Rolfe et al. 2016; Schwelm et al. 2015). Secreted proteins described to play a role in plant pathogenic organisms in suppressing the plant defense responses and modifying the host metabolism. *P. brassicae* genome also contains genes potentially involved in host hormone manipulation, such as the auxin-responsive Gretchen Hagen 3, isopentenyl-transferases, methyltransferase, and cytokinin oxidase. The clubroot genome also appears to be lacking several metabolic pathways, a characteristic of the eukaryotic biotrophic plant pathogens. The missing genes encode proteins involved in sulfur and nitrogen uptake, as well as the arginine, lysine, thiamine, and fatty acid biosynthesis pathways. In addition, only a few carbohydrate-active enzymes (CAZymes) involved in the synthesis, metabolism, and transport of carbohydrates were observed (Spanu et al. 2010). Schwelm et al. (2016) summarized the current life stage-specific transcriptomics data related to the molecular events in the life cycle of *P. brassicae*. During spore germination, and at the primary zoospore stage, enzymes involved in chitinous cell wall digestion are highly expressed. Several metabolism pathways such as the starch, citrate cycle, pentose phosphate pathway, as well as pyruvate, and trehalose metabolism were also very active. Although the molecular mechanisms involved in Primaryroot-hair infection are unknown, the secondary phase has been better described. During re-infection by the Secondaryzoospores in root hairs or the root cortex, genes involved in basal and lipid metabolism were highly expressed. The plant clubroot symptoms (hyperplasia and hypertrophy of the infected tissues leading to the gall phenotype) were found to result from the deregulation of the host plant Primaryand Secondarymetabolism as well as from the modification of plant hormone homeostasis by the pathogen. The main plant hormone pathways modified were cytokinin biosynthesis, auxin homeostasis, salicylic acid, and jasmonic acid metabolism. However, the mechanisms by which the pathogen may drive plant hormone homeostasis are still not completely described. At the end of the *P. brassicae* life cycle, Secondaryplasmodia transform into resting spores. In these structures, trehalose could play an important role for their long-term survival capacity. This description of the current transcriptomics data in *P. brassicae* shows that some important steps in *P. brassicae* developmental stages are still not well-characterized at the molecular level.

To gain further structural and functional knowledge of the *P. brassicae* genome, Daval et al. (2019) generated a de novo-sequence from a second European isolate. It was reported the genome sequence of the European eH *P. brassicae* isolate belonging to a pathotype widespread in France but different to e3, the reference clubroot isolate. The mitochondrial genome of this clubroot pathogen was described. The aim was to use this improved genome annotation to predict new metabolic pathways, and to develop Clubroot Cycle, a metabolic pathway database for *P. brassicae*. This new tool gives researchers access to a repertoire of *P. brassicae* metabolic and transport pathways as well as to an Omics Viewer to analyze expression data, which can be updated with new data. The genome of eH, a second European isolate, has been sequenced. Refinement of the annotation of the eH genome led to the identification of the mitochondrial genome sequence, which was found to be bigger than that of *Spongospora subterranea*, another plant parasitic Plasmodiophorid phylogenetically related to *P. brassicae*. New pathways were also predicted, such as those for the synthesis of spermidine, a polyamine up-regulated in clubbed regions of roots. *P. brassicae* pathway genome database was created to facilitate the functional study of metabolic pathways in transcriptomics approaches (Fig. 1.1). These available tools can help in understanding of the regulation of *P. brassicae* metabolism during infection, and in response to diverse constraints (Daval et al. 2019).

1.9 Historical Research Development

The research developments since the first discovery of the disease and establishment of causal organism have been briefly described on host–pathogen interactions and disease management strategies accomplished from initial scientific studies period to the present era of technical and scientific advances through modern era as follows.

1.9.1 Studies on Clubroot Disease Discovery

In the ancient world, there was increasing interest in observing agricultural crop problems, and conditions similar to clubroot were recorded. The Romans were aware of root symptoms which they described as spongy fungus-like roots (*Radices fungosae*). Bohner (1922) suggests avoiding stable manure, and chaff so early, as a connection had been established between animals eating malformed roots, and spreading a disease of crops through their infested feces. Subsequent early records come from Albert the Great of Cologne (AD 1196/1206 to 1280) who had widely traveled in the Spanish Low Countries (now parts of Belgium and the Netherlands) and Italy. In both locations, *Brassicas* were increasingly important sources of food for man, and his animals. Albertus noted spongy, fungus-like roots which have been interpreted as a record of clubroot. Later infections in Spain were

Fig. 1.1 Assignment of genes in different categories [*GO* gene ontology term (Blast2GO)], *EC* enzyme commission number (Blast2GO and EC Domain Miner), *IP* InterPro term (Blast2GO and gene ontology database). The term in which at least 80 genes were classified is represented (Daval et al. 2019)

recorded on cabbages in the fifteenth century where cabbages are described as being syphilitic (Chupp 1934)—translation of Woronin (1878) and Lancereaux (1869). Prominence was given to the disease in England and Scotland through the eighteenth. and nineteenth centuries co-incidentally with the agricultural revolution (Anderson 1855), and possibly following the development of swedes and turnips as parts of the Norfolk four-course rotation. Causes ascribed to conditions similar to clubroot include infection by "*worms,*" unsatisfactory soil, and poor crop nutrition, particularly the use of unbalanced fertilizer applications (Karling 1968). Karling also describes the subsequent spread of records of the incidence of what could be interpreted as "*clubroot disease*" emerging across northern Europe, and into lands which were colonized by European migrants in the nineteenth and twentieth centuries. Karling (1968) also recorded the multiplicity of vernacular names given to clubroot in European countries. This multiplicity indicates the widespread nature of the disease, and its importance for the general population as a cause of devastation to an important commodity of food crops.

1.9.2 Initial Scientific Studies Period (1830–1930)

The first person who reported having seen in 1736 the "*Turnip Disease, called in Norfolk and Suffolk, Anbury*" (Ellis 1742). He believed it to be a contagious, and probably due to an excess of barnyard manure, especially "*the long undigested ranker sort.*" In Scotland from 1829 to 1851, Farquharson (1831), Abbay (1831), and Birnie (1831) described the disease, and attributed it to unsatisfactory soil conditions or unbalanced fertilizing practices. Abbay stated that he saw the disease first in 1801. In 1855, Anderson asserted that the trouble first appeared about 1813. At about the time Anderson wrote, American and English students, among them Curtis (1843), and later Slingerland (1894) studied the trouble, and concluded that it might be due at least in part to insects. Henderson, in his gardening book published in 1867, discussed observations made on clubroot in the northeastern part of the United States many years previously. He believed that the disease was caused by the attack of the cabbage maggot. In 1874, Sorauer (1874) attributed the disease in part to insects. In one of paper by Ravn (1908) wrote of the history of the disease, and included in his bibliography citations of 18 articles published before Woronin's final description of the nature of the causal organism. Woronin began his studies of the malady in 1873 and in 1875, published a preliminary report on the organism, but did not name it at that time. The article in which the nature of the organism and its host relationships was carefully described and illustrated did not appear until 1878 (Woronin 1878). He published a total of eight papers on this disease, of which three were in Russian, and five in German. It is fact that for nearly two centuries, botanists have been studying the nature as well as control of clubroot disease. Economically, the effects of a plant disease are hard to gauge, and because of the nature of the trouble, it is peculiarly so in this case. Woronin (1878) estimated that in 1869, in the vicinity of St. Petersburg, Russian gardens lost approximately half their cabbage plants, and in 1918, it was reported that in the United States, New York sustained a loss of several thousand tons of cabbage. The important fact is that the disease spreads readily, and that, once established in a field, it may completely destroy, for an indefinite number of years, the usefulness of the plot as ground on which to grow crucifers. Geographically, the disease is very widely distributed. In the Old World, it occurs in nearly all regions where cruciferous crops are important. In the United States, it has been reported as occurring in 36 States, and as an important in twenty one states. It also occurs in Alaska. No trucking sections growing crucifers intensively appear to be incapable of becoming infested with the trouble (Wellman 1930).

In a technical bulletin No. 181 of USDA, Washington DC, Wellman (1930) revealed research work carried out on clubroot of crucifers in quite details as summarized up to 1930. The temperature range for spore germination occurs at a minimum of about 6 °C, and a maximum of about 27°. Spores germinate well at temperatures ranging from 18° to a little above 25°, with the peak at 25°. Disease development occurs over a range of 12° to 27°, with the optimum from 18° to 25° and with most malignant disease development at about 25°. The temperature range for disease development and spore germination is practically identical, whereas, as shown by Tisdale (1923) that the temperature effect upon the growth of the host is different. This indicates that probably the temperature range of the disease development is the direct result of the action of temperature on the pathogen. It is shown that infection of the host occurs quite readily after 18 h of exposure to infested soil held at sufficient moisture content. This indicates that upon the occurrence of a heavy rain even the most adequate system for drainage could not necessarily be expected to inhibit infection by *Plasmodiophora brassicae*. A survey of clubroot-infested fields in three states showed in some cases seriously infested soils which were naturally high in lime. A H-ion concentration survey of 116 disease-infested soils showed that the disease occurred at a range of pH 5 to 7.8, which was also found to be the range covered by samples of cabbage-growing soils selected without respect to disease condition. The action of certain alkaline chemicals was studied by adding them at different rates to infective soils. In a soil with a pH of 7.2, the addition of K_2CO_3 produced a pH of 8.1 without inhibiting the disease, while by the application of $Ca(OH)_2$,, the disease was completely prevented at a pH of 7.3. Data from field experiments have also shown that the H-ion concentration could not be considered a limiting factor in disease control. Experiments carried out on with different limes in pots, and in the field confirm the above results. Limes consisting of $CaCO_3$ and $CaSO_4.2H_2O$ are not good clubroot inhibitors. The limes which are of CaO or $Ca(OH)_2$ composition controlled the disease well in plants grown in the clubroot-infested soils used. In thoroughly infested seed beds, it was found necessary to apply at least 1500 pounds of hydrated lime per acre for satisfactory clubroot control. Unusually large quantities of raw ground limestone applied in the field did not inhibit the disease. Air-slaked limes were found to be of questionable value in their inhibitory effects. This form was, therefore, considered as an unsatisfactory control material. Hydrated lime applied at the rate of one-half ton per acre was found to check the disease noticeably, but it was not until 2 tons/acre were used that a commercially satisfactory control was obtained on the soil in question.

In the mid-nineteenth century, these disease conditions interested the microbiologist Michael Woronin to initiate work on this disease who was working in St. Petersburg, Russia. In northern parts of Europe and western Asia, considerable reliance was placed on the production and storage of cabbage as vital winter supplies of human food and animal fodder. In the nineteenth century, losses due to a disease which turned out to be clubroot were escalating and Woronin took up its study. Previously, he had traveled widely in Western Europe studying botany and mycology under authorities such as Anton de Bary in Freiburg and Holle in Heidelberg. Consequently, he was knowledgeable and had a network of proficient mentors with whom he could correspond and consult. In a 3-year period (1873–1876), Woronin established the identity of the causal agent of clubroot which he classified as *Plasmodiophora brassicae*, made a description of symptoms, identified key hosts and salient parts of the life cycle. He also established the taxonomic relationships ascribing *P. brassicae* as belonging to the Protista (as then defined by E. Hackel) (Chupp 1934, p. 20). Woronin's detailed cytology and taxonomic work established the scientific credentials of this organism, and its relationship with clubroot symptoms. His advice for its control included sanitation involving removal of infested plants, the elimination of diseased seedlings prior to transplanting, the use of crop rotation, and dipping transplant roots in a suspension of soot. This latter probably contained some heavy metals which are now known to be antagonistic to *P. brassicae*. Woronin (1878) as translated by Chupp (1934) identified *P. brassicae* as a protista which resembled the Myxomycetes but differed in the absence of a true sporangial membrane. *Plasmodiophora brassicae* has subsequently been reclassified several times (Cook and Schwartz 1930). The movement of *P. brassicae* through the Proteomyxa of the Protozoa, Mycetozoa, Chytridiales, Archimycetes, and many other classifications has been cataloged (Karling 1968; Buczacki 1983). The organism is now placed in the order Plasmodiophorales, family Plasmodiophoraceae (Braselton 1995).

1.9.3 Modern Era of the Twentieth Century

During this period, research results achieved in the last century have been retested, and validated with the cooperation and coordination of *Brassica* scientists at national and International levels with the availability of modern technology and fast communication facilities. Information has been generated on host-pathogen biology, ecology, and disease management.

i. Establishing Pathogen Biology, Host Responses, and Clubroot Management

Research over the following three-quarters of a century has been summarized in the monographs produced by Colhoun (1958) and Karling (1968). Both provide substantial accounts of the life history, and reproduction of the pathogen and host reactions. Damage caused by clubroot was substantial in some geographical areas. For instance, three-quarters of the Swede crops in the north-east of Scotland, an essential component of the agricultural system which demanded overwintering fodder for sheep, were considered as infected to some degree which resulted in the breeding of Swede "Bruce" that exhibited resistance to clubroot in that region. While in Wisconsin, USA substantial disease on cabbage in the 1930s resulted in the selection of the resistant cultivar "Badger Shipper" (Walker and Larson 1960). Generally, more research focused on establishing the distribution of resistance genes in hosts. It was soon evident that monogenic resistance is concentrated in *B. rapa* while *B. napus* and *B. oleracea* contained numbers of genes of small effect providing forms of field or polygenic resistance. Assessments of resistance in cultivars, and strains of host Brassicas are listed by both Colhoun (1958) and Karling (1968), but it was not possible at that time to provide logical commentary on reasons for the distribution of genes within host lines. Since then, there has been substantial progress in understanding the genetic basis for resistance (Diederichsen et al. 2009).

Attempts at defining the life cycle of *P. brassicae* while following on from Woronin's classical work were hampered technically by the non-axenic nature of the pathogen, and consequently the protracted time required for completion of studies using seedling plants. Ingram (1969) and Ingram and Tommerup (1972) added substantially to an understanding of the life cycle. They intended using tissue culture and callus culture in order to take this work further. But it was realized that the ploidy of the cultures becomes unstable, and hence could influence the resemblance of results to the nature of events in soil. At the same time, Butcher et al. (1974) achieved considerable success in understanding the intimate relationship between host and pathogen by the use of electron microscopy. Studies in the Netherlands (Dekhuizen and Overeem 1971) initiated the contention that symptoms reflect disrupted host growth hormone metabolism. Gustafsson and Falt (1986) also observed that as the pathogen passed through the cortical cells, the hypertrophy of the infected cells occurred. This is support for the view that *P. brassicae* is well evolved, and once established inside host tissue, unlikely to re-emerge, and be exposed to the rigors of the soil environment as suggested in some putative life cycles. Collaborative European research evolved the European Clubroot Differential (ECD) series which replaced a multitude of sets of differentials phenotypes which had been used worldwide by many different laboratories, making comparability of results difficult to establish, evaluate, and compare. Using this set, it was possible to undertake a worldwide study of resistance reactions (Toxopeus et al. 1986). That made for opportunities for breeding resistant cultivars, and in the 1980s, interest in the disease was stimulated because of epidemics developing in European spring-sown oilseed rape. The ECD series helped resolve problems relating to the possible existence and importance or otherwise of physiological races of *P. brassicae*. Jones (1981) and Jones et al. (1982a, b) using ECD demonstrated the highly variable nature of the pathogen micro-geographically, while Toxopeus et al. (1986) established that macro-geographically variation is related to cropping sequences. Following that, Voorips and Visser (1993) refined the use of the ECD series and single-spore isolates for plant breeding purposes.

Asian researchers especially in Japan produced a stream of work which identified further details of pathogen reproduction (Kageyama and Asanoo 2009). This was stimulated by the important uses of *Asian Brassicas* (*B. rapa*) in the diet, and growing concern for the environmental consequences of applying agrochemicals, which contained heavy metals such as mercury and chlorinated hydrocarbon compounds. The evaluation of resistance became part of the regular assessment of cultivar characteristics. The USA established the cytological and biochemical details of the secondary phase of the life cycle in considerable detail (Williams 1966; Williams et al. 1971; Aist and Williams 1971). Host metabolism is badly disrupted, and it became evident that galling symptoms resulted from disrupted host growth regulator metabolism. They also defined in considerable detail the sequence of events taking place when the contents of primary zoospores are physically injected into root hairs. Once attached to the root hair, the zoospore flattens, and the flagella coil around the zoospore which at this stage is filled with ribosomes, lipid bodies, mitochondria, and various vesicles. The flagella are retracted into the zoospore body, and then absorbed. The zoospore body "rounds up," and along a tubular vesicle, the "*rohr*" develops which is bound by a cyst plasma membrane. The open end of the "*rohr*" is directed towards the host cell wall. Contained within the "*rohr*" is a sharp pointed rod, "*the statchel.*" The vacuole enlarges, filling half of the cyst, and the end of the "*rohr*" swells, forming an adhesorium, which enables the cyst to attach onto the root hair. Pressure is created inside the cyst by the enlargement of its vacuole, and that forces "*the statchel*" through the adhesorium, and into the host cell wall. The zoospore protoplasm is injected into the host cell upon evagination of the "*rohr*," forming a Primaryplasmodium within the host root hair. The infection process is considered to be physical as no enzyme involvement has yet been detected.

Defining the environmental conditions conducive or suppressive to clubroot forms a major part of research interest linked with studies of the effects of soil structure, texture, and nutrient composition (Rouxel 1991). This was based on the seminal studies by Garrett in the 1940s (Samuel and Garrett 1945; Garrett 1956) which tracked root-hair infection, and outlined the environments conducive to invasion and colonization. Samuel and Garrett (1945) found that the level of root-hair infection increased with greater resting spore densities, and that at high alkalinity (pH 7.2), infection only occurred at the higher resting spore concentrations of 10^6 and 10^7. At a pH of 6.2, however, infection occurred at all spore concentrations. This was later supported by the work of Colhoun (1952, 1953) who also concluded that the level of clubroot infection increases with increasing resting spore concentration. His studies provided a comprehensive analysis of environmental factors which interact with pathogen, and host characteristics which lead to symptom expression. In support of agricultural and horticultural practice, very considerable efforts were invested in attempts at finding means of chemical control of clubroot. Many of the compounds evaluated failed to enter commercial use because of their potentially damaging effects on the environment. The impetus underpinning the interest of agrochemical companies resulted from the increasing worldwide importance of clubroot disease. However, the results of such initiatives have largely failed in Europe (Dixon and Wilson 1984; Dixon et al. 1994). The chemicals such as flusulphamide have considerably eased the problems which intensive growers of vegetable *Brassicas* face with this disease, and flusulphamide is now widely used in Japan and New Zealand (Donald and Porter 2009).

ii. Epidemiology, Biochemistry, and Physiology

Studies from the 1980s to the present day have been increasing scientific interest in *P. brassicae* possibly because of the huge effects of infection on host growth regulator metabolism. At the same time, recognition of clubroot disease as a major agricultural problem (Dixon 2009c) was identified. This possibly reflects recognition of its worldwide impact on a larger range of *Brassica* crops, including some oil seeds and vegetables destined for export trades. Stimulated by disease development in spring-sown oilseed rape in Europe, breeding programs built around the use of major genes taken from *B. rapa* developed in Germany and Scandinavia (Diederichsen et al. 2006; Wallenhammar et al. 2000). An active program charting field epidemiology in oil rape crops in Sweden determined the half-life, and capacities for disease spread. In these studies, Wallenhammar (1996) established that the half-life of *P. brassicae* resting spores is 3.6 years. It would, therefore, take 18 years, in the absence of a suitable host, for a field population of *P. brassicae* to decrease to 3% of the original spore population. This finding added much needed scientific clarity to estimations of the time needed for crop rotations which have been long advocated as one means of control.

Clubroot has been a scourge of overwintered cauliflower production in France, and during this period, studies demonstrated that suppressiveness could be identified in some soils in Brittany, the major area of production. Extensive programs over several decades studied potential resistance across cruciferous genomes (Delorme et al. 1998; Jubault et al. 2008). Major studies conducted under Japan conditions on the life cycle, understandings of interactions with other soil-borne organisms, the effects of using crop rotation and non-host crops, and detailed aspects of host–pathogen physiology which is now being expressed in considerable interest in biological and integrated control (Narisawa et al. 2005). Breeders produced a stable of new cultivars of Chinese cabbage with varying levels of resistance to clubroot disease (Yoshikawa 1993). There was a long-term interest in the host–pathogen physiology in the Free University in Berlin which produced more details relating to the life cycle and other Clubroot—agricultural and biological challenge worldwide aspects of *P. brassicae*. This also led to a major program concerned with the effects of the pathogen on host physiology. The overwhelming of host defense in the first 12 h after infection results in the down-regulation of reactive oxygen species (ROS) metabolism. It appears that host defenses then recover after 1–2 days, and ROS cell death is minimal during the entire progression of gall formation. Cytokinin biosynthesis is up-regulated after 3 days in combination with the parallel down-regulation of adenosine kinase, catalyzing the conversion (inactivation) of active cytokinin ribosides to the corresponding ribotides. Auxin synthesis is gradually increased by inducing myrosinases and nitrilases. Along with the host defense, metabolism (S-adenosylmethionine (SAM) synthetase and glycolysis), protein degradation, and signaling mechanisms (more specific proteasome proteins and chaperonins) are overwhelmed in the first 12 h after infection, and up-regulated thereafter. Because SAM is also a precursor of ethylene and polyamines, it is possible to predict that these plant growth regulators might also have a similar pattern. Cellular defense (via chitinase, lignin biosynthesis, phenol synthesis, tubulins, and peroxidases) is down-regulated during the entire primary stage of the infection process. Finally, energy production is gradually increased from 4 to 5 days after infection. This latter corresponds with the functioning of the plasmodia as metabolic sinks' (Keen and Williams 1969; Kavanagh and Williams 1981).

1.9.4 Present Era of the Twenty-First Century

Advances in molecular biology generally produced research technology and tools which enabled previously unattainable results to be achieved in the present century. The intriguing relationship between *P. brassicae* and *Brassica* hosts offers molecular biologists fertile areas for study. Since the model plant for much of molecular biology research, *A. thaliana*, is a *Brassica* species, this adds impetus to such studies. Much progress has been achieved in identifying clubroot resistance genes (CR). Over 20 quantitative trait loci (QTL) have been mapped in different linkage groups of *B. oleracea*, along with the dominant genes already recognized in *B. napus* (Piao et al. 2009). The impetus for this work in South Korea has stemmed from the economic and social importance of Chinese cabbage as a vital part of their diet. Similar comparable work has developed in Europe where the emphasis has focused more predominantly on understanding virulence in the pathogen. Chromosomes of *P. brassicae* varying from 2.2 Mb to 680 kb have been characterized, and the total genome is estimated at 20 Mb (Siemens et al. 2009), which has recently been increased to more than 25 Mb (Schwelm et al. 2015, 2016). The genomic gene structure and cDNA structure of several genes have been elucidated, and the expression of these genes linked with the development of clubroot. The sequence data support the inclusion of *P. brassicae* in particular, and the plasmodiophorids more generally in the Cercozoa (Siemens et al. 2009). Combining knowledge of resistance in *Brassicas* with that of virulence in the pathogen provides a robust platform from which the development of resistant cultivars may proceed. These are vital and increasingly important tools needed for the control of the disease, especially where it causes destruction in broad hectare crops, such as oilseed rape, in which the intensive mitigation measures used on vegetable *Brassicas* cannot be deployed for practical reasons.

i. Host Metabolism and Clubroot Development

One of the most significant symptoms of *P. brassicae* infection is the formation of large galls or swellings on the roots of susceptible hosts. Studies of pathogen-induced changes in host metabolism, and symptom development have been hampered, however, by the difficulties associated with working with this obligate parasite. Early work on the metabolic changes resulting from infection by *P. brassicae* revealed that sugars, such as glucose, fructose, mannose, and trehalose are found at much higher levels in infected hypocotyls than in healthy plants, whereas the concentration of galactose is lower (Keen and Williams 1969).

Transcriptomic and proteomic approaches have been employed to elucidate the mechanisms of *P. brassicae* infection and gall formation. Devos et al. (2006) and Siemens et al. (2006) analyzed changes in gene expression and protein content, respectively, whereas in *Arabidopsis*, Cao et al. (2008) examined changes in protein composition in *B. napus*. Siemens et al. (2006) choose two time points to analyze *P. brassicae*-induced changes in *Arabidopsis* gene expression: 10 days after inoculation (dai), when colonization by the pathogen had occurred but symptoms had not developed, and 23 dai, when root morphology had been extensively altered, and galls had formed. Using a full genome Affymetrix chip, more than 1000 *Arabidopsis* genes are found to be differentially expressed at each time point in infected vs. non-infected roots. Differentially expressed genes included those associated with defense, sugar phosphate metabolism, plant growth, and the cell cycle. Genes involved in auxin homeostasis are up-regulated in response to infection, whereas those involved in cytokinin homeostasis are down-regulated, providing support for the involvement of plant growth regulators in clubroot development (Siemens et al. 2006). A key role for cytokinins was further suggested by the observation that cytokinin oxidase/dehydrogenase overexpressing lines of

Arabidopsis are resistant to *P. brassicae*. The proteomic studies employed two-dimensional gel electrophoresis and quantitative image analysis in conjunction with mass spectrometry-based protein identification to analyze changes in protein abundance at 4 dai in *Arabidopsis* (Devos et al. 2006), and at 0.5, 1, 2, and 3 dai in *B. napus* (Cao et al. 2008). In *Arabidopsis*, 12% of the visualized proteins exhibited an altered abundance in infected vs. non-infected roots, including proteins involved in cell defense, cell differentiation, and detoxification of reactive oxygen species (ROS) (Devos et al. 2006). In *B. napus*, changes in the levels of 20 proteins can be detected after infection by *P. brassicae*. These included proteins with roles in lignin biosynthesis, cytokinin metabolism, glycolysis, intracellular calcium homeostasis, as well as the detoxification of ROS (Cao et al. 2008). Interestingly, the ROS detoxification system appears to be overwhelmed in the first hours (0.5 dai) after infection, but recovered by 1–2 dai. Collectively, these transcriptomic and proteomic studies, together with other research, revealed that clubroot infection is associated with the initial promotion of root growth, and auxin production, before the growth of secondary plasmodia (Devos et al. 2005). As galls begin to form, it becomes a major metabolic sink for auxins and cytokinins. In addition, the expression of auxins is reduced, and that of auxin hydrolases is increased during gall development (Schuller and Ludwig-Muller 2006). These disturbances in the levels of plant growth regulators that are largely responsible for the hypertrophy and hyperplasia observed in infected root tissues. Ultimately, the root galls function as strong metabolic sinks in infected plants, drawing nutrients and other resources away from the shoots and leaves, and contributing to above-ground symptom development (Ludwig-Müller et al. 2009).

ii. Genetic of Host–Parasite Interaction

With the exception of *B. juncea* and *B. carinata*, genotypes with resistance to one or more pathotypes of *P. brassicae* can be found in all of the major *Brassica* crops (Diederichsen et al. 2009). Both qualitative (Crute et al. 1980; Wit and van de Weg 1964) and quantitative (Chiang and Crete 1970; Figdore et al. 1993; Grandclement and Thomas 1996; Voorrips 1996) types of resistance have been reported. Most of these sources of resistance, however, are pathotype specific.

Inheritance of Resistance

Early studies indicated that at least three dominant, pathotype-specific resistance genes were located in the *Brassica* A genome. These were termed A, B, and C by Wit and van de Weg (1964), and *Pb1*, *Pb2*, and *Pb3* by James and Williams (1980), however, the correspondence of their gene positions cannot be demonstrated. The studies on *B. rapa* for resistance revealed that between one and three major genes, and one quantitative trait locus (QTL) are involved in conferring resistance to *P. brassicae*. The major genes are named *CRa* by Matsumoto et al. (1998), *CRb* by Piao et al. (2004), *Crr3* by Hirai et al. (2004), and *Crr1* and *Crr2* by Suwabe et al. (2003, 2006). Furthermore, an additional QTL was identified by Suwabe et al. (2006), and termed *Crr4*. Therefore, a total of eight resistance loci were mapped to the A genome, however, whether these genes that have been identified to date are homologous is unknown.

Some accessions of *B. oleracea* with pathotype-independent resistance to *P. brassicae* have also been reported (Voorrips 1996). Indeed, most studies on the C genome have revealed that clubroot resistance in *B. oleracea* is quantitative, and under polygenic control by one or two major QTLs, and some QTLs with minor effects (Figdore et al. 1993; Grandclement and Thomas 1996; Landry et al. 1992; Moriguchi et al. 1999; Nomura et al. 2005; Rocherieux et al. 2004; Voorrips et al. 1997). A few studies, however, have indicated that clubroot resistance in *B. oleracea* is qualitative, and controlled by either dominant (Chiang and Crete 1983) or recessive (Yoshikawa 1993) genes. It is possible that both quantitative and qualitative resistance mechanisms may be at play in this species.

In *B. napus*, most studies have found oligogenic control of resistance to *P. brassicae* (Crute et al. 1980). If this is true, the pyramiding of resistance genes in *B. napus* genotypes will be more practical than in other species. Gustafsson and Falt (1986) proposed models based on three, four, and five resistance genes, where the most favored model is based on four genes. A complex type of inheritance, with dominant genes from *B. rapa,* and recessive genes from *B. oleracea*, can be expected in a resynthesized *B. napus*, with resistance from both ancestral species (Diederichsen et al. 1996). Segregation analysis has suggested that resistance in resynthesized *B. napus* is controlled by at least two dominant and unlinked genes (Diederichsen and Sacristan 1996). Manzanares-Dauleux et al. (2000) located one major gene (*Pb-Bn1*) for resistance against two *P. brassicae* isolates on chromosome N03, as well as one additional minor QTL for each isolate on chromosomes N12 and N19.

iv. Pathotype Virulence Versus Host Resistance

Populations of *P. brassicae* that can overcome most of the commercially available sources of resistance have been present in Europe for decades. Such pathogen populations appeared in Chinese cabbage crops in Japan only a few years after the introduction of clubroot-resistant cultivars (Kuginuki et al. 1999). European populations of *P. brassicae* display great genetic and pathogenic variation, and therefore can quickly overcome resistance sources from *B. rapa* and *B. oleracea*. Resistance genes from stubble turnips (*B. rapa*) have been used in resistance breeding of various *Brassica* crops, including Chinese cabbage, oilseed rape, and *B. oleracea*. Although most turnip lines carry more than one resistance gene, cultivars of the other *Brassica* crops with resistance derived from turnip generally carry a single, dominant resistance gene that is pathotype specific. Breeding programs for clubroot resistance in oilseed rape in Canada in the 1980s utilized non-specific resistance from *B. rapa* (syn. *B. campestris*), which was backcrossed into cultivars of *B. napus* possessing pathotype-specific resistance genes (Vigier et al. 1989).

The assessment of the impact of genetic resistance on the frequency of primary (root-hair infection) and secondary (root) colonization of Chinese cabbage demonstrated that root-hair infection occurred more frequently with a compatible isolate (susceptible reaction) compared with an incompatible isolate (resistant reaction) and, subsequently, fewer cells within the root became infected when the host was resistant (Tanaka et al. 2006). The compatible isolate formed Secondaryplasmodia with many nuclei and, eventually, resting spores in the host root tissue, whereas plasmodia formed by the incompatible isolate remained immature with only a small number of nuclei and no production of resting spores. A similar response was observed in resistant and susceptible canola lines (Hwang et al. 2011). These results suggest that resistance in these host species is associated with the suppression of infection and subsequent plasmodial development during both Primaryand Secondarycolonization.

In contrast, Primaryand Secondaryinfection and development were observed in both resistant and susceptible lines of *B. oleracea*. Symptoms of cortical invasion by *P. brassicae* in resistant and susceptible hosts included cell wall breaks, the presence of vesicles or inclusion bodies within the cell walls, cell wall thickening in association with plasmodesmata, and enlarged and/or disorganized host nuclei (Donald et al. 2008). The main difference between the resistant and susceptible host reactions is that, in the resistant hosts, the Secondarythickening and cell walls of the xylem are not degraded. This study supports previous reports of an amoeboid form of the pathogen, in addition to the recognized two-phase life history of *P. brassicae*. Furthermore, it indicates that resistance in *B. oleracea* does not prevent the development of this amoeboid form. Instead, the reduced number of cell wall breakages in the resistant line suggests that movement of the amoeboid form is restricted, but not prevented, in the resistant host (Donald et al. 2008).

v. Molecular Mechanism of Pathogenicity

An improved understanding of the mechanisms of *P. brassicae* pathogenesis can contribute to the development of novel sources of resistance, and other control measures. However, the nature of *P. brassicae* as an obligate parasite hampers the application of the majority of techniques for the study of molecular mechanisms of pathogenesis. To date, there have been only a few reports on the molecular characterization of *P. brassicae* genes expressed during growth of the pathogen in host tissues (Bulman et al. 2006, 2007; Feng et al. 2010; Siemens et al. 2009). Although the plasmodiophorales are an important pathogen group, information on their genomic makeup is limited. To date, approximately 100 partial cDNA fragments or full-length clones of *P. brassicae* have been isolated. Graf et al. (2004) isolated a few partial cDNAs that had no significant homology to GenBank sequences. Suppression subtractive hybridization (SSH) has been conducted between RNA from *P. brassicae*-infected and uninfected *Arabidopsis* [*Arabidopsis thaliana* (L.) Heynh] tissue, and from the resulting library, 232 clones were screened. In addition, an oligo-capping procedure was used to screen 305 full-length cDNA clones from the infected tissue (Bulman et al. 2006). A total of 76 new *P. brassicae* gene sequences are identified, the majority of which are extended to full length by rapid amplification of cDNA ends (RACE). Many of the unisequences are predicted to contain signal peptides for translocation of the encoded proteins. Again using SSH, but starting with *B. rapa* rather than *Arabidopsis* clubroot galls, Sundelin (2008)

Table 1.1 Identified genes with a potential function in the pathogenicity of *Plasmodiophora brassicae* (Hwang et al. 2012)

Gene Name	GenBank accession number	Description	Reference
Y10	AB009880	Expression exclusively correlated with the vegetative plasmodial stage	Ito et al. (1999a)
PbTPS	Unknown	A trehalose-6-phosphate synthase gene with expression correlated with an accumulation of trehalose in resting spores	Brodmann et al. (2002)
PbSTKL1	AB231687	Expression increased strongly beginning 30 days after inoculation and coincident with resting spore formation	Ando et al. (2006)
PbBrip9 PbCC249	EU345432 AF539801	Strongly expressed at disease stages corresponding to the occurrence of sporulating plasmodia	Siemens et al. (2009)
PbO1	GU082362	A serine protease that stimulates resting spore germination	Feng et al. (2010)

isolated about 140 genes. Half of these gene sequences originated from the pathogen. Of these, 10 clones are newly characterized *P. brassicae* genes. The remainder of the *P. brassicae* genes had been identified previously (Bulman et al. 2006).

Only a few genes have been postulated or demonstrated to be related to the pathogenicity of *P. brassicae* (Table 1.1). Among others, a gene encoding a serine protease (*PRO1*) has been proven experimentally to be important for resting spore germination (Feng et al. 2010). *PRO1* is a single-copy gene present in a broad range of *P. brassicae* pathotypes. *PRO1* transcript was detected during plant infection, and the quantity of transcript fluctuates according to the stage of pathogenesis. *PRO1* exhibited high activity at a temperature of 25 °C and a pH of 6.0–6.4, whereas Dixon (2009a) indicated that the optimal temperature for germination of resting spores was 24 °C and the optimal pH was 6.0–6.7. *PRO1* protein enhances the stimulating effect of the root exudates on resting spore germination. This indicates that *PRO1* may play a role in clubroot pathogenesis by stimulating resting spore germination through its proteolytic activity. Because of this apparent role in pathogenesis, *PRO1* is a potential target for clubroot management, as inhibition of the expression of this single-copy gene can eliminate the corresponding function without interference from other gene copies (Feng et al. 2010).

vi. Induction of Gene in *Brassica* by *P. brassicae*

Orphan genes are genes that are unique to a particular taxon which exhibit no significant sequence similarity outside their species. These genes are also known as taxonomically restricted genes (TRGs) or lineage-specific genes (LSGs), and they are also commonly referred to as "ORFans" (opening reading frames ORFs) with no detectable sequence similarity to any other ORF in a targeted database. With the development of sequencing technologies, and genome analysis methods, research has shown that orphan genes are common in all kingdoms of life, and orphan genes represent 5–15% of the genomes of several species. Orphan genes can be determined by comparative genomic analysis across various species using BLAST program. The identification and characterization of orphan genes have been analyzed in several taxa including *Arabidopsis*. (Donoghue et al. 2011; Neme and Tautz 2013; Arendsee et al. 2014; Guo 2013; Li et al. 2009; Lin et al 2010; Yang et al. 2010; Jiang et al. 2018).

Despite the wildly found orphans in kingdoms of life, only a small number of them have been well-characterized as a result of the absence of functional motifs, identifiable folds, and recognizable domains. Qua-quine starch (QQS; TAIR; *AT3G30720*) is the first plant orphan gene with a biochemically characterized function, and it is known to regulate carbon and nitrogen partitioning to lipid, starch, and protein in seeds, and leaves across species through NFYC (a component of the nuclear factor Y complex is conserved across eukaryotes) interactions. A wheat *Triticum aestivum*, *Fusarium* resistance orphan gene (*TaFROG*) contributes to disease resistance in wheat. The Brassicaceae-specific *Enhancer of vascular Wilt Resistance* 1 (EWR1) in *Arabidopsis* and *Brassica oleracea* is responsible for the enhancement of resistance to Verticillium wilt pathogens. Gene profiling and mutant screening also provide some clues to the functions of orphans. In *Arabidopsis*, rice, and sweet orange, the result of mutant screening or qRT-PCR analysis indicated that orphan genes were associated with biotic and abiotic stress responses. Orphan genes in *Brassica* are associated with clubroot stress, and responsible for species-specific adaptive processes (Li et al. 2009, 2015; Yadeta et al. 2014; Jiang et al. 2018).

Orphan gene also called lineage-specific genes (LSGs) are important for responses to biotic and abiotic stresses, which are associated with lineage-specific structures, and biological functions. A total of 1540 *Brassica* specific genes (BSGs), and 1824 cruciferae-specific genes (CSGs) have been identified based on the genome of *Brassica rapa*. The genetic features analysis indicated that BSGs and CSGs possessed a lower percentage of multi-exon genes, higher GC content, and shorter gene length than evolutionary-conserved genes (ECGs). In addition, five types of BSGs were obtained and 145 out of 529 real A sub-genome-specific BSGs were verified by PCR in 51 species. In silico and semi-qPCR, gene expression analysis of BSGs suggested that BSGs are expressed in various tissues, and can be induced by *P. brassicae*. Moreover, an A/C sub-genome-specific BSG, BSGs1 was specifically expressed during the heading stage, indicating that the gene might be associated with leafy head formation. The results provide valuable biological information for studying the molecular function of BSGs for *Brassica* specific phenotypes, and biotic stress in *B. rapa* (Jiang et al. 2018).

1.10 Challenging Aspects of Clubroot Disease Management

It is now possible to research opportunities which offer both avenues for deeper understanding of the relationship of *P. brassicae* with its environment and increase capacities for controlling the clubroot disease. Even before Woronin, it was realized that *P. brassicae* or clubroot disease is favored by high concentrations of hydrogen ions in the soil, i.e. those with acidic pH. Soils with high calcium content and pH

>7.0 are antagonistic to the pathogen and the disease. However, other members of the Plasmodiophorales react quite differently (e.g., *S. subterranean* which causes powdery scab of potatoes).

1.10.1 Interactions of Physical, Chemical, and Nutritional Factors

Throughout the history of clubroot research, there has been a long trail of field studies attempting to understand the effects of different forms of calcium. Much of these researches are of limited value since it had been done with little regard to many uncontrolled biological and agricultural variables. In early studies, the optimum pH for clubroot infection was considered as approximately 6.0 (slightly acidic soils) (MacFarlane 1952; Colhoun 1953). Alkaline soils were thought to reduce symptom development, and it was recommended for good control of clubroot that the pH of the soil should be raised to between 7.3 and 7.5 (Anon 1984). The debate is as to whether the alkalinity of the soil is responsible for decreased infection or whether this is partly due to the increase in calcium ions (Dixon and Webster 1988; Webster and Dixon 1991a, b). Earlier work by Samuel and Garrett (1945), in which soil alkalinity was altered using nutrient solutions which did not contain calcium, suggested that it was the direct effect of pH which decreased the level of infection. A soil pH greater than 7.2 was found by Myers and Campbell (1985) to decrease Primaryinfections due to abortion of Primarythallus before the release of secondary zoospores. At a soil pH greater than 8.0, thalli that did develop zoospores were misshapen, and aborted, and no clubs developed. This supports the contention that Secondaryzoospores are necessary for gall development. Infection can occur in alkaline soils if the soil inoculum levels are sufficiently high (Colhoun 1953; Karling 1968). The importance of soil moisture content has long been hinted at in the literature, but Thuma et al. (1983) established that there is a direct relationship between increased soil moisture content, and the level of Primaryinfection, such that as the moisture content of the soil rose, the level of infection increased.

It was clearly demonstrated that elevated calcium concentrations in the soil environment retarded the rate of maturation of *P. brassicae* in root hairs, reducing the speed, and quantities of plasmodia which matured into sporangia. Detailed studies quantified these effects, showing that the quantity of calcium in the rhizosphere can be related to changes in the reproductive rate, and maturation of *P. brassicae*. There had been debate as to whether raising the soil pH is responsible for the decrease in *P. brassicae* infection or the presence of calcium ions. This was resolved by some elegant experiments involving the use of organic buffers by Webster (1986), who determined that calcium and pH effects were separate but complementary in relation to the promotion and reduction of the growth of the pathogen. It was also determined that boron had similar but more intense effects compared with calcium, and that the forms of nitrogen present can also regulate the success or failure of the host–pathogen interaction. Boron at elevated concentrations also appeared to be associated with the reduction of the secondary phase of the *P. brassicae* life cycle, and symptom expression was reduced by high concentrations of boron in the root environment. Craig and Dixon (1993) reinforced these findings with studies of root-hair infection rates in the presence of elevated boron concentrations. Subsequently, Page (2001) clearly demonstrated that a combination of calcium and nitrate nitrogen was antagonistic to the growth and reproduction of *P. brassicae*. By contrast, ammonical-nitrogen encouraged the life cycling of *P. brassicae*. It was observed that repeated field applications of calcium compounds could encourage the development of soil suppressiveness which inhibited the development of disease. Dixon (2012) has reported that calcium cyanamide fertilizer is considered to stimulate the development of soil microbes which are antagonistic to *P. brassicae*. It is evident from these studies that calcium, boron, and nitrate nitrogen have suppressive effects which may act synergistically. This evidence provides some scientific basis for the long-term practice of liming soil, changing soil chemistry so that hydrogen ion concentration is reduced, and there are increased quantities of calcium and other elements in the rhizosphere. In all probability, the relative concentrations of these molecules in soil moisture have a retarding effect on the ability of primary zoospores of *P. brassicae* to reach the root hair, and support changes to the flora associated with the rhizosphere, and root-hair surfaces, such as penetration becomes more difficult, and importantly, the internal environment which *P. brassicae* Primaryzoospores experience once inside the host becomes less conducive for growth and reproduction. Changing soil chemistry is a well-established farming practice, and there are now some good biological reasons underpinning its efficacy. There is a dynamic and integrated relationship between calcium and boron to affect the Primaryzoospore, and the attendant invasion, and colonization stages of *P. brassicae*. Boron is additionally involved in the Secondarysymptom development, and expression stages. Forms of nitrogen may also be involved in some of these events. Pathogenesis results from a sequence of interactive events firstly in the soil, and then within the host has been reported (Dixon 2014).

1.10.2 Interactions of Biological Factors

Biological control was defined by Garrett (1956) as any conditions under which, or practice whereby, survival or activity of a pathogen is reduced through the agency of any other living organism (except man himself), with the result that there is a reduction in the incidence of the disease caused by the pathogen. Biological control not only includes individual which reduce disease but also husbandry methods, such as intercropping, flooding, rotations, and the application of manure. These measures which influence the biological components of the habitat are referred to as indirect biological control (Campbell 1989).

In the last decade, research has re-focused on the opportunities for direct biological control of *P. brassicae*. Research into epoxydon (5-hydroxy-3-(hydroxymethyl)-7-oxolate bicyclo [4.1.0] hepta-3-en-2-one) may pave the way for a new range of agrochemicals. Epoxydon was obtained from a culture of *Phoma glomerata* (No.324 = JCM 9972) from the leaves of *Viola* spp. Comparisons of the effectiveness of epoxydon with several anti-auxins indicated that the anti-auxin 2, 3, 5 triiodo benzoic acid was the active ingredient (Arie et al. 1998). More recently, *Bacillus subtilis* has shown promise in commercial preparations as a biocontrol agent (Lahlali et al. 2011; Li et al. 2012).

Some soils which are suppressive to *P. brassicae* have been identified. In some instances, the soils did not remain suppressive after sterilization, indicating that a biological agent could be responsible but that the physical and chemical properties of the soil were potentially also responsible for the reduction in disease intensity (Jaw-Fen 1986; Tsushima et al. 2010; Workhu and Gerhardson 1996). It seems likely that suppressiveness could result from a combination of microbial action conditioned by the physical and chemical environment present in the soil. Rouxel (1984) and Rouxel et al. (1988) studied the receptiveness of soils in Brittany to *P. brassicae,* and concluded that biotic suppressive soils were present and their performance was influenced by soil pH. This indicated the presence of a specific type of microflora but the biocontrol agent was not identified. Factors which influence the suppressiveness of a soil or compost include soil texture, pH, moisture content, organic matter content, and mineral or nutrient composition. Suppressiveness may, therefore, be encouraged by altering these factors (Broadbent and Baker 1974; Boehm and Hoitink 1992). The actions of physical or chemical factors which enhance the soil suppressiveness are also likely to increase the capacities of microorganisms for biological control (Campbell 1989). Therefore, the addition of organically derived factors, such as chitin and seaweed extracts, to the soil environment can alter its suppressive qualities, and help to decrease disease development. These findings suggest that a more holistic approach to clubroot disease control may be the most fruitful.

1.10.3 Soil Suppressiveness and Conduciveness

The identification of the soils having suppressive characteristics along with factors governing this trait may be very helpful in integrated management of clubroot. There is occurrence of soils in some areas which are suppressive to *P. brassicae* but not due to biotic factors. Such soils have been identified in Taiwan. These soils had a pH greater than 7.4, and a calcium content of 1210 ppm (Hseith and Wang 1986). The latter used suppressive soil samples with a pH of 7.3, and a calcium content of 1460 ppm. When the suppressive soil was diluted with conducive soil, the pH was altered to 6.7, and the calcium content to 1210 ppm. Suppressiveness was lost from soils when they were acidified with sulfuric acid but returned when sodium carbonate was applied. Acidification of the soil increased the level of calcium by the dissolution of calcium compounds. As a result, the availability of calcium decreased due to the higher concentrations of hydrogen ions. The fertilizer calcium cyanamide provides a classic example of the manner by which soil suppressiveness to *P. brassicae* may be encouraged by the manipulation of soil chemistry, which then encourages the development of microbial flora which is antagonistic to the pathogen. This fertilizer is well established, and environmentally benign (Dixon 2012). The activity of calcium cyanamide in relation to *P. brassicae* infection was recognized as long ago as 1928 by Kindshoven. Since that date, there have been numerous studies which related the application of this fertilizer with enhanced microbial activity, and the suppression of clubroot disease.

1.10.4 Host Resistance

Resistant cultivars of vegetable *Brassicas* and oilseed rape have emerged (Diederichsen et al. 2006; Hirai 2006), and these formed important parts of an integrated disease control program. Indeed, the use of a range of strategies for control of clubroot disease is essential in order to preserve and protect resistant cultivars from erosion due to the changes in the spectrum of virulence dominating in populations of *P. brassicae*. This is especially important because Diederichsen et al. (2009) found that isolates of the pathogen display great variation, and show tendencies for overcoming resistance sources from either *B. rapa,* and *B. oleracea*. At present, resistance genes from *B. rapa* (stubble turnip types) are the most effective, and widely used. The biological progress made in mapping resistance and virulence genes should be used in practice by breeding programs, with the aim of broadening the genetic basis for clubroot resistance by using both natural variation and transgenic strategies (Diederichsen et al. 2009).

1.10.5 Use of Molecular Techniques

It is essential that robust diagnostic tools are utilized in order to quantify the inoculum potential (sensu Garrett), and track the efficacy or otherwise of remedial mitigation husbandry. Molecular biology offers methods based on the polymerase chain reaction (PCR) which provides opportunities for assaying soil, tissue, and water samples (Faggian and Strelkov 2009). In addition, serological-based tests could, if available, offer a straightforward field testing system. Robust diagnostic tools (Rennie et al. 2011) have been developed, and put into service to evaluate disease development and spread. These substantial studies of disease epidemiology have shown new and previously under-appreciated means of spread, produced solid basic research results on the disease, and pathogen cycles, and helped lay the foundations for genetically modified cultivars which are coming on stream from private and public sources (Hasan et al. 2012). Systems for the amelioration of soils as part of integrated control have been investigated (Hwang et al. 2011). This research program has investigated the effects of boron on reducing the growth and reproduction of *P. brassicae* (Deora et al. 2011). Alongside all this work runs a strand of more fundamental studies looking at host–pathogen relationships (Deora et al. 2012; Feng et al. 2012; Gossen et al. 2012; Hwang et al. 2012). Currently, clubroot is recognized as of major significance in both *Brassica* vegetables and oilseeds in China, Canada, Australia, and India. The problem there is so huge that special measures are being taken to boost research and attain methods for control.

1.10.6 Complexities of Soil Environmental Factors

The approach which preserves, and enhances soil integrity linked with capacities for increased agricultural productivity, and profitability, that offers means by which some of the serious problems of mankind in increasing food supplies while conserving biodiversity, and reducing adverse impacts on the environment may be achieved. This requires a detailed understanding of the complexities of the soil environment in which the hosts of *P. brassicae* grow, and in which clubroot disease develops (Fig. 1.2). The pathogen *P. brassicae* and its hosts have been subjected to some detailed scientific studies over the last one and a half centuries. As a result, some of the biological challenges are more clearly understood, and some of the agricultural challenges have been reduced. But the combination of *P. brassicae* and clubroot disease still demands substantial further investigation. This is a biologically basically microbial pathogen with a novel life cycle well adapted to the colonization of hosts which then provide energy supplies for use by its future generations. Understanding the biology of this host–pathogen interaction potentially offers routes for deeper knowledge of cellular metabolism, and the manner by which it can be altered in beneficial as well as harmful directions. Some of the interactions between *P. brassicae* and its hosts in the soil environment have been unraveled over the last two decades. There is now a much enhanced understanding of the effects of soil chemistry on the early vulnerable stage of the life cycle of *P. brassicae* when naked primary zoospores are released. Additionally, substantial knowledge has been gained of the effects of soil chemistry on the Primaryinvasion and colonization of root hairs. This is summarized in Fig. 1.3. The practical objective is achieving the suppression of growth by *P. brassicae* as described in Fig. 1.4. Agriculturally, there

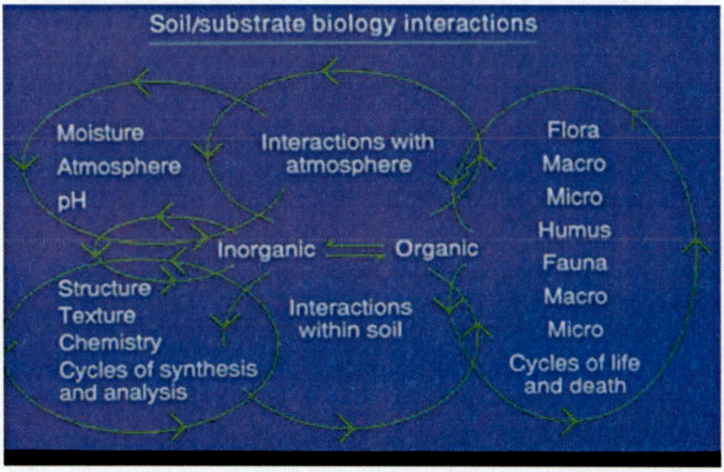

Fig. 1.2 Summary of the interactions of physical, chemical, and biological factors which form the environment in which the interaction between *Plasmodiophora brassicae* and *Brassica* hosts takes place (Dixon 2014)

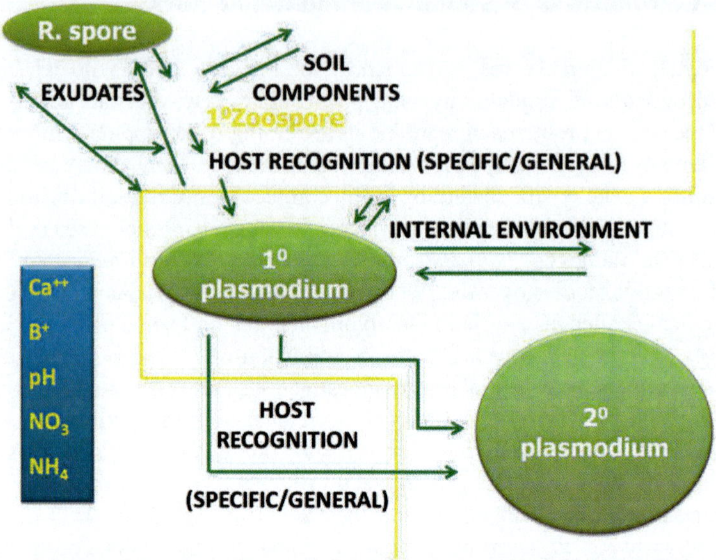

Fig. 1.3 Summary of the soil chemical factors which impinge on successful movement of primary zoospores of *Plasmodiophora brassicae* from resting spores to invasion, and colonization of host root hairs (Dixon 2014, 2009c)

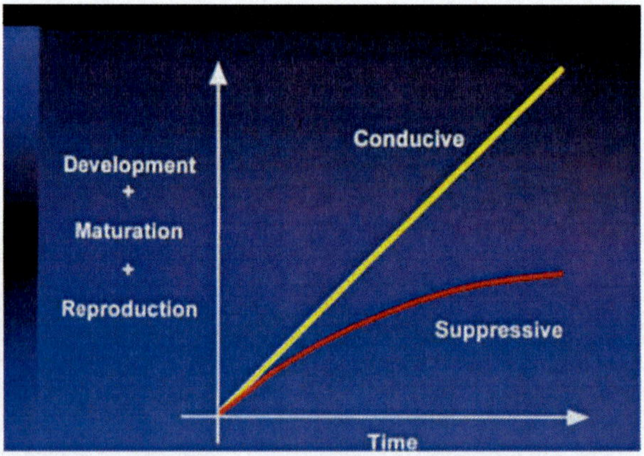

Fig. 1.4 Summary of conducive and suppressive conditions for *Plasmodiophora brassicae* (Dixon 2014, 2009c)

are opportunities for developing far greater understandings of the mechanisms of genetic resistance, and how products of plant breeding may be integrated with other controls. Most interestingly in that respect are studies of how rotational cropping alters the microbial flora that it is antagonistic to *P. brassicae*.

The phenomenon of suppressiveness is now accepted scientifically as a valid area for study, and engineering means for developing this are key for sustainable and holistic integrated control strategies. Potentially interweaving crop resistance, rotational cropping, and the resultant suppressiveness with biological control, and husbandry which tailors soil chemistry, and physics towards limiting the disease causing potential of *P. brassicae* are likely to be the most effective and cost-beneficial agricultural strategy (Dixon 2014). Now this disease has become very serious and devastating in one of the world's key traded commodity crops canola produced in Canada, and has been admitted to be a scourge in China, which is the world's most populous country, needs intensive research and extensive efforts to manage at global level.

1.11 Pathogen Diagnosis and Quantification

The challenges in *P. brassicae* research and clubroot disease management have been the absence of rapid and reliable technique for the detection and quantification of the pathogen in soil and host material. Clubroot infestation has historically been detected by means of plant bioassays, in which bait plants are grown in the suspect soil for a 5–6-week period, and then examined for the presence of root galls (Faggian and Strelkov 2009). Although this approach represents a reliable method to detect moderate to high levels of viable *P. brassicae* inoculum, it is time-consuming, labor intensive, prone to the effects of environmental conditions, and requires fairly large amounts of glasshouse space. Moreover, although bioassays may provide an indication of the relative amount of pathogen in soil samples, that cannot provide a quantitative measure of infestation, nor can they be applied to host tissue or in circumstances in which soil material is Limited (Hwang et al. 2012).

Alternative diagnostic approaches have included microscopic examination of the root hairs to assess the incidence of root-hair infection (Hwang et al. 2011; MacFarlane 1952), as well as fluorochrome staining of the *P. brassicae* resting spores to distinguish them from soil particles (Takahashi and Yamaguchi 1988, 1989). These techniques, however, can also be time-consuming, and must be conducted by trained research personnel. Serological methods to detect *P. brassicae* have also been developed (Lange et al. 1989; Wakeham and White 1996), but, if based on polyclonal antiserum, the results obtained can be influenced by variation in the specificity, and sensitivity of the antiserum used. The development of serological assays based on monoclonal antibodies may represent a desirable alternative to polyclonal-based assays, as monoclonal antibodies show improved specificity, and are potentially available in almost infinite quantities (Faggian and Strelkov 2009).

Various other diagnostic tests based on the polymerase chain reaction (PCR) have been developed for the detection of *P. brassicae*. Most of these protocols have targeted the ribosomal genes (rDNA), and internal transcribed spacer (ITS) regions of the pathogen genome (Cao et al. 2007; Chee et al. 1998; Faggian et al. 1999; Wallenhammar and Arwidsson 2001), although one test was based on the amplification of an isopentyl transferase-like gene (Ito et al. 1999b). In PCR-based detection protocols, *P. brassicae*-specific primers, and reaction conditions are developed to ensure that they amplify DNA only from the pathogen. If appropriately validated, these tests can provide a rapid, and reliable method to determine whether or not *P. brassicae* DNA is present in a variety of samples. In western Canada, one such PCR-based protocol developed by Cao et al. (2007) has been commercialized by a diagnostic laboratory, and is used for the routine detection of *P. brassicae* in plant, and soil materials submitted by farmers, municipalities, and industry personnel. This test, which represents the first one-step PCR-based protocol for the detection of the pathogen, can consistently detect 100 fentogram (fg) or less of *P. brassicae* DNA, which corresponds to 1 x 10^3 resting spores per gram of soil, or an index of disease of 11% or lower, when the soil is bioassayed (Cao et al. 2007). Although providing consistent results, this protocol, like all other earlier PCR-based tests, provides an indication only of the presence or absence of the pathogen, rather than the amount of inoculum present in a particular sample.

In contrast, the advent of quantitative PCR (qPCR) technologies has enabled the quantification of *P. brassicae* in a manner that was not previously possible (Wallenhammar et al. 2012). In a study, a qPCR assay was used, in conjunction with the analysis of signature fatty acids, for the *in planta* quantification of *P. brassicae* (Sundelin et al. 2010). The amounts of arachidonic acid (20:4) from whole-cell fatty acid analysis, and DNA from qPCR were found to be well correlated in infected plants, suggesting that these parameters may represent useful tools for studies requiring the quantification of the pathogen. Similarly, a qPCR-based protocol has been developed for the quantification of *P. brassicae* resting spores that occur as external contaminants of seeds, and tubers of various crops (Rennie et al. 2011), in order to assess the risk posed by such inoculum. Assays for the quantification of *P. brassicae* based on qPCR methodologies have the potential to greatly facilitate many aspects of clubroot related research, as primers, and protocols could be modified as necessary to meet the specific requirements of particular studies (Hwang et al. 2012). With technologies used in precision farming, such as Global Positioning System (GPS), and Geographic Information System (GIS), it is possible to track, and visualize infestation levels of soil-borne inoculum of the pathogen.

1.12 Detection of Clubroot Pathogen

The detection of clubroot pathogen has traditionally been relied on plant bioassays, either by visual assessment of root symptoms or by microscopic examination of root-hair infection (Toxopeus and Janssen 1975). Fluorescence microscopy, serology, and DNA-based methods have all been used to test soil, water, or plant samples for clubroot. The current, and historical approaches that have been used to detect, and quantify *P. brassicae* are as follows (Faggian and Strelkov 2009).

1.12.1 Bait Plants

The use of bait plants is the most reliable diagnostic method for assessing soils for the presence of *P. brassicae,* and is still heavily relied upon today despite the availability of more modern techniques. Following problematic early attempts to examine soil microscopically for the presence of *P. brassicae* resting spores (Fedotova 1933), bait plant methods were developed. Samuel and Garrett (1945) developed the infected root-hair count in which cabbage seedlings were grown for one week in the suspect soil sample. A 2-cm length of root was stained with acetocarmine, and root hairs were examined microscopically for the presence of zoosporangia. MacFarlane (1952) used this method to assess resting spore survival in soil. However, questions relating to the relationship between root-hair infection, and subsequent gall development led to further refinements in the method. Therefore, Colhoun (1957) grew plants under controlled conditions for five weeks in suspect soil before examining the roots visually for clubroot symptoms. Melville and Hawken (1967) adapted the method further after encountering problems, including the poor growth of test plants, the prevalence of other pathogens such as *Rhizoctonia solani* Kuhn, and difficulty in maintaining the maximum water holding capacity of the test soil. As a result, a variety of bait plant techniques (or bioassays) are currently in use to provide a reliable means of establishing the presence of viable *P. brassicae* inoculum in soils (Wallenhammar 1996), or to provide the only means of validating newly developed diagnostic procedures (Cao et al. 2007).

These methods are useful, however, only when inoculum reaches levels of greater than 1000 spores per gram dry soil, which is the generally accepted threshold for symptom development depending on soil type and environmental conditions. Bait plant techniques are also time-consuming (typically taking five weeks or longer), labor intensive, or both. Five weeks is also an unacceptably long time to wait for a test result, whereas examining root hairs microscopically is a laborious task requiring trained personnel. Such methods also require large amounts of glasshouse space (Melville and Hawken 1967), and, therefore, for practical and economic reasons, they cannot form the basis of a routine diagnostic test for large numbers of samples (Wakeham and White 1996).

1.12.2 Fluorescent Microscopy

Microscopic examination of soil was revisited by Takahashi and Yamaguchi (1987), who developed a fluorescence assay for *P. brassicae* detection. Soil suspensions were stained with a single fluorochrome, and fluorescing resting spores were counted under a fluorescent microscope. The method was modified after fluorescing soil particles prevented unambiguous counts (Takahashi and Yamaguchi 1988). The updated assay involved the use of two fluorochromes, and an altered soil preparation protocol that improved spore detection, and provided a means of determining spore viability. One fluorochrome, Calcofluor White M2R, bound to chitin (a major component of the resting spore wall) in undamaged cells, whereas the second fluorochrome, ethidium bromide, penetrated damaged and nonviable cells. This enabled the differential staining of viable (blue fluorescence) and nonviable (red fluorescence) resting spores. A correlation was later found between blue fluorescing spores and disease severity (Takahashi and Yamaguchi 1989; Takahashi 1991). However, Donald et al. (2002) later determined that the correlation between spore viability and disease severity, as predicted by the fluorescent staining method, was less clear. Again, although potentially useful for observing, and counting spores in soil, it is unlikely that microscopic examination of soil would form the basis of a routine diagnostic assay. Fluorescing soil particles and other artifacts can make it difficult to identify *P. brassicae* resting spores, so sample throughput is restricted by operator skill.

1.12.3 Serology

A serology based detection method for *P. brassicae*, developing an immunofluorescence assay for use with soil samples was used by Arie et al. (1988). However, the cross-reactivity of the antiserum with other organisms was not tested, and detection limits were not discussed. Lange et al. (1989) prepared a polyclonal antiserum for use in a dot immunobinding protocol to detect *P. brassicae*. The antiserum did not cross-react with *Polymyxa graminis*, another plasmodiophorid, or the common root pathogens *R. solani*, *Pythium ultimum*, and *Fusarium oxysporum*. The technique allowed the detection of *P. brassicae* in infected plant tissue up to a dilution of 1 in 2048, prompting the authors to state that the "sensitivity obtained was within the range permissible for a routine test." It was suggested that a fluorescent antibody technique be developed to detect *P. brassicae* in soil samples. Wakeham and White (1996) prepared several polyclonal antisera against *P. brassicae*, and used them in the development of soil diagnostic tests in the form of Western blots, dip-sticks, dot blots, immunoblotting assays, indirect enzyme-linked immunosorbent assays (ELISA), and indirect immunofluorescence assays. A range of soil fungi, including the closely related plasmodiophorid *Spongospora subterranean*, generally showed low cross-reactivity with the antisera. A detection limit of 100 spores per gram of soil was achieved with one of the polyclonals, *P. brassicae* 15/2, when used in the dip-stick method, indirect ELISA, and immunofluorescence. Orihara and Yamamoto (1998) developed another polyclonal-based assay, but the detection limit in artificially inoculated soil was 10,000 spores per gram, whereas in field soil, some non-specific reactions were observed. It was suggested that improvements in the soil preparation method might help to increase the detection sensitivity of the assay.

In general, the specificity of antisera raised against fungal plant pathogens can vary greatly (Savage and Sall 1981; Hardham et al. 1986), and as polyclonal antisera cannot be reproduced, and are therefore limited in quantity; their usefulness as diagnostic tools is limited. However, serologic assays based on monoclonal antibodies are likely to be a significant advance, improving specificity, and providing an infinite source of antibodies. Serologic assays are also inexpensive and amenable to high-throughput sample processing. A number of monoclonal antibodies have been developed for the detection of *P. brassicae*, and preliminary experiments, in a range of formats that include a lateral flow device, indicate that the monoclonal antibodies are highly specific, and can detect *P. brassicae* in naturally infested soil (Faggian and Strelkov 2009).

1.12.4 Polymerase Chain Reaction (PCR)

Polymerase chain reaction (PCR)-based assays are now routine components of plant pathogen diagnostic procedures, and because it require only DNA, and are ideally suited to the detection of obligate pathogens such as *P. brassicae*. Also, PCR is applicable to soil, and small tissue samples, making it an attractive alternative to more limited and time-consuming methods such as bioassays. Several PCR-based diagnostic assays have been developed for the detection of *P. brassicae*. Buhariwalla et al. (1995) developed PCR primers for the specific amplification of polymorphic DNA from *P. brassicae*. This provided a rapid means of studying the genetic diversity of the pathogen. The primers, however, also facilitated the amplification of host *Brassica* DNA, making them of little use for diagnostic purposes. Ito et al. (1997) developed a semi-nested PCR for specific amplification of an isopentyl transferase-like gene from purified *P. brassicae* DNA. These primers were later used to develop a detection protocol for *P. brassicae* in soil (Ito et al. 1999b). The PCR protocol could detect as little as one resting spore per gram of artificially infested soil. However, detection was difficult in naturally infested soil, necessitating a "double" PCR that involved another round of PCR with the nested set of primers. PCR results were not correlated with disease severity.

A number of PCR assays have been developed that were directed at the ribosomal genes rDNA) or internal transcribed spacer (ITS) regions of *P. brassicae*. Chee et al. (1998) developed a PCR detection protocol directed at the rDNA of *P. brassicae*, and were able to detect the organism in infected roots of Chinese cabbage and radish. Faggian et al. (1999) also developed a nested PCR detection protocol directed at the rDNA and ITS regions. The assay was used with artificially inoculated soil, and potting mix, naturally infested soil, and with plant tissue, enabling detection down to 10 spores per gram potting mix or fentogram (fg) levels of *P. brassicae* DNA, although PCR results were not correlated with disease severity. Wallenhammar and Arwidsson (2001) developed an rDNA-directed nested PCR protocol that enabled detection of *P. brassicae* in a range of naturally infested soils. Positive test results were obtained with soils containing inoculum that corresponded with a disease severity index higher than 21% (35% of bioassay plants showing symptoms). In some of the soils tested, however, plant bioassays returned positive results below the PCR detection limit (down to inoculum levels corresponding to a disease severity index of 8%), and also on one occasion what appeared to be a false negative for the PCR assay (inoculum level corresponding to a disease severity index of 88%). Cao et al. (2007) developed an rDNA-directed, one-step PCR protocol that enabled consistent detection of *P. brassicae* in naturally infested soil down to inoculum levels corresponding to a disease index of 11% (14% of plants infected), or fg

Table 1.2 Summary of PCR-based methods for the detection of *P. brassicae* DNA (Faggian and Strelkov 2009)

Author	Target	Sensitivity	Validated in soil?	Correlated with disease?
Chee et al. (1998)	rDNA/ITS region	–	X	X
Ito et al. (1999a, b)	Isopentyl transferase-like gene	1 resting spore per gram of soil	√	X
Faggian et al. (1999)	rDNA/ITS region	0.1 fg *P. brassicae* DNA	√	X
Wallenhammar and Arwidsson (2001)	rDNA/ITS region	260 pg DNA from infected roots	√	√
Cao et al. (2007)	rDNA/ITS region	100 fg *P. brassicae* DNA	√	√

quantities of DNA. The assay also allowed the detection of *P. brassicae* DNA on/in symptomless roots three days after inoculation. The PCR assay could still reliably detect *P. brassicae* in naturally infested soil that had been diluted up to five times with "clean" soil of the same type. However, as with Wallenhammar and Arwidsson (2001), some variation between soils was observed. It was suggested that this resulted from the patchy distribution of the pathogen in soil, and from sampling inadequacies. Table 1.2 summarizes all the PCR-based methods for the detection of *P. Brassicae* DNA. None of the PCR-based assays mentioned above reported any non-specific amplification, and progressive improvements in sensitivity and improved correlation with disease severity indicate that PCR may be a useful foundation to a routine diagnostic assay. Problems with sampling, sample size, and consistency between soil types require further research. Ren et al. (2012, 2014) used PCR amplification to confirm pathogens identity using quantitative PCR, and DNA of *P. brassicae* was detected and quantified on canola seeds (Rennie et al. 2011). A loop mediated isothermal DNA amplification technique (LAMP) has been developed to detect *P. brassicae* (Kaczmarek et al. 2014).

1.13 Major Sources of Information on Clubroot

The information on clubroot of crucifers is spread over in several publications of National and International importance in the form of research papers, reviews, bulletins, books, book chapters, extension articles, farmer's advisory notes, and news papers all over the world. However, the following major sources give an overall picture of research developments which took place since its first report in the year 1736. More than 283 years ago, the disease was recognized which has now spread over more than 88 countries of the world, and causing quantitative loss varied from 10-100 per cent along with qualitative deterioration of the valuable *Brassica* vegetables and oil yielding crops as well.

The International clubroot working group (ICWG) was founded in 1975, and use to publish newsletter until 1985. The reports from such meetings had been contained in the Cruciferae Newsletter. Meetings of ICWG are usually arranged in conjunction with *Brassica* symposia of the International Society for Horticultural Science (ISHS), and Crucifer Genetics Workshops, International Horticultural Congress, or International Congress of Plant Pathology. Apart from this, the group meetings of *Brassica* researchers in different countries, workshops, symposia, special meetings of vegetables, and Plant Pathology societies are platforms where information is shared on clubroot of crucifers.

Abstracts of the Turin Meeting have been published in Plant Protection Science (44, 1, 2009). Seed of European Clubroot Differentials (ECD) series which was developed by ICWG in 1975 is available on request from the GeneBank, Warwick University Horticulture Research International (Warwick HR1), Wallesbourne, Warwickshire CV 35 9EF, UK. Following sources have contributed immensely in clubroot comprehension with future priority areas of research to manage this menace. Early history (1742–1930), importance, geographical distribution, life history of the pathogen, spore germination, disease development, and control measures have been covered in quite details by Wellman (1930) in a technical Bulletin No. 181 "*Clubroot of Crucifers*" published by USDA. The first monographic treatment to clubroot was by Colhoun (1958), Clubroot disease of crucifers caused by *P. brassicae,* a monograph. Phytopathological papers No. 3 Kew, Surrey: The Common wealth Mycological Institute. Other significant publications are as under:

1. A monograph by Karling (1968). The Plasmodiophorales. Hafner Pub. Co. New York covers all kinds of clubroot research up to 1968.
2. Special issue of Canadian Journal of Plant Pathology 36, 2014 includes reviews on disease development, epidemiology, and disease management.
3. Several Publications of Dixon G A, published from 1976 to date on all aspects of clubroot with very useful, and motivating suggestions for researchers to fill the gaps in overall comprehension of this serious disease of worldwide concern. Dixon is the person who had been contributing continuously on the progress of research through his up-to-date reviews for the benefit of clubroot researchers to refine their technology to get better results for the management of this serous malady of crucifers.
4. Special issue of European Journal of Plant pathology 2016 on clubroot, and black leg diseases of *Brassicas*. It highlights advances in host–pathogen interaction at population, molecular, genomics, and management strategies.
5. Workshops and meetings
   a. European foundation for Plant Pathology conference in Cracow, Poland in 2014.
   b. Clubroot meeting at Chungnam Univ. Daejeon, South Korea, 2015. Reflect the major progress in clubroot research.
6. Journal of Plant Growth Regulation special issues on clubroot research, 2009.

7. Comprehensive reviews of the Molecular Basis of the Crucifers-Clubroot Pathosystem by Ludwig-Müller et al. (2009), Piao et al. (2009), Siemens et al. (2009), Hwang et al. (2012), and Feng et al. (2014).
8. The Journal Acta Horticulturae (No. 706) contains series of reviews papers, and Plant Protection Science 45(1) 2009 contains the abstracts from a conference held in 2008.

References

Abbay J (1831) Report on the disease in turnips called finger and toe. Highland Agric Soc Scot Trans 2:238–240

Adams D (2000) News feature –now the hard ones. Nature 408:792–793

Aist JR, Williams PH (1971) The cytology and kinetics of cabbage root hair penetration by *Plasmodiophora brassicae*. Can J Bot 49:2023–2034

Anderson (1855) Report on the disease of finger and toe in turnips. Highland Agric Soc Scot Trans 6(3):118–140

Ando S, Yamada T, Asano T, Kamachi S, Tsushima S, Hagio T, Tabei Y (2006) Molecular cloning of PbSTKL1 gene from *Plasmodiophora brassicae* expressed during club root development. J Phytopathol 154:185–189

Anon (1984) Agricultural Development and Advisory Service (ADAS).Clubroot Leaflet no 10276, Her Majesty's Stationery Office, London

Anon (2000) (The *Arabidopsis* genome initiative partnership). Analysis of the genome sequence of the flowering plant *Arabidopsis thaliana*. Nature 408:796–815

Arendsee ZW, Li L, Wurtele ES (2014) Coming of age: orphan genes in plants. Trends Plant Sci 19:698–708. https://doi.org/10.1016/j.tplants.2014.07.003

Arie T, Namba S, Yamashita S, Doi Y (1988) Detection of resting spores of *Plasmodiophora brassicae* Woron. from soil and root by fluorescent antibody technique. Ann Phytopathol Soc 54:242–245

Arie T, Kobayashi Y, Okada G, Kono Y, Yamaguchi I (1998) Control of soil borne clubroot disease of cruciferous plants by epoxydon from *Phoma glomerata*. Plant Pathol 47:743–748

Bi K, He Z, Gao Z, Zhao Y, Fu Y, Cheng J, Xie J, Jiang D, Chen T (2016) Integrated omics study of lipid droplets from *Plasmodiophora brassicae*. Sci Rep 6:36965. https://doi.org/10.1038/srep36965

Bi K, Chen T, He Z, Gao Z, Zhao Y, Liu H, Fu Y, Xie J, Cheng J, Jiang D (2019) Comparative genomics reveals the unique evolutionary status of *Plasmodiophora brassicae* and the essential role of GPCR signaling pathways. Phytopathol Res 1:12

Birnie M (1831) Report on the disease in turnips called fingers and toes. Highland Agric Soc Scot Trans 2:241–242

Boehm MJ, Hoitink HA (1992) Sustenance of microbial activity in potting mixes and its impact on the severity of Pythium root rot of Poinsettia. Phytopathology 82:259–264

Bohner K (1922) Vom Kohlkroof. Frankischer Kurier 8:111

Braselton JP (1995) Current status of the Plasmodiophorids. Crit Rev Microbiol 21:263–275

Broadbent D, Baker KF (1974) Behaviour of *Phytophthora cinnamomi* in soils suppressive and conducive to root rot. Austr J Agric Res 25:121–127

Brodmann A, Schuller A, Ludwig-Muller J, Aeschbacher RA, Wiemken A, Boller T, Wingler A (2002) Induction of trehalase in arabidopsis plants infected with the trehalose producing pathogen *Plasmodiophora brassicae*. Mol Plant-Microbe Int 15:693–700

Buczacki ST (1983) *Plasmodiophora*: an inter-relationship between biological and practical problems. In: Buczacki ST (ed) Zoosporic plant pathogens –a modern perspective. Academic, London, pp 161–192

Buhariwalla H, Greaves S, Magrath R, Mithen R (1995) Development of specific primers for the amplification of polymorphic DNA from the obligate root pathogen *Plasmodiophora brassicae*. Physiol Mol Plant Pathol 47:83–94

Bulman S, Siemens J, Ridgway HJ, Conner AJ (2006) Identification of genes from the obligate intracellular plant pathogen, *Plasmodiophora brassicae*. FEMS Microbiol Lett 264:198–204

Bulman S, Ridgway HJ, Eady C, Conner AJ (2007) Intron-rich gene structure in the intracellular plant parasite *Plasmodiophora brassicae*. Protist 158:423–433

Butcher DN, El-Tigani S, Ingram DS (1974) The roles of indole glucosinolates in the clubroot disease of the cruciferae. Physiol Plant Pathol 4:127–141

Campbell R (1989) Biological control of microbial plant pathogens. Cambridge University Press, London

Cao T, Tewari J, Strelkov SE (2007) Molecular detection of *Plasmodiophora brassicae*, causal agent of clubroot of crucifers, in plant and soil. Plant Dis 91:80–87

Cao T, Srivastava S, Rahman MH, Kav NNV, Hotte N, Deyholos MK, Strelkov SE (2008) Proteome-level changes in the roots of *Brassica napus* as a result of *Plasmodiophora brassicae* infection. Plant Sci 174:97–115

Chee HY, Kim WG, Cho WD, Jee HJ, Choi YC (1998) Detection of *Plasmodiophora brassicae* by using polymerase chain reaction. Korean J Plant Pathol 14:589–593

Cheng F, Mandakova T, Wu J, Xie Q, Lysak MA, Wang X (2013) Deciphering the diploid ancestral genome of the mesohexaploid *Brassica rapa*. Plant Cell 25:1541–1554

Chiang MS, Crete R (1970) Inheritance of clubroot resistance in cabbage (*Brassica oleracea* L. var. *capitata*). Can J Genet Cytol 12:253–256

Chiang MS, Crete R (1983) Transfer of resistance to race 2 of *Plasmodiophora brassicae* from *Brassica napus* to cabbage (*B. oleracea* sp. *capitata*). V. The inheritance of resistance. Euphytica 32:479–483

Chupp C (1934) A translation of *Plasmodiophora brassicae*–the cause of cabbage hernia by MS Woronin 1878, Phytopathological Classics No. 4, American Phytopathological Society, St. Paul, Minnesota, USA

Colhoun J (1952) Factors affecting the incidence of club root disease of Brassicae. Nature 169:21–22

Colhoun J (1953) A study of the epidemiology of club root disease of Brassicae. Ann Appl Biol 40:262–283

Colhoun J (1957) A technique for examining soil for the presence of *Plasmodiophora brassicae* Woron. Ann Appl Biol 45:559–565

Colhoun J (1958) Club root disease of crucifers caused by *Plasmodiophora brassicae* Woron. Commonwealth Mycological Institute, Kew

Cook WRI, Schwartz EJ (1930) The life-history, cytology and method of infection of *Plasmodiophora brassicae* Woron., the cause of finger and toe disease in cabbages and other crucifers. Phil Trans R Soc A 218:283–314

Craig MA, Dixon GR (1993) Alterations to the growth of *Plasmodiophora brassicae* Wor. (Clubroot) associated with boron. In: Proceedings of the 6th international congress of plant pathology, National Council for Research Canada, Montreal, Canada

Crute IR, Gray AR, Crisp P, Buczacki ST (1980) Variation in *Plasmodiophora brassicae* and resistance to clubroot disease in *Brassicas* and allied crops –a critical review. Plant Breed Abstr 50:91–103

Curtis J (1843) Observations on the natural history and economy of various insects affecting the turnip-crops, including the surface caterpillars, the turnip-gall weevil, and the dipterous flies and rove beetles infesting anbury. J Roy Agric Soc 4:100–138

Daval S, Belcour A, Gazengel K, Legrand L, Gouzy J, Cottret L, Lebreton L, Aigu Y, Mougel C, Manzanares-Dauleux MJ (2019) Computational analysis of the *Plasmodiophora brassicae* genome: mitochondrial sequence description and metabolic pathway database design. Genomics 111(6):1629–1640

Dekhuizen HM, Overeem JC (1971) The role of cytokinins in clubroot formation. Physiol Plant Pathol 1:151–162

Delorme R, Foissart N, Horvais R, Barrett P, Champagne G, Cheung WY, Landry BS, Renard M (1998) Characterisation of the radish introgression carrying the *Rfo* restorer gnes for the *Ogu*-INRA cytoplasmic male sterility in rapeseed (*Brassica napus* L). Theor Appl Gen 97:129–134

Deora A, Gossen BD, Walley F, McDonald MR (2011) Boron reduces development of clubroot in canola. Can J Plant Pathol 33:475–484

Deora A, Gossen BD, McDonald MR (2012) Infection and development of *Plasmodiophora brassicae* in resistant and susceptible canola cultivars. Can J Plant Pathol 34:239–247

Devos S, Vissenberg K, Verbelen JP, Prinsen E (2005) Infection of Chinese cabbage by *Plasmodiophora brassicae* leads to a stimulation of plant growth: impacts on cell wall metabolism and hormone balance. New Phytol 166:241–250

Devos S, Laukens K, Deckers P, Van Der Straeten D, Beeckman T, Inze D, Van Onckelen H, Witters E, Prinsen E (2006) A hormone and proteome approach to picturing the initial metabolic events during *Plasmodiophora brassicae* infection on *Arabidopsis*. Mol Plant Microbe Interact 19(12):1431–1443

Diederichsen E, Sacristan MD (1996) Disease response of resynthesized *Brassica napus* L. lines carrying different combinations of resistance to *Plasmodiophora brassicae*. Plant Breed 115:5–10

Diederichsen E, Wagenblatt B, Schallehn V, Deppe US, Sacristan MD (1996) Transfer of clubroot resistance from resynthesised *Brassica napus* into oilseed rape—identification of race-specific interactions with *Plasmodiophora brassicae*. Acta Hortic 407:423–430

Diederichsen E, Beckman J, Jorg Schondelmeier J, Dreyer F (2006) Genetics of clubroot resistance in *Brassica napus* Mendel. Acta Hortic 706:307–311

Diederichsen E, Frauen M, Linders EGA, Hatakeyama K, Hirai M (2009) Status and perspectives of clubroot resistance breeding in cruicifer crops. J Plant Growth Regul 28:264–281

Dixon GR (1980) Variation in *Plasmodiophora brassicae*. Ann Appl Biol 94:278–280

Dixon GR (1981) Vegetable crop diseases. Macmillan Publishers Ltd., London

Dixon GR (1996) Repression of the morphogenesis of *Plasmodiophora brassicae* Worn. by boron -a review. Acta Hortic 407:393–402

Dixon GR (2002) Interactions of soil nutrient environment, pathogenesis and host resistance. Plant Prot Sci 38:87–94

Dixon GR (2006) The biology of *Plasmodiphora brassicae* Worn., A review of recent advances. Acta Hortic 706:271–282

Dixon GR (2009a) *Plasmodiophora brassicae* in its environment. J Plant Growth Regul 28:212–228

Dixon GR (2009b) Husbandry –the sustainable means of controlling soil borne pathogens:-a synoptic review. Acta Hortic 817:233–242

Dixon GR (2009c) The occurrence and impact of *Plasmodiophora brassicae* and clubroot disease. J Plant Growth Regul 28:194–202

Dixon GR (2010) Calcium and pH as elements in a coherent control system for clubroot. Acta Hortic 867:151–155

Dixon GR (2012) Calcium cyanamide-a synoptic review of an environmentally benign fertiliser which enhances soil health. Acta Hortic 938:211–217

Dixon GR (2014) Clubroot (*Plasmodiophora brassicae* Woronin): an agricultural and biological challenge worldwide. Can J Plant Pathol 36(Supp 1):518

Dixon GR, Page LV (1998) Calcium and nitrogen eliciting alterations to growth and reproduction of *Plasmodiophora brassicae* (clubroot). Acta Hortic 459:343–349

Dixon GR, Tilston EL (2010) Soil-borne pathogens and their interactions with the soil environment. In: Dixon GR, Tilston EL (eds) Soil microbiology and sustainable crop production. Springer, Dordrecht, pp 197–271

Dixon GR, Webster MA (1988) Antagonistic effects of boron, calcium and pH on pathogenesis caused by *Plasmodiophora brassicae* (clubroot) - a review of recent work. Crop Res 28:83–95

Dixon GR, Wilson F (1984) Field evaluation of WL 105 305 (NK 483) for control of clubroot (*Plasmodiophora brassicae*). Tests Agrochem Cultivars No 5. Ann Appl Biol 104:34–35

Dixon GR, Craig MA, Burgess PJ, Thomas J (1994) MTF 651: a new soil-applied fungicide for the control of plasmodial fungi. In: Proceeding of brighton crop protect conf pests and diseases (session 5A-6), vol 2. Br Crop Prot Council, Farnham, pp 541–548

Donald EC, Porter I (2009) Integrated control of clubroot. J Plant Growth Regul 28:289–303

Donald EC, Lawrence JM, Porter IJ (2002) Evaluation of a fluorescent staining technique as an indicator of pathogenicity of resting spores of *Plasmodiophora brassicae*. Austr Plant Pathol 31:373–379

Donald EC, Jaudzems G, Porter IJ (2008) Pathology of cortical invasion by *Plasmodiophora brassicae* in clubroot resistant and susceptible *Brassica oleracea* hosts. Plant Pathol 57:201–209

Donoghue MTA, Keshavaiah C, Swamidatta SH, Spillane C (2011) Evolutionary origins of Brassicaceae specific genes in *Arabidopsis thaliana*. BMC Evol Biol 11:47

Einhorn G, Bochow H, Huber J, Krebs B (1991) Methodological studies to detect anatagomnists of the clubroot pathogen, *Plasmodiophora brassicae* Worn. Arch Phytopathol Plant Protect 27:205–208

Ellis W (1742) The modern husbandman, or, the practice of farming. 4 v. London

Faggian R, Strelkov SE (2009) Detection and measurement of *Plasmodiophora brassicae*. J Plant Growth Regul 28:282–287

Faggian R, Bulman SR, Lawrie AC, Porter IJ (1999) Specific polymerase chain reaction primers for the detection of *Plasmodiophora brassicae* in soil and water. Phytopathology 89:392–397

Farquharson J (1831) Report on the disease of Finger and Toe in Turnips. Highland Agric Soc Scot Trans 2:233–238

Fedotova T (1933) Contribution to the evolution of a method for the evaluation of soil infection with clubroot (*Plasmodiophora brassicae* Worn.). Trudjjw Zashchite Rastenii Ser 2(3):51–81

Feng J, Hwang R, Hwang SF, Strelkov SE, Gossen BD, Zhou Q, Peng G (2010) Molecular characterization of a serine protease Pro1 from *Plasmodiophora brassicae* that stimulates resting spore germination. Mol Plant Pathol 11:503–512

Feng LQ, Guoru X, Zichao M, Yixin W, Yueqiu H (2012) Analyses for the colonization ability of *Bacillus subtilis* XF-1 in the rhizosphere. Acta Phytophy Sin 39:425–430

Feng J, Cao T, Rennie DC, Strelkov SE, Hwang SF (2014) Host-parasite interactions in clubroot of crucifers. Can J Plant Pathol 36(Suppl 1):113–321

Figdore SS, Ferreira ME, Slocum MK, Williams PH (1993) Association of RFLP markers with trait loci affecting clubroot resistance and morphological characters in *Brassica oleracea* L. Euphytica 69:33–44

Garrett SD (1956) Biology of root invading fungi. Cambridge University Press, Cambridge

Gossen BD, Adhikari KKC, McDonald MR (2012) Effects of temperature on infection and subsequent development of clubroot under controlled conditions. Plant Pathol 61:593–599

Graf H, Faehling M, Siemens J (2004) Chromosome polymorphism of the obligate biotrophic parasite *Plasmodiophora brassicae*. J Phytopathol 152:86–91

Grandclement C, Thomas G (1996) Detection and analysis of QTL based on RAPD markers for polygenic resistance to *Plasmodiophora brassicae* Wor. in *Brassica oleracea* L. Theor Appl Genet 93:86–90

Guo YL (2013) Gene family evolution in green plants with emphasis on the origination and evolution of *Arabidopsis thaliana* genes. Plant J 73:941–951. https://doi.org/10.1111/tpj.12089

Gustafsson M, Falt AS (1986) Genetic studies on resistance to clubroot in *Brassica napus*. Ann Appl Biol 108:409–415

Hardham AR, Suzaki E, Perkin JL (1986) Monoclonal antibodies to isolate species and genus specific components on the surface of zoospores and cysts of the fungus *Phytophthora cinnamomi*. Can J Bot 64:311–325

Hasan MJ, Strelkov SE, Howard RJ, Rahman H (2012) Screening of *Brassica* germplasm for resistance to *Plasmodiophora brassicae* pathotypes prevalent in Canada for broadening diversity in clubroot resistance. Can J Plant Sci 92:501–515

Hasse I, May-DeMio LL, Lima Neto Vda C (2007) The effect of pre-plantation with medicinal plants in the *Plasmodiophora brassicae* control. Summa Phytopathol 33:74–79

Henderson P (1867) Gardening for profit; a guide to the successful cultivation of the market and family garden. 243 p, Illinois, USA, New York

Hirai M (2006) Genetic analysis of clubroot resistance in *Brassica rapa*. Breed Sci 56:223–229

Hirai M, Harada T, Kubo N, Tsukada M, Suwabe K, Matsumoto S (2004) A novel locus for clubroot resistance in *Brassica rapa* and its linkage markers. Theor Appl Genet 108:639–643

Hseith WH, Wang J (1986) Investigation on suppressive soils of clubroot crucifers in Taiwan. Plant Prot Bull 28:353–362

Hwang SF, Ahmed HU, Zhou Q, Strelkov SE, Gossen BD, Peng G, Turnbull GD (2011) Influence of cultivar resistance and inoculum density on root hair infection of canola (*Brassica napus*) by *Plasmodiophora brassicae*. Plant Pathol 60(5):820–829. https://doi.org/10.1111/j.1365-3059.2011.02457.x

Hwang SF, Strelkov SE, Feng J, Gossen BD, Howard RJ (2012) *Plasmodiophora brassicae*: a review of an emerging pathogen of the Canadian canola (*Brassica napus*) crop. Mol Plant Pathol 13:105–113. https://doi.org/10.1111/j.1364-3703.2011.00729.x

Ingram DS (1969) Growth of *Plasmodiophora brassicae* in host callus. J Gen Microbiol 56:455–467

Ingram DS, Tommerup IC (1972) The life history of *Plasmodiophora brassicae* Woron. Proc R Soc Lond Ser B 180:103–112

Ito S, Maehara T, Tanaka S, Kameya-Iwaki M, Yano S, Kishi F (1997) Cloning of a single-copy DNA sequence unique to *Plasmodiophora brassicae*. Physiol Mol Plant Pathol 50:289–300

Ito S, Ichinose H, Yanagi C, Tanaka S, Kameya-Iwaki M, Kishi F (1999a) Identification of an in planta-induced mRNA of *Plasmodiophora brassicae*. J Phytopathol 147:79–82

Ito S, Maehara T, Maruno E, Tanaka S, Kameya-Iwaki M, Kishi F (1999b) Development of a PCR-based assay for the detection of *Plasmodiophora brassicae* in soil. J Phytopathol 147:83–88

James RV, Williams PH (1980) Clubroot resistance and linkage in *Brassica campestris*. Phytopathology 70:776–779

Jaw-Fen W (1986) Studies on the suppressive factors and characteristics of suppressive soils of clubroot in crucifers. Plant Prot Bull 28:363–370

Jiang M, Dong X, Lang H, Pang W, Zhan Z, Li X, Piao Z (2018) Mining of *Brassica*-specific genes (BSGs) and their induction in different developmental stages and under *Plasmodiophora brassicae* stress in *Brassica rapa*. Int J Mol Sci 19(7):E2064. https://doi.org/10.3390/ijms19072064

Jones DR (1981) Differential pathogenicity of *Plasmodiophora brassicae* Worn. [dissertation]. University of Cambridge

Jones DR, Ingram DS, Dixon GR (1982a) Characterisation of isolates derived from single resting spores of *Plasmodiophora brassicae* and studies of their interactions. Plant Pathol 31:239–245

Jones DR, Ingram DS, Dixon GR (1982b) Factors affecting tests for differential pathogenicity in populations of *Plasmodiophora brassicae*. Plant Pathol 31:229–238

Jubault M, Hamon C, Gravot A, Lariagon C, Delorme R, Bouche-Reau A, Manzanzares-Dauleux MJ (2008) Differential regulation of root arginine catabolism and polyamine metabolism in clubroot-susceptible and partially resistant *Arabidopsis* genotypes. Plant Physiol 146:2008–2019

Kaczmarek J, Latunde-Dada AO, Irzykowski W, Cools HJ, Stonard JF, Brachaczek A, Jedryczka M (2014) Molecular screening for avirulence alleles *AvrLm1* and *AvrLm6* in airborne inoculum of *Leptosphaeria maculans* and winter oilseed rape (*Brassica napus*) plants from Poland and the UK. J Appl Genet 5:529–539

Kagale S, Robinson SJ, Nixon J, Xiao R, Huebert T, Condie J, Kessler D, Clarke WE, Edger PP, Links MG, Sharpe AG, Parkin IAP (2014) Polyploid evolution of the Brassicaceae during the Cenozoic Era. Plant Cell 26(7):2777–2791

Kageyama K, Asanoo T (2009) Life cycle of *Plasmodiophora brassicae*. J Plant Growth Regul 28:203–211

Karling JS (1968) The plasmodiophorales. Hafner Publishing Company, New York

Kavanagh JA, Williams PH (1981) Indole auxins in *Plasmodiophora* infected cabbage roots and hypocotyls. Mycol Res 77:125–129

Keen NT, Williams PH (1969) Translocation of sugars into infected cabbage tissues during clubroot development. Plant Physiol 44:748–754

Kindshoven J (1928) Entseuchhung des bodens und bekampfung der kohlhernie und kalkstickstoff. Mitt Deut Landw Gesell 43:522–523

Kowalski K, Bochow H (1996) Observations on the behaviour of resting spores of *Plasmodiophora brassicae* in the presence of cruciferous and non-cruciferous plant roots. Acta Hortic 407:419–422

Kuginuki Y, Hiroaki Y, Hirai M (1999) Variation in virulence of *Plasmodiophora brassicae* in Japan tested with clubroot-resistant cultivars of Chinese cabbage (*Brassica rapa* L. sp. *pekinensis*). Eur J Plant Pathol 105:327–332

Lahlali R, Peng G, McGregor L, Gossen BD, Hwang SF, McDonald M (2011) Mechanisms of the biofungicide Serenade (*Bacillus subtilis* QST713) in suppressing clubroot. Biocon Sci Technol 21:1351–1362

Lancereaux E (1869) A treatise on syphilis, historical and practical. Trans. by G Whitley, New Sydenham Society, London, vol 2, p 363

Landry BS, Hubert N, Crete R, Chiang MS, Lincoln SE, Etoh T (1992) A genetic map for *Brassica oleracea* based on RFLP markers detected with expressed DNA sequences and mapping of resistance gene to race 2 of *Plasmodiophora brassicae*. Genome 35:409–420

Lange L, Heide M, Hobolth L, Olson LW (1989) Serological detection of *Plasmodiophora brassicae* by dot immunobinding and visualization of the serological reaction by scanning electron microscopy. Phytopathology 79:1066–1071

Li L, Foster CM, Gan Q, Nettleton D, James MG, Myers AM, Wurtele ES (2009) Identification of the novel protein QQS as a component of the starch metabolic network in *Arabidopsis* leaves. Plant J 58:485–498

Li X, Mao Z, Wang Y, Wu Y, He Y, Long C (2012) ESI LC-MS and MS/MS characterization of antifungal cyclic lipopeptides produced by *Bacillus subtilis* XF-1. J Mol Microbiol Biotechnol 22:83–93

Li L, Zheng W, Zhu Y, Ye H, Tang B, Arendsee ZW, Jones D, Li R, Ortiz D, Zhao X, Du C, Nettleton D, Scott MP, Salas-Fernandez MG, Yin Y, Wurtele ES (2015) QQS orphan gene regulates carbon and nitrogen partitioning across species via NF-YC interactions. Proc Natl Acad Sci U S A 112:14734–14739

Lin H, Moghe G, Ouyang S, Iezzoni A, Shiu SH, Gu X, Buell CR (2010) Comparative analyses reveal distinct sets of lineage-specific genes within *Arabidopsis thaliana*. BMC Evol Biol 10:41

Ludwig-Müller J, Prinsen E, Rolfe S, Scholes J (2009) Metabolism and plant hormone action during the clubroot disease. J Plant Growth Regul 28:229–244

MacFarlane I (1952) Factors affecting the survival of *Plasmodiophora brassicae* Worn. in the soil and its assessment by a host test. Ann Appl Biol 39:239–256

Manzanares-Dauleux MJ, Delourme R, Baron F, Thomas G (2000) Mapping of one major gene and of QTL involved in resistance to clubroot in *Brassica napus*. Theor Appl Genet 101:885–891

Matsumoto E, Yasui C, Ohi M, Tsukada M (1998) Linkage analysis of RFLP markers for clubroot resistance and pigmentation in Chinese cabbage. Euphytica 104:79–86

Melville IE, Hawken RW (1967) Soil testing for club root in Devon and Cornwall. Plant Pathol 16:145–147

Moriguchi K, Kimizuka-Takagi C, Ishii K, Nomura K (1999) A genetic map based on RAPD, RFLP, isozyme, morphological markers and QTL analysis for clubroot resistance in *Brassica oleracea*. Breed Sci 49:257–265

Moxham SE, Buczacki ST (1983) Chemical composition of the resting spore wall of *Plasmodiophora brassicae*. Mycol Res 80:17–28

Myers DE, Campbell RN (1985) Lime and the control of clubroot of crucifers: effects of pH, calcium, magnesium and their interactions. Phytopathology 75:670–673

Nagaharu U (1935) Genome analysis with special reference to the experimental formation of *B. napus* and particular mode of fertilization. Jpn J Bot 7:389–452

Narisawa K, Shimura M, Usuki F, Fukuhara S, Hashiba T (2005) Effects of pathogen density, soil moisture and soil pH on biological control of clubroot of Chinese cabbage by *Heteroconium chaetospora*. Plant Dis 89:285–290

Neme R, Tautz D (2013) Phylogenetic patterns of emergence of new genes support a model of frequent de novo evolution. BMC Genomics 14:117

Neuhauser S, Kirchmair M, Gleason FH (2011) Ecological roles of the parasitic phytomyxids (Plasmodiophorids) in marine ecosystems –a review. Mar Freshw Res 62:365–371

Nomura K, Minegishi Y, Kimizuka-Takagi C, Fujioka T, Moriguchi K, Shishido R, Ikehashi H (2005) Evaluation of F2 and F3 plants introgressed with QTL for clubroot resistance in cabbage developed by using SCAR markers. Plant Breed 124:371–375

Orihara S, Yamamoto T (1998) Detection of resting spores of *Plasmodiophora brassicae* from soil and plant tissues by enzyme immunoassay. Ann Phytopathol Soc 64:569–573

Page LV (2001) Studies of components for a potential integrated control system for *Plasmodiophora brassicae*. [dissertation]. University of Strathclyde, Glasgow

Piao ZY, Deng YQ, Choi SR, Park YJ, Lim YP (2004) SCAR and CAPS mapping of CRb, a gene conferring resistance to *Plasmodiophora brassicae* in Chinese cabbage (*Brassica rapa* ssp. *pekinensis*). Theor Appl Genet 108:1458–1465

Piao Z, Ramchiary N, Lim YP (2009) Genetics of clubroot resistance in *Brassica* species. J Plant Growth Regul 28:252–264

Rakow G (2004) Species origin and economic importance of *Brassica*. In: Pua EC, Douglas CJ (eds) Biotechnology in agriculture and forestry, vol 54, 1st edn. Springer, Berlin, pp 3–11

Ravn FK (1908) Kaalbroksvampen. Landbr Planteavl 15:5271–5620

Ren L, Jia JG, Li M, Liu F, Cheng YG, Zhang MH, Chen KR, Fang XP (2012) Distribution of rapeseed clubroot disease in Hubei province and evaluation of yield loss. Agric Sci Technol 13 (4):775–777

Ren L, Fang X P, Sun C C, Chen K R, Liu F, Li M, Xu L (2014) First Report of Clubroot on *Capsella bursa-pastoris* caused by *Plasmodiophora brassicae* in Sichuan Province of China. https://doi.org/10.1094/PDIS-04-13-0395-PDN

Rennie DC, Manolii VP, Cao T, Hwang SF, Howard RJ, Strelkov SE (2011) Direct evidence of surface infestation of seeds and tubers by *Plasmodiophora brassicae* and quantification of spore loads. Plant Pathol 60:811–819

Rocherieux J, Glory P, Giboulot A, Boury S, Barbeyron G, Thomas G, Manzanares-Dauleux MJ (2004) Isolate-specific and broad-spectrum QTLs are involved in the control of clubroot in *Brassica oleracea*. Theor Appl Genet 108:1555–1563

Rolfe SA, Strelkov SE, Links MG, Clarke WE, Robinson SJ, Djavaheri M, Malinowski R, Haddadi P, Kagale S, Parkin IAP, Taheri A, Borhan MH (2016) The compact genome of the plant pathogen *Plasmodiophora brassicae* is adapted to intracellular interactions with host *Brassica* spp. BMC Genomics 17:272

Rouxel F (1984) Studies on soil receptivity to clubroot. In: Methods for studying the biological control of diseases caused by soil borne pathogens. Rothamsted Experiment Station, London

Rouxel F (1991) Natural suppressiveness of soils to plant diseases. In: Beemster ABR, Bolen GL, Gerlag M, Ruissen MA, Tempel B, Schippers A (eds) Biotic interactions and soil borne diseases. Elsevier, Amsterdam

Rouxel F, Briard M, Lejeune B (1988) Studies of soil receptiveness to clubroot caused by Plasmodiophora brassicae: experiments on responses of a series of vegetable soils in Brittany. In: Progress on pest management in field vegetables. INRA, Rennes, pp 145–152

Samuel G, Garrett SD (1945) The infected root-hair count for estimating the activity of *Plasmodiophora brassicae* Woron. in the soil. Ann Appl Biol 32:96–101

Savage SD, Sall MA (1981) Radio immunosorbent assay for *Botrytis cinerea*. Phytopathology 71:411–415

Schuller A, Ludwig-Muller J (2006) A family of auxin conjugate hydrolases from *Brassica rapa*: characterization and expression during clubroot disease. New Phytol 171:145–158

Schwelm A, Fogelqvist J, Knaust A, Jülke S, Lilja T, Bonillarosso G, Karlsson M, Shevchenko A, Dhandapani V, Choi SR, Kim HG, Park JY, Lim YP, Ludwig-Muller J, Dixelius C (2015) The *Plasmodiophora brassicae* genome reveals insights in its life cycle and ancestry of chitin synthases. Sci Rep 5:11153

Schwelm A, Dixelius C, Ludwig-Muller J (2016) New kid on the block – the clubroot pathogen genome moves the plasmodiophorids into the genomic era. Eur J Plant Pathol 145:531–542

Siemens J, Keller U, Sarx J, Kunz S, Schuller A, Nagel W, Schmülling T, Parniske M, Ludwig-Müller J (2006) Transcriptome analysis of *Arabidopsis* clubroots and disease resistance of cytokinin oxidase/dehydrogenase gene overexpressing plants indicate a key role for cytokinin in disease development. Mol Plant-Microbe Interact 19:480–494

Siemens J, Bulman S, Rehn F, Sundelin T (2009) Molecular biology of *Plasmodiophora brassicae*. J Plant Growth Regul 28:245–251

Slingerland MV (1894) The cabbage maggot with notes on the onion maggot and allied insects. N Y Cornell Agric Exp 78:481–577

Sorauer P (1874) Handbijch der pflanzenkrankheiten fur tandwirthf. Gartner, Irnd Forstleute, p 406

Spanu PD, Abbot JC, Amselem J, Burgis TA, Soanes DM, Stüber K, van Themaat EVL, Brown JKM, Butcher SA, Gurr SJ, Lebrun MH, Ridout CJ, Schulze-Lefert P, Talbot NJ, Ahmadinejad N, Ametz C, Barton GR, Benjdia M, Bidzinski P, Bindschedler LV, Both M, Brewer MT, Cadle-Davidson L, Cadle-Davidson MM, Collemare J, Cramer R, Frenkel O, Godfrey D, Harriman J, Hoede C, King BC, Klages S, Kleemann J, Knoll D, Koti PS, Kreplak J, López-Ruiz FJ, Lu X, Maekawa T, Mahanil S, Micali C, Milgroom MG, Montana G, Noir S, O'Connel RJ, Oberhaensli S, Parlange F, Pendersen C, Quesneville H, Reinhardt R, Rott M, Sacristán S, Schmidt SM, Schön M, Skaminoti P, Sommer H, Stephens A, Takahra H, Thordal-Christensen H, Vigouroux M, Weßling R, Wicker T, Panstruga R (2010) Genome expansion and gene loss in powdery mildew fungi reveal tradeoffs in extreme parasitism. Science 330:1543–1546

Sundelin T (2008) Identification of expressed *Plasmodiophora brassicae* genes during plant infection and molecular diagnosis of plant pathogens. PhD Thesis, University of Copenhagen, Copenhagen

Sundelin T, Christensen CB, Larsen J, Moller K, Lübeck M, Bodker L, Jensen B (2010) In planta quantification of *Plasmodiophora brassicae* using signature fatty acids and real-time PCR. Plant Dis 94:432–438

Suwabe K, Tsukazaki H, Iketani H, Hatakeyama K, Fujimura M, Nunome T, Fukuoka H, Matsumoto S, Hirai M (2003) Identification of two loci for resistance to clubroot (*Plasmodiophora brassicae* Woronin) in *Brassica rapa* L. Theor Appl Genet 107:997–1002

Suwabe K, Tsukazaki H, Iketani H, Hatakeyama K, Kondo M, Fujimura M, Nunome T, Fukuoka H, Hirai M, Matsumoto S (2006) Simple sequence repeat-based comparative genomics between *Brassica rapa* and *Arabidopsis thaliana*: the genetic origin of clubroot resistance. Genetics 173:309–319

Takahashi K (1991) Correlation between fluorescent staining reaction and germination in resting spores of *Plasmodiophora brassicae*. Ann Phytopathol Soc Jpn 57:160–164

Takahashi K, Yamaguchi T (1987) An improved method for estimating the number of resting spores of *Plasmodiophora brassicae* in soil. Ann Phytopathol Soc 53:507–515

Takahashi K, Yamaguchi T (1988) A method for assessing the pathogenic ability of resting spores of *Plasmodiophora brassicae* by fluorescence microscopy. Ann Phytopathol Soc 54:466–475

Takahashi K, Yamaguchi T (1989) Assessment of pathogenicity of resting spores of *Plasmodiophora brassicae* in soil by fluorescence microscopy. Ann Phytopathol Soc 55:621–628

Tanaka S, Mido H, Ito S (2006) Colonization by two isolates of *Plasmodiophora brassicae* with differing pathogenicity on a clubroot-resistant cultivar of Chinese cabbage (*Brassica rapa* L. subsp. *pekinensis*). J Gen Plant Pathol 72:205–209

Thuma BA, Rowe RC, Madden LV (1983) Relationships of soil temperature and moisture to clubroot (*Plasmodiophora brassicae*) severity on radish in organic soil. Plant Dis 67:758–762

Tisdale WB (1923) Influence of soil temperature and roil moisture upon the Fusarium disease in cabbage seedlings. J Agric Res 24:55–86

Tommerup IC, Ingram DS (1971) The life-cycle of *Plasmodiophora brassicae* Worn. in *Brassica* tissue cultures and intact roots. New Phytol 70:327–332

Toxopeus H, Janssen AMP (1975) Clubroot resistance in turnip-II. The 'slurry' screening method and clubroot races in the Netherlands. Euphytica 1975:751–755

Toxopeus H, Oost EH, Reuling G (1984) Current aspects of the taxonomy of cultivated *Brassica* species. Cruciferae Newslett 9:55–58

Toxopeus H, Dixon GR, Mattusch P (1986) Physiological specialisation in *Plasmodiophora brassicae* –an analysis by international experimentation. Mycol Res 87:279–287

Tsushima S, Murakami H, Akimoto T, Katahira M, Kuroyanagi Y, Shishido Y (2010) A practical estimating method of the dose-response curve between inoculum density of *Plasmodiophora brassicae* and the disease severity for long-term IPM strategies. JARQ 44:383–390

Vigier B, Chiang MS, Hume DJ (1989) Source of resistance to clubroot (*Plasmodiophora brassicae* Wor.) in triazine-resistant spring canola (rapeseed). Can Plant Dis Surv 69:113–115

Voorips RE, Visser DL (1993) Examination of resistance to clubroot in accessions of *Brassica oleracea* using a glasshouse seedling test. Eur J Plant Pathol 99:269–276

Voorrips RE (1996) Production, characterization and interaction of single-spore isolates of *Plasmodiophora brassicae*. Eur J Plant Pathol 102:377–383

Voorrips RE, Jongerius MC, Kanne HJ (1997) Mapping of two genes for resistance to clubroot (*Plasmodiophora brassicae*) in a population of doubled haploid lines of *Brassica oleracea* by means of RFLP and AFLP markers. Theor Appl Genet 94:75–82

Wakeham AJ, White JG (1996) Serological detection in soil of *Plasmodiophora brassicae* resting spores. Physiol Mol Plant Pathol 48:289–303

Walker JC, Larson RH (1960) Development of the first clubroot resistant cabbage variety. Wisconsin Agric Exp State Bull 547:12–16

Wallenhammar AC (1996) Prevalence of *Plasmodiophora brassicae* in a spring oil seed rape growing area in central Sweden and factors influencing soil infestation levels. Plant Pathol 45:710–719

Wallenhammar AC, Arwidsson O (2001) Detection of *Plasmodiophora brassicae* by PCR in naturally infested soils. Eur J Plant Pathol 107:313–321

Wallenhammar AC, Johnsson L, Gerhardson B (2000) Agronomic performance of partly clubroot-resistant spring oilseed turnip rape lines. J Phytopathol 148:495–499

Wallenhammar AC, Almquist C, Soderstrom M, Jonsson A (2012) In field distribution of *Plasmodiophora brassicae* measured using quantitative real-time PCR. Plant Pathol 61:16–28

Wang Y, Wu F, Bai J, He Y (2014) *BrpSPL9* (*Brassica rapa* ssp. *pekinensis* SPL9) controls the earliness of heading time in Chinese cabbage. Plant Biotechnol J 12:312–321

Webster MA (1986) pH and nutritional effects on infection by *Plasmodiophora brassicae* Wor. and clubroot symptoms [dissertation]. Scotland: University of Aberdeen

Webster MA, Dixon GR (1991a) Boron, pH and inoculum concentration influencing colonisation by *Plasmodiophora brassicae*. Mycol Res 95:74–79

Webster MA, Dixon GR (1991b) Calcium, pH and inoculum concentration influencing colonisation by *Plasmodiophora brassicae*. Mycol Res 95:64–73

Wellman FL (1930) Clubroot of crucifers. Technical bulletin no. 181, United States Department of Agriculture Washington, DC
Williams PH (1966) Acytological study of hypertrophy in clubroot of cabbage. Phytopathology 56:521–524
Williams PH, Aist SJ, Aist JR (1971) Response of cabbage root hairs to infection by *Plasmodiophora brassicae*. Can J Bot 49:41–47
Wit F, van de Weg M (1964) Clubroot resistance in turnips (*Brassica campestris* L.). I. Physiological races of the parasite and their identification in mixtures. Euphytica 13:9–18
Workhu Y, Gerhardson B (1996) Suppressiveness to clubroot, pea root rot and fusarium wilt in Swedish soils. J Phytopathol 144:143–146
Woronin M (1878) *Plasmodiophora brassicae* Urheber der Kohlpflanzen-Hernie. Jahrb Wiss Bot 11:548–574
Yadeta KA, Valkenburg DJ, Hanemian M, Marco Y, Thomma BPHJ (2014) The Brassicaceae-specific EWR1 gene provides resistance to vascular wilt pathogens. PLoS ONE 9:e88230
Yang YF, Hou S, Cui GH, Chen SL, Wei JH, Huang LQ (2010) Characterization of reference genes for quantitative real-time PCR analysis in various tissues of Salvia miltiorrhiza. Mol Biol Rep 37:507–513
Yoshikawa H (1993) Studies on breeding of clubroot resistance in cole crops. Bull Natl Res Inst 7:1–165
Zhang CW, Wei YP, Xiao D, Gao LW, Lyu SW, Hou XL, Bouuema G (2016) Transcriptomic and proteomic analyses provide new insights into the regulation mechanism of low-temperature-induced leafy head formation in Chinese cabbage. J Proteome 144:1–10

The Disease: Clubroot

2

Abstract

The disease is known by the name clubroot due to its characteristics symptoms of infected hypertrophied root cortex cells leading to the formation and development of root galls. The importance of the disease can be judged by its distribution in more than 88 countries of the world infecting 330 genera and more than 3700 species of plants, with more than 60 vernacular names causing 10–100 damage to vegetables and oil yielding crops. The pathogen can survive in soil for more than 20 years in the form of resting spores in the soil, and is difficult to eradicate. The disease which was discovered on turnip as early as about 284 years ago (1742–1744) in England as a serious and contagious form has now been spread on all the continents of the world wherever cruciferous vegetation are available. Clubroot is a potential threat to cruciferous crops covering 1384.8 m ha areas with a production of 147.69 mt at global level which is 2.69% of total cropped area of the world. There is direct and strong relationship between disease incidences, degree of soil infestation, and crop yield losses. Host resistance helps to reduce yield losses. All cultivated *Brassica* vegetables and oil yielding crops including several cruciferous weeds are susceptible to this disease. The pathogen can survive in the root cortex even on non-crucifers plants also. Symptoms of the disease may vary depending upon the growth stage of the crop when infection occurs. The characteristics symptoms of the disease appear on the roots as spindle like shredded, knobby or club shaped swellings. The clubroot has been assessed in the form of incidence and intensity using different rating scales, descriptive as well as pictorial to assess yield losses, host resistance, and efficacy of technology. The disease is usually monocyclic, and zoospore movement in the soil is limited. However, any activity that results in transport of soil from one point to another has the potential to disseminate clubroot from one area to another in the same vicinity and even to long distances carrying resting spores of the pathogen.

G. S. Saharan et al., *Clubroot Disease of Crucifers*,
https://doi.org/10.1007/978-981-16-2133-8_2

Keywords

Disease · History · Introduction · Geographical distribution · Economic importance · Host range · Symptomatology · Disease assessment · Dissemination

2.1 Introduction

The vernacular nomenclature of the disease indicates that it had a well-established importance in agriculture and horticulture since at least the middle Ages onwards in Europe and probably earlier. Subsequently the pathogen probably spread worldwide as a result of transfer on and in fodder taken by colonists as livestock feed. It is a moot point, however, whether there was much earlier spread of *Plasmodiophora brassicae* into China, and subsequently Japan as *Brassica rapa* (Chinese cabbage and many variants) colonized those lands in archeological time. Clubroot disease has been known on cultivated *Brassicas* in Europe at least as far back as the thirteenth century, and quite possibly much earlier to Roman times, and has been ascribed to many causes. During early nineteenth century, in view of the Scottish disease resulted from "*unsatisfactory soil conditions or unbalanced fertilizer practices*," in the absence of microbiological knowledge. Recent considerations of the ecology of the organism make an association of epidemics of *P. brassicae* with unbalanced host nutrition a quite tenable view. In 1873, the Russian biologist M. S. Woronin commenced studies of the disease in St. Petersburg, and five years later identified *Plasmodiophora brassicae* Woronin as the causal agent of clubroot disease (Woronin 1878). Woronin's interest was quite possibly stimulated by the ravages of clubroot disease in cabbage crops in that part of Russia. More than a century of subsequent studies has unraveled much concerning the biology, and host–pathogen relationships of *P. brassicae*, and the Brassicaceae (Dixon 2006, 2009a). The disease clubroot of crucifers known by more than 60 different vernacular names worldwide has very wide host range (>3700 genera) with damage ranging from 10–100%. Its pathogen (*Plasmodiophora brassicae*) is an obligate parasite with very complex life cycle partitioning into three stages: survival in the soil for more than 20 years as resting spores, primary infection of root hairs, and secondary infection within root cortex cells. The production of primary and secondary zoospores and their infection nature make the process of pathogenesis into more complicated. The intracellular development of plasmodia in the root cortex cells induces hypertrophy and hyperplasia leading to the formation and development of root galls known as clubroot.

2.2 Disease Synonymous

The disease is known by several common names in different countries of the world (Table 2.1). In countries such as Belgium and Germany, the multiplicity of names probably indicates a considerable degree of regionality and localization reflecting the distinctly separate dialects used by farmers and growers living in historically

Table 2.1 Common names for clubroot disease (Karling 1968; Dixon 2009b)

Country	Common names
Australia, New Zealand	Clubroot
Belgium	Bosse, gross pied, kanker, klinger, knobbel, knoll, knoop, knotze, knuist, kwab, kwabbe, kwabbel, kwabbeziekte, kwadevoeten, oolen, tol, wratten, verrue
Denmark	Kaalbrok, kaolbrok
Finland	Mohojuuri
France	Hernie du chou, gros pied, maladie digitoire, maladie du chou
Germany, Switzerland, Austria	Fingerkrankheit, galle, herniekrankheit, huas, kelch, klumpenfuss, knotesucht, kohlhernie, kohlkropf, krof des kohles, kropfkrankheit des kohles, kuss, nolle
Great Britain	Clubroot, finger-and-toe, anbury, banbury, clubbing
Italy	Ernia, mal degosso dei cavoli, tuberculosi dei cavoli
Netherlands	Knolvoet
North America	Clubroot, club foot, clump foot, finger-and-toe
Norway	Klumprot
Russia	Kapoustnaja
South Africa	Clubroot, finger- and- toe, club foot, dik voet
Spain	Hernia, hernia de la col, potra
Sweden	Klumprotsjuka

separate states. The naming also reflects the relative importance of the crops affected. In northern Europe, cabbage was the major crop grown, and hence diseased by clubroot, whereas further in southern France, cauliflower was of greater importance. Colonists from Europe took their names for this disease with them to the New World, and at the same time the pathogen was exported in animal fodder such as swedes (*Brassica napus*) and turnips (*B. rapa*) used as nutrition for their beasts. It is possible that *P. brassicae* reached China and Japan much earlier as ancestral forms of *B. rapa* moved from southeastern Europe, establishing Chinese cabbage and all its associated variant forms (Dixon 2007). In southern Europe the disease was viewed as a form of plant "*hernia*," whereas further north as "*club foot*," a term exported by German and Scandinavian colonists with their crops. In recent years, the disease has generally been referred to as "*clubroot*" as opposed to "*club foot*." This reflects a desire for uniformity in the nomenclature of diseases stemming from European Union rules relating to crop cultivar assessment.

2.3 History of Clubroot Introduction and Distribution

The first historical mention of the disease is by Ellis (1742) who observed in England on turnip as a serious and contagious disease in 1736. In Scotland from 1829 to 1831, Farquharson (1831), Abbay (1831), and Birnie (1831) described the disease. Henderson (1867) made observations on clubroot in the North eastern parts of the USA. Ravn (1908) wrote of the history of the disease before Woronin's final description of the nature of the causal organism as *P. brassicae*. In USA, it was reported in 36 states (Wellman 1930).Kunkel (1918) reviewed the work done on *P. brassicae* infection process from 1878 to 1918. In a technical Bulletin No. 181 Published by USDA, Wellman (1930) covered all details of clubroot of crucifers since its first report in 1742–1744 by Ellis. Clubroot was first described in Russia in the fourteenth century (Woronin 1878), and has since spread through eastern and northern Europe, including Finland (Rastas et al. 2012), Poland (Wesołowska 2014), Ukraine (Jedryczka et al. 2014), and Germany (Diederichsen et al. 2014). Clubroot may have been carried to North America together with the large influx of immigrants from eastern Europe that arrived in the 1800s (Minogue 2013). However, clubroot was not detected on oilseed rape (canola) in western Canada until 2003 (Tewari et al. 2005).

Information concerning how *P. brassicae* arrived in Latin America is limited. However, a report of clubroot like symptoms on cabbage in Spain by Diaz de Isla in 1539 (Howard et al. 2010; Dixon 2014) indicates that the pathogen was probably present in Spain during the period of Spanish colonization of Latin America. It is possible or even probable that the introduction of *P. brassicae* to Latin America coincides with the introduction of cruciferous crops from Spain. Alternatively, more than 13 million Europeans, mainly Germans, Scandinavians, and eastern Europeans, migrated to Latin America between 1879 and 1930, and probably carried the pathogen with them in infested food and fodder (Sanchez-Alonso 2007).

In Latin America, most research on clubroot has focused on cruciferous vegetables (Garcıa et al. 2006; Botero-Ramırez et al. 2015; Telenchana 2015). The Spanish for cabbage is *col* or *repollo* and clubroot is known in South America as *potra de la col* or *hernia de la col*, which translates as "*cabbage tumour*." *Plasmodiophora brassicae* has been reported in Mexico (Romero-Cova 1990; Robles Sanchez 1991), in Costa Rica on broccoli and cabbage (Cortes 2015), in Guatemala on broccoli (Hernandez 2002), in Bolivia on cauliflower (Guerrero 2016), in Venezuela on broccoli (Garcıa et al. 2006; Labrador-Morales et al. 2014), in Colombia on broccoli, cauliflower, and cabbage (Torres 1972; Castillo and Guerrero 2008; Botero-Ramırez et al. 2015), in Brazil on cauliflower, Chinese cabbage, broccoli, arugula, and cabbage (Lima et al. 2004; Cruz et al. 2008), in Ecuador on broccoli (Sanchez-Alonso 2013; Telenchana 2015), and in Peru (Pachari 2011), and Chile on canola (Galdames et al. 2014). Few studies on clubroot from each of these countries have been published, often as a student thesis or in Spanish or Portuguese. In addition, *P. brassicae* has been reported in Guatemala, Puerto Rico, Trinidad and Tobago, Argentina and Guyana, as identified in the European and Mediterranean Plant Protection Organization database (EPPO 2019; Dixon 2009a, b). The severity and economic losses caused by *P. brassicae* in the region are not available through online sources.

2.4 Geographical Distribution

The disease is likely to be distributed throughout the world wherever cruciferous vegetation is available. However, its published records on occurrence, intensity, and distribution are very limited (Table 2.2). At present, it is reported to be distributed in more than 88 countries of the world. In some of the countries, it has spread widely covering large areas with significant yield losses in cruciferous vegetables and oil yielding crops. Some of the countries have not paid required attention on this disease which may become a threat to the cruciferous crops in the near future. Above all, it is such a disease which cannot be eradicated once introduced in an area. Its management is very essential right from very early stages of infection.

The published record is available only from the countries of continents at Africa, Angola; South Africa (Natal, S.W. Cape); Asia, Brunei, China (E), Hong Kong, India (general), Israel, Japan, Korea, Peninsular Malaysia, Philippines, Sri Lanka, Taiwan, Turkey; USSR (Omsk, Irkutsk, Kemerovo, Yakutsk Republic, Siberia); Australasia and Oceania, Australia, Hawaii, New Zealand; Europe, Austria, Belgium, Britain and Northern Ireland (Jersey) (Shetland Islands), Bulgaria, Czechoslovakia, Denmark (Faeroe Islands), Finland, France, Germany, Greece, Hungary, Iceland, Irish Republic, Italy (Sardinia), Netherlands, Norway, Poland, Portugal (Azores), Romania, Spain, Sweden, Switzerland; USSR (Byelorussia, Estonia, Latvia, Lithuania, Russia), Yugoslavia; NORTH AMERICA, Canada (general); Mexico; USA (general incl. Alaska); Central America and West Indies, Puerto Rico, Trinidad; South America, Argentina (Buenos Aires), Brazil (Rio Grande do sul, Sao Paulo), Chile, Guyana, Venezuela (Table 2.2). The distribution given in Table 2.2 is based on all the information available to date. When several references are cited, they may give conflicting information on the status. Further, details may be available for individual references in Table 2.2 and details section which can be selected by going to generate report of its presence. The pictorial presentation of the clubroot in the world among the various continents revealed that the disease is prevalent in almost all the parts of the world (Fig. 2.1).

2.4.1 Geographical Distribution of Clubroot in China

Before China's Liberation in 1949, little information about clubroot in China was available. There were, however, documented discoveries of clubroot being found in Taiwan in 1912 and 1936, as well as Fujian Province (Fig. 2.2) in 1947 (Wang 1962). In 1953, clubroot was recognized as one of the important quarantine targets in the first National Plant Quarantine Conference (Wang 1962). The frequency of reports on clubroot increased substantially in the 1950s–1960s as agricultural activity increased in order to feed the burgeoning populations serving the Socialist Industrialization. Clubroot disease was observed in 1950 in Nanchang City, Jiangxi Province, and 5 years later, the disease had spread to almost all areas of Jiangxi Province, with the general incidence above 70% (Huang et al. 1955). Clubroot prevailed in Zhejiang Province, including Zhuji, Huangyan, Yongjia, and Jinhua

Table 2.2 Geographical distribution of clubroot of crucifers (Anon 1987, 1996; Buczacki 1979; Dixon 2009a, b; CABI/EPPO 2011; EPPO 2014; IMI Herbarium 2019)

Continent/country/region	Distribution	Reference
Asia		
Brunei Darussalam	Present	CABI/EPPO (2011), EPPO (2014)
China	Present	CABI/EPPO (2011), EPPO (2014)
– Anhui	Present	CABI/EPPO (2011), EPPO (2014)
– Beijing	Present	Chai et al. (2014)
– Fujian	Present	CABI/EPPO (2011), EPPO (2014)
– Gansu	Present	CABI/EPPO (2011), EPPO (2014)
– Guangdong	Present	CABI/EPPO (2011), EPPO (2014)
– Guangxi	Present	CABI/EPPO (2011), EPPO (2014)
– Hong Kong	Present	CABI/EPPO (2011), EPPO (2014)
– Hubei	Present	CABI/EPPO (2011), EPPO (2014)
– Hunan	Present	CABI/EPPO (2011), EPPO (2014)
– Jiangsu	Present	CABI/EPPO (2011), EPPO (2014)
– Jiangxi	Present	CABI/EPPO (2011), EPPO (2014)
– Jilin	Present	CABI/EPPO (2011)
– Liaoning	Present	CABI/EPPO (2011)
– Shandong	Present	CABI/EPPO (2011)
– Sichuan	Present	Wang et al. (2002); CABI/EPPO (2011); Ren et al. (2014)
– Yunnan	Present	CABI/EPPO (2011), EPPO (2014)
– Zhejiang	Present	CABI/EPPO (2011), EPPO (2014)
India	Widespread	CABI/EPPO (2011), EPPO (2014)
– Manipur	Present	Dixon (2009b); Bhattacharya et al. (2014)
– Meghalaya	Present	Dixon (2009b); Bhattacharya et al. (2014)
– Nagaland	Present	Dixon (2009b); Bhattacharya et al. (2014)
– Orissa	Present	Dixon (2009b); Bhattacharya et al. (2014)
– Sikkim	Present	Dixon (2009b); Bhattacharya et al. (2014)
– Tamil Nadu	Present	Loganathan et al. (2001); CABI/EPPO (2011)
– West Bengal	Present	Sen (2005); CABI/EPPO (2011)
Indonesia	Present	Yamada et al. (2004); CABI/EPPO (2011)
– Java	Present	CABI/EPPO (2011)
Iraq	Present	CABI/EPPO (2011)
Iran	Present	Dixon (2009a, b)
Israel	Present	CABI/EPPO (2011); EPPO (2014)
Japan	Present	Yoshikawa and Buczacki (1978); CABI/EPPO (2011); EPPO (2014)
– Hokkaido	Present	CABI/EPPO (2011)
– Honshu	Present	CABI/EPPO (2011)
– Kyushu	Present	CABI/EPPO (2011)
– Ryukyu Archipelago	Present	CABI/EPPO (2011)
Korea	Present	Kim et al. (2009)
– Gyeonggi	Present	Kim et al. (2009)

(continued)

Table 2.2 (continued)

Continent/country/region	Distribution	Reference
Korea, DPR	Present	CABI/EPPO (2011); EPPO (2014)
Korea, Republic of	Present	CABI/EPPO (2011); EPPO (2014)
Kyrgyzstan	Present	CABI/EPPO (2011)
Malaysia	Present	CABI/EPPO (2011); EPPO (2014)
– Peninsular Malaysia	Present	IMI Herbarium, undated; CABI/EPPO (2011); EPPO (2014)
Nepal	Widespread	CABI/EPPO (2011)
– Bhaktapur	Present	Timila et al. (2008)
– Kathmandu	Present	Timila et al. (2008)
– Lalitpur	Present	Timila et al. (2008)
– Palung valley	Present	Timila et al. (2008)
Pakistan	Present	Muhammad and Khan (2008); CABI/EPPO (2011)
Philippines	Present	CABI/EPPO (2011), EPPO (2014)
Singapore	Eradicated	AVA (2001)
Sri Lanka	Present	CABI/EPPO (2011), EPPO (2014)
Taiwan	Present	CABI/EPPO (2011), EPPO (2014)
Thailand	Present	Dixon (2009a, b)
Turkey	Present	CABI/EPPO (2011); EPPO (2014)
Vietnam	Present	Dixon (2009a, b)
Africa		
Angola	Present	CABI/EPPO (2011), EPPO (2014)
Malawi	Present	Mazuma et al. (2008); CABI/EPPO (2011)
Sao Tome and Principe	Present	CABI/EPPO (2011)
South Africa	Present	CABI/EPPO (2011); EPPO (2014)
Uganda	Present	Dixon (2009a, b)
Zambia	Present	Dixon (2009a, b)
North America		
Canada	Widespread	CABI/EPPO (2011), EPPO (2014)
– Alberta	Present	Tewari et al. (2004), CABI/EPPO (2011), EPPO (2014)
– Atlantic Provinces	Present	Dixon (2009a, b)
– British Columbia	Present	Rimmer et al. (2003), Xue et al. (2008), CABI/EPPO (2011)
– Manitoba	Present	CABI/EPPO (2011), EPPO (2014)
– New Brunswick	Present	CABI/EPPO (2011)
– Newfoundland and Labrador	Present	CABI/EPPO (2011)
– Nova Scotia	Present	CABI/EPPO (2011)
– Ontario	Present	Xue et al. (2008), CABI/EPPO (2011)
– Ottawa	Present	Dixon (2009a, b)
– Prince Edward Island	Present	CABI/EPPO (2011)
– Quebec	Present	Pageau et al. (2006), CABI/EPPO (2011)

(continued)

Table 2.2 (continued)

Continent/country/region	Distribution	Reference
– Saskatchewan	Present	CABI/EPPO (2011), EPPO (2014)
Mexico	Present	CABI/EPPO (2011), EPPO (2014)
USA	Widespread	CABI/EPPO (2011); EPPO (2014)
– Alaska	Present	CABI/EPPO (2011), EPPO (2014)
– California	Present	CABI/EPPO (2011)
– Colorado	Present	CABI/EPPO (2011)
– Connecticut	Present	CABI/EPPO (2011)
– Delaware	Present	CABI/EPPO (2011)
– Florida	Present	CABI/EPPO (2011)
– Hawaii	Present	CABI/EPPO (2011), EPPO (2014)
– Illinois	Present	CABI/EPPO (2011)
– Iowa	Present	CABI/EPPO (2011)
– Louisiana	Present	CABI/EPPO (2011)
– Maine	Present	CABI/EPPO (2011), EPPO (2014)
– Maryland	Present	CABI/EPPO (2011)
– Massachusetts	Present	Porth et al. (2003), publ. 2003; CABI/EPPO (2011)
– Michigan	Present	CABI/EPPO (2011)
– Minnesota	Present	CABI/EPPO (2011)
– Missouri	Present	CABI/EPPO (2011)
– New Hampshire	Present	CABI/EPPO (2011)
– New Jersey	Present	CABI/EPPO (2011)
– New York	Present	CABI/EPPO (2011)
– North Carolina	Present	CABI/EPPO (2011)
– North Dakota	Present	CABI/EPPO (2011), Chittem et al. (2014)
– Ohio	Present	CABI/EPPO (2011)
– Oklahoma	Present	CABI/EPPO (2011)
– Oregon	Present	CABI/EPPO (2011)
– Pennsylvania	Present	CABI/EPPO (2011)
– Rhode Island	Present	CABI/EPPO (2011)
– South Carolina	Present	CABI/EPPO (2011)
– Texas	Present	CABI/EPPO (2011)
– Vermont	Present	CABI/EPPO (2011)
– Virginia	Present	CABI/EPPO (2011)
– Washington	Present	CABI/EPPO (2011)
– West Virginia	Present	CABI/EPPO (2011)
– Wisconsin	Present	CABI/EPPO (2011)
Central America and Caribbean		
Costa Rico	Present	Dixon (2009a, b)
Ecuador	Present	Dixon (2009a, b)
Guatemala	Present	CABI/EPPO (2011)
Puerto Rico	Present	CABI/EPPO (2011), EPPO (2014)
Trinidad and Tobago	Present	CABI/EPPO (2011), EPPO (2014)

(continued)

Table 2.2 (continued)

Continent/country/region	Distribution	Reference
South America		
Argentina	Present	CABI/EPPO (2011), EPPO (2014)
Bolivia	Present	Dixon (2009a, b)
Brazil	Present	CABI/EPPO (2011), EPPO (2014)
– Minas Gerais	Present	CABI/EPPO (2011)
– Parana	Present	Lima et al. (2004), CABI/EPPO (2011)
– Rio Grande do Sul	Present	CABI/EPPO (2011), EPPO (2014)
– Sao Paulo	Present	CABI/EPPO (2011), EPPO (2014)
Chile	Present	CABI/EPPO (2011), EPPO (2014)
Colombia	Present	CABI/EPPO (2011)
Guyana	Present	CABI/EPPO (2011), EPPO (2014)
Venezuela	Present	CABI/EPPO (2011), EPPO (2014)
Europe		
Austria	Widespread	CABI/EPPO (2011); EPPO (2014)
Belarus	Present	CABI/EPPO (2011), EPPO (2014)
Belgium	Present	CABI/EPPO (2011), EPPO (2014)
Bosnia-Herzegovina	Present	CABI/EPPO (2011)
Bulgaria	Widespread	CABI/EPPO (2011), EPPO (2014)
Channel Islands	Present	Dixon (2009a, b), Anon (1987, 1996)
Croatia	Present	Cvjetkovic (2008), CABI/EPPO (2011)
Czech Republic	Present	CABI/EPPO (2011), EPPO (2014)
Czechoslovakia (Former)	Widespread	EPPO (2014)
Denmark	Widespread	CABI/EPPO (2011), EPPO (2014)
Estonia	Present	CABI/EPPO (2011); EPPO (2014)
Faroe Islands	Present	CABI/EPPO (2011); EPPO (2014)
Finland	Restricted distribution	CABI/EPPO (2011); EPPO (2014)
France	Present	CABI/EPPO (2011); EPPO (2014)
– France (mainland)	Present	CABI/EPPO (2011)
Germany	Widespread	CABI/EPPO (2011); EPPO (2014)
Greece	Present	CABI/EPPO (2011); EPPO (2014)
Hungary	Restricted distribution	CABI/EPPO (2011); EPPO (2014)
Iceland	Present	CABI/EPPO (2011); EPPO (2014)
Ireland	Widespread	CABI/EPPO (2011); EPPO (2014)
Italy	Present	CABI/EPPO (2011); EPPO (2014)
– Italy (mainland)	Present	CABI/EPPO (2011)
– Sardinia	Present	CABI/EPPO (2011); EPPO (2014)
Latvia	Present	CABI/EPPO (2011); EPPO (2014)
Lithuania	Present	CABI/EPPO (2011); EPPO (2014)
Luxembourg	Present	CABI/EPPO (2011)
Montenegro	Present	CABI/EPPO (2011)
Netherlands	Present	CABI/EPPO (2011); EPPO (2014)

(continued)

Table 2.2 (continued)

Continent/country/region	Distribution	Reference
Norway	Present	CABI/EPPO (2011); EPPO (2014)
Poland	Present	CABI/EPPO (2011); EPPO (2014)
Carpathian Foot hill region		
– West Pomerania	Present	Robak (1991)
– Varnia	Present	Robak (1991)
– Mazuria	Present	Robak (1991)
– Kuyavia Pomerania	Present	Robak (1991)
– Lower Silesia	Present	Robak (1991)
– Lubuskie	Present	Robak (1991)
– Wielkopolska (Great Poland)	Present	Robak (1991)
Portugal	Restricted distribution	CABI/EPPO (2011); EPPO (2014)
– Azores	Present	CABI/EPPO (2011); EPPO (2014)
– Portugal (mainland)	Restricted distribution	CABI/EPPO (2011)
Romania	Present	CABI/EPPO (2011); EPPO (2014)
Russian Federation	Present	CABI/EPPO (2011); EPPO (2014)
– Central Russia	Present	CABI/EPPO (2011)
– Eastern Siberia	Present	CABI/EPPO (2011); EPPO (2014)
– Russia (Europe)	Restricted distribution	EPPO (2014)
– Western Siberia	Present	CABI/EPPO (2011), EPPO (2014)
Serbia	Present	Todorovic (2007), CABI/EPPO (2011), EPPO (2014)
Slovakia	Present	CABI/EPPO (2011)
Spain	Present	CABI/EPPO (2011), EPPO (2014)
– Spain (Mainland)	Present	CABI/EPPO (2011)
Sweden	Widespread	CABI/EPPO (2011), EPPO (2014)
Switzerland	Widespread	CABI/EPPO (2011), EPPO (2014)
UK	Widespread	Buczacki and Ockendon (1979), CABI/EPPO (2011), EPPO (2014)
– Channel Islands	Present	CABI/EPPO (2011), EPPO (2014)
– England and Wales	Present	CABI/EPPO (2011), EPPO (2014)
– Northern Ireland	Present	CABI/EPPO (2011), EPPO (2014)
– Scotland	Widespread	CABI/EPPO (2011), EPPO (2014)
Ukraine	Present	CABI/EPPO (2011)
Yugoslavia (Serbia and Montenegro)		CABI/EPPO (2011)
Australia and New Zealand		
Australia	Present	Khangura and Wright (2012), CABI/EPPO (2011), EPPO (2014)
– New South Wales	Present	Hind-Lanoiselet and Parker (2005); CABI/EPPO (2011), EPPO (2014)

(continued)

Table 2.2 (continued)

Continent/country/region	Distribution	Reference
– Queensland	Present	CABI/EPPO (2011), EPPO (2014)
– South Australia	Present	CABI/EPPO (2011), EPPO (2014)
– Tasmania	Present	CABI/EPPO (2011), EPPO (2014)
– Victoria	Present	CABI/EPPO (2011), EPPO (2014)
– Western Australia	Present	CABI/EPPO (2011), Khangura and Wright (2012), EPPO (2014)
New Zealand	Present	CABI/EPPO (2011), EPPO (2014)
Papua New Guinea	Restricted distribution	CABI/EPPO (2011), EPPO (2014)

Fig. 2.1 The presence of clubroot in various parts of the world (CABI/EPPO 2011)

counties, where the incidence ranged between 55% and 70% (Zhu 1956). Clubroot on turnip was first recorded in Jiamusi City in Heilongjiang Province in 1956, and 3 years later, it had spread to most cruciferous vegetable fields, with an incidence of about 10–42% (Wang 1962). Clubroot disease was also a serious problem on cruciferous vegetables in Altai, Fuhai, and Burqin counties in Xinjiang Province (Wang 1962). Additionally, severe and widespread outbreaks of clubroot were also reported in many southern provinces, such as Guangdong, Guangxi, Jiangsu, Hunan, Sichuan, and Yunnan. By 1962, clubroot had spread through 13 Chinese provinces (Wang 1962). Since then, records of clubroot gradually decreased (Yang et al. 1992; Li et al. 1993) until the 1990s, especially the end of the 1990s, when disease outbreaks were reported in the southwest, and northeast regions of China (Li 1990; Tang et al. 1990; Chen et al. 1997; Li and Pan 1999).

By 2010, clubroot was recognized as a major constraint reducing the production of cruciferous vegetables throughout the country. Yunnan Province was the most seriously affected region. The infected areas in Eshan County of Yunnan Province

Fig. 2.2 The distribution of clubroot-infested provinces in China (Chai et al. 2014)

increased from 1056 ha to 2388 ha from 2005 to 2012, respectively, more than doubling in the past 8 years (Fan et al. 2013). At present, clubroot is distributed throughout China although it is more severe in the southwest, northeast, and central regions than in other areas. The disease affects 3.2–4.0 million ha of cruciferous crops annually, accounting for more than a third of the area in which cruciferous crops are grown. The average rate of yield loss was reported to be 20–30% (Wang et al. 2011). Clubroot infection is much more severe in Yunnan, Sichuan, Guizhou, Tibet, Chongqing, Hunan, Hubei, Henan, Shandong, Liaoning, Jilin, Heilongjiang and Shanxi Provinces of China. These regions are mostly mild, moist, and warm mountainous areas with acidic soils. The damage is not as severe in Guangxi, Guangdong, Jiangxi, Fujian, Anhui, Zhejiang, Jiangsu, Beijing, Shanghai, Xinjiang, and Taiwan provinces. To date, no instances of clubroot have been reported in Gansu, Qinghai, Neimenggu, Ningxia, and Shanxi Provinces. These regions with low rainfall commonly have alkaline soils. The current distribution of clubroot in China is illustrated in Fig. 2.2.

As the climate differences are significant between the south and the north, clubroot occurs at different times in different areas of China. The disease mainly occurs during July to August in the Northwest, while it develops during April to June and September to October in the Southwest. The time when clubroot occurs in the middle and lower streams of the Yellow River is from April to June and October to November. It affects the middle and lower reaches of the Yangtze River around from July to August and October to November and the occurrence times for the Southern coast and the Northeast are from January to March and October to December, respectively (Table 2.3).

Table 2.3 Geographic regions and season of clubroot occurrence in China (Chai et al. 2014)

Geographical region	Province	Season of occurrence
Northwest China	Xinjiang, Shanxi	July–August
Southwest China	Yunnan, Guizhou, Chongqing, Xizang, Sichuan	April–June, September–October
Middle and lower streams of the yellow river	Henan, Shandong, Beijing	April–June, October–November
Middle and lower reaches of the Yangtze River	Hunan, Hubei, Jiangsu, Zhejiang, Anhui, Jiangxi, Shanghai	July–August, October–November
Southern coast	Guangdong, Guangxi, Hainan, Fujian, Taiwan	January–March
Northeast China	Liaoning, Jilin, Heilongjiang	October–December

2.5 Economic Importance

The adverse effects on the economy of the world can be judged through geographical distribution and yield losses including quality of the crops affecting health of people and animals and reduced value of clubroot-infected fields. *Plasmodiophora brassicae* is widespread throughout the world. It is known to occur in more than 88 countries and results in a 10–100% reduction in yields on a global scale (Dixon 2009a, b). The disease is especially prevalent in mild, moist, temperate areas; these areas may be mountainous parts of tropical countries. Epidemics are developing quickly as the dietary, and industrial significance of *Brassica* crops is increasingly appreciated, and concomitantly the intensity of cultivation rises. It is thought that movement of the pathogen from Europe resulted from transport of diseased animal fodder taken by colonists traveling to America, Australasia, and other similar centers of settlement. Occurrence of *P. brassicae* in countries such as China and Japan may be more ancient. The pathogen might have accompanied ancestral *B. rapa* types moving from the Fertile Crescent in what is now known as Turkey, Iraq, and Iran to Asia (Dixon 2009a, b).

Disease incidence surveys have been published by the Commonwealth Mycological Institute (CMI) and more recently by the European Plant Pathology Organization (EPPO). The perusal of data in Table 2.2 indicates geographical distribution of the disease at present. In those countries where *Brassicas* are popular crops, clubroot disease has been identified for many decades. One of the problems associated with the historical record of clubroot incidence is its relative simplicity of diagnosis. Quite frequently the incidence goes formally unrecorded, hence its geographical distribution is much more then published reports.

There are several European reports of what appears to be clubroot disease in the sixteenth and seventeenth centuries and some illustrations of cruciferous plants by noted herbalists, and artists appear to have root swellings that might represent symptoms elicited by *P. brassicae* on the roots. The Agricultural Revolution during the eighteenth century encouraged the use of root crops such as the turnip, and as part of the Norfolk four-course rotation, and their use possibly increased the incidence of disease. The frequency of reports increased substantially in nineteenth century in Europe as agriculture increased to feed burgeoning populations serving the Industrial Revolution. Urbanized Victorian populations demanded both *Brassica* vegetables and meat from animals sustained over winter with cruciferous fodder. During the latter part of the nineteenth century, disease incidence was being noted quiet widely in North America and Australasia, presumably as a result of spread along with colonization of the mid-west, and west coast of North America, and movement from coastal centers into the inland fertile areas in Australasia. In northeastern Europe, where cabbage represented a significant part of the diet, and was conserved in a pickled form (sauerkraut) to carry populations through the winter, the disease appeared to have reached severe epidemic proportions. In the first half of the twentieth century, clubroot was recognized as a major constraint to the production of cruciferous vegetables on all continents. In rural areas where sheep still formed a major segment of the agrarian economy, the effects of clubroot disease were devastating. In northeastern Scotland, for instance, swede and turnip formed the backbone of winter fodder supplies and Morrison (1977) estimated that 50% of the acreage was infested with *P. brassicae*. The social significance of clubroot in that area led to early attempts in the 1930s to breed for resistance. This resulted in swede (*B. napus*) cvs. Wallace, and Bruce which even today have good field resistance. In Asia, crops of Chinese cabbage in all its variant forms, which are of major importance in the diet of local people, have been, and still are at serious risk from this pathogen. The importance of this pathogen is demonstrated by the intensive breeding of resistant Chinese cabbage cultivars from the 1960s onward at the National Institute for Vegetables, Ornamentals, and Tea (NIVOT) in Japan. Breeding has been supplemented by a successful joint industry-government search in Japan for agrochemicals competent to control the pathogen. In the Republic of Korea, breeding year-round Chinese cabbage cultivars has been accompanied by the search for clubroot resistance.

There is a significant parallel between Chinese cabbage (*B. rapa*) and European heading cabbage (*B. oleracea* var. *capitata*) in the Korean pickle kimchi, which was originally used to carry the population through the winter months. Clubroot disease in China is recognized as a major cause of depressed yield of their vast *Brassica* crops such as Chinese cabbage (Daowang et al. 2004). Jing et al. (2008) quantified losses of oilseed rape in China; there is 17% loss of young growing plants, and 15% at maturity with 10.2% losses of yield. The report indicates significant losses of plant height, numbers of siliquae, and seed production. Clubroot is now found throughout the world wherever *Brassica* crops are grown, and is assessed as the major source of disease-induced loss. The severity of infestation and symptom expression increases with the intensity and frequency of crop production. The only scientifically valid worldwide attempt to establish the frequency of infestation was made by Crete (1981). His survey looked at infection in individual *Brassica* crops. High levels of infestation (>10%) were recorded in *B. oleracea* in Australia, Canada, Czechoslovakia, Finland, Germany, Ireland, the Netherlands, Norway, Poland, Scotland, and Wales; in *B. rapa*, in Germany, New Zealand, and the USA; and in *B. napus*, in Finland, New Zealand, Scotland, and Wales. In total, Crete's survey

Table 2.4 Historic estimate of yield and financial losses to cauliflower crops caused by clubroot disease (Dixon 2009b)

Treatment	Yield	Overall return (£ sterling)
Return/ha in the absence of clubroot	1635 crates at 40p	654.00
Return/ha if clubroot present	832 crates at 40p	332.95
Return/ha clubroot present and controlled (cost of control with mercurous chloride = £46.48/ha)	1180 crates at 40p	472.00

covered 6 million hectares of *Brassica,* and related crops, and found an overall mean infection of 11%. This concurs with the general estimate of in-field losses to individual crop pathogens varying from 10–15%. Worldwide crop losses caused by clubroot in *Brassica* crops, and historical estimates of yield, and financial loss caused by clubroot in cauliflower are given in Table 2.4. Clubroot disease is the main constraint for oilseed rape and mustard cultivation in Eastern part of India mainly in West Bengal bordering Bangladesh and sporadic occurrence in Orissa states where yellow Sarson *Brassica campestris* (*B. rapa*) var. yellow sarson has been found to be highly susceptible.

Clubroot disease is also a serious problem on cruciferous vegetables in Darjeeling Hills in the eastern Himalayan region, and the northeastern region states like Manipur, Meghalaya, Mizoram, and Arunachal Pradesh bordering China. Bangladesh and China are now both considered high risk zones for clubroot disease (Bhattacharya et al. 2014). In Nepal, clubroot disease has been observed since 1993 (Timila et al. 2008). Severe and widespread outbreaks have been seen since 2004 in Bhaktapur, Kathmandu, Lalitpur, and the Palung Valley. The disease is especially severe in the Kathmandu Valley, and the Palung/Dama area of Makwanpur District. Cauliflower appears to be the crop most at risk, with 40% overall loss due to clubroot. In some parts of the country, production was reduced from 5–6 mt/household plot of 1500 m^2 prior to 2004 to 300 kg/plot. A report from Indonesia (Cicu 2006) comments on the serious nature of clubroot disease but fails to quantify it.

This disease has been known about for many years in the vegetable crops near Toronto, Montreal, and Newfoundland (McDonald et al. 2004; Belec et al. 2004). Reports of infection in oilseed rape from the area near Quebec (Pageau et al. 2006) indicated the susceptibility of the canola crop. Testing cultivars indicated that losses due to clubroot in grain yield were 80, 91, and 85% in an Argentinian-derived cultivar over the 1998–2000 seasons, and 69, 96, and 89% in Polish cultivars in the same period. Clubroot has now reached the Great Plains which has 6 million ha of canola (oilseed rape) (mostly *B. napus* with *B. rapa* to a lesser extent). Appreciable losses in yield of canola have been reported by Tewari et al. (2005) in Alberta. Donald and Porter (2003) indicated the upsurge in importance of oilseed rape (*B. napus*) to Australia's broad acre agricultural economy ($Au 545 million), indicating the risks posed by clubroot based on the losses sustained in that country's

Table 2.5 Worldwide crop losses caused by clubroot disease of crucifers

Country	Infection (%)	References
Australia	30–60	Donald and Porter (2014)
Canada	30–100	Hwang et al. (2011)
China	20–30 14–72 80	Chai et al. (2014) Ren et al. (2016) Feng et al. (2013)
Czechoslovakia (Czech Republic)	10–100	Ricarova et al. (2016a)
Denmark	5	Crete (1981)
England (UK)	6	Crete (1981)
Finland	5–16	Rastas et al. (2012)
France	3	Crete (1981)
Germany	8	Crete (1981)
India	30–70	Bhattacharya et al. (2014)
Ireland	17	Crete (1981)
Japan	5	Crete (1981)
Nepal	35–40	Timila et al. (2008)
Netherlands	10	Crete (1981)
New Zealand	15	Crete (1981)
Norway	12	Crete (1981)
Poland	25	Jedryczka et al. (2014)
Scotland	48	Crete (1981)
Sweden	82	Wallenhammar et al. (2014)
USA	10	Crete (1981)
Wales	45	Crete (1981)
UK	50	McGrann et al. (2016)

field vegetable *Brassicas*. Clubroot causes losses of at least 50% (1.6 t ha^{-1}) in diseased crops of oilseed rape in Australia (Donald and Porter 2003). The crop loss estimates from different countries of the world are given in Table 2.5.

Disease may also result in crop rejection because of blemishes such as residual and untrimmed tumors on the roots of swede or turnip. A crop may also be rejected because disease has delayed maturity, and hence, it is out of phase with the requirements of the supermarket buyer. Disease may also cause loss of quality in terms of appearance and size leading to downgrading. The incidence of clubroot in a field may kill out a portion of the plants resulting in luxury growth by those remaining and as a consequence crop heads mature erratically, and are of unacceptably variable sizes. Once established, this soil-borne pathogen reduces the asset value of the land that previously was highly prized for field *Brassica* production, causing a direct reduction in the wealth of a holding. Where land is rented for crop production, the presence of clubroot reduces the owners' opportunities for leasing to other farmers and growers (Dixon 2009a, b).

In field experimental studies carried out under Sweden conditions by Wallenhammar et al. (2014) where increasing amounts of *P. brassicae* soil inoculum were applied to the soil, yield loss was quantified. Strong relationship was found between disease incidence, and yield ($R^2 = 0.90$), and between natural soil inoculum, and yield ($R^2 = 0.94$) of spring oilseed rape (*Brassica napus* L.). A clear relationship ($R^2 = 0.87$) was also found between disease incidence and degree of soil infestation as tested in a bioassay. In another series of field experiment testing effects of partly resistant lines of summer oilseed turnip rape (*B. campestris*), multiplication of clubroot was moderate, 2.1–10% plants were infected after harvest, when initial soil inoculum levels were high 47–72% of the test plants infected in the bioassay. The yield of resistant lines was 5–10% higher than the yield of the susceptible cultivars SW Kulta. In a third series of field experiment placed as natural infested soils, yield reduction in summer oilseed rape was 1.62 t/ha^{-1} (50%) at the highest level of infection (91% plants infected at harvest), while at less than <20% plants infected, yield reductions were estimated to about 10%. On fields with low levels of soil infestation, <20% plants infected in the bioassay, the risk of heavy losses from clubroot could be avoided by including partly resistant cultivars in the crop rotation.

Risk potential of clubroot disease on winter oilseed rape under German conditions was determined using a model. According to cumulative link mixed models, disease rating is positively correlated with the amount of inoculum. Linear regression analysis revealed a negative correlation between seed yield and inoculum density. Yield losses of 60% appeared at the lowest inoculum density, plant losses, and reduced seed per pod were accountable for yield losses although the resistance cultivar showed clubroot symptoms only. Seed yield was not affected by the pathogen. The green house experiment revealed that clubroot severity in subsequent oilseed rape (OSR) was reduced after cropping the resistant cultivar. This study showed significant yield damage of *P. brassicae* at low infestation levels (Strehlow et al. 2015). In the recent past, clubroot distribution and intensity have increased at an alarming rate and have become persistent threat to the trade, commerce, and economy of some countries.

2.6 Host Range

All 330 genera and 3700 species of the family Brassicaceae are thought to be potential hosts for *Plasmodiophora brassicae* and within them, pathogen is able to complete both the root hair and cortical stages of its life cycle (Kageyama and Asano 2009; Howard et al. 2010). Cultivated crops appear to be especially susceptible. This includes all varieties (Bailey 1961) of *B. oleracea*, the Occidental Cole vegetables (Brussels sprout, cabbages, calabrese/green broccoli, cauliflower, culinary, and fodder kale, kohlrabi); *B. rapa* (syn. *B. campestris*), including turnip, turnip rape, *Sarson*, and the enormous range of oriental variants which provide leaf and root vegetables such as *Brassica rapa* var. *pekinensis*, and *B. rapa* var. *chinensis* (Chinese cabbages); *B. napus*, including swede (rutabaga), oilseed rape, and fodder rape; and seed, condiment (mustard), and vegetable crops derived from *B. carinata*, *B. nigra*, and *B. juncea*. Related genera such as radish (*Raphanus sativus*), cruciferous weeds, *Sinapis alba*, and decorative ornamentals including stocks (*Matthiola* spp), and wallflower (*Cheiranthus cheiri*) can be infected (Anderson 1855).

Cruciferous weeds such as Shepherd's purse (*Capsella bursa-pastoris*), *Barbarea vulgaris*, *B. kaber*, *B. hirta*, and Stinkweed (*Thlaspi arvense*) are susceptible to *P. brassicae* (Buczacki and Ockendon 1979; Howard et al. 2010; Chai et al. 2014).

The scientific model and rock garden plant *Arabidopsis thaliana* are also susceptible. Very few studies of infection have been done outside the genera *Brassica*, *Raphanus*, and *Arabidopsis* in the past 50 years (Colhoun 1958; Karling 1968; Dixon 2009b; Chai et al. 2014). *Raphanus* is credited with possessing a greater degree of resistance compared with *Brassica* but this may solely reflect a lower frequency of virulences in current pathogen populations. Certainly, where radish crops are grown intensively, clubroot disease rapidly becomes a major problem for crop producers. Glasshouse radish crops grown in what was East Germany during the 1960s and 1970s were reported as being extensively diseased, and field crops in Florida in the USA are also susceptible to the pathogen. Despite intensive searching in the Mediterranean centers of origin of several *Brassica* hosts, the pathogen is difficult to locate. These results lead to speculation that it may be a "*Disease of cultivation*." This concept suggests that increasingly intensive cultivation of *Brassicas* and related crops provided conditions under which the pathogen could thrive and where it is less subject to predation by antagonistic organisms.

In Japan, clubroot has been recorded on five species of cruciferous weeds, *C. bursa-pastoris*, *Rorippa indica*, *B. islandica*, *Cardamine flexuosa*, and *C. sculata* in addition to *Brassica* vegetable crops and oil yielding crops (Tanaka and Ito 2013). *Eutrema wasabi* and *Raphanus sativus* subsp. *longipinnatus* are also susceptible (Horikoshi and Tairako 2002).

In Germany, *P. brassicae* has been detected through SEM in the roots of (cortex) *Tropaeolum majus*, *Carica papaya*, *Reseda alba*, and *Beta vulgaris* (Ludwig-Muller et al. 1999). Structures resembling *P. brassicae* have been observed in the root cortex of *Tropaeolum majus, Carica papaya, Reseda alba*, and *Beta vulgaris* through scanning electron microscopy. In Brazil, clubroot disease has been reported on *Eruca sativa* by Lima et al. (2004).

In China *P. brassicae* is reported to infect and sporulate in roots of 16 genera of crops, ornamental, and weeds under artificial inoculation conditions. The disease incidence and disease indices (DI) of clubroot on different species of *Brassica* in the field were assessed which ranged from 23.6 to 85.9% and 13.6 to 71.5%, respectively (Ren et al. 2016; Tables 2.6 and 2.7). Chai et al. (2014) reported about distribution of *P. brassicae* races on *Brassica* species in China.

2.7 Symptomatology

Very often the symptoms of clubroot infection initiation are not seen above ground until after the host passes the seedling stage. The infection initiates below-ground, while above-ground symptoms are expressed only when gall formation interrupts movement of water and nutrients in the host plant foliage. Unless the roots of such infected plants are examined carefully after pulling, disease diagnosis is ignored. Recently infected roots quite often appear to be healthy which at later stage show

Table 2.6 Cruciferous plants surveyed or tested in pots for susceptibility to clubroot (Ren et al. 2016)

Common name	Botanical name	N[a]	Source	Clubroot[b]		
				Field	Pot	PCR
Oilseed rape	*Brassica napus*	>10	Hubei	+	+	+
Chinese cabbage	*B. rapa pekinensis*	3	Jingyan Yinong, Beijing	+	+	+
Brown mustard	*B. juncea*	>10	Hubei	+	+	+
Chinese kale	*B. alboglabra*	3	LuHeng Seeds, Beijing	+	+	+
Cabbage	*B. oleracea*	6	LuHeng Seeds, Beijing	+	+	+
Kale	*B. oleracea* var. *acephala*	1	Bejo Seeds, the Netherlands	+	+	+
Radish	*Raphanus sativus*	8	Syngenta and Tezier	+	+	+
Kohlrabi	*B. oleracea* var. *gongylodes*	2	Dayu Seeds, Hebei	+	ND	ND
Tuber mustard	*B. oleracea* var. *tumida*	2	Chengdu, Sichuan	+	ND	ND
Sprouting broccoli	*B. oleracea* var. *italica*	2	Xinye, Henan	+	ND	ND
Watercress	*Nasturtium officinale*	1	Luhe Seeds, Jinan, Shandong	ND	+	+
Arugula	*Eruca sativa*	1	Lujinlan Seeds, Beijing	ND	+	+
Woad	*Isatis tinctoria*	2	Yunnan	ND	+	+
Sweet alyssum	*Lobularia maritima*	2	Takii seed, Kyoto, Japan	ND	+	+
Wallflower	*Cheiranthus cheiri*	1	Takii seed, Kyoto, Japan	ND	+	+
Siberian wall—flower	*C. allionii*	1	Guomei Horticulture Co. Ltd, Zhejiang	ND	+	+
Wallflower	*C. maritimus*	1	Guomei Horticulture Co. Ltd, Zhejiang	ND	+	+
Dames rocket	*Hesperis matronalis*	1	Guomei Horticulture Co. Ltd, Zhejiang	ND	+	+
Globe candytuft	*Iberis umbellata*	1	Takii seed, Kyoto, Japan	ND	+	+
Violet orychophragmus	*Orychophragmus violaceus*	2	Guomei Horticulture Co. Ltd, Zhejiang	+	+	+
Virginia stock	*Malcolmia maritime*	1	Qinghai	–	+	+
Stock	*Matthiola incana*	8	Xinnongfeng Seed and Pan American Seed	–	–	+
White mustard	*Sinapis alba*	3	Qinghai	+	+	+
Shepherd's purse	*Capsella bursa-pastoris*	3	Sichuan	+	+	+

(continued)

Table 2.6 (continued)

Common name	Botanical name	N[a]	Source	Clubroot[b]		
				Field	Pot	PCR
Herb Sophia	*Descurainia saphia*	1	Qinghai	–	+	+
Pepper weed	*Lepidium apetalum*	1	Qinghai	–	+	+
Saltwater cress	*Thellungiella salsuginea*	1	Qinghai	–	+	+

[a]Number of varieties
[b]Symbols: + visible clubs on the roots or specific fragment present via polymerase chain reaction (PCR), – is no visible clubs on the roots, and *ND* not determined

characteristically swollen, malformation, and clubs. Above-ground symptoms consist of a flogging or wilting which may be confused with any other soil-borne disease or deficiency or abiotic stress till it is confirmed from observation of clubroots. It is a very typical disease which requires experience and knowledge of identification by the observer for proper diagnosis of the disease.

Clubroot galls are a nutrient sink, so they tie up nutrients, and severely infected roots of crucifers cannot transport sufficient water and nutrients for above-ground plant parts. Symptoms may vary depending on the growth stage of the crop when infection occurs. Early infection at the seedling stage can result in pale green to yellowish leaves than wilting, stunting, and yellowing of infected plants in the late rosette to early podding stages. Such symptoms may be wrongly attributed to heat stress during periods with high temperatures or to other diseases such as blackleg or Fusarium wilt or other root rots or even to water and salt stresses. In such cases, proper diagnosis includes digging up wilted plants to check for gall formation on roots (Plates 2.1, 2.2 and 2.3).

Infection that occurs at later crop stages may not show plant wilting, stunting, or yellowing. However, infected plants will ripen prematurely, and seeds will shrivel. Thus, yield and quality (oil content) are reduced. Patches of prematurely ripening of crops due to clubroot infection (Plate 2.4) can be confused with other diseases such as Sclerotinia stem rot, blackleg, and Fusarium wilt. In such cases, proper diagnosis should include digging up affected plants to check for gall formation on roots.

The characteristic symptoms of the disease appear on the roots as spindle like shredded, knobby, or club shaped swelling. The swellings may be few, and isolated or they may coalesced, and cover the extensive root system. The suspected plants if are not sampled until after swathing, the root galls may have decayed already, and typical whitish galls will no longer be present (Plate 2.5). Instead, the decayed galls cause roots to have a brown peaty appearance rather than a healthy white color, which should be a signal to carefully dig up more roots for closer inspection. Hybridization nodules on crucifers roots (Plate 2.6), although rare, can be confused with clubroot galls, but they appear as small, round nodules located at root nodes. The interior texture of a clubroot gall is spongy or marbled while hybridization nodules are uniformly dense inside, like healthy roots. Furthermore, hybridization

Table 2.7 Disease incidence and disease indices (DI) of clubroot on different species of *Brassica* in the field[a] (Ren et al. 2016)

Species			Disease incidence (%)		Disease indices		DI ranges (N^b)		
Common name	Botanical name	N^c	Range	Average	Range	Average	<30	30–50	>50
Oilseed rape	*Brassica napus*	100	26.1–100	81.9	20.3–96.3	61.2	3	24	63
	B. rapa	100	52.8–100	81.0	40.5–96.5	64.4	0	19	81
	B. juncea	97	45.5–100	85.1	25.8–100	71.5	1	10	86
Turnip	*B. rapa* var. *rapa*	2	83.8–88.0	85.9	59.5–85.7	68.2	–	–	–
Flowering stalk	*B. rapa ssp. Chinensis* var. *utilis*	2	39.6–77.3	58.5	44.7–72.7	54.0	–	–	–
Chinese kale	*B. alboglabra*	2	55.0–56.5	55.8	36.2–41.7	39.9	–	–	–
Tuber mustard	*B. juncea* var. *tumida*	2	39.3–66.7	53.0	27.8–55.6	37.0	–	–	–
Potherb mustard	*B. juncea* var. *crispifolia*	2	23.6–58.3	41.0	16.7–38.9	24.1	–	–	–
Sprouting broccoli	*B. oleracea* var. *italica*	2	28.0–47.1	37.5	22.7–23.5	22.9	–	–	–
Kohlrabi	*B. oleracea* var. *gongylodes*	2	9.1–38.1	23.6	6.1–28.6	13.6	–	–	–

[a]Trials were laid in randomized complex block design with two replications. Results were means of two experiments
[b]Number of accessions with DI ranges shown
[c]Number of accessions

Plate 2.1 Very severe symptoms of clubroot on crucifers (Hartman 2007)

Plate 2.2 Severe clubroot galls or clubs on crucifer roots (Hartman 2007)

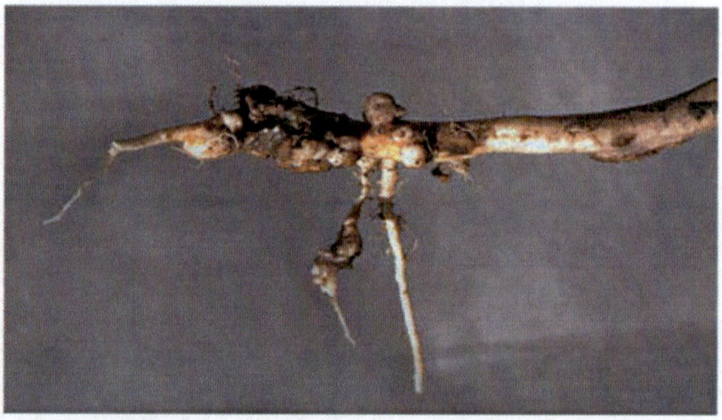

Plate 2.3 Moderately infected crucifers root (Hartman 2007)

nodules will not decay rapidly to a peaty appearance like clubroot galls do (Hartman 2007).

2.7.1 Anatomical Symptoms

Through SEM clubroot symptoms were observed in diseased root cells of turnip and Chinese cabbage by Ikegami (1978). In the early stage of infection, *P. brassicae* in

Plate 2.4 Patchy premature ripening field of crucifers (Hartman 2007)

Plate 2.5 Decayed clubroot galls and whitish stem appearance (Hartman 2007)

Plate 2.6 Hybridization nodules on canola root (Hartman 2007)

an infected cell can be observed as nearly to completely spherical plasmodial bodies of 3–5 μm in diameter. The young plasmodia assume various shapes, such as spherical bodies with or without strand like material, those with smooth or rough surfaces, and cluster like bodies. The strand like material connecting the young plasmodia seems to be the same material as reported by Dekhuijzen (1976) with transmission electron micrographs. The morphological diversity of young plasmodia may be due to the nutritional condition in a host cell and the root portion from which the cell is sampled. As some young plasmodial bodies grow a large multinucleate plasmodium may possibly be formed by their fusions. The vegetative plasmodia grow progressively in an infected cell as observed to form a large pathogen mass in the cell.

In observations of sectioned tissue of clubs from 45–60 days old plants, plasmodia in the infected cells showed clearly different growth rates. The various cells including a young plasmodium, a vegetative plasmodium, or a resting spore mass are observed in the same section. An infected cell seems to influence the growth of pathogen in the contiguous cell. A young plasmodium or a vegetative plasmodium, and a mature spore mass are recognized in an infected cell. It is presumed that these plasmodia are finally unable to mature into resting spores. The mature resting spores in the infected cells failed to show any membrane enveloping them. In the process by which resting spores mature from vegetative plasmodium, the plasmodial membrane is presumably disintegrated. Williams and McNabola (1967) described that as sporogenesis is initiated, the outer membrane of the plasmodial envelop disintegrates. A great number of resting spores is easily distributed into the soil by rotting of clubbed roots, where they become a formidable inoculum of clubroot disease. The details of fine structures of plasmodium and resting spores observed through scanning electron microscopy (SEM) are given in Chap. 4.

2.8 Disease Assessment

To assess the clubroot infection under field and controlled conditions (Laboratory, and Green house) in the form of incidence, intensity several rating scales have been devised, and being used all over the world. These rating scales are very important parameters for the clubroot researchers to assess losses caused by the pathogen, identify occurrence of pathotypes along with its virulence spectrum, assess host resistance, know efficacy of disease management strategies, determine mechanisms of host–pathogen interaction, and to test various hypothesis at macro- and microlevel (molecular aspects). The various rating scales proposed by different researchers are as follows:

2.8.1 Disease Severity Index of Watson (1965)

He has expressed the severity of clubbing by an indexing method similar to that used by Colhoun (1953). Each root was pulled out from the ground, and the amount of clubbing was graded (Fig. 2.3) according to the following scales: 0 = no visible clubbing, ½ = one or two small clubs on the tips of lateral roots, 1 = several small clubs on the lateral roots, 2 = large clubs on lateral roots, 3 = slight clubbing of the taproot, many large clubs on the lateral, and slight reduction of root system, 4 = considerable clubbing of the whole taproot and virtually no lateral roots, and 5 = complete clubbing of the whole taproot and virtually no lateral roots.

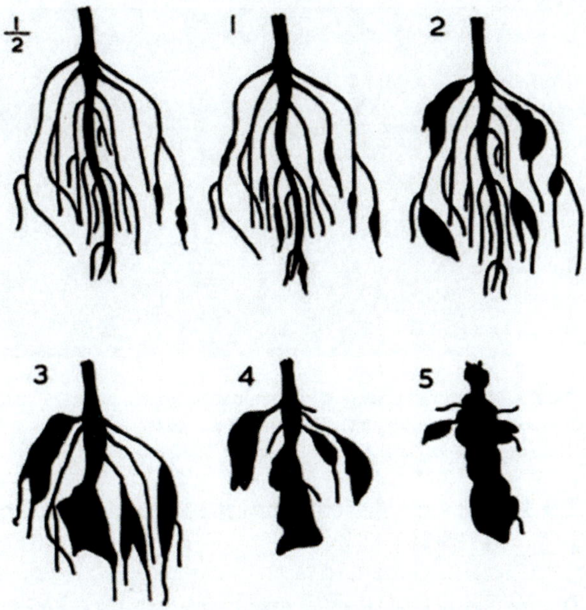

Fig. 2.3 The grades of clubbing used for obtaining the disease indices (Watson 1965)

2.8.2 Disease Assessment Scale of Ricarova et al. (2016b)

To assess the variation in *P. brassicae* infection on differential hosts, 5 weeks after inoculation, plants were removed from the pots, and washed with water. The roots were evaluated for clubroot severity on a 0–3 scale (Kuginuki et al. 1999), where0 = no galling, 1 = a few small galls, 2 = moderate galling on the main and lateral roots, and 3 = severe galling, the root is totally deformed.

An index of disease (ID) was calculated for each differential host using the formula of Horiuchi and Hori (1980) and modified by Strelkov et al. (2006):

$$\mathrm{ID}(\%) = \frac{\sum(n \times 0 + n \times 1 + n \times 2 + n \times 3)}{N \times 3} \times 100$$

where Σ is the total sum, *n* is the number of plants in class, *N* is the total number of plants, 0, 1, 2, and 3 arc the symptom severity classes.

The results shown represent the mean ID of each host over the three replicates for each *P. brassicae* population. Two thresholds were compared; ID 25% as described by Some et al. (1996), and ID <50% with the 95% confidence interval (CI) not exceeding 50%, as proposed by LeBoldus et al. (2012). Statistical analysis was carried out using Statistical v.7.0 (Stat Soft Inc., USA) and Microsoft Excel 2010 (Microsoft, Inc., USA) (Plate 2.7).

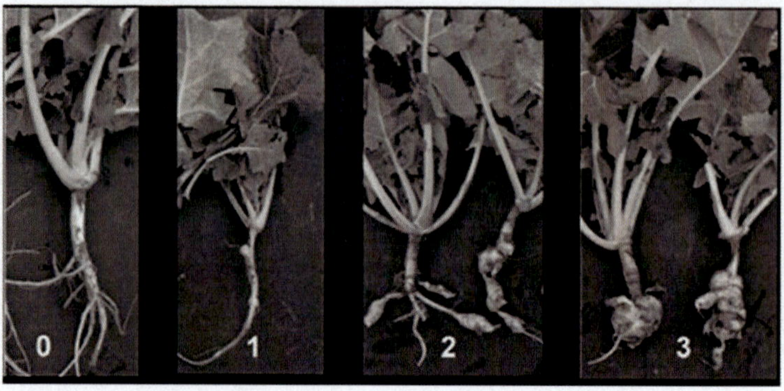

Plate 2.7 Pictorial disease rating scale (0 = no symptoms, 1 = minor clubs, 2 = clubs on lateral roots, 3 = severe clubbing) (Ricarova et al. 2016a, b)

2.8.3 Disease Assessment scale of Linnasalmi and Toiviainen (1991)

The following disease assessment scale was used for identification of pathotypes of *P. brassicae* in Finland by Linnasalmi and Toiviainen (1991). The degree of clubroot infection of each plant was assessed on a scale of 0–3, and the clubroot index was calculated according to Williams (1966):

$$\text{Clubroot index} = \frac{\sum(0 \times n_0 + 10 \times n_1 + 60 \times n_2 + 100 \times n_3)}{n_0 + n_1 + n_2 + n_3}$$

where n_0 to n_3 indicates n_0 no clubs, n_1 a few small clubs on the secondary roots, n_2 considerable clubbing on the lateral roots, and n_3 severe clubs on the primary and secondary roots.

2.8.4 Disease Assessment scale of Kim et al. (2016)

Following scale has been used to classify *P. brassicae* pathotypes on clubroot cultivars of Chinese cabbage. The severity of the club development on each plant was rated using a scale of 0–4, where 0 = no Clubbing, 1 = one or a few small clubs on the lateral roots, 2 = a few medium, and separate globular clubs on the lateral roots, 3 = medium-sized clubs on the main roots, and 4 = severe clubbing on the lateral roots, and main roots (Kuginuki et al. 1999; Suwabe et al. 2003).

The experimental units consisted of 20 individual plants, and the experiments were repeated three times independently. Host responses developing disease severity (DS) ≤ 1 were regarded as "*resistant*," while those developing a DS ≥ 2 were regarded as "susceptible."

2.8.5 Disease Assessment scale of Porth et al. (2003)

Disease evaluation was made according to 0-3 scale for identification of pathotypes in Turkey by Apaydin et al. (2010). Scale used was as follows: 0 = there was no gall formation on the roots, 1 = there was only gall formation on the lateral roots, 2 = there was less than 50% gall formation on the main roots, and 3 = there was more than 50% gall formation on the main roots.

Disease index (DI) was calculated by means of disease scale values to identify races (Dobson et al. 1983). The formula used to calculate disease index was as follows:

$$\mathrm{DI}: \{(n_0 \times 0) + (n_1 \times 1) + (n_2 \times 2) + (n_3 \times 3)\}/\{n_0 + n_1 + n_2 + n_3\} \times (100/3)$$

In this formula: where, n_0 Plant number having 0 value according to 0–3 scale, n_1 Plant number having 1 value according to 0–3 scale, n_2 Plant number having 2 value according to 0–3 scale, and n_3 Plant number having 3 value according to 0–3 scale.

If DI was equal to 0, reaction was accepted as resistant. If DI was between the 0 and 33 values, reaction was accepted unknown, and if DI was more than 33, reaction was sensitive. Race identification was done according to obtained reaction categories (Donald et al. 2006; Some et al. 1996). Race numbers were calculated according to race identification codes in European Differential Set and host rape (*Brassica napus*) reactions.

2.8.6 Disease Assessment scale of Strelkov et al. (2018)

For the assessment of clubroot development and disease reactions, plants were evaluated for clubroot symptom development at 6 weeks after inoculation. The roots of each plant were gently dug out of the potting mix, washed with water, and rated for clubroot severity on a 0–3 scale (Kuginuki et al. 1999), where: 0 = no galling, 1 = A few small galls (small galls on less than one-third of the roots), 2 = moderate galling (small to medium galls on one-third to two-thirds of the roots), and 3 = severe galling (medium to large galls on more than two-thirds of the roots).

The severity ratings were then used to calculate an index of disease (ID) for each experimental unit according to the formula of Horiuchi and Hori (1980) as modified by Strelkov et al. (2016):

$$\mathrm{ID}(\%) = \frac{\sum(n \times 0 + n \times 1 + n \times 2 + n \times 3)}{N \times 3} \times 100$$

where n is the number of plants in a class; N is the total number of plants; and 0, 1, 2, and 3 are the symptom severity classes. The mean ID was calculated by averaging the ID for each host–pathogen combination across repetitions. Genotypes were considered resistant if the mean ID was <50%, and its associated 95% confidence interval (CI) did not overlap 50% (LeBoldus et al. 2012; Strelkov et al. 2016).

2.8.7 Disease Assessment scale in Relation to Virulence

The overall virulence of the pathogen populations was measured by Cao et al. (2009) on a scoring scale of 0-3. To compare the overall virulence of the *P. brassicae* populations, a disease score was calculated for each differential genotype according to the following formula:

$$\text{Score} = \sum(n \times 0 + n \times 1 + n \times 2 + n \times 3)$$

where n is the number of plants in a class; and 0, 1, 2, and 3 are the symptom severity classes (Strelkov et al. 2007). The sum of the disease scores induced by each population on the 13 differential hosts was then utilized to test the homogeneous null hypothesis by means of the Savage test of the NPAR1WAY procedure; the Wilcoxon rank sum test was used for pair wise comparisons of virulence between populations (SAS Institute Inc. 1999).

2.8.8 Disease Assessment scale of Xue et al. (2008)

Granaat plants were assessed for symptom development eight weeks after inoculation with single spores while the differential hosts were examined six weeks after inoculation for pathotype designation. The plants were pulled from the soil, and then the roots were washed free of soil mix, and visually evaluated for disease development using a 0-to-3 scale, where: 0 = no galling, 1 = a few small galls (small galls on less than one-third of the roots), 2 = moderate galling (small to medium-sized galls on one-third to two-thirds of the roots), and 3 = severe galling (medium to large-sized galls on more than two-thirds of the roots).

An index of disease (ID) was calculated using the formula of Horiuchi and Hori (1980) as modified by Strelkov et al. (2006):

$$\text{ID}(\%) = \frac{\sum(n \times 0 + n \times 1 + n \times 2 + n \times 3)}{N \times 3} \times 100$$

where n is the number of plants in a class; N is the total number of plants in an experimental unit; and 0, 1, 2, and 3 are the symptom severity classes. The mass of the clubbed roots (total root mass minus mass of non-affected roots) also was recorded for each infected plant.

2.8.9 Scoring Scale of Wu et al. (2012)

Disease index (ID) and incidence of clubroot (IC) on three *B. rapa* lines inoculated with *P. brassicae* isolate for statistical analysis, plants were scored for disease reaction using the system of Kuginuki et al. (1999), the symptoms of each plant were rated on a scale of 0–3, where 0 = no clubs, 1 = few small clubs, 2 = moderate clubbing, and 3 = severe clubbing.

The disease index (ID) was calculated for each line as follows:

$$\text{Disease Index} = (n_0 \times 0 + n_1 \times 1 + n_2 \times 2 + n_3 \times 3)/(\text{number of inoculated plants}),$$

n_0–n_3 are the numbers of plants in each disease class. The mean ID for each line was calculated from the results of three tests. The degree of resistance of each line was classified into three categories according to Hatakeyama et al. (2004): mean ID <1.0, resistant; $1.0 \leq$ mean ID <2.0, partially resistant; mean ID ≥ 2.0, susceptible. The incidence of clubroot (IC) was calculated for each line as follows: (Number of diseased plants)/(number of inoculated plants) × 100.

The combinations of each line/isolate group (Granaat/e3, Line C09604/e3, Line CR701/e3, Granaat not infected as control) were studied in three replicates using a minimum of 30 control, and infected plants for each time point analysis.

2.8.10 Disease Assessment Scales of Deora et al. (2012)

secondary infection In order to assess host cultivars interactions with *P. brassicae* pathotypes, a trial was conducted under the conditions with the secondary inoculum in a factorial randomized complete block design with four replicates and 10 plants per experimental unit. The two factors were cultivar and pathotype, plus a non-inoculated control. At 6- weeks after inoculation, the plants were uprooted, the roots were washed, and clubroot severity was assessed using a 0–3 scale, where plants were divided into classes according to the extent of clubbing; 0 = No clubbing, 1 = <1/3 of root clubbed, 2 = 1/3 to 2/3 of roots clubbed, and 3 = >2/3 of roots clubbed (Xue et al. 2008 based on Kuginuki et al. (1999). The trial was repeated.

Clubroot incidence was calculated as the proportion (%) of plants showing clubroot symptoms. A disease severity index (DSI) was calculated using the following equation (Kobringer and Hagedorn 1983):

$$\text{DSI} = \frac{\{(\text{Class no.})(\text{ No. of plants in each class})]}{(\text{Total no.of plants per sample})(\text{No. classes} - 1)} \times 100$$

The height (cm) of each plant at 6 week after inoculation was measured from hypocotyl to plant apex, and the mean height per replicate was calculated. The developmental stage of each plant was assessed as follows: vegetative (no reproductive structures), buds, flowers or pods, and the proportion of plants in each stage were calculated for each experimental unit.

2.8.11 Disease Assessment Scale for Root-Hair Infection of He et al. (2019)

Clubroot severity was investigated and scored on a six-point scale (McGrann et al. 2016). A disease severity index (DI and DIR) was calculated as follows:

$$\text{Disease Index (DI)} = \frac{\sum(\text{Rating} \times \text{diseased plant number})}{5 \times \text{Total plant number}} \times 100$$

Disease incidence rate (DIR) = Infected plant number/total plant number. In some experiments, plant height, total fresh weight, and clubroot gall fresh weights were also measured in plants. In other, disease control effect was calculated using the following formula:

$$\text{Control effect} = (1 - \text{DI in treatment}/\text{DI in control}) \times 100.$$

The data presented are the mean ± standard errors (SE) of at least 30 replicates and were analyzed by SPSS software 22.0.

2.8.12 Scoring Scale to Classify Genotypes for Resistance to *P. brassicae* (Liu et al. 2018)

Roots of each accession were assessed for clubroot disease severity at 5 weeks after inoculation using a standard 0 to 3 scale where: 0 = no clubbing, 1 = small clubs only, 2 = moderate clubs, and 3 = severe clubbing.

For statistical analysis, two indicators, disease incidence and disease index (DI), were used. The disease incidence of each accession was calculated according to the formula: disease incidence = Number of susceptible plants/total number of investigated plants. The DI was calculated using the following formula (Strelkov et al. 2006):

$$\text{ID}\ (\%) = \frac{\sum(\text{rating class}) \times (\#\text{plants in rating class})}{(\text{total}\#\text{plants in treatment}) \times 3} \times 100$$

The criteria for resistance classification according to the DI were as follows: DI = 0, highly resistant (HR); DI $<$ 10, resistant (R); 10 $\leq$ DI $\leq$ 20, moderately susceptible (MS); 20 $\leq$ DI $\leq$ 50, susceptible (S); DI $>$ 50, highly susceptible (HS). The plants with DI values of 0–10 were identified as resistant.

2.9 Dissemination of Clubroot

The soil-borne pathogen, intra- and inter-field spread of *P. brassicae* has generally been regarded as slow. The disease is usually monocyclic, and zoospore movement in the soil is limited (Howard et al. 2010). However, any activity that results in transport of soil from one point to another has the potential to disseminate clubroot. An examination of the distribution of *P. brassicae*-infected canola plants within clubroot-infested fields revealed that the highest frequency of infection occurs at the field entrances, and drops significantly at sampling points 150 m and 300 m from the entrance. This finding suggests that pathogen inoculum is introduced on farm or other machinery as it entered the field, presumably in the form of resting spores present in soil transported from clubroot-infested fields (Cao et al. 2009; Strelkov et al. 2011) .

While the dissemination of clubroot on field equipment appears to be the primary method of pathogen spread, other mechanisms have also been implicated. There is some evidence of the movement of clubroot via soil erosion although this is largely anecdotal and/or preliminary in nature. In addition, *P. brassicae* resting spores have been detected as external contaminants of seeds and tubers of various field crops grown in clubroot-infested fields (Rennie et al. 2011). Although the numbers of spores (as measured by quantitative PCR) were generally much lower than what can be carried on machinery, seed-borne inoculum may have the potential to initiate new infestations under favorable conditions. Hence, the dissemination of clubroot on propagative materials may represent another secondary mechanism of field-to-field spread. It is suspected that passenger vehicles, transport trucks, and heavy equipment associated with activities such as oil and gas well drilling and servicing, pipelining, earthmoving, and custom agricultural work also have the potential to move clubroot-infested soil both short and long distances, but this risk has not been critically assessed. Nevertheless, many companies engaged in these activities have adopted risk mitigation strategies by ACMC (Alberta Clubroot Management Committee) 2010; Strelkov et al. (2017).

Since *P. brassicae* is a soil-borne pathogen, any process that moves infested soil from one place to another can contribute to its dissemination. Therefore, the transportation of machinery from one field to another, such as during routine farming operations, can be expected to transport *P. brassicae* as well, if this machinery is contaminated with infested soil. Indeed, approximately 50 kg and 150 kg of soil could be recovered from a cultivator and tractor, respectively, in farm equipment sanitation trials in southern Alberta (Hwang et al. 2014), suggesting that the amounts of infested soil that can be moved on machinery are significant. The important role of machinery in the spread of clubroot was confirmed in a survey that examined the distribution of *P. brassicae*-infected canola plants within clubroot-infested fields (Cao et al. 2009). This study revealed that the highest frequency of infection occurred at the field entrances, and dropped significantly at sampling points 150 m and 300 m from the entrance, consistent with the introduction of *P. brassicae* inoculum in soil carried on contaminated farm machinery. This is also consistent with what has been observed in Europe, where farm and field gateways are traditional infection foci. The movement of heavy equipment associated with activities such as oil and gas well drilling and servicing, earthmoving, and custom agricultural work may also contribute to the dispersal of clubroot (Strelkov et al. 2011). Nonetheless, while the dissemination of *P. brassicae* on farm and other machinery appears to be its primary means of spread, other mechanisms also may be contributing to clubroot dispersal although to varying degrees.

Early studies (Warne 1943) and numerous anecdotal reports suggested that the transmission of *P. brassicae* as an external contaminant on seeds and tubers is possible. To evaluate this possibility, particularly within a western Canadian context, a quantitative PCR-based assay was developed to detect and measure the amounts of *P. brassicae* resting spores found on seeds of common field crops and potato tubers grown in clubroot-infested fields (Rennie et al. 2011). With this assay, quantifiable levels of infestation were found on seven of 46 seed and tuber samples tested, although at amounts that were generally much less than what can be carried on machinery. Nonetheless, the identification of seed- and tuber-borne *P. brassicae* inoculum suggests that clubroot may spread on propagative materials, and that this possibility should not be ignored in the development of clubroot-containment strategies (Rennie et al. 2011). Practices such as commercial seed cleaning and seed treatments with products containing fungicides such as flusulfamide appear to be highly effective at reducing seed-borne inoculum of *P. brassicae* (Rennie et al. 2011; Hwang et al. 2012), and therefore should be routinely employed by farmers to help prevent clubroot spread.

Anecdotal evidence also suggests that *P. brassicae* can be disseminated through wind- and water-mediated erosion. Indeed, surveys in Alberta have revealed the occurrence of clubroot along water runs and the edges of otherwise disease-free fields, consistent with the movement of resting spores in water or wind-borne dust (Strelkov et al. 2011). A preliminary study was conducted to assess the significance of wind-borne dispersal of *P. brassicae* in commercial canola production (Rennie et al. 2012). Dust samplers were deployed at five fields in central Alberta and one field in southern Alberta, which were under various tillage regimes. The samplers collected dust at five different heights at four or five different locations at each site, with the accumulated dust retrieved at biweekly intervals and analyzed for both dust loads and resting spore presence/quantity. The resulting data indicated that P. brassicae resting spores can be moved in wind-borne dust, and that dust loads varied according to sampler height (Rennie et al. 2012). Resting spore concentrations as high as 2.2×10^5 spores g^{-1} soil were found in some of the dust samples, but in preliminary studies no clear relationships were established between dust quantity, sampler height or position, and resting spores (Rennie et al. 2012). Nonetheless, the results support the suggestion that soil erosion may contribute to the dissemination of clubroot, with the magnitude of this contribution dependent on the amount of infested dust moved and the distance that it travels. The various mechanisms implicated in the dissemination of *P. brassicae* in western Canada are summarized in Fig. 2.4.

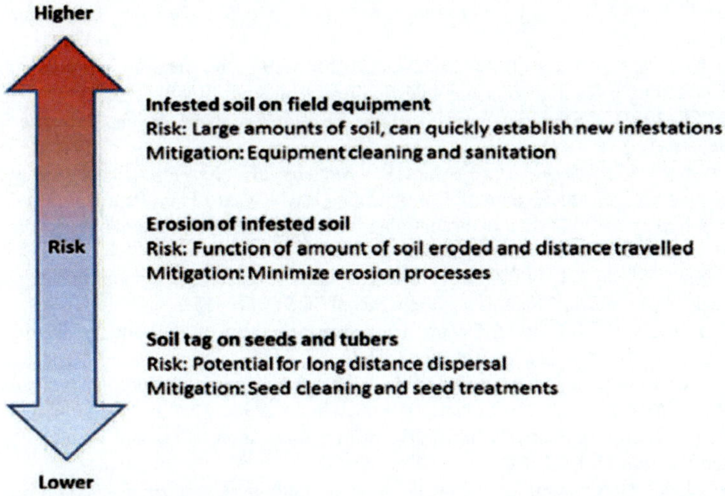

Fig. 2.4 Possible *Plasmodiophora brassicae* dissemination mechanisms in western Canada. The relative risk and possible mitigation strategies associated with each mechanism are indicated (Strelkov and Hwang 2014)

References

Abbay J (1831) Report on the disease in turnips called finger and toe. Trans Highland and Agri Soc Scot 2:238–240

ACMC (Alberta Clubroot Management Committee) (2010). Available from http://www1.agric.gov.ab.ca/

Anderson A (1855) Report on the disease of finger and toe in turnips. Trans Highland Agric Soc Scot 1855:118–140

Anon (1987) Insect pests of economic significance affecting major crops of the countries in Asia and the Pacific region. APPPC, Technical Document No. 135. Bangkok, Thailand: Regional FAO Office for Asia and the Pacific (RAPA)

Anon (1996) European plant protection organization (EPPO) PQR database. France, Paris

Apaydin A, Deligoz I, Kar H, Kibar B, Karaagac O (2010) An investigation on clubroot disease (*Plasmodiophora brassicae* Wor.) races in the Black Sea Region of Turkey. Zirat Fabrikasi Dergisi 28:57 60

AVA (2001) Diagnostic records of the plant health diagnostic services, plant health centre. Agri-Food Veterinary Authority, Singapore

Bailey LH (1961) Manual of cultivated plants. The MacMillan Company, New York

Belec C, Tremblay N, Coulombe J (2004) Liming and calcium cyanamide for clubroot control in cauliflower. Acta Hortic 635:41–46

Bhattacharya I, Dutta S, Mondal S, Mondal B (2014) Clubroot disease on *Brassica* crops in India. Canadian J. Plant Pathol 36(1):154–160

Birnie M (1831) Report on the disease in turnips called fingers and toes. Trans Highland Agric Soc Scot 2:241–242

Botero-Ramırez A, Gomez I, Benıtez E, Garcıa C (2015) Liming with dolomite reduces the efficacy of the biocontrol fungus *Trichoderma koningiopsis* against cabbage clubroot. Agron Colomb 33:49–57

Buczacki ST (1979) IMI descriptions of fungi and bacteria (No. 63), sheet 621. CAB International, Wallingford

Buczacki ST, Ockendon JG (1979) Preliminary observations on variation in susceptibility to clubroot among collections of some wild crucifers. Ann Appl Biol 92(1):113–118

CABI/EPPO (2011) *Plasmodiophora brassicae*. Distribution maps of plant diseases. CABI, Wallingford

Cao T, Manolii VP, Hwang SF, Howard RJ, Strelkov SE (2009) Virulence and spread of *Plasmodiophora brassicae* [clubroot] in Alberta, Canada. Can J Plant Pathol 31:321–329

Castillo J, Guerrero O (2008) Efecto de controladores biologicos sobre la hernia de las crucıferas en Tabio, Cundinamarca. Inventum 5:30–40

Chai AL, Xie XW, Shi YX, Li BJ (2014) Research status of clubroot (*Plasmodiophora brassicae*) on cruciferous crops in China. Can J Plant Pathol 36(S1):142–153

Chen SF, Liu XB, Hu BS (1997) Occurrence and management of clubroot disease on Chinese cabbage in Aletai, Xingjiang. Plant Prot Tech Ext 17:21–22

Chittem K, Mansouripour SM, Mendoza LEdel R (2014) First report of clubroot on canola caused by *Plasmodiophora brassicae* in North Dakota. Plant Dis 98(10):1438

Cicu (2006) Clubroot disease (*Plasmodiophora brassicae*) on crucifers and its control. J Penel Pengemb Pertan 25(1):16–21

Colhoun J (1953) Observations on the incidence of club root disease of *Brassicae* in limed soils in relation to temperature. Ann Appl Biol 40:639–644

Colhoun J (1958) Club root disease of crucifers caused by plasmodiophora brassicae Woron. A monograph. Phytopathological paper No 3. Surrey, Commonwealth Mycological Institute

Cortes MR (2015) Impacto de la Hernia de las Crucıferas (*Plasmodiophora brassicae* Woronin) en la Horticultura de Costa Rica: Epidemiologıa y Manejo Organico. Mexico: Colegio de Postgraduados (COLPOS), PhD thesis

Crete R (1981) Worldwide importance of clubroot. Clubroot Newslett 11:6–7

Cruz JCS, de Souza NL, Nakatani AK, Rosa DD, Basseto MA, Padovani CR, Furtado EL (2008) Caracterizacao patogenica e molecular de *Plasmodiophora brassicae*. Trop Plant Pathol 33:415–424

Cvjetkovic B (2008) Less frequent diseases of oilseed rape. (RjeÐe ucestale bolesti uljane repice.). Glasilo Biljne Zatite 8(5):318–320

Daowang S, Jialuan Y, Mingying Y, Weizhong Y (2004) Decreasing yield loss and residual analysis using 75% Dacotech in controlling cabbage clubroot. Southwest Chin J Agric Sci 17 (2):189–191

Dekhuijzen HM (1976) Microscopical studies on the cortex and medullary rays of *Brassica campestris* var. *rapa* infected with *Plasmodiophora brassicae*. J Phytopathol 87(2):171–186

Deora A, Gossen BD, McDonald MR (2012) Infection and development of *Plasmodiophora brassicae* in resistant and susceptible canola cultivars. Can J Plant Pathol 34:239–247

Diederichsen E, Frauen M, Ludwig-Muller J (2014) Clubroot disease management challenges from a German perspective. Can J Plant Pathol 36(S1):85–98

Dixon GR (2006) The biology of *Plasmodiophora brassicae* Worn., -a review of recent advances. Acta Hortic 706:271–282

Dixon GR (2007) Vegetable brassicas and related crucifers. Crop production science in horticulture series No 14. Crop production science in horticulture CABI Publishing, Wallingford, Oxfordshire

Dixon GR (2009a) *Plasmodiophora brassicae* in its environment. In: Dixon GR (eds) *Plasmodiophora brassicae* (clubroot) –a plant pathogen that alters host growth and productivity. J Plant Growth Regul 28:212–228

Dixon GR (2009b) The occurrence and economic impact of *Plasmodiophora brassicae* and clubroot disease. J Plant Growth Regul 28:194–202

Dixon GR (2014) Clubroot (*Plasmodiophora brassicae* Woronin): an agricultural and biological challenge worldwide. Can J Plant Pathol 36(1):5–18

Dobson RL, Robak J, Gabrielson RL (1983) Pathotypes of *Plasmodiophora brassicae* in Washington, Oregon and California. Plant Dis 67:269–271

Donald C, Porter I (2003) Clubroot (*Plasmodiophora brassicae*) an imminent threat to the Australian canola industry. In: Proceedings of the 13th biennial research assembly on brassicas. New South Wales Agriculture, pp 114–118

Donald EC, Porter IJ (2014) Clubroot in Australia: the history and impact of *Plasmodiophora brassicae* in *Brassica* crops and research efforts directed towards its control. Can J Plant Pathol 36:66–84

Donald EC, Porter IJ, Faggian R, Lancaster RA (2006) An integrated approach to the control of clubroot in vegetable *Brassica* crops. Acta Hortic 706:283–300

Ellis W (1742The modern husbandman, or, the practice of farming. 4 v. London

EPPO (2014) PQR database. Paris: European and Mediterranean Plant Protection Organization. http://www.eppo.int/DATABASES/pqr/pqr.htm

EPPO (2019) EPPO global database. https://www.eppo.int/RESOURCES/eppo_databases/global_database

Fan GP, Yang QH, Ning JC, Qian CP, Li AW, Li ZS (2013) Occurrence and management of clubroot on rape in Eshan city during 2005–2012. Chin Plant Prot 8:46–48

Farquharson J (1831) Report on the disease of finger and toe in Turnips. Highland Agric Soc Scot Trans 2:233–238

Feng J, Hwang SF, Strelkov SE (2013) Studies into primary and secondary infection processes by *Plasmodiophora brassicae* on canola. Plant Pathol 62:177–183

Galdames R, Aguilera N, Mera M (2014) Outbreak risk of clubroot caused by *Plasmodiophora brassicae* on oilseed rape in Chile. Plant Dis 98:1437

Garcıa R, Riera R, Zambrano C, Gutierrez L (2006) Desarrollo de un fungicida biologico a base de una cepa del hongo *Trichoderma harzianum* proveniente de la region andina venezolana. Fitosanidad 10:115–121

Guerrero SLA (2016) Evaluacion del Comportamiento Productivo de Dos Variedades de Coliflor Hibrido (Brassica oleracea var. botrytis), a Diferentes Densidades de Plantaci_on, Bajo Ambiente Atemperado en el Centro Experimental de Cota. Cota, Bolivia: Universidad Mayor de San Andres, Agronomist thesis

Hartman M (2007) Clubroot disease of canola and mustard. Alberta agriculture and rural development. http://www1.agric.gov.ab.ca/department/deptdocs.nsf/all/agdex8593

Hatakeyama K, Fujimura M, Ishida M, Suzuki T (2004) New classification method for *Plasmodiophora brassicae* field isolates in Japan based on resistance of F1 cultivars of Chinese cabbage (*Brassica rapa* L.) to clubroot. Breed Sci 54:197–201

He P, Cui W, Munir S, He P, Li X, Wu Y, Yang X, Tang P, He Y (2019) *Plasmodiophora brassicae* root hair interaction and control by *Bacillus subtilis* XF-1 in Chinese cabbage. Biol Control 128:56–63

Henderson P (1867) Gardening for profit; A guide to the successful cultivation of the market and family garden, New York, p 243

Hernandez REC (2002) Efecto Biofumigante de Diversas Fuentes de Materia Organica en el Cultivo de Brocoli(Brassica oleracea var. italica). Chimaltenango, Guatemala: Universidad Rafael Landıvar, Guatemala, Agronomist thesis

Hind-Lanoiselet T, Parker P (2005) Clubroot of canola and mustard. Primefact 115:1–2

Horikoshi N, Tairako K (2002) Occurrence of clubroot in Japanese radish caused by *Plasmodiophora brassicae* Woronin in Fukushima (in Japanese with English summary). Ann Rep Prot Plant 53:58–60

Horiuchi S, Hori M (1980) A simple greenhouse technique for obtaining high levels of clubroot incidence. Bull Chugoku Nat Agric Exp Stn 17:33–55

Howard RJ, Strelkov SE, Harding MW (2010) Clubroot of cruciferous crops – new perspectives on an old disease. Can J Plant Pathol 32:43–57

Huang QW, Ouyan L, Wang YJ (1955) Occurrence and prevention of clubroot disease on cruciferous crops in Jiangxi. Bull Plant Prot 8:1–4

Hwang SF, Ahmed HU, Strelkov SE, Gossen BD, Turnbull GD, Peng G, Howard RJ (2011) Seedling age and inoculum density affect clubroot severity and seed yield in canola. Can J Plant Sci 91:183–190

Hwang SF, Strelkov SE, Feng J, Gossen BD, Howard RJ (2012) *Plasmodiophora brassicae*: a review of an emerging pathogen of the Canadian canola (*Brassica napus*) crop. Mol Plant Pathol 13(2):105–113

Hwang SF, Howard RJ, Strelkov SE, Gossen BD, Peng G (2014) Management of clubroot (*Plasmodiophora brassicae*) on canola (*Brassica napus*) in western Canada. Can J Plant Pathol 36(S1):49–65

Ikegami H (1978) Research trends of clubroot on cruciferous vegetables in recent years (in Japanese). Shokubutsu Boeki 32:53–61

IMI Herbarium (2019) Herbarium specimen. CABI Bioscience, Egham

Jedryczka M, Kasprzyk I, Korbas M, Jajor E, Kaczmarek J (2014) Infestation of Polish agricultural soils by *Plasmodiophora brassicae* along the Polish-Ukrainian border. J Plant Prot Res 54:238–241

Jing W, Yun H, Xiaoling H, Yingze N, Xiaolan L, Yong L (2008) Study of symptoms, yield loss of clubroot and modality of *Plasmodiophora brassicae* in rape. Chin J Oil Seed Sci 30(1):112–115

Kageyama K, Asano T (2009) Life cycle of *Plasmodiophora brassicae*. J Plant Growth Regul 28 (3):203–211

Karling JS (1968) The plasmodiophorales. Hafner Publishing Company Inc., New York

Khangura RK, Wright DG (2012) First report of club root caused by *Plasmodiophora brassicae* on canola in Australia. Plant Dis 96(7):1075–1076

Kim WG, Moon MH, Kim JH, Choi HW, Hong SK (2009) Occurrence of clubroot on Pak-Choi caused by *Plasmodiophora brassicae*. Mycobiology 37(1):69–71

Kim H, Jo EJ, Choi YH, Jang KS, Choi GJ (2016) Pathotype classification of *Plasmodiophora brassicae* isolates using clubroot-resistant cultivars of Chinese cabbage. Plant Pathol J 32 (5):423–430

Kobringer KM, Hagedorn DJ (1983) Determination of bean root rot potential in vegetable production fields of Wisconsin's Central Sands. Plant Dis 67:177–178

Kuginuki Y, Hiroaki Y, Hirai M (1999) Variation in virulence of *Plasmodiophora brassicae* in Japan tested with clubroot-resistant cultivars of Chinese cabbage (*Brassica rapa* L. spp. *pekinensis*). Eur J Plant Pathol 105:327–332

Kunkel LO (1918) Tissue invasion by *Plasmodiophora brassicae*. J Agric Res 14:543–572

Labrador-Morales M, del Pozo-Nunez E, Garcıa-Cruz I (2014) Efecto de *Trichoderma harzianum* Rifai sobre *Plasmodiophora brassicae* Woronin, en brocoli, en la localidad de Escaguey, municipio Rangel, estado Merida. Centro Agrıcola 41:85–90

LeBoldus JM, Manolii VP, Turkington TK, Strelkov SE (2012) Adaptation to Brassica host genotypes by a single-spore isolate and population of *Plasmodiophora brassicae* [clubroot]. Plant Dis 96:833–838

Li MY (1990) Occurrence and prevention of clubroot disease of cruciferous crops. Beijing Agric Sci 4:11–13

Li HM, Pan YH (1999) Occurrence rules and control technology of clubroot disease on cruciferous crops. J Changjiang Veg 9:15–16

Li JK, Bai M, Xia XY, Zhaxi CR, Ou Z, Wei J (1993) Research on the soil causes of clubroot on cruciferous vegetables in Yadong City, Tibet. J Tibet Univ 8:70–72

Lima MLP, CaféFilho AC, Nogueira NL, Rossi ML, Schuta LR (2004) First report of clubroot of Eruca sativa caused by *Plasmodiophora brassicae* in Brazil. Plant Dis 88(5):573

Linnasalmi A, Toiviainen A (1991) Occurrence of clubroot and *Plasmodiophora brassicae* Wor. races in Finland. J Agric Sci Finl 63:415–434

Liu C, Yang Z, He P, Munir S, Wu Y, Ho H, He Y (2018) Deciphering the bacterial and fungal communities in clubroot-affected cabbage rhizosphere treated with *Bacillus Subtilis* XF-1. Agric Ecosyst Environ 256:12–22

Loganathan M, Swarnakumari N, Sivakumar M, Prakasam V, Mohan L, Ramaraj B, Samiyappan R (2001) Biological suppression of fungal nematode complex diseases of major cruciferous vegetables. South Indian Hortic 49:296–300

Ludwig-Muller J, Bennett RN, Kiddle G, Ihmig S, Ruppel M, Hilgenberg W (1999) The host range of *Plasmodiophora brassicae* and its relationship to endogenous glucosinolate content. New Phytol 141(3):443–458

Mazuma ED, Theu MP, Likongwe MC (2008) Participatory on-farm evaluation of Ridomil Gold MZ 68 WP for its efficacy for clubroot disease control in cabbage. In: Plant protection progress report for the 2007/2008 season, presented at the Department of Agricultural Research Services Planning and Review Meeting, Andrews Hotel, Mangochi, 14-20 September, 2008. Lilongwe, Malawi: Ministry of Agriculture and Food Security, pp 127–134

McDonald MR, Kornatowski B, McKeown AW (2004) Management of clubroot in Asian brassica crops grown on organic soils. Acta Hortic 635:25–30

McGrann GRD, Gladders P, Smith JA, Burnett F (2016) Control of clubroot (*Plasmodiophora brassicae*) in oilseed rape using varietal resistance and soil amendments. Field Crop Res 186:146–156

Minogue L (2013) The history of clubroot. Grainews

Morrison D (1977) Preliminary report on turnip and swede surveys, 1975/1976. In: Proceedings of the *Brassica* Fodder Crops Conference, Scottish Agricultural Development Council, Dundee published by Scottish Plant Breeding Station (now Scottish Crop Research Institute), Dundee

Muhammad S, Khan SH (2008) Effect of abiotic factors on incidence and severity of diseases on cabbage (Brassica oleracea L.) in Dera Ghazi Khan. Pak J Phytopathol 20(1):108–112

Pachari L (2011) Trichoderma harzianum y Bacillus subtilis en el manejo integrado de la ‘potra de la col’ (*Plasmodiophora brassicae*) en brocoli (*Brassica oleracea* L. var. *italica*) cv. Legacy en Chilina-Arequipa. Arequipa, Ecuador: Universidad Nacional de San Agustın, Agronomist thesis

Pageau D, Lajeunesse J, Lafond J (2006) Impact of clubroot [*Plasmodiophora brassicae*] on the yield and quality of canola. (Impact de l'hernie des cruciferes [*Plasmodiophora brassicae*] sur la productivité et la qualité du canola.). Can J Plant Pathol 28(1):137–143

Porth G, Mangan F, Wick R, Autio W (2003) Evaluation of management strategies for Clubroot (*Plasmodiophora brassicae* Woron) in Brassicaceae. In: Proceedings of the Inter-American Society for Tropical Horticulture [48th Annual Meeting of the Inter American Society for Tropical Horticulture, Tegucigalpa, Honduras, 7–11 October, 2002.], vol. 46, pp 78–80

Rastas M, Latvala S, Hannukkala A (2012) Occurrence of *Plasmodiophora brassicae* in Finnish turnip rape and oilseed rape fields. Agric Food Sci 21:141–158

Ravn FK (1908) Kaalbroksvampen. Tidsskr Landbr Planteavl 15:5271–5620

Ren L, Fang XP, Sun CC, Chen KR, Liu F, Li M, Xu L (2014) First report of clubroot on *Capsella bursa-pastoris* caused by *Plasmodiophora brassicae* in Sichuan Province of China. Plant Dis 98 (5):687

Ren L, Xu L, Liu F, Chen K, Sun C, Li J, Fang X (2016) Host range of *Plasmodiophora brassicae* on cruciferous crops and weeds in China. Plant Dis 100:933–939

Rennie DC, Manolii VP, Cao T, Hwang SF, Howard RJ, Strelkov SE (2011) Direct evidence of surface infestation of seeds and tubers by *Plasmodiophora brassicae* and quantification of spore load. Plant Pathol 60:811–819

Rennie DC, Turkington TK, Leboldus JM, Howard RJ, Kutcher HR, Strelkov SE (2012) Dust monitoring for clubroot [*Plasmodiophora brassicae*] in Alberta. Can J Plant Pathol 35:99–132

Ricarova V, Kazda J, Baranyk P, Rysanek P (2016a) Clubroot – an emerging disease faced by Czech oilseed rape growers. Sci Agric Bohem 47(3):105–112

Ricarova V, Kazda J, Singh K, Rysanek P (2016b) Clubroot caused by *Plasmodiophora brassicae* Wor.: a review of emerging serious disease of oilseed rape in the Czech Republic. Plant Prot Sci 52:71–86

Rimmer SR, Kutcher HR, Morrall RAA (2003) Diseases of canola and mustard. In: Bailey KL, Gossen BD, Gugel RK, Morrall RAA (eds) Diseases of field crops in Canada. Canadian Phytopathological Society, Saskatoon, pp 129–146

Robak J (1991) Zmiennosc patotypow *Plasmodiophora brassicae* Wor. występujących w Polsce i ich patogeniczność w stosunku do odmian i linii hodowlanych *Brassica oleracea*. Praca habilitacyjna. [Variability of the pathotypes of *Plasmodiophora brassicae* Wor. present in Poland and their pathogenicity to the cultivars and breeding lines of *Brassica oleracea*. Habilitation monograph]. Instytut Warzywnictwa, Skierniewice, Poland, p 59

Robles Sanchez R (1991) Produccion de oleaginosas y textiles, 3rd edn. Limusa, Cuidad de Mexico

Romero-Cova S (1990) Identification of several phytopathogenic fungi which are new or little known in Mexico. Rev Mex Fitopatol 8:1–8

Sanchez-Alonso B (2007) The other Europeans: immigration into latin America and the international labour market, 1870-1930. J Iberian Latin Am Econ Hist 25(3):395–426

Sanchez-Alonso BL (2013) Respuesta del Brocoli (*Brassica oleracea* var. *italica*) Hıbrido Fantastico a la Fertilizaci_on con Cinco Fuentes de Azufre Aplicadas a Dos Dosis. Quito, Ecuador: Universidad Central del Ecuador, Agronomist thesis

SAS Institute, Inc. (ed) (1999) SAS/STAT user's guide, version 9.2, 2nd edn. SAS Institute, Inc., Cary

Sen P (2005) Antagonistic effect of Ca, B and Mo on club-root disease of rape-mustard. Indian Agric 49(1/2):13–16

Some A, Manzanares MJ, Laurens F, Baron F, Thomas G, Rouxel F (1996) Variation for virulence on *Brassica napus* L. amongst *Plasmodiophora brassicae* collections from France and derived single spore isolates. Plant Pathol 45:432–439

Strehlow B, de Mol F, Struck C (2015) Risk potential of clubroot disease on winter oilseed rape. Plant Dis 99:667–675

Strelkov SE, Hwang SF (2014) Clubroot in the Canadian canola crop: 10 years into the outbreak. Can J Plant Pathol 36(S1):27–36

Strelkov SE, Tewari JP, Smith-Degenhardt E (2006) Characterization of *Plasmodiophora brassicae* populations from Alberta, Canada. Can J Plant Pathol 28:467–474

Strelkov SE, Manolii VP, Cao T, Xue S, Hwang SF (2007) Pathotype classification of *Plasmodiophora brassicae* and its occurrence in *Brassica napus* in Alberta. J Phytopathol 105:706–712

Strelkov SE, Hwang SF, Howard RJ, Hartman M, Turkington TK (2011) Progress towards the sustainable management of clubroot [*Plasmodiophora brassicae*] of canola on the Canadian Prairies. Prairie Soils Crops J 4:114–121

Strelkov SE, Hwang SF, Manolii VP, Cao T, Feindel D (2016) Emergence of new virulence phenotypes of *Plasmodiophora brassicae* on canola (*Brassica napus*) in Alberta, Canada. Eur J Plant Pathol 145:517–529

Strelkov SE, Manolii VP, Harding MW, Hwang SF, Manolii E, Zuzak K, Rennie DC, Feng J, Rahaman M, Daniels GC, Burke DA, Hill TB, Zhar K, Feindel D (2017) Occurrence and spread of clubroot on canola in Alberta in 2016. Can Plant Dis Surv 97:164–167

Strelkov SE, Hwang SF, Manolii VP, Cao T, Fredua-Agyeman R, Harding MW, Peng G, Gossen BD, Mcdonald MR, Feindel D (2018) Virulence and pathotype classification of *Plasmodiophora brassicae* populations collected from clubroot resistant canola (*Brassica napus*) in Canada. Can J Plant Pathol 40(2):284–298

Suwabe K, Tsukazaki H, Iketani H, Hatakeyama K, Fujimura M, Nunome T, Fukuoka H, Matsumoto S, Hirai M (2003) Identification of two loci for resistance to clubroot (*Plasmodiophora brassicae* Woronin) in *Brassica rapa* L. Theor Appl Genet 107:997–1002

Tanaka S, Ito S (2013) Pathogenic and genetic diversity in *Plasmodiophora brassicae* (clubroot) from Japan. J Gen Plant Pathol 79:297–306

Tang WH, Li H, Sun J (1990) Ecological conditions and management of clubroot disease on Chinese cabbage in Beijing. Beijing Agric Sci 6:6–9

Telenchana PMY (2015) Aplicacion de Productos Sello Verde en el Manejo de la Hernia de las Crucıferas (*Plasmodiophora brassicae*) en el Cultivo de Brocoli (*Brassica oleracea* var. Avenger), en las Condiciones Agroecol_ogicas de Izamba. Cevallos, Ecuador: Universidad Tecnica de Ambato, Agronomist thesis

Tewari JP, Orchard D, Hartman M, Lange R, Turkington TK, Strelkov S (2004) First report of clubroot of canola caused by *Plasmodiophora brassicae* in the Canadian Prairies. Can J Plant Pathol 26:228–229

Tewari JP, Strelkov SE, Orchard D, Hartman M, Lange RM, Turkington TK (2005) Identification of clubroot of crucifers on canola (*Brassica napus*) in Alberta. Can J Plant Pathol 27:143–144

Timila RD, Correll JC, Duwadi VR (2008) Severe and widespread clubroot epidemics in Nepal. Plant Dis 92:317

Todorovic D (2007) Quarter-century of presence and spread of *Plasmodiophora brassicae*, cabbage hernia in Leskovac Region. (Cetvrt veka prisustva i irenja kile kupusa (*Plasmodiophora brassicae*) u okolini Leskovca.). Biljni Lekar 35(5):505–511

Torres E (1972) Reaccion de algunas crucıferas al ataque de *Plasmodiophora brassicae* Woronin en Manizales, Colombia. Acta Agron 22:185–207

Wallenhammar AC, Almquist C, Schwelm A, Roos J, Marzec-Schmidt K, Jonsson A, Dixelius C (2014) Clubroot, a persistent threat to Swedish oilseed rape production. Can J Plant Pathol 36 (S1):135–141

Wang HF (1962) Discussion of clubroot disease on cruciferous vegetables in China - from the disease found in Jiamusi City. J Northeast Agric Univ 2:47–48

Wang XY, Peng HJ, Gao MQ, Han HB, Zhang JH, Xiao CG (2002) Preliminary identification of the pathogen of root tumour of stem mustard and factors influencing the incidence of the disease. Southwest Chin J Agric Sci 15(4):75–78

Wang J, Huang Y, Li XL, Li HZ (2011) Research progress in clubroot of crucifers. Plant Prot 37:153–158. (in Chinese)

Warne LGG (1943) A case of club-root of swedes due to a seed-borne infection. Nature 152:509

Watson AG (1965) Clubroot control in Pukekohe. N Z J Agric Res 8(4):988–996

Wellman FL (1930) Clubroot of crucifers. Technical Bulletin No. 181, United States Department of Agriculture Washington, DC

Wesołowska M (2014) Identification of *Plasmodiophora brassicae* Wor. isolate suppressing clubroot resistance in 'Kilaxy' F_1 white cabbage. Fol Hortic 26:57–62

Williams PH (1966) A system for the determination of races of *Plasmodiophora brassicae* that infect cabbage and rutabaga. Phytopathology 56(6):624–626

Williams PH, McNabola SS (1967) Fine structure of *Plasmodiophora brassicae* in sporogenesis. Can J Bot 45:1665–1669

Woronin M (1878) *Plasmodiophora brassicae*, the Cause of Cabbage Hernia. Jahrb Wiss Bot 11:548-574 [translated from German by Chupp C (1934) Phytopathological classics no 4. American Phytopathological Society, St. Paul]

Wu L, Siemens J, Li S, Ludwig-Müller J, Gong Y, Zhong L, He J (2012) Estimating *Plasmodiophora brassicae* gene expression in lines of *B. rapa* by RT-PCR. Sci Hortic 133:1–5

Xue S, Cao T, Howard RJ, Hwang SF, Strelkov SE (2008) Isolation and variation in virulence of single-spore isolates of *Plasmodiophora brassicae* from Canada. Plant Dis 92(3):456–462

Yamada M, Asandhi AA, Purwati E, Dianawati M (2004) Control of cabbage clubroot disease by employing one-year crop rotation with three vegetable combinations in the West Java highlands of Indonesia. JIRCAS Working Report, No. 43:175-184. [Evaluation of vegetable-based farming systems and improvement of vegetable and fruit cultivation in highland regions of West Java, Indonesia.]

Yang YH, Zhang XF, Zou LJ (1992) Studies of the occurrence of Chinese cabbage clubroot in Changsha. J Hunan Agric Univ 18:854–858

Yoshikawa H, Buczacki ST (1978) Clubrot in Japan: research and problems. Rev Plant Pathol 57 (7):253–257

Zhu GN (1956) Clubroot on cruciferous vegetables. Bull Agric Sci 11:677–678

The Pathogen: *Plasmodiophora brassicae* 3

Abstract

The pathogen *Plasmodiophora brassicae* Woronin is a very serious soil borne, intracellular, obligate, non-axenic, endoparasitic, wide spread, and destructive to cruciferous crops all over the world causing clubroot disease which is nutrient sick to the host plant. The pathogen is highly variable, and exists as mixture of pathotypes in the soil and even in single-clubbed roots. The isolates differ in preferences and aggressiveness towards different cruciferous hosts. The survival of the pathogen in the soil leads to infection, development, and multiplication in host tissues as well as in soil. It produces morphological structures like resting spores, zoospores, and plasmodia. The pathogen has a very complex life cycle consisting of different zoosporic stages (primary and secondary), formation of primary and secondary plasmodia, and resting spores inside the host cells. Plasmodia provoke hypertrophy and hyperplasia leading to the development of galls that obstruct nutrients and water transport to the infected plants. The only host-free stages are the short-lived biflagillated zoospores and resting spores. Resting spores are extremely resilient to harsh environmental conditions, and survive in the soil for more than 20 years. Although, traditionally, *Plasmodiophora* was classified as fungus but now molecular phylogenetic analysis of the small subunit (SSU) ribosomal RNA, actin, and polyubiquitin genes suggests that it belongs to the protists phylum *Cercozoa* in the kingdom *Rhizaria* and Plasmodiophorida order. The sexual reproduction and meiosis appear to occur in the plasmodium before resting spore formation but how and when karyogamy takes place is not exactly known so far. The *P. brassicae* genome is compact with small size (24.2–25.5 Mb), high gene density, and with scarcity of repeats. In its genome, there is reduction and overlapping of intergenic distances. High degree of variation in field populations indicates sexual recombination. The molecular biology techniques have revealed the *P. brassicae* genome size; recombination pattern; chromosome number; number of genes; gene location;

G. S. Saharan et al., *Clubroot Disease of Crucifers*,
https://doi.org/10.1007/978-981-16-2133-8_3

gene structure; gene functions during pathogenesis, disease developmental stages, pattern of gene expression, and signaling molecules for metabolic pathways.

Keywords

Pathogen · *Plasmodiophora brassicae* · Morphology · Classification and taxonomy · Life cycle · Molecular biology · Pathogenesis · *P. brassicae* genome · Genome characterization · Factors affecting host-pathogen interactions

3.1 Introduction

Members of Plasmodiophoromycota are usually necrotrophic endoparasites of vascular plants and stramenopiles. These members cause an abnormal enlargement of host cells called hypertrophy, and often an abnormal multiplication of host cells called hyperplasia. The result is the enlargement of infected portion of the host, and in vascular plants, disruption of vascular elements. General stunting of the host may occur, and may be more dramatic when plants are not flowering. This fact has been attributed to reduce competition for resources by the pathogen in the absence of the nutrient sink required in flowering development. Some species parasitize stramenopiles, including freshwater algae such as *Vaucheria,* and water molds including *Saprolegnia*, *Achlya,* and *Pythium*. Other species parasitize aquatic and amphibious vascular plants in the genera *Callitriche*, *Isoetes*, *Zostera,* and *Juncus* or plants such as cabbage, sugar beet, grasses, potato, *Nasturtium,* and *Veronica*.

Only a few species are of real economic importance, *Plasmodiophora brassicae* is the widespread cause of clubroot or finger, and toe disease of cabbage, and related crucifers, both cultivated and wild, whereas *Spongospora subterranean* is the causal agent of powdery scab of potato. Its special form *S. subterranean* f. sp. *nasturtii* causes crook root disease of watercress. Various species of plasmodiophorids also may serve as vector for many plant viruses. Plasmodiophorids produce multinucleate, unwalled protoplasts typically referred to as plasmodia. These structures are, however, significantly different from the plasmodia of Myxomycota as they are incapable of translocational movement, lack the ability to phagocytize food materials, and exist wholly within the cells. In some species, the plasmodium appears to be intracellular, separated from host cytoplasm only by the plasma membrane of the plasmodium, while in *P. brassicae,* the closely appressed plasma membranes of both the host cell and the plasmodium appear to form the host–pathogen interface.

The plasmodiophorid life cycle of *P. brassicae* involves the production of two different plasmodial phases. One phase is termed the primary or sporangial plasmodium, and gives rise to thin walled zoosporangia, while the other is called as the secondary or sporogenic plasmodium which produces thick walled (apparently chitinous) resting spores capable of surviving desiccation to germinate by zoospores. During their growth both primary and secondary plasmodia are characterized by distinctive mitotic divisions, which presumably have associated centrioles. During mitosis, an intra-nuclear spindle is formed on which the chromosomes are arranged in a ring around the nucleolus. As the chromosomes separate, a ring of chromosomes passes to each pole. The nucleolus persists for some time, and elongates. Because of the elongated nucleolus which is surrounded by a ring of chromosomes appears like a cross when viewed from the side, the name cruciform has been given to this type of division.

3.2 Morphology of *Plasmodiophora brassicae* in Host–Pathogen Interface

In the early stage of infection, *Plasmodiophora brassicae* in an infected cell appeared as nearly too completely spherical bodies of 3-5μm in diameter. The young plasmodia assume various shapes, such as spherical bodies with or without strand-like material, those with smooth or rough surfaces, and cluster-like bodies. The strand-like material connecting the young plasmodia seems to be the same material (Dekhuijzen 1976) as observed with transmission electron micrographs (TEM). The morphological diversity of young plasmodia is presumably due to the nutritional condition in a host cell, and the root portion from which the cell is sampled. Thus, the initial phase of infection requires more detailed investigation by scanning and transmission electron microscopes. As some young plasmodial bodies grew a large multinucleate plasmodium may possibly be formed by their fusions. The vegetative plasmodia which grew progressively in an infected cell are observed to form a large pathogenic mass in the cell.

In observations of sectioned tissue of clubs from 45 to 60 days old plants, plasmodia in the infected cells show clearly different growth rates. The various cells including a young plasmodium, a vegetative plasmodium, or a resting spore mass were observed in the same section. An infected cell seems to influence the growth of a pathogen in the contiguous cell. A young plasmodium or a vegetative plasmodium, and a mature spore mass are recognized in an infected cell. It is presumed that these plasmodia are finally unable to mature into resting spores. The mature resting spores in the infected cells fail to show any membrane enveloping them. In the process by which resting spores mature from vegetative plasmodium, the plasmodial membrane is presumably disintegrated. Williams and McNabola (1967) described that as sporogenesis is initiated, the outer membrane of the plasmodial envelope disintegrates during maturity. A large number of resting spores is easily distributed into the soil by rotting of clubbed roots, where they become a formidable inoculum of clubroot disease. The number of resting spores on an average per g of fresh club 4×10^8 in both turnip, and Chinese cabbage (Ikegami et al. 1978) has been recorded. The pathogen produces following morphological structures for its survival, infection, development, and multiplication in the infected host tissues of crucifers.

3.2.1 Resting Spores

The resting spore is about 3 μm in size, and sub-spherical to spherical (Buczacki and Cadd 1976). The surfaces of each resting spores are covered with spines (Williams and McNabola 1967; Ikegami et al. 1978). Young resting spores are connected by fine filamentous material, and mature resting spores have spines (Ikegami et al. 1978).

3.2.2 Zoospores

Zoospores are spindle-shaped or pyriform, 2.8–5.9 μm long, and biflagellate (Ayers 1944). The flagellum has two shapes: a shorter flagellum with a blunt end, and a longer flagellum with a whiplash or tail piece. According to McDonald et al. (2014), zoospores are 9.6–14.4 μm in size and uninucleate. Primary and secondary zoospores cannot be differentiated visually; both are ovate, with two whiplash flagella of unequal length.

3.2.3 Plasmodium

Young plasmodium is spherical in shape with smooth surface and 3.7 μm in diameter, ranging from 2.7 to 5 μm strands like materials connected to nearly spherical bodies. These increase in size ranging from 3.3 to 4.5 μm with an average size of 4.2 μm. Each of them is enveloped by a membranous material with a smooth surface. Several young round plasmodia with partially rough surface and an average diameter of 8 μm can be seen in host tissues. Vegetative plasmodia are considerably large mass in infected cells. At this stage, its size is about 45 μm in diameter. The vegetative plasmodia almost fill the host cells looking like sponge-like structure (Ikegami et al. 1978).

3.2.4 Morphology of Pathogen into Host Cell

The details of morphological features of plasmodium, zoospores, and resting spores from early stages of infection and to maturity have been observed in the diseased root cells through scanning electron microscopy (SEM). The details have been presented in Chap. 4.

3.2.5 Chemical Composition of Resting Spore Wall

The walls of the resting spores of *Plasmodiophora brassicae* were isolated mechanically and analyzed quantitatively for their major components. At least 88 % of the spore wall dry weight was accounted for and spore walls dry weight (88%)

Table 3.1 Distribution of *P. brassicae* spore wall components (Moxham and Buczacki 1983)

Component	Analytical method	Fraction	Walls dry Wt (%)
Reducing sugar	Nelson-Somogyi	% ethanol	0.3
		1.5M-HCl	14.4
		0.5% oxalic acid	0.7
		6M-HCI	5.2
		Total (sequential hydrolysates)	20.6
		Complete hydrolysate	22.4
Non-nitrogenous sugar	Anthrone	Complete hydrolysate	2.5
Hexosamine	Elson and Morgan	Complete hydrolysate	19.4
Glucosamine[a]	g.c	Complete hydrolysate	22.1
Glucose	g.c	Complete hydrolysate	0.5
Protein	Lowry	1M-NaOH 25°	13.7
		1M-NaOH 100°	11.2
		1M-acetic acid 100°	2.6
		Residue	6.1
		Total	33.6
Lipid	Bartnicki-Garcia and Nickerson/ Kessler and Nickerson	Readily extractable	16.0
		Bound	6.0
		Total	22.0[b]
Nitrogen	Micro Kjeldahl	Unhydrolyzed walls	6.8

[a]Glucosam initol was measured in complete hydrolysates by Gas Chromatography (GC) with glucosamine-HCI as standard and the weight adjusted by subtracting the value for HCl
[b]Of which not more than 4.5% could have been impurity derived from the solvents used

comprised of chitin (25.1%), carbohydrate (≥2.5%), protein (33.6%), lipid (≥17.5%), and residual ash equivalent to 10.5% of the total weight. The chitosan and callose are absent from the spore walls. The distribution of spore wall components is presented in Tables 3.1 and 3.2 (Moxham and Buczacki 1983). The resting spore wall is the interface of the pathogen with soil environment.

3.3 Classification and Taxonomy

Historically Plasmodiophoromycota is well defined as a monophyletic group by several characters including its cruciform nuclear division. The best evidence available suggests that the group is related to ciliate protists (Barr 1983; Castlebury et al. 1994). The hypothesis is supported by analysis of small subunit rDNA (Castlebury

Table 3.2 Amino acid composition of *P. brassicae* spore wall proteins (Moxham and Buczacki 1983)

Amino acid	Wall amino acid content (moles %)	Amino acid	Wall amino acid content (moles %)
Cysteic acid	3.0	Methionine	1.0
Aspartic acid	8.9	Isoleucine	2.6
Threonine	4.6	Leucine	4.9
Serine	9.8	Tyrosine	4'5
Glutamic acid	6.7	Phenylalanine	3.3
Proline	5.6	y-amino butyric acid	4.7
Glycine	13.5	Histidine	1.5
Alanine	14.0	Lysine	3.1
Cystine	1.2	Arginine	5.9
Valine	3.9		

and Domier 1998). The phylum Plasmodiophoromycota contains the single class Plasmodiophoromycetes with single order Plasmodiophorales (Karling 1968). Keys to genera can be found in Karling (1968); Waterhouse (1973), and Olive (1975). Brief descriptions of genera and species have also been provided by Dylewski (1990), who recognized a total of 29 species in 10 genera including *Plasmodiophora*, *Sorosphaera*, *Spongospora*, *Sorodiscus*, *Octomyxa*, *Polymyxa*, and *Woronin*. These genera are delimited primarily on the basis of morphology and arrangement of resting spores (Sporosori) even though doubts have been expressed about the reliability of such characters (Palm and Burk 1933; Miller 1959). Resting spores may be free from one another, as in *Plasmodiophora*, or adhere together to form sporosori, which may be large, consisting of many resting spores or small, consisting of only a few species. The sponge-like prosorus of *S. subterranean* f. sp. *subterranean* can be seen in a potato tuber. The present taxonomical status of *Plasmodiophora* is as under based on molecular and phylogenetic analysis.

3.3.1 Taxonomy

Since the establishment of *P. brassicae* to cause clubroot disease of crucifers (Woronin 1878), considerable research effort has been directed towards understanding the taxonomy, life history, and pathology of this pathogen. Recent taxonomic status has been that the group is included in the Order "Plasmodiophorida," Class "Phytomyxea," within the infra Kingdom or super group "Rhizaria" (Cavalier-Smith 1993, 2002; Goeche et al. 2012). Previously, the group has been included in the Kingdoms Protoctista (Margulis et al. 1989) and Protozoa (Barr 1992) as well as being described as primitive fungi by early workers (Karling 1968). Debate over the taxonomic status of the Plasmodiophorids remains current, arising largely from an incomplete knowledge of the life cycles of many of the Plasmodiophorids. Braselton (1995) comments that the most obvious gap in understanding of this group of pathogen is the identification of the point at which karyogamy occurs in the life cycle, or, if it does not occur, how nuclei become capable of meiosis. Zoospore fusion (Kole and Gielink 1961; Tommerup and Ingram 1971) or fusion of motile myxamoebae (Mithen and Magrath 1992) has been suggested as a means of genetic recombination in *Plasmodiophora*; however, neither of these processes has been extensively studied in *Brassica* spp. and the evidence is currently far from conclusive.

Once penetration of host tissue has occurred, *P. brassicae* induces in its host a suite of changes in primary and secondary metabolism which eventually cause gross physical distortion of the infected root. Relatively recent advances in technologies available to study changes in gene expression and protein composition have contributed significantly to understanding the pathology of *P. brassicae*, including changes in host nutrition, signaling, and the role of key plant hormones in defense responses.

The phylogeny and inter-relationships of the zoosporic plant pathogens in terms of morphological, biochemical, ultra-structural, and physiological attributes which suggested their possible protozoan origin was examined by Barr (1983). There is also evidence that the plasmodium of *P. brassicae* ingests material from the cytoplasm of the host cell (Karling 1968; Buczacki 1983; Mithen and Magrath 1992). This holozoic form of nutrition tends to be associated with free-living organisms rather than obligate biotrophs. Buczacki (1983) suggested that the Plasmodiophorales may be in the process of losing phagotrophic in favor of absorptive nutrition, having adopted the parasitic habit. The rRNA analysis of the Myxomycetes *Physarum polycephalum* and *Dicfyostelium discoideum* clearly separates them from the true fungi and places them among the protozoa (Cavalier-Smith 1993). Based on molecular data, taxonomists have reclassified *P. brassicae* in to the new taxon Rhizaria, which is a diverse group of amoeboid protists with the differentiated flagellae (Nikolaev et al. 2004). It is neither a fungus, nor a slime mold as earlier known. Although traditionally it was classified as a fungus but recently molecular phylogenetic analysis of the small subunit (SSU) ribosomal RNA, actin, and polyubiquitin genes suggests that it belongs to the protists phylum Cercozoa in the kingdom Rhizaria, and plasmodiophorida as order.

Phylogenetically, *P. brassicae* belongs to the order Plasmodiophorida, a plant pathogenic group of protists in the Phytomyxea within the eukaryotic super group Rhizaria (Bass et al. 2009; Burki et al. 2010; Neuhauser et al. 2014). *P. brassicae* isolates differ in preferences and aggressiveness towards different cruciferous hosts (Buczacki et al. 1975).

According to the database Index Fungorum (Kirk 2014; Cavalier-Smith and Chao 2003) the taxonomy of *P. brassicae* is as under:

Kingdom: Protozoa
Phylum: Cercozoa

Class: Phytomyxea
Order: Plasmodiophorida
Family: Plasmodiophoraceae
Genus: *Plasmodiophora*
Species: *brassicae*

According to Neuhauser et al. (2011), Burki and Keeling (2014), the taxonomy of the *P. brassicae* is as under:

Kingdom: Rhizaria
Phylum: Cercozoa
Class: Phytomyxea
Order: Plasmodiophorida
Family: Plasmodiophoraceae
Genus: *Plasmodiophora*
Species: *brassicae*

3.4 The Life Cycle of *Plasmodiophora brassicae*

Plasmodiophora brassicae is distinct from other plant pathogens, such as fungi or oomycetes, as it is an obligate biotrophic protist in the Plasmodiophorids within the eukaryote super group Rhizaria. Rhizaria share a common ancestor with Stramenopiles and Alveolata forming the S-A-R super group, which together with Plantae build the SARP mega group (Burki et al. 2010; He et al. 2014). Plasmodiophorids are parasites of plants, oomycetes, and a sister group to Phagomyxids, pathogens of sea grass, diatoms, and brown algae. Other agriculturally important Plasmodiophorid pathogens are *Spongospora subterranea*, the causal agent of potato powdery scab, and root galls, and vector for the Potato Mop Top Virus, where the virus transmitting *Polymyxa betae*, and *P. graminis* infecting sugar beets and cereals. Life cycle events of *P. brassicae* have been described by Schwelm et al. (2015) in quite details.

Plasmodiophorids have a complex life cycle consisting of different zoosporic stages, formation of plasmodia inside host cells, and resting spore formation (Fig. 3.1). The haploid resting spore releases a zoospore, which infect the root hairs, and form multinucleate plasmodia. Those develop several single-nucleate secondary zoospores, which are released into the soil. A fusion of these secondary zoospores is assumed to occur occasionally. Zoospores invade the root cortex, in which secondary multinucleate plasmodia develop. How and when karyogamy is taking place is not exactly known, but meiosis appears to occur in the plasmodia before resting spore formation. Plasmodia provoke abnormal cell enlargement (hypertrophy) and uncontrolled cell division (hyperplasia) leading to the development of galls that obstruct nutrients and water transport. In infected root tissue, different developmental stages of the Plasmodiophorids occur simultaneously. Host cells containing plasmodia in neighbor cells are filled with resting spores. The only host-free stages are the short-lived bi-flagellated zoospores and resting spores. Therefore molecular and genomic data for Plasmodiophorids are rare as they only grow inside living host cells and remain uncultivable on their own.

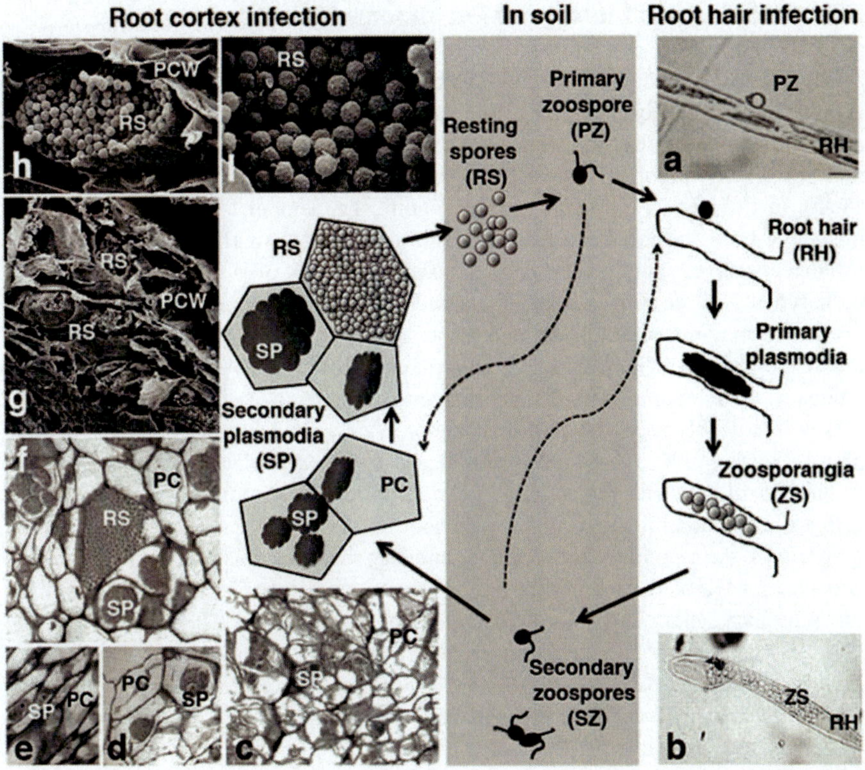

Fig 3.1 The life cycle of *Plasmodiophora brassicae*. A biflagellate primary zoospore (PZ) is released from each resting spore (RS). When the PZ reaches the surface of a root hair (RH) (**a**), it penetrates the cell wall, and forms primary plasmodia in the root hairs. After a number of nuclear divisions the plasmodia cease into zoosporangia (ZS), develop, and release secondary zoospores (SZ) (**b**). Occasional fusion of SZ has been suggested. PZ or SZ can penetrate root cortex cells, in which the pathogen develops into secondary plasmodia (SP) (**c–f**). Proliferation of SP is associated with cellular hypertrophy, causing a swollen root or club phenotype. After a number of nuclear divisions, the secondary plasmodia develop into multinuclear plasmodia, and finally into resting spores (**f–i**), which are released into the surrounding soil when plants decay. *RS* resting spore, *PZ* primary zoospore, *RH* root hair, *ZS* zoosporangia, *SZ* secondary zoospore, *SP* secondary plasmodia, *PC* plant cell, *PCW* plant cell wall (Schwelm et al. 2015)

Resting spores of *P. brassicae* are extremely resilient to harsh environmental conditions and contaminate arable land for decades. This feature makes it impossible to eradicate the organism via any known chemical or alternative soil treatment. The resting spores which are formed at the end of the life cycle inside the root contain chitin in their cell walls. Besides being an important component of the plasmodiophorid survival structure, chitin and other glycans are known as important signatures for plant immune activation or for the establishment of beneficial symbioses (Schwelm et al. 2015).

3.5 Molecular Biology of *Plasmodiophora brassicae*

The genus *Plasmodiophora* is a monophyletic group with uncertain systematic affinities. The species belonging to this genus possess unique features such as cruciform nuclear division, parasitism, obligate nature, biflagella, heterokont zoospores, environmentally resistance, and long-living resting spores (Braselton 1995). In this genus, the economically significant member is *P. brassicae*, which hosts *Brassica* species to cause clubroot disease. The pathogen shows a wide biological range, and its populations usually consist of a mixture of different pathotypes. Soil environmental factors such as physical, chemical, and biological properties may differentially influence the survival of some physiological races of the pathogen. Field isolates of *P. brassicae* display great variation and show a tendency to overcome different resistance sources from *Brassica* species (Dixon 2009). It is widely accepted that life cycle of *P. brassicae* consists of two infectious phases (Karling 1968; Ingram and Tommerup 1972). Inside the host cell, cytoplasm of the *P. brassicae* amoeba undergoes endomitotic division forming a multinucleate primary plasmodium. The primary plasmodium cleaves into sporoangiosori which gives rise to the secondary zoospores. Secondary zoospores fuse and form binucleate zoospores. Meiotic division in the multinucleate secondary plasmodium takes place just after karyogamy and before sporangiogenesis (Braselton 1981).

In order to understand the pathogenicity factors of *P. brassicae* causing clubroot disease on different *Brassica* hosts, several molecular techniques and tools have been employed to determine *P. brassicae* genome size, structure, and number of possible functional genes in the whole genome. Several studies have reported the use of pulse-field gel electrophoresis (PFGE) to determine the karyotypes for *P. brassicae*. Ito et al. (1994) used sheroplasts and differentiated 13 chromosomal bands in the range of 1.9 Mb to 750 kb. Bryan et al. (1996) used isolated plasmodia, and differentiated six chromosomal bands in the range of 1.7 Mb to 680 kb. Similarly, Graf et al. (2001) distinguished 16 chromosomal bands in the range of 2.2 Mb to 680 kb. Based on these studies, it is estimated that the *P. brassicae* total genome size can be 18–20.3 Mb (Graf et al. 2004). Several molecular marker techniques are employed to investigate virulent pattern of the *P. brassicae* population derived from single-spore isolate or field isolates, however, the number of distinguishing patterns are very low and that could not correlate with virulence patterns. In continuous efforts, two RAPD markers and one SCAR marker have been identified, which correlated to isolates of group pathotype 1. Yet, there are no sets of molecular markers that can distinguish other pathotypes from field isolates which make clubroot resistance breeding intriguing (Buhariwalla et al. 1995; Moller and Harling 1996; Manzanares-Dauleux et al. 2000, 2001).

3.5.1 The Size of the *P. brassicae* Genome

Plasmodiophora brassicae has small chromosomes that are impossible to distinguish by conventional cytogenetics. Braselton (1981) estimated 20 synaptonemal complexes in *P. brassicae* using serial sections for electron microscopy. Pulse-field gel electrophoresis (PFGE) can determine karyotypes (chromosome size and number) of organisms that cannot be studied using cytological methods or classical linkage analysis (Howlett 1996). Ito et al. (1994) used spheroplasts as starting material, and distinguished 13 chromosomal bands in the range of 1.9 Mb to 750 kb. Bryan et al. (1996) started with isolated plasmodia, and can be separated six bands in the range of 1.7 Mb to 680 kb. Graf et al. (2001) described 16 chromosomal bands of *P. brassicae* in the range of 2.2 Mb to 680 kb that corresponded with the chromosomal sizes estimated by Ito et al. (1994), but with three additional chromosomal bands. When three single-spore isolates of *P. brassicae* were analyzed by PFGE, chromosome polymorphisms between isolates were identified, with no isolates having more than 16 chromosomal bands (Graf et al. 2004). These chromosome polymorphisms might explain the differences in results between various isolates of *P. brassicae* (Ito et al. 1994; Graf et al. 2001). The number of chromosomes estimated by PFGE, however, was less than the 20 synaptonemal complexes estimated by electron microscopy (Braselton 1981), suggesting that some chromosomal bands might not be completely separated by PFGE. Taking the chromosome number and size from the PFGE studies into account, the total genome size of *P. brassicae* was estimated to be 18–20.3 Mb (Graf et al. 2004).

Southern hybridization technique has been used to map several *P. brassicae*-specific DNA fragments with chromosomes separated by PFGE. Repetitive elements were suitable for visualizing most of the chromosomes and were localized on 14 (H4), 8 (E56), or 5 (RFM8) chromosomes of the single-spore isolate "e3" (Graf et al. 2001, 2004; Fahling et al. 2004). The DNA-encoding ribosomal subunits were localized to a number of chromosomal bands of *P. brassicae* (Graf et al. 2004), and may well be involved in chromosome rearrangement because tracts of rRNA genes are frequently associated with chromosome variations (Daviere et al. 2001). Chromosome polymorphism between single-spore isolates may well hinder gene localization to specific chromosomal bands, but studies of such polymorphisms may also reveal new information behind the high variations in pathogenicity between *P. brassicae* isolates. In some organisms, chromosomal rearrangements have been associated with changes in pathogenicity, such as loss of toxin production, the loss of pathogenicity in the apple pathogen due to *Alternaria alternata* (Johnson et al. 2001) have been reported.

3.5.2 Recombination in *P. brassicae*

Synaptonemal complexes (Braselton 1981) and re-infection by dikaryotic plasmodia (Kobelt 2000) have been taken as evidence of sexual recombination in *P. brassicae*. *P. brassicae* exists as mixtures of pathotypes in the soil, and even in single-clubbed roots (Jones et al. 1982a, b; Fahling et al. 2003). Sexual recombination could explain the high degree of variation in field populations. In order to investigate the genetics of *P. brassicae*, Fahling et al. (2004) attempted to establish a cross between *P. brassicae* lines. Plants were inoculated with two different single-spore isolates,

"e_3," and "e_6," and then new single-spore isolates were extracted from the resulting new generation of clubs. No sexual recombination was observed by examining repetitive DNA from the progeny, but a chromosome with rearrangement was identified in one of the new single-spore isolates. This appeared to be a recent event occurring during the infection cycle (Fahling et al. 2004). This suggests an ability of *P. brassicae* to rearrange chromosomes without sexual recombination, and question the concept of whether genetically stable lines can be established by single-spore production. New sexual recombinants might have been missed in these experiments because of a lower propagation potential in the ECD05 host plant, as with almost all *P. brassicae*.

3.5.3 Genes Discovery of *P. brassicae*

In order to discover genes, several DNA detection assays for *P. brassicae* have used PCR primers designed in the sequence of ribosomal internal transcribed spacer (ITS) sequences (Faggian et al. 1999; Wallenhammar and Arwidsson 2001; Kageyama et al. 2003). These standard PCR-based assays tests have been found highly sensitive for spore detection but are not quantitative. Recently, a real-time PCR test based on the ribosomal ITS sequences has been designed in combination with measurements of the fatty acid biomarker arachidonic acid, quantification of *P. brassicae* infection is now possible (Sundelin 2008).

Many DNA detection studies have been combined with surveys of molecular diversity among *P. brassicae* field or single-spore isolates. In particular, there is the hope that molecular markers will eventually allow the rapid identification of virulence patterns among isolates. In some of the earliest experiments of this type, distantly related isolates of *Plasmodiophora* with divergent virulence were characterized by PCR amplification patterns using arbitrary primers (Buhariwalla et al. 1995; Moller and Harling 1996; Yano et al. 1997), and sequence-generated primers (Buhariwalla et al. 1995; Ito et al. 1999; Klewer et al. 2001). The number of distinguishing patterns was low; however, virulence patterns and PCR patterns could not be correlated unequivocally. Early successes in the cloning of repetitive DNA elements of the pathogen were achieved when small *P. brassicae* DNA libraries were established (Buhariwalla and Mithen 1995; Buhariwalla et al. 1995; Ito et al. 1998; Klewer et al. 2001). Difficulties in obtaining high-quality genomic *P. brassicae* DNA limited the library size in these studies between 27 and 150 clones.

Manzanares-Dauleux et al. (2001) applied 19 random primers and 9 sequenced-derived primers to obtain 103 polymorphic bands among 46 related isolates. This demonstrated the very high level of genetic diversity among isolates from one field (Manzanares-Dauleux et al. 2001). Two RAPD markers (Manzanares-Dauleux et al. 2001) and a sequence-characterized amplified region (SCAR) marker (Manzanares-Dauleux et al. 2000) were correlated to isolates of pathotype group 1. In a similar set of experiments, Klewer et al. (2001) examined the molecular diversity of single-spore isolates originating from a German field isolate "e." Reliable differentiation of isolates was produced with sequence-derived primers, and RFLP analysis using repetitive elements as hybridization probes. Using three highly repetitive *P. brassicae* probes and combinations of restriction enzymes, the number of polymorphic bands in these fingerprint patterns was increased to more than 100, and the patterns of different single-spore isolates showed fewer common bands (Fahling et al. 2003). Isolate-specific banding patterns have been obtained using a combination of sequence-derived primers from the borders of highly repetitive elements of *P. brassicae* (sequence-anchored primer), and primers for genomic walking (random primer). Overall, the molecular techniques to characterize isolates are now established but there is as yet no standard set of markers to describe an isolate are available. A widespread screening of pathotypes by standardized molecular techniques would make it easier to describe pathotypes and map their distribution. Molecular techniques that rapidly describe the virulence characteristics of *P. brassicae* field populations would be highly valuable for studies of population structure, and dynamics in *Brassica* crops which thereby benefit the rapeseed growers and farmers (Siemens et al. 2009a, b).

3.5.4 Evolutionary Position of *P. brassicae* and Other Plasmodiophorids

One of the most significant advances coming from the study of the molecular biology of *P. brassicae* has been a clarification of plasmodiophorid evolution. As with many other eukaryotic microbes, the first DNA target amplified and sequenced from *P. brassicae* was the ribosomal small subunit gene (SSU) (Castlebury and Domier 1998). Unlike all other plasmodiophorids evaluated, the *P. brassicae* SSU was found to contain self-splicing type I introns. The SSU sequence was used by Cavalier-Smith and Chao (1996/97) to provide the first robust phylogenetic evidence of linkage between the plasmodiophorids and cercozoan protists. To understand cercozoan evolution, further, Archibald and Keeling (2004) used degenerate PCR to amplify sequences of plasmodiophorid actin and polyubiquitin genes, including from *P. brassicae*. Phylogenies constructed with the actin amino acid sequences again showed a relationship between the plasmodiophorids and cercozoans. Even more convincingly, *P. brassicae* was shown to share with other cercozoans a characteristic amino acid insertion between ubiquitin monomers (Archibald et al. 2003). Subsequent analyses of these sequences using larger data sets have reinforced the inclusion of the plasmodiophorids within *Cercozoa* (Bass et al. 2005). The systematic placement of *P. brassicae* has been unclear for a long time, and the organism is still often referred to historically as a fungus. The more accurate evolutionary positioning of *P. brassicae* provides an essential basis for understanding many aspects of *P. brassicae* biology. It will be more rational to compare the *P. brassicae* genome with those of other cercozoans as they become available than with the currently available completely sequenced fungal genomes.

3.5.5 cDNA Cloning Approaches to Identify *P. brassicae* Genes

The isolation of a single-copy DNA sequence is unique to *P. brassicae* (Ito et al. 1997). While from subtractive cDNA libraries, Graf et al. (2004) isolated a few partial cDNAs that had no significant homology to GenBank sequences, whereas Ito et al. (1998) found a single-copy fragment with high similarity to the yeast RNA polymerase II gene. Brodmann et al. (2002) isolated a short partial sequence for a trehalose phosphate synthase (PbTPS). Ando et al. (2006) designated a full-length *P. brassicae* cDNA clone of a single-copy gene PbSTKL1 due to a serine/threonine kinase-like domain at the C-terminal region. An acceleration of the hunt for *Plasmodiophora* genes was achieved when more than 70 cDNA clones of the pathogen were isolated by suppression subtractive hybridization (SSH) or by direct EST sequencing (Bulman et al. 2006). More than 40 of these cDNA sequences were full-length. The study showed that while working with a pool containing both plant, and *P. brassicae*, cDNA was an acceptable starting point for plasmodiophorid gene discovery. Again using SSH, but starting with *Brassica rapa* rather than *Arabidopsis* clubroot galls. Sundelin (2008) isolated about 140 genes, half of these gene sequences originated from the pathogen, of these, ten clones were newly characterized *P. brassicae* genes. The remainder of the *P. brassicae* genes had been identified by Bulman et al. (2006), showing the consistency of results between these two SSH studies. Moreover, multiple copies of some *P. brassicae* genes (PbGST1, PbSUR2, and PbSHSP1) were sequenced in both studies, indicating a high transcription of these genes or perhaps preferential amplification of the transcripts (Siemens et al. 2009a, b).

3.5.6 *Plasmodiophora brassicae* Gene Structure

The presence of introns in plasmodiophorid genes was apparent during cloning of short *P. brassicae* gene fragments (Brodmann et al. 2002). The first full structure of *P. brassicae* protein-coding gene was provided when Ando et al. (2006) sequenced both the cDNA and the genomic DNA of the PbSTKL1 gene. By comparing these sequences, the PbSTKL1 gene was observed to contain 14 intron sequences. Subsequently, Bulman et al. (2007) obtained a genomic DNA sequence corresponding with 24 *P. brassicae* cDNAs, whereas Sundelin (2008) confirmed the genomic DNA sequence corresponding with five newly characterized genes. These included the cDNA and genomic sequence of the first ATPase gene (PbATP1) cloned from a plant parasitic protozoan species. The PbATP1 protein contained at least eight transmembrane helixes and a number of protein domains characteristic for P-type ATPases. Although the total number of genomic DNA sequences is small, the structure of *P. brassicae* genes appears to be consistent between these studies. On the whole, *P. brassicae* genes are intron-rich. The great majority of the introns were bordered by the consensus border sequences 5′ GT and AG 3′ which contained consensus branch-point sequences, indicating that the splicing event in *P. brassicae* proceeds in a similar way to that in the majority of eukaryotes. Examination of the promoter and transcription starter sites of genes indicated that *P. brassicae* transcription is likely to begin from initiator elements, as no TATA-box containing promoters were found. Where neighboring genes were confirmed, intergenic distances were found to be short, ranging from 44 to 470 bp, but a number of larger DNA fragments of up to 2.2 kb were also found, and these contained no obvious genes (Bulman et al. 2007). Calculations were made using a mean gene size of 1500 bp, and a mean intersection size of 500 bp mean that a *Plasmodiophora* genome of approximately has 20 Mbp with the potential to contain 10,000 genes. Consequently, 99% of the coding sequences on the genetic map of *P. brassicae* remain unknown (Siemens et al. 2009a, b).

3.5.7 Gene Functions at the Host–Pathogen Interaction

Genomic functions of *P. brassicae* are hampered by both the small number of *P. brassicae* genes isolated, and the limited homology of many *Plasmodiophora* sequences with known genes. The approximately 100 partial cDNA fragments or full-length clones of *P. brassicae* have been isolated up to now. About half of these *P. brassicae* genes, a biochemical function is strongly indicated by DNA or amino acid similarities with characterized genes of other organisms, but until now, no function for a *P. brassicae* gene has been proven experimentally. The remaining DNA fragments from *P. brassicae* reveal little or no similarity to sequence data in the databases, reflecting the paucity of genomic data available for protists from the Cercozoa (Keeling and Slamovits 2005). Nucleotide or amino acid sequence similarity with known genes from other organisms can give a hint as to gene function, but the presence of a protein domain in a sequence might also be misleading.

The small roots of *A. thaliana* facilitate the observation of clubroot symptoms in this host plant, revealing a reasonably reproducible pattern of disease development under controlled environmental conditions (Kobelt 2000; Kobelt et al. 2000; Siemens et al. 2002, 2006a). Correlating transcription with distinct stages of symptom development caused by the pathogen will be an opportunity to define landmarks and milestones of disease development with a possibility for gaining insights into the function of *P. brassicae* genes. A prominent actin cytoskeleton has been detected by immune-histochemistry over that part of the life cycle after secondary infection (Kobelt 2000). The PbActin and PbGST products appeared strongly expressed and well detectable over the whole infection cycle within *A. thaliana* roots, and can therefore be used as indicators for an actively growing pathogen (Siemens et al. 2009a, b).

Two unidentified single-copy genes, PbBrip9 and PbCC249, were strongly expressed at disease stages corresponding with the occurrence of sporulating plasmodia (Siemens et al. 2009b), whereas PbSTKL1 transcription coincided with resting spore formation (Ando et al. 2006). The expression of gene fragment Y10 was exclusively correlated with the vegetative plasmodial stage (Ito et al. 1999), and the transcription of PbTPS correlated with an accumulation of trehalose (Brodmann et al. 2002). Finally, preliminary evidence indicates that the genes PbPP2A (protein phosphatase 2A) and PbHMG (DNA binding protein with high-mobility group box) are preferentially expressed in the early phases of infection in *B. rapa* roots (Sundelin 2008). The gene Bj N1T1 expression in *B. juncea* during infection promotes over production of auxins to form galls in infected roots (Liu et al. 2018). Altogether, the very first molecular landmarks of clubroot disease development in *Arabidopsis* are established. The whole genome transcription data from the host plant *A. thaliana* has revealed new insights into clubroot development (Siemens et al. 2006a, b; Jubault et al. 2008; Ludwig-Muller et al. 2009). Combining *Arabidopsis* microarray data with expression analyses of *P. brassicae* genes will provide a strong new approach to clubroot research in the future.

As with other biotrophic pathogens, it is likely that *P. brassicae* secretes proteins into the host cell to effect modifications of cell structure. Predicted proteins with signal peptides from *P. brassicae* are thus of much interest (Sundelin 2008). The *P. brassicae* sequences with signal peptides known to date, two matched with functionally characterized genes with roles in pathogenicity, namely, PbPPI1, which contains an FKBP family peptidyl-prolyl cis-trans isomerase domain, and PbPDA1, which contains a "NodB-type" chitooligosaccharide deacetylase domain (Bulman et al. 2006). Surface-exposed or secreted PPIase domain-containing proteins have been found in several intracellular pathogens (Engleberg et al. 1989; Moro et al. 1995). The chitooligosaccharide deacetylase family includes members known to deacetylate substrates, including xylan and chitin, which might play roles in cell wall penetration during myxamoeboid movement of secondary plasmodia. The intracellular pathogen *P. brassicae* lives in close contact with the host, therefore, using a plant-derived RNAi strategy might work when an important parasitism gene of *P. brassicae* is identified (Siemens et al. 2009a, b).

3.6 Molecular Biology of *Plasmodiophora brassicae* Pathogenesis

A search of the National Center for Biotechnology Information (NCBI) database for *P. brassicae*—originated accessions yielded 180 nucleotide sequences, 138 expressed sequence tags, and 53 proteins. Of these sequences, only a small number have a potential importance in pathogenicity, largely based on sequence similarity with genes from other species. Most of these sequences are derived from cDNA libraries constructed by suppression subtractive hybridization (SSH). Among them, there are 76 *P. brassicae* genes isolated from a SSH library constructed between RNA from *P. brassicae*-infected and uninfected *Arabidopsis* tissue (Bulman et al. 2006). Many of the sequences are predicted to contain signal peptides for translocation of the encoded proteins. Using *Brassica rapa* L. clubroot galls, Sundelin et al. (2011) isolated about 140 clones. Half of these clones, which represented 24 unisequences (10 of which are newly characterized), originated from *P. brassicae*. Through the use of next generation sequencing (NGS) technologies, Burki et al. (2010) generated 3167 *P. brassicae* specific contigs from infected *B. rapa*. These sequences, however, are very short, and their functions or relevance to clubroot development are not known. Feng et al. (2010) constructed a SSH library at the early stages of infection (7 days after inoculation) of canola roots by *P. brassicae*. The library consisted of 797 clones that represented 439 unigenes. Thirty-two of these genes are demonstrated to be of a *P. brassicae* origin. Differential expression of a subset of these genes in the resting spores and at 7 days after inoculation was observed by quantitative PCR (qPCR), suggesting that some of these genes may play a role during pathogenesis. In addition to these libraries, a few other genes have been identified from *P. brassicae,* and postulated to be related to its pathogenicity. Ito et al. (1997) identified a gene (Y10) whose expression is exclusively correlated with the vegetative plasmodial stage of *P. brassicae*. Brodmann et al. (2002) found a trehalose-6-phosphate synthase gene with expression correlated with an accumulation of trehalose in resting spores. Ando et al. (2006) demonstrated that the expression of PbSTKL1 increases strongly beginning 30 days after inoculation and is coincident with resting spore formation. Two cDNAs, PbBrip9 and PbCC249, are found to be strongly expressed at stages of disease development corresponding to the occurrence of sporulating plasmodia (Siemens et al. 2009b). All these genes may have a potential role in clubroot pathogenesis.

Until now, only one *P. brassicae* gene, a serine protease (PRO1), has been functionally studied, and proven experimentally to be important for pathogenesis (Feng et al. 2010). Southern analysis and specific PCR amplification indicated that PRO1 is a single-copy gene present in a broad range of *P. brassicae* pathotypes. Northern analysis revealed expression of PRO1 during plant infection, as well as in dormant, and germinating resting spores. The expression pattern assessed by Northern analysis was further confirmed by qPCR in a later study (Feng et al. 2013). The optimal temperature of Pro1 is 25 °C, and it exhibits maximum activity at pH 6.2–6.4, and minimal activity at pH 6.8–7.6. These values coincide with the temperature and pH conditions favorable for *P. brassicae* resting spore germination in the field. When Pro1 is used to treat canola root exudates, it enhances the stimulatory effect of the root exudates on *P. brassicae* resting spore germination, indicating that Pro1 plays a role during clubroot pathogenesis by stimulating resting spore germination through its proteolytic activity (Feng et al. 2014). Bulman et al. (2007) identified DNA sequences spanning or flanking 24 *P. brassicae* genes, eventually sequencing close to 44 kb of genomic DNA. Some general characters of *P. brassicae* genes emerged from this preliminary genome survey: (1) *P. brassicae* genes are rich in spliceosomal introns with consensus splice sites, and branch-point sequences similar to those found in other eukaryotes, (2) as a consequence of being intron-rich, splicing is an important feature of gene expression, (3) *P. brassicae* transcription is likely to begin from initiator elements rather than TATA-box containing promoters, and (4) intergenic distances are short, ranging from 44 to 470 bp. Feng et al. (2013) reported a gene expression study in which all database-available *P. brassicae* genes (118) were investigated by dot-blot hybridization and qPCR. The expression of these genes in primary and secondary zoospores was compared. Both dot blot and qPCR identified up- and down-regulated genes, and the correlation between these two techniques was confirmed. Quantitative

PCR indicated that 58 genes are up-regulated in the secondary zoospores relative to the primary zoospores, whereas 55 are down-regulated.

3.7 The *Plasmodiophora brassicae* Genome and Life Stage-Specific Transcriptomics

Plasmodiophora brassicae genomes are available, including one reference genome from a European isolate, and additional re-sequencing of five Canadian, and one Chinese isolates. The pathotype classification of these isolates is difficult to compare because these have been characterized using different differential hosts sets. The main characteristics of the *P. brassicae* genome were described, such as its small size (24.2–25.5 Mb) and high gene density. Thirteen chitin synthases were described, suggesting an important role for this gene family in resting spore formation. Secreted proteins described to play a role in plant pathogenic organisms in suppressing the plant defense responses and modifying the host metabolism were previously predicted in silico in *P. brassicae*, but for most of them, no putative function could be assigned. The *P. brassicae* genome also contains genes potentially involved in host hormone manipulation, such as the auxin-responsive Gretchen Hagen 3, isopentenyl-transferases, methyltransferase, and cytokinin oxidase. The clubroot genome also appears to be lacking several metabolic pathways, a characteristic of the eukaryotic biotrophic plant pathogens. The missing genes encode proteins involved in sulfur, and nitrogen uptake, as well as the arginine, lysine, thiamine, and fatty acid biosynthesis pathways. In addition, only a few carbohydrate-active enzymes (CAZymes), involved in the synthesis, metabolism, and transport of carbohydrates, have been observed (Bi et al. 2016; Rolfe et al. 2016; Spanu et al. 2010).

The current life stage-specific transcriptomics data related to the molecular events in the life cycle of *P. brassicae* was summarized by Schwelm et al. (2016). During spore germination, and at the primary zoospore stage, enzymes involved in chitinous cell wall digestion are highly expressed. Several metabolism pathways such as the starch, citrate cycle, pentose phosphate pathway, as well as pyruvate and trehalose metabolism were also very active. Although the molecular mechanisms involved in primary root-hair infection are unknown, the secondary phase has been better described. During re-infection by the secondary zoospores in root hairs or the root cortex, genes involved in basal and lipid metabolism were highly expressed. The plant clubroot symptoms (hyperplasia and hypertrophy of the infected tissues leading to the gall phenotype) were found to result from the deregulation of the host plant primary, and secondary metabolism as well as from the modification of plant hormone homeostasis by the pathogen. The main plant hormone pathways modified were cytokinin biosynthesis, auxin homeostasis, salicylic acid, and jasmonic acid metabolism. However, the mechanisms by which the pathogen may drive plant hormone homeostasis are still not completely worked out. At the end of the *P. brassicae* life cycle, secondary plasmodia transform into resting spores. In these structures, trehalose could play an important role for their long-term survival capacity. The description of the current transcriptomics data in *P. brassicae* shows that some important steps in *P. brassicae* developmental stages are still not well-characterized at the molecular level (Siemens et al. 2006a; Wagner et al. 2012; Ludwig-Muller et al. 2009, 2015). Daval et al. (2018) presented the genome of eH, a second European isolate. Refinement of the annotation of the eH genome led to the identification of the mitochondrial genome sequence, which was found to be bigger than that of *Spongospora subterranea*, another plant parasitic Plasmodiophorid phylogenetically related to *P. brassicae*. New pathways were also predicted, such as those for the synthesis of spermidine, a polyamine up-regulated in clubbed regions of roots. A *P. brassicae* pathway genome database was created to facilitate the functional study of metabolic pathways in transcriptomics approaches. These available tools can help in understanding the regulation of *P. brassicae* metabolism during infection, and in response to diverse constraints.

3.8 Genome Characteristics of Pathotypes

A draft genome of *P. brassicae* pathotypes 3 and 6 (Pb3 and Pb6) differs in their host range. Pb3 is highly virulent on *B. napus* (also infects other *Brassica* species) while Pb6 infects only vegetable *Brassica* crops. Both the Pb3 and Pb6 genomes are highly compact, each with a total size of 24.2 Mb, and contain less than 2% repetitive DNA. Clustering of genome-wide single nucleotide polymorphisms (SNP) of Pb3, Pb6, and three additional re-sequenced pathotypes (Pb2, Pb5, and Pb8) shows a high degree of correlation of cluster grouping with host range. The Pb3 genome features significant reduction of intergenic space with multiple examples of overlapping un-translated regions (UTRs). Dependency on the host for essential nutrients is evident from the loss of genes for the biosynthesis of thiamine, and some amino acids, as well as the presence of a wide range of transport proteins, including some unique to *P. brassicae*. The annotated genes of Pb3 include those with a potential role in the regulation of the plant growth hormones cytokinin and auxin. The *P. brassicae* genome sequence reveals a compact genome, a dependency of the pathogen on its host for some essential nutrients, and a potential role in the regulation of host plant cytokinin and auxin. Genome annotation supported by RNA sequencing reveals significant reduction in intergenic space which, in addition to low repeat content, has likely contributed to the *P. brassicae* compact genome (Rolfe et al. 2016). In common with many pathogens, *P. brassicae* has been shown to synthesize trehalose. The Pb3 genome contained three genes encoding trehalose phosphate synthase (TPS) (PbPT3Sc00016_A_8.299_1, PbPT3Sc00018_Sm_4.147_1, PbPT3Sc00059_A_1.289_1), and single genes encoding trehalose phosphate phosphatase (PbPT3Sc00030_A_6.312_1), and trehalase (PbPT3Sc00031_A_12.282_1). The TPS gene PbPT3Sc00059_A_1.289_1 matched the fragment isolated by Brodmann et al. (2002), and the corresponding complete gene reported (Bulman et al. (2007).

The *P. brassicae* genome differ due to the small size and scarcity of repeats in comparison with most of other obligate biotrophs such as rust and powdery mildew fungi along with the downy mildew oomycete *Hyaloperonospora arabidopsidis*. Transposable and other repetitive elements are factors that contribute to genome expansion, evolution, and possible diversification of effector genes in biotrophic fungi. Another unique feature of the *P. brassicae* genome is the reduction (and in some occasions overlapping UTRs) of intergenic distances. This resembles the reduction of intergenic intervals in *Microsporidia*, another group of obligate intracellular pathogen in which the reduction of gene space has forced promoters and terminators into adjacent genes, and consequently the presence of mRNAs that encode sequences from more than one gene (Spanu 2012).

Comparison of the SNP profile of Pb3 with four other Canadian pathotypes showed a clear difference between Pb3 and Pb6 while the SNP profile for Pb2, Pb5, and Pb8 were very similar. Interestingly, this SNP-based grouping correlated with host specificity of these isolates. Despite the marked differences between the genome-wide SNP profile of Pb3 and Pb6, these two pathotypes were very similar with respect to genome structure and their effector profile.

As a biotrophic pathogen, *P. brassicae* relies on host plant carbon sources for growth. Among the Pb3 predicted transporters were a monosaccharide, sucrose, and several major facilitator super family (MSF) transporters that transport different compounds including simple sugars. A novel group of transporters among the predicted proteins of Pb3 with substrate binding motifs only found in prokaryotic ABC transporters. Seven proteins with these bacterial transporter motifs had 7 transmembrane sweet-taste receptor of 3 GCPR (7tm3) joined with either a bacterial extracellular solute-binding protein (SBP) or a bacterial periplasmic-binding (PBP) domain. Bacterial SBP motifs are essential for transmembrane transport irrespective of solute concentration, as they restrict substrate diffusion away from the cell, and improve transporter affinity. The evolutionary origin of these transporters remains to be established but they are likely to represent examples of horizontal gene transfer from prokaryotes to eukaryotes which have been documented in other protist genomes. Schwelm et al. (2015) also reported that a number of *P. brassicae* GH18 chitinases were closely related to bacterial chitinases which have not been reported in other eukaryotes before. Plant biotrophic pathogens are efficient in evading host recognition, and suppressing plant defense through secretion of effector proteins. A total of 221 small-secreted proteins evaluated among the predicted in Pb3 proteins, out of which 60% (132 SSPs) showed no homology to the NCBI nr database. Twenty-eight SSPs that were differentially expressed during infection consisted of proteins with chitin-binding domains, proteinase activity, and peptidase inhibitors. Salicylic acid (SA) is the main plant hormone for activating defense against biotroph pathogens. A putative Pb3 effector with homology to plant protein BSMT that function as SA methyltransferase has been identified. *In planta*, BSMT generates MeSA at the site of pathogen infection. MeSA acts as a SAR signal in distal leaves of infected plants. A BSMT from *P. brassicae* was recently reported by Ludwig-Muller et al. (2015). This gene function as induction of SAR, and it was differentially expressed during infection. One possible role for *P. brassicae* BSMT is to dampen the effect of SA in local tissue by converting it to MeSA with the potential benefit of protecting above-ground parts of the plant from other pathogens, and thus sustaining the life of the host plant which is required for obligate biotrophic pathogens. SA mediated plant defense is mainly described for foliar plant pathogens, and there is little information about involvement of the SA dependent defense pathway in roots. The possible involvement of SA against root colonizing pathogens is the induction of a SA marker gene encoding pathogenesis-related protein 1 (PR-1) in barley roots during challenge with *Bipolaris sorokiniana*, a root pathogen of barley. Lemarie et al. (2015) reported that in *Arabidopsis* plants with partial resistance to clubroot SA was induced. Two *B. napus* orthologues of *Arabidopsis* PR1 (At2g14610) were expressed in *B. napus* roots infected with Pb3 but only one of these orthologues was differentially expressed. Expression of this gene started at week 3, and then increased at weeks 4, and was significantly up-regulated at weeks 5 and 6 (7-fold at week 5, and 12-fold at week 6). This pattern of PR1 gene expression suggests the suppression of the SA pathway at earlier stages of infection, as was also supported by HORMONOMETER analysis (Rolfe et al. 2016).

3.9 Pathogen Dissemination

The clubroot pathogen (*P. brassicae*) disseminates from one location to another location through various agencies. However, its seed borne and/or seed contamination are both important modes of dissemination to long distances even across continents. The other sources of pathogen dissemination also have very significant contribution in spread and transport of pathogen inoculum from an infested area to new sites as described below.

3.9.1 Seed

Seed contaminated with resting spores of *P. brassicae* may represent a mode for the long-distance transmission of clubroot. Eriksson (1930) described an account of clubroot on turnips (*B. rapa* var. *rapa*) in Sweden that was reportedly caused by seed-borne inoculum, while Gibbs (1931) also attributed several cases of clubroot to seed-borne infestations. Using quantitative PCR to measure DNA of *P. brassicae*, Rennie et al. (2011) conclusively showed that seed-borne dissemination of this pathogen is possible. Wang (1981) reported that the occurrence of clubroot on cabbage (*B. oleracea*) in Henan Province, China, resulted from infested seeds obtained from a province where clubroot is prevalent. Similarly, it was suggested that seeds of white mustard (*Sinapis alba* L.) imported from Canada might have been the source of two cases of clubroot in New South Wales, Australia (Hind-Lanoiselet and Parker 2005).

3.9.2 Infected Plant Material

Long-distance transmission of *P. brassicae* may also occur via the movement of resting spores in infected plant material. Dissemination of clubroot is often associated with the spread of infested soil from a clubroot-infested field to an uninfested field on root vegetables. The planting time of cruciferous crops varies in different regions of the world. Domestic transportation of infected plants is the main route of long-distance transmission of *P. brassicae* (Li et al. 2010). Out of season vegetables infested with clubroot are dispatched among different provinces and countries, which create opportunities for the spread of *P. brassicae*. Transplanting of cole vegetables from clubroot-infested fields onto non-infested fields could introduce the pathogen to new areas or augment existing inoculum levels in the soil.

3.9.3 Soil

As a soil-borne pathogen, the main way that *P. brassicae* spreads is via the movement of infested soil on farm and other machinery (Howard et al. 2010). Soil becomes infested when *P. brassicae* resting spores are released from decomposing host root tissue. These resting spores can also be dispersed to uninfected plants through the movement of infested soil and water (Dixon 2009; Kageyama and Asano 2009), or perhaps even by windborne dust (Rennie et al. 2012). Wind disperses spores that are picked up together with light, dry, dusty soil particles over even greater distances. Earthworms and possibly moles, root nematodes, and other insects may be vectors for *P. brassicae* in the soil (Dixon 2009). Clubroot resting spores exhibit extreme longevity, which contributes to the severity of this disease. The half-life of resting spores has been estimated to be 3.6 to 4.4 years, while these can survive in the soil for nearly 20 years (Wallenhammar 1996; Dixon 2009). Using real-time PCR for direct detection and quantification of *P. brassicae* in soil samples, a population density of 10^4–10^6 resting spores g^{-1} soil was found in naturally infested soil samples (Wallenhammar et al. 2011). Based on the correlation between the population density of resting spores and gall formation, the minimum population density for gall formation was calculated as 3.5 resting spores g^{-1} soil (Naiki et al. 1978). There are, however, several reports of single resting spores found to induce gall formation (Scott 1985; Narisawa et al. 1996). Considering that primary infection results in millions of resting spores, the population density of *P. brassicae* increases substantially after continuous cropping with susceptible plants.

3.9.4 Rain and Irrigation Water

Rain and flood water disseminate *P. brassicae* over substantial distances, especially on slopping land. Irrigation water contaminated with resting spores of *P. brassicae* can potentially infest large areas of fields. *Plasmodiophora brassicae* has been detected in irrigation pond sediments (Datnoff et al. 1984), dams, and bores (Faggian et al. 1999). Resting spores of *P. brassicae* remain viable in water for 34 months and repeated irrigation with water containing as few as 10 spores ml^{-1} result in root galling (Donald 2005). Resting spores settle in undisturbed columns of water at a rate of 25 cm day^{-1}. In a farm dam, most of the resting spore population can be expected to be concentrated in the sediment at the bottom of the dam. The risk of spreading disease through irrigation using contaminated farm dam water can be reduced by locating the irrigation intake pipe in the stillest part of the dam, mounted on a float to collect water from near the surface (Donald and Porter 2009). Water should not be applied to cruciferous vegetables if it has been drawn from ponds or creeks contaminated with *P. brassicae* spores resulting from runoff from infested fields. In southern China, most areas of Yunnan, Hubei, Hunan, Sichuan, and other provinces are mountainous regions. In these areas, cruciferous crops are planted in slopping or terraced fields, and fields at lower elevations can easily be infested by rain or irrigation water contaminated with *P. brassicae* resting spores coming from infested fields located at higher elevations (Chai et al. 2014).

3.9.5 Animal Fodder and Manure

Clubroot can also be spread on animal fodder or manure from livestock fed with *P. brassicae*-infested fodder (Gibbs 1931). Spores are spread in manure and on farm animals themselves, being capable of withstanding the highly acidic gut environment (Li 2013). It is thought that the movement of the pathogen from Europe resulted from transporting diseased fodder turnips or more likely swedes used to feed the livestock taken by early European settlers to America, Australasia, and other similar centers of settlement (Dixon 2009). One of the earliest suggestions that clubroot was an agricultural problem coming from the fourth century AD, when the Roman Pallidus described the development of spongy roots on rape, turnip, and radishes grown in soil fertilized with manure in Italy (Watson and Baker 1969). This observation is believed to be the first record of what might have been clubroot transmission by livestock manure (Howard et al. 2010). In 1964, an outbreak of clubroot in previously uninfested fields was linked to the spread of manure from animals fed on diseased roots (Creelman 1965). In 2007–2012, clubroot spreads rapidly in HuoShaoPing, Hubei Province of China. More than 70% of the fields were infested within 5 years, with 60% yield loss (Li 2008). The survey found that livestock were raised free-range, and were feeding on *P. brassicae*-infected material. Local farmers often fertilized the fields with livestock manure directly without composting. Resting spores of *P. brassicae* can survive through the digestive tracts of cattle, pigs, and chickens, and consequently can also be spread through the movement of infested animals and their manure (Li 2013).

3.10 *In Vitro* Growth of *Plasmodiophora brassicae*

The possibility that *P. brassicae* may have a culturable amoebal-like growth stage has been investigated. Resting spores isolated from diseased roots were plated on to a medium strongly selective for amoebal growth. The first amoebal isolate obtained (isolate 9) was subcultured once and was in culture for a total of only 14 days, prior to inoculation of Savoy and Chinese cabbage plants. Control plants inoculated with clubroot resting spores were all severely galled and none of the plants "sham inoculated" with *E. coli* or Ringer's showed any disease symptoms. Of the plants inoculated with amoebal cysts, all 10 replicate of Chinese cabbage plants showed galling with a disease index of 1.8 (on a scale of 0 to 3), and five out of 10 Savoy cabbage plants showed galling with a disease index of 0.8. Microscopic examination of the galls produced in the Savoy cabbage plants showed apparently identical resting spores in plants inoculated with *P. brassicae* and the amoebal isolate. The disease levels on Chinese cabbage as expected were higher than Savoy cabbage highly susceptible to *P. brassicae* (Buczacki 1983). The lower levels of symptoms associated with the amoebal isolate compared with resting spores of *P. brassicae* may partly be due to the lower concentration used as an inoculum, i.e., $>1 \times 10^6$ ml^{-1} resting spores as opposed to 0.5×10^5 ml^{-1} amoebae and amoebal cysts. In subsequent tests, nine out of 26 individual amoebal isolates produced disease symptoms. The level of disease was lower than that observed with isolate 9, possibly due to the fact that the original isolate was only subcultured once while the subsequent isolates were subcultured twice prior to inoculation of host plants. When these putative positive amoebal isolates were subcultured again and the resulting amoebal cysts used to inoculate plants, no disease symptoms developed.

In order to investigate further the cultivability of *P. brassicae*, and to avoid the problems of cross-contamination seen in some of the plant inoculations, a molecular approach was used. Molecular probes of *P. brassicae* were produced and used to probe dot blots of amoebal and control DNA. Strong hybridization to three out of 12 isolates was observed under high stringency conditions although the degree of hybridization varied depending on the probe used. This hybridization was lost after a successive subculture. It is unlikely that the pathogenicity and positive hybridization of amoebal isolates were caused by carryover of *P. brassicae* resting spores. A more likely explanation for the loss of pathogenicity/hybridization is that the original amoebal cultures were mixed, and the subculturing led to the overgrowth of the putative amoebal form of *P. brassicae* by other saprotrophic amoebae. This explanation would account for both the simultaneous loss of pathogenicity and hybridization of the amoebal isolates and the continued culture of amoebae (Arnold et al. 1996).

3.11 Factor Affecting Host–Pathogen Interactions

The host–pathogen interaction initiates after germination of resting spores through zoospores which cause primary root-hair infection. The resting spores survive in a very complex soil environment where there is interaction of host root exudates to induce germination process of resting spores. In this process, pH plays very great role in the percent spore germination. The other events of pathogenesis have been explained in Chap. 7.

3.11.1 Resting Spore Germination

In soil environment, host–pathogen interactions begin at the early seedling stage when host plant root exudates are present, which induces germination of resting spores and releasing of primary zoospores. The root exudates act as stimulants for resting spore germination. It is not confined to the specific host of *P. brassicae*. Root exudates from *Brassica* host (broccoli) and non-*Brassica* host (ryegrass), both stimulates spore germination. Some specific stimulants such as caffeic acid, coumalic acid, corilagin, and others can stimulate resting spore germination in Chinese cabbage. *Brassica* species have unique root characteristics which permit pathogen invasion and subsequent infection for the disease development. Resting spore germination is observed stimulated by root exudates in other species, but zoospores could not establish primary infection (MacFarlane 1970; Friberg et al. 2005; Craig 1989; Ohi et al. 2003).

3.11.2 Effects of Root Exudates on Resting Spore Germination

The percentage of resting spore germination was close to zero on the first day of the experiment in all of the exudate treatments. The spores began to germinate one day after incubation in all treatments. The percentage of germination was significantly higher ($P \geq 0.05$) for those spores that had been incubated with root exudates prepared in nutrient solution compared to those that had been incubated in nutrient solution alone or in sterile distilled water (Fig.3.2). This trend continued throughout the 6th-day time of the experiment. Among the root exudates tested, the exudate preparation from perennial ryegrass produced a relatively higher rate of spore germination compared to the other two root exudate preparations. After 6 days, the germination for resting spores incubated with root exudates from perennial ryegrass, Chinese cabbage, and canola was 12%, 10%, and about 9.5%, respectively. In contrast, spore germination was only 4.0% and 3.0% for spores incubated in nutrient solution and sdH_2O, respectively (Fig. 3.2). Chemical composition differed among root exudates prepared from different plant species (Table 3.3). The exudate preparation from Chinese cabbage had a relatively higher concentration of N, K, S, Ca, and Mg compared to exudate preparations from perennial ryegrass or canola. The highest concentration of nutrient elements was detected in the nutrient solution used

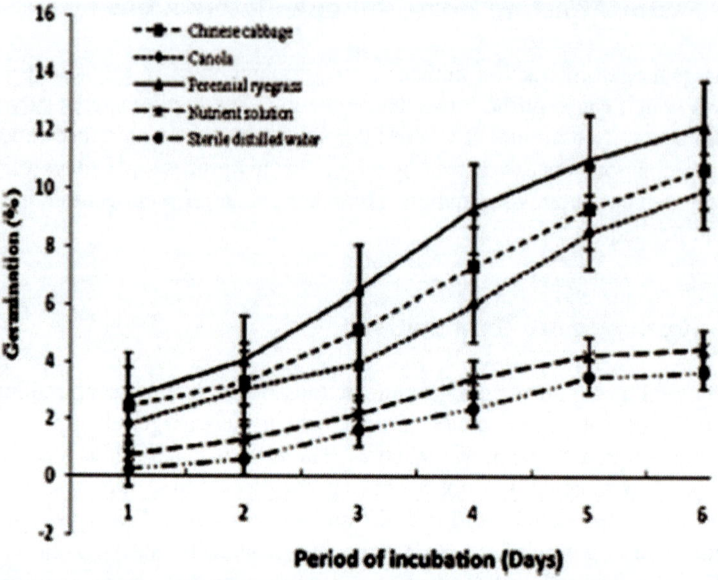

Fig. 3.2 Germination of *Plasmodiophora brassicae* resting spores in root exudates prepared from canola, Chinese cabbage, and perennial ryegrass in nutrient solution. Nutrient solution alone and sdH_2O served as controls. Data represent the means of eight replicates from two combined trials. The bars indicate standard errors (Rashid et al. 2013)

Table 3.3 Chemical composition and pH of root exudates preparations (Rashid et al. 2013)

	Concentration[a] ($mg\ L^{-1}$)							
Exudate preparation	N	K	S	Ca	Mg	Na	B	pH
Chinese cabbage	35.5	49	11	11	8	4	0.14	6.34
Canola	2.35	25	3	5	1	5	0.17	6.92
Perennial ryegrass	0.05	7	8	13	3	4	0.11	7.05
Nutrient solution	42.1	57	14	41	9	4	0.12	6.26
sdH_2O	0.01	4	3	2	1	4	0.08	7.39

[a]As determined by the inductively coupled plasma (ICP) method (Exova Canada Inc., Edmonton, AB, Canada); N ¼ nitrogen; K ¼ potassium; S ¼ sulfur; Ca ¼ calcium; Mg ¼ magnesium; Na ¼ sodium; B ¼ boron

as a control. The exudate preparation from perennial ryegrass had the lowest concentrations of boron and nitrogen. The pH values of root exudates prepared from Chinese cabbage, canola, and perennial ryegrass were 6.34, 6.92, and 7.07, respectively. The pH values of the nutrient solution and sdH_2O were 6.28 and 7.39, respectively (Rashid et al. 2013).

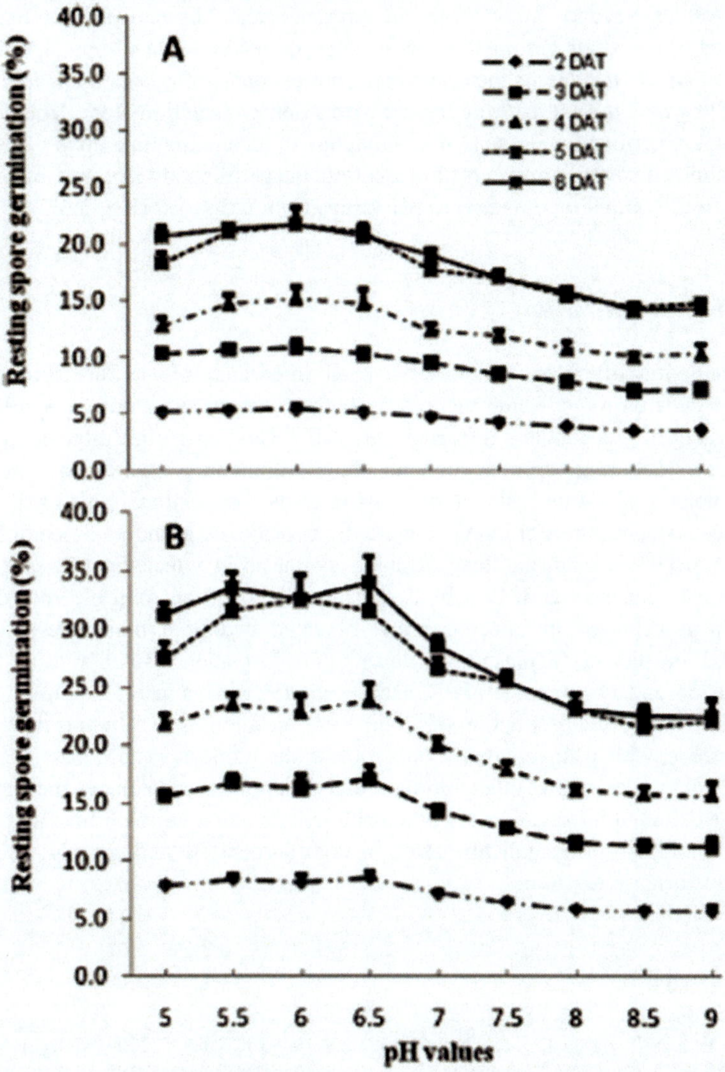

Fig. 3.3 Effect of a wide range of pH values on *Plasmodiophora brassicae* resting spore germination. The resting spores were incubated at 25 °C with nutrient solution alone (**a**) or with canola root exudates prepared in nutrient solution (**b**). Germination was recorded as a function of length incubation in days (DAT, days after treatment) over 6 days. Data represent the means of three replicates from two combined trials. The bars indicate the standard errors (Rashid et al. 2013)

3.11.3 Effects of pH on Resting Spore Germination

The percentage of germination of *P. brassicae* resting spores was maximal at pH values ranging from 5.0 to 7.0 in both incubation media (i.e., root exudate and nutrient solution) (Fig. 3.3). The percentage of spore germination gradually declined at pH values beyond 7.0 in both the presence and absence of root exudates. However, the overall germination percentage of resting spores was significantly higher in all pH treatments that contained root exudates of canola compared to the control in which the spores were treated with a nutrient solution alone. The percentage of spore germination increased as a function of incubation time up to 5 days, and then stabilized. After 5 or 6 days of incubation, the germination rate was lower at pH values from 7.5 to 9.0 compared to pH values from 5.0 to 7.0 (Fig. 3.3).

3.11.4 Gall Formation

The regulation of phytohormones plays an important role in the formation of massive galls on roots. Rapid increase in both cytokinin and auxin biosynthesis is observed during secondary infection and gall formation in the infected roots of *Brassicas*. *Brassica* species contain high aliphatic, indole, and aromatic glucosinolates which may play a vital role in disease development, and gall formation because conversion of indole-3-methyl glucosinolate to indole-3-acetonitrile is thought to be the main pathway of auxin synthesis in infected root tissues. The induction in nitrilase activity which cleavages indole-3-acetonitrole to indole-3-acetic acid occurred in infected roots. Elevated cytokinin biosynthesis is also observed in secondary plasmodia during gall formation (Butcher et al. 1974; Dekhuijzen and Overeem 1971; Rausch et al. 1981; Ugajin et al. 2003). Pedras et al. (2008) reported production of 45 different metabolites in *B. napus* infected by *P. brassicae*, which suggest that canola roots under biotic stress produce a complex blend of phytoalexins and other antimicrobial metabolites as defensive mechanisms. However, limited information about metabolic interaction between host and pathogen is available during gall formation in both susceptible and complete clubroot resistance disease reactions.

References

Ando S, Yamada T, Asano T, Kamachi S, Tsushima S, Hagio T, Tabei Y (2006) Molecular cloning of PbSTKL1 gene from *Plasmodiophora brassicae* expressed during clubroot development. J Phytopathol 154:185–189

Archibald JM, Keeling PJ (2004) Actin and ubiquitin protein sequences support a cercozoan/foraminiferan ancestry for the plasmodiophorid plant pathogens. J Eukaryot Microbiol 51:113–118

Archibald JM, Longet D, Pawlowski J, Keeling P (2003) A novel polyubiquitin structure in *Cercozoa* and *Formanifera*: evidence for a new eukaryotic super group. Mol Biol Evol 20:62–66

Arnold DL, Blakesley D, Clarkson JM (1996) Evidence for the growth of *Plasmodiophora brassicae in vitro*. Mycol Res 100:535–540

Ayers GW (1944) Studies on the life history of the club root organism, *Plasmodiophora brassicae*. Can J Res Sect C 32:143–149

Barr DJS (1983) The zoosporic grouping of plant pathogens — entity or non-entity. In: Buczacki ST (ed) Zoosporic plant pathogens — a modern perspective. Academic, London, pp 43–84

Barr DJS (1992) Evolution and kingdoms of organisms from the perspective of a mycologist. Mycologia 84:1–1

Bass D, Moreira D, Lopez-Garcia P, Polet S, Chao EE, Von der Heyden S, Pawlowski J, Cavalier-Smith T (2005) Polyubiquitin insertions and the phylogeny of Cercozoa and Rhizaria. Protist 156:149–161

Bass D, Chao EEY, Nikolaev S, Yabuki A, Ishida KI, Berney C, Pakzad U, Wylezich C, Cavalier-Smith T (2009) Phylogeny of novel naked filose and reticulose *Cercozoa*: Granofilosea cl. n. and Proteomyxidea revised. Protist 160:75–109

Bi K, He Z, Gao Z, Zhao Y, Fu Y, Cheng J, Xie J, Jiang D, Chen T (2016) Integrated omics study of lipid droplets from *Plasmodiophora brassicae*. Sci Rep 6:36965

Braselton JP (1981) Karyotypic analysis of *Plasmodiophora brassicae* based on serial thin sections of pachytene nuclei. Can J Bot 60:403–408

Braselton JP (1995) Current status of the Plasmodiophorids. Crit Rev Microbiol 21:263–275

Brodmann D, Schuller A, Ludwig-Muller J, Aeschbacher RA, Wiemken A, Boller T, Wingler A (2002) Induction of trehalase in *Arabidopsis* plants infected with the trehalose-producing pathogen *Plasmodiophora brassicae*. Mol Plant-Microbe Interact 15(7):693–700

Bryan RJ, Trese AT, Braselton JP (1996) Molecular karyotypes for the obligate, intracellular, plant pathogens, *Plasmodiophora brassicae* and *Spongospora subterranea*. Mycologia 88 (3):358–360

Buczacki ST (1983) *Plasmodiophora*. In: Buczacki ST (ed) Zoosporic plant pathogens. Academic, London, pp 161–191

Buczacki ST, Cadd SE (1976) Size distribution of the clubroot organism, *Plasmodiophora brassicae*. Trans Br Mycol Soc 67:133–136

Buczacki ST, Toxopeus H, Johnston P, Dixon TD, Hobolt LA (1975) Study of physiological specialization in *Plasmodiophora brassicae*: proposals for attempted rationalization through an international approach. Trans Br Mycol Soc 65:295–303

Buhariwalla H, Mithen R (1995) Cloning of a *Brassica* repetitive DNA element from resting spores of *Plasmodiophora brassicae*. Physiol Mol Plant Pathol 47:95–101

Buhariwalla S, Greaves S, Magrath R, Mithen R (1995) Development of specific PCR primers for the amplification of polymorphic DNA from the obligate root pathogen *Plasmodiophora brassicae*. Physiol Mol Plant Pathol 47:83–94

Bulman S, Siemens J, Ridgway HJ, Eady C, Conner AJ (2006) Identification of genes from the obligate intracellular plant pathogen, *Plasmodiophora brassicae*. FEMS Microbiol Lett 264 (2):198–204

Bulman S, Ridgway HJ, Eady C, Conner AJ (2007) Intron-rich gene structure in the intracellular plant parasite *Plasmodiophora brassicae*. Protist 158(4):423–433

Burki F, Keeling PJ (2014) *Rhizaria*. Curr Biol 24:R103–R107

Burki F, Kudryavtsev A, Matz MV, Aglyamova GV, Bulman S, Fiers M, Keeling PJ, Pawlowski J (2010) Evolution of Rhizaria: new insights from phylogenomic analysis of uncultivated protists. BMC Evol Biol 10:377

Butcher DN, El-Tigani S, Ingram DS (1974) The role of indole glucosinolates in the clubroot disease of the cruciferae. Physiol Plant Pathol 4:127–140

Castlebury LA, Domier LL (1998) Small subunit ribosomal RNA gene phylogeny of *Plasmodiophora brassicae*. Mycologia 90(1):102–107

Castlebury LA, Maddox JV, Glawe DA (1994) A technique for the extraction and purification of viable *Plasmodiophora brassicae* resting spores from host tissue. Mycologia 86:458–460

Cavalier-Smith T (1993) Kingdom protozoa and its 18 phyla. Microbiol Rev 57:953–994

Cavalier-Smith T (2002) The phagotrophic origin of eukaryotes and phylogenetic classification of Protozoa. Int J Syst Evol Microbiol 52:297–354

Cavalier-Smith T, Chao EE (1996/97) Sarcomonad ribosomal RNA sequences, rhizopod phylogeny, and the origin of euglyphid amoebae. Arch Protistenkd 147(3–4):227–236

Cavalier-Smith T, Chao EE (2003) Phylogeny and classification of phylum Cercozoa (Protozoa). Protist 154:341–358

Chai AL, Xie XW, Shi YX, Li BJ (2014) Research status of clubroot (*Plasmodiophora brassicae*) on cruciferous crops in China. Can J Plant Pathol 36:142–153

Craig MA (1989) Resting spores of *Plasmodiophora brassicae* as affected by root exudates of hosts and non-hosts. B. Sc. Honours Horticulture Dissertation, University of Strathclyde, Glasgow, p 71

Creelman DW (1965) Diseases of vegetable crops. Can Plant Dis Surv 45:52–63

Datnoff LE, Lacy GH, Fox JA (1984) Occurrence and populations of *Plasmodiophora brassicae* in sediments of irrigation water sources. Plant Dis 68:200–203

Daval S, Belcour A, Gazengel K, Legrand L, Gouzy J, Cottret L, Lebreton L, Aigu Y, Mougel C, Manzanares-Dauleux Maria J (2018) Computational analysis of the *Plasmodiophora brassicae* genome: mitochondrial sequence description and metabolic pathway database design. BMA Genomics. https://doi.org/10.1016/j.ygeno.2018.11.013

Daviere JM, Langin T, Daboussi MJ (2001) Potential role of transposable elements in the rapid reorganization of the *Fusarium oxysporum* genome. Fungal Genet Biol 34:177–192

Dekhuijzen HM (1976) Microscopical studies on the cortex and medullary rays of *Brassica campestris* var. *rapa* infected with *Plasmodiophora brassicae*. Phytopathol Z 87:171–186

Dekhuijzen HM, Overeem JC (1971) The role of cytokinins in clubroot formation. Physiol Plant Pathol 1:151–161

Dixon GR (2009) The occurrence and economic impact of *Plasmodiophora brassicae* and clubroot disease. J Plant Growth Regul 28:194–202

Donald EC (2005) The influence of abiotic factors and host plant physiology on the survival and pathology of *Plasmodiophora brassicae* of vegetable *Brassicas* [dissertation]. Melbourne: University of Melbourne

Donald C, Porter I (2009) Integrated control of clubroot. J Plant Growth Regul 28:289–303

Dylewski DP (1990) Phylum Plasmodiophoromycota. In: Margulis L, Corliss JO, Melkonian M, Chapman DJ (eds) Handbook of protoctista. Jones and Bartlett Publication, Boston, pp 399–416

Engleberg NC, Carter C, Weber DR, Cianciotto NP, Eisenstein BI (1989) DNA sequence of mip, a Legionella pneumophila gene associated with macrophage infectivity. Infect Immun 57:1263–1270

Eriksson J (1930) Fungous diseases of plants, 2nd edn. Bailliere Tindall and Co, London

Faggian R, Bulman SR, Lawrie AC, Porter IJ (1999) Specific polymerase chain reaction primers for the detection of *Plasmodiophora brassicae* in soil and water. Phytopathology 89(5):392–397

Fahling M, Graf H, Siemens J (2003) Pathotype separation of *Plasmodiophora brassicae* by the host plant. J Phytopathol 151:425–430

Fahling M, Graf H, Siemens J (2004) Characterization of a single-spore isolate population of *Plasmodiophora brassicae* resulting from a single club. J Phytopathol 152:438–444

Feng J, Hwang R, Hwang SF, Strelkov SE, Gossen BD, Zhou Q, Peng G (2010) Molecular characterization of a serine protease Pro1 from *Plasmodiophora brassicae* that stimulates resting spore germination. Mol Plant Pathol 11:503–512

Feng J, Hwang SF, Strelkov SE (2013) Assessment of gene expression profiles in primary and secondary zoospores of *Plasmodiophora brassicae* by dot blot and real-time PCR. Microbiol Res 168:518–524

Feng J, Cao T, Rennie DC, Strelkov SE, Hwang SF (2014) Host–parasite interactions in clubroot of crucifers. Can J Plant Pathol 36:113–121

Friberg H, Lagerlof J, Ramert B (2005) Germination of *Plasmodiophora brassicae* resting spores stimulated by a non-host plant. Eur J Plant Pathol 113:275–281

Gibbs JG (1931) Club-root in cruciferous crops. N Z J Agric 42:1–17

Goeche F, Wiese J, Núnez A, Labes A, Imhoff JF, Neuhauser S (2012) A novel phytomyxean parasite associated with galls on the bull-kelp *Durvillaea antarctica* (Chamisso) Hariot. PLoS ONE 7:e45358

Graf H, Sokolowski F, Klewer A, Diederichsen E, Luerssen H, Siemens J (2001) Electrophoretic karyotype of the obligate biotrophic parasite *Plasmodiophora brassicae* Wor. J Phytopathol 149:313–318

Graf F, Fahling M, Siemens J (2004) Chromosome polymorphism of the obligate biotrophic parasite *Plasmodiophora brassicae*. J Phytopathol 152:86–91

He D, Fu CJ, Fehling J, Tsai CC, Baldauf SL (2014) An alternative root for the eukaryote tree of life. Curr Biol 24:465–470

Hind-Lanoiselet T, Parker P (2005) Clubroot of canola and mustard. Primefact 115:1–2

Howard RJ, Strelkov SE, Harding MW (2010) Clubroot of cruciferous crops-new perspectives on an old disease. Can J Plant Pathol 32:43–57

Howlett BJ (1996) Application of pulsed field gel electrophoresis to the study of genomes of fungal pathogens of plants. Australian Plant Pathol 25(4):219–225

Ikegami H, Mukobata H, Naiki T (1978) Scanning electron microscopy of *Plasmodiophora brassicae* in diseased root cells of turnip and Chinese cabbage (Studies on the clubroot of cruciferous plants III). Ann Phytopathol Soc 44:456–464

Ingram DS, Tommerup IC (1972) The life history of *Plasmodiophora brassicae* Woron. Proc R Soc Lond B 180:103–112

Ito S, Yano S, Tanaka S, Kameya-Iwaki M (1994) The use of resting spore spheroplasts in the DNA analysis of *Plasmodiophora brassicae*. Ann Phytopathol Soc 60:491–495

Ito S, Maehara T, Tanaka S, Kameya-Iwaki M, Yano S, Kishi F (1997) Cloning of a single-copy DNA sequence unique *to Plasmodiophora brassicae*. Physiol Mol Plant Pathol 50:289–300

Ito S, Nakamura T, Matsumoto T, Maehara T, Tanaka S, Kameya-Iwaki M, Kishi F (1998) Characterization of DNA sequences cloned from resting spores of *Plasmodiophora brassicae*. J Phytopathol 146(2–3):143–147

Ito S, Maehara T, Maruno E, Tanaka S, Kameya-Iwaki M, Kishi F (1999) Development of a PCR-based assay for the detection of *Plasmodiophora brassicae* in soil. J Phytopathol 147:83–88

Johnson LJ, Johnson RD, Akamatsu H et al (2001) Spontaneous loss of a conditionally dispensable chromosome from the *Alternaria alternata* apple pathotype leads to loss of toxin production and pathogenicity. Curr Genet 40:65–72

Jones DR, Ingram DS, Dixon GR (1982a) Factors affecting tests for differential pathogenicity in populations of *Plasmodiophora brassicae*. Plant Pathol 31:229–238

Jones DR, Ingram DS, Dixon GR (1982b) Characterization of isolates derived from single resting spores of *Plasmodiophora brassicae* and studies of their interaction. Plant Pathol 31:239–246

Jubault M, Lariagon C, Simon M, Delourme R, Manzanares-Dauleux M (2008) Identification of quantitative trait loci controlling partial clubroot resistance in new mapping populations of *Arabidopsis thaliana*. Theor Appl Genet 117(2):191–202

Kageyama K, Asano T (2009) Life cycle of *Plasmodiophora brassicae*. J Plant Growth Regul 28:203–211

Kageyama K, Komatsu T, Suga H (2003) Refined PCR protocol for detection of plant pathogens in soil. J Gen Plant Pathol 69(3):153–160

Karling JS (1968) The plasmodiophorales, 2nd edn. Hafner, New York

Keeling PJ, Slamovits CH (2005) Causes and effects of nuclear genome reduction. Curr Opin Genet Dev 15:601–608

Kirk P (2014) Index fungorum. Available at http://www.indexfungorum.org/Names/NamesRecord.asp?Recor-dID=214750 (accessed April 22, 2014)

Klewer A, Luerben H, Graf H, Siemens J (2001) Restriction fragment length polymorphism markers to characterize *Plasmodiophora brassicae* single spore isolates with different virulence patterns. J Phytopathol 149(3–4):121–127

Kobelt P (2000) Die Verbreitung von sekundaren Plasmodien von *Plasmodiophora brassicae* (Wor.) im Wurzelgewebe von *Arabidopsis thaliana* nach immunhistologischer Markierung des plasmodialen Zytoskeletts. Ph.D. thesis, Free University Berlin, Berlin, Germany

Kobelt P, Siemens J, Sacristan MD (2000) Histological characterization of the incompatible interaction between *Arabidopsis thaliana* and the obligate biotrophic pathogen *Plasmodiophora brassicae*. Mycol Res 104(Pt 2):220–225

Kole AP, Gielink AJ (1961) Electron microscope observations on the flagella of the zoosporangial zoospores of *Plasmodiophora brassicae* and *Spongospora subterranea*. Proc Acad Sci 64:157–161

Lemarie S, Robert-Seilaniantz A, Lariagon C, Lemoine J, Marnet N, Jubault M, Manzanares-Dauleux MJ, Gravot A (2015) Both the jasmonic acid and the salicylic acid pathways contribute to resistance to the biotrophic clubroot agent *Plasmodiophora brassicae* in *Arabidopsis*. Plant Cell Physiol 56(11):2158–2168

Li BJ (2008) Cruciferous crop root disease in Changyang, Hubei Province. Chin Veg 7:55

Li JP (2013) Studies on the detection technology of *Plasmodiophora brassicae* and animal manure as a transmission route of the pathogen [dissertation]. Beijing: Chinese Acad Agri Sci

Li JP, Zhu YQ, Li BJ (2010) Transmission of clubroot of cruciferous vegetables. Chin Veg 5:21–22

Liu Y, Xu A, Liang F, Yao X, Wang Y, Liu X, Zhang Y, Dalelhan J, Zhang B, Qin M, Huang Z, Shaolin L (2018) Screening of clubroot-resistant varieties and transfer of clubroot resistance genes to *Brassica napus* using distant hybridization. Breed Sci 68:258–267

Ludwig-Muller J, Prinsen E, Rolfe SA, Scholes JD (2009) Metabolism and plant hormone action during clubroot disease. J Plant Growth Regul 28:229–244

Ludwig-Muller J, Julke S, Geiß K, Richter F, Mithöfer A, Sola I, Rusak G, Keenan S, Bulman S (2015) A novel methyltransferase from the intracellular pathogen *Plasmodiophora brassicae* methylates salicylic acid. Mol Plant Pathol 16(4):349–364

MacFarlane I (1970) Germination of resting spores of *Plasmodiophora brassicae*. Trans Br Mycol Soc 55:97–112

Manzanares-Dauleux MJ, Barret P, Thomas G (2000) Development of a pathotype specific SCAR marker in *Plasmodiophora brassicae*. Eur J Plant Pathol 106(8):781–787

Manzanares-Dauleux MJ, Divaret I, Baron F, Thomas G (2001) Assessment of biological and molecular variability between and within field isolates of *Plasmodiophora brassicae*. Plant Pathol 50:165–173

Margulis L, Corliss JO, Melkonian M, Chapman DJ (eds) (1989) Handbook of protoctista. Jones and Bartlett, Boston

McDonald MR, Sharma K, Gossen BD, Deora A, Feng J, Hwang SF (2014) The role of primary and secondary infection in host response to *Plasmodiophora brassicae*. Phytopathology 104:1078–1087

Miller CE (1959) Studies on the life cycle and taxonomy of *Ligniera verrucosa*. Am J Bot 46:725–729

Mithen R, Magrath R (1992) A contribution to the life history of *Plasmodiophora brassicae*: secondary plasmodia development in root galls of *Arabidopsis thaliana*. Mycol Res 96:877–885

Moller M, Harling R (1996) Randomly amplified polymorphic DNA (RAPD) profiling of *Plasmodiophora brassicae*. Lett Appl Microbiol 22:70–75

Moro A, Ruiz-Cabello F, Fernandez-Cano A, Stock RP, Gonzalez A (1995) Secretion by Trypanosoma cruzi of a peptidyl-prolyl cis-transisomerase involved in cell infection. EMBO J 14:2483–2490

Moxham SE, Buczacki ST (1983) Chemical composition of the resting spore wall of *Plasmodiophora brassicae*. Trans Br Mycol Soc 80(2):297–304

Naiki T, Kageyama K, Ikegami H (1978) The relation of spore density of *Plasmodiophora brassicae* Wor. to the root hair infection and club formation in Chinese cabbage (studies on the clubroot of cruciferous plant II). Ann Phytopathol Soc 44:432–439

Narisawa K, Kageyama K, Hashiba T (1996) Efficient root infection with single resting spores of *Plasmodiophora brassicae*. Mycol Res 100:855–858

Neuhauser S, Kirchmair M, Gleason FH (2011) Ecological roles of the parasitic Phytomyxea (plasmodiophorids) in marine ecosystems – a review. Mar Freshw Res 62:365–371

Neuhauser S, Kirchmair M, Bulman S, Bass D (2014) Cross-kingdom host shifts of phytomyxid parasites. BMC Evol Biol 14:33

Nikolaev SI, Berney C, Fahrni JF, Bolivar I, Polet S, Mylnikov AP, Aleshin VV, Petrov NB, Pawlowski J (2004) The twilight of heliozoa and rise of Rhizaria, an emerging super group of amoeboid eukaryotes. Proc Natl Acad Sci U S A 101:8066–8071

Ohi M, Kitamura T, Hata S (2003) Stimulation by caffeic acid, coumalic acid, and corilagin of the germination of resting spores of the clubroot pathogen *Plasmodiophora brassicae*. Biosci Biotechnol Biochem 67:170–173

Olive LS (1975) The Mycetozoans. Academic, New York

Palm BT, Burk M (1933) The taxonomy of the Plasmodiophoraceae. Arch Protistenkd 79:263–276

Pedras MS, Zheng QA, Strelkov S (2008) Metabolic changes in roots of the oilseed canola infected with the biotroph *Plasmodiophora brassicae:* phytoalexins and phytoanticipins. J Agric Food Chem 56(21):9949–9961

Rashid A, Ahmed HU, Xiao Q, Hwang SF, Strelkov SE (2013) Effects of root exudates and pH on *Plasmodiophora brassicae* resting spore germination and infection of canola (*Brassica napus* L.) root hairs. Crop Prot 48:16–23

Rausch T, Butcher DN, Hilgenberg W (1981) Nitrilase activity in clubroot diseased plants. Physiol Plant 52:467–470

Rennie DC, Manolii VP, Cao T, Hwang SF, Howard RJ, Strelkov SE (2011) Direct evidence of surface infestation of seeds and tubers by *Plasmodiophora brassicae* and quantification of spore loads. Plant Pathol 60:811–819

Rennie DC, Turkington TK, Leboldus JM, Howard RJ, Kutcher HR, Strelkov SE (2012) Dust monitoring for clubroot [*Plasmodiophora brassicae*] in Alberta. Can J Plant Pathol 35:99–132

Rolfe SA, Strelkov SE, Links MG, Clarke WE, Robinson SJ, Djavaheri M, Malinowski R, Haddadi P, Kagale S, Parkin IAP, Taheri A, Borhan MH (2016) The compact genome of the plant pathogen *Plasmodiophora brassicae* is adapted to intracellular interactions with host *Brassica* spp. BMC Genomics 17:272

Schwelm A, Fogelqvist J, Knaust A, Juelke S, Lilja T, Bonilla-Rosso G, Karlsson M, Shevchenko A, DhandapaniV CSR, Kim HG, Park JY, Lim YP, Ludwig-Mueller J, Dixelius C (2015) The *Plasmodiophora brassicae* genome reveals insights in its life cycle and ancestry of chitin synthases. Sci Reptr 5:11153

Schwelm A, Berney C, Dixelius C, Bass D, Neuhauser S (2016) The large subunit rDNA sequence of *Plasmodiophora brassicae* does not contain intra-species polymorphism. Protist 167:544–554

Scott ES (1985) Production and characterization of single-spore isolates of *Plasmodiophora brassicae*. Plant Pathol 34:287–292

Siemens J, Nagel M, Ludwig-Muller J, Sacristan MD (2002) The interaction of *Plasmodiophora brassicae* and *Arabidopsis thaliana*: parameters for disease quantification and screening of mutant lines. J Phytopathol 150(11–12):592–605

Siemens J, Keller I, Sarx J, Kunz S, Schuller A, Nagel W, Schmulling T, Parniske M, Ludwig-Muller J (2006a) Transcriptome analysis of *Arabidopsis* clubroots indicate a key role for cytokinins in disease development. Mol Plant-Microbe Interact 19(5):480–494

Siemens J, Keller I, Sarx J, Kunz S, Schuller A, Nagel W, Schmulling T, Parniske M, Ludwig-Muller J (2006b) Transcriptome analysis of *Arabidopsis thaliana* clubroots. Acta Hortic 706:301–306

Siemens J, Bulman S, Rehn F, Sundelin T (2009a) Molecular biology of *Plasmodiophora brassicae*. J Plant Growth Regul 28:245–251

Siemens J, Graf H, Bulman S, In O, Ludwig-Muller J (2009b) Monitoring expression of selected *Plasmodiophora brassicae* genes during clubroot development in *Arabidopsis thaliana*. Plant Pathol 58:130–136

Spanu PD (2012) The genomics of obligate (and Non obligate) biotrophs. Annu Rev Phytopathol 50:91–109

Spanu PD, Abbott JC, Amselem J, Burgis TA, Soanes DM, Stüber K, Brown JK, Butcher SA, Gurr SJ, Lebrun MH, Ridout CJ, Schulze-Lefert P, Talbot NJ, Ahmadinejad N, Ametz C, Barton GR, Benjdia M, Bidzinski P, Bindschedler LV, Both M, Brewer MT, Cadle-Davidson L, Cadle-

Davidson MM, Collemare J, Cramer R, Frenkel O, Godfrey D, Harriman J, Hoede C, King BC, Klages S, Kleemann J, Knoll D, Koti PS, Kreplak J, López-Ruiz FJ, Lu X, Maekawa T, Mahanil S, Micali C, Milgroom MG, Montana G, Noir S, O'Connell RJ, Oberhaensli S, Parlange F, Pedersen C, Quesneville H, Reinhardt R, Rott M, Sacristán S, Schmidt SM, Schon M, Skamnioti P, Sommer H, Stephens A, Takahara H, Thordal-Christensen H, Vigouroux M, Wessling R, Wicker T, Panstruga R (2010) Genome expansion and gene loss in powdery mildew fungi reveal tradeoffs in extreme parasitism. Science 330:1543–1546

Sundelin T (2008) Identification of expressed *Plasmodiophora brassicae* genes during plant infection and molecular diagnosis of plant pathogens. Ph.D. thesis, University of Copenhagen, Copenhagen

Sundelin T, Jensen DF, Lubeck M (2011) Identification of expressed genes during infection of Chinese cabbage (*Brassica rapa* subsp. *pekinensis*) by *Plasmodiophora brassicae*. J Eukaryot Microbiol 58:310–314

Tommerup IC, Ingram DS (1971) The life cycle of *Plasmodiophora brassicae* Woron. in *Brassica* tissue cultures and intact roots. New Phytol 70:327–332

Ugajin T, Takita K, Takahashi H, Muraoka S, Tada T, Mitsui T, Hayakawa T, Ohyama T, Hori H (2003) Increase inindole-3-acetic acid (IAA) level and nitrilase activity in turnips induced by *Plasmodiophora brassicae* infection. Plant Biotechnol 20:215–220

Wagner G, Charton S, Lariagon C, Laperche A, Lugan R, Hopkins J, Frendo P, Bouchereau A, Delourme R, Gravot A, Manzanares-Dauleux MJ (2012) Metabotyping: a new approach to investigate rapeseed (*Brassica napus* L.) genetic diversity in the metabolic response to clubroot infection. Mol Plant-Microbe Interact 25:1478–1491

Wallenhammar AC (1996) Prevalence of *Plasmodiophora brassicae* in a spring oilseed rape growing area in central Sweden and factors influencing soil infestation levels. Plant Pathol 45:710–719

Wallenhammar AC, Arwidsson O (2001) Detection of *Plasmodiophora brassicae* by PCR in naturally infested soils. Eur J Plant Pathol 107(3):313–321

Wallenhammar AC, Almquist C, Soderstrom M, Jonssona A (2011) In-field distribution of *Plasmodiophora brassicae* measured using quantitative real-time PCR. Plant Pathol 61:16–28

Wang SZ (1981) Changes of main crop diseases in Henan province during resent thirty years. J Henan Agric Univ 4:127–133

Waterhouse G (1973) Plasmodiophoromycetes. In: Ainsworth GC, Sparrow FK, Sussman AS (eds) The fungi, vol 4B. Academic, New York, pp 75–82

Watson AG, Baker KF (1969) Possible gene centers for resistance in the genus *Brassica* to *Plasmodiophora brassicae*. Econ Bot 23:245–252

Williams PH, McNabola SS (1967) Fine structure of *Plasmodiophora brassicae* in sporogenesis. Can J Bot 45:1665–1669

Woronin M (1878) *Plasmodiophora brassicae*, Urheber der Kohlpflanzen-Hernie. Jb Wissen Bot 11:548–574

Yano S, Tanaka S, Ito SI, Kameya-Iwaki M (1997) Variations of random amplified polymorphic DNA (RAPD) patterns among field populations of *Plasmodiophora brassicae*. Ann Phytopathol Soc 63(3):179–182

Electron Microscopy and Ultra-Structures 4

Abstract

During infection process, small galls emerge from epidermal cells in the elongation zone of host roots, and abundant secondary plasmodia, and resting spores can be seen within the parenchymal cells of such infected galls. At early infection stages in infected host cells, primary plasmodia contain complex cellular components including nuclei, lipid droplets, and an endomembrane system which is visible in epidermal cells in the root elongation zone. Cortical invasion by *P. brassicae* shows disruption of cell wall, presence of vesicles or inclusion bodies inside the cell wall, thickening of cell wall in combination with plasmodesmata, as well as elongation and/or disorganization of host nuclei. Multinucleate, intracellular plasmodia are visible within the clubs in the cortex and pericycle cells. Gall development amplifies merismatic activity within the vascular cambium and phloem parenchyma cells in the region of the hypocotyl. It also perturbs vascular development with reduction in xylem, and increase in phloem parenchyma cells in infected plants. Scanning electron microscopy of infected host galls revealed the initial plasmodia of the pathogen as spherical, smooth surfaces with an average diameter of 3.7 μm with an average size of 4.2 μm in diameter. Young plasmodia are round with partial rough surfaces and an average 8 μm in diameter. Vegetative plasmodia are smooth surface, spherical, and 45 μm in diameters which almost fill the host cells. Young resting spores are connected by fine filamentous materials, and mature resting spores have spiny surfaces. In resting sporangia during primary zoosporogenesis, single zoospores are differentiated within each sporangium being initiated by the emergence of two flagella in the tight space formed by invagination of plasma membrane within the sporangium. A deep groove is formed on the zoospore cell body by further invagination of the plasma membrane. Two flagella appear to coil around the zoospore cell body in parallel along this groove. A nucleolus like region is visible in the mature zoospores. A partial opened germ pore is seen in the sporangium

G. S. Saharan et al., *Clubroot Disease of Crucifers*,
https://doi.org/10.1007/978-981-16-2133-8_4

containing the mature zoospores. Zoospores have electron dense mitochondria and well-developed endoplasmic reticulum. Mitotic nuclear division is characterized by a persistent nucleolus, bipolar centrioles paired end-to-end, densely stained chromatin, and a complex array of membranes that surrounded and ramify through the spindle. Chromatin begins to condense in prophase, and is aligned at metaphase in a reticulate plate on the nuclear equator. Large amounts of membrane cisternae within the spindle are visible at metaphase. Several concentric sheets of perinuclear endoplasmic reticulum surround the spindle and give the mitotic nucleus the superficial appearance of having an intact nuclear envelope.

Keywords

Infection · Host tissues · Electron-microscopy · Fine structures · Anatomical changes · Cellular events in gall formation · Clubroot SEM · Electron microscopy of zoosporogenesis · Cytological changes and ultra-structures

4.1 Introduction

To comprehend life cycle of the pathogen, and host–pathogen interaction, mechanisms in the host tissues, electron microscopy, and observations of ultra-structures are very crucial, and important in association with anatomical, histological, cytological, and genetical studies. Through the use of electron microscopy, fine structures of *P. brassicae* have been revealed during infection process; hairy root infection; gall formation, and development of morphological structures like plasmodia, zoospores, flagella formation, resting spores, mitosis; anatomical and histological changes in the diseased host plant cells, galled sections, and callus cultures through axenic culture system.

4.2 Electron Microscopy of Infection Process

Resting spore suspensions of *P. brassicae* from calli were freshly prepared, and used to inoculate both *Arabidopsis* and rapeseed seedlings grown in Petri dishes, and proliferation of the pathogen within the infected roots was observed by microscopy. Approximately six weeks post-inoculation, small galls emerged from epidermal cells in the elongation zone of *Arabidopsis* roots, and abundant resting spores were observed within the parenchymal cells of these galls (Plate 4.1a, b). At 60 dpi, sections from these small galls were stained with CFW, and examined under confocal microscopy. Resting spores stained with CFW for cell wall depositions were clearly visible (Plate 4.1c). Similarly, at 30 dpi with the resting spore suspension freshly prepared from callus tissues, small galls were observed under light microscopy, on the surface of the elongation zone of canola roots grown in Petri dishes (Plate 4.1d). At this stage, abundant secondary plasmodia, and resting spores in the parenchymal cells of small galls were observed (Plate 4.1e). At 45 dpi (Plate

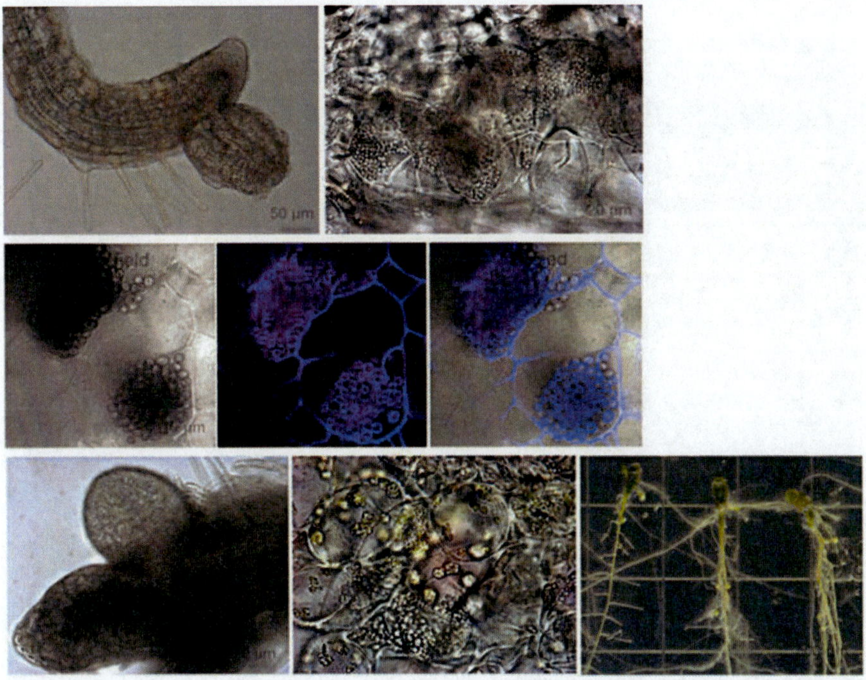

Plate 4.1 *Plasmodiophora brassicae* successfully proliferated and completed its life cycle in both *Arabidopsis* and rapeseed roots grown in Petri dishes. (**a**) A small gall emerging from the epidermis of the elongation zone of *Arabidopsis* roots at 45 days post-inoculation (dpi). (**b**) Microscopic view of a thin section of gall tissues showing resting spores formed in parenchyma cells of the infected gall on *Arabidopsis* roots at 60 dpi. (**c**) Formation of mature resting spores in infected galls of *Arabidopsis* roots. Cell walls of mature resting spores in infected galls at 60 dpi were stained by CFW (blue), and reviewed under confocal microscopy. (**d**) Microscopic view of a young gall emerging from the epidermis of the root elongation zone of rapeseed root at 30 dpi. (**e**) Microscopic view of a thin section of gall tissue showing plasmodia, and resting spores of *P. brassicae* at 30 dpi. *SP* secondary plasmodia; *RS* resting spores. (**f**) Development of visible clubroot galls on rapeseed roots at 45 dpi. Small galls (arrowheads) appeared on the surface of both main and lateral roots (Tu et al. 2018)

4.1f), galls along the surface of rapeseed roots were clearly visible by eye, and 3–10 visible galls appeared on the root surface of each infected seedling (Tu et al. 2018).

Live cell imaging of *Plasmodiophora brassicae–Arabidopsis* interactions during early infection stages has been performed by Tu et al. (2018) using transmission light microscopy. *P. brassicae* derived from callus tissue retains full virulence with the production of mature resting spores in *Arabidopsis* root tissues grown on medium in Petri dishes. However, to establish a live cell imaging system to study *P. brassicae–Arabidopsis* interactions, and pathogenesis during early infection stages remains largely unknown in the infection cycle (Schuller and Ludwig-Muller 2016). With a high inoculation pressure (5 × 10^7–10^8 spores/ml), inoculated *Arabidopsis* roots were shorter than control roots, and multiple primary plasmodia colonized epidermal

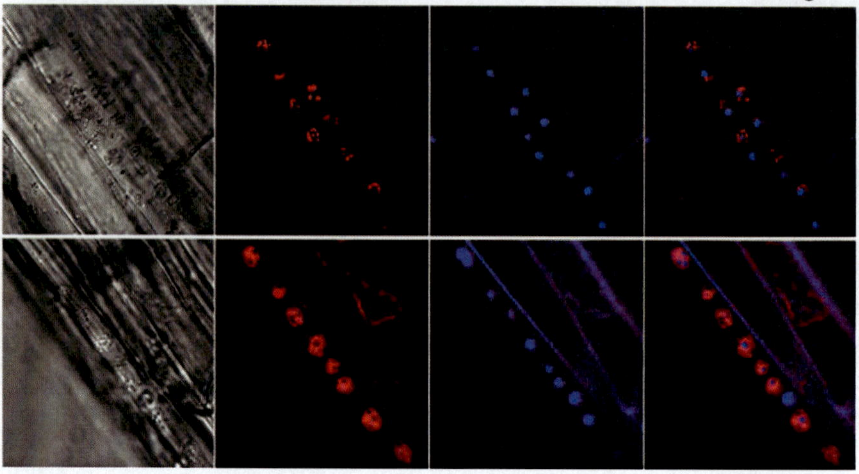

Plate 4.2 Confocal images of primary plasmodia in root epidermal cells of *Arabidopsis* seedlings grown on MS plates. *Arabidopsis* seedlings grown on MS plates for 7 days were inoculated with inoculum derived from the axenic culture of *Plasmodiophora brassicae* in rapeseed root tissues. Infected root epidermal cells of *Arabidopsis* seedlings at 2 days post-inoculation were stained with Nile red and DAPI and viewed under a confocal microscopy. Top panel: Intracellular lipid droplets and nuclei of primary plasmodia stained by Nile red and DAPI were visualized with EX/EM wave lengths 488/585–615 nm (red), and 405/420–480 nm (blue), respectively. Bottom panel: The endomembrane and nuclei of primary plasmodia stained by Nile red and DAPI were visualized with alternative wavelengths 543/>650 nm (red) and 405/420–480 nm (blue), respectively. Arrowheads indicate primary plasmodia in epidermal cells of the root elongation zone. Scale bars: 10 μm (Tu et al. 2018)

cells of the root elongation zone could be observed at 2 dpi (Plate 4.2). The *P. brassicae* primary plasmodia were spherical and approximately 5 μm in diameter. To confirm these spherical structures as the pathogen rather than host cell organelles, infected *Arabidopsis* roots were double-stained with Nile red and DAPI. Nile red-stained intracellular lipid droplets were observed at 585–615 nm after excitation at 488 nm (Plate 4.2, top panel), whereas the endomembrane compartments of primary plasmodia were observed at >650 nm after excitation with 543 nm (Plate 4.2, bottom panel). The single nucleus of each primary plasmodium was fluorescently labeled blue with DAPI, indicating primary plasmodia were uninucleate at initial formation after infection (Plate 4.2). Previously, as identified by TEM in an infected cabbage root hair, this stage had been referred to as uninucleate amoeba (Aist and Williams 1971). As protists in the eukaryotic kingdom, primary plasmodia of *P. brassicae* at early infection stages in host cells contain complex cellular components including nuclei, lipid droplets, and an endomembrane system.

Advantages of live cell imaging prompted Tu et al. (2018) to examine the interaction of *P. brassicae* with the host cellular responses by using transgenic *Arabidopsis* plants expressing a GFP fusion of the nuclear protein, AT4G19150-GFP, and the plasma membrane (PM) protein, and PMA4-GFP, respectively.

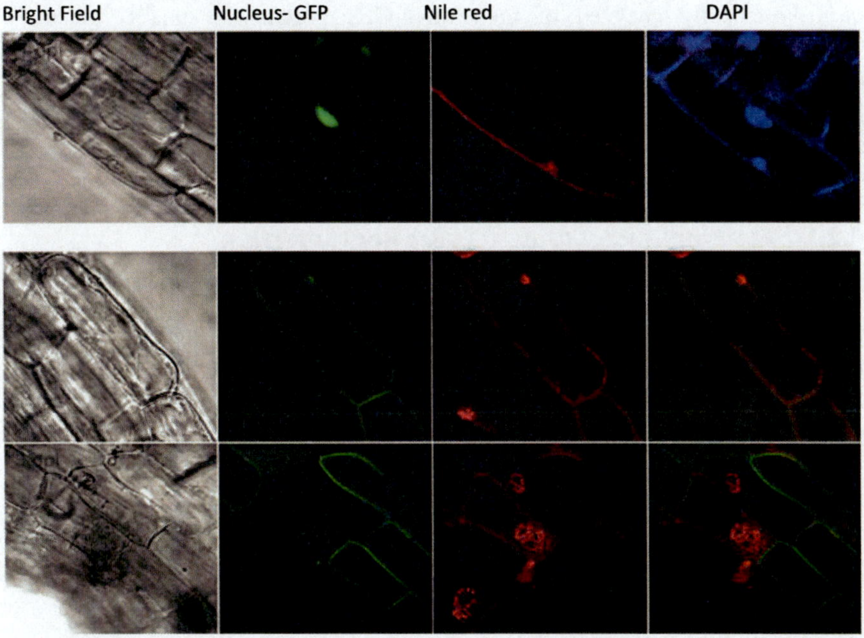

Plate 4.3 Live cell imaging of primary infection of *Plasmodiophora brassicae* in *Arabidopsis* roots. *Arabidopsis* seedlings expressing a 35S: AT4G19150-GFP (CS84731) construct for labeling the nucleus (**a**) or a 35S: PMA4-GFP (N799380) construct for labeling plasma membrane (**b**) were grown on MS plates for 7 days. Roots of *Arabidopsis* seedlings were inoculated with inoculum derived from the axenic culture of *P. brassicae* in rapeseed root tissues. Infected *Arabidopsis* roots were stained with Nile red and/or DAPI, and viewed under confocal microscopy. Inserts show images of single plasmodia with abundant lipid droplets. Arrowheads indicate primary plasmodia in epidermal cells of the root elongation zone. *Nu* host nucleus; *nu P. brassicae* nucleus. Scale bars: 10 μm (Tu et al. 2018)

Arabidopsis seedlings grown on MS medium were inoculated with *P. brassicae* inoculum from the dual culture, and then examined by confocal microscopy after staining with Nile red and DAPI. At 1 dpi, a root epidermal cell of the AT4G19150-GFP-expressing plant contained a primary plasmodium of *P. brassicae* that was located at the host cytoplasmic periphery and clearly labeled by Nile red and DAPI (Plate 4.3a). The infected epidermal cell displayed a nuclear GFP signal colocalizing with the DAPI staining. Similarly, an epidermal cell of the PMA4-GFP-expressing plant with a primary plasmodium at the cytoplasmic periphery showed the green fluorescence of the PM marker (Plate 4.3b). These images confirmed that *Arabidopsis* cells hosting primary plasmodia at the initial infection stage maintained cellular integrity. At 3 dpi, intracellular primary plasmodia labeled by Nile red were observed in infected root epidermal cells of the PMA4-GFP-expressing plant. In these infected cells, the green fluorescence of the PM marker disappeared compared to adjacent epidermal cells where no pathogen was visible, and the PMA4-GFP signal was retained (Plate 4.3b). These observations suggest that either the PM has been specifically depleted of the PMA4-GFP proteins or that it generally disintegrated when intracellular primary plasmodia were formed within the infected cells. In either case, whether this is actively triggered by the pathogen during pathogenesis or constitutes a defense response of the host cell remains to be elucidated.

4.2.1 Electron Microscopy of Hairy Root Infection

Infection of hairy roots was observed in all seven strains of hairy roots, but the frequencies differed among the host plant species (Table 4.1). Chinese cabbage and turnip hairy roots showed up to 80.0% incidence of root-hair infection. In contrast, the percentages of root-hair infection in cabbage and rapeseed hairy roots were moderate and low, respectively. Similar trends were observed with the number of zoosporangial groups. Chinese cabbage and turnip hairy roots developed galls 8 weeks after inoculation, while cabbage and rapeseed hairy roots showed no gall formation (Table 4.1). The number and fresh weight of galls formed in turnip hairy roots were both greater than those of Chinese cabbage hairy roots. The galls in turnip hairy roots were spherical or spindle-shaped and were scattered throughout the hairy roots (Plate 4.4) (Asano et al. 1999).

4.2.2 Microscopic Observations

There were no differences in the structure of zoosporangia among the seven hairy root strains. All stages in root-hair infection were observed, including the

Table 4.1 Root-hair infection and gall formation in hairy roots inoculated with *Plasmodiophora brassicae*[a] (Asano et al. 1999)

Root-hair infection			Gall formation		
Hairy root strains[b]	Infection (%)	Zoosporangial groups/infected hairy root (no)	Infection (%)	Galls/ infected hairy root (no)	Gall fresh weight/ infected hairy roots (mg)
BO 123	6.7a[c]	0.3a	0.0a	0.0a	0.0a
BP123	73.3b	18.5ab	33.3ab	39.9a	14.8a
BP1724	80.0b	22.0ab	13.3ab	18.3a	3.8a
BR123	53.3ab	8.5a	46.7b	276.6b	312.3b
BR13	80.0b	32.5b	20.0ab	75.2a	164.3ab
BN123	26.7ab	1.9a	0.0a	0.0a	0.0a
BN1725	46.7ab	4.1a	0.0a	0.0a	0.0a

[a]Values are means of three trials, each with five replicates. Resting spores were obtained from the galls 8 weeks after inoculation
[b]BO., BP., BR., and BN. = cabbage, Chinese cabbage, turnip, and rape, respectively
[z]Means within a column followed by the same letter are not significantly different ($P < 0.05$) according to Duncan's multiple range test

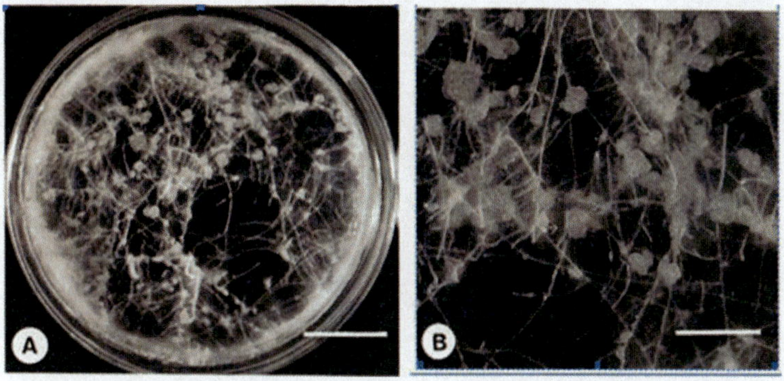

Plate 4.4 Gall formation on turnip hairy roots eight weeks after inoculation with resting spores of *Plasmodiophora brassicae*. (**a**) *In vitro* mass propagation of galls on Murashige and Skoog's medium. Bar = 20 mm. (**b**) Galls shown at high magnification. Bar = 10 mm (Asano et al. 1999)

development of immature zoosporangia, and complete evacuation of zoosporangia (Plate 4.5). The morphology of zoosporangia and the growth stages of root-hair infection in hairy root cultures were identical to the morphology and growth stages in infected intact plants. The secondary plasmodia and immature resting spores within whitish gall tissues were observed by light and electron microscopy 6 weeks after inoculation. Eight weeks after inoculation, however, galls became brownish, and a number of mature resting spores were observed in the gall tissues (Plate 4.6). The cortical infection phase of the pathogen was also identical in hairy roots, and diseased plants grown in infested soil.

1. *Pathogenicity of resting spores from diseased hairy roots:* The pathogenicity of resting spores from 8-week-old galls in hairy roots was tested on turnip seedlings by the slurry method (Table 4.2). Resting spores from galls developed in hairy root cultures showed the same infectivity as those from infected intact plants. All inoculated plants had shown severe clubroot symptoms.
2. *Susceptibility of the intact plant:* Root-hair infection was observed in all host species (Table 4.3). The percentage of infected plants and the number of zoosporangial groups per root were not significantly different among host plants. All four plant species also formed galls with high disease severity (Asano et al. 1999).

4.3 Fine Structures of *Plasmodiophora brassicae* in Host Tissues

Fine structures of *P. brassicae* were revealed by Riascos et al. (2011) through histopathology of cabbage root galls in the tissue sections of roots infected by *P. brassicae*. Plasmodia are observed in different stages of development in cabbage (Plate 4.7a) and turnip roots (Plate 4.7b). The observed structures are part of the life cycle of primary and secondary disease development in the host tissues during

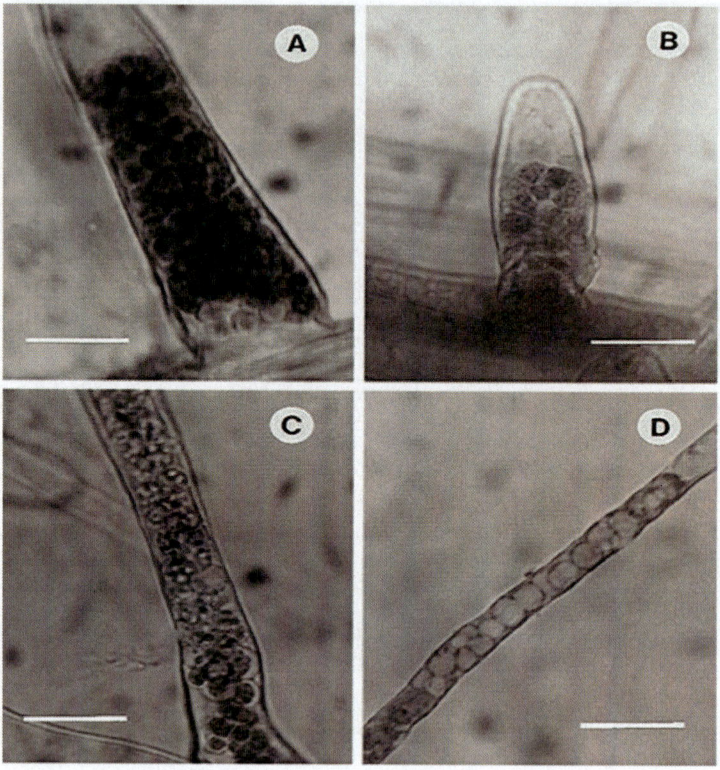

Plate 4.5 Root hair of a turnip hairy root 12 days after inoculation with *Plasmodiophora brassicae*. (**a**) Immature zoosporangia in the root hair; (**b**) zoosporangia containing zoospores in the root hair; (**c**) partially evacuated zoosporangia in the root hair, and (**d**) completely evacuated zoosporangia in the root hair. Bars = 20 μm (Asano et al. 1999)

pathogenesis (Suwabe et al. 2003). It can be seen by elongation and/or increased cell size associated with hypertrophy, and rupture processes of the cells containing the pathogen (Plate 4.8a), which is consistent with what was mentioned by Donald et al. (2008), and Kobelt et al. (2000), this being a mechanism of spread of the pathogen within the host tissue.

Symptoms of cortical invasion by *P. brassicae* included disruption of the cell wall, presence of vesicles or inclusion bodies inside the cell wall, cell wall thickness in combination with plasmodesmata, and elongation and/or disorganization of host nuclei. Amoebal shape of pathogen has been reported in root cortical cells, and it is often found in association with disruption of cell walls, and is presumed that it is the structure of the parasite that penetrates and invades the cortical cells (Donald et al. 2008; Kobelt et al. 2000). The observations carried out can be seen in the cell wall thickening (Plate 4.8b). At the level of epidermis, periderm and cortex become visible due to loss of tissue differentiation (Plate 4.9a, b).

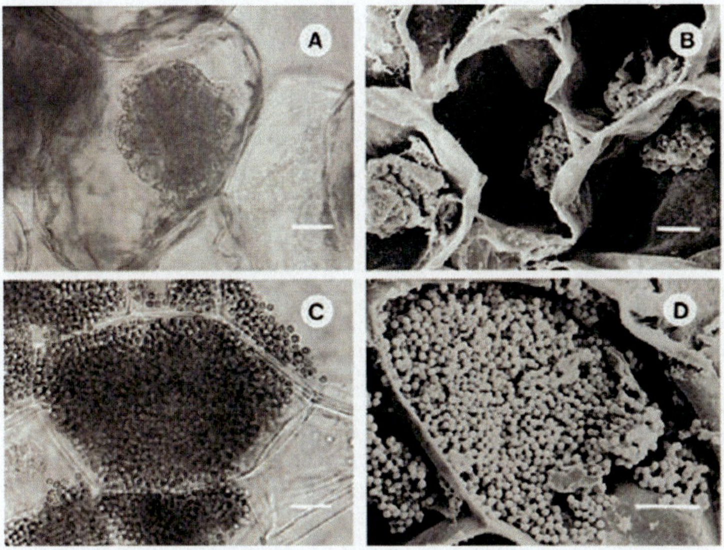

Plate 4.6 (**a**, **c**) Light micrographs and (**b**, **d**) scanning electron micrographs of gall tissue of turnip hairy roots inoculated with *Plasmodiophora brassicae*. (**a**, **b**) Six weeks after inoculation. Secondary plasmodia can be seen in the gall tissue. (**c**, **d**) Eight weeks after inoculation. Resting spores have developed in the gall tissue. Bars = 20 μm (Asano et al. 1999)

Table 4.2 Pathogenicity of resting spores of *Plasmodiophora brassicae* to turnip seedlings (Asano et al. 1999)

		Clubroot	
Resting spore source[a]	Viability of resting spores[b] (%)	Incidence (%)	Disease severity[c]
Hairy root	79.3a[d]	100.0b	100.0b
Intact plant	77.5a	100.0b	100.0b
Control		0.0a	0.0a

[a]Resting spores were obtained from the galls formed on hairy roots or diseased intact plants 8 weeks after inoculation

[b]Viability of resting spores was assessed by Nomarski interference optic microscopy

[c]Disease severity = $[(N0 \times 0) + (N1 \times 20) + (N2 \times 80) + (N3 \times 100)]/(N0 + N1 + N2 + N3)$. $N0$, $N1$, $N2$, and $N3$ are the numbers of seedlings in disease index classes 0 to 3, respectively. Disease index classes: 0 = no clubroot; 1 = slight swelling; 2 = moderate swelling on lateral or taproots; and 3 = severe swelling on lateral or taproots

[d]Means within a column followed by the same letter are not significantly different ($P < 0.05$) according to Duncan's multiple range test

The observations of artificial infection of *P. brassicae* in cabbage seedlings revealed anatomical changes of host plant. At 15 dpi, plants inoculated at a concentration of 1×10^8 spores ml^{-1} have a lower longitudinal growth of the root system (4–6.7 cm), compared with plants inoculated at a concentration of 1×10^7 spores ml^{-1} (3–8.5 cm) as compared with control (5.5–8.5 cm). Regardless of inoculation method, there is increased lateral growth (1–1.5 cm) as shown in Plate 4.10, which

Table 4.3 Root-hair infection and gall formation on several *Brassica* plants[a] (Asano et al. 1999)

Root-hair infection			Gall formation		
Host plant	Infection (%)	Zoosporangial groups (no)	Incidence (%)	Disease severity[b]	Fresh weight (g)
Cabbage	100.0a[c]	100.2a	100.0a	100.0a	0.9a
Chinese cabbage	100.0a	53.8a	100.0a	100.0a	1.4b
Turnip	100.0a	86.6a	100.0a	100.0a	2.2c
Rape	100.0a	81.2a	100.0a	100.0a	1.8bc

[a]Values are means of two trials. Two replicates were used for the root-hair infection experiment, and three replicates were used for the gall formation experiment

[b]Disease severity = $[(N0 \times 0) + (N1 \times 20) + (N2 \times 80) + (N3 \times 100)]/(N0 + N1 + N2 + N3)$. $N0$, $N1$, $N2$, and $N3$ are the numbers of seedlings in disease index classes 0 to 3, respectively. Disease index classes: 0 = no clubroot; 1 = slight swelling; 2 = moderate swelling on lateral and/or taproots; and 3 = severe swelling on lateral and/or taproots

[c]Means within a column followed by the same letter are not significantly different ($P < 0.05$) according to Duncan's multiple range test

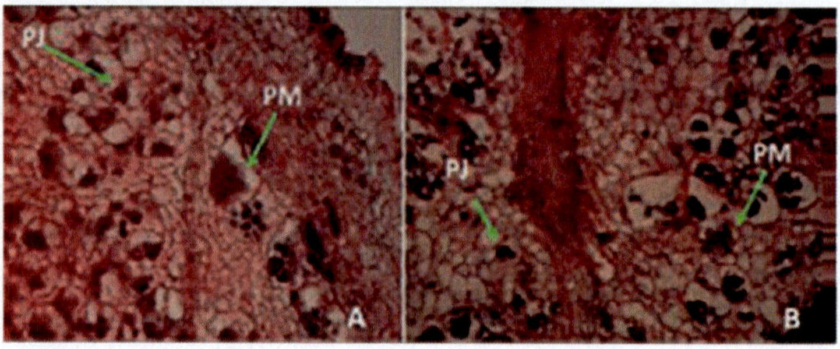

Plate 4.7 Detail of ruptured cells containing plasmodia: PJ, young plasmodia; PM, mature plasmodia 40×. **(a)** roots of cabbage, **(b)** turnip roots (Riascos et al. 2011)

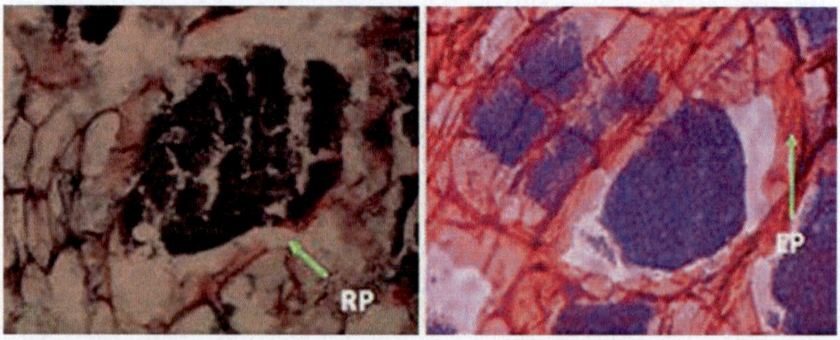

Plate 4.8 Damage to the cell wall level: **(a)** RP, breakage of cell wall in *B. rapa* (40×). **(b)** EP, cell wall thickening (40×) (Riascos et al. 2011)

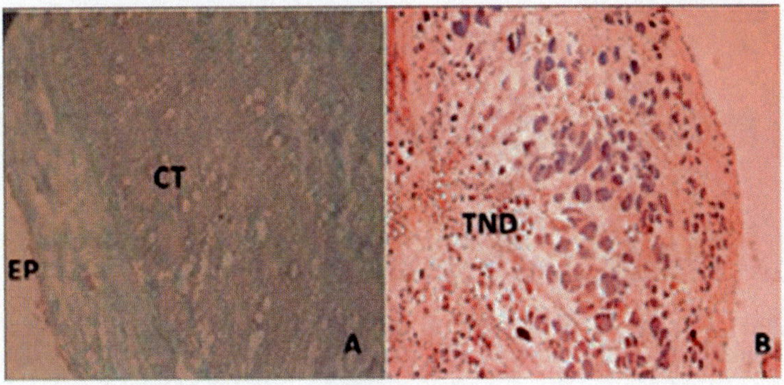

Plate 4.9 Contrast healthy and diseased tissues: (**a**) Differentiated tissues healthy root of *B. rapa*: EP, epidermis; CT: cortex (10×). (**b**) Undifferentiated tissue (TND) diseased root of *B. oleracea* (4×) (Riascos et al. 2011)

Plate 4.10 Plants infected with *P. brassicae*, 15 dpi. T1: direct addition of inoculum to 1 × 10^7spores ml^{-1}, T2: immersion 1 × 10^7spores ml^{-1}, T3: direct addition of inoculum to 1 × 10^8spores ml^{-1}, T4: immersion 1 × 10^8 spores ml^{-1}, T5: control (Riascos et al. 2011)

can be related to an imbalance of growth regulators in host–pathogen interaction, such as auxins and cytokinins, (Ludwig-Muller and Schuller 2008). Epicotyl length is higher in control plants compared to inoculated plants, suggesting that *P. brassicae* affects the growth and development of infected plants. Samples taken at 21 dpi revealed that plants inoculated at a concentration of 1 × 10^8 spores ml^{-1} by direct addition and immersion (T3 and T4) showed abundant lateral growth compared with control plants (T5), with values of 0.4 cm, 0.5 cm, and 0.3 cm, respectively. As seen in the stereoscope, it was evident in the axis tissue loss and darkening of the main root in the inoculated treatments, mainly in T4 (Plate 4.11). At this time, young plasmodia are found in treatments of immersion.

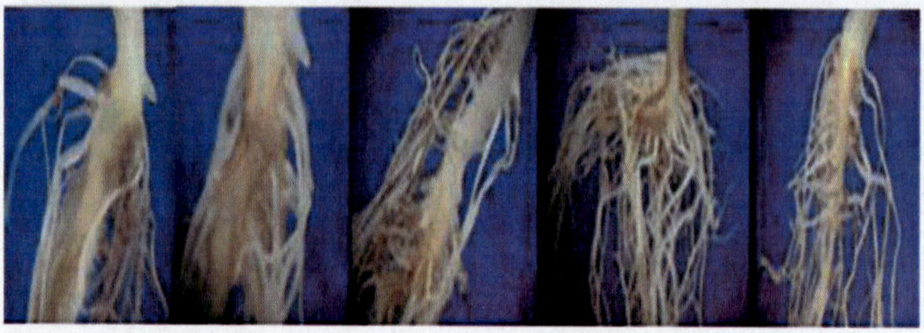

Plate 4.11 Cabbage roots of plants infected with *P. brassicae* (28 dpi): view of the top of the taproot; from left to right. T1: addition of inoculums 1×10^7 spores ml^{-1}; T2: immersion 1×10^7 spores ml^{-1}; T3: addition of inoculums 1×10^8 spores ml^{-1}; T4: immersion 1×10^8 spores ml^{-1}; T5: control (Riascos et al. 2011)

The results at 28 dpi were consistent with those obtained for the first and second sampling, for showing lower growth and higher longitudinal lateral growth for treatments T3 and T4. According to Devos et al. (2005) and Siemens et al. (2006), infection with *P. brassicae* results in the formation of adventitious roots, resulting in swelling, due to the involvement of auxins, cytokinins, and xyloglucan enzyme transglycosylase/hydrolase (XTH).

According to Voorrips and Kanne (1997) symptoms in infected plants may have variations from small swellings with ovoid or globular in the roots to the formation of large galls on the entire root system, which is consistent with that observed by Riascos et al. (2011). The growth rate of infected plants is delayed, and chlorosis occurs because the presence of large galls usually causes a reduction in nutrient uptake and water. As a result, infected plants grow abnormally and production is reduced (Karling (1968) and Voorrips (1995)). Discolored roots showed the presence of young plasmodia, whereas in the beginning of infection resulted in association with loss of main root axis, and increased lateral root growth, being more noticeable at a concentration of 1×10^8 spores ml^{-1} by immersion. The advanced stages were observed with a higher number of multinucleated plasmodia with different sizes and shapes.

4.4 Symptoms of Anatomical Changes

Symptoms caused by *P. brassicae* include abnormal enlargement (clubs) of the roots with severe swelling of main and lateral roots, and below-ground stems. In severely affected plants, clubs coalesce and cover most of the root system (Plate 4.12b). As a consequence, plants show reduced growth and stunting (Plate 4.12a). The histological sections showed that *P. brassicae* infected the root cortex, reached the vascular tissue, and promoted an abnormal increase in cabbage cell division (hyperplasia), and enlargement (hypertrophy) that resulted in extensive disorganization of tissue structure (Plate 4.12c, d). Infected cells occurred in clusters of variable size throughout diseased tissues surrounded by uninfected healthy looking cells (Plate 4.12b). Multinucleate, intracellular plasmodia were easily observed within the clubs in the

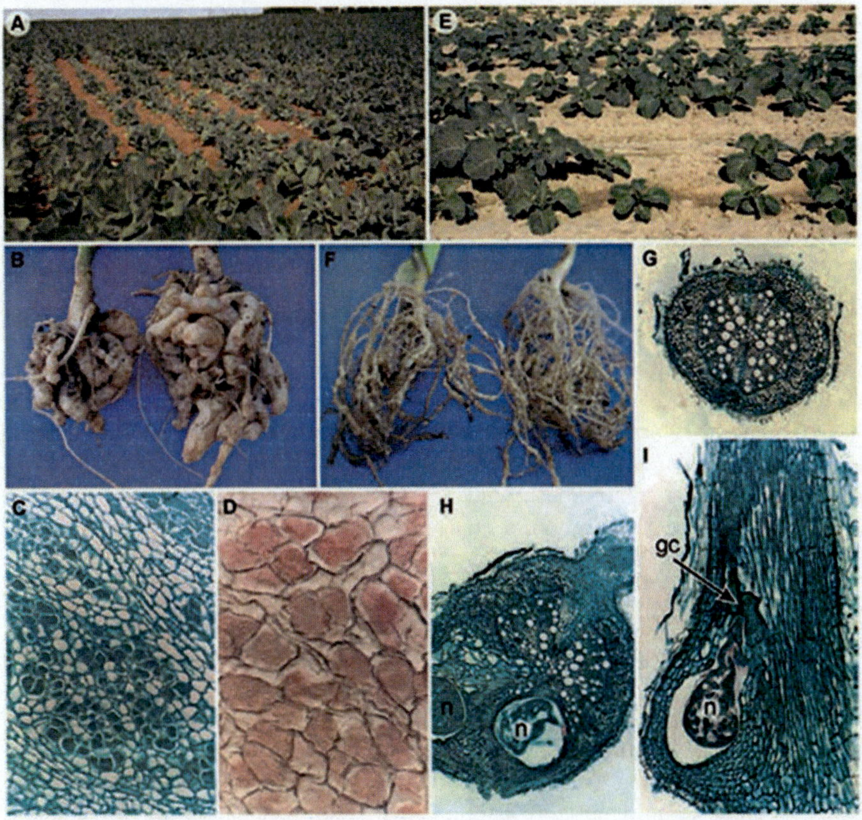

Plate 4.12 Cabbage cv. Lupini infected by *Plasmodiophora brassicae* (**a–d**), and by *Meloidogyne javanica* (**e–i**) in a field in southern Italy. (**a, e**) Field showing patches with stunted plants. (**b**) Root systems showing clubroot symptoms with extensive galling. (**c, d**) Cross-sections of roots showing the plasmodia in hypertrophied cortical cells. (**f**) Root system severely infected by *M. javanica* showing large and numerous galls. (**g**) Cross-section of a healthy cabbage root. (**h, i**) Cross and longitudinal sections of a root showing the nematode (n) feeding on multinucleate giant cells (gc) with hypertrophied nuclei (Navas-Cortés et al. 2010)

cortex and pericycle cells. Immature plasmodia were also present in the affected tissue. Plasmodia ultimately develop into resting spores (Plate 4.12c, d). These histological changes and pathogen structures are similar to those described previously for the second infection cycle of *P. brassicae* in susceptible hosts (Ingram and Tommerup 1972; Mithen and Magrath 1992; Kobelt et al. 2000; Ando et al. 2006). In Plate 4.12e–i depicts nematode infection and has been shown just for comparison with *P. brassicae* infection.

4.5 Cellular Events During Gall Formation

Gall development involves the amplification of existing meristematic activities within the vascular cambium (VC) and phloem parenchyma (PP) cells in the region of the hypocotyl. During gall formation, cellular events are initiated by activation of host cell division stimulating cambial activity by secondary plasmodia of the pathogen. Proliferation of cambium is a major cause of gall formation. The presence of secondary plasmodia in the host tissues reduces xylem formation and changes vascular system differentiation. Phloem parenchyma (PP) and cells within the phellogen are consistent with the activation of existing meristematic activity rather than the formation of new meristematic regions. In the infected roots, plasmodia are present in cortex and pericycle cells. Both infected and neighboring uninfected cells undergo extreme hyperplasia (cell division) and hypertrophy (cell expansion) that leads to the formation of root galls. Gall formation occurs as a result of perturbations in the activity of meristems in their regions and subsequent cell differentiation. Perturbations in phytohormones content, particularly auxin and cytokinin plays significant role in gall formation of infected plants.

4.5.1 *Plasmodiophora brassicae* Increases Cell Division within the Hypocotyl by Exploiting Existing Meristematic Activities

In *Arabidopsis*, *P. brassicae*-induced gall formation occurs largely in the hypocotyl, and only in the uppermost part of the root system—regions where secondary thickening is occurring. Visualization of cell divisions using a CYCB1: GUS construct showed an increase in mitotic activity and maintenance of meristematic (undifferentiated) cell division in the vicinity of the vascular cambium (VC) in the hypocotyl of infected plants 12 DPI. No difference in the patterns of cell division in the root or hypocotyl of infected plants was seen at 4 or 7 DPI. The timing and location of this activation of host cell division is consistent with the stimulation of existing cambial activity by secondary plasmodia at the onset of gall formation rather than the initiation of new meristems. The timing of this induction is similar to that observed by Mithen and Magrath (1992) who reported an increase in cell division 20 DPI that was associated with secondary infection of *Arabidopsis* by *P. brassicae* plasmodia. Devos et al. (2006) reported *P. brassicae*-induced cell division in the root cortex from 4 DPI using the same clubroot isolate as used by Malinowski et al. (2012) and a CYCB1: GUS construct in a *Ler* background.

Cross-sections through the hypocotyl of infected plants 16 DPI showed that ANT promoter activity became more widespread than in uninfected controls, but was still restricted to VC, phloem parenchyma (PP), and cells within the phellogen (PL), consistent with the activation of existing meristematic activity, rather than the formation of new meristematic regions. ANT activity overlapped with the regions of cell proliferation observed with the CYCB1: GUS marker. At later stages of infection, when the VC had become discontinuous and distorted, ANT expression was maintained in cambial regions, and was still associated with the localized islands of mitotic activity observed. These observations suggest that *P. brassicae*-induced cell division usurps the endogenous program of secondary thickening in the hypocotyl providing a reservoir of proliferating cells for further infection thus driving gall formation rather than de novo meristem formation.

Malinowski et al. (2012) tested the hypothesis that VC activity was necessary for gall formation in two ways. Firstly when used chemically induced expression of the cell cycle inhibitor ICK1/KRP1 to microscopic analysis revealed that cle41-1 mutants were unable to maintain the VC activity normally observed in infected Col-0 plants. As a result, cle41-1 mutants were unable to maintain a population of undifferentiated cells later during infection and this led to the early death of infected plants. Most infected cle41-1 mutants had died 32 DPI while the majority of Col-0 plants were still alive at this time (typically infected Col-0 plants can survive beyond 40 DPI). Despite this situation, spores collected from infected cle41-1 plants showed a similar viability to spores collected from infected Col-0. A comparison of the pOpON:KRP1 plants and the cle41-1 mutant is informative as although both approaches blocked gall formation without impairing *P. brassicae* development, only infected cle41-1 plants showed accelerated host mortality. It is speculated that a decrease in developmental capacity to generate phloem and xylem cells in the hypocotyl during the later stages of disease development affected the ability of the cle41-1 plants to tolerate infection as these plants still had large rosettes relative to induced pOpON: KRP1 plants placing significant demands on the remaining vascular system for the transport of water and assimilates. The increased proliferation of cambium is a major cause of gall formation in *Arabidopsis*.

Gall formation was not impaired in the wox4 mutant, although WOX4 has been identified as an important regulator of cambial activity, a number of lines of evidence point to the existence of additional regulatory pathways. The morphology of hypocotyls of 40-day-old uninfected wox4 mutants (26 DPI), when secondary thickening was well advanced, did not differ from that of uninfected Col-0 plants of the same age, indicating that the wox4 mutation had relatively little impact on cambial activity and subsequent development of this tissue at this age, in marked contrast to the cle41 mutation. Similarly, the morphology of hypocotyls of clubroot-infected wox4 mutants was similar to that of Col-0, again in contrast to cle41, leading to hypothesize that other additional regulatory pathways operate in the mature hypocotyls of both uninfected and clubroot-infected plants. The existence of additional regulatory pathways has also been proposed by Suer et al. (2011) who note that CLE41 acts upstream of WOX4, and the cle41 mutation has a wider impact on cambium activity than the wox4 mutation which still retains cambial function. Suer et al. (2011) propose that WOX4 is involved in auxin-mediated cambial proliferation, which supports the hypothesis that clubroot-induced gall formation is driven by changes in cytokinin rather than auxin, however, the regulation of WOX4 by other plant growth regulators remains to be established. Hirakawa et al. (2010) also support both WOX4-dependent and WOX4 independent pathways to regulate cambial activity.

4.5.2 Changes in Vascular System Differentiation, and Implications of the Host–Pathogen Interaction

P. brassicae is an obligate biotrophic pathogen that establishes a strong sink for carbohydrate during gall formation (Keen and Williams 1969; Mitchell and Rice 1979; Evans and Scholes 1995). Gustafsson and Falt (1986) reported an increased proportion of phloem bundles in developing galls, a process that would facilitate carbohydrate transport towards the rapidly growing tissue. Using *CLE44*: GUS as a marker of phloem tissue demonstrated that in infected *Arabidopsis* plants there was no increase in phloem bundle number 16 DPI, and even a small decrease 26 DPI, presumably as a consequence of the mechanical destruction of a proportion of cells by expanding neighboring cells infected with plasmodia. However, there were significant increases in the total number of PP cells, and a corresponding increase in the expression of the APL gene that regulates phloem development, which suggested that *P. brassicae* infection induces phloem formation via endogenous signaling pathways. PP cells act as conduits for carbohydrate transport with sieve elements (Lalonde et al. 2003), and an increased number of PP cells in infected tissue would influence carbohydrate distribution within the plant. This hypothesis is supported by the increased expression of the phloem-specific *SUC2* sucrose-H+ symporter observed 26 and 32 DPI. Siemens et al. (2011) found that expression of an extracellular invertase inhibitor from a root-specific promoter reduced the size of the gall in the roots (but not hypocotyl) of *P. brassicae*-infected Arabidopsis and demonstrated a requirement for sink-associated sucrolytic activity in gall formation.

Plasmodiophora brassicae infection had a profound effect on xylem development. Xylem formation was strongly repressed from 16 DPI, and only fragmented patches of xylem developed thereafter. As increased expression or mis-expression of CLE41 has been demonstrated to shift phloem/xylem balance and distribution (Etchells and Turner 2010), and CLE peptides are known to be tracheary element differentiation inhibitory factors (TDIFs; Ito et al. 2006), the alterations in *CLE41/44* expression pattern observed upon infection might contribute to reduced xylem formation. However, although CLE expression patterns were altered upon infection there was no significant change in expression level. Also, the characteristic loss of xylem was still observed in infected *cle41*-1 plants, which indicated that alterations in *CLE* signaling alone were unlikely to be the sole cause of disrupted xylogenesis. Transcripts of positive regulators of xylem cell specification VND6 (Kubo et al. 2005) and VND7 (Kubo et al. 2005; Yamaguchi et al. 2008, 2011) decreased upon infection. These results are consistent with a programmed down-regulation of the xylogenesis regulatory pathway. However, the expression of the negative regulator VNI2 (Yamaguchi et al. 2010) was not changed significantly upon infection, while the expression of a second negative regulator of xylogenesis, XND1, (Zhao et al. 2008) was strongly reduced at 16 and 26 DPI. It has been suggested (Yamaguchi et al. 2010) that XND1 may be an element of another parallel mechanism that represses expression of vessel-related genes, and thus the repression upon infection may result from a plant response to alterations in the *VND6/7* pathway. Downstream components of the xylogenesis pathway were strongly repressed upon infection at all time points examined by Malinowski et al. (2012). The expression of transcripts of genes involved in secondary cell formation, and programmed cell death (Funk et al. 2002), MYB46, XCP1, and XCP2 decreased. Microscopic observation of the *XCP2: GUS* line, a marker of late xylogenesis, showed that despite a drastic overall decline in xylem formation upon infection, expression was still maintained in a few cells that surrounded regions of active cambium islands in late stages (formation of resting spores) of gall development. It shows that in these localized regions xylem differentiation was restored. However, a role for cytokinins is likely. Mahonen et al. (2006) showed that cytokinin negatively regulates protoxylem formation while cytokinin content is elevated in the initial proliferative stages of gall development (Dekhuijzen and Overeem 1971; Ingram and Tommerup 1972; Devos et al. 2006). Ingram and Tommerup (1972) described a correlation between the presence of secondary plasmodia and reduction in xylem formation. Both host and *P. brassicae* have been proposed to contribute to this increase as a consequence of alterations in cytokinin biosynthesis and degradation (Dekhuijzen 1981; Muller and Hilgenberg 1986; Ando et al. 2005; Devos et al. 2006). Siemens et al. (2006) demonstrated that overexpression of cytokinin oxidase/dehydrogenase genes results in disease resistance. Given the importance of cytokinin in disease development, it would be interesting to test if elevated cytokinins were directly responsible for the decrease in xylogenesis.

4.5.3 Implications of Reduced Gall Size on the Disease Progression

As innate resistance to *P. brassicae* infection in *Brassicas* is limited, an alternative strategy is to explore how anatomical or physiological features of the host could significantly decrease the infection rate or progress of the disease. Malinowski et al. (2012) have demonstrated that *P. brassicae*-induced gall formation results from perturbations in host physiology and development but that the regulation of these processes is complex. It has been shown that manipulations (*pOpON: KRP1* and *cle41-1*) that have a profound impact on gall formation do not significantly affect pathogen development as both plants produced infective spores. However, extensive host cell proliferation during the development of the gall created an opportunity for the infection of a large number of cells in which resting spores eventually developed. Thus, large galls are, therefore, agronomically relevant as they facilitate the amplification of the pathogen, and the development of a large spore bank within the soil.

Although many studies have used gall size as an indicator of infection and hence host resistance or tolerance to the disease, these measurements do not account for differences in host or pathogen development and as such have the potential to be misleading. Therefore, Malinowski et al. (2012) propose that molecular markers of both host and pathogen development (Siemens et al. 2009) must be combined to understand the progress of the disease. The manipulation of gall size can provide a viable route to decreasing the number of resting spores produced in infected tissue, but whether this outcome can be achieved without compromising host viability remains to be established.

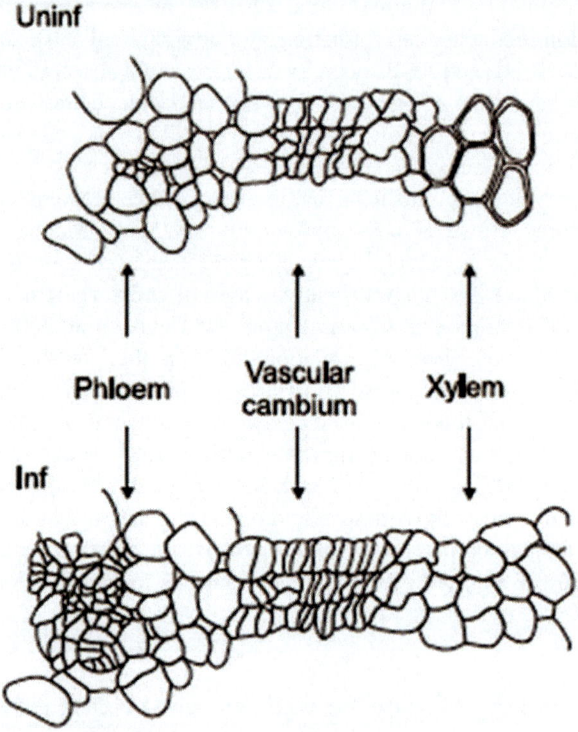

Fig. 4.1 A cellular model of the developmental changes that occur in hypocotyls of *P. brassicae*-infected *Arabidopsis*. In uninfected plants, vascular cambium proliferation generates progeny cells that differentiate to form phloem and xylem. Another meristematic activity is located in the phloem parenchyma cells and helps to maintain the function and integrity of phloem bundles. These meristematic activities are stimulated in infected plants and lead to the production of a greater number of cells. While phloem production is maintained in infected plants there is an increase in the proliferative state and a disruption of xylogenesis with less mature xylem created (Malinowski et al. 2012)

The emerging picture is that the gall forms as a consequence of pathogen-driven reprogramming of existing meristematic activity. Proliferation of the VC is the main factor that influences gall size; however, PP cells also proliferate. A shift in the developmental program in favor of proliferation also leads to disruption of xylem differentiation (Fig. 4.1). A key challenge is now to understand how the pathogen elicits these developmental responses in the host. This approach will require an integrated study of the regulatory networks that underlie gall formation, and alterations in hormone metabolism that drive them. Central to this process will be the requirement for understanding of *P. brassicae* metabolism and the extent to which it actively participates in these changes.

4.6 Scanning Electron Microscopy (SEM) of Clubroot Cells

Clubroot gall development was observed using scanning electron microscopy (SEM) of infected turnip and Chinese cabbage to reveal pathogens' morphological developments from very young plasmodium to the fully matured resting spores (Ikegami et al. 1978).

4.6.1 Early Young Plasmodia

Early young plasmodia in host cells are shown in Plate 4.13-figures 2–4. The initial plasmodia had a spherical shape with smooth surfaces and an average of 3.7 μm in diameter, ranging from 2.7 to 5.0 μm (Plate 4.13-figure 2). The strand-like materials connecting the nearly spherical bodies are shown in Plate 4.13-figure 3. The latter

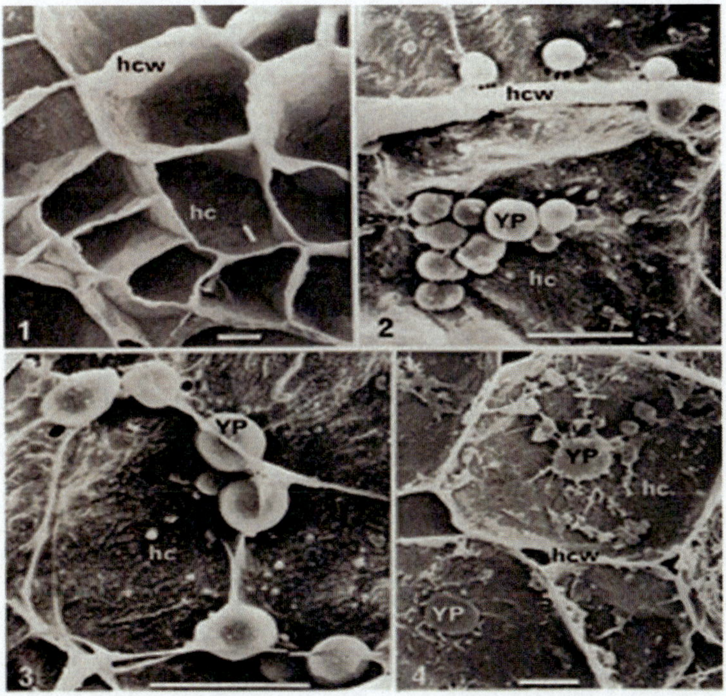

Plate 4.13 Scanning electron micrographs at the early stage of plasmodial formation in the infected root cells of Chinese cabbage. Each scale line=10μm. Figure 1 Healthy root cells of turnip, Taibyohikari-kabu at 35 days after sowing and cell organelles (arrow). Glutaraldehyde fixation figures 2–4 Infected root cells of Muso-hakusai at 25 (figure 4) and 35 days (2 and 3) after sowing. Very young plasmodia were observed as spherical bodies with or without strand-like material. Glutaraldehyde fixation. Abbreviations in capital letters refer to parasite, and in lower case to host. *RS* resting spore *YP* young plasmodium *YRS* young resting spore *VP* vegetative plasmodium *hc* host cytoplasm, *hcw* host cell wall (Ikegami et al. 1978)

bodies are with an average of 4.2 μm in diameter, ranging from 3.3 to 4.5 μm. Each of them is enveloped by a membranous material with a smooth surface. Some cell organelles of small particles can be recognized inside healthy host cells as depicted in Plate 4.13-figure 1 and those appeared also in the infected cells.

4.6.2 Young Plasmodia

Cluster-shaped plasmodia were observed also in an infected cell, as shown in Plate 4.14-figures 1 and 2. A small mass consisting of various-sized spherical bodies with smooth surfaces can be seen. Several young round plasmodia with partially rough surfaces and an averaging 8 μm in diameter are seen in Plate 4.14-figure 3. Low magnification of cut tissue revealed a young plasmodium in a cell and an actively growing plasmodium in a contiguous cell (Plate 4.14-figure 4).

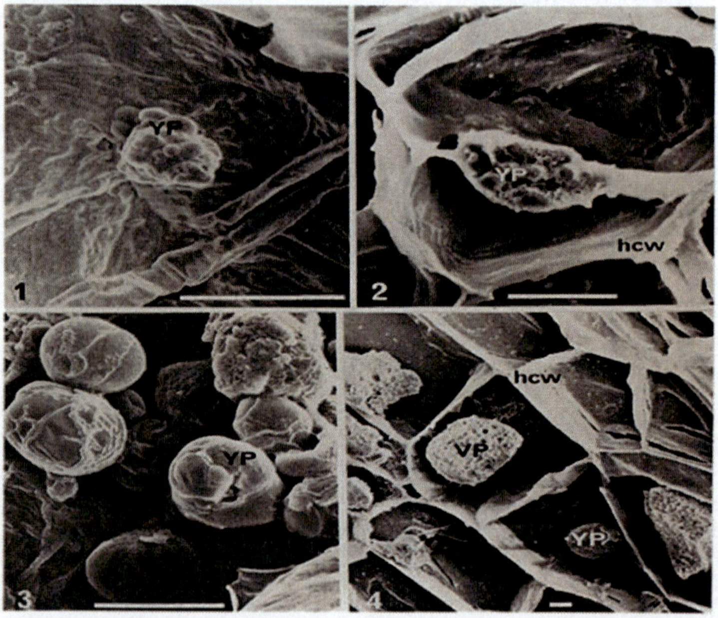

Plate 4.14 Scanning electron micrographs of young plasmodia growing in the infected root cells of turnip (figures 1, 3, and 4) and Chinese cabbage (2). Each scale line=10ƒÊm. (figures 1 and 2) Cluster-shaped plasmodia growing in an infected cell of Taibyohikari-kabu on 37 (figure 1) and 32 days (2) after sowing. Glutaraldehyde fixation. (figure 3) Spherical young plasmodia with partially rough surface in an infected cell of Waseo-kabu at 50 days after sowing. FAA fixation. (figure 4) Young and vegetative plasmodia in the infected cells of Waseo-kabu at 32 days after sowing. Glutaraldehyde fixation (Ikegami et al. 1978)

4.6.3 Vegetative Plasmodia

Vegetative plasmodia developing into a considerably large mass in an infected cell were observed in Plate 4.14-figure 4 at a low magnification; a higher magnification was shown in Plate 4.15-figure 4. The plasmodium is about 45 μm in diameter. The vegetative plasmodia, which consisted of several or many spherical bodies with smooth surfaces, are shown in Plate 4.15-figures 1, and 2. A plasmodium showing a sponge-like structure in figure 2, and an irregular shaped one in figure 3. The vegetative plasmodia almost filled the host cells (figures 2 and 3).

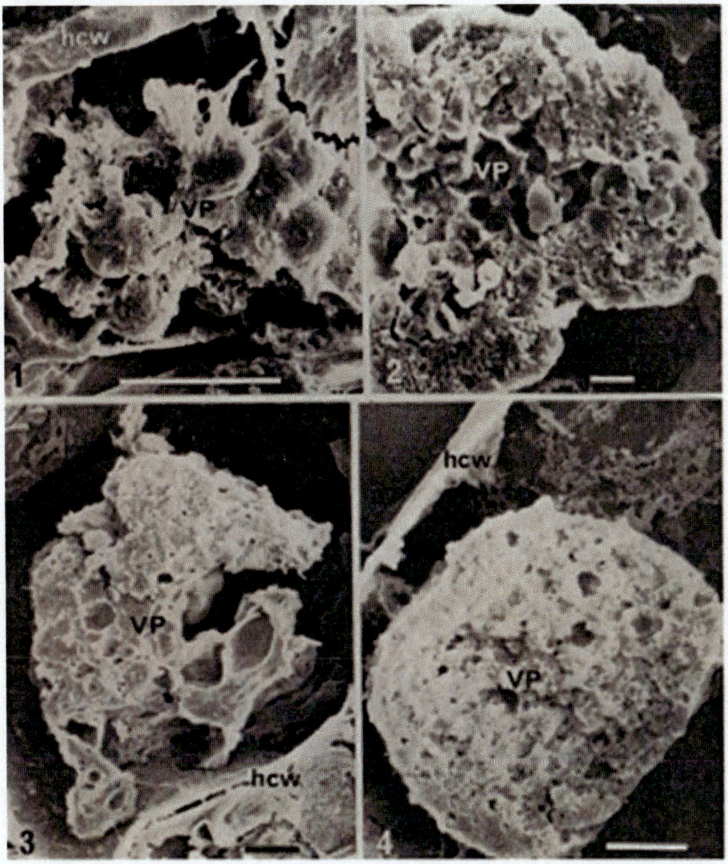

Plate 4.15 Scanning electron micrographs of different vegetative plasmodia showing rapid growth in an infected cell of Chinese cabbage (figure 1) and turnip (figures 2–4). Each scale line = 10 μm. (figure 1) A vegetative plasmodium with several spherical bodies observed in an infected cell of Harumakigokuwase-hakusai at 50 days after sowing. Glutaraldehyde fixation. (figure 2) A vegetative plasmodium of sponge-like structure in an infected cell of Waseo-kabu at 32 days after sowing. Glutaraldehyde fixation. (figure 3) A vegetative plasmodium growing irregularly in an infected cell of Waseo-kabu at 37 days after sowing. Glutaraldehyde and osmic acid fixations. (figure 4) Enlarged detail of a vegetative plasmodium in Plate 4.15-figure 4 (Ikegami et al. 1978)

4.6.4 Resting Spores

SEM in the samples of clubbed roots taken at 45 to 60 days after sowing revealed the abundant resting spores in an infected cell as observed in figures 1–4 of Plates 4.16 and 4.17. In Plate 4.16-figure 1, numerous resting spores (left cell) and many young plasmodia (right cell) are recognized. Also in an infected cell, a young plasmodium, and resting spores or vegetative plasmodial mass and resting spores are observed (Plate 4.16-figure 2). No membrane enveloping any resting spore masses is seen in Plates 4.16 and 4.17. Young resting spores are connected by fine filamentous materials (Plate 4.16-figure 3), and mature resting spores have spines in two infected cells (figure 4). Infected and non-infected cells in a club-cut tissue at low (figure 1)

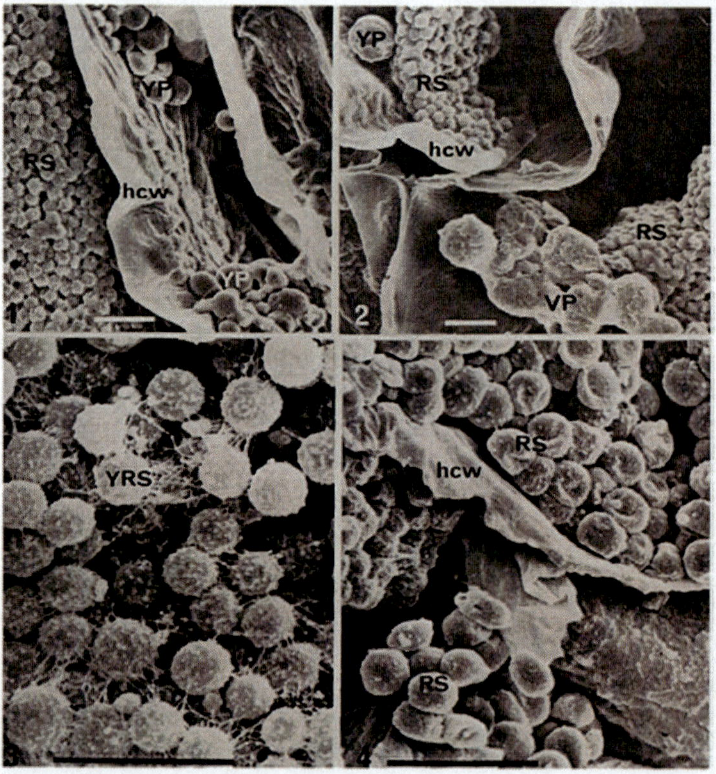

Plate 4.16 Scanning electron micrographs of plasmodia and resting spores in infected turnip cells. Each scale line=10 μm. (figure 1) Many young spherical plasmodia in an infected cell and numerous resting spores in its contiguous cell of Waseo-kabu on 57 days after sowing. FAA fixation. (figure 2) A young spherical plasmodium and a resting spore mass in an infected cell; a vegetative plasmodium and resting spores in its contiguous cell of Waseo-kabu at 57 days after sowing. FAA fixation. (figure 3) Young resting spores connected by fine filamentous materials in an infected cell of Taibyohikari-kabu at 45 days after sowing. Glutaraldehyde fixation. (figure 4) Mature resting spores with spines in two infected cells of Taibyohikari-kabu at 60 days after sowing. FAA fixation (Ikegami et al. 1978)

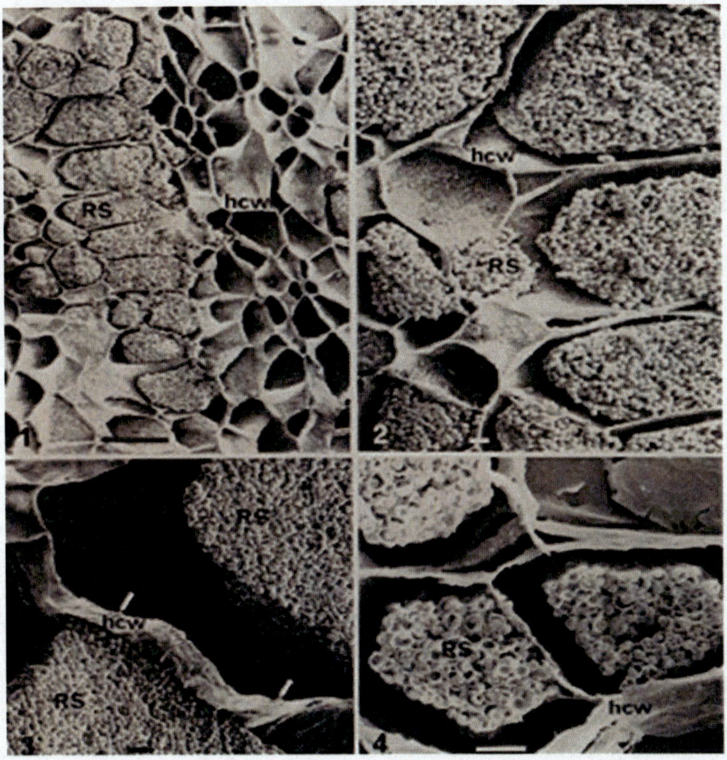

Plate 4.17 Scanning electron micrographs of mature resting spores in the infected cells of Chinese cabbage (figures 1 and 2) and turnip (figures 3 and 4). Each scale line except of 100 μm in figure 1 = 10 μm. (figures 1 and 2) Resting spore masses (fig. 1) and their higher magnifications (2) in an infected tissue of Muso-hakusai at 50 days after sowing. Glutaraldehyde fixation. (figure 3) Numerous resting spores in two infected cells of Waseo-kabu at 50 days after sowing, and cell organelles (arrow). FAA fixation. (figure 4) Resting spores showing definite cell walls in a paraffin sectioned tissue of clubroot of Waseo-kabu at 60 days after sowing FAA fixation (Ikegami et al. 1978)

and high magnification (figure 2) are depicted in Plate 4.17. A mass of innumerable resting spores filled in each infected cell. Plate 4.17-figure 3 shows a mass of myriad resting spores in two infected cells. Small particles of some organelles are found on the host cell wall. Clusters of resting spores in a paraffin section with well-defined spore cell walls are observed in figure 4.

4.7 Electron Microscopy of Zoosporogenesis

There are several reports of fine structures of the *P. brassicae* zoospores. Most of these are on secondary zoospores (Dekhuijzen 1979; Buczacki and Clay 1984) while others on germinating primary zoospores dried and shadowed for preparation (Kole

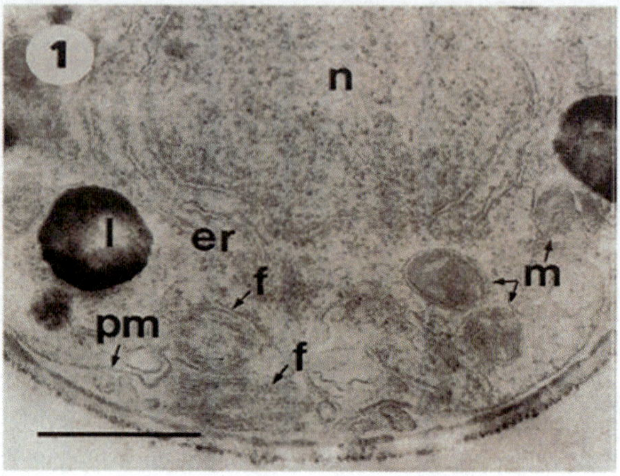

Plate 4.18 Differentiating primary zoospores within resting sporangium (fig. 1). Transversely sectioned flagella (f) are seen in the tight space between cell wall and invaginated plasma membrane (pm). Bar = 0.5 μm (Tanaka et al. 2001)

and Gielink 1962), and on primary zoospores adhering to the surface of host root hairs for infection (Aist and Williams 1971). Little detailed information is available on primary zoosporogenesis, although this knowledge is essential for understanding the biology of the primary zoospores and germination mechanisms of the resting sporangium. The process of zoosporogenesis was observed by Tanaka et al. (2001) through electron microscopy during resting spores germination. When resting sporangia of *P. brassicae* were incubated for 14 days in the culture solution containing turnip seedlings, the differentiation of primary zoospores was not synchronized among them. Primary zoospores could be observed only in a portion of sporangia, while other sporangia had released zoospores or had not yet differentiated them. Zoospores showed morphologically different features within incubated sporangia. The process of primary zoosporogenesis can be divided into four or five stages on the basis of the stage of the cell body, intracellular aspects, and other morphological features in zoospores observed within sporangia. Information on secondary zoosporogenesis in *P. brassicae* and other plasmodiophorids (Dekhuijzen 1979; Buczacki and Clay 1984; Talley et al. 1978; Miller et al. 1985) was also referred for identification of these stages.

A single zoospore differentiated in each sporangium. Differentiation of the primary zoospores was initiated by emergence of two flagella within the sporangium. Transversely sectioned flagella were seen in the tight space formed between the cell wall and the invaginated plasma membrane in the sporangium (Plate 4.18-figure 1). The zoospore cell was similar in intracellular aspects to mature sporangia within clubroot galls described by Yukawa and Tanaka (1979) but had electron dense mitochondria and well-developed endoplasmic reticulum. The zoospore at the stage of early development exhibited complicated intracellular aspects with the further invagination of the plasma membrane within the sporangium (Plate 4.19-figure 2). Transversely sectioned flagella were clearly observed in the invaginated portions of the plasma membrane (Plate 4.19-figures 2 and 3). The zoospore cell body appeared to be constricted in the middle by the invagination of the plasma

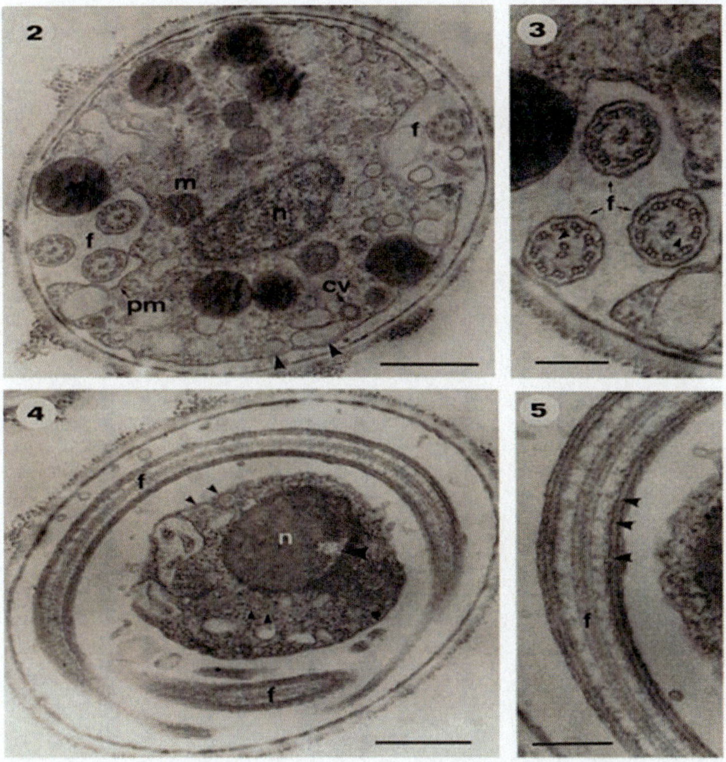

Plate 4.19 Primary zoospores at early stage of development and their flagella (figures 2–5). (figure 2) A zoospore showing transversely sectioned flagella (f) in a deep groove formed on the cell body by invagination of plasma membrane (pm). Electron dense nucleus (n), mitochondria (m) cored vesicle (cv), and other kinds of unidentified vesicle (arrowheads) are seen in the cell. Bar= 0.5 μm; (figure 3) Transversely sectioned flagella (f) showing radial spokes (arrowheads) between central pairs and other outer doubles of microtubules Bar= 0.5 μm; (figure 4) A zoospores sectioned at the constricted middle portion of the cell body. Longitudinally sectioned flagella (f) coil around the cell body at the constricted portion Bar = 0.5 μm. (figure 5) A longitudinally sectioned flagellum showing the arrangement of radial spokes (arrowheads) between central pairs and outer doubles of microtubules Bar = 0.5 μm (Tanaka et al. 2001)

membrane, and flagella were arranged in parallel along the inside of a deep groove formed on the cell body. This feature corresponded well in three-dimensional configurations with observation of the zoospores shown in Plate 4.19 (figure 4). Longitudinally sectioned flagella coiled around the zoospore cell body, which appeared to be small, indicating that the cell body was sectioned at its constricted middle portion (Plate 4.19-figure 4). Intracellularly, these zoospores were characterized by an electron dense nucleus without nucleolus, electron dense mitochondria, lipid droplets, and cored vesicles (Plate 4.19, figures 2 and 4). Unidentified vesicles containing electron opaque and dense materials were also observed in these zoospores (Plate 4.19, figures 2 and 4). An electron translucent

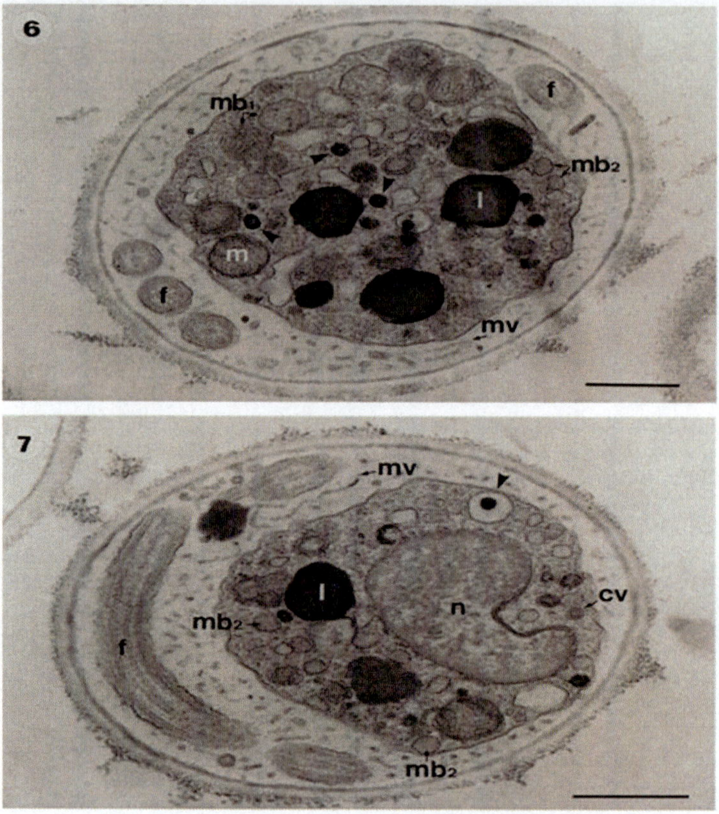

Plate 4.20 Primary zoospores at stage of late development within resting sporangium (figures 6 and 7). (figure 6) A zoospore showing a spherical cell body without the groove. Flagella (f) coil around the body in parallel. Numerous microvillus like vesicles (mv) are present in the periplasmic space. Large and small microbodies (mb_1 and mb_2, respectively) and small electron dense bodies (arrowheads) are rich in the cell. Bar = 0.5 μm; (figure 7) A small microbody (mb2) rich zoospores cell. A vesicle containing an electron dense granules (arrowheads), several cored vesicles (cv), and a few lipid droplets (l) are seen in the cell. Bar = 0.5 μm (Tanaka et al. 2001)

area was sometimes seen in the nucleoplasm (Plate 4.19, figure 4) as described for free-swimming primary zoospores of *Polymyxa graminis* (Barr and Allan 1982), and secondary zoospores of *Ligniera verrucosa* (Miller et al. 1985). Flagella were the whiplash type, and showed the typical 9+2 configuration of microtubules (Plate 4.19-figures 3 and 5). Radial spokes were seen unclearly between central pairs and outer doubles of microtubules in these flagella (Plate 4.19-figures 3 and 5).

Zoospores at the stage of late development are shown in Plate 4.20 (figures 6 and 7). The process of protoplasmic condensation (contraction of cell body) had begun in these zoospores, and a wide periplasmic space had developed in the sporangia containing them. Numerous microvillus like structures were observed within this space (Plate 4.20-figures 6 and 7). Although flagella coiled in parallel around the cell

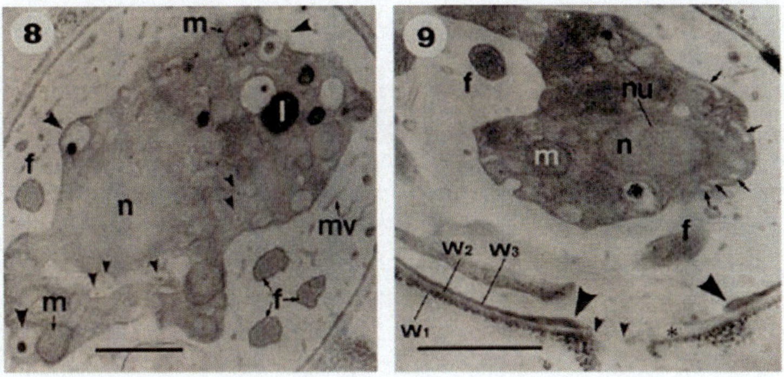

Plate 4.21 Mature primary zoospores within resting sporangium (figures 8 and 9). (figure 8) An irregular shaped condensed cell body of the zoospores. Many vesicles contain electron dense granules (large arrowheads) and small particles (small arrowheads). Bar = 0.5 μm; (figure 9) A zoospore and partially opened germ pore in the resting sporangium. The zoospore cell has a nucleus (n) with a nucleolus like region (nu), and a cluster of flattened vesicles (arrows) in the cell. A partially opened germ pore (small arrowheads) can be seen in a thickening (asterisk) of the middle (w2) wall layer in the area surrounded by the thickened ring (large arrowheads) of the innermost (W3) wall layer. Bar = 0.5 μm (Tanaka et al. 2001)

body, the groove was not seen on the cell body (Plate 4.20-figures 6 and 7). Intracellularly these zoospores were characterized by an electron dense nucleus without nucleolus, spherical mitochondria, relatively few lipid droplets, abundant microbodies (large and small), small electron dense bodies, cored vesicles and unidentified vesicles containing electron dense granules (Plate 4.20-figures 6 and 7).

The mature zoospores had an irregular shaped and fully condensed cell body (Plate 4.21-figures 8 and 9) as described the mature secondary zoospores of *L. verrucosa* (Miller et al. 1985). Intracellular features were complicated, characterized by an electron dense nucleus with a nucleolus like region (Plate 4.21-figure 9), unidentified vesicles containing electron dense granules and small particles (figure 8), and a cluster flattened vesicles (figure 9). Zoospores shown in figures 8 and 9 differed slightly in the degree of maturation. The zoospore shown in Fig. 8 had abundant microvillus like structures as well as developing zoospores (Plate 4.20-figures 6 and 7) in the periplasmic space, while the zoospores shown in fig. 9 had few. In addition, flagella coiled around the cell body in parallel in fig. 8 but no such arrangement of flagella was seen in fig. 9. These two zoospores appeared to be at early and late stages of maturation, respectively.

A partially opened germ pore was also observed in the sporangium containing the mature zoospores (fig. 9). This pore was located in the area thickening of the middle (W_2) wall layer surrounded by the thickened ring of the inner (W_3) wall layer, which has been reported for mature sporangia within clubroot galls by Yukawa and Tanaka (1979).

4.8 The Ultra-Structure of Mitosis in *Plasmodiophora brassicae*

Mitosis was examined in plasmodia of *Plasmodiophora brassicae* within artificially inoculated cabbage roots, using light and electron microscopy. Mitotic nuclear divisions are characterized by a persistent nucleolus, bipolar centrioles paired end-to-end, densely staining chromatin, and a complex array of membranes that surround and ramify through the spindle. Chromatin begins to condense in prophase, and is aligned at metaphase in a reticulate plate on the nuclear equator. The chromatin is not resolvable into distinct chromosomes at metaphase, and a chromosome count is not possible. Large amounts of membrane cisternae within the spindle are most clearly visible at metaphase, and apparently represent the remains of the nuclear envelope. The nuclear envelope is disrupted during prometaphase, and may become entangled in the spindle when centriolar microtubules enter the nucleus. Several concentric sheets of perinuclear endoplasmic reticulum surround the spindle and give the mitotic nucleus the superficial appearance of having an intact nuclear envelope. This interpretation of the identity of nucleus-associated membranes differs from those previously reported for other protists, including members of the Plasmodiophorales. Interesting features of mitosis in the Plasmodiophorales, including a persistent nucleolus, an apparently persistent nuclear envelope, and during metaphase, a dense equatorial plate of chromatin, were observed by light microscopy (Nawaschin 1899). The diagrammatic interpretation of mitotic nuclear division in *P. brassicae* is presented in Fig. 4.2 (Garber and Aist 1979). However, electron micrographs of mitotic nuclei have been published for only two members of the Plasmodiophorales, *Polymyxa betae* Keskin (Keskin 1971), and *Sorosphaera veronicae* Schroet (Braselton et al. 1975; Dylewski et al. 1978). Its ultra-structures have been revealed by Garber and Aist (1979).

4.9 Cytological Changes and Ultra-Structures of *Plasmodiophora brassicae* in *Brassica napus* Susceptible and Resistant Cultivars

Ultra-structures of *P. brassicae* and cellular changes in susceptible and resistance cultivars of *B. napus* were observed by Deora et al. (2013) through TEM and SEM.

1. *Transmission electron microscopy*: Similar ultra-structural features were observed across the root sections of all of the treatments in a reaction type (compatible and incompatible) at both 18 and 22 DAI. The most common features in the compatible interactions are represented in Plate 4.22, from the roots sections of "46A76"× P3 interaction at 22 DAI. Initial colonization of *P. brassicae* was characterized by the presence of uninucleate plasmodia completely immersed in the host cytoplasm that contained numerous, prominent lipid droplets (Plate 4.22a, b). Broken cell walls were already present at the beginning of colonization (arrow, Plate 4.22a). Outgrowths of the plasmodia (lobules) within the host cytoplasm were frequently observed (Plate 4.22b). Infected cells developed an electron dense (dark) ground cytoplasm, and the central vacuole

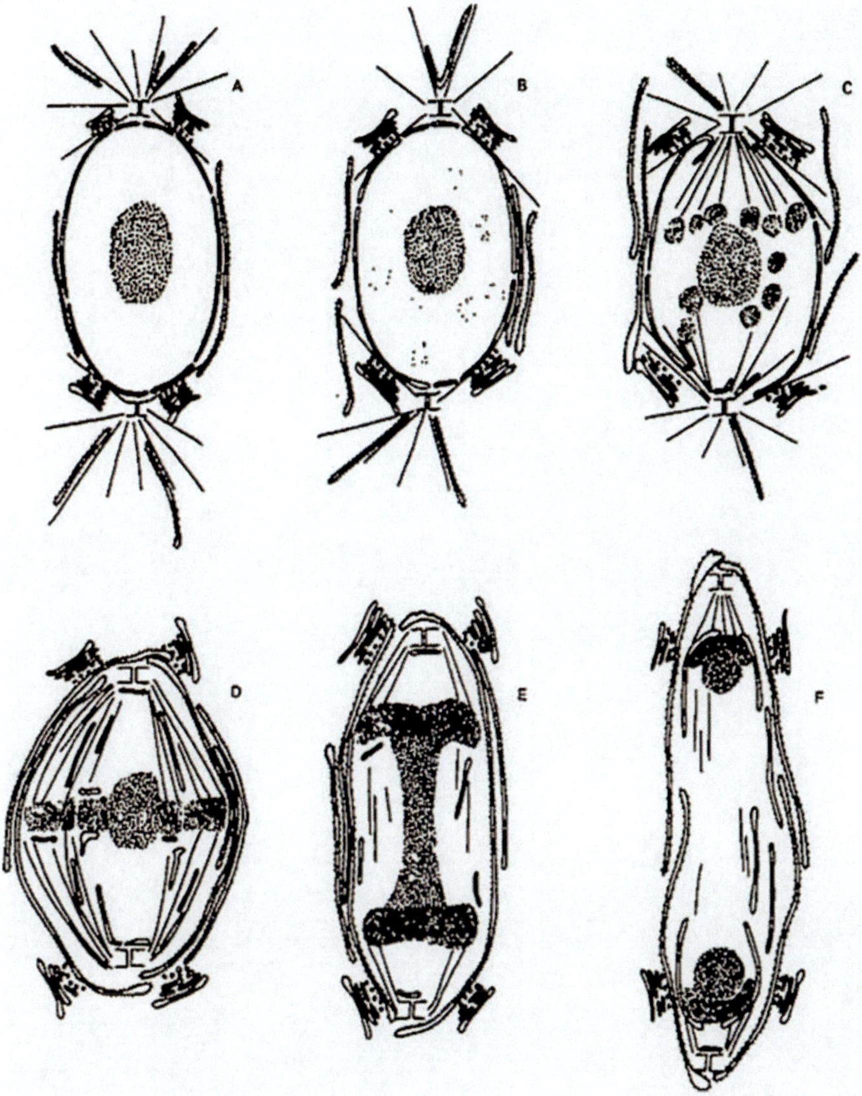

Fig. 4.2 Diagrammatic interpretation of mitotic nuclear division in *P. brassicae*. (**a**) interphase, (**b**) prophase, (**c**) prometaphase, (**d**) metaphase, (**e**) anaphase, (**f**) telophase (Garber and Aist 1979)

disintegrated or was inconspicuous (asterisks, Plate 4.22c). Cells adjacent to infected cells had thinner cell walls than other nearly cells, indicating that rapid cell division was occurring (arrows, Plate 4.22c). Increased mitochondrial populations were present in some of these affected cells, indicating that these cells were highly metabolic. Degradation of the secondary thickening of xylem cells walls (arrows, Plate 4.22d) and the presence of inclusion bodies in cell walls

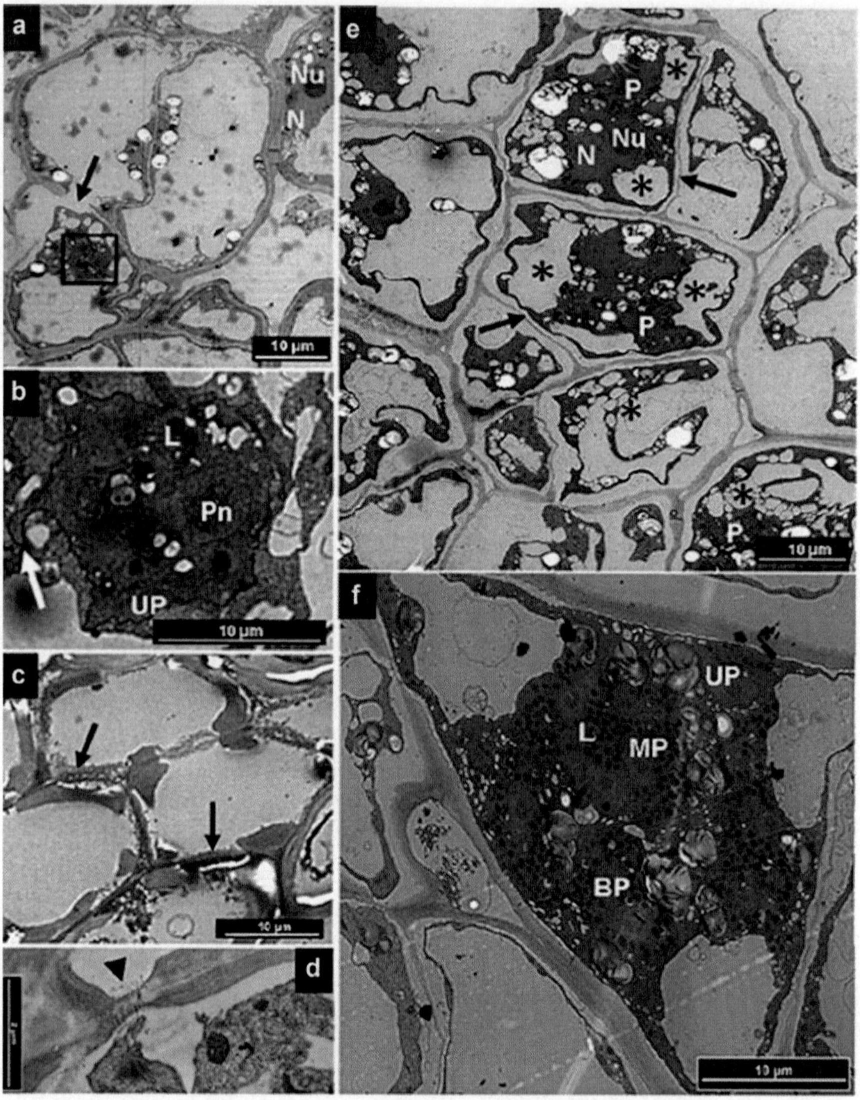

Plate 4.22 Transmission electron micrographs showing characteristic features of a compatible interaction of *Plasmodiophora brassicae* isolate P3 with canola cv. "46A76" at 22 days after inoculation. (**a**) A uninucleate plasmodium (rectangular portion) associated with a break in the cell wall (arrow). (**b**) Magnified portion of rectangular region from (**a**) showing uninucleate plasmodium with abundant lipid bodies. Arrow indicates an often observed phenomenon of outgrowth of pathogen (a lobule). (**c**) Thin cell walls (arrows) indicate formation of new wall following cell division. Disintegration of vacuole was commonly seen (asterisks). (**d**) Degradation of secondary thickening of xylem cell walls (arrows). (**e**) Inclusion bodies in cell wall (arrowheads). (**f**) Infected cell with uni-, bi- and multi-nucleated plasmodia. *BP* binucleated plasmodium, *L* lipid body, *MP* multinucleated plasmodium, *N* host nucleus, *Nu* host nucleolus, *Pn* plasmodial nucleus, *SG* starch granule, *UP* uninucleated plasmodium. *Note*: Isolate P3 was obtained from field where pathotype 3 (based on the Williams system) was found to predominate in 2006 (Deora et al. 2013)

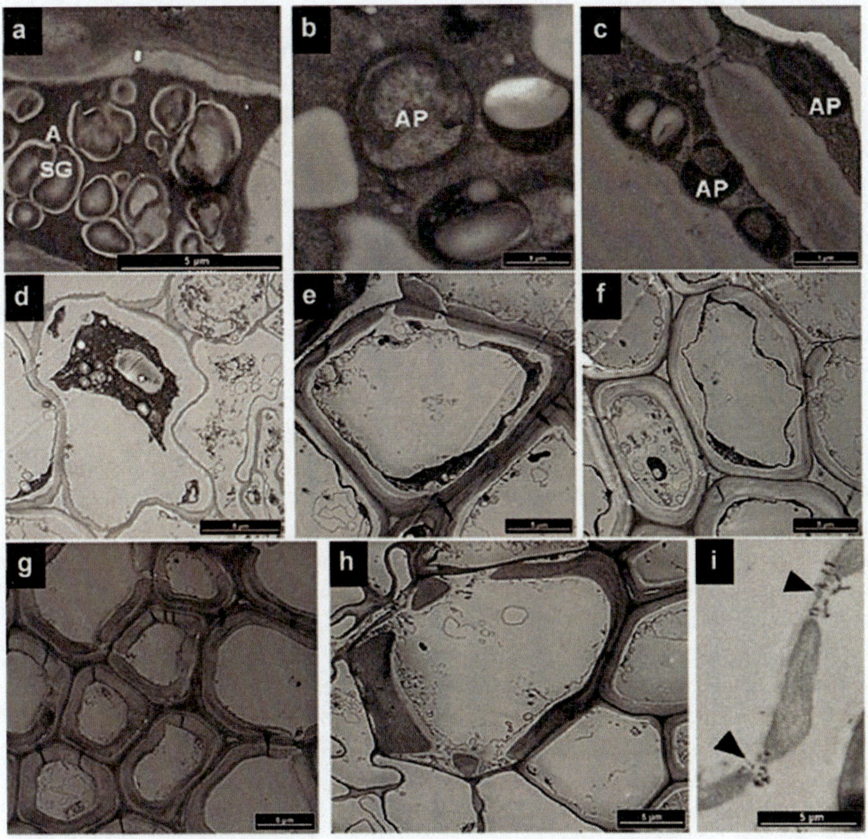

Plate 4.23 Transmission electron micrographs exhibiting characteristic features of a non-inoculated (**a**, **b**) and inoculated (isolate P3) resistant cv. "45H29" at 18 days after inoculation. (**a**) Abundant amyloplasts containing starch granules. (**b**) Amyloplast precursors were seen in non-inoculated roots. (**c**) Amyloplast precursors occurred in abundance in inoculated roots. (**d**) Cortical cell with shrunken cytoplasm. (**e**, **f**) The cells in neighborhood of shrunken cytoplasm cell looked normal. (**g**) Lignified xylem cells in stele. (**h**) A xylem cell with intact cell wall. (**i**) Inclusion bodies in cell wall: *A* amyloplasts, *AP* amyloplast precursor, *SG* starch granule. *Note*: Isolate P3 was obtained from field where pathotype 3 (based on the Williams system) was found to predominate in 2006 (Deora et al. 2013)

was frequently observed (arrowhead, Plate 4.22e). Heavily infected cells were completely occupied with one or more plasmodia at various stages of development, for example, uni-, bi- or multi-nucleated (Plate 4.22f). Also, lipid bodies were prominent, and other cell organelles were inconspicuous in these heavily infected cells (Plate 4.22f).

The features of the incompatible interactions were also similar at both 18 and 22 DAI. The most common features are represented in Plate 4.23, from the root sections of "45H29" × P3 interaction at 22 DAI. Amyloplasts containing starch granules (Plate 4.23a) and amyloplast precursors (Plate 4.23b) were abundant in the non-inoculated controls, but starch granules were less frequently observed in inoculated plants (Plate 4.23c). No pathogenic structures were observed in the assessments of resistant cultivars at 18 and 22 DAI, even though young plasmodia had been observed (although rarely) in the resistant cultivars using light microscopy. A few cortical cells with shrunken cytoplasm were observed (Plate 4.23d), but adjacent cells appeared normal (Plate 4.23e, f). Similarly, the lignified xylem cells (Plate 4.23g) and xylem vessels (Plate 4.23h) within the stele were normal. Cell wall breakage and inclusion bodies were present (arrows, Plate 4.23i), but at a much lower frequency than in the compatible interactions.

2. *Localization of reactive oxygen species and lignin*: In the incompatible interactions, no new points of ROS accumulation were observed in the cortical tissues at 18 or 35 DAI in response to inoculation. Instead, a high density of ROS accumulation, visible as a dense brown ring in sections stained with DAB, was present in the endodermis, pericycle, and vascular cambium of each of the incompatible interactions (Plate 4.24a), with an increasing density gradient towards the center of the root (Plate 4.24b). However, the ring formation was not seen in the root tissues in the cross-sections obtained beyond 1–2 mm from the cut end of the root piece. Similar ring formation was detected in roots of compatible interaction at 18 DAI, and in the non-inoculated controls of each cultivar at 18 and 35 DAI (not shown). In contrast, no ring was present at 35 DAI in any compatible interaction (Plate 4.24c, d).

   The location of pathogen colonization within the root was tracked over time in samples stained with methylene blue from the cortical infection experiment. In these sections, pathogen tissue stained a dark blue. Xylem cell walls with lignified secondary cell walls stained dark blue, and the non-lignified endodermis, pericycle, vascular cambium, cortical, and epidermal cells stained light blue (Plate 4.24e). In the compatible interactions at 24 DAI, pathogen colonization progressed from the cortex into the stele. The plasmodia preferentially colonized xylem parenchyma cells and stimulated cell division and enlargement of this cell type (Plate 4.24f, arrow). As colonization of the stele progressed, rapidly dividing xylem parenchyma cells (arrows) pushed into and between areas of xylem cells with lignified secondary walls (arrowheads, Plate 4.24g). No lignified points (indicative of localized hypersensitive reaction) were detected in cortical tissues stained with TBO at 18 or 35 DAI in the incompatible interactions. Lignin, which stains blue in TBO, was localized primarily in the secondary cell walls of xylem cells in the stele region (Plate 4.24h). The other tissues in the root stained pink because they are not lignified (Plate 4.24h). Similar staining patterns were observed in the non-treated controls at 18 and 35 DAI, and in the compatible interactions at 18 DAI. In the compatible reaction at 35 DAI, the area occupied by cells with walls containing lignin was substantially reduced during pathogen colonization. At an early stage of pathogen colonization into the stele, the area occupied by cells with walls that stained for lignin in the stele decreased (Plate 4.24i). When the pathogen subsequently colonized the pith, the cells with walls that stained for lignin declined to near zero (Plate 4.24j).

Plate 4.24 Light microscopic localization of reactive oxygen species (ROS) accumulation and lignification in a compatible ("46A76" × P3) and incompatible interaction ("45H29" × P3) of *Plasmodiophora brassicae* in canola roots. (**a–d**) Roots stained with diamino-benzidine (DAB). (**a**) In the incompatible interaction, staining with DAB reveals ROS accumulation (brown ring) in outer stele region of root section. (**b**) Phase contrast image of the section from (**a**) showing ROS accumulation in endodermis (E), pericycle (P), and vascular cambial (VC) region. (**c**) Compatible interaction; no ROS accumulation in outer stele region. (**d**) Phase contrast image of the section from (**c**) showing cells free of ROS accumulation. (**e–g**) Roots stained with methylene blue. (**e**) Incompatible interaction; lignified xylem cell walls stain dark blue in stele region and non-lignified walls stain light blue. (**f**) Compatible interaction; the pathogen colonizing xylem parenchyma cells in the stele region (arrow). (**g**) Compatible interaction, advanced stage; the xylem parenchyma cells (arrows) push the lignified xylem cells apart (arrowheads). (**h–j**) Roots stained with toluidine blue O. (**h**) Incompatible interaction; blue color indicates lignified xylem cells in stele region and purple-pink color reflects non-lignified (pectic) cell walls. (**i**) Compatible interaction; arrow indicates lignified stele cells changed from blue to purple-pink. (**j**) Compatible interaction; advanced stage, most lignified stele cells stained purple-pink instead of blue, arrow indicates pathogen cells. Magnification (**a**) and (**c**) ×20, (**b**, **d**, **e**, **g–j**) ×40, and (**f**) ×250. (**a**)–**d**, **h–j**) 35 DAI. (**e–g**) 24 DAI. *Note*: Isolate P3 was obtained from a research site where pathotype 3 had predominated in 2006 (Deora et al. 2013)

3. *Scanning electron microscopy:* Observations made using scanning electron microscopy supported the pattern of development in the compatible interaction that had been observed using light microscopy. It also demonstrated that P3 and P6 induced similar anatomical changes in a compatible reaction. The common features among these interactions are represented in Plate 4.25. In infected roots,

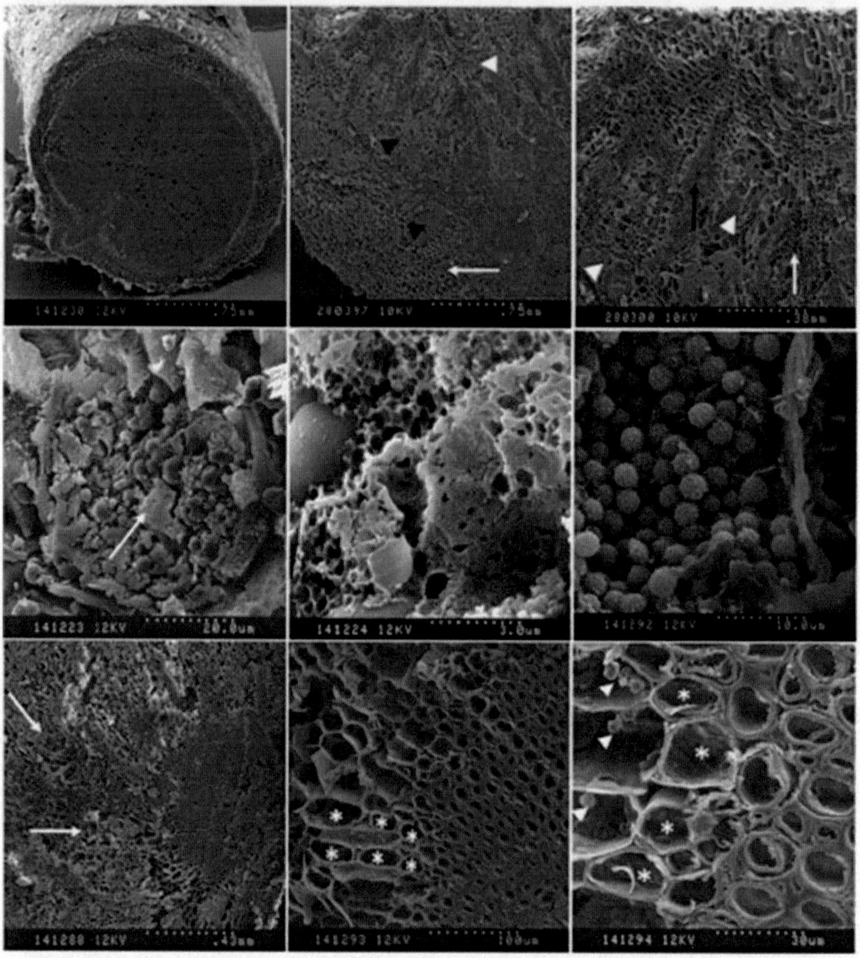

Plate 4.25 Scanning electron micrographs of a compatible interaction between *Plasmodiophora brassicae* isolate P3, and roots of cv. 46A76 at 35 DAI. (**a**) Non-inoculated control. (**b**) A portion of a root with infected and hypertrophied cortex (arrow) and stele cells (white arrowhead). A mass of resting spores occupying whole lumen of a group of cells (black arrowheads). (**c**) In the stele region, xylem parenchyma cells undergo division and enlargement (arrowheads). *De novo* synthesis of xylem parenchyma (white arrow) and vascular strands (phloem) (black arrow) took place. (**d**) New cell wall material (arrow) was often seen in the cells infected with secondary plasmodia. (**e**) A high magnification of wall material. (**f**) A mass of mature resting spores occupying a whole cortical cell. (**g**) A root showing advanced stage of interaction. Clusters of rapidly dividing xylem parenchyma cells pushed into and between (arrows) areas of xylem cells with lignified secondary walls. (**h**) Pathogen also induces cell wall expansion resulting in enlargement of lignified xylem cells (asterisks). (**i**) Pathogen (plasmodia, arrowheads) induces enlargement in the neighboring lignified xylem cells before invasion (asterisks). *Note*: Isolate P3 was obtained from a research site where pathotype 3 had predominated in 2006 (Deora et al. 2013)

the cortical and stele cells became disorganized (arrow and white arrowhead, respectively), and almost half of the root cells were already filled with masses of resting spores at 35 DAI (black arrowheads, Plate 4.25b). Within the stele, the plasmodia preferentially colonized xylem parenchyma cells, and induced cell division and enlargement (arrowheads, Plate 4.25c). In addition, the pathogen also stimulated de novo synthesis of parenchyma cells (white arrow) and vascular strands (phloem cells) (black arrow, Plate 4.25c). Thin cell walls were often noted in infected cells, indicating deposition of new cell wall material (Plate 4.25d, e). Cortical cells were filled with masses of mature resting spores (Plate 4.25f). At an advanced stage of clubroot symptom development, clusters of rapidly dividing xylem parenchyma cells pushed into and between areas of lignified xylem cells (arrows, Plate 4.25g). As the pathogen progression continued in the stele region, even the lignified xylem cells became enlarged (asterisks, Plate 4.25h). The pathogen (arrowhead) induced the hypertrophy of lignified cells even before colonizing the cells (asterisks, Plate 4.25i) (Deora et al. 2013).

References

Aist JR, Williams PH (1971) The cytology and kinetics of cabbage root hair penetration by *Plasmodiophora brassicae* Wor. Can J Bot 49:2023–2034

Ando S, Asano T, Tsushima S, Kamachi S, Hagio T, Tabei Y (2005) Changes in gene expression of putative isopentenyltransferase during clubroot development in Chinese cabbage (*Brassica rapa* L.). Physiol Mol Plant Pathol 67:59–67

Ando S, Tsushima S, Tagiri A, Kamachi S, Konagaya KI, Hagio T, Tabei Y (2006) Increase in *BrAO1* gene expression and aldehyde oxidase activity during clubroot development in Chinese cabbage (*Brassica rapa* L.). Mol Plant Pathol 7:223–234

Asano T, Kageyama K, Hyakumachi M (1999) Surface disinfestations of resting spores of *Plasmodiophora brassicae* used to infect hairy roots of *Brassica* spp. Phytopathology 89:314–319

Barr DJS, Allan PM (1982) Zoospore ultrastructure of *Polymyxa graminis* (Plasmodiophoromycetes). Can J Bot 60:2496–2504

Braselton JP, Miller CE, Pechak DG (1975) The ultrastructure of cruciform nuclear division in *Sorosphaera veronicae* (Plasmodiophoromycete). Am J Bot 62:349–358

Buczacki ST, Clay CM (1984) Some observations on secondary zoospore development in *Plasmodiophora brassicae*. Trans Br Mycol Soc 82:339–382

Dekhuijzen HM (1979) Electron microscopic studies on the root hairs and cortex of a susceptible and a resistant variety of *Brassica campestris* infected with *Plasmodiophora brassicae*. Neth J Plant Pathol 86:1–17

Dekhuijzen HM (1981) The occurrence of free and bound cytokinins in plasmodia of *Plasmodiophora brassicae* isolated from tissue cultures of clubroots. Plant Cell Rep 1:18–20

Dekhuijzen HM, Overeem JC (1971) The role of cytokinins in clubroot formation. Physiol Plant Pathol 1:151–161

Deora A, Gossen BD, McDonald MR (2013) Cytology of infection, development and expression of resistance to *Plasmodiophora brassicae* in canola. Ann Appl Biol 163:56–71

Devos S, Vissenberg K, Verbelen JP, Prinsen E (2005) Infection of Chinese cabbage by *Plasmodiophora brassicae* leads to a stimulation of plant growth: impacts in cell wall metabolism and hormone balance. New Phytol 166(1):241–250

Devos S, Laukens K, Deckers P, Van Der Straeten D, Beeckman T, Inze D, Van Onckelen H, Witters E, Prinsen E (2006) A hormone and proteome approach to picturing the initial metabolic events during *Plasmodiophora brassicae* infection on *Arabidopsis*. MPMI 19(12):1431–1443

Donald E, Jaudzems G, Porter IJ (2008) Pathology of cortical invasion by *Plasmodiophora brassicae* in clubroot resistant and susceptible *Brassica oleracea* hosts. Plant Pathol 57:201–209

Dylewski DP, Braselton JP, Miller CE (1978) Cruciform nuclear division in *Sorosphaera veronicae*. Am J Bot 65:258–267

Etchells JP, Turner SR (2010) The PXY-CLE41 receptor ligand pair defines a multifunctional pathway that controls the rate and orientation of vascular cell division. Development 137:767–774

Evans JL, Scholes JD (1995) How does clubroot alter the regulation of carbon metabolism in its host? Asp Appl Biol 42:125–132

Funk V, Kositsup B, Zhao C, Beers EP (2002) The *Arabidopsis* xylem peptidase XCP1 is a tracheary element vacuolar protein that may be a papain ortholog. Plant Physiol 128:84–94

Garber RC, Aist JR (1979) The ultrastructure of mitosis in *Plasmodiophora brassicae* (Plasmodiophorales). J Cell 40:89–110

Gustafsson M, Falt AS (1986) Genetic studies on resistance to clubroot in *Brassica napus*. Ann Appl Biol 108:409–415

Hirakawa Y, Kondo Y, Fukuda H (2010) TDIF peptide signaling regulates vascular stem cell proliferation via the *WOX4* homeobox gene in *Arabidopsis*. Plant Cell 22:2618–2629

Ikegami H, Mukobata H, Naiki T (1978) Scanning electron microscopy of *Plasmodiophora brassicae* in diseased root cells of turnip and Chinese cabbage (studies on the clubroot of cruciferous plants III). Ann Phytopathol Soc 444:456–464

Ingram D, Tommerup I (1972) The life history of *Plasmodiophora brassicae* Woron. Proc R Soc Lond B 180:103–112

Ito Y, Nakanomyo I, Motose H, Iwamoto K, Sawa S, Dohmae N, Fukuda H (2006) Dodeca-CLE peptides as suppressors of plant stem cell differentiation. Science 313:842–845

Karling JS (1968) The plasmodiophorales, 2nd edn. Hafner Publishing Company, New York, p 256

Keen NT, Williams PH (1969) Translocation of sugars into infected cabbage tissues during clubroot development. Plant Physiol 44:748–754

Keskin B (1971) Beitrag zur Protomitose bei Polymyxa betae Keskin. Arch Mikrobiol 77:344–348

Kobelt P, Siemens J, Sacristan M (2000) Histological characterization of the incompatible interaction between *Arabidopsis thaliana* and the obligate biotrophic pathogen *Plasmodiophora brassicae*. Mycol Res 104(2):220–225

Kole AP, Gielink AJ (1962) Electron microscope observations on the resting-spore germination of *Plasrnodiophora brassicae* Konikl. Nederl Acad Vet 65:117–121

Kubo M, Udagawa M, Nishikubo N, Horiguchi G, Yamaguchi M, Ito J, Mimura T, Fukuda H, Demura T (2005) Transcription switches for protoxylem and metaxylem vessel formation. Genes Dev 19:1855–1860

Lalonde S, Tegeder M, Throne-Holst M, Frommer WB, Patrick JW (2003) Phloem loading and unloading of sugars and amino acids. Plant Cell Environ 26:37–56

Ludwig-Muller J, Schuller A (2008) What can we learn from clubroots: alterations in host roots and hormone homeostasis caused by *Plasmodiophora brassicae*. Eur J Plant Pathol 121:291–302

Mahonen AP, Bishopp A, Higuchi M, Nieminen KM, Kinoshita K, Tormakangas K, Ikeda Y, Oka A, Kakimoto T, Helariutta Y (2006) Cytokinin signaling and its inhibitor AHP6 regulate cell fate during vascular development. Science 311:94–98

Malinowski R, Smith JA, Fleming AJ, Scholes JD, Rolfe SA (2012) Gall formation in clubroot-infected Arabidopsis results from an increase in existing meristematic activities of the host but is not essential for the completion of the pathogen life cycle. Plant J 71(2):226–238

Miller CE, Martin RW, Dylewski DP (1985) The ultrastructure of plasmodia, sporangia, and cystosori of *Ligniera verrucosa* (Plasmodiophoromycetes). Can J Bot 63:263–273

Mitchell DT, Rice KA (1979) Translocation of 14C-labelled assimilates in cabbage during club root development. Ann Appl Biol 92:143–152

Mithen R, Magrath R (1992) A contribution to the life history of *Plasmodiophora brassicae*: Secondary plasmodia development in root galls of *Arabidopsis thaliana*. Mycol Res 96:877–885

Muller P, Hilgenberg W (1986) Isomers of zeatin and zeatin riboside in clubroot tissue: evidence for trans-zeatin biosynthesis by *Plasmodiophora brassicae*. Physiol Plant 66:245–250

Navas-Cortés JA, Vovlas N, Trisciuzzi N, Castillo P, Troccoli A (2010) Anatomical changes induced by two soil-borne pathogens (*Plasmodiophora brassicae* and *Meloidogyne javanica*) in cabbage. Nematol Mediterr 38:39–43

Nawaschin S (1899) Beobachtungen Uber Den Feineren Bau Und Umwandlungen Von *Plasmodiophora brassicae* Woron. Im Laufe Ihres Intracellularen Lebens. Heft 5:404–427

Riascos D, Ortiz E, Quintero D, Montoya L, Hoyos-Carvajal L (2011) Histopathological and morphological alterations caused by *Plasmodiophora brassicae* in *Brassica oleracea* L. Agron Colomb 29(1):57–61

Schuller A, Ludwig-Muller J (2016) Histological methods to detect the clubroot pathogen *Plasmodiophora brassicae* during its complex life cycle. Plant Pathol 65:1223–1237

Siemens J, Keller I, Sarx J, Kunz S, Schuller A, Nagel W, Schmulling T, Parniske M, Ludwig-Muller J (2006) Transcriptome analysis of *Arabidopsis* clubroots indicate a key role for cytokinins in disease development. Mol Plant-Microbe Interact 19(5):480–494

Siemens J, Graf H, Bulman S, In O, Ludwig-Muller J (2009) Monitoring expression of selected *Plasmodiophora brassicae* genes during clubroot development in *Arabidopsis thaliana*. Plant Pathol 58:130–136

Siemens J, Gonzalez MC, Wolf S, Hofmann C, Greiner S, Du Y, Rausch T, Roitsch T, Ludwig-Muller J (2011) Extracellular invertase is involved in the regulation of clubroot disease in *Arabidopsis thaliana*. Mol Plant Pathol 12:247–262

Suer S, Agusti J, Sanchez P, Schwarz M, Greb T (2011) WOX4 imparts auxin responsiveness to cambium cells in *Arabidopsis*. Plant Cell 23:3247–3259

Suwabe K, Tsukazaki H, Iketani H, Hatakeyama K, Fujimura M, Nunome T, Fukuoka H, Matsumoto S, Hirai M (2003) Identification of two loci for resistance to clubroot (*Plasmodiophora brassicae* Woronin) in *Brassica rapa* L. Theor Appl Genet 107:997–1002

Talley MR, Miller CE, Braselton JP (1978) Notes on the ultrastructure of zoospores of *Sorosphaera veronicae*. Mycologia 70:1241–1247

Tanaka S, Ito S, Kameya-Iwaki M (2001) Electron microscopy of primary zoosporogenesis in *Plasmodiophora brassicae*. Mycoscience 42:389–394

Tu J, Bush J, Bonham-Smith P, Wei Y (2018) Live cell imaging of *Plasmodiophora brassicae*—host plant interactions based on a two step axenic culture system. Microbiol Open 2018:e765

Voorrips RE (1995) *Plasmodiophora brassicae*: aspects of pathogenesis and resistance in *Brassica oleracea*. Euphytica 83:139–146

Voorrips R, Kanne J (1997) Genetic analysis of resistance to clubroot (*Plasmodiophora brassicae*) in *Brassica oleracea*. II. Quantitative analysis of root symptom measurements. Euphytica 93:41–48

Yamaguchi M, Kubo M, Fukuda H, Demura T (2008) Vascular-related NAC-domain7 is involved in the differentiation of all types of xylem vessels in *Arabidopsis* roots and shoots. Plant J 55:652–664

Yamaguchi M, Ohtani M, Mitsuda N, Kubo M, Ohme-Takagi M, Fukuda H, Demura T (2010) VND-interacting2, a NAC domain transcription factor, negatively regulates xylem vessel formation in *Arabidopsis*. Plant Cell 22:1249–1263

Yamaguchi M, Mitsuda N, Ohtani M, Ohme-Takagi M, Kato K, Demura T (2011) Vascular-related NAC-domain7 directly regulates the expression of a broad range of genes for xylem vessel formation. Plant J 66:579–590

Yukawa Y, Tanaka S (1979) Scanning electron microscope observations on resting sporangia of *Plasmodiophora brassicae* in clubroot tissues after alcohol cracking. Can J Bot 57:2528–2532

Zhao C, Avci U, Grant EH, Haigler CH, Beers EP (2008) XND1, a member of the NAC domain family in *Arabidopsis thaliana,* negatively regulates lignocellulose synthesis and programmed cell death in xylem. Plant J 53:425–436

Pathogenic Variability

5

Abstract

Pathogenic variability in *P. brassicae* was detected long back during 1930s on the basis of differential interactions with cruciferous species. First generation widely used host differentials all over the world included genotypes from *B. napus* var. *napobrassica*, *B. napus* var. *rapifera*, *B. oleracea* var. *capitata*, *B. rapa* var. *pekinensis*, *B. rapa* ssp. *rapifera,* and *B. napus* var. *napus*. The occurrence, distribution, and pattern of *P. brassicae* virulence/pathotypes have been characterized from Australia, Canada, China, Czech Republic, Finland, France, GDR, Iceland, India, Japan, Korea, New Zealand, Norway, Poland, Sweden, Turkey, UK, USA, and USSR. Molecular techniques like PCR, RAPD, and RFLP/AFLP have also been used to characterize *P. brassicae* isolates. Although, these studies have revealed high degree of polymorphisms among as well as within *P. brassicae* populations, but so far no correlations have been established with pathogenicity. Modified differentials sets have been used at Canada, China, Japan, Korea, and Turkey to get clearer picture of pathotypes in their countries by including genotypes of *Brassica* having local/national importance. Numerous pathotypes of *P. brassicae* have been identified from more than 20 countries of the world with designations as serial numbers (1, 2. . .n), triplicates (16/14/15, 16/2/14. . .n), and letters (A, B, C, . . .X. . .n); (P1, P2. . .n); (Pb: 0, Pb: 1, Pb: 2, Pb: 3). However, the characterization of pathotypes with respect to virulence spectrum, and erosion of clubroot cultivars is from very limited countries. For the rapid identification, and detection of pathotypes infecting *Brassica* species, molecular markers have been developed. A *CR 811* gene has been identified in pathotype 3 and 5 which plays an important role in host resistance, and *P. brassicae* pathogenicity. Isolate/pathotype-specific molecular markers are very useful tools to identify genes to test, and differentiate unknown pathotypes. Differential sets development, and used so far have several limitations to detect, and identify new pathotypes as well as the distribution of their virulence all over

G. S. Saharan et al., *Clubroot Disease of Crucifers*,
https://doi.org/10.1007/978-981-16-2133-8_5

the world. The differential sets consist of genetically heterogeneous *Brassica* species which is against the International standard differentials host system with near isogenic lines. Similarly, isolates used so far are composed of genetically heterogeneous populations instead of genetically pure single-spore isolates. Similarly, designation and nomenclature of pathotypes have limitations since different systems have been followed to name the pathotypes as serial number, triplets (binary notation), and letters which makes difficult to compare, analyze, and detection of new pathotypes. Following this system, similar pathotypes have been reported from different countries which is not consistent with the biology of the pathogen. The intermediate and fluctuating results of certain deferential hosts following inoculation with the population of *P. brassicae* indicates importance of single-spore isolates for virulence analysis prevalent world over. It is inevitable to get deeper insight into the life cycle of *P. brassicae* through pathotypes genomics, proteomics, and transcriptomics analysis governing virulence pattern and relationship between pathogenic and genetic variations.

Keywords

Virulence · Identification and characterization of pathotypes · Differential sets · Nomenclature of pathotypes · Virulence pattern of pathotypes · Pathotype variation · Pathogenic and genetic variation · Global distribution of pathotypes · Pathotypes genome · Intraspecific biodiversity · Phylogenetics of pathotypes · Polymorphisms · Production of pure genotype isolates · Virulence pattern on cruciferous weeds

5.1 Introduction

Physiologic specialization in *Plasmodiophora brassicae* has been determined in different countries of the world since its first report in 1931 (Honig 1931; Karling 1968). Various host differential systems have been proposed, and used to identify pathotypes (or races) of the clubroot pathogen. These include the differentials of Williams (1966), Some et al. (1996), European clubroot differential (ECD) set (Buczacki et al. 1975), and Canadian clubroot differential set (CCD) (Strelkov et al. 2018). All the first three systems have been used by the researchers all over the world to assess the occurrence of pathotypes in *P. brassicae* in their respective countries. However, it has been felt that these differentials sets are not fully applicable to detect complete virulence in each country because of variations in geographical conditions, range of genotypes of cruciferous crops grown, and history of clubroot occurrence. Researchers now have started using modified differential sets with addition of supplementary and complementary genotypes of their regional importance and suitability to detect full range of *P. brassicae* virulence. The occurrence, distribution, and pattern of virulence of clubroot pathotypes in different countries of the world have been analyzed.

5.2 Occurrence, Distribution, and Pattern of *Plasmodiophora brassicae* Virulence

The occurrence, distribution, and pattern of pathogenic variability of *P. brassicae* have been characterized from different countries of the world using three most common differential sets to designate pathotypes at early phase of investigations (Table 5.1). However, widely used differential sets are given in Table 5.19. The information generated so far is presented in different sections as follows:

5.2.1 Australia

The Australian pathogen population has been analyzed by Donald and Porter (2014). Variation in the Australian pathogen population has been studied using the Williams (1966) and Chambers (1977), and European Clubroot Differential (ECD) series (Donald et al. 2006; Kong Kaw Wa 2009), and using molecular methods (Badi et al. 2010). All of these studies indicate that physiological variants of *P. brassicae* exist in Australia. Bioassay work using differential host plants indicates that the Australian populations appear more similar to populations reported from the USA than those from Europe. There is no variation in the reaction of the Australian pathogen collections to the *Brassica rapa* differential hosts (01–05) with only one causing a susceptible host reaction in a *B. rapa* host other than host 05 ("Granaat"), assumed to be universally susceptible (Donald et al. 2006). This is unsurprising since stubble turnips (*B. rapa*) are not frequently grown in mainland of Australia, hence strong virulence towards this group of hosts is not likely to have segregated in mainland pathogen collections. A wide range of virulence toward the *B. napus* differential hosts (06–10) exist worldwide (Toxopeus et al. 1986). Donald et al. (2006) reported that 80.5% of the 41 Australian collections screened failed to produce a susceptible reaction on any of the *B. napus* hosts or only caused a susceptible reaction on *B. napus* hosts 06 and/or 07. Screening conducted by Kong Kaw Wa (2009) produced a similar result for the *B. napus* differential hosts. Chambers (1977) reported the presence of races 6 and 7 (Williams' differentials) corresponding to ECD race numbers 16/2/30, 16/0/30, and 16/2/31, 16/0/31, respectively. It indicates the presence of high levels of virulence to *B. oleracea* differentials. Australian collections sent to the USA from New South Wales were identified as races 3, 6, and 7. A collection from Victoria was identified as race 6 (Williams 1966). These studies conducted using differential hosts indicate a lack of virulence towards *B. napus* hosts 08, 09 and, to a lesser extent, 10 in Australian pathogen populations. Differential variation in host reaction was greatest for the *B. oleracea* group of hosts, however, the poor differentiating ability of hosts 13 and 14 reported elsewhere (Toxopeus et al. 1986) was confirmed with susceptible reactions being recorded from these hosts for 93 or 86% of the collections screened by Donald et al. (2006) and Kong Kaw Wa (2009), respectively. There have been several Australian reports of different ECD codes being assigned to the same pathogen collection when that collection has been in used in multiple screening

Table 5.1 Most common differential sets used for designation of *P. brassicae* pathotypes

Name of the set	ECD genotype	ECD No./score	Differential name
ECD- European Clubroot Differential set (Buczacki et al. 1975)	*B. rapa* ssp. *rapifera* Polish rape	ECD01/1	Line aaBBCC
		ECD02/2	Line AAbbCC
		ECD03/4	Line AABBcc
		ECD04/8	Line AABBCC
	B. rapa var. *pekinensis*	ECD05/16	Chinese cabbage cv. Granaat
	B. napus var. *napus* (Fodder rape)	ECD06/1	Dc 101 Nevin
		ECD07/2	Dc 119 Giant rape
		ECD08/4	Dc 128 selection from Dc 119
		ECD09/8	New Zealand resistant rape
	B. napus var. *rapifera*	ECD10/16	Dc 130 Wilhelmsburger
	B. oleracea var. *capitata* (cabbage)	ECD11/1	Badger Shipper
		ECD12/2	Bindsachsener
		ECD 13/4	Jersey Queen
		ECD 14/8	Septa
		ECD 15/16	Verheul (Kale)
Williams (1966)		4 plants	Laurentian (*B. napus* var. *napobrassica;* ECD 10,11,13
Some et al. (1996)		3 plants	Brutor (*B. napus* var. *napus*). ECD 06, 10

bioassays (Donald et al. 2006; Agarwal et al. 2009). This lack of reproducibility, the ECD analysis may result from the large proportion of indeterminate host reactions, the presence of mixed populations of resting spores present in the galls used for inoculum preparation has been reported (Crute et al. 1980; Jones et al. 1982a, b; Scholze et al. 2002) or may be due to variable environmental conditions during the bioassay. A microsatellite-based molecular technique was proposed by Faggian et al. (1999) as a first step towards fast, accurate, and reliable identification of *P. brassicae* pathotypes. This approach is appealing as it avoids the need for a bioassay based-system, and the associated problems of indeterminate host reactions, and environmental influences on the host reaction. However, using RAPD, and microsatellite primers to investigate polymorphism in Australian collections of *P. brassicae*, RAPD primers OPA 1, 8, 11; OPB 3, 7, 20; OPM 2, 13, 16, and microsatellite primers (GACA)4, HKB 17/9, and HKB 23/52 have revealed key differences between the *P. brassicae* isolates (Badi et al. 2010).

5.2.2 Canada

In Canada, differential hosts presented in Tables 5.2 and 5.3 were used to designate pathotypes (Tables 5.4 and 5.5), and for characterization of virulence (Table 5.2) of *P. brassicae* populations prevalent under field conditions. Physiological characterization of *P. brassicae* was initiated during third decade of 1900 (Honig 1931), and numerous sets of differential hosts were proposed for the assessment of virulence in the pathogen (Ayers 1957; Williams 1966; Buczacki et al. 1975; Some et al. 1996). The studies of pathogenic variability from various countries, the differential set of Williams (1966), which consists of two cabbage, and two rutabaga cultivars, has been most widely used (Tables 5.1 and 5.6). In his original paper, Williams (1966) identified pathotypes (races) 2 and 6 in populations of *P. brassicae* obtained from Quebec and British Columbia, respectively. Pathotype 6 was also reported from another population from British Columbia (Strelkov et al. 2006b). Similarly, pathotype 6 was determined to be the predominant pathotype in cole crops from Ontario (Reyes et al. 1974), a finding that was confirmed by Strelkov et al. (2006b). In populations of *P. brassicae* from Nova Scotia, pathotype 3 was found to be predominant (Hildebrand and Delbridge 1995). In Alberta, a total of seven populations have been characterized: five populations obtained from a commercial canola field near St Albert were identified as pathotype 3, as was a population from a market garden near Leduc, while another population from experimental field plots located in northeast Edmonton was given a pathotype 5 designation (Strelkov et al. 2006b; 2007a). The virulence of the Alberta populations was also evaluated on the hosts of the European Clubroot Differential set (ECD) (Buczacki et al. 1975), which consists of 15 genotypes representing three subsets (*Brassica rapa* L., *Brassica napus* L., and *Brassica oleracea* L. subset). On the hosts of the ECD set, the populations from St Albert and Leduc were designated as ECD 16/15/12, while the population from northeast Edmonton was designated as ECD 16/15/0 (Strelkov et al. 2006b). The population of *P. brassicae* from Alberta was classified as ECD-15/12 on the hosts of ECD set or as pathotypes 3 or P2 on the differentials of Williams or Some et al. by Strelkov et al. (2006b). Although *P. brassicae* populations in Alberta appeared to be fairly homogenous (Strelkov et al. 2006b), most of the populations tested came from the same field. While this was a justified approach for evaluating pathogen diversity when only 12 tightly clustered fields were known to be infested with clubroot (Strelkov et al. 2005), the disease has since been found to be much more widespread in central Alberta (Strelkov et al. 2006a), and the diversity of *P. brassicae* over this larger region needs to be assessed. Such information is

Table 5.2 Differential hosts used to identify *Plasmodiophora brassicae* pathotypes in Alberta, Canada (Strelkov et al. 2006b)

ECD No.[a]	Differential host
ECD 05	*Brassica rapa* var. *pekinensis* cv. Granaat
ECD 06	*B. napus* var. *napus* cv. Nevin
ECD 07	*B. napus* var. *napus* cv. giant rape
ECD 08	*B. napus* var. *napus* selection ex. giant rape
ECD 09	*B. napus* var. *napus* New Zealand clubroot-resistant rape
ECD 10	*B. napus* var. *napobrassica* cv. wilhemsburger
ECD 11	*B. oleracea* var. *capitata* cv. badger shipper
ECD 12	*B. oleracea* var. *capitata* cv. bindsachsener
ECD 13	*B. oleracea* var. *capitata* cv. jersey queen
ECD 14	*B. oleracea* var. *capitata* cv. septa
ECD 15	*B. oleracea* var. *capitata* subvar. *laciniata* cv. verheul
Na	*B. napus* var. *napobrassica* cv. laurentian
Na	*B. napus* cv. brutor (spring oilseed rape)

[a]European Clubroot Differential (ECD) numbers denote the differential hosts of the ECD set (Buczacki et al. 1975); hosts ECD 10, 11, 13, and Laurentian represent the differentials of Williams (1966), whereas hosts ECD 06, ECD 10, and Brutor represent the differentials of Some et al. (1996). *Na* not applicable

important in the development of canola cultivars with resistance to the predominant pathotypes of *P. brassicae*, as well as for the appropriate management of this resistance once it is obtained.

In Canada, the virulence of seven *P. brassicae* populations from Alberta as well as of one population from British Columbia, and another from Ontario was characterized on the differentials of Williams (1966) and ECD set (Buczacki et al. 1975) by Strelkov et al. (2006b). The reaction of some hosts to these populations was clearly identified as resistant or susceptible, other differentials showed intermediate indices of disease. This caused difficulty in classifying *P. brassicae* populations into pathotypes, as neither differential system refers to intermediate ID values (Williams 1966; Buczacki et al. 1975). The inoculations were conducted using field populations of *P. brassicae*. Therefore, it is possible that intermediate ID values were a result of the heterogeneity of the pathogen populations, as these may consist of a mixture of pathotypes (Tinggal and Webster 1981; Jones et al. 1982a; Scott 1985). To preclude the possibility of mixed infections, Williams (1966) suggested that resting spores should be recovered from each differential host and used to re-inoculate the complete set of differentials. Similarly, Buczacki et al. (1975) stated that a population cannot be considered homogenous until at least two separate spore extractions and re-inoculations have not produced a change in host reaction.

In Japanese populations of *P. brassicae*, Kuginuki et al. (1999) also observed intermediate scores on several differentials, and suggested that indistinct reactions were due to genetic heterogeneity of the hosts rather than the pathogen. In an effort to accommodate intermediate ID values, Toxopeus et al. (1986) suggested that host reactions with an ID = 20% should be regarded as resistant, and those with an ID =

Table 5.3 Virulence of *Plasmodiophora brassicae* populations from Alberta, Canada, on the differential hosts of Williams (1966), Some et al. (1996) and the European Clubroot Differential (ECD) set (Strelkov et al. 2006b)

	Index of disease (%) induced by populations of *Plasmodiophora brassicae*[a]									
Differential host	Field 1	Field 30	Field 44	Field 52	Field 64	Field 90	Field 92	Field 111	Field 112	Field 113
ECD05	100 ± 0	100 ± 0	100 ± 0	100 ± 0	99 ± 1	100 ± 0	100 ± 0	100 ± 0	100 ± 0	100 ± 0
ECD06	95 ± 2	94 ± 3	95 ± 2	88 ± 4	92 ± 3	94 ± 5	86 ± 6	95 ± 4	80 ± 6	94 ± 4
ECD07	100 ± 0	100 ± 0	100 ± 0	99 ± 1	99 ± 1	99 ± 1	100 ± 0	100 ± 0	98 ± 2	99 ± 1
ECD08	99 ± 1	100 ± 0	100 ± 0	97 ± 3	95 ± 2	98 ± 2	99 ± 1	100 ± 0	99 ± 1	100 ± 0
ECD09	99 ± 1	96 ± 4	98 ± 2	94 ± 4	90 ± 4	99 ± 1	99 ± 1	97 ± 2	93 ± 6	97 ± 2
ECD10	10 ± 1	18 ± 3	33 ± 13	28 ± 5	16 ± 4	20 ± 8	29 ± 5	27 ± 4	22 ± 2	9 ± 2
ECD11	31 ± 6	40 ± 14	39 ± 3	45 ± 2	8 ± 3	41 ± 2	23 ± 5	21 ± 3	20 ± 4	44 ± 5
ECD12	27 ± 2	24 ± 3	36 ± 12	31 ± 4	42 ± 13	33 ± 2	22 ± 6	18 ± 4	16 ± 5	42 ± 7
ECD13	71 ± 8	72 ± 8	84 ± 4	86 ± 8	65 ± 5	83 ± 7	63 ± 11	62 ± 4	77 ± 6	82 ± 5
ECD14	94 ± 0	95 ± 2	99 ± 1	99 ± 1	86 ± 9	93 ± 3	89 ± 4	94 ± 4	90 ± 2	96 ± 2
ECD 15	40 ± 4	34 ± 7	34 ± 3	30 ± 3	34 ± 6	39 ± 14	26 ± 6	23 ± 1	21 ± 7	43 ± 2
Laurentian	94 ± 3	93 ± 4	100 ± 0	99 ± 1	89 ± 4	92 ± 3	93 ± 3	91 ± 3	93 ± 2	99 ± 1
Brutor	100 ± 0	100 ± 0	100 ± 0	97 ± 2	94 ± 3	98 ± 2	100 ± 0	100 ± 0	99 ± 1	99 ± 1

[a]Means of four replicates ±1 SE. Indices of disease were calculated using the formula: $ID(\%) = \sum(n \times 0 + n \times 1 + n \times 2 + n \times 3 \times 100\% / N \times 3$, where n is the number of plants in each class; N is the total number of plants; and 0, 1, 2, and 3 are the symptom severity classes

Table 5.4 Pathotype designations for field populations of *Plasmodiophora brassicae* collected from commercial fields in Alberta, Canada in 2005 (Strelkov et al. 2007b)

	Pathotype designation[a]		
Populations	ECD	Williams	Some et al.
Field 1	–/15/12	3	P2
Field 20	–/15/12	3	P2
Field 44	–/15/12	3	P2
Field 52	–/15/12	3	P2
Field 64	–/15/12	3	P2
Field 90	–/15/12	3	P2
Field 92	–/15/12	3	P2
Field 111	–/15/12	3	P2
Field 112	–/15/12	3	P2
Field 113	–/15/12	3	P2

[a]As determined on the differential hosts of the European Clubroot Differential set (Buczacki et al. 1975), Williams (1966), or Some et al. (1996). For classification into pathotypes, an index of disease of 50% was used as the cutoff between a resistant and a susceptible reaction (Strelkov et al. 2007b)

Table 5.5 Pathotype designations for field populations of *Plasmodiophora brassicae* collected in Alberta, British Columbia, and Ontario, using a disease index of 50% as the cut off between a resistant and susceptible reaction (Strelkov et al. 2006b)

		Pathotype designation[a]	
Population	Origin	ECD	Williams
SACAN03-1	St. Albert Alta (canola)	16/15/12	3
SACAN03-2	St. Albert Alta (canola)	16/15/12	3
SACAN03-3	St. Albert Alta (canola)	16/15/12	3
SACAN03-4	St. Albert Alta (canola)	16/15/12	3
SACAN03-5	St. Albert Alta (canola)	16/15/12	3
CDCN04-1	CDCN[b] Edmonton Alta (canola)	16/15/0	5
Leduc-1	Leduc, Alta (soil)	16/15/12	3
Abot JE04-1	Abbottsford, BC (soil)	16/2/12	6
ORCA04	Orton, Ont. (cabbage)	16/0/14	6

[a]As determined on the differential hosts of Williams (1966) or the European Clubroot Differential (ECD) set (Buczacki et al. 1975)
[b]Crop diversification centre north

80% should be regarded as susceptible. Although this criteria would allow classification of most of the *B. rapa* (ECD 01–ECD 05) and *B. napus* (ECD 06 to ECD 10, "Laurentian") hosts as either resistant or susceptible, the reactions of the majority of the *B. oleracea* hosts would still be indistinct (Toxopeus et al. 1986), since these often developed IDs higher than 20% but lower than 80%. Therefore, to be able to assign pathotype designations to all populations tested, the range of values corresponding with a resistant and a susceptible reaction were expanded: hosts with an ID from 0% to 49% were regarded as resistant, and those with an ID from 50% to 100% were regarded as susceptible. Under this scheme, the populations

Table 5.6 Differential hosts used to characterize the *Plasmodiophora brassicae* populations in Canada (Strelkov et al. 2006b)

Cultivar or line	ECD No.[a]
Polish rape (*Brassica rapa* subsp. *rapifera*)	
Line aaBBCC	01
Line AAbbCC	02
Line AABBcc	03
Line AABBCC	04
Chinese cabbage (*Brassica rapa* var. *pekinensis*)	
Granaat	05
Fodder rape *(Brassica napus* var. *napus)*	
Nevin	06
Giant rape	07
Giant rape selection	08
New Zealand resistant rape	09
Rutabaga *(Brassica napus* var. *rapifera)*	
Withemsburger	10
Cabbage *(Brassica oleracea* var. *capitata)*	
Badger shipper	11
Bindsachsener	12
Jersey queen	13
Septa	14
Kale *(Brassica oleracea* var. *capitata* subvar. *laciniata*	
Verheul	15
Rutabaga *(Brassica napus* var. *napobrassica)*	
Laurentian	NA

[a]European clubroot differential (ECD) number denote the differential hosts of the ECD set (Buczacki et al. 1975), hosts ECD 10, 11, 13, and *B. napus* var. *napobrassica* represent the differentials of Williams (1966)

from St. Albert and Leduc were classified as ECD 16/15/12, and the population from CDCN was classified as ECD 16/15/0 on the differentials of Buczacki et al. (1975). On Williams' (1966) hosts, the St. Albert and Leduc populations were classified as pathotype 3, and the CDCN population was classified as pathotype 5. Pathotype 3 was previously found to be pre-dominant on cruciferous vegetables in Nova Scotia (Hildebrand and Delbridge 1995), while the population from CDCN represents the first reported occurrence of pathotype 5 in Canada. The different populations and their pathotype classifications are summarized in Table 5.5. The pathogen population from Abbotsford, British Columbia was classified as ECD 16/2/12 or pathotype 6 (Table 5.5). This pathotype six classification was consistent with a prior report (Williams 1966), which also identified a population from British Columbia as pathotype 6. The population from Orton, Ontario, was designated as ECD 16/0/14 or pathotype 6 (Table 5.5). This finding was also consistent with a previous report indicating that pathotype 6 is predominant in cole crops from Ontario (Reyes et al. 1974).

All *P. brassicae* populations examined were avirulent or of low virulence on the *B. rapa* hosts ECD 01 to ECD 04 (Fig. 5.1). This observation was consistent with a survey of 299 ECD tests performed worldwide, in which these *B. rapa* hosts were found to be infected only and rarely by the clubroot pathogen (Toxopeus et al. 1986). The remaining *B. rapa* host, ECD 05, is considered to be the universal susceptible (Toxopeus et al. 1986) and was indeed susceptible to all populations characterized by Strelkov et al. (2006b) (Fig. 5.1). Being given the similar virulence patterns of all populations tested, the *B. rapa* subset did not appear useful for differentiating *P. brassicae* populations from Canada. In contrast, a complete range of reactions were observed on the *B. napus* hosts, indicating greater differentiating capacity in this subset. Interestingly, *P. brassicae* populations from Alberta were highly virulent on the fodder rape hosts ECD 06 to ECD 09 (Fig. 5.1). In the case of the populations from St. Albert and CDCN, originally collected from canola, this may reflect a high degree of adaptation to a *B. napus* var. *napus* host. This relationship is not so clear for the Leduc population, which originated from infested soil from a market garden. Nevertheless, the populations from Ontario and British Columbia, obtained from regions in which clubroot occurs mainly on cruciferous vegetables, were avirulent on all or all but one of the fodder rape varieties (Fig. 5.1). Seaman et al. (1963) reported that continuous cropping of a resistance source led to an erosion of its effectiveness, which was restored after a break of two years. This suggests that the pathotype composition of a population can shift quite rapidly, and is consistent with the increased virulence on *B. napus* observed in Alberta.

On the *B. oleracea* hosts, all pathogen populations produced at least some disease (Toxopeus et al. 1986; Voorrips 1995). However, the CDCN population caused less disease on this susceptible as compared to all other populations evaluated (Fig. 5.1). The reduced virulence of the CDCN population on the *B. oleracea* subset (and "Laurentian") resulted in its unique pathotype designation, relative to the other populations from Alberta, under both classification systems used. As these differences in virulence were largely based on the reactions of the *B. oleracea* hosts, which produced intermediate reactions against all populations, these results should be treated with caution. Nevertheless, they seem to indicate the presence of genetic diversity in the pathogenicity of *P. brassicae* populations in Alberta. Moreover, if the intermediate ID values observed on some hosts were caused by pathotype mixtures, then presumably different pathotypes are present, even within single plants (Jones et al. 1982a).

The occurrence and composition of physiological specialization in *P. brassicae* from populations and single-spore isolates has been analyzed by Strelkov and Hwang (2014) in Canada. They have reported that in Canada, work to characterize the strains of *P. brassicae* began in the 1950s, when the virulence patterns of parasite collections from the Maritimes were assessed on a differential set consisting of wild and cultivated crucifers (Ayers 1957). In later decades, the differential hosts of Williams (1966), Some et al. (1996), and the European Clubroot Differential (ECD) set (Buczacki et al. 1975) have been most widely used to analyze populations

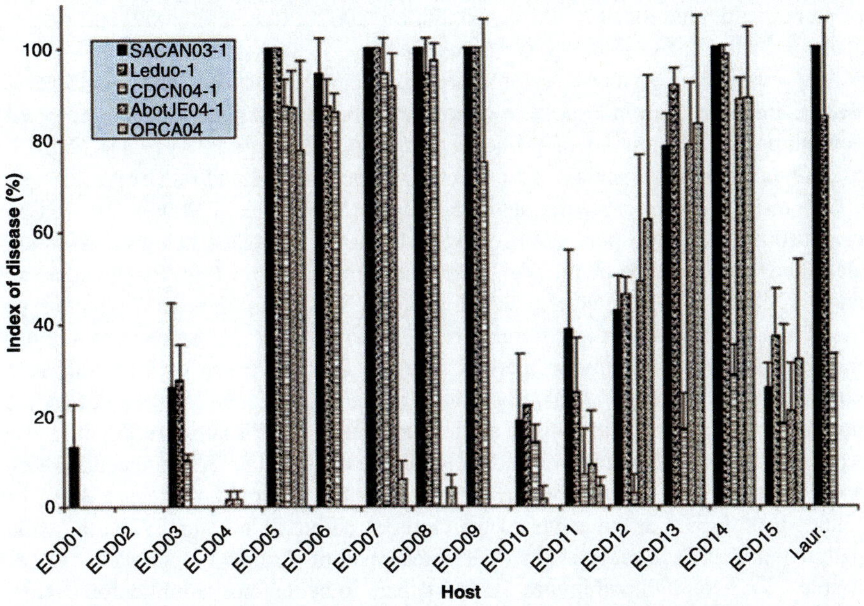

Fig. 5.1 Virulence of *Plasmodiophora brassicae* populations on the differential hosts of Williams (1966), and the European Clubroot Differential (ECD) set. SACAN03-1 was collected near St. Albert, Alberta; Leduc-1 was collected near Leduc, Alberta; CDCN04-1 was collected in northeast Edmonton, Alberta; AbotJE04-1 was collected near Abbotsford, British Columbia, and ORCA04 was collected near Orton, Ontario. Tests were repeated three to five times, and the mean indices of disease (ID) are shown as bars. Lines indicate standard deviations. ECD 01, *Brassica rapa* subsp. *rapifera* line aaBBCC; ECD 02, *B. rapa* subsp. *rapifera* line AAbbCC; ECD 03, *B. rapa* subsp. *rapifera* line AABBcc; ECD 04, *B. rapa* subsp. *rapifera* line AABBCC; ECD 05, *B. rapa* var. *pekinensis* "Granaat"; ECD 06, *Brassica napus* var. *napus* "Nevin"; ECD 07, *B. napus* var. *napus* "Giant Rape"; ECD 08, *B. napus* var. *napus* selection ex. "Giant Rape"; ECD 09, *B. napus* var. *napus* New Zealand clubroot-resistant rape; ECD 10, *B. napus* var. *napobrassica* "Wilhemsburger"; ECD 11, *Brassica oleracea* var. *capitata* "Badger Shipper"; ECD 12, *B. oleracea* var. *capitata* "Bindsachsener"; ECD 13, *B. oleracea* var. *capitata* "Jersey Queen"; ECD 14, *B. oleracea* var. *capitata* "Septa"; ECD 15, *B. oleracea* var. *acephala* subvar. *laciniata* "Verheul"; Laur., *B. napus* var. *napobrassica* "Laurentian." Hosts ECD 10, 11, and 13 are also members of Williams (1966) differential set, but "Laurentian" is unique to this set (Strelkov et al. 2006b)

of *P. brassicae* (Table 5.1). Until the outbreak of clubroot on canola, however, there had been limited evaluation of the pathotype ("race") composition of Canadian populations of the pathogen. Williams (1966) identified pathotypes 2 and 6 in populations of *P. brassicae* from Quebec and British Columbia, respectively. Pathotype 6 also was found to be predominant in cole crops in Ontario (Reyes et al. 1974), while pathotype 3 was most common in an analysis of *P. brassicae* populations from Nova Scotia (Hildebrand and Delbridge 1995). The identification of clubroot on canola in Alberta led to a concerted effort to characterize the virulence of pathogen populations (Strelkov et al. 2006b, 2007b; Cao et al. 2009) and single-spore isolates (Xue et al. 2008) from this crop.

The evaluation of more than fifty *P. brassicae* populations, and single-spore isolates recovered from canola in central Alberta indicates that pathotype 3, as defined on the differentials of Williams (1966), or pathotype P2 or ECD 16/15/12, as defined on the differentials of Some et al. (1996) or the ECD set (Buczacki et al. 1975), respectively, is predominant in the region (Strelkov et al. 2006b, 2007b; Cao et al. 2009). A single population recovered from soil in Saskatchewan also was classified as pathotype 3 on the differentials of Williams (1966). By contrast, analysis of two populations from canola in southern Alberta suggests that pathotype 5 or P3, as classified on the differentials of Williams (1966) or Some et al. (1996), respectively, is most common there (Cao et al. 2009). Testing of a *P. brassicae* population from canola in Manitoba and another from Quebec also revealed the presence of pathotype 5, as defined on the differentials of Williams (1966), but these were classified as P2 and P3, respectively, on the differentials of Some et al. (1996) and Cao et al. (2009). Collectively, the results point to the predominance of Williams (1966) pathotype 3 at the center of the clubroot outbreak in central Alberta, while pathotype 5 has been associated with isolated cases of clubroot on canola elsewhere (Table 5.7). Although pathotypes 3 and 5 appear to be the most common on canola, additional strains of *P. brassicae* also have been recovered, particularly when single-spore isolates rather than populations of the pathogen were examined. Other pathotypes collected from soil and canola in central Alberta include pathotypes 2, 6, and 8, as classified on the differentials of Williams (1966), Strelkov et al. (2006b), Xue et al. (2008), and Cao et al. (2009). While pathotype 6 is rare in Alberta, but it is still the predominant pathotype on cruciferous vegetables in British Columbia and Ontario (Strelkov et al. 2006b; Xue et al. 2008; Cao et al. 2009), just as it was decades ago (Williams 1966; Reyes et al. 1974). In Ontario, however, analysis of single-spore isolates versus populations of *P. brassicae* indicates a more complex pathotype composition than previously thought (Xue et al. 2008) (Table 5.7). The fairly diverse pathotype structure of *P. brassicae* in Canada suggests that clubroot resistance in canola will have to be well-managed, since the virulence of pathogen populations can shift quickly in response to the selection pressure imposed by the cropping of resistant hosts (Seaman et al. 1963; LeBoldus et al. 2012).

5.2.3 China

The studies on pathogen populations from China, the differential set of Williams (1966), consisting of two cabbage, and two rutabaga cultivars 'Jersey Queen', 'Badger Shipper', 'Laurentian', and 'Wilhelmsburger' has been used (Shen et al. 2009; Ding et al. 2013; Ji et al. 2013; Peng et al. 2013; Wang et al. 2013a, b; Zhao et al. 2013). The race (pathotype) 4 predominates in China. A total of 15 populations in 2009, were characterized, out of which 11 populations obtained from Liaoning, Shandong, Yunnan, Sichuan, and Jilin Provinces were identified as race 4, while two

Table 5.7 *Plasmodiophora brassicae* pathotype composition in Canada (Strelkov and Hwang 2014)

Province	Pathotypes (s)[a]		References
	Populations[b]	Single-spore isolates[c]	
Alberta	3, 5, 2	3, 8, 2, 6	Strelkov et al. (2006a, 2007b), Xue et al. (2008), Cao et al. (2009)
British Columbia	6	6	Strelkov et al. (2006a), Williams (1966), Xue et al. (2008)
Manitoba	5	n/a[d]	Cao et al. (2009)
Nova Scotia	3, 1, 2	n/a[d]	Hildebrand and Delbridge (1995)
Ontario	6	3, 5, 8	Reyes et al. (1974), Strelkov et al. (2006a), Xue et al. (2008), Cao et al. (2009)
Quebec	2, 5	n/a[d]	Williams (1966), Cao et al. (2009)
Saskatchewan	3	n/a[d]	Strelkov and Hwang (2014)

[a]Pathotype designated are based on the differentials of Williams (1966), since this was only system used in some of the references. Pathotypes are listed in order from most to least frequently reported in a province, in cases where two pathotypes are equally common, they are listed in numerical order
[b]A population refers to a collection of resting spores recovered from infested soil or galls of an infected plant (Voorrips 1995; Xue et al. 2008)
[c]A single-spore isolate refers to a populations derived from a gall inoculated with a single-spore isolate of *P. brassicae* and maintained in isolation (Voorrips 1995; Xue et al. 2008)
[d]*n/a* not available (i.e. single-spore isolate were not characterized)

other populations from Liaoning Province were identified as races 2 and 11, whereas another two populations from Sichuan Province were identified as races 7 and 10 (Shen et al. 2009). Race 4 was also characterized from 11 populations collected from Heilongjiang, Jilin, and Liaoning Provinces (Zhao et al. 2013). In populations of *P. brassicae* from canola, race 4 was found to be predominant at 16 locations, including Hubei, Anhui, Sichuan, and Yunnan Provinces (Ji et al. 2013). Similarly, race 4 was determined to be the predominant race in 16 populations from Yunnan, Hubei, Shanxi, Sichuan, Shanghai, and Liaoning Provinces (Ding et al. 2013). Four isolates collected from fields of Chinese cabbage, European cabbage, and radish in Hubei Province also were classified as race 4 (Wang et al. 2013b). Race 4 appears to be predominant on cruciferous vegetables and is spreading throughout China, resulting increasing damage.

In populations of *P. brassicae* from Hunan Province, races 1, 4, 9, and 13 were identified in 12 populations collected from clubroot-infected crops of Chinese cabbage, tuber mustard (*B. juncea* var. *tumida* Tsen, and Lee), purple flowering stalk (*B. rapa* ssp. *chinensis* var. *purpurea* Hort.), and canola (Peng et al. 2013). Races 7 and 2 were detected, respectively, in populations from Sichuan and Yunnan Provinces (Ding et al. 2013). Therefore, although race 4 represents 78.5% of the populations characterized from China. However, so far, races 1 (2.5%), 2 (3.8%),

7 (2.5%), 9 (2.5%), 10 (1.3%), 11 (1.3%), and 13 (7.6%) have also been identified (Table 5.8).

5.2.4 Czech Republic

Pathotypes in the Czech Republic were determined by Ricarova et al. (2016). Isolates collected from vegetable-producing areas were tested. A total of 92 isolates from the Czech Republic and Slovakia were tested. It revealed 6 pathotypes according to Williams (1966) with the most prevailing pathotype 7 (almost 50% of samples). Other samples revealed pathotypes 6, 4, 10, 2, 3, and pathotype 1 in one isolate. In some isolates pathotypes 2 and 3 were detected together. Pathotype 3 was revealed only in Slovakia. According to ECD classification, 35 pathotypes were identified, but 22 of them occurred only once. The most common was the pathotype ECD 16/15/13 (28% cases) as well as ECD 16/02/28 and 16/14/31 (9% cases). During 2014, out of 17 samples, six pathotypes according to Williams (1966) were detected with the most prevalent pathotype 7 (41% of cases). Other identified pathotypes were 6, 2, 3, 5, and 9. According to Some et al. (1996), 3 pathotypes were identified with the most frequent pathotype P3 (82% of cases) along with pathotypes P4 and P1. The ECD set identified 10 pathotypes and the most common was ECD 16/14/15 (24% cases), other pathotypes, which occurred more than once, were 16/14/13, 16/14/31, and 16/2/14.

Vegetable isolates and oilseed rape isolates have different "preference" levels to these species. Pathotypes 9 and 3 according to Williams (1966) seem to be unique for the infection of oilseed rape, because none of these pathotypes were recorded on vegetables during screenings performed by Rod (1991, 1994, 1996) in the Czech Republic. This is an interesting finding, because pathotype 3 is one of the most widespread pathotypes on oilseed rape especially in Canada (Strelkov et al. 2006a, 2007b). On the other hand, pathotypes 1 and 4 seem to be unique for vegetables.

The geographic pattern of *P. brassicae* distribution is not probably connected with any special environmental feature like soil type or climate conditions. The spread of clubroot is mainly historical, because of a long tradition of cabbage cultivation, which was dominant in all infested localities (Chytilova and Dusek 2007). The oilseed rape cultivation does not have such a long continuance as the first fields were recorded in the 20s of the last century (5900 ha), after Second World War, there were 32,605 ha and cropping on a larger area came only in the 80s of the last century (102,160 ha). Since then the cropped area was growing and a real breakthrough came with the new millennium, when the production areas of oilseed rape reached 344,117 ha and the peak was in 2013 (418,808 ha) (Czech Statistical Office). As the area of vegetable cultivation was gradually decreasing in the Czech Republic, fields used before for vegetable production started to be cropped with oilseed rape and further spreading of clubroot infection was therefore easier and faster (Ricarova et al. 2016).

Table 5.8 Distribution of *Plasmodiophora brassicae* races in China (Chai et al. 2014)

Province	City	Host	No. of samples		Reference
			Race[a]	Examined	
Anhui	Xiuning	*B. napus*	2	1	Ji et al. (2013)
	He county	*B. napus*	4	1	Ji et al. (2013)
	Xiuning	*B. napus*	13	2	Ji et al. (2013)
Chongqing	Wulong	*B. pekinensis*	4	1	Ding et al. (2013)
Heilongjiang	NanCheng, BaiCheng, Ningan	*B. pekinensis*	4	3	Zhao et al. (2013)
Hubei	Changyang	*B. pekinensis*	4	4	Wang et al. (2013b)
	Changyang	*B. pekinensis, R. sativus*	4	4	Ding et al. (2013)
	Zhijiang, Dangyang	*B. napus*	4	2	Ji et al. (2013)
Hunan	Huaihua	*B. pekinensis, B. juncea*	1	2	Peng et al. (2013)
	Yiyang, Zhuzhou	*B. pekinensis, B. chinensis var. purpurea, B. napus*	4	4	Peng et al. (2013)
	Changsha	*B. pekinensis, B. Chinensis* var. *purpurea*	9	2	Peng et al. (2013)
	Huaihua, Zhuzhou	*B. pekinensis, B. napus*	13	4	Peng et al. (2013)
Jilin	Wulajie	*B. pekinensis*	4	2	Shen et al. (2009)
	Lishu	*B. pekinensis*	4	1	Shen et al. (2009)
	Songyuan, Liuhe, Renping	*B. pekinensis*	4	4	Zhao et al. (2013)
Liaoning	Benxi	*B. pekinensis*	2	1	Shen et al. (2009)

(continued)

Table 5.8 (continued)

Province	City	Host	No. of samples		Reference
			Race[a]	Examined	
	Dalian	*B. pekinensis*	4	1	Ding et al. (2013)
	Shenyang, Dalian, Dandong	*B. pekinensis*	4	4	Shen et al. (2009)
	Shenyang	*B. pekinensis*	11	1	Shen et al. (2009)
	Xinmin, Fushun, Zhuanghe, Pulandian	*B. pekinensis*	4	4	Zhao et al. (2013)
Shanghai	Jiading	*Nigella damascene* × B. *oleracea* var. *botrytis*	4	1	Ding et al. (2013)
Shandong	Qingdao, Changshan	*B. pekinensis*	4	2	Shen et al. (2009)
	Qingdao	*B. pekinensis*	4	1	Ding et al. (2013)
Shanxi	Taibai	*B. pekinensis*	4	1	Ding et al. (2013)
Sichuan	Pixian	*B. pekinensis*	4	1	Shen et al. (2009)
	Pengzhou, Pixian, Shenfang	*B. pekinensis*, *B. oleracea*, *B. juncea* var. *tumida*	4	4	Ding et al. (2013)
	Chengdu, Mianyang, Qionglai, Meishan, Guanghan	*B. napus*	4	5	Ji et al. (2013)
	Pengzhou	*B. pekinensis*	7	1	Shen et al. (2009)
	Ganzi	*B. pekinensis*	7	1	Ding et al. (2013)
	Xichang	*B. pekinensis*	10	1	Shen et al. (2009)
Yunnan	Tonghai	*B. pekinensis*	2	1	Ding et al. (2013)

(continued)

Table 5.8 (continued)

Province	City	Host	No. of samples		Reference
			Race[a]	Examined	
	Jiangchuan	*B. pekinensis*	4	1	Shen et al. (2009)
	Weishan, Panlong, Yuxi	*B. pekinensis*	4	3	Ding et al. (2013)
	Yuxi, Baoshan, Tengchong Dali	*B. napus*	4	8	Ji et al. (2013)

[a]Physiological races of *P. brassicae* were determined with the differential set of Williams (1966) (Jersey Queen, "Badger Shipper," Laurentian and Wilhelmsburger)

Czech Republic and Poland

Pathotypes of *Plasmodiophora brassicae* causing damage to oilseed rape in the commercial fields of Czech Republic and Poland were determined by Ricarova et al. (2016). There were considerable differences between the populations of *P. brassicae* from both countries, and the number of pathotypes varied depending on the evaluation system and the threshold used to distinguish susceptible vs. resistant plant reactions. Using an index of disease (ID) of 25% to distinguish susceptible vs. resistant reactions, seven pathotypes were identified based on the differentials of Williams, five with the system of Some et al. (1996), and 18 with the European Clubroot Differential (ECD) set. However, based on a threshold of 50%, there were nine pathotypes according to the evaluation system by Williams (1966), four based on the differentials of Some et al. (1996), and 15 with the ECD set. Changing of the thresholds led to the reclassification of some pathotypes. Several pathotypes were common in both countries. High amounts of pathogen DNA were found in many of the field soils analyzed by quantitative PCR. There was a weak correlation between soil pH and infestation of *P. brassicae* for the Polish soils (Table 5.9 and 5.10).

5.2.5 Finland

In Finland, *P. brassicae* race determinations were performed using 90 isolates. The classification system of Williams (1966) was used. The races that were isolated were 1, 2, 3, 4, 6, and 7 (Table 5.11). Race 2 was by far the most common, being found in 32 communes; races 3,4,6, and 7 were each found in 9–12 communes; race 1 only in one commune. No clear differences in the occurrence of the races in the various parts of the country could be observed. A comparison was made between Williams' and the ECD differential set (Buczacki et al. 1975) classification systems. In addition, the pathotypes in clubroot material from Norway and Iceland were also determined (Linnasalmi and Toiviainen 1991). The regional distribution of the races is shown in

Table 5.9 Pathotype classifications for populations of *Plasmodiophora brassicae* collected from commercial oilseed rape fields in the Czech Republic, and Poland, 2010–2012. Threshold for reaction susceptible vs. resistant 25% ID (Some et al. 1996; Ricarova et al. 2016)

Isolate	Locality	Region	Pathotypes designation for different system			Soil pH
			Williams[a]	Some[b]	ECD[c]	
Czech Republic						
1	Holany	Liberecky	7	P2	16/15/15	7.3
2	Hradek n. Nisou		6	P4	16/14/4	6.8
3	Bily Kostel		2	P2	16/15/15	6.8
4	Modlibohov		7	P3	16/14/31	7.1
5	Horka u Bakova	Stredocesky	7	P3	16/14/31	6.9
6	Kbelnice	Kralovehradecky	6	P3	16/14/14	6.6
7	Miletin		2	P2	16/15/31	6.6
8	Trebnouseves		7	P2	16/15/31	6.5
9	Pohledy	Pardubicky	1	P1	16/31/12	6.6
10	Hrdejovice Ves	Jihocesky	7	P2	16/15/15	6.5
11	Horusice		4	P5	16/22/31	6.4
12	Zirovnice	Vysocina	1	P5	16/22/30	6.5
13	Kozmice	Moravskoslezsky	4	P1	16/31/15	6.6
14	Klokocov		7	P2	16/3/15	6.7
Poland						
1	Siemysl	West Pomerania	4	P1	17/31/31	7.6
2	Ramlewo		7	P3	16/7/31	6.7
3	Karlin		4	P1	17/31/31	5.8
4	Tuczno		9	P1	16/31/8	4.3
5	Karsibor		4	P1	17/31/31	6.4
6	Jablonowo		4	P1	17/31/31	7.3
7	Bielnik	Varmia	7	P3	16/31/31	5.4
8	Wegorzewo	Masuria	6	P3	16/14/30	6.9
9	Wrzesiny	Lubuskie	7	P3	16/14/31	7.6
10	Krotoszyn	Great Poland	9	P1	16/31/8	5.9
11	Lubań	Lower Silesia	4	P1	16/31/14	5.0
12	Ocice		3	P3	16/6/31	6.2
13	Bolkow		7	P3	16/14/15	6.8
14	Dobromierz		7	P3	16/14/15	5.0
15	Opole	Opole	7	P3	16/14/31	4.4
16	Kiczyce	Upper Silesia	7	P3	16/14/15	6.8

[a]Williams (1966)
[b]Some et al. (1996)
[c]European Clubroot Differential (ECD) Buczacki et al. (1975)

Table 5.11. Of the six races isolated, race 2 was by far the most common. It was found in 32 communes, i.e. in 46% of the communes with clubroot. Races 3, 4, 6, and 7 were each found in about ten communes (9–12 communes, i.e. 14–17%);

Table 5.10 Pathotype classifications for populations of *Plasmodiophora brassicae* collected from commercial oilseed rape fields in the Czech Republic and Poland, 2010–2012. Threshold for susceptible vs. resistant reaction of differential plant 50% ID with 95% confidence interval (LeBoldus et al. 2012; Ricarova et al. 2016)

Isolate	Locality	Region	Pathotypes designation for different system			Soil pH
			Williams[a]	Some[b]	ECD[c]	
Czech Republic						
1	Holany	Liberecky	7	P3	16/14/15	7.3
2	Hradek n. Nisou		6	P4	16/10/4	6.8
3	Bily Kostel		2	P3	16/14/13	6.8
4	Modlibohov		7	P3	16/14/31	7.1
5	Horka u Bakova	Stredocesky	7	P3	16/14/15	6.9
6	Kbelnice	Kralovehradecky	6	P3	16/2/14	6.6
7	Miletin		2	P3	16/14/15	6.6
8	Trebnouseves		7	P3	16/14/31	6.5
9	Pohledy	Pardubicky	9	P1	16/31/8	6.6
10	Hrdejovice Ves	Jihocesky	6	P3	16/14/12	6.5
11	Horusice		4	P5	16/18/15	6.4
12	Zirovnice	Vysocina	1	P3	16/2/14	6.5
13	Kozmice	Moravskoslezsky	7	P3	16/14/15	6.6
14	Klokocov		7	P3	16/2/15	6.7
Poland						
1	Siemysl	West Pomerania	4	P1	16/31/31	7.6
2	Ramlewo		6	P3	16/7/28	6.7
3	Karlin		10	P1	16/31/31	5.8
4	Tuczno		9	P1	16/31/8	4.3
5	Karsibor		4	P1	16/31/31	6.4
6	Jablonowo		4	P1	16/31/31	7.3
7	Bielnik	Varmia	7	P3	16/31/29	5.4
8	Wegorzewo	Masuria	6	P3	16/14/12	6.9
9	Wrzesiny	Lubuskie	6	P3	16/14/30	7.6
10	Krotoszyn	Great Poland	16	P1	16/31/8	5.9
11	Luban	Lower Silesia	12	P1	16/31/14	5.0
12	Ocice		3	P3	16/2/14	6.2
13	Bolkow		7	P3	16/14/15	6.8
14	Dobromierz		7	P3	16/14/15	5.0
15	Opole	Opole	6	P3	16/14/12	4.4
16	Kiczyce	Upper Silesia	7	P3	16/14/15	6.8

[a]Williams (1966)
[b]Some et al. (1996)
[c]European Clubroot Differential (ECD) Buczacki et al. (1975)

race 1 occurred only once. Two different races were found in only a few communes: races 2 and 3 in Nurmijarvi (N) and Joutseno (Sa); "races 2," and 6 in Haukivuori (Sb), races 6, and 7 in Hartola (Ta), Kangasala (Ta), and Oulu (Ob). No clear

Table 5.11 Distribution of clubroot and occurrence of *Plasmodiophora brassicae* Wor. races in Finland, 1974–1978 (Linnasalmi and Toiviainen 1991)

	Communes		Farms inspected		Samples		*P. brassicae* races						
Biological province[a]	Total no.	With clubroot (%)	Total no.	With clubroot (%)	Total no.	With club- root (%)	1	2	3	4	6	7	1–7
							No. of isolates						Total
Ab-Varsinais-Suomi	13	100	19	100	46	100	0	6	1	2	2	1	12
N-Uusimaa	6	100	11	100	28	100	0	3	1	2	0	1	7
Ka-Etela-Karjala	4	100	9	56	14	71	0	3	1	0	0	0	4
St-Satakunta	12	75	21	48	26	42	0	4	0	4	0	0	8
Ta-Etela-Hame	20	95	37	73	88	77	0	11	0	1	8	3	23
Sa-Etela-Savo	8	50	13	54	20	55	0	1	4	0	0	1	6
Oa-Etela-Pohjanmaa	12	58	27	37	44	32	0	2	0	2	0	2	6
Tb-Pohjois-Hame	9	56	25	24	46	24	0	1	0	1	0	2	4
Sb-Pohjois-Savo	7	100	10	80	19	84	1	1	4	0	2	0	8
Kb-Pohjois-Karjala	6	83	10	60	23	52	0	7	1	0	0	0	8
Om-Keski-Pohjanraaa	2	50	5	60	6	50	0	1	0	0	0	0	1
Ob -Pohjois-Pohjanmaa	2	100	5	100	15	100	0	0	0	0	2	1	3
total/mean	101	81	192	61	375	65	1	40	12	12	14	11	90

[a]Heikinheimo and Raatikainen (1971)

prevalence of races was demonstrated from different parts of Finland. All races were distributed fairly evenly over the country, with the exception of the rare 1. However, race 4 seems to show some prevalence in the western parts of the country, while race 3 occurs mostly in the east. In the preliminary studies conducted in 1971–1972 on the occurrence of the *P. brassicae* races in Finland (Linnasalmi and Palonen 1974), the races 2, 4, 6, and 7 were isolated. In two test places the races that were found in 1972 were still the same, viz. race 4 in the test field of the Institute of Plant Pathology in Tikkurila (Vantaa), and race 7 in the test field of the Institute of Horticulture in Piikkio. In 1964, Williams identified race 7 from Piikkio and race 2 from Tikkurila (Williams 1966).

The races displayed the following distribution pattern on different plant species: races 2, 3, 4, 6, and 7 were isolated from cabbage (60 isolates) and cauliflower (17 isolates); races 2, 3, and 7 from broccoli (4 isolates); race 4 from red cabbage, and marrow stem kale (1 isolate from each); races 1, 2, and 4 from swede (5 isolates), and race 2 from turnip (2 isolates). Of the cabbage cultivars grown in Finland, (cf. App. 2) only the Norwegian Resista, and Respla (Weisaeth 1977) are partly resistant to some *P. brassicae* races. These cultivars were accepted for marketing in Norway in 1973, and in Finland in 1975. Both cultivars were found to have some resistance to races 2, 4, and 7 in Finland when compared with the severely contaminated Blatopp cultivar as follows:

	Cultivar, clubroot index (0–3)		
Race (isolate)	Resista	Respla	Blatopp (Control)
2 (F118)	1.80	1.47	2.88
3 (F147)	2.36	2.49	2.96
4 (F155)	1.95	1.14	2.78
7 (F171)	1.86	0.84	2.52

ZAt some farms where these cultivars were grown in field sectors situated next to each other, severe contamination and consequently weaker growth of Blatopp were evident, whereas Resista and Respla were only sparsely and slightly contaminated and headed well.

5.2.6 France

The populations of *P. brassicae* in Brittany are highly heterogeneous for both virulence and DNA pattern. Among the field isolates analyzed, four different pathotypes have been identified. The 37 single-spore isolates (SSIs) revealed seven pathotypes. The eight possible pathotypes that can be identified with the differential set used were detected in the restricted sample (Table 5.12). The identification of the new pathotype was achieved with SSIs. There is high degree of variation for virulence found previously among French *P. brassicae* collections (Some et al. 1996), and found by others around the world (Crute et al. 1980; Linnasalmi and Toiviainen 1991; Voorrips 1995; Kuginuki et al. 1999). As

Table 5.12 Characteristics and pathotype designation of *Plasmodiophora brassicae* from France (Manzanares-Dauleux et al. 2001)

Type of isolate	Source field isolate[a] (pathotype)	Host	French district	Year of collection	Isolate	Pathotype[b]
Field		Rapeseed	Cotes d' Armor	1993	PbM	P1
		Cauliflower	Finistere	1991	Pb137	P5
			Ille-et-Vilaine	1992	K92	P2
				1992	SJ92	P5
				1993	SJ93	P1
		Chinese cabbage	Finistere	1993	K93	P1
				1994	K94	P7
				1995	K95	P7
				1996	K96	P7
SSI[c]	PbM (P1)	Rapeseed	Cotes d'Armor	1993	Ms6	P1
					M651	P1
					M727	P1
					M746	P1
					M774	P1
	K92 (P2)	Cauliflower	Finistere	1992	K92-42	P1
					K92-33	P3
					K92-18	P3
					K92-16	P4
					K92-19	P4
					K92-118	P4
					K92-8	P6
					K92-24	P7
					K92-32	P7
	Pb137 (P5)	Cauliflower	Finistere	1991	Pb137-436	P4
					Pb137-437	P4
					Pb137-514	P4
					Pb137-538	P4
					Pb137-560	P4

(continued)

Table 5.12 (continued)

Type of isolate	Source field isolate[a] (pathotype)	Host	French district	Year of collection	Isolate	Pathotype[b]
					Pb137-524	P5
					Pb137-418	P7
					Pb137-434	P7
					Pb137-439	P7
					Pb137-499	P7
					Pb137-522	P7
					Pb137-532	P7
					Pb137-543	P7
					Pb137-555	P7
					Pb137-566	P7
					Pb137-574	P7
					Pb137-576	P7
					Pb137-483	P8
	SJ92 (P5)	Cauliflower	Ille-et-Vilaine	1992	SJ92-225	P1
					SJ92-336	P1
					SJ92-256	P4
					SJ92-233	P5
					SJ92-370	P7

[a]Field isolate from which single-spore isolates (SSIs) were derived
[b]Pathotype classification according to the disease reaction of three differential *B. napus* cultivars to the *P. brassicae* isolates (Some et al. 1996)
[c]For the SSIs, host of origin and year of sampling are those of the original field isolate

pathotype identification depends on host genotype, isolates identified as a determined pathotype can be separated into more than one pathotype when more differential cultivars are used. Therefore, natural *P. brassicae* variation for virulence is probably much greater than the variation reported. The molecular markers distinguished many more genotypes than did differential hosts, each isolate having a unique molecular genotype. The RAPD and SG primers were selected to obtain as many polymorphic amplicons specific to one isolate or group of isolates as possible. It is obvious that selection of polymorphic bands leads, in contrast to the virulence study, to an over estimation of genetic dissimilarities between *P. brassicae* isolates; such amplicons would be of interest for the development of molecular epidemiological tools. Amplification of *P. brassicae* with arbitrary primers has proved to be successful for characterizing genetic diversity (Moller and Harling 1996). Accurate choice of Operon primers and standardized conditions of *P. brassicae,* DNA extraction produced simple amplification patterns that could be clearly interpreted, and with selected bands different from the host plant patterns, as with the SG primers (Buhariwalla and Mithen 1995).

Virulence and DNA polymorphisms were detected within and between field isolates, and among SSIs from different pathotypes, hosts, and geographic origins. General mean genetic dissimilarity among 46 isolates of *P. brassicae* was high compared with genetic dissimilarity found in other obligate biotrophic fungi (Delye et al. 1997; Jennings et al. 1997), and in pathogens for which no sexual stage has been observed under field conditions (Gonzalez et al. 1998; Hsiang and Mahuku 1999); comparable levels of genetic dissimilarity were observed among populations of the facultative stem canker pathogen *Leptosphaeria maculans* (Mahuku et al. 1997). The relatively high level of genetic diversity was similar among and within-field isolates, i.e. genetic distances among SSIs derived from a single-club field isolate were similar to genetic distances among field isolates from different host and geographical origins. The degree of global variability (virulence and DNA patterns) assessed in the populations of *P. brassicae* studied, and the high level of genetic diversity observed at a very small scale (among SSIs from a single-club field isolate), support the assumption that recombination plays an important role in the life cycle of the pathogen. Progress in identifying and monitoring variability in *P. brassicae* is possible by the use of molecular markers that are isolate-specific or pathotype-specific. Overall, the heterogeneous group of isolates could not be classified according to pathotype, host or geographical origin. The lack of correlation between virulence and/or DNA patterns and host origin (*B. oleracea*, *B. rapa* or *B. napus*) confirmed earlier studies showing no evidence for a pathogenic specialization of *P. brassicae* to a particular host species (Crute et al. 1983). The low correlation between virulence and RAPD patterns may be due to the high degree of DNA polymorphism among isolates within pathotypes. The evolution of the pathotype seems to be independent of the evolution of neutral DNA markers, as shown for other pathogens (Chen et al. 1993). Isolates classified by Manzanares-Dauleux et al. (2001) as pathotype P1 (virulent on all hosts) were clustered, although at low levels of similarity. Furthermore, two primers (OPL14 and OPA13) gave polymorphic patterns that matched the biological classification. A SCAR±PCR marker from the OPL14-1200 RAPD fragment differentiating the RAPD pattern of the isolates belonging to pathotype P1 has recently been developed (Manzanares-Dauleux et al. 2000).

5.2.7 Germany

Determination of pathogen variation in *P. brassicae* is based on the monitoring for virulent pathotypes against "Mendel" on demand by growers who experience clubroot infections in their "Mendel" crop. These cases are brought to the attention of field services of either the breeding company NPZ or the marketing company (Rapool-Ring) or to local institutions of the official advisory service. The identity of diseased plants was confirmed by testing for "Mendel"-specific markers, and the virulence of the local *P. brassicae* isolate towards "Mendel" was evaluated in a greenhouse assay. A new breeding material (NPZ-CR 21), and a susceptible genotype Chinese cabbage was included for comparison. In many cases, the occurrence of clubroot in a crop of "Mendel" can be attributed to infected susceptible volunteers from previous crops. Confirmed cases of virulent isolates attacking "Mendel" have been recorded regularly, however, the number of these cases – although slowly increasing – is not as high as originally anticipated for a monogenic, race-specific resistance trait. Table 5.13 provides an overview of the geographic distribution of

Table 5.13 Clubroot reactions (disease index, DI) of the resistant cultivar "Mendel," in comparison to new breeding material (NPZ CR21) and the universally susceptible host Chinese cabbage with selected *Plasmodiophora brassicae* field populations or single-spore isolates (Diederichsen et al. 2014)

P. brassicae isolate name	Origin	Population type[a]	Disease index (DI)		
			Mendel	NPZ CR21	Chinese cabbage
Leduc[b]	CA (Edmonton)	SSI	67	40	100
Wusterhusen	D (MVP)	FP	98	92	100
Dersekow	D (MVP)	FP	73	21	100
b	D (Westphalia)	FP	10	6	100
Pol4	PL (Silesia)	FP	71	16	100
k	D (SH)	FP	63	5	100
HRO1	D (MVP)	FP	27	3	100
Schwaan	D (MVP)	FP	71	31	100
N18	D (SH)	FP	67	7	100
SS2[c]	F (Brittany)	SSI	0	0	98
L1	D (SH)	FP	67	42	100
CZ1	CZ (NE Bohemia)	FP	77	24	100
Qsch	D (Saarl)	FP	39	0	100
Visschl	D (SH)	FP	3	0	99
BB1	UK (Scotland)	FP	3	0	100
WY1	UK (Scotland)	FP	0	0	100

[a]Single resting spore-derived isolate = SSI, field population = FP
[b]Provided by S. Strelkov, Univ. of Alberta, Canada
[c]Provided by G. Thomas, INRA Le Rheu, France

field isolates which are virulent against "Mendel." Most virulent isolates originated from the northeast area of Germany, close to Greifswald, Stralsund or the island of Rugen, where 18 out of 27 field isolates collected turned out to be virulent on "Mendel." Confirmed cases from Lower Saxony, Saxony, Hessia, and Bavaria had to be added to the map since 2005, while virulent races from the West German Emsland and Westphalia regions were known before the cultivar was released.

Outside of Germany, virulent isolates have been identified so far in Poland, and in a few cases in the UK. Not all field isolates from the highlighted areas are virulent for "Mendel," and based on regular reports from advisory services, the "Mendel" resistance still seems to give some protection in the field even when there are virulent pathotypes in the local population, particularly when environmental conditions are less favorable for clubroot. A study on the occurrence of clubroot in susceptible oilseed rape crops has been conducted in the federal state of Rhineland-Palatinate. U. Preiss (DLR-RLP, Bad Kreuznach) collected 453 soil samples from south-western German rapeseed crops with no known history of clubroot incidence, of which 16% tested positive for the presence of *P. brassicae* in bioassays; two of the recovered field isolates were virulent against "Mendel." There are differentiation between north and middle to southern isolates when using the set of Some et al. (1996). Differentiation based on molecular variation in different regions in northern Germany with those from the south-western region are based on AFLP-marker analysis, isolates from these two regions were separated into different clusters; however, also within the regions and between fields in the immediate vicinity, there was significant variation in molecular patterns (Diederichsen et al. 2014).

5.2.8 India

There is distinct physiological specialization in *P. brassicae* in different agro-climatic regions of India (Bhattacharya 2003). All the popular and economically important cultivars of cauliflower, cabbage, and oilseed *Brassicas* belonging to the *B. oleracea*, *B. rapa,* and *B. juncea* groups showed high disease severity measured as Per cent Disease Index (PDI). The field screening and *in vivo* growth chamber studies in Kalimpong using field isolates of rayosak showed that the *B. juncea* and *B. oleracea* groups of vegetables including Indian mustard developed high PDIs, whereas the *B. rapa* group of oilseeds and vegetable *Brassicas* are moderately susceptible hosts, including Chinese cabbage "Granaat," which is recognized as universally susceptible. There is a statistically significant difference in the disease reactions between different *Brassica* species, such as *B. juncea*, *B. oleracea,* and *B. rapa* including toria Sarson, and yellow Sarson. The results of screening of *Brassica* hosts in the plains of South Bengal revealed that *B. napus* "Tower," and "Midas"; *B. carinata* HC-1, HC-4, and HC-5, and line ACCBN-479 of the *B. nigra* group, were either resistant or highly tolerant to *P. brassicae* isolates from India (Chattopadhyay et al. 1991). All the indigenous cultivars of *B. juncea*, yellow Sarson, and toria Sarson are moderately or highly susceptible to *P. brassicae*. The variability in the virulence of *P. brassicae* populations was observed since a toria

Sarson cultivar (B-54) was resistant in Odisha State (Das et al. 1987), and highly susceptible in West Bengal State. Similarly, Benoy (B-9), a yellow Sarson cultivar, is highly susceptible in the plains of South and North Bengal but the same genotype showed tolerant reactions in the Darjeeling Hills. It indicates the existence of different pathotypes of *P. brassicae* in the different agro-climatic zones of West Bengal (Bhattacharya et al. 2014). However, standard procedure and differential sets recommended for identifying pathotypes are lacking.

5.2.9 Japan

In *P. brassicae* populations from Japan, races 1, 2, 3, 4, 6, 7, 8, and 9 have been identified by the Williams' method, with 1, 2, 3, and 4 as major races (Yoshikawa 1990, 1993). Tanaka et al. (1998) also identified races 1, 4, and 9 using the Williams' method, with race 4 as the most major race. On the other hand, widely varying races in populations were identified using the European Clubroot Differential (ECD) hosts (Buczacki et al. 1975); another report also identified races 16 and 20 using only 01–05 ($n = 10$ group) of ECD hosts, with 16 as the major race (Yoshikawa 1993). A notable problem, however, with these race differentiation systems is that they are unable to distinguish between pathogenic and nonpathogenic populations on clubroot cultivars of Chinese cabbage bred in Japan (Hatakeyama et al. 2004; Kuginuki et al. 1999; Tanaka et al. 1998; Yano et al. 1996, 1997; Yoshikawa 1990). As per Williams' proposal, race 9 populations tested are pathogenic on clubroot cultivars of Chinese cabbage, whereas Williams' race 4 populations are divided into two groups pathogenic and nonpathogenic on the cultivars (Tanaka et al. 1998; Yano et al. 1996, 1997). These differential systems involve a fundamental problem in which some differential hosts often yield intermediate and fluctuating responses (disease index) to most populations, and interpretation of the results is often ambiguous (Kuginuki et al. 1999). Therefore, Kuginuki et al. (1999) proposed a new race-differential system using two clubroot cultivars of Chinese cabbage as hosts. Because these clubroot cultivars show a clear response (resistant or susceptible) to *P. brassicae* populations from Japan, and interpretation of the data obtained is extremely simple and easy. Of these clubroot cultivars, CR Ryutoku shows population resistance identical to many other clubroot cultivars of Chinese cabbage (CR Kukai 65, CR Fukutakara, and CR Kanko), while CR Utage 70 has a unique resistance and clearly differs in resistance from CR Ryutoku (Kuginuki et al. 1999; Osaki et al. 2008b). This system classifies *P. brassicae* populations into four groups on the basis of the responses of CR Ryutoku, and CR Utage 70. Osaki (2008) and Osaki et al. (2008b) modified the system of Kuginuki et al. (1999) by using CR Kukai 65 (same response as CR Ryutoku), and CR Utage 70 as differential hosts and assigned four pathotypes (pathotypes A: pathogenic on both cultivars; B: pathogenic on CR Kukai 65 but not on CR Utage 70; C: nonpathogenic on CR Kukai 65 but pathogenic on CR Utage 70; D: nonpathogenic on both cultivars). On the basis of this system, pathotypes A, B, C, and D have been identified in 5, 5, 13, and 12 populations, respectively, from various areas of Japan (Osaki 2008; Osaki et al. 2008b). These differential systems using the

CR cultivars of Chinese cabbage are advised for practical use to directly ascertain the effectiveness of the cultivars in the fields, as these cultivars are widely grown in Japan. Chinese cabbage cv. CR Utage 70 is considered to have a very unique, and interesting genetic background (Kuginuki et al. 1999), and is an important determinant in the pathotype classification system. Unfortunately, this cultivar is not currently for sale, and cannot be obtained. Therefore, a modified differential system has been proposed using Chinese cabbage cv. Super CR Hiroki as an alternative for CR Utage 70 (Hatakeyama et al. 2004).

Variations in virulence among populations from crops have also been recorded from Japan (Tanaka and Ito 2013). The levels of virulence (aggressiveness) on cruciferous crops also often vary with the populations of *P. brassicae*. Highly virulent populations of *P. brassicae* that can severely affect a common cultivar of Chinese cabbage even at a low inoculum density (1×10^2 spores/g dry soil) are known, while weakly virulent populations that can severely affect the cultivar only at a high inoculum density (1×10^6 spores/g dry soil) have also been observed (Tanaka et al. 1990, 1991, 1997). However, the virulence on common cultivars does not always parallel that on clubroot cultivars among populations. The three populations (Williams' race 9) that are relatively a weakly virulent on a common cultivar of Chinese cabbage, can severely affect various clubroot cultivars of the crop, whereas three other populations (Williams' races 1 and 4), which were relatively highly virulent on the common cultivar, did not affect many clubroot cultivars (Tanaka et al. 1990, 1991). The levels of effectiveness of fungicides and other various control measures vary not only with the inoculum densities but also with the variations in virulence of populations (Horiuchi and Hori 1983; Tanaka et al. 1997). The efficacy of fungicides (trichlamide, flusulfamide, and fluazinam) is high against the weakly virulent population but lower against the highly virulent populations (Tanaka et al. 1997). The effect of day length on clubroot incidence also varies depending on the virulence of populations and is relatively low against highly virulent populations (Horiuchi and Hori 1983; Tanaka et al. 1997). Most CR cultivars of cabbage are considered to have partial resistance to *P. brassicae* (Voorrips 1995; Yoshikawa 1993). However, the levels of resistance of the CR cultivars of cabbage and broccoli bred in Japan also vary depending on the populations (Tanaka et al. 1998). Populations may be divided into two groups on the basis of the levels of virulence on cabbage cvs. YCR Rinen, and YCR Tae (Tanaka and Ito 2013).

5.2.10 Korea

In Korea, it has been reported that 14 of the 16 races exist based on the Williams' differential set, which consists of cabbage (*Brassica oleracea* L. var. *capitata* L. "Jersey Queen" and "Badger Shipper"), and rutabaga (*B. napobrassica* Mill. "Laurentian" and "Wilhelmsburger") (Cho et al. 2003; Williams 1966). Although, the Williams' differential set has been primarily used as a differential host set. Hatakeyama et al. (2004) suggested that two CR F_1 cultivars of Chinese cabbage, Super CR Hiroki and Ryutoky, can be used as differential hosts to determine the races of *P. brassicae* isolates rather than those of the Williams' differential set. Thus, more efficient systems are required for the identification of pathotypes of *P. brassicae* isolates, which is useful in breeding. Since clubroot disease of Chinese cabbage (*B. rapa*) was first reported at Korea in 1928, it has emerged as a significant concern in Chinese cabbage cultivation with spreading rapidly in cultivated areas (Cho et al. 2003). In order to overcome this problem, many Korean seed companies have bred clubroot-resistant cultivars of Chinese cabbage using clubroot-resistant genes identified from European fodder turnip (*B. rapa* subsp. *rapifera*), and their clubroot resistances have been evaluated using *P. brassicae* isolates that were classified by the Williams' host responses (Jo et al. 2011; Piao et al. 2009; Zhang et al. 2015). Despite considerable efforts, clubroot disease by *P. brassicae* has been observed on Chinese cabbage in Korea. To obtain a better understanding of the pathogenic diversity of the *P. brassicae* isolates Kim et al. (2016) investigated the pathotypes of Korean *P. brassicae* field isolates with various cultivars of Chinese cabbage, including 22 cultivars from Korea, China, and Japan, and 15 inbred lines. It was found that four differential hosts can distinguish four pathotype groups of *P. brassicae* field isolates in Korea, which provides a new foothold for breeding strategies for CR Chinese cabbage.

The pathotypes of 12 Korean *P. brassicae* field isolates were determined using various Chinese cabbage including 22 commercial cultivars from Korea, China, and Japan, and 15 inbred lines. All *P. brassicae* isolates exhibited the typical clubroot disease on non-clubroot-resistant cultivar, indicating that the isolates were highly pathogenic. According to the reactions on the Williams' hosts, the 12 field isolates were initially classified into five races. However, when these isolates were inoculated onto clubroot-resistant (CR) cultivars of Chinese cabbage, several isolates led to different disease responses even though the isolates have been assigned to the same race by the Williams' host responses. Based on the pathogenicity results, the 12 field isolates were reclassified into four different groups: pathotype 1 (GN1, GN2, GS, JS, and HS), 2 (DJ and KS), 3 (HN1, PC, and YC), and 4 (HN2 and SS). In addition, the CR cultivars from Korea, China, and Japan exhibited distinguishable disease responses to the *P. brassicae* isolates, suggesting that the 22 cultivars used in this study, including the non-CR cultivars, are classified into four different host groups based on their disease resistance. Combining these findings, the four differential hosts of Chinese cabbage, and four pathotype groups of *P. brassicae* might provide an efficient screening system for resistant cultivars and a new foundation of breeding strategies for CR Chinese cabbage (Table 5.14).

5.2.11 Turkey

The races of the cabbage clubroot disease (*Plasmodiophora brassicae* Wor.) in the Black Sea region of Turkey were determined by Apaydın et al. (2010). Some species belonging to cruciferous family, European Clubroot Differential set, isolates of the cabbage clubroot disease, and cabbage areas in Ordu province were used as material. According to studies by using European Clubroot Differential Set, and host rape

Table 5.14 Race determination of the 12 field isolates of *Plasmodiophora brassicae* from Korea using Williams' system (Kim et al. 2016)

Isolate ID	Location	Host	Williams' differential host[a]				Race	Reference
			JQ	BS	LR	WB		
DJ	Daejeon	Chinese cabbage	0.0 (R)[b]	0.0 (R)	0.7 (R)	0.1 (R)	5	Jo et al. (2011)
GN1	Gangneung	Chinese cabbage	0.7 (R)	0.0 (R)	3.5 (S)	2.7 (S)	9	Jo et al. (2011)
GN2	Gangneung	Chinese cabbage	3.6 (S)	0.5 (R)	4.0 (S)	4.0 (S)	1	Kim et al. (2016)
GS	Goesan	Cabbage	0.5 (R)	0.0 (R)	3.9 (S)	2.9 (S)	9	Jo et al. (2011)
HN1	Haenam	Chinese cabbage	4.0 (S)	4.0 (S)	4.0 (S)	4.0 (S)	4	Jo et al. (2011)
HN2	Haenam	Chinese cabbage	4.0 (S)	3.5 (S)	4.0 (S)	4.0 (S)	4	Jo et al. (2011)
HS	Hoengseong	Chinese cabbage	4.0 (S)	3.8 (S)	4.0 (S)	0.6 (R)	2	Jo et al. (2011)
JS	Jeongseon	Chinese cabbage	0.7 (R)	0.0 (R)	2.9 (S)	3.1 (S)	9	Jo et al. (2011)
KS	Keumsan	Chinese cabbage	0.0 (R)	0.0 (R)	0.9 (R)	0.3 (R)	5	Kim et al. (2016)
PC	Pyeongchang	Chinese cabbage	0.9 (R)	0.1 (R)	2.9 (S)	3.2 (S)	9	Jo et al. (2011)
SS	Seosan	Chinese cabbage	0.3 (R)	0.0 (R)	0.7 (R)	0.4 (R)	5	Jo et al. (2011)
YC	Yeoncheon	Chinese cabbage	3.9 (S)	3.7 (S)	3.8 (S)	0.2 (R)	2	Jo et al. (2011)

[a]*JQ* jersey queen, *BS* badger shipper, *LR* Laurentian, *WB* wilhelmsburger, *R* resistance; *S* susceptibility
[b]The numbers indicate average disease rating of 20 replications and three independent experiments showed similar results

(*Brassica napus*), it was found that there were ECD 16/31/31 race, and pathotype 1, respectively, in the Black Sea region of Turkey. Eight clubroot isolates were collected from different parts of the Kabaduz country of Ordu province in 2006. The differential set consists of 7 white cabbages, 1 black cabbage (*Brassica oleracea* var. *nigra*), 1 rape (*Brassica napus*), 1 Chinese cabbage (*Brassica campestris* subsp. *pekinensis*), and 1 cress (*Lepidium sativum*) cultivars, total 11 cultivars to test the differences among the isolates. The reaction of the cruciferous hosts against isolates was observed, and it was established that all tested materials but cress were sensitive against all tested isolates. Cress gave resistant reaction against all isolates. There was no difference in virulence among the isolates brought from Kabaduz country while using European Clubroot Differential set and host rape (*Brassica napus*) cultivars. The ECD 16/31/31 race and pathotype 1 were identified in the Black Sea region of Turkey (Apaydın et al. 2010).

5.3 Identification and Characterization of *Plasmodiophora brassicae* Pathotypes

The different isolates of *P. brassicae* cannot be distinguished by any other phenotype except virulence, and therefore, they have been identified through phytopathological tests using differential host sets (Williams 1966; Buczacki et al. 1975; Kuginuki et al. 1999). The European Clubroot Differential (ECD) set is widely accepted as a method to characterize isolates using species of the genus *Brassica* as differential hosts (Buczacki et al. 1975). However, when using species of *Brassica,* the tests need time and space in the greenhouse. Smaller crucifers, such as *Capsella bursa pastoris* (Ikegami and Naiki 1981), *Cardamine flexuosa,* and *Cardamine scutata* (Tanaka et al. 1993) or *Arabidopsis thaliana* (Koch et al. 1991; Mithen and Magrath 1992; Fuchs and Sacristan 1996) have been shown to be the host plants of *P. brassicae*, but these species have not been used as test plants. Distantly related isolates of *Plasmodiophora* were also characterized by PCR-amplification patterns using arbitrary primers (Buhariwalla et al. 1995; Moller and Harling 1996; Yano et al. 1997) and sequence-generated primers (Buhariwalla and Mithen 1995; Faggian et al. 1999). However, the number of distinguishing patterns is low, and virulence patterns, and polymerase chain reaction (PCR) patterns could not be unequivocally correlated, although a correlation between ECD classification and one pattern of a random amplified polymorphic DNA (RAPD) primer for three isolates was described (Moller and Harling 1996).

Further molecular studies by restriction fragment length polymorphism (RFLP) or amplified fragment length 8polymorphism (AFLP) have been limited by difficulties in obtaining sufficient amounts of DNA of the pathogen suitable for restriction analysis (Buhariwalla and Mithen 1995). The intracellular life cycle of the pathogen is such that DNA preparations from plasmodia were strongly contaminated by host DNA. Spores of the pathogen can be isolated free of host cell material, but contamination of host DNA was shown within the spores, and the yield of DNA was low (Bryngelsson et al. 1988; Buhariwalla and Mithen 1995). The methodological improvements have facilitated the study of closely related isolates of *P. brassicae*. Large amounts of DNA of the pathogen can be isolated with the modified CTAB-method (Cetyl-trimethyl ammonium bromide). This DNA is suitable for RFLP studies in combination with the cloned *P. brassicae*-specific DNA fragments. A small plant tester set using ecotypes of *A. thaliana* has been introduced allowing phytopathological tests under a controlled environment with large numbers of plants (Klewer et al. 2001).

Polyvinylpyrrolidone has been used for other species (Lodhi et al. 1994; Porebski et al. 1997) to eliminate polyphenols, tannins or polysaccharides in extracted DNA. The modified DNA extraction methods using either polyvinylpyrrolidone (PVP) purification during DNA extraction with CTAB or column purification after DNA extraction yielded DNA free of factors inhibiting restriction enzymes. The presented method of DNA extraction from young clubs is simpler, and resulted in a higher yield in comparison with the method starting with lyophilized spores (Ito et al. 1997) or spheroplasts (Ito et al. 1994). However, this method revealed a high contamination of host DNA. Therefore, DNA extracted from spores is necessary to increase the size of partial genomic libraries. *Plasmodiophora brassicae*-specific DNA fragments have been cloned. Comparable partial genomic libraries were established by Buhariwalla and Mithen (1995) and Ito et al. (1998). The number of clones was always small with 27, and 44 clones, respectively. Buhariwalla and Mithen (1995) selected clones containing repetitive elements by inverse Southern analysis and have used these clones to establish sequence-generated primers for *P. brassicae*. The low number of specific fragments found in the partial genomic libraries might be the result of high homology between pathogen and host. Earlier isolates were studied by PCR to avoid large-scale DNA extraction. However, PCR markers are suitable to distinguish distantly related isolates (Buhariwalla and Mithen 1995; Moller and Harling 1996; Yano et al. 1997; Faggian et al. 1999) but are less useful for more closely related isolates, such as e3 and e6, because the number of reproducible differentiating PCR markers is low. Sequence-generated primers seem to be necessary for reliable markers. However, the primer pairs RFM8-1/8-2 and RFM9-1/9-2 might be useful for screening of large populations of isolates.

Establishing a small tester set with *A. thaliana* lines, which needs less time and space than a set with hosts of the genus *Brassica* allows more plants within a test under a controlled environment. However, the *Arabidopsis*-set is a supplement but not a substitute for the ECD characterization, because it is restricted to a subpopulation of the field isolate "e." All other tested isolates produced clubs in all tested *A. thaliana* lines (Fuchs and Sacristan 1996). In contrast to isolate e6, isolate e3 shows the resistance pattern described by Fuchs and Sacristan (1996) for the original isolate "e." Furthermore, the isolates e3 and e6 revealed high genetic similarity according to molecular markers. Therefore, this set is a useful tester set to search for correlation between molecular markers, and virulence, and for recombination in a progeny of e3 and e6. RFLP fingerprints with highly repetitive elements of *P. brassicae* can characterize isolates. By this method, the number of genetic markers for the pathogen can be increased and the virulence characteristics of an isolate can be supplemented by a molecular fingerprint. Furthermore, completely retractable DNA of the pathogen is also a precondition for AFLP-studies to generate more markers for establishing a genetic map of the pathogen (Klewer et al. 2001). Identification of the geographic populations/isolates of *P. brassicae* are important to understand the distribution, infection route, and mechanism of host resistance. Therefore, development of reliable molecular markers that enable rapid identification of the pathotypes, and geographic populations is required, and for this purpose, molecular characterization of the genome is essential. Bulman et al. (2007) compared the nucleotide sequences of 24 genes in the *P. brassicae* genome with those of the corresponding cDNA and found that the genome was rich in spliceosomal introns.

Random amplified polymorphic DNA (RAPD) analysis on *P. brassicae* populations that differed in pathogenicity revealed that the genome was highly polymorphic among the populations, although correlations between the pathogenicity and RAPD patterns were not clear (Manzanares-Dauleux et al. 2001; Osaki et al. 2008b). Niwa et al. (2011) studied the complete structure of the rDNA repeat unit of

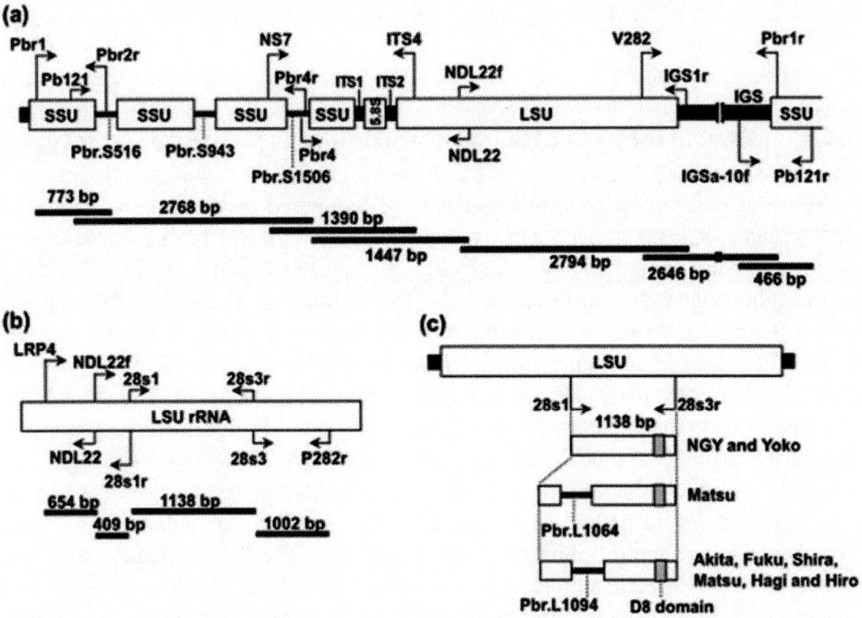

Fig. 5.2 (**a**) Structures of the ribosomal RNA gene (rDNA) transcription unit of *Plasmodiophora brassicae* population NGY, and relative locations, and directions of PCR primers used to amplify the gene. Lengths of the PCR products are indicated. The lines and open boxes indicate introns (Pbr. S516, Pbr.S943, and Pbr.S1506) and exons, respectively. *SSU* small subunit rDNA; *5.8S* 5.8S rDNA, *LSU* large subunit rDNA, *ITS* internal transcribed spacer, *IGS* intergenic spacer. (**b**) Relative locations, and directions of the primers for reverse transcription-PCR to amplify the large subunit ribosomal RNA (LSU rRNA) of population NGY. (**c**) Locations of Pbr.L1064, and Pbr.L1094 introns and divergent 8 (D8) domain in the LSU rDNA of *P. brassicae* field populations. Pbr.L1064 was found only in the Matsuzaka (Matsu) population, whereas Pbr.L1094 was found in Akita (Akita), Fukushima (Fuku), Shiraoka (Shira), Matsuzaka (Matsu), Hagi, and Hiroshima (Hiro). The NGY and Yokosuka (Yoko) populations possessed no intron in the gene. The relative locations and directions of the 28s1/28s3r primer set are indicated (Niwa et al. 2011)

P. brassicae with particular emphasis on polymorphisms among field populations of different geographic origin in Japan. The nuclear ribosomal RNA gene (rDNA) repeat unit of *P. brassicae* was determined, with particular emphasis on the structure of large subunit (LSU) rDNA, in which polymorphic regions were expected to be present. The complete rDNA complex was 9513 bp long, which included the small subunit, 5.8S and LSU rDNAs as well as the internal transcribed spacer and intergenic spacer regions. Among eight field populations collected from throughout Honshu Island, Japan, a 1.1 kbp region of the LSU rDNA, including the divergent 8 domain, exhibited intraspecific polymorphisms that reflected geographic isolation of the populations. Two new group I introns were found in this region in six out of the eight populations, and the sequences also reflected their geographic isolation (Fig. 5.2). The polymorphic region found may have potential for the development of molecular markers for discrimination of field populations/isolates of this pathogen (Niwa et al. 2011).

5.3.1 Molecular Marker for Identification of *P. brassicae* Pathotypes

Breeding of clubroot-resistant plants has been hampered by the large variation of pathogenicity in *P. brassicae* and by the lack of an efficient means for detecting specific isolates. To improve the practicality of *P. brassicae* pathotype identification, a molecular approach was developed. RAPD profiles of 37 single-spore-derived isolates belonging to seven different pathotypes were compared. A RAPD marker, OPL141200, was found in the molecular pattern of all the isolates belonging to one particular pathotype (P1), pathogenic on all differential hosts tested. The DNA band corresponding to this marker was cloned and sequenced. No significant homology to previously characterized nucleotide sequences was found. Primers were designed to specifically amplify the OPL141200 band. The SCAR marker was observed in all isolates belonging to pathotype P1 and was absent in isolates belonging to other pathotypes and in the different plant hosts analyzed. The SCAR marker was also generated from direct amplification of DNA from clubs (mixture of host and pathogen DNA) developed after infection by P1 isolates. This molecular marker may be a valuable tool for rapid and reliable identification of *P. brassicae* P1 isolates in areas where resistant varieties are cultivated (Manzanares-Dauleux et al. 2000).

To investigate genetic diversity and geographic differentiation of *P. brassicae* populations from different regions, Strehlow et al. (2014) compared three regions of Germany that differ in oilseed rape cropping history, oilseed rape acreage, and incidence of clubroot. These regions were either spatially separated or separated by the former inner German border. *Plasmodiophora* isolates were collected from 59 fields (29, 17, and 13 fields per region, respectively), and 174 amplified fragment length polymorphism (AFLP) markers were analyzed. Every field isolate showed a unique genotype pattern; that is, no genotype was shared among the regions and different fields. The mean gene diversity was 0.27, suggesting that *P. brassicae* is a genetically diverse species. The comparison of indexes (gene diversity, genotypic diversity, and linkage disequilibrium) between the regions does not support the hypotheses that cropping history, oilseed rape acreage, and incidence of clubroot affect these estimates. Principal component analysis (PCA), fixation index (FST), and generalized linear model (GLM) were suitable to specify regional differences. PCA revealed two clusters of isolates based on the geographic origin of the isolates, and FST showed that these clusters were highly differentiated. Hypotheses about association of genotypes with different spatial scales were tested with GLM: the region, reflecting the cropping history, and the individual field had a significant effect on the AFLP pattern. Strehlow et al. (2014) proposed that individual field isolates represent a discrete population and that geographic differentiation results from low levels of gene flow due to the limited dispersal of this soil-borne pathogen and from localized selection pressure as unifying force on the genotypes. By using PCR with gene-specific primers, 117 non-housekeeping *P. brassicae* genes were investigated for their presence/absence in isolates representing pathotypes 2, 3, 5, 6 or 8 (Williams' system), and in new strains of the pathogen recently identified from Alberta, Canada, which are highly virulent on clubroot-resistant canola (*Brassica napus*) cultivars. One gene, named *Cr811*, was present in isolates of pathotype 5 and in a new strain of *P. brassicae* that also is classified as pathotype 5 on the system of Williams, but causes severe clubroot on resistant canola. Three additional newly identified virulent populations, similar to pathotype 3, also were tested, and found to carry *Cr811*. In contrast, the *Cr811* gene was absent in all isolates representing the original pathotypes 2, 3, 6, and 8. Quantitative real-time PCR analysis indicated greater copy numbers of *Cr811* in the new virulent pathotype 3-like and pathotype 5-like populations versus the original pathotype 5, indicating a potential method to identify new pathotypes of *P. brassicae* before the breakdown of resistance could be detected in a canola crop. The association of *Cr811* with infection by *P. brassicae* of previously resistant canola cultivars, as revealed by qPCR analysis, suggests that *Cr811* plays an important role in the interaction between host resistance and *P. brassicae* pathogenicity (Feng et al. 2016).

5.3.2 Identification of Geographical Isolates of *P. brassicae* by Ribosomal DNA Sequence Polymorphisms

In plant disease diagnostics, the application of marker-based detection of isolates of a pathogen has become essential. Polymerase chain reaction (PCR)-marker-based detection is advantageous for accurate identification of non-culturable obligate parasites. PCR-based techniques for detection of *P. brassicae* have targeted both a single-copy sequence, and the multicopy ribosomal DNA. Ribosomal DNA (rDNA) is a nuclear DNA sequence that codes for ribosomal RNA (rRNA), which is necessary for the synthesis of proteins, cellular growth, and development of organisms. About 80% of the rRNA present in a growing cell takes part in modulating approximately half of all cellular translation. An rDNA repeat unit in eukaryotes generally consists of several coding regions including a smaller subunit, a larger subunit, a comparatively larger intergenic spacer (IGS), and two comparatively larger internal transcribed spacers, ITS1 and ITS2, on either side of a 5.8s subunit. Ribosomes translate mRNA molecules to produce proteins. Figure 5.3 reveals that the rDNA of *P. brassicae* consists of a small subunit (SSU), ITS1, 5.8s, ITS2, and a large subunit (LSU). Designing, testing, and confirming molecular markers that can quickly identify *P. brassicae* isolates collected from different geographical locations may improve efforts to halt the dissemination, and better control this pathogen. *P. brassicae* isolates with geographical variation (Swedish versus UK) had single-nucleotide polymorphisms (SNP) in the ITS, and 18s smaller subunits. Such variation can be exploited for accurate detection of a particular isolate with SNP markers. To design these markers, molecular characterization of the rDNA is essential. Niwa et al. (2011) published the complete rDNA structure of *P. brassicae* of Japanese origin (accession number AB526843; clone NGY; 9513 base pairs (bp)). They also sequenced partial rRNA sequences of the LSU from

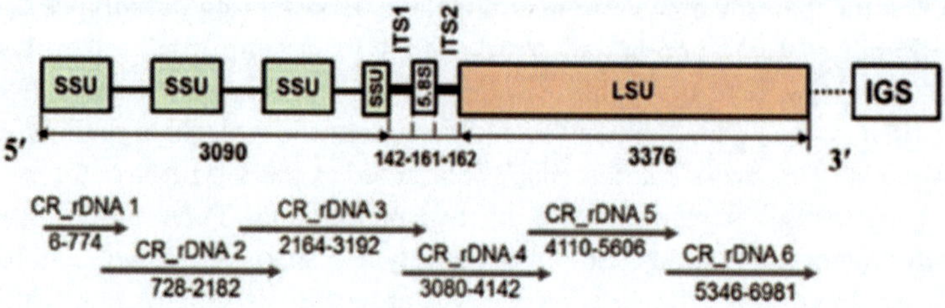

Fig. 5.3 Schematic structure of the ribosomal DNA (rDNA) of *P. brassicae*. SSU, small subunit; LSU, large subunit; ITS, internal transcribed spacer; IGS, intergenic spacer based on literature; CR DNA (from 1 to 6) indicates marker sequences developed for cloning the target regions. Numbers in the figure are lengths in base pairs (bp) of each subunit cloned and sequenced. The dotted line in-between LSU and IGS at the 3′ end indicates uncompleted sequences. Arrows from 5′ to 3′ direction indicate length of primers (Laila et al. 2017)

25 clones from eight Japanese geographical isolates (field populations); these sequences were between 1131 and 1479 bp, and were quite diverse in nucleotide sequence compared to the NGY isolate. Schwelm et al. (2016) published nearly complete rDNA sequences of two isolates of *P. brassicae* of European origin, AT (KX011115, 7909 bp) and e3 (KX0111135, 7382 bp), which were nearly identical to each other at their 28s LSU. They also published partial sequences for 28s rRNA (1143 bp) from 19 other *P. brassicae* isolates originating from Germany (KX011116), Sweden (KX011117–KX011123), New Zealand (KX011124), Canada (KX011125–KX011132), and South Korea (KX011133–KX011134). These 19 partial rDNA sequences originating from five different countries were identical to AT and e3 isolates originating from Germany and Austria, respectively (Schwelm et al. 2016). In Korea, a number of geographical isolates of *P. brassicae* have been collected (Jo et al. 2011). Laila et al. (2017) determined the complete structure of the rDNA repeat unit of *P. brassicae* and identified the polymorphisms among Korean field populations of different geographic origin in order to develop molecular markers for characterization of the isolates. Understanding *P. brassicae* genetics, and developing efficient molecular markers, is essential for effective detection of harmful races of this pathogen. Samples from 11 Korean field populations of *P. brassicae* (geographic isolates) collected from nine different locations in South Korea were used in this study. Genomic DNA was extracted from the clubroot-infected samples to sequence the ribosomal DNA. Primers and probes for *P. brassicae* were designed using a ribosomal DNA gene sequence from a Japanese strain available in GenBank (accession number AB526843; isolate NGY). The nuclear ribosomal DNA (rDNA) sequence of *P. brassicae*, comprising 6932 base pairs (bp), was cloned and sequenced and found to include the small subunits (SSUs) and a large subunit (LSU), internal transcribed spacers (ITS1 and ITS2), and a 5.8s. Sequence variation was observed in both the SSU and LSU (Fig. 5.3). Four markers showed useful differences in high-resolution melting (HRM) analysis to identify nucleotide polymorphisms including single-nucleotide polymorphisms (SNPs), oligonucleotide polymorphisms, and insertions/deletions (InDels). A

combination of three markers was able to distinguish the geographical isolates into two groups (Laila et al. 2017).

5.3.3 Race and Isolate-Specific Molecular Marker

In Korea, there are at least five races that belong to four pathotype groups. A recent study conducted in Korea attempted to develop molecular markers based on ribosomal DNA polymorphism to detect *P. brassicae* isolates, but none of those markers was either race-specific or pathotype-specific. Jeong et al. (2018) developed race and isolate-specific markers by exploiting genomic sequence variations. A total of 119 markers were developed based on unique variation exists in genomic sequences of each of the races. Only 12 markers were able to detect *P. brassicae* strains of each isolate or race. Ycheon14 markers were specific to isolates of race 2, Yeoncheon, and Hoengseong. Ycheon9 and Ycheon10 markers were specific to Yeoncheon isolate (race 2, pathotype 3), ZJ1-3, ZJ1-4, and ZJ1-5 markers were specific to Haenam2 (race 4) isolate, ZJ1-35, ZJ1-40, ZJ1-41, and ZJ1-49 markers were specific to Hoengseong isolate and ZJ1-56, and ZJ1-64 markers were specific to Pyeongchang isolate (race 4, pathotype 3). The PCR-based sequence characterized amplified region (SCAR) markers developed are able to detect five Korean isolates of *P. brassicae*. These markers can be utilized in identifying four Korean *P. brassicae* isolates from different regions.

5.3.4 Pathotype-Specific Molecular Marker

Pathotype differentiation via virulence on sets of differential hosts is time consuming, labor intensive, and can be affected by environmental conditions. Further, due to the fact that different pathotypes may be present in single galls, hence complicated results may be obtained from such tests. Development of molecular markers specific to particular pathotypes would be a valuable tool to identify and monitor *P. brassicae* populations. A random amplified polymorphic DNA (RAPD) derived sequenced characterized amplified region (SCAR) marker specific to P1 as defined on the differential of Some et al. (1996) has been developed, and proven to be useful for the identification of this pathotype from field isolates. The molecular determinants of a pathotype are most likely related to the presence/absence or differential expression of genes related to pathogenicity. In GenBank, there are approximately 120 non-housekeeping genes of *P. brassicae* represented by expressed sequence tags (EST) or predicted proteins. 118 primers pairs specific to most of these genes were used by Zhang et al. (2015) to evaluate the presence of the corresponding genes in genomic DNA samples prepared from single-spore isolates belonging to five pathotypes as classified on the differential set of Williams (1966). The members of these 118 genes may be specific to certain pathotypes. To identify gene(s) that is deferentially present in pathotypes, and to molecularly characterize the identified gene(s), and to test the usefulness of these gene(s) as markers to differentiate unknown pathotypes such markers are very useful. One gene, designated *Cr811*, was identified to be exclusively present in P5. The in planta expression of this gene was studied using an infection time course on both a resistant, and a susceptible canola genotype. Primers specific to this gene were used to identify P5 from field isolates.

Isolates of the pathogen can be classified into pathotypes according to their pathogenicity on differential hosts. The presence or absence of all database available non-housekeeping *P. brassicae* genes (118 in total) were assessed by polymerase chain reaction (PCR) analysis in isolates belonging to five *P. brassicae* pathotypes (2, 3, 5, 6, and 8 according to Williams differential set). One gene designated *Cr811* was present exclusively in the isolate of pathotype 5 (Fig. 5.4). This was further confirmed by dot blot hybridization, and by PCR using alternate DNA preparations, and primers. Reverse transcription quantitative PCR analysis indicated that in planta, expression of *cr811* was up-regulated during canola infection, especially in the stage of secondary plasmodia. Primers specific to *cr811* could distinguish a field isolate of *P. brassicae* belonging to pathotype 5 from two other field isolates representing pathotypes 3 and 8. These findings suggest that *cr811* is a gene that is potentially involved in clubroot pathogenesis, and that it also might serve as a molecular marker for differentiation of pathotype 5 from other pathotypes (Zhang et al. 2015). Differential expression of genes between pathotypes 4(P4) and pathotype 7 (P7) was assessed according to transcriptome data of molecular marker screening. Among the pathotypes (P2, P4, P7, P9, P10, and P11) tested, six genes were exclusive to P4, dividing the isolates into three types: *PBRA_003263* and *PBRA_003268* were present in all P4 isolates. *PBRA_000003/Novel 1512* were found in a type of P4 (P4-1) and Novel 137/PBRA_005772 were found in another P4 type, P4-2 Amplicons for all six genes were produced for only one isolate, which was named P4-3. The genes identified might serve as molecular markers for differentiation of P4 from other pathotypes and may also distinguish different types of P4 (Zheng et al. 2018/19).

5.4 Differential Sets

Differential sets consisting of *Brassica* species can separate pathotypes according to disease symptoms on roots of differential hosts (Table 5.1) (Hwang et al. 2012). In a differential set, Williams (1966) used four differential plants (hosts) which can identify 16 pathotypes. It was created mainly for North America, but this technique has been used worldwide. A differential set according to Buczacki et al. (1975) is called ECD set (European Clubroot Differential set), and has been used mainly in Europe. It encloses 15 differential plant species which are divided into 3 genomic subsets. The first group of 5 hosts covers the genotype of *Brassica rapa*; the second group contains the genotype of *Brassica napus*; and, the third subset includes hosts from the *Brassica oleracea* genotype. In the case of infection each host gets a special score. The pathotype designation is arranged in triplets (e.g. 16/30/24), each genotype is counted separately. This set can theoretically identify 48 pathotypes. A differential system developed by Some et al. (1996) contains just 3 differentials

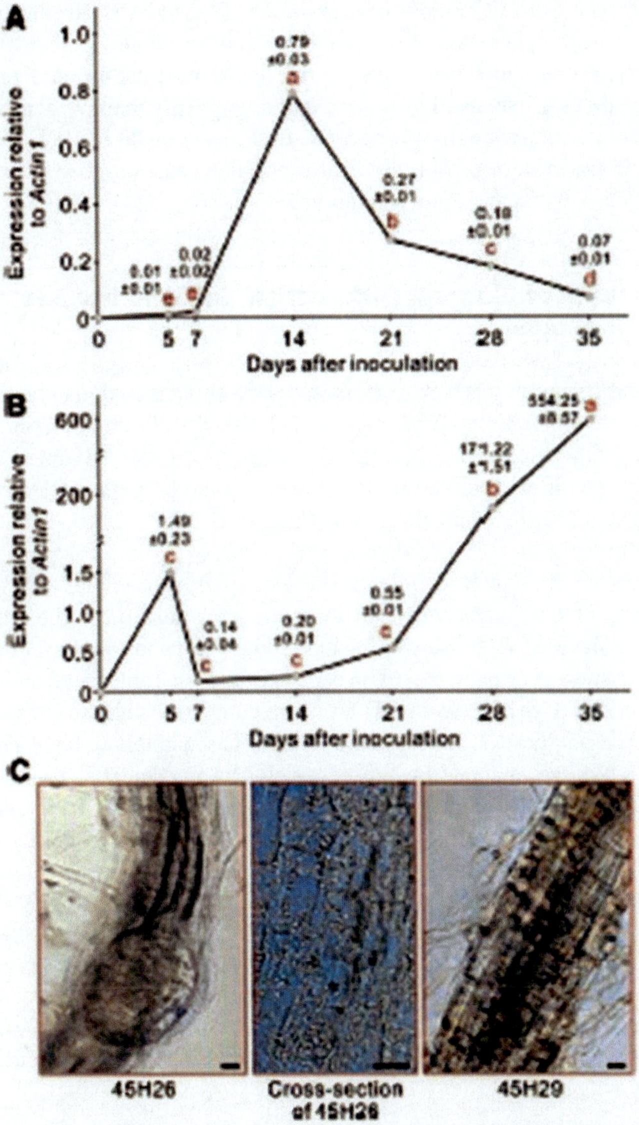

Fig. 5.4 Expression of Cr 811 during plant infection, canola cultivars (**a**) 45H26, and (**b**) 45H29 were inoculated with pathotype 5 of *P. brassicae* and the infected root samples were collected at 0–35 days after inoculation. The primers used for quantitative polymerase chain reaction analysis were ActF/ActR for the *actin 1* gene, and h/r for *Cr811*. Means ± standard deviation in the plot annotated by the same letter do not differ based on Fisher's least significant difference test at $P \leq 0.01$ ($n = 3$). (**c**) Root cortex of canola cultivars 45H26 and 45H29 at 35 days after inoculation. Bars = 10 μm (Zhang et al. 2015)

with two of them from ECD classification – ECD 06 and ECD 10. This system can distinguish 8 pathotypes. Some et al. (1996) used a single-spore isolate, which means that just one spore was used for the hosts' inoculation, and results bring information about a genetically uniform pathotype. This method demands lots of time and work, because single-spore inoculation is successful in 10% of cases only. Molecular designation of pathotypes is also possible since number of primers have been designed, which can identify pathotype.

5.4.1 Validity of Clubroot Differentials Sets and Isolates of Pathogen

All clubroot differential sets lack exact information about the number, and identity of resistance genes in the hosts. In *B. rapa* ECD hosts, there is combinations of two to three resistance genes (A, B, C) (Wit and van de Weg 1964; Toxopeus and Jansen 1975). There are different models of the genetics of the *B. napus* ECD hosts that are based on at least four race-specific genes (Crute et al. 1983; Crute 1986; Gustafsson and Falt 1986).

It is assumed that most differential hosts represent complex CR genotypes. Combinations of CR genes will lead to an underestimation of the frequency of compatible isolates of *P. brassicae* for individual CR genes, and moreover, of the degree of variation in *P. brassicae*. The perusal of data in Tables 5.15 and 5.16, race-specific reactions of the *B. rapa* ECD hosts are compared with those from *B. napus* lines that represent the *B. rapa* CR genes A, B, or C separately or have a single QTL according to markers mapped by Werner et al. (2008). The *B. rapa* hosts showed much fewer compatible interactions than did the monogenic differentials, and even when two genes were overcome individually, their respective combination in a

Table 5.15 Percent clubroot differential sets (Diederichsen et al. 2009)

Set	No. of hosts	Species represented	Known resistance genes	No. of pathotypes to be identified
ECD set (Buczacki et al. 1975)	18 (12 resistant)	5 × *B. rapa*	A, B, C in *B. rapa* hosts	48 theoretical
		5 × *B. napus*		23 realistic[a]
		5 × *B. oleracea*		
Williams (1966)	4	2 × *B. napus*	Not identified	16
		2 × *B. oleracea*		
Some et al. (1996)	3	*B. napus*	Not identified	8
Kuginuki et al. (1999)	2	*B. napus*	Not identified	4

[a]Difference between theoretical and realistic numbers is due to complex genotypes and shared R genes between different hosts

Table 5.16 Race specificity of CR in *Brassica* hosts displaying CR genes from *B. rapa* (Diederichsen et al. 2009)

Host	Postulated loci	Disease index with *P. brassicae* isolate			
		China 1	Schwaan	H1	Polen 1
B. rapa ECD hosts					
ECD-01	B+C	42	9	44	2
ECD-02	A+C	5	2	29	0
ECD-03	A+B	0	19	48	0
ECD-04	A+B+C	0	0	9	0
B. napus hosts					
55-6	A	23	17	91	40
140-4	B	82	109	96	99
102-4	C	83	0	41	100
914-8	(A03 65.3)[a]	39	0	86	100
914-48	(A08 57.8)[a]	10	8	93	8

[a]Map position according to Werner et al. (2008); susceptible reactions (DI[33]) in italic; testing procedures as in Diederichsen and Sacristan (1996)

B. rapa host still provided at least a quantitative level of CR. At least from a European perspective, the number of *P. brassicae* pathotypes seems to be limited more by the number of differential hosts rather than by a capacity for adaptation.

An important problem for race differentiation is the genetic heterogeneity present on both sides of this host–pathogen interaction. An important problem for race differentiation is the genetic heterogeneity present on both sides of this host pathogen interaction as well as due to lack of standard sets of host differentials (isogenic lines), and use of mixed population of pathogenic isolates instead of single spore isolates. Many host species, such as *B. rapa*, *B. oleracea*, and *R. sativus*, are out breeders, and display heterogenic resistance reactions in crop cultivars as well as in differential hosts. These reactions lead to speculation about the genetic identity of the differential host, and to uncertain interpretations of compatibility. Heterogeneity has to be expected even more for *P. brassicae* isolates. In accordance with most clubroot literature, the term "isolate" is used here for a heterogeneous *P. brassicae* field population in the natural state in which it was collected from the field as infected roots, and maintained subsequently in susceptible host genotypes. In Europe, such a field collection has to be regarded as heterogenic. Different pathotypes can be found even inside a root tumor club (Jones et al. 1982b). Single resting spore-derived isolates (SSI) have been generated (Buczacki 1977; Jones et al. 1982a; Schoeller and Grunewaldt 1986; Voorrips 1996; Manzanares-Dauleux et al. 2000; Fahling et al. 2004; Xue et al. 2008), and often these differ in pathogenicity in comparison to the original collection. Single-spore isolates (SSIs) are most valuable for genetic analysis. The breeding, and selection purposes, specific virulence properties are of greater importance. Clubroot disease reactions in resistance tests frequently give quantitative reactions; this can also apply to the interaction between monogenic CR hosts, and SSI. Race-specific CR genes can yield quantitative disease phenotypes, which should not be interpreted as a criterion for race-independent resistance as in other pathosystems.

5.4.2 Limitations of the Current Host Differential Sets

All the present systems have limitations with respect to which pathotype has been detected from which *Brassica* species, and what are its virulence spectrum/ pattern on *Brassica* genotypes. Earlier studies of physiological specialization in *P. brassicae* have focused on its virulence on the differentials of Williams (1966), Some et al. (1996), and the ECD set (Buczacki et al. 1975). Although these systems have been fairly effective for identifying the predominant strains of the pathogen, each has its own limitations, particularly with respect to the applicability of pathotype designations to the *P. brassicae-Brassica* pathosystem. The differential set of Williams (1966) consists of two rutabagas (*B. napus* L. var. *napobrassica*), and two cabbages (*B. oleracea* L. var. *capitata*), while the system is straight forward, and consists of a small number of hosts (thereby reducing greenhouse space requirements when conducting pathotype assessments). It was originally developed to identify strains of *P. brassicae* from cabbage and rutabaga. Therefore, a pathotype designation based on the system of Williams (1966) may not necessarily reflect virulence patterns on all *Brassicas*. For instance, most of the *P. brassicae* isolates, and populations from the Canadian prairies share an identical virulence pattern on many canola cultivars, but are classified as pathotype 3 or 5 based on the distinct reactions of the cabbage "Jersey Queen," and the rutabaga "Laurentian," which are two of Williams' differential hosts. Conversely, some pathogen populations that are classified into the same pathotype based on this system possess distinct virulence patterns on other hosts (Table 5.7 Strelkov and Hwang 2014). An additional challenge associated with the use of Williams' differential set is the occurrence of host reactions that are inconsistent, and (or) intermediate between resistant, and susceptible (Kuginuki et al. 1999; Strelkov et al. 2006a). In some cases at least, this appears to reflect heterogeneity within the differential host genotypes, which are older, open-pollinated cultivars. These intermediate and fluctuating reactions can make the pathotype classification of some *P. brassicae* populations difficult or may lead to erroneous results or different from different countries. Such results may also lead to inconsistent interpretations. Similar limitations have been observed with the differential set of Some et al. (1996). While this set consists of only three *B. napus* hosts that yield clear reactions to the pathogen, its capacity to differentiate pathotypes is fairly low, at least as assessed with Canadian populations of *P. brassicae* (Strelkov et al. 2007b; Xue et al. 2008; Cao et al. 2009). The ECD set (Buczacki et al. 1975) consists of three subsets (*B. rapa*, *B. napus,* and *B. oleracea*) of five hosts each. This enables comparisons between multiple genera, and genotypes, and can provide significant information on the virulence patterns of a particular isolate or population. The large number of hosts, however, results in a requirement for a large amount of greenhouse space when characterizing *P. brassicae* populations. It should, nevertheless, be borne in mind that where there is interest in pathogenesis in a particular host genus, then only the differentials representing that genus need to be used. This reduces the amount of work, and bench space needed by two-thirds. Importantly, the differentials in the *B. napus* subset were identified for a swede and forage rape breeding program in the 1960s at the then

Welsh Plant Breeding Station (WPBS). Moreover, not all of the hosts are effective differentials (Toxopeus et al. 1986) and the *B. rapa* subset has not been useful for distinguishing pathogen populations from Canada (Strelkov et al. 2006b). The binary notation system of nomenclature used to designate ECD pathotypes also is viewed as cumbersome and confusing by researchers resulting in limited acceptance of the ECD system despite the fact that this system allows users to deduce which differentials are resistant or susceptible to a given population of *P. brassicae* based on its ECD designation.

5.4.3 Proposed Novel Clubroot Differential Set

The limitations associated with the currently available differential sets have stimulated interest in the development of a novel system(s) to more effectively use, and identify strains of *P. brassicae* in the Canadian canola crop and others from Japan, Korea, China New Zealand, and Turkey. Any new differential set developed within this context, however, would have to meet four main criteria: (1) it must have a good differentiating capacity; (2) it must have relevance to crucifers production; (3) it must provide clear, and consistent results; and (4) it must consist of plant hosts for which the seeds are readily available. Research aimed at evaluating the suitability of various crucifer genotypes as clubroot differentials is underway, as described below.

The reactions of five lines from the Rapid Cycling *Brassica* Collection (RCBC; Wisconsin Fast Plants, Madison, WI), and 84 lines of *Arabidopsis thaliana* (L.) Heynh. were assessed against pathotypes 2, 3, 5, and 6 of the pathogen, as classified on the system of Williams' (Sharma et al. 2013) revealed that some of the lines were found to effectively differentiate these four pathotypes, but they could not detect any new virulence patterns among the limited populations tested. These may be more appropriate for the identification of strains of the pathogen from cruciferous vegetables, since they served to confirm selected pathotype designations of Williams (1966). Some of the RCBC, and *A. thaliana* lines also may have significant potential as model crops for the study of clubroot, particularly with respect to resistance reactions (Sharma et al. 2013), and thereby contribute to an improved understanding of this pathosystem.

In a different approach, 11 *B. napus* and two *B. rapa* genotypes were tested for their suitability as potential hosts in a putative Canadian clubroot differential set. These included some of the genotypes from the ECD set (Buczacki et al. 1975), the differentials of Some et al. (1996), and several commercially available Canadian cultivars of canola. The hosts from the ECD set, and Some et al. (1996) were selected because, in earlier studies, it gave clear reactions, and/or possessed good differentiating capacity against Canadian populations of the pathogen; their inclusion in a new differential system also would allow direct comparisons with data obtained previously using the older systems. Clubroot-resistant canola cultivars also were included as putative differentials because they could help to monitor changes in the virulence profiles of pathogen populations from the Canadian canola crop. This collection of hosts was tested against 11 populations, and seven single-spore isolates of *P. brassicae*, representing collections of the pathogen from canola in the Prairies, canola in Quebec, and cruciferous vegetables from Canada, the USA, and China. The putative differentials were able to distinguish five strains of *P. brassicae*, with some isolates, and populations formerly classified as pathotypes 5 and 6 on the differentials of Williams (1966) separated into distinct groups. Several of the hosts gave similar reactions, and can be represented by a single genotype, thereby reducing the number of genotypes that would need to be included in a new differential set. The data indicate that these hosts may constitute an effective foundation for the development of a Canadian clubroot differential set, aimed at a better characterization of *P. brassicae* populations. Ultimately, a coordinated approach will be needed to ensure that a robust, widely accepted system of differentials is developed. This perspective may also be desirable to explore the development of near-isogenic *B. napus* lines to differentiate strains of the pathogen (Strelkov and Hwang 2014).

5.5 Nomenclature of Pathotypes

In terminology of clubroot pathogen, the term "pathotype" of *P. brassicae* is used instead of the term "race" (Crute et al. 1980; Voorrips 1995). The term "population" is used for a collection of spores obtained from infested soil or infected plant material, which is used for the inoculation of differential set of plants (Buczacki et al. 1975). "Differential set" means the number of plant species with specific reaction to *P. brassicae* infection being used for pathotype identification (Buczacki et al. 1975). The term used as "race" of *P. brassicae* cannot be used because it designates a genetically uniform population which does not occur in *P. brassicae* populations (Parlevliet 1985). *P. brassicae* populations are combinations of several pathotypes (Diederichsen et al. 2009). More than one pathotype can even occur in one gall (Jones et al. 1982b; Fahling et al. 2003). Pathotypes can be distinguished on the basis of differential reaction of pathogen isolates on well-defined national and International differential sets. The term isolate is used for a single spore or population of pathogen collected from different geographical area to test on differential set for pathotypes identified. Numerical system has been used to name a pathotype of *P. brassicae* all over the world during early phases of investigation. However, in subsequent studies letters have been used to name pathotype.

5.6 Proposed Novel Nomenclature System of Pathotypes

There is a need to develop a sound system for designation and nomenclature of pathotypes of *P. brassicae* at International level like nomenclature of microbes as par International code of binomial botanical nomenclature. The name of a pathotype should indicate its host species/genotype along with pathotypes identity number, e.g. pathotypes BoPb1. In this system, first two letters (Bo) indicate host species as *Brassica oleracea* which recognized Pb1 pathotypes of *P. brassicae* (second two letters Pb). The consistency should be maintained in the number of pathotypes identified earlier with new ones for keeping parity and contiguity in the system. Adaptation of uniform and sound system of nomenclature of pathotypes will be useful for researchers to compare results and will also avoid unnecessary confusion to name a new pathotype as well as confirm already known or discovered earlier by others. This system can be extended up to genotypes level also as per convenience. The other details have been given at Sect. 5.10 in this chapter.

5.7 Virulence Pattern of *Plasmodiophora brassicae* Pathotypes

Recently virulence and pathotype classification of *P. brassicae* populations collected from clubroot-resistant canola (*Brassica napus*) in Canada was characterized for virulence on seven clubroot-resistant (CR) canola cultivars. Sixty-one of these populations could overcome resistance in at least one CR cultivar, and were evaluated further by inoculation on 13 *Brassica* hosts termed the Canadian Clubroot Differential (CCD) set (Table 5.17). The CCD Set included the differentials of Williams (1966), and Some et al. (1996), selected hosts of the European Clubroot Differential Set, and the *B. napus* cultivars "Brutor," "Mendel," "Westar," and "45H29." Each unique virulence pattern on the CCD Set represented a distinct pathotype, and was identified with a letter. Five reference isolates, obtained prior to the introduction of CR canola, also were assessed. A total of 17 pathotypes were detected using the CCD Set, compared with five pathotypes using the system of Williams (1966), and two with the system of Some et al. (1996), suggesting that the CCD Set has a greater differentiating capacity. Pathotype A, a variant of pathotype 3 (as per Williams) which is able to overcome the resistance in CR *B. napus*, was predominant. The original pathotype 3, which is avirulent on CR canola, was classified as CCD pathotype H (Table 5.17). An integrated strategy, combining other tools in addition to resistance, will be needed for the sustainable management of clubroot (Strelkov et al. 2018)

5.7.1 Virulence and Spread of *Plasmodiophora brassicae* (Clubroot) in Alberta, Canada

The composition of pathotypes and virulence determined from different provinces of Canada is very complex (Table 5.7) as analyzed using a single-spore isolate verses populations of *P. brassicae* (Xue et al. 2008; Strelkov and Hwang 2014). Virulence and spread of *P. brassicae* inciting clubroot was determined by Cao et al. (2009), within-field surveys suggested that the main mechanism of clubroot spread is through the movement of infested soil on farm machinery, as the frequency of infestation was highest at the field entrances, and decreased sharply at distances of 150 and 300 m from the entrance. Characterization of *P. brassicae* populations from Alberta on the differentials of the European Clubroot Differential (ECD) series, P.H. Williams (1966) and A. Some et al. (1996) revealed that pathotype ECD–/15/

Table 5.17 Pathotype classification scheme on the hosts of the Canadian Clubroot Differential Set (Strelkov et al. 2018)

Pathotype designation[a,b]																	
CCD	A	B	C	D	E	F	G	H	I	J	K	L	M	N	O	P	X
Williams	3	2	5	3	8	2	5	3	5	8	5	5	6	8	3	8	5
Some et al.	P_2	P_2	P_2	P_2	P_2	P_2	P_3	P_2	P_2	P_3	P_3	P_3	P_2	P_2	P_3	P_2	P_3
Differential host[c]									Reaction[d]								
ECD 02	−	−	−	−	−	−	−	−	−	−	−	−	−	−	−	−	−
ECD 05	+	+	+	+	+	+	+	+	+	+	+	+	+	+	+	+	+
ECD 06	+	+	+	+	+	+	−	+	+	−	−	−	+	+	−	+	−
ECD 08	+	+	+	+	+	+	+	+	+	+	−	+	+	+	+	+	+
ECD 09	+	+	+	+	+	+	−	+	+	−	−	−	+	+	+	+	−
ECD 10	−	−	−	−	−	−	−	−	−	−	−	−	−	−	−	−	−
ECD 11	−	+	−	−	−	+	−	−	−	−	−	−	−	−	−	−	−
ECD 13	+	+	−	+	−	+	−	+	−	−	−	−	+	−	−	−	−
Brutor	+	+	+	+	+	+	+	+	+	+	+	+	+	+	+	+	+
Laurentian	+	+	−	+	+	+	−	+	−	+	−	−	−	+	+	+	−
Mendel	+	+	−	−	−	−	−	−	−	−	−	−	−	−	−	+	+
Westar	+	+	+	+	+	+	+	+	+	+	+	+	+	+	+	+	+
45H29	+	+	+	+	+	−	+	−	−	+	+	−	−	−	+	+	+

[a]Pathotype designations are based on the systems of the Canadian Clubroot Differential (CCD) Set, Williams (1966), and Some et al. (1996); ECD 10, 11, and 13 are included in the differentials of Williams (1966), along with "Laurentian," while the differentials of Some et al. (1996) consist of ECD 06, "Laurentian," and "Brutor." All hosts designated ECD are also members of the European Clubroot Differential Set (Buczacki et al. 1975), but ECD pathotype designations could not be obtained since not all ECD differentials were included in the test

[b]CCD pathotypes F, H, I, M, and N correspond to the original pathotypes 2, 3, 5, 6, and 8, as defined on the differentials of Williams (1966), which cannot overcome resistance in any clubroot-resistant *B. napus* host; CCD pathotype L corresponds to the *P. brassicae* population D-G3 reported by Strelkov et al. (2016), while CCD pathotype X corresponds to the populations L-G1, L-G2, and L-G3 reported by Strelkov et al. (2016)

[c]European Clubroot Differential (ECD) 02 = *Brassica rapa* ssp. *rapifera* line AAbbCC; ECD 05 = *B. rapa* var. *pekinensis* "Granaat"; ECD 06 = *Brassica napus* "Nevin"; ECD 08 = *B. napus* "Giant Rape" selection; ECD 09 = *B. napus* New Zealand resistant rape; ECD 10 = *B. napus* var. *napobrassica*

"Wilhemsburger"; ECD 11 = *Brassica oleracea* var. *capitata* "Badger Shipper"; ECD 13 = "Jersey Queen"; Brutor = *B. napus* "Brutor"; Laur. = *B. napus* var. *napobrassica* "Laurentian"; Westar = *B. napus* "Westar"; and 45H29 = *B. napus* "45H29"

[d]A plus (+) sign denotes a susceptible host reaction, while a minus (−) sign denotes a resistant reaction. Host reactions were classified as + or − based on the index of disease (ID) that developed 6 weeks following inoculation; for a reaction to be considered resistant, the mean ID must be <50% and the 95% confidence interval must not exceed 50%

12, 3, or P2, respectively, is predominant in the central part of the province. However, novel virulence phenotypes were also observed. Two populations from southern Alberta were classified as pathotype 5 or P3 on the hosts of Williams (1966) and Some et al. (1996), respectively, or as pathotypes ECD–/6/8 and –/4/0 on the ECD set, suggesting independent introductions of *P. brassicae* to this region or greater diversity than previously thought. Pathogen populations collected from infected canola plants from Quebec and Manitoba also yielded unique virulence patterns but were still virulent on most *B. napus* hosts, and an Ontario population induced significant disease only on a few of the vegetable differentials. Virulence levels overall were highest for the *P. brassicae* populations from central Alberta, which may help to explain the rapid increase in clubroot observed (Cao et al. 2009).

5.8 Pathotype Variation Analysis Through Single Spore

Variation in virulence of single-spore isolates (SSI) of *Plasmodiophora brassicae* from Canada was determined by Xue et al. (2008). Populations of the pathogen often consist of a mixture of different pathotypes. Therefore, a simple and efficient method to isolate single resting spores of *P. brassicae* was developed, based on serial dilution of spore suspensions. The virulence of 24 single-spore isolates, representing five populations of the pathogen from Alberta, Ontario, and British Columbia, was characterized on the differentials of Williams (1966) and Some et al. (1996). Symptoms were rated 6 weeks after inoculation, and Fisher's least significant difference ($P < 0.05$) was used to differentiate resistant from susceptible host reactions. The pathotype composition of *P. brassicae* in Canada appeared more diverse when single-spore isolates were examined rather than populations of the pathogen. In Alberta, at least three, and possibly four pathotypes were identified among the 14 isolates tested, whereas a maximum of only two pathotypes had been reported previously when populations of the pathogen were examined. Pathotype 3 or P2, as classified on the differentials of Williams (1966), and Some et al. (1996), respectively, was found to be predominant in the region. The occurrence of other pathotypes at lower frequencies suggests that caution should be used in any breeding strategy, because rare pathotypes of *P. brassicae* may quickly become predominant if susceptible host genotypes are continuously grown (Table 5.18). The intermediate and fluctuating results of certain differential hosts following inoculation with the populations of *P. brassicae* indicates importance of single-spore isolates for virulence analysis. This phenomenon, which was also reported by others, may result from heterogeneity in the pathogen populations or from genetic heterogeneity in the differential hosts themselves. The use of single-spore-derived isolates of *P. brassicae* has been suggested as a way to more accurately assess virulence in the pathogen, and a number of agarose-based methods for the isolation of single resting spores have been developed. However, these agarose-based techniques are generally time consuming, and inoculation efficiencies can be very low.

Table 5.18 Reactions of six *Brassica* differential hosts in response to inoculation with single-spore-derived isolates of *Plasmodiophora brassicae* from Canada[a] (Xue et al. 2008)

Single-spore isolate	*Brassica napus* cvs.				*B. oleracea* cvs.		Pathotype[b]
	Brutor	Laurentian	Nevin	Wilhelms burger	Badger Shipper	Jersey Queen	
SACAN-ss1	+	+	+	−	−	+	$3/P_2$
SACAN-ss2	+	+	+	−	−	+	$3/P_2$
SACAN-ss3	+	+	+	−	−	+	$2/P_2$[c]
SACAN-ss4	+	+	+	−	−	+	$3/P_2$
SACAN-ss5	+	+	+	−	−	+	$3/P_2$
AbotJE-ss1	+	−	−	−	−	+	$6/P_3$
AbotJE-ss2	+	−	−	−	−	+	$6/P_3$
AbotJE-ss3	+	−	−	−	−	+	$6/P_3$
AbotJE-ss4	+	−	−	−	−	+	$6/P_3$
AbotJE-ss5	+	−	−	−	−	+	$6/P_3$
CDCN-ss1	−	+	+	−	−	−	$8/P_6$
CDCN-ss2	+	+	+	−	−	+	$3/P_2$
CDCN-ss3	+	+	+	−	−	−	$8/P_2$
CDCN-ss4	+	+	+	−	−	+	$3/P_2$
CDCN-ss5	+	+	+	−	−	+	$3/P_2$
Leduc-ss1	+	+	+	−	−	+	$3/P_2$
Leduc-ss2	+	−	−	−	−	+	$6/P_3$
Leduc-ss4	+	+	+	−	−	+	$3/P_2$
Leduc-ss5	+	+	+	−	−	+	$3/P_2$
ORCA-ss1	+	+	+	−	−	+	$3/P_2$
ORCA-ss2	+	+	+	−	−	−	$8/P_2$
ORCA-ss3	+	−	[illegible]	[illegible]	[illegible]	−	$5/P_3$
ORCA-ss4	+	−	−	−	−	−	$5/P_3$
ORCA-ss5	+	+	+	−	−	+	$3/P_2$

[a]Susceptible (compatible) and resistant (incompatible) host reactions are indicated by + and −, respectively. Least significant difference (LSD; $P < 0.05$) was used to differentiate resistant from susceptible reactions; genotypes developing disease indices (IDs) that were not significantly different from the genotype with the lowest ID in the same treatment were regarded as resistant, and all others were regarded as susceptible

[b]As determined on the differential hosts of Williams (1966) and Some et al. (1996)

[c]SACAN-ss3 was classified as pathotype 2 on the differentials of Williams based on LSD groupings, although analysis of variance indicated no significant interactions between the differential hosts and SACAN-ss1 to SACAN-ss5

5.9 Relationship Between Pathogenic and Genetic Variation in *Plasmodiophora brassicae*

Recently, molecular markers have been used to analyze the relationship between genetic variation and pathogenicity or host preference in populations and single-spore isolates of *P. brassicae* (Buhariwalla et al. 1995; Fahling et al. 2003; Ito et al. 1997, 1998; Manzanares-Dauleux et al. 2001; Moller and Harling 1996; Osaki et al. 2008b; Yano et al. 1997). These findings are contributing to understanding the genetic background involved in the pathogenic variation of *P. brassicae* populations. In Japan, Yano et al. (1996, 1997) have shown that two populations (Williams' race 9) are clearly distinguishable from 14 other populations (Williams' races 1, and 4) by random amplified polymorphic DNAs (RAPDs), and the result corresponds well with results using isozyme patterns in the same populations. However, their methods could not distinguish between pathogenic and nonpathogenic populations on clubroot resistance cultivars of Chinese cabbage bred in Japan. Thereafter, Osaki et al. (2008b) analyzed the relationship between RAPDs, and pathogenic variation (pathotypes determined using clubroot resistance cultivars of Chinese cabbage, and Williams' races) among additional populations, including those from the weed, *C. flexuosa*, from Japan. The cluster analysis based on RAPD data has indicated that three populations (Williams' race 9) are clearly distinguishable from other populations (Williams' races 1 and 4), as described by Yano et al. (1997) and that partial correspondence exists between pathotypes and genetic variations in populations as follows: all pathotype B and D populations were distinguishable from all pathotype C populations except one (Fig. 5.5). However, their method could not distinguish pathotype B populations from those of pathotype D or pathotype A populations from other locations.

Pathotype A populations are extremely heterogenic, and are presumed to have arisen through plural evolutionary processes (Osaki et al. 2008b). Genetic variation among populations of *P. brassicae* from Japan is relatively high, but the correlation between pathogenic, and genetic variations in these populations is not very high (Yano et al. 1997; Osaki et al. 2008b). Heterogeneity in each population may have affected the results of the authors' analyses of pathogenicity and DNA polymorphisms. When Manzanares-Dauleux et al. (2001) used single-spore isolates for similar analyses, the correlation between pathogenic, and genetic variations was relatively low, even among these isolates. Specialization of races and pathotypes seems to reflect only a limited aspect within the high level of genetic diversity in this organism. In a dendrogram based on RAPD analysis, *P. brassicae* populations from *C. flexuosa* grouped into a single sub-cluster, and were located in a single cluster with all Williams' race 9 populations from crops (Osaki et al. 2008b, Fig. 5.5). The results indicate that these two groups of populations are genetically closely related. The race 9 populations are minor within *P. brassicae* from Japan, and have unique pathogenicity (avirulent on many common cultivars of cabbage, and have relatively lower virulence than race 4 populations on common cultivars of Chinese cabbage, but are virulent on many clubroot resistance cultivars of Chinese cabbage) (Tanaka et al. 1991, 1997, 1998). The significance of the genetically close relationship

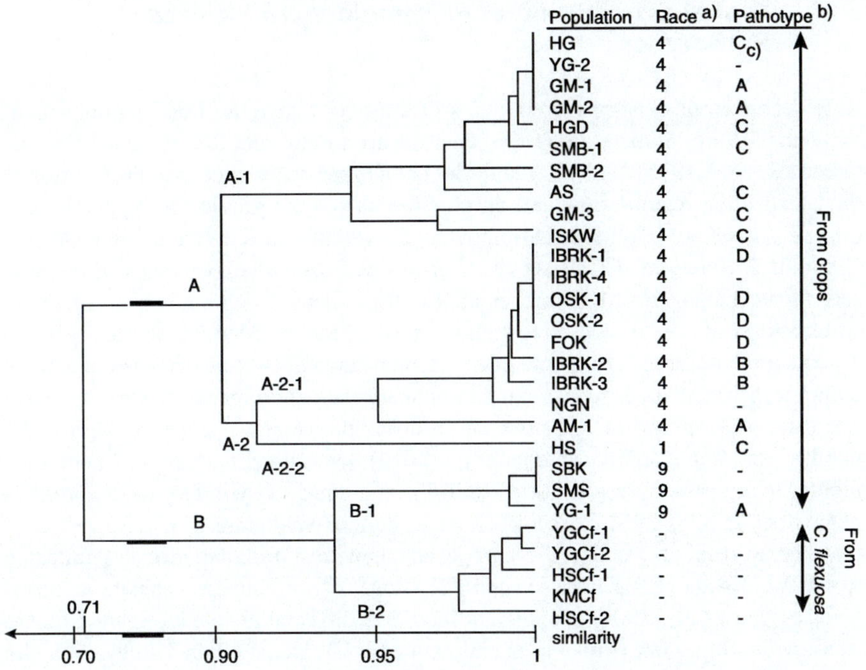

Fig. 5.5 Dendrogram based on RAPD data for 28 populations of *Plasmodiophora brassicae* from cruciferous crops, and the wild crucifer *Cardamine flexuosa*. The dendrogram was constructed by the unweighted paired group method with arithmetic averaging from the matrix of coefficients of simple matching similarity. a Williams' race, b pathotype determined on the basis of pathogenicity on two clubroot-resistant cultivars of Chinese cabbage, CR Kukai 65, and CR Utage 70, pathotype A pathogenic on both cultivars, pathotype B pathogenic on CR Kukai 65 but nonpathogenic on CR Utage 70, pathotype C nonpathogenic on CR Kukai 65 but pathogenic on CR Utage 70, pathotype D nonpathogenic on both cultivars, c not tested (Tanaka and Ito 2013)

between the two groups of populations from crops, and *C. flexuosa* is not clear, however, available results may point to an evolutionary process of the pathogen in Japan. *Cardamine flexuosa* is thought to have originated in Eurasia, and become naturalized in Japan following the introduction of rice cultivation (Takematsu and Ichizen 1993). On the other hand, clubroot on cruciferous crops was first recognized at the end of the nineteenth century in Japan (Ikegami 1992), so clubroot on *C. flexuosa* may precede clubroot on cruciferous crops. *Plasmodiophora brassicae*, an obligate parasite, may thus have a long history of co-evolution with *C. flexuosa* in Japan. Because of its wide distribution in Japan (Tanaka et al. 1993, 2006), clubroot of *C. flexuosa* may be more common than the disease on crops in Japan and may be one source of the pathogenic genotypes on cruciferous crops. Additional studies are thus necessary for a thorough understanding of the origin of *P. brassicae* and the genetic, and epidemiological relationship between the populations from *C. flexuosa*, and cruciferous crops in Japan (Tanaka and Ito 2013).

5.10 Global Distribution of *Plasmodiophora brassicae* Pathotypes

Researchers from most of the countries of the world have used differential sets of Williams (1966), Some et al. (1996), and European clubroot differential (ECD) set (Buczacki et al. 1975) to identify clubroot pathotypes occurrence, and distribution in their respective region/areas, and on cruciferous species grown for vegetable, and oilseed *Brassicas*. However, Strelkov et al. (2018) used a modified Canadian Clubroot Differential (CCD) set of 13 *Brassica* hosts which includes differential sets of Williams (1966), Some et al. (1996), selected eight hosts of ECD set (Buczacki et al. 1975), and three genotypes of *B. napus* (Mendel, Wester, 45H29) as explained in Table 5.19. Some other attempts have been made to develop a novel clubroot differential set which can be applicable widely with fair system to detect spectrum of virulence of *P. brassicae* on different species of crucifers with wider validity, and applicability. Sharma et al. (2013) determined reaction of lines of the rapid cycling *Brassica* collection (RCBC), and lines of *Arabidopsis thaliana* to pathotypes 2, 3, 5, and 6 classified on the system of Williams.

In Japan Kuginuki et al. (1999) proposed a new race differentiation system using two CR cultivars (CR Kukai 65, and CR Utage 70) of Chinese cabbage as hosts. Hatakeyama et al. (2004) proposed a modified differential system using Chinese cabbage cv. Super CR Hiroki as an alternative for CR Utage 70. In Turkey, Apaydın et al. (2010) used 11 hosts of cruciferous species to identify pathotypes of *P. brassicae*. A new race designation system for clubroot with four clubroot races and differentials from several *B. oleracea* species was developed to enable a relevant and reproducible description of clubroot-resistant *B oleracea* varieties (Table 5.20; International Seed Federation 2019). However, so far no sound system for identification of pathotypes of *P. brassicae* has been developed at International level which can detect virulence on different species of crucifers. There is a need to develop an ideal differential set with isogenic lines of crucifers for each species separately so that right spectrum of virulence on each species can be identified with number of pathotypes occurring in a country. Global distribution of *P. brassicae* is given in Table 5.21. It is clear from the Table that similar virulence is distributed all over the world. It indicates limitation of differential sets used in each country (same number of differential) which could not detect any new pathotypes. Addition of others (more) number of genotypes could have detected new or more number of pathotypes/virulence. The other limitation is in most of the cases instead of single spore (resting spore), populations of isolates have been used consisting of mixture of resting spores which may be unable to detect new pathotype. Single resting spore may also consist of more than one race.

No uniform system has been followed to designate or name in nomenclature of pathotypes. The pathotypes have been designated as serial numbers (1, 2, 3. . .n), as triplicates (16/2/15, 16/22/30. . .n), and as letters (P1, P2, P5,. . .n); (A, B, C, . . .X. . .n). It does not indicate which pathotype has been detected by which host genotype. The nomenclature of pathotype should indicate the host species which have been recognized by a specific pathotype. It is proposed that a uniform

Table 5.19 Widely used differential sets[a] to identify pathotypes of *P. brassicae*

Brassica species/varieties constituting differential set			
B. napus var. *napobrassicae*[b]	*B. rapa* ssp. *rapifera*[c]	*B. napus* var. *napus* brutor[d]	ECD 02 AAbbCC[e]
Laurentian	Polish rape	Nevin	ECD 05 Granaat
B. napus var. *rapifera*	Line aaBBCC ECD 01	*B. napus* var. *rapifera*	ECD 06 Nevin
Wilhelmobuger	Line AAbbCC ECD 02	Wilhelmsburger	ECD 08 Giant rape selection
B. oleracea var. *capitata*	Line AABBcc ECD 03		ECD 09 New Zealand resistant rape
Badger Shipper	Line AABBCC ECD 04		ECD 10 Wilhemsburger
Jersey Queen	*B. rapa* var. *pekinensis* Chinese cabbage cv. Granaat ECD 05		ECD 11 Badger Shipper
	B. napus var. *napus*		ECD 13 Jersey Queen
	Fodder rape DC 101 Nevin ECD 06		Bruter
	DC119 Giant Rape ECD 07		Laurentian
	DC 128 selection from DC 119 ECD 08		Mendel
	New Zealand resistant rape ECD 09		Wester
	B. napus var. *rapifera* DC 130, Wilhelmsburger ECD 10		45H29
	B. oleracea var. *capitata* Cabbage, Badger Shipper ECD 11		
	Bindasachsener ECD 12		
	Jersey Queen, ECD 13		
	Septa ECD 14		
	Verheul (Kale) ECD 15		

[a]Number of differential in each set 1 = 4, 2 = 15, 3 = 3, and 4 = 13
[b]Differential set of Williams (1966)
[c]European clubroot differential (ECD) set (Buczacki et al. (1975)
[d]Differential set of Some et al. (1996)
[e]Canadian clubroot differential (CCD) set of 13 *Brassica* hosts include differentials of Williams (1966), Some et al. (1996), selected hosts of ECD set (Buczacki et al. 1975), and three genotypes of *B. napus* (Strelkov et al. 2018)

nomenclature should be followed which can indicate from which species it has been identified. The pathotypes may be designated as BoPb1, BoPb2… from *B. oleracea*, BrPb1, BrPb2… from *B. rapa*, BnPb1, BnPb2, …from *B. napus,* AtPb1, AtPb2 from *A. thaliana,* and so on. In this scheme, first two letters indicate host species, and second two letters indicate pathotypes number. It is more benign and expedient for

Table 5.20 Identification of *Plasmodiophora brassicae* races using *Brassica oleracea* varieties (International Seed Federation 2019)

Differential hosts as *B. oleracea* varieties		Pathotypes/races			
		Pb: 0	Pb: 1	Pb: 2	Pb: 3
Bartolo[a]	White cabbage	S	S	S	S
Lodero[a]	Red cabbage	HR	HR	HR	S
051632[a]	White cabbage	HR	S	S	HR
Kilaton[a]	White cabbage	HR	HR	S	HR
Monclano[a]	Broccoli	HR	HR	S	HR
Clapton[a]	Cauliflower	HR	HR	S	HR
Crispus[a]	Brussel sprouts	HR	HR	S	HR

S-susceptible, HR-highly resistant
[a]Differential hosts and isolates that are used by seed sector

the nomenclature of pathotypes of *P. brassicae*. Although binary notation system of nomenclature used to designate ECD pathotypes allows users to deduce which differentials are resistant or susceptible to a given isolate of *P. brassicae* but it is viewed as cumbersome, and confusing. Recently, race and isolate-specific molecular markers have been developed through genome realignment which enables detection of *P. brassicae* isolates (Jeong et al. 2018). However, genomic characterization of pathotypes has not been found to correlate with pathogenesis.

5.11 Analysis of *Plasmodiophora brassicae* Pathotypes Genome

To obtain insights into the biology of *P. brassicae*, Rolfe et al. (2016) have sequenced, using a whole genome shotgun approach, the genomes of pathotypes 3 and 6 that differ in their host range and virulence, and re-sequenced pathotypes 2, 5, and 8. Comparisons were made among these novel genome sequences, and also with the recently published data from a European pathotype, e3 (Schwelm et al. 2015). The combined data with a transcriptomic approach in two host species, *B. napus* and *A. thaliana* allowed gaining insights into the processes that underline pathogenicity, the manipulation of host development and mechanisms of nutrient acquisition by this pathogen.

5.11.1 Genome Sequencing and Assembly

Plasmodiophora brassicae pathotype 3 (Pb3) is the predominant strain of the clubroot pathogen in the province of Alberta in western Canada, and is highly virulent on *B. napus* but also infects other *Brassica* species. By contrast, pathotype 6 (Pb6) is the predominant strain of *P. brassicae* in British Columbia and Ontario, and has a host range limited to vegetable Brassicas.

To gain insight into the genome of *P. brassicae* and genome diversity of various pathotypes, Rolfe et al. (2016) generated *de novo* assembled genomes of Pb3 and

Table 5.21 Global distribution of *Plasmodiophora brassicae* pathotypes. The differential system used was: Williams (1966), Some et al. (1996), ECD set (Buczacki et al. 1975), and Canadian clubroot differential set (Strelkov et al. 2018)

Countries	Pathotypes	References
Australia	3, 6, 7	Williams (1966)
Canada	1, 2, 3, 5, 6, 8	Strelkov and Hwang (2014)
	16/0/14,16/2/12	Strelkov et al. (2006b, 2007b)
	16/15/0, 16/15/12, –/15/12	
	P2, P3	Xue et al. (2008)
	A, B, C, D, E, F, G, H, I, J, K, L, M, N, O, P, X	Strelkov et al. (2018)
China	1, 2, 4, 7, 9, 10, 11, 13	Chai et al. (2014)
Czech Republic	1, 2, 4, 6, 7, 10	Ricarova et al. (2016)
	16/2/15, 16/2/14, 16/10/4, 16/14/13	
	16/14/15, 16/14/31, 16/18/15, 16/14/12	
	16/31/8, 16/15/15, 16/14/4, 16/14/14	
	16/15/31, 16/31/12, 16/22/31, 16/22/30	
	16/31/15, 16/3/15	
	P1, P2, P3, P4, P5	
Finland	1, 2, 3, 4, 6, 7	Linnasalmi and Toiviainen (1991)
	16/31/12, 16/3/31, 16/19/31, 16/18/30	
	16/31/31, 16/00/30, 16/00/31/, 16/02/31	
	16/03/31	
France	P1, P2, P3, P4, P5, P6, P7, P8	Manzanares-Dauleux et al. (2001)
GDR	1, 3, 4, 6, 7, 9	Williams and Seidel (1968)
Iceland	7	Linnasalmi and Weisaeth (1978)
India	Pathotype present	Bhattacharya et al. (2014)
Japan	1, 2, 3, 4, 5, 6, 7, 8, 9, 16, 20	Tanaka and Ito (2013)
	A,B,C,D	Osaki (2008), Osaki et al. (2008a, b)
Korea	1, 2, 3, 4, 5, 9	Kim et al. (2016)
New Zealand	1, 2, 4, 5, 6, 7, A, B, C, D, E, F, G, H	Williams (1966), Lammerink (1965, 1986)
Norway	1, 3, 4, 5, 6, 7, 9	Linnasalmi and Weisaeth (1978)
Poland	2, 3, 4, 6, 7, 9, 10, 12, 16	Ricarova et al. (2016)
	17/31/31, 16/7/31, 16/31/8, 16/31/31	
	16/14/30, 16/14/31, 16/31/14, 16/6/31	
	16/14/15, 16/7/28, 16/31/29, 16/14/12	
	16/2/14, P1, P3	
Sweden	1, 2, 3, 4, 6, 7, 15	Jonsson (1981)
Turkey	1, 16/31/31	Apaydın et al. (2010)

(continued)

Table 5.21 (continued)

Countries	Pathotypes	References
UK	1	Williams (1966)
USA	6, 7, 16/02/31, 16/03/31	Williams and Walker (1963), Dobson et al. (1983)
USSR	1, 2, 6, 7	Krivchenko et al. (1982)

Pb6 and re-sequenced three additional pathotypes: Pb2, Pb5, and Pb8. Pb3 is the most virulent strain of these pathotypes. Single-spore isolates of Pb3, and Pb6 were propagated in the roots of their respective hosts, *B. napus* cv. DH12075 and *B. rapa* var. *pekinensis* cv. Granaat. DNA from partially purified spores was used for whole genome shotgun sequencing using a Roche 454 GS FLX Titanium platform. In addition, an 8 kb mate-paired library was sequenced for each pathotype (using a Roche GS-FLX+ platform). After removal of the host plant DNA sequences, the remaining sequence reads were assembled into 109 scaffolds for Pb3 and 356 scaffolds for Pb6. Two Pb3 scaffolds [5 (2.522 kbp) and 38 (4.049 kbp)] were considered as potential host contamination based on 99% identity to *Brassica* ribosomal RNA sequence (NCBI blastn search, E value 0) and were excluded. The size of the assembled genome for both pathotypes was similar and approximately 24.2 Mbp, which is close to pre-existing estimates for the genome size for *P. brassicae* (Siemens et al. 2009), and to the genome sequence of *P. brassicae* isolate e3 reported by Schwelm et al. (2015). The estimated repeat content for Pb3 and Pb6 was less than 2%, characteristic of a small genome with few repetitive elements. The average GC content of the genome was 59% for both Pb3 and Pb6.

The total number of predicted genes for Pb3 and Pb6 was 10,851 and 10,070, respectively. Analysis of the single-copy Core Eukaryotic Genes (CEGs) via CEGMA revealed that the Pb3 genome contained 92% complete CEGs and 93% as partial sequences. The assembly of Pb6 yielded lower results, with 85% of the CEGs being identified as complete and 86% as partial. Close examination of predicted genes and RNA sequences (from infected *B. napus* and *A. thaliana* tissue sampled at different time points after infection) mapped to the Pb3 genome revealed that the majority of the genes were located in close proximity to each other along the compact genome of *P. brassicae*. The intergenic distance for the majority of the genes was less than 800 bp. Analysis of predicted ORFs supported by RNASeq data revealed highly reduced intergenic regions with overlapping UTRs indicating shared regulatory elements for at least some of the *P. brassicae* genes.

The overall features reported for the Pb3 genome are similar to the published genome of *P. brassicae* isolate e3 by Schwelm et al. (2015). However, the genome assembly of Pb3 was less fragmented than that of e3 (107 against 165 scaffolds) and a higher number of genes were predicted in Pb3 (10,851) than in e3 (9730). A comparison of the two proteomes identified 1070 Pb3 predicted proteins that did not have an orthologue among those reported for e3. A search of Pb3 RNA sequences expressed at various stages of infection supported the presence of transcripts for 943 of these 1070 predicted genes. In addition, only five of the 1070 Pb3 genes did not have a match in the e3 genome sequence, confirming that almost all of the additional 1070 genes were present as un-annotated sequences in the e3 genome assembly. Due to the importance of Pb3 as a highly prevalent and virulent pathotype infecting various *Brassica* crops, and also because of a more complete genome assembly (compared with Pb6 and e3), Pb3 was selected as the reference genome for further analysis.

5.11.2 Genome Comparison of *P. brassicae* Pathotypes

Alignment of the Pb3 and Pb6 genomes revealed a close to perfect synteny with rare examples of inversion of small segments of the Pb6 genome. The majority of predicted Pb6 genes had an orthologue in the Pb3 genome except for 14 genes that were unique to Pb6, with six of these having no homologue in the NCBI non-redundant (nr) protein database and eight that matched hypothetical proteins (E value <0.001). To gain insight into genome variation and its correlation with host specificity, Rolfe et al. (2016) re-sequenced the genomes of Pb2, Pb5, and Pb8. A total of 173, 134 SNPs were discovered by alignment of the genome sequence of Pb6 and the re-sequenced pathotypes against the Pb3 reference. Paired-end Illumina reads generated for Pb3 were also aligned to the Pb3 genome. After filtering (on allele frequency, SNP positions with heterozygous genotypes and indels) 132, 429 SNPs remained and were clustered into 90 haplotype blocks. One haplotype block was found to result from a single SNP between the reference genome and Pb3 Illumina reads; investigation of this SNP indicated that it resulted from a sequencing error. Out of the remaining 89 haplotype blocks, 86.5% contained more than one SNP. Comparison of whole genome SNP profiles revealed that Pb2, 3, 5, 8 were highly similar and clustered as one group, while the SNP profile of Pb6 was more divergent. The haplotypes correlate with the host range and geographical distribution of these pathotypes, with Pb2, 3, 5, 8 being virulent on *B. napus* and present in the prairie provinces and Pb6 failing to cause disease on *B. napus* and being prevalent in British Columbia and Eastern provinces of Canada. Without access to read data, analyses of the e3 genome could not be completed as with the other pathotypes. However, by direct sequence comparison 48,578 out of the identified 132, 429 SNPs were found to differentiate the genome of e3 reported by Schwelm et al. (2015) and Pb3. These SNPs were grouped into 58 haplotypes and showed that e3 clustered with Pb6.

5.11.3 The *Plasmodiophora brassicae* Pathotype 3 Proteome

To annotate and assign GO functions to the predicted genes, predicted Pb3 proteins were searched for homology to entries in the NCBI nr, GO and InterPro databases using Blast2GO-PRO (Conesa et al. 2005). Out of the 10,851 total predicted proteins for Pb3, 3460 did not have hits in the NCBI database (E value <10 − 3). An Ortho-MCL search of protein families identified 341 paralogous groups in Pb3 with the largest groups having up to 37 paralogues. A Blast search of the NCBI nr database identified matches for only 51 of the paralogous groups. A search for orthologous proteins identified *P. ramorum* as the organism with the highest number of orthologues matching Pb3 proteins (755 Pb3 proteins). Among these were genes involved in protein and nucleic acid binding, ATP and metal ion binding, oxidoreductases, kinases and ATPase activity. Rolfe et al. (2016) searched the genome of Pb3 for potential secreted proteins, which might act as effectors. The total number of predicted secreted proteins was 590, with 431 of these proteins having a probability score (D score) of greater than 0.7. The 221 proteins from this list were defined as small (less than 300 aa) secreted proteins (SSPs). Nearly 60% of the Pb3 SSPs (132 out of 221) had no homology with any protein in the NCBI nr database. Sixty-three of the 590 Pb3 SSPs did not have any homologue among the e3 predicted proteins (E value $<10-3$). A blastn search confirmed the presence of all but one (PbPT3Sc00048_Am_3.106_1) of these genes in the e3 genome. Comparing Pb3 SSPs with those predicted for Pb6 revealed that 119 of the Pb3 SSPs did not have a homologue in the Pb6 genome. Further study is required to validate the uniqueness of these 119 SPs for Pb3 and to reveal their possible role in determining host specificity. Rolfe et al. (2016) searched the SSPs of Pb3 for the presence of RXLR motifs. RXLR motifs identified in the effectors of oomycete pathogens are believed to function in translocation of effector proteins into the plant host cells. RXLR type effectors are highly prevalent in *Phytophthora* and downy mildew species. A potential RXLR motif was found in only 13 of the 221 predicted SSPs indicating that this motif is not prevalent among the effectors of *P. brassicae*. The clubroot pathogen is evolutionary distinct from oomycetes and the lack or underrepresentation of RXLR like motifs might be expected; however, this pathogen resides within the host cell and in this regard it resembles malaria parasites (plasmodium species). Malaria parasites are also evolutionary distinct from oomycetes but have adapted a similar mechanism for translocation of effector proteins (Sperschneider et al. 2015). Therefore, it is possible that a similar mechanism for translocation of virulent proteins operates in *P. brassicae*.

5.12 Intraspecific Biodiversity of *Plasmodiophora brassicae*

According to Schwelm et al. (2016), there is no genetic diversity in the LSU sequence of *P. brassicae* in isolates from around the world. Therefore, the large subunit (LSU) is not suitable to distinguish between different isolates or pathotypes of *P. brassicae*. Specific fluorescence in-situ hybridization (FISH) analyses confirmed that the LSU sequences belong to the causal agent of clubroot disease, *P. brassicae*. The phylogenetic analyses performed on different parts of the LSU show that downstream of the binding site of primers NDL22/NDL22F, the sequences of the NGY-type (AB525843) describing differences in the "*P. brassicae*" LSU sequences most likely originated from a glissomonad *Cercozoa*. The lack of overlap with the sequence, they had obtained upstream of that binding site prevented them from identifying the different biological origin of these sequences. The specific primers designed by Niwa et al. (2011) for their diversity study are situated in the downstream region of that point, and as a result instead of amplifying the LSU of *P. brassicae* isolates, the LSU of glissomonads that were present in the rhizosphere of the clubroot-infected plants was amplified. Glissomonad *Cercozoa* are highly abundant, and diverse organisms in soil (Howe et al. 2009), it is therefore not surprising that they could be found in most/all rhizosphere samples. Glissomonads are also like phytomyxids known to have introns in their ribosomal gene clusters (Howe et al. 2009). The presence of an intron in the downstream fragment obtained by Niwa et al. (2011) was therefore in concordance with a plasmodiophorid origin of the sequence. Schwelm et al. (2016) found that one long intron in the downstream sequence of *P. brassicae* that made amplification, and sequencing of that fragment difficult, but not in the same position as that present in the glissomonad sequence. At the time when the study of Niwa et al. (2011) was published only a small number of LSU sequences were available for *Cercozoa*. *P. brassicae* and glissomonads both belong to *Cercozoa*, the chimeric nature of the *P. brassicae* sequence did result in a plausible phylogenetic tree, and did not arouse suspicion. However, with an enlarged dataset as the one presented here, it became possible to highlight it (Fig. 5.6). This emphasizes the need to generate more sequence data for less studied eukaryotic groups to increase the accuracy of phylogenetic analyses. It is suggest that the described differences in the LSU sequences of the Japanese "*P. brassicae*" NGY-type isolates do actually represent genetic diversity in the LSU between different glissomonad species. However, Schwelm et al. (2016) could not identify any differences in the LSU sequences of *P. brassicae* from cosmopolitan isolates, and therefore, this gene is not suitable to distinguish between different isolates or pathotypes of *P. brassicae* model in Mega 6.06 software between the 11 Korean field isolates, and the reference isolates of *P. brassicae*, based on variations in the nucleotide sequences of rDNA. Complete rDNA, if available or LSU sequences were used. Accessions with "KX4304" were obtained in this study. Purple, red, and blue colors indicate *P. brassicae* rDNA sequences obtained from Schwelm et al. 2016; Niwa et al. (2011) and this study. Black colors indicate rDNA sequences from other *Cercozoa* species (Laila et al. 2017).

5.13 Phylogenetic Analysis of Pathotypes

The *P. brassicae* isolates collected from Korea have been compared with the isolates from Japan and Europe for their phylogenic relationship. Two separate clusters were formed in the phylogenetic tree based on the rDNA sequences that separated *P. brassicae* isolates from other *Cercozoa* species according to the Maximum Likelihood method (Fig. 5.7). Similar phylogenetic separation was also found in a circular cladogram constructed following the unweighted pair-group method with arithmetic mean (UPGMA) method, and Kimura 80 nucleotide distance measure method. The 11 Korean isolates, and three published sequences AB526843, KX011115, and KX011135 were in the same cluster (Fig. 5.7). However, all

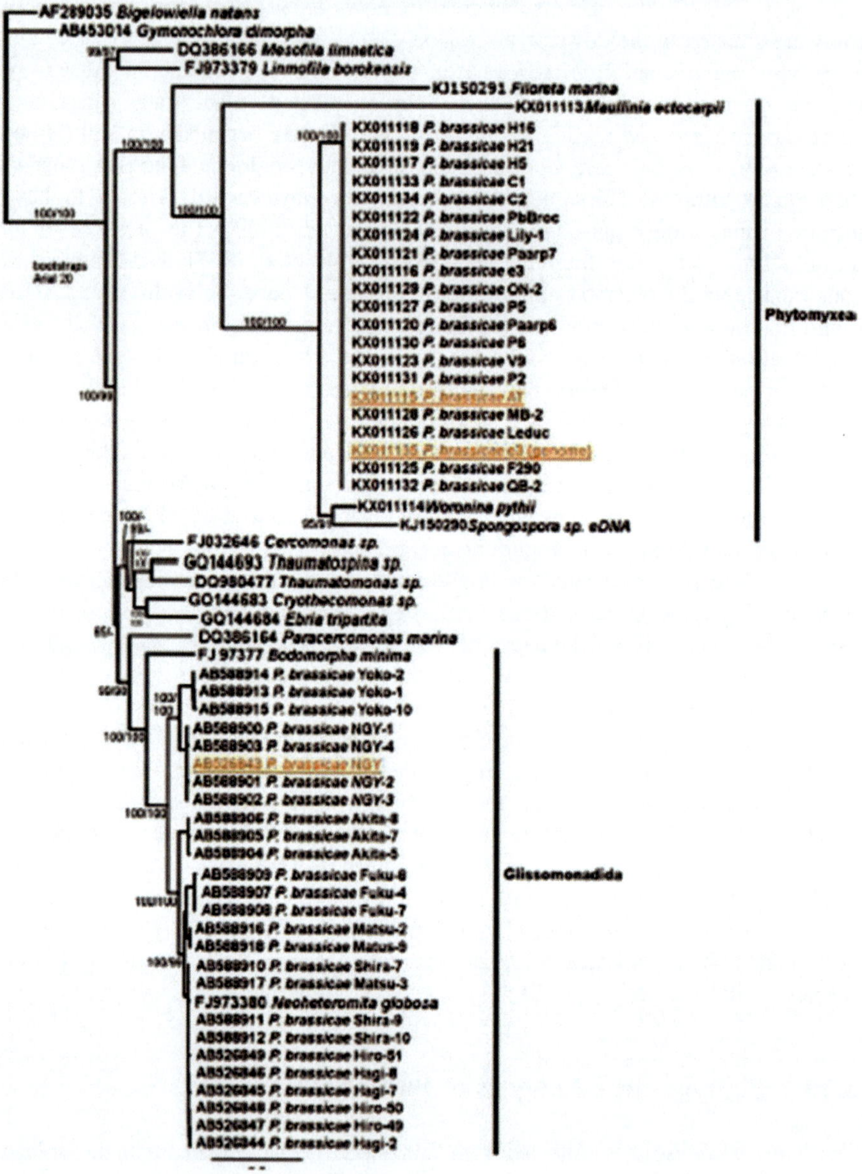

Fig. 5.6 Intraspecific diversity of *Plasmodiophora brassicae* isolates of worldwide origin. Sequences generated in this study (KX011115-KX011135) form a well-supported clade with sequences of other Phytomyxea from this study (KX011113, KX011114), and GenBank (KJ150290) while the NGY-type sequences (AB526844–AB526849; AB588900–AB588918) form a clade with Glissomonadida sequences from GenBank (FJ97377, FJ973380). Sequences, PhyML tree, support values are given Chi2 support values (%)/posterior probabilities (%) (Schwelm et al. 2016)

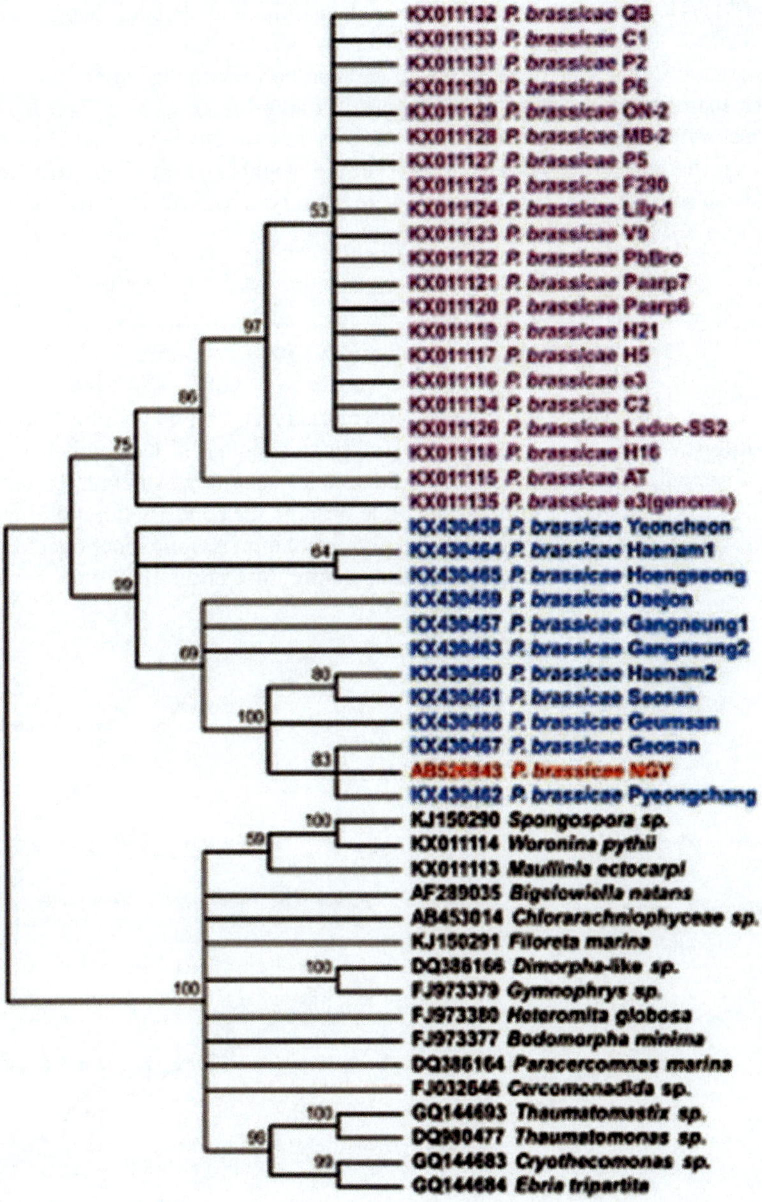

Fig. 5.7 Phylogenetic tree constructed following the Maximum Likelihood method, and Jukes–Cantor

11 Korean isolates formed a separate sub-cluster with the Japanese NGY isolate (Fig. 5.7). The Korean isolates having higher dG values between −131, and −163 kcal·mol^{-1} in their predicted secondary structures, due to sequence variations in

region 2 (Geosan, Geumsan, Seosan, Haenam2, and Pyeongchang), further sub-clustered together. The other six Korean isolates with smaller than -168 kcal·mol^{-1} dG clustered together, indicating that the variation in region 2 accounted for the majority of variation in phylogenetic clustering (Fig. 5.7). The Japanese reference isolate had a closer sequence similarity with the Korean isolate Geosan that predicted the largest dG value (Fig. 5.7). In the neighbor-joining tree, two Korean isolates, Haenaml and Hoengseong, were closely clustered with the European sequences KX011115 and KX011135 (Laila et al. 2017).

5.14 Polymorphisms in *Plasmodiophora brassicae*

A number of single-nucleotide polymorphisms (SNPs), oligonucleotide olymorphisms, and insertions-deletions (InDels) among the 11 Korean field isolates were observed. Such variation was predominant in the large subunit. To develop specific markers based on single-nucleotide or oligonucleotide polymorphisms, ten probes were designed and tested through high-resolution melting. Four out of the ten probes resulted in significant melting temperature differences (Fig. 5.8), however,

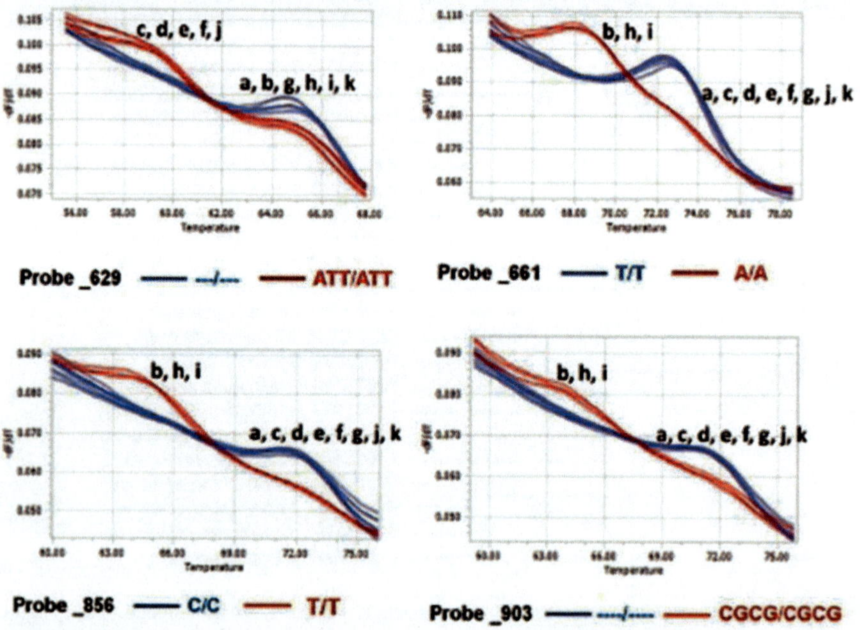

Fig. 5.8 Variation due to polymorphism between Korean *Plasmodiophora brassicae* isolates. The melting temperatures are shown for probes (single-nucleotide, and oligonucleotide polymorphism, and In Del) after high-resolution melting analysis in a Roche light cycler. (**a**) Gangneung1; (**b**) Yeoncheon; (**c**) Daejon; (**d**) Haenam2; (**e**) Seosan; (**f**) Pyeongchang; (**g**) Gangneung2; (**h**) Haenam1; (**i**) Hoengseong; (**j**) Geumsan, and (**k**) Goesan (Laila et al. 2017)

the remaining six probes showed no differences (Fig. 5.8). Probes 661, 856, and 903 clearly separated Yeoncheon, Haenam1, and Hoengseong (red peaks) from the other eight isolates. The melting temperature difference between the two groups for probes 661, 856, and 903 were 5, 8, and 7°C, respectively (Fig. 5.8). Probes 661 and 856 separated the Korean isolates based on SNPs, whereas probe 903 separated the two groups based on an InDel of four nucleotides. Probe 629 had about a 6 °C melting temperature difference between peaks produced by Daejon, Haenam2, Seosan, Pyeongchang, Geumsan (red peaks, Fig. 5.8), and the other six isolates (Gangneung1, Yeoncheon, Gangneung2, Haenam1, Hoengseong, Goesan, blue peaks in Fig. 5.8) due to variation in three adjacent nucleotides (Laila et al. 2017).

5.15 Production of Pure Genotype Isolates of *Plasmodiophora brassicae*

Production of pure genotype isolates of *Plasmodiophora brassicae* are hampered by the lack of amenability for axenic culture, and the high degree of variation pathogen is exhibiting for its pathogenicity. Its races can vary in their pathogenicity towards different cruciferous species, and cultivars as well (Crute et al. 1980; Toxopeus et al. 1986). *P. brassicae* is assumed to exist as mixtures of pathotypes in the soil, and even in single clubbed roots different pathotypes have been detected (Jones et al. 1982a). Therefore, results from resistance tests can lack reproducibility. Resistance breeding the most promising strategy for clubroot control is hampered by the lack of well defined, and pure pathotypes which are needed in selection procedures, and for the characterization of host resistance as well (Diederichsen et al. 2009). Many attempts have been made to produce single-spore-derived isolates of *P. brassicae* by inoculating susceptible host plants with single resting spores (Buczacki 1977; Tinggal and Webster 1981; Jones et al. 1982b; Scott 1985; Xue et al. 2008). Basically, in these studies single resting spores were detected in small droplets of paraffin, water, buffer, or on water agar before they were transferred to a seedling root of a susceptible *Brassica* host (cv. Granaat). Due to the small size of the resting spores, their diameter is 2–3μm, the practicability of these methods is poor, and their infection rates are low, and unreliable. According to the models of *P. brassicae* life cycle (Ingram and Tommerup 1972; Kageyama and Asano 2009), primary zoospores emerge from resting spores and enter the cytoplasm of rhizodermal cells, mainly root hairs. Undergo mitotic divisions to develop primary plasmodia which subsequently form clusters of zoosporangia, the so called sporangiosorus (Karling 1968).This multiplication leads to a strong increase in the number of nuclei, and genetically uniform secondary zoospores, that are subsequently released to initiate the cortical infection. Another problem when producing single-spore isolates of *P. brassicae* is the uncertainty of the spore's viability. Recently, infected root hairs containing only a single sporangiosorus were used as inoculum to obtain isolates that are finally also derived from a single resting spore. To compare the efficacy of this new method, single resting spore inoculations were also performed.

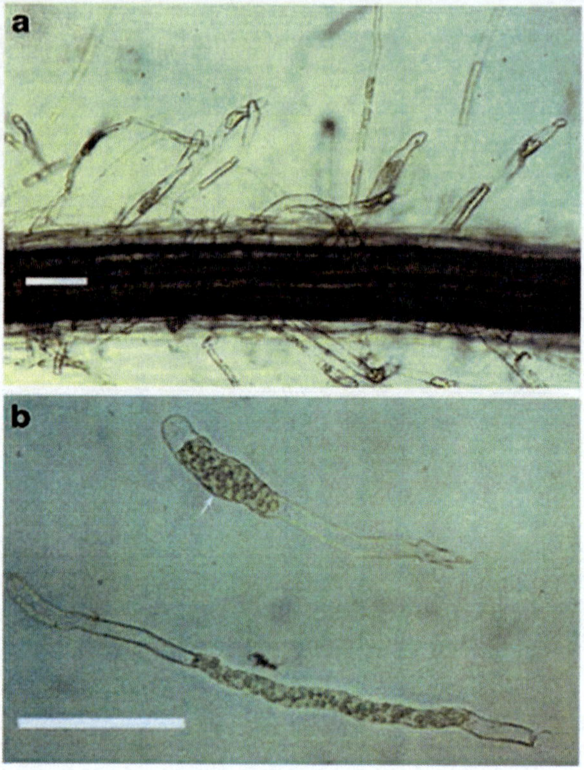

Plate 5.1 Micrograph of root hairs containing mature sporangiosori of *Plasmodiophora brassicae* before (**a**) and after detachment (**b**). Infected root hairs often showed deformed tips (arrow). White bar indicates 100 μm (Diederichsen et al. 2016)

Inoculations with detached, single infected root hairs containing a single sporangiosorus of *P. brassicae* have been successfully used by Diederichsen et al. (2016) to generate single pathotype isolates (Plate 5.1). Some of the single infected root hair derived isolates have been used to unravel the genetics of clubroot resistance in *Arabidopsis* (isolates "e2" and "e3"; Arbeiter et al. 2002; Jubault et al. 2008), and to clone a resistance gene from this host. In particular isolates "e2" and "e3" have proven their uniform reactions in different studies with *Arabidopsis* and *Brassica* hosts, and isolate "e3" has been used to generate the first genome sequence of a plant pathogenic species belonging to the kingdom of Rhizaria (Schwelm et al. 2015). Compared to the established methods of transferring single resting spores to *Brassica* seedlings, this technique is a useful alternative to produce single-spore-derived isolates. Basically, this method for inoculation of *Brassica* seedlings with single resting spores of *P. brassicae* resembled those applied in previous studies by Buczacki (1977), Tinggal and Webster (1981), Jones et al. (1982b), Scott (1985), Manzanares-Dauleux et al. (1994), and Xue et al. (2008). The major obstacles of single resting spore inoculations are:

1. It is difficult to proof the presence of only one single, tiny resting spore in a small droplet. Without the possibility of a direct microscopical comparison other small

and round objects like starch kernels can be mistaken for resting spores. The surface tension of the droplet impedes the identification of resting spores. Small droplets tend to evaporate while it is very difficult to use higher magnifications with bigger droplets. For these reasons, the microscopical determination of single resting spores in droplets is more time consuming than the detection of infected root hairs with a single sporangiosorus. The transfer of resting spores to the roots of a host plant is only indirectly feasible, and is difficult to be controlled.

2. Another disadvantage of single resting spore inoculations can be expected to arise from differences in the germination, and infection capacity of the isolated resting spore which cannot be estimated before inoculation. This should lead to an unknown proportion of inoculations with non-infectious spores.
3. The inoculum load is so low that it seems to be unlikely that a single resting spore derived isolate really is the result of a single resting spore infection, and some clubroot workers still have their doubts. Uncontrolled clubroot contaminations are difficult to be completely excluded, and consist probably of higher spore numbers of unknown origin.

These disadvantages can be overcome to some extent by the use of infected root hairs as inoculum source instead of resting spores. Merely low magnification light microscopy is required to identify sporangiosori without any need for dye. There is no need of micro-manipulative devices because all steps can be done by hand, and are visible with the inverted microscope. With some experience, root hairs can rapidly and easily be separated, and individually transferred to the root of a seedling. The success of the root hair's transfer to the susceptible host can be controlled microscopically by its absence in the droplet after the transfer. Furthermore, the infectivity of fresh secondary zoospores is probably higher than that of resting spores, because the first obstacle-the primary infection has already been passed, which means a selection for viable individuals. Single resting spore infection rates of established methods are in general low. According to Ingram and Tommerup (1972), single sporangiosori can consist of up to 100 zoosporangia each giving rise to 4–16 secondary zoospores. Thus, the inoculum amount could be up to 1600-fold higher compared to single resting spore inoculations. The infection rates after root-hair inoculation were higher than after single resting spore inoculation, though a further increase in the infection rate is still desirable. Infection rates were not increased by inoculating seedlings with very low inoculum densities (2–100 resting spores). A successful infection seems to depend mainly on the infection capability of the individual resting spore or of the zoospores from the sporangiosorus. In the other microorganisms, the fusion of gametes is a prerequisite for the continuation of their life cycle. Whether such sexual processes are a part of *P. brassicae* life cycle is still not clear. Fusion of secondary zoospores has been proposed, and Tommerup and Ingram (1971) reported the observation of such an event in their microscopical studies. If the fusion of secondary zoospores would be required for a successful cortical infection, it should be impossible to generate single genotype isolates of *P. brassicae*. Considering the number of successful reports on the creation of single resting spore isolates, it seems unlikely that the fusion of different mating types is needed for the initiation of the secondary infection phase. The absence of sexual recombination in *P. brassicae* is also supported by molecular studies from Graf et al. (2001), and Fahling et al. (2004), who characterized chromosomal polymorphisms or tried to detect recombinant marker patterns in subpopulations derived from mixed inoculations of isolates "e3," and "h1." Instead of recombinant patterns they could only recover the parental marker patterns again. One objection to root-hair inoculations as a means to create single pathotype isolates can be the idea that primary plasmodia might fuse before they form zoosporangia (Diederichsen et al. 2016).

5.16 Virulence Pattern of *Plasmodiophora brassicae* Populations on Cruciferous Weeds

Capsella bursa-pastoris is one of the most common cruciferous weeds worldwide and is well known as a wild host of *P. brassicae* in many countries. Although populations from cruciferous crops are often pathogenic on the weed, the level of compatibility varies depending on the specific combination of weed collection and *P. brassicae* population in Japan (Tanaka et al. 1983; Tanina et al. 1985). On the other hand, *P. brassicae* populations from the weed have been confirmed to be highly virulent on cruciferous crops in Fukushima (Tsushima 2003) and Fukuoka prefectures. Populations from *C. bursa-pastoris* collected in three fields of Fukuoka, Fukuoka Prefecture were highly virulent on cabbage and Chinese cabbage and were identical in pathotype (determined on the CR cultivars of Chinese cabbage) with populations from cruciferous crops collected in the three fields. The presence of *C. bursa-pastoris* in fields possibly perpetuates *P. brassicae* in soil, and reduces the effectiveness of crop rotation, and decoy crops for clubroot control in cruciferous crop production in certain areas of Japan (Tanaka and Ito 2013).

Clubroot of *C. flexuosa* has been recorded exclusively in Japan. Populations from cruciferous crops are nonpathogenic on *C. flexuosa* (Murai et al. 1986; Osaki et al. 2008b; Tanaka et al. 1984, 1994; Umehara and Tamura 1968; Yoshikawa 1993). However, *C. flexuosa* collected from a field of Hiroshima Prefecture was found to be highly susceptible to many populations from cruciferous crops. Collections of *C. flexuosa* are also thought to vary in susceptibility to populations from cruciferous crops. Some populations from this weed are pathogenic on Chinese cabbage and turnip (Murai et al. 1986; Osaki et al. 2008b; Tanaka et al. 1984). In many cases, however, although these populations suppress the growth of common cultivars of these crops by affecting their entire taproots, the infected roots frequently have slight or moderate swelling and necrotic root tissues. None to very few resting spores develop in infected roots with such symptoms. Therefore, these populations have often been assessed as moderately pathogenic on Chinese cabbage and turnip (Murai et al. 1986; Osaki et al. 2008b; Tanaka et al. 1984), but the symptoms may reflect incomplete resistance of these crops against populations from the weed. However, typical clubroot galls are also sometimes found on a part of their infected plants. Murai et al. (1986) reported that a population obtained primarily from such typical clubroot galls on Chinese cabbages was more virulent than an original population from the weed on Chinese cabbage. Clubroot on a cruciferous weed, *Rorippa islandica*, has been recorded only in extremely limited areas of Japan (Ikegami 1978; Umehara and Tamura 1968). Therefore, although populations from cruciferous crops have also been reported to be nonpathogenic on *R. islandica* (Osaki 2008; Umehara and Tamura 1968; Yoshikawa 1993), there had been no reports on the pathogenicity of *P. brassicae* from this weed. However, clubroot of *R. islandica* has been recently found in Fukuoka Prefecture, and in a pathogenicity test of a population from this weed, has been examined using inoculation tests (Osaki 2008). The results showed that the population from *R. islandica* often induced incomplete (slight or moderate) swelling with necrotic root tissues on most of the infected roots of Chinese cabbage and turnip as did some populations from *C. flexuosa*, but the population was nonpathogenic on a collection of *C. flexuosa* (Osaki 2008). The population from *R. islandica* may differ fundamentally in pathogenicity from those from cruciferous crops and *C. flexuosa*.

Different pathotypes within the field populations can be separated by repeated propagation in resistant host plants (Fahling et al. 2003). The process through which minor pathogenic genotypes develop into major ones through selective propagation on the crops may in fact act within each population from the weeds in the fields. The pathogenicity of *P. brassicae* from these weeds has been studied only in a limited number of populations. The study of clubroot susceptibility has been limited to a few collections. Variation in pathogenicity may be present within *P. brassicae* from the weeds as well as within that from crops. In addition, the level of clubroot susceptibility may vary among weed collections and may also be heterogenic even within a collection (Tanaka and Ito 2013).

References

Agarwal A, Kaul V, Faggian R, Cahill DM (2009) Development and use of a model system to monitor clubroot disease progression with an Australian field population of *Plasmodiophora brassicae*. Austr Plant Pathol 38:120–127

Apaydın A, Deligoz I, Kar H, Kibar B, Karaagac O (2010) An investigation on clubroot disease (*Plasmodiophora brassicae* Wor.) races in the black sea region of Turkey. GOU Ziraat Fakultesi Dergisi 27(2):57–60

Arbeiter A, Fahling M, Graf H, Sacristan MD, Siemens J (2002) Resistance of *Arabidopsis thaliana* to the obligate biotrophic parasite *Plasmodiophora brassicae*. Plant Prot Sci 38:519–522

Ayers GW (1957) Races of *Plasmodiophora brassicae*. Can J Bot 35:923–932

Badi A, Lawrie AC, Donald EC (2010) Genetic diversity of *Plasmodiophora brassicae* in Australia. In Stirling GR (ed) Proc. 6th Australasian Soil Borne Diseases Symposium, 9-11 August 2010, p 32

Bhattacharya I (2003) Study of race variation and characterization of *Plasmodiophora brassicae* causing clubroot disease of crucifers in West Bengal (India). Mohanpur, Bidhan Chandra Krishi Viswavidyalaya, Final Report, ICAR (Indian Council of Agricultural Research) Adhoc Project. New Delhi, India

Bhattacharya I, Dutta S, Mondal S, Mondal B (2014) Clubroot disease on *Brassica* crops in India. Can J Plant Pathol 36:154–160

Bryngelsson T, Gustafsson M, Gréen B, Lind-Halldén C (1988) Uptake of host DNA by the parasitic fungus *Plasmodiophora brassicae*. Physiol Mol Plant Pathol 33(2):163–171

Buczacki ST (1977) Root infections from single resting spores of *Plasmodiophora brassicae*. Trans Br Mycol Soc 69:328–329

Buczacki ST, Toxopeus H, Mattusch P, Johnston TD, Dixon GR et al (1975) Study of physiological specialization in *Plasmodiophora brassicae*: proposals for attempted rationalization through an international approach. Trans Br Mycol Soc 65:295–303

Buhariwalla S, Mithen R (1995) Cloning of a *Brassica* repetitive DNA element from resting spores of *Plasmodiophora brassicae*. Physiol Mol Plant Pathol 47:95–101

Buhariwalla S, Greaves S, Magrath R, Mithen R (1995) Development of specific PCR primers for the amplification of polymorphic DNA from the obligate root pathogen *Plasmodiophora brassicae*. Physiol Mol Plant Pathol 47:83–94

Bulman S, Ridgway HJ, Eady C, Conner AJ (2007) *Intron*-rich gene structure in the intracellular plant parasite *Plasmodiophora brassicae*. Protist 158:423–433

Cao T, Manolii VP, Hwang SF, Howard RJ, Strelkov SE (2009) Virulence and spread of *Plasmodiophora brassicae* [clubroot] in Alberta, Canada. Can J Plant Pathol 31:321–329

Chai AL, Xie XW, Shi YX, Li BJ (2014) Research status of clubroot (*Plasmodiophora brassicae*) on cruciferous crops in China. Can J Plant Pathol 36:142–153

Chambers SC (1977) Studies on some populations of Plasmodiophora brassicae (Woron.) in Victoria. Plant Research Institute, Department of Agriculture, Victoria

Chattopadhyay AK, Bagchi BN, Roychoudhury UK (1991) Reactions of some cultures of rapeseed mustard against clubroot disease. Indian Phytopathol 1991:238–239

Chen X, Line RF, Leung H (1993) Relations between virulence variation and DNA polymorphism in *Puccinia striiformis*. Phytopathology 83:1489–1497

Cho WD, Kim WG, Takahashi K (2003) Occurrence of clubroot in cruciferous vegetable crops and races of the pathogen in Korea. Plant Pathol J 19:64–68

Chytilova V, Dusek K (2007) Metodika testováni odolnosti brukvovitých plodin k nádorovitosti. Metodika pro praxi (in Czech). Výzkumný ústav rostlinné výroby, Praha

Conesa A, Gotz S, Garcia-Gomez JM, Terol J, Talon M, Robles M (2005) Blast2GO: a universal tool for annotation, visualization and analysis in functional genomics research. Bioinformatics 21(18):3674–3686

Crute IR (1986) The relationship between *Plasmodiophora brassicae* and its hosts: the application of concepts relating to variation in inter-organismal associations. Adv Plant Pathol 5:1–52

Crute IR, Gray AR, Crisp P, Buczacki ST (1980) Variation in *Plasmodiophora brassicae* and resistance to clubroot disease in Brassicas and allied crops - a critical review. Plant Breed Abstr 50:91–104

Crute IR, Phelps K, Barnes A, Buczacki ST, Crips P (1983) The relationship between genotypes of three *Brassica* species and collections of *Plasmodiophora brassicae*. Plant Pathol 32:405–420

Das SN, Mishra SK, Swain PK (1987) Reactions of some toria varieties to *Plasmodiophora brassicae*. Indian Phytopathol 40:120

Delye C, Laigret F, Corio-Costet MF (1997) RAPD analysis provides insight into the biology and epidemiology of *Uncinula necator*. Phytopathology 87:670–677

Diederichsen E, Sacristan MD (1996) Disease response of resynthesized *Brassica napus* L. lines carrying different combinations of resistance to *Plasmodiophora brassicae* Wor. Plant Breed 115:5–10

Diederichsen E, Frauen M, Linders EG, Hatakeyama K, Hirai M (2009) Status and perspectives of clubroot resistance breeding in crucifer crops. J Plant Growth Regul 28:265–281. https://doi.org/10.1007/s00344-009-9100-0

Diederichsen E, Frauen M, Ludwig-Muller J (2014) Clubroot disease management challenges from a German perspective. Can J Plant Pathol 36:85–98

Diederichsen E, Wagenblatt B, Schallehn V (2016) Production of pure genotype isolates of *Plasmodiophora brassicae* Wor. comparison of inoculations with root hairs containing single sporangiosori or with single resting spores. Eur J Plant Pathol 145:621–627

Ding YH, Jian YC, Yu YJ, Wang WH, Geng LH, Kang JG (2013) Identification of pathotype of *Plasmodiophora brassicae* on crucifer vegetables in eight provinces of China. Chin Veg 16:85–88

Dobson RL, Robak J, Gabrielson RL (1983) Pathotypes of *Plasmodiophora brassicae* in Washington, Oregon and California. Plant Dis 67:269–271

Donald EC, Porter IJ (2014) Clubroot in Australia: the history and impact of *Plasmodiophora brassicae* in *Brassica* crops and research efforts directed towards its control. Can J Plant Pathol 36:66–84

Donald EC, Cross SJ, Lawrence JM, Porter IJ (2006) Pathotypes of *Plasmodiophora brassicae*, the cause of clubroot, in Australia. Ann Appl Biol 148(3):239–244

Faggian R, Bulman SR, Lawrie AC, Porter IJ (1999) Specific polymerase chain reaction primers for the detection of *Plasmodiophora brassicae* in soil and water. Phytopathology 89(5):392–397

Fahling M, Graf H, Siemens J (2003) Pathotype-separation of *Plasmodiophora brassicae* by the host plant. J Phytopathol 151:425–430

Fahling M, Graf H, Siemens J (2004) Characterization of a single spore isolate population of Plasmodiophora brassicae resulting from a single club. J Phytopathol 152:438–444

Feng J, Jiang J, Feindel D, Strelkov SE, Hwang SF (2016) The gene *Cr811* is present exclusively in pathotype 5 and new emerged pathotypes of the clubroot pathogen *Plasmodiophora brassicae*. Eur J Plant Pathol 145:615–620

Fuchs H, Sacristan MD (1996) Identification of a gene in *Arabidopsis thaliana* controlling resistance to clubroot and characterization of the resistance response. Mol Plant-Microbe Interact 9:91–97

Gonzalez M, Rodroguez R, Zavala ME, Jacobo JL, Hernandez F, Acosta J, Martonez O, Simpsom J (1998) Characterization of Mexican isolates of *Colletotrichum lindemuthianum* by using differential cultivars and molecular markers. Phytopathology 88:292–299

Graf H, Sokolowski F, Klewer A, Diederichsen E, Luerssen H, Siemens J (2001) Electrophoretic karyotype of the obligate biotrophic parasite *Plasmodiophora brassicae* Wor. J Phytopathol 149:313–318

Gustafsson M, Falt AS (1986) Genetic studies on resistance to clubroot in *Brassica napus*. Ann Appl Biol 108(2):409

Hatakeyama K, Fujimura M, Ishida M, Suzuki T (2004) New classification method for *Plasmodiophora brassicae* field isolates in Japan based on resistance of F_1 cultivars of Chinese cabbage (*Brassica rapa* L.) to clubroot. Breed Sci 54:197–201

Heikinheimo O, Raatikainen M (1971) Paikan ilmoittaminen Suomesta tallennetuissa biologisissa aineistoissa. Ann Ent Fenn 37(1a):1–30

Hildebrand PD, Delbridge RW (1995) Race survey of *Plasmodiophora brassicae* in Nova Scotia. Can Plant Dis Surv 75:170

Honig F (1931) Der Kohlkropferreger (*Plasmodiophora brassicae* Wor.). Gartenbauwissenschaft 5:116–225

Horiuchi S, Hori M (1983) Control of clubroot bisease of crucifers, with reference to the soil solarization technique. JARQ 17(I):1–5

Howe AT, Bass D, Vickerman K, Chao EE, Cavalier-Smith T (2009) Phylogeny, taxonomy, and astounding genetic diversity of *Glissomonadida* ord. nov., the dominant gliding zooflagellates in soil (Protozoa: Cercozoa). Protist 160:159–189

Hsiang T, Mahuku GS (1999) Genetic variation within and between southern Ontario populations of *Sclerotinia homoeocarpa*. Plant Pathol 48:83–94

Hwang SF, Strelkov SE, Feng J, Gossen BD, Howard RJ (2012) *Plasmodiophora brassicae*: a review of an emerging pathogen of the Canadian canola (*Brassica napus*) crop. Mol Plant Pathol 13:105–113

Ikegami H (1978) Research trends of clubroot on cruciferous vegetables in recent years (in Japanese). Plant Prot 32:53–61

Ikegami H (1992) Proliferation and pathogenicity of *Plasmodiophora brassicae* in infected callus tissue (studies on the clubroot of cruciferous plant X). Proc Kansai Plant Prot Soc 34:17–28

Ikegami H, Naiki T (1981) The number of Plasmodiophora brassicae resting spores in club and infected callus. Ann Phytopathol Soc 48:101–102

Ingram DS, Tommerup IC (1972) The life history of *Plasmodiophora brassicae* Wor. Proc R Soc Lond 80:103–112

International Seed Federation (2019) Nyon, Switzerland, July 2019

Ito S, Yano S, Tanaka S, Kameya-Iwaki M (1994) The use of resting spore spheroplasts in the DNA analysis of *Plasmodiophora brassicae*. Ann Phytopathol Soc 60:491–495

Ito S, Maehara T, Tanaka S, Kameya-Iwaki M, Yano S, Kishi F (1997) Cloning of a single-copy DNA sequence unique to *Plasmodiophora brassicae*. Physiol Mol Plant Pathol 50:289–300

Ito S, Nakamura T, Matsumoto T, Maehara T, Tanaka S, Kameyaiwaki M, Kishi F (1998) Characterization of DNA sequences cloned from resting spores of *Plasmodiophora brassicae*. J Phytopathol 146(2-3):143–147

Jennings JM, Newton AC, Buck KW (1997) Detection of polymorphism in *Puccinia hordei* using RFLP and RAPD markers, differential cultivars, and analysis of the intergenic spacer region of rDNA. J Phytopathol 145:511–519

Jeong JY, Robin AHK, Natarajan S, Laila R, Kim HT, Park JI, Nou I-S (2018) Race- and isolate-specific molecular marker development through genome-realignment enables detection of Korean *Plasmodiophora brassicae* isolates, causal agents of clubroot disease. Plant Pathol J 34(6):506–513

Ji HW, Ren L, Chen KR, Xu L, Liu F, Sun CC, Li J, Liu SY, Fang XP (2013) Identification of physiological races of club root and resistance of rape cultivars to *Plasmodiophora brassicae*. Chin J Oil Crop Sci 35:301–206

Jo SJ, Shim SA, Jang KS, Choi YH, Kim JC, Choi GJ (2011) Resistance of cultivars of Chinese cabbage to *Plasmodiophora brassicae* isolates of several races collected in Korea. Korean J Hort Sci Technol 29:610–616

Jones DR, Ingram DS, Dixon GR (1982a) Characterization of isolates derived from single resting spores of *Plasmodiophora brassicae* and studies of their interaction. Plant Pathol 31 (3):239–246

Jones DR, Ingram DS, Dixon GR (1982b) Factors affecting tests for differential pathogenicity in populations of *Plasmodiophora brassicae*. Plant Pathol 31:229–238

Jonsson R (1981) Breeding for improved resistance to clubroot in oil rape. Proc Brassica Conf 1981:44–46

Jubault M, Lariagon C, Simon M, Delourme R, Manzanares-Dauleux MJ (2008) Identification of quantitative trait loci controlling partial clubroot resistance in new mapping populations of *Arabidopsis thaliana*. Theor App Genet 117:191–202

Kageyama K, Asano T (2009) Life cycle of *Plasmodiophora brassicae*. J Plant Growth Regul 28:203–211

Karling JS (1968) The plasmodiophorales. Hafner Publishing Co, New York

Kim H, Jo EJ, Choi YH, Jang KS, Choi GJ (2016) Pathotype classification of *Plasmodiophora brassicae* isolates using clubroot-resistant cultivars of Chinese cabbage. Plant Pathol J 32 (5):423–430

Klewer A, Luerben H, Graf H, Siemens J (2001) Restriction fragment length polymorphism markers to characterize *Plasmodiophora brassicae* single-spore isolates with different virulence patterns. J Phytopathol 149:121–127

Koch E, Cox R, Williams PH (1991) Infection of *Arabidopsis thaliana* by *Plasmodiophora brassicae*. J Phytopathol 132:99–104

Kong Kaw Wa S (2009) Discovering clubroot resistance genes in *Brassica* vegetable crops. Ph.D. Thesis, RMIT University, Melbourne, Australia

Krivchenko VI, Boos GV, Surmava ME (1982) A characteristic of the gene bank of cabbage as regards resistance to *Plasmodiophora brassicae* Woron. Trudy po Prikladnoi Botanike, Genetike i Selecktsii 72:113–120

Kuginuki Y, Yoshikawa H, Hirai M (1999) Variation in virulence of *Plasmodiophora brassicae* in Japan tested with clubroot-resistant cultivars of Chinese cabbage (*Brassica rapa* L. ssp. *pekinensis*). Eur J Plant Pathol 105(4):27–332

Laila R, Robin AH, Yang K, Choi GJ, Park JI, Nou IS (2017) Detection of ribosomal DNA sequence polymorphisms in the protist *Plasmodiophora brassicae* for the identification of geographical isolates. Int J Mol Sci 18:84

Lammerink J (1965) Six pathogenic races of *Plasmodiophora brassicae* Wor. in New Zealand. N Z J Agric Res 13:156–164

Lammerink J (1986) Identification of an eighth race of *Plasmodiophora brassicae*, the cause of clubroot in New Zealand. N Z J Agric Res 29:101–104

LeBoldus JM, Manolii VP, Turkington TK, Strelkov SE (2012) Adaptation to *Brassica* host genotypes by a single-spore isolate and population of *Plasmodiophora brassicae* (clubroot). Plant Dis 96(6):833–838

Linnasalmi A, Palonen S (1974) Om *Plasmodiophora brassicae* rasernas identifiering och kartering Finland. Nord Jordbr Forskn 56:6–10

Linnasalmi A, Toiviainen A (1991) Occurrence of clubroot and *Plasmodiophora brassicae* Wor. races in Finland. J Agric Sci 63:415–434

Linnasalmi A, Weisaeth G (1978) Om klumprotraser i Trondelag Plasmodiophorarase 1, 7 og 9. Summary: races of clubroot in Trondelag. Norway Res Norway Agric 29:223–239

Lodhi MA, Ye GN, Weeden NF, Reisch BI (1994) A simple and efficient method for DNA extraction from grapevine cultivars and *Vitis* species. Plant Mol Biol Report 12(1):6–13

Mahuku GS, Goodwin PH, Hall R, Hsiang T (1997) Variability in the highly virulent type of *Leptosphaeria maculans* within and between oilseed rape fields. Can J Bot 75:1485–1492

Manzanares-Dauleux MJ, Laurens F, Baron F, Thomas G (1994) Production of single spore isolates of *Plasmodiophora brassicae*. Cruciferae Newslett 16:132–133

Manzanares-Dauleux MJ, Delourme R, Baron F, Thomas G (2000) Mapping of one major gene and of QTLs involved in resistance to clubroot in *Brassica napus*. Theor Appl Genet 101:885–891

Manzanares-Dauleux MJ, Divaret I, Baron F, Thomas G (2001) Assessment of biological and molecular variability between and within field isolates of *Plasmodiophora brassicae*. Plant Pathol 50:165–173

Mithen R, Magrath R (1992) A contribution to the life history of *Plasmodiophora brassicae*: secondary plasmodia development in root galls of *Arabidopsis thaliana*. Mycol Res 96:877–885

Moller M, Harling R (1996) Randomly amplified polymorphic DNA (RAPD) profiling of *Plasmodiophora brassicae*. Lett Appl Microbiol 22:70–75

Murai K, Tanaka S, Katumoto K, Nishi Y (1986) Variation in pathogenicity of a population from *Cardamine flexuosa* by propagation in Chinese cabbage (abstract in Japanese). Ann Phytopathol Soc 52:132

Niwa R, Kawahara A, Murakami H, Tanaka S, Ezawa T (2011) Complete structure of nuclear rDNA of the obligate plant parasite *Plasmodiophora brassicae*: intra-specific polymorphisms in the exon and Group I intron of the large subunit rDNA. Protist 162:423–434

Osaki K (2008) Studies on diversity of pathogenicity within *Plsamodiophora brassicae* (in Japanese with English summary). PhD dissertation, Tottori Univ, Tottori, Japan

Osaki K, Tanaka S, Ito S (2008a) Pathogenicity of *Plasmodiophora brassicae* populations from small, spheroid, resistant-type clubroot galls on roots of clubroot-resistant cultivars of Chinese cabbage (*Brassica rapa* L. subsp. *pekinensis*). J Gen Plant Pathol 74:242–245

Osaki K, Fujiyama S, Nakayama A, Shimizu Y, Ito S, Tanaka S (2008b) Relation between pathogenicity and genetic variation within *Plasmodiophora brassicae*. J Gen Plant Pathol 74:281–288

Parlevliet JE (1985) Race and pathotype concepts in parasitic fungi. OEPP Bull 15:145–150

Peng SS, Ren ZH, Huang XL, Liu MJ, Sun LF, Liu EM (2013) Physiological race identification on clubroot of cruciferous crops caused by *Plasmodiophora brassicae* in Hunan. J Changjiang Veg 6:46–49

Piao Z, Ramchiary N, Lim YP (2009) Genetics of clubroot resistance in *Brassica* species. J Plant Growth Regul 28:252–264

Porebski S, Bailey LG, Baum BR (1997) Modification of a CTAB DNA extraction protocol for plants containing high polysaccharide and polyphenol components. Plant Mol Biol Report 15 (1):8–15

Reyes AA, Davidson RT, Marks CF (1974) Races, pathogenicity and chemical control of *Plasmodiophora brassicae* in Ontario. Phytopathology 64:173–177

Ricarova V, Kaczmarek J, Strelkov SE, Kazda J, Lueders W, Rysanek P, Manolii V, Jedryczka M (2016) Pathotypes of *Plasmodiophora brassicae* causing damage to oilseed rape in the Czech Republic and Poland. Eur J Plant Pathol 145:559–572

Rod J (1991) Výskyt a rozšíření patotypů nádorovitosti (*Plasmodiophora brassicae* Wor.) v Československu. In: Proceedings XII. Czechoslovak Plant Protection Conference, Sept 17–19, 1991, Prague, Czech Republic, pp 237–238

Rod J (1994) The occurrence and distribution of pathotypes of *Plasmodiophora brassicae* Wor. in the Czech Republic and Slovakia. Ochrana Rostlin 30:171–182

Rod J (1996) Reports – agent of clubroot of crucifer. Brno: UKZUZ 37:1–45

Rolfe SA, Strelkov SE, Links MG, Clark WE, Robinson SJ, Djavaheri M, Malinowski R, Haddadi P, Kagale S, Parkin IAP, Ali T, Borhan MH (2016) The compact genome of the plant pathogen *Plasmodiophora brassicae* is adapted to intracellular interactions with host *Brassica* spp. BMC Genomics 17:272–286

Schoeller M, Grunewaldt J (1986) Production and characterization of single spore derived lines of *Plasmodiophora brassicae* Wor. Cruciferae Newslett 11:110–111

Scholze P, Malorny M, Dippe R (2002) A comparative study of reaction to clubroot (*Plasmodiophora brassicae* Wor.) by inoculating plants with single races or a race mixture. J Plant Dis Prot 109(3):217–226

Schwelm A, Fogelqvist J, Knaust A, Julke S, Lilja T, Bonilla-Rosso G, Karlsson M, Shevchenko A, Dhandapani V, Choi SR, Kim HG, Park JY, Lim YP, Ludwig-Muller J, Dixelius C (2015) The *Plasmodiophora brassicae* genome reveals insights in its life cycle and ancestry of chitin synthases. Sci Rep 2015:11153. https://doi.org/10.1038/srep11153

Schwelm A, Berney C, Dixelius C, Bass D, Neuhauser S (2016) The large subunit rDNA sequence of *Plasmodiophora brassicae* does not contain intra-species polymorphism. Protist 167:544–554

Scott ES (1985) Production and characterization of single-spore isolates of *Plasmodiophora brassicae*. Plant Pathol 34:287–292

Seaman WL, Walker JC, Larson RH (1963) A new race of *Plasmodiophora brassicae* affecting Badger Shipper cabbage. Phytopathology 53:1426–1429

Sharma K, Gossen BD, Greenshields D, Selvaraj G, Strelkov SE, McDonald MR (2013) Reaction of lines of the rapid cycling *Brassica* collection and *Arabidopsis thaliana* to four pathotypes of *Plasmodiophora brassicae*. Plant Dis 97:720–727

Shen XQ, Nie K, Wu Q, Zhang YG, Meng XH (2009) Initial research report on differentiation identification of Chinese cabbage clubroot main physiological races. Chin Veg 8:59–62

Siemens J, Bulman S, Rehn F, Sundelin T (2009) Molecular Biology of *Plasmodiophora brassicae*. J Plant Growth Regul 28(3):245–251

Some A, Manzanares MJ, Laurens F, Baron F, Thomas G, Rouxel F (1996) Variation for virulence in *Brassica napus* L. amongst *Plasmodiophora brassicae* collections from France and derived single-spore isolates. Plant Pathol 45(3):432–439

Sperschneider J, Dodds PN, Gardiner DM, Manners JM, Singh KB, Taylor J (2015) Advances and challenges in computational prediction of effectors from plant pathogenic fungi. PLoS Pathol 11 (5):e1004806. https://doi.org/10.1371/journal.ppat.1004806

Strehlow B, de Mol F, Struck C (2014) History of oilseed rape cropping and geographic origin affect the genetic structure of *Plasmodiophora brassicae* populations. Phytopathology 104:532–538

Strelkov SE, Hwang SF (2014) Clubroot in the Canadian canola crop: 10 years into the outbreak. Can J Plant Pathol 36:27–36

Strelkov SE, Tewari JP, Hartman M, Orchard D (2005) Clubroot on canola in Alberta in 2003 and 2004. Can Plant Dis Surv 85:72–73

Strelkov SE, Cao T, Manolii VP, Lange MR, Smith-Degenhardt E, Orchard D, Tewari JP (2006a) Incidence of clubroot on canola in Alberta in 2005. Can Plant Dis Surv 86:91–93

Strelkov SE, Tewari JP, Smith-Degenhardt E (2006b) Characterization of *Plasmodiophora brassicae* populations from Alberta, Canada. Can J Plant Pathol 28(3):467–474

Strelkov SE, Manolii VP, Cao T, Hwang SF, Orchard D (2007a) Incidence of clubroot on canola in Alberta in 2006. Can Plant Dis Surv 87:109–111

Strelkov SE, Manolii VP, Cao T, Xue S, Hwang SF (2007b) Pathotype classification of *Plasmodiophora brassicae* and its occurrence in *Brassica napus* in Alberta. Can J Phytopathol 105:706–712

Strelkov SE, Hwang SF, Manolii VP, Cao T, Feindel D (2016) Emergence of new virulence phenotypes of *Plasmodiophora brassicae* on canola (*Brassica napus*) in Alberta, Canada. Eur J Plant Pathol 145:517–529

Strelkov SE, Hwang SF, Manolii VP, Cao T, Fredua-Agyeman R, Harding MW, Peng G, Gossen BD, Mcdonald MR, Feindel D (2018) Virulence and pathotype classification of *Plasmodiophora brassicae* populations collected from clubroot resistant canola (*Brassica napus*) in Canada. Can J Plant Pathol 40(2):284–298

Takematsu T, Ichizen N (1993) Sekai no Zassou II. Zenkoku Noson Kyoiku Kyokai, Tokyo, pp 415–417

Tanaka S, Ito S (2013) Pathogenic and genetic diversity in *Plasmodiophora brassicae* (clubroot) from Japan. J Gen Plant Pathol 79(5):297–306

Tanaka S, Tanina K, Muramoto K, Katumoto K, Nishi Y (1983) Ecological studies on clubroot of crucifers (III) Clubroot susceptibility of *Capsella bursa-pastoris* from various areas (abstract in Japanese). Ann Phytopathol Soc 49:390

Tanaka S, Inoue T, Murai K, Katumoto K, Nishi Y (1984) Ecological studies on clubroot of crucifers (IV) Pathogenicity of *Plasmodiophora brassicae* populations from *Cardamine flexuosa* (abstract in Japanese). Ann Phytopathol Soc 50:396

Tanaka S, Negoro M, Ota T, Katumoto K, Nishi Y (1990) Clubroot of spring Chinese cabbage in Nagasaki Prefecture, Kyushu (in Japanese with English summary). Bull Fac Agric Yamaguchi Univ 38:33–45

Tanaka S, Sakamoto Y, Kajima K, Fujieda K, Katumoto K, Nishi Y (1991) Pathogenicity of three isolates of clubroot fungus attacking clubroot-resistant cultivars of Chinese cabbage (in Japanese with English summary). Bull Fac Agric Yamaguchi Univ 39:113–122

Tanaka S, Ito S, Kameya-Iwaki M, Katumoto K, Nishi Y (1993) Occurrence and distribution of clubroot disease on two cruciferous weeds, *Cardamine flexuosa* and *C. scutata*, in Japan. Trans Mycol Soc 34:381–388

Tanaka S, Murai K, Ito S, Katumoto K, Nishi Y (1994) The occurrence clubroot disease of wasabi (*Eutrema wasabi* Maxim.) and its possible source of infection (in Japanese with English summary). Ann Phytopathol Soc 60:257–259

Tanaka S, Yoshihara S, Ito S, Kameya-Iwaki M (1997) The influence of virulence of *Plasmodiophora brassicae* populations on epidemiology of Chinese cabbage clubroot and efficacy of fungicides (in Japanese with English summary). Ann Phytopathol Soc 63:183–187

Tanaka S, Fujiyama S, Shigemori S, Nakayama A, Ito S, Kameya-Iwaki M (1998) Pathogenesis of isolates of *Plasmodiophora brassicae* from Japan (1) Race and pathogenesis in clubroot resistant cultivars (in Japanese with English summary). Kyushu Plant Prot Res 44:15–19

Tanaka S, Mizui Y, Terasaki H, Ito S (2006) Distribution of clubroot disease of a cruciferous weed, *Cardamine flexuosa*, in major isolated islands, Hokkaido and Okinawa in Japan. Mycoscience 47(2):72–77

Tanina K, Tanaka S, Katumoto K, Nishi Y (1985) Ecological studies on clubroot of crucifers (V) variation in susceptibility to *Plasmodiophora brassicae* populations from Chinese cabbage among collections of *Capsella bursa-pastoris* (abstract in Japanese). Ann Phytopathol Soc 51:77

Tinggal SH, Webster J (1981) Technique for single spore infection by *Plasmodiophora brassicae*. Trans Br Mycol Soc 76:187–190

Tommerup IC, Ingram DS (1971) The life-cycle of *Plasmodiophora brassicae* Woron. in *Brassica* tissue cultures and in intact roots. New Phytol 70:327–332

Toxopeus H, Jansen AMP (1975) Clubroot resistance in turnip II. The 'slurry' screening method and clubroot races in the Netherlands. Euphytica 24:751–755

Toxopeus H, Dixon G, Mattusch P (1986) Physiological specialization in *Plasmodiophora brassicae*: an analysis by international experimentation. Trans Br Mycol Soc 87:279–287

Tsushima S (2003) Clubroot disease. Plant Prot 57:233–236

Umehara Y, Tamura M (1968) Ecological studies on the clubroot of cruciferous crops (in Japanese with English summary). Bull Ishikawa ken Agric Exp Stat 5:1–18

Voorrips RE (1995) *Plasmodiophora brassicae* – aspects of pathogenesis and resistance in *Brassica*. Euphytica 83(2):139–146

Voorrips RE (1996) Production, characterization and interaction of single-spore isolates of *Plasmodiophora brassicae*. Eur J Plant Pathol 102:377–383

Wang MQ, Xiang YY, Zhou FZ (2013a) Application effect of lime in aquatic vegetable bases. J Changjiang Veg 4:63–65

Wang WH, Yu YJ, Ding YH, Zhang FL, Yu SC, Zhang DS, Lu GX (2013b) Physiological race identification of clubroot (*Plasmodiophora brassicae* Wor.) from Changyang county, Hubei province and resistant resource screening for Chinese cabbage breeding. J Changjiang Veg 12:55–60

Weisaeth G (1977) Hodekalsortene 'Respla' og 'Resista' foredlinger med klumprotresistens. Inst, for gronnsakdyrking. Norges Landbr hogsk Meld 74:937–939

Werner S, Diederichsen E, Frauen M, Schondelmaier J, Jung C (2008) Genetic mapping of clubroot resistance genes in oilseed rape. Theor Appl Genet 116:363–372

Williams PH (1966) A system for the determination of races of *Plasmodiophora brassicae* that infect cabbage and rutabaga. Phytopathology 56:624–626

Williams PH, Seidel D (1968) Zum vorkommen von *Plasmodiophora brassicae*-rassen in der Deutschen Demokratischen Republik. Arch Pflanzenschutz 4:31–36

Williams PH, Walker JC (1963) Races of clubroot in North America. Plant Dis Rep 47:608–611

Wit F, van de Weg M (1964) Clubroot resistance in turnips (*Brassica campestris* L.). I. physiological races of the parasite and their identification in mixtures. Euphytica 13:9–18

Xue S, Cao T, Howard RJ, Hwang SF, Strelkov SE (2008) Isolation and variation in virulence of single-spore isolates of *Plasmodiophora brassicae* from Canada. Plant Dis 92:456–462

Yano S, Tanaka S, Ito S, Kameya-Iwaki M (1996) Variations of isozyme and soluble proteins among field populations of *Plasmodiophora brassicae*. Ann Phytopathol Soc 62:365–371

Yano S, Tanaka S, Ito S, Kameya-Iwaki M (1997) Variations of random amplified polymorphic DNA (RAPD) patterns among field populations of *Plasmodiophora brassicae*. Ann Phytopathol Soc 63:179–182

Yoshikawa H (1990) Races of *Plasmodiophora brassicae* on crucifer crops. Plant Prot 44:295–298

Yoshikawa H (1993) Studies on breeding of clubroot resistance in cole crops (in Japanese with English summary). Bull Natl Res Inst Veg Ornam Plants Tea Jpn Ser A 7:1–165

Zhang H, Feng J, Zhang S, Zhang S, Li F, Strelkov SE, Sun R, Hwang SF (2015) Resistance to *Plasmodiophora brassicae* in *Brassica rapa* and *Brassica juncea* genotypes from China. Plant Dis 99:776–779

Zhao Y, Bai YJ, Miao ZY, Li Y, Zhao KH (2013) Identification of physiological races of *Plasmodiophora brassicae* causing clubroot in Chinese cabbage from Northeast China. J Hum Agric Univ 39:176–178

Zheng J, Wang X, Xiao Y, Wei S, Wang D, Huang Y (2018/19) Specific genes identified in pathotype 4 of the clubroot pathogen *Plasmodiophora brassicae*. Plant Dis 103(3):1–9

Perpetuation and Survival of the Pathogen 6

Abstract

Plasmodiophora brassicae causing clubroot disease of crucifers is highly evolved and very successful to perpetuate and survive to complete its life cycle in all kinds of ecological niche it occupies. It produces robust, well-protected soil-borne resting spores which can withstand adverse conditions to survive for a very long period of more than 20 years and respond speedily once comes in a contact with compatible host and cause infection through primary zoospores emerging out from the perennating spores. Various physical, chemical, and biological components of the soil environment have effect on the survival, growth, and reproduction of this pathogen being solely soil borne in nature. Resting spores germination is triggered as a result of signals initiating by the root exudates of compatible hosts in the soil environ. The resultant motile zoospores move rapidly to the root hairs surface to penetrate and colonize. The short period between germination and penetration is one of the greatest vulnerable stages in the life cycle of the pathogen. At this stage, perpetuation and survival of the pathogen is affected by numerous factors in the soil environment. Various factors influence the inoculum potential, viability, and invasion capacity of the pathogen. The different factors interacting with resting spores and zoospores of the pathogen are soil moisture, temperature, light intensity, soil texture and structure, spore load, pathotypes, host resistance, soil pH, acidity and alkalinity, amount of various chemicals like calcium, boron, nitrogen, magnesium, molybdenum, calcium cyanamide, calcareous seaweed, soil fauna and flora, plant growth regulators, extracts of compost, rhizosphere components, residue of the pesticides applied, host and non-host plants, and cultural activity to manage the pathogen. Interactions with the molecular and biochemical cellular environment are also a factor of recent origin to develop resistant cultivars of crucifers.

G. S. Saharan et al., *Clubroot Disease of Crucifers*,
https://doi.org/10.1007/978-981-16-2133-8_6

Keywords

Perpetuation and survival · Factors affecting survival · Effect on resting spores · Effect on zoospores · Effect of soil moisture · Temperature · Light intensity · Soil texture and structure · Spore load · Pathogenic variability · Host resistance · pH · Calcium and magnesium · Acidity · Alkalinity · Boron · Nitrogen · Calcium cyanamide · Molybdenum · Biological constituents · Host and non-host plants

6.1 Introduction

Plasmodiophora brassicae Wor. causes clubroot disease of crucifers. It is very well fitted for successful life to perpetuate and survive for long period of more than 20 years through its robust, well-protected, and apparently long-lived, soil-borne resting spores. These spores withstand adverse conditions, and yet these dormant structures appear to be capable of responding speedily once a compatible host arrives. When host is available, the primary zoospores emerging out from the perennating spores possess efficient and effective locomotion, penetration, and invasive capacities. These features enable *P. brassicae* to exploit the soil environment and its interface, the rhizosphere, with the host to best for its advantage. Once within the host environment, the reproductive cycles of *P. brassicae* are shielded from adverse external conditions, allowing the production of multitudinous new resting spores that eventually rebuild the soil inoculum potential. During this phase, the pathogen appears to have the capability of altering the host's metabolic activities to its own advantage. Only for short duration and across minute distances in the soil are the primary zoospores exposed to hostile and adverse conditions. While in this soil phase, the delicate and vulnerable single walled, zoospores equipped with twin flagellae swim through the soil moisture films from germinated resting spores to the outer surfaces of root hairs. This is the singularly most vulnerable part of the entire life cycle of *P. brassicae* which is very hard to notice, detect, and manage with present techniques.

6.2 Factors Affecting Survival of Pathogen

There are several physical, chemical, and biological factors which influence *P. brassicae* spores (resting spores, zoospores) perpetuation, survival, germination, and movement in the soil environment to reach to the host to continue their life cycle. Above all, interaction of host plants with other vegetation (host and non-host), agricultural operations, human, animals, soil fauna and flora may also influence clubroot infection and development under field conditions. However, experimental evidences on such interactions will strengthen our knowledge to manage clubroot pathogen more efficiently.

6.2.1 Effect on Resting Spores

The resting spores of *P. brassicae* are the first important link in the life cycle of clubroot to commence considering environmental interactions. These robust spores have the purpose of providing long-term survival, and perennation for *P. brassicae* and consequently they have evolved to retain viability in the soil despite exposure to many seasons of adverse weather conditions. Field studies indicate that these spores have a half-life of at least 3.6 years, and some spores may exist for at least 18 years in the absence of suitable hosts before spore populations are eroded to undetectable levels (Wallenhammar 1996). Apparently long-term resting spore longevity perplexed early researchers (Gibbs 1939), not least because of reports that cruciferous crops grown on land previously carrying permanent grassland leys (lays is a English agricultural term for grassland pasture providing grazing over an extended period of time) could become common with clubroot disease after one or two years of arable cropping with *Brassicas*, frequently swedes (*Brassica napus*) grown for stock feed. Temperature, moisture content, and position in the soil profile will influence spore longevity (Monteith 1924; Fedorintschik 1935). Soil pH apparently affects the rate of production of primary zoospores, numbers increase in acidic compared with alkaline soils (Bochow 1961) but without much change in total germination. Spore dormancy, and the need for external stimulants form elements in the initial relations between *P. brassicae* and the environment (Honig 1931). Only a few spores on release from rotten roots which germinate immediately (Ogawa et al. 2001), forms of external stimulus (Ohi et al. 2003; Hata et al. 2002) can be needed to initiate the process. The readiness for germination of spores released from host roots was examined (Humphrey 1892; Chupp 1917; Naumov 1925; Honig 1931; Colhoun 1958). It was concluded that bacteria and other organisms disintegrate the diseased host tissues, and "conditions" the spores for more effective germination. But these secondary microorganisms are not essential for the germination process itself. Unknown mechanisms present within the resting spore initiate germination and control its speed. It appears that these mechanisms within individual spores operate quite separately from other spores since not all spores germinate in synchrony.

Rain and flood water disseminate *P. brassicae* over quite substantial distances especially on sloping land. Wind disperses spores that are collected with light, dry, dusty soil particles over even greater distances. Earthworms (Gleisberg 1922) and possibly moles, root nematodes, and insects may be vectors (Chupp 1924; Eriksson 1936) for *P. brassicae* in soil. Spores are spread in manure (Gibbs 1931b) and on farm animals themselves being capable of withstanding the gut environment. Farm animals and their food supplies sailing with European colonists to the New World and Australasia were probably vehicles for *P. brassicae* infesting virgin territory. The worldwide survey of physiological races by Toxopeus et al. (1986) indicated a predominance of virulences for *B. napus* in these regions. Dirty machinery, wheels, boxes, and tillages all provide potential means for the spread of *P. brassicae*. Wild and weedy members of the Brassicaceae and infested crop transplants harbor, spread, and perpetuate the pathogen. Once established in a soil profile, subsequent distribution is related to soil textural and structural properties, and the frequency and intensity of husbandry operations. Soil compaction and panning by rototilling reduces the movement of spores into the subsoil as did a large "A" horizon of top soil with an active rhizosphere (Murakami et al. 2003). The population density of resting spores decreases at increasing soil depths, more than 97% of the total of *P. brassicae* inoculum is present in the surface soil (0–5 cm depth) and few resting spores are found below 40 cm deep soil (Kim et al. 2000). Since the density of resting spores is affected by soil type, pH, and host susceptibility, a combination of these factors determines intensity of the inoculum potential at a particular site. It follows that after germination in a specific environment, the inoculum potential of *P. brassicae* produces dose–response curves (Murakami et al. 2002b).

At germination, the resting spore volume increases as vacuoles enlarge and the walls thicken becoming more transparent (Woronin 1878; Favorski 1910; Chupp 1917). A single swarm zoospore is liberated from each resting spore leaving behind residual cytoplasm. Germination is characterized by a loss of refractile globules characteristic of stored reserves in dormant spores and probably indicating the enzymic mobilization of these resources. Immersion of spores in water may encourage germination (Chupp 1917; Bawden 1948). Ayers (1944) obtained germination in 1–10 days using tap water and speed of germination appeared to be dependent on spore maturity. Optimal temperature for resting spore germination was established as 24 °C, and pH 6.0–6.7 with an upper lethal temperature of 45 °C, and with visible light inhibiting germination. Spores may be stored as dense suspensions at 3–4 °C for 3 years without loss of viability (MacFarlane 1958), apparently withstanding anaerobic conditions and are not killed by exposure to −20 °C for 3 days. It is standard practice to store galls at −20 °C for several years as stock inoculum (Dixon 1976a, b).

Influences of environmental factors such as temperature, soil moisture, and soil pH on the viability of resting spores of *Plasmodiophora brassicae* incubated in sterile soil were examined by using the direct method based on the fluorescent staining reaction in spores (Takahashi 1994). When resting spores are incubated in acidic soil, the frequency of nonviable spores increases readily and profusely after incubation and reaches approximately 70 to 80% after 30 days of incubation (Fig. 6.1). The rate of the increase was not appreciably affected by temperatures between 15 and 30 °C and by soil moisture contents between 35 and 100% during the 30-day incubation, though affected slightly by low temperature and moisture content (Fig. 6.2). The temperatures and soil moistures in the range tested were not so important factor as affecting the viability of resting spores. On the other hand, soil pH distinctly affected spore viability. A remarkable increase in the nonviable spore frequency was found in acidic soils with pH values of approximately 4.5–6.5. In alkali soils, the rate of increase was conspicuously low in the beginning of incubation period as compared with those in acidic soils, but then the nonviable spore frequency increased continuously at an almost constant rate during 7-month incubation (Fig. 6.3). The soil pH is one of the important factors affecting the viability of resting spores, but may not be the determinate factor in their survival due to lack of ability to suppress independently the decline in the viability during a long period of time. Evidence for an impact of the host on resting spore germination is provided

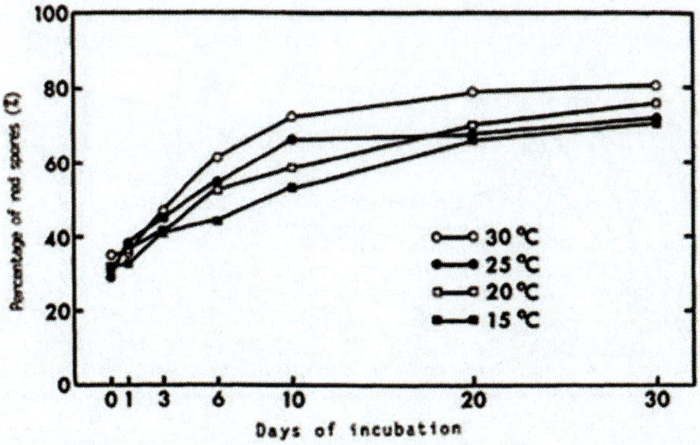

Fig. 6.1 Effect of temperature on the viability of resting spores of *Plasmodiophora brassicae* incubated in sterile soil. Nonviable spores are shown as red spores in figure. The infested soils with a pH of 4.7, and 55% moisture content were incubated at 15, 20, 25, or 30 °C for 30 days. Data are means of four replicated experiments (Takahashi 1994)

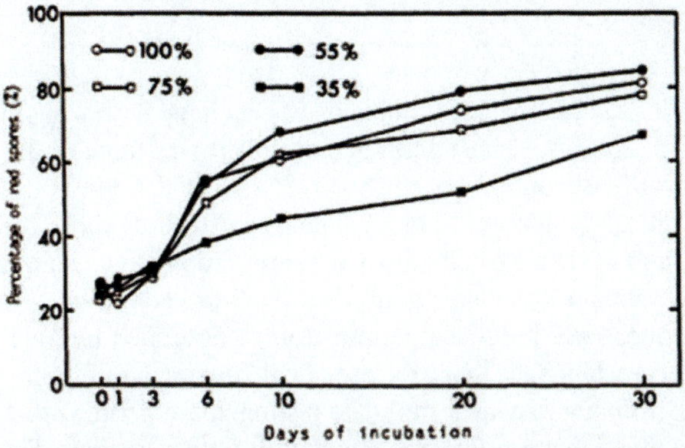

Fig. 6.2 Effect of soil moisture content on the viability of resting spores of *Plasmodiophora brassicae* incubated in sterile soil. Nonviable spores are shown as red spores in figure. The infested soils with a pH of 4.7, and moisture contents of 35, 55, 75, or 100% were incubated at 25 °C for 30 days. Data are means of four replicated experiments (Takahashi 1994)

by Niwa et al. (2008) who reported a significant increase in the percentage of germinated spores (lacking a nucleus) in rhizosphere where the host *B. rapa* var. *perviridis* (neep greens) was present. The involvement of root exudates as stimulants for resting spore germination was postulated and eventually confirmed (Chupp 1917; Hooker et al. 1945; MacFarlane and Last 1957; Bochow 1963, 1965; Macfarlane 1970). Substantial studies by Kowalski and Bochow (1996) concluded that the stimulant effect for germination is non-specific and can come from exudates emanating of many species and is not confined solely to those from hosts of *P. brassicae*. This was supported by the evidence of Craig (1989) who found that root exudates from both calabrese (green broccoli) and perennial ryegrass stimulated

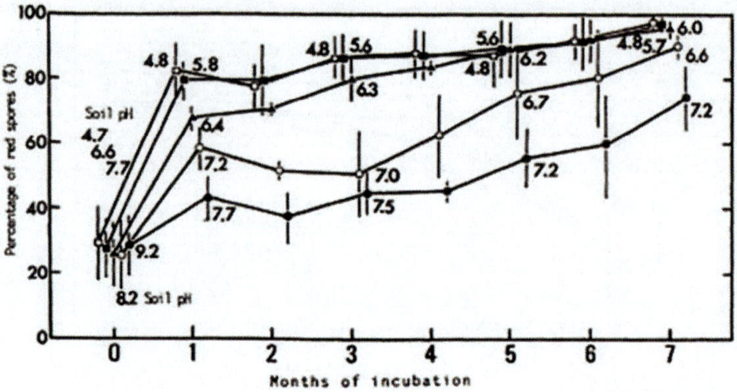

Fig. 6.3 Viability of resting spores of *Plasmodiophora brassicae* incubated for a long period of time in sterile soil. Nonviable spores are shown as red spores in figure. The infested soils with a 55% moisture content, and pH values adjusted by the addition of calcium hydroxide at various concentrations or non-amended, were incubated at 25 °C for 7 months. Numerals in column represent soil pH values. The pH value of non-amended infested soil as control was 4.7 at the beginning of incubation. Data are means of four replicated experiments. Bars indicate standard errors (Takahashi 1994)

spore germination. Some specific stimulants effects have been suggested (Ohi et al. 2003; Hata et al. 2002). Greatest germination (75%) was found to be induced with root exudates from susceptible cabbage hosts. Suzuki et al. (1992) established that an abiotic stimulant can be present in root exudates particularly those from susceptible and resistant Chinese cabbage cultivars. Complex carbohydrate compounds found in the exudates of cabbage stimulated germination in vitro conditions. Possibly several factors may sequentially influence germination as described by Yun et al. (2007). Yano et al. (1991) established that a release of calcium ions from spores induces their germination. Host plant exudates stimulate resting spore germination which in turn releases a second stimulatory factor, encouraging further activity. The environment in which the host plant grows affects the composition of exudates, and drought encourages a release of amino acids. Page (2001) implicated calcium as a factor in generating soil suppressiveness to *P. brassicae* and hence adversely affecting germination but recognized that this element does not operate in isolation from the effects of soil microbial flora. Similar findings were reported by Stewart (2007) who used a comparable range of calcium sources. The number of resting spores present is adversely affected by adding highly calcareous converter furnace slag to soils in Japan (Shinoda et al. 2005). Direct evidence that the inhibition of spore germination is a primary cause of pathogen suppression under neutral pH was observed by Niwa et al. (2008).

The number of germinated resting spore in soil correlates with levels of root hair invasion. As soil calcium declined, it reduces the number of germinated spores and level of root hair invasion. It seems, therefore, that not only do host exudates affect germinability, but the numbers of spores available for germination in the first instance relates in some way to calcium availability in soil. Potentially calcium and pH may affect the longevity and viability of the resting spore in situ in soil. Calcium-rich compost or calcium carbonate changing soil pH from 6.0 to 6.9, and 6.2 to 7.1 respectively, significantly reduces the percentage of germinated spores in the rhizosphere and the number of root-hair infections. Spore germination and subsequent root-hair colonization are retarded by the presence of calcium and alkaline pH values. Earlier, Niwa et al. (2007) found that addition of large amounts of organic matter over 15 years raises soil calcium concentration, changes pH to alkaline values resulting in previously clubroot disease conducive soil becoming suppressive. Organic matter suppresses infection by *P. brassicae,* and the finer particle fractions (<5 mm) changed pH most effectively. Calcium hydroxide, calcium carbonate, and potassium hydroxide also suppress infection with potassium hydroxide being the least effective (Webster 1986). Adding sulfuric acid to a suppressive soil promoted infection by acidifying it. It is concluded that soil pH has major influence on the processes of infection and calcium contributes separately to these influences with both factors operating in unison. Resting spores from "unnatural" sources such as callus cultures are less capable of germination than those from galls from whole plants (Matsumiya 1989). This may result from such spores differing physiologically from those grown under natural conditions perhaps as a result of the callus culturing system. Resting spore numbers per diseased plant increase with low values of disease severity but thereafter remain almost constant for plants with category "3" symptoms and beyond (Dixon and Doodson 1970, 1971; Dixon 1977, 1984b, 1985). Mean numbers of resting spores per diseased plant ranged from 1×9^3 to 1×10^9 regardless of the value of the disease index having apparently passed a saturation threshold, even when resting spore load in soil reaches at modest concentrations as the disease severity increases (Murakami et al. 2002a, b).

6.2.2 Effect on Zoospore

The movement of zoospores is very limited after liberation from the resting spore to the point of encystment on root surfaces within soil. This can be termed Colhoun's Dilemma (Colhoun 1958) since in vivo studies are obscured by soil and in vitro studies are limited by the problems associated with culturing a minute biotrophic microbe which defies axenic culture. Possibly flagellae are those parts of the zoospores primarily affected by soil pH, moisture, calcium, temperature, and interactions with other microbes. The motility of other flagellate organisms is known to be affected by such factors, but little information is available for *P. brassicae*. A combined function of locomotion and location to the flagellae of *P. brassicae* is supported by Dick (1997). Less environmental specialization was suggested in the primary zoospores as compared with secondary stages of zoospores (Dixon 1984a, b, 1985).

6.2.3 Effect of Soil Moisture

Soil moisture is viewed empirically as the medium by which the pathogen reaches to its host. Although clubroot is regarded as a disease of wet soils, there are many reports of its severity increasing during dry seasons or on dry sites. Clubroot disease is, however, considered to be associated with low lying, poorly drained soils and it is most severe after wet weather. Due to this reason, soil moisture is classed as a dominant environmental factor in interactions with *P. brassicae*. Colhoun obtained infection up to pH 8.2 where moisture content was at 70% maximum water holding capacity (Colhoun 1952, 1953). Thereby, it was demonstrated that plants grown under alkaline conditions are vulnerable to clubroot disease when other environmental factors are weighted heavily in favor of the pathogen. In support of this contention, infection developed in 10 to 18 hours with an excessively moist soil (Wellman 1930). Hence, when soil moisture content rises above 50% soil water holding capacity, disease develops very quickly demonstrating indirectly the speed at which primary zoospores travel. Soil texture can affect the motility of *P. brassicae* zoospores (Samuel and Garrett 1945) because in sand–soil mixtures, highest levels of infection have been observed. Infection develops at moisture levels as low as 9% in mineral soils whereas 60% is necessary with organic soils (Hamilton and Crete 1978). Where soil moisture rises from 50% of maximum water holding capacity up to saturation (Dixon 1981, 1984b) disease severity escalates. Lange and Olson (1983) emphasized the dependence of zoosporic microbes on free water existing between the soil crumbs for the movement of zoospores. Free water is critically important for the formation, discharge, and dispersal of zoospores and may influence the encystment and penetration processes at the root-hair surface. The distances traveled by soil-borne zoospores are relatively short, probably between 10–20 mm. Invasion of root hairs occurred up to 75mm from the source of *P. brassicae* infection in soils where water mass movements are minimized (Watson 1967). Mathematical modeling by Yang et al. (2004) demonstrated a relationship between soil moisture and host invasion.

6.2.4 Effect of Temperature

Temperature has been regarded as a factor of lesser importance than soil moisture affecting the successful movement and invasion of *P. brassicae*. Severe infection develops in acidic soils at air temperatures of 16.6 °C, when alkaline soils are used, disease expression is less severe. Conditions for disease development are more favored when working with alkaline soils and air temperature of 23 °C and fluctuations around the mean (Colhoun 1953). Temperatures below 20 °C are thought to present a barrier to clubroot disease development (Chupp 1917; Gibbs 1931a). Monteith (1924) showed that symptoms developed throughout the range of 9–30 °C and were supported in this contention by Wellman (1930). In New Zealand, Ayers (1944) identified the minimum temperature for root-hair infection of swede at 12–14 °C. The early stages of root-hair infection immediately following inoculation require temperatures above 22.5 °C. Once root-hair invasion is completed, then it is believed that lower and fluctuating temperatures are sufficient to support symptom formation. Growth analysis studies by Buczacki et al. (1978) suggested that temperature is most significant as a regulatory factor in the second week after inoculation, which is when root-hair colonization reaches its peak and zoosporangia are forming.

It is possible that the predominance of individual environmental factor changes according to the stage in the life cycle under consideration. Webster (1986) postulated: "that when one factor limits disease expression (such as pH) another may significantly modulate their levels (such as temperature)"; this implies that one factor sets an actual limit while another interacts establishing a frequency or intensity. Wallenhammar (1999), citing C. Williamson, suggested that galling can develop at 7 °C with diurnal fluctuations in temperature and increases of 8–12 h in day length. Certainly, rising spring temperatures are associated with infection, and disease expression in Swedish oilseed rape (*B. napus*) crops, Wallenhammar claimed as day length increases, and host growth accelerates. Interactions may go further and relate to effects on the constitution of the host, since Roback and Gabrielson (1988) found that temperature influences the expression of host resistance. The cauliflower cultivars resistant at 15 °C can show susceptibility at 20 °C with high inoculum potentials.

6.2.5 Effect of Light Intensity

Plasmodiophora brassicae is a soil-borne pathogen so light has no significant impact on growth. Colhoun (1961) showed that light intensity had marked effects on the relationship between spore load and the number of plants that develop clubroot disease. Consequently, Grainger (1962) was able to relate clubroot development with his C_P/R_S ratio system, where C_P = the weight of total carbohydrate in the whole plant, and R_S = the residual dry weight of the shoot. Webster (1986) commented that "light may influence disease expression via an effect on host photosynthetic efficiency, and hence on energy reserves available to fuel clubbing." Light may also regulate the balance between shoot and root growth. Rausch et al. (1981) showed that low light intensities gave a greater reduction in root growth in infected compared with control plants.

6.2.6 Effect of Soil Texture and Structure

Clubroot disease expression is associated with all kinds of soils. Palm and Mc New (1956) demonstrated infection happens more readily in mixtures of soil and sand or of clay and sand or in undiluted soil as compared with pure sand cultures. Soil compaction and consequent water-logging caused by animals or machinery is associated with increased disease severity around headlands and gateways (Anderson 1855; Russell 1859; Somerville 1895). In contrast to these, Anderson (1855) showed the disease with loose open soils, and Somerville (1895) associated the disease on turnips (*B. rapa*) with soil aeration caused by hoeing during winter, and Larson and Walker (1934) contended that aerating alkaline soil favors infection. Light, sandy, humus-rich, and clayey soils are thought most disease prone. As with the interaction of clubroot and acidity or alkalinity, there is a dearth of rigorously tested science-based evidence concerning disease development and soil characteristics. Soil type was shown by SBH (1980) to influence the disease causing abilities of physiological races of *P. brassicae*. Clay and loam soils are prone to compaction and may also be calcareous, resulting in interactions between conducive and suppressive effects moderated by factors such as inoculum potential thereby determining disease expression. Soils of high water holding capacity, such as silts, are likely to encourage this pathogen. Soil type interacts with calcium and pH is considered by Campbell and Greathead (1996) as factors determining disease intensity.

6.2.7 Effect of Spore Load

A relationship between root-hair infections and cortical clubbing can be expected from epidemiological theory (Vander Plank 1975) and was investigated by Naiki et al. (1984) using the ECD series of genotypes and 54 common Japanese crucifers. Despite detailed recordings of root-hair infection in susceptible and resistant types, disease frequency could not be related to subsequent clubbing. Nonetheless, spore load is accepted as seminally important in determining the intensity of subsequent disease especially at low concentrations. Samuel and Garrett (1945) found that number of infected root hairs increases concomitantly with spore density/spore load with percentage of clubbed plants. There are interactions with spore maturity. Older spores appear more capable of causing infection compared with younger ones (Colhoun 1958) termed as "infective power." Under less than optimal environments, spore load is related with the number of diseased plants, thus alkaline soils have a direct relationship between spore load and number of diseased plants.

The conditions under which resting spores are stored affected their viability (MacFarlane and Last 1957) whereas germination is encouraged and the proportion of germinated spores increases by the presence of host root exudates. Interacting factors such as moisture, temperature, pH, light intensity, and internal factors including spore size, age, and nutritional status affect the overall outcome of the interaction between host and pathogen. "Inoculum pressure is modified by environment" as noted by Webster (1986) and observed that under environmentally unlimiting conditions, and below the threshold level of infection needed for maximum disease expression, severity of clubbing is proportional to increasing inoculum concentration and total root-hair infection. Above this threshold level, however, increasing spore concentration may generate higher root-hair infection levels but not relate to increased disease severity. Webster's "threshold" is effectively a saturation point beyond which the physiological and biochemical processes governing symptom development within the host cannot be affected by inoculum load. Saturation itself is not a fixed and immutable value because Webster's work supported the view that only a low percentage of spores in an inoculum are capable of causing successful infection or invasion at any one time. Saturation of the root-hair space within an observed section of root, distribution of spores around susceptible root hairs, and distances over which they can travel are all factors influencing the chances of an additional spore being able to establish an infection and proceeding to cause clubroot disease. The two stage life cycle of *P. brassicae* inevitably means that only a limited number of invading zoospores can ultimately proceed to incite symptoms.

6.2.8 Effect of Pathogenic Variability

There may be competition for root-hair space by different physiological races of *P. brassicae* as suggested by Jones (1980). Both antagonistic and synergistic relationships between races of *P. brassicae* may affect relationships between physiological forms (Dixon 1979, 1980, 1979, 1980; Jones et al. 1981). Within any population of *P. brassicae* spores, there may be a range of vigor or infective capacities such that some on establishing infection proceed more rapidly through the life cycle than others. Jones (1980) reported that more than one physiological race of *P. brassicae* may occur within a population or within a spore suspension prepared from a single gall. Jones et al. (1981) and Dixon et al. (1981) further demonstrated competition between physiological races in experiments using inocula composed of mixed races. Races varied in "vigor" or "aggressiveness" as indicated by the intensity of differential reactions between races and hosts using the ECD series. It has been observed that inoculum from *B napus* cv. *marian* clubs generates a higher disease index on cv. Wilhelmsburger than on cv. Nevin, but the reverse was the case for inoculum from cv. Acme clubs (Dixon 1981). Tanaka et al. (2006b) suggested that there is a suppression of plasmodial development during secondary colonization in *B. rapa* ssp *pekinensis* cv. Kubai 70 which occurs differentially when using a range of isolates of *P. brassicae*. How differences in vigor, or aggressiveness, or infective capacity are derived, and when they come into action are questions yet to be answered (Dixon 2009c).

6.2.9 Effect of Host Resistance

Host resistances may be considered as an environmental component affecting the success of *P. brassicae*. In that perspective, it becomes an additional sink for the energy expended by invading spores. The process of breaching host resistance may be a function of the biological fitness of successive waves of invasions by zoospores, both primary and secondary, and diminishing general resistance in the host. Ultimately, successful infections are established in the root hairs of resistant cultivars. It appears that specific resistance is expressed against the secondary phase of the *P. brassicae* life cycle. Hence, during that phase, more energy may be expended by *P. brassicae*, fruitlessly where robust resistance is present but more successfully when this is not the case. In view of the highly polygenic nature of some forms of resistance to *P. brassicae,* especially in *B. oleracea,* these events might go some way towards explaining the lags in time between invasions which result in less advanced states of infections on assessment days where plants are subjected to lower inoculum concentrations. As a result, infection numbers fall below an observable threshold and in practice are not counted in assays used to determine the value of resistant genotypes. These phenomena have generated much more interest in what constitutes observable or phenotypic resistance to *P. brassicae* (Toxopeus et al. 1975).

6.2.10 Effect of pH, Calcium, and Magnesium

The *P. brassicae* and its interaction with the environment of calcium content and the associated hydrogen ion content (pH) of soil are very important for successful life cycle of the pathogen. Calcium emerges as a fundamental factor in the life cycles of both *P. brassicae* and its hosts. Datnoff et al. (2007) summarized the involvement of calcium in host metabolism, physiology, and signaling of many host–pathogen interactions indicating a relationship with expression of resistance. The disease is associated with acidic soils and is alleviated by the use of various forms of agricultural lime. The incidence of clubroot is not limited at pH 7.0 as is still claimed especially in much farm advisory and home gardening literature. Colhoun (1958) stated "results obtained by field experimentation show the difficulty encountered in determining the exact upper limit of the soil pH at which infection can (still) occur." The chemical and physical forms, and quantities of calcium used also affect the results and add additional variables. The effects of pH and of calcium can be separated and quantified individually as factors influencing the environmental success of *P. brassicae*. The impact of the balance of nutrients in the soil is significant while the actual content of individual ions is still important. Myers and Campbell (1985) suggested that clubroot disease expression depends on the balance between pH and the amounts of calcium and magnesium in the soil. Fletcher et al. (1982) achieved greatest effects of clubroot disease with field applications of calcium carbonate and calcium nitrate which increased pH to 7.9 and 8.3, respectively. Although pH is a major factor in reducing disease expression, some other factor than pH possibly the calcium itself is involved (Hamilton and Crete 1978). There is a tendency to assume that these factors affect the microbe while in the soil, but because *P. brassicae* spends most of its life cycle within the host, it can be fair to suggest that calcium and pH also affect the environments. A role for calcium in the post-infection development of *P. brassicae* is supported by the demonstration that incorporation into roots is pH-dependent (Myers and Campbell 1985). Campbell and Greathead (1996) contended that *P. brassicae* is affected by pH and calcium concentration at more than one point in the life cycle between spore germination and the completion of resting spore formation in the cortical cells as confirmed by Webster (1986), Dixon and Webster (1988), Webster and Dixon (1991a), Dixon and Page (1998), and Page (2001). It is evident that calcium has the greatest impact when it is present in the period between spore germination to post-penetration of root hairs. The latter appears to be when root-hair infection has the biggest impact on subsequent gall formation. Apparently, there may be separate mechanisms involved because the periods 0–3 and 0–7 days post-penetration seem to be different with respect to the extent of their influence on subsequent disease development. The expression of effect seems to be cumulative because it took longer when a 30 mel^{-1} Ca^{++} solution was used as compared with 55 mel^{-1} Ca^{++} solution to achieve similar final results. The host–pathogen response varies also with pH, however, that is a separate factor. It is worth stating here that calcium at pH 7.2 needed to be present by day 14 to suppress root-hair infection or alter the progress of galling. The pathogen may be affected by the calcium environment in the root hair, and this alters subsequent behavior in the cortical cells. The major significance is that high concentrations of calcium at pH 6.2 or 7.2 reduce total numbers of root-hair infections, and the rate of maturation through plasmodial, sporangial, and zoosporangial stages as compared with the controls. Raised concentrations of calcium completely inhibit the later stages of *P. brassicae* development in the root hair even where high inoculums doses are applied. The calcium effect commences in the soil because (Dixon and Page 1998), the germination of resting spores, motility of zoospores, and the composition of benign microbial flora around roots are altered. High concentrations of calcium can possibly reduce flagellar action. Satir (1982), and Sleigh and Barlow (1982) reported that changes in calcium of the order of 10^{-6} to 10^{-4} M affected the action of demembranated flagellae.

6.2.11 Effect of pH, Acidity, and Alkalinity

The uneven distribution of acidic and alkaline areas of soil in individual fields with pH ranging from 5.73–8.45 in localized patches may influence disease development (Wallenhammar 1999). Mattsson (1995) showed that pH values of the subsoil are frequently more alkaline than the upper horizons in Sweden especially in the calcareous glacier clay region near Uppsala in eastern central Sweden. This modernizes aspects of *Colhoun's Dilemma* related to pH. Earlier, Palm (1958, 1963) had concluded that the effect of pH is not restricted solely to the establishment of *P. brassicae* as a pathogen because the rate of gall proliferation was markedly suppressed by an alkaline condition of the medium after infection in the host tissues. The changes in the soil reaction may have more drastic effects on gall development than on the number of infections by zoospores. Using organic buffers, Myers and Campbell (1985) adjusted pH and calcium content separately from each other and showed that at 10 mel^{-1} Ca^{++} and a pH of above 7.1 reduced the numbers of primary zoosporangia in root hairs thus inhibiting galling. Webster and Dixon (1991b) demonstrated that the effects of pH are independent of calcium concentration and found that alkaline pH reduces total root-hair infection number and retards the maturation of plasmodia, sporangia, and zoosporangia. The pH effect on the maturation of root-hair infections is activated by exposure to alkaline pH within 3 days of penetration. Prolonged exposure beyond 3 days gives no additional effect.

There may be a dual effect that alkaline pH increases sensitivity of the host and/or *P. brassicae* to calcium effects as well as increasing the efficiency of calcium uptake.

The effects of pH and calcium are remarkably similar but this does not necessarily mean they are one and the same as has been suggested by some workers. They may regulate the pathogenic potential of an inoculum quite separately. Since pH regulates the response to calcium, intracellular function may be modified. A high concentration of H^+ ions in plant tissues is potentially antagonistic to calcium. Membrane permeability is decreased by both alkaline pH and by high calcium. This environment can affect the growth and reproduction of *P. brassicae* as it proliferates within the host root-hair and epidermal cells or within the cortical cells. Alkaline environments can affect primary and secondary invasions, cortical migration, and cell hypertrophy. An involvement of Ca^{++} in the growth and reproduction of *P. brassicae* ultimately leading to induced cell death or hypersensitivity is suggested by Takahashi et al. (2006). At the agronomic level, promotion of high alkalinity linked with continuous cropping is suggested by Shinoda et al. (2005) as a means of reducing the soil inoculum load.

6.2.12 Effect of Boron

Boron is associated with affecting the activities of *P. brassicae* (O'Brian and Dennis 1936). Palm (1963) investigated the effect of boron on *P. brassicae* in sand cultures and recorded maximum root-hair infection at 0.3 mel^{-1} or less. In the absence of boron, the inhibitory effect of calcium on root-hair infection is suppressed. Lime may fail to diminish clubroot disease in boron deficient soils. Dixon and Wilson (1983, 1984a, b, 1985) have achieved significant reductions in disease index with sodium tetraborate applied to acidic granitic soils. In environments with elevated boron concentration, there are significant effects both in the root-hair and cortical phases of *P. brassicae*. Throughout the *in planta* stages of the life cycle of *P. brassicae*, boron has an impact on the pathogen. There appears also to be a relationship between the quantity of boron in the plant which is moderated by uptake over time and space as determined by the size of the plant root system and its capacity to absorb boron. It is likely that there are interactions with other ions. Lime applications in the forms of calcium carbonate or oxide may alter the nutrient environment in soil to the detriment of *P. brassicae* and therefore, make the host–pathogen association more affected by other factors such as boron. Alternatively, boron may have a primary effect because Webster and Dixon (1991a, b) found that the effects of boron interact with both the primary and secondary stages of development of *P. brassicae* ultimately affecting the intensity of symptom expression. The environment induced by boron in cells where membrane permeability and wall structure are altered may be to the detriment of *P. brassicae*. It can also make for conditions less conducive for nuclear division by the pathogen. Quite possibly, boron effects are distinct from those of calcium and pH. Dixon (1991) identified that boron affects the progress of *P. brassicae* by retarding the rate of sporangial maturation. The correlation of diminished intensity of disease expression and boron suppression of root-hair infection and gall formation appears related to host exposure. Long exposures to low concentrations seem to equate with the effects of shorter exposures to higher concentrations. Field and controlled laboratory studies (Craig and Dixon 1993a, b) identified that boron has a substantial effect on the ability of *P. brassicae* to invade root hairs and establish colonization in the field. Raising the boron content of the rhizosphere prior to the availability of a susceptible host to infested soil limited the subsequent ability of *P. brassicae* zoospores to penetrate, colonize root hairs, and cause symptoms.

6.2.13 Effect of Nitrogen, Calcium Cyanamide, Molybdenum, Calcareous Seaweed, and Plant Growth Regulators

High concentrations of nitrate (4–6 times standard) consistently suppress disease symptoms. Webster (1986) postulated a two-phase response with low concentrations enhancing and high concentrations retarding symptoms, as found with other biostimulants (Dixon 1991). Adding nitrate nitrogen results in the stimulation of cellular free amino acid pools. If this stimulated arginine or lysine-rich histones, then this can possibly lead to repression of RNA polymerases in the pathogen preventing its access to the gene products needed for pathogenesis. As nitrate concentration increases enzyme sites become saturated with consequent substrate and/or product inhibition and the amino acid moieties being diverted towards forming an environment inhibitory to *P. brassicae*. Nitrate metabolism is regulated by the availability of reduced cofactors NAD (P) H (nicotinamide adenine dinucleotide (phosphate)-reduced) for conversion to the ammonium form (Hewitt 1970). Under conditions of high nitrate supply, all available NAD (P) H are used up, and then a shortage of such cofactors can influence the activity of *P. brassicae in planta*. Raising nitrate levels above 20 mel^{-1} reduces symptom expression and numbers of infected plants. Results obtained in controlled conditions using split root techniques (Dixon 2009c) demonstrated the effects of nitrate ions in influencing the rhizosphere environment to the detriment of *P. brassicae*. These results were then supported by glasshouse and subsequent field experiments (Dixon 2009a). Bhattacharya and Mandal (2006) found that calcium ammonium nitrate and calcium nitrate significantly reduce the intensity of clubroot disease. Page (2001) showed by detailed laboratory and field studies that calcium nitrate is associated with decreases in *P. brassicae* infection with a subsequent reduction in the severity of symptom expression. During the soil phase of *P. brassicae*, the viability of resting spores and the ability of primary zoospores to invade the host are reduced by the presence of calcium nitrate. There may be changes to the fitness of *P. brassicae* as a result of the presence of calcium nitrate. In this compound, calcium is available in a highly soluble form linked to the nitrate ion. Page (2001) also concluded that the presence of calcium nitrate in the rhizosphere may also have been associated with changes to the dominant physiological race of *P. brassicae*.

Interactions between the fertilizer calcium cyanamide, *P. brassicae,* and other soil microbes have also been demonstrated (Dixon 2009b). The calcium cyanamide and the products of its degradation, calcium and nitrate nitrogen are associated with the reduced viability of *P. brassicae*. Additional research is required to elucidate the means by which this effect is achieved, but it is now considered likely that calcium cyanamide alters the balance of biological components in the soil environment surrounding *P. brassicae*. Thereby, the growth and reproduction of soil-borne microbes antagonistic to *P. brassicae* are encouraged. The adding of calcium cyanamide to soil infested with *P. brassicae* results ultimately in the reduction in the intensity of clubroot disease. Other calcareous substances such as calcified seaweed (a form of coral), or true extracts of algal seaweed which contain inorganic nutrient ions and organic compounds including plant growth regulators and extracts of composts (Tilston et al. 2002) have been associated with changes to the biological environment of soil to the detriment of growth and reproduction of *P. brassicae*. Sen (2005) found effects of molybdenum along with calcium and boron in the root environment in relation to pathogenesis of *P. brassicae* on rapeseed-mustard in West Bengal, India. Interactions of the nutrient environment, pathogenesis, and resistance are described in the other chapters of this book.

6.2.14 Effect of Biological Soil Constituents

There are relationships between *P. brassicae* and the macro- and microflora and fauna in soil. The free-swimming zoospores of *P. brassicae* are undoubtedly at risk from the predatory habits of soil inhabitants. Instances of "disease suppression" may well relate to the presence of such organisms which can increase in unquantified amounts either naturally or following husbandry activities. Soil suppressiveness to pathogenic organisms resulting from the activities of saprophytic microflora is a well-accepted phenomenon (Alabouvette et al. 1996). Organic or inorganic soil amendments that stimulate the microflora have a significant effect on the survival of *P. brassicae*. Bacteria such as *Bacillus* spp. and fluorescent *Pseudomonas* spp. are recognized in affecting the growth of *P. brassicae* (Einhorn et al. 1991). Biotic suppressive soils were identified in Taiwan with pH of above 7.4, and a calcium content of 1210 ppm (Hseith and Wang 1986). Because the resting spore walls contain chitin, it is likely that chitinolytic bacteria can be major antagonists of *P. brassicae* reducing the inoculum potential (Anonymous 2008). Antibiosis resulting from microbial sources has usually been approached as a means for the biological control of *P. brassicae* as opposed to developing an understanding of the ecological relationships between organisms. Soil suppressiveness relating to *P. brassicae* has been found to be more conducive to *P. brassicae* than low-humic andosols even when high spore concentrations were present in the latter. It was suggested that the suppressiveness of low-humic andosols relates to the presence of biological antagonists (Murakami et al. 2000). Biotic suppression of *P. brassicae* in presence of Chinese cabbage (*B. rapa*) host plants results from the presence of the soil endophytic fungus *Heteroconium chaetospira* (Narisawa et al. 2005). Soil moisture content, pH, and spores density significantly affected the level of repression of *P. brassicae*. Crop rotations, particularly those containing maize (*Zea mays*) depressed the activities of *P. brassicae* (Yamada et al. 2003). This may be expressed as ecological interaction and biological control (Dixon 2003a, b).

Primary plasmodia are found in the root cultures of both susceptible and resistant cultivars (Takahashi et al. 2006), but secondary plasmodia proliferate only in cultures of susceptible hosts. The alkalization of the root culture of resistant cultivars is responsible for this difference. In the rhizosphere (Nicholas 1965), saprophytic species thrive supported by nutrients in plant root exudates. This is the environment in which the primary zoospores of *P. brassicae* are actively attempting penetration and colonization of the host root hairs. This is a dynamic situation in constant flux as the microbial flora changes under the influence of substantial alterations to root activity. It is observed that root hairs themselves last for only a few hours, and the range of microbial species alters in some instances almost hourly. This situation must have a major impact on the inoculum potential of *P. brassicae*. Ultimately, this potential includes the supply of biological energy needed for penetration and colonization of a host. It is a function of inoculum density or intensity (mass or units of inoculum per unit of soil), available nutrient (both internal and external to the propagules), environmental factors, and genetic capacity of *P. brassicae* itself (Martinson 1963). The levels of genetic resistance for a particular host species, which may vary during its life cycle and interactions with the environment, are key factors determining the outcome of encounters with *P. brassicae*. Combining all these factors of host and pathogen interacting with their respective environments offers a predictive index for the success of growth and reproduction by *P. brassicae*. Dixon and Walsh (1998) showed that while *P. brassicae* only interacts with the soil environment for a short period both in time and space, this period is critical for the success of *P. brassicae* in establishing subsequent growth and reproduction within the host. The rhizosphere is not a rigid entity either in its parameters of shape, content, or time. The propagules of *P. brassicae* experience an environment in which there is an irregular distribution of nutrients, water, and oxygen. Elements such as calcium which move towards the root surface by mass flow may accumulate in the rhizosphere in quantities larger than required for uptake by the plant root. As a result, surplus ions accumulate in the immediate environs of the rhizosphere and can have immense impact on *P. brassicae*. Although rhizosphere around the root apex is acidic that further reduces can be alkaline had effect on the colonization, and invasion of these areas by this pathogen. Important aspects for the survival of soil-borne pathogens include the conditions prevailing at the time of arrival at the root surface, and in the surrounding rhizosphere affecting establishment and penetration of the pathogen (Bowen and Rovira 1999). Page (2001) showed that under some circumstances, the presence of calcium and nitrate nitrogen can be associated with changes in the virulence spectrum of the pathogen population begins to indicate the powerful forces present in the rhizosphere which impinge on the primary zoospores of *P. brassicae*. It appears that signals can pass out from the host root to the resting spores of *P. brassicae,* triggering germination, and then the primary zoospore proceed with location finding and flagella motion towards the root surface (Chap. 1). It has been suggested that sugars and/or carbohydrates in root exudates are involved in triggering the germination of *P. brassicae* and subsequent processes. It is quite possible these compounds also offer an external source of energy for the pathogen. Location and direction finding by the zoospore is most likely a result of some form of general or specific chemotaxis such as a gradient of compounds such as carbon dioxide, oxygen, or a redox gradient. Alternatively, the zoospores may simply be passively swept through the soil water films by physical flow. That might explain why high levels of soil moisture are prerequisite for successful colonization. Because temperature is also an important factor, it possibly regulates the rates of energy release within the zoospore, and chemical interactions with soil components. While *P. brassicae* is germinating, and in motion, the primary zoospores are subject to attack by other soil inhabitants such as *Bacillus* species. Decoy crops such as the Japanese leafy daikon (*Raphanus sativus*) appear to operate by encouraging resting spore germination and root-hair colonization but without successful secondary stage colonization and subsequent symptom expression (Murakami et al. 2000). Presumably such hosts offer sources of energy for germination and colonization but do not have an internal environment conducive to the growth and reproduction of the secondary stages of *P. brassicae*. The degree of decomposition of organic matter apparently influences its suppressive effects in soil and can indirectly also influence these processes.

The fungus *Heteroconium chaetospira* inhibited the activities of *P. brassicae* even where soil physical conditions, soil moisture, and soil pH would have otherwise been conducive (Narisawa et al. 2005). Chinese cabbage (*B. rapa*) roots became colonized by hyphae of *Heteroconium chaetospira*, but there is no visible evidence of host cell wall degradation, host reactions, or invigilation of the host plasma membrane around the hyphae (Yonezawa et al. 2004). Those soils with negative charges matching similar potential on spores of *P. brassicae* are conducive for development of the pathogen while those with positive charges are suppressive (Murakami et al. 2004a, b). Other members of the soil microflora such as *Bacillus* spp., *Pseudomonas* spp., and *Trichoderma* spp. are recorded as reducing the activity of *P. brassicae* (Yeoung et al. 2003). The biotic basis for soil suppressiveness to *P. brassicae* was shown to be supported by abiotic factors (Murakami et al. 2000, 2007). The effect of rhizosphere components on the success of *P. brassicae* was emphasized by Belec et al. (2004).

6.2.15 Effect of Host and Non-host Plants

Soil environments created by non-host and host plants such as leek (*Allium porrum*), winter rye (*Secale cereale*), and perennial ryegrass (*Lolium perenne*) tended in glasshouse studies to reduce the growth of *P. brassicae* but such effects have been less dramatic in the field. There was no host species-specific interaction between *P. brassicae* and non-host types (Friberg et al. 2006). Root exudates from *Lolium perenne* stimulated more spore germination than was obtained from other plants (Friberg et al. 2005). These differences cannot be explained by variations in the composition of the exudates or differences in root activity. But alternatively environment can mitigate against *P. brassicae* so that the pathogen fails to invade otherwise susceptible hosts such as *Cardamine flexuosa* (Tanaka et al. 2006a). Break or rotational crops may alter the soil environment in a manner suppressive to *P. brassicae* (Cheah et al. 2006). Studies of the association of *P. brassicae* and hosts other than cruciferous types have attempted to construct rotations which are antagonistic to *P. brassicae*. The pathotypes interact and are influenced in these interactions by the surrounding environments (SBH 1980; Jones 1980). The extent to which such interactions are influenced by host plants remains to be determined.

The success of *P. brassicae* is dependent on the density of resting spores, soil type, soil pH, and host susceptibility. Dose–response curves may vary even where soils are of similar pedological type (Murakami et al. 2002b). The impact of edaphic chemistry including boron, calcium, nitrogen concentrations, and pH on the growth and reproduction of *P. brassicae* within host cells is known (Dixon 2002; Dixon and Page 1998; Webster and Dixon 1991a, b). The implications for the manner by which resting spores germinate, the motility of primary zoospores, growth, and reproductive efficiency of *P. brassicae in planta,* and for the expression of host resistance (Chap. 1) have been reviewed by Dixon (2009a).

The death of *Plasmodiophora brassicae* apparently occurs logarithmically in soil so that a few propagules persist for a very long time (MacFarlane 1952). This conforms to Wallenhammar's findings. Possibly even the lowest level of inoculum can be significant for the survival of *P. brassicae* because, as Ayers (1944) described, infection commences from a single zoospore. That being so then *Plasmodiophora brassicae* is indeed superbly well evolved and fitted for survival in hostile soil environments (Dixon 2009c).

6.2.16 Effect of Traditional and Modern Technology

For the management of clubroot under field conditions, various traditional and modern techniques have been developed, and are in use which can influence the pathogens survival and perpetuation. The details are given in respective chapters of this book.

6.2.17 An Over View

Perpetuation and survival of *P. brassicae* is influenced by all the biotic and abiotic factors interacting in the soil environ for its infection, pathogenesis, and completion of life cycle including disease management strategies employed. All the factors affecting different process have been described in detail in subsequent chapters of this book.

References

Alabouvette C, Lemanceau P, Steinberg C (1996) Biological control of *Fusarium* wilts: opportunities for developing a commercial product. Chapter 9. In: Hall R (ed) Principles and practice of managing soil borne plant pathogens. American Phytopathol Soc, St Paul, pp 192–212

Anderson A (1855) Report on the disease of finger and toe in turnips. Trans Highland Agric Soc 1855:118–140

Anonymous (2008) Chitin against clubroot. Gemuse 44(2):26–27

Ayers GW (1944) Studies on the life history of the clubroot organism *Plasmodiophora brassicae*. Can J Res 1944:143–149

Bawden FC (1948) Department of plant pathology. Rep Rothamsted Exp 1947:49–52

Belec C, Tremblay N, Coulombe J (2004) Managing soil – borne pathogens: a sound rhizosphere to improve productivity in intensive horticultural systems. Acta Hortic 635:41–46

Bhattacharya TK, Mandal NC (2006) Management of clubroot (*Plasmodiophora brassicae*) of rapeseed and mustard by nitrogenous fertilisers. Ann Plant Prot Sci 14(1):260–261

Bochow H (1961) Uber die Beeinflessung von *Plasmodiophora brassicae* Wor. durch Mineralsalze. Phytopath Zeitschr 40:420–421

Bochow H (1963) Untersuchungen zur Ökologie und indirekten Bekämpfung von *Plasmodiophora brassicae* Worn. Ein Beitrag zur Klärung von Fragen der Bodenhygiene. Habilitationschrift 1963:155

Bochow H (1965) The effect of root diffusates of host and non-host plants on the spore germination of *Plasmodiophora brassicae* Worn., Plant microbe relationships. In: Proceedings of a symposium on the relationships between soil micro-organisms and plant roots. Czechoslovak (now the Czech Republic) Academy of Science, Prague, pp 296–299

Bowen GD, Rovira AD (1999) The rhizosphere and its management to improve plant growth. Adv Agron 66:1–102

Buczacki ST, Ockendon JG, Freeman GH (1978) An analysis of some effects of light and soil temperature on clubroot. Ann Appl Biol 88:229–238

Campbell RN, Greathead AS (1996) Control of clubroot of crucifers by liming. In: Engelhard AW (ed) Soil borne pathogens: management of diseases with macro – and microelements. Am Phytopathol Soc (APS), St Paul, pp 90–101

Cheah LH, Gowers S, Marsh AT (2006) Clubroot control using *Brassica* break crops. Acta Hortic 706:329–332

Chupp C (1917) Studies on clubroot of cruciferous plants. Bull Cornell Agric 387:419–452

Chupp C (1924) Cabbage diseases. Plant Dis Rep 8:125–126

Colhoun J (1952) Factors affecting the incidence of club root disease of *Brassica*. Nature 169:21–22

Colhoun J (1953) Observations on the incidence of club root disease of *Brassica* in limed soils in relation to temperature. Ann Appl Biol 40:639–644

Colhoun J (1958) Club root disease of crucifers caused by *Plasmodiophora brassicae* Woron. – A monograph. The Commonwealth Mycological Institute, Kew, Surrey, Phytopathological Paper No 3

Colhoun J (1961) Spore load, light intensity and plant nutrition as factors influencing the incidence of club root disease of *Brassicas*. Trans Br Mycol Soc 44:593–600

Craig MA (1989) Resting spores of *Plasmodiophora brassicae* as affected by root exudates of hosts and non–hosts. B.Sc. (Hons) Horticulture Dissertation, University of Strathclyde, Glasgow, 71 p

Craig MA, Dixon GR (1993a) The influence of boron on the growth of *Plasmodiophora brassicae* in Summer Cabbage. Proc Crop Prot 1993:277–282

Craig MA, Dixon GR (1993b) Alterations to the growth of *Plasmodiophora brassicae* Wor. (Clubroot) associated with boron. Proc 6th Intl Cong Plant Pathol Montreal Poster 3.4.12

Datnoff LE, Elmer WH, Huber DM (2007) Mineral nutrition and plant disease. The American Phytopathol Soc, St Paul

Dick MW (1997) Fungi, flagella and phylogeny. Mycol Res 101:385–394

Dixon GR (1976a) Assessment keys for the evaluation of seedling and adult plant symptoms of clubroot. Keys 3.1.2 and 3.1.3. In: Manual of plant growth stages and disease assessment keys. Ministry of Agriculture, Fisheries & Food, London

Dixon GR (1976b) Methods for testing seedling resistance to clubroot (*Plasmodiophora brassicae*) used in Western Europe and U.S.A. Plant Pathol 25:129–134

Dixon GR (1977) Manual of plant growth stage and disease assessment keys - contains keys developed for the assessment of clubroot of root crops, clubroot-seedling key, Department for Environment, Food & Rural Affairs (Defra) (previously Ministry of Agriculture, Fisheries and Food MAFF), Plant Pathology Laboratory, Harpenden, Hertfordshire now at Central Science Laboratory, York

Dixon GR (1979) Interactions between host cultivar and *Plasmodiophora brassicae* populations. In: Proceedings of Woronin + 100 International Conference on Clubroot. Published by University of Wisconsin, USA, pp 93–96

Dixon GR (1980) Variation in *Plasmodiophora brassicae*. Ann Appl Biol 94:278–280

Dixon GR (1981) *Plasmodiophora brassicae* (clubroot) and epidemic development and the measurement of disease levels. In: Vegetable crop diseases. MacMillan, London, pp 137–142

Dixon GR (1984a) Vegetable Husbandry. Research investigations and field trials, horticulture division, Aberdeen School of Agriculture, Aberdeen

Dixon GR (1984b) Galls caused by fungi and bacteria. In: Wood RKS, Jellis GJ (eds) Chapter 15 in plant diseases: infection, damage and loss. Blackwell Scientific Publications, Oxford

Dixon GR (1985) Vegetable Husbandry. Research investigations and field trials, horticulture division, Aberdeen School of Agriculture (University of Aberdeen Department of Agriculture and North of Scotland College of Agriculture), Aberdeen

Dixon GR (1991) Primary and secondary stages of *Plasmodiophora brassicae* (clubroot) as affected by metallic cations and pH. In: Beemster ABR, Bollen GJ, Gwerlagh M, Ruissen MA, Schippers B, Tempel A (eds) Biotic interactions and soil borne diseases, no 23 in developments in agricultural and managed forestry ecology. Elsevier, Amsterdam, pp 381–386

Dixon GR (2002) Interactions of soil nutrient environment, pathogenesis and host resistance. Plant Prot Sci 38(1):87–94

Dixon GR (2003a) Clubroot, a case for integrated control. Plantsman 2(3):179–183

Dixon GR (2003b) Suppressing *Plasmodiophora brassicae*, the cause of clubroot disease by bio-manipulation. In: 8th International Congress of Plant Pathology: Solving Problems in the Real World. February 2nd to 7th 2003, Christchurch, New Zealand. Delivered as Oral Poster on Thursday 6th February; Poster Number 20.17, page 275 Abstracts volume 2 - Offered Papers

Dixon GR (2009a) Husbandry– the sustainable means of controlling soil borne pathogens: a synoptic review. Acta Hortic 817:233–242

Dixon GR (2009b) Calcium cyanamide -100 years of successful integrated control. Plant Prot Sci 45:37–38

Dixon GR (2009c) *Plasmodiophora brassicae* (clubroot) – a plant pathogen that alters host growth and productivity. J Plant Growth Regul 28:212–228

Dixon GR, Doodson JK (1970) Clubroot of *Brassica* crops. Agriculture 77:500–503

Dixon GR, Doodson JK (1971) Assessment keys for some diseases of vegetable, fodder and herbage crops. J Natl Inst Agric Bot 12:299–307

Dixon GR, Page LV (1998) Calcium and nitrogen eliciting alterations to growth and reproduction of *Plasmodiophora brassicae* (clubroot). Acta Hortic 459:343–349

Dixon GR, Walsh UF (1998) Suppression of plant pathogens by organic extracts - A Review. Acta Hortic 69:383–390

Dixon GR, Webster MA (1988) Antagonistic effects of boron, calcium and pH on pathogenesis caused by *Plasmodiophora brassicae* (clubroot) - a review of recent work. Crop Res 28:83–95

Dixon GR, Wilson F (1983) Evaluation of calcium cyanamide for control of *Plasmodiophora brassicae* (clubroot). Tests of agrochemicals and cultivars no 4. Ann Appl Biol 102:50–51

Dixon GR, Wilson F (1984a) Field evaluation of chemicals for control of clubroot (*Plasmodiophora brassicae*). In: Proceedings of the Crop Protection in Northern Britain Conference, Dundee, pp 400–405

Dixon GR, Wilson F (1984b) Field evaluation of WL 105305 (NK483) for control of clubroot (*Plasmodiophora brassicae*). Tests of Agrochemicals and Cultivars no. 5. Ann Appl Biol 104:34–35

Dixon GR, Wilson F (1985) Fungicides applied to propagation blocks and the field to control *Plasmodiophora brassicae*, clubroot on cabbage. Ann Appl Biol 106:44–45

Dixon GR, Jones DS, Ingram DS (1981) Studies on the populations of *Plasmodiophora brassicae*. In: Proceedings Eucarpia Cruciferae 1981 Conference, Published by Agricultural University of Norway, Aas, pp 41–44

Einhorn G, Bochow H, Huber J, Krebs B (1991) Methodological studies to detect antagonists of the clubroot pathogen *Plasmodiophora brassicae* Worn. Arch Phytopath Plant Prot 27(3):205–208

Eriksson J (1936) Die Pilzkranwheiten der Kulturge-wäche. Hanbuch fur Pflanzenbauer und Studierende. I. Teil, 2 auf., pp 38–42

Favorski W (1910) Nouvelle recherché sur le developpement et la cytologie de *Plasmodiophora brassicae* Worn. (In Russian). Mém Soc Nat Kieff 20:149–183

Fedorintschik NS (1935) Life history of club – root of cabbage (*Plasmodiophora brassicae* Woron.). (In Russian) Summary Science Research Wk. Institute of Plant Prot, Leningrad, pp 69–70

Fletcher JT, Hims MJ, Archer FC, Brown A (1982) Effects of adding calcium and sodium salts to field soils on the incidence of clubroot. Ann Appl Biol 100:245–251

Friberg H, Lagerlof J, Ramert B (2005) Germination of *Plasmodiophora brassicae* resting spores stimulated by a non–host. Eur J Plant Pathol 113(3):275–281

Friberg H, Lagerlof J, Ramert B (2006) Usefulness of non-host plants in managing *Plasmodiophora brassicae*. Plant Pathol 55(5):690–695

Gibbs JG (1931a) Clubroot in cruciferous crops. N Z J Agric 42:1–17

Gibbs JG (1931b) Dissemination of clubroot in the dung of farm stock. N Z J Agric 42:193–198

Gibbs JG (1939) Factors influencing the control of club root. N Z J Sci Technol 20:409–412

Gleisberg W (1922) Das Ratsel der Hernieverbreitung. Nachr Bl Pfl'sch'dienst Berlin 2:89–90

Grainger J (1962) The host plant as a habitat for fungal and bacterial parasites. Phytopathology 52:140–150

Hamilton HA, Crete R (1978) Influence of soil moisture, soil pH and liming source on the incidence of clubroot, the germination and growth of cabbage produced in mineral and organic soils under controlled conditions. Can J Plant Sci 58:45–53

Hata S, Sumi Y, Ohi M (2002) Dry powder and extract of *Posidonia australis* Hook. F, a species of seagrass, stimulate the germination of the pathogen *Plasmodiophora brassicae* and control clubroot of Chinese cabbage. J Jpn Soc Hort Sci 71(2):197–202

Hewitt EJ (1970) Physiological and biochemical factors controlling the assimilation of inorganic nitrogen supplies by plants. In: Nitrogen nutrition of the plant. Waverley Press, Leeds, pp 78–103

Honig F (1931) Der Kohlkropfereger (*Plasmodiophora brassicae* Wor.). Gartenbauwissenschaft 5:116–225

Hooker WJ, Walker JC, Link KP (1945) Effects of two mustard oils on *Plasmodiophora brassicae* and their relation to resistance to clubroot. J Agric Res 70:63–78

Hseith WH, Wang J (1986) Investigation on suppressive soils of clubroot of crucifers in Taiwan. Plant Prot Bull 28:353–362

Humphrey JE (1892) Report on plant diseases etc. With observations in the field and in the vegetation house. Publ. Document Massachusetts Agricultural Experimental Station, no 33, pp 218–248 (Abstract in Bot Zbl 307-309)

Jones DR (1980) Differential pathogenicity of *Plasmodiophora brassicae* Worn, Ph.D thesis. University of Cambridge

Jones DR, Ingram DS, Dixon GR (1981) Differential pathogenicity of *Plasmodiophora brassicae*. Cruciferae Newslett 6:54–56

Kim CH, Cho WD, Kim HM (2000) Distribution of *Plasmodiophora brassicae* causing clubroot disease of Chinese cabbage in soil. Plant Dis Res 6(1):27–32

Kowalski K, Bochow H (1996) Observations on the behaviour of resting spores of *Plasmodiophora brassicae* in the presence of cruciferous and non-cruciferous plant roots. Acta Hortic 407:419–421

Lange L, Olson LW (1983) Zoosporic plant pathogens – a modern perspective. Academic, London

Larson RH, Walker JC (1934) Soil treatment in relation to clubroot of cabbage. J Agric Res 48:749–759

MacFarlane I (1952) Factors affecting the survival of *Plasmodiophora brassicae* Worn. in the soil and its assessment by a host test. Ann Appl Biol 39:239–256

MacFarlane I (1958) A solution–culture technique for obtaining root–hair or primary infection by *Plasmodiophora brassicae*. J Gen Microbiol 18:720–732

MacFarlane I (1970) Germination of resting spores of *Plasmodiophora brassicae*. Trans Br Mycol Soc 55:97–112

MacFarlane I, Last FT (1957) Club root of cruciferous plants. Rep Rothamsted Exp 1956:117–119

Martinson CA (1963) Inoculum potential relationships in *Rhizoctonia solani* measured with microbiological sampling tubes. Phytopathology 53:634–638

Matsumiya E (1989) *In vitro* system for an obligate parasite *Plasmodiophora brassicae*. Dissertation for Master of Agriculture degree Graduate School of Agriculture, Hokkaido University, Japan, p 63

Mattsson L (1995) Skordevariationer inom enskilda fält. Storlek och tankbara orsaker. Swedish University of Agricultural Sciences, Department of Soil Sciences, Division of Soil Fertility. Report 196

Monteith J (1924) Relation of soil temperature and soil moisture to infection by *Plasmodiophora brassicae*. J Agric Res 28:549–561

Murakami H, Tsushima S, Shishido Y (2000) Soil suppressiveness to clubroot disease of Chinese cabbage caused by *Plasmodiophora brassicae*. Soil Biol Biochem 32(11/12):1637–1642

Murakami H, Tsushima S, Kuroyanagi Y, Shishido Y (2002a) Reduction in resting spore density of *Plasmodiophora brassicae* and clubroot severity by liming. Soil Sci Plant Nutr 48(5):685–691

Murakami H, Tsushima S, Shishido Y (2002b) Factors affecting the pattern of dose response curve of clubroot disease caused by *Plasmodiophora brassicae*. Soil Sci Plant Nutr 48(3):421–427

Murakami K, Shinoda H, Maruta R, Goto I (2003) Relationship between physical and chemical properties of soil and resting spore density of *Plasmodiophora brassicae* and clubroot disease index in production areas of cruciferous vegetables. Jpn J Soil Sci Plant Nutr 74(6):781–786

Murakami H, Tsushima S, Akimoto T, Kuoryanagi Y, Shishido Y (2004a) Quantitative studies on the relationship between plowing into soil of clubbed roots of proceeding crops caused by *Plasmodiophora brassicae* and disease severity in succeeding crops. Soil Sci Plant Nutr 50 (8):1307–1311

Murakami K, Shinoda H, Nakamura F, Goto I (2004b) Influence of soil types and soil pH on the incidence of clubroot disease caused by *Plasmodiophora brassicae*. Jpn J Soil Sci Plant Nutr 75 (3):339–345

Murakami K, Nakamura F, Goto I (2007) Effects of *Bacillus subtilis* NB 22 addition to nursery soils on incidence of clubroot disease. Jpn J Soil Sci Plant Nutr 78(4):387–390

Myers DF, Campbell RN (1985) Lime and the control clubroot of crucifers: effects of pH, calcium, magnesium and their interaction. Phytopathology 75:670–673

Naiki T, Tanahashi K, Kageyama K (1984) The relationship between root hair infection with *Plasmodiophora brassicae* Worn. and subsequent club formation among cruciferous species. Ann Phytopathol Soc Jpn 50:211–215

Narisawa K, Shimura M, Usuki F, Fukuhara S, Hashiba T (2005) Effects of pathogen density, soil moisture, and soil pH on biological control of clubroot in Chinese cabbage by *Heteroconium chaetospira*. Plant Dis 89(3):285–290

Naumov NA (1925) Contribution to the study of club root of cabbage II. Morbi Plant 14:49–73

Nicholas DJD (1965) Influence of the rhizosphere on the mineral nutrition of the plant. In: Baker KF, Synder WC (eds) Ecology of soil-borne plant pathogens – prelude to biological control. John Murray, London, pp 210–251

Niwa R, Kumei T, Nomura Y, Yoshida S, Osaki M, Ezawa T (2007) Increase in soil pH due to Ca-rich organic matter application causes suppression of clubroot disease of crucifers. Soil Biol Biochem 39:778–785

Niwa R, Nomura Y, Osaki M, Ezawa T (2008) Suppression of clubroot disease under neutral pH caused by inhibition of spore germination of *Plasmodiophora brassicae* in the rhizosphere. Plant Pathol 57:445–452

O'Brian DG, Dennis RWG (1936) Further information relating to control of raan in swedes. Scottish J Agric 19:1–18

Ogawa S, Takahashi H, Hayakawa T, Ito K, Mitsui T, Hori H, Kiso A (2001) Enhancement of germination of spores from obligatory plant pathogen, *Plasmodiophora brassicae* which causes clubroot disease. Bull Fac Agric Niigata Univ 54(1):35–43

Ohi M, Kitamura T, Hata S (2003) Stimulation by caffeic acid, coumalic acid, and corilagin of the germination of resting spores of the clubroot pathogen *Plasmodiophora brassicae.* Biosci Biotechnol Biochem 67(1):170–173

Page LV (2001) Studies of components for a potential integrated control system for *Plasmodiophora brassicae*, Ph.D thesis. University of Strathclyde, Glasgow

Palm ET (1958) Effect of mineral nutrition on invasiveness of *Plasmodiophora brassicae* Worn. and the development of club root. Dis Abstr 19:425–426

Palm ET (1963) Effect of mineral nutrition on the invasion and response of turnip tissue to *Plasmodiophora brassicae* Worn. Cont Boyce Thompson Inst 22:91–112

Palm ET, Mc New GL (1956) A method for determining the incidence of club root infection in nutrient cultures. Cont Boyce Thompson Inst 18:333–337

Rausch T, Mattusch P, Hilgenberg W (1981) Influence of clubroot disease on the growth kinetics of Chinese cabbage. Phytopathol Z 102:28–33

Roback J, Gabrielson RL (1988) The effects of inoculums potential in screening for resistance to *Plasmodiophora brassicae* in greenhouse trials. Acta Agrobot 41:217–224

Russell R (1859) The cause of finger and toe in turnips. J Agric For 9:529–544

Samuel G, Garrett SD (1945) The infected root hair count for estimating the activity of *Plasmodiophora brassicae* Woron. in the soil. Ann Appl Biol 32:96–101

Satir P (1982). Mechanisms and controls of microtubule sliding in cilia. In: Amos WB, Duckett JG (eds) Prokaryotic and Eukaryotic Flagella, Cambridge University Press (Symposium of the Society for Experimental Biology, Symposium no XXXV), pp 179–201

SBH T (1980) In: Devon and Cornwall (ed) Physiologic populations of *Plasmodiophora brassicae* Woron, Ph.D Thesis. University of Exeter, Exeter, p 294

Sen P (2005) Antagonistic effect of Ca, B and Mo on club-root disease of rape-mustard. Indian Agric 49(1/2):13–16

Shinoda H, Murakami K, Gotto I (2005) Effect of amelioration of soil acidity and continuous cropping of cruciferous vegetables on the incidence of clubroot disease and resting spore density in soil. Jpn J Soil Sci Plant Nutr 76(6):891–896

Sleigh MA, Barlow DI (1982) How are different ciliary beat patterns produced? In: Amos WB, Duckett JG (eds) Prokaryotic and Eukaryotic Flagella, Cambridge University Press (Symposium of the Society for Experimental Biology, Symposium no XXXV), pp 139–157

Somerville W (1895) Further infection experiments with finger and toe. J R Agric Soc 6:749–759

Stewart K (2007) Conventional and novel treatments for control of clubroot of *Brassicas*, Ph.D thesis. University of Edinburgh, Edinburgh

Suzuki K, Matsumiya E, Ueno Y, Mizutani J (1992) Some properties of germination-stimulating factor from plants for resting spores of *Plasmodiophora brassicae.* Ann Phytopathol Soc Jpn 60:699–705

Takahashi K (1994) Influences of some environmental factors on the viability of resting spores of *Plasmodiophora brassicae* Worn. incubated in sterile soil. Ann Phytopathol Soc Jpn 60:658–666

Takahashi H, Ishikawa T, Kaido M, Takita K, Hayakawa T, Okazaki K, Itoh K, Mitsui T, Hori H (2006) *Plasmodiophora brassicae*–induced cell death and medium alkalization in clubroot-resistant cultured roots of *Brassica rapa*. J Phytopathol 154:156–162

Tanaka S, Mizui Y, Terasaki H, Yasuaki S, Ito SI (2006a) Distribution of clubroot disease of a cruciferous weed, *Cardamine flexuosa*, in major isolated islands Hokkaido and Okinawa in Japan. Mycoscience 47(2):72–77

Tanaka S, Mido H, Ito S (2006b) Colonization by two isolate of *Plasmodiophora brassicae* on a clubroot-resistant cultivar of Chinese cabbage (*Brassica rapa* L. spp *pekinensis*). J Gen Plant Pathol 72(4):205–209

Tilston EL, Pitt D, Groenhof AC (2002) Composted recycled organic matter suppresses soil-borne diseases of field crops. New Phytol 154:731–740

Toxopeus H, Gowers S, Crete R (1975) The problem of 'cut-off points'. Clubroot Newslett 4:6–8

Toxopeus H, Dixon GR, Mattusch P (1986) Physiological specialization in *Plasmodiophora brassicae,* an analysis by international experimentation. Trans Br Mycol Soc 87:279–287

Vander Plank JE (1975) Principles of plant infection. Academic, New York, p 216

Wallenhammar AC (1996) Prevalence of *Plasmodiophora brassicae* in a spring oilseed rape growing area in central Sweden and factors influencing soil infestation levels. Plant Pathol 45:710–719

Wallenhammar AC (1999) Monitoring and control of *Plasmodiophora brassicae* in spring oilseed *Brassica* crops. Acta Universitatis Agriculturae Sueciae, Agraria 183, Swedish University of Agricultural Sciences, Uppsala, p 53

Watson AG (1967) The movement of *Plasmodiophora brassicae* in soil. Phytopathology 57:608

Webster MA (1986) pH and nutritional effects on infection by *Plasmodiophora brassicae* Worn. and on clubroot symptoms. PhD thesis, University of Aberdeen, p 273

Webster MA, Dixon GR (1991a) Calcium, pH and inoculum concentration influencing colonization by *Plasmodiophora brassicae*. Mycol Res 95:64–73

Webster MA, Dixon GR (1991b) Boron, pH and inoculum concentration influencing colonization by *Plasmodiophora brassicae*. Mycol Res 95:74–79

Wellman FL (1930) Club root of Crucifers. Tech Bull US Dept Agric 181:331

Woronin M (1878) *Plasmodiophora brassicae* Urheber der Kohlpflanzen – Hernie. JB. Wiss. Bot., 11, 548-574. (Translation by Chupp, C. Phytopathological Classics, no 4, 1934, American Phytopathol Soc)

Yamada M, Asandhi AA, Purwati E (2003) Employing one–year rotations with three vegetable combinations to control clubroot damage in the West Java Highlands. Research Highlights published by the Japan International Research Center for Agricultural Sciences (JIRCAS), Tsukuba, Japan, pp 16–17

Yang MY, Yanf JL, Sun DW, Yan WZ (2004) Effects of soil moisture on the incidence of clubroot. J Agric Sci 17(4):482–483

Yano S, Tanaka S, KameyaIwaki M, Katumoto K (1991) Relation of Ca^{2+} efflux to germination of resting spores of clubroot fungus. Bull Fac Agric 39:105–112

Yeoung YR, Kim JH, Kim BS, Young JJ, Yoon CS (2003) Effects of beneficial antagonists *Bacillus* sp., *Pseudomonas* sp. and *Trichoderma* sp. on the control of clubroot of Chinese cabbage. Korean J Hort Sci Technol 21(3):194–198

Yonezawa M, Usuki F, Narisawa K, Takahashi J, Hashiba T (2004) Anatomical study on the interaction between the root endophytic fungus *Heteroconium chaetospira* and Chinese cabbage. Mycoscience 45(6):367–371

Yun H, Quing MS, Qin LX, Jing W, Ling HX (2007) Morphology of *Plasmodiophora brassicae* and biological characteristic of the pathogenic resting spores in rapeseed. Sci Agric Sin 40 (7):1388–1394

7 Infection and Pathogenesis

Abstract

The infection process by *P. brassicae* is initiated soon after the germination of resting spores triggered by calcium ions and host exudates under humid and favorable temperature conditions of 20–25 °C. Resting spores in the soil germinate and release primary zoospores that penetrate root hairs to form primary plasmodia. The primary plasmodia form multinucleate zoosporangia which produce uninucleate secondary zoospores. These secondary zoospores are released into the soil and infect root cortical cells and develop into motile secondary plasmodia. Cortical infection results in hyperplasia and hypertrophy of the roots cells which produces swollen, rounded clubroot symptoms. Clubroot development and severity of the pathogenesis is induced and regulated by several metabolites, growth regulators, sugars, hormones, enzymes, proteins, lipids, glucosinolates, etc., during host–pathogen interactions. The involvement of auxin and cytokinin as pathogenicity factors causes increase in cell division and cell elongation during late infection stages. Molecular mechanism of pathogenesis revealed a serine protease (Pro1) gene important for resting spore germination. During infection, and pathogenesis of *P. brassicae* members of the TF families, MYB, bHLH, and APZ/EREBP are highly up- or down-regulated in both shoot and root of host plants. *P. brassicae* infection triggers active sugars translocation between the sugars producing tissues and the clubbed tissues, and SWEET family genes are involved in this process. *Brassica* specific genes (BSGs) play important roles in *P. brassicae* stress responses. Genes Pb-Bripq and PbPSA are expressed in early stages of infection. miRNAs are expressed during the process of clubroot initiation and development. G protein-coupled receptors (GPCR) signal transduction pathways play important roles in the growth, development, and pathogenesis of *P. brassicae* to their cruciferous host species extensively.

G. S. Saharan et al., *Clubroot Disease of Crucifers*,
https://doi.org/10.1007/978-981-16-2133-8_7

Keywords

Infection and pathogenesis · Role of metabolites · Host hormones metabolism · Auxin metabolism · Cytokinin metabolism · Host resistance · Role of Nitrilase · Serine protease · Lipids · Endogenous glucosinolate · Auxin signaling pathways · Molecular mechanisms · Induction of gene · Gene expression · Identification of effector gene · Differential expression · Expression of hypoxia genes · Expression SWEET genes · RT-PCR analysis · Induction of microRNAs · Induction of signaling molecules and genes · Role of G-protein-coupled receptors (GPCR) signal transduction pathways

7.1 Introduction

The disease cycle of *P. brassicae* consists of three phases; first phase is survival in the soil through resting spores, a secondary phase, root-hairs infection by primary zoospores is restricted to root hairs and occasionally epidermal cells, and a third phase, cortical infection by secondary zoospores which involves pathogen infection and proliferation within the root and subsequent symptom development. In the primary phase, resting spores (RS) in the soil germinate and release primary zoospores that penetrate root hairs to form primary plasmodia. The primary plasmodia form multinucleate zoosporangia which produce uninucleate secondary zoospores (SZ). This process takes more than 5 days at the optimum temperature of 25 °C. The SZ are released into the soil solution, infect root cortical cells, and develop into motile secondary plasmodia. After a mobile phase within the root, the young plasmodia coalesce into sessile vegetative plasmodia that divide to form long-lived RS. Cortical infection results in hyperplasia and hypertrophy of the root cells, which produces the characteristic swollen, rounded clubbed root symptoms. Clubbed roots are readily visible within 5–6 weeks of inoculating at temperatures of 20 to 25 °C. Inoculating with RS initially produces only primary (root hair) infection but inoculating with SZ results in both primary and secondary infection in susceptible hosts. Primary infection occurs in both susceptible and resistant host cultivars. Primary infection can also occur in several non-host species but secondary infection and development of small numbers of RS have been reported in only a few species that are generally considered non-hosts.

7.2 Process of Infection and Pathogenesis

In primary infection, each of the inoculating results in root-hair infection (RHI) (Table 7.1). At 13 and 15 days after seeding (DAS) (3 and 5 days after inoculating), root-hair infection is higher following inoculating with secondary zoospores (SZ-P3) compared with SZ-P6. Substantial levels of infection are first observed at 1 day after inoculating in plants inoculated with RS-P3 + SZ-P3 (63%), RS-P3 (46%), and RS-P6 + SZ-P3 (35%). At 15 DAS (5 days after inoculating), RHI is relatively high

Table 7.1 Incidence of primary infection (root-hair infection) on canola "Zephyr" resulting from inoculating with resting spores (RS) and secondary zoospores (SZ) of pathotypes 3 and 6 (P3 and P6, respectively) of *Plasmodiophora brassicae,* assessed at 11, 13, and 15 days after seeding (DAS) (1, 3, and 5 days after inoculating with secondary zoospores) (McDonald et al. 2014)

	Root-hair infection (%)[b]		
Spore type, pathotype[a]	11 DAS	13 DAS	15 DAS
RS-P6 (10 DAS)	2 a	15 a	26 a
RS-P3 (10 DAS)	3 a	26 b	48 b
SZ-P6	4 a	32 c	55 c
SZ-P3	10 a	47 e	74 e
RS-P6	22 b	41 d	50 b
RS-P6+SZ-P3	35 c	55 f	60 d
RS-P3	46 d	64 g	72 e
RS-P3+SZ-P3	63 e	73 h	80 f
Standard error	0.44	1	0.7

[a]The 10 DAS treatments were inoculated with RS spores at the same time as the SZ treatments were applied. All other RS treatments were applied 5 DAS

[b]Values are the means of eight replications, with one plant per rep. Means within a column followed by the same letter do not differ at $P < 0.05$ based on Tukey's test

for all of the treatments except RS-P6 at 10 DAS (26%), although differences among treatments are still evident. Root-hair infection is higher for RS-P3 than RS-P6 and higher for SZ-P3 than SZ-P6. Root-hair infection with RS-P3 + SZ-P3 is higher than RS-P3 alone. However, root-hair infection is higher with RS-P3 + SZ-P3 than RS-P6 + SZ-P3. The incidence of root-hair infection increases over time. There is a small inoculating treatment sampling date interaction ($F = 148.5$, $P = 0.0001$, compared with $F = 5308$ for sampling date and 2249 for treatment) but inoculating treatment generally continues to exhibit a pattern of response at the later assessment dates similar to that observed at 11 DAS.

During the secondary infection, most important differences are observed in infection of the root cortex (Fig. 7.1) and subsequent symptom expression (Table 7.2). Substantial numbers of plasmodia in the root cortex are observed at 11 DAS in plants inoculated with RS-P3 + SZ-P3 (15 secondary plasmodia) and RS-P3 (10 secondary plasmodia). Plasmodia are first observed at 13 days after sowing (DAS) in plants inoculated with resting spore with pathotypes 3 (RS-P3) and secondary zoospores pathotype 6 (SZ- P6) at 10 DAS whereas at 15 DAS in plants inoculated with pathotype 6 (RS-P6) at 10 DAS. Plasmodia produced from primary and SZ are similar in morphology (Plate 7.1a, b). By 15 DAS, the differences among treatments are even more distinct than at the earlier assessment dates (Fig. 7.1). As expected, inoculating with RS-P3 results in many more secondary plasmodia per field (mean of 24 plasmodia per field) than RS-P6 (12 plasmodia). Inoculating with SZ alone results in fewer plasmodia than RS of the same pathotype; 12 plasmodia for RS-P6 versus 9.5 plasmodia for SZ-P6, and 24 plasmodia for RS-P3 versus 16 plasmodia for SZ-P3. RS-P3 + SZ-P3 resulted in more plasmodia (30 plasmodia) than either SZ-P3 alone (15 plasmodia) or RS-P3 alone

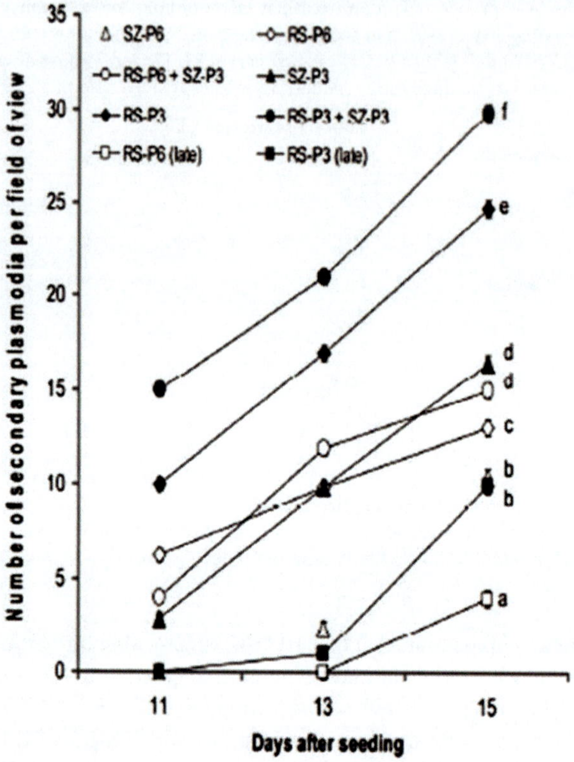

Fig. 7.1 Secondary infection (number of plasmodia per field of view) of canola "Zephyr" at 1, 3, and 5 days after inoculating with resting spores (RS) or secondary zoospores (SZ). Regression slopes among the treatments were similar and, therefore, are not presented. Data points are the means of observed values of eight replications (one plant per rep). Means at 5 days after inoculating with the secondary zoospores followed by the same letter do not differ based on Tukey's test at $P \leq 0.05$. Capped lines = standard error (McDonald et al. 2014)

(24 plasmodia). The treatments RS-P6 + SZ-P3 results in fewer plasmodia than the treatment RS-P3+SZ-P3 (15 versus 30). There are no differences in the number of plasmodia when SZ-P3 was applied alone or in combination with RS-P6.

The symptom development pattern and the extent and development of cortical infection were evaluated at 52 DAS by McDonald et al. (2014). The area of cortical infection, clubroot severity, and the number of cells containing RS all showed the same pattern of response to inoculating treatment (Table 7.2). The area of cortical colonization and the DSI are highly correlated ($r = 0.92$, $P < 0.0001$). These trends are similar to that of early (15 DAS) cortical infection above, except that no symptoms developed on plants inoculated with RS-P6, even though some plasmodia had developed by 15 DAS and young plasmodia are still visible at 47 DAS (22 infected cells per field).

In both the compatible and incompatible interactions, inoculating with RS elicits a different pattern of response than inoculating with SZ (Table 7.2). Inoculating with

Table 7.2 Clubroot severity on harvested roots, area of cortical infection (Area of CI), and number of infected cells with young plasmodia, mature plasmodia, or resting spores in five fields of view in sections of canola roots; and mean plant height (cm), shoot weight (g), and root weight (g) of canola "Zephyr," assessed at 52 days after seeding, which was 47 days after inoculating with resting spores (RS), and 42 days after inoculating with secondary zoospores (SZ) of pathotypes 3 and 6 (P3 and P6, respectively) of *Plasmodiophora brassicae*[a] (McDonald et al. 2014)

Spore type, pathotype	Number of infected cells						Plant growth parameters		
	Incidence (%)	Severity (%)	Area of CI (%)	Young plasmodia	Mature plasmodia	Resting spores	Plant height (cm)	Shoot weight (g)	Root weight (g)
RS-P6	0 a	0 a	0.1 a	22 d	0 a	0 a	49 c	4.0 c	1.1 a
SZ-P6	67 b	31 b	4 b	13 c	6 b	2 b	54 c	4.0 c	1.4 a
SZ-P3	78 c	67 c	12 c	6 b	12 d	12 c	28 b	3.0 b	2.2 b
RS-P6 + SZ-P3	85 c	84 d	18 d	1 a	15 e	26 d	28 b	2.6 b	2.2 b
RS-P3	100 d	100e	33 e	1 a	10 c	32 e	15 a	1.3 a	3.6 c
RS-P3 + SZ-P3	100 d	100 e	34 e	1 a	9 c	32 e	10 a	1.0 a	4.2 c
Control	0	0	0	0	0	0	50 c	4.6 c	1.1 a
Standard error	0.97	1.8	0.40	0.44	0.48	0.60	1.1	0.68	0.97

[a]Values are the means of eight replications (10 plants per rep for incidence, severity, and plant growth parameters and 1 plant and five fields of view per rep for data on number of infected cells). Means in the columns followed by the same letter do not differ based on Tukey's test at $P \leq 0.05$

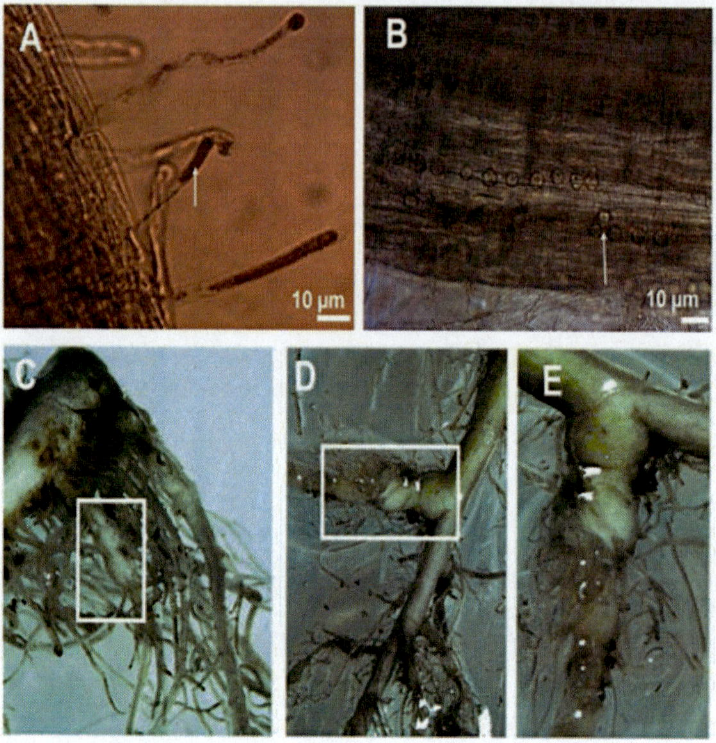

Plate 7.1 (**a**) Infected root hair (arrow) and (**b**) young secondary plasmodia (arrow) on canola root inoculated with secondary zoospores of pathotype 6 (avirulent on "Zephyr") of *Plasmodiophora brassicae*. (**c–e**) Bead-shaped clubs from inoculating with secondary zoospores of pathotype 6 (McDonald et al. 2014)

RS-P3 results in the highest values for both area of cortical infection (33%) and number of cells with RS (32 cells), and 100% clubroot incidence and severity. SZ-P3 produced less cortical infection (12%), and lower incidence and severity (78 and 67% DSI, respectively) than RS-P3. The opposite pattern is observed in the incompatible interaction; inoculating with RS-P6 produced no clubroot incidence (no symptoms) but inoculating with SZ-P6 produced low levels of cortical infection (4%) and development of RS (2 cells). Clubroot symptoms developed following inoculating with SZ-P6, with a low incidence and severity consisting solely of small, bead-shaped clubs.

The treatment RS-P3 + SZ-P3 produced 100% incidence and severity with highest cortical infection and number of cells with RS, the same as RS-P3 alone (Table 7.2). However, RS-P6 + SZ-P3 results in reduced levels of pathogen development in the root cortex, incidence and severity are lower (85 and 86% DSI, respectively), as is the area of cortical infection (18%) and number of cells with RS (26 cells). The only unusual result is the comparison of SZ-P3 with RS-P6 +

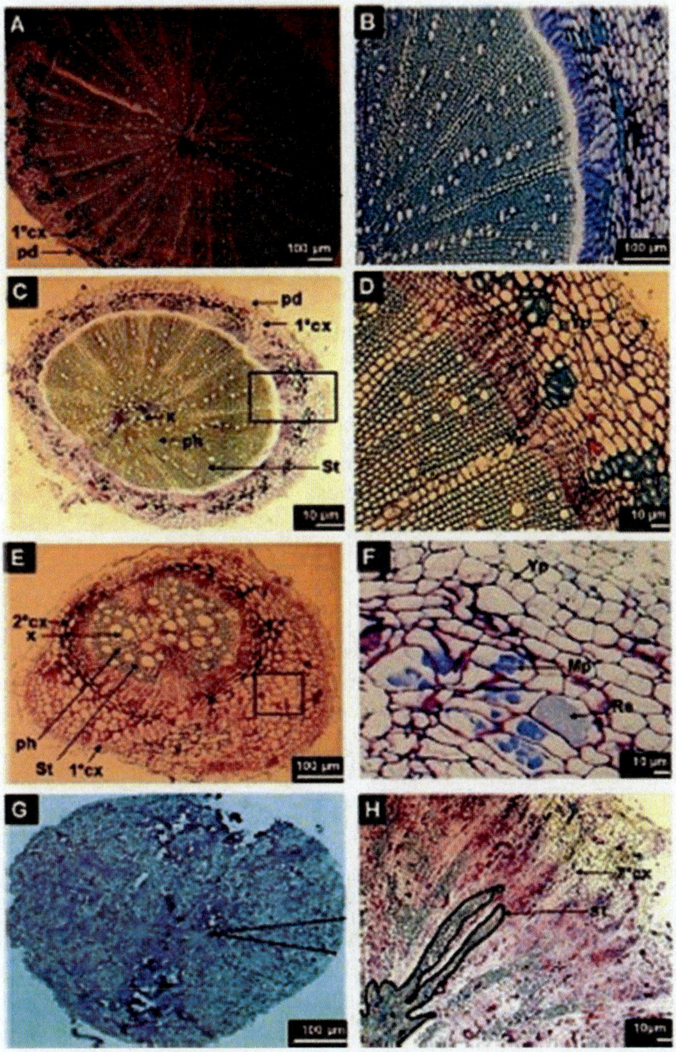

Plate 7.2 Cross section of roots of healthy and *Plasmodiophora brassicae* infected canola plants, stained with toluidine blue O. (**a**) Healthy root (control). Lignified xylem cell walls stain dark blue, non-lignified walls are pink. (**b**) Magnified region from A showing intact stele, cortex, and periderm. (**c**) Roots of plants inoculated with resting spores of pathotype 6 (avirulent on "Zephyr"); (**d**) Magnified region from C showing intact stele, cortex, and periderm (young plasmodia are marked with arrow). (**e**) Bead-shaped clubs produced after inoculating with secondary zoospores of pathotype 6. Note the proliferation of cortical tissue outside of the stele, and no proliferation within the stele. (**f**) Magnified region from E showing young plasmodia, mature plasmodia, and resting in the cortical cells. (**g**) Typical clubs. Stele region is disrupted. (**h**) Magnified region from G showing invasion and expansion of stele (marked with black boundaries). Annotations: *x* xylem, *ph* phloem, *1°cx* primary cortex, *2°cx* secondary cortex, *pd* periderm, *St* stele, *Yp* young plasmodia, *Mp* mature plasmodia, and *Rs* resting spores (McDonald et al. 2014)

SZ-P3. Inoculating with RS-P6 +SZ-P3 resulted in higher area of cortical infection, more RS, and higher clubroot severity than SZ-P3 alone.

No clubs develop on plants inoculated with RS-P6 or on the uninoculated control (Table 7.2). Typical large clubs developed on all of the other treatments. The only exception is SZ-P6, where tiny, bead-shaped clubs developed in strings or clusters along the root (Plate 7.1c–e). RS were present in the infected cortical cells of the bead-shaped clubs (Plate 7.2f). In all, more than 20 plants were assessed after an additional 2 weeks of growth (8 weeks after inoculating), and the size and shape of the clubs had not changed. Plant height and shoot weight are generally highest, and root mass and clubroot incidence and severity are lowest, in plants with no clubbing (inoculated with RS-P6 or controls) or with bead-shaped clubs (SZ-P6), intermediate with RS-P6 + SZ-P3 and SZ-P3, and lowest and highest in plants with large clubs (RSP3 and RS-P3 + SZ-P3) (Table 7.2). Similarly, the root biomass of plants inoculated with RS-P6 (1.1 g) and SZ-P6 (1.4 g) was similar but only SZ-P6 produced clubroot symptoms, although the bead-shaped clubs are considerably smaller in size and weight than typical clubs (Table 7.2). Proliferation of *P. brassicae* results in severe distortion of root growth and development (Plate 7.2). In cross-sections of the roots of control plants, the periderm, cortex, and stele region are organized and continuous (Plate 7.2a, b). In plants inoculated with RS-P6, young plasmodia are observed in the cortical cells and stele but the pathogen did not develop further (Plate 7.2c, d). As a result, plants inoculated with RS-P6 did not differ morphologically from the controls. Typical clubs developed from inoculating with RS-P3, RS-P3 + SZ-P3, SZ-P3, and RS-P6 + SZ-P3 but the clubs from inoculating with SZ-P3 and RS-P6 + SZ-P3 are smaller than normal. In all of the typical clubs, cortical cells of the infected roots became enlarged and intermingled with the secondary phloem, and the distinct organization of tissues into cortex, endodermis, and stele is lost (Plate 7.2g, h). The area occupied by cells with walls that stained for lignin in the stele decreased rapidly during pathogen colonization (Plate 7.2g, h). In the bead-shaped clubs, cortical cells are highly plasmolyzed and the stele is continuous (Plate 7.2e, f). The pathogen did not penetrate into the stele. As a result, there is limited cell proliferation in these tissues (Plate 7.2f). Plants inoculated with RS from the bead-shaped clubs became heavily infected and produced large clubs.

In nuclear visualization, zoosporangia are present in most of the infected root hairs at 12 days after inoculating. In mature zoosporangia, the plasmodial cytoplasm is cleaved to produce individual zoosporangia (Plate 7.3a). CLSM observation showed that these zoosporangia are multinucleate and each nuclei reflected production of uninucleate incipient SZ within the root hairs. Depending on the size, one to six nuclei are present in a zoosporangium ($n = 20$), and the nuclear size varied from 1.0 to 1.5 μm ($n = 50$) (Plate 7.3a). The flagellate SZ collected in deionized water had a mean length of 14.4 ± 0.19 μm and width of 9.6 ± 0.32 μm ($n = 55$), and all had two anterior whiplash flagella (Plate 7.3b). The mean diameter of encysted zoospores (Plate 7.3d) obtained by centrifugation is 21.7 ± 0.37 μm ($n = 50$). All of the flagellate and encysted zoospores are uninucleate (Plate 7.3c, d). SZ that are released into phosphate buffer or physiological saline were slightly smaller than

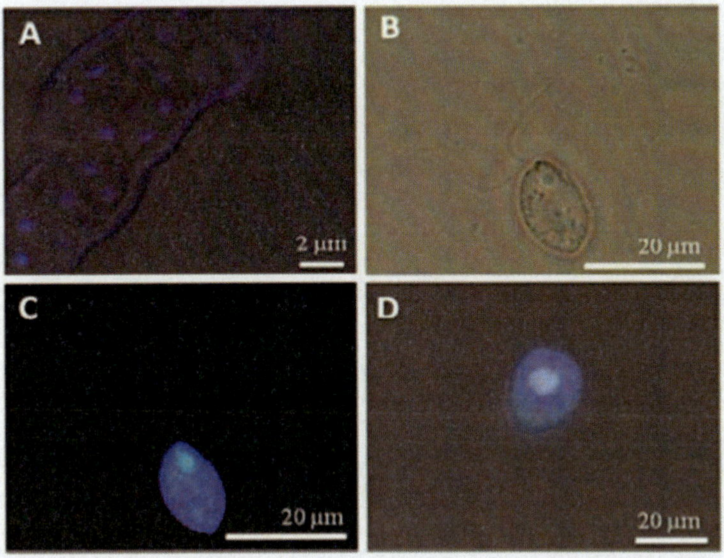

Plate 7.3 Developmental stages and nuclear visualization in secondary zoospores. (**a**) Conf °Cal laser scanning microscope micrograph (overlaid fluorescence and differential interference contrast image) of an infected root hair stained with Hoechst 33,258 showing the plasmodial mass cleaved into zoosporangia. Each zoosporangium contains a varying number (1–6) of nuclei that represent incipient secondary zoospores. (**b–d**) Epifluorescence micrographs. (**b**) Phase contrast of a biflagellate zoospore produced by mechanical shaking of root hairs. (**c**)-Fluorescence image of 4′-6-diamidino-2-phenylindole staining of a single nucleus in the flagellated zoospore. (**d**) Overlaid image (fluorescence and phase contrast) of an encysted uninucleate zoospore (produced by centrifugation) (McDonald et al. 2014)

those produced in deionized water. The length and width of flagellate zoospores in phosphate buffer are 12.5 ± 0.23 by 8.7 ± 0.25 µm, respectively, and in physiological saline are 11.8 ± 0.29 by 8.3 ± 0.30 µm, respectively. The diameters of encysted zoospores in phosphate buffer and physiological saline are 17.8 ± 0.28 and 17.3 ± 0.44 µm, respectively (McDonald et al. 2014).

7.2.1 Root-Hair Infection Process

During the life cycle of the *P. brassicae*, pathogen can survive through the winter and summer in soil or animal manure with non-rotted diseased plants. Under appropriate soil and environmental conditions, these resting spores release primary zoospores; when a zoospore swims to the surface of a root hair, it penetrates through the cell wall. This stage is termed the root-hair infection stage, and inside root hairs, the pathogen forms primary plasmodia (Agrios 2005). Sporangia are produced 4 days after inoculating (Kageyama and Asano 2009). According to Ji et al. (2014), the primary plasmodia inside the root hair of susceptible lines began to

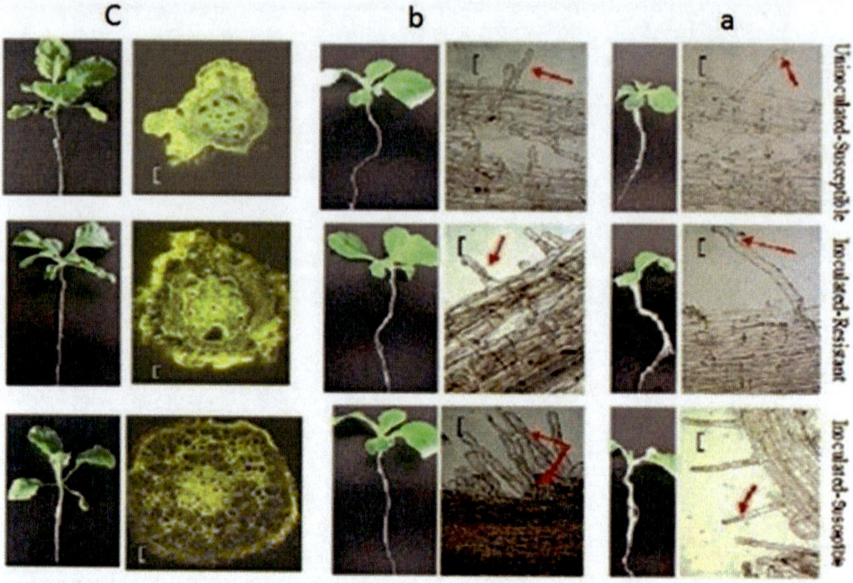

Plate 7.4 External morphology and microscopic observation of Cabbage roots after inoculating with *Plasmodiophora brassicae*. (**a**) The 7th day after inoculating. (**b**) The 14th day after inoculating, (**c**) The 30th day after inoculating. Scale bars = 50 μm (Ji et al. 2014)

multiply, increasing gradually, and more than a quarter of the lines are occupied on the 7th day after inoculating (Plate 7.4a). This stage is referred to as the root-hair infection stage (Agrios 2005). Kageyama and Asano (2009) reported that at the root-hair infection stage, there were no visible symptoms, which was confirmed by Ji et al. (2014) (Plate 7.4b). Later secondary zoospores release, the secondary plasmodia contain two nuclei in the early stages of growth, which develop into multinuclear plasmodia during the cortical infection or secondary infection stage (Kageyama and Asano 2009). Fourteen days after inoculating, the stage varied gradually from root-hair infection to cortex infection, but no external differences were visible (Plate 7.4b). These plasmodia undergo a series of cell divisions and eventually form multinucleate secondary plasmodia that invade cells in the cortex and stele. Both infected and neighboring uninfected cells undergo extreme hyperplasia (cell division), and hypertrophy (cell expansion), leading to the formation of characteristic galls at 30 days (Plate 7.4c); these results agree with those of Malinowski et al. (2012) in normal cells, and greater rates of cell division were evident in infected plants on 16, 22, and 26 days after inoculating. Cell division largely ceases by 32 days after inoculating, while cell division in the hypocotyls of infected plants appears as a diffuse ring in the vascular cambium (VC) and neighboring progeny cells. Ji et al. (2014) clearly determined the life cycle of *P. brassicae*, the dynamic changes of Chinese cabbage roots, and specified the time of each life cycle stage of *P. brassicae* infection in Chinese cabbage. IAA acts as signaling molecular that putatively stimulates root hair infection in the early stages of *B. napus* to *P. brassicae* infection (Xu et al. 2016)

7.2.2 Cortical Infection Process

The observations of cortical infection revealed that young plasmodia are small and roughly spherical in shape (Plate 7.5a). The young plasmodia proliferate and form clusters, followed by the development of mature plasmodia (Plate 7.5b). Mature plasmodia are generally much larger and irregular in shape. Resting spores develop from the cleavage of mature plasmodia. Resting spores were round and distinct (Plate 7.5c). Microscopic study of stained root sections revealed that temperature had a large effect on the development of *P. brassicae* within the root cortex. Infection and development of the pathogen in the root cortex was followed in roots grown at 15 to 30 °C (Fig. 7.2). No cortical infection was present at 10 °C. The extent of cortical infection increases as the temperature increases above 15 °C. Infected cells were first observed at 10 DAI in plants at 20 to 30 °C, and at 22 DAI in plants at 15 °C. The increase in root area infected was roughly linear over time but occurred at different rates for different temperatures. In plants at 25 °C, cortical infection was 2% at 10 DAI and increased to 37% by 42 DAI. There was initially little or no difference in cortical infection between the 20 and 30 °C treatments, but cortical infection at 20 °C was slightly higher from 34 to 42 DAI (Fig. 7.2a). Cortical infection at 42 DAI was 30% at 20 °C and 24% at 30 °C. In contrast, cortical infection remained very low at 15 °C, reaching a maximum of 9% at 42 DAI. The effect of temperature on cortical infection at 42 DAI was quadratic, with an optimum temperature of 24.4 °C ($y = -0.04x2 + 1.84x - 16.2$, where y = cortical infection and x = temperature; $R^2 = 0.88$; $P < 0.001$) (Fig. 7.3).

Young plasmodia, mature plasmodia, and resting spores were observed in the root cortex of plants grown at temperatures from 15 to 30 °C, but were not present at 10 °C (Fig. 7.2b). The numbers of cells per field of view containing the developmental stages of secondary infection (young plasmodia, mature plasmodia, and resting spores) of *P. brassicae* at 10 DAI were highest at 25 °C (48 cells), lower at 20 °C (34 cells), and lowest at 30 °C (29 cells) at 10 DAI, but the number of infected cells do not differ after 22 DAI in the temperature range of 20 to 30 °C (41 cells at 20 °C, 44 cells at 25 °C, and 45 cells at 30 °C) (Fig. 7.2b). In contrast, the number of infected cells per field of view at 15 °C increased slowly, appearing first at 22 DAI (7 cells), and increasing overtime to a maximum at 34 DAI (49 cells) that was higher than the number of infected cells at other temperature (Fig. 7.2b). Young plasmodia were first observed at 10 DAI at 20 to 30 °C, and at 22 DAI at 15 °C (Fig. 7.2b). At 15 and 30 °C, young plasmodia were still present in roots until 42 DAI. In contrast, young plasmodia were present only until 22 DAI at 25 °C, after which time only resting spores could be seen. Mature plasmodia were first detected in the root cortex at 10 DAI at 25 °C (16 cells per field of view), at 14 DAI at 20 °C (15 cells), and 30 °C (7 cells), and at 26 DAI at 15 °C (3 cells). Differentiation of mature

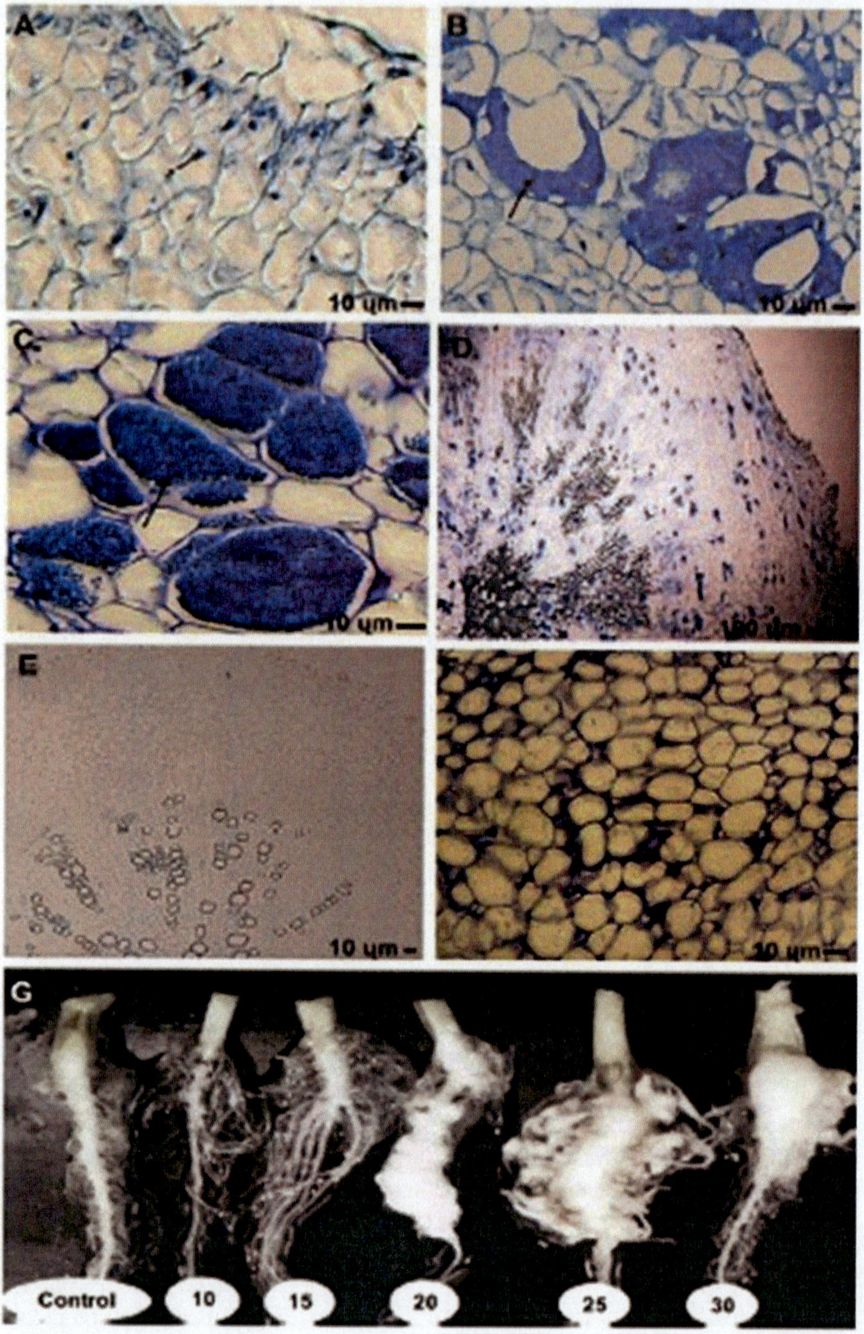

Plate 7.5 Stages of secondary infection of *Plasmodiophora brassicae* in cross-sections of the root cortex of Shanghai pak choy stained with methylene blue: (**a**) young plasmodia (arrow), (**b**) mature plasmodia in hypertrophied cell (arrow), (**c**) resting spores in cortical cells (arrow), (**d**) root sampled

plasmodia into resting spores was not affected by temperature and occurred 12 days after the first occurrence of mature plasmodia at temperatures from 15 to 30 °C. In the period from 26 to 42 DAI, the number of cells with resting spores increased from 8.5 to 36.5 at 20 °C, and 4.2 to 22.4 at 30 °C. The transition from mature plasmodia to resting spores was complete at 34 DAI at 25 °C, and 42 DAI at 20 °C, but at 15 and 30 °C the transition was still not completed when the trial was terminated at 42 DAI. The gross anatomy of root tissues in the non-inoculated controls was not affected by temperature but as expected, infection by *P. brassicae* resulted in severe distortion of root growth and development. At 10 °C, both non-inoculated and inoculated roots had distinct regions of cortical cells, endodermis, and stele. In inoculated roots where swelling symptoms were just beginning to develop, cortical cells looked similar to those in the non-inoculated roots, although the cortical cells were highly plasmolyzed. In contrast, the endodermis and stele exhibited signs of disruption and abnormal cell division (Plate 7.5d–f). The cellular organization within infected roots broke down as symptoms developed (Plate 7.5f). The cortical cells of the infected roots became enlarged and intermingled with the secondary phloem, and the distinct organization of tissues into cortex, endodermis, and stele was lost. Based on visual observation, infected cells were much larger than the corresponding cells of the non-inoculated roots, and even cells without plasmodia in the galled roots tended to be larger than the cells of the non-inoculated root (Sharma et al. 2011).

7.2.3 Symptom Development and Clubroot Severity

Temperature also has an effect on symptom development and clubroot severity. Clubroot symptoms (swelling and abnormal growth of root tissues) develop earlier and are more severe in plants grown at temperatures above 15 °C than at 10 or 15 °C. Swelling of the taproot is visible at 10 DAI in plants at 25 °C, at 14 DAI at 20 and 30 °C, and at 28 DAI at 15 °C. No symptoms of clubroot are present at 10 °C when the trial was terminated at 42 DAI (Fig. 7.4). Clubroot severity increased quickly at 25 °C and more slowly at 20 and 30 °C, with a DSI rating of 100 first observed at 22 DAI in plants at 25 °C, at 30 DAI at 30 °C, and at 32 DAI at 20 °C. In contrast, clubroot severity at 15 °C developed very slowly to a maximum DSI of 47 at 42 DAI (Fig. 7.4). There was a quadratic relationship between temperature and severity at 22 DAI (when DSI first reached 100 at 25 °C): $y = -0.025 \times 2 + 1.54x - 14.3$, where $y =$ DSI and $x =$ temperature ($R^2 = 0.80$; $P < 0.01$) with an optimum temperature of 25.6 °C (Fig. 7.5). The area of cortical infection was strongly and

Plate 7.5 (continued) at gall initiation; plasmodia are distributed throughout the tissue and there is no distinct cortex or stele. Note the extensive cell division and cells enlargement. (**e**) Healthy root showing distinct regions of cortex and stele divided by the endodermis, (**f**) healthy root showing small, uniform cells, and (**g**) effect of temperature (10–30 °C) on clubroot severity at 22 days after inoculating (Sharma et al. 2011)

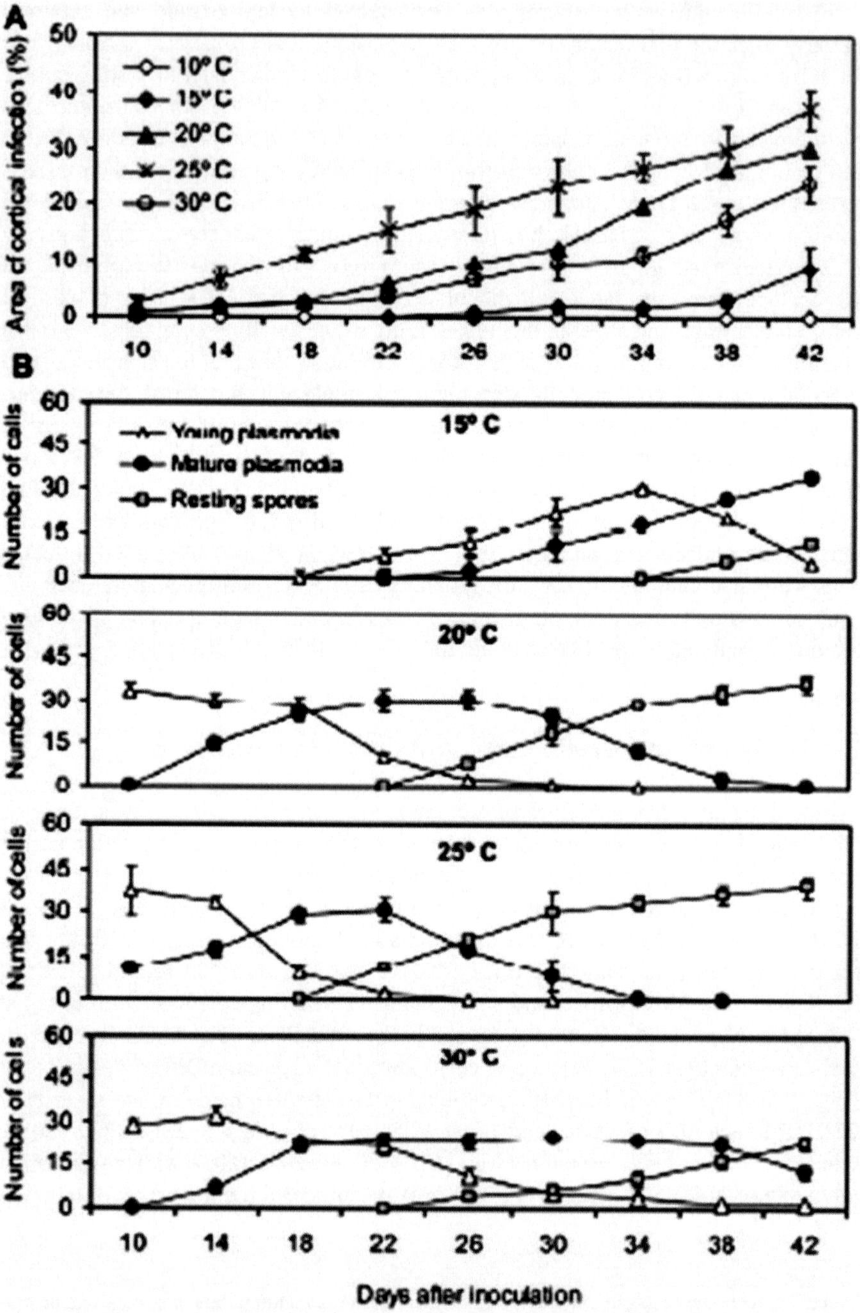

Fig. 7.2 Effect of temperatures from 10 to 30 °C on Shanghai pak choy inoculated with P*lasmodiophora brassicae*. (**a**) Proportion of the area of the root cortex occupied by the pathogen (cortical infection, %) and (**b**) stages of pathogen development in root tissue over time. Capped lines = standard error (Sharma et al. 2011)

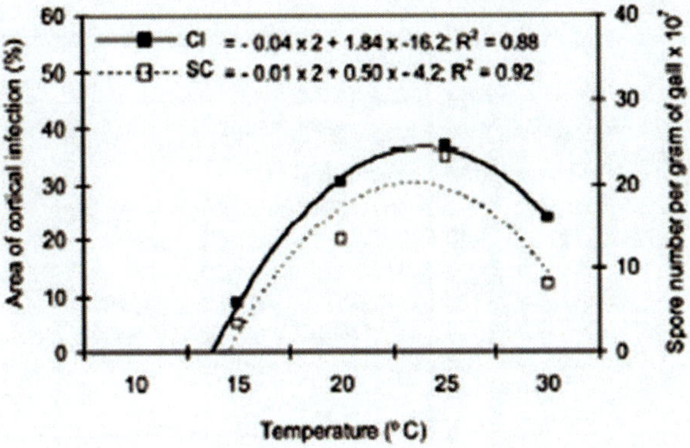

Fig. 7.3 Effect of temperature on the proportion of the area of the root cortex occupied by the pathogen (cortical infection, %) and number of spores per gram of gall at 42 days after inoculating. Curves are only meant to show responses, in their general pattern (Sharma et al. 2011)

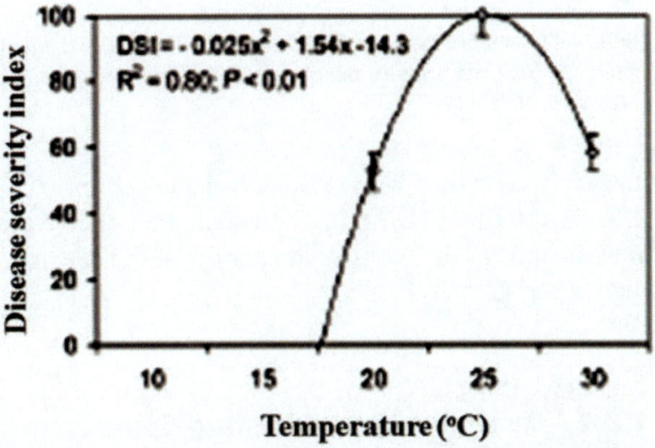

Fig. 7.4 Effect of temperatures from 10 to 30 °C on clubroot severity (disease severity index) on Shanghai pak choy assessed at 22 days after inoculating. Capped lines = standard error (Sharma et al. 2011)

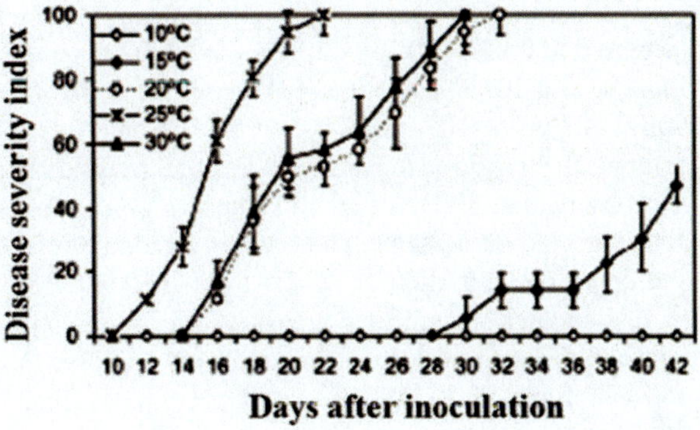

Fig. 7.5 Effect of temperatures from 10 to 30 °C on clubroot severity (disease severity index) on Shanghai pak choy over time. Capped lines = standard error (Sharma et al. 2011)

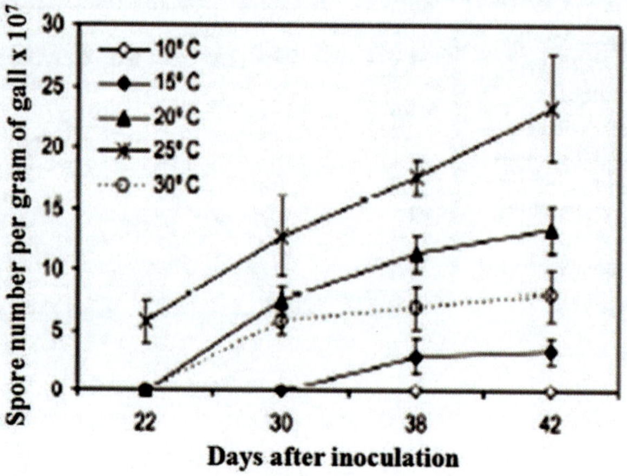

Fig. 7.6 Relationship between (**a**) disease severity index (DSI) and proportion of the area of the root cortex occupied by the pathogen (cortical infection, CI), (**b**) spores per gram of gall weight (SC), and CI, and (**c**) SC, and DSI from 10 to 30 °C. For the uniformity of presentation, four data points for each temperature treatment, each representing 22, 30, 38, and 42 DAI were selected (Sharma et al. 2011)

positively correlated with DSI but had a lower linear coefficient of regression ($R^2 = 0.60$; $P < 0.01$) (Fig. 7.6a). The maximum possible clubroot severity (DSI of 100) was obtained at 10% cortical infection; and the rate of cortical infection increase was high at 25 °C.

7.2.4 Quantification of Resting Spores

Temperature affected the number of resting spores per gram of gall. Resting spore concentration increased linearly over time at the optimum temperature of 25 °C, and progressively more slowly at 20, 30, and 15 °C (Fig. 7.7). At 25 °C, substantial quantities of resting spores (5.7×10^7 spores g^{-1}) were first observed at 22 DAI and increased to 23.2×10^7 by 42 DAI. At 30 DAI, spore concentration was 7.5×10^7 spores g^{-1} at 20 °C, and 5.7×10^7 spores g^{-1} at 30 °C. At 42 DAI, spore concentration was 13.3×10^7 spores g^{-1} at 20 °C, and 7.9×10^7 spores g^{-1} at 30 °C. In contrast, the transition to resting spores occurred slowly at 15 °C; resting spores were first observed at 38 DAI (2.9×10^7spores g^{-1}) and increased very slowly to 3.4×10^7 spores g^{-1} at 42 DAI. There was a quadratic relationship between temperature and spore concentration at 42 DAI ($R^2 = 0.92$; $P < 0.01$) (Fig. 7.3). Similarly, regression analysis indicated a positive linear relationship between spore concentration and the area of cortical infection ($R^2 = 0.89$; $P < 0.01$), and between spore concentration and the DSI ($R^2 = 0.59$; $P < 0.01$) (Fig. 7.6b, c). The main difference was that no resting spores were detected from the gall at DSI of 53 and 58 from plants at 20 and 30 °C at 22 DAI, respectively. In contrast, substantial levels of

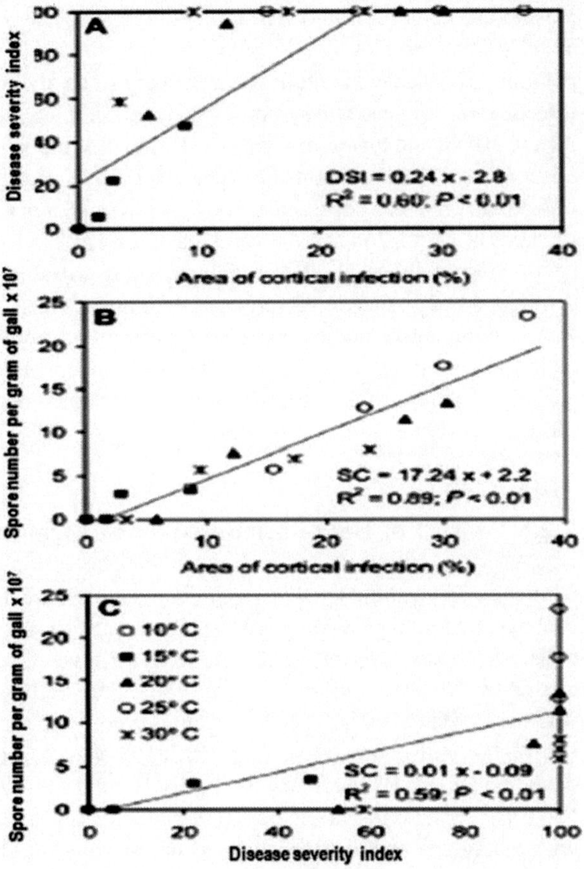

Fig. 7.7 Effect of temperatures from 10 to 30 °C on number of spores per gram of gall over time. Capped lines = standard error (Sharma et al. 2011)

resting spores (2.9×10^7 spores g^{-1}) were observed at DSI of 22 from plants at 15 °C at 38 DAI (Sharma et al. 2011).

7.2.5 Effect of Temperature on Cortical Infection by *P. brassicae* and Clubroot Severity

To assess the effect of temperature on infection and development of *Plasmodiophora brassicae* in the root cortex of Shanghai pak choy (*Brassica rapa* subsp. *chinensis*) and on subsequent clubroot severity, ten-day-old seedlings were grown individually, inoculated with resting spores, and maintained in growth cabinets at 10, 15, 20, 25, and 30 °C (Fig. 7.2). Seedlings were harvested at 2-day intervals, starting 8 days after inoculating (DAI) and continuing until 42 DAI. Roots were assessed at 4-day intervals for the incidence of cortical infection and stage of infection (young plasmodia, mature plasmodia, and resting spores), at 2-day intervals for symptom development and clubroot severity, and at 8-day intervals for the number of spores per gram of gall. Temperature affected every stage of clubroot development. Cortical infection was highest and symptoms were observed earliest at 25 °C, intermediate at 20 and 30 °C, and lowest and latest at 15 °C. No cortical infection or symptoms were observed at 42 DAI in plants grown at 10 °C. A substantial delay in the development of the pathogen was observed at 15 °C. Resting spores were first observed at 38 DAI in plants at 15 °C, 26 DAI at 20 and 30 °C, and 22 DAI at 25 °C. The yield of resting spores from galls was higher in galls that developed at 20 to 30 °C than those that developed at 15 °C over 42 days of assessment. These results support the observation in companion studies that cool temperatures result in slower development of clubroot symptoms in *Brassica* crops, and demonstrate that the temperature has a consistent pattern of effect throughout the life cycle of the pathogen (Sharma et al. 2011).

7.2.6 Effect of Host Resistance on Cortical Infection

Cortical colonization by *Plasmodiophora brassicae* in susceptible and resistant cabbage cultivars was assessed by Gludovacz et al. (2014). Three cultivars of cabbage (*Brassica oleracea* L. var. *capitata*) that differed in reaction to clubroot caused by *Plasmodiophora brassicae* were identified in field trials. Cultivar Bronco was susceptible to pathotype 6 (all plants severely clubbed), "Kilaherb" was resistant (no plants with symptoms), and "B-2819" was intermediate (small clubs). The development of *P. brassicae* in the roots of these cultivars was assessed under controlled conditions at 28 days after inoculating using root sectioning, staining with methylene blue, and bright field microscopy. Plasmodia of *P. brassicae* were present in the root tissue of each cultivar. There was extensive cortical colonization

Table 7.3 Clubroot incidences and severity on green and red cabbage cultivars grown in naturally infested soil at sites with high and low levels of resting spores of *Plasmodiophora brassicae* at the Muck Crops Research Station, Holland Marsh, ON in 2011 and 2012 (Gludovacz et al. 2014)

	High inoculum site (2011, 2012)		Low inoculum site (2012 only)	
Cultivar	Clubroot incidence (%)	Disease severity index	Clubroot incidence (%)	Disease severity index
Klimaro (S)	100 a[a,b]	100 a	72 a	27 a
Bronco (S)	100 a	100 a	71 a	24 a
B-2819 (MS)	98 b	53 b	11 a	4 a
Kilaherb (R)	0 c	0 c	0 b	0 b

[a]Values in a column with the same letter do not differ at $P = 0.05$, based on Tukey's multiple means comparison test

[b]Arcsine transformation used prior to analyses to achieve a closer approximation to normal distribution

Table 7.4 Percent area of cortical colonization and incidence of cortical cells containing selected developmental stages on cabbage at 28 days after inoculating with pathotype 3 of *Plasmodiophora brassicae*[a] (Gludovacz et al. (2014)

Cultivar	Area infected (%)	Number of cells with young plasmodia	Number of cells with mature plasmodia	Number of cells with resting spores	Total cells infected
Bronco (S)	26 a[b]	69 a	32 a	34 a	135 a
B-2819 (MS)	9 b	66 a	26 b	3 b	95 b
Kilaherb (R)	7 c	42 b	9 c	0 c	51 c

[a]Data combined across three repetitions of the study, with 76 pictures in total of cross-sections of root below the hypocotyl from 16 plants

[b]Values in the same column followed by the same letter do not differ at $P = 0.05$, based on Tukey's multiple means comparison test

(26%) and some resting spore formation in "Bronco," and lower colonization (9%) and only trace levels of resting spore formation in "B-2819." In the clubroot-resistant "Kilaherb," root colonization was similar (7%) to the intermediate "B-2819," but no resting spores formed and no clubbing symptoms developed (Tables 7.3 and 7.4). This pattern of response differed substantially from that observed previously in commercial cultivars of spring canola (*B. napus*), and may indicate that the mechanism of resistance in "Kilaherb" is different from that currently available in canola. Also, the intermediate phenotype of "B-2819" may represent a quantitative resistance reaction that may also differ from that in canola.

7.2.7 Induction and Role of Metabolites in Clubroot Host–Parasite Interaction

The clubroot symptoms clearly give evidence that plant growth regulators are involved in disease development. Hyperplasia and hypertrophy indicate the involvement of auxins (Ludwig-Muller et al. 1993) and cytokinins (Dekhuijzen and Overeem 1971). Although cytokinins are most likely synthesized by plasmodia of the pathogen (Dekhuijzen 1981; Muller and Hilgenberg 1986), it appears that the disease stimulates production of the auxin indole-3-acetic acid in the host (Grsic et al. 1999; Grsic-Rausch et al. 2000). Other symptoms of clubroot disease include changes in carbohydrate metabolism, such as an accumulation of starch in the roots and hypocotyls of infected plants (Evans and Scholes 1995; Williams et al. 1968). In addition, the disaccharide trehalose (α-D-glucopyranosyl-[1, 1]-α-D-glucopyranoside) has been shown to accumulate in the hypocotyls of cabbage plants affected by clubroot disease (Keen and Williams 1969). Trehalose, which is usually synthesized by trehalose-6-phosphate synthase (TPS) and trehalose-6-phosphate phosphatase, is a typical microbial sugar accumulating in a variety of symbiotic or pathogenic interactions of microorganisms with plants (Muller et al. 1995a, b). The synthesis of trehalose could affect the carbon metabolism of infected plants in two ways—by converting plant photosynthates into trehalose, microorganisms would divert carbon away from plant metabolism and thereby create a carbon sink. In addition, trehalose that is released into the plant could affect plant metabolism by interfering with the plant's sugar sensing system. Trehalose has been shown to induce enzymes of fructan synthesis in barley (Muller et al. 2000; Wagner et al. 1986) and sucrose synthase activity in soybean (Muller et al. 1998). In *Arabidopsis thaliana*, growth on trehalose-containing medium results in an inhibition of root growth, an increased accumulation of starch in the shoots, and an increased activity of ADP-glucose pyrophosphorylase, the first enzyme of starch synthesis, accompanied by an induction of the ADP-glucose pyrophosphorylase gene *ApL3* (Fritzius et al. 2001; Wingler et al. 2000).

Plants such as *A. thaliana* are themselves capable of synthesizing small amounts of trehalose (Blázquez et al. 1998; Muller et al. 2001; Vogel et al. 1998, 2001); therefore, trehalose synthesis also may be an important intrinsic component in the regulation of plant metabolism (Goddijn and Smeekens 1998; Muller et al. 1999). In fact, an *Arabidopsis* mutant with a T-DNA insertion disrupting *AtTPS1* is embryo lethal, indicating that trehalose synthesis plays an essential role in plants metabolism (Eastmond et al. 2002). By releasing trehalose into the plant, pathogens such as *P. brassicae* might be able to exploit the plant's trehalose-sensing system and alter carbohydrate metabolism to their favor.

The concentration of trehalose in plant cells is probably regulated by trehalase, the enzyme catalyzing the hydrolysis of trehalose into two glucose molecules. By preventing an over accumulation of trehalose that would interfere with the regulation of plant metabolism, trehalase could play an important role in plant–microbe interactions. Trehalase activity has been found to be increased in symbiotic relationships of plants with trehalose-producing microorganisms, such as arbuscular mycorrhizal fungi (Schubert and Wyss 1995), and rhizobia (Muller et al. 1994). The role of trehalose metabolism in the interaction of *A. thaliana* with *P. brassicae*, focusing on the amounts of trehalose accumulating in various plant organs and on the activity and expression of plant trehalase was assessed by Brodmann et al. (2002). The trehalose accumulated strongly in the infected organs (i.e., the roots and hypocotyls) and, to a lesser extent, in the leaves and stems of infected plants. This accumulation pattern of trehalose correlated with the expression of a putative trehalose-6-phosphate synthase (EC 2.4.1.15) gene from *P. brassicae*, *PbTPS1*. Clubroot formation also resulted in an induction of the *Arabidopsis* trehalase gene, *ATTRE1*, and in a concomitant increase in trehalase (EC 3.2.1.28) activity in the roots and hypocotyls, but not in the leaves and stems of infected plants. Thus, induction of *ATTRE1* expression was probably responsible for the increased trehalase activity. Trehalase activity increased before trehalose accumulated; therefore, it is unlikely that trehalase was induced by its substrate. The induction of trehalase may be part of the plant's defense response and may prevent excess accumulation of trehalose in the plant cells, where it could interfere with the regulation of carbon metabolism (Fig. 7.8) (Brodmann et al. 2002).

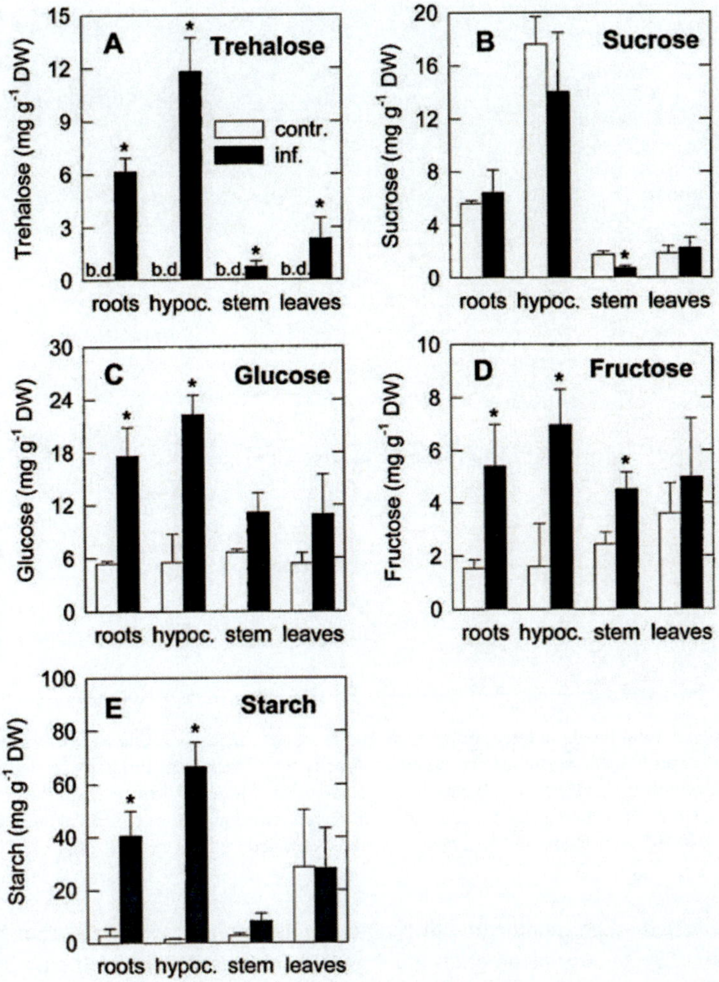

Fig. 7.8 An amount of (**a**) trehalose, (**b**) sucrose, (**c**) glucose, (**d**) fructose, and (**e**) starch in the roots, hypocotyls, stems, and cauline leaves of control plants (control = open bars) and of infected plants (infected = black bars) 38 days after inoculating with *Plasmodiophora brassicae*. Data are means of between four and five plants. Standard error. Asterisks (*) indicate significant differences between the control and the infected plants (analysis of variance; $P = 0.05$). *b.d.* below detection limit (Brodmann et al. 2002)

7.2.8 Role of Host Hormone Metabolism in Pathogenesis

Infection by *P. brassicae* results in the misregulation of plant-derived hormonal pathways leading to alterations in the root phenotype, called clubs or clubroots, as well as reduced growth in the aerial parts of the infected plants. Assimilates from leaves are redirected towards the colonized roots, thereby reducing the overall plant performance (Ludwig-Muller et al. 2009; Siemens et al. 2011). The root gall size can

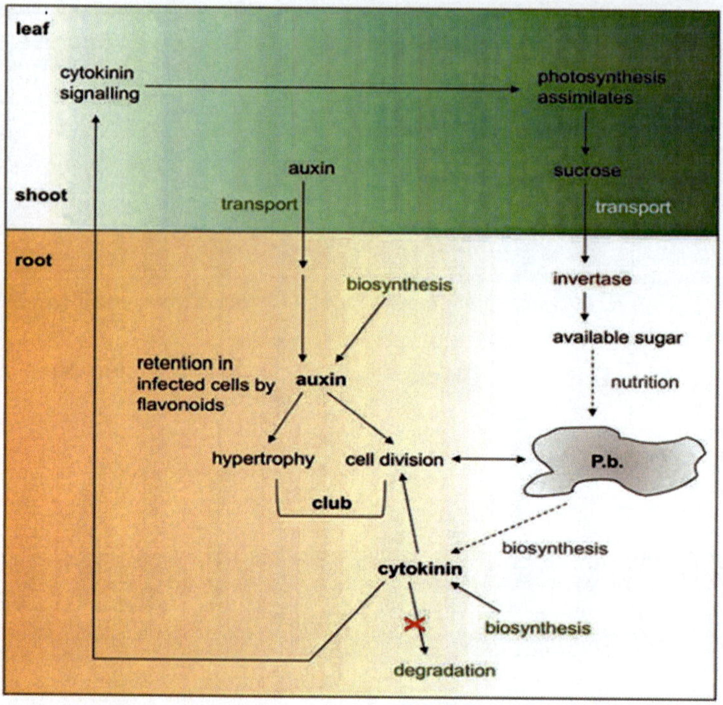

Fig. 7.9 A model for the interaction of roots and leaves in a *Plasmodiophora brassicae*-infected plant. Roots provide the hormonal signals either directly by *P. brassicae* (Pb) or by induction within the host, which then affect the above-ground plant parts. These in turn provide the roots with nutrients, which are an essential feature for an obligate biotrophic pathogen. Hormones affect and are also affected by primary and secondary metabolites (Diederichsen et al. 2014)

be associated with plant growth-promoting hormones, especially auxins and cytokinins. Secondary metabolites are, however, also associated with club development, either directly through the action of growth hormones or indirectly via their general bioactive properties. Lastly, the hormones also control primary metabolism, such as sugar availability for the pathogen.

The role of auxins in the development of typically hypertrophied root tissues (clubroots) has been partially elucidated. The model plant *A. thaliana* is a good host for *P. brassicae*, so its metabolism was studied using genetic, biochemical, and molecular tools. A simple model (Fig. 7.9) indicates that *P. brassicae* causes synthesis of cytokinins (CK) (Muller and Hilgenberg 1986) and blocks the CK degradation pathway in the host plant (Siemens et al. 2006), increasing the CK levels in infected tissues. In addition, the host CK synthesis is also up-regulated (Ludwig-Muller et al. 2009). CKs could then be transported to the upper part of the plant, leading to an increase in assimilate production. Sucrose and other sugars are transported back to the metabolic sink in the roots created by the elevated CK levels (Fig. 7.9; Table 7.5). This is indicated by the large numbers of starch grains present

Table 7.5 Selected changes in hormone status and metabolism, and corresponding effects on the clubroot phenotype in *Arabidopsis thaliana* (Siemens et al. 2002)

Hormone/ compound	Genes or pathways altered	Type of alteration	Effect on clubroot	Reference
Auxin	Biosynthesis	Mutant nit1	Reduction	Grsic-Rausch et al. (2000)
		Antisense nit2	Delayed gall formation	Neuhaus et al. (2000)
		Overexpression NIT2	None	Neuhaus et al. (2000)
	Conjugate hydrolysis	Mutant ilr1	None	Siemens et al. (2002)
	Conjugate synthesis	Single mutant various GH3 genes	None	Jahn et al. (2013)
	Conjugate synthesis	Double mutant gh3.5 gh3.17	Slightly more susceptible	Jahn et al. (2013)
Auxin/ glucosinolate	Biosynthesis	Mutant CYP79B2/B3	None	Siemens et al. (2008)
Glucosinolate	tu8 (heterochromatin protein, lower IGL levels)	Mutant	Reduction	Ludwig-Muller et al. (1999)
Camalexin	Biosynthesis	Mutant pad3	None	Siemens et al. 2002, 2008)
Cytokinin	Degradation	Overexpression cytokinin oxidase	Reduction	Siemens et al. (2006)
Brassinosteroid	Biosynthesis	Mutant *det2*	Some reduction	Siemens et al. (2002)
	Signaling pathway	Mutant *bri1*-6	Reduction	Schuller and Ludwig-Muller (unpublished results)
Gibberellin	Biosynthesis	Chemical inhibition	None	Pasold and Ludwig-Muller (2013)
Jasmonic acid	Conjugation/ activation	Mutant *jar1*	Higher susceptibility	Siemens et al. (2002)
Ethylene	Signaling pathway	Different receptor mutants (*etr1*)	Higher susceptibility	Knaust and Ludwig-Müller (2013)
Flavonoid	Biosynthesis	Different mutants	Slight reduction	Pasold et al. (2010)
	Root-specific MYB transcription factor	Overexpression	Some reduction	Pasold et al. (2010)
Sucrose	Invertase	Repression of enzyme activity	Reduction	Siemens et al. (2011)

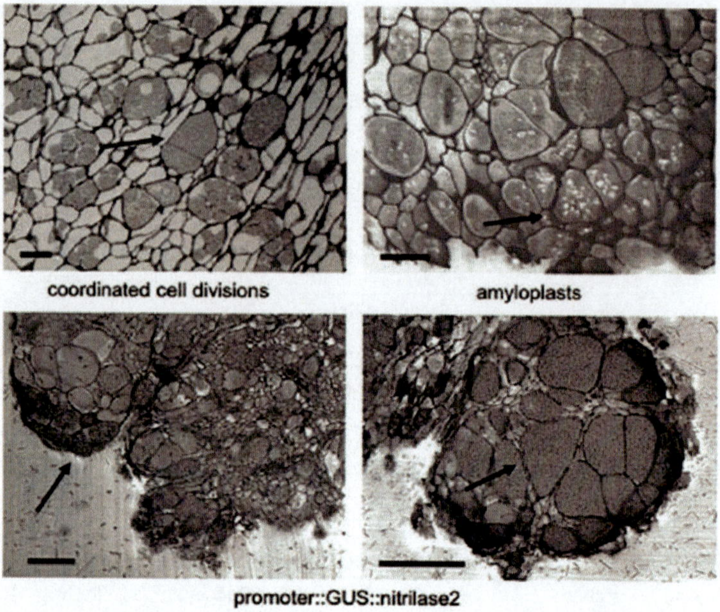

Plate 7.6 *Plasmodiophora brassicae* might induce synchronized cell divisions together with host cells (arrow). The metabolic sink also results in the accumulation of starch grains/amyloplasts (arrow). These can be seen by the white appearance in the staining of the sections with methylene blue-azur II and basic fuchsin to differentiate pathogen structures from host tissue. The two lower pictures show promoter: GUS activity for nitrilase 2 transgenic lines, the ß-glucuronidase activity is indicated by dark parts in the sections. In the left photo, the arrow points to such stained regions; in the right photo, the arrow indicates large hypertrophied cells filled with resting spores of *P. brassicae*. Bars represent 10 μm (upper two panels) and 20 μm (lower two panels) (Ludwig-Muller et al. 2009)

in infected roots with well-developed clubs (Plate 7.6). Simultaneously, host auxin synthesis is induced by *P. brassicae* and perhaps the indole glucosinolate pathway is hijacked by the pathogen, producing larger amounts of free IAA via the nitrilase pathway (Ludwig-Muller 2009a). The auxin produced in *P. brassicae*-colonized host cells (Grsic-Rausch et al. 2000; Pasold et al. 2010) might be retained in the infected cells by increased flavonoid accumulation in infected host roots, because flavonoids are believed to prevent auxin efflux from a cell (Pasold et al. 2010). This indicates a role for auxin transport in clubroots. Together, auxin (indole-3-acetic acid; IAA) and CK increase the cell division rate and plasmodia of *P. brassicae* might divide together with the host cells. This hypothesis is supported by the observation that in highly infected roots, the cells harboring the protists are connected to each other in large patches (Plate 7.6). In addition, during later stages of infection, the high auxin levels in cells containing plasmodia of the protists cause hypertrophy, so that the number of resting spores formed increases (Plate 7.6). Another plant growth-promoting hormone class, gibberellins, does not seem to be involved in club development. Application of a gibberellin biosynthesis inhibitor (chlorocholine chloride, CCC) to *P. brassicae*-infected plants did not alter the size of a club, so it was concluded that the formation of clubroots is independent of GA synthesis (Pasold and Ludwig-Muller 2013), even though some genes involved in GA signaling were found to be elevated in a microarray experiment (Siemens et al. 2006). On the other hand, there is some evidence for the involvement of brassinosteroids, because the det2 mutant showed a more tolerant phenotype (Siemens et al. 2002). The gene DET2 encodes a protein from the biosynthetic pathway leading to brassinolide (BR) in *Arabidopsis*. In addition, a BR receptor mutant bri1-6 exhibited smaller root galls compared with the wild type after infection with the protists (Table 7.5) (Diederichsen et al. 2014).

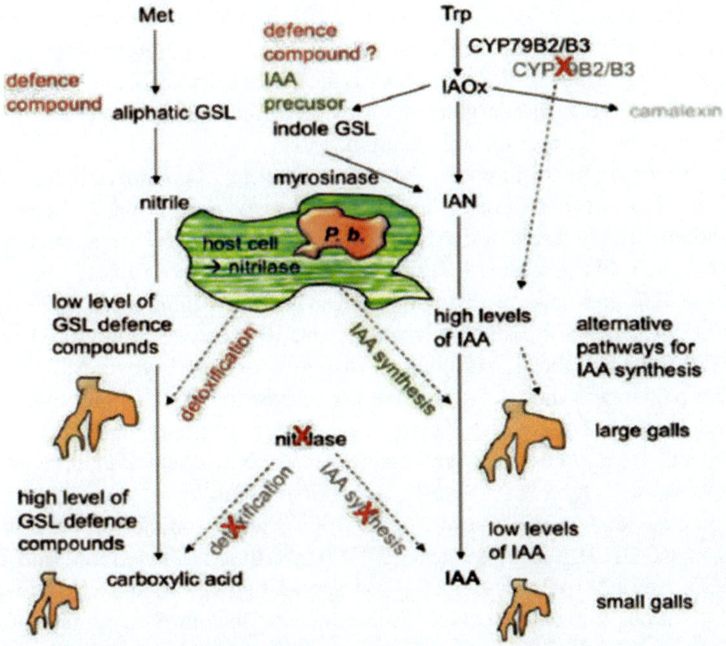

Fig. 7.10 A model explaining the possible dual roles for glucosinolates in defense or gall growth induced by *Plasmodiophora brassicae* (*P.b*). Aliphatic glucosinolates are viewed as defense compounds, while indole glucosinolates could play dual roles. The hatched (green) compartment symbolizes a hypertrophied host cell, harboring plasmodia of *Plasmodiophora brassicae*, which expresses high nitrilase transcript/protein. Nitrilase would convert the nitriles to carboxylic acids, yielding on the left-hand side less toxic defense compounds, whereas indole-3-acetonitrile (IAN) is converted to the auxin IAA. Both would result in large galls. If nitrilase is blocked, then either more toxic degradation products are formed from the glucosinolates or less IAA is synthesized, both leading to smaller galls. Small galls symbolize reactions leading to resistance or tolerance, large gall symbols stand for reactions leading to gall growth. A red cross symbolizes the reduction of the compound or pathway. The involvement of the IAA biosynthesis pathway and the side branch to the phytoalexin camalexin are shown together with their effects on gall size and IAA levels. When the first step in the IAA pathway is genetically blocked, then the IAA levels and gall size are still the same as in the wild type, indicating that additional biosynthetic pathways have to be activated (Piotrowski 2008)

The so-called omics approaches have helped to elucidate the role of plant hormones in clubroot development. Siemens et al. (2006) compared the transcriptome of clubroots at two different points after inoculating with somewhat defined developmental stages of the pathogen for changes in gene expression profiles. Agarwal et al. (2011) used early stages during pathogenesis for a similar experiment, and Devos et al. (2006) analyzed the proteome in infected roots during these early stages of colonization. These data have all been obtained with the model plant *Arabidopsis thaliana*, which is a good host for *P. brassicae*. Additionally, proteome data were supplied for *Brassica* species by Cao et al. (2008).

7.2.8.1 Auxin Metabolism and Signaling

Auxin can be synthesized in *Arabidopsis* via several different pathways, presumably occurring in parallel to each other (Fig. 7.10). While the involvement of some pathways has only been inferred to be of importance based on microarray studies (Siemens et al. 2006), the pathway typical for Brassicaceae via indole-3-acetaldoxime (IAOx), indole glucosinolates (IG), and indole-3-acetonitrile (IAN) has received a lot of experimental attention.

There seem to be differences, however, between *Arabidopsis* and *Brassica* species (Ludwig-Muller 2009a). In an *Arabidopsis* mutant with altered indole glucosinolate (IGL) levels resulting from a mutation in the gene encoding the heterochromatin-like protein 1 (tu8), the lower levels of indole-3-acetonitrile (IAN), and IGL correlated with smaller clubs (Ludwig-Muller et al. 1999; Bennett et al. 2005). Early work pointed to many correlations between glucosinolate levels (especially IGL) and the size of clubs in *Brassica* species (Ludwig-Muller 2009a). Since the pathway is thought to involve the conversion of IGL via myrosinase to IAN, and further to IAA by the activity of nitrilase, the activity, and transcription of nitrilase was monitored in several plant species. Nitrilase (NIT) transcripts are up-regulated in clubs of *Arabidopsis* (Grsic-Rausch et al. 2000). Using immunolocalization (Grsic-Rausch et al. 2000), and promoter-reporter analyses (Pasold et al. 2010), the presence of NIT transcriptional activation, and protein localization in cells with plasmodia, and sporulating plasmodia close to resting spore formation was demonstrated. Consequently, the nitrilase mutant nit1, and antisense plants for the NIT1 and NIT2 genes with reduced levels of NIT expression had reduced gall size or the development was slower (Grsic-Rausch et al. 2000; Neuhaus et al. 2000) (Table 7.5). In this context, it is interesting to note that NIT2 was the predominantly expressed isoform. By contrast, when a mutant was investigated, which is blocked in the first step of the pathway devoted to the synthesis of IAA, camalexin and indole glucosinolates via indole-3-acetaldoxime (IAOx), no reduction in club size was detected, and the levels of endogenous IAA in galls were similar to wild-type plants (Siemens et al. 2008). It is suggested that blocking early steps in the IAA pathway can induce alternative pathways, while blocking later steps, such as nitrilase, resulted in a reduction of IAA and thus club formation was also inhibited (Plate 7.6). The importance of increased nitrilase transcription within the enlarged clubs has been demonstrated in *Brassica* species, but additional work has provided evidence for the up-regulation of other genes involved in auxin biosynthesis, such as aldehyde oxidase and amidase (Ludwig-Muller et al. 2009).

Since glucosinolates such as indole glucosinolate (IGL) are involved as defense compounds in other plant–pathogen systems, another possibility for the localization of nitrilase in plasmodia-containing cells might be important. In addition to the conversion of IGL to IAN, and IAA, thereby leading to the hypertrophy of infected host cells (Fig. 7.10), nitrilase could be a defense mechanism to degrade the glucosinolates to less toxic nitriles as suggested by Piotrowski (2008). In the presence of epithiospecifier proteins, and myrosinase, the glucosinolates are degraded to toxic compounds, the (epi) thi °Cyanates. To avoid this, *P. brassicae* could also induce host nitrilase in the cells where the protists are residing to prevent the production of the toxic compounds (Fig. 7.10). Thus, the induction of nitrilase could have a twofold effect for the pathogen: first, induction of IAA synthesis and second, alteration of the defense metabolite spectrum provided by the host. Both hypotheses fit with the observation of nitrilase being closely connected to host cells with the presence of pathogen plasmodia. While the biosynthesis to IAA is better understood, the signaling pathways have so far not been elucidated. There are two different auxin receptor types discussed (Christian et al. 2006), the auxin binding protein 1 (ABP1) and the TIR F-box protein class of receptors. Both classes of receptors were up-regulated in transcription during disease progression as shown by promoter: GUS analysis for ABP1 and transcript levels for TIR1 (Jahn et al. 2013). In addition, TIR-class receptor double mutants showed less susceptibility to the clubroot pathogen, indicating that hormone signaling must be intact (Jahn et al. 2013). As targets for auxin perception via the TIR pathway, the GH3 genes could be required, as these are also involved in regulating the auxin homeostasis by conjugating IAA to various amino acids (Staswick et al. 2005). Consequently, it might be expected that several GH3 genes were transcriptionally up-regulated during clubroot development as shown by microarray data (Siemens et al. 2006) and transcript analysis (Jahn et al. 2013). Thus, with the GH3 genes, a target group for the auxin pathway could be identified (Table 7.5). Other target genes for IAA in the formation of hypertrophied cells could be those involved in cell wall expansion (Ludwig-Muller 2009b; Diederichsen et al. 2014)). In *Brassica rapa* L., it has been shown at an enzymatic level that the IAA conjugates with amino acids were hydrolyzed back more strongly in infected roots compared with controls, also contributing to the free IAA levels in clubs (Ludwig-Muller et al. 1996). An *Arabidopsis* mutant in one of the conjugate hydrolase genes ilr1 has been recognized, however, formed clubs comparable with wild-type galls (Table 7.5) (Siemens et al. 2002).

7.2.8.2 Cytokinin Metabolism

The involvement of cytokinins in club formation was demonstrated for *Brassica* species, and *Arabidopsis* (Ludwig-Muller et al. 2009). It has been (Dekhuijzen and Overeem 1971; Dekhuijzen 1980, 1981) shown that clubs contained increased amounts of cytokinins, which was confirmed by later studies using more modern analytical techniques. In the 1980s, it was demonstrated that *P. brassicae* can synthesize small amounts of cytokinins (Muller and Hilgenberg 1986). In addition, plants are contributing to increased cytokinin levels by induction of their own biosynthesis, or by repression of degradation. The first possibility seems to be present in *Brassica*, where the induction of *ipt* genes was found. In *Arabidopsis*, the induction of cytokinins seems to involve the second possibility, i.e., repression of metabolic steps, also leading to increased cytokinin levels (Siemens et al. 2006). The reduction of transcripts encoding two root-specific cytokinin oxidase/dehydrogenase genes (*CKX*) was described. Consequently, plants overexpressing *CKX* genes were much more tolerant against the clubroot pathogen, but since they also show a dwarf phenotype, which cannot be used to control the disease symptoms. Solutions to this problem might include the control of gene expression under the control of root-specific promoters, as was used in reducing invertase (Siemens et al. 2011). Since disease development is also dependent on cell divisions in the host root cortex, the role of cytokinins might be to stimulate cell divisions together with auxin.

Cytokinins not only regulate tissue growth, can also be involved in the redirection of assimilates and possibly also other nutrients from the aerial parts of an infected plant towards the root galls (Ludwig-Muller et al. 2009; Siemens et al. 2011). Several sugars were found to be increased in *P. brassicae*-infected roots (Brodmann et al. 2002), an indication of either higher de novo synthesis and/or relocation. Together with the observation that primary metabolism was strongly induced in microarray experiments; this led to a hypothesis connecting the presence of cytokinins with sugar metabolism. It was shown that invertase gene expression and protein amount can be increased by addition of cytokinins (Ehness and Roitsch 1997), so that the role of these proteins for carbon partitioning was investigated (Siemens et al. 2011). It could be demonstrated that several invertase promoters showed strong activity in root galls as determined by promoter: reporter fusions, which increased during disease progression. Mutants in a single invertase gene, however, did not alter the clubroot phenotype. Therefore, a different approach was used, namely to inhibit invertase activity by overexpression of a gene for a proteinaceous invertase inhibitor. The results showed that in transgenic plants, the invertase activity was reduced and the infected roots were smaller (Siemens et al. 2011). Thus, invertases can be regarded as targets for cytokinins, which increase the available soluble sugars for uptake into the plasmodia of the pathogen.

7.2.8.3 Role of Other Plant Hormones

No clear role for several other plant hormones has yet been identified so far in case of *P. brassicae* infection. Some observations, however, from microarray data have led to the assumption that during the later developmental stages of a clubbed root, drought stress signaling might be important. A large number of abscisic acid (ABA) related transcripts were strongly up-regulated once the pathogen became established in the host root (Ludwig-Muller 2009b). During this period, the vascular tissue is heavily affected, so the infected plant is also suffering from low water and nutrient transport rates. Whether the levels of ABA are indeed increased has yet to be demonstrated, but it is reasonable to hypothesize that the plant is defending itself against drought stress (Fig. 7.11). This action would be also beneficial for the

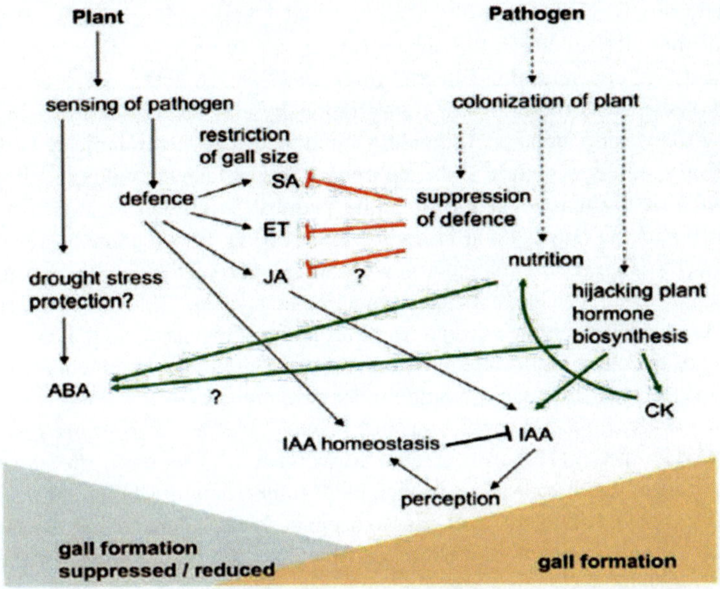

Fig. 7.11 A model compiling the putative roles for several plant hormones in clubroot disease development. Green arrows stand for inducing reactions, red blocks for inhibition. The different pathways and hormonal signals result in a homeostasis between the pathogen and the host plant, which can result in different outcomes such as gall formation or suppression of gall growth/ formation with intermediate states possible (Siemens et al. 2008)

clubroot pathogen itself, which is dependent on obtaining nutrition from the host. Some experiments results have provided evidence that the clubroot pathogen can down-regulate (partially) plant defense mechanisms or avoid them. Camalexin is increased in clubroots (Siemens et al. 2008), but a mutant in the camalexin biosynthetic pathway did not show altered clubroot phenotypes, indicating that this phytoalexin is not important in the plant's defense against *P. brassicae* (Fig. 7.10). Many plant defense genes in microarray analyses have been shown not to be regulated or even down-regulated. Several mutant analyses support this idea. In the jasmonate-triggered defense response, one critical step is the conjugation of jasmonic acid (JA) to the amino acid isoleucine (Staswick and Tiryaki 2004), which results in the active component recognized by the F-box protein COI1 involved in the signal transduction. In line with observations that in several plant–pathogen interactions, the mutant in the gene encoding the conjugating enzyme (jar1) was more susceptible towards the tested pathogens, the jar1 mutant was also more susceptible to *P. brassicae* (Siemens et al. 2002). On the other hand, a positive connection between jasmonic acid and nitrilase has been shown (Grsic et al. 1999), indicating possibly a two-fold role for JA in clubroot pathogenesis (Fig. 7.11): the plant may use it as defense signal, while JA can increase indirectly IAA levels, which is of advantage to *P. brassicae*. The ethylene (ET) signal is also involved in different defense responses. In the case of the interaction of *Arabidopsis* with *P. brassicae*, the role as a defense molecule is not clear (Knaust and Ludwig-Müller 2013). While the levels of the intermediate in the biosynthetic pathway aminocyclopropane carboxylic acid (ACC) decreased in infected roots, one gene, ACO2, encoding a biosynthetic enzyme, ACC oxidase, was transcriptionally elevated in clubroots compared with controls. Using mutants in signaling components, however, indicated that these were clearly more susceptible to the clubroot pathogen than the wild type, leading to the result that the functional ET signaling pathway is needed to restrict root gall growth (Fig. 7.11) (Knaust and Ludwig-Müller 2013). In this context, it should be noted that the ET receptor mutant etr1, when tested with high inoculum densities, did not show an aberrant clubroot phenotype (Siemens et al. 2002). Finally, salicylic acid (SA) as a defense molecule against biotrophic pathogens could play a role in the outcome of the clubroot disease syndrome. It was shown that treatments with SA at an early stage after inoculating could render *Arabidopsis* (Agarwal et al. 2011) and *B. oleracea* var. *italica* Plenck (broccoli) (Lovelock et al. 2016) more resistant to *P. brassicae* infection. However, treatments with SA at a later stage after inoculating did not result in the increased resistance of *B. rapa* (Ludwig-Muller et al. 1995). It can be concluded that the treatment has to occur early during the infection process, or otherwise *P. brassicae* is somehow capable of suppressing this plant defense pathway. Nevertheless, an increase in SA levels, together with increased biosynthesis from chorismic acid, has been shown in infected *Arabidopsis* plants. These findings have stimulated the idea that the plant defense mechanisms might be suppressed by the pathogen. Using an endophytic fungus (*Acremonium alternatum* Link), it was possible to induce tolerance against *P. brassicae* in *Arabidopsis,* and *B. rapa* (Chinese cabbage) (Doan et al. 2010; Jaschke et al. 2010). The effect was dependent on the application of the fungus in a period between 3 and 0 days before inoculating with *P. brassicae*, and the best effect was achieved when both organisms were administered at the same time. Autoclaved spores were, however, also able to induce the tolerance phenomenon (Jaschke et al. 2010). Analysis of gene expression from a set of genes from *P. brassicae* together with histological examinations indicated that the protists were not inhibited, but rather development in the roots slowed down. Indeed, preliminary experiments results showed that the fungus *A. alternatum* is able to induce some defense-related genes from the JA and SA signaling pathways. Despite these findings, which have been obtained recently, there is still a very incomplete understanding of the hormonal signals leading to the homeostasis between the clubroot pathogen and its host plants (Diederichsen et al. 2014).

7.2.8.4 Hormonal Responses to *P. brassicae* Infection in *Brassica* Cultivars Differing in Resistance

The pathogen *P. brassicae* requires the modulation of host metabolism including hormones pools for successful infection and pathogenesis. Gall formation is associated with the stimulation of cell division, which requires cytokinin (CK) and auxin activity. Moreover, CKs enhance the sink strength; thus, their elevation in galls is important for the attraction of nutrients to infected roots, as the pathogen is dependent on nutrients from the host, such as carbohydrates, amino acids, and lipids. Enhanced CK levels in roots after *P. brassicae* infection have been reported

(Dekhuijzen and Overeem 1971; Dekhuijzen 1981). Muller and Hilgenberg (1986) demonstrated that *P. brassicae* is able to convert adenine to trans-zeatin, and its riboside, which indicates that the pathogen is able to form CKs in limited amounts. Further evidence for the involvement of CKs came from Siemens et al. (2006), who showed that CK-depleted *Arabidopsis thaliana* plants were more tolerant to the clubroot pathogen. When the early response of *A. thaliana* to *P. brassicae* infection was followed, plasmodial-produced CKs (mainly isopentenyladenine and isopentenyladenosine) triggered the CK response (expression of response regulator ARR5) three days after inoculating (dai), which resulted in the 1 °Cal re-initiation of cell division in the root cortex 4 dai. Even during the later stages, CKs were not differentially regulated in *A. thaliana*, it was suggested that CKs can also regulate the development of the plasmodial stage of *P. brassicae* (Devos et al. 2006; Malinowski et al. 2016). At 6 dai, elevation of auxin (indole-3-acetic acid, IAA) content, as well as the stimulation of auxin-inducible genes is observed, promoting morphological changes in the roots including cell elongation. Polar auxin transport seems to be indispensable for early clubroot symptom development. After early up-regulation, auxin content may be maintained, even at the later stages of infection. The up-regulation of nitrilases and myrosinases seems to be responsible for this effect via the indole-3-acetonitrile pathway. The other growth-promoting hormones that were reported to be involved in *P. brassicae* infection are brassinosteroids (Grsic et al. 1999; Schuller et al. 2014). The pathogen imposes alteration of the levels of other hormones too. Abscisic acid (ABA), hormone predominantly associated with abiotic stress responses, is found to be elevated in the host plant at the later stages of *P. brassicae* infection, at least partially due to the restriction of water transport by roots and the resulting water deficit in the above-ground parts (Ludwig-Muller et al. 2009).

The key hormone in the defense against biotroph infection is salicylic acid (SA). Its level is found to be strongly enhanced by *P. brassicae* infection in resistant Bur-0 ecotype of *A. thaliana*, in contrast to an only mild elevation of jasmonic acid (JA) levels and signaling pathway. A substantial JA increase was detected in Bur-0 after infection with another, more aggressive pathotype, as well as in the sensitive ecotype *A. thaliana* Col-0. Furthermore, mutants with a constitutive SA-response and PR1-expression (cpr1) showed enhanced resistance to *P. brassicae* (Lemarie et al. 2015; Lovelock et al. 2016). Elevation of JA biosynthesis and JA-responsive gene expression in sensitive Col-0 may raise a question whether this effect is a part of the host plant defense or a part of the pathogen strategy to overcome plant defense by suppression of the SA pathway. A similar effect was observed in the case of coronatine produced by *Pseudomonas syringae*. The pathogen responses may differ among different species or even cultivars (*A. thaliana* and *Brassica* species). Prerostova et al. (2018) compared two cultivars of *Brassica napus*—the more resistant cultivar Alister and the sensitive cultivar Hornet—in order to characterize responses associated with enhanced resistance to *P. brassicae*. The responses have been determined at the levels of hormones (CK, auxin, SA, JA, and ABA) as well as the expression of hormone-related genes. Taking into account the communication between roots and shoots, the effect of *P. brassicae* infection was followed not only in roots and galls, but also in leaves.

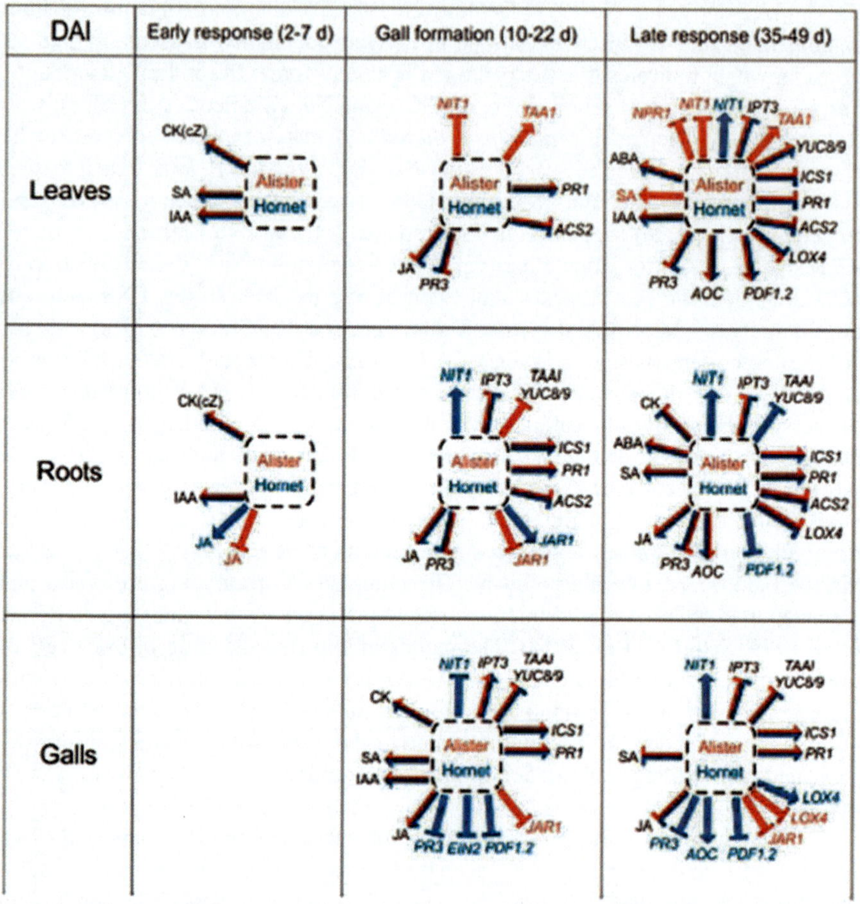

Fig. 7.12 Summary of the *Plasmodiophora brassicae* impact on *Brassica napus* plants during infection progression. Up- or down-regulation of phytohormone contents and gene expression levels in Alister (specific response—red letters and arrows), and Hornet (specific response—blue letters and arrows) leaves, roots, and galls detected during early phase (2–7 dai), gall formation (10–22 dai), and late phase (35–49 dai). Response common for both cultivars—black letters (Prerostova et al. 2018)

Hormonal dynamics after *P. brassicae* infection was compared in two *Brassica napus* cultivars—more resistant SY Alister, and more sensitive Hornet, in order to elucidate responses associated with efficient defense. Both cultivars responded to infection by the early transient elevation of active cytokinins (predominantly cis-zeatin), and auxin indole-3-acetic acid (IAA) in leaves and roots, which was longer in Hornet. Moderate IAA levels in Hornet roots coincided with a high expression of biosynthetic gene nitrilase NIT1 (contrary to TAA1, YUC8, and YUC9). Alister had a higher basal level of salicylic acid (SA), and it stimulated its production (via the expression of is °Chorismate synthase (ICS1)) in roots earlier than Hornet (Fig. 7.12). Gall formation stimulated cytokinin, auxin, and SA levels—

with a maximum 22 days after inoculating (dai). SA marker gene *PR1* expression was the most profound at the time point where gall formation began, in leaves, roots, and especially in galls. Jasmonic acid (JA) was higher in Hornet than in Alister. To investigate SA and JA function, SA was applied before infection, and twice (before infection and 15 dai), and JA at 15 dai. Double SA application diminished gall formation in Alister, and JA promoted gall formation in both cultivars. Activation of SA/JA pathways reflects the main differences in clubroot resistance (Prerostova et al. 2018).

7.2.9 Role of Nitrilase in Clubroot Development

he development of clubroot disease symptoms are correlated with an increase of auxin and cytokinin, thus resulting in increased cell division and cell elongation. The high IAA content was attributed to the conversion of indole glucosinolates to IAN by the enzyme myrosinase, which is compartmented against its substrate in healthy tissue, and further conversion by nitrilase to IAA (Butcher et al. 1974). Several components of IAA synthesis via the indole glucosinolate pathway are induced during clubroot in Chinese cabbage (*Brassica rapa* subsp. *pekinensis*). The activity of the enzyme, which converts Trp to indole-3-acetaldoxime, presumably the first step in indole glucosinolate biosynthesis (Ludwig-Muller and Hilgenberg 1988), was enhanced in infected roots (Ludwig-Muller et al. 1997), presumably leading to the elevated glucosinolate levels found in clubs (Ludwig-Muller et al. 1993, 1997). The increase in myrosinase expression was also demonstrated in infected roots of Chinese cabbage (Grsic et al. 1999). The activity of nitrilase was enhanced in hypertrophied roots of *Brassica napus* (substrate 3-cyanopyridine; Rausch et al. 1981), and *B. rapa* (substrate IAN; Grsic et al. 1999), and *in vivo* studies showed elevated IAN levels in infected roots after feeding of labeled Trp (Rausch et al. 1983).

The expression of nitrilase during clubroot disease in *Arabidopsis* was investigated by using transgenic plants carrying promoter-reporter gene fusions for each nitrilase isoform, as well as by RNA gel-blot analysis during symptom development. While no differences between nitrilase expression in younger infected and control roots were found in RNA gel-blot analysis, in 32 dpi roots, corresponding to the exponential growth phase of clubs, the expression of nitrilase was prominently enhanced in infected roots (Fig. 7.13). This has been confirmed by western-blot analysis using polyclonal antibodies against the 38-kD protein from *Arabidopsis*. At later stages no differences were visible on the mRNA level, but the amount of nitrilase protein was still higher, indicating that it is more stable than nitrilase mRNA. Two GUS activity peaks were observed, one at 21 dpi and a second one at 32 dpi. While the first peak was only the result of *NIT1* expression, the second peak was predominantly due to expression of *NIT2*. The increase in *NIT* expression 21 dpi can bc correlated with the growth of plasmodia, the increase 32 dpi with sporulating plasmodia and development of sporangia. Hist °Chemical GUS staining in infected roots of *NIT1::uidA* and *NIT2::uidA* plants 32 dpi showed a correlation

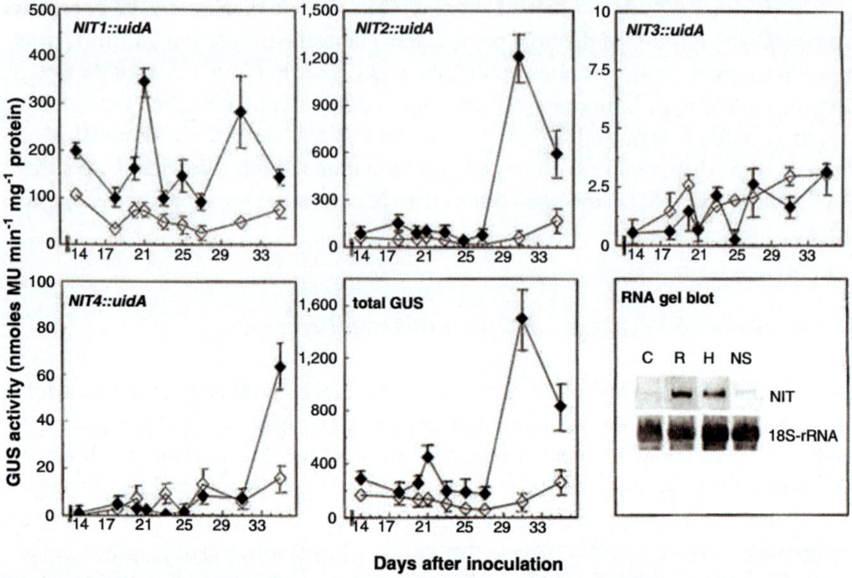

Fig. 7.13 Expression of different nitrilase genes during clubroot development in *Arabidopsis*. Control (L) and infected (l) roots of transgenic *Arabidopsis* plants harboring the *uidA* gene under the control of NIT1-4 promoters were analyzed at different time points (14–35 dpi) for GUS activity. Labeling of the y-axes (GUS activity) is meant for all panels except RNA gel-blot. Values are means ± SE from different plants of the same experiment. Total GUS activity was calculated by addition of the activities obtained for NIT1-4:: *uidA* plants. At 32 dpi a RNA gel-blot analysis was conducted with control roots (C), separately harvested root (R), and hypocotyl (H) clubs, as well as roots from infested soil showing no symptoms (NS). Nitrilase was detected with a NIT1-cDNA probe. Equal sample loading was confirmed by hybridizing the same blot with an 18S-rRNA probe. MU, Methylumbelliferone (Grsic-Rausch et al. 2000)

between GUS expression and sporangia containing cells, confirming the results of the immunolocalization of nitrilase protein. The increased *NIT2* expression after pathogen infiltration in leaves (Bartel and Fink 1994) points to an additional role for *NIT1* in the development of the hypertrophy.

Rausch et al. (1983) proposed that only very small amounts of auxins are needed for club formation, and therefore, enhancement of auxins in a few cells would be sufficient to induce symptom development. Therefore, the cellular localization of nitrilase in clubs using polyclonal antibodies against nitrilase was investigated by Grsic-Rausch et al. (2000). Only occasional nitrilase signals occurred in control roots, and noninvaded cells of clubs, but at certain stages of *P. brassicae* development, high amounts of nitrilase signals were observed, which correlated tightly with sporulating plasmodia. Only weak signals were found in cells harboring secondary plasmodia, and no signals in cells with mature resting spores. A time course study showed that cells containing strong nitrilase signals were first visible 29 dpi, with the number of these cells peaking 35–42 dpi, and then declining again. This confirms the

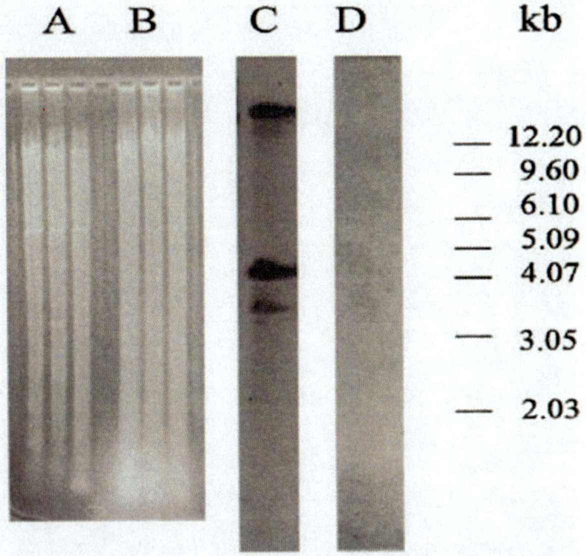

Plate 7.7 Detection of nitrilase from genomic DNA of *Arabidopsis* and *Plasmodiophora brassicae* by DNA gel-blot analysis. Lanes A and B show the digested genomic DNA of *Arabidopsis* and *P. brassicae*, respectively, prior to blotting after separation on a 0.8% (w/v) agarose gel. From left to right: BamHI, EcoRI, and HindIII digest. Southern blot from genomic DNA of *Arabidopsis* (lane C) and *Plasmodiophora brassicae* (lane D) digested with Eco RI was performed with a NIT1-cDNA probe. No signals are visible in the lane with *P. brassicae* DNA (Grsic-Rausch et al. 2000)

western analysis results, which showed increased nitrilase protein 35 and 42 dpi, although *NIT* mRNA was only increased 32 dpi, and then declined again.

Several lines of evidence suggest that the nitrilase detected during club development is of plant, and not pathogen origin. First, Southern analysis with a plant-specific *NIT1* cDNA probe (Bischoff et al. 1995) revealed that no nitrilase sequence similar to plant nitrilase is present in *P. brassicae* (Plate 7.7). Similarities between *NIT1* and the other nitrilase genes *NIT2*, *NIT3*, and *NIT4* sequences were 95%, 87%, and 69%, respectively, according to sequence alignment on the nucleotide level. The *NIT1* cDNA probe used clearly cross-hybridizes with *NIT1-3*, but not very strongly with *NIT4*. Therefore, it is highly unlikely that an antibody against the plant protein derived from this DNA would recognize a distantly related putative nitrilase from *P. brassicae*. However, it may be still formally possible that a nitrilase with different primary amino acid sequence could be structurally similar enough to cross-react with the NIT1 antibody. Since, transgenic *Arabidopsis* plants overexpressing *NIT4* (the least homologous *Arabidopsis* nitrilase) do not show an increased signal with the NIT1 antibody compared with wild type which is highly unlikely. Second, isolated resting spores from *P. brassicae* did not show immune signals with the NIT1 antibody. Third, in *NIT::uidA* lines, the immune signal obtained with an antibody against GUS is located in the same cells as the signal obtained with the NIT1 antibody. Measurement of GUS activity in clubs of untransformed plants has shown that *P. brassicae* does not possess GUS activity; therefore, the immune signal obtained with the GUS antibodies must also derive from the plant promoter. It was hypothesized that the increase in IAA is derived from the decompartmentalization of host cells by *P. brassicae* (Butcher et al. 1974), and subsequently, indole glucosinolates stored in the vacuole (Helmlinger et al. 1983) can be converted by

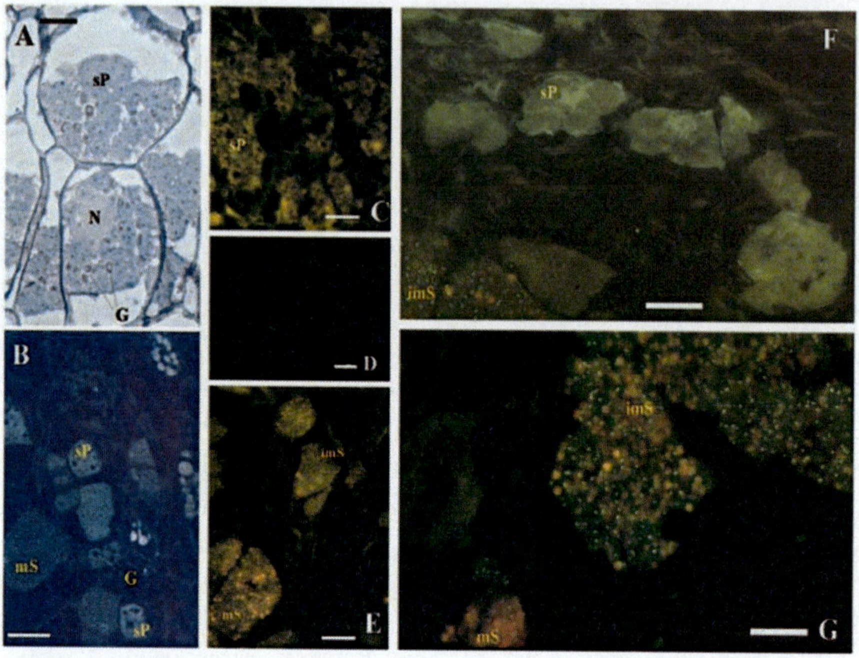

Plate 7.8 Immunolocalization of nitrilase in longitudinal sections of infected roots of *Arabidopsis* using a specific antibody against NIT1. Secondary plasmodia (sP), immature spores (imS), and mature (mS) resting spores of *P. brassicae* as well as nuclei (N) and starch granules (G) of the cortical host plant cells are labeled. Bars indicate 20 μm. A and B, Light microscopic pictures of large secondary plasmodia (23 dpi) stained with toluidine blue and basic fuchsin (**a**), and plasmodia, and resting spores (29 dpi) stained with ethidium bromide, and DAPI using epifluorescence optics specific for DAPI (**b**). Using different staining techniques, different parts of the host cell and the pathogen can be distinguished. In A the plasmodia, and the enlarged host cells are clearly visible using toluidine blue, and basic fuchsin staining, while in (**b**) the multinucleate plasmodia and the nuclei of the resting spores are stained with DAPI and ethidium bromide. Immunosignals of nitrilase in infected cells at 21 dpi (**f**) and 35 dpi (**g**) using FITC-specific epifluorescence optics, and the corresponding controls only using the secondary antibody for staining at 21 dpi (**c**), and 35 dpi (**e**). Immunolocalization of nitrilase in non-infected roots of a nitrilase-overproducing line (35SNIT2) 28 days after germination is also shown (**d**) (Grsic-Rausch et al. 2000)

myrosinase to IAN, which is then converted to IAA by nitrilase. The data presented in Plate 7.8 reveal the host cells are still intact during the development of secondary plasmodia, whereas sporangia containing cells are enlarged compared with uninfected cells, and the compartmentation is disrupted. This correlates with an increase in nitrilase signals. At later stages of sporangia development, the host cell might be completely destroyed, so that no protein synthesis can take place, which is reflected in reduced nitrilase signals. Since nitrilase can be induced by its own substrate (Allard 1995; Grsic et al. 1998), one possible regulation mechanism for increased nitrilase expression could be the increased amounts of IAN synthesized by myrosinase from indole glucosinolates. However, other regulatory mechanisms cannot be ruled out. Overexpression of nitrilase does not result in more severe symptoms because nitrilase activity was already induced above a sufficient level. However, the club size and infection rate were reduced in *nit1-3* mutants, showing that loss of one *NIT* gene results in a less severe phenotype of club formation, although it was previously shown that one *NIT* isoform can substitute for another mutant (Normanly et al. 1997). The mechanism that leads to the induction of nitrilase during clubroot formation (signals from the pathogen or from the host plant) needs attention to researchers.

7.2.10 Transcriptome Analysis of *Arabidopsis* Clubroots Indicates a Key Role for Cytokinins in Disease Development During the Pathogenesis of *P. brassicae*

The morphological description of the disease in *A. thaliana* revealed two overlapping processes in the root cortex during the development of the club. At the beginning of club growth, cell division is the predominant process, whereas cell enlargement dominated the later developmental stages (Kobelt et al. 2000). This correlates with cytokinin production by the vegetative plasmodia of the pathogen during earlier disease stages (Dekhuijzen 1981) and increased levels of auxin during later disease stages, when plasmodia were forming sporangia and resting spores (Grsic-Rausch et al. 2000). Club growth might be an overall result of the ratios of these two classes of hormones during the disease, whereas cytokinin dominated the earlier and auxin the later phase. To understand disease development on the molecular level, several approaches were followed to isolate genes differentially expressed during gall formation. RNA fingerprinting with arbitrary primed polymerase chain reaction (PCR) (Ito et al. 1999), a random priming approach to isolate cDNAs (Schuller and Ludwig-Müller 2002), and subtractive cDNA libraries (Arbeiter et al. 2002; Graf et al. 2004) were used to isolate gene fragments that could be related to the development of a club. However, all these studies revealed a mixture of plant and protists genes, and so far, only one unknown gene from the protists has shown correlation to pathogenesis (Ito et al. 1999). In order to overcome these limitations in the molecular analysis of clubroot, host gene expression was monitored during the development of the disease (10 and 23 days after inoculating [dai]), using the *Arabidopsis* ATH1 genome array. The transcriptome analysis has further elucidated the role of cytokinin regulation during clubroot development, and has refined the decisive steps in establishing a strong metabolic sink in the root tissue by the pathogen. A resistance phenotype to clubroot of cytokinin deficient *AtCKX*-overexpressing plants has been identified. To investigate host gene expression during the development of the disease, Siemens et al. (2006) have used the *Arabidopsis* ATH1 genome array. Two time points were chosen, an early time point at which the pathogen has colonized the root but has induced only very limited change of host cell and root morphology and a later time point at which more than 60% of the host root cells were colonized and root morphology was drastically altered. At both time points, more than 1000 genes were differentially expressed in

Table 7.6 Down- (<0.5×) and up-regulated genes (>2×) related to auxin synthesis, metabolism, transport, and gene regulation[a] (Siemens et al. 2006)

AGI No.	Gene annotation	Expression level[b]				Mean ratio of induction	
		Time point 1		Time point 2			
		Control (24 dag)	Inoc. (10 dpi)	Control (37 dag)	Inoc. (23 dpi)	Time point 1[c]	Time point 2[d]
At2g23170	Auxin-responsive *GH3* family protein	1.8	162.5	1.7	229.2	92.0 ± 61.5	137.5 ± 40.5
At1g72540	Dual specificity cytosolic kinase	0.8	38.1	6.1	124.9	47.6 ± 8.2	20.5 ± 13.1
At5g13360	Auxin-responsive *GH3* family protein	19.8	67.1	16.4	164.2	3.4 ± 0.9	10.0 ± 3.8
At4g34760	Auxin-induced protein X15	19.6	65.6	43.1	64.1	3.4 ± 1.2	1.5 ± 19.8
At1g77690	Amino acid permease, similar to *AUX1*	92.7	290.9	225.7	27.6	3.1 ± 0.5	0.1 ± 0.3
At1g73590	Auxin efflux carrier protein. putative (*PIN1*)	56.2	176.5	123.8	76.0	3.1 ± 0.5	0.6 ± 0.3
At2g32410	Auxin-resistance protein (*AXR1*)	23.7	61.2	35.5	28.1	2.6 ± 0.4	0.8 ± 0.1
At4g34750	Auxin-responsive protein	23.4	58.2	38.2	21.3	2.5 ± 0.6	0.6 ± 20.2
At1g28130	Auxin-responsive *GH3* family	37.2	91.9	43.7	70.9	2.5 ± 0.4	1.6 ± 19.7
At2g33860	Auxin-responsive factor (*ARF3*)	70.7	161.7	99.7	245.8	2.3 ± 0.3	2.5 ± 1.0
At5g37020	Auxin-responsive factor (*ARF8*)	84.7	178.4	134.8	206.3	2.1 ± 0.3	1.5 ± 0.4
At1g70940	Auxin transport protein. putative (*PIN3*)	83.2	173.5	138.7	200.9	2.1 ± 0.4	1.4 ± 0.5

At4g02980	Auxin binding protein 1 (*ABP1*)	73.3	150.8	90.2	166.7	2.1 ± 0.2	1.8 ± 0.4
At2g14960	Auxin-responsive *GH3* family	21.4	43.4	14.2	17.1	2.0 ± 1.5	1.2 ± 0.6
At1g43630	Activated protein kinase C receptor	514.7	1021.0	583.5	262.7	2.0 ± 0.2	0.5 ± 0.2
At3g12830	Auxin-responsive family protein	84.6	154.4	35.9	137.5	1.8 ± 0.4	3.8 ± 2.9
At3g44310; At3g44300	AT3g44310. nitrilase 1 (*NIT1*) AT3g44300. nitrilase 2 (*NIT2*)[e]	410.7	632.1	348.4	2565.7	1.5 ± 0.3	7.4 ± 2.8
At5g01990	Auxin efflux carrier family protein	46.3	65.1	37.1	178.5	1.4 ± 0.1	4.8 ± 1.6
At4g37390; At1g59500	AT4g37390. auxin-responsive *GH3* family AT1g59500. auxin-responsive *GH3* family protein	27.8	36.1	2.3	70.9	1.3 ± 0.5	31.3 ± 9.1
At1g77850	Transcriptional factor B3 family protein	34.9	43.0	51.7	22.6	1.2 ± 0.5	0.4 ± 0.3
At4g38840	Auxin-responsive protein	16.6	19.5	18.1	38.4	1.2 ± 0.3	2.1 ± 19.5
At4g14550	Auxin-responsive protein (*IAA14*)	21.4	24.3	15.3	76.1	1.1 ± 0.2	5.0 ± 1.9
At4g03400	Auxin-responsive *GH3* family	17.1	18.9	13.3	97.4	1.1 ± 0.3	7.3 ± 2.9
At5g35580	Protein kinase	243.9	250.2	399.6	41.5	1.0 ± 0.3	0.1 ± 0.4
At1g19850	Transcription factor MONOPTEROS (*IAA24*)	183.8	181.6	225.0	83.9	1.0 ± 0.0	0.4 ± 0.3
At4g00880	Auxin-responsive family protein	154.1	141.7	355.0	52.3	0.9 ± 0.2	0.1 ± 0.3

(continued)

Table 7.6 (continued)

AGI No.	Gene annotation	Expression level[b]				Mean ratio of induction	
		Time point 1		Time point 2			
		Control (24 dag)	Inoc. (10 dpi)	Control (37 dag)	Inoc. (23 dpi)	Time point 1[c]	Time point 2[d]
At2g21050	Amino acid permease, similar to *AUX1*	93.1	87.6	246.8	54.7	0.9 ± 0.1	0.2 ± 0.3
At3g23050	Auxin-responsive protein/indoleacetic acid-induced protein 7 (*IAA7*)	45.9	37.7	26.7	135.6	0.8 ± 0.2 ± 0.1	5.1 ± 1.7
At2g22670	Auxin-responsive protein/indoleacetic acid-induced protein 8 (*IAA8*)	1319.9	1038.7	1503.3	641.4	0.8 ± 0.1	0.4
At4g17280	Auxin-responsive family protein	46.1	36.1	51.8	16.6	0.8	0.3
At4g27260	Auxin-responsive *GH3* family protein	66.1	48.8	34.2	190.0	0.7 ± 0.2	5.6 ± 2.0
At1g51780	Indole-3-acetic acid (IAA)-amino acid hydrolase 5/auxin conjugate hydrolase (*ILL5*)	236.2	173.2	288.0	104.1	0.7 ± 0.0	0.4 ± 0.3
At4g36110	Auxin-responsive protein	29.1	21.2	15.4	38.3	0.7 ± 0.2	2.5 ± 0.8
At2g33310	Auxin-responsive protein/indoleacetic acid-induced protein 13 (*IAA13*)	409.8	277.3	669.0	323.6	0.7 ± 0.1	0.5 ± 0.2
At5g20730	Auxin-responsive factor (*ARF7*)	170.3	108.4	178.7	80.4	0.6 ± 0.1	0.4 ± 0.2
At4g27450	Expressed protein similar to auxin down-regulated protein *ARG10*	690.5	426.3	385.0	1969.9	0.6 ± 0.2	5.1 ± 2.5
At1g04250	Auxin-responsive protein/indoleacetic acid-induced protein 17 (*IAA17*)	441.4	249.3	466.9	221.7	0.6 ± 0.1	0.5 ±0.2

At3g16500	Auxin-responsive Aux/IAA family protein	254.9	139.5	284.4	123.6	0.5 ± 0.1	0.4 ± 0.2
At2g04850	Auxin-responsive protein-related	57.5	29.2	133.1	6.9	0.5 ±0.2	0.1 ± 0.0
At2g46370	Auxin-responsive *GH3* family protein	357.1	179.7	313.7	142.0	0.5 ±0.2	0.5 ± 0.2
At1g56220	Dormancy/auxin associated family protein	551.7	264.1	205.9	705.6	0.5 ± 0.1	3.4 ±1.7
At2g46690	Auxin-responsive family protein	533.7	251.1	941.6	76.4	0.5 ± 0.1	0.1 ± 0.4
At5g53590	Auxin-responsive family protein	1787.8	813.1	1497.7	405.9	0.5 ± 0.0	0.3 ±0.3
At4g12980	Auxin-responsive protein	133.1	60.0	179.9	36.2	0.5 ± 0.1	0.2 ± 0.3
At3g15460;	AT3G15460. brix domain-containing protein						
At3g15450	AT3g15450. expressed protein	2935.7	1246.0	1650.0	472.0	0.4 ±0.1	0.3 ± 0.3
At1g75590	Auxin-responsive family protein	29.9	11.8	23.6	19.1	0.4 ± 0.1	0.8 ± 0.5
At1g04100	Auxin-responsive protein/indoleacetic acid-induced protein 10 (*IAA10*)	160.7	56.0	115.2	69.3	0.3 ± 0.0	0.6 ± 0.5
At5g25890	Auxin-responsive protein/indoleacetic acid-induced protein 28 (*IAA28*)	319.6	101.7	288.6	80.7	0.3 ±0.1	0.3 ± 0.3
At5g20820	Auxin-responsive protein-related	70.4	21.7	82.1	48.5	0.3 ± 0.1	0.6 ± 0.6
At1g04240	Auxin-responsive protein/indoleacetic acid-induced protein 3 (*IAA3*)	131.2	37.8	130.1	39.6	0.3 ± 0.1	0.3 ± 0.3
At3g23030	Auxin-responsive protein/indoleacetic acid-induced protein 2 (*IAA2*)	529.6	150.5	587.9	218.7	0.3 ± 0.0	0.4 ± 0.3

(continued)

Table 7.6 (continued)

AGI No.	Gene annotation	Expression level[b]				Mean ratio of induction	
		Time point 1		Time point 2			
		Control (24 dag)	Inoc. (10 dpi)	Control (37 dag)	Inoc. (23 dpi)	Time point 1[c]	Time point 2[d]
At4g29080	Auxin-responsive Aux/IAA family protein	959.6	262.1	1469.5	99.1	0.3 ± 0.0	0.1 ± 0.4
At1g51950	Auxin-responsive protein/indoleacetic acid-induced protein 18 (*IAA18*)	441.3	117.9	675.5	137.8	0.3 ±0.1	0.2 ± 0.3
At1g15580	Auxin-responsive protein/indoleacetic acid-induced protein 5 (*IAA5*)	32.2	8.1	69.8	4.1	0.3 ± 0.2	0.1 ± 0.1
At3g47620	TCP family transcription factor	475.2	110.5	364.4	374.1	0.2 ± 0.0	1.0 ± 0.2
At5g19140	Auxin/aluminum-responsive protein	898.3	196.7	875.7	925.6	0.2 ± 0.1	1.1 ± 0.1
At4g22620	Auxin-responsive family protein	172.3	35.7	180.7	41.0	0.2 ± 0.1	0.2 ± 0.3
At3g15540	Auxin-responsive protein/indoleacetic acid-induced protein 19 (*IAA19*)	276.2	56.2	330.8	56.3	0.2 ± 0.1	0.2 ± 0.3
At3g04730	Auxin-responsive protein/indoleacetic acid-induced protein 16 (*IAA16*)	765.4	98.8	501.8	188.8	0.1 ± 0.0	0.4 ± 0.3
At3g60690	Auxin-responsive family protein	673.5	86.7	190.5	49.5	0.1 ± 0.0	0.3 ± 0.5
At2g45210	Auxin-responsive protein	314.1	36.0	58.7	62.4	0.1 ± 0.0	1.1 ± 0.6

[a]Differentially influenced by *Plasmodiophora brassicae* in infected and not inoculated plants of ecotype Col-0 at two time points after inoculating. All values are normalized to the expression level of not inoculated Col-0 plants at the corresponding time point. Mean values of expression levels of three independent experiments and the mean values of induction and standard deviations are given
[b]*dag* days after germination, *dpi* days post-infection
[c]10 dpi/24 dag ± SD (standard deviation)
[d]23 dpi/37 dag ± SD
[e]*NIT1* and *NIT2* cannot be distinguished on the microarray by the oligonucleotides used

Table 7.7 Down- (<0.5×) and up-regulated genes (>2×) related to cytokinin metabolism and signaling[a] (Siemens et al. 2006)

		Expression level[b]				Mean ratio of induction	
		Time point 1		Time point 2			
		Control	Inoculated	Control	Inoculated		
AGI No.	Gene annotation	(24 dag)	(10 dpi)	(37 dag)	(23 dpi)	Time point 1[c]	Time point 2[d]
At4g31920	Two-component response regulator protein (*ARR10*)	28.9	136.5	54.3	87.4	4.7 ± 0.5	1.6 ± 0.5
At2g01830	Histidine kinase (*CRE1/AHK4*)	71.7	202.5	183.4	143.5	2.8 ± 0.3	0.8 ± 0.3
At3g48100	Response regulator 5 (*ARR5*)	75.3	175.3	295.4	94.2	2.3 ± 0.6	0.3 ±0.3
At1g10470	Response regulator 4 (*ARR4*)	283.0	340.1	382.5	120.5	1.2 ± 0.1	0.3 ± 0.3
At1g74890	Response regulator 15 (*ARR15*)	23.6	25.5	34.3	15.1	1.1 ±0.2	0.4 ± 0.2
At3g55950	Protein kinase family	234.5	251.6	292.0	95.8	1.1 ± 0.2	0.3 ± 0.3
At4g29740	Cytokinin oxidase (*CKX4*)	40.6	42.4	36.5	73.9	1.0 ± 0.1	2.0 ±0.5
At1g28230	Purine permease (*PUP1*)	39.6	32.7	29.9	103.4	0.8 ± 0.2	3.5 ± 1.8
At4g16420	Transcriptional adaptor (*ADA2b*)	93.4	73.0	84.0	169.3	0.8 ± 0.2	2.0 ±0.4
At3g19160	Cytokinin synthase (*IPT8*)	22.1	12.1	10.4	35.6	0.5 ± 0.1	3.4 ± 2.6
At3g57040	Response regulator 9 (*ARR9*)	79.2	42.2	70.6	25.8	0.5 ± 0.1	0.4 ±0.3
At5g62920	Response regulator 6 (*ARR6*)	42.6	19.4	72.6	16.9	0.5 ±0.1	0.2 ±0.2
At4g16110	Response regulator (similar to *ARR2*)	85.8	38.2	43.8	37.6	0.4 ± 0.1	0.9 ± 0.2
At1g67710	Response regulator (*ARR11*)	53.3	20.3	53.3	3.6	0.4 ± 0.1	0.1 ± 0.1
At2g41510	Cytokinin oxidase (*CKX 1*)	36.7	13.8	42.8	10.7	0.4 ±0.1	0.2 ±0.2
At1g19050	Response regulator 7 (*ARR7*)	611.2	160.8	524.2	141.7	0.3 ±0.1	0.3 ± 0.3
At3g63110	Cytokinin synthase (*IPT3*)	194.4	45.5	260.3	24.0	0.2 ± 0.1	0.1 ± 0.4
At5g05440	Cytokinin-specific binding protein	1171.6	259.4	979.9	280.5	0.2 ± 0.1	0.3 ± 0.3
At3g63440	Cytokinin oxidase (*CKX6*)	364.6	36.5	336.6	9.4	0.1 ± 0.1	0.1 ± 0.1

[a]Influenced by *Plasmodiophora brassicae* in infected and not inoculated plants of ecotype Col-0 at two time points after inoculating. All values are normalized to the expression level of not inoculated Col-0 plants at the corresponding time point. Mean values of expression levels of three independent experiments and the mean values of induction and standard deviations are given

[b]*dag* days after germination, *dpi* days post-infection

[c](10 dpi/24 dag) ± SD (standard deviation)

[d](23dpi/37 dag) ± SD

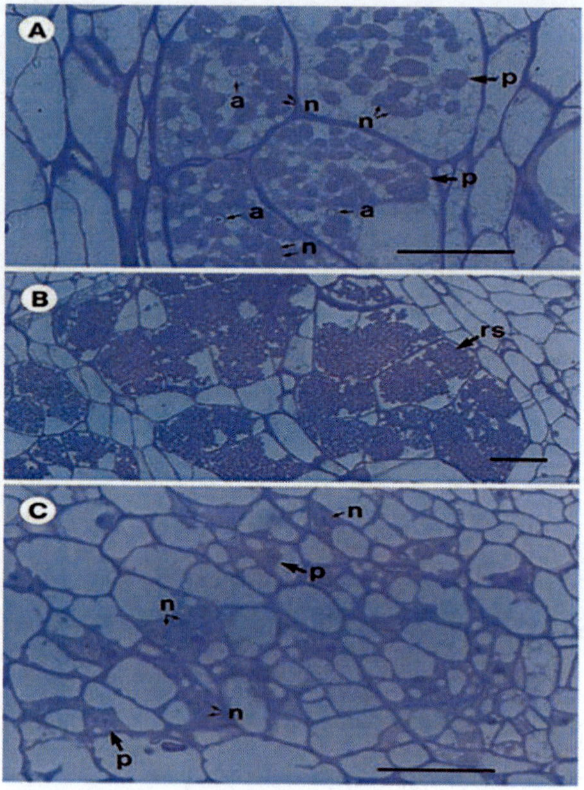

Plate 7.9 Light micrographs of plasmodia and resting spores within infected cells of root tissues of clubroot-resistant cultivar Kukai 70 grown in pots of soil amended with either isolate Yamaguchi (YGM) or Hagi (HG) of *Plasmodiophora brassicae*. (**a**) Plasmodia (*p*) of YGM with many nuclei (*n*) and starch-accumulating amyloplasts (*a*) in infected cells at 30 days after inoculating. (**b**) Resting spores (*rs*) of YGM at 40 days after inoculating. (**c**) Small plasmodia (*p*) of HG with a small number of nuclei (*n*) at 40 days after inoculating. Bars = 50 μm (Tanaka et al. 2006a)

infected versus control roots. These included genes associated with growth and cell cycle, sugar phosphate metabolism, and defense. The involvement of plant hormones in club development was further supported; genes involved in auxin homeostasis, such as nitrilases and members of the *GH3* family, were up-regulated, whereas genes involved in cytokinin homeostasis (cytokinin synthases and cytokinin oxidases/dehydrogenases) were already strongly down-regulated at the early time point (Tables 7.6 and 7.7). Cytokinin oxidase/dehydrogenase overexpressing lines were disease resistant, clearly indicating the importance of cytokinin as a key factor in clubroot disease development.

7.2.11 Role of Host Resistance in Modulating Colonization of *P. brassicae*

A commercial clubroot-resistant F_1 cultivar of Chinese cabbage (*Brassica rapa* L. subsp. *pekinensis*), Kukai 70, is resistant to an isolate of populations of *Plasmodiophora brassicae* from Hagi (HG) city and is susceptible to another from Yamaguchi (YMG) city, Japan. The degree and frequency of primary and secondary cycle colonization by the isolates in the root hairs and root tissues of cv. Kukai 70 were compared. Seedlings of cv. Kukai 70 were grown in soils amended with inoculum of either HG or YMG and harvested 10 days after inoculating to observe the primary cycle (number of root-hair infections) and 20, 30, and 40 days after inoculating to observe of the secondary cycle (frequency of infected cells and degree of plasmodial development based on the number of nuclei in infected cells). Although more root hairs were infected in HG than in YMG, fewer cells in root tissues including the cortex and medullary rays were infected in HG than in YMG. In addition, YMG developed plasmodia with many nuclei and formed resting spores, whereas plasmodia remained immature with a small number of nuclei in HG and did not form resting spores even by 40 days after inoculating (Plate 7.9). These results suggest that suppression of plasmodial development during secondary colonization is associated with resistance mechanisms to HG in cv. Kukai 70. Starch did not accumulate (i.e., development of amyloplasts) in HG-infected cells. This may be involved in the suppression of secondary colonization of *P. brassicae* in the cultivar (Tanaka et al. 2006a).

7.2.12 Role of a Serine Protease pro 1 in *P. brassicae* Pathogenesis

Proteases are enzymes that cleave proteins by the catalysis of peptide bond hydrolysis. It can be classified into five main classes on the basis of the reactive nucleophile in the active site of the enzyme: cysteine proteases, aspartate proteases, metalloproteases, threonine proteases, and serine proteases. Enzymes in each class are further classified into evolutionarily distinct clans, and each clan is subdivided into families on the basis of sequence homology and the order of the catalytic triad (Rawlings and Barrett 1993). Serine proteases are one of the most important and widely distributed families of proteases, which utilize a uniquely activated serine residue in the substrate binding pocket to catalytically hydrolyze peptide bonds. These are found in all organisms, and carry out a diverse array of physiological functions, such as digestion, defense, and the production of active peptides (Yousef et al. 2003). The pathogenicity function of serine protease has been well studied in protozoans that are pathogenic to humans. Serine proteases have also been reported to be important for pathogenicity in plant pathogens. A serine protease and a metalloprotease from *Xanthomonas campestris* pv. *campestris* play important roles for pathogenesis in black rot on turnip leaves (Dow et al. 1990).

A serine protease gene (*PRO1*) was cloned from *P. brassicae* and its molecular characteristics were investigated by Feng et al. (2010). Southern analysis and specific polymerase chain reaction (PCR) amplification indicated that *PRO1* is a single-copy gene present in a broad range of *P. brassicae* pathotypes. Northern analysis revealed that the expression of *PRO1* was induced during plant infection, and that the quantity of transcript fluctuated according to the stage of pathogenesis. Amino acid sequence analysis suggested that the encoded protein (Pro1) belongs to the S28 family of proteases, with a predicted signal peptide, and a theoretical molecular mass of 49.4 kDa. The open reading frame (ORF) of *PRO1* was transferred into *Pichia pastoris,* and Pro1 was heterologously produced. Pro1 showed proteolytic activity on skimmed milk and *N*-succinyl-Ala-Ala-Phe-7-amido-4-methylcoumarin, and the activity could be inhibited by serine protease inhibitors with the chelating agent ethylenediaminetetraacetic acid. The optimal temperature of Pro1 was 25 °C, and it exhibited high activity at pH 6.0–6.4. These values coincide with the temperature and pH conditions favorable for *P. brassicae* resting spore germination in the field. When Pro1 was used to treat *Brassica* root exudates, it enhanced the stimulating effect of the root exudates on *P. brassicae* resting spore germination, indicating that Pro1 may play a role during clubroot pathogenesis by stimulating resting spore germination through its proteolytic activity (Feng et al. 2010).

7.2.13 Role of Lipids in *P. brassicae* Pathogenesis

In pathogen dormancy, carbohydrate and lipid composition of spores play a determinant role. Plasmodiophorid spores contained an abundance amount of lipid bodies. The most abundant fatty acid in resting spores of *P. brassicae* is Arachidonic acid (ARA, 20:4). Transcriptional data suggest that fatty acids are synthesized in the plasmodia and degraded in resting spore. Lipid droplet is a cellular organelle that stores neutral lipids [triacylglycerol (TAG) and cholesterol esters] as a source of energy and carbon. They serve as the energy reservoir for the entire organism. Lipid droplets have been shown to be physiological important organelles in lipid metabolism rather than just storage organelles. Numerous proteins have been shown by immune detection or proteomic approaches to be associated with lipid droplets and to change under different physiological conditions. Lipid droplets biogenesis and function have been studied in many organisms from archaea, bacteria, plant to mammalian cells. Lipid with lipid droplets and their associated proteins will facilitate to improve the understanding of their functional roles for *P. brassicae*. In addition to comprising cellular membranes, fatty acids are stored within cells as energy-rich TAGs in lipid droplets. In response to metabolic signals, mobilized fatty acids and other precursors derived from stored neutral lipids are used for a wide variety of functions within lipoproteins. Fatty acids are essential metabolites used for energy production via β-oxidation. Because an excess of free fatty acids in the cytoplasm is toxic to cells, cells esterify fatty acids into neutral lipids, and package them into lipid droplets. Bi et al. (2016) found that lipid droplets are present in the resting spores, and secondary plasmodia of *P. brassicae* by transmission electron microscopy (TEM). Nile red staining showed that different zoosporic stages of *P. brassicae* contained lipid droplets. Two hundred and ninety-five lipid droplet-associated proteins were identified by proteomic study, and seven types of fatty acid compositions were chemically characterized using GC-MS in lipid droplets. Genome integrated transcriptome, proteome of its lipid synthesis, storage, and metabolism are very essential aspect in *P. brassicae* life cycle.

Lipid droplets are observed in the resting spores of *P. brassicae*. Lipid droplet-associated proteins were identified and categorized into nine groups. Transcriptome analysis of these proteins during three different zoosporic stages revealed differences in gene expression pattern. GO enrichment analysis revealed that these proteins associated with lipid droplets were mainly linked to biosynthesis and metabolism. GC-MS analysis revealed that lipid droplets contain seven types of free fatty acids: saturated fatty acids C16:0 and C18:0 and unsaturated fatty acids C18:1Δ9, C18:1Δ11, C18:2, C20:4, and C20:5. *P. brassicae* accumulated a large amount of triacylglycerols (TAGs). Bi et al. (2016) systematically analyzed the putative proteins involved in TAG biosynthesis, and its metabolic pathway. KEGG pathway analysis defined 3390 genes, including 167 genes involved in lipid metabolism. Transcriptome analysis revealed that 162 candidate enzymes involved in lipid metabolism were differential expressed.

7.2.14 Relationship of Endogenous Glucosinolate Content to *P. brassicae* Infection

The infection by *P. brassicae* may occur in non-*Brassica* species, leading to the potential formation of resting spores. Structures resembling *P. brassicae* are found in the root cortex of *Tropaeolum majus* (Nasturtium) *Carica papaya* (Pawpaw), *Reseda alba* (White mignonette or White upright mignonette), and *Beta vulgaris* (Beet) as demonstrated by scanning electron microscopy (SEM). Inoculating of *Brassica rapa* roots with spores extracted from either *T. majus* or *B. vulgaris* roots which had been previously inoculated with *P. brassicae* led to development of clubroot in the roots of *B. rapa*. It is also observed that the development of the symptom might be correlated with glucosinolate content, although other host factors are implicated in the *B. vulgaris* interaction with *P. brassicae*. In the glucosinolate-containing non-*Brassicas*, *T. majus,* and *C. papaya*, the concentrations of benzyl-glucosinolate increased markedly in roots inoculated with *P. brassicae*, compared with the controls. There were also increases in concentrations of benzyl-glucosinolate in leaves of *T. majus* after *P. brassicae* infection. However, in *R. alba* roots, the total glucosinolate content decreased after inoculating with *P. brassicae* compared with the controls. High root concentrations of 2-OH-2-phenylethylglucosinolate (glucobarbarin) compared with low root indole glucosinolates in this species might limit *P. brassicae* infection and development. The importance of cultivation of non-*Brassica* species on fields infested with *P. brassicae* needs careful selection of non-host species (Ludwig-Muller et al. 1999).

7.2.15 Auxin Signaling Pathways for Clubroot Formation

The infection of plants by *P. brassicae* in the secondary phase, which occurs in the cortex and stele of roots and hypocotyl region leads to abnormal development. Especially during this later phase, the host root responds to infection by increased cell division rates followed by hypertrophy of infected cells. These harbor first the plasmodia of *P. brassicae*, probably dividing together with the host cells. Later, the

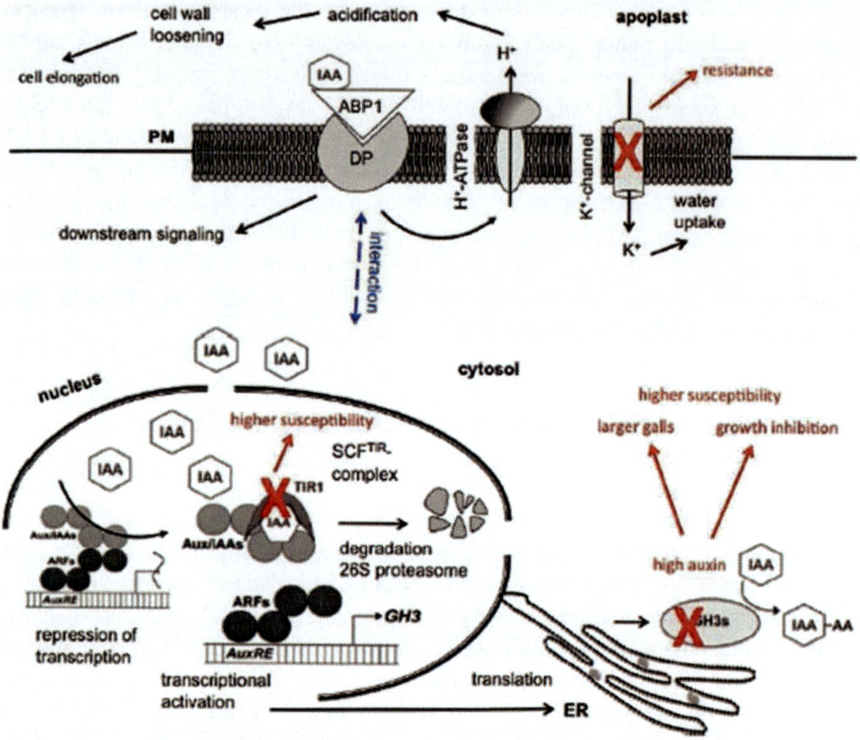

Fig. 7.14 Model for the auxin-dependent degradation of Aux/IAA proteins via the TIR receptor family to induce gene expression by auxin in *Arabidopsis* and the possible involvement of *Plasmodiophora brassicae* in this process (indicated in red). The role for a second auxin receptor, ABP1, might lie in processes at the plasma membrane, leading in turn to cell elongation (Jahn et al. 2013)

plasmodia grow and the host cell increases concomitantly in size. These enlarged host cells can reach at least ten times the size of uninfected cells. While cell division has been attributed to the action of auxins and cytokinins, cell enlargement has been so far linked exclusively to higher auxin concentrations and synthesis. The induction of auxin in *Arabidopsis* correlates with an increase in seedling growth and Xyloglucan-Endo-Transferase/Hydrolase leading to cell elongation. In addition, a microarray revealed that genes involved in cell division and expansion such as cell cycle genes and expansins are up-regulated, especially at the first analyzed time point of the disease (10 days after inoculating). Consequently, altering hormone concentrations has led to reduced clubroot symptoms (Siemens et al. 2006; Devos et al. 2005). Despite a long lasting research on hormonal events, the exact signaling and control mechanisms are still not known. Therefore, Jahn et al. (2013) investigated, which auxin signaling pathway(s) contribute to clubroot formation.

Auxin signaling is regulated by two types of receptors: the nuclear-l °Calized TIR/AFB family and the plasma membrane associated auxin binding protein 1 (ABP1) (Fig. 7.14). The nuclear signaling pathway leads to the activation of the transcriptional response via auxin perception by the F-box protein TIR1 (transport inhibitor response 1). F-box proteins function as substrate recognition modules for the multisubunit complex of ubiquitin ligases (also called Skp1-Cullin1-F-box protein (SCF); here SCFTIR). The F-box protein TIR1 is the receptor, which recruits, by binding to auxin, the protein target that is designated for degradation by ubiquitination. Polyubiquitinated proteins are then transferred into the 26S proteasome and degraded (Dharmasiri et al. 2005; Christian et al. 2006). Auxin-mediated transcriptional response is controlled by negative (Aux/IAA) and mainly positive (ARF) regulators. ARFs (auxin-response-factors) bind to the auxin-responsive elements (AuxRe) in respective promoters of auxin-inducible genes for transcriptional activation, but some ARFs are also inhibitors. Both Aux/IAAs and ARFs are present as large families. The Aux/IAA proteins bind to the ARFs through homologous domains in both proteins and thereby repress auxin-regulated transcription (Guilfoyle and Hagen 2007; Santner and Estelle 2009).

The Aux/IAA proteins are short-lived, because they are degraded via the SCFTIR pathway and their degradation is promoted by auxin. Many mutations in *Aux/IAA* genes are stabilizing the resulting protein, because degradation domains are affected and thus they constitute gain-of-function mutations. Among the targets of ARFs are genes encoding proteins involved in the regulation of auxin homeostasis. The IAA amino acid conjugate synthetases (GH3) have been first recognized as auxin-induced genes. Later, it was discovered that a large gene family and that some proteins are involved in the conjugation of IAA to various amino acids. GH3 proteins are involved in various responses of plants to abiotic and biotic stresses. Especially the protein GH3.5 seems to play various roles in addition to the synthesis of IAA amino acid conjugates. It also conjugates the plant defense signal salicylic acid to amino acids and is involved in the synthesis of the *Arabidopsis* phytoalexin camalexin. One family member, GH3.11 (JAR1), catalyzes the formation of the isoleucine conjugate of jasmonic acid (JA). Contrary to the inactivation of IAA by conjugation, the isoleucine conjugate of JA is active, because it alone binds to the JA receptor COI1. ABP1 has been shown to control events at the plasma membrane by the regulation of membrane fluxes leading in consequence to cell elongation. ABP1 has an ER retention sequence, but it is hypothesized that for action at the plasma membrane some protein "escapes"|| its intracellular location. In addition, there is no transmembrane sequence present in ABP1, which led to the speculation that a d °Cking protein could be responsible for anchoring ABP1 in the membrane (Fig. 7.14). ABP1 is involved in protoplast swelling, which is an indication for its role *in planta* in cell elongation. In addition, many growth processes seem to rely on ABP1 activity, which is an ABP1 knockout mutant is embryo lethal. Downstream of ABP1, the action of ATPases and ion channels was postulated. Both a H^+-ATPase and K^+-channels are needed for the auxin-mediated cell elongation response. K^+-channels necessary for the osmotic changes occurring during cell elongation can be induced by auxin in maize and *Arabidopsis*. In contrast, the TIR pathway can be responsible for the increased gene expression for these channels in response to auxin, linking the two pathways. Initial evidence for the involvement of the TIR pathway in the clubroot disease came from Alix et al. (2007) who reported the partial resistance of the *Arabidopsis axr3-1* mutant. The *AXR3* gene encodes the transcriptional repressor IAA17, a member of the Aux/IAA family. The transcript is stabilized in the mutant, so that the transcriptional activation cannot take place. Jahn et al. (2013) investigated the contribution of both auxin signaling pathways mentioned above to the clubroot symptom development. In addition, targets of both pathways were functionally investigated for their roles during disease progression.

Clubroot of crucifers is characterized by root swellings as symptoms, which are dependent on the alteration of auxin and cytokinin metabolism. Two different classes of auxin receptors, the TIR family and the auxin binding protein 1 (ABP1) in *Arabidopsis thaliana* are transcriptionally up-regulated upon gall formation. Mutations in the *TIR* family resulted in more susceptible reactions to the root pathogen. As target genes for the different pathways were investigated for the transcriptional regulation of selected transcriptional repressors (*Aux/IAA*) and transcription factors (*ARF*). As the TIR pathway controls auxin homeostasis via the up-regulation of some auxin conjugate synthetases (GH3), the expression of selected *GH3* genes was also investigated, showing in most cases up-regulation. A double *gh3* mutant showed also slightly higher susceptibility to *P. brassicae* infection, while all tested single mutants did not show any alteration in the clubroot phenotype. As targets for the ABP1-induced cell elongation the effect of potassium channel blockers on clubroot formation was investigated. Treatment with tetraethylammonium (TEA) resulted in less severe clubroot symptoms. This research provides evidence for the involvement of two auxin signaling pathways in *Arabidopsis* needed for the establishment of the root galls by *P. brassicae*.

7.2.16 Effect of Host Resistance on Pathogens Infection and Development

Infection of root hairs by *P. brassicae* occurs in a broad range of plants, not only in susceptible host cultivars but also in resistant cultivars (Diederichsen et al. 2009) and even non-host plants (Feng et al. 2012b). Root hairs serve as a niche for pathogen increase prior to invasion of the root cortex, but also provide an opportunity for close interaction between host and pathogen, which facilitates the evolution of mechanisms by which *P. brassicae* can break down basal resistance to cortical infection (Feng et al. 2013). In susceptible plants, both root-hair infection and clubroot severity increase with higher inoculum density (Hwang et al. 2011a, b, c). Root-hair infection occurs more frequently with a compatible isolate (susceptible reaction) compared with an incompatible isolate (resistant reaction) and, subsequently, fewer cells within the root become infected when the host is resistant (Tanaka et al. 2006b). The compatible isolate forms secondary plasmodia with many nuclei and eventually resting spores are formed in the host root tissue. In contrast, plasmodia formed by the incompatible isolate remain immature with only a small number of nuclei and do not produce resting spores (Hwang et al. 2011b; Deora et al. 2013). It suggest that resistance in host species is associated with a small amount of suppression of infection during primary infection, but a much greater suppression of secondary infection and subsequent plasmodial development in the root cortex. One important difference between the resistant and susceptible host reactions is that in a resistant host, the secondary thickening of the cell walls in the xylem is not degraded and there are fewer cell wall breakages (Donald et al. 2008). The influence of cultivar resistance on root-hair infection was studied across a range of inoculum densities in canola in one resistant and one susceptible cultivar (Hwang et al. 2011a, b). Subsequently it was observed that root-hair infection in two resistant and three susceptible canola cultivars (Hwang et al. 2012). Infection and clubroot severity were higher and plant height was reduced in the susceptible cultivars relative to the resistant cultivars. Root-hair infection and the amount of *P. brassicae* DNA (assessed using qPCR analysis) increased in both resistant and susceptible cultivars over time, but the increases were greater in the susceptible cultivars. Also, there was a strong linear relationship between root infection and the amount of DNA of *P. brassicae* in the root hairs. The slow increase in the amount of pathogen DNA and root-hair infection observed in the resistant cultivars, and the sharp increases observed between 4 and 10 days after inoculating in the susceptible cultivars, indicate that secondary infection and pathogen development are occurring quickly in the susceptible cultivars. However, sharp declines in detectable levels of pathogen DNA in the seedling roots occurred during the transition from zoospore development in the root hairs to secondary infection (Hwang et al. 2012). These levels quickly rebounded as the pathogen spread into the root cortex after secondary infection (Gludovacz 2013).

The rate of infection and pathogen development is most rapid in susceptible canola lines (Deora et al. 2012, 2013). However, resistant plants showed reduced growth and delayed development when inoculated with resting spores of an avirulent pathotype (Hwang et al. 2012; Deora et al. 2013), which indicates that resistance is an active process. In at least one vegetable *Brassica* line, microscopic analysis has revealed substantial levels of pathogen infection in inoculated roots of symptomless plants (Gludovacz 2013; Gludovacz et al. 2013). The mechanism for this response has not yet been determined, but it is possible that it involves genes for tolerance to *P. brassicae*. Successful management of *P. brassicae* through the deployment of genetically resistant cultivars will require very careful control of susceptible weeds and canola volunteers to maximize the impact of resistant cultivars on reducing resting spore levels in the soil and to prevent the build-up of virulent populations (Hwang et al. 2012).

7.3 Metabolic Events During *Plasmodiophora brassicae* Infection

At an advanced stage of the infection, host plants suffer from permanent wilting, because the galls obstruct the normal water and nutrient transport in the root or destroy the root tissue. The intrinsic characteristics of clubroot disease point to an involvement of plant hormones in clubroot symptom development. The role for cytokinins in gall development has already been observed. The availability of

Table 7.8 Plant hormone levels of control and infected *Arabidopsis thaliana* Col-0 measured 4 days after inoculating (Devos et al. 2006)

	Levels (p mol g^{-1} fresh weight)[a]	
Hormone compound	Control	Infected
Zeatin (Z)	0.35 ± 0.34	0.82 ± 0.34
Dihydro Z	BD	BD
Z-N^9-glucosides	BD	BD
Z riboside	BD	BD
Dihydro (DH) Z riboside	BD	BD
DH Z-N^9-glucosides	BD	BD
Z riboside monophosphate	1.29 ± 0.27	0.80 ± 0.27
DH Z riboside monophosphate	BD	BD
Isopentenyl (iP) adenine	0.68 ± 0.31[b]	1.52 ± 0.31[b]
iP adenine N^9-glucoside	BD	BD
iP adenosine	BD[b]	0.88 ± 0.29[b]
iP adenosine monophosphate	1.04 ± 0.20	1.23± 0.19
Indole-3-acetic acid (IAA)	6.0± 0.5	6.2± 0.5
IAA conjugates	10.7± 0.3	11.4± 0.4

[a]BD = below detection limit (±100 fmol), $n = 3$
[b]$P < 0.05$

Arabidopsis thaliana mutants has been exploited to investigate a set of biochemical and morphological mutant lines that might affect the interaction with *P. brassicae*. Mutants affected in glucosinolate metabolism and altered plant hormone response were screened for a tolerant reaction to clubroot (Ludwig-Muller et al. 1999; Siemens et al. 2002).

Due to its exclusive intracellular life cycle, it remains difficult to distinguish between plant and pathogen genes as well as the proteins that control the confined phases in the life cycle and disease development. However, some information about proteins involved in *P. brassicae* host's response is available. During *P. brassicae* infection, the expression of a glutamine-rich domain (Pyk20) coupled to β-glucuronidase (GUS) is up-regulated (Puzio et al. 2000). Twenty hours after inoculating with *P. brassicae,* the activity of phenylalanine ammonia-lyase (PAL) was transiently increased in clubroot-resistant calli (Takahashi et al. 2002). Hansen et al. (1994) observed a differential protein pattern in *P. brassicae*-infected roots at a late stage of infection (4 weeks after inoculation). The overall protein content in the extract of infected roots is higher than in control roots. However, the level of some specific host proteins is reduced in the infected roots (Hansen et al. 1994). Ito et al. (1996) analyzed the protein patterns related to a susceptible and a resistant response in Chinese cabbage roots infected with two different *P. brassicae* isolates, and showed that a protein that was increased in the susceptible response revealed homology to a pathogenesis-related protein. Ludwig-Muller et al. (1994) investigated the activity of pathogen related proteins in clubroot infected roots of a susceptible and a resistant cultivar, and found that a chitinase isoenzyme is induced in the susceptible interaction (Ludwig-Muller et al. 1994). The proteome is the full complement of proteins expressed by a genome at a specific point in time (Wasinger et al. 1995). An analysis of the protein complement of a biological system, done through protein profiling, provides the opportunity to directly monitor the functional readout of gene expression. Differential proteome expression studies enable to compare the proteome between systems and have proven to be useful in the search for stress-specific proteins.

A combined differential hormone and proteome approach of *Arabidopsis* roots inoculated with *P. brassicae* elicits a susceptible response in *Arabidopsis* (Colhoun 1958). Devos et al. (2006) reported on the early response of *Arabidopsis thaliana* to *Plasmodiophora brassicae* at the hormone and proteome level. Using a *CYCB1*;1:: GUS construct, the re-initiation of infection-related cell division is shown from 4 days after inoculating. Sensitivity to cytokinins and auxins as well as the endogenous hormone levels was evaluated (Table 7.8). Both an enhanced cytokinin gene response and an accumulation of isopentenyl adenine and adenosine precede this re-initiation of cell division, whereas an enhanced auxin gene response is observed from 6 days after inoculating. The *alh*1 mutant, impaired in the cross-talk between ethylene and auxins, is resistant to *P. brassicae*. A differential protein analysis of infected versus non-infected roots and hypocotyls was performed using two-dimensional gel electrophoresis and quantitative image analysis, coupled with matrix-assisted laser desorption ionization time of flight/time of flight mass spectrometry-based protein identification. The visualized proteins analysis 12% showed altered abundance compared with the non-infected plants, including proteins involved in metabolism, cell defense, cell differentiation, and detoxification (Table 7.9). Combining the hormone and proteome data, it postulates that, at the very first stages of *Plasmodiophora* infection, plasmodial-produced cytokinins trigger a l °Cal re-initiation of cell division in the root cortex. Consequently, a de novo meristematic area is established that acts as a sink for host derived indole-3-acetic acid, carbohydrates, nitrogen, and energy to maintain the pathogen which triggers gall development (Fig. 7.15).

7.3.1 The Role of Cytokinins in Clubroot Disease Progression

Gall formation is thought to involve perturbations in phytohormones homeostasis, particularly of auxin and cytokinin and more recently brassinosteroids, contributing to the altered morphogenic processes within the root and hypocotyl (Ludwig-Muller et al. 2009; Schuller et al. 2014). Studies of *P. brassicae*-infected *Arabidopsis* have shown that in the early stages of infection, elevated cytokinins are associated with increased cell division. This is thought to be occurring as a result of a complex interaction between host and pathogen metabolism and signaling. At later stages of gall formation, the expression of host cytokinin biosynthetic genes is repressed, but so too is the expression of host cytokinin oxidases and dehydrogenases (Devos et al. 2006; Siemens et al. 2006). However, the measurement of many CK species, isomers, conjugates, and degradation products is technically difficult and many components are either not resolved (Devos et al. 2006) or high numbers of biological replicates are required (Devos et al. 2005). The pathogen may also have the

Table 7.9 Identification of proteins demonstrating significant relative abundance level changes during *Plasmodiophora brassicae* infection (ratio >1.5)[a] (Devos et al. 2006)

MIPS fun cat[b]	Ratio	Accession	Locus	Description	Spot
1	4.1	Q8LE20	At4g13940	S-adenosyl homocysteinase	17
1	2.0	Q6KCR2	At5g17920	Cobalamin-independent methionine synthase	20
01.03.01	−1.8	Q9LZG0	At5g03300	Adenosine kinase 2	39
2	5.5	gil 30421658	AtCg00490	RuBisCO large subunit	4
2	1.5	Q8H1Y0	At1g24180	Dehydrogenase E1a-like subunit (IAR4)	23
2	1.5	Q9M5K3	At1g48030	Lipoamide dehydrogenase precursor	28
02.01	−2.0	Q9ZU52	At2g01140	Fructose biphosphate aldolase	42
32	7.3	O24433	At3g09260	Myrosinase	1
32	7.8	Q75L84	At3g09260	Myrosinase	2
32	7.8	O24433	At3g09260	Myrosinase	7
32.07	8.0	Q96350	At1g20620	Catalase (cat3)	8
32	19.1	Q9C8Y9	At1g66280	Myrosinase	9
32	6.1	O23656	At1g66270	Myrosinase	12
32.07	1.6	O23626	At2g30860	Glutathione S-transferase	16
32	11.6	Q9SR37	At3g09260	Myrosinase	21
32.07	3.5	Q42957	At2g15620	Ferredoxin-nitrite reductase	24
32.07	2.7	gil 11134159	At1g77510	Thioredoxin	33
32.07	−3.7	Q6JAC2	sAt1g21750	Disulfide isomerase precursor	37
32.07	−5.7	Q949U7	At3g52960	Peroxiredoxin	41
40.02	−3.6	Q9M4R9	At5g44340	β-Tubulin 4	38
40.02	−5.6	Q39412	At5g44340	Tubulin β-1/-4 chain	44
40.02	−5.1	Q39808	At5g62700	Tubulin β-2/β-3 chain	45
70.04	−9.2	Q9M3T2	At4g14960	α-tubulin	40
98	32.5	Q9ZVF3	At2g01520	Major latex protein	3
98	1.9	Q9SUR0	At4g23670	Major latex protein	19
98	16.4	Q9LVT8	At5g47210	Hypothetical protein	27
98	6.6	Q9LVT8	At5g47210	Hypothetical protein	31
99	16.9	Q94C39	At1g11580	Pectin methylesterase	6
99	3.5	Q6PNZ2	At5g06860	Polygalacturonase inhibiting protein1 (PIPG1)	13
99	4.0	Q8HOX6	At5g12140	Putative cysteine protease inhibitor	14
99	7.2	O04317	At3g16400	Myrosinase binding protein	18
99	7.0	Q9FPJ8	At5g54900	Putative RNA-binding protein	25
99	1.6	Q940A4	At1g11840	Putative lactoylgluthathione lyase	26
99	2.5	Q9LIE2	At3g22200	Putative aminobutyrate aminotransferase	29
99	3.0	Q94EG6	At5g02240	Hypothetical protein	30

(continued)

Table 7.9 (continued)

MIPS fun cat[b]	Ratio	Accession	Locus	Description	Spot
99	2.3	gil 21592716	At2g21870	Putative ATP synthase	32
99	4.1	Q9CAT7	At1g73230	Putative transcription factor BTF3	34
99	2.8	Q9SU14	At4g12720	A MutT/nudix family protein	35
99	−3.3	Q96269	At1g79930	Heat shock protein70	36
99	−7.9	Q8LCI7	At2g40010	Putative 60S acidic ribosomal protein P0	46
4	−1.9	O23716	At4g31300	20S proteasome β subunit A	43
12.04	1.6	Q9ASR1	At1g56070	Elongation factor EF-2	22
14	23.3	gil 21592448	At2g16600	Cytosolic cyclophilin R °C3	15
16	5.2	Q8LDR5	At1g30580	Putative GTP-binding protein	11
20	25	P41916	At5g20010	GTP-binding protein RAN1A	5
20.03	7.9	gil 21537313	At5g67500	Putative porin	10

[a]Positive ratio values indicate higher protein abundance in infected plants as compared with non-infected plants, whereas negative values indicate the opposite
[b]MIPS functional categories (Schoof et al. 2002)

capability to synthesize cytokinins from nucleotide precursors (Muller and Hilgenberg 1986). Interfering with the balance of CK synthesis and degradation has the potential to impact disease progression. Plants constitutively overexpressing cytokinin oxidase show reduced gall formation (Siemens et al. 2006). Cytokinins and auxins act in concert to regulate tissue development. In later stages of infection, alterations in the concentrations and transport of auxin and auxin-related metabolites such as indole glucosinolates are thought to be important. It has been proposed that *P. brassicae* plasmodia act as a sink for IAA and that host auxin biosynthetic reactions are stimulated. Elevated auxins are thought to lead to increases in cell wall extensibility associated with cell expansion (Devos et al. 2005; Ludwig-Muller et al. 2009). The completion of the *P. brassicae* genome (Schwelm et al. 2015) allows the transcriptional response of both the host and pathogen to be studied simultaneously. The development of ultrasensitive methods for the quantification of CKs and their metabolites in milligram amounts of tissue has allowed developing a complete CK and auxin profile in control and infected tissues.

The completion of the *P. brassicae* genome has allowed the identification of pathogen CK biosynthetic genes and their expression in infected hosts. The pathogen hijacks host vascular cambium development, and cytokinins are implicated in this process. Transcriptomics and metabolomics were used to investigate changes in cytokinin metabolism during gall formation of clubroot-infected *Arabidopsis thaliana*. RNA-Seq analysis of infected tissue showed that host cytokinin metabolism was strongly down-regulated both at the onset and late stages of gall formation. Expression of host genes associated with cytokinin biosynthesis, signaling,

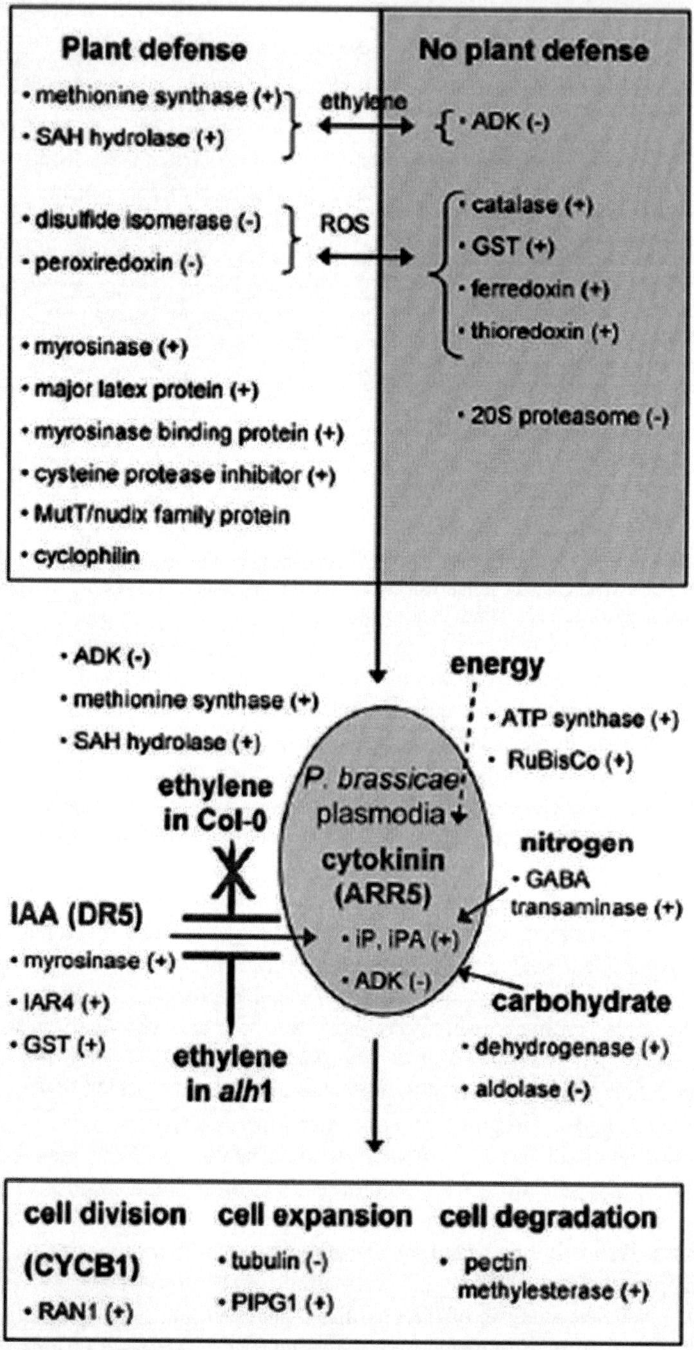

Fig. 7.15 Hypothesis of plant processes upon *Plasmodiophora brassicae* infection of *Arabidopsis thaliana* Col-0 at 4 days after inoculating. Proteins that indicate plant defense (see white box) and proteins that point to a suppression of plant defense (see gray box) control the defense against

degradation, and conjugation was strongly repressed. Analysis of cytokinin precursors, active components, and conjugates by micro-analytical techniques was consistent with these transcriptional responses. Two isopentenyl-transferase genes associated with cytokinin biosynthesis are present in the *P. brassicae* genome and are expressed throughout gall formation. The impact of pathogen-derived cytokinins on the total cytokinin content of infected tissue and host gene expression was minimal in wild-type plants. However, infection of ipt1; 3; 5; 7 mutants that are severely restricted in their ability to synthesize active cytokinins led to an increase in expression of host cytokinin responsive genes. *P. brassicae* can synthesize small amounts of cytokinin, but this has little impact on the host plant as the ipt1; 3; 5; 7 phenotype is not rescued. Intriguingly, plasmodial development was slowed and spore viability reduced in these mutants indicating a potential role for cytokinins in plasmodial development (Malinowski et al. 2016).

7.3.2 The Role of IAA During the Infection of *P. brassicae*

The involvement of auxin and cytokinin as pathogenicity factors has been demonstrated by the correlation of clubroot disease symptoms with an increase in the levels of these plant hormones, which cause increased cell division and cell elongation during the late infection stage (Devos et al. 2005; Devos and Prinsen 2006; Jubault et al. 2013; Siemens et al. 2006). It was reported that free indole-3-acetic acid (IAA) content fluctuated in infected roots at all the time points tested (10–42 days after inoculating [DAI]), although some studies have shown conflicting results. IAA is generally higher in infected roots than in control roots of the same age plants, which suggests the involvement of IAA accumulation during the late stage of gall formation (Ludwig-Muller 2009a). However, the level of IAA in infected cabbage roots was much lower than uninfected roots when the pathogen produced resting spores at 28 DAI (Kavanagh and Williams 1981). These complicated results may be explained by differences in the interactions between different hosts and the pathogen.

Fig. 7.15 (continued) P. brassicae. The plant is unable to defend itself and plasmodial development proceeds (*P. brassicae* plasmodia = gray circle). The plasmodia synthesize cytokinins that induce a sink for chlorophyll, nitrogen, carbohydrates, and indole-3-acetic acid (IAA). This process requires energy (dashed arrow). IAA transport that is necessary for accurate plasmodial development is not blocked by ethylene. When the cross-talk of ethylene and auxin is altered, as in the *alh*1 mutant, clubroot development is impaired. Upon a susceptible interaction, the newly formed plasmodial sink induces cell division, cell expansion, and, eventually, cell degradation. The proteins, reporter genes, and hormone compounds that are related to each process are presented (up-regulated [+] and down-regulated [–] upon clubroot disease). Abbreviations: reactive oxygen species (ROS), S-adenosylhomocysteine (SAH), adenosine kinase (ADK), glutathione *S*-transferase (GST), dehydrogenase E1a-like subunit (IAR), 4-aminobutyrate aminotransferase (GABA), GTP-binding nuclear protein (RAN), polygalacturonase inhibiting protein (PIPG), isopentenyl adenine (iP), and isopentenyl adenosine (iPA) (Devos et al. (2006)

Transcriptomic and proteomic experiments have provided evidence for the involvement of IAA pathway related genes during gall formation in clubroot disease (Cao et al. 2008; Siemens et al. 2006, 2009; Sundelin et al. 2011). The expression levels of nitrilase (Ando et al. 2008; Grsic-Rausch et al. 2000; Siemens et al. 2006), myrosinase (Searle et al. 1982; Sundelin et al. 2011), aldehyde oxidase (Ando et al. 2006), auxin conjugate hydrolase (Ludwig-Muller et al. 1996; Schuller and Ludwig-Muller 2006), and other enzymes involved in auxin homeostasis are induced during the late stage of the infection process. The importance of nitrilase in gall development was shown with a nit1 mutant of the model plant *Arabidopsis*, and antisense plants have shown reduced gall size that correlated with reduced gall IAA content (Grsic-Rausch et al. 2000; Neuhaus et al. 2000). Thus, the involvement of IAA in gall formation during the secondary phase of the pathogen's life cycle was demonstrated.

By treating *B. napus* with exogenous IAA or N-1-naphthylphthalamic acid (NPA), which inhibits the polar transport of IAA in the plant, the effect of IAA on clubroot disease has been revealed. RT-PCR and qPCR were used to examine the expression of IAA-biosynthesis-related genes to determine the gene expression patterns in the roots and leaves of *B. napus* soon after *P. brassicae* inoculating (3, 7, and 10 DAI), defined as the "root hair infection stage." Treatment of infected *B. napus* seedlings with exogenous IAA promoted the development of clubroot disease, including an increase in the number and size of galls, whereas treatment with N-1-naphthylphthalamic acid (NPA) attenuated the disease and altered the location and size of the root nodules, resulting in a reduced disease index. An increased level of free IAA was detected in the roots at 3 and 7 days after inoculating (DAI) compared with the control, suggesting the activation of IAA signaling in the early stage of *P. brassicae* infection. Moreover, IAA treatment caused no significant change in the germination rate of the resting spores but resulted in enhanced root-hair infection. RT-PCR and quantitative real-time PCR (qPCR) were used to analyze the expression of IAA homeostasis-related genes (3, 7, and 10 DAI) to identify gene expression patterns in the roots and leaves of *B. napus* after inoculating. Five IAA-biosynthesis-related genes were induced in the roots after inoculating with *P. brassicae* but not in the leaves. The rapid induction of BnAAO4 expression at 3 DAI may be responsible for the over production of IAA during the early stages of infection. Xu et al. (2016) suggests that IAA acts as a signaling molecule that putatively stimulates root-hair infection in the early response of *B. napus* to *P. brassicae* infection (Fig. 7.16).

7.4 Molecular Mechanisms of Pathogenesis

The Information on the genomic makeup of *P. brassicae* and the molecular basis of host resistance/susceptibility is very limited, largely because of the nature of this organism as an obligate parasite. Graf et al. (2004) isolated a few partial cDNAs that had no significant homology to GenBank sequences. Suppression subtractive hybridization (SSH) has been conducted between RNA from *P. brassicae*-infected

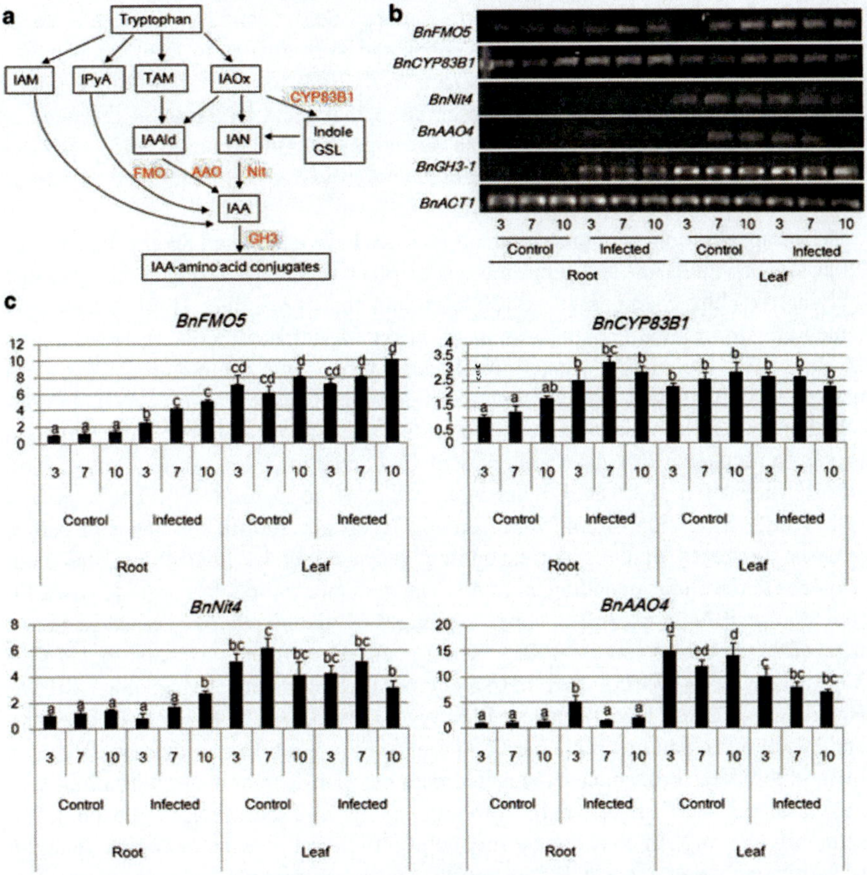

Fig. 7.16 Summary of the proposed IAA biosynthetic pathway in plants (**a**), and the expression patterns of genes involved in this pathway after infection with *Plasmodiophora brassicae* by RT-PCR (**b**), and qPCR (**c**). Enzyme names shown in bold indicate genes induced by *Plasmodiophora brassicae* infection. Different letters above the bars indicate that the differences are significant ($P < 0.05$). *IAM* indole-3-acetamide, *IPyA* indole-3-pyruvic acid, *TAM* tryptamine, *IAOx* indole-3-acetaldoxime, *IAAld* indole-3-acetaldehyde, *IAN* indole-3-acetonitrile, *GSL* glucosinolates, *CYP83B* cytochrome P450 monooxygenase, *FMO* flavin monooxygenase-like, *AAO* aldehyde oxidase, *Nit* nitrilase, *GH3* Gretchen Hagen 3 (Xu et al. 2016)

and uninfected *Arabidopsis* tissue, and a total of 76 new *P. brassicae* gene sequences have been identified (Bulman et al. 2006). Many of the sequences are predicted to contain signal peptides for translocation of the encoded proteins. Again using SSH, but starting with *B. rapa* rather than *Arabidopsis* clubroot galls, Sundelin et al. (2011) isolated about 140 clones. Half of these clones, which represented 24 unisequences, originated from *P. brassicae*. Out of these, 10 unisequences are newly characterized *P. brassicae* genes. By using next generation sequencing (NGS) technologies, Burki et al. (2010) generated 27,333 contigs assembled from 454 reads from *P. brassicae*-infected *B. rapa*. Among these contigs, 3167 are from *P. brassicae*. A few genes have been postulated to be related to the pathogenicity of *P. brassicae* (Ito et al. 1999; Brodmann et al. 2002; Ando et al. 2006; Siemens et al. 2009), mainly based on the observation that the expression of the gene is correlated with certain stage of disease development. One gene, a serine protease (*PRO1*), has been heterologously expressed and proven experimentally to be important for resting spore germination (Feng et al. 2010).

The transcriptomic and proteomic approaches have elucidated the mechanisms of *P. brassicae* infection and coordination in plant reactions. Using a full genome Affymetrix chip, Siemens et al. (2006) found that more than 1000 *Arabidopsis* genes are differentially expressed in *P. brassicae*-infected roots at either 10 or 23 days after inoculating versus non-infected roots. Two proteomic studies, both using two-dimensional gel electrophoresis and quantitative image analysis in conjunction with mass spectrometry-based protein identification, illustrated changes in protein abundance at 4 days after inoculating (dai) in *Arabidopsis* (Devos et al. 2006), and at 0.5, 1, 2, and 3 dai in *B. napus* (Cao et al. 2008). The observed differential expression on mRNA or protein levels at the different stages of pathogenesis suggested that the corresponding genes might be involved in the plant response to the clubroot pathogen. Functional genomic approaches provide powerful tools for identifying expressed genes. Among these techniques, expressed sequence tags (ESTs, Adams et al. 1991), serial analysis of gene expression (SAGE, Velculescu et al. 1995), and massively parallel signature sequencing (MPSS, Brenner et al. 2000) have been successfully employed. Due to its relative simplicity and inexpensive nature, single pass EST sequencing has been the most widely used method to characterize genes associated with cellular development, with biotic and abiotic stress in plant research. Although many EST clones lack the complete sequences of mRNAs, they can be very helpful to investigate the biological function of particular genes, especially in poorly annotated genomes. Additionally, it is possible to obtain full-length clones starting from ESTs. EST resources are also important for comparative transcriptomics because gene expression profiles can be obtained by analyzing ESTs generated under varying conditions.

The whole transcriptome sequencing using next generation sequencing (NGS) technologies or RNA sequencing (RNA-seq) has been started to be utilized to reveal the complex landscape of the transcriptome from various organisms at an unprecedented level of sensitivity and accuracy (Ozsolak and Milos 2011). Compared with traditional EST sequencing and other high-throughput technologies such as microarrays, NGS and RNA-seq offer a near-complete snapshot of a transcriptome with a higher dynamic range of expression level. However, common NGS platforms including Illumina, SOLiD, and 454 suffer from the disadvantage that the sequence reads are often very short (35–500 bp). As a consequence, high technologies on transcriptome assembly are required after NGS.

Despite the advantages of NGS and RNA-seq, EST cloning incorporated with SSH (Diatchenko et al. 1996) has been used to identify genes involved in host–pathogen interactions. This method reduces, by subtractive hybridization, the cloning of abundantly expressed housekeeping genes or genes commonly expressed in both control and treated plants, thereby normalizing expressed cDNA profiles during library construction. Moreover, it significantly enhances, by suppression PCR, the chances of cloning differentially expressed genes. This is particularly important because many pathogenesis-related genes are expressed at low levels, and can be limited to a particular tissue or cell type. This technique has been used to isolate plant and pathogen genes that are expressed during infection (Huang et al. 2007). Both of the two cDNA libraries of *P. brassicae*-infected plants, one using *Arabidopsis* (Bulman et al. 2006) and the other *B. rapa* (Sundelin et al. 2011), were constructed by using the SSH technique.

The construction of a SSH cDNA library followed by EST analysis was used to achieve a better insight into the genes expressed during the compatible interaction between a susceptible *Brassica* cultivar and *P. brassicae*. The infection of plants by pathogenic and the subsequent establishment of disease involves substantial changes in the biochemistry and physiology of both partners. Analysis of genes that are expressed during these interactions represents a powerful strategy to obtain insights into the molecular events underlying these changes. Feng et al. (2012a) have used suppression subtractive hybridization (SSH) and expressed sequence tag (EST) analysis to investigate gene expression during the early stages of colonization of *Brassica* roots by *P. brassicae*. A cDNA library was constructed by SSH which consisted of 797 clones that represented 439 unigenes. Thirty-two of these genes were demonstrated to be of *P. brassicae* origin. The remaining 407 genes, which were of a *Brassica* origin, were subjected to gene ontology and *in silico* analyses. Real-time PCR analysis of ten *P. brassicae* and seven *Brassica* genes indicated that seven of the former and five of the latter were up-regulated at 7 days after infection, suggesting the importance of these genes in pathogenesis or host resistance.

Molecular mechanisms of *P. brassicae* infection during primary and secondary zoospores cannot be differentiated morphologically. When inoculated onto healthy *Brassica* roots, secondary zoospores cause primary infections similar to those caused by primary zoospores (Feng et al. 2012b). In contrast, when inoculated onto *Brassica* roots that are already under primary infection, primary zoospores can cause secondary infections without going through a primary infection phase (Feng et al. 2013). Thus, from a developmental biology standpoint, neither morphogenesis nor functional differentiation can clearly distinguish between these two kinds of zoospores. The primary and secondary zoospores may be essentially same or they undergo a divergence. No matter the state of divergence between the spore types, the "arms race" with the host would induce changes in the secondary zoospores. These changes are most likely at the gene expression level, and can be both genetic and epigenetic in nature. Knowledge of the molecular differences between primary and secondary zoospores may be useful for rational clubroot resistance breeding efforts as well as for the development of pesticides effective against *P. brassicae*. If these two kinds of zoospores are the same with respect to gene expression, a gene or quantitative trait loci (QTL) for resistance against either primary or secondary infection will be also effective against the other. Thus, the selection for resistance and subsequent genetic studies can reliably be conducted at the primary infection stage, which would save time and resources. This scenario is also applicable to fungicide development. Despite this importance, data on gene expression in these two kinds of zoospores is limited with information on the genetic makeup of *P. brassicae*, with only a small number of expressed sequence tags (ESTs) and protein sequences available in GenBank. Similarly, very few *P. brassicae* genes have been functionally characterized (Feng et al. 2010). Most of the database-available genes are derived from cDNA libraries that were constructed using the suppression subtractive hybridization techniques (Bulman et al. 2006; Feng et al. 2012b; Sundelin et al. 2011), which are enriched in genes up-regulated during pathogenesis. Thus, these genes are all good candidates for gene expression studies during the different stages of pathogenesis. All database-available of *P. brassicae* genes were analyzed in silico. To compare the gene regulation profiles of the primary and secondary zoospores, these genes were subjected to both dot-blot and real-time PCR analyses to understand clubroot pathogenesis and provide gene candidates for gene functional studies, which will in turn facilitate breeding for resistance and the identification of potential targets for the development of novel pesticides.

To get information on the molecular basis of pathogenicity of the clubroot pathogen *Plasmodiophora brassicae*-specific primers were designed and used to amplify genomic fragments of *P. brassicae* genes that represent all database-available proteins and ESTs. The PCR products were blotted on membranes and hybridized with digoxigenin-labeled double-stranded cDNA, derived from either primary or secondary zoospores of *P. brassicae*. The same primers were also used in real-time PCR against the single-stranded cDNA synthesized from the two types of zoospores. Both dot-blot and real-time PCR identified up- and down-regulated genes provide the correlation between these two techniques. Real-time PCR indicated that 58 genes were up-regulated in the secondary zoospores relative to the primary zoospores, whereas 55 were down-regulated. These data suggest that different mechanisms are utilized by the pathogen in causing primary and secondary infections. The expression patterns of genes with known or putative functions suggest the relative importance of these genes during pathogenesis. In contrast, highly expressed or regulated genes with unknown function can be further studied in identifying pathogenicity factors (Feng et al. 2013).

7.4.1 Induction of Gene During the Primary Phase of *P. brassicae* Infection

The infection of *P. brassicae* occurs in two distinct phases: the primary phase, which occurs predominantly in root hairs, and the secondary phase, which occurs predominantly in the root cortex (Ingram and Tommerup 1972). It is the secondary phase that produces the characteristic disease symptoms of root tissue hypertrophy and hyperplasia, or "clubbing" (Dekhuijzen 1979; Takahashi et al. 2006; Donald et al. 2008). In a compatible (susceptible) interaction with *Arabidopsis thaliana* (L.) Heynh., the pathogen can proceed through its entire life cycle from 3 to 28 days (Siemens et al. 2002). *P. brassicae* attachment and penetration of the root hair can occur from 4 days after inoculating onwards (Agarwal et al. 2009). This occurs after a resting spore releases a single motile zoospore that attaches to the wall of the root hair and encysts by means of a tubular structure (the rohr and the stachel), injecting its protoplast into the host cytoplasm (Aist and Williams 1971). Between 5 and 12 days after inoculating, plasmodia and then zoosporangia (producing secondary zoospores) develop within the root hairs with little visual sign of damage to the host cells. Between 10 and 13 days after inoculating, *P. brassicae* undergoes a transition from the primary phase to the secondary phase of its life cycle as observed by Devos et al. (2005) and Siemens et al. (2006) depending on the isolate used (single spore or field population). Within 28 days after inoculating, the secondary plasmodia multiply into resting spores within the cortical cells proliferating into galls in *A. thaliana* (Agarwal et al. 2009). Siemens et al. (2006) have reported the use of microarray analysis to examine gene expression following inoculating of a susceptible *A. thaliana* ecotype Col-0 with *P. brassicae* at 10 and 23 days after inoculating (DAI) using a single-spore isolate. These two time points were within the so-called secondary infection stage, which takes place in the root cortex, the 10-day time point being described as the early secondary plasmodial stage and the 23-day time point as the late secondary stage, by which time the disease has progressed substantially in the roots to develop galls. Their study found wide-scale changes in the expression of genes in several important functional groups including those involved in growth, sugar phosphate metabolism, defense, and plant hormone synthesis. The two time points analyzed, however, were relatively late in the interaction for studying the potentially important events that take place during initial pathogen invasion. Proteomic-based approaches have also been reported (Devos et al. 2006; Cao et al. 2008). Devos et al. (2006) examined the susceptible response in *A. thaliana* roots at 4 days after inoculating (DAI) with a field population of *P. brassicae*. Most of the differentially regulated proteins (12%) were involved in plant defense, cell differentiation, hormone metabolism, and detoxification. Cao et al. (2008) examined the susceptible response in canola (*Brassica napus* L.) roots 12, 24, 48, and 72 h after inoculating using the field population of *P. brassicae* collected from central Alberta, Canada. Only 20 protein spots in total were differentially regulated including proteins involved in lignin biosynthesis, cytokinin metabolism, glycolysis, intracellular calcium homeostasis, and the detoxification of reactive oxygen species (ROS). Further, a review of the interactions of *P. brassicae* with its hosts (Ludwig-Muller and Schuller 2008) emphasized that although the components of early signaling in the interaction between *A. thaliana* and leaf pathogens are well understood, there are relatively few studies with root pathogens like *P. brassicae*.

Analyses of microarray conducted at 4, 7, and 10 days after inoculating (DAI) revealed significant induction or suppression of a relatively low number of genes in a range of functional categories. At 4 DAI, there was induced expression of several genes known to be critical for pathogen recognition and signal transduction in other resistant host–pathogen interactions. As the pathogen further colonized root tissue and progressed through the primary plasmodium stage for production of zoosporangia at 7 and 10 DAI, respectively, fewer genes showed changes in expression. The microarray results were validated by examining a subset of induced genes at 4 DAI by quantitative real-time reverse transcriptase PCR (RT-qPCR)

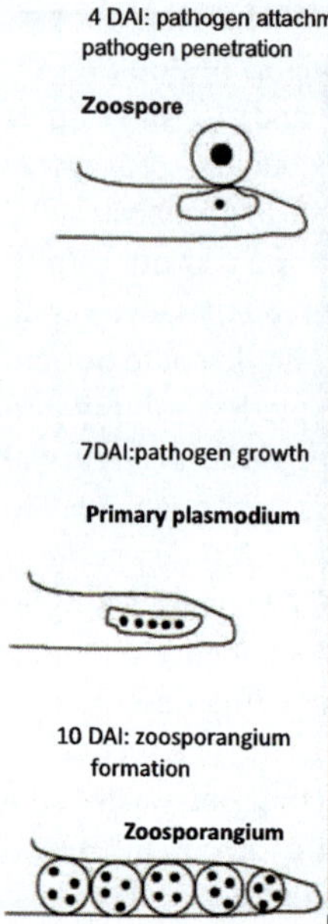

UP	DOWN
Signaling pathways/ transcription and transduction Leucine –rich repeat protein CDPK protein TIR-NBS –LRR protein MAPKK protein Serine/threonine protein kinase WARKY transduction factor (WRKY18) **Cell wall modification** Expansion protein Xyloglucan:xyloglucosyl transferase Pectinesterase protein **Phenylproponoid pathway lignin biosynthesis** 4-coumarate-CoA ligase (*4CL1*) **Jasmonic acid biosynthesis** Lipoxygenase (*LOX4*) **Lipid metabolism** Phytoalexin-defcient 4 protein (*PAD4*)	**Oxidative burst/stress** Peroxidase 72 (*PER72*) Glutathione S-transferase Heat shock protein 101 (*HSP101*) NADP oxidoreductase **Cell wall modification** Hydroxyproline- rich glycoprotein protein Arabinogalactan protein **Phenylproponoid pathway ligninbiosynthesis** Cinnamoyl-CoA reductase (*CCR1*) 4-coumarate-CoA ligase (*4CL1*) Cinnamyl-alcohol dehydrogenase (*CAD5*) **Cytokinin biosynthesis** UDP-glucoronosyl/UDP- glucosyl transferase protein **Salicylic acid biosynthesis** Phenylalanine ammonia-lyase (*PAL4*) **Ethylene biosynthesis** 1-aminocyclopropane-1- carboxylate synthase 2 (*ACS2*) Ethylene receptor (*ETR2*) **Antimicrobial proteins** Chitinase Class IV chitinase
Signaling pathways/transcription and transduction protein kinase Oxidative burst/stress NADPH oxidase **Cell wall modification** Pectinesterase Endo-a-4 beta glucanase	
Signaling pathways/ transcription and transduction Zinc finger protein Leucine –rich repeat protein Serine/threonine protein kinase Protein phosphatase 2C Rapid alkalinization factor protein **Oxidative burst/stress** Glutathione peroxidase **Cytokinin biosynthesis** UDP-glucoronosyl/UDP- glucosyl transferase protein **Ethylene biosynthesis** 1-aminocyclopropane-1-caroxylate oxidase (*ACC oxidase*)	**Oxidative burst/stress** L-ascorbate peroxidase

Fig. 7.17 Left panel: the life cycle stages of *Plasmodiophora brassicae* (primary root-hair phase) at 4, 7, and 10 days after inoculating (DAI). Right panel: summary table of up- and down-regulated genes and functional groups at 4, 7, and 10 DAI (Agarwal et al. 2011)

analysis all of which correlated positively with the microarray data. The two *A. thaliana* mutants *jar1* and *coi1* tested were found to be susceptible to *P. brassicae*. The involvement of defense-related hormones in the interaction was further investigated (Agarwal et al. 2011) and the findings indicate that addition of salicylic acid can suppress clubroot disease in *A. thaliana* plants (Fig. 7.17).

7.4.2 Expression of Genes During Infection of *P. brassicae*

Microarray chips, two-dimensional electrophoresis, and high-throughput sequencing technology have been used to study gene expression in the host plant in response to *P. brassicae* infection. Siemens et al. (2002, 2006, 2009) explored the gene expression of *A. thaliana* Col-0 inoculated with *P. brassicae* after 10 and 23 days using a microarray chip. They found that more than 1000 host genes were differentially expressed in the infected roots as compared to the control roots. Starch, sulfur, and secondary metabolism, auxin, and cytokinin synthesis, and the expression of transport-related genes changed significantly, whereas that of genes associated with defense and lignin synthesis did not. In addition, some flavonoid genes were also differentially expressed (Siemens et al. 2006). Devos et al. (2006) and Cao et al. (2008) discovered protein changes in *Arabidopsis* following inoculating with *P. brassicae* using two-dimensional electrophoresis. Devos et al. (2006) observed 35 up-regulated and 11 down-regulated proteins 4 days after *P. brassicae* inoculating, which were mainly associated with defense, cell metabolism, cell differentiation, and active oxygen activity. Cao et al. (2008) observed changes in the expression of 20 proteins (with 13 spots increasing and seven spots decreasing) 12, 24, and 48 h post-inoculating, including lignin synthesis, cytokinin synthesis, calcium steady-state, glycolysis, and oxygen activity. Agarwal et al. (2011) detected 147, 27, and 37 differentially expressed genes (DEGs) in *A. thaliana* after 4, 7, and 10 days, respectively, using a microarray chip. They further deduced that changes observed at 4 days post-infection (dpi) were mainly related to host and pathogen recognition and signal transduction. Using transcriptome analysis, Chen et al. (2015) observed major changes between sensitive and resistant varieties of *B. rapa* 0–96 h post-infection in metabolism, transport, and signal transduction. Now a days, new high-through put technologies are gradually being used to study the interaction between *P. brassicae* and the host, such as metabotyping, laser microdissection coupled to transcriptional profiling and miRNA sequencing (Wagner et al. 2012; Schuller et al. 2014; Verma et al. 2014). The *P. brassicae* genome was sequenced, which will be very convenient for studies on the interaction between host plants and *P. brassicae* in the future (Schwelm et al. 2015; Rolfe et al. 2016). In order to clarify the early events taking place between the pathogen and the host, Zhao et al. (2017) analyzed differentially expressed genes and pathways in *A. thaliana* 24 and 48 h following infection with *P. brassicae*, using transcriptome analysis. In addition, the molecular mechanism of *A. thaliana's* response to *P. brassicae* in the early infection phase was described. Zhao et al. (2017) demonstrated changes on gene expression and pathways in *A. thaliana* during early infection with *P. brassicae* using transcriptome analysis, and identified 1903 and 1359 DEGs at 24 and 48 h post-inoculating (hpi), respectively. Flavonoids and the lignin synthesis pathways were enhanced, glucosinolates, terpenoids, and proanth °Cyanidins accumulated and many hormonal- and receptor-kinase related genes were expressed, caused by *P. brassicae* infection during its early phase (Fig. 7.18). Therefore, the early interaction between *A. thaliana* and *P. brassicae* plays an important role in the entire infection process. In order to confirm the results of the RNA-seq analysis,

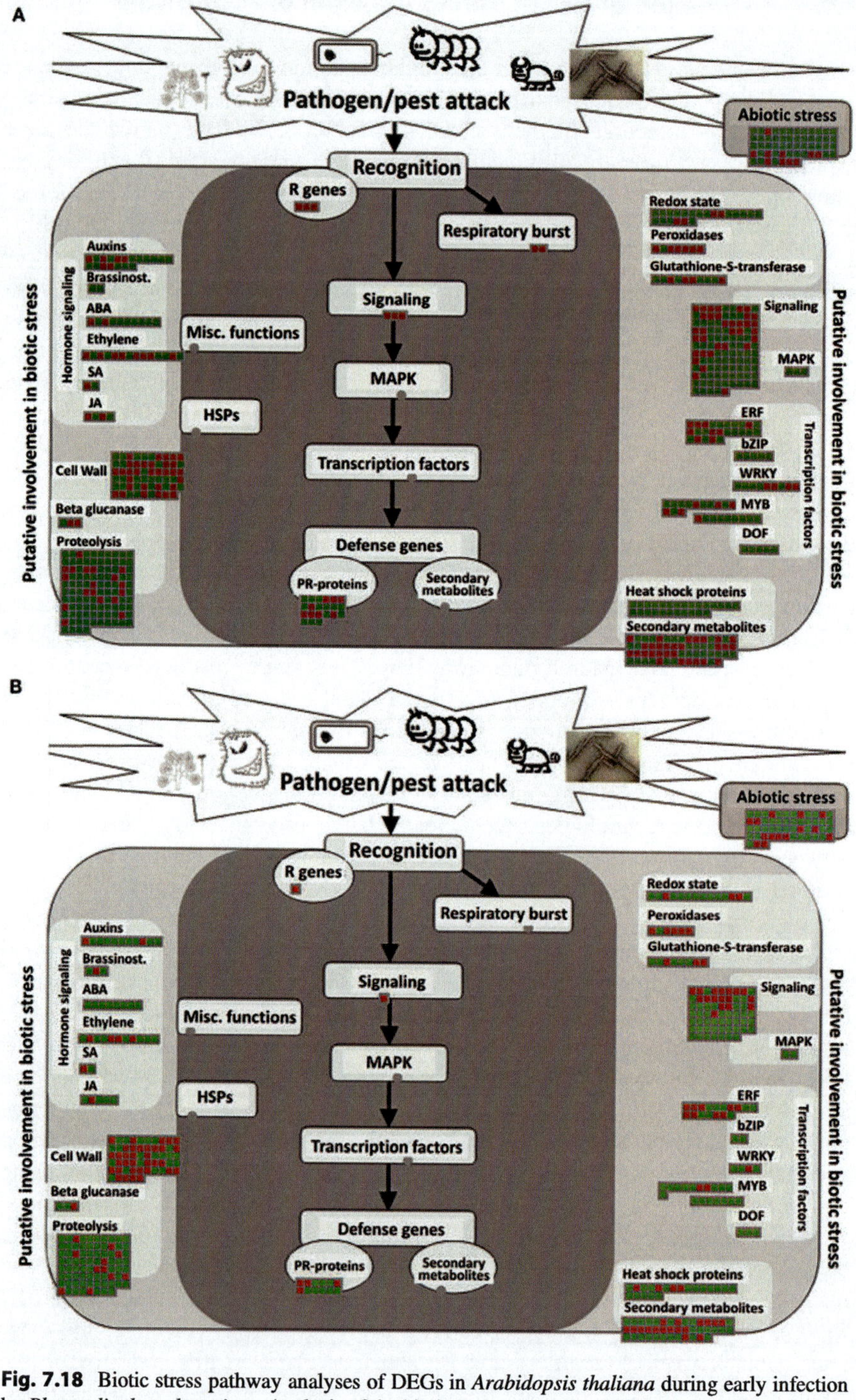

Fig. 7.18 Biotic stress pathway analyses of DEGs in *Arabidopsis thaliana* during early infection by *Plasmodiophora brassicae.* Analysis of the biotic stress pathway of DEGs was performed using MAPMAN software. Red boxes mean up-regulated genes and green mean down-regulated genes.

40 DEGs were selected for qRT-PCR, including 10 PR genes, three MYB transcription factor genes, one WRKY gene, 11 genes involved in the IAA, SA, JA, and cytokinin signaling pathways, 11 genes associated with secondary metabolism, and four genes related to the cell wall.

7.4.3 Identification of Clubroot Effectors Genes

In order to manipulate plant defenses and enable pathogen colonization, biotrophic plant pathogens like *P. brassicae* have evolved advanced strategies to deliver effector proteins (genes) into the host cell during infection. Successful infection relies primarily on the success of the release of the effectors, which in many cases are responsible for the suppression of plant immunity. The initial recognition of conserved microbial features, known as pathogen-associated molecular patterns (PAMPs), leads to PAMP-triggered immunity (PTI) in the host. PAMP-triggered immunity is a first level of immune response that can be overcome by effector proteins produced by adapted pathogens. Resistance (R) to adapted pathogens is achieved through specific recognition of effectors, also known as avirulence proteins, by corresponding R proteins produced by the plant host. This effector-R protein recognition constitutes the second level of immune response, effector triggered immunity (ETI). *P. brassicae* is a well adapted pathogen of *Brassica* hosts; though indirect evidence suggests lack of either response, both PTI and ETI have not been well-characterized in the host-*P. brassicae* pathosystem.

Understanding and identifying the proteins that comprise the secretome of *P. brassicae* is an important step towards identifying the complementary R proteins in the plant host. Putative effectors from biotrophic plant pathogens such as oomycetes and fungi are emerging from the sequencing and assembly of their genomes or transcriptomes, followed by comparative analysis with candidate effector genes. Schwelm et al. (2015) and Rolfe et al. (2016) identified putative effectors released by *P. brassicae* during the infection process. However, the obligate parasitic nature of *P. brassicae* has made it impossible to obtain a complete genome, with six draft genomes barely annotated for five Canadian pathotypes and one European pathotype. Transcriptomes of *P. brassicae* infecting *B. napus* and *Arabidopsis thaliana* in the Canadian study, and infecting *B. rapa, B. napus*, and *B. oleracea* var. *capitata* in the European study, have identified some candidate effectors.

The overexpression of a benzoic acid/salicylic acid methyltransferase-encoding gene [PBRA_T000444 in *P. brassicae* European pathotype 3(Pbe3), PbPT3Sc00 026_A_1.308_1 in *P. brassicae* pathotype 3 (Pb3) draft genomes; GenBank accession number AFK13134] during the second week of infection, with expression peaking three and four weeks after infection has been observed (Schwelm et al.

Fig. 7.18 (continued) (**a**) 24 h after inoculating; (**b**) 48 h after inoculating. The pathway frames are from the MAPMAN software database (Zhao et al. 2017)

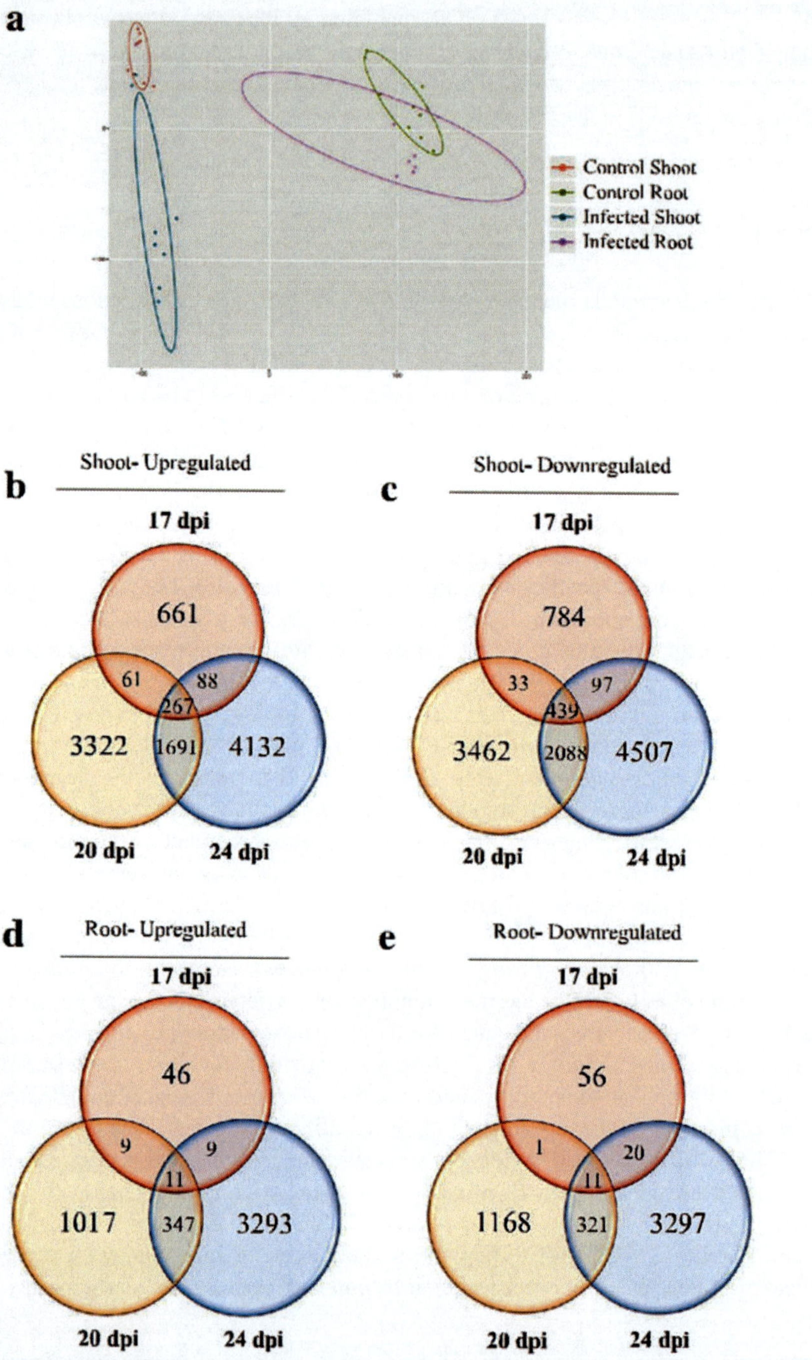

Fig. 7.19 (**a**) PCA plot displaying biological variation between infected and control samples in shoot and root. (**b–e**) Venn diagrams showing the total number of significantly DEGs (p-value ≥ 0.05) at 17, 20, and 24 dpi in infected shoot and root compared to mock-infected control samples.

2015; Rolfe et al. 2016; Ludwig-Muller et al. 2015). Salicylic acid (SA) is essential for the activation of plant defense. In the plant cell, regulation of active SA is managed through the maintenance of different inactive forms of SA, such as methyl salicylate (MeSA). The methyltransferase identified in *P. brassicae* has been characterized in detail by Ludwig-Muller et al. 2015, identifying its role in the methylation of salicylic, benzoic, and anthranilic acids, thereby contributing to the suppression of the salicylic acid-induced defense in a plant host (Pérez-Lópeza et al. 2018).

7.4.4 Differential Expression of Host Shoot and Root Genes During *P. brassicae* Infection

A broad-spectrum view of the *Arabidopsis* transcriptomic response to *P. brassicae* inoculating, in above- and below-ground tissues of the host plant was provided by Irani et al. (2018). They observed more DEGs at later stages of the disease establishment, with the largest transcriptomic changes occurring in the 24 dpi samples of both shoot and root. Although the root is the infected organ, the number of DEGs and the fold change in the DEGs, across all three time points sampled was higher in the shoot (Fig. 7.19a–e), suggesting that metabolic and cellular processes of the shoot are in fact more affected by clubroot disease progression than those of the infected root. A higher number of DEGs in the shoot is probably a result, in some part, of a general stress response to the disease, as shown by the down-regulation of photosynthesis and light reaction genes, whereas gene expression in diseased roots would predominantly be a specific response to the pathogen attack. A similar down-regulation of genes involved in photosynthesis, together with an up-regulation of pathogen protection and oxidative stress genes, has been reported in shoots in response to inoculating of roots with rhizobacterium (Rojas et al. 2014). The functional classification of highly DEGs showed that the expression of many of these genes was co-regulated or oppositely regulated in shoot and root. In the shoot, as expected, DEGs involved in photosynthesis were down-regulated after root inoculating with *P. brassicae* and subsequent clubroot progression. Furthermore, several genes involved in the degradation of starch and triacylglycerols (lipid storage molecules) and the Calvin cycle were also down-regulated in the shoot. Following decreased water and nutrient uptake by the host plant, the general down-regulation of these genes in the shoot of infected plants is predictable. A decrease in starch and soluble carbohydrates in the leaves of *Arabidopsis*, five weeks after infection with *P. brassicae*, has previously been reported. These findings were compatible with a reduced rate of photosynthesis in *Arabidopsis* leaves. Contrary to these decreases,

Fig. 7.19 (continued) The overlapping regions correspond to the number of DEGs present at more than one time point. (**b**) Up-regulated DEGs in shoot. (**c**) Down-regulated DEGs in shoot. (**d**) Up-regulated DEGs in root. (**e**) Down-regulated DEGs in root (Irani et al. 2018)

tryptophan (shikimate pathway metabolite) biosynthesis genes were highly induced in infected shoot, especially at 24 dpi. Infected *Arabidopsis* plants exhibited impaired growth and induced senescence, particularly at 20 and 24 dpi. Over this time period, the genes encoding cell wall structural proteins were mainly down-regulated in infected shoot. In addition, expression of many cell wall modification and degradation genes were turned off in the shoot over the 17, 20, and 24 dpi time points.

In infected root, expression of starch and sucrose biosynthesis and glycolysis pathway genes was mainly up-regulated. An increase in various sugars and starch has been previously reported in *P. brassicae*-infected *Arabidopsis* roots, whereas the suppression of invertase expression in transgenic plants resulted in enhanced tolerance to *P. brassicae*. An increased availability of carbon sources at the infection sites is a way to resource for the metabolic activities of the pathogen as well as for the increased cell division of the host plant, resulting in galls at these sites. Further, DEGs involved in lipid biosynthesis and elongation is up-regulated at 17 and 20 dpi in root. The higher abundance of these genes might be a response for the pathogen lipid demand. An up-regulation of lipid transfer genes was found in infected *Arabidopsis* roots when galls and resting spores are present, suggesting the formation of a lipid reserve. This is consistent with the observation that accumulation of abundant lipid droplets in the pathogen cytoplasm. Increased expression of phospholipases in infected root, mainly at 24 dpi, may be attributed to their roles in the synthesis of JA, oxylipins, and other important plant defense signaling molecules. However, since phospholipases hydrolyze phospholipids, an increased expression may indicate an increased level of catabolization of phospholipids by the host plant (Canonne et al. 2011).

A group of cell wall modification proteins, including expansins, xyloglucan endotrans-glucosylase/hydrolases (XTHs), and endo-β-1, 4-glucanases (EGases), facilitate cell wall expansion and function in regulating the plasticity and rheology of the cell wall. The up-regulation of cell wall modification genes in the root is likely linked to cell wall loosening and cell expansion in infected root. However, several DEGs encoding cell wall degradation enzymes were strongly down-regulated in infected root at 24 dpi. *P. brassicae* induces regulation of cell wall modification proteins involved in an alteration of cell wall structure during infection. Moreover, in root, an up-regulation of *SAUR* genes that are an early auxin-responsive gene family is consistent with the suggestion that IAA induces cell division and elongation during gall formation in *P. brassicae*-infected *Arabidopsis*. An involvement of auxin in the host–*P. brassicae* interaction has been observed by Ludwig-Muller et al. (2015).

In both shoot and root at 24 dpi, the expression of several genes in major branches of the shikimate pathway is up-regulated. At 17 dpi, genes involved in the biosynthesis of sinapate esters through the phenylpropanoid branch of the shikimate pathway were up-regulated in both tissues. Both a higher level of transcription of shikimate pathway genes, together with an increased production of the pathway's metabolites in plants, have been reported in response to pathogenic attacks. In *P. brassicae*-infected roots, an accumulation of phenylpropanoid pathway

Table 7.10 DEGs belonging to TF families with $\log_2$ fold change $\geq \pm 2$ at three tested time points (Irani et al. 2018)

TF family	Total number	Common expression	Opposite expression	Change	Shoot	Root
AP2/ EREBP	17	0	0	Up	5	4
				Down	7	1
Aux/IAA	2	0	0	Up	0	0
				Down	2	0
B3 (DBP)	6	1 (up) At1g49475	0	Up	1	2
				Down	3	0
bHLH	36	1 (up) At4g29930	0	Up	9	7
				Down	17	3
bZIP	8	0	0	Up	4	1
				Down	2	1
C2C2-CO-like	8	1 (up) At3g21890	1 (up in shoot, down in root)	Up	4	1
			At3g21150	Down	2	1
C2H2	9	1 (up) At3g53600	0	Up	2	1
				Down	6	0
C3H	5	2 (down) At5g44260	0	Up	1	0
		At2g25900		Down	2	2
CCAAT	8	2 (up) At3g05690	0	Up	2	5
		At5g06510		Down	1	0
G2-like	4	1 (down) At4g37180	0	Up	0	0
				Down	2	2
HB	11	0	0	Up	0	4
				Down	6	1
HSF	4	0	0	Up	3	1
				Down	0	0
LOB	8	1 (down) At4g37540	0	Up	2	1
				Down	4	1
MYB	39	1 (up) At5g56840	0	Up	20	9
				Down	7	3
Orphans	4	0	0	Up	2	0
				Down	2	0
Putative	8	0	0	Up	2	1
				Down	3	2
Unclassified	21	2 (down) At1g32540	1(up in shoot, down in root)	Up	3	4
		At4g01335	At2g05160	Down	11	3
WRKY	6	0	0	Up	6	0
				Down	0	0

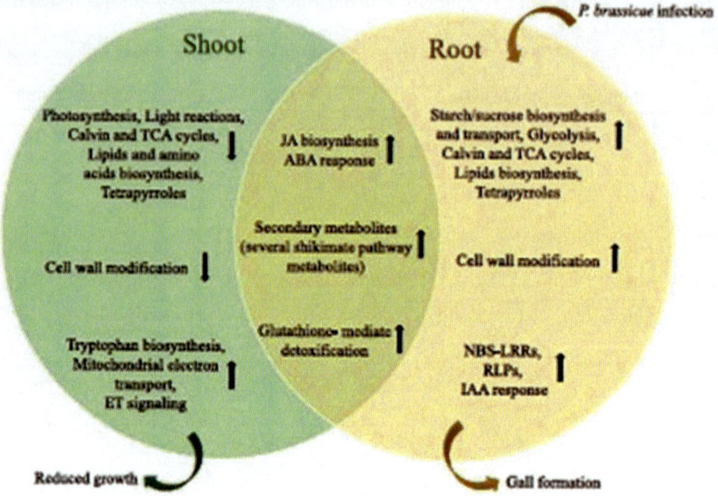

Fig. 7.20 Simplified model to describe the major transcriptional regulation in shoot and root of *Plasmodiophora brassicae*-infected *Arabidopsis*. The overlapping region corresponds to common DEGs in shoot and root. The upward black arrows show up-regulated DEGs and the downward arrows show down-regulated DEGs. *ABA* abscisic acid, *ET* ethylene, *IAA* indole acetic acid (auxin), *JA* jasmonic acid, *RLPs* receptor-like proteins (Irani et al. 2018)

compounds or a higher expression of related genes has been reported. The shikimate pathway metabolites have important roles in both root and shoot responses in infected host plants (Irani et al. 2018).

Transcriptome levels for genes in the JA biosynthesis pathway are elevated at all three time points in shoot and at 24 dpi in root. However, several genes involved in the JA response pathway are strongly down-regulated in root at 24 dpi, perhaps by the pathogen, indicating that the transcript regulation of JA response genes might be manipulated by the pathogen. *P. brassicae* is able to manipulate a number of hormone pathways and can alter hormone homeostasis in the host plant to its benefit. On the other hand, JA response genes may be down-regulated once the pathogen is established and having successfully disarmed the plant immune response by 24 dpi. An enhanced JA level was reported, in Chinese cabbage roots, 3–5 weeks after infection with *P. brassicae*. High levels of jasmonate-related compounds were also reported in *Arabidopsis* ecotype Col-0 roots, 21 days after inoculating with *P. brassicae*. In addition, during secondary infection by *P. brassicae,* the JA pathway is induced in susceptible *Arabidopsis* Col-0 but not in the partially resistant ecotype, Bur-0. The JA pathway is involved in clubroot disease during *P. brassicae* secondary infection (Rolfe et al. 2016; Leman et al. 2015).

A number of DEGs in the ABA response pathway are up-regulated in both shoot and root at 24 dpi. The level of ABA has been reported to increase in *P. brassicae*-infected *B. rapa* plants during gall formation. Moreover, up-regulation of ABA response and signaling genes during late time points after *P. brassicae* infection has been reported. Up-regulation of ABA response genes might be a response to abiotic stress conditions in infected tissues that are subsequently under water and nutrient stress. However, it has also been proposed that ABA plays a role in the induction of plant defense response genes and JA-induced defense genes (Adie et al. 2007).

Members of the TF families, MYB, bHLH, and AP2/EREBP, were highly up- or down-regulated in both shoot and root (Table 7.10). In *Arabidopsis*, the functions of MYB TFs are very diverse, including anthocyanin and flavonol biosynthesis, terminal differentiation and cell cycle regulation. Similarly, members of the bHLH family play various roles in the regulation of ABA and gibberellins signaling pathways, stomata development and root development in *Arabidopsis*. The comparison of expression profiles of TFs, in shoot and root tissues, highlights the major differences in response of each tissue to the infection. As shown in Table 7.10, only a few TFs are commonly up- or down-regulated in both infected shoot and root (Feller et al. 2011). *Arabidopsis* has a complex transcriptomic response to *P. brassicae* infection, and in general, the plant response to pathogen infection of the root is different between shoot and root. Irani et al. (2018) proposed a simplified model of the major molecular changes in shoot and root in response to *P. brassicae* infection and clubroot disease progression (Fig. 7.20). In the root, there is a general induction of genes related to cell wall modification, starch/sucrose biosynthesis and transport to provide energy. In contrast, in the shoot, and possibly as a result of the impairment of nutrient and water uptake, a down-regulation of photosynthesis, chlorophyll synthesis, starch/sucrose biosynthesis, and an inhibition of synthesis of proteins and enzymes involved in cell wall formation and its modification. As a result, the leaves show symptoms of stress such as reduced growth and necrosis. However, the shoot and root also responded to the pathogen in a number of common ways, such as an induction of JA biosynthesis, ABA response and the production of shikimate pathway metabolites.

7.4.5 Expression of Hypoxia Genes in *Arabidopsis* Roots During Secondary Infection of *P. brassicae*

Hypoxia responses in *Arabidopsis* roots during infection by the clubroot pathogen *P. brassicae* were assessed by Gravot et al. (2016). The hypoxia-related gene markers Pyruvate decarboxylase 1 (PDC1), Pyruvate decarboxylase 2 (PDC2), and Alcohol dehydrogenase 1 (ADH1) were induced during secondary infection by two isolates of *P. brassicae*, eH and e2. The PDC2 was highly induced as soon as 7 days post-inoculating (dpi), i.e., before the development of gall symptoms, and GUS staining revealed that ADH1 induction was localized in infected cortical cells of root galls at 21 dpi. Clubroot symptoms were significantly milder in the pdc1 and pdc2 mutants compared with Col-0, but a null T-DNA insertional mutation of ADH1 did not affect clubroot susceptibility. The Arg/N-end rule pathway of ubiquitin-mediated proteolysis controls oxygen-sensing in plants. Mutants of components of this pathway, *ate1ate2* and *prt6*, that both exhibit constitutive hypoxia responses, showed enhanced clubroot symptoms. In contrast, gall development was reduced in quintuple and sextuple mutants where the activity of all oxygen-sensing group VII ethylene response factor transcription factors (ERFVIIs) is absent (erfVII, and prt6 erfVII). The induction of PDC1 and PDC2 during the secondary infection of roots by *P. brassicae* contributes positively to clubroot development, and which is controlled by oxygen-sensing through ERFVIIs. The absence of any major role of ADH1 in symptom development may also suggest that PDC activity could contribute to the formation of galls through the activation of a PDH bypass.

7.4.6 Expression of Sweet Genes During *P. brassicae* Infection

After infection of the plant roots, *P. brassicae* colonization leads to swelling roots and gall formation, eventually inhibiting roots to uptake nutrients and water from the soil. Sugars are the main carbon source for the living organisms including pathogens and their host plants. In plants, sugars are synthesized in the photosynthetic organs and allocated into the sink organs via long-distance transport. Sugars produced in the living plants not only play key roles for plant growth and development, but also are transferred or secreted into the plant surface. Sugars secreted by plants could be used as carbon supply for pathogen growth (Patrick 1989; Chen et al. 2010). Especially, biotrophic microbes like *P. brassicae* obtain nutrients from host cells by extracting sugars (Manck-Götzenberger and Requena 2016; Roman and Rathjen 2017). In *P. brassicae*, it has been reported that 2% sucrose solution can promote germination of resting spores (Guo 2001). In addition, accumulation of soluble sugars (hexoses and sucrose) is found in the galls of *P. brassicae*-infected *Arabidopsis thaliana* (Evans and Scholes 1995; Brodmann et al. 2002) and *Brassica* plants (Williams et al. 1968; Keen and Williams 1969), indicating that the galls might act as a newly established carbon sink during clubroot development. Sugar transporters including well-known gene families such as MSTs (monosaccharide transporters) (Toyofuku et al. 2000), SUTs (sucrose transporters) (Reinders et al. 2008, 2011), and SWEETs (sugar will be eventually exported transporters) (Chen et al. 2012) are responsible to transl °Cate monosaccharides or disaccharides.

Transcriptome studies revealed that several genes involved in sugar transport and metabolism strongly up-regulated during gall formation in clubroot-infected *Arabidopsis* (Siemens et al. 2006; Malinowski et al. 2012). A transcriptome profiling performed on both shoot and root tissues of *A. thaliana* challenged by *P. brassicae* also pointed that genes involved in sucrose and starch biosynthesis are differently expressed in shoot and root (Irani et al. 2018). These transcriptomic and metabolic experiments provided insights that sugar accumulation is important for gall formation, probably depending on some sugar transport genes. Moreover, during the process of plant–pathogen co-evolution, the pathogens have evolved a mechanism to compete for sugars with infected host cells to sustain their life cycle (Doidy et al. 2012; Lemoine et al. 2013). Competition for sugars at the plant–pathogen interface is likely controlled by the sugar transporters (Toyofuku et al. 2000; Doidy et al. 2012), thus regulation of those transporters would give a new insight into disease control.

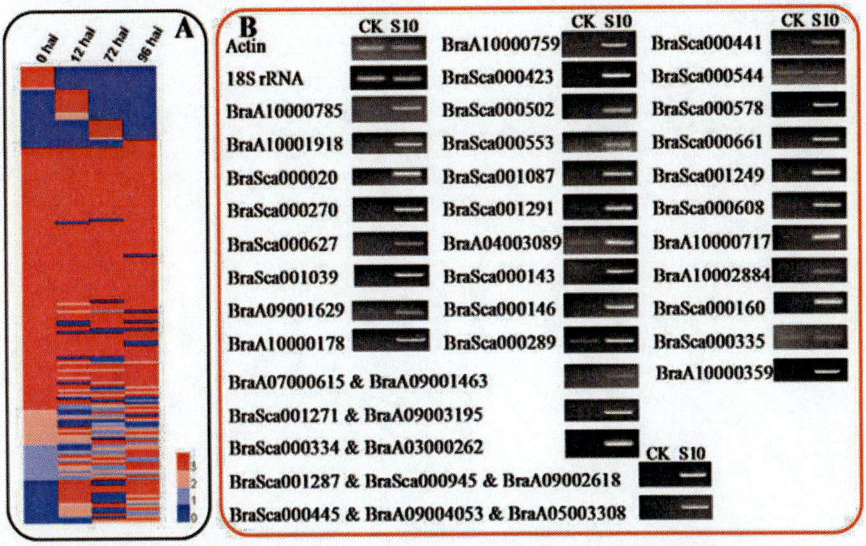

Fig. 7.21 Expression patterns of BSGs in *Brassica rapa* after *Plasmodiophora brassicae* infection. **(a)** Heat map of BSGs after *P. brassicae* treatment of Chinese cabbage (FPKM > 2). Zero, 12, 72, and 96 hai (hai: hours after inoculating) represents clubroot-susceptible Chinese cabbage (BJN3-2) that was infected by *P. brassicae.* Gene expressions were analyzed by semi-qPCR. *B. rapa* 18S rRNA and Actin genes were used as the internal control. **(b)** *P. brassicae* treatment (Jiang et al. 2018)

The role of SWEETs during *P. brassicae*-induced gall formation in Chinese cabbage was examined for sugar contents in roots and leaves of the Chinese cabbage clubroot resistant (CR) and susceptible (CS) near-isogenic lines (NILs) after infection of *P. brassicae*. In addition, the response of *B. rapa* SWEET (hereafter BrSWEET) sugar transporters to *P. brassicae* infection as well as clubroot disease development in *Arabidopsis* sweet11 mutant was examined. The pathogen lives inside roots and hijacks nutrients from the host plants. It is suggested that clubroot galls created an additional nutrient sink in infected roots. Sugar contents in leaves and roots before and after *P. brassicae* infection using a pair of Chinese cabbage near-isogenic lines (NILs), carrying either a clubroot resistant (CR) or susceptible (CS) allele at the CRb Locus were assessed. *P. brassicae* infection caused significant increase of glucose and fructose contents in the root of CS-NIL compared to CR-NIL, suggesting that sugar translocation and *P. brassicae* growth are closely related. Among 32 *B. rapa* SWEET homologs, several BrSWEETs belonging to Clade I and III were significantly up-regulated, especially in CS-NIL upon *P. brassicae* infection. Moreover, *Arabidopsis* sweet11 mutant exhibited slower gall formation compared to the wild-type plants. *P. brassicae* infection probably triggers active sugar translocation between the sugar producing tissues and the clubbed tissues, and the SWEET family genes are involved in this process (Li et al. 2018).

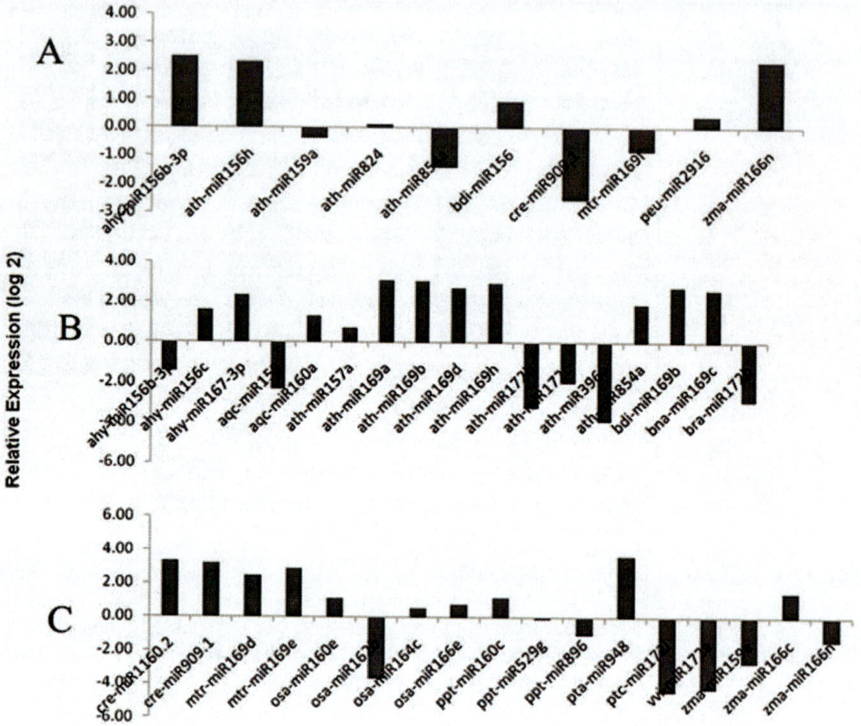

Fig. 7.22 miRNA-microarray expression of *Plasmodiophora brassicae*-responsive miRNAs exhibiting differential expression at 10- (**a**), and 20- dpi (**b, c**) following pathogen infection (Verma et al. 2014)

7.4.7 Expression Patterns of *Brassica* Specific Genes (BSGs) in *B. rapa* After *P. brassicae* Infection

Orphan genes are often associated with biotic and abiotic stress, and they usually function in stress-response signal transduction pathways. Using RNA sequencing, transcriptome profiling was compared between clubroot-resistant and clubroot-susceptible *B. rapa* genotypes after infection with *P. brassicae* during early infection (Chen et al. 2015). Jiang et al. (2018) reassembled respective RNA sequencing data to analyze the BSGs expression patterns, and the results indicated that 180 unique BSGs were supported by RNA-Seq data (FPKM > 2 for least one sampling stage (Fig. 7.21). Further, 24 BSGs exhibited high expression abundance at least in one sampling stage (FPKM > 50). Seventy-eight BSGs displayed constitutive expression at every point (FPKM > 2), and 32 BSGs showed period-specific expression (FPKM > 2). Deora et al. (2012) indicated that *P. brassicae* is blocked at middle or late stages of infection in resistant *Brassica* crops. However, BSGs may play more important roles in middle or late periods (4–20 days after inoculating (dai)). To confirm this hypothesis, roots from *B. rapa* plants were collected at 10 dai with

P. brassicae (referred to CK and S10), and 52 successfully amplified BSGs in Chiifu were analyzed using semi-qPCR. As expected, 78.85% (41 genes) of analyzed genes responded to *P. brassicae* treatments, and 39 out of 41 BSGs were up-regulated (Fig. 7.21b). However, only 8 out of these 52 BSGs have transcript support (FPKM > 2) (Fig. 7.21). The results strongly suggest that BSGs play important roles in *P. brassicae* stress responses (Jiang et al. 2018).

7.4.8 *Plasmodiophora brassicae* Gene Expression During Pathogenesis of *Brassica rapa* Assessed by RT-PCR Analysis

The crucifer *Arabidopsis thaliana* has been used to characterize the expression of several *P. brassicae* genes during pathogenesis (Siemens et al. 2009). *A. thaliana* is a host of *P. brassicae* and the small size of the roots facilitates the observation of disease development. However, clubroot development in *Brassica* takes a long time and the roots show more homogeneous developmental stages of plasmodia in comparison to *A. thaliana* (Gustafsson and Falt 1986).

Therefore, the short life span of this small weedy species might prevent researchers from identifying genes that are expressed in a stage-specific manner during the interaction. The expression of several *P. brassicae* genes over the time course of the host–pathogen interaction in *Brassica rapa* was analyzed by semi-quantitative RT-PCR in order to find molecular markers for the actively growing pathogen, and for distinct developmental stages of *P. brassicae*, early developmental stages in root hairs or young vegetative plasmodia shortly after secondary infection. The gene expression was analyzed in three *B. rapa* host lines with a virulent isolate, to test whether semi-quantitative RT-PCR can be used to characterize host–pathogen interactions in crop plants. To find landmarks for specific clubroot development stages, expression of several *P. brassicae* genes was analyzed during the infection process in a universally susceptible *Brassica rapa* line (cv. Granaat, ECD 05). The results revealed that the genes *Pb-YPT, Pb-Brip9*, and *Pb-PSA* were regulated during infection of *P. brassicae, Pb-Brip9* and *Pb-PSA* were expressed in early stages of infection. Wu et al. (2012) provide evidence that semi-quantitative RT-PCR could be used for discriminating developmental stages of *P. brassicae* in *B. rapa*. Stage-specific gene expression of *P. brassicae* should be studied in *B. rapa* roots due to longer disease development in comparison to *Arabidopsis thaliana* (Wu et al. 2012).

7.4.9 Induction of MicroRNAs in *Brassica* During *P. brassicae* Infection

The discovery of miRNAs in plants has shed new light on the understanding of a wide range of biological processes, such as growth development and plant morphology (Ori et al. 2007; Carlsbecker et al. 2010; Lelandais-Briere et al. 2010), cell proliferation (Schommer et al. 2014), flowering and sex determination (Chuck et al. 2009; Wu et al. 2013), and abiotic stress tolerance (Barrera-Figueroa et al. 2012;

Yang et al. 2013). Further, plant miRNAs are extensively involved in immune responses against various invading pathogens, such as bacteria, fungi, and viruses (Jones and Dangl 2006; Navarro et al. 2006; Padmanabhan et al. 2009; Dunoyer et al. 2015). In the defense against pathogen attacks in *Arabidopsis*, it was found to negatively regulate auxin signaling by targeting the auxin receptors TIR1 (transport inhibitor response 1), AFB2 (auxin signaling F-box protein 2), and AFB3 (Navarro et al. 2006). It was revealed that miRNAs function indirectly by targeting protein-coding genes. Therefore, the identification of miRNAs and their targets can contribute to the further understanding of metabolic processes in plants. The next generation sequencing (NGS) technology, also known as small RNA sequencing, degradome sequencing, and transcriptome sequencing, all of which can globally identify miRNAs and their targets, has been successfully applied in *Brassica napus* (Xu et al. 2012), *Glycine max* (Soybean) (Song et al. 2011), and *Gossypium hirsutum* L. (Cotton) (Zhang et al. 2015).

Xiaochun et al. (2016) constructed and sequenced small RNA and degradome libraries derived from the roots using Illumina/Solexa sequencing. The miRNAs is responsive to *P. brassicae* infection and validated the target transcripts of *P. brassicae*-responsive miRNAs by degradome sequencing in Chinese cabbage. It will provide valuable information for elucidating the miRNA-guided regulatory networks of Chinese cabbage during its stress response to *P. brassicae*. Increasing evidence has revealed that microRNAs play a pivotal role in the post-transcriptional regulation of gene expression in response to pathogens in plants. However, there is little information available about the expression patterns of miRNAs and their targets in Chinese cabbage (*Brassica rapa* ssp. *pekinensis*) under *P. brassicae* infection. Using deep sequencing and degradome analysis, a genome-wide identification of miRNAs and their targets during *P. brassicae* stress was performed. A total of 221 known and 93 potentially novel miRNAs were successfully identified from two root libraries of one control (635-10CK) and *P. brassicae*-treated Chinese cabbage samples (635-10T). Of these, 14 known and 10 potentially novel miRNAs were found to be differentially expressed after *P. brassicae* treatment. Degradome analysis revealed that the 223 target genes of the 75 miRNAs could be potentially cleaved. KEGG (Kyoto Encyclopedia of Genes and Genomes) pathway analysis suggested that the putative target genes of the miRNAs were predominately involved in selenocompound metabolism and plant hormone signal transduction. Then the expression of 12 miRNAs was validated by quantitative real-time PCR (qRT-PCR). These results provide insights into the miRNA-mediated regulatory networks underlying the stress response to the plant pathogen *P. brassicae* (Xiaochun et al. 2016).

MicroRNAs are a highly conserved class of small non-coding RNAs that regulate gene expression by post-transcriptional repression. Emerging evidence indicates that hosts' endogenous small RNAs represent an important mechanism of control in plant immune responses and hormone signaling at the time of stress. RNAs like ath-miR160 and ath-miR167 are involved in pathogenesis and target the auxin-response-factor (ARF). Another microRNA, ath-miR164, has been implicated in auxin homeostasis and lateral root development, which may have a bearing on clubroot development. Therefore, it is conceivable that miRNAs may be involved in mediating important plant processes following infection with *P. brassicae*, which results in the development of root galls, leading to stunting and decreased productivity. Now, microarrays of miRNAs have become available for high-throughput global analysis of miRNAs expression patterns in response to pathogens. This valuable tool has been further exploited in a number of plants to study the miRNA: mRNA interactions in tomato, rice, and *Arabidopsis* which may lead to the identification of miRNAs that are involved in plant responses to specific stresses, including *P. brassicae* challenge (Li et al. 2010; Mallory et al. 2005; Guo et al. 2005; Liu et al. 2008). Clubroot disease of Brassicaceae family has potential to substantially limit *Brassica* production. A miRNA microarray-based approach was used to identify miRNA expression and regulation during the disease initiation and progression (Verma et al. 2014). Results revealed the useful basis towards understanding the dynamics of miRNAs related to *P. brassicae* infection of *Brassica*. The differential expression analysis of miRNA during pathogenesis of *P. brassicae* was carried out at two developmentally distinct time points. The expression of various miRNAs was observed to be modulated during the process of clubroot initiation and development. Many of these miRNAs were involved in the regulation of gene activity by targeting, among others, TFs, ARFs, MYB, MADS box, AP2, and ERFs. Many of these are known to be involved in stress-responses and developmental events related to pathogenesis. A diverse array of miRNAs and their target genes are modulated during the pathogenesis of *Brassica* by *P. brassicae*. It enhances the understanding related to the biological processes involved in *P. brassicae–Brassica* interactions and may lead to unique ways to generate disease resistant phenotypes (Fig. 7.22)

7.4.10 Induction of Signaling Molecules and Genes During Gall Formation

Plasmodiophora brassicae establishes an intricate interaction with their hosts during at least some parts of the infection process, because of their dependence on host carbon sources. They influence host physiology and alter host regulatory networks over a wide range of its genome. Especially the plant's hormonal balance is altered by this interaction. The changes in host hormone metabolism are connected to the intracellular life style of these protists (Ludwig-Muller et al. 2009). The infection process of plants by *P. brassicae* consists of two phases: the primary phase, which is restricted to root hairs and the secondary phase, which occurs in the cortex and stele of roots and hypocotyl and leads to abnormal development. Especially during this later phase, the host root responds to infection by increased cell division rates followed by hypertrophy of infected cells. These harbor first the plasmodia of *P. brassicae*, probably dividing together with the host cells. Later, the plasmodia grow and the host cell increases concomitantly in size. These enlarged host cells can reach at least ten times the size of uninfected cells. While cell division has been attributed to the action of auxins and cytokinins, cell enlargement has been so far linked exclusively to higher auxin concentrations and synthesis. The induction of auxin in *Arabidopsis* correlates with an increase in seedling growth and Xyloglucan-

Endo-Transferase/Hydrolase leading to cell elongation. In addition, a microarray revealed that genes involved in cell division and expansion such as cell cycle genes and expansins are up-regulated, especially at the first analyzed time point of the disease (10 days after inoculating). Consequently, altering hormone concentrations has led to reduced clubroot symptoms. Auxin signaling pathway(s) contribute to clubroot formation (Siemens et al. 2006; Devos et al. 2005; Pasold et al. 2010; Neuhaus et al. 2000) Auxin signaling is regulated by two types of receptors: the nuclear-localized TIR/AFB family and the plasma membrane associated auxin binding protein 1 (ABP1) (Dharmasiri et al. 2005; Spartz and Gray 2008; Christian et al. 2006) (Fig. 7.14). The nuclear signaling pathway leads to the activation of the transcriptional response via auxin perception by the F-box protein TIR1 (transport inhibitor response 1). F-box proteins function as substrate recognition modules for the multisubunit complex of ubiquitin ligases (also called Skp1-Cullin1-F-box protein (SCF); SCFTIR). The F-box protein TIR1 is the receptor, which recruits, by binding to auxin, the protein target that is designated for degradation by ubiquitination. Polyubiquitinated proteins are then transferred into the 26S proteasome and degraded.

Auxin-mediated transcriptional response is controlled by negative (Aux/IAA), and mainly positive (ARF) regulators. ARFs (auxin-response-factors) bind to the auxin-responsive elements (AuxRe) in respective promoters of auxin-inducible genes for transcriptional activation, but some ARFs are also inhibitors (Guilfoyle and Hagen 2007). Both Aux/IAAs and ARFs are present as large families. The Aux/IAA proteins bind to the ARFs through homologous domains in both proteins and thereby repress auxin-regulated transcription (Reed 2001).

The Aux/IAA proteins are short-lived, because they are degraded via the SCFTIR pathway and their degradation is promoted by auxin. Many mutations in *Aux/IAA* genes are stabilizing the resulting protein, because degradation domains are affected and thus constitute gain-of-function mutations (Reed 2001). Among the targets of ARFs are genes encoding proteins involved in the regulation of auxin homeostasis. The IAA amino acid conjugate synthetases (GH3) have been first recognized as auxin-induced genes. Later, it was discovered that it is a large gene family and that some proteins are involved in the conjugation of IAA to various amino acids. GH3 proteins are involved in various responses of plants to abiotic and biotic stresses. Especially the protein GH3.5 seems to play various roles in addition to the synthesis of IAA amino acid conjugates. It also conjugates the plant defense signal salicylic acid to amino acids and is involved in the synthesis of the *Arabidopsis* phytoalexin camalexin. One family member, GH3.11 (JAR1), catalyzes the formation of the isoleucine conjugate of jasmonic acid (JA). Contrary to the inactivation of IAA by conjugation, the isoleucine conjugate of JA is active, because it alone binds to the JA receptor COI1 (Hagen and Guilfoyle 2002; Park et al. 2007; Zhang et al. 2007).

ABP1 has been shown to control events at the plasma membrane by the regulation of membrane fluxes leading in consequence to cell elongation. ABP1 has an ER retention sequence, but it is hypothesized that for action at the plasma membrane some protein—escapes|| its intracellular location. In addition, there is no transmembrane sequence present in ABP1, which led to the speculation that a d °Cking protein could be responsible for anchoring ABP1 in the membrane (Fig. 7.14). ABP1 is involved in protoplast swelling, which is an indication for its role *in planta* in cell elongation. In addition, many growth processes seem to rely on ABP1 activity, for example, an ABP1 knockout mutant is embryo lethal (Chen et al. 2001). Downstream of ABP1, the action of ATPases and ion channels was postulated. Both H^+-ATPase and K^+-channels are needed for the auxin-mediated cell elongation response. K^+-channels necessary for the osmotic changes occurring during cell elongation can be induced by auxin in maize and *Arabidopsis* (Fuchs et al. 2006). In contrast, the TIR pathway can be responsible for the increased gene expression for these channels in response to auxin, linking the two pathways. Initial evidence for the involvement of the TIR pathway in the clubroot disease came from Alix et al. (2007) who reported the partial resistance of the *Arabidopsis axr3-1* mutant. The *AXR3* gene encodes the transcriptional repressor IAA17, a member of the Aux/IAA family. The transcript is stabilized in the mutant, so that the transcriptional activation cannot take place. The contribution of both auxin signaling pathways to the clubroot symptom development and their roles during disease progression have been revealed by Jahn et al. (2013).

Two different classes of auxin receptors, the TIR family and the auxin binding protein 1 (ABP1) in *Arabidopsis thaliana* are transcriptionally up-regulated upon gall formation. Mutations in the *TIR* family resulted in more susceptible reactions to the root pathogen. As target genes for the different pathways have been investigated for the transcriptional regulation of selected transcriptional repressors (*Aux/IAA*), and transcription factors (*ARF*), since the TIR pathway controls auxin homeostasis via the up-regulation of some auxin conjugate synthetases (GH3), and the expression of selected *GH3* genes also show in most cases of up-regulation. A double *gh3* mutant showed also slightly higher susceptibility to *P. brassicae* infection, while all tested single mutants do not show any alteration in the clubroot phenotype. As targets for the ABP1-induced cell elongation the effect of potassium channel blockers on clubroot formation was investigated. Treatment with tetraethylammonium (TEA) resulted in less severe clubroot symptoms. This research provides evidence for the involvement of two auxin signaling pathways in *Arabidopsis* needed for the establishment of the root galls by *P. brassicae*.

7.4.11 Expression of Immunophilins During *P. brassicae* Infection

Searches in the *P. brassicae* genome resulted in 20 putative immunophilins comprising of 11 cyclophilins (CYPs), 7 FK506-binding proteins (FKBPs), and 2 parvulin-like proteins. RNA-seq data showed that immunophilins were differentially regulated in enriched life stages such as germinating spores, maturing spores, and plasmodia, and infected *Brassica* hosts (*B. rapa*, *B. napus*, and *B. oleracea*). *PbCYP3* was highly induced in all studied life stages and during infection of all three *Brassica* hosts, and hence was selected for further analysis. *PbCYP3* was heterologously expressed in *Magnaporthe oryzae* gene-inactivated Δ*Cyp1* strain. The new strain Δ*Cyp1+* overexpressing *PbCYP3* showed increased virulence on rice compared to the Δ*Cyp1* strain. These results suggest that the predicted immunophilins and particularly *PbCYP3* are activated during plant infection. *M. oryzae* is a well-studied fungal pathogen and could be a valuable tool for future functional studies of *P. brassicae* genes, particularly elucidating their role during various infection phases (Singh et al. 2018).

7.4.12 Role of G-protein-Coupled Receptors (GPCR) Signal Transduction Pathways in *P. brassicae* Pathogenesis

Unlike other obligate biotrophic pathogens with diverse infection structures, *P. brassicae* is an intracellular pathogen that spends most of its life cycle in the intracellular environment of the host, similar to the life cycle of endophytic microbes (Kageyama and Asano 2009). An effective molecular mechanism is, therefore, required for *P. brassicae* to adapt to the intracellular environment of its plant hosts. To date, draft genome sequences from single-spore isolates of Pbe3 (Europe strain), Pb3, and Pb6 (Canadian strain) of *P. brassicae* have been obtained, and the genomes of three other pathotypes (Pb2, Pb5, and Pb8) have been re-sequenced (Schwelm et al. 2015; Rolfe et al. 2016). Although the genomes of these strains of *P. brassicae* have been sequenced and analyzed, the evolutionary status of *Plasmodiophora* and their relatives have not yet been investigated in detail.

The ability to sense and transduce extracellular signals into intracellular responses using a sophisticated molecular signaling system is crucial for living organisms. G-protein-coupled receptors (GPCRs), characterized by a seven transmembrane domain structure, are involved in a wide variety of physiological processes. Signals from a diverse range of extracellular signaling molecules, including hormones, neurotransmitters, ions, photons, and odorants, can be relayed by GPCRs to intracellular downstream signaling pathways to ultimately elicit cellular responses (Hilger et al. 2018). Based on their sequence homology and functional similarity, the super family of GPCRs can be generally divided into 6 classes (A, B, C, D, E, and F) (Qiu et al. 2009; Alexander et al. 2017). The signal transduction of GPCRs is dependent upon receptor-mediated activation of heterotrimeric G proteins consisting of three subunits, Gα, Gβ, and Gγ (Kamp et al. 2016). As one of the main signaling pathways in eukaryotes, the signaling system of GPCRs is involved in numerous processes including developmental control, cell growth, migration, density sensing, and neurotransmission (Rosenbaum et al. 2009; de Mendoza et al. 2014). As important central sensors that activate eukaryotic signaling, GPCRs are the primary targets of human drugs (van den Hoogen et al. 2018). Despite this, the specific roles of these receptors and their downstream signaling pathways in protists remain poorly understood. Bi et al. (2019) conducted a comparative genomics analysis to determine the role of the GPCR signaling pathway in *P. brassicae* pathogenesis of crucifers. It was observed that *Plasmodiophora brassicae* is an important biotrophic eukaryotic plant pathogen, and a member of the rhizarian protists. This biotrophic pathogen causes clubroot in cruciferous plants via novel intracellular mechanisms that are markedly different from those of other biotrophic organisms. To date, genomes from six single-spore isolates of *P. brassicae* have been sequenced. An accurate description of the evolutionary status of biotrophic protists, however, remains lacking. Bi et al. (2019) determined the draft genome of the *P. brassicae* ZJ-1 strain. A total of 10,951 protein-coding genes were identified from a 24.1 Mb genome sequence. After a comparative genomics approach to prove the Rhizaria super group is an independent branch in the eukaryotic evolutionary tree. It was found that the GPCR signaling pathway, the versatile signal transduction to multiple intracellular signaling cascades in response to extracellular signals in eukaryotes, is significantly enriched in *P. brassicae*-expanded and *P. brassicae*-specific gene sets. Additionally, treatment with a GPCR inhibitor relieved the symptoms of clubroot and significantly suppressed the development of plasmodia. These findings suggest that GPCR signal transduction pathways play important roles in the growth, development, and pathogenicity of *P. brassicae* to the crucifers at large.

References

Adams MD, Kelley JM, Gocayne JD, Dubnick M, Polymeropoulos MH, Xiao H (1991) Links complementary DNA sequencing: expressed sequence tags and human genome project. Science 252:1651–1656

Adie BAT, Perez-Perez J, Perez-Perez MM, Godoy M, Sanchez-Serrano JJ, Schmelz EA, Solano R (2007) ABA is an essential signal for plant resistance to pathogens affecting JA biosynthesis and the activation of defenses in *Arabidopsis*. Plant Cell 19:1665–1681

Agarwal A, Kaul V, Faggian R, Cahill DM (2009) Development and use of a model system to monitor clubroot disease progression with an Australian field population of *Plasmodiophora brassicae*. Austr Plant Pathol 38:120–127

Agarwal A, Kaul V, Faggian R, Rookes JE, Ludwig-Muller J, Cahill DM (2011) Analysis of global host gene expression during the primary phase of the *Arabidopsis thaliana–Plasmodiophora brassicae* interaction. Funct Plant Biol 38(6):462–478

Agrios GN (2005) Plant pathology. Elsevier, Amsterdam

Aist JR, Williams PH (1971) The cytology and kinetics of cabbage root hair penetration by *Plasmodiophora brassicae* Worn. Can J Bot 49:2023–2034

Alexander SP, Christopoulos A, Davenport AP, Kelly E, Marrion NV, Peters JA, Faccenda E, Harding SD, Pawson AJ, Sharman JL, Southan C, Davies JA, Collaborators CGTP (2017) The concise guide to pharmacology 2017/18: G protein-coupled receptors. Br J Pharmacol 174 (Suppl 1):S17–S129

Alix K, Lariagon C, Delourme R, Manzanares-Dauleux MJ (2007) Exploiting natural genetic diversity and mutant resources of *Arabidopsis thaliana* to study the *Arabidopsis thaliana–Plasmodiophora brassicae* interaction. Plant Breed 126(2):218–221

Allard R (1995) The functional analysis of plant nitrilases in transgenic plants. PhD thesis.1995 Göttingen, Germany

Ando S, Yamada T, Asano T, Kamachi S, Tsushima S, Hagio T, Tabei Y (2006) Molecular cloning of PbSTKL1 gene from *Plasmodiophora brassicae* expressed during clubroot development. J Phytopathol 154(3):185–189

Ando S, Tsushima S, Kamachi S, Konagaya K, Tabei Y (2008) Alternative transcription initiation of the nitrilase gene (BrNIT2) caused by infection with *Plasmodiophora brassicae* Woron. in Chinese cabbage (*Brassica rapa* L.). Plant Mol 68:557–569

Arbeiter A, Fahling M, Graf H, Sacristan MD, Siemens J (2002) Resistance of *Arabidopsis thaliana* to the obligate biotrophic parasite *Plasmodiophora brassicae*. Plant Prot Sci 38:519–522

Barrera-Figueroa BE, Gao L, Wu Z, Zhou X, Zhu J, Jin H, Liu R, Zhu JK (2012) High throughput sequencing reveals novel and abiotic stress-regulated microRNAs in the inflorescences of rice. BMC Plant Biol 12:132

Bartel B, Fink GR (1994) Differential regulation of an auxinproducing nitrilase gene family in *Arabidopsis thaliana*. Proc Natl Acad Sci U S A 91:6649–6653

Bennett RN, Wenke T, Freudenberg B, Mellon FA, Ludwig-Muller J (2005) The tu8 mutation of *Arabidopsis thaliana* encoding a heterochromatin protein 1 homolog causes defects in the induction of secondary metabolite biosynthesis. Plant Biol 7:348–357

Bi K, He Z, Gao Z, Zhao Y, Fu Y, ChengJ XJ, Jiang D, Chen T (2016) Integrated omics study of lipid droplets from *Plasmodiophora brassicae*. Sci Rep 6:36965

Bi K, Chen T, He Z, Gao Z, Zhao Y, Liu H, Fu Y, Xie J, Cheng J, Jiang D (2019) Comparative genomics reveals the unique evolutionary status of *Plasmodiophora brassicae* and the essential role of GPCR signaling pathways. Phytopathol Res 1:12

Bischoff M, Löw R, Grsic S, Rausch T, Hilgenberg W, Ludwig-Muller J (1995) Infection with the obligate biotroph *Plasmodiophora brassicae*, the causal agent of the clubroot disease, does not affect expression of NIT1/2-related nitrilases in roots of Chinese cabbage. J Plant Physiol 147:341–345

Blázquez MA, Santos E, Flores CL, Martínez-Zapater JM, Salinas J, Gancedo C (1998) Isolation and molecular characterization of the *Arabidopsis* TPS1 gene, encoding trehalose-6-phosphate synthase. Plant J 13:685–689

Brenner S, Johnson M, Bridgham J, Golda G, Lloyd DH, Johnson D (2000) Gene expression analysis by massively parallel signature sequencing (MPSS) on microbead arrays. Nat Biotechnol 18:630–634

Brodmann D, Schuller A, Ludwig-Müller J, Aeschbacher RA, Wiemken A, Boller T, Wingler A (2002) Induction of trehalase in *Arabidopsis* plants infected with the trehalose-producing pathogen *Plasmodiophora brassicae*. MPMI 15(7):693–700

Bulman S, Siemens J, Ridgway HJ, Eady C, Conner AJ (2006) Identification of genes from the obligate intracellular plant pathogen, *Plasmodiophora brassicae*. FEMS Microbiol Lett 264 (2):198–204

Burki F, Kudryavtsev A, Matz MV, Aglyamova GV, Bulman S, Fiers M, Keeling PJ, Pawlowski J (2010) Evolution of Rhizaria: new insights from phylogenomic analysis of uncultivated protists. BMC Evol Biol 10(1):377

Butcher DN, El-Tigani S, Ingram DS (1974) The role of indole glucosinolates in the clubroot disease of the Cruciferae. Physiol Plant Pathol 4:127–141

Canonne J, Marino D, Jauneau A, Pouzet C, Brière C, Roby D, Rivas S (2011) The *Xanthomonas* type III effector *XopD* targets the *Arabidopsis* transcription factor *MYB30* to suppress plant defense. Plant Cell 23:3498–3511

Cao T, Srivastava S, Rahman MH, Kav NNV, Hotte N, Deyholos MK, Strelkov SE (2008) Proteome-level changes in the roots of *Brassica napus* as a result of *Plasmodiophora brassicae* infection. Plant Sci 174(1):97–115

Carlsbecker A, Lee JY, Roberts CJ, Dettmer J, Lehesranta S, Zhou J, Lindgren O, Moreno-Risueno MA, Vaten A, Thitamadee S, Campilho A, Sebastian J, Bowman JL, Helariutta Y, Benfey PN (2010) Cell signalling by micro RNA165/6 directs gene dose-dependent root cell fate. Nature 465:316–321

Chen JG, Ullah H, Young JC, Sussman MR, Jones A (2001) ABP1 is required for organized cell elongation and division in *Arabidopsis* embryogenesis. Genes Dev 15:902–911

Chen LQ, Hou BH, Lalonde S, Takanaga H, Hartung ML, Qu XQ, Guo WJ, Kim JG, Underwood W, Chaudhuri B, Chermak D, Antony G, White FF, Somerville SC, Mudgett MB, Frommer WB (2010) Sugar transporters for intercellular exchange and nutrition of pathogens. Nature 468:527–532

Chen LQ, Qu XQ, Hou BH, Sosso D, Osorio S, Fernie AR, Frommer WB (2012) Sucrose efflux mediated by SWEET proteins as a key step for phloem transport. Science 335:207–211

Chen J, Pang W, Chen B, Zhang C, Piao Z (2015) Transcriptome analysis of *Brassica rapa* near-isogenic lines carrying clubroot-resistant and-susceptible alleles in response to *Plasmodiophora brassicae* during early infection. Front Plant Sci 6:1183

Christian M, Steffens B, Schenck D, Burmester S, Böttger M, Luthen H (2006) How does auxin enhance cell elongation? Roles of auxin binding proteins and potassium channels in growth control. Plant Biol 8:346–352

Chuck G, Candela H, Hake S (2009) Big impacts by small RNAs in plant development. Curr Opin Plant Biol 12:81–86

Colhoun J (1958) Club root disease of crucifers caused by *Plasmodiophora brassicae* Woron (Commonwealth Mycological Institute: Kew, UK)

de Mendoza A, Sebé-Pedrós A, Ruiz-Trillo I (2014) The evolution of the GPCR signaling system in eukaryotes: modularity, conservation, and the transition to metazoan multicellularity. Genome Biol Evol 6(3):606–619

Dekhuijzen HM (1979) Electron microscopic studies on the root hairs and cortex of a susceptible and a resistant variety of *Brassica campestris* infected with *Plasmodiophora brassicae*. Neth J Plant Pathol 86:1–17

Dekhuijzen HM (1980) The occurrence of free and bound cytokinins in clubroots and *Plasmodiophora brassicae* infected turnip tissue cultures. Physiol Plant 49:169–176

Dekhuijzen HM (1981) The occurrence of free and bound cytokinins in plasmodia of *Plasmodiophora brassicae* isolated from tissue cultures of clubroots. Plant Cell Rep 1:18–20

Dekhuijzen HM, Overeem JC (1971) The role of cytokinins in clubroot formatiosn. Physiol Plant Pathol 1:151–161

Deora A, Gossen BD, McDonald MR (2012) Infection and development of *Plasmodiophora brassicae* in resistant and susceptible canola cultivars. Can J Plant Pathol 34:239–247

Deora A, Gossen BD, McDonald MR (2013) Cytology of infection, development, and expression of resistance to *Plasmodiophora brassicae* in canola. Ann Appl Biol 163:56–71

Devos S, Prinsen E (2006) Plant hormones: a key in clubroot development. Commun Agric Appl Biol Sci 71:869–872

Devos S, Vissenberg K, Verbelen JP, Prinsen E (2005) Infection of Chinese cabbage by *Plasmodiophora brassicae* leads to a stimulation of plant growth: impacts on cell wall metabolism and hormone balance. New Phytol 166:241–250

Devos S, Laukens K, Deckers P, Straeten DV, Beeckman T, Inze D, Onckelen H, Witters E, Prinsen E (2006) A hormone and proteome approach to picturing the initial metabolic events during *Plasmodiophora brassicae* infection on *Arabidopsis*. Mol Plant-Microbe Interact 19 (12):1431–1443

Dharmasiri N, Dharmasiri S, Weijers D, Lechner E, Yamada M, Hobbie L, Ehrismann JS, Jurgens G, Estelle M (2005) Plant development is regulated by a family of auxin receptor F box proteins. Dev Cell 9:109–119

Diatchenko L, Lau YF, Campbell AP, Chenchik A, Moqadam F, Huang B, Lukyanov S, Lukyanov K, Gurskaya N, Sverdlov ED, Siebert PD (1996) Suppression subtractive hybridization: a method for generating differentially regulated or tissue-specific cDNA probes and libraries. Proc Natl Acad Sci U S A 93:6025–6030

Diederichsen E, Frauen M, Linders E, Hatakeyama K, Hirai M (2009) Status and perspectives of clubroot resistance breeding in crucifer crops. J Plant Growth Regul 28:265–281

Diederichsen E, Frauen M, Ludwig-Muller J (2014) Clubroot disease management challenges from a German perspective. Can J Plant Pathol 36:85–98

Doan TT, Jaschke D, Ludwig-Muller J (2010) An endophytic fungus induces tolerance against the clubroot pathogen *Plasmodiophora brassicae* in *Arabidopsis thaliana* and *Brassica rapa* roots. Acta Hortic 867:173–180

Doidy J, Grace E, Kuhn C, Simonplas F, Casieri L, Wipf D (2012) Sugar transporters in plants and in their interactions with fungi. Trends Plant Sci 17:413–422

Donald EC, Jaudzems G, Porter IJ (2008) Pathology of cortical invasion by *Plasmodiophora brassicae* in clubroot resistant and susceptible *Brassica oleracea* hosts. Plant Pathol 57:201–209

Dow JM, Clarke BR, Milligan DE, Tang JL, Daniels MJ (1990) Extracellular proteases from *Xanthomonas campestris* pv. *campestris*, the black rot pathogen. Appl Environ Microbiol 56:2994–2998

Dunoyer P, Himber C, Voinnet O (2015) Induction, suppression and requirement of RNA silencing pathways in virulent *Agrobacterium tumefaciens* infections. Nat Genet 47:847

Eastmond PJ, van Dijken AJH, Spielman M, Kerr A, Tissier AF, Dickinson HG, Jones JDG, Smeekens SC, Graham IA (2002) Trehalose-6-phosphate synthase 1, which catalyses the first step in trehalose synthesis, is essential for *Arabidopsis* embryo maturation. Plant J 29:225–235

Ehness R, Roitsch T (1997) Co-ordinated induction of mRNAs for extracellular invertase and a glucose transporter in *Chenopodium rubrum* by cytokinins. Plant J 11:539–548

Evans JL, Scholes JD (1995) How does clubroot alter the regulation of carbon metabolism in its host? Asp Appl Biol 42:125–132

Feller A, Machemer K, Braun EL, Grotewold E (2011) Evolutionary and comparative analysis of MYB and bHLH plant transcription factors. Plant J 66:94–116

Feng J, Hwang R, Hwang SF, Strelkov SE, Gossen BD, Zhou QX, Peng G (2010) Molecular characterization of a serine protease *Pro1* from *Plasmodiophora brassicae* that stimulates resting spore germination. Mol Plant Pathol 11(4):503–512

Feng J, Hwang SF, Strelkov SE (2012a) Analysis of expressed sequence tags derived from a compatible *Plasmodiophora brassicae*-canola interaction. Can J Plant Pathol 34:562–574

Feng J, Xiao Q, Hwang SF, Strelkov SE, Gossen BD (2012b) Infection of canola by secondary zoospores of *Plasmodiophora brassicae* produced on a nonhost. Eur J Plant Pathol 132:309–315

Feng J, Hwang SF, Strelkov SE (2013) Studies into primary and secondary infection processes by *Plasmodiophora brassicae* on canola. Plant Pathol 62:177–183

Fritzius T, Aeschbacher R, Wiemken A, Wingler A (2001) Induction of ApL3 expression by trehalose complements the starch-deficient *Arabidopsis* mutant adg2-1 lacking ApL1, the large subunit of ADP glucose pyrophosphorylase. Plant Physiol 126:883–889

Fuchs LS, Fuchs D, Compton DL, Powell SR, Seethaler PM, Capizzi AM, Schatschneider C, Fletcher JM (2006) The cognitive correlates of third-grade skill in arithmetic, algorithmic computation, and arithmetic word problems. J Educ Psychol 98:29–43

Gludovacz TV (2013) Clubroot development on Canola and Cabbage in relation to soil temperature and resistance. University of Guelph, Guelph

Gludovacz TV, Gossen BD, McDonald MR (2013) Evidence for tolerance to *Plasmodiophora brassicae* in a cabbage cultivar. Can J Plant Pathol 35:111

Gludovacz TV, Deora A, McDonald MR, Gossen BD (2014) Cortical colonization by *Plasmodiophora brassicae* in susceptible and resistant cabbage cultivars. Eur J Plant Pathol 140(4):859–862

Goddijn O, Smeekens S (1998) Sensing trehalose biosynthesis in plants. Plant J 14:143–146

Graf H, Faehling M, Siemens J (2004) Chromosome polymorphism of the obligate biotrophic parasite *Plasmodiophora brassicae*. J Phytopathol 152:86–91

Gravot A, Gautier R, Lime T, Lemarié S, Jubault M, Lariagon C, Lemoine J, Vicente J, Robert-Seilaniantz A, Holdsworth MJ, Manzanares-Dauleux MJ (2016) Hypoxia response in *Arabidopsis* roots infected by *Plasmodiophora brassicae* supports the development of clubroot. BMC Plant Biol 16:251

Grsic S, Sauerteig S, Neuhaus K, Albrecht M, Rossiter J, Ludwig-Müller J (1998) Physiological analysis of transgenic *Arabidopsis thaliana* plants expressing one nitrilase isoform in sense or antisense direction. J Plant Physiol 153:446–456

Grsic S, Kirchheim B, Pieper K, Fritsch M, Hilgenberg W, Ludwig-Muller J (1999) Induction of auxin biosynthetic enzymes by jasmonic acid and in clubroot diseased Chinese cabbage plants. Physiol Plant 105:521–531

Grsic-Rausch S, Kobelt P, Siemens JM, Bischoff M, Ludwig-Muller J (2000) Expression and localization of nitrilase during symptom development of the clubroot disease in *Arabidopsis*. Plant Physiol 122:369–378

Guilfoyle TJ, Hagen G (2007) Auxin response factors. Curr Opin Plant Biol 10:453–460

Guo XH (2001) Biological characteristics and pathogenesis of *Plasmodiophora brassicae*. Southwest Agriculture University, Chongqing

Guo HS, Xie Q, Fei JF, Chua NH (2005) MicroRNA directs mRNA cleavage of the transcription factor NAC1 to down regulate auxin signals for *Arabidopsis* lateral root development. Plant Cell 17:1376–1386

Gustafsson M, Falt AS (1986) Genetic studies on resistance to clubroot in *Brassica napus*. Ann Appl Biol 108:409–415

Hagen G, Guilfoyle T (2002) Auxin-responsive gene expression: genes, promoters and regulatory factors. Plant Mol Biol 49(3-4):373–385

Hansen CE, Flengsrud R, Kopperud C (1994) Two-dimensional gel electrophoresis of proteins in healthy roots and clubroots of *Brassica oleracea*. Acta Agric Soil Plant Sci 44:123–128

Helmlinger J, Rausch T, Hilgenberg W (1983) Localization of newly synthesized indole-3-methyl-glucosinolate (=glucobrassicin) in vacuoles from horse radish (*Armoracia rusticana*). Physiol Plant 58:302–310

Hilger D, Masureel M, Kobilka BK (2018) Structure and dynamics of GPCR signaling complexes. Nat Struct Mol Biol 25(1):4–12

Huang Y, Ma SQ, Li XQ, Wang J, Hu XL (2007) Morphology of *Plasmodiophora brassicae* and biological characteristic of the pathogenic resting spores in rapeseed. Sci Agric Sin 40:1388–1394

Hwang SF, Ahmed HU, Strelkov SE, Gossen BD, Turnbull GD, Peng G, Howard RJ (2011a) Seedling age and inoculum density affect clubroot severity and seed yield in canola. Can J Plant Sci 91:183–190

Hwang SF, Ahmed HU, Zhou Q, Strelkov SE, Gossen BD, Peng G, Turnbull GD (2011b) Influence of cultivar resistance and inoculum concentration on root hair infection of canola (*Brassica napus*) by *Plasmodiophora brassicae*. Plant Pathol 60:820–829

Hwang SF, Strelkov SE, Gossen BD, Turnbull GD, Ahmed HU, Manolii V, Howard RJ (2011c) Soil treatments and amendments for amelioration of clubroot of canola. Can J Plant Sci 91:999–1010

Hwang SF, Ahmed HU, Zhou Q, Strelkov SE, Gossen BD, Peng G, Turnbull GD (2012) Assessment of the impact of resistant and susceptible canola on *Plasmodiophora brassicae* inoculum potential. Plant Pathol 61:945–952

Ingram D, Tommerup I (1972) The life history of *Plasmodiophora brassicae* Woron. Proc R Soc Lond B 180:103–112

Irani S, Trost B, Waldner M, Nayidu N, Tu J, Kusalik AJ, Todd CD, Wei Y, Bonham-Smith PC (2018) Transcriptome analysis of response to *Plasmodiophora brassicae* infection in the *Arabidopsis* shoot and root. BMC Genomics 19:23

Ito S, Tanaka S, Miyanaga C, Takabayashi S, Yano S, Kameya-Iwaki M (1996) Constitutive and inducible proteins in the root of Chinese cabbage infected with *Plasmodiophora brassicae*. J Phytopathol 144:89–95

Ito S, Ichinose H, Yanagi C, Tanaka S, Kameya-Iwaki M, Kishi F (1999) Identification of an in planta-induced mRNA of *Plasmodiophora brassicae*. J Phytopathol 147:79–82

Jahn L, Mucha S, Bergmann S, Horn C, Staswick P, Steffens B, Siemens J, Ludwig-Muller J (2013) The clubroot pathogen (*Plasmodiophora brassicae*) influences auxin signaling to regulate auxin homeostasis in *Arabidopsis*. Plant Theory 2:726–749

Jaschke D, Dugassa-Gobena D, Karlovsky P, Vidal S, Ludwig-Müller J (2010) Suppression of clubroot development in *Arabidopsis thaliana* by the endophytic fungus *Acremonium alternatum*. Plant Pathol 59:100–111

Ji R, Zhao L, Xing M, Shen X, Bi Q, Peng S, Feng H (2014) Infection of *Plasmodiophora brassicae* in Chinese cabbage. Genet Mol Res 13(4):10976–10982

Jiang M, Dong X, Lang H, Pang W, Zhan Z, Li X, Piao Z (2018) Mining of *Brassica*-specific genes (BSGs) and their induction in different developmental stages and under *Plasmodiophora brassicae* stress in *Brassica rapa*. Int J Mol Sci 19:2064

Jones JD, Dangl JL (2006) The plant immune system. Nature 444:323–329

Jubault M, Lariagon C, Taconnat L, Renou JP, Gravot A, Delourme R, Manzanares-Dauleux MJ (2013) Partial resistance to clubroot in *Arabidopsis* is based on changes in the host primary metabolism and targeted cell division and expansion capacity. Funct Integr Genomics 13:191–205

Kageyama K, Asano T (2009) Life cycle of *Plasmodiophora brassicae*. J Plant Growth Regul 28:203–211

Kamp ME, Liu Y, Kortholt A (2016) Function and regulation of heterotrimeric G proteins during chemotaxis. Int J Mol Sci 17(1):90

Kavanagh JA, Williams PH (1981) Indole auxins in *Plasmodiophora* infected cabbage roots and hypocotyls. Trans Br Mycol Soc 77:125–129

Keen NT, Williams PH (1969) Translocation of sugars into infected cabbage tissues during clubroot development. Plant Physiol 44:748–754

Knaust A, Ludwig-Müller J (2013) The ethylene signalling pathway is needed to restrict root gall growth in *Arabidopsis* after infection with the obligate biotrophic protist *Plasmodiophora brassicae*. J Plant Growth Regul 32:9–21

Kobelt P, Siemens J, Sacristan MD (2000) Histological characterisation of the incompatible interaction between *Arabidopsis thaliana* and the obligate biotrophic pathogen *Plasmodiophora brassicae*. Mycol Res 104:220–225

Lelandais-Briere C, Sorin C, Declerck M, Benslimane A, Crespi M, Hartmann C (2010) Small RNA diversity in plants and its impact in development. Curr Genomics 11:14–23

Leman S, House L, Hoegh A (2015) Developing a new interdisciplinary computational analytics undergraduate program: a qualitative-quantitative-qualitative approach. The Am Stat 69 (4):397–408

Lemarie S, Robert-Seilaniantz A, Lariagon C, Lemoine J, Marnet N, Jubault M, Manzanares-Dauleux MJ, Gravot A (2015) Both the jasmonic acid and the salicylic acid pathways contribute to resistance to the biotrophic clubroot agent *Plasmodiophora brassicae* in *Arabidopsis*. Plant Cell Physiol 56:2158–2168

Lemoine R, Camera SL, Atanassova R, Dedaldechamp F, Allario T, Pourtau N, Bonnemain JL, Laloi M, Coutos-Thevenot P, Maurousset L, Faucher M, Girousse C, Lemonnier P, Parrilla J, Durand M (2013) Source-to-sink transport of sugar and regulation by environmental factors. Front Plant Sci 4:272

Li Y, Zhang Q, Zhang J, Wu L, Qi Y, Zhou JM (2010) Identification of microRNAs involved in pathogen-associated molecular pattern-triggered plant innate immunity. Plant Physiol 152:2222–2231

Li H, Li X, Xuan Y, Jiang J, Wei Y, Piao Z (2018) Genome wide identification and expression profiling of SWEET genes family reveals its role during *Plasmodiophora brassicae*-induced formation of clubroot in *Brassica rapa*. Front Plant Sci 9:207

Liu D, Wang P, Zhu W, Gu X, Zhou W, Zhou Z (2008) Enantio selective degradation of fipronil in Chinese cabbage (*Brassica pekinensis*). Food Chem 110:399–405

Lovelock DA, Sola I, Marschollek S, Donald CE, Rusak G, van Pee KH et al (2016) Analysis of salicylic acid-dependent pathways in *Arabidopsis thaliana* following infection with *Plasmodiophora brassicae* and the influence of salicylic acid on disease. Mol Plant Pathol 17:1237–1251

Ludwig-Muller J (2009a) Glucosinolates and the clubroot disease: defense compounds or auxin precursors? Phytol Chem Rev 8:135–148

Ludwig-Muller J (2009b) Plant defence – what can we learn from clubroots? Austr Plant Pathol 38:318–324

Ludwig-Muller J, Hilgenberg W (1988) A plasma membrane-bound enzyme oxidizes l-tryptophan to indole-3-acetaldoxime. Physiol Plant 74:240–250

Ludwig-Muller J, Schuller A (2008) What can we learn from clubroots: alterations in host roots and hormone homeostasis caused by *Plasmodiophora brassicae*. Eur J Plant Pathol 121:291–302

Ludwig-Muller J, Bendel U, Thermann P, Ruppel M, Epstein E, Hilgenberg W (1993) Concentrations of indole-3-acetic acid in plants of tolerant and susceptible varieties of Chinese cabbage infected with *Plasmodiophora brassicae* Woron. New Phytol 125(4):763–769

Ludwig-Muller J, Thermann P, Pieper K, Hilgenberg W (1994) Peroxidase and chitinase isoenzyme activities during root infection of Chinese cabbage with *Plasmodiophora brassicae* Woron. Physiol Plant 90:661–670

Ludwig-Muller J, Kasperczyk N, Schubert B, Hilgenberg W (1995) Identification of salicylic acid in Chinese cabbage and its possible role during root infection with *Plasmodiophora brassicae*. Curr Top Phytol Chem 14:39–45

Ludwig-Muller J, Epstein E, Hilgenberg W (1996) Auxin-conjugate hydrolysis in Chinese cabbage: characterization of an amidohydrolase and its role during the clubroot disease. Physiol Plant 97:627–634

Ludwig-Muller J, Schubert B, Pieper K, Ihmig S, Hilgenberg W (1997) Glucosinolate content in susceptible and tolerant Chinese cabbage varieties during the development of the clubroot disease. Phyochemistry 44:407–414

Ludwig-Muller J, Pieper K, Ruppel M, Cohen JD, Epstein E, Kiddle G, Bennett R (1999) Indole glucosinolate and auxin biosynthesis in *Arabidopsis thaliana* L. glucosinolate mutants and the development of the clubroot disease. Planta 208:409–419

Ludwig-Muller J, Prinsen E, Rolfe SA, Scholes JD (2009) Metabolism and plant hormone action during clubroot disease. J Plant Growth Regul 28:229–244

Ludwig-Muller J, Julke S, Geiß K, Richter F, Mithofer A, Sola I, Rusak G, Keenan S, Bulman S (2015) A novel methyltransferase from the intracellular pathogen *Plasmodiophora brassicae* methylates salicylic acid. Mol Plant Pathol 16(4):349–364

Malinowski R, Smith JA, Fleming AJ, Scholes JD, Rolfe SA (2012) Gall formation in clubroot-infected *Arabidopsis* results from an increase in existing meristematic activities of the host but is not essential for the completion of the pathogen life cycle. The Plant J 71(2):226–238

Malinowski R, Novak O, Borhan MH, Spıchal L, Strnad M, Rolfe SA (2016) The role of cytokinins in clubroot disease. Eur J Plant Pathol 145:543–557

Mallory AC, Bartel DP, Bartel B (2005) MicroRNA-directed regulation of Arabidopsis AUXIN RESPONSE FACTOR17 is essential for proper development and modulates expression of early auxin response genes. Plant Cell 17:1360–1375

Manck-Götzenberger J, Requena N (2016) *Arbuscular mycorrhiza* symbiosis induces a major transcriptional reprogramming of the potato *SWEET* sugar transporter family. Front Plant Sci 7:487

McDonald MR, Sharma K, Gossen BD, Deora A, Feng J, Hwang SF (2014) The role of primary and secondary infection in host response to *Plasmodiophora brassicae*. Phytopathology 104:1078–1087

Muller P, Hilgenberg W (1986) Isomers of zeatin and zeatin riboside in clubroot tissue: evidence for trans-zeatin biosynthesis by *Plasmodiophora brassicae*. Physiol Plant 66(2):242–250

Muller J, Xie ZP, Staehelin C, Mellor RB, Boller T, Wiemken A (1994) Trehalose and trehalase in root nodules from various legumes. Physiol Plant 90:86–92

Muller J, Boller T, Wiemken A (1995a) Trehalose and trehalase in plants: recent developments. Plant Sci 112:1–9

Muller J, Boller T, Wiemken A (1995b) Effects of validamycin A, a potent trehalase inhibitor, and phytohormones on trehalose metabolism in roots and root nodules of soybean and cowpea. Planta 197:362–368

Muller J, Boller T, Wiemken A (1998) Trehalose affects sucrose synthase and invertase activities in soybean [*Glycine max* (L.) Merr.] roots. J Plant Physiol 153:255–257

Muller J, Wiemken A, Aeschbacher R (1999) Trehalose metabolism in sugar sensing and plant development. Plant Sci 147:37–47

Muller J, Aeschbacher RA, Sprenger N, Boller T, Wiemken A (2000) Disaccharide-mediated regulation of sucrose: fructan-6-fructosyl transferase, a key enzyme of fructan synthesis in barley leaves. Plant Physiol 123:265–273

Muller J, Aeschbacher RA, Wingler A, Boller T, Wiemken A (2001) Trehalose and trehalase in *Arabidopsis thaliana*. Plant Physiol 125:1086–1093

Navarro L, Dunoyer P, Jay F, Arnold B, Dharmasiri N, Estelle M, Voinnet O, Jones JD (2006) A plant miRNA contributes to antibacterial resistance by representing auxin signaling. Science 312:436–439

Neuhaus K, Grsic-Rausch S, Sauerteig S, Ludwig-Muller J (2000) *Arabidopsis* plants transformed with nitrilase 1 or 2 in antisense direction are delayed in clubroot development. J Plant Physiol 156(5/6):756–761

Normanly J, Grisafi P, Fink GR, Bartel B (1997) *Arabidopsis* mutants resistant to the auxin effects of indole-3-acetonitrile are defective in the nitrilase encoded by the *NIT1* gene. Plant Cell 9:1781–1790

Ori N, Cohen AR, Etzioni A, Brand A, Yanai O, Shleizer S, Menda N, Amsellem Z, Efroni I, Pekker I, Alvarez JP, Blum E, Zamir D, Eshed Y (2007) Regulation of LANCEOLATE by miR319 is required for compound-leaf development in tomato. Nat Genet 39:787–791

Ozsolak F, Milos PM (2011) RNA sequencing: advances, challenges and opportunities. Nat Rev Genet 12:87–98

Padmanabhan C, Zhang X, Jin H (2009) Host small RNAs are big contributors to plant innate immunity. Curr Opin Plant Biol 12:465–472

Park SW, Kaimoyo E, Kumar D, Mosher S, Klessig DF (2007) Methyl salicylate is a critical mobile signal for plant systemic acquired resistance. Science 318:113–116

Pasold S, Ludwig-Muller J (2013) Reduction of clubroot (*Plasmodiophora brassicae*) formation in *Arabidopsis thaliana* after treatment with prohexadione calcium, an inhibitor of oxoglutaric acid-dependent dioxygenases. Plant Pathol 62:1357–1365

Pasold S, Siegel I, Seidel C, Ludwig-Muller J (2010) Flavonoid accumulation in *Arabidopsis thaliana* root galls caused by the obligate biotrophic pathogen *Plasmodiophora brassicae*. Mol Plant Pathol 11:545–562

Patrick JW (1989) Solute efflux from the host at plant-microorganism interfaces. Funct Plant Biol 16:53–67

Pérez-Lópeza E, Waldnerb M, Hossaina M, Kusalikb AJ, Weia Y, Bonham-Smitha PC, Todd CD (2018) Identification of *Plasmodiophora brassicae* effectors—a challenging goal. Virulence 9 (1):1344–1353

Piotrowski M (2008) Primary or secondary? Versatile nitrilases in plant metabolism. Phytochemistry 69:2655–2667

Prerostova S, Dobrev PI, Konradyova V, Knirsch V, Gaudinova A, Kramna B, Kazda J, Ludwig-Muller J, Vankova R (2018) Hormonal responses to *Plasmodiophora brassicae* infection in *Brassica napus* cultivars differing in their pathogen resistance. Int J Mol Sci 19:4024

Puzio PS, Newe M, Grymaszewska G, Ludwig-Muller J, Grundler FMW (2000) *Plasmodiophora brassicae*-induced expression of *pyk20*, an *Arabidopsis thaliana* gene with glutamine-rich domain. Physiol Mol Plant Pathol 56:79–84

Qiu BL, Harvey JA, Raaijmakers CE, Vet LEM, van Dam NM (2009) Nonlinear effects of plant root and shoot jasmonic acid application on the performance of *Pieris brassicae* and its parasitoid *Cotesia glomerata*. Funct Ecol 23:496–505

Rausch T, Butcher DN, Hilgenberg W (1981) Nitrilase activity in clubroot diseased plants. Physiol Plant 52:467–470

Rausch T, Butcher DN, Hilgenberg W (1983) Indole-3-methylglucosinolate biosynthesis and metabolism in clubroot diseased plants. Physiol Plant 58:93–100

Rawlings ND, Barrett AJ (1993) Evolutionary families of peptidases. Biochem J 290:205–218

Reed JW (2001) Roles and activities of Aux/IAA proteins in *Arabidopsis*. Trends Plant Sci 6:420–425

Reinders A, Sivitz AB, Starker CG, Gantt J, Stephen W, John M (2008) Functional analysis of LjSUT4 a vacuolar sucrose transporter from *Lotus japonicus*. Plant Mol Biol 68:289–299

Reinders A, Sivitz AB, Ward JM (2011) Evolution of plant sucrose uptake transporters. Front Plant Sci 3:22

Rojas CM, Senthil-Kumar M, Tzin V, Mysore KS (2014) Regulation of primary plant metabolism during plant-pathogen interactions and its contribution to plant defense. Front Plant Sci 5:17

Rolfe SA, Strelkov SE, Links MG, Clarke WE, Robinson SJ, Djavaheri M, Malinowski R, Haddadi P, Kagale S, Parkin IAP, Taheri A, Borhan MH (2016) The compact genome of the plant pathogen *Plasmodiophora brassicae* is adapted to intracellular interactions with host *Brassica* spp. BMC Genomics 17:272

Roman RV, Rathjen JP (2017) Apoplastic sugar extraction and quantification from wheat leaves infected with biotrophic fungi. Methods Mol Biol 1659:125–134

Rosenbaum DM, Rasmussen SG, Kobilka BK (2009) The structure and function of G-protein-coupled receptors. Nature 459(7245):356–363

Santner A, Estelle M (2009) Recent advances and emerging trends in plant hormone signalling. Nature 459(7250):1071–1078

Schommer C, Debernardi JM, Bresso EG, Rodriguez RE, Palatnik JF (2014) Repression of cell proliferation by miR319-regulated TCP4. Mol Plant 7:1533–1544

Schoof H, Zaccaria P, Gundlach H, Lemcke K, Rudd S, Kolesov G, Arnold R, Mewes HW, Mayer KFX (2002) MIPS *Arabidopsis thaliana* database (MAtDB): an integrated biological knowledge resource based on the first complete plant genome. Nucleic Acids Res 30:91–93

Schubert A, Wyss P (1995) Trehalase activity in mycorrhizal and non-mycorrhizal roots of leek and soybean. Mycorrhiza 5:401–404

Schuller A, Ludwig-Müller J (2002) Isolation of differentially expressed genes involved in clubroot disease. Plant Prot Sci 38(2):483–486

Schuller A, Ludwig-Muller J (2006) A family of auxin conjugate hydrolases from Brassica rapa: characterization and expression during clubroot disease. New Phytol 171:145–158

Schuller A, Kehr J, Ludwig-Muller J (2014) Laser micro-dissection coupled to transcriptional profiling of *Arabidopsis* roots inoculated by *Plasmodiophora brassicae* indicates a role for brassinosteroids in clubroot formation. Plant Cell Physiol 55:392–411

Schwelm A, Fogelqvist J, Knaust A, Julke S, Lilja T, Bonilla-Rosso G, Karlsson M, Shevchenko A, Dhandapani V, Choi SR, Kim HG, Park JY, Lim YP, Ludwig-Muller J, Dixelius C (2015) The *Plasmodiophora brassicae* genome reveals insights in its life cycle and ancestry of chitin synthases. Sci Rep 5:11153

Searle LM, Chamberlain K, Rausch T, Butcher DN (1982) The conversion of 3-indole-methylgluco- sinolate to 3-indoleacetonitrile by myrosinase and its relevance to the clubroot disease of the cruciferae. J Exp Bot 33:935–942

Sharma K, Gossen BD, McDonald MR (2011) Effect of temperature on cortical infection by *Plasmodiophora brassicae* and clubroot severity. Phytopathology 101:1424–1432

Siemens J, Nagel M, Ludwig-Muller J, Sacristan MD (2002) The interaction of *Plasmodiophora brassicae* and *Arabidopsis thaliana*: parameters for disease quantification and screening of mutant lines. J Phytopathol 150:592–605

Siemens J, Keller I, Sarx J, Kunz S, Schuller A, Nagel W, Schmulling T, Parniske M, Ludwig-Muller J (2006) Transcriptome analysis of Arabidopsis clubroots and disease resistance of cytokinin oxidase/ dehydrogenase gene overexpressing plants indicate a key role for cytokinin in disease development. Mol Plant-Microbe Interact 19:480–494

Siemens J, Glawischnig E, Ludwig-Müller J (2008) Indole glucosinolates and camalexin do not influence the development of the clubroot disease in *Arabidopsis thaliana*. J Phytopathol 156:332–337

Siemens J, Graf H, Bulman S, In O, Ludwig-Muller J (2009) Monitoring expression of selected *Plasmodiophora brassicae* genes during clubroot development in *Arabidopsis thaliana*. Plant Pathol 58:130–136

Siemens J, González MC, Wolf S, Hofmann C, Greiner S, Du Y, Rausch T, Roitsch T, Ludwig-Muller J (2011) Extracellular invertase is involved in the regulation of clubroot disease in *Arabidopsis thaliana*. Mol Plant Pathol 12(3):247–262

Singh K, Tzelepis G, Zouhar M, Rysanek P, Dixelius C (2018) The immunophilin repertoire of *Plasmodiophora brassicae* and functional analysis of *PbCYP3* cyclophilin. Mol Genet Genomics 293(2):381–390

Song QX, Liu YF, Hu XY, Zhang WK, Ma B, Chen SY, Zhang JS (2011) Identification of miRNAs and their target genes in developing soybean seeds by deep sequencing. BMC Plant Biol 11:5

Spartz AK, Gray WM (2008) Plant hormone receptors: new perceptions. Genes Dev 22:2139–2148

Staswick PE, Tiryaki I (2004) The oxylipin signal jasmonic acid is activated by an enzyme that conjugates it to isoleucine in *Arabidopsis*. Plant Cell 16:2117–2127

Staswick PE, Serban B, Rowe M, Tiryaki I, Maldonado MT, Maldonado MC, Suza W (2005) Characterization of an *Arabidopsis* enzyme family that conjugates amino acids to indole-3-acetic acid. Plant Cell 17:616–627

Sundelin T, Jensen DF, Lübeck M (2011) Identification of expressed genes during infection of Chinese cabbage (*Brassica rapa* subsp *pekinensis*) by *Plasmodiophora brassicae*. J Eukaryot Microbiol 58(4):310–314

Takahashi H, Takita K, Kishimoto T, Mitsui T, Hori H (2002) Ca2+ is required by clubroot resistant turnip cells for transient increases in PAL activity that follow inoculating with *Plasmodiophora brassicae*. J Phytopathol 150:529–535

Takahashi H, Ishikawa T, Kaido M, Takita K, Tayakawa T, Okazaki K, Itoh K, Mitsui T, Hori H (2006) *Plasmodiophora brassicae*-induced cell death and medium alkalization in clubroot-resistant cultured roots of *Brassica rapa*. J Phytopathol 154:156–162

Tanaka S, Mizui Y, Terasaki H, Ito S (2006a) Distribution of clubroot disease of a cruciferous weed, *Cardamine flexuosa*, in major isolated islands, Hokkaido and Okinawa, in Japan. Mycoscience 47:72–77

Tanaka S, Mido H, Ito S (2006b) Colonization by two isolates of *Plasmodiophora brassicae* with differing pathogenicity on a clubroot resistant cultivar of Chinese cabbage (*Brassica rapa* L. subsp. *pekinensis*). J Gen Plant Pathol 72:205–209

Toyofuku K, Kasahara M, Yamaguchi J (2000) Characterization and expression of monosaccharide transporters (*OsMSTs*) in rice. Plant Cell Physiol 41:940–947

van den Hoogen DJ, Meijer HJG, Seidl MF, Govers F (2018) The ancient link between Gprotein-coupled receptors and C-terminal phospholipid kinase domains. MBio 9(1):e02119–e02117

Velculescu VE, Zhang L, Volgelstein B, Kinzler KW (1995) Serial analysis of gene expression. Science 5235:484–487

Verma SS, Rahman MH, Deyholos MK, Basu U, Kav NN (2014) Differential expression of miRNAs in *Brassica napus* root following infection with *Plasmodiophora brassicae*. PLoS ONE 9:e86648

Vogel G, Aeschbacher RA, Müller J, Boller T, Wiemken A (1998) Trehalose-6-phosphate phosphatases from *Arabidopsis thaliana*: Identification by functional complementation of the yeast tps2 mutant. Plant J 13:673–683

Vogel G, Fiehn O, Jean-Richard-dit-Bressel L, Boller T, Wiemken A, Aeschbacher RA, Wingler A (2001) Trehalose metabolism in *Arabidopsis*: occurrence of trehalose and molecular cloning and characterization of trehalose-6-phosphate synthase homologues. J Exp Bot 52:1817–1826

Wagner W, Wiemken A, Matile P (1986) Regulation of fructan metabolism in leaves of barley (*Hordeum vulgare* L. cv. *Gerbel*). Plant Physiol 81:444–447

Wagner G, Charton S, Lariagon C, Laperche A, Lugan R, Hopkins J (2012) Metabotyping: a new approach to investigate rapeseed (*Brassica napus* L.) genetic diversity in the metabolic response to clubroot infection. Mol Plant-Microbe Interact 25:1478–1491

Wasinger VC, Cordwell SJ, Cerpa-Poljak A, Yan JX, Gooley AA, Wilkins MR, Duncan MW, Harris R, Williams KL, Humphery-Smith I (1995) Progress with gene-product mapping of the Mollicutes: *Mycoplasma genitalium*. Electrophoresis 16:1090–1094

Williams PH, Keen NT, Strandberg JO, McNabola SS (1968) Metabolite synthesis and degradation during clubroot development in cabbage hypocotyls. Phytopathology 58:921–928

Wingler A, Fritzius T, Wiemken A, Boller T, Aeschbacher RA (2000) Trehalose induces the ADP-glucose pyrophosphorylase gene, ApL3, and starch synthesis in *Arabidopsis*. Plant Physiol 124:105–114

Wu L, Siemens J, Li S, Ludwig-Muller J, Gong Y, Zhong L, He J (2012) Estimating *Plasmodiophora brassicae* gene expression in lines of *B. rapa* by RT-PCR. Sci Hortic 133:1–5

Wu L, Liu D, Wu J, Zhang R, Qin Z, Li A, Fu D, Zhai W, Mao L (2013) Regulation of *flowering locus T* by a microRNA in *Brachypodium distachyon*. Plant Cell 25:4363–4377

Xiaochun W, Xu W, Yuan Y, Yao Q, Zhao Y, Wang Z, Jiang W, Zhang X (2016) Genome-wide investigation of microRNAs and their targets in *Brassica rapa* ssp. *pekinensis* root with *Plasmodiophora brassicae* infection. Hortic Plant J. https://doi.org/10.1016/j.hpj.2016.11.004

Xu MY, Dong Y, Zhang QX, Zhang L, Luo YZ, Sun J, Fan YL, Wang L (2012) Identification of miRNAs and their targets from *Brassica napus* by high-throughput sequencing and degradome analysis. BMC Genomics 13:421

Xu L, Ren L, Chen K, Liu F, Fang X (2016) Putative role of IAA during the early response of *Brassica napus* L. to *Plasmodiophora brassicae*. Eur J Plant Pathol 145:601–613

Yang C, Li D, Mao D, Liu X, Ji C, Li X, Zhao X, Cheng Z, Chen C, Zhu L (2013) Over expression of microRNA319 impacts leaf morphogenesis and leads to enhanced cold tolerance in rice (*Oryza sativa* L.). Plant Cell Environ 36:2207–2218

Yousef GM, Kopolovic AD, Elliott MB, Diamandis EP (2003) Genomic overview of serine proteases. Biochem Biophys Res Commun 305:28–36

Zhang Z, Li Q, Li Z, Staswick PE, Wang M, Zhu Y, He Z (2007) Dual regulation role of GH3.5 in salicylic acid and auxin signaling during *Arabidopsis–Pseudomonas syringae* interaction. Plant Physiol 145:450–464

Zhang Y, Wang W, Chen J, Liu J, Xia M, Shen F (2015) Identification of miRNAs and their targets in cotton inoculated with *Verticillium dahliae* by high-throughput sequencing and degradome analysis. Int J Mol Sci 16:14749–14768

Zhao Y, Bi K, Gao Z, Chen T, Liu H, Xie J, Cheng J, Fu Y, Jiang D (2017) Transcriptome analysis of *Arabidopsis thaliana* in response to *Plasmodiophora brassicae* during early infection. Front Microbiol 8:673

Disease Cycle

8

Abstract

Plasmodiophora brassicae causing clubroot disease of cruciferous crops very successfully completes its disease cycle through three stages of life cycle; first long-time survival of resting spores in the soil, second root hairs infection by the primary zoospores, and third cortical infection by secondary zoospores. In the infected root cells, the pathogen develops into secondary plasmodia which cause hypertrophy followed by gall formation, and generation of new resting spores. Calcium ions and root exudates trigger the germination of the resting sores. The germination of mature resting spores is affected by pH, humidity, temperature, and other inorganic ions as well as biological and environs factors of the soil. The zoospores released from resting spores survive for short periods of time. In dual culture, the development of *P. brassicae* in hairy roots is similar to that found in intact plants. The secondary plasmodia in the inner cortical cells are distributed by cytoplasmic streaming and host cell division. The role of effectors/gene has been determined in the disease cycle. The serine protease Pro1 is involved in the germination of resting spores. During root-hair infection by primary zoospores, primary plasmodia formation, secondary zoospore formation, and secondary plasmodia formation, protease inhibitors such as cys-rich proteases, and proteases and methyltransferases are produced to suppress plant immunity for the success of infection process and disease cycle.

Keywords

Disease cycle · Factors affecting · Root hair infection · Cortical infection · Life cycle of the pathogen · Growth of *P. brassicae* · Secondary plasmodia growth · Identification and characterization of genes in disease cycle · Role of effector genes in life cycle · Bioinformatics analysis to identify effectors

G. S. Saharan et al., *Clubroot Disease of Crucifers*,
https://doi.org/10.1007/978-981-16-2133-8_8

8.1 Introduction

Plasmodiophora brassicae is a soil-borne obligate pathogen. It has three stages in its life cycle: survival in soil, root-hair infection, and cortical infection. Resting spores of *P. brassicae* have a great ability to survive in soil (>20 years). These resting spores release primary zoospores. When a zoospore reaches the surface of a root hair, it penetrates through the cell wall. This stage is termed the root-hair infection stage. Inside root hairs the pathogen forms primary plasmodia. A number of nuclear divisions occur synchronously in the plasmodia, followed by cleavage into zoosporangia. Later, 4–16 secondary zoospores are formed in each zoosporangium and released into the soil. Secondary zoospores penetrate the cortical tissues of the main roots, a process called cortical infection. Inside invaded roots cells, the pathogen develops into secondary plasmodia which are associated with cellular hypertrophy, followed by gall formation in the tissues. The plasmodia finally develop into a new generation of resting spores, followed by their release back into soil as survival structures. *In vitro* dual cultures of *P. brassicae* with hairy root culture and suspension cultures have been developed to provide a way to non-destructively observe the growth of this pathogen within host cells. The development of *P. brassicae* in the hairy roots was similar to that found in intact plants. The observations of the cortical infection stage suggest that swelling of *P. brassicae*-infected cells and abnormal cell division of *P. brassicae*-infected and adjacent cells will induce hypertrophy, and that movement of plasmodia by cytoplasmic streaming increases the number of *P. brassicae* infected cells during cell division.

8.2 Disease Cycle

The clubroot of crucifers caused by *Plasmodiophora brassicae* is a soil-borne obligate pathogen. The primary stage of life cycle has been recorded in root hairs and epidermal cells of some non-cruciferous hosts in the family's Papaveraceae, Poaceae, and Rosaceae (Colhoun 1958; Karling 1968). The pathogen has three stages in its life cycle: survival in soil, root-hair infection, and cortical infection (Fig. 8.1) (Ayers 1944; Ingram and Tommerup 1972; Naiki and Dixon 1987; Kageyama and Asano 2009). The primary inoculum is in the form of resting spores dispersed from infected rotten host tissue into the surrounding soil of field. The resting spore is about 3 μm in size and sub-spherical to spherical (Fig. 8.1a) (Buczacki and Cadd 1976). The surface of each resting spore is covered with spines (Fig. 8.1b) (Williams and McNabola 1967; Ikegami et al. 1978). A primary zoospore is released from each resting spore, spindle-shaped or pyriform, 2.8–5.9 μm long, and biflagellate (Fig. 8.1b) (Ayers 1944). The flagellae have two shapes: a shorter flagellum with a blunt end and a longer flagellum with a whiplash or tail piece. When the zoospore reaches the surface of a root hair, it penetrates the cell wall. This stage is termed the root-hair infection stage or primary infection stage. In root hairs, the pathogen forms primary plasmodia (Fig. 8.1c). A number of nuclear divisions occur synchronously in the plasmodia, followed by cleaving into zoosporangia. The

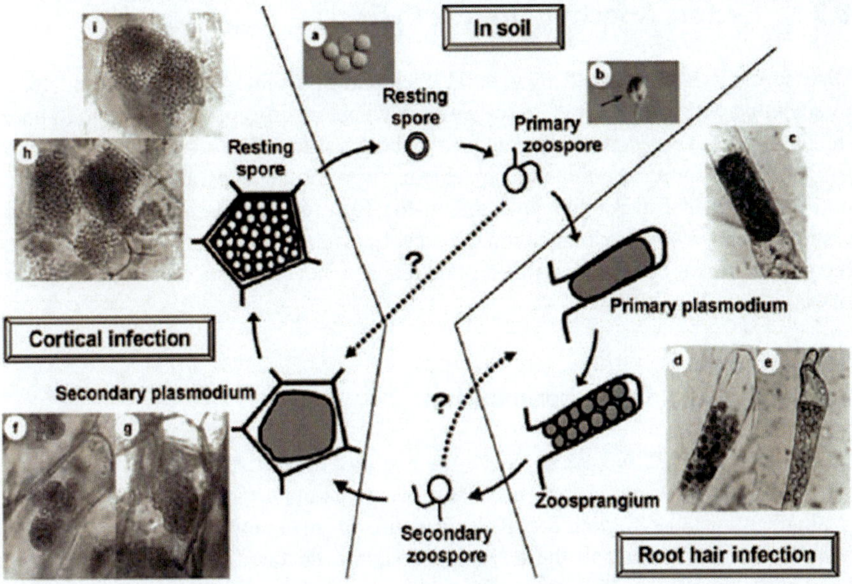

Fig. 8.1 Life cycle of *Plasmodiophora brassicae*. (**a**) Resting spore, (**b**) primary zoospore, (**c**) primary plasmodium in root hair, (**d**) zoosporangial cluster in root hair, (**e**) empty zoosporangium, (**f, g**) secondary plasmodia in cortical cells, (**h, i**) resting spores in cortical cells (Kageyama and Asano 2009)

zoosporangia form clusters in the root hair (Fig. 8.1d) and sometimes in epidermal cells. Later, 4–16 secondary zoospores are formed in each zoosporangium. After releasing these zoospores, the empty zoosporangia remain in the root hairs (Fig. 8.1e).

The secondary zoospores cannot be visually differentiated from the primary zoospores. Binucleate zoospores are sometimes found and interpreted as having formed by the fusion of two distinct zoospores, not from division within nuclei (Tommerup and Ingram 1971; Ingram and Tommerup 1972). The secondary zoospores penetrate the cortical tissues, a process called cortical infection or secondary infection stage. Inside infected cells the pathogen develops into secondary plasmodia which proliferate and are associated with cellular hypertrophy (Fig. 8.1f, g), followed by gall formation in root tissues. After a number of nuclear divisions, the secondary plasmodia contain two nuclei in the early stages of growth and then develop into multinuclear plasmodia (Garber and Aist 1979). In plasmodia with haploid nuclei, the nuclei may fuse forming diploid nuclei. Further, meiotic cleavage may occur in the diploid plasmodia, indicating that the plasmodia return to the haploid state again (Buczacki 1983). This hypothesis is not universally accepted. The plasmodia finally develop into resting spores (Fig. 8.1h, i) (Ikegayami et al. 1982), followed by their release into soil as survival structures. During these complex cleavages, the pathogen produces resting spores and increases its genetic diversity (Kageyama and Asano 2009).

8.3 Factors Affecting Disease Cycle

Disease cycle is influenced by several chemical and biological factors and their interactions with the pathogen in soil environ where resting spores survive to initiate infection process and development in the host tissues. The major factors which trigger resting spores' germination are calcium ions, host root exudates, temperature, and moisture. The inoculum potential in the form of inoculum load (number of resting spores) also contribute in infection cycle. After germination of resting spores, the process of root hairs and cortical infection is achieved for successful completion of disease cycle as follows.

8.3.1 Resting Spore Germination

Resting spores of *P. brassicae* have a great ability to survive in soil. Gibbs (1931) reported that resting spores may survive without host plants for 5 years. Wallenhammar (1996) found that the half-life of an inoculum was 3–6 years in heavily infested fields and the level of infestation declined to below a detectable level after a period of 17.3 years. The resting spores remained active when held at 40 °C for 24 h but were inactivated by treatment at 30 °C for 14 days (White and Buczacki 1979).

Resting spore germination is the first step in the life cycle. The environmental conditions that affect germination are key to controlling clubroot disease. The release of calcium ions reputedly triggers the germination of the resting spores (Yano et al. 1991). A re-absorption of calcium ions is necessary for successful germination. MacFarlane (1970) found that the resting spores from old rotten galls had higher germinability than those extracted from young hard galls. Scanning electron microscopy (SEM) demonstrated that the young spores are covered with fibrous materials (Plate 8.1a), whereas the mature spores possess a number of spines but the fibrous material is not present (Plate 8.1b). Mature spores germinate regardless of the presence of calcium ions, whereas immature young spores require the presence of calcium ions (Fig. 8.2). The populations of resting spores have different levels of maturation and their ability to germinate is associated with the level of maturation of the spores. The germination of mature resting spores is affected by environmental factors such as pH, humidity, temperature, and other inorganic ions and by biological factors (Takahashi 1994a; Friberg et al. 2005).

The zoospores released from resting spores survive for relatively short periods of time (Suzuki et al. 1992; Takahashi 1994b), suggesting that induction of spontaneous germination would provide a valuable opportunity to reduce the inoculum potential of this pathogen. Germination may be triggered by host root exudates as well as those from non-hosts (Narita and Nishiyama 1955; Bochow 1965; Kroll et al. 1983; Ikegami 1985; Suzuki et al. 1992; Friberg et al. 2005). Root-hair infection has also been observed in non-host plants such as *Lolium perenne*, *Reseda odorata*, and *Tropæolum majus* but apparently does not result in cortical infection (MacFarlane 1952).

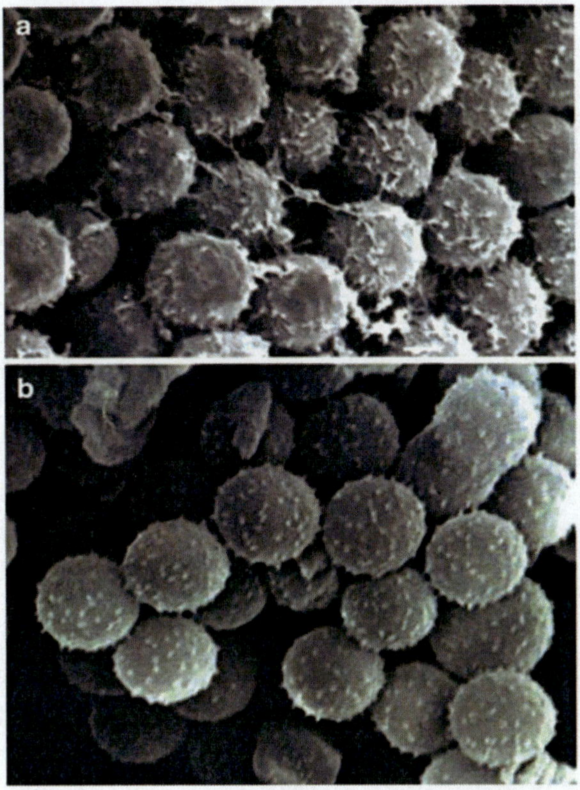

Plate 8.1 Scanning electron micrographs of immature and mature resting spores in clubroot of turnip. (**a**) Immature resting spores, (**b**) Mature resting spores. Scale bar = 1 μm (Kageyama and Asano 2009)

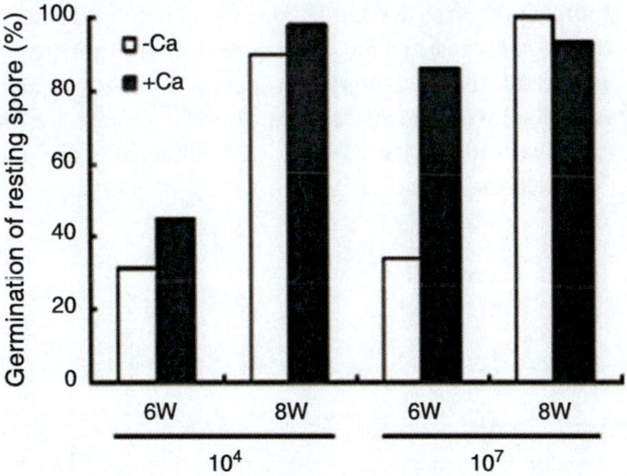

Fig. 8.2 Effect of maturation level of resting spores and calcium ion concentration on the germination of resting spores. The resting spores were collected from clubroots of turnips grown in soils infested with 10^4 and 10^7 spores/g soil for 6 and 8 weeks (Kageyama and Asano 2009)

Therefore, the use of decoy or trap plants can be a valuable tactic for disease control. Murakami et al. (2000, 2001) demonstrated that leafy daikon (*Raphanus sativus*), spinach, and oats when grown on infested land reduced the pathogen population density and results in a parallel decline in disease severity.

8.4 Root-Hair Infection

Although the primary zoospores infect root hairs, zoosporangial clusters with many zoospores are formed within the root hairs. Naiki et al. (1984a) found that secondary zoospores released from zoosporangia in root hairs re-infected them. These events suggest that the pathogen may proliferate in a short cycle around root hairs and epidermal cells and this enhances the infection of cortical tissues during the root-hair infection stage of the life cycle. Root-hair infection is observed in non-host plants as well as host plants. It is not clear whether the secondary zoospores from non-host plant roots will infect the roots of *Brassica* hosts. Host specificity may be less in the root hairs compared with the cortical cells. Further study is necessary to understand the role of root-hair infection of non-host plants.

The estimation of infestation levels in fields was based on the relationship between population densities of the pathogen, and the frequencies of root-hair infections (Samuel and Garrett 1945). Naiki et al. (1978) reported, however, that no root-hair infection was observed when there were fewer than 1×10^3 resting spores/g^{-1} soil. Clubroot formation occurs above this concentration (Fig. 8.3a). This level of sensitivity is not enough for practical diagnosis.

Calcium concentration may be related to root-hair infection as well as spore germination. Webster and Dixon (1991) found that increasing calcium concentration around roots reduces the development of zoosporangia in root hairs and slowed the release of the secondary zoospores from zoosporangia. These effects can be associated with successful control of the disease by liming the soil. Liming could have two effects: the addition of calcium ions and the increase of soil alkalinity (Webster and Dixon 1991).

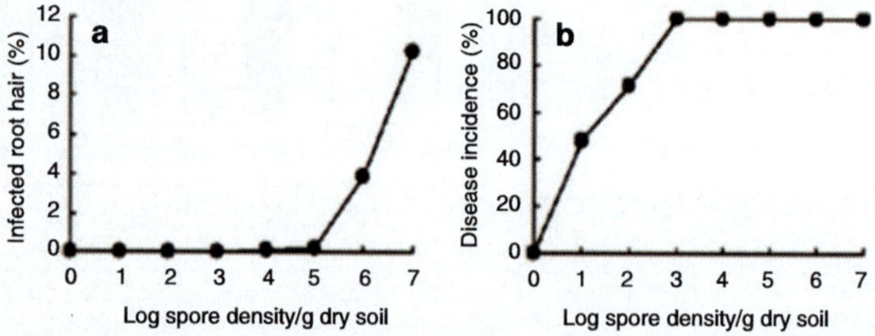

Fig. 8.3 Relationship of spore density to root-hair infection and gall formation of Chinese cabbage (**a**) root-hair infection (**b**) gall formation (Kageyama and Asano 2009)

8.5 Cortical Infection

The cortical infection stage is induced by secondary zoospores released from zoosporangia in root hairs. The primary zoospores cannot infect cortex tissues (Dobson and Gabrielson 1983). The infection processes resulting in cortical infection are still not clear. Whether the zoospores are released from the root hair into the soil or move inside the root hair or both has yet to be clarified. Stages of movement within the host have been demonstrated in a dual culture of the host and pathogen as described below. The relationship between root-hair infection and cortical infection leading to gall formation is complex (Naiki et al. 1984b). The susceptibility of root hairs could not be correlated with that of cortical cells and resultant gall formation. The data in Fig. 8.4 reveal that the cultivar represented by the symbol (•) showed high levels of root hair infection but no gall formation when exposed to the pathogen physiologic race ECD20/15/12. In addition, the balance of root-hair infection and gall formation varied when different pathogen races were used. The cultivar represented by (open square) showed moderate root-hair infection and low gall formation when exposed to the race ECD20/15/12, whereas the cultivar showed low root-hair infection and high gall formation when exposed to the race ECD20/31/31. Using highly susceptible cultivars of Chinese cabbage, the population density of resting spores was significantly correlated with gall formation (Fig. 8.3b) (Naiki et al. 1978).

Based on the regression curve of this relationship, the minimum population density for gall formation was calculated as 3.5 resting spores/g^{-1} soil. There are two pathogenic strategies in each life cycle: root-hair infection forming zoosporangial clusters with many zoospores and cortical infection forming a number of resting spores. Therefore, even if the infestation level is low and does not affect the yield of a susceptible crop, the population density will increase substantially for the next cropping season. Actually, there are several reports that single resting spores

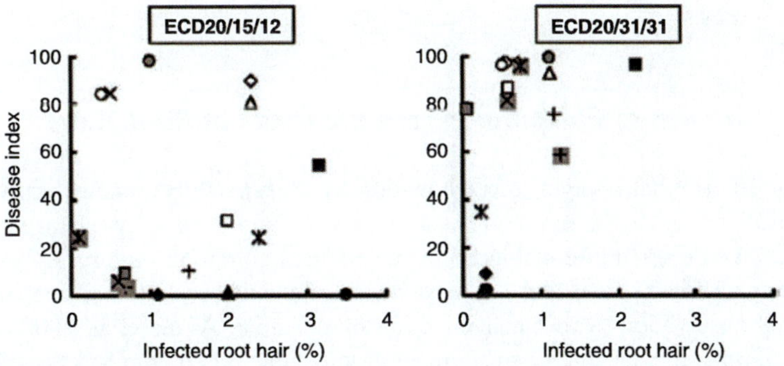

Fig. 8.4 Relationship between root-hair infection and gall formation in 16 cultivars of crucifers using different physiologic races of *Plasmodiophora brassicae*, ECD20/15/12 and ECD20/31/31 (Kageyama and Asano 2009)

induced gall formation (Tinggal and Webster 1981; Jones et al. 1982; Scott 1985; Kageyama et al. 1995; Narisawa et al. 1996).

8.6 Life Cycle of Pathogen in Dual Culture

Several investigations have been carried out with dual cultures of *P. brassicae* using callus culture (Strandberg et al. 1966; Ingram 1969; Tommerup and Ingram 1971; Dekhuijzen 1975; Buczacki 1980; Ikegami 1992), with difficulties in observing this pathogen within host tissues. In vitro dual cultures of *P. brassicae* with hairy root culture (Asano et al. 1999, 2000) and suspension cultures (Asano and Kageyama 2006) are developed to provide a way to non-destructively observe the growth of this pathogen within host cells. The life cycle of *P. brassicae* has been revealed from these cultural techniques under controlled conditions. The details have been given in Chap. 3

8.7 Germination of Surface-Disinfected Resting Spores

Satisfactory surface disinfection of resting spores is one of the most important techniques needed when establishing a dual culture of *P. brassicae* in hairy root cultures. Resting spores of *P. brassicae* are surface-disinfected with 2% (w/v) cloramine-T solution for 20 min, and an antibiotic solution (1μg mL^{-1} of colistin sulfate, 1μg mL^{-1} of vancomycin hydrochloride, and 6 μg mL^{-1} of cefotaxime sodium) for 1 day (Asano et al. 1999). The germination of surface-disinfected resting spores starts 2 days after incubation, increase steadily for 6 days, and then stops. On the other hand, the germination of non-disinfected resting spores is delayed by 2 days, and the percentage of successful germination is lower. The germination rate of surface-disinfected resting spores (12.0%) is significantly higher than that of non-disinfected resting spores (6.7%). The germination process is identical with one for intact resting spores but other biological factors will affect the germinability of resting spores.

8.8 Growth of *Plasmodiophora brassicae* in Root Hairs

The age of root hairs might affect infection by *P. brassicae* (Samuel and Garrett 1945; Naiki et al. 1978). Serial observation of root-hair infection by resting spores is difficult when using traditional methods of soil and hydroponic culture. By contrast, dual culture of hairy roots and *P. brassicae* under axenic conditions allows successful serial observation of root-hair infection in a sample (Asano et al. 2000) over a period of time. In dual culture, the number of infections is higher in 5–6-day-old root hairs, indicating that the age of root hairs is influencing successful infection. Infection start about 3 days after inoculation, increases up to 6 days, and continued increasing more slowly until 10–12 days after inoculation. The germination of

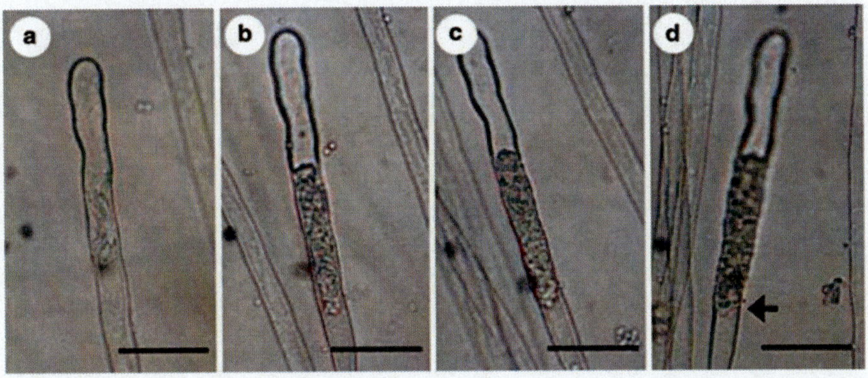

Plate 8.2 Growth of *Plasmodiophora brassicae* within a single root hair of turnip hairy root photographed **(a)** 4, **(b)** 5, **(c)** 6, and **(d)** 8 days after inoculation with resting spores. **(a)** Amorphous primary plasmodium. **(b)** Young zoosporangia developed from primary plasmodium. **(c)** Mature zoosporangia containing secondary zoospores. **(d)** Partly evacuated zoosporangia. Arrows indicate empty zoosporangium. Scale bar = 50 μm (Kageyama and Asano 2009)

resting spores and root-hair infections occurred synchronously with a delay of 2 days between these events. This might be due to a time lag in infection by primary zoospores following resting spore germination.

The growth of *P. brassicae* in a single root hair in turnip hairy root culture was serially observed for 8 days after inoculation. An amorphous primary plasmodium (Plate 8.2a) expanded to fill as much as one-quarter of the volume of the root hair in 1 day (Plate 8.2b). The protoplasm of the plasmodium became tuberculate in shape, and each tubercle subsequently became bounded with a membrane and developed zoosporangia (Plate 8.2c). The cytoplasm within these zoosporangia cleaved and then differentiated into mature zoosporangia containing secondary zoospores. Secondary zoospores were released from zoosporangia (Plate 8.2d). In dual cultures of *P. brassicae,* and turnip hairy roots, swimming secondary zoospores were observed 6–8 days after inoculation. The process of secondary infection in dual culture could not be observed. The development of *P. brassicae* in hairy roots was apparently similar to that found in intact plants (Katsura et al. 1970; Ingram and Tommerup 1972), despite the fact that hairy roots without shoots may differ physiologically and biochemically from those of intact plants.

Asano et al. (1999) also established gall formation in hairy root cultures. Therefore, dual culture appears to have some advantages over traditional methods of using soil or other substrates. The dual-culture method allows for the investigation of *P. brassicae* under carefully defined chemical and physical conditions such as temperature, humidity, pH, and nutrients without the effects of other microorganisms being present. This method can be used to study the effect of environmental and biological factors on the infection and growth of P. brassicae (Dixon 2014).

8.9 Growth of Secondary Plasmodia

Dual cultures of *P. brassicae* and turnip was developed by Asano and Kageyama (2006) using suspension cell cultures. The suspension culture of *P. brassicae*-infected turnip cells was derived from *P. brassicae* infected callus in MS liquid medium with 0.1 mg, 4 D L^{-1} and 0.02 mg kinetin L^{-1}. Propagated suspension cells were spherical to cylindrical or filamentous in shape (Plate 8.3a–d). *Plasmodiophora brassicae*-infected cells were significantly larger than uninfected cells (Williams et al. 1969; Gustafsson et al. 1986). Abnormally shaped cells (Plate 8.3d) were sometimes observed among the infected cells, but not among uninfected cells. These observations led to speculation that *P. brassicae*-infected cells are induced to produce auxin following invasion by this pathogen (Ludwig-Muller et al. 1999, 2015). Stimulated auxin production was thought to be a cause of the swelling of infected cells and result in hypertrophy of infected tissues. The fact that uninfected cells were not significantly hypertrophied suggests that the cell-stimulating activity of auxin was localized within infected cells. Because secondary plasmodia of *P. brassicae* may be able to synthesize cytokinins (Muller and Hilgenberg 1986), *P. brassicae* may promote cell division in gall tissue, leading to hyperplasia

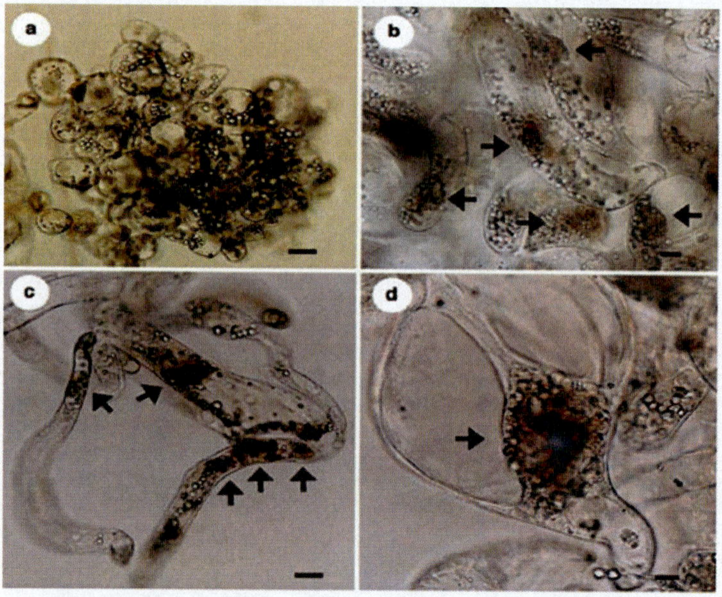

Plate 8.3 *Plasmodiophora brassicae*-infected or non-infected turnip suspension cell morphology. (**a**) *P. Brassicae* non-infected suspension cells. (**b**) Cylindrical cells containing secondary plasmodia (arrows). (**c**) Filamentous cells containing secondary plasmodia (arrows). (**d**) An abnormally growing cell containing secondary plasmodia (arrow). Scale bars = 20 μm (Kageyama and Asano 2009) other microorganisms being present. This method can be used to study the effect of environmental and biological factors on the infection and growth of *P. brassicae* (Dixon 2014)

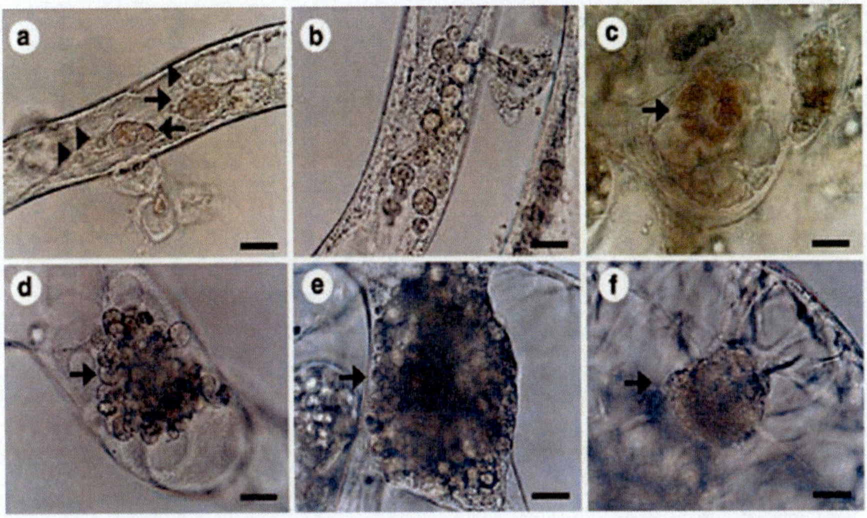

Plate 8.4 Growth stages of *Plasmodiophora brassicae* in transformed turnip suspension cells. (**a**) Very small (arrowheads) and small secondary plasmodia (arrows), (**b**) Proliferation of small secondary plasmodia, (**c**) Small cluster of secondary plasmodia (arrow), (**d**) Large cluster of secondary plasmodia with numerous subunits (arrow), (**e**) Vegetative plasmodium with several spherical bodies (arrow), (**f**) Vegetative plasmodium (arrow), Scale bars = 20 μm (Kageyama and Asano 2009)

(Gustafsson et al. 1986; Kobelt et al. 2000). Although, the concentrations of kinetin added to the medium were low or even nil, both in infected and uninfected cells propagated vigorously under these experimental conditions. These results suggest that swelling of *P. brassicae*-infected cells and abnormal cell division of *P. brassicae*-infected and adjacent cells induce the hypertrophy and hyperplasia required to form gall tissue.

Plasmodiophora brassicae grew into secondary plasmodia in suspension-culture cells (Asano and Kageyama 2006). The characteristics of secondary plasmodia in the cells were identical with those in naturally infected galls. All the growth stages of secondary plasmodia were observed (Plate 8.4a–f). Spherical or sub-spherical young plasmodia (Plate 8.4a) divided into numerous small, spherical plasmodia (Plate 8.4b). The small plasmodia developed (Plate 8.4c), and then formed a cluster by repeated division (Plate 8.4d). Plasmodia fused with each other (Plate 8.4e), followed by development of vegetative plasmodia (Plate 8.4f). Such observations have been made with the scanning electron microscopy (SEM) by Ikegami et al. (1978) also.

Observations made of a single suspension cell over time using light microscopy with Nomarski optics revealed that secondary plasmodia moved in transformed suspension cells (Plate 8.5a–d). Spherical plasmodia and starch grains separated from a cluster of plasmodia (Plate 8.5a) and in 20 min, they had moved to the opposite sides of host cells by means of cytoplasmic streaming (Plate 8.5b–d). Light

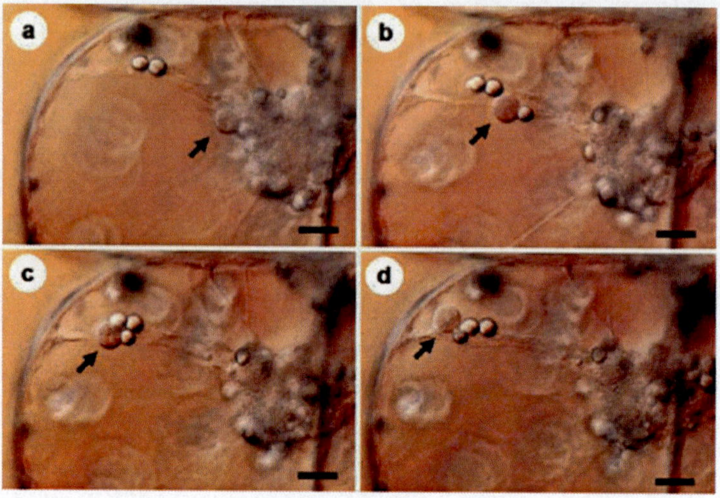

Plate 8.5 Time-lapse microscopy of plasmodium movement by transformed turnip suspension cell cytoplasmic streaming using Nomarski optics. Secondary plasmodium moved across the cell in 20 min (arrows). Scale bars = 20 μm (Kageyama and Asano 2009)

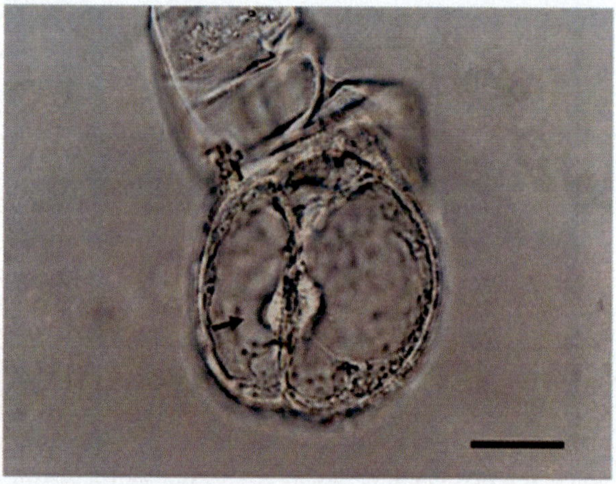

Plate 8.6 Division of secondary plasmodia (arrow) of *Plasmodiophora brassicae* concurrent with the mitotic division of transformed turnip suspension-culture cells. Scale bar = 20 μm (Kageyama and Asano 2009)

microscopy revealed that spreading mechanisms of secondary plasmodia within host cells were accompanied by division of the latter (Plate 8.6) as reported by Buczacki (1983). It was suggested that vegetative plasmodia were clustered at the time when infected cells divided and gave rise to two daughter cells. The probability of two daughter cells forming with plasmodia was low. The plasmodia were actively moved during host cell division into infected cells (Plate 8.5a–d), then the probability of two daughter cells containing plasmodia would be much higher. This suggests that the movement of plasmodia by cytoplasmic streaming may increase the number of *P. brassicae*-infected cells during cell division. Mithen and Magrath (1992) and Kobelt et al. (2000) showed cell-to-cell movement of myxamoebae via cell wall breakage by pseudopodia-like structures in the outer cortex of host roots. On the other hand, distribution of secondary plasmodia in the inner cortex and medullar rays increased following host cell division (Dekhuijzen 1976). Asano and Kageyama (2006) made continuous nondestructive observations of intracellular plasmodial movement and its division. Their observations led to the proposition that there are several stages in the growth of *P. brassicae* during the cortical infection phase. Secondary zoospores infect cortical cells, become myxamoebae, and then invade internal root tissues. Myxamoebae penetrate and migrate from the epidermis to the vascular stele, then become secondary plasmodia and grow in the cortical cells. The secondary plasmodia in the inner cortical cells are distributed by cytoplasmic streaming and host cell division rather than by disruption of the intact cell wall. The similarity of growth characteristics in in vitro culture to those in intact *P. brassicae*-infected plants strongly suggests that the use of hairy root culture and suspension culture is a realistic method for growth studies of *P. brassicae* and brassicaceous plants (Kageyama and Asano 2009).

8.10 Identification and Characterization of Genes in Disease Cycle

Many eukaryotic biotrophic plant pathogens have evolved advanced strategies to deliver effector proteins (genes) into the host cell during infection to manipulate plant defenses and enable pathogen colonization. Successful infection relies primarily on the success of the release of the effectors, which in many cases are responsible for the suppression of plant immunity. The initial recognition of conserved microbial features, known as pathogen-associated molecular patterns (PAMPs), leads to PAMP-triggered immunity (PTI) in the host. PAMP-triggered immunity is a first level of immune response that can be overcome by effector proteins produced by adapted pathogens. Resistance (R) to adapted pathogens is achieved through specific recognition of effectors, also known as avirulence proteins, by corresponding R proteins produced by the plant host. This effector-R protein recognition constitutes the second level of immune response, effector triggered immunity (ETI). *P. brassicae* is a well adapted pathogen of *Brassica* hosts; though indirect evidence suggests lack of either response, both PTI and ETI have not been well-characterized in the host-*P. brassicae* pathosystem (Dodds and Rathjen 2010; Jones and Dangl 2006; Siemens et al. 2011).

Identifying the proteins that comprise the secretome of *P. brassicae* is an important step towards understanding the complementary R proteins in the plant host. Putative effectors from biotrophic plant pathogens such as fungi are emerging from the sequencing and assembly of their genomes or transcriptomes, followed by comparative analysis with candidate effector genes. This strategy has been followed by Schwelm et al. (2015) and Rolfe et al. (2016) to identify putative effectors released by *P. brassicae* during the infection process. However, the obligate parasitic nature of *P. brassicae* has made it impossible to obtain a complete genome, with six draft genomes barely annotated for five Canadian pathotypes and one European pathotype. Transcriptomes of *P. brassicae* infecting *B. napus* and *Arabidopsis thaliana* in the Canadian study, and infecting *B. rapa*, *B. napus*, and *B. oleracea* var. *capitata* in the European study, have identified some candidate effectors. However, while the information obtained in both studies is a good start, the effectors specifically responsible for *P. brassicae* infection and subsequent disease progression have still to be identified (Perez-Lopez et al. 2018).

The overexpression of a benzoic acid/salicylic acid methyltransferase-encoding gene [PBRA_T000444 in *P. brassicae* European pathotype 3(Pbe3), PbPT3Sc00 026_A_1.308_1 in *P. brassicae* pathotype 3 (Pb3) draft genomes; GenBank accession number AFK13134] during the second week of infection, with expression peaking three and four weeks after infection has been observed. Salicylic acid (SA) is essential for the activation of plant defense. In the plant cell, regulation of active SA is managed through the maintenance of different inactive forms of SA, such as methyl salicylate (MeSA). The methyltransferase identified in *P. brassicae* has been characterized in detail by Ludwig-Muller et al. (2015), identifying its role in the methylation of salicylic, benzoic, and anthranilic acids, thereby contributing to the suppression of the salicylic acid-induced defense in a plant host.

8.10.1 Host-Pathosystem

The basic criteria to identify candidate secreted effector proteins are: proteins with a signal peptide (within the initial 60 amino acids at the N-terminus), no transmembrane domains, small size between 300 and 450 amino acids, and mostly species-specific. These parameters have been used in the identification of putative *P. brassicae* effectors. In addition, several other characteristics, motifs, and domains have been associated with effector proteins and have been used to improve identification and functional characterization of *P. brassicae* effectors/genes, i.e., (1) Cysteine-rich proteins, (2) Rxlr motif, (3) C-chitin binding domains, (4) D-protease/protease inhibitors, (5) E-nuclear localization domains, and (6) Pexal motif. The details have been covered in the Chap. 12.

8.10.2 Effector Proteins in Other Plant Pathogenic Plasmodiophorids

Plasmodiophora brassicae is not the only plant pathogenic protists, nor is it the only plasmodiophorid affecting economically important crops: the group includes

Spongospora subterranea, the causal agent of powdery scab on potato tubers, and *Polymyxa* spp., *Polymyxa graminis* and *Polymyxa betae*, which affect graminaceous plants and Chenopodiaceae plants, respectively. Genomic sequences, although limited, are available from *S. subterranea* and comprehensive *S. subterranea* transcriptomic datasets are available from root galls. These data suggest intron-rich genes and an enrichment of chitin-related enzymes in the *S. subterranean* transcriptome. Transposable elements are more expressed in *S. subterranea* than in *P. brassicae*, but evidence for recombination in *S. subterranea* is limited and there is little understanding of sexual recombination in phytomyxids. A study of secreted proteins in *S. subterranea* has been carried out, identifying the benzoic acid/salicylic acid methyltransferase-encoding gene overexpression (Schwelm et al. 2015, 2017).

8.10.3 Role of Effectors (Genes) in Clubroot Life Cycle

Clubroot has three main stages to its life cycle, (1) survival in the soil and germination of resting spores, (2) root hair or epidermal cell infection, and (3) cortical infection, although the pathogen has also been observed in phloem (Kageyama and Asano 2009). The serine protease, Pro1, is thought to be involved in step 1, the germination of resting spores (Feng et al. 2010). It is from step 2 to 5 of the life cycle, it is expected that other proteases and protease inhibitors, such as cys-rich proteases and methyltransferases will be produced to suppress plant immunity,

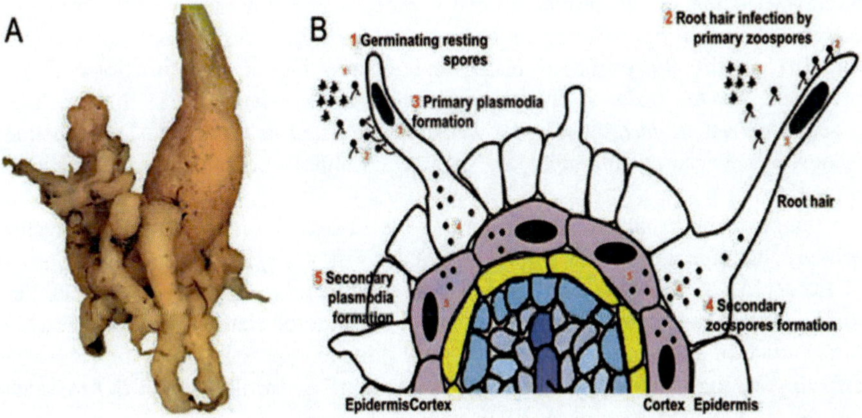

Fig. 8.5 *Brassica* plant root affected by *Plasmodiophora brassicae*. (**a**) Canola root with typical galls after 1 month of inoculation with *P. brassicae* resting spores. (**b**) Life cycle of *P. brassicae* showing the steps involved in infection through to the production of secondary plasmodia in the host plant cortical cells. Scheme based on that of Kageyama and Asano (2009), representing spindle-shaped resting spores, biflagellate primary zoospores, zoospores, and primary and secondary plasmodia (oval black figure in root hairs and cortical cells, respectively). Further steps in *P. brassicae*´s life cycle, such as the formation of resting spores in cortical cells and its ejection to the soil, are not shown in this scheme (Perez-Lopez et al. 2018)

thereby preventing the host plant from mounting responses such as programed cell death and increasing the probability of success of the infection process (Siemens et al. 2011). During the formation of primary and secondary plasmodia, infected root tissues develop into swollen galls; it is expected that *P. brassicae* will secret an array of effector proteins triggering growth, expansion and differentiations of infected host cells. During zoospore maturation, as expected, *P. brassicae* effectors with chitin-binding domains will express to remove chitin fragments that otherwise could trigger PAMP-associated immunity during late stages of the *P. brassicae* life cycle (Fig. 8.5). Based on previous studies, might be also expected over expression of effectors implies on manipulating plant defense such as PbBSMT (Malinowski et al. 2012; Ludwig-Muller et al. 2015) and effectors disturbing plant meristematic activity during the formation of secondary zoospore and secondary plasmodia. While there are still steps in the life cycle of *P. brassicae* that remain to be clarified, the study of effectors and their roles in life cycle completion and clubroot disease progression is an arena waiting to be explored (Perez-Lopez et al. 2018).

8.10.4 Bioinformatics Analysis to Identify Effectors in Future

Computational prediction is an excellent starting point to screen for putative effectors, to identify functional domains, and to help us understand the evolution, distribution, and characteristics of effectors. Utilizing the information available, Perez-Lopez et al. (2018) have designed a coherent pipeline aimed at identifying putative effectors involved in the infection process by *P. brassicae,* and subsequent clubroot disease progression in its plant host (Fig. 8.6). The pipeline makes use of tools to identify the origin of reads, and identify motifs, and functions of the predicted secreted proteins. The bioinformatics tools referred to in the pipeline, together with their descriptions and websites, are listed in Table 8.1. This pipeline is only a suggestion, although key parameters should be set in order to run it properly.

Bioinformatics analysis is the route to the identification of candidate effector proteins, but laboratory confirmation of function will always be required. Validation of the RNA-seq data through qPCR or ddPCR, identification of the sub-cellular localization of the candidate effector proteins through transient expression or stable transformation and identification of the plant proteins interacting with the pathogen effectors are some of the logical next steps towards the identification of *P. brassicae* effector proteins (Bulman et al. 2019; Irani et al. 2018).

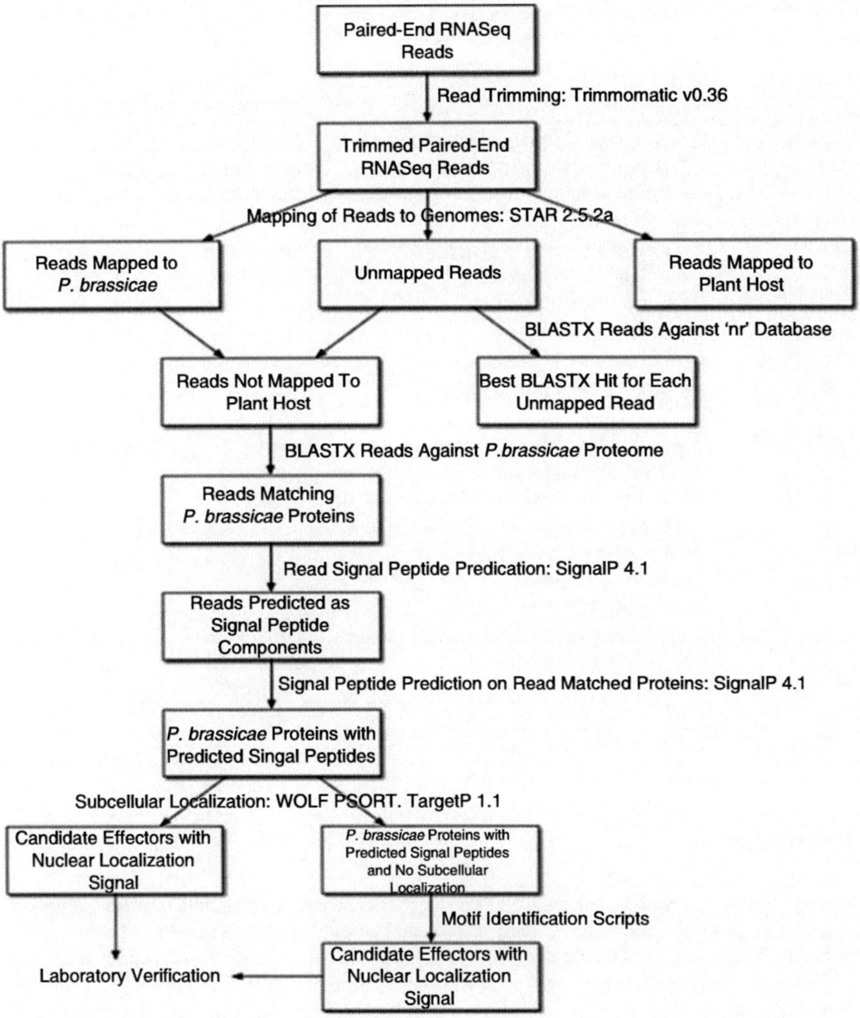

Fig. 8.6 Coherent pipeline to identify putative effector proteins of *Plasmodiophora brassicae*. The pipeline assumes: (1) Researchers are starting with RNA-Seq reads from a host plant infected with *P. brassicae*; (2) The draft genomes available for *P. brassicae*, other plasmodiophorids, oomycetes, and other plant pathogens are used; (3) Motifs mentioned in this chapter or structural similarities with previously described effector proteins were identified (Perez-Lopez et al. 2018)

Table 8.1 Bioinformatics tools used in a coherent pipeline to identify putative effectors of *P. brassicae* (Perez-Lopez et al. 2018)

Bioinformatics tool	Description	Website	Reference
Trimmomatic	A fast, multithreaded command line tool that can be used to trim and crop Illumina data as well as to remove adapters	http://www.usadellab.org/trimmomatic	Bolger et al. (2014)
STAR	An algorithm for aligning high-throughput long and short RNA-seq data to a reference genome	http://code.google.com/p/rna-star	Dobin et al. (2013)
SignalP	Software that predicts signal peptide cleavage sites in proteins from eukaryotes and prokaryotes	http://www.cbs.dtu.dk/services/SignalP	Petersen et al. (2011)
WoLF PSORT	An algorithm that predicts the sub-cellular localization sites of proteins based on their amino acid sequences. It makes predictions based on both known sorting signal motifs and some correlative sequence features such as amino acid content	www.wolfpsort.org	Horton et al. (2007)
TargetP	Tool that predicts sub-cellular location of eukaryotic proteins	http://www.cbs.dtu.dk/services/TargetP	Emanuelsson et al. (2000)

References

Asano T, Kageyama K (2006) Growth and movement of secondary plasmodia of *Plasmodiophora brassicae* in turnip suspension-culture cells. Plant Pathol 55(1):145–151

Asano T, Kageyama K, Hyakumachi M (1999) Surface disinfestations of resting spores of *Plasmodiophora brassicae* used to infect hairy roots of *Brassica* spp. Phytopathology 89:314–319

Asano T, Kageyama K, Hyakumachi M (2000) Germination of surface-disinfected resting spores of *Plasmodiophora brassicae* and their root hair infection in turnip hairy roots. Mycoscience 41:49–54

Ayers GW (1944) Studies on the life history of the club root organism, *Plasmodiophora brassicae*. Can J Res Sect C 32:143–149

Bochow H (1965) The effect of root diffusates of host and non-host plants on the spore germination of *Plasmodiophora brassicae* Worn., plant microbe relationships. In: Proceedings of symposium on the relationships between soil micro-organisms and plant roots. Czechoslovak Academy of Science, Prague, pp 296–299

Bolger AM, Lohse M, Usadel B (2014) Trimmomatic: a flexible trimmer for illumina sequence data. Bioinformatics 30(15):2114–2120

Buczacki ST (1980) Culture of *Plasmodiophora brassicae* in host callus tissue. In: Ingram DS, Helgeson JP (eds) Tissue culture methods for plant pathologists. Blackwell Scientific, Oxford, pp 145–150

Buczacki ST (1983) *Plasmodiophora*. In: Buczacki ST (ed) Zoosporic plant pathogens. Academic, London, pp 161–191

Buczacki ST, Cadd SE (1976) Size distribution of the clubroot organism, *Plasmodiophora brassicae*. Trans Br Mycol Soc 67:133–136

Bulman S, Richter F, Marschollek S, Benade F, Jülke S, Ludwig-Muller J (2019) *Arabidopsis thaliana* expressing PbBSMT, a gene encoding a SABATH-type methyltransferase from the plant pathogenic protist *Plasmodiophora brassicae*, show leaf chlorosis and altered host susceptibility. Plant Biol 21(1):120–130

Colhoun J (1958) Club root disease of crucifers caused by *Plasmodipophora brassicae* Woron. – A monograph. The Commonwealth Mycological Institute, Kew, Surrey, Phytopathological Paper No 3

Dekhuijzen HM (1975) The enzymatic isolation of secondary vegetative plasmodia of *Plasmodiophora brassicae* from callus tissue of *Brassica campestris*. Physiol Plant Pathol 6:187–192

Dekhuijzen HM (1976) Microscopical studies on the cortex and medullary rays of *Brassica campestris* var. *rapa* infected with *Plasmodiophora brassicae*. J Phytopathol Z 87:171–186

Dixon GR (2014) Clubroot (*Plasmodiophora brassicae* Woronin): an agricultural and biological challenge worldwide. Can J Plant Pathol 36(Supp 1):518

Dobin A, Davis CA, Schlesinger F, Drenkow J, Zaleski C, Jha S, Batut P, Chaisson M, Gingeras TR (2013) STAR: ultrafast universal RNA-seqaligner. Bioinformatics 29:15–21

Dobson RL, Gabrielson RL (1983) Role of primary and secondary zoospores of *Plasmodiophora brassicae* in the development of clubroot in Chinese cabbage. Phytopathology 73:559–561

Dodds PN, Rathjen JP (2010) Plant immunity: towards an integrated view of plant–pathogen interactions. Nat Rev Genet 11(8):539–548

Emanuelsson O, Nielsen H, Brunak S, von Heijne G (2000) Predicting sub-cellular localization of proteins based on their N-terminal amino acid sequence. J Mol Biol 300:1005–1016

Feng J, Hwang R, Hwang SF, Strelkov SE, Gossen BD, Zhou Q, Peng G (2010) Molecular characterization of a serine protease *Pro1* from *Plasmodiophora brassicae* that stimulates resting spore germination. Mol Plant Pathol 11:503–512

Friberg H, Lagerlof J, Ramert B (2005) Germination of *Plasmodiophora brassicae* resting spores stimulated by a non – host. Eur J Plant Pathol 113(3):275–281

Garber RC, Aist JR (1979) The ultrastructure of mitosis in *Plasmodiophora brassicae* (Plasmodiophorales). J Cell Set 40:89–110

Gibbs JG (1931) Club root in cruciferous crops. N Z J Agric 42:1–17

Gustafsson M, Liljeroth E, Gunnarsson M, Lundborg T (1986) Effects of infection by *Plasmodiophora brassicae* on root anatomy of rape. J Phytopathol 117:144–151

Horton P, Park KJ, Obayashi T, Fujita N, Harada H, Adams-Collier CJ, Nakai K (2007) WoLF PSORT: protein localization predictor. Nucleic Acids Res 35:W585–W587

Ikegami H (1985) Disease of clubroot fungus by cultivation of different crops in heavily infested soil. Res Bull Fac Agric 50:19–32

Ikegami H (1992) Studies on the clubroot of cruciferous plants. X. Proliferation and pathogenicity of *Plasmodiophora brassicae* in infected callus tissue. Proc Kansai Plant Prot 34:17–28

Ikegami H, Mukobata H, Naiki T (1978) Scanning electron microscopy of *Plasmodiophora brassicae* in diseased root cells of turnip and Chinese cabbage (Studies on the clubroot of cruciferous plants III). Ann Phytopathol Soc Jpn 44:456–464

Ikegayami H, Naiki T, Ito T, Imuro Y (1982) Ultra-structural growth process of Plasmodiophora brassicae in infected cells of Chinese cabbage and turnip (studies on the clubroot of cruciferous plants IV). Res Bull Fac Agric 46:9–19

Ingram DS (1969) Growth of *Plasmodiophora brassicae* in host callus. J Gen Microbiol 55:9–18

Ingram DS, Tommerup IC (1972) The life history of *Plasmodiophora brassicae* Woron. Proc R Soc Lond Ser B 180:103–112

Irani S, Trost B, Waldner M, Irani S, Trost B, Waldner M, Nayidu N, Tu J, Kusalik AJ, Todd CD, Wei Y, Bonham-Smith PC (2018) Transcriptome analysis of response to *Plasmodiophora brassicae* infection in the *Arabidopsis* shoot and root. BMC Genomics 19:23

Jones JDG, Dangl JL (2006) The plant immune system. Nature 444:323–329

Jones DR, Ingram DS, Dixon GR (1982) Characterization of isolates derived from single resting spores of *Plasmodiophora brassicae* and studies of their interaction. Plant Pathol 31 (3):239–246

Kageyama K, Asano T (2009) *Plasmodiophora brassicae* (clubroot)–a plant pathogen that alters host growth and productivity. J Plant Growth Regul 28:203–211

Kageyama K, Kamimura Y, Hyakumachi M (1995) A simple inoculation method with a single resting spore of *Plasmodiophora brassicae*. Ann Phytopathol Soc Jpn 61:415–418

Karling JS (1968) The plasmodiophorales, 2nd edn. Hafner Publishing Company, New York, p 256

Katsura K, Tamura M, Yamaguchi N (1970) Some observations on the infection of *Plasmodiophora brassicae* Worn. in root hair of Cruciferae plants. Proc Kansai Plant Prot 12:23–29

Kobelt P, Siemens J, Sacristan M (2000) Histological characterization of the incompatible interaction between *Arabidopsis thaliana* and the obligate biotrophic pathogen *Plasmodiophora brassicae*. Mycol Res 104(2):220–225

Kroll TK, Lacy GH, Moore LD (1983) A quantitative description of the colonization of susceptible and resistant radish plants by *Plasmodiophora brassicae*. Phytopathol Z 108:97–105

Ludwig-Muller J, Pieper K, Ruppel M, Cohen JD, Epstein E, Kiddle G, Bennett R (1999) Indole glucosinolate and auxin biosynthesis in *Arabidopsis thaliana* L. glucosinolate mutants and the development of the clubroot disease. Planta 208:409–419

Ludwig-Muller J, Julke S, Geiß K, Richter F, Mithofer A, Sola I, Rusak G, Keenan S, Bulman S (2015) A novel methyltransferase from the intracellular pathogen *Plasmodiophora brassicae* methylates salicylic acid. Mol Plant Pathol 16(4):349–364

MacFarlane I (1952) Factors affecting the survival of *Plasmodiophora brassicae* Worn. in the soil and its assessment by a host test. Ann Appl Biol 39:239–256

MacFarlane I (1970) Germination of resting spores of *Plasmodiophora brassicae*. Trans Br Mycol Soc 55:97–112

Malinowski R, Smith JA, Fleming AJ, Scholes JD, Rolfe SA (2012) Gall formation in clubroot-infected *Arabidopsis* results from an increase in existing meristematic activities of the host but is not essential for the completion of the pathogen life cycle. Plant J 71(2):226–238

Mithen R, Magrath R (1992) A contribution to the life history of *Plasmodiophora brassicae*: secondary plasmodia development in root galls of *Arabidopsis thaliana*. Mycol Res 96:877–885

Muller P, Hilgenberg W (1986) Isomers of zeatin and zeatin riboside in clubroot tissue: evidence for trans-zeatin biosynthesis by *Plasmodiophora brassicae*. Physiol Plant 66:245–250

Murakami H, Tsushima S, Shishido Y (2000) Soil suppressiveness to clubroot disease of Chinese cabbage caused by *Plasmodiophora brassicae*. Soil Biol Biochem 32(11/12):1637–1642

Murakami H, Tsushima S, Akimoto T, Shishido Y (2001) Reduction of spore density of *Plasmodiophora brassicae* in soil by decoy plants. J Gen Plant Pathol 67:65–88

Naiki T, Dixon GR (1987) The effects of chemicals on developmental stages of *Plasmodiophora brassicae* (clubroot). Plant Pathol 36(3):316–327

Naiki T, Kageyama K, Ikegami H (1978) Studies on the clubroot of cruciferous plants. II. The relation of spore density of *Plasmodiophora brassicae* Worn. to the root hair infection and club formation in Chinese cabbage. Ann Phytopathol Soc Jpn 44:432–439

Naiki T, Kawaguchi C, Ikegami H (1984a) Root hair reinfection in Chinese cabbage seedlings by the secondary zoospores of *Plasmodiophora brassicae* Worn. Ann Phytopathol Soc Jpn 50:216–220

Naiki T, Tanahashi K, Kageyama K (1984b) The relationship between root hair infection with *Plasmodiophora brassicae* Worn. and subsequent club formation among cruciferous species. Ann Phytopathol Soc Jpn 50:211–215

Narisawa K, Kageyama K, Hashiba T (1996) Efficient root infection with single resting spores of *Plasmodiophora brassicae*. Mycol Res 100:855–858

Narita T, Nishiyama Y (1955) Some observation on the germination of the resting spore of *Plasmodiophora brassicae* Worn. In: Jubilee publication in commemoration of the sixtieth birthday of Prof. Yoshihiko Tochinai and Prof. Teikichi Fukushi. Sapporo, Japan, pp 309–315

Perez-Lopez E, Waldner M, Hossain M, Kusalik AJ, Wei Y, Bonham-Smith PC, Todd CD (2018) Identification of *Plasmodiophora brassicae* effectors a challenging goal. Virulence 9 (1):1344–1353

Petersen TN, Brunak S, Von Heijne G, Nielsen H (2011) SignalP 4.0: discriminating signal peptides from transmembrane regions. Nat Methods 8:785–786

Rolfe SA, Strelkov SE, Links MG, Clarke WE, Robinson SJ, Djavaheri M, Malinowski R, Haddadi P, Kagale S, Parkin IAP, Taheri A, Borhan MH (2016) The compact genome of the plant pathogen *Plasmodiophora brassicae* is adapted to intracellular interactions with host *Brassica* spp. BMC Genomics 17:272

Samuel G, Garrett SD (1945) The infected root-hair count for estimating the activity of *Plasmodiophora brassicae* Woron. in the soil. Ann Appl Biol 32:96–101

Schwelm A, Fogelqvist J, Knaust A, Juelke S, Lilja T, Bonilla-Rosso G, Karlsson M, Shevchenko A, DhandapaniV CSR, Kim HG, Park JY, Lim YP, Ludwig-Mueller J, Dixelius C (2015) The *Plasmodiophora brassicae* genome reveals insights in its life cycle and ancestry of chitin synthases. Sci Rep 5:11153

Schwelm A, Badstober J, Bulman S, Schwelm A, Badstober J, Bulman S, Desoignies N, Etemadi M, Falloon RE, Gachon CMM, Legreve A, Lukes J, Merz U, Nenarokova A, Strittmatter M, Sullivan BK, Neuhauser S (2017) Not in your usual top 10: protists that infect plants and algae. Mol Plant Pathol 19(4):1029–1044

Scott ES (1985) Production and characterization of single-spore isolates of *Plasmodiophora brassicae*. Plant Pathol 34:287–292

Siemens J, Gonzalez MC, Wolf S, Hofmann C, Greiner S, Du Y, Rausch T, Roitsch T, Ludwig-Muller J (2011) Extracellular invertase is involved in the regulation of clubroot disease in *Arabidopsis thaliana*. Mol Plant Pathol 12:247–262

Strandberg JO, Williams PH, Yukawa Y (1966) Monoxenic culture of *Plasmodiophora brassicae* with cabbage tissue (Abstr.). Phytopathology 56:903

Suzuki K, Matsumiya E, Ueno Y, Mizutani J (1992) Some properties of germination-stimulating factor from plants for resting spores of *Plasmodiophora brassicae*. Ann Phytopathol Soc Jpn 60:699–705

Takahashi K (1994a) Influences of some environmental factors on the viability of resting spores of *Plasmodiophora brassicae* Worn. incubated in sterile soil. Ann Phytopathol Soc Jpn 60:658–666

Takahashi K (1994b) Biological agents affecting the viability of resting spores of *Plasmodiophora brassicae* Worn. in soil without host roots. Ann Phytopathol Soc Jpn 60:667–674

Tinggal SH, Webster J (1981) Technique for single spore infection by *Plasmodiophora brassicae*. Trans Br Mycol Soc 76:187–190

Tommerup IC, Ingram DS (1971) The life-cycle of *Plasmodiophora brassicae* Woron. in *Brassica* tissue cultures and in intact roots. New Phytol 70:327–332

Wallenhammar AC (1996) Prevalence of *Plasmodiophora brassicae* in a spring oilseed rape growing area in central Sweden and factors influencing soil infestation levels. Plant Pathol 45:710–719

Webster MA, Dixon GR (1991) Calcium, pH and inoculum concentration influencing colonization by *Plasmodiophora brassicae*. Mycol Res 95:64–73

White JG, Buczacki ST (1979) Observations on suppression of clubroot by artificial or natural heating of soil. Trans Br Mycol Soc 73:271–275

Williams PH, McNabola SS (1967) Fine structure of *Plasmodiophora brassicae* in sporogenesis. Can J Bot 45(9):1665–1669

Williams PH, Reddy MN, Strandberg JO (1969) Growth of non-infected and *Plasmodiophora brassicae* infected cabbage callus in culture. Can J Bot 47(8):1217–1221

Yano S, Tanaka S, KameyaIwaki M, Katumoto K (1991) Relation of Ca^{2+} efflux to germination of resting spores of clubroot fungus. Bull Fac Agric 39:105–112

Epidemiology and Disease Forecasting

9

Abstract

Clubroot of crucifers is strictly a disease of soil environment; therefore, its development is influenced by environmental parameters of soil and their interactions with all abiotic and biotic soil inhabitants. The disease development is affected by many factors such as temperature, soil pH, soil type, rainfall, drainage, micronutrients like boron, calcium, and nitrogen, inoculum potential, host resistance, weeds, cultural practices, and interactions among all these factors in a soil complex environment. Temperature below 17 °C inhibits the development of clubroot at all stages of its disease cycle. Alkaline pH also reduces infection and symptoms development. However, alkaline pH does not eliminate clubroot when other conditions are conducive for infection. Nutrients cannot limit the clubroot development when soil moisture in the rhizosphere during primary and secondary infection favors disease development. Both calcium and nitrate forms of nitrogen and calcium cyanamide inhibit the development of plasmodium into sporangia in root hairs, and reduces the viability of resting spores as well as the ability of primary zoospores to invade the host. Root-hair infection, cortical infection, and disease severity are high at 25 °C temperature and pH of 5.0–6.0 but decreases as pH increases, and temperature decreases. Clubroot severity is positively correlated with inoculum density in the field. Host resistance inhibits cortical infection by the pathogen. Disease forecasting is based on congenial factors for disease development.

Keywords

Epidemiology and disease forecasting · Factors affecting disease development · Effect of temperature · soil pH · Soil type · Rain fall and drainage · Effect of boron · Calcium · Nitrogen · Movement of inoculum · Root exudates and pH on resting spores · Effect of factors interacting in soil on disease development · Temperature · pH · Soil type · Moisture and soil compaction · Mechanism of

G. S. Saharan et al., *Clubroot Disease of Crucifers*,
https://doi.org/10.1007/978-981-16-2133-8_9

clubroot spread · Effect of host age · Inoculum density · Disease development in relation to host resistance · Role of weeds in clubroot epidemiology · Role of land elevation · Cropping sequence · Soil conditions · Risk factors of disease spread

9.1 Introduction

Several factors have an important effect on infection and development of *P. brassicae*, including temperature, pH, cation, and micronutrient concentration, moisture and inoculum concentration, pathogenic variability, host resistance, cultural operation, sowing time, soil type and interaction of host–pathogen with various abiotic and biotic factors in soil environment. The influence of pH on clubroot development has been reported on many *Brassica* spp. The severe clubbing in broccoli was observed at pH values up to 7.2, even at a low inoculum concentration, but no clubbing occurred at pH 8.0. Application of cations to raise soil pH above 7.2 under greenhouse conditions reduced root-hair infection and subsequent symptom development. Clubbing developed irrespective of pH when spore density was high and moisture was optimum, and infection of root hairs was not affected by high pH in the absence of calcium. Raising the soil pH to 7.2 or above is a standard recommendation for clubroot management in vegetable crops, but this treatment is not always effective. Even heavy liming may not prevent clubroot in cabbage if moisture and spore load are sufficiently high. The optimum temperature for development of *P. brassicae* lies between 20 and 26 °C, based on controlled environment.

Successful infection by *P. brassicae* depends on several environmental factors. Uneven distribution of these environmental conditions in the field can lead to a scattered pattern of infested patches. One major factor that affects disease incidence is soil pH. Infections occur mainly where the pH is below 7. Reports of disease incidence at pH values around 8 are rare but do exist (Colhoun 1953). This may reflect the capacity of *P. brassicae* to adapt, especially when other factors are optimal. Soil humidity is another key factor; the zoospore stages of *P. brassicae* life cycle depend on soil water for movement (Dobson et al. 1982) and this can explain the prevalence of infections in those patches of a field where drainage is not sufficient. Soil temperature between 20 and 22 °C is optimum for disease severity in the field (Thuma et al. 1983; Einhorn and Bochow 1990) although the geographical distribution of the disease leads to the suggestion that *P. brassicae* can adapt itself to different temperature regimes. Infections can be expected when resting spores, young host roots, low pH, and wet and warm weather coincide. *Plasmodiophora brassicae* depends on the movement of contaminated material for long-distances. Opposite to leaf pathogens, wind dispersal is only of minor relevance and is expected to depend on the drift of contaminated soil. Medium to long-distance spread is assisted by infected transplants of cole crops, farm machinery, animals, and water runoff. Spore dispersal over short distances by active movement in the rhizosphere seems of limited relevance for spreading because both primary and secondary zoospores appear to be incapable of swimming more than a few millimeters.

Zoospore movements should be sufficient to allow attachment to the nearest root surface. An evolutionary strategy to cope with poor dispersal is longevity and a high reproduction rate, which applies to *P. brassicae*. Resting spores are capable of remaining viable in soil for at least 17 years, with an estimated half-life time of 3.6 years (Wallenhammar 1996). One gram of infected root tissue contains at least 1–2 × 10^8 spores, and a single host plant can produce root tumors weighing more than 20 g. On the assumption that the plant as example was initially infected by 10^4 spores would mean a 100,000-fold increase in one generation, lasting approximately 6 weeks. Due to the non-specificity of the primary infection phase, it has been assumed that some non-host plants as well as resistant plants reduce the inoculum level in contaminated soils by causing the resting spores to germinate without allowing their multiplication (Yamagishi 1987; Harling and Kennedy 1991; Friberg et al. 2006) but the experimental evidence is unclear.

9.2 Factors Affecting Clubroot Development

The epidemiology of clubroot is entirely different from other foliar and root rot diseases of crops. The clubroot pathogen survives in the soil as resting spores and intracellular obligate pathogen in the host tissues during infection, pathogenesis, and multiplication leading to gall formation in host roots in the soil environ. The above-ground environ factors have minor role in clubroot development. However, below-ground environ factors and their interactions have very significant role in clubroot epidemics. The host rhizosphere biology and abiotic and biotic factors like temperature, soil moisture, soil type, soil texture, soil pH, soil fauna and flora, rainfall, drainage, calcium, nitrogen, boron, organic matter, inoculum load, inoculum potential, pathogen virulence, mechanism of clubroot spread, host resistance, weeds, cropping sequence, and interactions of all these factors in the soil have a great bearing on clubroot epidemiology with significant risk factors in a particular cultivation situation in an area. The role of all individual biotic and abiotic factors on disease development is discussed below:

9.2.1 Effect of Temperature

A minimum temperature of 14 °C is required for germination of the resting spores of *P. brassicae* (Chupp 1917). Symptom development is also affected by temperature. The minimum temperature required to produce severe clubroot symptoms on *Brassica* vegetable crops is estimated to be 18–19.5 °C (Colhoun 1953; Buczacki et al. 1978), and clubroot levels increase as soil temperature increases, with very low levels at or below 14 °C (Thuma et al. 1983). Similarly, clubroot severity is highest in short-season *Brassica* vegetable crops harvested in July and August, when soil temperatures ranges between 20 and 22 °C during the growing period, and lowest for crops harvested in October, with mean air temperatures during the final 10 days before harvest are below 12 °C (McDonald and Westerveld 2008).

Plasmodiophora brassicae (primary infection, secondary infection, colonization, and symptom development) exhibits a similar response to *B. rapa* subsp. *chinensis* at different stages of pathogenesis. The development is very slow at temperatures below 17 °C, increases quickly to a maximum at about 23–26 °C, and then decline as temperature increases to 30 °C (Sharma et al. 2011a, b). This same pattern of response to temperature was also demonstrated for primary infection and symptom severity on canola under controlled conditions (Gossen et al. 2012a) and for Shanghai pak choy in field trials (Gossen et al. 2012b). The daily fluctuations of up to 10 °C have little or no effect on infection success or subsequent pathogen development (Gludovacz 2013). However, when conditions were maintained near the optimum for infection, root-hair infection is highly correlated with symptom development in susceptible *Brassica* crops (Hwang et al. 2011a, b). This indicates that the incidence of root-hair infection is a useful indicator of inoculum potential (levels of resting spores in soil) when assessed under controlled conditions (Hwang et al. 2011a), but is likely to be much less useful under the much more variable conditions found in the field.

Clubroot severity declines as the initial infection of seedlings is delayed by as little as one week (Hwang et al. 2011b). Therefore, seeding into cool soils to delay infection may be a useful approach to clubroot management. Early (or late) seeding to avoid periods of high soil temperature has been shown to provide effective management of clubroot in short-season *Brassica* vegetable crops (Gossen et al. 2012b). In canola, early seeding reduces clubroot severity and increases seed yield slightly (Hwang et al. 2012a).

9.2.2 Effect of Soil pH

Germination of resting spores occurs readily in acidic soil, but is slower in alkaline soil (Bramer 1924). As a result, the standard recommendation for clubroot management has been to raise the soil pH to 7.2 or above by applying lime (Colhoun 1953; Hamilton and Crête 1978; Fletcher et al. 1982). Reduction in root-hair infection and subsequent reduction in clubroot development in Brassicas at alkaline pH have been reported in many crops (Samuel and Garrett 1945; Hamilton and Crête 1978; Crute et al. 1981; Myers and Campbell 1985; Webster and Dixon 1991b; Donald and Porter 2004). Development of clubbing symptoms on turnip is inhibited under alkaline conditions (Palm 1963) and severe clubbing in broccoli is observed at pH values up to 7.2, but no clubbing occur at pH 8.0 (Myers and Campbell 1985). Similarly, clubroot levels in fields surveyed in Sweden (Wallenhammar 1996) and Finland (Rastas et al. 2012) are highest in soils with acidic pH. The application of organic matter plays an important role in clubroot suppression by increasing soil alkalinity (Niwa et al. 2007).

Application of calcium, magnesium, or boron to raise soil pH above 7.2 under controlled conditions reduces root-hair infection and subsequent symptom development (Webster and Dixon 1991a, b; Donald and Porter 2004; Rashid et al. 2013). However, infection of root hairs is not affected by high pH in the absence of calcium (Myers and Campbell 1985). Reduction in primary infection and clubroot incidence at pH ≥7.2 has been attributed to inhibition of primary plasmodia prior to the release of secondary zoospores (Myers and Campbell 1985).

The slightly alkaline soil conditions reduce but did not completely stop root-hair infection, secondary colonization by the pathogen, or development of clubroot symptoms in susceptible *Brassica* seedlings when conditions are highly conducive for the pathogen (moderate concentration of resting spores, and warm [20–27 °C] wet soil). Similarly, assessment of soil pH in 267 commercial canola fields in Alberta infested with *P. brassicae* showed that soil pH was weakly correlated with clubroot severity (Gossen et al. 2013a). This indicates that other factors may also have an important impact on clubroot levels in many fields.

9.2.3 Effect of Soil Type, Rainfall, and Drainage

Clubroot incidence and severity may be influenced by the soil organic matter content, moisture, pH, and physical and chemical properties. High levels of clubroot are reported in heavy, acidic loamy soils relative to lighter soils (Colhoun 1953). However, light soil, sandy soil, humus-rich soil, and clay soil also favor clubroot development (Karling 1968). High levels of clubroot have been associated with low organic matter content (Wallenhammar 1996; Murakami et al. 2000), and adding organic amendments to field soils has been shown to reduce clubroot (Niwa et al. 2007). Organic matter improves soil texture and increases soil microbial activity and nutrient availability, which may be associated with pathogen suppression (Dixon and Tilston 2010). In contrast, high levels of clubroot developed each year in soils with very high (70%) organic matter content (Thuma et al. 1983; McDonald and Westerveld 2008; Gossen et al. 2012b).

Soil moisture has an important and consistent influence on clubroot levels. However, the use of different methods of measuring and expressing soil moisture (gravimetric vs. volumetric; World Meteorological Organization 2008) has made comparisons difficult. High soil moisture maintained for a short period is sufficient for clubroot infection to occur (Wellman 1930), whereas Colhoun (1953) indicated that 60–70% soil moisture is most favorable for clubroot development at optimum temperature, but low levels of infection occur even when soil moisture is low. An increase in soil moisture increases clubroot severity on turnip (Ayers 1944). When the impact of soil type on clubroot was examined under low soil matric potential, almost 100% infection occurred in silty loam soil, sandy loam soil, and muck soil (Dobson et al. 1982). This indicated that clubroot is most severe in soil with a high capacity for moisture retention. Clubroot developed on cabbage at soil moisture content as low as 60% (w/w) in muck soil and 9% in mineral soil, which was equivalent to about 25% moisture-holding capacity in both soils (Hamilton and Crête 1978). Clubroot incidence increases as soil moisture increases. Similarly, rainfall is positively correlated with clubroot severity on radish, *Brassica* vegetables, and *Brassica* on muck soil (Thuma et al. 1983; Adhikari 2010; Gludovacz 2013).

Avoiding compacted soils is a standard recommendation for clubroot management (McDonald et al. 2004). Soil compaction has been associated with root disease in many host–pathogen systems (Raghavan et al. 1982; Moots et al. 1988; Carter and Johnston 1989). In general, clubroot development is reduced in well-drained soil, but heavily waterlogged and compacted soils favor the disease (Dixon and Tilston 2010). This may be related to the sizes, frequency, and continuity of soil pores, which are crucial for zoospore movement (Cook and Papendick 1972). To examine the impact of soil texture on infection and development of clubroot on canola (Kasinathan 2012), canola was grown in tall pots filled with muck soil (pH 6.2, organic matter 68%), mineral soil (pH 6.8, organic matter 3.5%), autoclaved non-calcareous coarse sand (pH 6.5, organic matter 0%), or soil-less mix (pH 6.0). Mineral and muck soil were selected to represent field soils, and sand and soil-less mix to represent growth media used. The concentration of inoculum used in this experiment was 5×10^6, which reflects the concentration of resting spores in infested commercial canola fields in Alberta (Ahmed et al. 2011). The low severity on soil-less mix may be associated with its larger pore space, which made it difficult to maintain saturation in the root zone. Results obtained supported that severity of clubroot can develop on both muck and mineral soil when moisture is adequate. Mineral soil has a higher bulk density than muck soil or soil-less mix, and pore volume is larger in soil-less mix than in the other growth media, so soil-less mix presumably has the greatest moisture-holding capacity. Clubroot incidence and severity are highest on mineral soil, intermediate on muck soil and lowest in soil-less mix (Kasinathan 2012). There was a negative correlation between bulk density and micro-pore volume, and a positive correlation between bulk density and clubroot severity. This may indicate that soil pore volume or pore size affects the success of primary infection, possibly via an effect on soil moisture content (Dobson et al. 1982).

9.2.4 Effect of Boron

Application of mineral nutrients can be an effective tool for management of many crop pathogens (Datnoff et al. 2007). One of the most successful applications of this approach has been the use of boron, calcium, and nitrate nitrogen to suppress clubroot in *Brassica* vegetable production (Stangoulis and Graham 2007; Dixon 2009). Canola and other *Brassica* crops require large amounts of boron to achieve optimum growth (Asad et al. 2000, 2002). A deficiency of boron can reduce foliar and root growth, flower retention, and pollen germination (Asad et al. 2000; Stangoulis et al. 2001). Although crop tolerance to excess boron is higher than in cereals and pulse crops (Gupta 1993), the range of concentration between deficiency and phytotoxicity is narrow (Tanaka and Fujiwara 2008). Excess boron can cause phytotoxic effects such as necrosis on leaf edges and upward cupping of older leaves (Brown and Shelp 1997; Brown and Hu 1998). Boron application in *Brassica* vegetables generally reduced clubroot incidence by up to 50% and increased yield by 30% (Dixon 1996; Nott et al. 1999; Ruaro et al. 2009). Application rate and timing can be important; prolonged exposure to a low concentration of boron has an equivalent effect on symptom development as short exposure to a high concentration (Webster and Dixon 1991b).

Application of boron inhibited the development of primary plasmodia in root hairs into sporangia, and maturation of secondary plasmodia in the root cortex. This inhibition reduces clubroot severity in Chinese cabbage (*B. rapa* ssp. *pekinensis*) under field conditions (Webster and Dixon 1991b). Boron also has a role in strengthening plant cell walls. Up to 90% of cellular boron is localized in the cell walls, where it forms cross-links with pectin polymers that enhanced cell wall stability and rigidity (Loomis and Durst 1992) and stabilizes the structure of plasma membranes (Cakmak et al. 1995). Stabilization of these structural components may interfere with penetration of zoospores or development of *P. brassicae* (Dixon 1996), or may retard the movement of the pathogen through the root cortex (Graham and Webb 1991). The boron concentration may also influence the cellular environment. This may occur through induction of secondary metabolites that are toxic to a wide range of pathogens, such as phenolic and peroxidase compounds (Ruiz et al. 1998; Camacho-Cristobal et al. 2008), or through an increase in the pH of cytoplasm to create conditions that are less favorable for *P. brassicae*.

In controlled environment on canola seedlings grown in sand, primary infection in root hairs and secondary development in the root cortex are inhibited with increasing rates of boron, irrespective of boron formulation (Deora et al. 2011). The timing of boron application (pre- or post-emergence) has no effect on clubroot development. This indicates that the rate of boron applied is more important than the timing of application. The application of boron at the rates >2 kg ha^{-1} resulted in phytotoxicity, but the seedlings grew out of the symptoms within about 10 days (Deora et al. 2011).

High rates of boron (16 and 32 kg ha^{-1}) reduce clubroot severity by 25–35% on muck soil, but even high rates of boron application did not reduce severity on mineral soil. The difference in response may reflect the difference in organic matter content (~3% vs. 70%). Organic matter is a storehouse of boron because of the binding ability of dihydroxy compounds of humus, the major component of organic matter in soil (Parks and White 1952; Goldberg 1997). More boron is available to the plant for a longer period during the growing season in soil with high organic matter than in mineral soil, where boron leaches out more quickly (Parks and White 1952; Yermiyahu et al. 1988). The longer period of availability of boron on muck soil was likely an important factor in the higher efficacy against clubroot compared with mineral soil (Gupta and Cutcliffe 1978; Deora et al. 2013).

9.2.5 Effect of Calcium and Nitrogen

Application of lime has been used since Roman times for management of clubroot. Liming makes soils more alkaline, and also increases levels of available calcium. Alkaline pH reduces root-hair infection and the rate of maturation of the pathogen, and calcium also inhibits the pathogen (Webster and Dixon 1991a). The efficacy of lime is affected by the type of lime applied, the interval before planting, and frequency of application (Wellman 1930; Campbell et al. 1985; Campbell and Greathead 1996). It is also affected by environmental factors and inoculum density (Colhoun 1958).

Calcium applied as calcium sulfate ($CaSO_4$) or calcium carbonate ($CaCO_3$), or in combination with nitrogen as calcium cyanamide ($CaCN_2$), calcium nitrate ($CaNO_3$), or calcium ammonium nitrate (Ca $(NO_3)2$ NH_4 NO_3 $10H_2O$), reduces clubroot severity on *Brassica* vegetables (Fletcher et al. 1982; Dixon and Wilson 1983; Klasse 1999; Donald et al. 2002, 2004; Porth et al. 2003) and rapeseed-mustard under field conditions (Bhattacharya and Mandal 2006). High levels of calcium salts increased spore dormancy (MacFarlane 1970). The salts that increased calcium and raised pH are also associated with a larger reduction in clubroot symptoms than those that affected pH alone. The efficacy of several calcium and nitrogen fertilizers reduces clubroot on *Brassica* vegetables and has been assessed and substantiated in Canada (Belec and Tremblay 2004; McDonald et al. 2004; Tremblay et al. 2005; Abbasi and Lazarovits 2006) and on canola in Australia (Donald et al. 2002, 2004). Both calcium and nitrate nitrogen inhibit the development of plasmodia into sporangia in root hairs (Webster and Dixon 1991a; Dixon and Page 1998), similar to the effect of boron. In addition to an effect on pathogen development, calcium fertilizers, alone or in combination with nitrogen as calcium nitrate, reduce the viability of *P. brassicae* resting spores and the ability of primary zoospores to invade the host (Naiki and Dixon 1987; Webster and Dixon 1991a; Page 2001). Application of calcium cyanamide, which degrades into calcium and nitrate nitrogen in the soil, has a similar effect.

Several mechanisms for suppression of *P. brassicae* by calcium and nitrogen have been suggested. A high calcium concentration decreases membrane permeability of root cells, which can affect the growth and reproduction of *P. brassicae* as it proliferates within the host. Alternatively, the alkaline environment created by the addition of large amounts of calcium could affect primary and secondary infection, cortical migration, and cell hypertrophy. Calcium ions may also have a role in induced cell death in response to infection by *P. brassicae* (Takahashi et al. 2002, 2006). Nitrate nitrogen is known to stimulate arginine—or lysine-rich histones in plant cells. These histones may inhibit RNA polymerases in the pathogen, and ultimately block the pathogen's access to gene products needed for pathogenesis. Another possible mechanism is that a high supply of nitrate nitrogen influences the activity of *P. brassicae* via a shortage of cofactors such as NADPH (nicotinamide adenine dinucleotide phosphate) that are needed for conversion of nitrate to ammonium (Webster 1986).

9.2.6 Effect of Movement of Inoculum

The potential avenue of long-distance movement of *P. brassicae* is with the seed of crops grown in infested fields. A low potential for seed-to-seedling transmission from contaminated canola seed has been demonstrated under controlled conditions. This low risk can be reduced even further by commercial seed cleaning and seed treatment with fungicides (Rennie et al. 2011). Therefore, contaminated seed likely represents a very small risk for long-distance dispersal of the pathogen. In contrast, movement of resting spores with soil from infested fields by wind and water may be an important means of dispersal. When dust traps were positioned adjacent to infested fields, spores were routinely collected at more than 30 cm above the soil surface (Strelkov and Hwang 2014). Air-borne spores and soil are readily moved upward into the air column and have the potential to be moved tens or even hundreds of kilometers before being re-deposited. There is evidence that transmission of wind-borne spores deposited in water used to irrigate seedlings for transplant results in rapid spread of clubroot in parts of Australia (Donald et al. 2006). Given the large hectare of fields infested and the high concentration of resting spores ($>10^6$ spore g^{-1}) in some fields in Alberta (Cao et al. 2009), the number of spores that can potentially be moved by a single windstorm is staggering. It appears likely that at least some of the infestations of *P. brassicae* at widely dispersed sites across the region are the result of movement of spores by wind.

9.2.7 Effect of Root Exudates and pH on Resting Spore Germination and Infection

Resting spores germination induced by the root exudates (REs) may reduce soil inoculum loads due to mortality of the zoospores in the absence of an appropriate living host (Kroll et al. 1983; Ikegami 1985; Murakami et al. 2001; Friberg et al. 2005, 2006), and consequently reduce disease severity on the following crops. It has been suggested that exudate preparation from the roots of perennial ryegrass (*Lolium perenne* L.) stimulated the germination *of P. brassicae* RSs in an aqueous solution (Friberg et al. 2005). MacFarlane (1970) observed that the germination percentage of *P. brassicae* resting spores was increased in the presence of cabbage (*Brassica oleracea* L. var. *capitata* cv. King of Hearts) plants or a diffusate from cabbage roots. However, in field and greenhouse trials, RS populations were found to be only slightly reduced following two rounds of cruciferous crops (canola or Chinese cabbage) relative to non-cruciferous host crops (red clover, perennial ryegrass, orchard grass, bent grass) and non-host crops (barley, wheat) (Ahmed et al. 2011). Based on the observations of Ahmed et al. (2011), it can be indirectly concluded that the effects of root exudates/diffusates from non-host plants (bait crops) on the germination of RSs and consequently on the severity of clubroot on susceptible crucifers were insignificant.

Clubroot severity under field conditions has also been suggested to be influenced by soil pH values (Myers and Campbell 1985; Takahashi 1994). The liming of soil by the application of Ca-rich compost which neutralizes acidity has been reported to be effective in reducing clubroot disease severity in cruciferous crops (Karling 1968; Myers and Campbell 1985; Webster and Dixon 1991a). Suppression of clubroot infection in Chinese cabbage (*Brassica rapa* var. *pekinensis* cvs. Satokaze and Tomikaze) at neutral pH values was suggested to be related to the inhibition of RS germination (Niwa et al. 2008). However, the exact mechanism of this inhibition is unclear (Asano et al. 2000) although it was previously suggested that at elevated pH levels, the abortion of primary thalli (primary plasmodia) before release of the secondary zoospores might be the likely reason (Myers and Campbell 1985). Acidic soils can increase the rate of production of primary zoospores, relative to alkaline soils (Dixon 2009). However, variable clubroot severity was also reported when the soil pH exceeded 7.2 or when pH was below 5.5 (Myers and Campbell 1985; McDonald et al. 2004; Rashid et al. 2011). It has been suggested that high concentration of calcium, derived from the application of Ca-rich compost, may provide clubroot control even at pH values <7.2 and low concentration of calcium may induce clubroot development even at pH values >7.2 (Myers and Campbell 1981, 1985; Campbell et al. 1985). Young resting spores required the presence of Ca-ions for germination, while mature spores germinate regardless of whether calcium was present or not (Kageyama and Asano 2009). Resting spores from old, decaying galls germinated more rapidly than spores from young galls (MacFarlane 1970). Clubroot infection and gall development were found to occur even at alkaline pH values if other conditions such as temperature, moisture, and inoculum concentration were optimal for disease development (MacFarlane 1952; Colhoun 1973; Fletcher et al. 1982). It is unclear from the above reports, whether it is the pH values or the level of calcium in the soil that influences clubroot severity under field conditions.

Rashid et al. (2013) examined under controlled conditions: (1) the effects of root exudates from host (*Brassica*, *B. napus*, Chinese cabbage, *B. rapa* var. *pekinensis*, cv. Granaat) and non-host (perennial ryegrass, *L. perenne*) plants and a wide range of pH values on resting spore germination of Canadian populations of the clubroot pathogen, and (2) the effects of a wide range of pH values either alone or in conjunction with a calcium level on root-hair infection and clubroot severity on canola.

Germination of soil-borne RSs, followed by their infection of susceptible canola plants are important requirements for the establishment of clubroot in the field. Root exudates/diffusates from non-host plants have been suggested to positively influence RS germination (MacFarlane 1970; Friberg et al. 2005, 2006). However, a study conducted under field and greenhouse conditions reported no significant effects of bait crops on clubroot severity (Ahmed et al. 2011). Elevation of soil pH values by application of Ca-rich compost was suggested to reduce clubroot severity (Karling 1968; Myers and Campbell 1985). However, it was not clear whether elevated pH values or elevated calcium levels in the soil were responsible for the negative influence on clubroot development (Rashid et al. 2013).

The study conducted under controlled conditions showed that the percentage of RS germination was significantly higher when incubated with REs from host or non-host plants compared to controls that did not contain REs. Among the RE preparations tested, those from the non-host species, perennial ryegrass produced a higher percentage of spore germination than those from canola or Chinese cabbage. Previous reports showed that the non-host perennial ryegrass stimulated RS germination more than the host species, *B. rapa* var. *pekinensis* or three other non-host species (Friberg et al. 2005, 2006). However, contrasting results were also reported earlier showing that perennial ryegrass had no effect on the germination of clubroot

RSs (Robak 1996). Similarly bait crops were also shown to be ineffective in preventing clubroot severity on canola (Ahmed et al. 2011). The possible reason for these apparently contradictory findings might be related to different experimental conditions used in the different studies. Ahmed et al. (2011) conducted experiments under field and greenhouse conditions using naturally clubroot-infested soil. In contrast, Rashid et al. (2013) conducted study under laboratory conditions using test tubes and in vitro preparation of root exudates. Thus, the concentration, soil-binding-related inactivation, and the rate of microbial degradation of root exudates were different in two experiments. This might have resulted in differential effects of REs on clubroot RS germination.

To understand if there were any correlations between nutrient elements present in the RE preparations and the percentage of stimulation of RS germination, Rashid et al. (2013) conducted chemical analysis of the RE preparations. Analysis of the results suggests that no consistent correlation between nutrient elements and the percentage of RS germination was present in the different RE preparations. Friberg et al. (2005) also reported poor correlations between spore germination and nutrient levels in the exudates. However, Dixon (2009) showed from field and controlled laboratory studies that boron had a significant effect on the ability of *P. brassicae* to colonize host plants, and suggested that by raising the boron content in the rhizosphere, it may be possible to limit the ability of *P. brassicae* zoospores to infect the host roots. Similarly, elevated level of nitrate was shown to suppress symptoms of many plant diseases, but there have been limited investigations into its effect on clubroot (Dixon 2009).

In addition to various mineral nutrient elements discussed above, REs may also contain a diverse array of chemical compounds such as amino acids, organic acids, polysaccharides, and proteins (Flores et al. 1999). These chemicals which serve important roles as chemical attractants and/or repellents in the rhizosphere (Walker et al. 2003) may have also contributed to differential induction of RS germination. Suzuki et al. (1992) identified a putative germination-stimulating factor (GSF), described as a heat stable, fairly polar, and low molecular mass compound. GSF was not considered to be related to clubroot resistance or host recognition, since it was not specific to the root exudates of clubroot-susceptible crucifers, but instead was also present in exudates of a resistant crucifer and the non-host lettuce, *Lactuca sativa* L. (Suzuki et al. 1992). Feng et al. (2010) studied the stimulatory effect of host REs on *P. brassicae* RS germination. It was found that treatment of these exudates with a serine protease (Pro1) from the pathogen further increased germination rates. They hypothesized that the GSFs consist of two components: (1) innate compounds in the REs containing caffeic acid, coumalic acid, and corilagin and (2) hydrolytes produced by the enzyme activities of Pro1 and/or other enzymes. In the case of later component, the involvement of hydrolytic enzymes not only generates stimulatory chemicals, but also provides a potential pathway for the pathogen to recognize its host (Feng et al. 2010). Soil pH can also either positively or negatively influence clubroot disease development and severity (Myers and Campbell 1985; Takahashi 1994). Rashid et al. (2013) studied the effects of a wide range of pH values either alone or in conjunction with canola REs on the germination of RSs, a pre-requisite for plant infection. Spore germination was enhanced at pH values above 5.0 and up to 7.0 and maximal at pH 6.5 and then gradually declined at pH values above 7.0 in both the presence and absence of REs although percentage of RS germination was always higher in the presence of REs than in their absence. Takahashi (1994) also reported mixed spore germination rates in response to pH differences. It was found that the RSs incubated in acid soil with pH values from 4.5 to 6.5 had a significantly lower spore viability after a 30-day incubation period than RSs incubated in alkaline soil (pH 7.8–8.2) for a similar length of time. In contrast, in alkaline soil, the spore viability declined at a more or less constant rate over a 7-month incubation period (Takahashi 1994). Niwa et al. (2008) reported a significant reduction in clubroot RS germination as a result of neutralization of soil by the application of Ca-rich compost or calcium carbonate into the rhizosphere. Feng et al. (2010) reported the optimal activity of *P. brassicae* serine protease (Pro1), an enzyme that stimulates RS germination ranged from pH 6.0 to 6.4. Other studies however reported optimal activity of serine protease at pH values ranging from 6.0 to 12.0 (Hachem et al. 2005; Saurabh et al. 2007; George-Okafor and Odibo 2011). It is likely that the highest percentage of RS germination at pH value of 6.5 might be linked to higher serine-protease activity at this pH. However, this result was not consistently reflected on disease severity.

Rashid et al. (2013) determined the effect of a wide range of pH values on root-hair infection and clubroot severity caused by *P. brassicae*. Variable results were obtained when the experiment was conducted in sand culture versus soil culture conditions. In the sand culture experiment, this was conducted in the presence of a Ca-containing NS, root-hair infection was higher at relatively lower pH values (<7.0), whereas in soil culture experiment, where water instead of NS was used, root-hair infection was higher at elevated pH values (pH 6.5–9.0). The above observations indicate that calcium might also play roles in root-hair infection and clubroot severity caused by *P. brassicae*. As mentioned earlier, application of Ca-rich compost that increases soil pH values was shown to reduce clubroot severity on crucifers (Karling 1968; Myers and Campbell 1985). However, it was not clear whether the negative impacts of Ca-rich compost on clubroot development reported earlier were related to elevated pH values or increased concentrations of Ca^{+2}, or a combination of both factors in the soil. It was also suggested that high concentrations of divalent cations such as Ca^{+2} and/or Mg^{+2} can inhibit clubroot development (Myers and Campbell 1985; Dixon and Page 1998). Furthermore, Ca^{+2} availability and/or uptake by plants were shown to be pH dependent, and directly proportional to the pH value of the growing medium (Maas 1969; Niwa et al. 2007; Merino-Gergichevich et al. 2010). Alkaline soil was shown to contain higher concentration of available calcium than soil with acidic pH values (Maas 1969; Merino-Gergichevich et al. 2010). It can be suggested that acidic pH can exert an antagonistic effect on Ca^{+2} absorption thereby enhancing clubroot infection, whereas alkaline pH stimulates the process, thereby reducing clubroot infection in canola plants. The above might have resulted in differential susceptibility of canola plants to root-hair infection and consequently clubroot severity caused by *P. brassicae* observed under two growth conditions. Further research is needed to determine the interaction of calcium and pH on clubroot severity on canola. The studies revealed that (1) root exudates of both host and non-host plant species tested stimulated *P. brassicae* resting spore germination, (2) the resting spore germination of *P. brassicae* was higher at low pH values than at high pH values, and (3) in the presence of an added calcium salt, root-hair infection of canola by resting spores of *P. brassicae* was more frequent at low pH values, however, in the absence of added calcium salt, root-hair infection and disease severity were greatest at the higher pH levels. Further research is necessary to determine the interaction of different pH values and varying concentrations of calcium on root-hair infection and disease severity caused by *P. brassicae*.

9.2.8 Effect of Factors Interacting in Soil on Disease Development

In soil environment, there is impact of several factors of physical, chemical, and biological origin and their complex interactions under different types of soils on disease development and pathogen life cycle.

9.2.8.1 Temperature and pH Interaction in Soil

An early study on the effects of temperature indicated that clubroot developed on cabbage at soil temperatures ranging from 9 °C to 30 °C, with an optimum of 25 °C (Monteith 1924). A subsequent study found that the most favorable air temperatures for clubroot development were 18–23 °C although severity was high when temperatures fluctuated near 23 °C (Colhoun 1953). Severity on cabbage, Chinese cabbage, mustard, and radish showed a bell-shaped response curve with maximum severity at 21–22 °C under controlled conditions (Thuma et al. 1983). In a field study, rainfall in the first 2–3 weeks after seeding was associated with high severity of clubroot on radish, but severity increased as soil temperature increased (Thuma et al. 1983). Similar results were reported on Shanghai pak choy and flowering Chinese cabbage grown on a muck soil, where temperatures around 22 °C (the highest average air temperature recorded) were associated with the highest clubroot severity (McDonald and Westerveld 2008). However, in contrast with the study on radish, temperature in the 10 days prior to harvest had the strongest influence on severity. Studies under controlled conditions have shown that the optimum temperature for each stage of clubroot development, from primary infection to symptom development, is consistently near 25 °C on short-season Brassica vegetables (Sharma et al. 2011a, b). Root hairs of Shanghai pak choy became infected at 2 days after inoculation (DAI) at 25 °C, but infection was not observed until 6 DAI in plants at 10 °C. Root-hair infection peaked at 26 °C. At 10 °C, the pathogen did not develop beyond the primary plasmodial stage, while many zoosporangia had matured and dehisced by 10 DAI at 25 °C and 93% of the root hairs had been infected (Sharma et al. 2011a). Similar temperature effects were observed for cortical (secondary) infection. The infected area of the root cortex was highest at 25 °C, intermediate at 20°C and 30 °C, and lowest and last to develop at 15 °C. No cortical infection was observed in plants at 10 °C (Sharma et al. 2011b).

Canola grown at 25/20 °C day/night produced a similar result (Deora et al. 2012). Root-hair infection was 68% at 14 DAI, and there was 38% cortical infection at 28 DAI. Inoculated plants developed 100% clubroot incidence and very high severity (94–100%). Trials comparing the temperature response of *P. brassicae* on canola and Shanghai pak choy under controlled conditions confirmed that they were similar (Gossen et al. 2012a). Clubroot incidence on canola was slightly higher than on Shanghai pak choy and approached 100% at temperatures from 15 °C to 30 °C, with maximum symptom severity at around 25 °C. Similarly, in controlled environment trials to evaluate the effect of soil type on canola and Shanghai pak choy, both crops developed 100% incidence of clubroot when grown in saturated soil at 18–23 °C (Kasinathan 2012). Application of lime to raise soil pH for control of clubroot on Brassica vegetables has been recommended for decades but does not always reduce clubroot severity (Webster and Dixon 1991a; McDonald and Westerveld 2008). An early study reported that clubroot could develop on cabbage in both acidic and alkaline soil. Infection in alkaline soil was favored by high moisture content, high temperature, and high spore load. The application(s) of lime might not be effective if temperature and moisture were not limiting (Colhoun 1953). On canola under controlled conditions high pH inhibited infection and development of the pathogen, but this negative impact was largely negated when other conditions for infection were highly conducive for the pathogen. Clubroot incidence and severity were 100% on canola grown in sand at pH 5.0–6.5 under optimum temperatures (20–25 °C) plus high moisture and high inoculum pressure. Severity at pH 8.0 was substantially reduced, but was still 40% (Kasinathan et al. 2011; Kasinathan 2012). Similarly, a large survey of clubroot-infested canola fields in Alberta (267 fields with soil pH ranging from 4.8 to 7.8, assessed over 6 years) found only a weak negative association ($r = 0.33$) between clubroot severity and pH (Gossen et al. 2013a). This demonstrates that pH is only one of the factors that can influence the severity of clubroot in a field.

Although clubroot severity is lower at alkaline pH (>7.2), alkaline pH alone did not prevent clubroot development when inoculum pressure and soil moisture and temperature were optimum. Controlled environment studies have shown that the temperature optimum for clubroot development is about 25 °C. The lower estimates of optimum temperature (21–22 °C) reported from field studies (Thuma et al. 1983; McDonald and Westerveld 2008) likely indicate that soil temperature under the crop canopy in these trials did not reach 25 °C. However, clubroot incidence can reach 100% across a wide range of temperatures when soil moisture is high and there is a high concentration of inoculum. The impact of temperature and pH on clubroot was very similar on canola and Shanghai pak choy.

9.2.8.2 Soil Type, Moisture, and Compaction of Soil

There are conflicting reports on the influence of any, soil type on clubroot severity. The clubroot was most severe in soils with low organic matter in a survey of oilseed rape (*B. napus*) crops growing on mineral soil in northern Europe (Wallenhammar 1996), but there are reports of high levels of clubroot on soils with high organic matter (60–70%) on radish (Thuma et al. 1983), Shanghai pak choy (McDonald and

Westerveld 2008) and other *Brassica* vegetables (Adhikari et al. 2012) and canola (Deora et al. 2012). In a trial where cabbage was grown on mineral (Fox sandy loam, 2–3% organic matter) and muck (74% organic matter) soil sites in the same season and region, clubroot was severe on both soil types (Saude et al. 2012). In a study that directly compared soil types under controlled environment conditions, a high incidence of clubroot (85–100%) developed on canola and Shanghai pak choy grown in muck soil (60–70% organic matter), mineral soil (4%), or sand (0% organic matter) (Kasinathan 2012). Taken together, these results indicate that organic matter does not have a substantial effect on clubroot severity.

Compacted soils often have poor drainage and correspondingly long periods of high moisture levels. High levels of soil moisture provide optimum conditions for the spread of primary and secondary zoospores by increasing the free moisture between the soil particles. Compacted soils also have a low pore volume, which may facilitate movement of zoospores towards a nearby root.

A soil compaction, clubroot incidence, and severity were higher on canola grown on mineral soil (bulk density 1.1 g cm^{-3}, organic matter 4%) than muck soil (bulk density 0.4 g cm^{-3}, organic matter 60–70%), and higher in compacted soil-less mix than in loose soil-less mix in a controlled environment study (Kasinathan 2012). When soil-less mix was added to mineral or muck soil to increase pore volume and decrease bulk density, incidence and severity were lower in soil plus soil-less mix than in the original soil sample. However, pore volume was not the only factor that affected clubroot levels, because clubroot incidence and severity were higher in canola grown in the amended mineral soil than in the amended muck soil, even though both amended soils had the same pore volume (Kasinathan 2012).

9.3 Mechanism of Clubroot Spread

Factors affecting clubroot of crucifers in canola have been assessed by Gossen et al. (2015). Accordingly there is circumstantial evidence that there have been at least two separate introductions of *P. brassicae* inoculum pathogenic on canola to the prairie region, with one introduction near Edmonton and a second one in southern Alberta. Multiple introductions require consideration because it has implications for the genetic and pathogenic variability of the pathogen population. High variability may in turn affect the rate that host resistance can be overcome. There is compelling evidence that occasional long-distance dispersal of the pathogen is occurring in the prairie region, with the most likely candidates being movement of soil on equipment and air-borne dispersal of spores with infested soil. Short-distance movement of spores via wind has been confirmed in Alberta, and it is simply a matter of time for long-distance dispersal of air-borne resting spores. Seed transmission does not appear to be an important mechanism for transmission of the pathogen. Similarly, movement of spores in water is likely not a major mechanism of pathogen dispersal in the prairie region, but that could change if clubroot was to become established in the Red River Valley, where large-scale flooding is a frequent occurrence, such situation may operate in other areas of the world.

Use of resistant cultivars can reduce the rate of clubroot spread by reducing or eliminating almost all crop infection from low concentrations of resting spores, e.g., wind-borne spores. Use of resistant cultivars alone is unlikely to halt the advance of *P. brassicae*, but can reduce the rate and impact of clubroot spread when used together with sanitation of equipment brought from infested fields, early seeding (Hwang et al. 2014), and a longer (>2 years) break from canola (Peng et al. 2013, 2015). Effective use of existing resistant cultivars and longer crop rotations are especially important to provide a much-needed reprieve to allow development of canola cultivars with improved resistance and development of alternative clubroot management strategies.

Plasmodiophora brassicae has a remarkable life cycle that involves two motile zoospore stages, a mobile amoeboid phase that can slip from cell to cell within the host, and small (~3-μm dia.), sub-spherical to spherical resting spores (Kageyama and Asano 2009). In terms of pathogen dissemination, however, the only important stage of the life cycle is the resting spores. The resting spores are roughly the size of silt particles and bacteria, but have a tough, multi-layered cell wall that makes them highly resistant to both microbial breakdown in soil and standard sanitation treatments (Hwang et al. 2014). Once a field is infested by *P. brassicae*, resting spores can build up rapidly in the presence of susceptible crops (Hwang et al. 2013). The half-life of resting spores was estimated to be 3.6 in Sweden (Wallenhammar 1996) and 4.4 years in central Alberta (Hwang et al. 2013).

The mechanism(s) of transmission of *P. brassicae* inoculum in canola on the Canadian prairies has been reviewed (Gossen et al. 2014; Strelkov and Hwang 2014). Transmission of the pathogen from field to field in infested soil on heavy equipment appears to be an important method of short-distance dissemination (Cao et al. 2009) and may also be a mechanism for long-distance dissemination. The spore concentrations observed in many infested fields are often $\geq 10^7$ spores g^{-1}, and hundreds of kilograms of infested soil can be carried from field to field on heavy equipment such as tractors and cultivators (Hwang et al. 2014). As a result, more than enough resting spores can be transported on equipment that is moved from an infested field to a clean field to initiate a new infestation. Also, recent studies on the importance of wind erosion for the dispersal of *P. brassicae* demonstrated that substantial amounts of inoculum are present in wind-blown dust from infested fields in Alberta (Rennie et al. 2015). This indicates that dispersal of inoculum by wind may be an important mechanism for both short- and long-distance transmission of *P. brassicae*.

Dispersal of resting spores on seed produced in infested fields is a potential mechanism of long-distance transmission that has not been studied in detail. A low but repeatable incidence of seed-to-seedling transmission was demonstrated under optimal conditions in a greenhouse study (Rennie et al. 2011). Seed cleaning or seed treatment with a fungicide reduced this transmission to trace levels. The risk of transmission directly on canola seed, which is routinely cleaned and treated prior to planting, was therefore considered small (Rennie et al. 2011). Seed of other crops, such as field pea and potato tubers, can also carry *P. brassicae*-infested soil as an external contaminant, and may thereby also serve to disperse the pathogen.

9.3.1 Mechanism of Clubroot Spread in India

In India, clubroot has spread on cabbage, cauliflower, and *Brassica* crops in the Eastern Himalayan Darjeeling hills of West Bengal and South Indian Nilgiri hills in Tamil Nadu. Since the early 1980s, this pathogen has spread rapidly on rape and mustard grown in red and lateritic soil regions of West Bengal. At present clubroot disease is spreading at an alarming rate in the plains of north and south of West Bengal and its endemic form in Darjeeling hills is causing substantial damage to *Brassica* cultivation. Chemical control of clubroot disease is difficult due to prolong survival of pathogen as resting spores in soil, high pathogenic variability of different races, and very few varieties of *Brassicas* have shown resistance to the pathogen. Development of resistant lines is difficult due to narrow range of genetic variability of *Brassica* hosts in India and wide variability of pathogen. No resistance source has been found among the cultivated oilseed *Brassicas* and vegetables. All the cultivated *Brassicas* are highly susceptible to clubroot disease in West Bengal. The plant resistance to disease although genetically controlled may also be affected by environmental factors. Prolong survival of the pathogen in soil as resting spore is unique features of this disease. Survival of resting spores in soil and subsequent germination and infection process are affected by soil texture and structure; its moisture; temperature; pH; calcium, boron, and nitrogen content; and the presence of active microbial antagonists (Bhattacharya and Mondal 2016).

9.3.2 Clubroot Occurrence and spread in India

In India, clubroot is restricted to very few states (Fig. 9.1) on *Brassica* vegetables and oil yielding crops. However, its intensity is increasing on all kinds of *Brassica* species owing to no sources of resistance and narrow genetical base. Clubroot disease was surveyed for two consecutive years during 1998–2000 (Bhattacharya 2003) in an area covering the Darjeeling (altitude 2100 m), Kalimpong (1200 m), and Sonada (2550 m) sub-divisions in the Darjeeling Hills, situated in the Eastern Himalayan region and also in the foothills (Terai region) of North Bengal and plains of South Bengal. Kalimpong in the Darjeeling Hills was the hot spot for clubroot disease where all of the *Brassica* plants were found to be susceptible to *P. brassicae*. Rayosak (*Brassica juncea* L. ssp. *rugosa* (Roxb), a local leafy vegetable, and cauliflower were found to be the most susceptible hosts at all locations surveyed. Among the oilseed Brassicas in the Darjeeling Hills, Indian mustard (*B. juncea*) was severely diseased (rated 2–3 on a 0–3 scale). In the plains of North and South Bengal districts (soil pH 5.3–6.7) such as Cooch behar, Jalpaiguri, Birbhum, Bankura, West Midnapore, and Purulia, the *B. rapa* group of oilseeds was found to be the most susceptible host. The sporadic occurrence of clubroot on cole vegetables was also found in the foothills of Darjeeling, including the Cooch behar and Jalpaiguri districts of North Bengal. A more recent survey for clubroot in the Darjeeling Hills and plains of North and South Bengal (Baskey 2013) showed that clubroot disease is spreading at an alarming rate in the plains of North Bengal, expanding its host range from oilseed *Brassicas* to vegetables. With the exception of the Nadia,

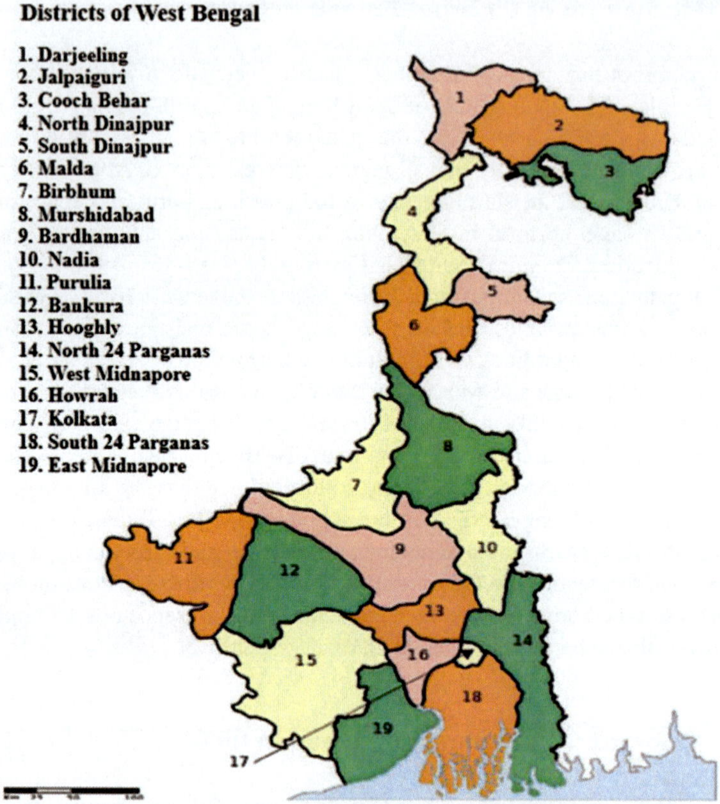

Fig. 9.1 Clubroot disease occurrence in West Bengal, India. The most severe symptoms have been observed in the Darjeeling Hills, Jalpaiguri, and Cooch behar districts of North Bengal and Birbhum, Bankura, Purulia, and West Midnapore on the red and lateritic soil regions of South Bengal. With the exception of Nadia, North 24 Parganas, and the coastal areas of South 24 Parganas and East Midnapore, clubroot disease is prevalent in all districts of West Bengal. Recently, clubroot has also been found in the Gangetic alluvial fertile and neutral soils of Howrah, Hooghly, Bardhaman, and Murshidabad districts (Bhattacharya et al. 2014)

North and South 24 Parganas and East Midnapore districts, clubroot was reported from all other districts at moderate to high disease severity (Fig. 9.1). Since 2003–2004, clubroot disease has started to appear on oilseed Brassicas in the most fertile part of West Bengal, the Gangetic alluvial region (soil pH 6.7–7.3), covering the Hooghly Bardhaman and Howrah districts of South Bengal. It is possible that in this potato-growing region, farmers applied large amounts of nitrogenous fertilizers such as urea to boost production. The urea, through ammonification and nitrification processes, made the soils more acidic (pH 5.7–6.7). Symptoms of clubroot started appearing when farmers replaced potato with *Brassica* oilseeds in their cropping sequences (Bhattacharya et al. 2014).

9.4 Clubroot Development in Relation to pH and Temperature Interaction

The interaction of a wide range of temperature and pH conditions on the development of clubroot on canola (or any other host) under controlled conditions was assessed by Gossen et al. (2013a). Soil pH above 7.0 reduced and delayed clubroot development, but moderately severe clubroot developed at pH 8.0 when temperatures were optimum for the pathogen (20–25 °C) and soil moisture was adequate. RHI, CI, and DSI were all at the maximum at 25 °C × pH 5.0–6.0, but decreased as pH increased, and temperature decreased. Similarly, soil pH was only weakly correlated with clubroot incidence and severity in 267 infested commercial canola fields in Alberta. In the growth cabinet studies, the high RHI observed at 25 °C × pH 6.0 was consistent with the maximum levels of RHI (Ayers 1944; Samuel and Garrett 1945; MacFarlane 1952; Sharma et al. 2011a). However, RHI decreased with increasing pH and decreasing temperature. RHI was low (40%) at 20 °C × pH 7.5–8.0. The development of *P. brassicae* in root hairs was influenced by both temperature and pH. At a combination of 25 °C × pH 5.0–5.5, and 20 °C × pH 5.5, the primary infection phase (primary plasmodia, differentiated plasmodia, and zoosporangia) had been completed by 12 DAI, leaving only empty sporangia in the root hairs after the zoospores had dehisced. Development was much slower at 15 °C × pH 6.0–8.0, such that only primary plasmodia and a few zoosporangia were present when the study was terminated at 12 DAI. Development was even more inhibited at 10 °C, where only a few primary plasmodia developed at the optimum pH of 6.0–6.5 and no pathogen development was observed in the other pH treatments.

Clubroot symptoms were not completely suppressed at slightly alkaline pH (7.5–8.0) although there was a substantial decline in severity (MacFarlane and Last 1959; Palm 1963; Myers and Campbell 1985). The development of moderate levels of clubroot at 15 °C and acidic pH indicates that clubroot can develop under less-than-ideal conditions in slightly acidic soils. RHI was strongly correlated with subsequent clubroot incidence and severity (Fig. 9.2). There was a strong correlation between primary infection and subsequent gall development under controlled conditions (MacFarlane 1952; Sharma et al. 2011a). In contrast, there was no correlation of RHI with clubroot symptoms in a greenhouse study (Voorrips 1992). It may be due to the complex factors that influence infection success by secondary zoospores, including moisture level, temperature, and pH. The weak negative correlation observed between soil pH and clubroot levels (CI and DSI) supports the report of a weak correlation between soil pH and clubroot levels on oilseed and turnip rape (*B. napus* and *B. rapa*) in Finland (Rastas et al. 2012). These results indicate that other factors, such as inoculum load (MacFarlane 1952; Hildebrand and McRae 1998), moisture (Colhoun 1952, 1953; Gossen et al. 2012b; Kasinathan 2012; Rastas et al. 2012), level of soluble calcium (Donald and Porter 2009), temperature (Thuma et al. 1983; McDonald et al. 2004; McDonald and Westerveld 2008; Gossen et al. 2012a, b), and cropping rotation (Rastas et al. 2012) likely also influence clubroot levels. The severe infestations of clubroot are likely to

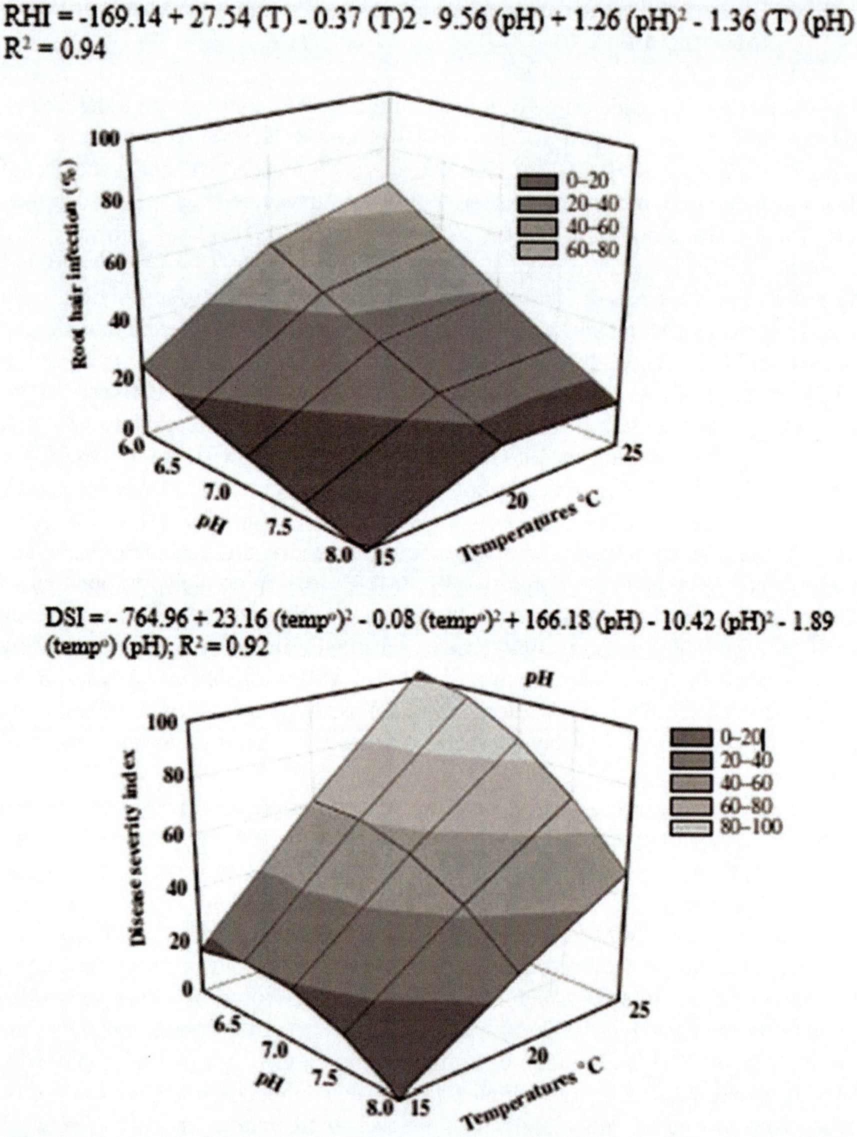

Fig. 9.2 The effect of temperature and pH on root-hair infection and subsequent clubroot severity in canola inoculated with *Plasmodiophora brassicae* under controlled conditions. Only those treatments that were assessed in both repetitions of these studies are presented (Gossen et al. 2013a, b)

develop more quickly and more consistently on acidic soils than alkaline soils, other factors being equal. This pattern of response almost certainly applies equally to canola, to other *Brassica* field crops, and to vegetable *Brassica* crops (Gossen et al.

2013b). Although application of lime to raise the soil pH is a common practice for reducing clubroot in susceptible vegetable crops, it does not always result in an effective reduction in clubroot levels (McDonald et al. 2004). While raising soil pH may afford some protection against clubroot, it is unlikely to be enough on its own to prevent the development of clubroot when inoculum pressure is high and environmental conditions are favorable for the pathogen. There may be potential for clubroot to spread into regions of the Canadian prairies with neutral or slightly alkaline pH if large numbers of resting spores are introduced through human activities such as movement of contaminated equipment or via movement by wind and water (Gossen et al. 2013b).

9.5 Effect of Host Age and Inoculum Density on Clubroot Development

The effect of inoculum density was studied by diluting heavily infested field soil with pathogen-free soil-less potting mix at seven increments, ranging from completely pathogen-free to 100% infested soil, and also by incorporating resting spores into the soil-less mix at concentrations of 1×10^5 to 1×10^8 spores cm^{-3}, along with a non-inoculated control. Seed of the susceptible canola cultivar 34 SS 65 was planted in soil of each treatment, grown to maturity, and rated for plant height, seed yield, and clubroot severity (0–3 scale) at harvest. Clubroot severity increased and plant height and seed yield decreased with increasing inoculum density. To assess the effect of seedling age on reaction to clubroot, resting spores of *P. brassicae* were inoculated onto roots of 1-, 2-, 3-, and 4-week-old seedlings of 34 SS 65. In addition, seed (0-week-old seedlings) was sown into infested soil. Inoculation of young seedlings resulted in higher clubroot severity (Figs. 9.3 and 9.4), shorter plants, and lower yield than inoculation of older seedlings (Hwang et al. 2011a).

Clubroot severity is positively correlated with inoculum density, disease severity (DS) increased from 0.6 to 0.9 with increasing inoculum density between 10^6 and 10^8 spores l^{-1} in the case of susceptible oilseed rape(OSR). In canola, DS increased from 0.4 to 1.0 with increasing inoculum density of 10^5 to 10^8 resting spores' cm^{-1} of soil (Hwang et al. 2011a). Disease severity of *B. rapa* increased from approximately 0.4 to 1.0 at spore conc. between 10^4 and 10^5 spores g^{-1} of soil, and DS of *B. oleracea* increased from 0.1 to 0.9 with increasing spore conc. between 10^4 and 10^6 spores g^{-1} of soil (Murakami et al. 2002). Murakami et al. (2002) observed that the role of inoculum density on symptom development changes with soil type, soil pH, plant species, cultivar resistance, and inoculum used. Therefore, estimating the damage potential of *P. brassicae* is difficult on the basis of the inoculum density in the soil alone. Little is known about the resting spore population in the soil of commercial fields. In Japan, the soil infestation level of the fields was usually below 10^6 spores g^{-1} of soil (Tsushima et al. 2010). On Swedish fields, soil infestation levels above 1.3×10^6 spores g^{-1} of soil were detected (Wallenhammar et al. 2012), and the conc. of *P. brassicae* spores on Canadian fields has been estimated to exceed

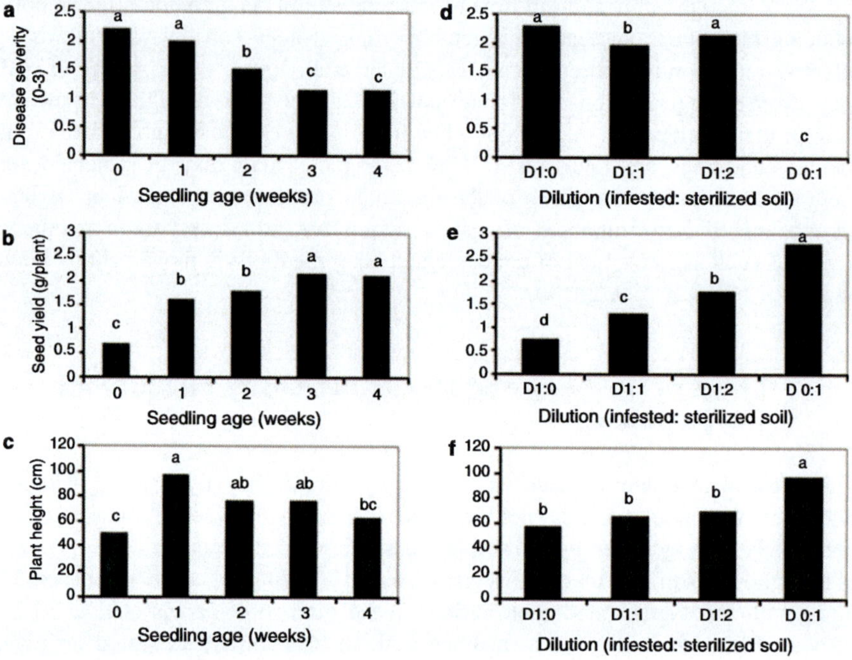

Fig. 9.3 Effect of seedling age (in weeks) at first exposure to infection and pathogen inoculum density [infested soil: sterilized soil (v/v)] on clubroot severity, plant height, and yield of canola under greenhouse conditions. No clubroot developed in the sterilized soil controls, so these data were dropped from the analysis. Bars with the same letter within a graph do not differ according to Fisher's protected least significant difference test at $P \leq 0.05$ (Hwang et al. 2011a)

10^8 spores g^{-1} of soil (Hwang et al. 2011b). These high soil infestation levels demonstrate the severe potential impact of *P. brassicae* on cruciferous crops on commercial fields.

Regression analysis revealed a significant influence of inoculum density and DS on seed yield of the susceptible OSR. Murakami et al. (2002) found that disease seldom occurs in an inoculum density of 10^3 spores' g^{-1} of soil in any crop or cultivar. Strehlow et al. (2015) recorded mean DS and seed yield losses high; yield losses of about 60% occurred at the generally accepted threshold level for symptoms development of 10^3 spores g^{-1} of soil (Faggian and Strelkov 2009). Hwang et al. (2011a) detected seed yield losses of about 80% at an inoculum density of 10^5 spores' cm^{-1} soil. However, all plants died due to severe clubroot infection at a similar spore concentration. This can be attributed to a high pathogenicity of the inoculum used. Hwang et al. (2011a) studied the effect of inoculum density of Canadian field isolate on canola; for similar DS by using at least 100× higher inoculum densities to get clubroot severity.

Field isolates of the new world seemed to be less virulent against *B. napus* compared with field isolates of Europe (Donald et al. 2006). Thus, already low

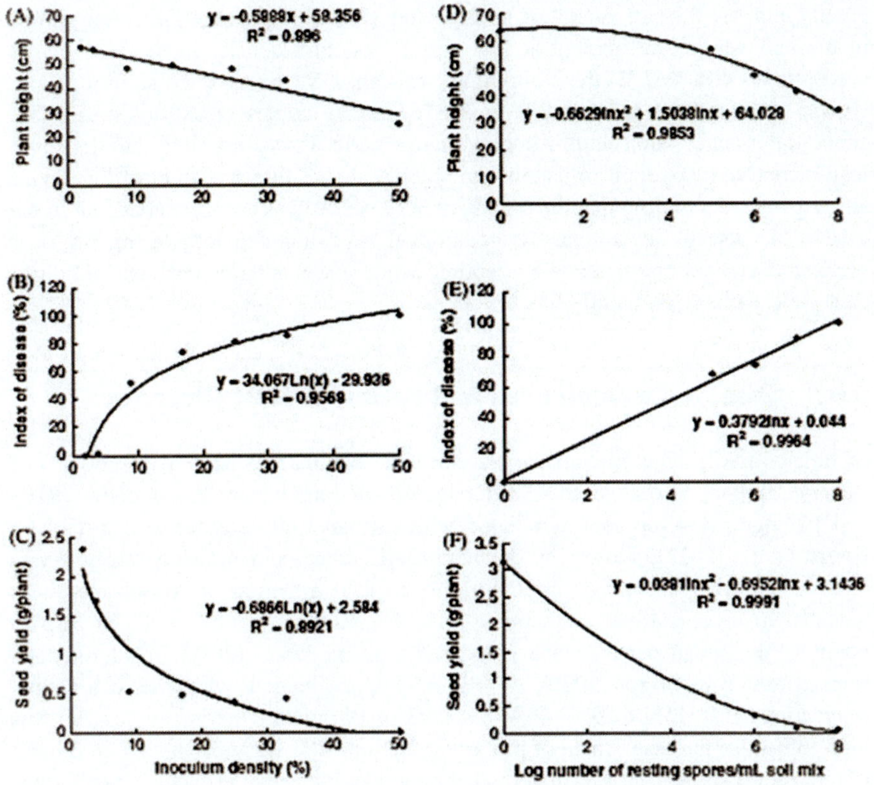

Fig. 9.4 Effect of inoculum density on plant height, clubroot severity (index of disease), and seed yield of canola ((**a–c**) dilution of naturally infested soil; (**d–f**) soil inoculated with known concentrations of resting spores extracted from root galls) (Hwang et al. 2011a)

levels of soil inoculum can become an important limiting factor for crop profitability especially in regions where *P. brassicae* possesses a high pathogenicity against *B. napus*. Strehlow et al. (2014) showed that *Plasmodiophora* field isolates in Germany are generally highly diverse. This indicates that *P. brassicae* displays great pathogenic variations. Therefore, information about the pathogenicity of the predominant field isolate is necessary when characterizing the damage potential of *P. brassicae*.

The green house experiment using the previous year soil revealed that mean DS was twice as high in the soil that was cultivated with the susceptible OSR compared with the soil that was cultivated with the resistant OSR (0.83 v/s 0.39). After cropping the resistant OSR, mean DS of the subsequent susceptible OSR was even lower compared with DS before repeated cropping of OSR at the lowest initial inoculum density. Mean DS did not differ significantly between the initial inoculum densities. All these findings suggest that (1) cropping susceptible cultivars will greatly increase the soil inoculum, as expected, (2) the propagation of resting spores in the soil was more efficient at low initial inoculum density in the case of the susceptible OSR, and (3) the cropping of resistant OSR reduced the soil inoculum. Hwang et al. (2013) counted the number of resting spores in club of susceptible canola after inoculation with different spores' conc. It was revealed that the spore load increased by over 60-fold at low and by less than 3-fold at high inoculum levels and supports the finding that the propagation of spores was density dependent in the case of the susceptible cultivars. In conclusion, inoculum density affected DS, seed yield, and yield components of susceptible winter OSR and demonstrated a significant yield damage potential of *P. brassicae*.

9.6 Disease Development in Relation to Host Resistance

In time-course studies of four canola cultivars inoculated with two pathotypes of *Plasmodiophora brassicae*, small differences in the rates of root-hair infection (RHI) and pathogen development associated with primary infection were identified by Deora et al. (2012). However, the major difference in resistance reaction was expressed during secondary infection. No cortical infection or colonization was detected in incompatible interactions ("45H21" × P6, "45H29" × P3 or P6). In contrast, the mean cortical area colonized was moderate (20%) in intermediate interactions ("In Vigor 5030" × P3 or P6) compared with the compatible interactions ("45H21 × P3," "46A76" × P3 or P6), where the area colonized was high (38%). In the assessment of primary infection, RHI was present at 4 DAI in all of the reaction types, but was lower in the incompatible and intermediate interactions than in the compatible interactions. However, total RHI in the incompatible interactions increased over time, so that by 12 DAI it was similar (64%) to the compatible interaction (68%) (Table 9.1). A similar pattern was observed for individual developmental stages of the pathogen: development was initially slower in the incompatible and intermediate interactions than in the compatible interactions, but development progressed slightly more quickly in the incompatible interaction over time. In canola (Hwang et al. 2011b) and Chinese cabbage (Tanaka et al. 2006), the primary infection developed later in resistant ("45H29") than susceptible cultivars. In comparison, the RHI in intermediate interactions remained lower (57%) at 12 DAI than in the compatible or incompatible interactions. Similar to RHI, fewer (9%) sporangia dehisced in the intermediate interactions. In several canola genotypes and two pathotypes interaction, RHI was slightly delayed in the incompatible interaction and slightly reduced in the intermediate interactions (Deora et al. 2012).

Even though differences in RHI in compatible and incompatible interactions were small, primary infection may play an important role in expression of resistance. It has been observed that high levels of cortical infection occurred in ryegrass inoculated with secondary zoospores of *P. brassicae*, but no cortical infection occurred in ryegrass where RHI occurred before exposure to secondary zoospores (Feng et al. 2012). Although primary infection is non-specific phenomenon that can

Table 9.1 Contrasts between the mean incidence (%) of developmental stages of *Plasmodiophora brassicae* during root-hair infection (RHI) for three classes of clubroot reaction†: compatible (C), intermediate (I), and incompatible (Inc), at 4 and 12 days after inoculation (DAI) (Deora et al. 2012)

	Means of reaction types (%)			Contrast estimates between means‡		
Developmental stages	C	I	Inc	C vs. I	C vs. Inc	I vs. Inc
Plasmodia						
4 DAI	43	35	33	−8***	−10***	−2
12 DAI	31	32	33	1	2*	1
4 vs. 12 DAI	−12***	−3	0	9***	12***	3
Sporangia						
4 DAI	10	5	4	−5***	−6***	−1
12 DAI	18	16	15	−2*	−3*	−1
4 vs. 12 DAI	8***	11***	11***	3*	3*	0
Dehisced sporangia						
4 DAI	6	3	3	−3*	−3*	0
12 DAI	19	9	15	−10***	−4*	6***
4 vs. 12 DAI	13***	6***	12***	−7***	−1	6***
Total RHI						
4 DAI	59	43	40	−16***	−19***	−3
12 DAI	68	57	63	−11***	−5*	6***
4 vs. 12 DAI	9***	14***	23***	5***	14***	9***

*-***Significant at $P < 0.05$ and 0.0001, respectively
†Reaction types: compatible ('45H21' × P3, '46A76' × P3 or P6); incompatible ('45H21' × P6, '45H29' × P3 or P6) and intermediate ('InVigor5030' × P3 or P6).
‡The comparisons are based on orthogonal single degree of freedomcontrasts.

occur in both host species and in plants that do not develop any clubroot symptoms (Ludwig-Muller et al. 1999), this indicates that primary infection may have a role in recognition and induction of resistance. Another study showed that RHI is related to the levels of indole acetic acid (IAA), a phytohormone that influences clubroot development. In cabbage, the level of IAA initially increased in susceptible cultivars and then declined, but continued to increase even after 14 DAI in resistant cultivars (Ludwig-Muller et al. 1993).

The cortical infection was completely inhibited in all of the incompatible interactions. Cultivar "45H29" was resistant to both P3 and P6, and "45H21" was resistant to only P6. This differential response to pathotype indicates that the genetics of resistance to P3 and P6 differ in these two cultivars. However, the timing of action of the resistance genes in these cultivars appears to be similar, because the pathogen was completely inhibited at the cortical infection stage. This indicates that high levels of resistance in canola are expressed in a manner similar to that observed in other hosts (Kroll et al. 1983; Kobelt et al. 2000; Tanaka et al. 2006; Donald et al. 2008). Morgner (1995) monitored the histology of *P. brassicae* infections in roots of resistant *B. napus* (ECD-06), *B. rapa* ("Chorus"), and *B. oleracea* ("Bohmerwaldkohl") and identified secondary stages, particularly the early secondary plasmodia, as present in the cortex 12 DAI, but no plasmodia or resting spores were detected at 22 DAI. Apparently, pathogen development was stopped before the completion of the life cycle. Similarly, Hwang et al. (2011b) reported the presence of secondary zoospores and secondary plasmodia in the resistant canola "45H29." Complete inhibition of secondary infection by *P. brassicae* has been reported in several host species, including radish (Kroll et al. 1983), Chinese cabbage (Tanaka et al. 2006), cauliflower (Donald et al. 2008), ECD-06 (Morgner 1995), and *Arabidopsis* (Kobelt et al. 2000). However, the mechanism of this resistance is not consistent across host species. Occurrence of a hypersensitive reaction to secondary infection has been observed in turnip (Dekhuijzen 1979) and *Arabidopsis* (Kobelt et al. 2000), but not in radish (Kroll et al. 1983), Chinese cabbage (Tanaka et al. 2006), or canola (Morgner 1995).

The area of cortical infection in the intermediate interactions was much lower than in the compatible interactions. This is consistent with a study on radish, where moderate resistance delayed the development of secondary infection (Kroll et al. 1983). In a study on *Arabidopsis*, there were differences in regulation of arginine catabolism and polyamine metabolism (associated with cell proliferation and hypertrophy) between a susceptible and a moderately resistant ecotype (Jubault et al. 2008). Additional studies to assess the role of IAA and arginine catabolism and polyamine metabolism in clubroot resistance in canola may provide insights into how resistance functions in this important field crop. Inoculation with *P. brassicae* reduced plant height and delayed plant development, irrespective of reaction type (compatible, intermediate, incompatible) or pathotype (P3, P6). This result supports a report of reduction in plant height and yield in both susceptible and resistant canola cultivars as a result of infection by *P. brassicae* (Hwang et al. 2011a). In general, P3 was more aggressive than P6, it infected more root hairs, colonized a larger proportion of the root cortex, and produced more severe clubroot symptoms than P6 in both the compatible and intermediate interactions. Pathotype 3 also reduced plant height and development more than P6. These results support a previous report that P3 is more aggressive than the other pathotypes in Canada (Cao et al. 2009).

9.6.1 Effect of Host Resistance on Resting Spore Populations in Soil

In principle, cropping of resistant cultivars should result in a reduction in the population of resting spores in the soil by stimulating spore germination. Also, few or no resting spores are produced in the resistant plants. However, fields cropped to resistant crops are not necessarily free of susceptible plants. Susceptible weed species are endemic, susceptible canola volunteers will continue to be present in infested fields for many years, and there may be a small percentage of susceptible siblings or off-types in seed lots. Each of these factors may offset the benefits that can be derived from a resistant cultivar (Hwang et al. 2012a). The effects of growing resistant and susceptible canola genotypes on resting spore populations were assessed under greenhouse and field conditions (Hwang et al. 2012a). One crop of susceptible canola contributed 1.4×10^8 resting spores ml^{-1} soil in a mini-plot experiment, and 1.0×10^{10} spores per g gall under field conditions. Repeated cropping of susceptible canola over four, 6-week cycles increased clubroot severity over time, while severity declined slightly in continuously fallow soil and where a resistant cultivar was repeatedly grown. When a resistant canola cultivar was included in the field study, the susceptible cultivar produced much larger galls. Moreover, galls in the resistant cultivar produced only 60% of the number of spores per g of gall mass than in the susceptible cultivar, which further reduced the relative contribution of inoculum by the resistant cultivar (Hwang et al. 2012a). Resting spore concentrations in the soil increased following the cultivation of both resistant and susceptible canola cultivars in heavily infested soil although the increase was greatest with a susceptible cultivar (Hwang et al. 2012a). In Sweden, partially resistant cultivars of oilseed turnip rape (*B. rapa* var. *oleifera*) grown on moderately infested soils produced a slight decrease in inoculum levels (Wallenhammar 1996). Taken together, these results indicate that production of resistant cultivars may reduce inoculum levels in lightly or moderately infested soils, but will not appreciably reduce clubroot infestations in heavily infested fields. Further research, including a long-term study, is required to confirm this view. A number of studies have examined the effect of susceptible weeds and volunteers on resting spore populations by growing mixtures of resistant and susceptible plants. In a recent trial, repeated cropping of a resistant cultivar reduced resting spore populations compared with a susceptible cultivar, and also reduced clubroot severity in the subsequent susceptible canola crop (Hwang et al. 2012a). However, soil inoculum loads were similar following resistant canola and continuous fallow. This indicates that while the resistant cultivar does not increase soil inoculum loads, it also does not act as a bait crop, i.e., it does not appear to significantly stimulate the germination of resting spores. However, the resistant crucifers reduced the numbers of residual resting spores after 4 years of continuous cultivation (Yamagishi et al. 1986). Colonization and gall formation also were reduced when susceptible and resistant radish (*Raphanus sativus* L.) cultivars were grown together, relative to the susceptible cultivar alone (Kroll et al. 1984). In Canada, clubroot severity was lower in a susceptible canola crop following a mixed crop (resistant + susceptible) relative to a susceptible cultivar only, but was higher than following fallow or a resistant cultivar only (Hwang et al. 2013). This indicates that susceptible volunteer canola may play an important role in the persistence of resting spore populations in infested fields. No empirical data are available, however, on the number of resting spores contributed by volunteers in infested fields where resistant canola is grown (Hwang et al. 2014).

9.7 Role of Weeds in Clubroot Epidemiology

Knowledge regarding clubroot hosts in the world is fragmentary. Theoretically, all species of the family Brassicaceae are potential hosts for *P. brassicae*. Karling (1968) described more than 300 clubroot hosts; however, isolates of the pathogen have different capacities to infect specific host genotypes (Honig 1931). Few infection studies have been performed outside the genera *Brassica*, *Raphanus*, and

Arabidopsis (Dixon 2009; Howard et al. 2010). Natural infections have been reported in genera *Brassica*, *Raphanus*, *Arabidopsis*, *Eruca* (Paz Lima et al. 2004), *Capsella* (Kim et al. 2011), *Cardamine* (Tanaka et al. 1993), *Matthiola* (Liu 2012), and *Isatis* (Xiong et al. 2012); among these, *Brassica, Raphanus* (Zhang and Zhao 2012), *Isatis, Matthiola*, and *Capsella* (Ren et al. 2014) are reported from various countries (see host range in chapter; subhead 2.5). Some non-cruciferous plants are also listed as hosts of *P. brassicae*, including nasturtium (*Tropaeolum majus*), papaya (*Carica papaya*), corn poppy (*Papaver rhoeas*), and clover (*Trifolium repens*) (Ludwig-Muller et al. 1999). The root-hair (primary) infection in these species has been observed with artificial inoculation under ideal environmental conditions.

Cruciferous weeds may play a role in clubroot epidemics because the disease has been reported in at least five species of cruciferous weeds (Kim et al. 2011; Ren et al. 2014; Tanaka et al. 1993). In Japan, clubroot on the cruciferous weed *Cardamine flexuosa* was distributed in 43 prefectures. Thus, this plant was considered to be the most suitable plant species to investigate the distribution of clubroot (Tanaka et al. 2006). In Britain, Buczacki and Ockendon (1979) observed clubroot infection on several common cruciferous weeds under controlled conditions, including wild cabbage (*Brassica oleracea*), shepherd's purse (*Capsella bursa-pastoris*), wall rocket (*Diplotaxis muralis*), treacle mustard (*Erysimum cheiranthoides*), white mustard (*Sinapis alba*), charlock (*S. arvensis*), hedge mustard (*Sisymbrium officinale*), and field penny cress (*Thlaspi arvense*). However, *C. bursa-pastoris, Sinapis alba*, herb sophia (*Descurainia sophia*), and pepper weed (*Lepidium apetalum*) are widespread cruciferous weeds in China. Saltwater cress (*Thellungiella salsuginea*) is also a common weed in saline fields in northern China. Knowledge of the susceptibility of these weeds to *Plasmodiophora brassicae* is limited but extremely important for management of clubroot on cultivated cruciferous crops. A field survey of natural infection and assays in pot cultured plants were conducted to gain a better understanding of the host range of *P. brassicae* in China. Ren et al. (2016) found 16 genera of crops, 9 ornamentals, and 5 weeds as potential sources of cruciferous epidemics.

9.8 Role of Land Elevation, Cropping Sequence, and Soil Conditions on Clubroot Epidemiology

Crop losses of about 30–80% were observed in the case of the *Brassica* leafy vegetable rayosak, which is cultivated under shade in moist and low land conditions in the Darjeeling district (Bhattacharya 2003). Resting spore populations of 1×10^7/g soil were recorded at Kalimpong in highly infested soils, as estimated by bioassays. The cultivation of more than one *Brassica* crop in a cropping sequence increases the risk of clubroot incidence. Inclusion of rice in a cropping sequence reduced disease incidence and severity in Kalimpong at Darjeeling Hills, possibly because the anaerobic conditions created in paddy/rice fields reduced the viability of *P. brassicae* resting spores (Bhattacharya 2003). Physical, chemical, and biological components of the soil environment affect the survival, growth, and reproduction of

P. brassicae (Dixon 2009). In the case of rapeseed-mustard cultivated under lowland moist conditions in the red and lateritic regions of West Bengal, clubroot development is favored when the temperatures fluctuate between 18 °C and 25 °C, the soil is sandy, the moisture level is above 50% of field capacity, and the pH is acidic conditions (5.3–6.5) (Sen 2005).

9.9 Disease Forecasting

Generation of site specific decision support system is essential component to forecast clubroot for integrated management of *P. brassicae* under a prevalent ecosystem of crucifers' cultivation. The ability to predict disease and expected yield loss is the foundation upon which an integrated clubroot control strategy can be implemented. In its simplest form, it involves a sound knowledge of site history, including the severity of disease in the most recent *Brassica* crop, the rotational history, site preparation, and treatments applied in previous crops. This information is readily available to most growers, but often not recorded. Molecular techniques offer rapid and reliable means for detecting and/or quantifying *P. brassicae* in the site. Sound local knowledge of factors such as soil type and climatic conditions and basic site history, these techniques can be powerful decision support tools for farmers. The data, information of the various experimental findings and knowledge assimilated thereof, on the *Plasmodiophora brassicae*—*Brassica* system involving Brassicas and cole vegetables, will lead to a simple, easy to use PC based decision support system. The proposed system will benefit the end users, researchers, and policy makers in agriculture towards forecasts regarding the outbreak of such disease in the various crops across the identified agroclimates and also be a useful tool towards planning for control of the pathogen. The congenial conditions for clubroot development are depicted in Fig. 9.5.

Preliminary forecasts of the potential distribution and severity of the disease in Canada have been developed using CLIMEX V2.0 software (Hearne Scientific Software Pty. Ltd., Melbourne, Australia). These models are based on temperature and moisture parameters obtained from previous reports, the occurrence of clubroot in epidemic areas, and an iterative process (Turkington et al. 2004). The resulting forecasts have predicted the occurrence of clubroot in regions where it is already a problem on cruciferous vegetables. They have also indicated the potential for losses in southeastern Manitoba and in central Alberta (Turkington et al. 2004). Adjustment of climatic data to account for above-average rainfall indicates that, in wetter years, clubroot could also be a problem over a wider region of the Prairies, including parts of Saskatchewan. Further refinement of the modeling parameters is ongoing and the risk maps produced may eventually serve as important tools in clubroot management (Howard et al. 2010).

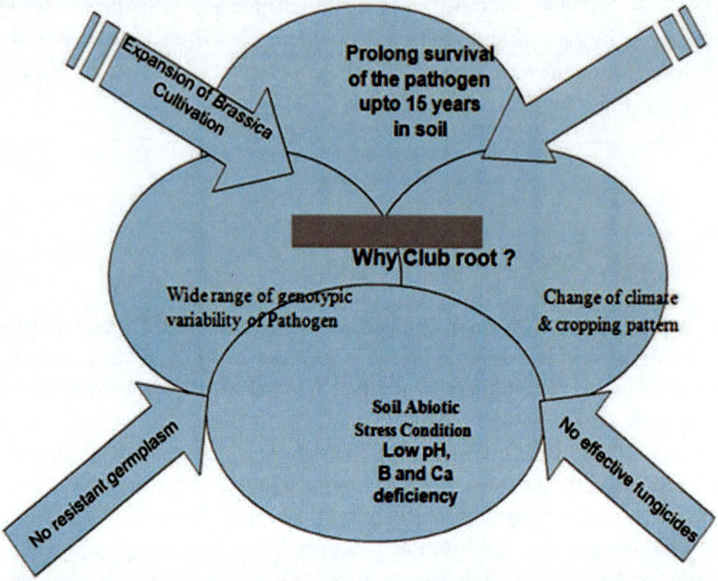

Fig. 9.5 Congenial factors for crucifer's clubroot disease (Bhattacharya and Mondal 2016)

9.10 The Risk Factors of Clubroot Spread

The rapid spread of clubroot is influenced by a combination of factors in the crucifers growing areas, viz. (1) a highly susceptible crop grown in large, often contiguous fields, (2) short intervals between successive susceptible crops, (3) movement of resting spores within and among fields by farm machinery, (4) development of high concentrations of resting spores in many fields before the disease was discovered, (5) lack of resistant cvs., and (6) lack of implementation of proper regulatory measures (quarantine) on movement of infected materials from clubroot endemic areas (Howard et al. 2010). This rapid spread is of particular concern because clubroot causes severe losses in susceptible oilseed crops, and many of the management methods used to reduce clubroot in high-value *Brassica* vegetable crops are not economical for these crops (Gossen et al. 2013b; Hwang et al. 2013).

Long-distance dispersal of the pathogen may occur in infested soil on farm machinery and miscellaneous vehicles/equipment, via wind or water erosion, with infested seed and inputs such as manure, and other mechanisms (Strelkov and Hwang 2014). The relative importance of wind- and waterborne dispersal has yet to be established definitively, but the wide geographic distribution of new sites of infestation indicated that at least some long-distance dispersal was occurring. If the pathogen has been dispersed by wind, low levels of resting spores are almost certainly already present at the soil surface across much of the region.

The impact of pH on development of clubroot on canola demonstrated that slightly alkaline soil pH alone did not eliminate the risk of clubroot when other conditions were conducive for pathogen infection and development (Gossen et al. 2013a). Similarly, assessments of soil type and organic matter under controlled conditions (Kasinathan 2012) indicated that these factors had a limited impact on development of clubroot when other conditions were conducive for the pathogen.

Soil moisture has been shown to consistently affect clubroot development (Thuma et al. 1983; Adhikari 2010; Kasinathan 2012; Gludovacz 2013). The interaction of soil moisture with soil pH, soil type, and micronutrient content may have an important impact on clubroot risk. Little is known about the nature of these interactions, but areas of low rainfall with well-drained soils are at much lower overall risk of clubroot than regions with high rainfall and heavy, compacted soils that drain slowly. However, even sites with low mean rainfall occasionally experience periods or even seasons of above-normal rainfall, and such fluctuations may have an important impact on the risk of clubroot establishment at a particular site. Also, high levels of clubroot can develop even after prolonged drought in a growing season. The heavy rains in mid-June ended a severe spring drought in central Alberta were associated with a severe epidemic of clubroot in field trials (Peng et al. 2011) and commercial fields (Strelkov et al. 2010, 2013).

Inoculum concentration is also an important factor in clubroot risk (MacFarlane 1952; Hildebrand and McRae 1998; Hwang et al. 2012b). Although susceptible plants can be infected by a single resting spore (Xue et al. 2008), a concentration of $\geq$1000 spores g^{-1} soil is needed to produce consistent clubroot symptoms under conditions that are highly conducive for infection (Hwang et al. 2011b). As conditions in the soil become increasingly unfavorable for the pathogen, based on combinations of low or fluctuating soil moisture, alkaline soil pH, high boron or calcium concentration, and other factors, the resting spore concentration required to produce consistent symptoms will also increase.

The need for a larger number of resting spores to initiate infection as conditions become less conducive for the pathogen is expected to have a substantial influence on the success of pathogen transmission associated with the various methods of dispersal. On the one hand, the minute quantities of resting spores that are introduced to a field in the dust on wheat seed harvested from an infested field and used to seed a subsequent wheat crop would almost certainly require optimum conditions to have any chance of infecting a susceptible crop grown in that field one or two years later. If conditions are even slightly less than ideal, there is little or no risk of successful establishment of the pathogen from this mode of transmission. In contrast, trillions of spores can be introduced to a field in each of the large clumps of infested soil that fall off contaminated field equipment. Over the potential lifetime of these long-lived resting spores, conditions would have to be highly unfavorable to reduce the risk of pathogen establishment if susceptible hosts (crops or weeds) are present over several years. The concentration of spores deposited by wind is likely to be much lower than the numbers of spores deposited in clumps of soil fallen from farm machinery. Weather systems occasionally pick up and deposit large volumes of soil at a considerable distance from the initial infested field. In general, however, deposition of windborne spores would be higher at sites close to a heavily infested field than at sites further out along the dispersal gradient. Therefore, conditions for infection would generally need to be highly favorable for the pathogen to establish at a site far from a large source of resting spores. Therefore, the risk of pathogen establishment from wind dispersal of spores would be higher at sites close to large sources of inoculum or in regions where conditions are highly conducive for the pathogen (high rainfall and heavy, acidic soils) and lower in areas where conditions are unfavorable (low rainfall, light or alkaline soils, high calcium or boron).

Long-distance dispersal of spores in water could occur in situations where runoff from infested fields is used to irrigate a susceptible crop. This was shown to be an important mechanism of dispersal in Australia (Donald et al. 2006) and in paddy irrigation systems in China (Yu et al. 2013). Rapid spread of the pathogen in an area by flooding has also been reported in Canada (Canadian Department of Agriculture 1956), but dispersal of resting spores by water on the Canadian prairies is more likely to occur over relatively short distances, associated with localized flooding (Gossen et al. 2014).

The clubroot spread risk factors have been analyzed in Canada by Gossen et al. (2014) which are eye opener for the crucifer growers to reduce the clubroot epidemics. These factors are applicable to clubroot-infected areas of the world with minor variations in the intensity of infection. Clubroot epidemic is closely tied to weather, climate, soil characteristics, and the mechanism that brings resting spores to a particular site. Large areas of Saskatchewan appear to be at a relatively low risk that *P. brassicae* will become established and develop into an important constraint to canola production in the near future. Low mean rainfall, the presence of unfavorable soil characteristics (especially neutral to alkaline pH and high calcium), and the physical distance from existing infested fields all contribute to this low risk. These analyses indicate that it would be very difficult for the pathogen to become established in these areas from wind- or seed-borne resting spores. However, it appears likely that introduction of large amounts of heavily infested soil from contaminated machinery into a field could allow the pathogen to establish. Once introduced, the pathogen has the potential to build up on susceptible crops and weeds whenever conditions are conducive for infection. Dissemination of the pathogen from an undetected site of infestation to other sites within the field and to adjacent fields would likely occur quickly on susceptible canola cultivars, based on the rate of expansion of infested sites observed in Alberta.

In large areas of Manitoba, conditions of precipitation, soil pH, and calcium concentration are much more conducive for the pathogen than many parts of Saskatchewan and southern Alberta. The risk of transmission of *P. brassicae* via wind and water is currently low in Manitoba because of the physical distance from existing infested fields, but *P. brassicae* appears to have the potential to be a serious constraint to canola production in Manitoba if it becomes established in that region. The development of resistant cultivars provides producers with options for clubroot management. Where populations of resting spores in a field are low, even a short break between canola crops has a positive impact on yield of resistant cultivars (Peng et al. 2013). Maintaining resting spore populations in the soil at low levels will also reduce the risk of a rapid breakdown of genetic resistance in these cultivars (Diederichsen et al. 2003; Peng et al. 2011). However, low populations of susceptible off-types, volunteer plants from susceptible cultivars, or even susceptible weeds will serve to maintain high levels of resting spores (Hwang et al. 2012b) in many infested fields. The best option in areas that are currently free of clubroot is to keep the pathogen out by ensuring that all equipment (including that of oil companies, utilities, custom contractors, and even hunters and other occasional visitors) is free of contaminated soil from infested areas.

References

Abbasi PA, Lazarovits G (2006) Effect of soil application of AG3 phosphonate on the severity of clubroot of Bok Choy and cabbage caused by *Plasmodiophora brassicae*. Plant Dis 90:1517–1522

Adhikari KKC (2010) Effect of temperature biofungicides and fungicides on clubroot of selected Brassica crops. University of Guelph Guelph ONT, Canada Master's Thesis

Adhikari KKC, McDonald MR, Gossen BD (2012) Reaction to *Plasmodiophora brassicae* pathotype 6 in lines of *Brassica* vegetables, Wisconsin Fast Plants, and canola. Hortic Sci 47:374–377

Ahmed HU, Hwang SF, Strelkov SE, Gossen BD, Peng G, Howard RJ, Turnbull GD (2011) Assessment of bait crops to reduce inoculum of clubroot (*Plasmodiophora brassicae*) of canola. Can J Plant Sci 91:545–551

Asad A, Bell RW, Dell B (2000) Uptake and distribution of boron in canola at vegetative and early flowering stages using boron buffered solution culture. Commun Soil Sci Plant Anal 31:2233–2249

Asad A, Blamey FPC, Edwards DG (2002) Dry matter production and boron concentrations of vegetative and reproductive tissues of canola and sunflower plants grown in nutrient solution. Plant Soil 243(2):243–252

Asano T, Kageyama K, Hyakumachi M (2000) Surface infestation of resting spores of *Plasmodiophora brassicae* used to infect hairy roots of *Brassica* spp. Phytopathology 89:314e319

Ayers GW (1944) Studies on the life history of the club root organism, *Plasmodiophora brassicae*. Can J Res 23:143–149

Baskey S (2013) Screening of *Brassica* hosts against *Plasmodiophora brassicae* causing club root disease of crucifers to identify sources of resistance [dissertation]. Bidhan Chandra Krishi Viswavidyalaya, Mohanpur, Nadia, West Bengal India

Belec C, Tremblay N (2004) Liming and calcium cyanamid for clubroot control in cauliflower. Acta Hortic 635:41–46

Bhattacharya I (2003) Study of race variation and characterization of *Plasmodiophora brassicae* causing clubroot disease of crucifers in West Bengal (India) Mohanpur, Bidhan Chandra Krishi Viswavidyalaya, Final Report ICAR (Indian Council of Agricultural Research) Adhoc Project. New Delhi, India

Bhattacharya TK, Mandal NC (2006) Management of clubroot (*Plasmodiophora brassicae*) of rapeseed and mustard by nitrogenous fertilizers. Ann Plant Prot Sci 14(1):260–261

Bhattacharya I, Mondal B (2016) Sustainable management of clubroot disease *Plasmodiophora brassicae* (Woronin) of Brassicas in West Bengal. J Bio-Resour Env Agric Sci 2(1):216–221

Bhattacharya I, Dutta S, Mondal S, Mondal B (2014) Clubroot disease on *Brassica* crops in India. Can J Plant Pathol 36(Sup1):154–160

Bramer H (1924) Unteesuchunqen über biologie und biacämpfuno de8 errboebh deb kohlhernib, plasmodiophora bbassicae wobonin. landw. Jahrb 59:227–243

Brown PH, Hu H (1998) Boron mobility and consequent management in different crops. Better Crops 82:28–31

Brown PH, Shelp BJ (1997) Boron mobility in plants. Plant Soil 193:85–101

Buczacki ST, Ockendon JG (1979) Preliminary observations on variation in susceptibility to clubroot among collections of some wild crucifers. Ann Appl Biol 92:113–118

Buczacki ST, Ockendon JG, Freeman GH (1978) An analysis of some effects of light and soil temperature on clubroot. Ann Appl Biol 88:229–238

Cakmak I, Kurz H, Marschner H (1995) Short-term effects of boron, germanium and high light intensity on membrane permeability in boron deficient leaves of sunflower. Physiol Plant 95:11–18

Camacho-Cristobal JJ, Rexach J, Gonzalez-Fontes A (2008) Boron in plants: deficiency and toxicity. J Integr Plant Biol 50:1247–1255

Campbell RN, Greathead AS (1996) Control of clubroot of crucifers by liming. In: Engelhard AW (ed) Soil borne pathogens: management of diseases with macro- and microelements. American Phytopathological Society (APS), St Paul, pp 90–101

Campbell RN, Greathead AS, Myers DF, DeBoer GJ (1985) Factors related to control of clubroot of crucifers in the Salinas Valley of California. Phytopathology 90:769–774

Cao T, Manolii VP, Hwang SF, Howard RJ, Strelkov SE (2009) Virulence and spread of *Plasmodiophora brassicae* [clubroot] in Alberta, Canada. Can J Plant Pathol 31:321–329

Carter M, Johnston H (1989) Association of soil macro porosity and relative saturation with root rot severity of spring cereals. Plant Soil 120:149–152

Chupp C (1917) Studies on clubroot of cruciferous plants. Bull Cornell Agric Expt Stat 387:419–452

Colhoun J (1952) Factors affecting the incidence of clubroot disease of *Brassica*. Nature 169:21–22

Colhoun J (1953) Observations on the incidence of club root disease of Brassicae in limed soils in relation to temperature. Ann Appl Biol 40:639–644

Colhoun J (1958) Club root disease of crucifers caused by *Plasmodipophora brassicae* Woron. A monograph. The Commonwealth Mycological Institute, Kew, Surrey, Phytopathological Paper No 3

Colhoun J (1973) Effects of environmental factors on plant disease. Annu Rev Phytopathol 11:343e364

Cook RJ, Papendick RI (1972) Influence of water potential of soils and plants on root disease. Annu Rev Phytopathol 10:349–374

Crute IR, Buczacki ST, Karen S (1981) A solution culture method for observing the development of clubroot symptoms on young *Brassica* plants. Ann Appl Biol 99:241–245

Datnoff LE, Elmer WH, Huber DM (2007) Mineral nutrition and plant disease. APS Press, St. Paul

Dekhuijzen HM (1979) Electron microscopic studies on the root hairs and cortex of a susceptible and a resistant variety of *Brassica campestris* infected with *Plasmodiophora brassicae*. Netherl J PIant Pathol 86:1–17

Deora A, Gossen BD, Walley F, McDonald MR (2011) Boron reduces development of clubroot in canola. Can J Plant Pathol 33:475–484

Deora A, Gossen BD, McDonald MR (2012) Infection and development of *Plasmodiophora brassicae* in resistant and susceptible canola cultivars. Can J Plant Pathol 34:239–247

Deora A, Gossen BD, Hwang SF, Pageau D, Howard RJ, Walley F, McDonald MR (2013) Effect of boron on clubroot of canola in organic and mineral soils and on residual toxicity to rotational crops. Can J Plant Sci 93:1

Diederichsen E, Deppe U, Sacristán MD (2003) Characterization of clubroot resistance in recent winter oilseed rape material. In: Proceedings of the 11th International Rapeseed Congress, Vol. 1, 2003 Jul 6; Copenhagen, Denmark, pp 68–70

Dixon GR (1996) Repression of the morphogenesis of *Plasmodiophora brassicae* Worn. by boron – a review. Acta Hortic 407:393–401

Dixon GR (2009) *Plasmodiophora brassiciae* in its environment. J Plant Growth Regul 28 (3):212–228

Dixon GR, Page LV (1998) Calcium and nitrogen eliciting alterations to growth and reproduction of *Plasmodiophora brassicae* (clubroot). Acta Hortic 459:343–349

Dixon GR, Tilston EL (2010) Soil-borne pathogens and their interactions with the soil environment. In: Dixon GR, Tilston EL (eds) Soil microbiology and sustainable crop production. Springer, Dordrecht, pp 197–271

Dixon GR, Wilson F (1983) Evaluation of calcium cyanamide for control of *Plasmodiophora brassicae* (clubroot). Tests of agrochemicals and cultivars no 4. Ann Appl Biol 102 (Suppl):50–51

Dobson R, Gabrielson RL, Baker AS (1982) Soil water matric potential requirements for root-hair and cortical infection of Chinese cabbage by *Plasmodiophora brassicae*. Phytopathology 72:1598–1600

Donald C, Porter IJ (2004) A sand solution culture technique used to observe the effect of calcium and pH on root hair and cortical stages of infection by *Plasmodiophora brassicae*. Aus Plant Pathol 33(4):585–589

Donald EC, Porter IJ (2009) Integrated control of clubroot. J Plant Growth Regul 28:289–303

Donald EC, Porter IJ, Lancaster RA (2002) Strategic application of lime, fertilizers and fungicides for improved control of clubroot. Cruciferae News Lett 24:81–82

Donald EC, Lawrence JM, Porter IJ (2004) Influence of particle size and application method on the efficacy of calcium cyanamide for control of clubroot of vegetable brassicas. Crop Prot 23:297–303

Donald EC, Porter IJ, Faggian R, Lancaster RA (2006) An integrated approach to the control of clubroot in vegetable *Brassica* crops. Acta Hortic 706:283–300

Donald EC, Jaudzems G, Porter IJ (2008) Pathology of cortical invasion by *Plasmodiophora brassicae* in clubroot resistant and susceptible *Brassica oleracea* hosts. Plant Pathol 57:201–209

Einhorn G, Bochow H (1990) The effects of temperature and pH value on the pathogenesis of *Plasmodiophora brassicae* Worn., the causal agent of clubroot. Arch Phytopathol Pflanzenschutz 26:131–138

Faggian R, Strelkov SE (2009) *Plasmodiophora brassicae* (clubroot) –a plant pathogen that alters host growth and productivity. J Plant Growth Regul 28:282–287

Feng J, Hwang R, Hwang SF, Strelkov SE, Gossen BD, Zhou Q, Peng G (2010) Molecular characterization of a serine protease Pro1 from *Plasmodiophora brassicae* that stimulates resting spore germination. Mol Plant Pathol 11:503–512

Feng J, Xiao Q, Hwang SF, Strelkov SE, Gossen BD (2012) Infection of canola by secondary zoospores of *Plasmodiophora brassicae* produced on a non-host (ryegrass). Eur J Plant Pathol 132:309–315

Fletcher JT, Hims MJ, Archer FC, Brown A (1982) Effects of adding calcium and sodium salts to field soils on the incidence of clubroot. Ann Appl Biol 100:245–251

Flores HE, Vivanco JM, Loyola-Vargas VM (1999) Radical biochemistry: the biology of root-specific metabolism. Trends Plant Sci 4:220–226

Friberg H, Lagerlof J, Ramert B (2005) Germination of *Plasmodiophora brassicae* resting spores stimulated by a non-host plant. Eur J Plant Pathol 113:275–281

Friberg H, Lagerlof J, Ramert B (2006) Usefulness of nonhost plants in managing *Plasmodiophora brassicae*. Plant Pathol 55:690–695

George-Okafor UO, Odibo FJC (2011) Purification and some properties of thermostable alkaline serine protease from thermophilic *Bacillus* sp. Gs-3. J Biol Sci 11:299–306

Gludovacz TV (2013) Clubroot in canola and cabbage in relation to soil temperature, plant growth and host resistance [dissertation]. University of Guelph, Guelph

Goldberg S (1997) Reactions of boron with soils. Plant Soil 193:35–48

Gossen BD, Adhikari KKC, McDonald MR (2012a) Effects of temperature on infection and subsequent development of clubroot under controlled conditions. Plant Pathol 61:593–599

Gossen BD, Adhikari KKC, McDonald MR (2012b) Effect of seeding date on development of clubroot in vegetable *Brassica* crops. Can J Plant Pathol 34:516–523

Gossen BD, Kasinathan H, Cao T, Manolii VP, Strelkov SE, Hwang SF, McDonald MR (2013a) Influence of pH and temperature on infection and symptom development of clubroot in canola. Can J Plant Pathol 35:294–303

Gossen BD, McDonald MR, Hwang SF, Strelkov SE, Peng G (2013b) Comparison of clubroot (*Plasmodiophora brassicae*) development and management on canola and *Brassica* vegetables. Can J Plant Pathol 35:175–191

Gossen BD, Deora A, Peng G, Hwang SF, McDonald MR (2014) Effect of environmental parameters on clubroot development and the risk of pathogen spread. Can J Plant Pathol 36:37–48

Gossen BD, Strelkov SE, Manolii VP, Rennie DC, Cao T, Hwang SF, Peng G, McDonald MR (2015) Spread of *Plasmodiophora brassicae* on canola in Canada, 2003–2014: old pathogen, new home. Can J Plant Pathol 37(4):403–413

Graham RD, Webb MJ (1991) Micronutrients and disease resistance and tolerance in plants. In: Mortvedt JJ, Cox FR, Schuman LM, Welch RM (eds) Micronutrients in Agriculture. Soil Sci Soc America, Madison, pp 329–370

Gupta UC (1993) Factors affecting boron uptake by plants. In: Gupta UC (ed) Boron and its role in crop production. CRC Press, Boca Raton, pp 87–104

Gupta UC, Cutcliffe JA (1978) Effects of methods of boron application on leaf tissue concentration of boron and control of brown-heart in rutabaga. Can J Plant Sci 58:63

Hachem JP, Man MQ, Crumrine D, Uchida Y, Brown BE, Rogiers V, Roseeuw D, Feingold KR, Elias PM (2005) Sustained serine proteases activity by prolonged increase in pH leads to degradation of lipid processing enzymes and profound alterations of barrier function and stratum corneum integrity. J Invest Dermatol 125:510–520

Hamilton HA, Crête R (1978) Influence of soil moisture, soil pH and liming source on the incidence of clubroot, the germination and growth of cabbage produced in mineral and organic soils under controlled conditions. Can J Plant Sci 58:45–53

Harling R, Kennedy SH (1991) Biological control of *Plasmodiophora brassicae* using a bait crop. In: Forty-third international symposium on crop protection, part 1. Mededelingen van de Faculteit, Rijksuniversiteit Gent Diergenseskunde vol. 56, pp 159–170

Hildebrand PD, McRae KB (1998) Control of clubroot caused by *Plasmodiophora brassicae* with non-ionic surfactants. Can J Plant Pathol 20(1):1–11

Honig F (1931) Der Kohlkropferreger (*Plasmodiophora brassicae* Wor.). Eine Monographie Die Gartenbauwissenschaft 5:116–225

Howard RJ, Strelkov SE, Harding MW (2010) Clubroot of cruciferous crops- New perspectives on an old disease. Can J Plant Pathol 32:43–57

Hwang SF, Ahmed HU, Strelkov SE, Gossen BD, Turnbull GD, Peng G, Howard RJ (2011a) Seedling age and inoculum density affect clubroot severity and seed yield in canola. Can J Plant Sci 91:183–190

Hwang SF, Ahmed HU, Zhou Q, Strelkov SE, Gossen BD, Peng G, Turnbull GD (2011b) Influence of cultivar resistance and inoculum concentration on root hair infection of canola (*Brassica napus*) by *Plasmodiophora brassicae*. Plant Pathol 60:820–829

Hwang SF, Ahmed HU, Zhou Q, Strelkov SE, Gossen BD, Peng G, Turnbull GD (2012a) Assessment of the impact of resistant and susceptible canola on *Plasmodiophora brassicae* inoculum potential. Plant Pathol 61:945–952

Hwang SF, Strelkov SE, Feng J, Gossen BD, Howard RJ (2012b) *Plasmodiophora brassicae*: a review of an emerging pathogen of the Canadian canola (*Brassica napus*) crop. Mol Plant Pathol 13:105–113

Hwang SF, Ahmed HU, Zhou Q, Rashid A, Strelkov SE, Gossen BD, Peng G, Turnbull GD (2013) Effects of cropping clubroot-resistant canola plants on *Plasmodiophora brassicae* resting spore populations in the soil. Plant Pathol 62:404–412

Hwang SF, Howard RJ, Strelkov SE, Gossen BD, Peng G (2014) Management of clubroot (*Plasmodiophora brassicae*) on canola (*Brassica napus*) in western Canada. Can J Plant Pathol 36(sup1):49–65

Ikegami H (1985) Decrease of clubroot fungus by cultivation of different crops in heavily infested soil. Res Bull Faculty Agric Gifu Univ 50:19–32

Jubault M, Hamon C, Gravot A, Lariagon C, Delorme R, Bouche-Reau A, Manzanzares-Dauleux MJ (2008) Differential regulation of root arginine catabolism and polyamine metabolism in clubroot-susceptible and partially resistant *Arabidopsis* genotypes. Plant Physiol 146:2008–2019

Kageyama K, Asano T (2009) Life cycle of *Plasmodiophora brassicae*. J Plant Growth Regul 28:203–211

Karling JS (1968) The plasmodiophorales. Hafner Publishing Company, New York

Kasinathan H (2012) Influence of pH, Temperature, and Biofungicides on Clubroot of Canola [dissertation]. University of Guelph, Guelph

Kasinathan H, Gossen BD, Mcdonald MR (2011) The influence of temperature and pH on clubroot (*Plasmodiophora brassicae*) symptom development in canola under controlled environment conditions. Can J Plant Pathol 33:260

Kim WG, Lee SY, Choi HW, Hong SK, Lee YK (2011) Occurrence of clubroot on shepherd's-purse caused by *Plasmodiophora brassicae*. Mycobiology 39:233–234

Klasse HJ (1999) Calcium cyanamide–a unique source of nitrogen promoting healthy growth and improving crop quality in vegetables. In: Improved crop quality by nutrient management. Kluwer Academic Publishers, Dordrecht, pp 233–235

Kobelt P, Siemens J, Sacristan M (2000) Histological characterization of the incompatible interaction between *Arabidopsis thaliana* and the obligate biotrophic pathogen *Plasmodiophora brassicae*. Mycol Res 104(2):220–225

Kroll TK, Lacy GH, Moore LD (1983) A quantitative description of the colonization of susceptible and resistant radish plants by *Plasmodiophora brassicae*. J Phytopathol 108:97–105

Kroll TK, More LD, Lacy GH (1984) Inter planting susceptible and resistant radish cultivars reduces colonization by *Plasmodiophora brassicae*. Hortic Sci 19:403–404

Liu WW (2012) Control of clubroot on *Matthiola incana*. Yunnan Agric 11:20

Loomis WD, Durst RW (1992) Chemistry and biology of boron. Biofactors 3:229–239

Ludwig-Muller J, Bendel U, Thermann P, Ruppel M, Epstein E, Hilgenberg W (1993) Concentrations of indole-3- acetic acid in plants of tolerant and susceptible varieties of Chinese cabbage infected with *Plasmodiophora brassicae* Woron. New Phytol 125:763–769

Ludwig-Muller J, Bennett RN, Kiddle G, Ihmig S, Ruppel M, Hilgenberg W (1999) The host range of *Plasmodiophora brassicae* and its relationship to endogenous glucosinolate content. New Phytol 141:443–458

Maas EV (1969) Calcium uptake by excised maize roots and interactions with alkali cations. Plant Physiol 44:988–989

MacFarlane I (1952) Factors affecting the survival of *Plasmodiophora brassicae* Worn. in the soil and its assessment by a host test. Ann Appl Biol 39:239–256

MacFarlane I (1970) Germination of resting spores of *Plasmodiophora brassicae*. Trans Br Mycol Soc 55:97–112

MacFarlane I, Last F (1959) Some effects of *Plasmodiophora brassicae* Woron. on the growth of the young cabbage plant. Ann Bot 23:547–570

McDonald MR, Westerveld SM (2008) Temperature prior to harvest influences the incidence and severity of clubroot on two Asian *Brassica* vegetables. Hortic Sci 43:1509–1513

McDonald MR, Kornatowska B, McKeown AW (2004) Management of clubroot of Asian *Brassica* crops grown on organic soils. Acta Hortic 635:25–30

Merino-Gergichevich C, Alberdi M, Ivanov AG, Reyes-Díaz M (2010) Al3þeCa2þ interaction in plants growing in acid soil: Al-phytotoxicity response to calcareous amendments. J Soil Sci Plant Nutr 10:217–243

Monteith J (1924) Relation of soil moisture and soil temperature to infection by *Plasmodiophora brassicae*. J Agric Res 28:549–562

Moots CK, Nickell CD, Gray LE (1988) Effects of soil compaction on the incidence of *Phytophthora megasperma* f. sp. *glycinea* in soybean. Plant Dis 72:896–900

Morgner M (1995) Mikroskopische Untersuchungen der Reaktion empfindlicher und resistenter Genotypen verschiedener *Brassica*-Arten gegenuber *Plasmodiophora brassicae* Worn. Ph.D. Thesis, Free University of Berlin, Berlin

Murakami H, Tsushima S, Shishido Y (2000) Soil suppressiveness to clubroot disease of Chinese cabbage caused by *Plasmodiophora brassicae*. Soil Biol Biochem 32:1637–1642

Murakami H, Tsushima S, Akimoto T, Shishido Y (2001) Reduction of spore density of *Plasmodiophora brassicae* in soil by decoy plants. J Gen Plant Pathol 67:85–88

Murakami H, Tsushima S, Kuroyanagi Y, Shishido Y (2002) Reduction in resting spore density of *Plasmodiophora brassicae* and clubroot severity by liming. Soil Sci Plant Nutr 48:685–691

Myers DF, Campbell RN (1981) Clubroot of crucifers in California: soils respond differentially to lime for clubroot control (Abstr.). Phytopathology 71:1005–1006

Myers DF, Campbell RN (1985) Lime and the control of clubroot of crucifers: effects of pH, calcium, magnesium and their interactions. Phytopathology 75:670–673

Naiki T, Dixon GR (1987) The effects of chemicals on developmental stages of *Plasmodiophora brassicae* (clubroot). Plant Pathol 36(3):316–327

Niwa R, Kumei T, Nomura Y, Yoshida S, Osaki M, Ezawa T (2007) Increase in soil pH due to Ca – rich organic matter application causes suppression of clubroot disease of crucifers. Soil Biol Biochem 39:778–785

Niwa R, Nomura Y, Osaki M, Ezawa T (2008) Suppression of clubroot disease under neutral pH caused by inhibition of spore germination of *Plasmodiophora brassicae* in the rhizosphere. Plant Pathol 57:445–452

Nott H, Falloon R, Cheah LH (1999) Clubroot control from safe chemicals. NZ Commer Grower 54:24–26

Page LV (2001) Studies of components for a potential integrated control system for *Plasmodiophora brassicae*. PhD thesis, University of Strathclyde, Glasgow UK

Palm ET (1963) Effect of mineral nutrition on the invasion and response of turnip tissue to *Plasmodiophora brassicae* Worn. Contribut Boyce Thompson Inst 22:91–112

Parks WL, White JL (1952) Boron retention by clay and humus systems saturated with various cations. Soil Sci Soc Am Proc 16:298–300

Paz Lima MLP, Cafe-Filho AC, Nogueira NL, Rossi ML, Schuta LR (2004) First report of clubroot of *Eruca sativa* caused by *Plasmodiophora brassicae* in Brazil. Plant Dis 88:573

Peng G, Lahlali R, Hwang SF, Hynes RK, McDonald MR, Pageau D, Gossen BD, Strelkov SE, Mc Grego L (2011) Control of clubroot on canola using the biofungicide Serenade plus cultivar resistance. In: Proceedings of the 13th international rapeseed congress. Czech Republic, Prague, pp 1134–1137

Peng SS, Ren ZH, Huang XL, Liu MJ, Sun LF, Liu EM (2013) Physiological race identification on clubroot of cruciferous crops caused by *Plasmodiophora brassicae* in Hunan. J Changjiang Veg 6:46–49

Peng G, Pageau D, Strelkov SE, Gossen BD, Hwang SF, Lahlali RL (2015) A more than 2-year break from susceptible canola reduces *Plasmodiophora brassicae* resting spores in heavily infested soil and maximizes canola yield. Eur J Agron 70:78–84

Porth G, Mangan F, Wick R, Autio W (2003) Evaluation of management strategies for clubroot (*Plasmodiophora brassicae* Woron) in Brassicaceae. Proc Int Am Soc Trop Hortic 46:78–80

Raghavan GSV, Taylor F, Vigier B, McKyes E, Gauthier L (1982) Effect of compaction and root rot disease on development and yield of peas. Can Agric Eng 24:31–34

Rashid A, Ahmed H, Hwang SF, Strelkov S (2011) Effect of a wide range of pH values on resting spore germination and clubroot (*Plasmodiophora brassicae* Woronin) severity on canola. In: Proceed. Plant Pathological Society of Alberta Conference, Edmonton, Alberta, Canada, November 7–9

Rashid A, Ahmed HU, Xiao Q, Hwang SF, Strelkov SE (2013) Effects of root exudates and pH on *Plasmodiophora brassicae* resting spore germination and infection of canola (*Brassica napus* L.) root hairs. Crop Prot 48:16–23

Rastas M, Latvala S, Hannukkala A (2012) Occurrence of *Plasmodiophora brassicae* in Finnish turnip rape and oilseed rape fields. Agric Food Sci 21:141–158

Ren L, Fang XP, Sun CC, Chen KR, Liu F, Li M, Xu L (2014) First report of clubroot on *Capsella bursa pastoris* caused by *Plasmodiophora brassicae* in Sichuan province of China. Plant Dis 98:687

Ren L, Xu L, Liu F, Chen K, Sun C, Li J, Fang X (2016) Host range of *Plasmodiophora brassicae* on cruciferous crops and weeds in China. Plant Dis 100:933–939

Rennie DC, Manoli VP, Cao T, Hwang SF, Howard RJ, Strelkov SE (2011) Direct evidence of surface infestation of seeds and tubers by *Plasmodiophora brassicae* and quantification of spore loads. Plant Pathol 60:811–819

Rennie DC, Holtz MD, Turkington TK, LeBoldus JM, Hwang SF, Howard RJ, Strelkov SE (2015) Movement of *Plasmodiophora brassicae* resting spores in windblown dust. Can J Plant Pathol 37:188–196

Robak J (1996) The effect of some crop rotations on decrease of clubroot, *Plasmodiophora brassicae*, in soils. In: Proceedings Brighton Crop Protection Conference: Pests and Diseases, 1996, pp 647–651

Ruaro L, Lima Neto VDC, Justiniano P, Júnior R (2009) Influence of boron, nitrogen sources and soil pH on the control of club root of crucifers caused by *Plasmodiophora brassicae*. Trop Plant Pathol 34:231–238

Ruiz JM, Bretones G, Baghour M, Ragala L, Belakbir A, Romero L (1998) Relationship between boron and phenolic metabolism in tobacco leaves. Phytochemistry 48:269–272

Samuel G, Garrett SD (1945) The infected root hair count for estimating the activity of *Plasmodiophora brassicae* Woron. in the soil. Ann Appl Biol 32:96–101

Saude C, Mckeown A, Gossen BD, Mc Donald MR (2012) Effect of host resistance and fungicide application on clubroot pathotype 6 in green cabbage and napa cabbage. Hortic Technol 22:311–319

Saurabh S, Jasmine I, Pritesh G, Rajendra Kumar S (2007) Enhanced productivity of serine alkaline protease by *Bacillus* sp. using soybean as substrate. Malaysian J Microbiol 3:1–6

Sen P (2005) Antagonistic effect of Ca, B and Mo on clubroot disease of rapeseed-mustard. Indian Agric 49(1/2):13–16

Sharma K, Gossen BD, McDonald MR (2011a) Effect of temperature on primary infection by *Plasmodiophora brassicae* and initiation of clubroot symptoms. Plant Pathol 60:830–838

Sharma K, Gossen BD, McDonald MR (2011b) Effect of temperature on cortical infection by *Plasmodiophora brassicae* and clubroot severity. Phytopathology 101:1424–1432

Stangoulis JCR, Graham RD (2007) Boron and plant disease. In: Datnoff LE, Elmer WH, Huber DM (eds) Mineral nutrition and plant disease. APS Press, St. Paul, pp 207–214

Stangoulis JCR, Brown PH, Bellaloui N, Reid RJ, Graham RD (2001) The efficiency of boron utilization in canola. Aust J Plant Physiol 28:1109–1114

Strehlow B, de Mol F, Struck C (2014) History of oilseed rape cropping and geographic origin affect the genetic structure of *Plasmodiophora brassicae* populations. Phytopathology 104:532–538

Strehlow B, de Mol F, Struck C (2015) Risk potential of clubroot disease on winter oilseed rape. Plant Dis 99:667–675

Strelkov SE, Hwang SF (2014) Clubroot in the Canadian canola crop: 10 years into the outbreak. Can J Plant Pathol 36(S1):27–36

Strelkov SE, Manolii VP, Marquez-Zequera I, Manolii E, Hwang SF (2010) Incidence of clubroot on canola in Alberta in 2009. Can Plant Dis Surv 90:123–125

Strelkov SE, Manolii VP, Rennie DC, Manolii EV, Fu H, Strelkov IS, Hwang SF, Howard RJ, Harding MW (2013) The occurrence of clubroot on canola in Alberta in 2012. Can Plant Dis Surv 93:145–148

Suzuki K, Matsumiya E, Ueno Y, Mizutani J (1992) Some properties of germination-stimulation factor from plants for resting spores of *Plasmodiophora brassicae*. Ann Phytopathol Soc Jpn 58:699–705

Takahashi K (1994) Influences of some environmental factors on the viability of resting spores of *Plasmodiophora brassicae*. Ann Phytopathol Soc 60:658–666

Takahashi H, Takita K, Kishimoto T, Mitsui T, Hori H (2002) Ca2+is required by clubroot–resistant turnip cells for transient increases in PAL activity that follows inoculation with *Plasmodiophora brassicae*. J Phytopathol 150(10):529–535

Takahashi H, Ishikawa T, Kaido M, Takita K, Hayakawa T, Okazaki K, Itoh K, Mitsui T, Hori H (2006) *Plasmodiophora brassicae* induced cell death and medium alkalisation in clubroot-resistant cultured roots of *Brassica rapa*. J Phytopathol 154(3):156–162

Tanaka M, Fujiwara T (2008) Physiological roles and transport mechanisms of boron: perspectives from plants. Eur J Phys 456:671–677

Tanaka S, Ito S, Kameya-Iwaki M, Katumoto K, Nishi Y (1993) Occurrence and distribution of clubroot disease on two cruciferous weeds, *Cardamine flexuosa* and *C. scutata* in Japan. Trans Mycol Soc 34:381–388

Tanaka S, Mido H, Ito S (2006) Colonization by two isolate of *Plasmodiophora brassicae* on a clubroot-resistant cultivar of Chinese cabbage *Brassica rapa* L.spp *pekinensis*. J Gen Plant Pathol 72(4):205–209

Thuma BA, Rowe RC, Madden LV (1983) Relationships of soil temperature and moisture to clubroot (*Plasmodiophora brassicae*) severity on radish in organic soil. Plant Dis 67:758–762

Tremblay N, Belec C, Coulombe J, Godin C (2005) Evaluation of calcium cyanamide and liming for control of clubroot disease in cauliflower. Crop Prot 24(9):798–803

Tsushima S, Murakami H, Akimoto T, Katahira M, Kuroyanagi Y, Shishido Y (2010) A practical estimating method of the dose-response curve between inoculum density of *Plasmodiophora brassicae* and the disease severity for long-term IPM strategies. JARQ 44:383–390

Turkington TK, Olfert OO, Weiss RM, Clear RM, Tewari JP, Strelkov SE (2004) Forecasting the potential distribution and abundance of plant diseases using CLIMEXTM modeling with historical and potential weather scenarios associated with climate change. In: Proceedings of Manitoba Agron. Conf. 2004, pp 99–110. http://www.umanitoba.ca/afs/agronomists_conf/proceedings/2004/turkington_forecasting_potential.pdf

Voorrips RE (1992) Root hair infection by *Plasmodiophora brassicae* in clubroot-resistant and susceptible genotypes of *Brassica oleracea*, *B. rapa* and *B. napus*. Netherl J Plant Pathol 98:361–368

Walker TS, Bais HP, Grotewold E, Vivanco JM (2003) Root exudation and rhizosphere biology. Plant Physiol 132:44–51

Wallenhammar AC (1996) Prevalence of *Plasmodiophora brassicae* in a spring oilseed rape growing area in central Sweden and factors influencing soil infestation levels. Plant Pathol 45:710–719

Wallenhammar AC, Almquist C, Soderstrom M, Jonsson A (2012) In-field distribution of *Plasmodiophora brassicae* measured using quantitative real-time PCR. Plant Pathol 61 (1):16–28

Webster MA (1986) pH and nutritional effects on infection by *Plasmodiophora brassicae* Worn. and on clubroot symptoms. PhD thesis, University of Aberdeen, p 273

Webster MA, Dixon GR (1991a) Calcium, pH and inoculum concentration influencing colonization by *Plasmodiophora brassicae*. Acta Hortic 95:64–73

Webster MA, Dixon GR (1991b) Boron, pH and inoculum concentration influencing colonization by *Plasmodiophora brassicae*. Mycol Res 95:74–79

Wellman FL (1930) Club root of crucifers. Technical Bulletin of the United States Department of Agriculture, no. 181, p 331

Xiong GR, Zheng JF, Wu YX, He YQ (2012) A new disease, clubroot of *Isatis tinctoria* L. J Agric Catastro 2:3–4

Xue S, Cao T, Howard RJ, Hwang SF, Strelkov SE (2008) Isolation and variation in virulence of single-spore isolates of *Plasmodiophora brassicae* from Canada. Plant Dis 92:456–462

Yamagishi H (1987) Cultivation of clubroot-resistant strains to suppress population density of resting spores of *Plasmodiophora brassicae* in the soil. Jpn Agric Res Qurt 21:36–41

Yamagishi H, Yoshikawa H, Ashizawa M, Hida K, Yui S (1986) Effects of resistant plants as a catch crop on the reduction of resting spores of clubroot (*Plasmodiophora brassicae* Worn.) in soil. J Jpn Soc Hort Sci 54:460–466

Yermiyahu U, Keren R, Chen Y (1988) Boron sorption on composted organic matter. Soil Sci Soc Am J 52:1309–1313

Yu XK, Wu YX, Mao ZC, He YQ (2013) Water-mediated dissemination and chemical control of cabbage clubroot disease. J Huazhong Agric Univ 1:48–53

Zhang ZM, Zhao YC (2012) Identification of new symptoms of radish clubroot. Chin Plant Prot 12:44–45

Biochemistry of Host–Pathogen Interaction 10

Abstract

The life cycle of *P. brassicae* is entirely different from other biotrophic pathogens. The intimate relationship between host and pathogen leads to changes in hormonal status, cellular development of pathogen, and sources sink relations of clubbed roots and shoot of host plant. In the infected plants growth-promoting hormones such as auxins, cytokinin, and changes in secondary metabolites are associated with club development which exhibits their effect either directly on growth hormones or indirectly via their bioactive properties. The changes in status, and metabolism of hormones, and compounds like auxins, cytokinin, glucosinolates, camalexin, brassinosteroids, gibberellin, jasmonic acid, ethylene, flavonoids, and sugars have effects on clubroot phenotypes. The metabolic sink results in the accumulation of starch grains in clubbed tissues. The different pathways and hormonal signals result in a homeostasis between the pathogen and the host plant resulting in galls formation and/or suppression of gall growth. A protein (PbGH3) can modify auxins and jasmonic acid produced in infected root galls. The resilient resting spores of the pathogen contain chitin in their cell walls. A trehalase encoding gene is primarily expressed in spores to release glucose as an energy source in the germination process. The *PLeff* genes are expressed in plasmodia, the *Heff* genes are expressed in the transcriptomes of the host, and the *Leff* genes are expressed late in the life cycle during spore formation. A complete pathway for the spermidine biosynthesis has been predicted in the *P. brassicae.*

Keywords

Biochemistry of host pathogen interaction · Pathogen and host hormones metabolism · Expression of hormones in cultivars · Auxin metabolism · Signaling · Cytokinin metabolism · Other plant hormones · Metabolic pathways ·

G. S. Saharan et al., *Clubroot Disease of Crucifers*,
https://doi.org/10.1007/978-981-16-2133-8_10

CAZymes · Prediction of new pathways · Host response through transcriptome · Genes regulating enzymes · Nutrient acquisition

10.1 Introduction

Biochemical and physiological events during *Plasmodiophora brassicae* infection, pathogenesis, and expression of host resistance are very essential parts in the life cycle of clubroot disease to plan disease management strategies with the co-operation and coordination of clubroot researchers from the disciplines of Plant Pathology, Biochemistry, Molecular biology, Genetics and Plant Breeding. Identification of genes responsible for infection, pathogenesis, and host resistance can be used to find out the ways of modification of clubroot infection and suppression in the host. Expression of an invertase inhibitors gene, altered trehalose metabolism, and arginine metabolism are the events which can suppress clubroot development.

10.2 *Plasmodiophora brassicae* and Host Hormone Metabolism

Infection by *P. brassicae* results in the misregulation of plant-derived hormonal pathways leading to alterations in the root phenotype, called clubs or clubroots, as well as reduced growth in the aerial parts of the infected plants. Assimilates from leaves are redirected towards the colonized roots, thereby reducing the overall plant performance (Ludwig-Muller 2009a, b; Siemens et al. 2011). The root gall size can be associated with plant growth-promoting hormones, especially auxins and cytokinins. Secondary metabolites are, however, also associated with club development, either directly through the action of growth hormones or indirectly via their general bioactive properties. Lastly, the hormones also control primary metabolism, such as sugar availability for the pathogen.

The role of auxins in the development of typically hypertrophied root tissues (clubroots) has been partially elucidated. The model plant *A. thaliana* is a good host for *P. brassicae*, so its metabolism was studied using genetic, biochemical, and molecular tools. A simple model (Fig. 7.9; Chap. 7) indicates that *P. brassicae* causes synthesis of cytokinins (CK) (Muller and Hilgenberg 1986) and blocks the CK degradation pathway in the host plant (Siemens et al. 2006), increasing the CK levels in infected tissues. In addition, the host CK synthesis is also up-regulated (Ludwig-Muller 2009b). CKs can then be transported to the upper part of the plant, leading to an increase in assimilate production. Sucrose and other sugars are transported back to the metabolic sink in the roots created by the elevated CK levels (Fig. 7.9, Chap. 7; Table 10.1). This is indicated by the large numbers of starch grains present in infected roots with well-developed clubs (Plate 7.6; Chap. 7). Simultaneously, host auxin synthesis is induced by *P. brassicae* and perhaps the indole glucosinolate pathway is hijacked by the pathogen, producing larger amounts of free IAA via the nitrilase pathway (Ludwig-Muller 2009a). The auxin produced in

Table 10.1 Selected changes in hormone status and metabolism, and corresponding effects on the clubroot phenotype in *Arabidopsis thaliana* (Diederichsen et al. 2014)

Hormone/ compound	Genes or pathways altered	Type of alteration	Effect on clubroot	Reference
Auxin	Biosynthesis	Mutant *nit1* antisense *nit2* overexpression NIT2	Reduction	Grsic-Rausch et al. (2000)
			Delayed gall formation	Neuhaus et al. (2000)
			None	Neuhaus et al. (2000)
	Conjugate hydrolysis	Mutant *ilr1*	None	Siemens et al. (2002)
	Conjugate synthesis	Single mutant various *GH3,* genes	None	Jahn et al. (2013)
	Conjugate synthesis	Double mutant *gh3*.5, *gh3*.17	Slightly more susceptible	Jahn et al. (2013)
Auxin/ glucosinolate	Biosynthesis	Mutant *CYP79B2*/ *B3*	None	Siemens et al. (2008)
Glucosinolate	tu8 (heterochromatin protein, lower IGL levels)	Mutant	Reduction	Ludwig-Muller et al. (1999)
Camalexin	Biosynthesis	Mutant *pad3*	None	Siemens et al. 2002, 2008)
Cytokinin	Degradation	Overexpression cytokinin oxidase	Reduction	Siemens et al. (2006)
Brassinosteroid	Biosynthesis signaling pathway	Mutant *det2* mutant *bri1*-6	Some reduction reduction	Siemens et al. (2002)
Gibberellin	Biosynthesis	Chemical inhibition	None	Pasold and Ludwig-Muller (2013)
Jasmonic acid	Conjugation/activation	Mutant *jar1*	Higher susceptibility	Siemens et al. (2002)
Ethylene	Signaling pathway	Different receptor mutants (*etr1*)	Higher susceptibility	Knaust and Ludwig-Müller (2013)
Flavonoid	Biosynthesis root-specific MYB transcription factor	Different mutants overexpression	Slight reduction some reduction	Pasold et al. (2010)
				Pasold et al. (2010)
Sucrose	Invertase	Repression of enzyme activity	Reduction	Siemens et al. (2011)

P. brassicae-colonized host cells (Grsic-Rausch et al. 2000; Pasold et al. 2010) might be retained in the infected cells by increased flavonoid accumulation in infected host roots, because flavonoids are believed to prevent auxin efflux from a cell (Pasold et al. 2010). This indicates a role for auxin transport in clubroots. Together, auxin (indole-3-acetic acid; IAA) and CK increase the cell division rate and plasmodia of *P. brassicae* might divide together with the host cells. This hypothesis is supported by the observation that in highly infected roots, the cells harboring the protist are connected to each other in large patches (Plate 7.6; Chap. 7). In addition, during later stages of infection, the high auxin levels in cells containing plasmodia of the protist cause hypertrophy, so that the number of resting spores formed increases (Plate 7.6; Chap. 7). Another plant growth-promoting hormone class, gibberellins, does not seem to be involved in club development. Application of a gibberellin biosynthesis inhibitor (chlorocholine chloride, CCC) to *P. brassicae*-infected plants did not alter the size of a club, so it was concluded that the formation of clubroots is independent of GA synthesis (Pasold and Ludwig-Muller 2013), even though some genes involved in GA signaling were found to be elevated in a microarray experiment (Siemens et al. 2006). On the other hand, there is some evidence for the involvement of brassinosteroids, because the *det2* mutant showed a more tolerant phenotype (Siemens et al. 2002). The gene *DET2* encodes a protein from the biosynthetic pathway leading to brassinolide (BR) in *Arabidopsis*. In addition, a BR receptor mutant *bri1-6* exhibited smaller root galls compared with the wild type after infection with the protist (Table 10.1) (Diederichsen et al. 2014).

The so-called omics approaches have helped to elucidate the role of plant hormones in clubroot development. Siemens et al. (2006) compared the transcriptome of clubroots at two different points after inoculation with somewhat defined developmental stages of the pathogen for changes in gene expression profiles. Agarwal et al. (2011) used early stages during pathogenesis for a similar experiment, and Devos et al. (2006) analyzed the proteome in infected roots during these early stages of colonization. These data have all been obtained with the model plant *Arabidopsis thaliana*, which is a good host for *P. brassicae*. Additionally, proteome data were supplied for *Brassica* species by Cao et al. (2008).

10.2.1 Expression of Hormones in Different Cultivars

The expression of hormones in the *P. brassicae*-infected resistant and susceptible cvs. has been characterized. In the early stages of infection, there is a change in growth-promoting hormones in response to level of host resistance. During later phase of infection for gall formation, the amount and rate of hormonal production is high in susceptible cvs. than the resistant ones. There is a differential expression of hormones in leaves versus roots and formation of galls in R and S cvs. indicating organ and host resistance specificity in production of hormones.

10.2.1.1 The Early Response is Characterized by Changes in Growth-Promoting Hormones

Colonized cells were demonstrated to exhibit a stimulation of cell division and elongation with the elevation of CKs in roots during the early response to *P. brassicae* infection. The impact of *P. brassicae* infection on hormone and gene expression data is shown in Fig. 7.12; Chap. 7. Active CKs are partially produced by the pathogen, as indicated by the conversion of a precursor to an active CK, and by the occurrence of two putative IPT genes in the genome. In *Brassica rapa* and *Arabidopsis thaliana,* Prerostova et al. (2018) detected the increase of active CKs, predominantly caused by cis-zeatin, already 2 dai in roots and leaves of both resistant and susceptible cultivars. In the disease-sensitive cultivar Hornet, the elevation of the CK content was maintained for 10 days in both organs, while in the resistant cultivar Alister, it is dropped to the control level 5 dai, and further decreased below the controls in leaves 7 dai, and in roots 10 dai. Transient CK elevation is in accordance with the up-regulation of several IPT genes during the first phase of infection in inoculated roots reported for *Brassica rapa* (Chinese cabbage (Ludwig-Muller 2009a, b). The CK down-regulation seems to be part of plant defense, as plants overexpressing CK degrading enzyme cytokinin oxidase/dehydrogenase are more resistant to the clubroot disease (Siemens et al. 2006). This effect can be interpreted as the reduction of the signals involved in cell division, because the elevated CKs trigger the local re-initiation of cell cycle progression in the root cortex, as indicated by up-regulation of cyclin CYCB1;1 expression described by Devos et al. (2006).

Stimulation of cell division also requires the presence of auxin. The elevated content of IAA was reported in *Brassica campestris* 10 dai. In *Brassica napus*, up-regulation of IAA was detected after 3 dai, to a higher extent in roots. In *Arabidopsis*, an enhanced response of auxin-related genes was observed from 6 dai. Due to the importance of active polar auxin transport, it seems that the host auxin needs to be moved to the site of infection. In addition, the signaling pathways responsible for auxin also need to be functioning. A transient elevation of IAA contents in leaves 2 dai in Alister, and 2–5 dai in Hornet was observed. In roots, IAA content was promoted between 2–7 dai in Alister, and from 2 dai to 15 dai in Hornet. The stronger induction in the susceptible cultivar indicates a prominent role for IAA in disease progression.

Not many changes in defense-related hormones were detected during the early response. The content of SA in roots was similar as in the controls, while the content of JA was transiently elevated in Hornet roots between 2–5 dai, and in Alister at 7 dai. Irani et al. (2018) showed that *P. brassicae* stimulates the expression of JA related genes in *A. thaliana*, but not genes related to SA. It seems that the onset of *P. brassicae* infection is associated with colonization of the root tissue, but still without visible symptoms (Ludwig-Muller 2009a, b).

10.2.1.2 Differences in the Gall Formation of the Susceptible Cultivar Hornet and Resistant Cultivar Alister

The later phase of *P. brassicae* infection can be characterized by gall formation, which was more profound in the susceptible cultivar Hornet. During this stage, the CK content in leaves was higher in Alister (at 15–42 dai) than in Hornet which may reflect the better physiological state of the more tolerant cultivar. CK down-regulation was found in the roots of both cultivars during the later stage. Malinowski et al. (2016) reported that CKs are diminished in *A. thaliana* after infection (16 dai). In galls of both cultivars, however, high CK levels were detected 22 dai.

In addition, auxin might also be responsible for enlarged cells during the development of large sporulating plasmodia and resting spores. In roots, the susceptible cultivar Hornet exhibited increased IAA content until 15 dai. In both cultivars, the IAA level was elevated in galls 22 dai (to a higher extent in Hornet), which might be related to cell expansion governed by auxin. In Hornet galls, NIT1 transcription was up-regulated 35 dai, and especially at 49 dai. Several reports describe the up-regulation of transcription of *nitrilase* isoforms in clubroot tissues compared to controls in different host plants such as *Brassica rapa* (turnip, Chinese cabbage) or *Brassica juncea*, and specific expression of *nitrilase 1* and *2* genes in developing galls of *Arabidopsis thaliana*, as well as with the up-regulation of myrosinases and glucosinolate metabolism. The difference between IAA levels and the expression of IAA-biosynthetic genes may be given by the up-regulation of the expression of GH3 genes (encoding auxin conjugating enzymes), as was shown in the case of *Arabidopsis thaliana*. It may indicate that plant defense includes IAA conjugation, i.e., the inactivation of the excess of IAA involved in gall formation (Jahn et al. 2013).

SA is the key hormone in plant combat with biotrophic pathogens. SA increase in roots was more profound in Alister (35–49 dai) than in Hornet, where it occurred later (42–49 dai). In Alister galls, SA level was enhanced at 22 and 42 dai, while its enhancement was delayed in Hornet (only between 42–49 dai). The expression of SA biosynthetic gene *ICS1* was higher in Alister roots between 35–42 dai, whereas its expression was lower in Hornet root galls. The induction of defense-related responses in susceptible infected plants is of limited extent and duration. In *A. thaliana*, an up-regulation of the corresponding *ICS1* transcript was reported at 21 but not 28 dai. The responses to *P. brassicae* infection were followed in the resistant genotype *Brassica macrocarpa* Guss. and in the susceptible genotype *Brassica oleracea* var. *italica*, the SA signal transduction was up-regulated mainly in the resistant genotype, in comparison with the susceptible one. Only minor differences were observed in the expression of the central component of the SA signaling pathway *NPR1*. Regulation of its activity seems to be realized predominantly by oligomerization of the resulting protein, and subsequent differential cellular localization. PR1 expression was up-regulated in Hornet roots 35–49 dai, and in Alister already 22–42 dai. It has a role as a marker for SA-dependent defense. Constitutive up-regulation of PR1 expression in the cpr1 mutant of *A. thaliana* resulted in more clubroot-resistant plants. The experiments with SA deficient/

insensitive and constitutively active mutants indicate that SA is very important, but not sufficient in defense to *P. brassicae* (Lovelock et al. 2016).

During the later stages, the JA content was elevated in the roots of both cultivars, which indicates that JA might not play a role in the defense of the resistant cultivar to the clubroot disease. Jasmonic acid content was especially high during the gall initiation, occurring around 22 dai. As JA is able to induce nitrilase activity, there could be a link between jasmonate and auxin metabolism, and not so much of a signal for plant defense. The gene for the rate-limiting enzyme of ethylene biosynthesis, ACS2, exhibited lower expression in roots and galls of the susceptible cultivar Hornet compared to the resistant Alister. Knaust and Ludwig-Müller (2013) reported for *A. thaliana* that the expression of another gene for the ethylene biosynthetic enzyme ACO2 was up-regulated throughout the whole experimental period of 10–28 dai. However, the expression of a crucial component of the ethylene signaling pathway EIN2 was not significantly affected in the two *Brassica* cultivars. Ethylene signaling mutants of *A. thaliana* were more susceptible to *P. brassicae* infection, thereby making larger galls at a lower infection pressure. The ethylene might be needed to restrict gall size under normal conditions (Prerostova et al. 2018).

10.2.1.3 The Impacts of Salicylic Acid and Jasmonic Acid Applications

Taking into account that SA is a crucial hormone in the response to biotrophic pathogens, Prerostova et al. (2018) tested the effect of exogenous SA applied before infection or twice before infection and 15 dai, when the gall formation starts. The impact of exogenous SA in the fight with the clubroot pathogen is dependent on the time point and method of administration as well as on the plant material. Agarwal et al. (2011) reported that a very early incubation with SA significantly reduced the clubroot symptoms of *A. thaliana*, which was confirmed for *Brassica oleracea* (broccoli). However, a later treatment of *A. thaliana* plants with SA during infection had no effect on disease severity. The exogenous SA was rather quickly metabolized, as already 2 days after its application of only a moderate amount of the hormone was left. At the time point of the second foliar spraying 15 dai, a very high peak of SA was detected 2 h after application. SA metabolism may be partly due to the host plant tendency to re-establish hormone homeostasis disturbed by the application of the hormone, or it may be a part of the strategy of the pathogen to down-regulate SA levels, and thus suppress plant defense. Indeed, *P. brassicae* was reported to have high SA-methyltransferase activity, creating an SA transport form, which can be translocated from the galls to the upper plant parts where it is emitted. Consequently, *A. thaliana* plants overexpressing the methyltransferase gene were more susceptible to clubroot. A similar mode of action was observed in the case of *Pseudomonas syringae*, which produces coronatine, a compound structurally similar to JA-isoleucine. Coronatine induces the JA signaling pathway, which leads to the suppression of the SA pathway. Also, SA glucosylation might be considered in the SA deactivation. SA application before infection had a negligible effect on gall formation. However, SA application at the onset of gall formation prevented disease development in the more resistant cultivar Alister, indicating that in cv. Alister, the defense is already partially working, and therefore, only small amounts of SA increase resistance. The fact that the JA-metabolic mutant jar1, which is unable to produce the active JA metabolite JA-isoleucine, is more sensitive to *P. brassicae* infection, indicates that JA signaling is also involved in the response to *P. brassicae* infection. Nevertheless, exogenous application of JA at 15 dai promoted gall formation in Hornet, but also in the more resistant cultivar Alister. Similarly, the partially resistant *Arabidopsis* ecotype Bur was reported to activate after infection of the SA pathway, while the sensitive *Arabidopsis* ecotype Col-0 activated the JA pathway. When more virulent pathotype of *P. brassicae* (isolate e2 instead of eH isolate) was used, even Bur did not activate the SA pathway. These data indicate a complex cross-talk among these two pathways which depends on the mutual ratio of their activities (Lemarie et al. 2015; Jubault et al. 2013).

10.2.1.4 Organ Specificity—Leaves Versus Roots and Galls

Shoot/root communication is fast and intensive within the plant. This can be demonstrated by changes of leaf hormone levels upon attack by soil pathogens as well as by elevation of SA or JA in roots after their foliar application. SA has been proven to be the key hormone in defense against biotrophs. In some cases, when exogenous SA did not decrease clubroot growth, it diminished at least the stress-induced inhibition of shoot growth. Thus, the (partial) preservation of active CK and auxin content in leaves as shown for both cultivars might be a measure of the shoot fitness, and not the consequence of the pathogen impact, as is seen in roots. In the latter, the regulation of CK and auxin levels (including the degradation with cytokinin oxidase/dehydrogenase CKX or conjugation with GH3) seems to reflect the combat between the pathogen and plant. Hormone homeostasis in leaves might either be maintained through transport, degradation, or conjugation, or through de novo synthesis. Up-regulation of transcripts for auxin biosynthesis genes in both cultivars, such as *NIT1* and the two *YUC* genes, indicates the stimulation of IAA synthesis (Lovelock et al. 2016).

The modulation of SA and JA content in leaves reflects the intensive activation of the defense, also in organs that are not directly exposed to the pathogen. The SA level was increased in Alister leaves between 35–49 dai, while in Hornet leaves, the increase occurred earlier, being maintained until the end of the experimental period (between 15–49 dai). In the roots, the SA level increased earlier, and to a higher extent in the resistant than in the susceptible cultivar. Methylation of SA by the protist's methyltransferase can result in a form that is more readily transported into the leaves of *A. thaliana*. The methyl ester of SA can be converted back to free SA in the leaves using specific esterases. The role of such esterases and methyltransferases for SA metabolism of *B. oleracea* during clubroot infection was demonstrated. Not only in the roots, but also in the leaves, differentially expressed genes were found that could be involved in SA metabolism (Manoharan et al. 2016).

In the case of JA, the situation might be a bit different. The JA levels were up-regulated in leaves of the cultivar Alister only (15–35 dai), the expression of the JA biosynthetic gene *LOX4* was down-regulated in leaves of both cultivars. Finally, the fitness of the plant is also dependent on the water status. In roots, the content of abscisic acid (ABA) was enhanced in both cultivars during later stages of the infection, a bit higher in the susceptible Hornet than in the resistant Alister cultivar, which confirmed earlier data where ABA was also increased in *Brassica* galls at a later stage of disease development. Ludwig-Muller (2009a) proposed, based on the up-regulation of many transcripts related to drought stress during later stages of the clubroot disease, a strong influence of ABA. The drought stress could be attributed to the destroyed root tissues that include the vasculature. The content of ABA was enhanced in leaves and roots of both cultivars at 42 and 49 dai. ABA has obviously a dual role in the regulation of stomata aperture and stimulation of the expression of stress-related genes. However, the ABA elevation in leaves seems to be rather a consequence of impaired water transport (leaf wilting) than modulation of the transcriptome (Prerostova et al. 2018).

10.3 Auxin Metabolism and Signaling

Auxin can be synthesized in *Arabidopsis* via several different pathways, presumably occurring in parallel to each other (Fig. 7.10; Chap. 7). While the involvement of some pathways has only been inferred to be important based on microarray studies (Siemens et al. 2006), the pathway typical for Brassicaceae via indole-3-acetaldoxime (IAOx), indole glucosinolates (IG), and indole-3-acetonitrile (IAN) has received a lot of attention. There seem to be differences, however, between *Arabidopsis* and *Brassica* species (Ludwig-Muller 2009a). In an *Arabidopsis* mutant with altered indole glucosinolate (IGL) levels resulting from a mutation in the gene encoding the heterochromatin-like protein 1 (tu8), the lower levels of indole-3-acetonitrile (IAN), and IGL correlated with smaller clubs (Ludwig-Muller et al. 1999; Bennett et al. 2005). Early work pointed to many correlations between glucosinolate levels (especially IGL) and the size of clubs in *Brassica* species (Ludwig-Muller 2009a). Since the pathway is thought to involve the conversion of IGL via myrosinase to IAN and further to IAA by the activity of nitrilase, the activity and transcription of nitrilase was monitored in several plant species. Nitrilase (*NIT*) transcripts are up-regulated in clubs of *Arabidopsis* (Grsic-Rausch et al. 2000).Using immunolocalization (Grsic-Rausch et al. 2000) and promoter (Pasold et al. 2010), the presence of *NIT* transcriptional activation and protein localization in cells with plasmodia and sporulating plasmodia close to resting spore formation was demonstrated. Consequently, the nitrilase mutant nit1 and antisense plants for the *NIT1* and *NIT2* genes with reduced levels of *NIT* expression had reduced gall size or the development was slower (Grsic-Rausch et al. 2000; Neuhaus et al. 2000) (Table 10.1). In this context, it is interesting to note that *NIT2* was the predominantly expressed isoform. By contrast, when a mutant was investigated, which is blocked in the first step of the pathway devoted to the synthesis of IAA, camalexin and indole glucosinolates via indole-3-acetaldoxime (IAOx), no reduction in club size was detected and the levels of endogenous IAA in galls were similar to wild-type plants (Siemens et al. 2008). That suggested that blocking early steps in the IAA pathway can induce alternative pathways, while blocking later steps, such as nitrilase, resulted in a reduction of IAA and thus club formation was also inhibited (Plate 7.6, Chap. 7).

The importance of increased nitrilase transcription within the enlarged clubs has been demonstrated in *Brassica* species, but additional work has provided evidence for the up-regulation of other genes involved in auxin biosynthesis, such as aldehyde oxidase and amidase (Ludwig-Muller et al. 2009).

Since glucosinolates such as indole glucosinolate (IGL) are involved as defense compounds in other plant–pathogen systems, another possibility for the localization of nitrilase in plasmodia containing cells might be important. In addition to the conversion of IGL to IAN and IAA, thereby leading to the hypertrophy of infected host cells (Fig. 7.10, Chap. 7), nitrilase could be a defense mechanism to degrade the glucosinolates to less toxic nitriles as suggested by Piotrowski (2008). In the presence of epithiospecifier proteins and myrosinase, the glucosinolates are degraded to toxic compounds, the (epi) thiocyanates. To avoid, *P. brassicae* could also induce host nitrilase in the cells where the protist is residing to prevent the production of the toxic compounds (Fig. 7.10, Chap. 7). Thus, the induction of nitrilase could have a twofold effect for the pathogen: first, induction of IAA synthesis and second, alteration of the defense metabolite spectrum provided by the host. Both hypotheses fit with the observation of nitrilase being closely connected to host cells with pathogen plasmodia present.

While the biosynthesis to IAA is better understood, the signaling pathways have so far not been elucidated. Two different auxin receptor types have been discussed (Christian et al. 2006): the auxin binding protein 1 (ABP1), and the TIR F-box protein class of receptors. Both classes of receptors were up-regulated in transcription during disease progression as shown by promoter: GUS analysis for ABP1, and transcript levels for TIR1 (Jahn et al. 2013). In addition, TIR-class receptor double mutants showed less susceptibility to the clubroot pathogen, indicating that hormone signaling must be intact (Jahn et al. 2013). As targets for auxin perception via the TIR pathway, the *GH3* genes can be required, as these are also involved in regulating the auxin homeostasis by conjugating IAA to various amino acids (Staswick et al. 2005). Consequently, it might be expected that several *GH3* genes were transcriptionally up-regulated during clubroot development as shown by microarray data (Siemens et al. 2006) and transcript analysis (Jahn et al. 2013). Thus, with the *GH3* genes a target group for the auxin pathway can be identified (Table 10.1). Other target genes for IAA in the formation of hypertrophied cells can be those involved in cell wall expansion (Ludwig-Muller 2009b). In *Brassica rapa* L., it has been shown at an enzymatic level that the IAA conjugates with amino acids were hydrolyzed back more strongly in infected roots compared with controls, also contributing to the free IAA levels in clubs (Ludwig-Muller et al. 1996). An *Arabidopsis* mutant in one of the conjugate hydrolase genes ilr1, however, formed clubs comparable with wild-type galls (Table 10.1) (Siemens et al. 2002).

10.4 Cytokinin metabolism

The involvement of cytokinins in club formation was demonstrated for *Brassica* species and *Arabidopsis* (Ludwig-Muller et al. 2009). Dekhuijzen and Overeem (1971) and Dekhuijzen (1980, 1981) showed that clubs contained increased amounts of cytokinins, which was confirmed by later studies using more modern analytical techniques. In the 1980s, it was demonstrated that *P. brassicae* can synthesize small amounts of cytokinins (Muller and Hilgenberg 1986). In addition, plants are contributing to increased cytokinin levels by induction of their own biosynthesis, or by repression of degradation. The first possibility seems to be present in *Brassica*, where the induction of *ipt* genes was found. In *Arabidopsis*, the induction of cytokinins seems to involve the second possibility, i.e., repression of metabolic steps, also leading to increased cytokinin levels (Siemens et al. 2006). The reduction of transcripts encoding two root-specific cytokinin oxidase/dehydrogenase genes (*CKX*) was described. Consequently, plants overexpressing *CKX* genes were much more tolerant against the clubroot pathogen, but since they also show a dwarf phenotype, this cannot be used to control the disease symptoms. Solutions to this problem might include the control of gene expression under the control of root-specific promoters, as was used in reducing invertase (Siemens et al. 2011). Since disease development is also dependent on cell divisions in the host root cortex, the role of cytokinins might be to stimulate cell divisions together with auxin. Cytokinins not only regulate tissue growth, they can also be involved in the redirection of assimilates, and possibly also other nutrients from the aerial parts of an infected plant towards the root galls (Ludwig-Muller et al. 2009; Siemens et al. 2011). Several sugars were found to be increased in *P. brassicae*-infected roots (Brodmann et al. 2002), an indication of either higher de novo synthesis and/or relocation. Together with the observation that primary metabolism was strongly induced in microarray experiments; this led to a hypothesis connecting the presence of cytokinins with sugar metabolism. It was shown that invertase gene expression and protein amount can be increased by addition of cytokinins (Ehness and Roitsch 1997), so that the role of these proteins for carbon partitioning was investigated (Siemens et al. 2011). It could be demonstrated that several invertase promoters showed strong activity in root galls as determined by promoter, reporter fusions, which increased during disease progression. Mutants in a single invertase gene, however, did not alter the clubroot phenotype. Therefore, a different approach was used, namely to inhibit invertase activity by overexpression of a gene for a proteinaceous invertase inhibitor. The results showed that in transgenic plants, the invertase activity was reduced and the infected roots were smaller (Siemens et al. 2011). Thus, invertases can be regarded as targets for cytokinins, which increase the available soluble sugars for uptake into the plasmodia of the pathogen.

10.5 Role of Other Plant Hormones

The role of several other plant hormones in disease development has yet to be identified. Some observations, however, from microarray data have led to the assumption that during the later developmental stages of a clubbed root, drought stress signaling might be important. A large number of abscisic acid (ABA) related transcripts were strongly up-regulated once the pathogen became established in the host root (Ludwig-Muller 2009b). During this period, the vascular tissue is heavily affected, so the infected plant is also suffering from low water and nutrient transport rates. Whether the levels of ABA are indeed increased has yet to be demonstrated, but it is reasonable to hypothesize that the plant is defending itself against drought stress (Fig. 7.11, Chap. 7). This action would be also beneficial for the clubroot pathogen itself, which is dependent on obtaining nutrition from the host.

Some experiments have provided evidence that the clubroot pathogen can down-regulate (partially) plant defense mechanisms or avoid them. The phytoalexin camalexin is increased in clubroots (Siemens et al. 2008), but a mutant in the camalexin biosynthetic pathway did not show altered clubroot phenotypes, indicating that this phytoalexin is not important in the plant's defense against *P. brassicae* (Fig. 7.10, Chap. 7). Many plant defense genes in microarray analyses have been shown to be not regulated or even down-regulated. Several mutant analyses support this idea. In the jasmonate-triggered defense response, one critical step is the conjugation of jasmonic acid (JA) to the amino acid isoleucine (Staswick and Tiryaki 2004), which results in the active component recognized by the F-box protein COI1 involved in the signal transduction. In line with observations that in several plant–pathogen interactions the mutant in the gene encoding the conjugating enzyme (*jar1*) was more susceptible towards the tested pathogens, the jar1 mutant was also more susceptible to *P. brassicae* (Siemens et al. 2002). On the other hand, a positive connection between jasmonic acid and *nitrilase* has been shown (Grsic-Rausch et al. 1999), indicating possibly a twofold role for JA in clubroot pathogenesis (Fig. 7.11, Chap. 7), the plant may use it as defense signal, while JA can increase indirectly IAA levels, which is of advantage to *P. brassicae.*

The ethylene (ET) signal is also involved in different defense responses. In the case of the interaction of *Arabidopsis* with *P. brassicae*, the role as a defense molecule is not clear (Knaust and Ludwig-Müller 2013). While the levels of the intermediate in the biosynthetic pathway aminocyclopropane carboxylic acid (ACC) decreased in infected roots, one gene, *ACO2*, encoding a biosynthetic enzyme, ACC oxidase, was transcriptionally elevated in clubroots compared with controls. Using mutants in signaling components, however, indicated that these were clearly more susceptible to the clubroot pathogen than the wild type, leading to the result that the functional ET signaling pathway is needed to restrict root gall growth (Fig. 7.11, Chap. 7) (Knaust and Ludwig-Müller 2013). In this context, it should be noted that the ET receptor mutant etr1, when tested with high inoculum densities, did not show an aberrant clubroot phenotype (Siemens et al. 2002).

Finally, salicylic acid (SA) as a defense molecule against biotrophic pathogens could play a role in the outcome of the clubroot disease syndrome. It was shown that treatments with SA at an early stage after inoculation could render *Arabidopsis* (Agarwal et al. 2011) and *B. oleracea* var. *italica* Plenck (broccoli) (Lovelock et al. 2013) more resistant to *P. brassicae* infection. However, treatments with SA at a later stage after inoculation did not result in the increased resistance of *B. rapa* (Ludwig-Muller et al. 1995). It can be concluded that the treatment has to occur early during the infection process, or otherwise *P. brassicae* is somehow capable of suppressing this plant defense pathway. Nevertheless, an increase in SA levels, together with increased biosynthesis from chorismic acid, has been shown in infected *Arabidopsis* plants (Diederichsen et al. 2014).

These findings have stimulated the idea that the plant defense mechanisms might be suppressed by the pathogen. Using an endophytic fungus (*Acremonium alternatum* Link), it was possible to induce tolerance against *P. brassicae* in *Arabidopsis* and *B. rapa* (Chinese cabbage) (Doan et al. 2010; Jaschke et al. 2010). The effect was dependent on the application of the fungus in a period between 3 and 0 days before inoculation with *P. brassicae*, and the best effect was achieved when both organisms were administered at the same time. Autoclaved spores were, however, also able to induce the tolerance phenomenon (Jaschke et al. 2010). Analysis of gene expression from a set of genes from *P. brassicae* together with histological examinations indicated that the protist was not inhibited, but rather development in the roots slowed down. Indeed, preliminary experiments showed that the fungus, *A. alternatum* is able to induce some defense-related genes from the JA, and SA signaling pathways. Despite these findings, which have been obtained, there is still a very incomplete understanding of the hormonal signals leading to the homeostasis between the clubroot pathogen and its host plants (Diederichsen et al. 2014).

10.6 Metabolic Pathways

Similar to other eukaryotic biotrophic plant pathogens, *P. brassicae* and *S. subterranea* are reduced in several metabolic pathways. Both Plasmodiophorids lack genes encoding for proteins involved in sulfur and nitrogen uptake as well as genes for histidine, tryptophan, and threonine biosynthesis. Arginine and lysine synthesis were also compromised in *P. brassicae*. Thiamine biosynthesis genes were also reduced in the Plasmodiophorids, suggesting a dependence of their host plants for those metabolites. No fatty acid synthase was identified in *P. brassicae*. The inability for fatty acid synthesis has been described for obligate biotrophic bacterial pathogens of insects and plants, and kinetoplast parasites such as *Trypanosoma* species. The *P. brassicae* gene content for fatty acid biosynthesis was similar as for the kinetoplasts, which use an alternative parasitic-specific fatty acid synthesis, the microsomal elongase pathway. The *P. brassicae* metabolizes fatty acids from the host by elongating and modifying molecules of four or more carbons. Protein and transcript data showed that fatty acids are synthesized in the plasmodia and degradation occurs in the spores. Seven putatively secreted lipases in the *P. brassicae* genome were among the highest expressed genes in the plasmodial stage providing the pathogen with fatty acid starter molecules (Kemen et al. 2011; Oshima et al. 2004; Lee et al. 2006). Schwelm et al. (2015) predicted carbohydrate conversion to mainly occur via glycolysis, glyoxylate, pyruvate metabolism, pentose phosphate pathway, and the TCA cycle. Proteins and transcripts of the glyoxylate cycle were highly abundant in the germinating spore sample indicating a high conversion of sugars and lipids in this active life stage. The *P. brassicae* genome contained trehalose biosynthesis genes expressed in all transcriptome sets, which corroborates to earlier findings of trehalose accumulation by *P. brassicae* in infected plants. The trehalose production by plant pathogens interferes with the sugar sensing pathways of their host and thereby manipulate plant metabolism to their favor or regulate energy and nitrogen supply during colonization. Besides affecting the host, trehalose in the resting spores can be important for their long-term survival as an energy resource or an osmo-protectant against the effects of desiccation. A trehalase encoding gene of *P. brassicae* was primarily expressed in spores indicating a release of glucose as an energy resource in the germination process (Brodmann et al. 2002; Wilson et al. 2010).

10.6.1 Plant Hormone Homeostasis

Plant tissue infected with *P. brassicae* shows a modified hormone balance that induces hypertrophy and cell divisions in host roots leading to gall formation. *P. brassicae* can potentially modify the plant hormone homeostasis of at least four plant hormones: cytokinins and salicylic acid (SA), auxins, and jasmonic acid (JA). The *P. brassicae* genome, but not the *S. subterranea* transcripts, contained an auxin-responsive Gretchen Hagen 3 (GH3) protein (PbGH3). GH3-proteins are known to maintain hormonal homeostasis by conjugating plant hormones to amino acids. The PbGH3 predicted protein structure resembles plant GH3-proteins (Fig. 10.1a). The heterologous expressed PbGH3 protein conjugated auxins and JA with proteinogenic amino acids *in vitro*, although it is phylogenetically not related to plant GH3-like proteins (Fig. 10.1b). Both Plasmodiophorids contained two isopentenyl-transferases (IPTs) belonging to DMAPP: tRNA-IPTs classes I and II (Ludwig-Muller et al. 2009, 2015; Siemens et al. 2006; Staswick et al. 2005). Class I DMAPP: tRNA-IPTs include members from *Physcomitrella patens* and *Arabidopsis thaliana* that are involved in *cis*-zeatin synthesis. Cytokinin homeostasis can also be inferred by a *P. brassicae* cytokinin oxidase, although the expression in transcript sets is low. *P. brassicae* produces *trans*-zeatin and the contribution of the Plasmodiophorid IPTs and the potential cytokinin oxidase to the host cytokinin homeostasis remains to be shown. The gall phenotype might be a result of pathogen-driven interference of the recently described secreted methyl transferase (PbBSMT) can reduce the accumulation of SA in infected roots. *PbBSMT* expression in spores and plasmodia was low, but high expression levels in clubroots of *Brassica napus*, *B. oleracea,* and *B. rapa* host plants highlight its important role (Lindner et al. 2014; Malinowski et al. 2012).

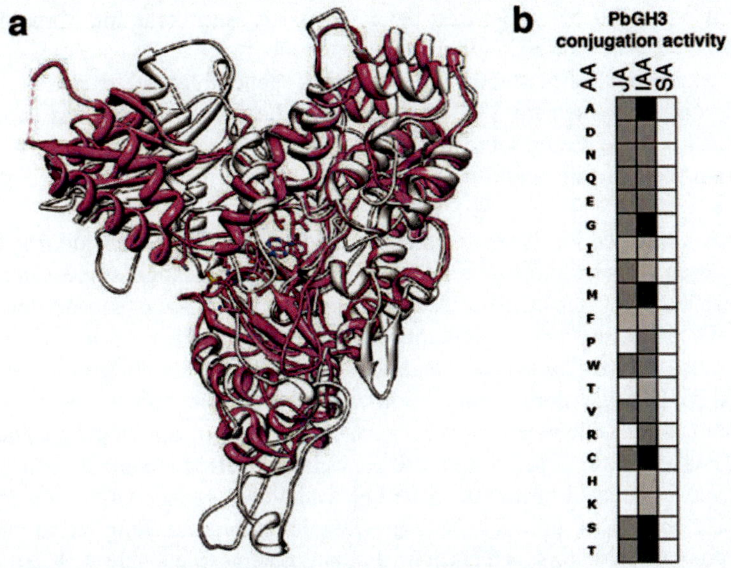

Fig. 10.1 Structure prediction and activity of the PbGH3 protein. (**a**) Modeling of PbGH3 on plant GH3-proteins for which crystal structures are available. An overlay of the obtained protein model for PbGH3 (gray) with a monomer of the *Arabidopsis thaliana* AtGH3.12 (PDB database entry 4EPM (http://www.rcsb.org/pdb/home/home.do) protein (magenta) is shown. The model was built in SWISS MODEL65 and the overlaid with the standard settings of Chimera66. (**b**) Conjugation activity of PbGH3 with IAA, SA, and JA, and amino acids (AA). Activity is indicated by black boxes (high activity) to white boxes (no activity) (Schwelm et al. 2015)

10.6.2 The Plasmodiophorid Secretome

Microbes secrete effector proteins to manipulate host processes in order to facilitate colonization. Schwelm et al. (2015) predicted 553 secreted proteins in *P. brassicae*, and 613 in *S. subterranea*. Both secretomes were smaller than that of *B. natans* but similar in numbers to other biotrophic fungi and oomycete plant pathogens. An OrthoMCL comparison of the two secretomes revealed that most secreted proteins were not shared between the two species but were enriched in ankyrin and protein domains detected in effectors of other plant pathogens, such as small cysteine-rich secreted proteins, chitin recognition proteins, proteases, and protease inhibitors. The secretome of *P. brassicae* was further enriched in leucine-rich repeat domains and pathogenesis-related (PR) proteins, such as PR-5 thaumatin like-proteins. A cysteine-rich secretory protein with BLASTP hits to PR-1-like proteins with no homologues in the non-plant pathogenic Rhizarians, was also present in *S. subterranea*, and identified in the Plasmodiophorid *P. betae* (Pbef2) (Okmen and Doehlemann 2014; Desoignies et al. 2014).

Oomycete effectors frequently code for host-targeting sequence motifs, such as RXLR, LXLFLAK, and CHXC34. The secreted proteins of *P. brassicae* and *S. subterranea* do not contain such over represented amino acid motifs. However,

Schwelm et al. (2015) categorized the *P. brassicae* secretome into three groups based on their gene expression to identify potential effectors: (1) the PLeff genes—highest expressed in plasmodia, (2) the Heff genes–highest expressed in the transcriptomes obtained from the hosts, and (3) the Leff genes–highest expressed late in the life cycle during spore formation. Each group contains proteins that could reprogram the plant metabolism and interfere with the plant defense on a protein level.

Highly expressed PLeff genes contained ankyrin domains, which are frequent in all Rhizarian, but are effector candidates for bacterial pathogens and symbionts. PLeff genes also included a secreted lipase, leucine-rich repeats proteins, PBRA_T005609, that shows similarities to effector candidates of biotrophic smut fungi in the Ustilaginomycetes, and RING finger domain-containing proteins. Secreted RING finger domain proteins interfere with the host ubiquitination system of the human intracellular pathogen *T. cruzi*, a strategy also seen for plant pathogens. *PbBSMT* is one of the highest expressed genes in the Heff group, likely reducing the SA content in infected host tissue. The Leff candidates include CE4-domain polysaccharide acetylases, proteases, and protease inhibitors including Kazal-like and papain protease inhibitors, and fasciclin domain proteins (Stephanie et al. 2012; Fan et al. 2012; Hashimoto et al. 2010; Laurie et al. 2012).

10.6.3 Carbohydrate-Active Enzymes (CAZymes)

Like other biotrophic plant pathogens, the Plasmodiophorids have a reduced set of CAZymes. The total number of CAZymes was similar to those identified in the genomes of *Hyaloperonospora arabidopsidis, Ustilago maydis* or the arbuscular mycorrhizal fungus *Rhizophagus irregularis.* However, the arsenal for both potential plant cell wall (PCW) modifying, and chitin cell wall modifying enzymes, differed substantially from fungal and oomycete species (Fig. 10.2). In the proteins models of *P. brassicae* and *S. subterranea* only a few domains were identified that can degrade hemicellulose. Predicted CAZymes with domains involved in cellulose, lignin, or pectin degradation or their modifications were also rare, compared to selected fungi, oomycetes, *B. natans* and *R. filosa* (Fig. 10.2a). Only one protein with a cellulose-binding motif (CBM1) was found. The Plasmodiophorids contain several members of the cellulase containing GH5-hydrolase family. However, the GH5 hydrolases of the Plasmodiophorids was based on phylogeny belonged to the ceramidase GH5_12 and GH5_27 subfamilies (Aspeborg et al. 2012).

The most prominent chitin-related CAZymes domains in the protein models belonged to the GH18 chitinases, CE4, and CBM18 (Fig. 10.2b). *P. brassicae* contained 16 GH18 domain-containing chitinases and their evolutionary history shows whole gene duplications and internal GH18 domain duplication events. GH18 domains can be separated into three phylogenetic clusters A, B, and C. The Plasmodiophorid GH18 chitinase domains grouped to cluster A, and the bacterial cluster C previously not reported to contain eukaryotic members. In the cluster A, the Plasmodiophorid GH18s formed a specific subgroup (Pd1). These Pd1-chitinase

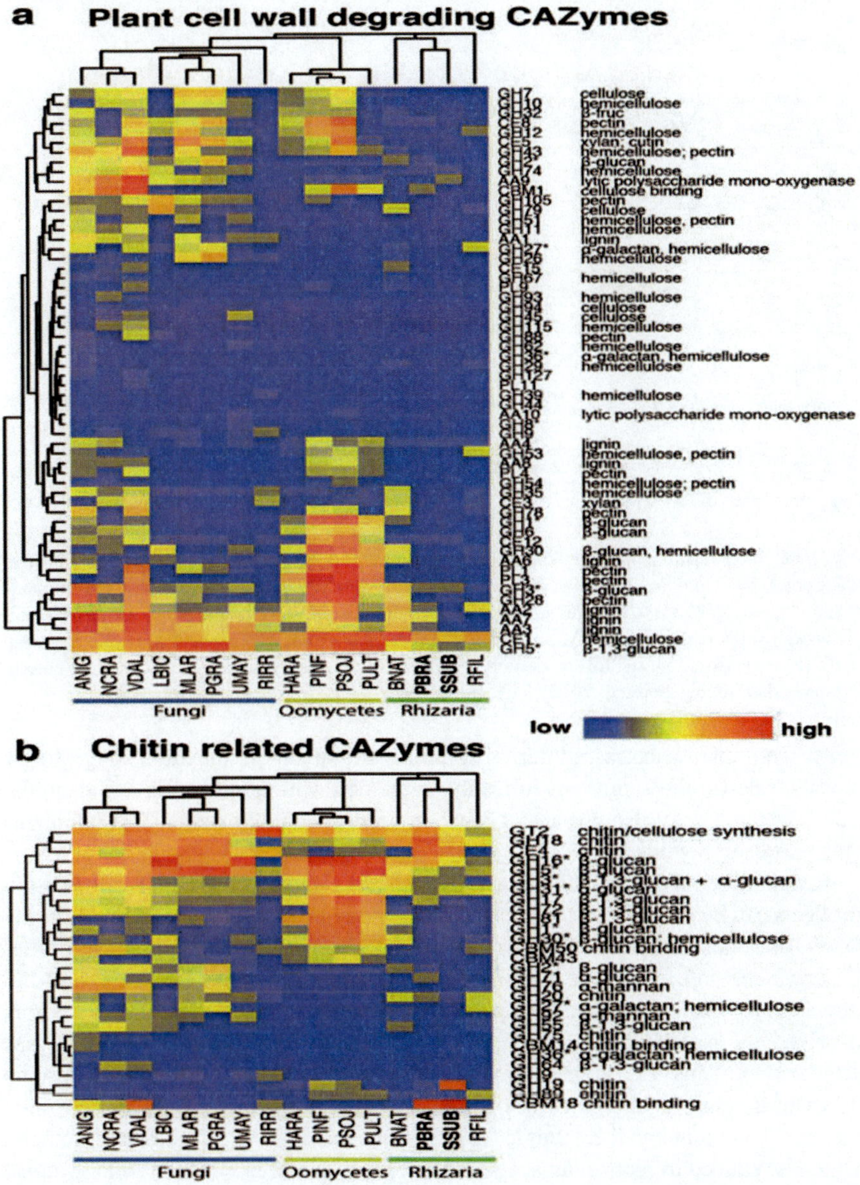

Fig. 10.2 Distribution of (**a**) plant cell wall degrading enzymes, and (**b**) chitin-related enzymes in Rhizarian and selected fungal and oomycete genomes. The heat map represents the abundance of CAZyme domains in each genome. Clustering of gene families is based on Pearson correlation of gene numbers in each family in each species. *Indicates CAZymes with a possible effect on polysaccharides contained in both chitinous and/or plant cell walls. Species abbreviations: HARA *Hyaloperonospora arabidopsidis*, PINF *Phytophthora infestans*, PSOJ *Phytophthora sojae*, PULT *Pythium ultimatum*, BNAT *Bigellowiellanatans*, PBRA *Plasmodiophora brassicae*, SSUB *Spongospora subterranea*, RFIL *Reticulomyxae filosa*, ANIG *Aspergillus nigra*, NCRA

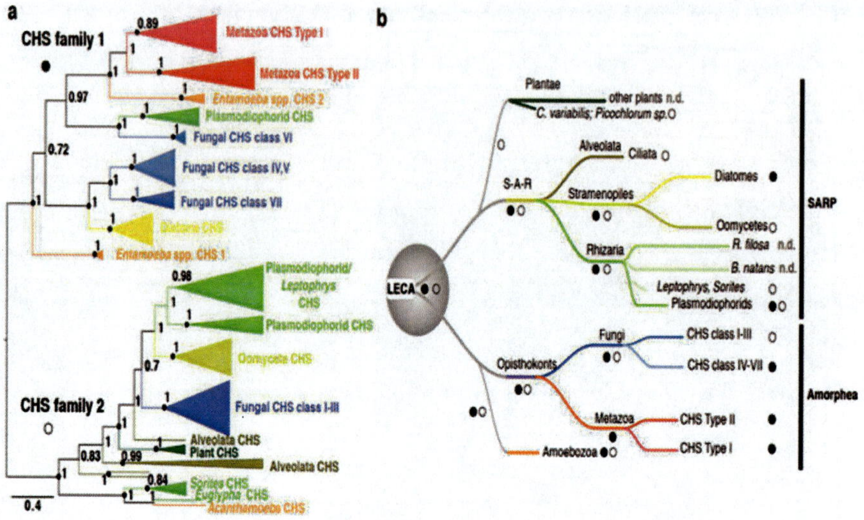

Fig. 10.3 Inter-relationships of eukaryotic chitin synthases (CHS) and model for CHS evolution in eukaryotes. (**a**) Unrooted Bayesian phylogenetic reconstruction of chitin synthase CHS domains (Chitin_synth_2; PF03142). Node support as represent posterior probabilities. Black circles indicate nodes supported of parallel ML analysis (RAxML, LG model, 500 replicates) with a bootstrap >70. (**b**) A proposed model for the evolution of CHS in the eukaryotic kingdom based on current available data (Schwelm et al. 2015)

genes were mainly expressed during spore formation and germination suggesting a specific role for those proteins in resting spore cell wall modification and degradation. The phylogeny also revealed a new oomycete-specific cluster of GH18 domains (Karlsson and Stenlid 2009).

Recognition of microbial cell walls including chitin and chitin degradation products can be recognized by plants and trigger defense responses. In a microarray assay, the plant receptor genes for this type of recognition were suppressed after *P. brassicae* infection. The CBM18 domain and the LysM harboring CBM50 domain are known to bind chitin. Plants contain LysM domains in receptors involved in microbe recognition. Fungi can escape host recognition via LysM effectors, proteins that consist of a signal peptide and multiple LysMs, but no catalytic domain. In the datasets of the two Plasmodiophorids, only *P. brassicae* contained one protein with a single CBM50 domain. However, the CBM18 domain, which is reduced in biotrophic fungal lineages and absent in *B. natans* and *R. filosa* and plant pathogenic oomycetes, is frequent in both Plasmodiophorids (Fig. 10.3b). CE4-domain proteins can convert chitin into its derivate chitosan and were frequent

Fig. 10.2 (continued) Neurospora crassa, VDAL *Verticillium dahliae*, LBIC *Laccaria bicolor*, MLAR *Melampsora larici-populina*, PGRA *Puccinia graminis* f. sp. *tritici*, UMAY *Ustilago maydis*, RIRR *Rhizophagus irregularis* (Schwelm et al. 2015)

in the Plasmodiophorids but absent in the other available Rhizaria genomes. Some CE4-domain proteins also included the chitin-binding CBM18 domain, suggesting that these proteins bind to the Plasmodiophorid chitin in order to promote modification into chitosan, a weaker inducer of immune responses than chitin in many plants (Sanchez-Vallet et al. 2014; Zhao et al. 2014; Soanes et al. 2008; Schwelm et al. 2015).

10.6.4 Plasmodiophorids Contain Two Ancient Chitin Synthase (CHS) Families

CHS are essential for the chitin synthesis. In *P. brassicae* 13 *CHS*-encoding genes and six in *S. subterranea* were identified. Whereas *B. natans* and *R. filosa* contained no predicted CHS, transcript data from *Leptophrys*, *Sorites*, and *Euglypha* species include CHS domains, suggesting that chitin is also synthesized by Rhizarian outside the Plasmodiophorids. CHS are divided into two paralogous families, with family 1 represented by fungal CHS classes IV–VII, and family 2 by fungal CHS classes I–III. Plasmodiophorid harbor both CHS families. No Plasmodiophorid CHS acquired additional protein domains, unlike Metazoa, and fungi CHS where fusion with myosin domains have occurred. The protein domain structure and phylogeny of the Plasmodiophorid family 1 CHS are related to fungal class VI CHS. A family previously only detected in fungal species (Fig. 10.3a). In contrast to recent analyses, animal CHS regrouped with fungal CHS classes IV–VII in family 1, together with SARP CHS of Plasmodiophorids and diatoms. The CHS family 2 comprised fungal classes I–III, and CHS of oomycetes, plants, Alveolata, and Rhizaria, including Plasmodiophorid in the SARP (Ruiz-Herrera et al. 2002; Zakrzewski et al. 2014).

It was proposed that a branch of β-glycosyl-transferases developed into both CHS families after the plant kingdom split from the evolutionary line of Metazoa and fungi. Schwelm et al. (2015) revealed a different evolutionary scenario where the two lineages had already diverged in the last common ancestor of the SARP and Amorphea. This makes the absence of family 2 in Metazoa and family 1 in Plants and Alveolata all secondary losses (Fig. 10.3b). Family 2 was lost in the Metazoa concurrent with the divergence from Fungi in the Amorphea. In the SARP, Family 1 was lost in the Alveolata when they diverged from Rhizaria and Stramenopiles. Subsequently in the Stramenopiles, the loss of family 2 in diatoms, and family 1 in oomycetes occurred when those groups separated. In the Rhizaria, both families represent the Plasmodiophorids. At least family 2 is also contained in *Sorites*, *Euglypha* and *Leptophrys*, whereas CHS appear to been lost in *R. filosa* and *B. natans*. The ancient origin of the CHS families suggests that synthesis of chitin or chitin-like structures has been a driving force in the evolution and diversification of early eukaryotes.

10.6.5 Metabolic Pathway Mediated by Clubroot Resistance (CR)

Several processes of primary metabolism were also differentially regulated in CR mediated by *Rcr1*, including enhanced MET and reduced glycolysis. Plant defense is an energy-consuming process resulting from enhanced demand on certain metabolism and cytological activities. Glycolysis and MET are two major pathways for plant respiration which can be stimulated during defense responses to generate ATP and carbon skeletons. It has been demonstrated that glycolysis is contributing to disease resistance in several plant–pathogen interactions. However, the expression of proteins related to glycolysis, especially the major regulator phosphofructokinase (PFK), was reduced while those related to MET were up-regulated in resistant samples (Song et al. 2016). This result suggests that MET serve as the major energy source required for defense responses mediated by *Rcr1* against clubroot. It is also possible that the down-regulation of glycolysis serves as another defense response by inhibiting certain pathogen-induced metabolism (carbohydrate or auxin, for example) favoring excessive cell growth and division. During a compatible interaction between *Arabidopsis* and *P. brassicae*, an increased carbon flow toward glycolysis was observed (Jubault et al. 2013; Schuller et al. 2013).The pathogen-induced arginine catabolism appeared to be inhibited in resistant samples. In *Arabidopsis*, proline deriving from the arginine catabolism accumulated substantially in a susceptible ecotype upon *P. brassicae* infection while such an accumulation was not observed in the partially resistant ecotype Bur-0 (Jubault et al. 2008). The abundance in 11-pyyroline-5-carboxylate synthase (Bra005012) was decreased in resistant samples. This enzyme catalyzes the rate-limiting step of proline biosynthesis responding to environmental stresses (Zheng et al. 2014). Event woof enzymes involved in arginine biosynthesis were up-regulated, the down-regulation of this key enzyme in proline biosynthesis may indicate that the *P. brassicae*-induced arginine catabolism and proline biosynthesis are both disrupted in resistant samples carrying *Rcr1* (Song et al. 2016).

10.7 Prediction of New Pathways

New biochemical pathways and annotation data has been predicted by Stephanie et al. (2019) using *P. brassicae* crucifers host-pathosystem (Fig. 1.1, Chap. 1). A complete pathway for the spermidine biosynthesis was predicted in eH *P. brassicae*. Particularly, the three enzymes (EC 2.7.2.11, EC 1.2.1.41, and EC 2.6.1.13) required for conversion of glutamate to ornithine were present in the eH genome although they were not found in the genomes described by Rolfe et al. (2016) and Schwelm et al. (2015). Ornithine is converted into putrescine, a precursor of the polyamine spermidine. The roles of polyamines synthesized by plants are well documented. They are involved in fundamental cellular processes, such as chromatin organization, cell proliferation, differentiation, and programmed cell death, but also in adaptive responses to abiotic and biotic stresses. Polyamines have been found being accumulated in the galls caused by *P. brassicae*. In turnip (*Brassica rapa* var. *rapa*), the levels of putrescine, spermidine, and spermine were reported to be higher in the roots infected with clubroot than in the non-infected normal roots. In cabbage, broccoli, and Komatsuna clubroot galls, the levels of putrescine, spermidine, and spermine also increased after infection with *P. brassicae*. The high level of polyamines measured in the roots may be contributed by the host plant, as reported by Cao et al. (2008) who showed the up-regulation of the *B. napus* polyamine biosynthetic enzyme spermidine synthase in response to *P. brassicae* infection. However, it could also be due to *P. brassicae* since it is difficult to distinguish between metabolites of plant or pathogen origin. Polyamine biosynthesis could be vital to the survival and pathogenesis of *P. brassicae* as described in other pathogens, such as bacteria, fungi, and protists.

The chorismate/Shikimate pathway, which was predicted by Schwelm et al. (2015) but lacking the chorismate mutase (EC 5.4.99.5), Stephanie et al. (2019) found to be complete in the eH *P. brassicae* genome suggesting that this pathway is operational in *P. brassicae*. This pathway is known in plants, bacteria, fungi, and apicomplexan parasites, and is central for the biosynthesis of carbocyclic aromatic compounds. Chorismate is the common branch point for the production of these metabolites, and some intermediates in these pathways also have important physiological relevance. Salicylate, which is derived from chorismate, is involved in plant defense against *P. brassicae*. Anthranilate, another derivative of chorismate, is the precursor of the *Pseudomonas quinolone* signal, which regulates numerous virulence factors in *Pseudomonas aeruginosa*. Further, the chorismate mutase enzyme of this pathway, newly identified in the eH genome, was described as being involved in gall formation in *Meloidogyne javanica* and in *Ustilago maydis*. The potential role of this enzyme in *P. brassicae* pathogenicity would require functional studies.

Clubroot Cyc also predicted the L-Dopachrome biosynthesis pathway, which is involved in melanin synthesis. Melanin is a broad term for a group of natural pigments found in most organisms and localized in cell walls of different structures of fungi. They represent well-known virulence factors for several pathogenic fungi and bacteria. So Clubroot Cyc, a freely accessible tool that can be updated with newly published results, provides a way to easily visualize omics data and represent metabolic pathways potentially involved in particular experimental conditions of interest. Metabolic pathway prediction in *P. brassicae* is still, however, incomplete and poorly annotated because there is still not enough information in the reference databases. Further progress in metabolic network reconstruction and modeling, appropriate gap-filling methods are required, and software is being developed to identify and add lacking pathways, such as Meneco, a tool dedicated to the topological gap-filling of genome-scale draft metabolic network (Prigent et al. 2017; Stephanie et al. 2019).

10.8 Analysis of Host Response Through Transcriptome Data

Two RNA-Seq experiments were performed to gain insight into the genes involved in host penetration, infection, and gall formation. RNA extracted from the roots of the clubroot-susceptible *B. napus* line DH12075 sampled at weekly intervals (from 1 to 6 weeks) after inoculation with Pb3 spores was sequenced. Sequences generated for RNA prepared from the roots of DH12075 plants mock inoculated with water at the same time points served as controls. In addition, Rolfe et al. (2016) sequenced RNA from control and infected roots or hypocotyls of *A. thaliana* plants at the onset (16 DPI) and at late stages (26 DPI) of gall formation with three independent biological replicates. This second experiment used a population of *P. brassicae* described by Malinowski et al. (2012). Analysis of differential expression in infected *B. napus* found significant changes in the expression of 617 Pb3 genes during the 6-week course of infection (genes in the *B. napus* data were considered to be differentially regulated if they showed a twofold or greater change in expression in at least 2 consecutive samples and an adjusted p value of $\leq$0.05). Within the differentially expressed genes there were genes involved in carbohydrate processing, amino-sugar and nucleotide sugar metabolism, transporters, cytoskeleton, motor proteins, and flagella-associated proteins. Of the differentially regulated transcripts, 41 were genes encoding for predicted secreted proteins with 14 of them being SSPs. Hydrolases and peptidases were the most common differentially regulated secreted proteins. The infected *A. thaliana* samples, ~25 million paired-end reads were obtained from each sample. For the 16 DPI samples, ~13% (3.5 million) of the paired reads matched the *P. brassicae* genome in both hypocotyl and root samples. For the 26 DPI samples, 48% of the paired reads (12 million) from hypocotyl tissue and somewhat less (34 %, 9 million) from root tissue matched the *P. brassicae* genome. The root and hypocotyl samples at each time point were very similar to each other. Only 53 and 32 genes showed differential expression (twofold or greater with a corrected probability of $p \leq 0.05$) between the 16 DPI and 26 DPI hypocotyl and root tissue, respectively. In contrast, gene expression differed markedly between the 16 and 26 DPI samples within each tissue (1661 genes for hypocotyl tissue and 957 genes for root tissue, of which 763 were in common). Genes that were differentially up- or down-regulated ($\geq$twofold, $p \leq 0.05$ after correction for false discovery) between 16 and 26 DPI samples in both tissue types were identified using DESeq2 and GO terms analyzed using Blast2GO (Rolfe et al. 2016).

Several GO terms associated with cytoskeletal and membrane activities (cytoskeletal protein binding, plasma membrane, ion transport, cytoskeleton, cell differentiation, protein transport) were down-regulated at 26 DPI compared with 16 DPI, indicative of a shift in metabolic processes associated with the change from plasmodial growth form to spore formation. Many genes associated with general cellular metabolism (carboxylic, lipid and organo-nitrogen metabolic processes, catalytic activity) showed the opposite pattern, with enrichment in the later stages of gall development and spore formation (Rolfe et al. 2016).

10.9 Identification of Genes Regulating Enzymes

As on obligate biotroph, *P. brassicae* relies entirely on a living host for completion of its life cycle. A common feature of biotrophic fungal and oomycete pathogens sequenced to date is the reduction of plant cell wall degrading enzymes and inability to synthesize important metabolites. Rolfe et al. (2016) used the Kyoto Encyclopedia of Genes and Genomes (KEGG) Automated Annotation server to identify biosynthetic pathways that are present in the *P. brassicae* transcript models and also searched manually, the DNA sequence of the genome in case the transcript models were incomplete. The *P. brassicae* Pb3 genome contains only one (PbPT3Sc00011_Am_2.145_1) which was expressed weakly during gall formation. Another feature of biotrophy is a reduction in plant cell wall degrading enzymes (CWDEs). A search of the carbohydrate-active enzymes (CAZy) database for CWDEs identified 113 CWDEs enzymes in Pb3, which is nearly half of the number reported for necrotrophic and hemi-biotrophic fungal plant pathogens. Figure 10.4 indicates a heat map of normalized expression for these genes in the *B. napus* and *Arabidopsis* transcriptomes. Pb3 lacks GH61 (the copper-containing lytic polysaccharide monooxygenases), GH78 (β-glucanases), and pectate lyases (PL) of groups PL1 and PL3, which are also reported to be absent from the genome of other biotrophic pathogens (Zhao et al. 2014). The only exception was the presence of one gene (PbPT3Sc00083_Am_0.81_1) encoding for a cellobiohydrolase (GH6). The major classes of CWDEs enzymes found in the Pb3 were that for xylan (hemicellulose) degradation such as carbohydrate esterase (CE) CE1-CE4 and glycoside hydrolase (GH) family 16. Two of the most abundant GH families were GH18 and GH114 that act on chitin and chitosan, respectively. Also a carbohydrate-binding module (CBM) family 18, which binds chitin, was the most prevalent of the CBM identified in Pb3 and Pb6. Although phylogenetically distant from fungi, chitin seems to be prevalent in *P. brassicae*. Chitin is a major component (25%) of the wall of *P. brassicae* resting spores (Moxham and Buczacki 1983). In their analysis of the e3 clubroot genome Schwelm et al. (2015) identified the importance of chitinases and chitin-binding proteins in infection.

One of the Pb3 genes with a predicted chitin-binding domain (PbPT3Sc00048_S_5.266_1, CBM18) was highly expressed in the first week of infection of *B. napus*, with the expression significantly reduced in week 2, and weakly expressed thereafter (weeks 3–6 after infection). A blastp search of NCBI identified a *Fusarium fujikuroi* (accession CCT72994) chitinase as the best hit (E value 2e-04). Fungal chitinases play a role in cell division, sporulation, and spore germination among several other functions (Duo-Chuan 2006). Given the timing of the expression of this Pb3 predicted chitinase, it is plausible that this gene is involved in the germination and formation of spores. In addition, there are several chitin synthases, GH family 18, and a chitin deacetylase (CDA), which were differentially expressed in *P. brassicae* during *B. napus* infection. One predicted CDA gene (PbPT3Sc00039_A_4.359_1, CE4) was significantly up-regulated in weeks 3–6 after inoculation (coinciding with the production and release of resting spores). Another gene (PbPT3Sc00039_ Sm_3.268_1, CE4) predicted as a

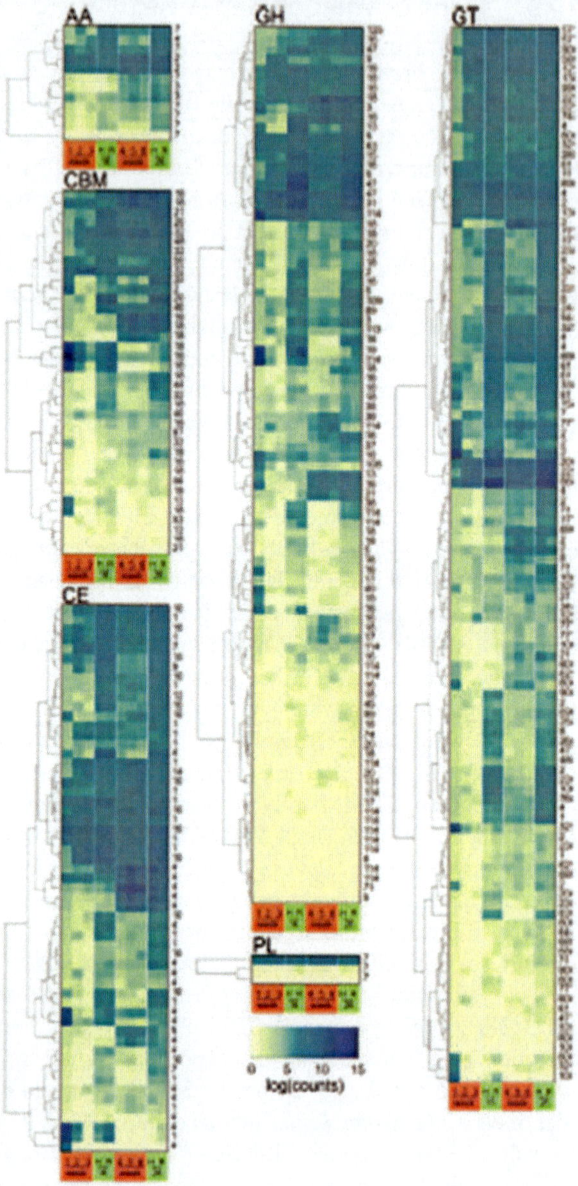

Fig. 10.4 A heat map of normalized gene expression (regularized log normal expression output of DESeq2) of carbohydrate-active enzymes in infected *Brassica napus* (red boxes: weeks 1–6) and *Arabidopsis thaliana* hypocotyl (green boxes) (H) and root (R) tissues at the onset (16 DPI) and late (26 DPI) stages of gall formation. Samples are ordered by similarity but sample similarity dendrograms have been omitted for clarity. *AA* auxiliary activities, *CBM* carbohydrate-binding modules, *CE* carbohydrate esterases, *GH* glycoside hydrolases, *GT* glycosyl transferases, *PL* pectate lyases. The numbers to the right refer to the class of carbohydrate-binding modules as listed (Rolfe et al. 2016)

CDA/polysaccharide deacetylase was significantly up-regulated in the first week of infection. In fungi, it has been suggested that modification of surface chitin by CDA can result in protection of the cell wall from enzymatic hydrolysis, and help to avoid generation of an auto-catalytic defense response system in the invaded host plant. Collectively, these data provide a picture of the fine regulation of genes involved in the synthesis, turn-over (hydrolysis), and modification (deacetylation to protect chitin containing cell walls from degradation) of chitin in *P. brassicae*, which correlates with the various life stages of the pathogen.

10.10 Nutrient Acquisition

The *P. brassicae* genome contains a complete molybdopterin biosynthetic pathway but the thiamine biosynthetic pathway is absent. Eleven putative thiamine transporters were identified within the Pb3 genome, consistent with the acquisition of this essential vitamin from its host. The genome encoded a sulfite oxidase but nitrate reductase was missing, with nitrogen acquired from the host in more complex forms.

The shikimic acid pathway to chorismate was complete with incomplete evidence supporting the synthesis of Tyr and Phe. However, the branch via anthranilate to Trp was absent. The His and Val/Leu/Ile pathways were also absent. Incomplete pathways were present for Thr, Met, Arg, and Lys. A complete understanding of the biosynthetic capacity of *P. brassicae* will require further investigation, but it is clear that this obligate biotroph has lost the capacity to synthesize a number of amino acids and thus, must acquire them from the host. The predicted transcriptome contained 10 putative amino acid transporters. Inspection of the *P. brassicae* genome showed a complete glycolytic and TCA cycle, indicating a capacity for the utilization of carbohydrate from the host for energy production. However, unlike many biotrophic pathogens (including *P. brassicae* infestans), there was no evidence for extracellular invertase genes that would degrade host apoplastic sucrose. Therefore, *P. brassicae* is likely to take up sugars directly. The genome contained 37 ABC transporters, some of which may be involved in nutrient acquisition. PbPT3Sc00087_Am_0.68_1 showed homology (28%, E value $= 6 \times 10^{-11}$) to a SWEET sugar permease from *Ascaris suum*, PbPT3Sc00016_A_10.333_1 was predicted to encode a sucrose proton transporter, while PbPT3Sc00069_Am_1.146_1 and PbPT3Sc00030_A_2.294_1 encode putative monosaccharide transporters. The predicted sucrose transporters were expressed most strongly in developing galls (weeks 4–6 in *B. napus* and 26 DPI in *A. thaliana*) indicating that these genes may play a role in the development of sink strength during gall formation.

During the annotation process, in which predicted gene sequences were compared with known proteins in the KEGG, a series of matches were found against proteins with prokaryotic substrate binding domains. Inspection of these sequences found that these encoded a hitherto unknown class of eukaryotic transporters in which a prokaryotic substrate binding domain is fused to a 7 alpha-helical transmembrane domain. The structure, predicted solutes transported and expression pattern are shown in Fig. 10.5. No such fusion proteins were found in any other eukaryotic genome suggesting that this class of transporter is unique to *P. brassicae*. The gene sequences are present within the e3 genome but either not annotated as genes or annotated as two separate genes. However, transcriptome analysis shows clearly that these are single coding regions in each case. Although characterization of the

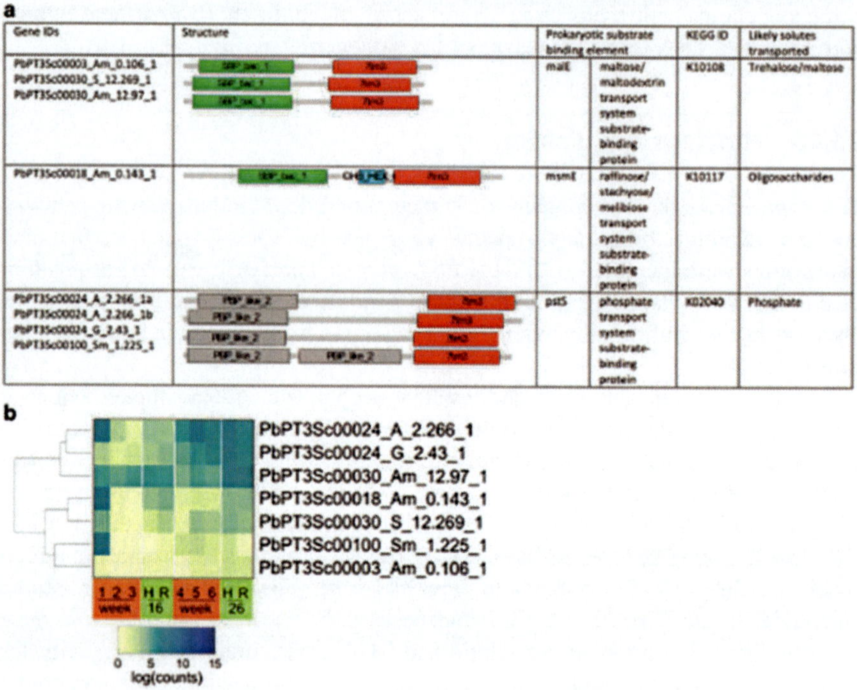

Fig. 10.5 (**a**) Novel putative transporters in the *Plasmodiophora brassicae* genome fuse a eukaryotic 7tm3 motif with prokaryotic substrate binding domains. PFAM domains: SBP_bac_1 bacterial extracellular solute-binding protein; CHB_HEX N-terminal domain of chitobiases, PBP_like_2 periplasmic binding protein. Homologies of these domains to known prokaryotic substrate binding elements are listed. (**b**) Expression in *P. brassicae*-infected *B. napus* (red boxes: week 1–6) and *Arabidopsis thaliana* (green boxes) (16 and 26 DPI hypocotyls (H) and roots (R)) (Rolfe et al. 2016)

substrates transported by these proteins requires experimental confirmation, sequence homology suggests that PbPT3Sc00003_Am_0.106_1, PbPT3Sc00030_S_12.269_1, PbPT3Sc00030_Am_12.97_1, and PbPT3Sc00018_Am_0.143_1 encode sugar (trehalose, maltose, or oligosaccharide) transporters, while PbPT3Sc00024_A_2.266_1a, PbPT3Sc00024_A_2.266_1b, PbPT3Sc00024_G_2.43_1, and PbPT3Sc00100_Sm_1.225_1 encode phosphate transporters. The majority of these transporters were expressed early in infection and during gall development, suggesting they play roles in nutrient acquisition at these key developmental stages (Rolfe et al. 2016). In common with many pathogens, *P. brassicae* has been shown to synthesize trehalose. The Pb3 genome contained 3 genes encoding trehalose phosphate synthase (TPS) (PbPT3Sc00016_A_8.299_1, PbPT3Sc00018_Sm_4.147_1, PbPT3Sc00059_A_1.289_1) and single genes encoding trehalose phosphate phosphatase (PbPT3Sc00030_A_6.312_1) and trehalase (PbPT3Sc00031_A_12.282_1) in agreement with The TPS gene PbPT3Sc00059_A_1.289_1 matched the fragment isolated by Brodmann et al. (2002) and the corresponding complete gene reported elsewhere (Bulman et al. 2007).

References

Agarwal A, Kaul V, Faggian R, Rookes JE, Cahill DM, Ludwig-Muller J (2011) Analysis of global host gene expression during the primary phase of the *Arabidopsis thaliana-Plasmodiophora brassicae* interaction. Funct Plant Biol 38:1–16

Aspeborg H, Coutinho PM, Wang Y, Brumer H, Henrissat B (2012) Evolution, substrate specificity and subfamily classification of glycoside hydrolase family 6 (GH5). BMC Evol Biol 12:186

Bennett RN, Wenke T, Freudenberg B, Mellon FA, Ludwig-Muller J (2005) The tu8 mutation of *Arabidopsis thaliana* encoding a heterochromatin protein 1 causes defects in the induction of secondary metabolite biosynthesis. Plant Biol 7:348–357

Brodmann A, Schuller A, Ludwig-Muller J, Aeschbacher RA, Wiemken A, Boller T, Wingler A (2002) Induction of trehalase in *Arabidopsis* plants infected with the trehalose-producing pathogen *Plasmodiophora brassicae*. Mol Plant-Microbe Interact 15:693–700

Bulman S, Ridgway HJ, Eady C, Conner AJ (2007) Intron-rich gene structure in the intracellular plant parasite *Plasmodiophora brassicae*. Protist 158(4):423–433

Cao T, Srivastava S, Rahman MH, Kav NV, Hotte N, Deyholos M, Strelkov SE (2008) Proteome-level changes in the roots of *Brassica napus* as a result of *Plasmodiophora brassicae* infection. Plant Sci 174(1):97–115

Christian M, Steffens B, Schenck D, Burmester S, Böttger M, Lüthen H (2006) How does auxin enhance cell elongation? Roles of auxin binding proteins and potassium channels in growth control. Plant Biol 8:346–352

Dekhuijzen HM (1980) The occurrence of free and bound cytokinins in clubroots and *Plasmodiophora brassicae* infected turnip tissue cultures. Physiol Plant 49:169–176

Dekhuijzen HM (1981) The occurrence of free and bound cytokinins in plasmodia of *Plasmodiophora brassicae* isolated from tissue cultures of clubroots. Plant Cell Rep 1:18–20

Dekhuijzen HM, Overeem JC (1971) The role of cytokinins in clubroot formation. Physiol Plant Pathol 1:151–161

Desoignies N, Carbonell J, Moreau JS, Conesa A, Dopazo J, Legrève A (2014) Molecular interactions between sugar beet and *Polymyxa betae* during its life cycle: molecular interactions between sugar beet and *Polymyxa betae*. Ann Appl Biol 164:244–256

Devos S, Laukens K, Deckers P, Van Der Straeten D, Beeckman T, Inzé D, Van Onckelen H, Witters E, Prinsen E (2006) A hormone and proteome approach to picturing the initial metabolic events during *Plasmodiophora brassicae* infection on *Arabidopsis*. Mol Plant-Microbe Interact 19:1431–1443

Diederichsen E, Frauen M, Ludwig-Muller J (2014) Clubroot disease management challenges from a German perspective. Can J Plant Pathol 36:85–98

Doan TT, Jaschke D, Ludwig-Muller J (2010) An endophytic fungus induces tolerance against the clubroot pathogen *Plasmodiophora brassicae* in *Arabidopsis thaliana* and *Brassica rapa* roots. Acta Hortic 867:173–180

Duo-Chuan L (2006) Review of fungal chitinases. Mycopathologia 161(6):345–360

Ehness R, Roitsch T (1997) Co-ordinated induction of mRNAs for extracellular invertase and a glucose transporter in *Chenopodium rubrum* by cytokinins. Plant J 11:539–548

Fan L, Reynolds D, Liu M, Stark M, Kjelleberg S, Webster NS, Thomas T (2012) Functional equivalence and evolutionary convergence in complex communities of microbial sponge symbionts. Proc Natl Acad Sci U S A 109:E1878–E1887

Grsic-Rausch S, Kirchheim B, Pieper K, Fritsch M, Hilgenberg W, Ludwig-Muller J (1999) Induction of auxin biosynthetic enzymes by jasmonic acid and in clubroot diseased Chinese cabbage plants. Physiol Plant 105:521–531

Grsic-Rausch S, Kobelt P, Siemens J, Bischoff M, Ludwig-Muller J (2000) Expression and localization of nitrilase during symptom development of the clubroot disease in *Arabidopsis thaliana*. Plant Physiol 122:369–378

Hashimoto M, Murata E, Aoki T (2010) Secretory protein with RING finger domain (SPRING) specific to *Trypanosoma cruzi* is directed, as a ubiquitin ligase related protein, to the nucleus of host cells. Cell Microbiol 12:19–30

Irani S, Trost B, Waldner M, Nayidu N, Tu JY, Kusalik AJ, Todd CD, Wei YD, Bonham-Smith WC (2018) Transcriptome analysis of response to *Plasmodiophora brassicae* infection in the *Arabidopsis* shoot and root. BMC Genome 19:23

Jahn L, Mucha S, Bergmann S, Horn C, Siemens J, Staswick P, Steffens B, Ludwig-Muller J (2013) The clubroot pathogen (*Plasmodiophora brassicae*) influences auxin signaling to regulate auxin homeostasis. Plan Theory 2:726–749

Jaschke D, Dugassa-Gobena D, Karlovsky P, Vidal S, LudwigMüller J (2010) Suppression of clubroot (*Plasmodiophora brassicae*) development in *Arabidopsis thaliana* by the endophytic fungus *Acremonium alternatum*. Plant Pathol 59:100–111

Jubault M, Hamon C, Gravot A, Lariagon C, Delourme R, Bouchereau A, Manzanares-Dauleux MJ (2008) Differential regulation of root arginine catabolism and polyamine metabolism in clubroot-susceptible and partially resistant *Arabidopsis* genotypes. Plant Physiol 146:2008–2019

Jubault M, Lariagon C, Taconnat L, Renou J-P, Gravot A, Delourme R, Manzanares-Dauleux MJ (2013) Partial resistance to clubroot in *Arabidopsis* is based on changes in the host primary metabolism and targeted cell division and expansion capacity. Funct Integr Genome 13:191–205

Karlsson M, Stenlid J (2009) Evolution of family 18 glycoside hydrolases: diversity, domain structures and phylogenetic relationships. J Mol Microbiol Biotechnol 16:208–223

Kemen E, Gardiner A, Schultz-Larsen T, Kemen AC, Balmuth AL, Robert Seilaniantz A, Bailey K, Holub E, Studholme DJ, MacLean D, Jones JDG (2011) Gene gain and loss during evolution of obligate parasitism in the white rust pathogen of *Arabidopsis thaliana*. PLoS Biol 9:e1001094

Knaust A, Ludwig-Müller J (2013) The ethylene signaling pathway is needed to restrict root gall growth in *Arabidopsis* after infection with the obligate biotrophic protist *Plasmodiophora brassicae*. J Plant Growth Regul 32:9–21

Laurie JD, Ali S, Linning R, Mannhaupt G, Wong P, Guldener U, Munsterkotter M, Moore R, Kahmann R, Bakkeren G, Schirawski J (2012) Genome comparison of barley and maize smut fungi reveals targeted loss of RNA silencing components and species-specific presence of transposable elements. Plant Cell 24:1733–1745

Lee SH, Stephens JL, Paul KS, Englund PT (2006) Fatty acid synthesis by elongases in trypanosomes. Cell 126:691–699

Lemarie S, Robert-Seilaniantz A, Lariagon C, Lemoine J, Marnet N, Jubault M, Manzanares-Dauleux MJ, Gravot A (2015) Both the jasmonic acid and the salicylic acid pathways contribute to resistance to the biotrophic clubroot agent *Plasmodiophora brassicae* in *Arabidopsis*. Plant Cell Physiol 56:2158–2168

Lindner AC, Lang D, Seifert M, Podlesakova K, Novak O, Strnad M, Reski R, Schwartzenberg K (2014) Isopentenyltransferase-1 (IPT1) knockout in *Physcomitrella* together with phylogenetic analyses of IPTs provide insights into evolution of plant cytokinin biosynthesis. J Exp Bot 65:2533–2543

Lovelock DA, Donald CE, Conlan XA, Cahill DM (2013) Salicylic acid suppression of clubroot in broccoli (*Brassicae oleracea* var. italica) caused by the obligate biotroph *Plasmodiophora brassicae*. Austr Plant Pathol 42:141–153

Lovelock DA, Sola I, Marschollek S, Donald CE, Rusak G, van Pee KH, Ludwig-Muller J, Cahill DM (2016) Analysis of salicylic acid-dependent pathways in *Arabidopsis thaliana* following infection with *Plasmodiophora brassicae* and the influence of salicylic acid on disease. Mol Plant Pathol 17:1237–1251

Ludwig-Muller J (2009a) Plant defence—what can we learn from clubroots? Austr Plant Pathol 38:318–324

Ludwig-Muller J (2009b) Glucosinolates and the clubroot disease: defense compounds or auxin precursors? Phytochem Rev 8:135–148

Ludwig-Muller J, Kasperczyk N, Schubert B, Hilgenberg W (1995) Identification of salicylic acid in Chinese cabbage and its possible role during root infection with *Plasmodiophora brassicae*. Curr Top Phytochem 14:39–45

Ludwig-Muller J, Epstein E, Hilgenberg W (1996) Auxin-conjugate hydrolysis in Chinese cabbage: characterization of an amidohydrolase and its role during the clubroot disease. Physiol Plant 97:627–634

Ludwig-Muller J, Pieper K, Ruppel M, Cohen JD, Epstein E, Kiddle G, Bennett R (1999) Indole glucosinolate and auxin biosynthesis in *Arabidopsis thaliana* (L.) Heynh. glucosinolate mutants and the development of clubroot disease. Planta 208:409–419

Ludwig-Muller J, Prinsen E, Rolfe SA, Scholes JD (2009) Metabolism and plant hormone action during clubroot disease. J Plant Growth Regul 28:229–244

Ludwig-Muller J, Julke S, Geiß K, Richter F, Mithofer A, Sola I, Rusak G, Keenan S, Bulman S (2015) A novel methyltransferase from the intracellular pathogen *Plasmodiophora brassicae* participates in methylation of salicylic acid. Mol Plant Pathol 16(4):349–364

Malinowski RS, Jody A, Fleming AJ, Scholes JD, Rolfe SA (2012) Gall formation in clubroot-infected *Arabidopsis* results from an increase in existing meristematic activities of the host but is not essential for the completion of the pathogen life cycle. Plant J 71(2):226–238

Malinowski R, Novák O, Borhan MH, Spichal L, Strnad M, Rolfe SA (2016) The role of cytokinins in clubroot disease. Eur J Plant Pathol 145:543–557

Manoharan R, Shanmugam A, Hwang I, Park JI, Nou IS (2016) Expression of salicylic acid-related genes in *Brassica oleracea* var. capitata during *Plasmodiophora brassicae* infection. Genome 59:379–391

Moxham SE, Buczacki ST (1983) Chemical composition of the resting spore wall of *Plasmodiophora brassicae*. Trans Br Mycol Soc 80(2):297–304

Muller P, Hilgenberg W (1986) Isomers of zeatin and zeatin riboside in clubroot tissue: evidence for trans-zeatin biosynthesis by *Plasmodiophora brassicae*. Physiol Plant 66:245–250

Neuhaus K, Grsic-Rausch S, Sauerteig S, Ludwig-Muller J (2000) *Arabidopsis* plants transformed with nitrilase 1 or 2 in antisense direction are delayed in clubroot development. J Plant Physiol 156:756–761

Okmen B, Doehlemann G (2014) Inside plant: biotrophic strategies to modulate host immunity and metabolism. Curr Opin Plant Biol 20:19–25

Oshima K, Kakizawa S, Nishigawa H, Jung H-Y, Wei W, Suzuki S, Arashida R, Nakata D, Miyata S, Ugaki M, Namba S (2004) Reductive evolution suggested from the complete genome sequence of a plant-pathogenic phytoplasma. Nat Genet 36:27–29

Pasold S, Ludwig-Muller J (2013) Reduction of clubroot (*Plasmodiophora brassicae*) formation in Arabidopsis thaliana after treatment with prohexadione-calcium, an inhibitor of oxoglutaric acid-dependent dioxygenases. Plant Pathol 62(6):1357–1365

Pasold S, Siegel I, Seidel C, Ludwig-Muller J (2010) Flavonoid accumulation in *Arabidopsis thaliana* root galls caused by the obligate biotrophic pathogen *Plasmodiophora brassicae*. Mol Plant Pathol 11:545–562

Piotrowski M (2008) Primary or secondary? Versatile nitrilases in plant metabolism. Phytochemistry 69:2655–2667

Prerostova S, Dobrev PI, Konradyova V, Knirsch V, Gaudinova A, Kramna B, Kazda J, Ludwig-Muller J, Vankova R (2018) Hormonal responses to *Plasmodiophora brassicae* infection in *Brassica napus* cultivars differing in their pathogen resistance. Int J Mol Sci 19(12):E4024

Prigent S, Frioux C, Dittami SM, Thiele S, Larhlimi A, Collet G, Gutknecht F, Got J, Eveillard D, Bourdon J, Plewniak F, Tonon T, Siegel A (2017) Meneco, a topology-based gap-filling tool applicable to degraded genome-wide metabolic networks. PLoS Comput Biol 13:e1005276

Rolfe SA, Strelkov SE, Links MG, Clarke WE, Robinson SJ, Djavaheri M, Malinowski R, Haddadi P, Kagale S, Parkin IAP, Taheri A, Borhan MH (2016) The compact genome of the plant pathogen *Plasmodiophora brassicae* is adapted to intracellular interactions with host *Brassica* spp. BMC Genomics 17:272

Ruiz-Herrera J, Prieto G, Ruiz MJ, Medrano R (2002) Evolution and phylogenetic relationships of chitin synthases from yeasts and fungi. FEMS Yeast Res 1:247–256

Sanchez-Vallet A, Mesters JR, Thomma BPHJ (2014) The battle for chitin recognition in plant-microbe interactions. FEMS Microbiol Rev 39:171–183

Schuller A, Kehr J, Ludwig-Muller J (2013) Laser Micro-dissection coupled to transcriptional profiling of *Arabidopsis* roots inoculated by *Plasmodiophora brassicae* indicates a role for brassino-steroids in clubroot formation. Plant Cell Physiol 55:392–411

Schwelm A, Fogelqvist J, Knaust A, Julke S, Lilja T, Bonilla Rosso G, Karlsson M, Shevchenko A, Dhandapani V, Choi SR, Kim HG, Park JY, Lim YP, Ludwig-Müller J, Dixelius C (2015) The *Plasmodiophora brassicae* genome reveals insights in its life cycle and ancestry of chitin synthases. Sci Rep 5:11153

Siemens J, Nagel M, Ludwig-Muller J, Sacristan M (2002) The interaction of *Plasmodiophora brassicae* and *Arabidopsis thaliana*: parameters for disease quantification and screening of mutant lines. Phytopathology 150:592–605

Siemens J, Keller I, Sarx J, Kunz S, Schuller A, Nagel W, Schmulling T, Parniske M, Ludwig-Muller J (2006) Transcriptome analysis of *Arabidopsis* clubroots indicate a key role for cytokinins in disease development. Mol Plant-Microbe Interact 19:480–494

Siemens J, Glawischnig E, Ludwig-Muller J (2008) Indole glucosinolates and camalexin do not influence the development of the clubroot disease in *Arabidopsis thaliana*. J Phytopathol 156:332–337

Siemens J, Gonzalez MC, Wolf S, Hofmann C, Greiner S, Du Y, Rausch T, Roitsch T, Ludwig-Muller J (2011) Extracellular invertase is involved in the regulation of clubroot disease in *Arabidopsis thaliana*. Mol Plant Pathol 12:247–262

Soanes DM, Alam I, Cornell M, Wong HM, Hedeler C, Paton NW, Rattray M, Hubbard SJ, Oliver SG, Talbot NJ (2008) Comparative genome analysis of filamentous fungi reveals gene family expansions associated with fungal pathogenesis. PLoS One 3:e2300

Song T, Chu M, Lahlali R, Yu F, Peng G (2016) Shotgun label-free proteomic analysis of clubroot (*Plasmodiophora brassicae*) resistance conferred by the gene *Rcr1* in *Brassica rapa*. Front Plant Sci. https://doi.org/10.3389/fpls.2016.01013

Staswick PE, Tiryaki I (2004) The oxylipin signal jasmonic acid is activated by an enzyme that conjugates it to isoleucine in *Arabidopsis*. Plant Cell 16:2117–2127

Staswick PE, Serban B, Rowe M, Tiryaki I, Maldonado MT, Maldonado MC, Suza W (2005) Characterization of an *Arabidopsis* enzyme family that conjugates amino acids to indole-3-acetic acid. Plant Cell 17:616–627

Stephanie D, Arnaud B, Kevin G, Ludovic L, Jerome G, Ludovic C, Lionel L, Yoann A, Christophe M, Wawra S, Belmonte R, Lobach L, Saraiva M, Willems A, West P (2012) Secretion, delivery and function of oomycete effector proteins. Curr Opin Microbiol 15:685–691

Stephanie D, Arnaud B, Kelvin G, Ludovic L, Jerome G, Ludovic C, Lionel L, Yoann A, Christophe M, Manzanares-Dauleux MJ (2019) Computational analysis of the *Plasmodiophora brassicae* genome: mitochondrial sequence description and metabolic pathway database design. Genomics 111(6):1629–1640. https://doi.org/10.1016/j.ygeno.2018.11.013

Wilson RA, Gibson RP, Quispe CF, Littlechild JA, Talbot NJ (2010) An NADPH-dependent genetic switch regulates plant infection by the rice blast fungus. Proc Natl Acad Sci U S A 107:21902–21907

Zakrzewski AC, Weigert A, Helm C, Adamski M, Adamska M, Bleidorn C, Raible F, Hausen H (2014) Early divergence, broad distribution and high diversity of animal chitin synthases. Genome Biol Evol 6:316–325

Zhao ZT, Liu HQ, Wang CF, Xu JR (2014) Correction: comparative analysis of fungal genomes reveals different plant cell wall degrading capacity in fungi. BMC Genome 15:6

Zheng L, Dang Z, Li H, Zhang H, Wu S, Wang Y (2014) Isolation and characterization of a D1 -pyrroline-5-carboxylate synthetase (NtP5CS) from *Nitraria tangutorum* Bobr. and functional comparison with its *Arabidopsis* homologue. Mol Biol Rep 41:563–572

Host Resistance

11

Abstract

Genetical and molecular mechanisms of host resistance in the form of qualitative and quantitative genes have been expressed against *P. brassicae* in the economically important and widely grown cruciferous species, viz., *B. rapa*, *B. oleracea*, *B. napus*, *R. sativus,* and *Arabidopsis thaliana*. Mapping of quantitative and qualitative traits loci (QTL's) have revealed several dominant loci, twelve in *B. rapa*, more than 22 QTL's in *B. oleracea* and more than 19 QTL's in *B. napus* for clubroot resistance to different pathotypes of *P. brassicae*. Clubroot resistance in *Brassica* species is inherited by many dominant, recessive, additive, and complementary genes. Substantial progress has been made in breeding clubroot resistance genotypes/cvs. of *B. rapa*, *B. oleracea*, *B. napus*, *R. sativus,* and *A. thaliana*. Molecular mechanisms to identify host resistance have been extensively exploited through genetic mapping of clubroot resistance (CR) loci, genomics, transcriptomics, and proteomics approaches. Multiple pathotypes specific QTL's conferring resistance to *P. brassicae* have been identified and functionally characterized by RNA sequencing and genotyping-by-sequencing to facilitate cloning of R genes. To counteract the role of mitogen in modulation of CR QTL's, new sources of resistance have been found among the oilseed rape accessions under low and high N conditions. Marker-assisted selection of CR genes have ensured accumulation of more than one R gene into a single CR cv. for more durability under field conditions against different pathotypes of *P. brassicae*. To trigger the pathogen-associated molecular pattern (PAMP's) immunity, thirty three chitinase genes have been identified in *B. rapa* with their structure, evolution, and chromosome location. Map-based cloning and characterization of *Crr1A* genes which confers resistance to *B. rapa* encodes a Toll-Interlinked-1 receptor/nucleotide-binding set/Leucine-rich repeat (TIR-NB-LRR) protein expressed in the stele and cortex of hypocotyl and root, where secondary infection of pathogen occurs but not in root hairs. Some NBS-LRR genes mediate

G. S. Saharan et al., *Clubroot Disease of Crucifers*,
https://doi.org/10.1007/978-981-16-2133-8_11

resistance in a more complex manner then the classical gene for gene interactions (One host R-gene-one pathogen gene). Differential R-genes expression and their functional mechanisms during life cycle of pathogen to confer resistance at different stages of infection and pathogen development have been elucidated. Phenylpropanoid pathway genes (Lignin biosynthesis) are up-regulated in CR *B. rapa*. Clubroot resistance has been transferred in *Brassica* through molecular marker-assisted selection, interspecific hybridization, distant hybridization, and mutational approaches. Effective management of R cvs. of *Brassica* can enhance durability, and cost-effective control of clubroot. The nomenclature of CR genes has not been standardized in *Brassica* species. A basal defense mechanism is activated by a virulent pathogen through induction of trehalase gene. Cellular resistance is expressed through accumulation of reactive oxygen species (ROS) or lignification of epidermal or cortical cells in the resistant cvs. Small spheroid galls of clubroot acts as an indicator of host resistance. Histological resistance is expressed through deposition of callose in response to pathogen invasion. Systematic acquired resistance is associated with up-regulation of pathogenesis-related genes PR-1 and PR-2.

Keywords

Host resistance · Genetics of CR genes · Inheritance of CR · Resistance breeding · Molecular mechanism of host resistance · Genomics of resistance genes · Characterization of genes · Proteomic approach identify resistance genes · Differential expression of R-genes · Role of phenylpropanoid · Transferring CR through MAS · Interspecific hybridization · Distant hybridization · Management of CR for effective utilization · Nomenclature of CR genes · Non-host resistance · Expression of cellular resistance · Mutational approach resistance · Histological resistance · Systemic acquired resistance · Sources of resistance · Deployment of resistance · Breakdown of resistance · Manipulation of resistance · Gene sequencing · Genome complexity

11.1 Introduction

Crucifer plants are always challenged by various biotic and abiotic stresses. In agriculture cropping system, plant protection is being delivered using different approaches such as chemical control, various agronomic practices, biological control, integrated pest management (IPMs), and cultivation of resistance cultivars. Among these approaches, resistance cultivars are the most economical, environmentally sustainable solution to control different diseases including clubroot in crucifers. Resistance to pathogen in plants is by two different mechanisms, viz., pathotypes specific resistance and basal resistance. Both types of resistance are important in the protection of plants against pathogens. The advances in molecular biology especially the genomic information such as development of molecular markers, comparative mapping between related species and transcriptomic analysis have great promises for identification, localizing, colonizing, and functionally characterizing genes of economic importance. Molecular mapping of quantitative and qualitative trait loci (QTL's) have revealed several dominant loci in *B. rapa*. More than 22 QTL's in *B. oleracea* and over 19 QTL's in *B. napus* which confer clubroot resistance to different pathotypes of *P. brassicae*. Long-term cultivation of clubroot disease resistance cultivars and hybrids may result in breakdown of resistance due to evolution of new virulence of pathogen through selection pressure. The inheritance of clubroot resistance is either qualitative or quantitative in *Brassica* species. Recently, mapped clubroot resistance genetic loci and closely linked molecular markers to these loci have been used for marker-assisted selection in clubroot resistance breeding of *Brassica* species. Extensive use of recently available resistance sources can be combined with molecular tools and new technologies such as genes/QTL's, mapping, fine mapping, gene cloning, comparative genomics, and analysis of transcriptomic profile through next generation sequencing can enhance our understanding of clubroot resistance mechanisms. Novel information can help controlling clubroot disease in an effective way, so yield losses can be reduced, and the quality of *Brassica* crop product can be improved.

11.2 Genetics of Clubroot Resistance Genes

The genetic basis of clubroot resistance (CR) has been determined in economically important cruciferous species: *Brassica rapa, B. oleracea, B. napus, R. sativus*, and *Arabidopsis thaliana*. In *B. rapa*, mainly in Chinese cabbage, one minor and seven major CR genes introduced from European fodder turnips have been identified. Mapping of these CR genes localized *Crr1* on R8, *Crr2* on R1, *CRc* on R2, and *Crr4* on R6 linkage groups of Chinese cabbage. Genes *Crr3, CRa, CRb*, and *CRk* mapped to R3, but at two separate loci, *CRa* and *CRb* are independent of *Crr3* and *CRk*, which are closely linked. Further, analysis suggested that *Crr1, Crr2*, and *CRb* have similar origins in the ancestral genome as in chromosome 4 of *Arabidopsis thaliana*. Genetic analysis of clubroot resistance genes in *B. oleracea* suggests that these are quantitative traits. Twenty-two quantitative trait loci (QTLs) have been mapped in different linkage groups of *B. oleracea*. In *B. napus*, genetic analysis of clubroot resistance was found to be governed by one or two dominant genes. The quantitative analysis approach, however, proved that these are polygenic. In total, at least 19 QTLs have been detected on eight chromosomes of *B. napus*, N02, N03, N08, N09, N13, N15, N16, and N19. The chromosomal location of the other six QTLs is not clear. Cloning of these QTLs or resistance loci is possible. Progress in genomics, particularly the techniques of comparative mapping and genome sequencing, supplements cloning, and allows improved characterization of CR genes. Further, the development of DNA markers linked to CR genes will in turn hasten the breeding of clubroot-resistant crucifer's cultivars.

11.2.1 Inheritance of Clubroot Resistance in Crucifers

The inheritance of resistance to clubroot, Walker and Larson (1951) found that susceptibility was dominant in crosses between resistant and susceptible cabbage cultivars and disease response appeared to be governed by a polygenic type of inheritance. Strandberg and Williams (1967) reported that Chinese cabbage (*B. pekinensis* Rupr.) cvs. Michihili possessed one dominant gene for resistance to races 6, and 7 of the clubroot pathogen. Chiang and Crete (1970) crossed resistant and susceptible cabbage lines, F_1 data showed that susceptibility was dominant over resistance and data from F_2 and backcross progenies supported a duplicate gene hypothesis. Lammerink (1967) crossed New Zealand clubroot resistant rape and New Zealand Wilhelmsburger rutabaga and found that resistance to races B and C was determined by single dominant non-allelic genes. Johnston (1970), working with a new racial isolate, crossed susceptible Giant Rape and resistant New Zealand rape and found that resistance was dominant and controlled by a single gene, similar findings were reported when susceptible Giant Rape was crossed with resistant Wilhelmsburger rutabaga.

The York and Wilhelmsburger rutabaga cultivars, both resistant to races 2, and 3 of the clubroot pathogen, were crossed with the race 2 and 3 susceptible Laurentian cultivar. Race 2-resistant Ditmars S2 rutabaga selection was also crossed with susceptible Laurentian. Race exposure of F_1 seedlings of resistant X susceptible crosses showed that resistance to races 2 and 3 was dominant. Susceptibility studies of F_2 and backcross progenies showed that resistance to race 2 in York was controlled by a single major gene, whereas resistance to this race in Wilhemsburger was controlled by two major genes. Single major gene resistance to race 2 was indicated for the Ditmars S2 selection. This crucifer line is susceptible to race 3. The F_2 progeny exposures to race 3 showed that resistance was controlled by one major gene in both the York, and Wilhelmsburger rutabaga cultivars (Ayers and Lelacheur 1972). In case of *A. thaliana,* resistance to *P. brassicae* isolate is conferred by a dominant allele of a single nuclear gene designated as RPB1. With the aid of morphological marker, this locus is mapped near the markers dis 2 to chromosomes 1 (Fuchs and Sacristan 1996).

In the major cultivated cruciferous species, *B. rapa*, *B. napus*, *B. oleracea*, *R. sativus* and model crucifer species *A. thaliana,* clubroot disease has become a cause of serious concern because of rising crop failures. To breed clubroot-resistant cultivars of these species, a number of germplasms have been evaluated. This allowed sources of resistance to be identified in crucifer's species. Inheritance of resistance to clubroot varies between these species up to molecular level.

(i) ***Brassica rapa***

The breeding of clubroot-resistant cultivars of Chinese cabbage (*B. rapa* ssp. *pekinensis*) has been impeded because most cultivars are highly susceptible to clubroot disease (Yoshikawa 1981; Cho et al. 2002). The identification of resistant sources in European fodder turnips (*B. rapa* ssp. *rapifera*) (Karling 1968; Buczacki et al. 1975; Crute et al. 1983; Crisp et al. 1989) allowed the transfer of CR genes from European fodder turnip to Chinese cabbage (Yoshikawa 1981). The introgression of CR genes from European fodder turnip cultivars, including Gelria R, Siloga, Debra, and Milan White, has provided and broadened the genetic diversity of Chinese cabbage. The European Clubroot Differential (ECD) hosts 01–04 represent the spectrum of resistance to virulent pathotypes of *P. brassicae* (Toxopeus and Janssen 1975; Toxopeus et al. 1986; Buczacki et al. 1975). These have been used as resistant sources in breeding clubroot-resistant cultivars either in Chinese cabbage or in *B. napus*.

In these studies, at least three independent dominant genes, which conferred differential (race specific or vertical) resistance to particular pathotypes of *P. brassicae*, is present in turnip genotypes (Wit and van de Weg 1964; Tjallingii 1965; Toxopeus and Janssen 1975; James et al. 1978; Crute et al. 1980). Yoshikawa (1993) demonstrated that clubroot resistance of European fodder turnips, including cv. Siloga, was controlled mainly by a major gene and a few minor genes. James et al. (1978) identified three independent dominant genes that conferred resistance in three *B. rapa* genotypes to race 6 of *P. brassicae*. Crute et al. (1980) also demonstrated that three genes controlled resistance in turnip. This suggests that clubroot resistance in *B. rapa* is controlled by several genes independently. Because clubroot-resistant cultivars of Chinese cabbage were released by introducing CR genes from clubroot-resistant European fodder turnip, it is believed that clubroot-resistant cultivars of Chinese cabbage have the potential for introducing one to several genes from turnip. The data obtained by marker trait analysis confirmed that at least eight CR loci are present in *B. rapa.* Two loci crr1 and crr2 have been identified (Suwabe et al. 2003, 2006; Hirai et al. 2004; Piao et al. 2004; Sakamoto et al. 2008).

(ii) *Brassica oleracea*

Various sources of clubroot resistance through the screening of germplasm have been identified (Crute et al. 1980; Dixon and Robinson 1986; Dixon et al. 1986; Dixon 1988; Crisp et al. 1989; Dias et al. 1993; Voorrips and Kanne 1997; Manzanares-Dauleux et al. 2000b; Carlsson et al. 2004). In contrast to *B. rapa*, completely resistant accessions have only rarely been identified in *B. oleracea*. Crisp et al. (1989) evaluated about 1000 *B. oleracea* accessions, resistant sources were confirmed in some kales, including the curly marrow stem and 1000-head types, and in cabbage, including Bohmerwaldkohl, Bindsachsener cabbage, Badger Shipper, and Ladoszkaya cabbage. Some open pollinated Brussels sprouts, including cvs. Cambridge, Continuity, Rubine, and Catskill, and forms of southern European cabbage, cauliflower, and broccoli exhibited lower levels of susceptibility. Among the kale, cabbage, and winter cauliflower accessions evaluated, only some kale accessions, mainly the leafy and short leafy kale groups, exhibited high levels of resistance to clubroot, whereas all cabbage and cauliflower accessions were susceptible (Manzanares-Dauleux et al. 2000b). Some of these resistant sources are widely used in breeding programmes for *B. oleracea*.

The detailed genetics of clubroot resistance were studied in *B. oleracea* using either diallel crossing methods or segregating population. Most of these studies

Table 11.1 Inheritance of clubroot resistance in *Brassica oleracea* (Piao et al. 2009)

Resistant source	Population	Test (pathotype)	Genes involved	Reference
Bohmerwaldkohl	F2, BC	Glasshouse (field isolate)	Four genes	Yoshikawa (1993)
Bohmerwaldkohl	F2, BC	Glasshouse (field isolate)	Additive and recessive	Crute and Pink (1989)
Kale accessions	Diallel	Glasshouse (field isolate)	Many dominant genes	Laurens and Thomas (1993)
Bohmerwaldkohl, Badger Shipper, Larson 8353 T, Resistant Detroit and Curly kale	F1	Glasshouse (field isolate)	Recessive	Voorips and Visser (1993)
Cabbage	F1, F2, BC1	Glasshouse (field isolate, single-spore isolate)	Two complementary genes	Vorrips and Kanne (1997)
Resistant Detroit			Two genes	
Bindsachsener			Two genes	
Curly kale			More than two genes	

concluded that inheritance of this trait in *B. oleracea* is polygenic (Table 11.1). Yoshikawa (1993), who worked with progenies of crosses between cv. Bohmerwaldkohl and a susceptible cabbage cv. demonstrated that at least one to four genes controlled this trait. Voorips and Visser (1993) found that the inheritance of resistance to clubroot was recessive based on the genetic analysis of the 11 F_1 populations derived from crosses between resistant sources, including three accessions of Bohmerwaldkohl, cv. Badger Shipper, two lines derived from "Larson 8353 T," cv. Resistant Detroit, and four accessions of curly kale and susceptible cabbage, respectively. In kale (Laurens and Thomas (1993), the inheritance of resistance is controlled by many dominant alleles with a predominance of additive genetic effects with incomplete dominance. Based on qualitative and quantitative analyses of the F_1, F_2, and backcross progenies of four crosses derived from four sources of resistance and one common susceptible doubled haploid line, Voorrips and Kanne (1997) suggested that one was controlled by two complementary genes. Chiang and Crete (1970) agreed that resistance was controlled by two loci.

(iii) ***Brassica napus***

Clubroot resistance in swedes (*B. napus*) suggested that genotypes, including cvs. Wilhelmsburger and Studsgaard Bangholm, exhibited resistance under certain environmental conditions (Karling 1968; Dixon et al. 1972). These sources are used to breed clubroot-resistant cultivars of *B. napus*. During genetic studies of resistance to clubroot in swedes, different sources of resistance have been used including Wilhelmsburger, clubroot-resistant rape, Wye swede group, York swede, and

Ditmar S2 swede (Lammerink 1967; Johnston 1970; Ayers 1972; Ayers and Lelacheur 1972). The inheritance of resistance in *B. napus* is controlled by one or two single independent dominant genes. At least one of these genes is shared by several sources. Two recessive genes with additive effect were also identified in Danish Giant swede. Gustafsson and Falt (1986) suggested that four resistance genes are present in the *B. napus* ECD hosts. Based on genetic analysis using partial or genome-wide marker surveys in several *B. napus*, over 20 quantitative trait loci for CR were identified (Manzanares-Dauleux et al. 2000a; Werner et al. 2008). Additive effects in genetic control of CR in *B. napus* loci were found by Li et al. (2016) who predicted the role of TIR-NBS gene family. Because *B. napus* is a natural amphidiploid between *B. oleracea* and *B. rapa*, however, it is not clear where these resistance genes are located and their origins. It is suggested that genes for resistance in *B. napus* are located in the A genome from *B. rapa* (Chiang et al. 1977).

11.2.2 Clubroot Resistance Breeding in Cruciferous Crops

Substantial progress has been made in breeding for clubroot resistance genotypes/cultivars in cruciferous crops like *B. rapa*, *B. oleracea*, *B. napus*, *Raphanus sativus*, and *Arabidopsis thaliana*. However, work on race-specific resistance cvs. needs more attention of researchers.

(i) *B. rapa*

The perusal of data in Table 11.2 summarizes the available information about eight CR loci or QTL that have been mapped in *B. rapa*. Four of these reside on chromosome A03, probably representing two complex regions that control CR. One region, originally identified by mapping *CRb* and its marker TCR05 (Piao et al. 2004), is assumed to be located in the top part of A03 because a QTL derived from cv. Gelria could be allocated around TCR05 and was mapped between 12.5 cM (BRMS124) and 30.2 cM (BRMS206) (Suwabe et al. 2006). Locus *CRa* was identified as being linked with the SCAR marker HC352b (Matsumoto et al. 1998, 2005; Hayashida et al. 2008), which has been mapped by Sakamoto et al. (2008) on A03 in the vicinity of markers that were mapped in this region. Therefore, *CRa* and *CRb* might be allelic or closely linked CR loci. So far, it is not known whether these differ in their race specificity, but they differ in their mode of inheritance.

Another region on A03 is highlighted by the *Crr3* locus, which is located between 34.6 cM (BRMS158) and 73.5 cM (BRMS058) (Suwabe et al. 2006; Saito et al. 2006). Sakamoto et al. (2008) reported a large QTL (*CRk*) around OPC11-2S, a marker of *Crr3*. The locus *CRk* was derived from Chinese cabbage cv. CR Kanko, and possibly it is identical, allelic, or closely linked to *Crr3*. Other QTL or major loci that have a strong impact on CR were found on chromosomes A02 [*CRc* (Sakamoto et al. 2008)] and A08 [*Crr1* (Suwabe et al. 2003)]. Because only a few resistant turnip cultivars have been used as CR sources, it is likely that additional CR loci will be identified in this crop type. The QTLs for CR residing in the *Brassica* A genome

Table 11.2 Mapped CR loci in *Brassica rapa* and their linked markers (Diederichsen et al. 2009)

CR locus[a]	Identified as	Linkage group	Linked markers	Inheritance	References
Cra[b]	Major gene	A03	HC352b-SCAR	Dominant	Matsumoto et al. (1998), Hayashida et al. (2008)
Crr1	Major gene	A08	BRMS-088	Codominant	Suwabe et al. (2003, 2006)
Crr2	Modifier gene	A01	BRMS-096	Codominant	Suwabe et al. (2003, 2006)
Crr3	Major gene	A03	OPC11-2S	Codominant	Hirai et al. (2004), Saito et al. (2006)
CRb	Major gene	A03	TCR05	Codominant	Piao et al. (2004)
Crr4	QTL	A06	BRMS-125	Codominant	Suwabe et al. (2003, 2006)
CRk[c]	QTL, possibly major gene	A03	OPC11-2S	Codominant	Sakamoto et al. (2008)
CRc	QTL, possibly major gene	A02	m6R	Dominant	Sakamoto et al. (2008)

[a]Suffixes, a, b, and c in the locus name do not correspond to A, B, and C genes of ECD set, but indicate chronological order of discovery
[b]Very near or identical to *CRb*
[c]Very near or identical to *Crr3*

and originating from turnip (ECD-04) were also mapped in *B. napus* (Werner et al. 2008). A QTL on A03 had a strong effect on CR for three *P. brassicae* isolates. This QTL was located at 65.3 cM in the map (Werner et al. 2008); it might correspond to *Crr3* or eventually to *Crb*. Map did not have enough markers in common with the previously mentioned maps, however, the identity of these loci/QTLs remains to be determined. Differential race-specific responses of the *CRk* and *CRc* loci were reported by Sakamoto et al. (2008). According to the available information, about race-specific effects of CR genes from *B. rapa*, it is most likely that the mapped CR loci confer race-specific resistance, as in Japanese CR cultivars. Kuginuki et al. (1999) and Hatakeyama et al. (2004) suggested the use of two commercial CR Chinese cabbage cultivars showing opposite interactions with most Japanese field isolates of *P. brassicae* for race designation.

Most mapping studies have used a CR cultivar or CR breeding line as a resistant parent. These breeding lines have already been selected and may have only a proportion of those CR genes that might have been present in the original CR source. Suwabe et al. (2003) in their study used a doubled haploid (DH) line, G004, as a resistant parent. This line was selected from DH lines derived from an F_1 between the CR turnip cv. Siloga and a susceptible Chinese cabbage line. Kuginuki et al. (1997) selected DH lines with CR similar to that of cv. Siloga. This selection was essential for finding *Crr2* in addition to *Crr1*. The effect of *Crr2* was visible only in the presence of *Crr1*. The *Crr2* locus was lost during the breeding of commercial CR Chinese cabbage, but it was shown to be essential for resistance toward highly virulent pathotypes (Suwabe et al. 2003). These CR loci will become important in breeding cultivars with more durable resistance (Diederichsen et al. 2009).

(ii) *B. oleracea*

The cabbage cv. Badger Shipper was one of the first CR cabbage cultivars introduced in North America in the 1960s (Crute et al. 1980). The resistance was thought to be based on a single-gene group with an unknown number of genes (Chiang and Crete 1976). This resistance, however, was readily broken after introduction in the field. However, in the absence of the resistant cultivar for 3 years, the pathogen population seemed to lose pathogenicity to cv. Badger Shipper, suggesting that there are adverse fitness costs associated with pathogenicity (Crute et al. 1980). Other attempts to introduce resistant *B. oleracea* cultivars include broccoli and cabbage lines developed in North America (Baggett 1983; Baggett and Kean 1985) that are resistant to various pathotypes present in North America. Resistant cabbage cultivars released in Norway based upon resistance from the primitive cabbage cultivar Bohmerwaldkohl (Weisaeth 1977), and a cabbage landrace from the Shetland Islands (Dixon 1988; Crute et al. 1980; Voorrips 1995).

In Canada, the *B. napus* swede cv. Wilhelmsburger was chosen as a means to introduce resistance into *B. oleracea* (Chiang et al. 1977; Chiang and Crete 1983), as the prevalent races from the USA and Canada seem to lack virulence to most of the *B. napus* differential hosts (Chiang and Crete 1989). A single resistant BC1 plant with $n = 26$ was selected for introgression of CR into the cabbage cultivar Badger Shipper and the cabbage line 8–41 (Chiang et al. 1977, 1980; Landry et al. 1992). They both possess resistance against race 6, similar to cv. Wilhelmsburger (Chiang and Crete 1976; Williams 1966). It resulted in the release of two CR cultivars, Acadie and Richelain (Chiang and Crete 1985, 1989). The first disease resistance QTLs that were mapped in *Brassica* was reported in 1992 by Landry et al., who found two QTLs for CR. Toxopeus et al. (1986) concluded that the turnip *B. rapa* ECD hosts had the strongest and broadest resistance. Within *B. oleracea,* many resistance sources have been described (Crute et al. 1980; Crisp et al. 1989; Hansen 1989; Voorips and Visser 1993; Manzanares-Dauleux et al. 2000a). Only a few lines or cultivars have reached the commercial state (Crute et al. 1980; Voorrips 1995). Linkage drag between CR and undesired traits such as reduced shelf-life has been reported for some *B. oleracea* CR sources (Crute et al. 1980). Although CR from *B. oleracea* is regarded as mainly non-differential, it is less used in breeding than race-specific CR from *B. napus* or *B. rapa*. The CR *B. oleracea* cultivars carry race-specific CR that originates from *B. rapa*. CR from *B. oleracea* is genetically more complex, mainly recessive, and therefore, difficult to use in breeding particularly in hybrid breeding. In addition, *P. brassicae* isolates that can completely overcome the resistance of all *B. oleracea* ECD hosts are very common (Crute et al. 1983; Toxopeus et al. 1986).

In different QTL analyses of CR in *B. oleracea*, major and minor QTLs were found (Voorrips et al. 1997; Moriguchi et al. 1999; Rocherieux et al. 2004; Werner et al. 2008). It is unknown whether some of these QTLs are allelic or identical because most of them have not been mapped onto a reference map. Rocherieux et al.

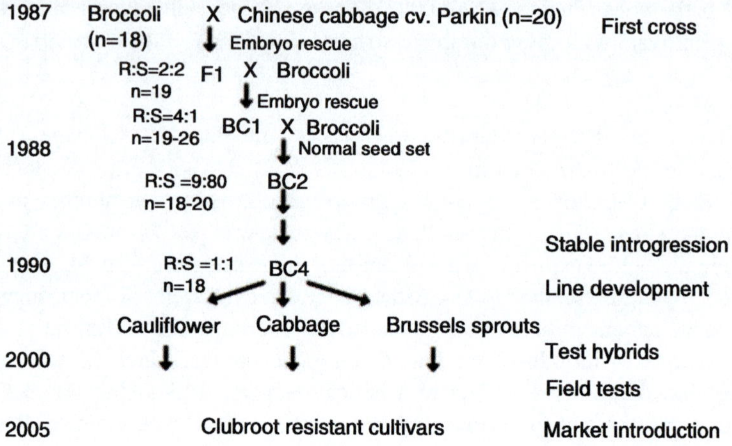

Fig. 11.1 Introgression of clubroot resistance from *Brassica rapa* cv. Parkin into *Brassica oleracea* crops at Syngenta Seeds B.V.; chromosome numbers and segregation ratios for resistance are given for initial generations (Diederichsen et al. 2009)

(2004) identified several isolate-specific QTLs and one major QTL that conferred resistance to all five *P. brassicae* isolates used in their study. Werner et al. (2008) mapped QTLs from *B. rapa* and *B. oleracea* after introduction into *B. napus* and identified eight QTLs in the *B. oleracea* C genome, all showing race specificity. Despite its non-differential appearance, it has to be assumed that CR from *B. oleracea* is likely to be based on a mixture of race-specific and race-independent CR loci. The first interspecific crosses were made in 1987 between a broccoli inbred line and Chinese cabbage cv. Parkin, which possesses a single dominant resistance gene (Fig. 11.1). Embryo rescue (Harbert 1969) was used to obtain F_1 and BC_1 plants. Nine resistant BC_2 plants were derived from a single BC_1 plant. The expected 1:1 segregations were found only in the BC_4, suggesting that the CR gene had been introgressed in the C genome. A backcross programme in cauliflower, cabbage, and Brussels sprouts was then initiated, it included after several years of backcrosses, selections, line development, and test crosses. Field tests evaluating the resistance in *B. oleracea* background under field conditions were carried out for several years at naturally infested locations in Europe. In all tests the CR entries remained resistant, whereas the susceptible controls were heavily infected. Trials performed in India, Australia (Donald et al. 2006a, b), and other countries yielded similar results. These results indicated that the resistance gene was as effective in the *B. oleracea* background as in the original *B. rapa* cultivar, and encouraged introduction of cultivars into the market. The CR cultivars were introduced to the market in 2005 (cauliflower cvs. Clapton and Clarify and cabbage cvs. Kilaton and Tekila), and well received by growers. The resistance from *B. rapa* does not cover all isolates of *P. brassicae* (Toxopeus et al. 1986), however, A small number of isolated infections were observed. Medium to severe attacks occurred at some locations in Germany, Poland, and Brittany, France. This situation had already been encountered with *B. rapa*

Table 11.3 Reproduction rate[a] of a standard isolate compared to an Indonesian isolate capable of infecting CR Chinese cabbage cv. Parkin on three cabbage entries with different levels of susceptibility (Diederichsen et al. 2009)

	Spores/g infected root		Spores/plant		Mean score (0–5)	
Entry	Stand.	Ind.	Stand.[b]	Ind.[c]	Stand	Ind.
Marathon	595	199b[d]	1809	347b	5.0	4.3b
Cab.1	563	248b	802	158a	4.9	3.4ab
Cab 2	468	68a	792	64a	4.6	2.5a
Mean	542b	172a	1134b	189a	4.8b	3.4a

[a]Reproduction rate = No. of spores × 10^6/g of infected root or per plant
[b]Stand. = race 4, Nord-Ditmar Germany
[c]Ind. Indonesian
[d]Different letters indicate significant differences between means according to two-way ANOVA and Tukey test after square root transformation; $n = 14, p = 0.05$. Scoring scale: 0 = no symptoms, 5 = fully diseased

cv. Parkin before introduction of the new cultivars. Often infections of the CR cultivars are quantitative with small clubs as reported for CR Chinese cabbage cultivars (Osaki et al. 2008a, b). Compatible isolates sometimes have a reduced fitness on susceptible *B. oleracea* hosts as compared with a standard isolate not capable of overcoming *B. rapa* genes. An isolate that was capable of infecting CR cv. Parkin from Indonesia did show a strongly reduced reproduction rate on three susceptible cabbage cultivars (Table 11.3). Donald et al. (2007) studied the histology of the Clapton-like resistance and showed that the pathogen was inhibited in the cortex; for some isolates this interaction may be more quantitative than qualitative. Fitness costs associated with certain pathogenicity factors, as has been observed with cv. Badger Shipper (Crute et al. 1980) or interaction between pathotypes may be present.

(iii) *B. napus*

Clubroot disease was originally most important to the storage-root-forming *B. napus* type spp. *napobrassica* (swede or rutabaga), and to fodder rape (*B. napus* ssp. *oleifera* var. annuum) compared with winter oilseed rape because their growing period better matches the pathogen's temperature requirements. As a result of the spread of oilseed rape into areas where soil conditions are more conducive, clubroot incidences are more frequent in this crop.

In a Swedish survey, 72% of 190 soil samples from oilseed rape fields tested positive for the presence of *P. brassicae* (Wallenhammar 1996). In naturally infested fields, Wallenhammar et al. (1999) found up to 50% yield reduction in spring oilseed rape. In Canada, the first incidences of clubroot in *B. napus* were recorded in 2003 in Alberta and the disease is still spreading in this region (Strelkov et al. 2005; Xue et al. 2008).

Breeding for CR in *B. napus* focused initially on swedes, where resistant types had been identified by Colhoun (1958) and Ayers and Lelacheur (1972). Swedes have been used to introduce CR into *B. oleracea* (Chiang et al. 1977) and received

Table 11.4 CR genotypes postulated for *Brassica napus* ECD hosts (Diederichsen et al. 2009)

B. napus ECD host	Gustafsson and Falt (1986)	Crute et al. (1983)
ECD-06	CR 2	R2+R3
ECD-08	CR 1+ CR 3	R1
ECD-09	CR 1+ CR 2	R1+R2
ECD-10	CR 1+ CR 4	R1+R2+R4

CR genes from *B. rapa* by interspecific hybridization (Lammerink 1969; Williamson and Bradshaw 1989; Bradshaw et al. 1997). Fodder rape cultivars with CR were bred after selection of resistant individuals (Lammerink 1964; Johnston 1970); some of these are included in the ECD set. The genetics of CR in the ECD *B. napus* differentials (ECD-06 to ECD-10) has been studied by Gustafsson and Falt (1986), who postulated that four race-specific genes are present in these hosts (Table 11.4). Based on differential pathogenicity patterns with many isolates, Crute et al. (1983) postulated a slightly different allocation of CR loci. Genes with the same numbers in both models may not be identical. The difference between both models is most probably due to the choice of *P. brassicae* isolates that were used in the Swedish study. By using an isolate with the capability to overcome the gene "R3" sensu Crute et al. (1983), the effect of this gene cannot be observed, this would explain the identification of only one gene in ECD-06 and two in ECD-10 (Gustafsson and Falt 1986).

Several attempts were made to introduce CR from both ancestral species into *B. napus*. Diederichsen and Sacristan (1994) combined different sources of CR from *B. rapa* and *B. oleracea*. The expression of CR in resynthesized *B. napus* forms was compared (Diederichsen and Sacristán 1996). Both CR sources were less efficient in the *B. napus* background compared with the original parent, in particular, CR from *B. oleracea* appeared to be strongly diluted, indicating the presence of epistatic factors in the *B. rapa* genome. Combining the most resistant ancestral parents led to resynthesized *B. napus* forms that were resistant against all isolates of *P. brassicae* that were available in their study, representing a level and broadness of CR that had not been present previously in *B. napus*. Bradshaw et al. (1997) crossed an auto tetraploid of the *B. rapa* host ECD-04 with a tetraploid kale and selected a CR swede after backcrossing, in 1996 the CR cultivar cv. Invitation was released in the UK seed market.

Two CR oilseed rape cultivars, cv. Mendel and cv. Tosca, were introduced to the European seed market in 2000. Both cultivars possess race-specific CR (Diederichsen et al. 2003) originating from a resynthesized *B. napus* form (Diederichsen et al. 2003; Happstadius et al. 2003). The hybrid cultivar Mendel has yields comparable with other cultivars from the same breeding period. In Germany, its market share was increased from 0.7% in 2001 to 2% in 2006. Its breeding process is illustrated in Fig. 11.2. Resistance was initially combined in a resynthesized *B. napus* by crossing two ECD hosts, *B. rapa* ECD-04 and *B. oleracea* ECD-15, with the aid of ovule culture (Diederichsen and Sacristan 1994). After backcrossing, di-haploidization, and selection with a highly virulent isolate of

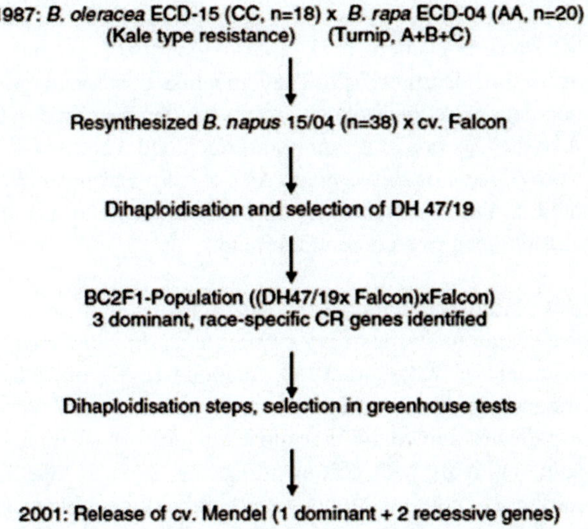

Fig. 11.2 Breeding of *Brassica napus* cv. mendel by Norddeutsche Pflanzenzucht H.G. Lembke KG (Diederichsen et al. 2009)

P. brassicae, a resistant DH line was identified, used for breeding and genetic analysis. The latter confirmed the presence of three dominant, race-specific CR genes from ECD-04 in a backcross population (Diederichsen et al. 1996), as expected. Further backcrosses and selections with undefined *P. brassicae* isolates originating from infested crops led to the resistant parent of the hybrid cultivar Mendel. Subsequent mapping of CR in cv. Mendel revealed one major locus plus two recessive genes (Diederichsen et al. 2006), indicating that two dominant CR genes were lost during breeding. The resistance of cv. Mendel is efficient in most cropping areas. Compatible populations of *P. brassicae* were known to exist before its release and occasionally infected crops of cv. Mendel have been observed. In many cases, the infected individuals were identified to be volunteers of previous crops. However, because cv. Mendel's resistance is based on only one major gene (recessive genes are unlikely to contribute in a hybrid cultivar). It is recommended only in those cases where clubroot infections have already been encountered and should not be grown as insurance against potential disease incidence. Control strategies involving the integrated use of pyramids of control methods will help to prolong the usefulness of this cultivar. Identification of race-independent CR genes from *B. oleracea* in *B. napus* was attempted by using as the CR source a resynthesized *B. napus* form that originated from a cross between ECD-04 and the cabbage landrace Bohmerwaldkohl. Nineteen QTLs were detected in a DH population (Werner et al. 2008), all showing race specificity. There was no indication of race-independent QTLs, which were assumed to be present in the *B. oleracea* C genome. The only strong QTL in the C genome that could be detected in this material originated from the susceptible parent cv. Express and conferred CR to only one isolate of *P. brassicae*. Some QTLs had overlapping confidence intervals on chromosomes A03, and A08 and conferred CR against more than one isolate. Based on this study, the QTLs with the broadest effects against *P. brassicae* pathotypes in *B. napus* could be expected to be located in the *B. rapa* A genome. Manzanares-Dauleux et al. (2000b) identified resistance against a SSI in the *B. napus* dwarf line Darmor *bzh*. They mapped one major gene and two QTLs in a DH population. As the linkage groups of their map had not been allocated to either the A or the C genome, it cannot be concluded whether CR from Darmor-*bzh* originated from *B. rapa* or *B. oleracea*. As CR from Darmor-*bzh* has not been observed in the field so far, it seems likely that the major gene has a race-specific effect that has already been overcome in the field.

(iv) Radish (*Raphanus sativus* L.)

Radish, a cruciferous crop closely related to *Brassica*, originated in the Mediterranean region (Kitamura 1958; Tsunoda 1980), but today is more important in East Asia where a wide variety of genotypes are found. Yoshikawa (1993) surveyed CR in radishes and found that most Japanese cultivars are highly resistant, except for some cultivars with bolting resistance, such as cvs. Tokinashi or Ninengo. Some northern Chinese cultivars were highly susceptible to this disease. European radish cultivars such as Comet and White Icicle showed mild symptoms or were resistant (Yoshikawa 1993). Rat-tail radish, which is grown in tropical Asia for the consumption of its pods, is resistant to the pathogen. Resistance to clubroot seems to be widespread in this species, but there are reports of clubroot outbreaks in radish in the USA (Thuma et al. 1983) and more recently in Japan (Horikoshi and Tairako 2002). In Japan, radish was believed for a long time to be totally immune to clubroot. Therefore, radish is often used as a cleaning crop for fields contaminated with *P. brassicae* (Murakami et al. 2000). The susceptible cultivars were newly bred, so it is possible that some CR genes were lost during breeding. Kamei et al. (2007) reported on a QTL analysis of radish CR using a Japanese radish as the resistant parent and a Chinese cultivar as the susceptible parent. They found that the trait is controlled by one locus or closely linked loci in a single linkage group. Using nine types of monosomic addition lines of radish, Akaba et al. (2008) showed that only one addition type had clear resistance, suggesting that CR is controlled predominantly by a single radish chromosome. Therefore, it seems likely that only a single locus controls a large part of radish CR. Considering the broadness of CR from radish, simple inheritance was unexpected. Resistant radishes can be bred by a simple backcross method starting with a Japanese radish as a donor plant. Mass selection is risky, however, when the trait is dominant. If CR from radish could be transferred into *Brassica* species, it might be valuable for breeding highly resistant *Brassica* cultivars. Some efforts to transfer CR to *Brassica* have been reported using sexual or somatic hybrids between radish and *Brassica* species (Hagimori et al. 1992; Hagimori 1995; Xing et al. 1989). Recombination between *Brassica* and *Raphanus* chromosomes would be a precondition for a successful introgression. Attempts to introgression other traits such as restorer genes for cytoplasmic male sterility into *Brassica* did show, however, this is a rare event (Sakai et al. 1996; Delourme et al. 1998).

11.2.3 Development of Resistant Cultivars

Resistant cultivars are the only solution for protection against *P. brassicae* in combination with suitable crop rotation and management strategies (Peng et al. 2014).

11.2.3.1 Genetics and Breeding

The clubroot host-plant resistance can be found in main cruciferous crops (except *Brassica juncea* and *B. carinata*) (Diederichsen et al. 2009). The resistance against *P. brassicae* is polygenic (quantitative) and also monogenic (qualitative) (Crute et al. 1980; Voorrips 1996). Many of the resistance types are pathotype specific and do not cover all pathotypes of *P. brassicae* (Diederichsen et al. 2009). The studies proved the existence of eight resistance loci, some of them are pathotype-specific major genes, and some are quantitative loci (QTL) with specific or non-specific resistance responses (Voorrips 1996; Hwang et al. 2012). Some studies also proved that resistance is qualitative and is controlled by dominant and recessive genes (Chiang and Crete 1983; Yoshikawa 1993). The oilseed rape resistant cultivar Mendel was bred from ECD hosts – resistant turnip (*B. rapa,* ECD-04) and kale (*B. oleracea* ECD-15) through backcross with oilseed rape (Diederichsen et al. 2009). The F_1 backcrosses were inoculated with *P. brassicae* and the most resistant ones were selected for further breeding and genetic analysis (Diederichsen et al. 2009). Later the presence of resistant genes was confirmed. There is one major locus and two recessive genes in the cultivar Mendel, whereas other genes were lost in the breeding process and negative selection (Diederichsen et al. 2009). This is why cv. Mendel is recommended to use wisely together with other management strategies (Diederichsen et al. 2009; Dixon 2009). The same situation exists in other resistant cultivars.

11.2.3.2 Mechanism of Resistance

The resistance mechanisms are based on suppressing the pathogen development in a host rather than on infection avoidance. The term resistance is used because the infection inhibition is based on resistance genes, not on tolerance genes. The zoospore penetration occurs in the root hairs of a resistant host (the primary phase of the pathogen life) (Hwang et al. 2012). The secondary phase of infection in the root cortex also follows, but formed plasmodia do not mature and resting spores are not created (Hwang et al. 2011; Deora et al. 2013). Slight suppression of primary infection and significant suppression of secondary infection occur during the resistance response, the disruption of the cell walls is lesser and the secondary thickening of the stem is not stopped (Donald et al. 2008). The immune response includes the activation of metabolic pathways for synthesis of phytoalexins, phenolic compounds, lignin, and other substances that block the pathogen invasion (Siemens et al. 2006).

11.2.3.3 Resistant Cultivars in the Field

The first resistant cultivars were Mendel and Tosca, marketed at the beginning of the new millennium. These cultivars are heterozygous hybrids with one gene of pathotype-specific resistance. Unfortunately, the Mendel host-plant resistance was broken in some parts of Germany and France, where it was grown since 2000 (Diederichsen et al. 2014). In Great Britain, resistance was broken quite rapidly in the field, where cultivars were grown under strong selection pressure (Diederichsen et al. 2003). These cultivars were also tested on local pathotypes in Canada. Cultivar Mendel showed resistance against pathotypes 3, 5, 6, and 8 (Williams set) (Rahman et al. 2014). Cultivar Tosca was susceptible to pathotype 3 and resistant to pathotype 2 (Williams set) (Peng et al. 2013). These results showed that the host-plant resistance is quite variable and some genes may be ineffective in some regions with specific pathotypes (Peng et al. 2014). New resistant cultivars of spring rape with genes from European cultivars were introduced into Canada in 2009 (Peng et al. 2014). In the European market, the available resistant cultivars which are derived from the first resistant cultivar Mendel (Diederichsen et al. 2014). It is expected that a wider range of resistant varieties will come to the market in the next years (Diederichsen et al. 2014). However, these varieties cannot be grown without suitable agronomic measures and pathogen-resistance management strategies (Diederichsen et al. 2014). Field experiments with resistant cultivars in infested fields were also conducted in the Czech Republic. Eight resistant cultivars were tested, some of them are normally available on the market and some were test cultivars in the breeding process (new breeds). Normally traded cultivars had stable performance in resistance results and good yield. Cultivars from new breeding showed some variations in resistance performance and yield (Ricarova et al. 2016).

11.3 Molecular Mechanism of Host Resistance

Molecular aspects of pathogens specific resistance have been extensively exploited which is known to be mediated by resistance (R) genes through the recognition of specific pathogen effectors. These effectors initially function as virulence/pathogenicity factors for the pathogen, but then become recognized by the host as signals of attempted infection. Most of the R-genes encode proteins carrying nucleotide-binding site (NM3) in the central region and a leucine-rich repeat (NBS-LRR) domain at the C-termінces. These genes form a large multigene family in the plant genome and can be separated into two subclasses, the toll-interleukin-1(TIR) class and the coiled-coil (CC) class. The NBS-LRR proteins are thought to recognize effectors or effector-specific signals from the pathogen and activate the plant immune system to provide hypersensitive responses (HR).

11.3.1 Mapping of Clubroot Resistance Genes

All the first generation of molecular markers such as restriction fragment length polymorphism (RFLP), amplified fragment length polymorphism (AFLP), random amplified polymorphic DNA (RAPD), and simple sequence repeats (SSR) are used to map clubroot resistance in various *Brassica* species. Landry et al. (1992) used RFLP markers and detected two QTLs in *B. oleracea*. Figdore et al. (1993) used RFLP markers and associated several linkage groups to clubroot resistance in *B. oleracea*. Grandclement and Thomas (1996) used RAPD markers, and analyzed QTL for clubroot resistance in broccoli and cauliflower, and identified some RAPD markers significantly linked to clubroot resistance. Voorrips et al. (1997) used RFLP and AFLP markers and detected two genes for clubroot resistance in *B. oleracea* doubled haploid lines. Rocherieux et al. (2004) performed QTL analysis in *B. oleracea* and detected isolate-specific and broad-spectrum QTLs, suggesting that the clubroot resistance in *B. oleracea* is genetically complicated and molecular marker-assisted selection might be not so effective. Nagaoka et al. (2010) performed QTL mapping using a DH line population between resistant cabbage and susceptible broccoli and detected two major QTL for clubroot resistance on chromosomes O2 and O5, and minor QTLs on chromosome O2, O3, and O7. Similar to the conclusion drowned from conventional genetic analysis, clubroot resistance in *B. oleracea* is most likely due to the effects of multiple minor genes, and molecular marker-assisted selection might be not as effective as that in Chinese cabbage which contains dominant resistance genes introduced from European turnips.

In Chinese cabbage, major clubroot resistance loci that are introduced from European turnips have been mapped (Table 11.5). These clubroot resistance loci are named as *Crr1*, *Crr2*, *Crr3*, and *Crr4* and also *CRa*, *CRb*, *CRc*, and *CRk* suggesting that there might be eight independent loci. Two clubroot resistance loci, *Crr1* and *Crr2* were mapped using SSR markers and a third locus, *Crr3* was identified using RAPD markers, which suggested that there are three independent clubroot resistance loci in Chinese cabbage. Using RFLP markers, a genetic map was constructed and a clubroot resistance locus, *CRa*, was mapped on linkage group 3, and SCAR and CAPS markers were used to map another locus, *CRb*, on chromosome R3. Moreover, three clubroot resistance loci *CRa*, *CRc*, and *CRk* have been added in the list of clubroot resistance through molecular marker-assisted selection (Saito et al. 2006; Hirai et al. 2004; Suwabe et al. 2006; Matsumoto et al. 2012).

Due to the complex genome structure of *B. napus*, QTL mapping for clubroot resistance is necessary in *B. napus*. It is quite common to find the clubroot resistance in *B. napus* that does not segregate as a typical Mendelian trait as in *B. rapa*. Although one major locus was mapped on chromosome N3, two minor QTL on chromosomes N12, and N19 were identified for clubroot resistance in *B. napus*. In another report, a DH line population derived from a cross of clubroot-resistant synthetic *B. napus* and susceptible canola was used to perform QTL mapping for clubroot resistance. The synthetic *B. napus* contains dominant clubroot resistance from European turnip ECD4 and may be medium resistance from *B. oleracea*. The

Table 11.5 Genetic mapping of clubroot resistance loci in *Brassica rapa* (Piao et al. 2009)

Resistant source	Population	Isolate[a]	Locus	Types of DNA marker	Flanking markers	Chromosome with interval (cM)	Reference
ECD02	F2	Race 2	*CRa*	RFLP	HC352b, HC181	R3 (15 cM)	Matsumoto et al. (1998)
				STS	HC352b-SCAR		Sakamoto et al. (2008)
Gelria R	F_2	Race 4	*CRb*	SCAR	TCR05, TCR09	R3 (3.0 cM)	Piao et al. (2004)
Siloga	F_2	Ano-o1 (race 2),	*Crr1*	SSR	BRMS-297,	R8 (1.6 cM)	Suwabe et al. (2006)
		Wakayama-01	*Crr2*	SSR	BRMS-088	R1 (2.2 cM)	
			Crr4	RFLP	BRMS-100, BRMS-096	R6 (2.7 cM)	
					BN288D, WE24-1		
Milan White	F_3	Ano-01	*Crr3*	STS	OPC11-1S, OPC11-2S	R3 (10 cM)	Hirai et al. (2004)
Debra	F_2	M85 (race 2), K04	*CRk*	STS	HC688, OPC11-2S	R3 (9.1 cM)	Sakamoto et al. (2008)
	F_2		*CRc*	AFLP	E14M3-02	R2 (5.1 cM)	
					E15M4-006		

[a]Isolate characterization based on the Williams' classification (Williams 1966)

identified nineteen QTLs on chromosomes N02, N03, N08, N13, N15, N16, and N19 for clubroot resistance, four QTLs with LOD values of over 11, of which three are located on chromosome N03, and one on N19, and the proportion of the phenotypic variance explained by each QTL is over 40%. The major QTLs might come from the C genome of *B. oleracea*, which is contradictory to previous reports where major clubroot resistance genes in *B. napus* come from the A genome of *B. rapa* (Hirani and Li 2015).

Identified clubroot resistance loci in Chinese cabbage have been fine mapped and some clubroot resistance genes have been eventually cloned. The *Crr3* locating on chromosome R3 was first mapped to a small genetic region between 0.35 cM genetic distance using 888 F2 individual plants. The clubroot resistance locus *CRa* has been further analyzed to identify the candidate gene. Over 1600 F_2 individual plants were used to select 80 recombinants using two closely linked molecular markers. Further, analysis of those recombinants allowed identifying one open reading frame located on chromosome R3, which belongs to a typical resistance gene family and encodes a TIR-NBS-LRR protein. There are two other independent reports that focused on fine mapping of clubroot resistance loci on chromosome R3. The *CRb* clubroot resistance locus which was described to be effective to *P. brassicae* isolates No. 14, a very aggressive isolate in Japan, has been fine mapped. Using over 2000 F_2 individual plants and F_3 progeny testing, 92 F_2 recombinants between two closely linked molecular markers were identified. The analysis of these 92 F_2 recombinants suggested that the *CRb* clubroot resistance locus might be the same as the *CRa* locus and the *CRa* and *CRb* clubroot resistance loci are different from the clubroot resistance locus *Crr3*. Similarly, gene mapping of five Chinese cabbage cultivars was performed and all these hybrid cultivars were found to contain the same clubroot resistance locus on chromosome R3. Fine mapping of the clubroot resistance locus in Chinese cabbage to a 187 kilo-base pair (kb) chromosomal region using a large segregating population with over 8,000 individual plants was done. Molecular markers which are closely linked to the mapped clubroot resistance locus have been developed and those molecular markers can be used in marker-assisted selection to breed Chinese cabbage with clubroot resistance (Saito et al. 2006; Ueno et al. 2012; Kato et al. 2012; Gao et al. 2014). Characterization of clubroot resistance genes offers opportunities for further understanding clubroot resistance and interactions of resistance genes and pathogens. Hatakeyama et al. (2013) cloned one clubroot resistance gene *Crr1a* on chromosome R8 and confirmed the resistance through plant transformation. Some transgenic *B. rapa* plants are resistant while others are susceptible, suggesting that the *Crr1a* gene may not explain the whole clubroot resistance in the original locus. The *Crr1a* and *Crr1b* are random repeats in the same locus and both genes encode typical resistance gene proteins with TIR-NBS-LRR structures. The clubroot resistance loci on chromosome 3 in *B. rapa* also contain multiple genes that encode TIR-NBS-LRR proteins. The complexity of these clubroot resistance loci needs to be investigated further. When a clubroot resistance locus contains multiplc genes encoding the similar proteins, it becomes challenging to know how each individual gene plays a role in the clubroot resistance and how they contribute to the differences of alleles from various resistant sources. It is

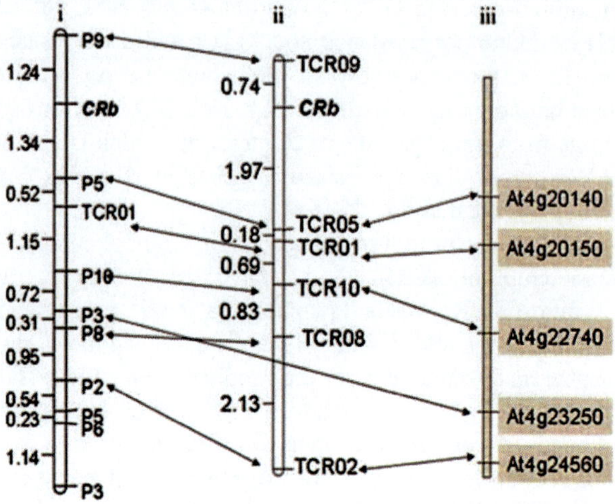

Fig. 11.3 Identification of the homologous region of *Brassica rapa* linkage group containing *CRb* locus with that of *Arabidopsis thaliana* chromosome 4 (Piao et al. 2004). (1) AFLP linkage group containing *CRb* locus, (2) Linkage group converted to SCARs and CAPS markers, (3) Homologous *Arabidopsis thaliana* chromosome 4 (Piao et al. 2009)

necessary to further dissect those complex clubroot resistance loci and investigate each individual gene to understand the functional properties of these loci. Therefore, gene functional analysis for clubroot resistance is an important in *Brassica* species (Fig. 11.3).

11.3.2 Mapping of Clubroot Resistance Genes in *Brassica* Species

With details of CR QTLs, their number, and location on specific chromosomes in each species are as under.

(i) *Brassica rapa*

Genetic analysis and genetic mapping of CR genes are well studied in *B. rapa*. All eight possible CR genes present in *B. rapa* have been identified through QTL mapping using a range of resistant sources and marker systems (Table 11.5). Kuginuki et al. (1997) identified three random amplified polymorphic DNA (RAPD) markers linked to a CR locus using turnip cv. Siloga as a source of resistance. Suwabe et al. (2003) identified *Crr1* and *Crr2*, two loci that originated from cv Siloga, using F_2 populations and simple sequence repeat (SSR) marker systems. It was concluded that these two loci are complementary for clubroot resistance. Inoculation of test plants with mild and virulent isolates of *P. brassicae* showed that plants with a homozygous resistance locus had greater resistance to clubroot than those having a heterozygous resistance locus. In addition to *Crr1* and *Crr2*, *Crr4*, a weak QTL, was also detected using a similar F_2 population (Suwabe et al. 2006). Five genes involved in glucosinolates biosysnthesis were mapped and GSL-BoELONG and GSL-BoPro were found to be linked to the Pb-Bo (Anju)1 and Bo (GC) 1 loci (Nagaoka et al. 2010). Hirai et al. (2004) identified and mapped a novel locus named *Crr3*, which originated from the clubroot-resistant turnip

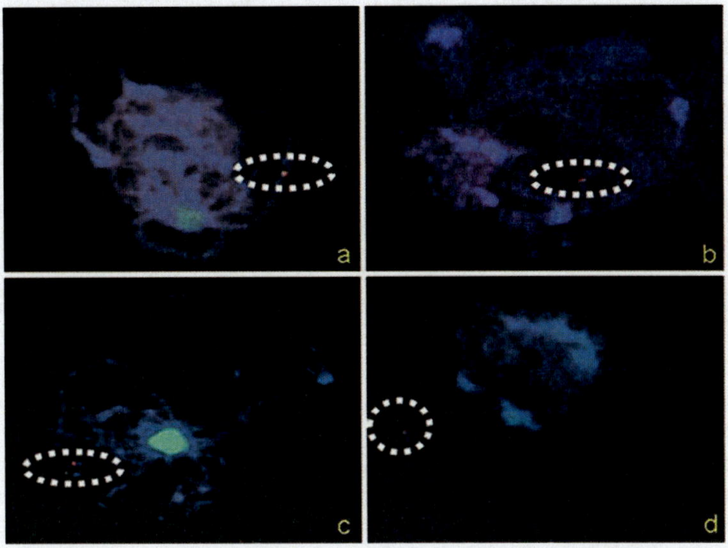

Plate 11.1 FISH localization of BAC clones containing SCAR markers (TCR05 and TCR09) linked to clubroot resistance locus in *Brassica rapa* CR Shinki genome. (**a**) BAC clones KBrH060E03 (green) and KBrH097J16 (red) from the same contig harboring TCR05 marker showing localization in the same position in *B. rapa* CR Shinki chromosome 1. (**b**) BAC clones KBrH083J12 (green) and KBrH103M15 (red) from the same contig harboring TCR09 marker showing localization in same region. (**c**) BAC clones KBrH97J16 (green, harboring TCR05), and KBrH88B11 (red, harboring TCR09) from different contig linked to TCR05 and TCR09 markers showing localization in different regions of the same chromosome. (**d**) BAC clones KBrH115F22 (green, harboring TCR05) and KBrH144K19 (red, harboring TCR09) from different contig linked to TCR05 and TCR09 markers showing localization in different regions of the same chromosome (Piao et al. 2009)

cv. Milan White. Another CR locus, *CRb*, derived from turnip cv. Gelria R, was mapped in an interval of 3 cM using sequence characterized amplified region (SCAR) markers converted from amplified fragment length polymorphism (AFLP) markers (Piao et al. 2004). Homology analysis of SCAR markers with that of *Arabidopsis* sequence database identified that the region containing the *CRb* locus is homologous to the central part of the *A. thaliana* chromosome 4 (Fig. 11.2). These are also homologs with the region of the *B. rapa* chromosome containing *Crr1* and *Crr2* CR QTL. *Brassica rapa* BAC clones containing these SCAR markers have been identified and located on a specific chromosome in clubroot resistance cv. CR Shinki through fluorescent in situ hybridization (FISH) (Plate 11.1). Co-localization of these BAC clones was found in *B. rapa* chromosome 1. This locus is independent of *Crr1*, *Crr2*, *Crr3*, and *Crr4* (Saito et al. 2006). Sakamoto et al. (2008) identified *CRk* and *CRc* loci derived from turnip cv. Debra using QTL analysis of two F_2 populations and two isolates of *P. brassicae*. In total, eight CR loci were mapped and allocated to five different chromosomes of *B. rapa*. The genes *Crr1, Crr2, Crr3*, *Crr4*, and *CRc* are mapped to R8, R1, R3, R6, and R2, respectively, of the internationally agreed upon *B. rapa* reference genetic map. It is noteworthy that *CRa, CRb*, and *CRk* with *Crr3* are mapped on the same linkage group of R3, but they are not located in the same chromosome region, except for *CRk* and *Crr3* (Piao et al. 2009). A clubroot resistance gene (*Rcr2*) was identified and fine mapped in Chinese cabbage cv. "Jazz" using single-nucleotide polymorphisms (SNP) markers identified from bulked segregant RNA sequencing (BSR-Seq) and molecular markers were developed for use in marker-assisted selection. In total, 203.9 million raw reads were generated from one pooled resistant (R) and one pooled susceptible (S) sample, and >173,000 polymorphic SNP sites were identified between the R and S samples. One significant peak was observed between 22 and 26 Mb of chromosome A03, which had been predicted by BSR-Seq to contain the causal gene *Rcr2*. There were 490 polymorphic SNP sites identified in the region. A segregating population consisting of 675 plants was analyzed with 15 SNP sites in the region using the Kompetitive Allele Specific PCR method, and *Rcr2* was fine mapped between two SNP markers, SNP_A03_32 and SNP_A03_67 with 0.1 and 0.3 cM from *Rcr2*, respectively. Five SNP markers co-segregated with *Rcr2* in this region. Variants were identified in 14 of 36 genes annotated in the *Rcr2* target region. The numbers of poly variants differed among the genes. Four genes encode TIR-NBS-LRR proteins and two of them *Bra019410* and *Bra019413*, had high numbers of polymorphic variants and so are the most likely candidates of *Rcr2* (Huang et al. 2017).

Eight Chinese cabbage germplasms were screened using published clubroot-resistant (CR) loci-/gene-linked markers. A CR gene *Crr3* potential carrier "85-74" was detected which linked to marker BRSTS61, however, "85–74" shows different responses to local pathogens "LAB-19," "LNND-2," and "LAB-10" from "CR-73" which harbors *Crr3*. Pang et al. (2018) used a next generation sequencing-based bulked segregant analysis approach combined with genetic mapping to detect CR genes in an F_2 segregant population generated from a cross between the Chinese cabbage inbred lines "85–74" (CR) and "BJN3-1" (clubroot susceptible). The "85–74" line showed resistance to a local pathogen "LAB-19" which was identified as race 4; a genetic analysis revealed that the resistance was conferred by a single dominant gene. The CR gene named *CRd* was mapped to a 60 kb (1 cM) region between markers yau389 and yau376 on chromosome A03. *CRd* is located upstream of *Crr3* which was confirmed based on the physical positions of *Crr3* linked markers. The identification of *CRd* linked markers can be applied to marker-assisted selection in the breeding of new CR cultivars of Chinese cabbage and other *Brassica* crops.

(ii) *Brassica oleracea*

A number of DNA markers linked to CR loci in *B. oleracea* have been developed. Two to nine QTLs have been identified (Table 11.6) (Landry et al. 1992; Figdore et al. 1993; Grandclement and Thomas 1996; Voorrips et al. 1997; Moriguchi et al. 1999; Rocherieux et al. 2004; Nomura et al. 2005). Two QTLs, *CR2a* and *CR2b*, showing resistance to race 2 of *P. brassicae* and contributing 58 and 15%, respectively, of the phenotypic variation, were identified using swede cv. Wilhelmsburger as a resistance source (Landry et al. 1992). Three QTLs showing resistance to race

Table 11.6 Genetic mapping of clubroot resistance loci in *Brassica oleracea* (Piao et al. 2009)

Resistant source	Population	Isclate	Loci	Types of DNA marker	Flanking markers	LG with interval (cM)	Reference
Wilhelmsburger	F_2	Race 2[a]	CR2a	RFLP	2NF11, 2ND3	LG6 (22)	Landry et al. (1992)
			CR2b		3NE4a, 3ND3	LG1 (12)	
Broccoli (OSU CR-7)[b]	F2	Race 7	3	RFLP	14a 48 177b	LG1 LG4 LG9	Figdore et al. (1993)
Kale (C10)	F2	16/31//31[c]	At least 2	RAPD	OPL6-780, OPB11-740, OPA18-14900, OPA4-700, OPE20-1250, OPA1-1880, OPA16-510		Grandclement et al. (1996)
Indsachsener	DH	Field isolate	*pb-3* *pb-4*	RFLP, AFLP	4NE11a 2NA8c	LG3 LG1	Voorrips et al. (1997)
Kale (K269)	F2	Race 1 and 3	1	RAPD, RFLP	WG6A1, WG1G5	LG3 (2.6)	Moriguchi et al. (1999)
Kale (K269)	F2	Three field isolates	QTL1 QTL3 QTL9	SCAR	SCA02a 2 SCB50b,SCB74c SOPT15a, SCA25	LG1 LG3 LG9	Nomura et al. (2005)
Kale (C10)	F2	P1 (Ms6, eH), P2 (K92), P4 (K92-16), P7 (Pb137-522)[d]	*Pb-Bo1*	RAPD, RFLP, ACGM	Ae05.8800, T2	LG1 (18.4)	Rocherieux et al. (2004)
			Pb-Bo2		PBB38a, r10.1200	LG2 (19.3)	
			Pb-Bo3		ae15.100, RGA8.450	LG3 (13.9)	
			Pb-Bo4		ELI3.983, aa9.983,	LG4 (3.1)	
					PBB7b, ae05.135		

(continued)

Table 11.6 (continued)

Resistant source	Population	Isolate	Loci	Types of DNA marker	Flanking markers	LG with interval (cM)	Reference
			Pb-Bo5a			LG5 (32.5)	
			Pb-Bo5b		ELI3.115, a18.1400	LG5 (12.0)	
			Pb-Bo8		c01.980, t16.500	LG8 (10.2)	
			Pb-Bo9a		aj16.570, W22B.400	LG9 (24.1)	
			Pb-Bo9b		a04.1900, ae03.136	LG9 (1.4)	

[a]Isolate characterization based on Williams' classification (Williams 1966)
[b]*OSU* Oregon State University
[c]Isolate characterization based on ECD set (Buczacki et al. 1975)
[d]Isolate characterization based on Somé et al. (1996)

7 were identified using broccoli (Figdore et al. 1993). Voorrips et al. (1997) identified two QTLs, *pb-3* and *pb-4*, and a minor QTL contained in cv. Bindsachsener using a multiple QTL mapping approach which analyzed the fresh weight of galls. The additive effects of two major loci were responsible for 68% of the difference between the parents and for 60% of the genetic variance among the means of DH lines. One QTL for resistance to clubroot disease was identified on linkage group 3 using resistant kale (K269) under conditions of natural infection (Moriguchi et al. 1999). Based on the quantitative analysis of F_3 family using controlled environments with four single-spore isolates and one field isolate from four *P. brassicae* isolates, Rocherieux et al. (2004) found two to five QTLs depending on the pathotype used. Of the nine QTLs fully identified, *Pb-Bo1* is common to all isolates and accounts for 20.7–80.7% of the phenotypic variation, whereas the rest was specific to one, two, or three isolates. Nomura et al. (2005) identified three QTLs, QTL1, QTL3, and QTL9, using a kale line (K-269) as the resistant parent, which was similar to that used by Moriguchi et al. (1999). Therefore, one of these QTLs is probably similar to the one QTL that is located at the end of LG3. The SCAR markers converted from RAPD and RFLP markers linked to these QTLs were evaluated for F_2 and F_3 plants. It was observed that F_2 individuals with three QTLs expressed very high clubroot resistance, similar to that of the kale parent, whereas the F_2 and F_3 plants carrying a single QTL expressed only intermediate resistance (Nomura et al. 2005).

At least 22 QTLs have been found in *B. oleracea* so far. The discovery of several CR QTLs indicates that the clubroot resistance in *B. oleracea* is controlled at several QTLs, further confirming the complex genetic basis of clubroot resistance in *B. oleracea*. Because these mapping studies used different CR sources and isolates, although the primer and marker sequence are disclosed, the comparison of these QTLs is impossible. To understand the genetics and genomics of CR loci in *B. oleracea* in detail, development of common PCR-based markers is required. In addition, comparative studies suggested that the genomes of *Brassica* species have evolved from a common ancestor. It is worthwhile investigating whether CR genes in *Brassica* species might have common origins and how their mechanisms of evolution are maintained. Comparative studies of CR genes or their linked markers should provide new insights into these processes (Piao et al. 2009).

(iii) *Brassica napus*

Twenty-two QTLs involving clubroot resistance have been identified in *B. napus* (Table 11.7). Manzanares-Dauleux et al. (2000a) mapped the major gene *Pb-Bn1*, which confers resistance to two single-spore isolates (SSI) of *P. brassicae*, onto linkage group DY4. Based on the quantitative resistance expressed against each SSI, they also found at least two additive QTLs on chromosomes DY4 and DY15, respectively. In addition, epistatic interactions between nine regions with or without additive effects have been located. The total phenotypic variation accounted for by additive and epistatic QTLs ranged from 62% to 81.4% depending on the *P. brassicae* isolate used. Analyses of double-haploid (DH) populations, in which CR genes are derived from ECD04, identified one major gene and two recessive

Table 11.7 Genetic mapping of clubroot resistance loci in *Brassica napus* (Piao et al. 2009)

Resistant source	Population	Isolate	Loci	Types of DNA marker	Flanking markers	LG with interval (cM)	Reference
Winter dwarf line	DH	P4 (K92-16)[a]	Pb-Bn1	RAPD	OPG03.960, OPV09.2100	DY4 (8.3)	Manzanares-Dauleux et al. (2000a)
		P7 (Pb137-522)	Two QTL		OPC18.1250, OPD20.760 OPQ01.930, OPG13.950	DY2 (11.8) DY15 (3.7)	
ECD04	DH	1	One dominant and at least two recessive	SSR	HMR337, HMR388, HMR307	MS06 (4.0)	Diederichsen et al. (2006)
Bohmer waldkohl	DH	01:60[b]	PbBn-01:60-1	AFLP	158_241, 128_330	N03 (13.6)	Werner et al. (2008)
ECD-04		0.1:07	PbBn-01:60-3		84_174, 1_103	N19 (6.4)	
		e4x04a	PbBn-01:60-4		107_370, 166_215	N19 (4.3)	
			PbBn-01:07-1		107_106, 128_330	N03 (7.7)	
			PbBn-01:07-2		84_258, 79_168	N08 (2.0)	
			PbBn-01:07-3		160_186, 160_193	N13 (2.0)	
			PbBn-e4x04-1		163_448, 159_296	N19 (4.4)	
			PbBn-a-1		84_258, 153_370	N08 (5.4)	
Bohmer waldkohl,	DH	*l*	PbBn-l-1	AFLP	107_106, 128_330	N03 (7.7)	Werner et al. (2008)
ECD-04		*k*	PbBn-l-2		84_258, 79_168	N08 (2.0)	
		Korporal	PbBn-k-1		154_103, 152_373	N02 (7.5)	
			PbBn-k-2		107_106, 128_330	N03 (7.7)	
			PbBn-k-3		79_75, HMR1382a	N15 (4.8)	
			PbBn-Korp-1		6_450, 165_156	N09 (3.1)	
			PbBn-Korp-2		107_366, 146_363	N09 (9.0)	
			PbBn-Korp-3		158_154, 148_158	N09 (1.3)	

			PbBn-Korp-4		19_155, 166_146	N16 (12.6)	
			PbBn-Korp-5			N16 (31.7)	

[a]Isolate characterization based on Somé et al. (1996)
[b]Isolate characterization based on ECD set (Buczacki et al. 1975)

genes (Diederichsen et al. 2006). A resynthesized *B. napus* was developed by crossing cv. Bohmerwaldkohl (*B. oleracea*) and ECD-04 (*B. rapa*). From this, the CR DH line "263/11" was obtained (Diederichsen et al. 1996). Werner et al. (2008) analyzed the DH population derived from a cross of 263/11, and the susceptible cv. Express using seven *P. brassicae* isolates. Nineteen QTLs expressing resistance to seven isolates were detected on eight chromosomes, N02, N03, N08, N09, N13, N15, N16, and N19. All QTLs were found to be race specific. The total phenotypic variation accounted for ranged from 20.8 to 79.6% depending on the pathogen isolate used. Among the 19 QTLs detected, four were closely linked to each other on chromosome NO3, three were linked also to chromosome N08. In *B. rapa*, genes *CRa*, *CRb*, *CRk*, and *Crr3* are located on chromosome R3, which corresponds with N03 in *B. napus*. Genes *CRk* and *Crr3* are located in the similar region of *PbBn-k-2*, *PbBn-1-1*, *PbBn-01:60-1* on N03. However, genes *CRa* and *CRb* are independent from them. *PbBn-01.07-2*, *PbBn-l-2*, and *PbBn-a-1* are linked to BRMS088 on chromosome N08 in *B. napus*, which is also linked with *Crr1* on R8 in *B. rapa*. Further, studies of these loci, using common markers, might explain whether they are identical. The QTLs located on N03 and N19 contribute strong effects and confer broad-spectrum resistance. The clubroot resistance QTL's/genes mapped/fine mapped in different *Brassica* species are given in Table 11.8.

11.3.3 Genomics Approaches for the Identification of Clubroot Resistance Genes

Plasmodiophora brassicae infects all *Brassica* species and the model plant *Arabidopsis thaliana*. It is possible to apply currently available genomic tools and techniques for the cloning and characterization of CR genes in addition to comparative mapping, and identification of candidate genes through transcriptomic analysis. The whole-genome sequence information available for *Arabidopsis* has been used for comparative genome analysis of *Brassica* species. Comparative mapping between *Brassica* and *Arabidopsis* has revealed a conservation of gene order in small chromosomal blocks despite inversions and large-scale deletions (Cavell et al. 1998; O'Neill and Bancroft 2000). This information has been used to align linkage groups containing CR genes in *B. rapa* with those of *Arabidopsis* chromosomes. Genes *Crr1, Crr2*, and *CRb* are in synteny with the central region of chromosome 4 of *A. thaliana* (Suwabe et al. 2006; Fig. 11.2). The studies by Jubault et al. (2008) of *A. thaliana* have identified one QTL for clubroot resistance in this region, suggesting the presence of functionally active candidate gene(s). This region of an *Arabidopsis* chromosome has clusters of the CR genes such as leucine-rich repeats (LRRs) and nucleotide-binding sites (NBSs). Similarly, RPP for resistance to *Hyaloperonospora parasitica* (downy mildew), RPS for a resistance to *Pseudomonas syringae* (bacterial blight), and ACD, which accelerates cell death in response to pathogen infection (Suwabe et al. 2006) have been identified.

It has been suggested that CR genes may be members of these clusters of resistance genes. The genes *Crr1*, *Crr2*, and *CRb,* however, are distributed on

Table 11.8 Clubroot resistances QTL/gene mapped/fine mapped in different *Brassica* species by different research studies (Hirani and Li 2015)

Brassica species (R sources)	Populations	LG	QTL/genes
B. rapa (Chinese cabbage)	BC1	A03	*Rcr1* fine mapped
B. rapa (Chinese cabbage)	F2	A03	*CRb* fine mapped
B. rapa (G004 line)	F2	A08	*Crr1a* fine mapped
B. rapa (Chinese cabbage)		A03	*CRa* fine mapped
B. oleracea (Anju)	DH	O2, O5,	pb-Bo(Anju)1, pb-Bo(GC)1
B. napus (synthetic line)	DH	N02, N03, N08, N13, N15, N16 and N19	Nineteen QTL identified on different LGs
B. oleracea (kale)	F2:3	LG1, 2, 5	Nine QTL (Pb-Bo1 to Pb-Bo9) with phenotypic variance 20-88%
B. rapa (Shinki)	F2	A03	*CRb*
Brassica oleracea (Bindsachsener)	DH	–	Two QTL (pb-3 and pb-4)
B. rapa (Chinese cabbage)	F2, BC1	A03	CR gene fined mapped
B. rapa (turnip line)	F2	A03, A08	Two major QTL (Pb-Br3, Pb-Br8 and
B. rapa (European turnip)	F2:3	A03	*Crr3*
B. rapa (G004)	F2	A06	*Crr4*
B. rapa (Chinese cabbage)	F2	A03 and A02	*CRk and CRc*

three different chromosomes in *B. rapa*, R8, R1, and R3, respectively (Saito et al. 2006; Suwabe et al. 2006). Based on this observation, it has been suggested that the evolution of CR genes occurs by one of two routes: first, clubroot resistance was originally controlled by a single major gene in the ancestral genome, which later differentiated and diverged as functionally duplicate genes during the course of evolution in the *Brassica* genome (Suwabe et al. 2006). Second, route might be that the resistance genes for clubroot were originally clustered in that region in the ancestral genome which was later distributed into different genomic regions following chromosomal rearrangement in *Brassica*. Current *Brassica* speciess, which diverged 17–18 million years ago from *Arabidopsis*, are evolutionarily believed to be derived from whole-genome triplication and rearrangement of one ancestral genome (Lagercrantz 1998; O'Neill and Bancroft 2000; Yang et al. 2006). This hypothesis explains why these CR genes are dispersed and located on different chromosomes in *B. rapa*. Saito et al. (2006) suggested that the genomic region around *Crr3* exhibits homology to the top of the long arm of *Arabidopsis* chromosome 3, and possibly also to *CRk* (Sakamoto et al. 2008). It was concluded that *Crr3* has a different origin from that of *Crr1, Crr2*, and *CRb*. Gene *CRk* is independent of the CR genes *Crr1*, *Crr2*, *CRa*, and *CRb* but has a similar QTL region with *Crr3* (Sakamoto et al. 2008). Another novel CR locus, *CRc*, which is independent of all other CR loci, is located on chromosome R2 (Sakamoto et al. 2008). Fuchs and Sacristan (1996) mapped a CR locus (RPB1) in *Arabidopsis* chromosome 1. Their mapping for partial clubroot resistance identified two QTLs in chromosome 5 in F_2 and four QTLs in the RIL population, one each in chromosome 1 and 4 and two in the chromosome 5 of *A. thaliana* (Jubault et al. 2008). The CR QTL region in chromosome 5 co-localized with that containing several resistance gene clusters. These resistance genes could be candidates for clubroot disease resistance. Fine mapping and detailed analysis of the levels of expression possessed by these genes would help to identify specific ones capable of conferring clubroot resistance. The complete genome sequence of only one species of Brassicaceae family, *Arabidopsis thaliana*, is available. It has been reported that gene sequence identity of *Brassica* and *Arabidopsis* varies from 75% to 90% (Quiros et al. 2001). Therefore, sequence information from the model species will greatly enhance cloning and characterization of CR genes in other *Brassica* species because the partial sequences of several accessions of *A. thaliana* showing degrees of response to the pathogen *P. brassicae* collected worldwide, are available now (Fuchs and Sacristan 1996; Siemens et al. 2002; Nordborg et al. 2005; Alix et al. 2007). Only a few studies in transcriptomic/ expression levels have been completed with respect to disease development. These include investigation of involvement of several metabolic pathways in disease pathogenesis, such as hormonal regulation by auxins (Grsic et al. 1999; Neuhaus et al. 2000), cytokinins (Devos and Prinsen 2006; Siemens et al. 2006), and trehalose synthesis (Brodmann et al. 2002). Grsic et al. (1999) observed that de novo indole-3-acetic acid (IAA) biosynthesis plays a role in symptom development during later stages of disease development; it was suggested that jasmonic acid, which increased during club development, may be involved in the up-regulation of enzymes involved in IAA synthesis. Neuhaus et al. (2000) further supported this hypothesis by showing a delay in clubroot development after transforming *A. thaliana* with an antisense construct of Nitrilase 1 and 2, enzymes required for auxin biosynthesis. Siemens et al. (2006) investigated host gene expression during clubroot development in *A. thaliana* using an ATH1 microarray at two points in time, an early initial stage of infection and a later stage at which 60% of the host root cells were colonized.

More than 1000 genes were observed as being associated with the growth and cell cycle, sugar phosphate metabolism, and defense genes that were differentially expressed between infected versus control plants were observed. Up-regulation of auxin biosynthesis genes such as nitrilases and members of the GH3 family and down-regulation of cytokinin homeostasis were observed. It was further observed that lines overexpressing cytokinin oxidase/hydrolases were resistant to clubroot thereby strongly suggesting cytokinin is a key factor in clubroot development. Of the 312 genes identified as defense and disease resistant related genes, only 5 and 7% were up-regulated at the first and the second points in time. This functional analysis of gene expression in clubroot-resistant and susceptible lines provided initial preliminary information. To strengthen further the understanding of genes and genetic networks involved in the mechanisms of clubroot disease and host interaction, more detail analyses at the transcriptomic level are needed (Piao et al. 2009).

11.3.4 The Role of Transcriptome and Proteomic Approaches in Host Resistance

With the use of transcriptomic and proteomic approaches, the mechanisms of *P. brassicae* infection, pathogenesis, host specificity, and expression of host resistance have been elucidated. To date, the complete genome sequence of only one species of Brassicaceae, *A. thaliana* is available. The gene sequence identity of *Brassica* and *Arabidopsis* varies from 75% to 90% (Quiros et al. 2001). Using a full genome Affymetrix chip, Siemens et al. (2006) found that more than 1000 *Arabidopsis* genes are differentially expressed in *P. brassicae*-infected roots at either 10 or 23 days after inoculation versus non-inoculated roots. These included genes associated with growth and cell cycle, sugar phosphate metabolism, and defense. The involvement of plant hormones in club development further supported, genes involved in auxin homeostasis, such as nitrilases and members of the GH3 family, are up-regulated, whereas genes involved in cytokinin homeostasis are strongly down-regulated at both time points. Cytokinin oxidase/dehydrogenase overexpressing lines are disease resistant, clearly indicates the importance of cytokinin as a key factor in clubroot disease development. Most known defense or resistance related genes are either not differentially expressed or down-regulated at 23 vs. 10 days after inoculation. Feng et al. (2012a, b) observed the expression of several canola genes homologous to *Arabidopsis* genes found to be down-regulated at 35 days after inoculation, are highly expressed at early stages (7 days after inoculation) and down-regulated at later stages (42 days after inoculation) of infection. At the later stages of infection, the galls were forming and the resting spores were mature. Thus, the host genes involved in resistance/susceptibility may no longer be as active as during the early infection stages. Another explanation is that during the later stages, the pathogen already has successfully suppressed the expression of most resistance-related genes.

Proteome-level analysis, a powerful tool for high-throughput global protein expression analysis using two-dimensional gel electrophoresis (2-DE) coupled with mass spectrometry (MS) and bioinformatics, has been extensively used in the identification of proteins that are differentially expressed in plants in response to various abiotic or biotic stresses including those caused by plant pathogens. In order to investigate whether the pattern of specific host proteins are changed during clubroot formation, Hansen et al. (1994) employed 2-DE to compare the root protein profiles of *Brassica oleracea* L. with *B. oleracea* infected with *P. brassicae* four weeks after inoculation. They found that several proteins present in non-infected roots were absent or strongly reduced in the infected roots. In a proteomic study of a Chinese cabbage cultivar and two *P. brassicae* isolates, Ito et al. (1996) were able to resolve 133 differentially abundant protein spots, relative to non-infected controls, in infected roots 3 to 4 weeks after inoculation. The 133 protein spots were further classified into 6 groups, including spots that were enhanced in the susceptible response (59 spots), unique to the susceptible response (6 spots), repressed in the susceptible response (9 spots), enhanced in the resistant response (35 spots), unique to the resistant response (5 spots), and repressed in the resistant response (19 spots). N-terminal amino acid sequencing revealed that one protein (25 kDa, pI 7.0) that was enhanced in the susceptible response group showed high homology with pathogenesis-related protein group 5. These results suggest that the patterns of gene expression are different in the susceptible and the resistant responses.

In a differential protein analysis of infected versus non-infected roots and hypocotyls of *Arabidopsis* with 2-DE and MS, Devos et al. (2006) reported that 12% (46/390) of the visualized *Arabidopsis* proteins showed an altered abundance 4 days after inoculation with *P. brassicae* compared with the non-inoculated plants, including proteins involved in metabolism, cell defense, cell differentiation, and detoxification. The metabolism related enzymes adenosine kinase 2 (ADK2) and fructose bisphosphate aldolase (FBA) were found to be down-regulated at 4 days after inoculation. Down-regulation of ADK2 during the early stages of *P. brassicae* infection may keep the elevated levels of active cytokinins high, which is consistent with the increased endogenous isopentenyl adenine content. Down-regulation of FBA upon *P. brassicae* infection suggests a flow toward the production of glucose, which correlates with the fact that galls of *P. brassicae* act as a carbon sink that may be induced by the high cytokine level present at the infection site (Devos et al. 2006). Myrosinase, an enzyme that can catalyzes indole-3-methyl glucosinolate degradation to form indole-3-acetic acid, was also found to be six times more abundant in infected *Arabidopsis* at 4 days after inoculation. During *P. brassicae* infection, a number of detoxification related proteins are found to be up-regulated, including catalase, glutathione S-transferase, thioredoxin, and ferredoxin-nitrate reductase. Out of the 46 differentially regulated proteins, 11 proteins including ADK2, FBA, peroxiredoxin, α-tubulin, and heat shock protein 70 are down-regulated, and 35 are up-regulated including myrosinase, glutathione S-transferase, ferredoxin-nitrite reductase, and pectin methyl-esterase (Devos et al. 2006).

The changes in the root protein profile were examined by 2-DE at 12, 24, 48, and 72 h after inoculation of a susceptible canola (*B. napus*) genotype with *P. brassicae* (Cao et al. 2008). A total of 20 protein spots were identified as either up-regulated (13 spots) or down-regulated (7 spots). Decreased abundance of adenosine kinase, which is involved in cytokine homeostasis, supported the previous reports that cytokinins play a key role in the early phases of *P. brassicae* infection (Devos et al. 2006). An approximately sixfold reduction in caffeoyl-CoA O-methyltransferase abundance suggested a reduction in host lignin biosynthesis after inoculation, and is consistent with the compatible nature of the *B. napus/P. brassicae* interaction. Levels of enzymes involved in the metabolism of reactive oxygen species, such as copper/zinc superoxide dismutase and cytochrome c-oxidase, declined sharply at 12 h after inoculation, but increased at 24–72 h. Protein identification by MS analysis revealed that the 20 differentially regulated proteins can be classified into proteins involved in lignin biosynthesis, cytokinin metabolism, glycolysis, intracellular calcium homeostasis, and the detoxification of reactive oxygen species.

11.3.5 Pathotype-Specific Resistance

This type of resistance is mediated by resistance (R) genes through the recognition of specific pathogen effectors. These effectors initially function as virulence/pathogenicity factors for the pathogen, but then become recognized by the host as signals of attempted infection (Bernoux et al. 2011). Most R genes encode proteins carrying a nucleotide-binding site (NBS) in the central region and a leucine-rich repeat (NBS-LRR) domain at the C-terminus. These genes form a large multigene family in the plant genome and can be separated into two subclasses, the toll-interleukin-1 (TIR) class and the coiled-coil (CC) class (Rafiqi et al. 2009). The NBS–LRR proteins are thought to recognize effectors or effector-specific signals from the pathogen and activate the plant immune system to provide hypersensitive responses (Bernoux et al. 2011).

Although many clubroot resistance (CR) loci have been identified through genetic analysis and quantitative trait loci (QTL) mapping (Piao et al. 2009), the resistance mechanisms associated with these genes remained unknown until when two R genes were cloned and characterized. The first gene is *CRa* (Ueno et al. 2012) from *B. rapa*, which confers specific resistance to *P. brassicae* isolate M85. Fine mapping of the *CRa* locus to the *A. thaliana* and *B. rapa* genomes revealed a candidate gene encoding a TIR-NBS-LRR protein. The cloned alleles of this gene in susceptible and resistant *B. rapa* lines have several structural differences, and *CRa* expression was observed only in the resistant line. Four mutant lines lacking clubroot resistance were obtained by the UV irradiation of pollen from a resistant line, and all of these mutant lines were found to carry independent mutations in this candidate gene. This genetic and molecular evidence strongly suggests that the identified gene is *CRa*. The second gene is *Crr1a* (Hatakeyama et al. 2013), which is also from *B. rapa*. The *Crr1a* is one of the two loci of *Crr1* that was previously considered as a single locus. This gene (named $Crr1a^{G004}$) was cloned from the resistant line G004 and encodes a TIR-NB-LRR protein. By comparison, the susceptible allele $Crr1a^{A9709}$, cloned from the same locus in the susceptible line A9709, encodes a truncated NB-LRR protein. The $Crr1a^{G004}$ protein is expressed in the stele and cortex of the hypocotyl and roots, but not in root hairs, suggesting that it controls resistance to secondary infection or later pathogen development. Gain-of-function analysis proved that $Crr1a^{G004}$ was sufficient to bestow resistance to isolate Ano-01 in susceptible *Arabidopsis* and *B. rapa* genotypes. Since resistance genes from different plant genomes often show similar structures because they are most likely to be derived from a common ancestral gene (Leister 2004), the information about *CRa* and *Crr1a* will facilitate the cloning and molecular characterization of other clubroot resistance genes (Feng et al. 2014).

11.3.5.1 Identification of Pathotype-Specific Clubroot Resistance Genes Through Bulked Segregant RNA Sequencing

A clubroot resistance gene, *Rcr1*, with efficacy against pathotype 3 of *P. brassicae*, was mapped to chromosome A03 of *B. rapa* in pak choy cultivar "Flower Nabana." The resistance to pathotypes 2, 5, and 6 was shown to be associated with *Rcr1* region on chromosome A03. Bulked segregant RNA sequencing was performed and short read sequences were assembled into 10 chromosomes of the *B. rapa* reference genome v1.5. The resistant (R) bulks, a total of 351.8 million (M) sequences, 30,836.5 million bases (Mb) in length, produced 120-fold coverage of the reference genome. The susceptible (S) bulks, 322.9 M sequences, 28,216.6 Mb in length, produced 109-fold coverage. In total, 776.2 K single-nucleotide polymorphisms (SNPs), and 122.2 K insertion / deletion (InDels) in R bulks, and 762.8 K SNPs and 118.7 K InDels in S bulks were identified, each chromosome had about 87% SNPs and 13% InDels, with 78% monomorphic and 22% polymorphic variants between the R and S bulks. Polymorphic variants on each chromosome were usually below 23%, but made up 34% of the variants on chromosome A03. There were 35 genes annotated in the *Rcr1* target region and variants were identified in 21 genes. The numbers of poly variants differed significantly among the genes. Four out of them encode Toll-Interleukin-1 receptor/nucleotide-binding site/leucine-rich-repeat proteins, *Bra019409* and *Bra019410* harbored the higher numbers of polymorphic variants, which indicates that they are more likely candidates of *Rcr1*. Fourteen SNP markers in the target region were genotyped using the Kompetitive Allele Specific PCR method and were confirmed to associate with *Rcr1*. Selected SNP markers were analyzed with 26 recombinants obtained from a segregating population consisting of 1587 plants, indicating that they were completely linked to *Rcr1*. Nine SNP markers were used for marker-assisted introgression of *Rcr1* into *B. napus* canola from *B. rapa*, with 100% accuracy (Yu et al. 2016).

11.3.5.2 GBS Identification of QTL's Conferring Resistance to Multiple Pathotypes

Lines with resistance to a broad range of pathotypes of *P. brassicae* have been identified in the canola progenitor species *B. rap*. This species can be used to broaden the genetic base of clubroot resistance in canola. Introgression of traits from *B. rapa* into canola through interspecific crosses is possible, so resistance to clubroot from *B. rapa* could be transferred into canola through conventional breeding. The identification and genetic mapping of clubroot resistance genes has been carried out in *B. rapa*, *B.oleracea*, and *B. napus*. Two resistance genes, *CRa* and *Crr1*, have been isolated from Chinese cabbage lines of *B. rapa*. They encode Toll-Interleukin-1 receptor/nucleotide-binding site/leucine-rich-repeat (TIR-NBS-LRR, TNL) proteins. Genotyping-by-sequencing (GBS) offers a new tool to explore the genetic control of complex traits. GBS analysis was used in the current study to: (1) characterize genome-wide variants in *B. rapa*; (2) identify SNP sites that could be used for genetic mapping; and (3) detect QTL effectively resistant to multiple pathotypes of *P. brassicae* identified in western Canada; and (4) identify possible candidate genes for each QTL (Yu et al. 2017).

F_1 progeny from the *Brassica rapa* lines T19 (resistant) × ACDC (susceptible) were backcrossed with ACDC, then self-pollinated to produce BC1S1 lines. From genotyping-by-sequencing (GBS) of the parental lines and BC_1 plants, about 1.32 M sequences from T19 were aligned into the reference genome of *B. rapa* with 0.4-fold coverage, and 1.77 M sequences with 0.5-fold coverage in ACDC. The number of aligned short reads per plant in the BC1 ranged from 0.07 to 1.41 M sequences with 0.1-fold coverage. A total of 1584 high quality SNP loci were obtained, distributed on 10 chromosomes. A single co-localized QTL, designated as *Rcr4* on chromosome A03, conferred resistance to pathotypes 2, 3, 5, 6, and 8. The peak was at SNP locus A03_23710236, where LOD values were 30.3 to 38.8, with phenotypic variation explained (PVE) of 85–95%. Two QTLs for resistance to a novel *P. brassicae* pathotype 5x, designated *Rcr8* on chromosome A02 and *Rcr9* on A08, were detected with 15.0 LOD and 15.8 LOD, and PVE of 36% and 39%, respectively. Bulked segregant analysis was performed to examine TIR-NBS-LRR proteins in the regions harboring the QTL (Yu et al. 2017).

11.3.6 Role of Nitrogen in Modulation of Clubroot Resistance QTL's

The positive or negative modulation of plant disease development resulting from variations in nitrogen supply has been reported for many pathosystems (Dordas 2008; Fagard et al. 2014; Walters and Bingham 2007). In case of clubroot disease, a few studies have suggested that a high-nitrogen supply tends to reduce the damage caused by *P. brassicae* infection (Dixon 2009; Gossen et al. 2014), while another drew the opposite conclusion (Pryor 1940). None of these reports, however, took into account the possible influence of plant genotype (mainly the presence of genetic resistance factors) on the observed interactions between disease development and nitrogen supply. Genetic sources of both qualitative and quantitative resistance to *P. brassicae* have been identified in *B. rapa, B. oleracea,* and *B. napus.* Nonetheless, only a few commercial varieties with clubroot resistance are available currently, and most of these ("Mendel," "Mendelsson," "Cracker," "Andromeda," "SY Alister") possess qualitative resistance under oligogenic or major gene control. The deployment of clubroot-resistant varieties based on qualitative resistance can impose strong selection pressure on pathogen populations, favoring strains capable of overcoming resistance. In oilseed rape or canola, resistance defeating strains of *P. brassicae* were detected within 3–4 years of the introduction of clubroot-resistant host cultivars (Strelkov et al. 2016). Indeed, many examples exist in which the effectiveness of major resistance genes has been broken, including various cases involving major genes for clubroot resistance. The pyramiding of qualitative and quantitative resistance is considered today as a key lever to increase the durability of plant resistance against rapidly evolving pathogen populations (Brun et al. 2010; Tomita et al. 2013). As with most quantitative traits, the effect of quantitative resistance is dependent on other biotic and abiotic environmental factors. The stability of quantitative resistance to clubroot under different fertilization regimes, including low-nitrogen input was determined. A preliminary assay was conducted with non-inoculated (healthy) plants grown under standard clubroot resistance screening conditions, in order to assess the effect of two different NO_3-levels on plant growth and leaf nitrogen content. A segregating host population and a collection of oilseed rape genotypes then were used to identify and compare clubroot resistance QTL under high- and low-nitrogen supply following inoculation with two different isolates of *P. brassicae.* To assess the variability of clubroot resistance under different nitrogen levels and to characterize the impact of nitrogen supply on genetic resistance factors a linkage analyses and a genome-wide association study were conducted to detect QTL for clubroot resistance and evaluate their sensitivity to nitrogen. The clubroot response of a set of 92 diverse oilseed rape accessions and 108 lines derived from a cross between "Darmor-*bzh*" (resistant) and "Yudal" (susceptible) was studied in the greenhouse under high- and low-nitrogen conditions, following inoculation with the *P. brassicae* isolates eH and K92-16. Resistance to each isolate was controlled by a major QTL and a few small-effects QTL. While the same QTL were detected under both high and low nitrogen, their effects were altered. Clubroot resistance to isolate eH, but not K92-16, was greater under a low-N supply versus a high-N supply. New sources of resistance were found among the oilseed rape accessions under both low and high-N conditions (Laperche et al. 2017).

11.3.7 Linkage Markers of Clubroot Resistance in *Brassica*

The loss of resistance genes during breeding process is a very common phenomenon. For the determination of such genes, Kuginuki et al. (1997) observed linkage markers for a CR locus in *Brassica rapa.* Two CR loci, *Crr1* and *Crr2*, were identified in a doubled haploid (DH) line, G004 (Suwabe et al. 2003). The former locus corresponds to Kuginuki's CR locus because the linkage marker RA12-75 is tightly linked to BRMS-088, a marker for *Crr1*. These are based on CR traits derived from a European fodder turnip, Siloga. The studied conducted by Yoshikawa (1993) on the CR turnips, Milan White, was found to have polygenic resistance. To elucidate the genetics of CR of this turnip, Hirai et al. (2004), analyzed an inbred line having the resistant trait from Milan White. An inbred turnip (*Brassica rapa* syn. *campestris*) line, N-WMR-3, which carries the trait of clubroot resistance (CR) from a European turnip, Milan White, was crossed with a clubroot-susceptible doubled haploid line, A9709. A segregating F_3 population was obtained by single-seed descent of F_2 plants and used for a genetic analysis. Segregation of CR in the F_3 population suggested that CR is controlled by a major gene. Two RAPD markers, OPC11-1 and OPC11-2, were obtained as candidates of linkage markers by bulked segregant analysis. These were converted to sequence-tagged site markers, by cloning and sequencing of the polymorphic bands, and named OPC11-1S and OPC11-2S, respectively. The specific primer pairs for OPC11-1S amplified a clear dominant band, while the primer pairs for OPC11-2S resulted in codominant bands. Frequency distributions and statistical analyses indicate the presence of a major dominant CR gene linked to these two markers. The present marker for CR was independent of the previously found CR loci, *Crr1* and *Crr2*. Genotypic distribution and statistical analyses did not show any evidence of CR alleles on *Crr1* and *Crr2* loci in N-WMR-3. *B. rapa* has at least three CR loci. Therefore, the new CR locus was named *Crr3*. The present locus may be useful in breeding CR Chinese cabbage cultivars to overcome the decay of present CR cultivars (Hirai et al. 2004).

11.3.8 Marker-Assisted Selection of Clubroot Resistance Genes

To ensure stable production, more than one resistance genes need to be accumulated into a single CR cultivar. Early studies on clubroot resistance found that at least 3 major resistance genes existed in turnips, and that cultivars with such qualitative CR genes showed highly stable resistance (Crute et al. 1980; Toxopeus and Janssen 1975; Wit 1964). These genes, which are probably contained in source materials of CR Chinese cabbage, differ in their protective action. Genetic accumulation can bring elevated resistance corresponding to the wide pathogenicity of clubroot fungus. Marker-assisted selection (MAS) is indispensable in CR breeding because it allows for the acquisition of higher resistance by combining different resistance genes (Hirai 2006). DNA markers for clubroot resistance have been developed in major *Brassica* crops. Seven major CR loci, *CRa, CRb, CRk, CRc, Crr1, Crr2*, and *Crr3*, have been identified for Chinese cabbage using various types of markers (Hirai et al. 2004; Matsumoto et al. 1998; Piao et al. 2004; Sakamoto et al. 2008; Suwabe et al. 2003), though mutual relations among them are somewhat ambiguous, because different resistant materials and pathogenic isolates are employed for detecting the loci. The complexity of the clubroot pathogen in particular makes an exact evaluation of resistance genes difficult. Matsumoto et al. (2012) attempted to accumulate 3 CR genes, *CRa, CRk*, and *CRc*, through marker-assisted selection. Five doubled haploid CR lines with an individual CR locus were used as breeding materials. The CR lines were crossed with each other. A subsequent selection for resistance was performed using sequence characterized amplified region markers in segregating generations. As a result, 4 homozygous lines for 3 resistance genes and the F_1 hybrids between them were developed. CR pyramiding lines were inoculated with 6 field isolates of *P. brassicae*. The homozygous lines for 3 CR genes, whether selfed or crossed, exhibited exceedingly high resistance against all of the isolates. Morphological characters of F_1 hybrids were comparable to those of a control cultivar, but the degree of heterosis was less than expected, which is probably because of genetic similarity of the parents. The results prove that clubroot resistance can be reinforced through the accumulation of varied resistance genes in *B. rapa*.

11.3.9 Role of Chitinase Genes in Clubroot Resistance

Chitinases (EC 3.2.1.14), a subgroup of pathogenesis-related (PR) proteins, catalyze the hydrolysis of chitin. The resulting chitin fragments act as potent pathogen-associated molecular patterns (PAMPs) that induce PAMP-triggered immunity. Therefore, chitinases are believed to function as defense-related genes against pathogens containing chitin, which is supported by several studies that indicated increased levels of chitinase in response to pathogen attacks. Chitinases are classified into seven classes (Class I–VII) and most of them belong to the first four classes. The multigene family chitinase genes involved in biotic and abiotic stresses such as temperature changes, high salt content as well as metal and wounding stresses. Chinese cabbage responds to *P. brassicae* infection by activating a variety of genes, an enhanced chitinase activity has been reported in both susceptible and tolerant Chinese cabbage lines after inoculation with *P. brassicae*. Chen et al. (2016) identified 10 chitinase genes regulated by *P. brassicae* infection using RNA-Seq. Ahmad et al. (2012) identified 30 chitinase-like genes in *B. rapa* and discussed their role in resistance to the biotic and abiotic stress. Thirty three chitinase genes have been identified in *B. rapa* and their structure, evolution, and chromosome location have been further analyzed. The responses of the chitinase family genes to *P. brassicae* infection were investigated and the symptoms of clubroot disease were compared between susceptible lines subjected to pre-treatments with or without chitin oligosaccharides (CTOS). Chitinases are believed to function as a guardian against chitin containing pathogens. A total of 33 chitinase genes have been identified and grouped into five classes based on their conserved domain. They were distributed unevenly across eight chromosomes in *B. rapa*, and 31 of them contained few introns ($\leq$2). In addition, the expression of these genes was organ-specific, and 14 genes were expressed differentially in response to *P. brassicae* challenge of clubroot-susceptible (CS NIL) and resistant (CR NIL) lines. Further, reduced pathogen DNA content and clubroot symptoms were observed in the CS NILs after their treatment with chitin oligosaccharides 24 h prior to inoculation with *P. brassicae*. The findings indicate that chitinases play a crucial role in pathogen resistance of the host plants. The results offer an insight into the role of chitinase in *B. rapa*–*P. brassicae* interaction (Chen et al. 2018).

11.3.10 Characterization of Genes for Resistance to Clubroot Disease in *Brassica rapa*

European fodder turnips (*Brassica rapa*) were identified as sources of resistance and have been used to transfer their clubroot resistance (CR) genes into Chinese cabbage, oilseed rape, and *Brassica oleracea*. The CR trait of European turnip cultivars such as Siloga, Gelria R, and Debra was considered at first to be controlled by a single dominant gene. Yet after the release of more than 50 CR F_1 cultivars of Chinese cabbage so far, the breakdown of resistance, caused by variation in the pathogenicity of *P. brassicae*, has been reported (Yoshikawa 1981; Kuginuki et al. 1999). Genetic analysis and quantitative trait locus (QTL) mapping studies have identified at least 8 CR loci in *B. rapa*, 22 QTLs in *B. oleracea*, and 16 QTLs in *Brassica napus*. In *B. rapa*, *Crr1* and *Crr2* were identified on chromosomes A08 and A01, respectively. These two loci were detected by using two *P. brassicae* isolates, the mild Ano-01 and the more virulent Wakayama-01. *Crr1* was necessary for the resistance to both isolates, but plants having *Crr1* alone were susceptible to Wakayama-01. *Crr2*, which by itself does not show any effect against either isolate, was necessary for resistance to Wakayama-01 in combination with *Crr1*. Therefore, *Crr1* may play a role in a common pathway of resistance, and *Crr2* may be a modifier locus for the resistance expressed by *Crr1*. Four CR loci–*CRa*, *CRb*, *Crr3*, and *CRk* were identified on chromosome A03. The relationships between common markers and the location of the CR loci suggest that *CRa* and *Crr3* are identical, allelic, or closely linked to *CRb* and *CRk*, respectively. *CRc* was mapped on A02, and a weak QTL, *Crr4*, on A06. Comparative mapping of CR loci between *B. rapa* and *Arabidopsis* revealed that *Crr1, Crr2*, and *CRb* are syntenic with the central region of *Arabidopsis* chromosome 4. Because this region is located within a disease resistance gene cluster, it has been suggested that CR genes are members of these clusters. Suwabe et al. (2012) fine mapped the *Crr1* locus and found that it was likely to comprise two gene loci, a major locus for clubroot resistance and another locus with minor effect named *Crr1a* and *Crr1b*. The map-based cloning and characterization of *Crr1a*, the major locus for resistance to Ano-01, derived from European fodder turnip "Siloga." $Crr1a^{G004}$ encodes a TIR-NB-LRR disease resistance protein and is expressed in the stele or cortex of hypocotyl and roots, where the secondary infection phase occurs. Transgenic *Arabidopsis* and susceptible *B. rapa* harboring $Crr1^{G004}$ showed resistance to *P. brassicae* isolates similar to that of the resistant *B. rapa* (Hatakeyama et al. 2013). Fine mapping of the *Crr1* locus, which was originally identified as a single locus, revealed that it comprises two gene loci, *Crr1a* and *Crr1b*. The map-based cloning and characterization of *Crr1a*, which confers resistance to clubroot in *Brassica rapa*. $Crr1a^{\mathrm{G004}}$, cloned from the resistant line G004, encodes a Toll-Interleukin-1 receptor/nucleotide-binding site/leucine-rich repeat (TIR-NB-LRR) protein expressed in the stele and cortex of hypocotyl and roots, where secondary infection of the pathogen occurs, but not in root hairs, where primary infection occurs (Fig.11.4). Gain-of-function analysis proved that $Crr1a^{G004}$ alone conferred resistance to isolate Ano-01 in susceptible *Arabidopsis* and *B. rapa*. In comparison, the susceptible allele $Crr1a^{A9709}$ encodes a truncated NB-LRR protein, which lacked more than half of the TIR domain on account of the insertion of a solo-long terminal repeat (LTR) in exon 1 and included several substitutions and insertion-deletions in the LRR domain.

11.3.11 Proteomic Approach to Identify Clubroot Resistance Genes

Proteomics has been used in studying plant responses to biotic stresses based on differentially accumulated proteins (DAPs) as well as their functional annotation (Mehta et al. 2008; Quirino et al. 2010). Two-dimensional electrophoresis (2-DE) used commonly to quantify and compare proteins in different samples (Gao et al. 2012; Wu et al. 2014a, b) but technical challenges have limited its application in the post-genomics era, especially for large-scale global profiling of proteome. Quantitative proteomics, especially the use of label-free shotgun techniques, has become a popular approach to replace the 2-DE in studying plant–pathogen interactions (Novo et al. 2014). Critical metabolic or signaling pathways may be identified via

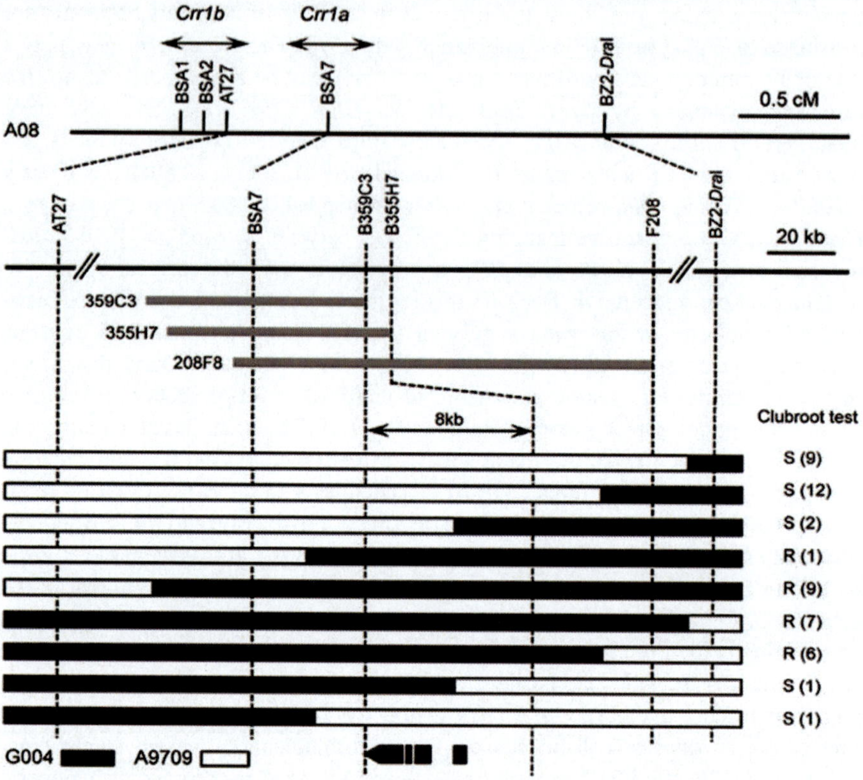

Fig. 11.4 Map-based cloning of *Crr1a*: a genetic map around the *Crr1* locus, and graphical genotypes of F_3 populations in which recombination occurred between AT27 and BZ2–DraI. The major locus for resistance to Ano-01 was predicted around BSA7 and another locus with minor effect was predicted around BSA2 (Suwabe et al. 2012). Named the former locus *Crr1a* and the latter *Crr1b*. Gray bars indicate BAC clones isolated by using BSA7 as a probe. Black bars, homozygous for resistant G004 allele; white bars, homozygous for susceptible A9709 allele. Phenotypes of resistance (R), and susceptibility (S) of F_3 populations to Ano-01 are indicated to the right of the graphical genotype with the number of F_3 populations of each in parentheses. The 3 F_3 populations in which recombination occurred between B359C3 and B355H7 were susceptible to Ano-01. The position of *Crr1a* was delimited to an 8-kb region between B355H7 and B359C3. The predicted ORFs of the *Crr1a* are presented (Hatakeyama et al. 2013)

functional annotation of DAPs, complementing the findings from other studies such as transcriptomic analysis. A label-free shotgun proteomic approach was used to profile and compare the proteomes of *Brassica rapa* carrying and not carrying the CR gene *Rcr1* in response to *P. brassicae* infection. A total of 527 differentially accumulated proteins (DAPs) have been identified between the resistant (with *Rcr1*) and susceptible (without *Rcr1*) samples, and functional annotation of these DAPs indicates that the perception of *P. brassicae* and activation of defense responses are triggered via an unique signaling pathway distinct from common modes of recognition receptors reported with many other plant–pathogen interactions.

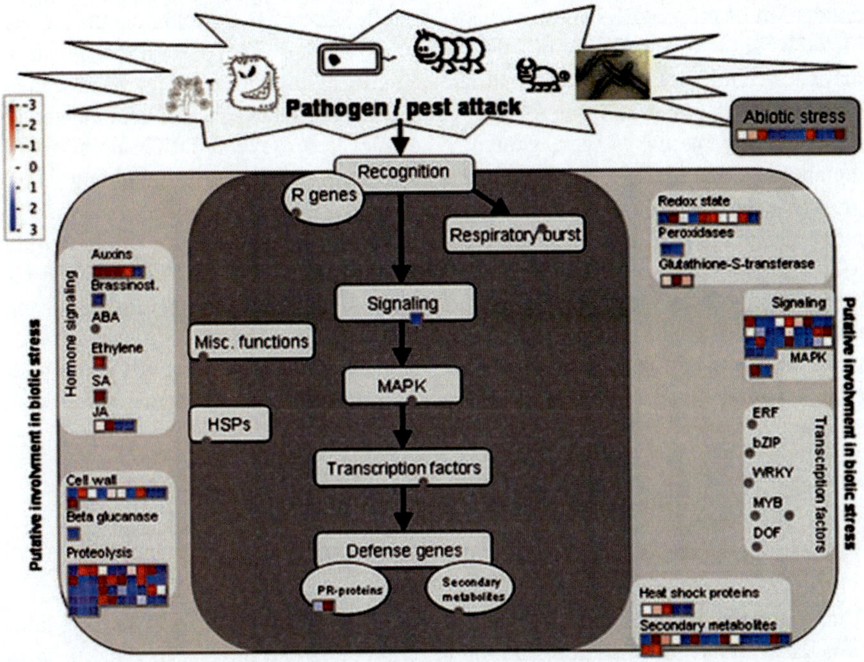

Fig. 11.5 Annotation of 145 DAPs in relation to signaling pathways involved in biotic/abiotic stress responses based on the analysis using the software Map man. The blue and red colors indicate up-and down-regulation, respectively, and gray circles indicate no DAPs identified in these categories (Song et al. 2016)

This pathway appears to act in a calcium-independent manner through a not-well-defined cascade of mitogen-activated protein kinases and may require the ubiquitin-26S proteasome found to be related to abiotic stresses, especially the cold-stress tolerance in other studies (Fig. 11.5). Both up-regulation of defense-related and down-regulation of pathogenicity-related metabolism were observed in plants carrying *Rcr1*, and these functions may all contribute to the CR mediated by *Rcr1*. These results, combined with those of transcriptomic analysis reported earlier, improved the understanding of molecular mechanisms associated with *Rcr1* and CR at large, and identified candidate metabolites or pathways related to specific resistance mechanisms. Deploying CR genes with different modes of action may help to improve the durability of CR (Song et al. 2016).

11.3.12 Differential Expression of R-genes

In the transcriptome analysis conducted using RNA sequencing on a segregating F_1 population inoculated with *P. brassicae*, 2,212 differentially expressed genes (DEGs) were identified between plants carrying and not carrying *Rcr1*. Functional annotation of these DEGs showed that several defense-related biological processes, including signaling and metabolism of jasmonate and ethylene, defensive deposition of callose and biosynthesis of indole-containing compounds, were up-regulated significantly in plants carrying *Rcr1* while genes involved in salicylic acid metabolic and signaling pathways were generally not elevated. Several DEGs involved in metabolism potentially related to clubroot symptom development, including auxin biosynthesis and cell growth/development, showed significantly lower expression in plants carrying *Rcr1*. The CR gene *Rcr1* and closely linked markers will be highly useful for breeding new resistant *Brassica* cultivars. The identification of DEGs between inoculated plants carrying and not carrying *Rcr1* is an important step towards understanding of specific metabolic/signaling pathways in clubroot resistance mediated by *Rcr1*. This information may help judicious use of CR genes with complementary resistance mechanisms for durable clubroot resistance (Chu et al. 2014).

11.3.13 Gene Expression and Functional Mechanisms

The use of microarray analysis to examine gene expression at 10 and 23 days after inoculation (DAI) in *Arabidopsis thaliana* ecotype Col-0 following inoculation with a single-spore isolate of *P. brassicae* was suggested by Siemens et al. (2006). Two methods have been used to study gene expression in partially resistant and susceptible *B. rapa*. The first used a low-cost "boutique" *Brassica* oligonucleotide array of 150 *Arabidopsis–Brassica* derived features constructed using nucleotide sequences of mostly defense associated and regulatory genes, from Gen Bank. This indicated a high number of constitutively and differentially expressed genes at 48 h after inoculation (HAI) as opposed to 24 or 72 HAI. It was observed that 48 h time point correlates with timing of penetration by primary zoospores. Key observations included the constitutive overexpression of a pathogenesis-related protein, chitinase, as well as the up-regulation of a lignin biosynthesis enzyme, caffeoyl-CoA 3-O-methyltransferase, in the partially resistant cultivars. A more sophisticated microarray platform, the Affymetrix *Arabidopsis* ATH1 GeneChip, was used for whole-genome transcriptional profiling of *B. rapa* cultivars at 48 HAI. This analysis showed three major salicylic acid (SA) dependent defense responses that were consistent with elevated levels of SA in partially resistant and resistant cultivars. Agarwal et al. (2011) used the Affymetrix *Arabidopsis* ATH1 Gene Chip to study host gene expression in *A. thaliana* (susceptible) ecotype Col-0 during the primary stages of infection (4, 7, and 10 DAI) by *P. brassicae*. The study revealed significant induction or suppression of a relatively low number of genes in a range of functional categories. Differential expression was greatest at DAI with 147 genes differentially expressed compared with 27 genes, and 37 genes at 7, and 10 DAI, respectively. This confirmed results obtained using a qPCR assay to detect and monitor the progression of *P. brassicae* infection in *A. thaliana* roots up to 23 DAI (Agarwal et al. 2009).

The *P. brassicae* was not detected in inoculated roots at 1, 2 or 3 days after inoculation. Pathogen DNA was first detected in roots at 4 DAI confirming pathogen penetration. Differentially expressed genes were classified into functional categories and related to the life cycle stage of the pathogen in *A. thaliana* for the time point of interest (Agarwal et al. 2011). In the susceptible reaction, salicylic acid, and ethylene biosynthesis were down-regulated at 4 DAI but jasmonic acid biosynthesis was induced (Agarwal et al. 2011). The elevated levels of SA in resistant and partially resistant *B. rapa/P. brassicae* interactions together with the down-regulation of SA biosynthesis in the compatible *Arabidopsis* Col-0/*P. brassicae* interaction (Agarwal et al. 2011) provide clear evidence for the role of SA in host defense responses to *P. brassicae.*

11.3.14 Role of Phenylpropanoid in Host Resistance

Phenylpropanoids are a group of important compounds, produced through the shikimate pathway, that contribute to biotic stress responses in plants. The phenylpropanoid pathway is the starting point for biosynthesis of several metabolites such as lignin, flavonoids, tannins, and anthocyanins. It has also been suggested that the pathway plays a role in the plant response to clubroot disease. The accumulation of phenolic compounds in root cell walls suggests that cell wall lignification is a defense response in clubroot-resistant ecotypes of *Arabidopsis.* Increased levels of naringenin and flavonols (quercetin and kaempferol) have been reported in *Arabidopsis* clubroot galls where they are thought to be associated with the regulation of auxin efflux from gall cells. Infected *Arabidopsis* roots with accumulated naringenin showed a decrease in gall size, and higher expression levels of phenylpropanoid pathway genes have been reported in several general, comparative transcriptomic studies of clubroot disease, while an accumulation of several lignin biosynthesis enzymes has been identified in clubroot-resistant (*Rc1*) *Brassica rapa* (Fuchs and Sacristan 1996; Pasold et al. 2010; Pasold and Ludwig-Müller 2013; Agarwal et al. 2011; Irani et al. 2018). Roots and leaves of two clubroot susceptible, and one resistant *Brassica napus* lines, inoculated with *P. brassicae* spores, were monitored for disease progression at four time points after inoculation. Quantitative gene expression analysis of phenylpropanoid pathway genes in these tissues showed increased expression of several genes involved in lignin biosynthesis in response to infection. In leaves, a higher expression of genes involved in flavonoid biosynthesis was also observed. In both roots and leaves of the resistant line, phenylpropanoid pathway genes were up-regulated at more time points and at higher levels than in susceptible lines. The differential expression of phenylpropanoid pathway genes between resistant and susceptible lines suggests that this pathway is instrumental in resistance to clubroot disease progression in the resistant line (Irani et al. 2019).

11.4 Transferring Clubroot Resistance Through Molecular Marker-Assisted Selection in *Brassica*

To develop resistant *Brassica* cultivars, several resistant sources such as European turnips, Chinese cabbage, and rutabaga cultivars are available and the resistance in these sources is dominant, which makes it easier to transfer clubroot resistance through interspecific and intraspecific hybridization. Rutabaga cultivars have been identified as clubroot-resistant sources. However, the genetics of clubroot resistance in rutabaga is complicated, so it is difficult to develop molecular markers that are closely linked to the dominant clubroot resistance genes (Ayers and Lelacheur 1972).

In Chinese cabbage, the dominant clubroot resistance from European turnips has been successfully used to develop clubroot-resistant Chinese cabbage. Since the gene mapping has been performed extensively in Chinese cabbage, molecular markers closely linked to clubroot resistance loci that are used in gene mapping can be easily selected to transfer clubroot resistance genes in the development of Chinese cabbage cultivars through molecular marker-assisted selection. Since the amphidiploid *B. napus* has a very complex genome, most of the molecular markers developed in *B. rapa* may not be polymorphic and cannot be directly used in *B. napus*. Additional efforts are required to develop molecular markers in *B. napus* when the mapped clubroot resistance loci in Chinese cabbage are transferred into *B. napus*. Currently, most clubroot resistance genes in European turnips have not been intensively investigated, and mapping and cloning of these clubroot resistance genes in European turnips will allow using these genes effectively and efficiently in *Brassica* breeding (Hirani and Li 2015).

11.4.1 Transfer of Clubroot Resistance Through Interspecific Hybridization

A dominant clubroot resistance gene from *B. rapa* (Chinese cabbage) was fine mapped, and molecular markers were developed in Chinese cabbage that could be used for marker-assisted selection (MAS) in other *Brassica* crops. To transfer this clubroot resistance gene to *B. napus* (canola), an interspecific hybridization was made between *B. napus* (canola) and *B. rapa* (Chinese cabbage) (Fig. 11.6). Subsequently, the F_1 was backcrossed to the canola recurrent parent for three generations to produce BC_1, BC_2, and BC_3 progenies (Table 11.9). Using these populations, simple sequence repeat (SSR) markers flanking the clubroot resistance gene were used to perform MAS in canola. These molecular markers were then evaluated in 13 different canola and rapeseed quality genotypes in *B. napus* and *B. rapa*. These markers exhibited high reliability in identifying clubroot resistance in this diverse set of *Brassica* genotypes. Clubroot resistance also co-segregated with the SSR markers flanking the clubroot resistance gene in the BC_3 and BC_3S1. The segregation ratio of resistant and susceptible individuals in the BC_3 supported the expected 1:1 ratio for the segregation of a single Mendelian gene. BC_3S1 families with homozygous

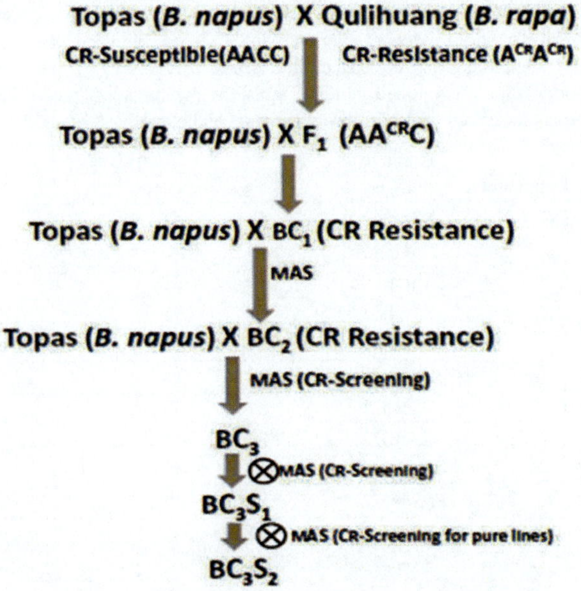

Fig. 11.6 Schematic diagram of interspecific backcross breeding strategy to introgress clubroot resistance gene from *Brassica rapa* (Chinese cabbage) to *Brassica napus* (canola) (Hirani et al. 2016)

clubroot resistance were developed during this process, and will be valuable sources of clubroot resistance in *B. napus* breeding activities (Hirani et al. 2016).

11.4.2 Transfer of Clubroot Resistance Through Distant Hybridization

The three clubroot-resistant Chinese cabbages (1003, 1007, and 1008) were used to transfer the clubroot resistance genes to *B. napus* by distant hybridization combined with embryo rescue. Three methods including morphological identification, cytology identification, and molecular marker-assisted selection were used to determine hybrid authenticity, and 0, 2, and 4 false hybrids were identified by these three methods. In total, 297 true hybrids were identified. Clubroot resistance markers and artificial inoculation were utilized to determine the source of clubroot resistance in the true hybrids. Two simple sequence repeat (SSR), and two intron polymorphic (IP) markers linked to clubroot resistance genes were identified. The clubroot resistance genes 1007 and 1008 were mapped to A03. At last, 159 clubroot-resistant hybrids were obtained by clubroot resistance markers and artificial inoculation. Four molecular markers associated with clubroot resistance were identified, which were located in a region from 23.77 Mb to 25.56 Mb on A03 of *B. rapa*. In the adjacent area of this region, two clubroot-resistant genes *CRa* (Ueno et al. 2012) and *CRb* (Hatakeyama et al. 2017) have been cloned, in which both genes encode the TIR-NBS-LRR (TNL) protein. Through analyzing the genes structure of the region surrounding the *CRa* and *CRb*, it was found that six genes *Bra012540*,

Table 11.9 Analysis of molecular markers for clubroot resistance gene introgression in canola using marker-assisted selection in a cross between the clubroot-resistant *Brassica rapa* cv. Qulihuang and the susceptible *Brassica napus* cv. Topas. The BC1, BC2, BC3, and BC3S1 populations developed using "Topas" as the recurrent parent and screened using a set of SSR markers linked with a clubroot resistance allele (Hirani et al. 2016)

Populations	Name of markers	Total[a]	CR resistance allele	Susceptible allele	χ^2	P[b]
BC_1	SScr062	69	41	28	2.45	0.12
	SScr045b	69	40	29	1.74	0.19
	SScr079b	69	31	38	0.71	0.4
BC_2	SScr062	27	15	12	0.33	0.56
	SScr045b	30	18	12	1.2	0.27
	SScr079b	29	18	11	1.69	0.19
BC_3	SScr071	87	38	49	1.39	0.24
	SScr072	87	44	43	0.01	0.91
	SScr082b	86	45	41	0.19	0.67
	SScr062	82	42	40	0.05	0.83
	SScr045b	82	43	39	0.2	0.66
	SScr079b	83	41	42	0.01	0.91
		AA	AB	BB	χ^2	P[c]
BC_3S_1	SScr071	7	16	9	0.25	0.88
	SScr072	7	15	7	0.31	0.86
	SScr045b	7	13	7	0.81	0.67
	SScr079b	7	16	8	0.13	0.94

[a]Indicates total number of individual plants excluding missing values
[b]Expected segregation ratio was 1:1
[c]Expected segregation ratio was 1:2:1 in BC_3S_1

Bra012541, Bra019409, Bra019410, Bra019412, and *Bra019413* have the structure of TNL (Liu et al. 2018).

11.4.3 Transfer of Clubroot Resistance to *B. rapa* from the European Clubroot Differential Set

European turnips are the most resistant sources and their resistance genes have been introduced into other crops such as oilseed rape (*Brassica napus* L.), Chinese cabbage, and other *Brassica* vegetables. The European clubroot differential (ECD) set contains four turnip accessions (ECD1–4). These ECD turnips exhibit high levels of resistance to clubroot when they are tested under controlled environmental conditions with Canadian field isolates. Gene mapping of the clubroot resistance genes in ECD1–4 were performed and three independent dominant resistance loci were identified. Two resistance loci were mapped on chromosome A03, and the third on chromosome A08. Each ECD turnip accession contained two of these three resistance loci. Some resistance loci were homozygous in ECD accessions while others showed heterozygosity based on the segregation of clubroot resistance in

20 BC_1 families derived from ECD1 to 4. Molecular markers were developed linked to each clubroot resistance loci for the resistance gene introgression in different germplasms (Hirani et al. 2018).

11.5 Management of Clubroot Resistance for Effective Utilization

The *Brassica* genomes (A, B, and C genome) are crucially important to provide novel genetic inheritance for economically important traits that can be used for the overall improvement of crop production and quality. The single A genome of diploid *B. rapa* (A genome) holds more than 230 R-gene sequences in 16 gene families, among which over 8 genetic loci have been identified which have functional properties for clubroot disease resistance. There are possibilities of the existence of more R-genes specific to the clubroot disease resistance and their allelic variations in different genetic pools or wild relative species. Effective utilization of resistance loci and their allelic variations may enhance the durability of resistance against clubroot disease in different *Brassica* species (Cheng et al. 2011).

As a long history of clubroot disease revealed relatively high evolutionary patterns of the pathogen, *P. brassicae*. In various cultivating geographical regions of *Brassica* crops, persistence of *P. brassicae* pathotypes with high levels of pathogenic virulence poses challenges to breed durable clubroot-resistant cultivars. The breakdown of clubroot-resistant cultivars has become a serious problem in Chinese cabbage and leafy cabbage in China, Korea, and Japan. Effective management of resistance genetic resources in breeding novel cultivars can enhance the performance of resistance loci in different *Brassica* species for sustainable, more durable, and cost-effective control of the clubroot disease (Hatakeyama et al. 2006; Osaki et al. 2008a).

11.6 Nomenclature of Clubroot Resistance (CR) Genes

The nomenclature of CR genes has not been standardized in *Brassica* species. In *B. rapa*, there are *Crr* series (*Crr1* to *Crr4*) and CR series (*CRa*, *CRb*, *CRc*, and *CRk*). In *B. oleracea*, there are *CR2a*, and *CR2b*, *Pb* series, *Pb-Bo* series (Piao et al. 2009). In *B. napus*, there are *Pb-Bn* and *PbBn* series. An international standardization of CR loci nomenclature is needed to avoid confusion between and within the different *Brassica* species as well as to differentiate the CR loci. Piao et al. (2009) recommended the use of *PbBr, PbBo*, and *PbBn* for clubroot resistance loci found in *B. rapa, B. oleracea*, and *B. napus*, respectively (*Pb* is derived from *P. brassicae*, *Br* from *B. rapa*, *Bo* from *B. oleracea*, and *Bn* from *B. napus*, respectively).

At least three discasc rcsistance loci of the ancestral genome, that is, *Crr1, Crr2*, and *CRb* (together considered as one locus), *Crr3,* and *CRkI* (together another locus), and *RPB1* (third locus), and three unclear loci, *CRa, CRc*, and *Crr4*, might be involved in the evolution of CR genes in cruciferous plants. According to the hypothesis of hexa-polyploidization applied to *Brassica* species (Lagercrantz and Lydiate 1996; Lagercrantz 1998), there is the possibility that they originated from one common gene. The cloning and comparison of these genes at the nucleotide level would help to unravel stages in the evolution of CR genes. The cloning of the *CRb* and *Crr3* genes is underway (Piao et al. 2004; Saito et al. 2006).

11.7 Non-host and Basal Resistance

Non-host resistance is the resistance shown by an entire plant species to all genetic variants of a pathogen (Mysore and Ryu 2004). In this context, the pathogen is referred to as non-adapted to the plant. Basal resistance is the first line of defense of a host to the adapted pathogen, which is achieved through a set of defined receptors, also referred to as pattern recognition receptors (PRRs), that recognize and conserved microbe-associated molecular patterns (MAMPs) (Jones and Dangl 2006; de Wit 2007). Upon MAMP recognition, primary defense responses are induced such as cell wall alterations, deposition of cellulose, and the accumulation of defense-related proteins (van Loon et al. 2006). A new conceptual framework has been recently proposed that non-host resistance and basal resistance to adapted pathogens may rest on similar principles (Niks and Marcel 2009; Fan and Doerner 2012). Based on studies of other pathosystems, plant primary metabolism has been demonstrated to play a significant role in non-host and basal resistance (Stuttmann et al. 2011).

The involvement of plant primary metabolism in clubroot pathogenesis has been studied, with the induction of sink metabolism regarded as a non-specific response of host roots to *P. brassicae* infection (Ludwig-Müller et al. 2009). In *P. brassicae*-infected *Arabidopsis* roots, expression of sucrose synthase and starch synthase is induced (Siemens et al. 2006). The lower rate of photosynthesis is associated with an increase in invertase activity in the roots, and an accumulation of hexoses with a down-regulation of photosynthetic gene expression is observed (Siemens et al. 2011). It is postulated that plant genotypes with down-regulated sucrose/starch synthase or invertase activity may carry a basal resistance to the pathogen (Gravot et al. 2011). The induction of plant trehalase during clubroot disease in *Arabidopsis* is a defense mechanism, which can help the plant to cope with the accumulation of pathogen-synthesized trehalose. This mechanism of resistance is also found in the susceptible *Arabidopsis* accession Col-0, and the inhibition of trehalase activity during clubroot infection results in enhanced disease symptoms, suggesting that trehalase induction is a basal defense mechanism and act as a disease resistance mechanism activated by a virulent pathogen on a susceptible host (Jones and Dangl 2006).

11.8 Response of Single Spore and Population of *Plasmodiophora brassicae* on Host Resistance

Clubroot-resistant *Brassica* genotypes are being deployed to reduce yield losses. Two experiments were conducted to examine the effect on *P. brassicae* virulence of repeated exposure of a population, and single-spore isolate of the pathogen to the same host. The first experiment examined revealed changes in the index of disease over five cycles of infection on seven *Brassica* hosts (European Clubroot Differential [ECD] 02, ECD 04, ECD 05, ECD 15, "45H26," "45H29," and 08N823R). The second experiment tested the virulence of five cycled populations ("45H29," 08N823R, ECD 05, and ECD 15) and three cycled single-spore isolates ("45H29," 08N823R, and ECD05) on four resistant canola genotypes ("73-77," "73-67," VT-SD-09, and "9558C"). Both single-spore isolates and populations of *P. brassicae* rapidly eroded the resistance present in the two canola genotypes, "45H29" and 08N823R. Although the index of disease increased on these two genotypes, the four resistant canola genotypes remained resistant to all the cycled populations, and single-spore isolates in the second experiment (Fig. 11.7) (LeBoldus et al. 2012).

11.9 Expression of Cellular Resistance to *Plasmodiophora brassicae*

In the resistance genotypes, there is lignification of epidermal and cortical cells of the host in addition to hypersensitive response to *P. brassicae* infection. The timing and expression of resistance to four pathotypes (2, 3, 5, and 6) of *P. brassicae* were assessed in four new commercial cultivars of canola (*Brassica napus*) with resistance to clubroot. Each of the resistant cultivars was highly resistant to all four isolates, and there was no difference in their response to infection. Root-hair infection occurred at high levels, but pathogen development occurred more slowly than in a susceptible cultivar. Secondary infection and development in cortical cells was severely inhibited in each of the resistant cultivars, only a few bi-nucleated plasmodia were observed at 12 days after inoculation (DAI), and plasmodia were rarely observed at 18 and 24 DAI. In contrast, development in the susceptible cultivar had progressed to resting spores by 24 DAI. A dense ring of accumulated reactive oxygen species (ROS) was observed in the endodermis, pericycle, and vascular cambium of non-inoculated controls and inoculated plants of the resistant cultivars. However, the ROS ring disappeared rapidly in infected plants of the susceptible cultivar. Plasmodia invaded the stele of susceptible roots by preferentially colonizing the xylem parenchyma cells. Expansion and enlargement of lignified xylem cells was observed by 35 DAI.

The absence of any specific points of ROS accumulation or lignification of epidermal or cortical cells in the resistant cultivars indicates that a hypersensitive response is not the main mechanism of resistance in these lines. The uniform response of these resistant cultivars to the four isolates of *P. brassicae* indicates

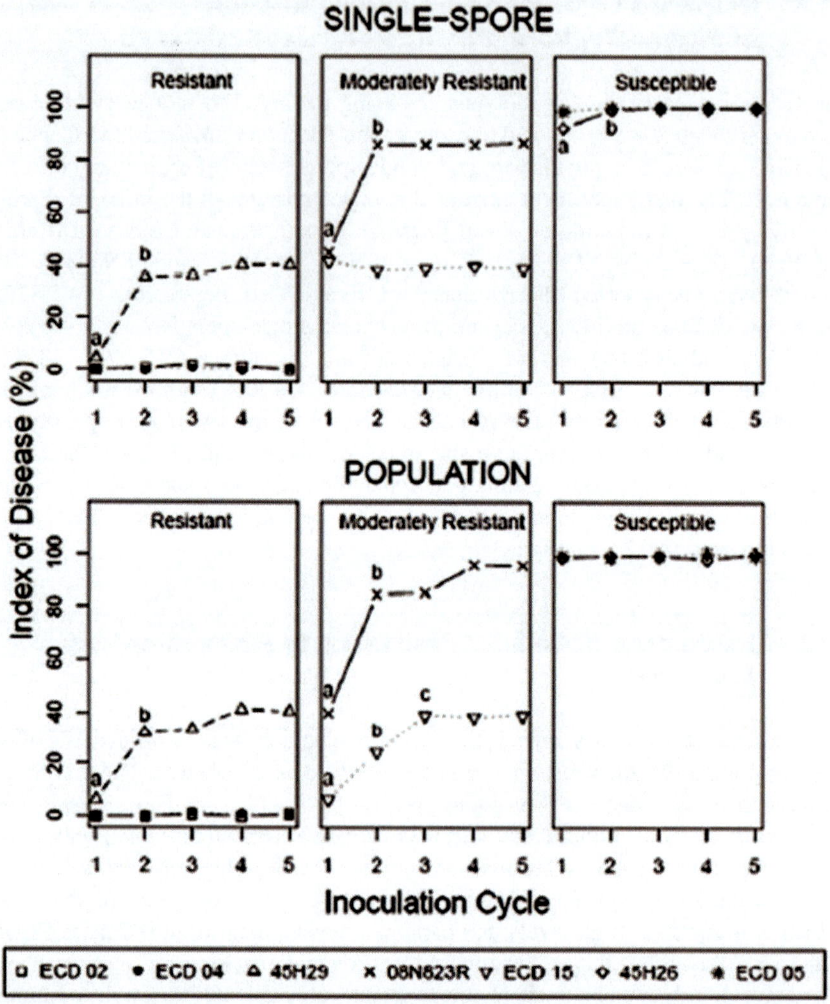

Fig. 11.7 Change in mean index of disease (ID) following five cycles of inoculation, isolation, and reinoculation on the same host with a single-spore isolate (SACAN03-ss1) or population (SACAN03-1) of *Plasmodiophora brassicae*. In total, seven different hosts exhibiting different levels of clubroot resistance were inoculated (resistant = European Clubroot Differential [ECD] 02, ECD 04, and "45H29"; moderately resistant = 08N823R and ECD 15; and susceptible = "45H26" and ECD 05). Mean ID values with different letters are significantly different for a particular host (LeBoldus et al. 2012)

that the resistance in each cultivar may be conditioned by a gene(s) from a single source that confers broad resistance, because most sources of resistance to *P. brassicae* are pathotype specific (Deora et al. 2013).

11.10 Small Spheroid Galls of Clubroot as an Indicator of Resistance

In the evaluation of resistance, small spheroid galls (SSGs), which are distinct from typical spindle galls, and are generally regarded as a host resistance reaction, were observed. The occurrence of SSGs was investigated when evaluating the host resistance of seven commercial clubroot-resistant (CR) Chinese cabbage (*Brassica rapa*) cultivars inoculated with twelve *P. brassicae* populations collected from different infested fields in Chuanyu region of China. All seven CR Chinese cabbage cultivars exhibited high resistance levels to the major populations tested, SSGs occurred in host-population combinations containing either Chinese cabbage cultivar "T1–145" or population "FL." This indicated that some CR cultivars or *P. brassicae* populations might be more likely to produce SSGs than others (Plate 11.2). Thus, recording the occurrence of SSGs separately in the resistance evaluation might be useful in mitigating the potential for selection of SSG producing hosts.

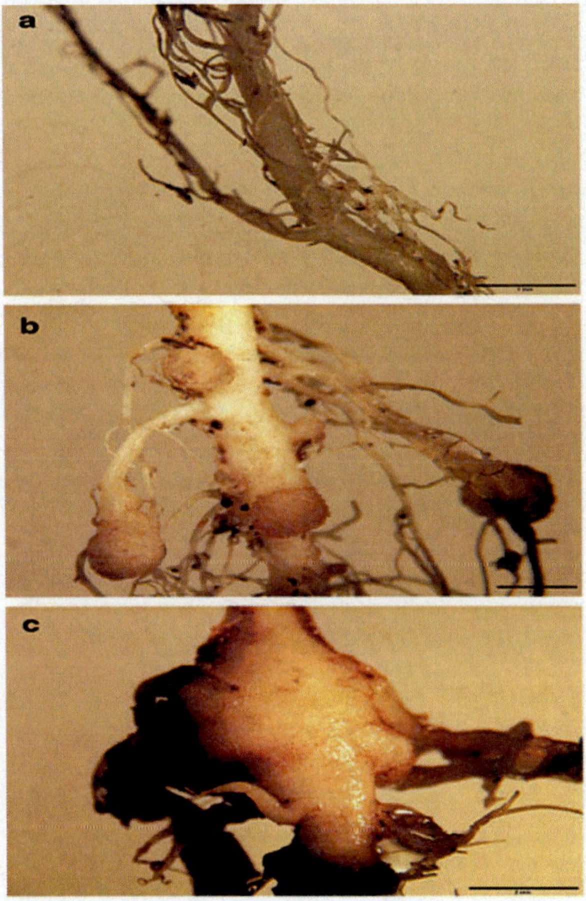

Plate 11.2 Morphology differences among Chinese cabbage plants with (**a**) healthy root (scale bar =1 mm), (**b**) root with small spheroid galls (SSGs) (scale bar =1 mm), (**c**) root with spindle galls (scale bar =2 mm) (Peng et al. 2016)

External morphology observations revealed that both fresh root weight and gall diameter of spindle galls were significantly higher than SSGs, and histological observations indicated that delayed pathogen development in cortical infection might contribute to the restricted development of SSGs. Additionally, the number of SSGs formed per root was low (2.2 in average). This observation further confirmed that some host-population combinations cannot form large numbers of SSGs on single roots (Peng et al. 2016).

11.11 Mutational Approach for Clubroot Resistance

Botrytis-induced kinase1 (BIK1), a receptor-like cytoplasmic kinase, plays an important role in resistance against pathogens and insects in *Arabidopsis thaliana*. Chen et al. (2016) investigated the potential roles of receptors FLS2, BAK1, and BIK1 in the infection of *P. brassicae* cruciferous plants. Wild-type plants, *fls2,* and *bak1* mutants showed typical symptom on roots, and the galls were filled with large quantities of resting spores, while *bik1* mutant plants exhibited strong resistance to *P. brassicae*. Compared with that of the wild-type plants, the root hair and cortical infection rate of *bik1* mutant were significantly reduced by about 40–50%. A considerable portion of *bik1* roots failed to form typical galls. Even if some small galls were formed, they were filled with multinucleate secondary plasmodia. The *bik1* plants accumulated less reactive oxygen species (ROS) at infected roots than other mutants and wild-type plants. Exogenous salicylic acid (SA) treatment alleviated the clubroot symptoms in wild-type plants, and the expression of the SA signaling marker gene *PR1* was significantly increased in *bik1*. Both *sid2* (salicylic acid induction-deficient2) and *npr1-1* [non-expresser of PR genes that regulate systemic acquired resistance (SAR)] mutants showed increased susceptibility to *P. Brassicae* compared with wild-type plants. The resistance of *bik1* to *P. brassicae* is possibly mediated by SA inducible mechanisms.

11.12 Mechanism of Resistance in Clubroot Resistance (CR) Cultivars

Most CR cultivars (highly resistant cultivars) of crucifers, including Chinese cabbage, suppress the formation of large and/or spindle-form clubroot galls, which affect the growth of plants. Alternatively, small, spheroids, resistant-type galls (SSGs) are often seen on the outer surface of roots of these cultivars (Osaki et al. 2008a; Seaman et al. 1963; Williams 1966). The SSGs have been considered to be a result of a resistant reaction to *P. brassicae* (Seaman et al. 1963). In roots of the CR cultivars of Chinese cabbage, populations of *P. brassicae*, which are avirulent on these cultivars, can commonly infect at the root hairs, form primary plasmodia within these cells, and produce secondary zoospores (primary infection phase) (Tanaka et al. 2006; Yamagishi et al. 1986). In addition, although these populations can infect the root cortex and medullary rays and form secondary plasmodia within the infected cells (secondary infection phase), these remain immature plasmodia with a small number of nuclei and cannot develop resting spores in root tissues, with the exception of SSGs (Tanaka et al. 2006). These populations, however, can develop mature resting spores within SSGs (Osaki et al. 2008a). Populations from these SSGs have characteristics of pathogenicity identical to those of original populations (Osaki et al. 2008a). Thus, populations that are avirulent on CR cultivars of Chinese cabbage can complete their life cycle only within SSGs in these cultivars.

The formation of SSGs indicates the restriction of the migration of *P. brassicae* in resistant root tissues (Seaman et al. 1963), and is considered to be one of the resistant mechanisms in the CR cultivars of Chinese cabbage. In addition, suppression of secondary colonization (i.e., suppression of plasmodial development) of *P. brassicae* within infected root cells may also be associated with resistance in the CR cultivars of Chinese cabbage (Tanaka et al. 2006). Such an incompatible response against *P. brassicae* has also been observed in the CR cultivars of radish (Kroll et al. 1983) and canola (Deora et al. 2012; Hwang et al. 2011) in other countries.

The details of the physiological mechanisms involved in the suppression of secondary colonization by *P. brassicae* in CR cultivars are unknown. However, hardly any starch accumulation is detected within infected root cells of the CR cultivars (Tanaka et al. 2006). Starch accumulation, which is abnormally stimulated within *P. brassicae*-infected root cells of susceptible plants, is very important for secondary colonization of the pathogen because the starch serves as its carbon source (Keen and Williams 1969; Williams et al. 1968). Morgner (1995) has also reported that starch accumulation is observed in infected root cells of susceptible lines but never in those of CR lines of *Brassica*. Perhaps CR cultivars of some cruciferous crops can selectively regulate starch accumulation in response to different populations of *P. brassicae*. A new commercial CR cultivar of Chinese cabbage, Akimeki, bred through selection using multiple DNA markers, was recently released in Japan (Matsumoto et al. 2012). This cultivar is highly resistant to various populations of *P. brassicae*, including those pathogenic on conventional CR cultivars of Chinese cabbage, therefore, the stability and persistence of their usefulness in Japan are widely expected. However, an effective method to avoid a breakdown in resistance should be considered for practical use of this cultivar in fields (Tanaka and Ito 2013).

11.13 Histological Resistance

The observations of root-hair infection stages and a comparison of the infection in a clubroot-resistant line of *B. oleracea* var. *botrytis* (cauliflower) with the susceptible cultivar "White Rock" revealed anatomical mechanisms of host resistance against clubroot pathogen through light and electron microscopy (Donald et al. 2008). Using this method, primary (root hair) and secondary (cortical) stages of *P. brassicae* were observed in both resistant and susceptible hosts. Cell wall breaks, the presence of vesicles or inclusion bodies within cell walls, cell wall thickening in association with

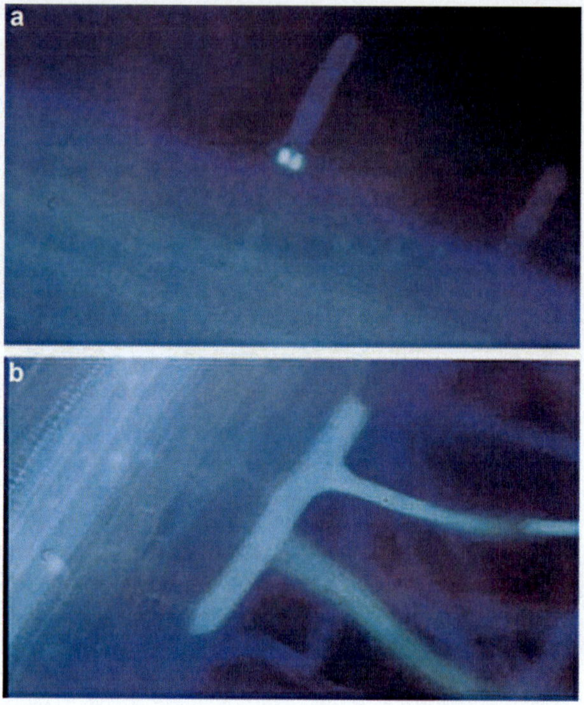

Plate 11.3 (a) "Localized" callose deposition at the base of a root hair, and (b) "generalized" callose deposition in root hairs and basal epidermal cells of clubroot-resistant cauliflower seedlings fluorescing after staining with Aniline blue (×400) (Donald and Porter 2014)

plasmodesmata and enlarged and/or disorganized host nuclei are observed in the cortex of inoculated resistant and susceptible hosts. The key difference observed between infected resistant and susceptible hosts is the absence of degradation of the secondary thickening and cell walls of the xylem in the resistant host (Donald et al. 2008). Differences between the resistant and susceptible host responses in cauliflower were studied using light microscopy to observe lignification after staining with 1% (w/v) phloroglucinol in 70% ethanol for 12 h. Additionally, fluorescence microscopy was used to observe callose deposition after staining with 0.5% (w/v) aniline blue in 1.15 M dipotassium hydrogen orthophosphate for 12 h (Donald 2005). No evidence of lignification in response to inoculation in either the resistant or susceptible roots was observed. The resistant host reaction, however, associated with production of α 1–3 glucan (callose) and two types of reactions were observed. The first type was a "local" reaction in which small deposits were usually, but not exclusively, observed at the base of infected root hairs (Plate 11.3a). This reaction occurred in 14% of resistant and only 1% of inoculated susceptible root hairs. The second type was a "general" reaction in which the whole root hair and/or basal epidermal cell would fluoresce (Plate 11.3b). This reaction occurred in 17% of the root hairs from inoculated resistant plants and only 3% from the inoculated susceptible plants. For both of these reaction types, the difference between the number of resistant and susceptible root hairs exhibiting "localized" or "generalized" fluorescence following inoculation was significant (Donald 2005). Callose is a glucan that can be rapidly deposited in stressed, wounded or pathogen-infected cells because of the presence of a pre-existing synthase enzyme present in the plasma membrane (Lamb et al. 1989). Deposition of callose is one of few host responses to pathogen invasion that does not require novel gene expression (Kobayashi et al. 1995), and, it can be the "first line" of defense against pathogen invasion.

11.14 Systemic Acquired Resistance

Salicylic acid (SA) biosynthesis is down-regulated in the compatible (susceptible) interaction, so exogenous application of SA will reduce the susceptibility of *A. thaliana* Col-0 to *P. brassicae* (Agarwal et al. 2011). The dipping of 14-day-old *A. thaliana* seedlings in 500 μM unbuffered SA for 60s before planting and inoculation with *P. brassicae* reduces the disease index from 81.5 (inoculated/untreated) to 20 (inoculated/SA treated) out of a possible 100 (Agarwal et al. 2011). Lovelock et al. (2013) reported systemic transportation of exogenously applied SA from roots to leaves in *B. oleracea*. This is associated with up-regulation of pathogenesis-related genes PR-1, and PR-2 in a "systemic acquired resistance like" response as early as 24 h after treatment. Application of concentrations greater than 5 mM significantly reduces root galling in between 25 and 65% measured 6 weeks after inoculation, but concentrations in excess of 20 mM are phytotoxic, significantly reducing root and shoot weights. The use of *Arabidopsis* mutants confirmed the role of SA in the host defense against *P. brassicae* (Lovelock 2013). The mutants cprl and dndl over-express SA. Both shows reduced root galling after inoculation with *P. brassicae* compared with the SA deficient NahG mutant and the wild-type Col-0. Elevated levels of PR-1 were measured in the over-expressor mutants and in the wild-type treated with 0.5 or 1 mM SA.

11.15 Sources of Resistance

Sources of resistance effective against clubroot pathogen, *P. brassicae* have been identified from cruciferous host species, viz., *B. napus, B. carinata, B. oleracea* var. *capitata, B. oleracea* var. *botrytis, B. oleracea* var. *gongylodes, Raphanus sativus, B. rapa* subsp. *pekinensis*, and *B. rapa* subsp. *rapifera* from Australia, Canada, China, Germany, India, Japan, and UK (Table 11.10). Pathotype-specific sources of resistance (Table 11.11) have been identified from *Arabidopsis thaliana* (Pathotype 2,5,6,e); *B. napus* (Pathotype 2,3,5,6,8); *B. oleracea* (Pathotypes 2,3,4,5); *B. nigra* (Pathotype 2,3,5,6,8), and *B. rapa* (Pathotype 2,3,5,6,8). The sources have been utilized to breed CR cvs. in different countries of the world which are providing protection to the crops from clubroot disease. Breeding disease resistance cvs. is a continuous process to replace breakdown of R-genes on the event of selection pressure of the pathogen and search of more stable sources of R-genes.

Table 11.10 Sources of resistance to *Plasmodiophora brassicae*

Host species/genotypes	Country	References
B. napus		
GSL-1, Culture-1, Culture-2, PCR -80, WW-1507, ISN-700, HNS-3	India	Chattopadhyay et al. (2001)
Mendel, NPZ CR-21	Germany	Diederichsen et al. (2014)
Wilhelmsburgar	Germany	Chiang and Crete (1983)
Pioneer 45H29, D 3152, Dekalb 73-67 RR, 73-77 RR, Canterra 1960, Proven 9558C	Canada	Strelkov et al. (2011)
B. carinata		
HC-1, HC-4, HC-5, HC-9001, PC-3, PCC-2, PPSC-1, PC-5, ACCBN-479	India	Chattopadhyay et al. (2001)
B. oleracea var. *capitata*		
KK Cross, Drumhead	India	Chattopadhyay et al. (2001)
Clapton	UK	Donald and Porter (2009)
Highfield	Australia	Donald and Porter (2009)
B. oleracea var. *botrytis*		
Danila, Snowball	India	Chattopadhyay et al. (2001)
Tekila, Kilatan	UK	Donald and Porter (2009)
Syngenta	Australia	Donald and Porter (2009)
Bindsachsenar, Bohmerwaldkohl	Germany	Crute et al. (1983)
B. oleracea var. *gongylodes*		
White viena	India	Chattopadhyay et al. (2001)
Raphanus sativus		
Novitas, Poem van, Zwijndrechat, Flevo, Robijn, Saxafire, Scharo, Verano		Rowe (1980)
B. rapa subsp. *pekinensis*		
Accession 1003, 1007, 1008	China	Liu et al. (2018)
B. rapa subsp. *rapifera*		
ECD-04	Germany	Diederichsen et al. (2006, 2009)
Milan white	Japan	Otani et al. (1982)
Golria, Siloga, Debra, Milan white	Japan	Hirai (2006)

11.15.1 Nature of Genetical Mechanism in Sources of Resistance

Genotypes with resistance to one or more of the pathotypes of *P. brassicae* have been reported in all of the major *Brassica* crops, except *B. juncea* (L.) Czern. and *B. carinata* Braun (Diederichsen et al. 2009). Both qualitative (Wit and van de Weg 1964; Crute 1986) and quantitative (Chiang and Crete 1970; Figdore et al. 1993; Grandclement and Thomas 1996; Voorrips et al. 1997) types of resistance have been reported. Most of these sources of resistance, however, are pathotype or race specific. In *B. napus*, most studies have reported oligogenic control of resistance to *P. brassicae* (Crute 1986).

Table 11.11 Pathotype-specific sources of resistance to *Plasmodiophora brassicae*

Host species/genotypes	Pathotypes	References
Arabidopsis thaliana		
Liue ct-1	2	Sharma et al. (2013)
Pu2-23	5	-do-
Ws-2, Sorbe	6	-do-
Tsu-0, Ze-0	e	Fuchs and Sacristan (1996)
B. napus		
(Rapid cycling Brassica collection (RCBC)	All	Sharma et al. (2013)
45H21, Invigor 5020 LL	6	Adhikari et al. (2012)
B. napus		
Rutabaga	2, 3, 5, 6, 8	Hasan et al. (2012)
B. oleracea		
RCBC lines	2, 3, 5	Sharma et al. (2013)
B. carinata		
RCBC lines	2, 5	Sharma et al. (2013)
B. rapa		
RCBC lines	2, 5	Sharma et al. (2013)
B. nigra		
BB	2, 3, 5, 6, 8	Hasan et al. (2012)
B. rapa subsp. *pekinensis* (Napa cabbage)		
Deneko, Bilko, Yuki	6	Adhikari et al. (2012)
Emiko, China Gold, Flower Nabana	6	Saude et al. (2012)
B. rapa var. *rapifera*		
Turnip	2,3,5,6,8	Hasan et al. (2012)
B. rapa var. *oleifera*		
Winter and spring rapeseed	2, 3, 5, 6, 8	Hasan et al. (2012)
B. oleracea var. *capitata*		
Kilaton, Tekila, Kilaxy, Kilaherb	6	Saude et al. (2012)
Xiangan 336, Verheul, Bindsachsener, Zhouyebai, 2358	4	Ning et al. (2018)
Pathotype-specific resistant species		
B. rapa (AA)		Hasan et al. (2012)
B. oleracea (CC)		
B. napus (AACC), Turnip, Rutabaga		

This would make the pyramiding of resistance genes in *B. napus* genotypes more practical than in other species. Models based on three, four, and five resistance genes have been proposed, and the most favored model was based on four genes (Gustafsson and Falt 1986). A complex type of inheritance, with dominant genes from *B. rapa* and recessive genes from *B. oleracea*, was expected in a resynthesized *B. napus*, with resistance from both ancestral species (Diederichsen and Sacristán 1996). Segregation analysis indicated that resistance in resynthesized *B. napus* was controlled by at least two dominant and unlinked genes (Diederichsen and Sacristán

1996). One major gene (*Pb-Bn1*) for resistance against two *P. brassicae* isolates is located on chromosome N03 and additional minor QTL for each isolate on chromosomes N12 and N19 (Manzanares-Dauleux et al. 2000a). Previous reports indicate that *P. brassicae* resistance in canola is controlled by a combination of major genes and quantitative trait loci (Matsumoto et al. 1998; Suwabe et al. 2003, 2006; Hirai et al. 2004; Piao et al. 2009).

Accessions of *B. oleracea* with pathotype-independent resistance to *P. brassicae* have also been reported (Voorrips 1996). Most studies on the C genome indicate that *P. brassicae* resistance in *B. oleracea* is quantitative and under the polygenic control of one or two major QTLs and some QTLs with minor effects (Landry et al. 1992; Figdore et al. 1993; Grandclement and Thomas 1996; Voorrips et al. 1997; Moriguchi et al. 1999; Rocherieux et al. 2004; Nomura et al. 2005). A few studies, however, indicated that *P. brassicae* resistance in *B. oleracea* is qualitative and controlled by either dominant (Chiang and Crete 1983) or recessive (Yoshikawa 1993) genes. It is possible that both quantitative and qualitative resistance mechanisms may be at play in this species.

Resistance genes from fodder turnip (*B. rapa*) have been used in resistance breeding of various *Brassica* crops, including Chinese cabbage, oilseed rape, and *B. oleracea*. Although most turnip lines carry more than one resistance gene, cultivars of the other *Brassica* crops with resistance derived from turnip generally carry a single, dominant resistance gene that is pathotype-specific. While important to clubroot management, genetic resistance has generally been race- or pathotype-specific (Diederichsen et al. 2009) and can break down when virulent races increase in the pathogen population. Therefore, genetic resistance should be carefully managed in combination with other methods of clubroot control (Hwang et al. 2012).

11.16 Deployment of Resistance

Clubroot-resistant canola cultivars have been the dominant component of the clubroot management strategy in western Canada in recent years. The first resistant cultivar for western Canada, Pioneer "45H29," was registered in 2009 and its release was quickly followed by several other resistant cultivars (Strelkov et al. 2011; Gossen et al. 2013). However, the durability of this resistance is not known. The genetic basis for the resistance in these cultivars also, is proprietary knowledge that is not available to the public. This has complicated efforts aimed at resistance stewardship, since it is not possible to develop rational strategies for the rotation of resistance sources in infested fields. The number of resistant genes, that are currently available to breeders are limited (Hirai 2006). Recent studies indicated that there were no detectable differences in pathotype reactions among clubroot-resistant canola cultivars available in western Canada (Deora et al. 2012, 2013). This represents one line of evidence that resistance genes in all these lines originated from a similar source. In contrast, another study by LeBoldus et al. (2012) revealed that pathogen populations that had been repeatedly cycled on and become adapted to one *Brassica* host did not exhibit a similar adaptation to other hosts, when the cycled populations were inoculated on the latter, indicating that distinct sources of resistance may occur in at least some genotypes. Regardless of the relationships between resistance sources, there is substantial genetic and pathotype variation present in the *P. brassicae* populations in western Canada (Strelkov et al. 2006; Xue et al. 2008; Cao et al. 2009), and the deployment of a cultivar with single-gene resistance against a genetically diverse pathogen on a large scale imposes a strong selection pressure for pathogen genotypes that are able to overcome this resistance (Hwang et al. 2012). Similar mechanisms may operate in other countries also.

11.17 Erosion or Breakdown of Resistance

Resistance to *P. brassicae* has broken down in various cruciferous crops, including oilseed rape and vegetable brassicas (Seaman et al. 1963; Kuginuki et al. 1999; Oxley 2007; Diederichsen et al. 2009). While *P. brassicae*-resistant cultivars of *B. napus*, *B. oleracea,* and *B. rapa* are available, their use is often limited by the short durability and the pathotype-specificity of the resistance (Voorrips 1995; Diederichsen et al. 2009). Each of the resistant canola cultivars available for production in western Canada will develop small galls under high inoculum pressure, and repeated cycles of inoculation can result in a rapid decline in the level of resistance (LeBoldus et al. 2012). The effective and sustainable deployment of genetically resistant cultivars to manage clubroot disease requires that the use of these cultivars be integrated with various other approaches for control (Diederichsen et al. 2009; Donald and Porter 2009). As such, regardless of the nature of the resistance, it is prudent to ensure that resistance is carefully managed to maximize its durability (Hwang et al. 2012).

Erosion and breakdown of clubroot-resistant (CR) cultivars to the old and evolution of new virulent pathogen has been assessed from very few countries of the world. The identification in 2013 of a few *P. brassicae* populations able to overcome the resistance in CR canola, only 3–4 years after the introduction of this trait on the Canadian prairies, underscored the continued vulnerability of *Brassica* to clubroot (Strelkov et al. 2016). This threat was further highlighted in the study conducted by Strelkov et al. (2018) with the identification of an additional 61 *P. brassicae* populations in Alberta between 2014 and 2016 capable of overcoming the clubroot resistance in the existing suite of CR cultivars. In a greenhouse study, rapid increases in clubroot severity were observed on some *Brassica* host genotypes after repeated cultivation in a *P. brassicae*-inoculated potting mix, which indicated that adaptation by the pathogen to selection pressure had occurred quickly (LeBoldus et al. 2012). Likewise, the loss or erosion of clubroot resistance was reported under field conditions in cabbage in the USA (Seaman et al. 1963), Dutch stubble turnip (*B. rapa* ssp. *rapifera*) in the UK, Chinese cabbage in Japan (Osaki et al. 2004; Hatakeyama et al. 2006), and “Mendel” oilseed rape in the UK and Germany (Oxley 2007; Diederichsen et al. 2014). In Canada, all of the CR canola (*Brassica*) cultivars evaluated were susceptible to at least some of the new, virulent *P. brassicae* field populations tested. Also, if a population of the pathogen could cause clubroot on one of these cultivars, it most often was virulent on all of them. Indeed, there was little differentiating capacity between individual CR canola cultivars, and all were susceptible to pathogen populations that could induce IDs >75% on at least one of the CR cultivars. Moreover, while a few pathogen populations were virulent on only one or a few hosts, such populations produced IDs <75% on even the most susceptible CR canola cultivars. This suggests that those populations were either less aggressive or represented a mix of virulent and avirulent resting spores (Xue et al. 2008). All of the pathogen populations caused IDs >90% on the universally susceptible ECD 05, which confirmed that these populations were highly virulent on a susceptible host.

It is interesting to note the burst of pathotype diversity observed in 2014 (nine pathotypes) and its apparent reduction in 2015 (five pathotypes) and 2016 (three pathotypes), with pathotype A becoming increasingly predominant. The phenomenon of diversifying selection in response to host resistance is not novel in agricultural pathosystems. Major resistance genes impose strong selection pressures on pathogen populations and, in combination with a number of factors, can act to cause "boom-and-bust" cycles of pathotype diversification within a population (Wolfe and McDermott 1994; Kolmer 1996; de Vallavieille-Pope et al. 2012; Lof and Van Der Werf 2017). Protection from infection due to a single major gene is short-lived and often brings on a measurable increase in pathotype diversity followed by a stabilizing shift toward a few dominant genotypes.

Additionally, genome environment/context and the "two speed" genome concept also can accommodate an effective, sustainable amount of allele diversity in a pathogen population under selection. Since, *P. brassicae* is an obligate parasite, elucidating the mechanisms of virulence diversification is more challenging and thus mechanistic details are not well-known or characterized in the *Brassica* clubroot pathosystem. Similarly, the amount of *P. brassicae* pathotype diversity prior to applying the R-gene selection is virtually impossible to measure. Nonetheless, *P. brassicae* clearly has mechanisms of generating diversity and maintaining low levels of rare pathotypes within field populations over extended periods, which can then respond to the varying selection pressures imposed by deployment of CR genes.

Given its prevalence and virulence on the CR cultivars, pathotype A would make a logical target for the resistance screening activities associated with canola development. Another novel pathotype, B (a variant of the original pathotype 2), also may be of interest in clubroot resistance breeding, since, it shares almost the same host range as pathotype A, but additionally is virulent on the *B. oleracea* host ECD 11, which was resistant to all of the other *P. brassicae* pathotypes characterized (Strelkov et al. 2018). Nonetheless, pathotype B was identified only once during 2014–2016. The second most common pathotype in this study, after A, was pathotype D. This pathotype, which was another variant of Williams' pathotype 3, shared an identical virulence pattern with pathotype A except that it was avirulent on var. "Mendel." Pathotype X (a variant of pathotype 5), which corresponded to the first populations found to be able to overcome resistance in 2013 (Strelkov et al. 2016), was not identified again in 2014–2016. Since, it was first identified, pathotype X has been one of the most commonly used for clubroot resistance screening, given that it represented the first pathotype found to overcome resistance in CR canola. In light of its relative rarity, however, and its inability to cause clubroot on all of the CCD hosts susceptible to pathotypes A, B, and D, its current prevalent use should be reconsidered.

Fields in which clubroot resistance appeared to be overcome often occurred as distinct clusters. This may indicate that new populations of *P. brassicae* with novel virulence patterns have emerged independently in multiple fields, via the selection pressure imposed by the planting of CR *Brassica*, and then spread locally by the movement of infested soil on farm and other machinery (Cao et al. 2009) and/or by the movement of wind-blown dust infested with the pathogen (Gossen et al. 2015; Rennie et al. 2015). Nonetheless, the occurrence of clusters of virulent clubroot pathotypes highlights the need for growers accessing *Brassica* fields to take precautions, such as cleaning farm machinery, to avoid spread of these new pathotypes.

The emergences of multiple new virulence phenotypes of *P. brassicae*, which are able to overcome genetic resistance in *Brassica,* have a serious impact on the long-term production of this crop. Collectively, CR *Brassica* cultivars represent the most important clubroot management tool available to farmers, but genetic resistance should be deployed carefully to maximize its longevity. As such, resistance must be used as part of an integrated clubroot management approach that includes longer rotations out of *Brassica* in infested fields, cleaning/sanitization of machinery, and potentially the use of liming or other strategies to reduce disease pressure.

11.18 Manipulation of Host Resistance

A number of lines of evidence support the view that *P. brassicae,* in common with many other biotrophic pathogens, manipulates the defense responses of its hosts. Analysis of differentially expressed putative Pb3 effectors within the first six weeks post-inoculation provided evidence for the manipulation of plant defense, by the induction of protease inhibitors and possibly suppression of salicylic acid (SA)-induced defense. Pair wise comparisons of *B. napus* genes differentially expressed in response to Pb3 infection using the application HORMONOMETER showed that salicylic acid mediated responses at the earliest stages of infection was negatively correlated with the *A. thaliana* gene expression profile in response to SA. In addition, the expression of a *P. brassicae* predicted secreted protein (PbPT3Sc00026_A_1.308_1) identified as a benzoic acid/salicylic acid methyltransferase (Pb3-BSMT) (NBCI accession number AFK13134) was extremely low in *B. napus* infected roots during the first week after inoculation, but increased in two week and peaked at weeks 3 and 4. Likewise, expression was higher in infected *A. thaliana* tissue at 16 DPI than at 26 DPI. A similar expression pattern for this gene was reported recently by Ludwig-Müller et al. (2015). Rolfe et al. (2016) confirmed the methyltransferase activity of this enzyme and suggested a possible role for this enzyme in inactivating plant defense by converting SA

(Salicylic Acid) to MeSA (Methyl Salicylate), which acts as a mobile signal in systemic acquired resistance (SAR) in plants.

11.19 Future Avenues and Scope of Clubroot Resistance

The great genetic diversity within and between the different *Brassica* crop species and other non-*Brassica* species within the Brassicaceae provides a rich resource for resistance (R) genes against the different major pathogens including *P. brassicae*. *B. napus* (AACC, n= 19) is an amphidiploids, formed by interspecific hybridization of the two diploid species *B. rapa* (AA, $n = 10$) and *B. oleracea* (CC, $n = 9$). During this hybridization, several rearrangement events occurred that resulted in extensive reordering among the diverged genomes (Truco et al. 1996). The genome sequences of *B. rapa* (Wang et al. 2011; Cai et al. 2017), *B. napus* (Chalhoub et al. 2014), *B. oleracea* (Parkin et al. 2014; Liu et al. 2014), *B. juncea,* and *B. nigra* (Yang et al. 2016) and the pan-genome of *B. oleracea* have been published. The A and C genomes in *B. napus* are highly homologous, due to their common ancestry. Evidence of the A and C genomes sharing high genome similarity is also provided from common ancestral blocks between *B. rapa* and *B. oleracea* have opened new era to develop clubroot-resistant cultivars (Parkin et al. 2005, 2014; Schranz et al. 2006; Liu et al. 2014; Golicz et al. 2016).

Plants recognize pathogens either through the pathogen's pathogen-associated molecular patterns (PAMPs) with the plant's pattern recognition receptors (PRRs) or the pathogen's virulence molecules, termed effectors, interacting with the plant R genes. Although different definitions of R- genes have been put forward but R-genes with the conventionally described version that is a single/major gene which results in qualitative resistance and highly specific. In contrast, quantitative resistance involves more than one gene, which often lie within a major QTL region that displays large phenotypic variance. The specificity between the host R gene and the pathogen effector happens only for pathogens that have adapted to the plant. This plant immunity is termed effector-triggered immunity (ETI), signified by localized cell death that constitutes a hypersensitive response (HR)—an "all-or-none" effect, which is often the exception rather than the rule (Dodds and Rathjen 2010). Mostly, plants respond to pathogens by mounting resistance called PAMP-triggered immunity (PTI), which is a non-host resistance. Plants also defend themselves by triggering defense-related hormone signaling pathways such as jasmonates (JA), salicylates (SA), ethylene (ET), abscisic acid (ABA), and brassinosteroids (BR). This type of resistance is called systemic acquired resistance (SAR) and often incurs broad-spectrum and long lasting immunity (Durrant and Dong 2004; Netea et al. 2011; Kazan and Lyons 2014)). Evidence shows that R gene mediated plant immunity and SAR involving defense response hormone signaling are tightly regulated revealing the complexity of the plant immunity (Thomma et al. 2011; Stotz et al. 2014; Larkan et al. 2013; Ma and Borhan 2015). In *Brassica* species, both major R genes and quantitative QTL have been identified for all key host pathogens. However, the number of these genes that have been characterized and cloned is very limited. This causes challenges for breeding for resistance and an increased reliance on chemical control. Current advances in genomics in *B. napus* have provided us with a good platform to search for candidate R genes using advanced molecular tools (Edwards et al. 2013; Zhang et al. 2016; Bayer et al. 2017) to explore stable R cvs. to *P. brassicae*. *P. brassicae*, causal agent for clubroot has a wide host range including non-cruciferous plants. The pathogen physiology, progress in efforts to combat disease, and an update on its epidemiology has recently been reviewed. The sources of *P. brassicae* R genes and genetic mapping studies have been reviewed by Piao et al. (2004) and Rahman et al. (2014). To date, at least 19 R-genes/QTL have been mapped for Clubroot resistance in *B. napus*, by Werner et al. (2008). However, only two Clubroot resistance genes have been cloned, *Crr1a* and *CRa*, both in *B. rapa* (Table 11.12 and Fig. 11.8).

Sources of *P. brassicae* R genes in *B. rapa* have been particularly promising as there are many race-specific, single, dominant R genes present, as compared with *B. oleracea* which has a more continuous resistance profile involving various combinations of both major and minor genes. Due to the different physiological resistance mechanisms toward *P. brassicae* in *B. rapa* and *B. oleracea*, complete resistance is often reported in the former but not in the latter (Tanaka et al. 2006; Donald et al. 2008; Hwang et al. 2012). Recent screening efforts in China have identified *B. rapa* to have good resistance against Canadian pathotype 3 (Zhang et al. 2015). Hence, it is not uncommon that breeding efforts make use of *B. rapa*, in particular using the European fodder turnip as the recurrent parent, to introduce resistance into *B. rapa*, *B. napus*, and *B. oleracea* (Diederichsen et al. 2009). The introgressed regions can be characterized using SNP genotyping and next generation sequencing in order to identify and characterize these resistance genes and determine if they are the same gene or distinct. They could then be deployed in a more targeted manner. In Canada, most oilseed *B. napus* shows little, or at best only moderate, resistance to Williams pathotypes 3 and 5, while *B. napus* rutabaga lines have better resistance. However, in Australia, the majority of *B. napus* was found to be resistant (Donald et al. 2006a), reflecting the influence of the local abiotic environments on the pathogenicity of the fungus or the origin of host and/or pathogen influencing the resistance outcome (Donald et al. 2006a; Donald and Porter 2014). Further QTL or Genome-Wide Association Study (GWAS) studies should be performed in *B. napus* in order to identify these resistance genes.

Traditionally, the AB genome amphidiploid *B. juncea* does not show high resistance. It often shows low, moderate (Peng et al. 2014) or intermediate (Sharma et al. 2012) resistance. However, one accession of *B. juncea* was found to be highly resistant to the local pathotype 4 under field conditions in China (Ren et al. 2016). Another good source of resistance from the B genome is diploid *B. nigra*, which has been reported to display high resistance against all local pathotypes in Canada. Therefore, there is the potential for all A, B, and C genomes of the *Brassica* species to harbor R genes against *P. brassicae* (Neik et al. 2017).

Table 11.12 *Brassica* crops and R genes/QTL identified and genetically mapped or cloned against *Plasmodiophora brassicae* (Neik et al. 2017)

Resistance gene/QTL mapped	References
B. napus (Canola lines "1CA0591.323" and "1CA0591.263" (Resistant, R) derived from cultivar "Mendel"× "A7-26NR" (Susceptible, S), using isolate SACAN-ss1, Williams pathotype 3): "Mendel" carries *CRa* and CRb^{Kato} on chromosome A3	Fredua-Agyeman and Rahman (2016)
CRa and CRb^{Kato} may be same allele also shown	Zhang et al. (2014)
B. napus (Inbred line "12-3" (R) × inbred line "12-1" (S) using pathotype 3): New CR gene, act singly or combined with *CRa*	Zhang et al. (2016)
B. napus (DH "263/11" (R) × Cultivar "Express" (S), 7 different isolates from Sweden, Germany, and France): 19 QTL mapped. Broad-spectrum resistance mapped on chromosome A2, A3, A8, A9, C3, C5, C6, and C9. QTL on A3 corresponds to the region for *CRk* and *Crr3*, some QTL have additive effects	Werner et al. (2008)
B. napus (Cultivar "Mendel"(R) × Breeding line (S), using isolate 1 that is highly virulent on *B. oleracea*): One single dominant locus from *B. rapa* ("ECD-04"), two recessive loci from *B. oleracea* ("ECD-15")	Diederichsen et al. (2006)
B. napus (DH progeny from Darmor-*bzh* (R) × Yudal (S), using single-spore isolate pathotypes 4 and 7 according to : Monogenic or polygenic depending on isolate, major resistance dominant gene *Pb- Bn1* toward isolate Pb137-522 on LG	Somé et al. (1996)
DY4, also show weaker partial resistance effect in association with other QTL. Additive and epistatic QTL were identified	Manzanares-Dauleux et al. (2000a, b)
B. napus (Natural lines *B. oleracea* "ECD-15" (R) × B. rapa" ECD-04" (R)), from which cultivar "Mendel" is produced, using virulent field isolates): Two dominant, unlinked genes	Diederichsen and Sacristán (1996)
B. rapa (Fine map *Rcr2* in Chinese cabbage cv. "Jazz" (R) × DH ACDC (S) using pathotype 3): *Rcr2* on 22 and 26 Mb of chromosome A3	Huang et al. (2017)
B. rapa (Breeding line "T19" originate from cultivar "Pluto" (R) × DH line "ACDC" (S), using Williams pathotype 2,3,5,6,8, 5x): GBS identified *Rcr4* on chromosome A3 and two novel QTL, *Rcr8* on A2 and *Rcr9* on A8	Yu et al. (2017)
B. rapa (Inbred lines Chinese cabbage "CCR13685" (R) × Pak choi "GHQ11021" (S), using isolate with unknown race/pathotype): single dominant gene	Chen et al. (2016)
B. rapa (Fine map *Rpb1* from Chu et al. 2013): Single dominant allele *Rcr1* (or also known as *Rpb1*) on LG A3, close position to both *CRa* and *CRb*, against Williams pathotype 3. Rcr1 is predicted to be TIR-NBS-LRR)	Chu et al. (2014), Yu et al. (2017)

(continued)

Table 11.12 (continued)

Resistance gene/QTL mapped	References
B. rapa (Pak choi "702-5" (S) × DH line "CR Shinki" (R), using Williams race 4): Mapped physical location of *CRb* on chromosome A3 on *B. rapa* with interval length of 83.5 kb with 15 candidate genes including NBS-LRR genes. *CRb* is not an allele of *CRa, but closely linked genes. CRa and* CRb^{Kato} *may be same allele*	Fredua-Agyeman and Rahman (2016), Zhang et al. (2014)
B. rapa (Inbred lines turnip "Siloga" (R) × Chinese cabbage "BJN3" (S), using isolate Williams race 4): One major locus *QS_B3.1* on chromosome A3 corresponding to *CRa* and *CRb*, and two minor loci *QS_B1.1* on chromosome A1 (homologous to *Arabidopsis* chromosome 3 independent of any other CR loci) and *QS_B8.1* on chromosome A8 (share same locus with *Crr1b* and *Crr1*). Additive effects and epistatic interactions were found in *B. rapa*	Pang et al. (2014)
B. rapa (Five Chinese cabbage cultivars (R) × *B. rapa* oilseed accession (S) using Canadian field isolate): linkage group N3 corresponding to chromosome A3	Gao et al. (2014)
B. rapa (Fine map *CRb* in DH line "CR Shinki" (R) × Chinese cabbage "702-5" (S) using pathotype 4): *CRb* gene was tightly linked to two other *CR* genes, *CRa* and CRb^{Kato}	Zhang et al. (2014)
B. rapa (DH "G004" (R) x DH "A9709" (S), using field isolates Ano-01, Wakayama-01, and Nos. 5, 7, 9, and 14 based on *Two* genes, *Crr1a* with major effect and *Crr1b* with minor effect at *Crr1* locus	Hatakeyama et al. (2004, 2013)
B. rapa (Inbred turnip line "ECD04" (R) × Inbred Chinese cabbage line "C59-1" (S) × using four different isolates Pb2, Pb4, Pb7, and Pb10): Partial resistance, *PbBa1.1 on A1, PbBa3.1, PbBa3.2 and PbBa3.3* on A3 (*PbBa3.1* and *PbBa3.3* on different region), *PbBa8.1* on A8	Chen et al. (2013)
B. rapa (Cultivar Pak choi "FN" (R) x DH "ACDC" (S), using isolate Williams pathotype 2, 3, 5, 6, 8): Single dominant gene *Rpb1* located on LG A3, close to *CRa*	Chu et al. (2013)
B. rapa (Cultivars "Akiriso" and "CR Shinki," using isolate No. 14 or pathotype group 3 according to : Single dominant gene linked to *CRb*, or is *CRb* found in "Akiriso" *CRb* and *CRa* are one and the same clubroot resistance gene	Hatakeyama et al. (2004, 2017) Kato et al. (2012)
B. rapa (Fine map *Crr1* in *Arabidopsis*): *Crr1* is likely consist of two genetic loci. The gene order is conserved except for one inversion in which insertion is found	Suwabe et al. (2012)
B. rapa (Fine map *Crr1* in *Arabidopsis*): *Crr1* is likely consisting of two genetic loci. The gene order is conserved except for one inversion in which insertion is found	
B. rapa [Mapping of *CRa* from (Matsumoto et al. 1998)]: RFLP marker, HC352b closely linked to *CRa*	Hayashida et al. (2008)

(continued)

Table 11.12 (continued)

Resistance gene/QTL mapped	References
B. rapa (Two F2 populations, one F2 progeny from DH "K10" (R) × DH "Q5" (S), the other DH "C9" (R) × DH "6R" (S), using isolate M85 Williams race 2 and K04 with unknown pathotype/race): two loci: *CRk* and *CRc*, broad-spectrum resistance. *CRk* is very near or identical to *CRb* on chromosome A3. *CRc* is mapped on chromosome A2 or R2	Sakamoto et al. (2008)
B. rapa (Doubled haploid, DH "G004" (R) × DH "A9709" (S), using isolate Wakayama-01 which has not been classified but is close to Williams race 4): Two loci, *Crr1* and *Crr2*, polygenic control, both loci needed for resistance. *Crr1* is closely linked to the major QTL toward Williams race 2	Kuginuki et al. (1997); Suwabe et al. (2003)
B. rapa (Map position of *Crr1* and *Crr2* on *A. thaliana* and *B. rapa*): *Crr1* is on linkage group LG7 (chromosome A8 or R8) and *Crr2* on LG6 (chromosome A1 or R1) of *B. rapa*, both aligned to same region of chromosome 4 in *Arabidopsis*	Suwabe et al. (2006)
B. rapa (Map position of *Crr3* on *B. rapa* and *Arabidopsis* from Hirai et al. 2004): *Crr3* is on R3, same LG as *CRb*, syntenic region on *Arabidopsis* chromosome 3	Saito et al. (2006)
B. rapa (DH Chinese cabbage "CR Shinki" (R) × "94SK" (S), using isolate Williams race 4): single dominant gene, CRb (Piao et al.2004) that is closely linked to *CRa* on chromosome A3 or R3 *CRb*, *Crr1*,and *Crr2* fall in the same syntenic region of *Arabidopsis* chromosome 4	Diederichsen et al. (2009) Saito et al. (2006)
B. rapa (DH "G004" (R) × DH "A9709" (S), using isolate Wakayama-01 and Ano-01): *Crr4* loci on LG 2 (chromosome A6 or R6), independent of *Crr1*, *Crr2*, and *Crr3*	Suwabe et al. (2006)
B. rapa (Inbred line "N-WMR-3" containing cultivar "Milan White" (R) × DH "A9709" (S), using isolate Ano-01Williams race 3): Crr3, major dominant gene, independent of *Crr1* and *Crr2*	Hirai et al. (2004)
B. rapa (DH "T136-8" (R) × DH "Q5" (S), both Chinese cabbage, using isolate Williams race 2): Single dominant major gene, *CRa* (chromosome A3 or R3)	Matsumoto et al. (1998)
B. rapa (European fodder turnip "Siloga S2" (R) × DH Chinese cabbage "Homei"(S), using isolate Williams race 2):Single dominant phenotype plus heterogeneous phenotype	Kuginuki et al. (1997)
B. oleracea (Korean inbred cabbage lines "C1220" (R) and "C1176" (S): Three QTL, *CRQTL-GN_1* (chromosome C2) and *CRQTL-GN_2* (chromosome C3), using GN isolate Williams race 9) and *CRQTL-YC* (chromosome C3), using YC isolate Williams isolate race 2): *pB-Bo (Anju)1* and *pB-Bo (Anju)3* overlap with *CRQTL-GN_1* and *CRQTL-GN_2* (or *CRQTL-YC*) respectively. *CRQTL-YC* is major QTL for race 2 and also minor QTL for race	Lee et al. (2016)

(continued)

Table 11.12 (continued)

Resistance gene/QTL mapped	References
9 isolates. Both closely linked QTL *CRQTL-GN_2* and *CRQTL-YC* have syntenic regions on *B. rapa* chromosome A3 (*Crr3*, *CRk*, *CRa*, and *CRb)* and A8 (*Crr1)*	
B. oleracea (DH cabbage "Anju" (R) × DH broccoli "GC" (S) using isolate Williams race 4): *pb-Bo(Anju)1* and *pb-Bo(Anju)2* on chromosome O2 (syntenic R2 where *CRc* is detected), *pb-Bo(Anju)3* on chromosome O3 (corresponds to R3 distal end), *pb-Bo(Anju)4* on chromosome O7 (closely linked to *CRb*, homologous to R3), *pb-Bo(GC)1* on chromosome O5. Polygenic control for "Anju" resistance with *pb-Bo (Anju)1* being major QTL while the rest is moderate and minor QTL. Reported that *CR2b* and *pb-3* are also on chromosome O3	Nagaoka et al. (2010)
B. oleracea (Inbred kale line "K269" (R) × Inbred cabbage line "Y2A" (S), using isolate Williams race 4): Three QTL namely QTL 1, 3, and 9 with complementary effect (Nomura et al. 2005). QTL 9 and *PbBo(GC)1* may be the same locus	Nagaoka et al. (2010)
B. oleracea (Landrace kale line "C10" (R) × DH broccoli "HDEM" (S), using isolate pathotype according to Ms6 (P1), K92-16 (P4), Pb137-522 (P7), eH (P1), K92 (P2)): nine QTL, major QTL *Pb-Bo1* (isolate specific) against three out of five isolates (Pb137-522, K92-16, and K92), minor QTL (non-specific) *Pb-Bo3, Pb-Bo4, Pb-Bo8, Pb-Bo9a, Pb-Bo9b for isolates* Pb137-522, K92-16, K92, eH, and Ms6, respectively	Somé et al. 1996, Rocherieux et al. (2004)
B. oleracea (Inbred kale "K269"(R) × Inbred cabbage "Y2A" (S), using field that is infected with isolate Williams race 1 and 3): One major QTL on LG 3	Moriguchi et al. (1999)
B. oleracea (German landrace "Bindsachsener" (R) × DH broccoli "Greenia" (S), using isolate ECD 16/3/30): *pb-3* and *pb-4* additive effect on LG 3 and 1 respectively	Voorrips et al. (1997)
B. oleracea (Kale line "C10" (R) × Cauliflower line "48.4.7" (S), using isolate Williams race 7): Polygenic control	Grandclement and Thomas (1996)
B. oleracea (Kale breeding lines, using isolate Williams race 4): Young and adult stage resistance was assessed. Dominant alleles, additive effects and incomplete dominance were found	Laurens and Thomas (1993)
B. oleracea (Inbred broccoli "CR7" (R) × Cultivar cauliflower "Early White" (S), using isolate Williams race 7): Mixed dominant QTL and polygenic QTL	Figdore et al. (1993)
B. oleracea (European Clubroot Differential or ECD host set including cabbage, broccoli, kale, using Dutch isolate): Display recessive character	Voorips and Visser (1993)
B. oleracea (Cabbage line "No. 86-16-5" (R) × rapid cycling line "CrGC No. 85", using isolate Williams race 2): Two dominant QTL, *CR2a* (derived from *B. napus*), *CR2b* on LG 6 and 1, respectively	Landry et al. (1992)

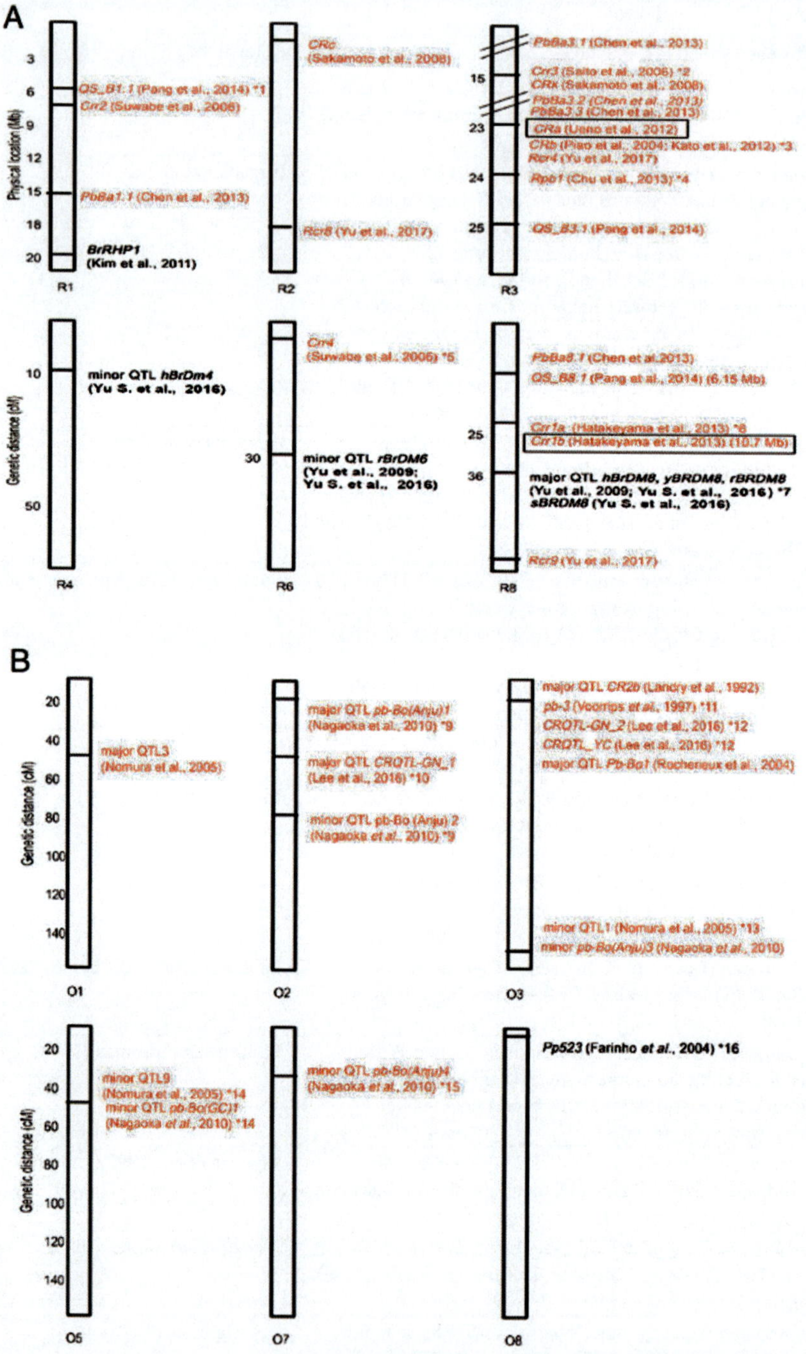

Fig. 11.8 *R* genes/QTL mapped on *Brassica rapa* (**a**), *B. oleracea* (**b**), and *B. napus* (**c**). Each vertical rectangular is a chromosome, with the chromosome name followed by chromosomenumber at the bottom. Black font denotes *R* genes/QTL mapped for *Sclerotinia*

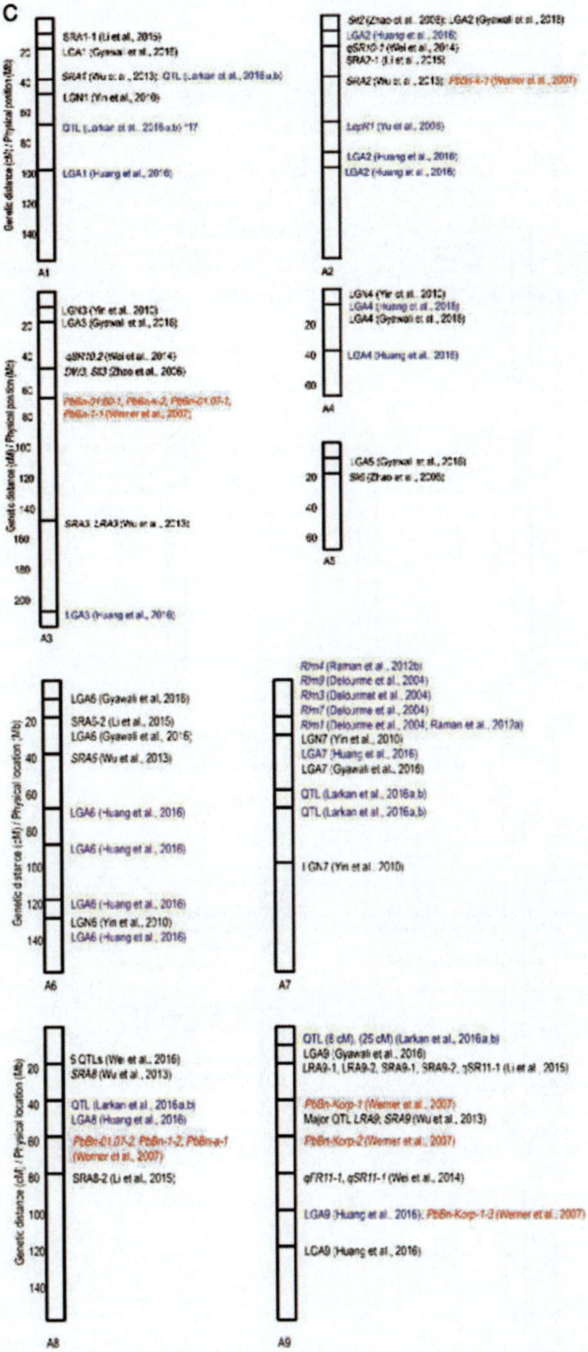

Fig. 11.8 (continued) sclerotiorum, blue font *Leptosphaeria maculans*, red font *P. brassicae,* and bold black font *Hyaloperonospora parasitica*. Boxed *R* gene is cloned. Where genetic distance applies, the chromosome position is stated in parentheses at the end of the locus. *1, Independent

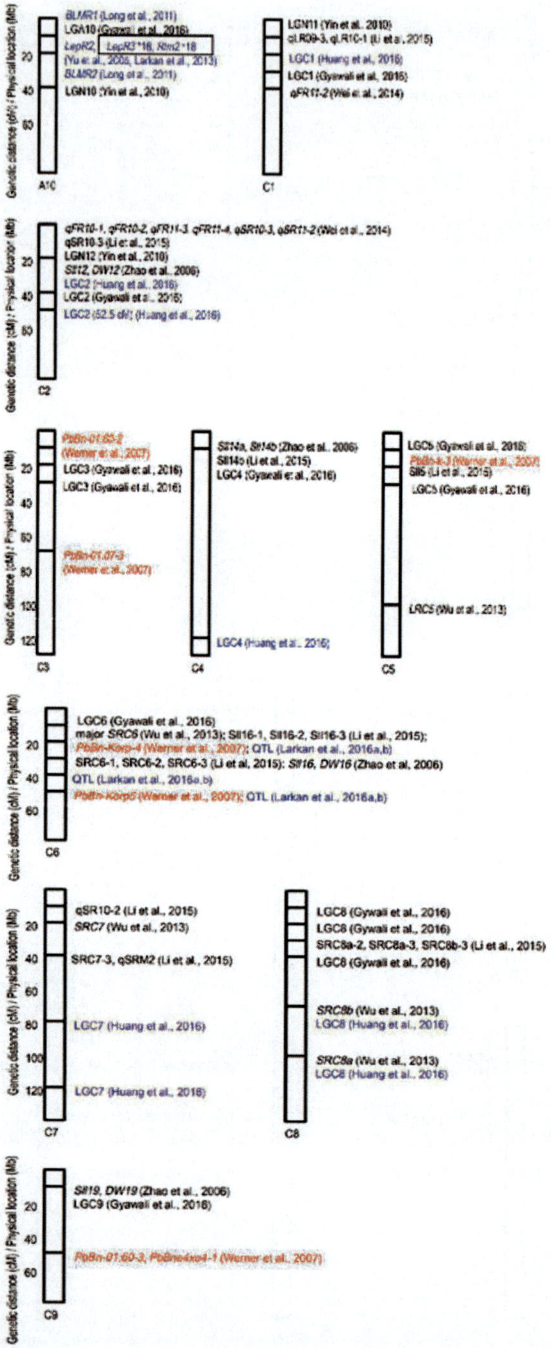

Fig. 11.8 (continued) minor loci, homologous to *A. thaliana* chromosome 3; *2, *Crr3* is independent of *Crr1* and *Crr2;* *3, Syntenic with *A. thaliana* chromosome 4. *Crr1* and *Crr2* also share the same syntenic region; *4, Close to *CRa* and *CRb*, also known as *Rcr1* (Chu et al. 2014), predicted as

11.19.1 Gene Sequencing

Recent advances in genome sequencing and assembly have allowed us to tap into the rich genetic resources from different crop species and their wild relatives. A reference genome is useful to align and compare all the different QTL mapping experiments. By having a reference genome, the physical location of the markers in QTL maps that are linked with candidate genes can be determined. The genome sequence of the *Brassica* diploids, *B. rapa* (Wang et al. 2011; Cai et al. 2017), *B. oleracea* (Parkin et al. 2014; Liu et al. 2014), and *B. nigra* (Yang et al. 2016) and the amphidiploids *B. napus* (Chalhoub et al. 2014; Bayer et al. 2017), *B. juncea* (Yang et al. 2016) have been published. This along with the advanced genotyping tools such as the 60K *Brassica Infinium* SNP array (Clarke et al. 2016) and Genotyping-by-Sequencing (GBS) (Bus et al. 2012; Bayer et al. 2015; Cheng et al. 2016; Scheben et al. 2017a), have allowed rapid identification of genome-wide QTL and GWAS regions for R genes for Clubroot (Lee et al. 2016; Li et al. 2016).

Genome sequencing in Brassicas has now been expanded to include the *B. oleracea* pan-genome where genomes of nine lines were assembled and compared for structural variation (Golicz et al. 2016). Pan-genome sequencing is an important advance, as the diversity of a species is not represented by a single cultivar, and in *B. oleracea* only 81% genes were present in all lines. Of the 19% variable genes, those associated with disease resistance were found to be among the most prevalent. Therefore, in identification of resistance genes, research should look at the variable genes, not only those in the reference genome sequence. Genome re-sequencing (Cheng et al. 2016) and pan-genomics (Golicz et al. 2016) approaches are promising tools to make identification, characterization, and cloning of candidate R genes more

Fig. 11.8 (continued) TIR-NBS-LRR (Yu et al. 2016); *5, Independent of *Crr1*, *Crr2*, and *Crr3*; *6, Two genes, *Crr1a* and *Crr1b* were identified at originally single *Crr1* locus; *7, All at the same locus and are identical to major QTL *BraDM* (Yu et al. 2009); *8, Candidate gene, *Bra016457*, is serine/threonine (S/T) kinase. The major QTL in *7 and *8 lie between 17.6 and 17.8 Mb on chromosome R8; *9, Syntenic with *CRc* on chromosome R2; *10, Strong collinearity with chromosome A2 in *B. rapa* where *CRc* gene is located. Also overlap with *pb-Bo (Anju)1*; *11, Identified by Voorrips et al. (1997) but mapped on chromosome O3 by Nagaoka et al. (2010); *12, *CRQTL-GN_2* (15.5–24.3 cM) and *CRQTL_YC* (8.5–38.1 cM) have similar location. Syntenic regions for these two QTL are on chromosome R3 (*Crr3*, *CRk*, *CRa*, and *CRb*), R8 (*Crr1*); *13, Same linkage group with QTL-LG1 (Moriguchi et al. 1999); *14, May be the same locus; *15, Closely linked to *CRb*, homologous to chromosome R3; *16, Adult resistance; *17, Field resistance for *L. maculans*. Candidate gene is cysteine-rich RLK; *18, *LepR3* and *Rlm2* are allelic variants (Neik et al. 2017) achievable and reliable (Fig. 11.9). The pan-genome approach to study structural variation of the genes in *Brassica* genomes, with whole-genome gene expression and methylation studies has already proved very useful toward uncovering the structure, function, and evolutionary origin of R genes. Copy number variation (CNV) has already been studied for the AP2/ERF super family involved in plant stress tolerance in *B. napus*, CNV of R genes in *Brassica* is only now being

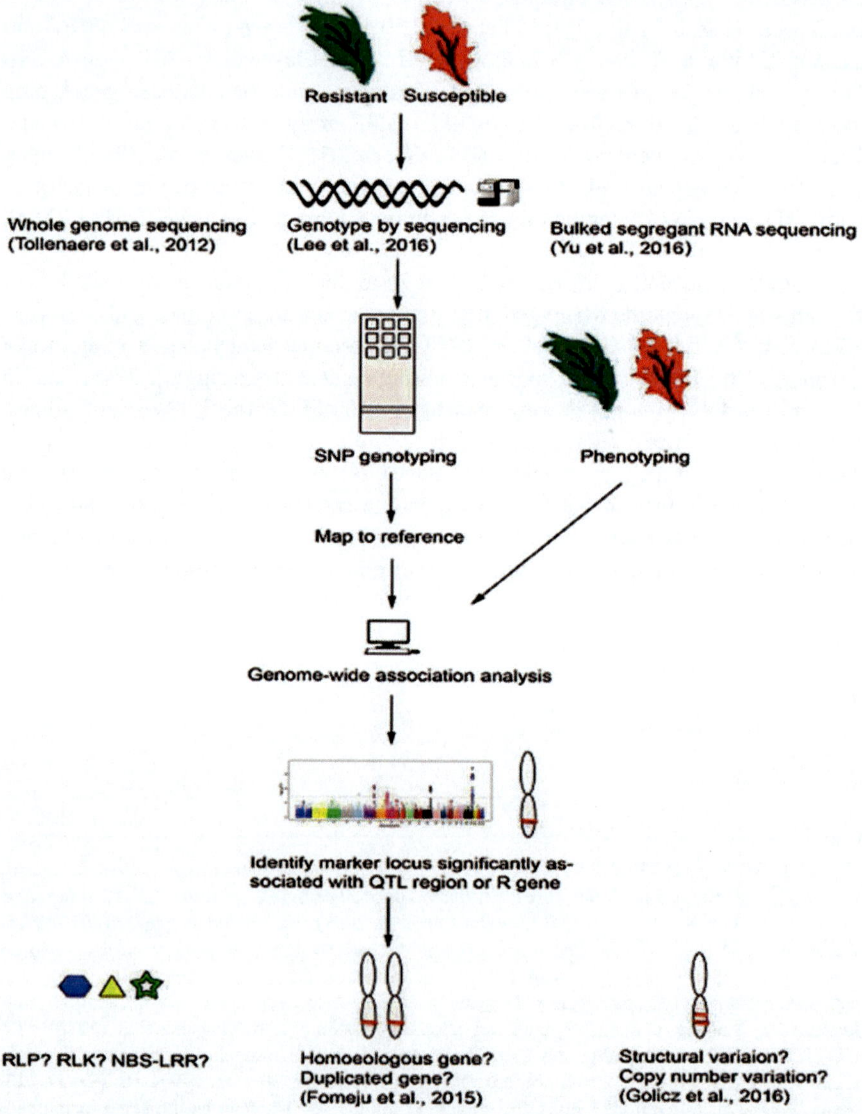

Fig. 11.9 Identification of *R* genes in *Brassica napus* using next generation sequencing method (Neik et al. 2017)

investigated (Batley et al. 2016) and has been implicated in important crops such as cereals (maize and rice), solanaceae, and soybean (Springer et al. 2009; McHale et al. 2012; Xu et al. 2012; Saxena et al. 2014; Wei et al. 2016). This gene expansion through structural variation allows the plant to adapt to rapid changes in the environment also (Cheung et al. 2009).

11.19.2 Genome Complexity

Brassica napus is a relatively young crop in evolutionary history that results from an accumulated series of polyploidization events. Prior to hybridization, the whole enome triplication in the diploids (*B. rapa* and *B. oleracea*) was preceded by three events of whole-genome duplication in the ancestor genome that have further contributed to complicated gene rearrangement, duplication, and loss. Despite the genome complexity, *Brassica* species have become models to study evolutionary events and how rich genetic resources can be exploited from the plant (Pires and Gaeta 2011; Liu et al. 2014) needs to be studied.

Domesticated *B. napus* has a lower genetic diversity compared to its diploid progenitors *B. rapa* and *B. oleracea*, evidenced by similar gene order and content between the three species due to strong genetic bottlenecks during cultivation. By contrast, homoeologous rearrangements/recombination occurs more frequently in resynthesized *B. napus*, thereby contributing greatly toward genetic diversity, which provides a good source of R genes for breeding resistant *B. napus* cultivars. The efforts were made to produce introgressed *B. napus* in a novel way, from *B. juncea* (A genome source) and *B. carinata* (C genome source) (Chatterjee et al. 2016) (rather than from *B. oleracea* and *B. rapa*) from where more sources of R genes can then be utilized and identified. While in *B. napus*, the homoeologous genes are often co-expressed (Chen 2007), there could also be bias in homoeolog expression, as found in *B. juncea* (Yang et al. 2016) and *B. rapa* (Cheng et al. 2012). How the different levels of gene expression affect the stability of R genes remains to be determined. The fate of homoeologous genes, whether they are silenced, expressed, or both copies diverge where the gene function may or may not change (sub- and neo-functionalization) is a result of the genetic and epigenetic interactions between redundant genes (Comai 2005), biased gene expression (Yang et al. 2016), gene fractionation (Subramaniam et al. 2013) or accelerated amino acid sequence evolution and positive selection (Liu and Adams 2010). These polyploidy events could potentially result in a selective advantage, the higher seed yield in *B. napus* (Osborn et al. 2003).

Although race-specific R genes are easier to manipulate for breeding resistant crops compared to polygenic resistance, which involves many genes and the effect of environmental factors, the identification of R genes in *B. napus* still remains a challenge due to the complexity of its genome. The success stories of cloning *Crr1a* and *CRa* (Clubroot R genes encoding TIR-NB-LRR) from *B. rapa* and *LepR3/Rlm2* in *B. napus*, revealed that the R genes are organized as members of allelic variants or tandem repeats. The clustering effect of the R genes for these two pathogens provides further evidence of complexity of the genome organization. Clustering of R genes is also prevalent in association with Clubroot, Blackleg, and Sclerotinia Stem Rot resistance on both *B. rapa* and *B. napus* Such gene clustering effects are common in plant genomes for quick adaptive response to pathogens via recombination events. Clustering of R genes may potentially enable recognition of different pathogen races carrying different avirulence genes, such as in clubroot R genes on chromosome A8 (Pang et al. 2014). Epistatic interaction was also observed for clubroot in *B. napus* (Manzanares-Dauleux et al. 2000a, b), where a single dominant resistance gene does not explain the resistance outcome in its entirety. It was also reported that the same genomic region in *B. napus* can act as both QTL and/or major gene based on different pathotypes of Clubroot (Manzanares-Dauleux et al. 2000b; Rocherieux et al. 2004). The clustering and duplication of R genes can also cause problems in R-gene identification as these genes tend to collapse in genome sequence assemblies. Reliance on the single reference genome sequence can therefore hamper efforts and future work should incorporate novel technologies such as isolated chromosome sequencing or targeted amplification of genomic regions of interest (Neik et al. 2017).

11.19.3 Pathotype, R-gene Classification, and Nomenclature

The usefulness of pathotype classification became obscured when research groups used unknown, different and/or universally un-obtainable sources of host differentials that prevented application of a universal and consistent set of host differentials at international level. Application of different host genotype differentials not commonly available across different locations and/or countries has meant findings in terms of definition of pathogen races/pathotypes is, at best, of limited use and then only to that particular locality. Use of pathogen isolates without appropriate pathotype/race classification, can easily lead to misleading R-gene identification when it comes to qualitative or quantitative resistance classifications and nomenclature. To overcome this, an additional differential/ host differential set was established for Clubroot in Japan using local Chinese cabbage cultivars and local field isolates (Kuginuki et al. 1999; Hatakeyama et al. 2006). A new pathogen isolates carrying new avirulence genes emerge as a normal consequence of pathogen evolutionary dynamics, host differential sets need to be updated and standardized to enable them to effectively delineate future new pathotypes/races. A standardized ideal host differential set with isogenic lines for each *Brassica* species should be developed and used internationally to get true picture of clubroot pathotypes. Application of different scoring systems can also lead to complications in the identification of R genes. It is also imperative that a robust race/pathotype differentiation system be developed to separate different and group similar races/ pathotypes. So far no standard system has been used to designate, and nominate pathotype and a resistant gene in *P. brassicae-Brassica* species host-patho system acceptable at international level. A robust system should indicate which pathotype has recognized which resistant gene from which *Brassica* species. At present,

P. brassicae pathotypes have been designated as serial numbers (1, 2, 3...), triplicates (2/3/15; 4/12/14...), and letters (P1, P2, P3, Pb: 1, Pb: 2, Pb: 3, Pb: 4 or A, B, C, D) without following any standard system. Simultaneously resistant genes have been designated as *CRa, CRb, RCr4, RCr* 9, *Crr1, Crr2, CRQTL-GN1, CRQTL–GN-2, QTL1, QTL* 2, *Pb-Bo3, PB-Bo9a,* etc. identified and mapped from different *Brassica* species (Table 11.12). The host resistant gene should indicate which pathotype has recognized which resistant gene. Therefore, it is proposed that a pathotype may be designated as BrPb1, BrPb2... from *B. rapa*, BnPb1, Bn Pb2, ... from *B. napus*, BjPb1, BjPb2, ... from *B. juncea*, BoPb1, BoPb2, ... from *B. oleracea* so on.

Similarly a resistant gene may be designated as BrCR1, BrCr2... from *B. rapa;* BnCR1, BnCr2... from *B. napus;* BjCR1, BjCR2... from *B. juncea,* and BoCR1, BoCR2 ... from *B. oleracea*. It is essential to follow a uniform system all over the world to determine. (1) Status of earlier reports races/pathotypes and R genes and their distribution in different *Brassica* species at regional level; (2) Emergence and evolution of new races/pathotypes; (3) Virulence spectrum of each race in terms of how many R genes are attacked by which pathotype; (4) Durability of R genes against each pathotype; and (5) Favorable combinations of R genes for effectiveness and durability of clubroot-resistant cultivars. With the advances in modern molecular techniques, now it has become easy and possible to detect, identity, map, and clone R genes and virulence genes which has facilitated breeding of *Brassica* cvs. with desired attributes and genes to increase productivity of crucifers. Moreover, techniques like gene editing and gene silencing allows the insertion of desired gene and deletion of undesirable gene to breed an ideal genotype of crucifer. With co-operation and international co-ordination including informatics and available worldwide database the desired goals can be achieved.

11.19.4 Breeding Resistant Cultivars

Continuous efforts are on-going to breed resistant *Brassica* species cultivars through introgression. *B. rapa* remains a good source of race-specific R genes that often result in high resistance against clubroot. For development of clubroot resistance, a single dominant gene from resistant *B. rapa* turnip morphotypes was successfully introduced into Chinese cabbage (Yoshikawa 1980). Transfer of a Clubroot resistance gene was also performed from *B. napus* into *B. oleracea*, with the resistance gene believed to also be carried in the A genome of *B. rapa* (Chiang et al. 1980). In Canada, efforts to transfer a Clubroot resistance gene from the *B. napus* hybrid "Mendel" into a local susceptible spring *B. napus* cultivar was promising (Rahman et al. 2011). Transfer of a clubroot R gene was also made from *B. rapa* Chinese cabbage "Qulihuang" into *B. napus* "Topas" using marker-assisted selection (Hirani et al. 2016). Within breeding programmes, it is important to make informed decisions about the source of novel R genes and which parental genomes to cross, because the R gene from the parents may be masked in the progenies of the resynthesized genomes due to one or more epistatic interactions, such as was found for a clubroot R gene (Werner et al. 2008). These challenges can be overcome by the latest genome editing techniques, such as CRISPR-Cas9 (Scheben et al. 2017b), as was recently reported in maize with precise mutation on the targeted allele (Svitashev et al. 2016; Liang et al. 2017).

To breed resistant cultivars, it is important to consider the fitness cost to pathogen isolates when they do overcome host resistances. The study on the pathogen avirulence/virulence genes in the field are adapted to the host cultivar and determine the genetic changes in the pathotype that cause heavy fitness consequences (Vera Cruz et al. 2000; Delmas et al. 2016). It is also important to consider the host R gene as selection pressure on the pathogen can affect the stability of the R gene. In terms of the linked genes in *L. maculans*, *AvrLm1*, and *AvrLm6*, high selection pressure on one *Avr* gene can cause a change in allele frequency of the other (Brun et al. 2010; Van De Wouw et al. 2010). Understanding the evolutionary potential between the pathogen and the host is essential in developing a durable resistant *Brassica* crop (McDonald and Linde 2002).

11.19.5 Molecular Identification of Pathotype and Pathogen Elicitors/Effectors

To control disease outbreaks in crops, it is imperative that pathogen population variability in the field can be quickly identified. This can be achieved by having a full set of genetic markers for different pathogen pathotypes, not only contained within the local fields but universally. The initiatives for *P. brassicae* showed that *Cr811* was a good candidate gene to differentiate pathotype 5 from the rest of the pathotypes (Feng et al. 2016). Molecular characterization of pathogen effectors is gaining momentum. With better genome assemblies and higher gene prediction accuracies, it will soon have better insights into the structures and functions of effector proteins, which allow more efficient screening and identification of R genes (Gibriel et al. 2016). There should be a focus on studies that provide additional molecular details at the gene level which would further our understanding of the factors leading to virulence and susceptibility associated with R-gene mediated resistance (Neik et al. 2017).

11.19.6 The Genetics of CR Genes and Marker-Assisted Breeding

To develop clubroot-resistant germplasm with a pool of CR genes for long-term management of this disease, diverse genetic sources must be found and characterized. This is important to enable rotation of cultivars with different CR genes or for pyramiding CR genes within a single cultivar. Pyramiding the major resistance genes in Chinese cabbage is known to confer high resistance against multiple pathotypes (Matsumoto et al. 2012). With advances in molecular technologies in the last two decades, at least nine CR loci in the A genome of *B. rapa* (Piao et al. 2004; Suwabe et al. 2003, 2006, 2012; Saito et al. 2006;

Sakamoto et al. 2008; Ueno et al. 2012; Hatakeyama et al. 2013), and several quantitative trait loci (QTL) in *B. oleracea* (C genome) (Figdore et al. 1993; Voorrips et al. 1997; Moriguchi et al. 1999; Rocherieux et al. 2004; Nagaoka et al. 2010) and *B. napus* (AC genome) (Manzanares-Dauleux et al. 2000b; Werner et al. 2008) have been identified. Chu et al. (2013) mapped a CR gene of *B. rapa* ssp. *chinensis* (Pak Choy) "Flower Nabana" using 318 SSR markers in a mapping population of 1299 backcross plants segregating for resistance to pathotype 3. A 1: 1 segregation ($\chi 2 = 2.98$, $P = 0.084$) in the mapping population suggested that the resistance in "Flower Nabana" is controlled by a dominant gene. This gene was designated as *Rpb1* and genetically mapped to *B. rapa* chromosome A3. Six molecular markers flanking this resistance gene, with three markers on each side, were identified. The markers sR6340I and sN8591 are the closest ones from each side, at a distance of 0.77 and 0.54 cM to *Rpb1*, respectively. Based on the location of these markers, and a sequence BLAST search of the four CR loci *CRa*, *CRb*, *CRk*, and *Crr3* reported to be in the linkage group A3 (Piao et al. 2004; Saito et al. 2006; Sakamoto et al. 2008; Piao et al. 2009), the *Rpb1* gene is seemingly located very close to *CRa* (Ueno et al. 2012) or *CRb* (Kato et al. 2012).

Often R genes afford resistance against diverse pathogens through interactions with pathogen-encoded proteins, but individual R genes can be pathogen/pathotype/race specific. The Toll/interleukin-1 receptor (TIR)-nucleotide-binding site (NBS)–Leucine-rich repeat (NBS-LRR) class is the largest among the five well-characterized classes of R genes (Pto; TIR-NBS-LRR; Coiled-Coil-NBS-NRR; Cf; Xa21 (Martin et al. 2003; Bernoux et al. 2011). Two *TIR-NBS-LRR* genes have been identified as the genes underlying two distinct CR loci in *B. rapa.* Ueno et al. (2012) cloned *CRa* present on chromosome A3. An intact *CRa* confers resistance to *P. brassicae* (isolate M85) in *B. rapa* while insertions/deletions that abrogate the gene also abolish the resistance. Even a single amino acid change in critical regions abolished the resistance trait. In a loss of function analysis, randomly mutagenized derivatives of the resistant line that were found to be sensitive to *P. brassicae* were also found to have undergone mutations in the *CRa* located *NBS-LRR* gene. The other resistant *B. rapa* accessions, such as "Debra" "Gelria R," and two commercially grown Chinese cabbage cultivars, "Ryutoku" and "CR Shinki," are likely to carry the functional *TIR-NBS-LRR* gene. The second *TIR-NBSLRR* associated with *Crr1a* in chromosome A8 was isolated by Hatakeyama et al. (2013). The transgenic expression of this gene confers clubroot resistance in an otherwise susceptible *B. rapa* line. There are some common features in the two functionally characterized CR genes of *B. rapa,* both are expressed at higher levels in the resistant lines than in the susceptible types, and interestingly, both genes showed alternative splicing. Alternative splicing of *TIR-NBS-LRR* genes, in addition to the production of normal transcripts, has been found to be necessary for complete resistance against Tobacco mosaic virus in tobacco (Dinesh-Kumar and Baker 2000) and against *Pseudomonas syringae* pv. tomato strain DC3000 in *Arabidopsis* (Zhang and Gassmann 2003).

As R genes are race or pathotype-specific, a pathotype can evolve to "defeat" an R gene until a new R gene(s) arises by mutation(s) to overcome the defeated gene. Some *NBS-LRR* genes mediate resistance in a more complex manner than the classical gene–gene interaction (one host R gene—one pathogen gene). An example is the requirement of two *NBS-LRR* genes for resistance against wheat leaf rust (*Puccinia triticina*) (Loutre et al. 2009).

Given that R genes are central players in adaptive resistance in plants, there are hundreds of *NBS-LRR* genes in a typical plant genome and these appear to have arisen by gene duplications and dispersions in the genome. It is very likely that additional CR loci will also be identified to carry *NBS-LRR*-encoding genes. This has implications for breeding resistance in *Brassica*, and especially for gene pyramiding. To manage clubroot in *Brassica* crops, there is requirement of specific R-gene alleles in various combinations. In this regard, it is useful to have a genome-wide view of the *TIR-NBS-LRR* genes as the landscape of potential sources of CR genes. Mun et al. (2009) identified 62 *TIR-NBS-LRR* genes in the *B. rapa* genome. Since then, additional *TIR-NBS-LRR* genes have been found in the genome (Rahman et al. 2014). Out of nine, two functionally known CR genes *CRa* and *Crr1a* are in Group I and Group IV, respectively. Group I contains nine of the 18 *TIR-NBS-LRR* genes found in chromosome A3. It is likely that additional NBS-LRR will be positionally matched to the other CR loci that have been mapped (*Crr2* at A1, *CRc* at A2, *Crr4* at A6 (Piao et al. 2009) but not yet characterized on a molecular level (Fig. 11.10).

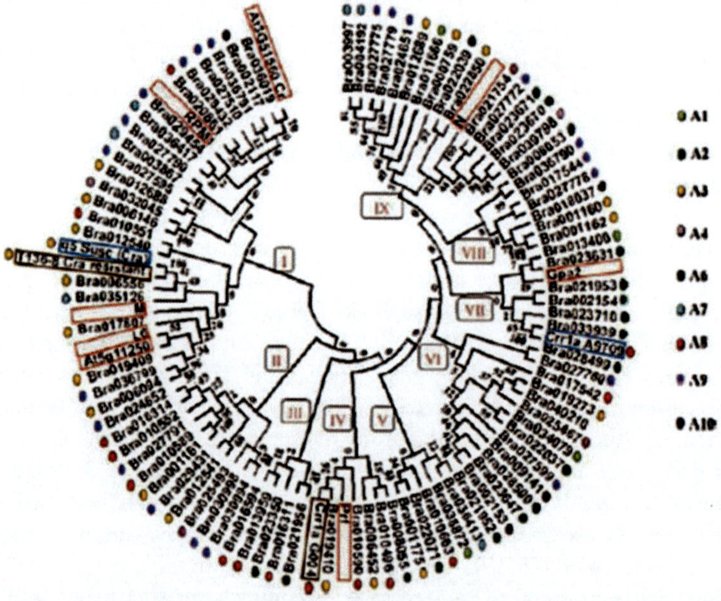

Fig. 11.10 Phylogenetic relationship of *TIR-NBS-LRR* gene products in the *Brassica rapa* genome (www.brassicadb.org) along with some R genes from other species: *AT3G51560, AT5G11250, N (U15605), L6 (U27081), M (U73916), Prf (U65391), Gpa2 (AF195939),* and *RPM1 (X87851).* Known R genes are marked by red rectangles. *CRa* and *Crr1a* are indicated by brown rectangles; their counterparts in susceptible lines are indicated by blue rectangles. Neighbor joining method in MEGA5 (Tamura et al. 2011) was used in the analysis. The chromosomal associations (A1 through A10) of the NBS-LRR of *B. rapa* are as indicated on the right (Rahman et al. 2014)

Arabidopsis is a well-established model plant for *Brassica* crops. The chromosomal regions are syntenic and often the deduced gene products appear very similar. Although most of the *Arabidopsis* accessions are susceptible to *P. brassicae*, some resistant ecotypes have been found (Fuchs and Sacristan 1996; Siemens et al. 2002; Alix et al. 2007; Agarwal et al. 2009; Sharma et al. 2013). Pathotype 3 is an aggressive race on *Arabidopsis* (Sharma et al. 2013). With reference to pathotype 1 resistance, Fuchs and Sacristan (1996) found that *Arabidopsis* ecotypes Tsu-0 and Ze-0 show pathotype-specific dominant resistance to *P. brassicae* isolate "e" and mapped a dominant gene (named *RPB1*) to chromosome 1. A major QTL for CR has been mapped on chromosome 5 along with four other QTL with minor effects. This region has many R genes which are already associated with resistance to other pathogens (*TTR1, RPS4,* and *RRS1*) (Jubault et al. 2008). Given the absence of information on *P. brassicae* effector and host R interactions, *Arabidopsis* is likely to be useful for discerning the molecular aspects of the clubroot pathosystem. This knowledge may in turn provide a direct source of diverse CR genes for transgene deployment or for allele identification in primary and secondary gene pools. In genetic management of diseases, qualitative loci (R genes) are most commonly deployed. Such resistance to blackleg or stem canker in *B. napus* was found to be more durable when combined with quantitative resistance afforded by small-effects loci that are not race specific for resistance (Brun et al. 2010). Alix et al. (2007) found partial resistance to pathotype 1 in Bur-0 and Kn-0 ecotypes and in some mutants which were affected in auxin metabolism. Pasold et al. (2010) reported an association between flavonoid content and disease, but it is unknown if auxin transport, a general stress response (anti-oxidant effects) or some other aspects of flavonoid functions are involved. Gravot et al. (2012) proposed that the enhanced arginase levels found in clubs is a form of negative control over hormone-induced hyperplasia. Transcriptome analysis has been conducted in pathotype 3-susceptible and resistant *B. napus* that showed expression modulation in clusters of biochemically related genes. The emerging patterns from *Arabidopsis* and *B. napus* are complex, underscoring the challenges in deciphering the genes contributing to resistance. Nevertheless, there are opportunities to learn about CR in *Arabidopsis*, and from the use of modern genomic approaches in *Arabidopsis* and *Brassica* crop plants, to find genetic networks that are pertinent to resistance at qualitative and quantitative levels and to build durable resistance in the long term (Rahman et al. 2014)

11.19.7 Pyramiding R-genes Against *P. brassicae* in *B. napus*

PbBa8.1 and *CRb* are two clubroot-resistant genes that are important for canola breeding. Shah et al. (2020) conducted comparative RNA sequencing (RNA-Seq) analysis between 618R and its parental lines (305R and 409R) in order to uncover the transcriptomic response of the superior defense mechanisms of 618R and to determine how these two different resistant genes coordinate with each other. It was elucidated that the number and expression of differentially expressed genes (DEGs) in 618R are significantly higher than in the parental lines, and *PbBa8*.1 shares more DEGs and plays a dominant role in the pyramided line. The common DEGs among the lines largely exhibit non-additive expression patterns and enrichment in resistance pathways. Among the enriched pathways, plant–pathogen interaction, plant hormone signaling transduction, and secondary metabolites are the key observation. However, the expressions of the salicylic acid (SA) signaling pathway and reactive oxygen species (ROS) appear to be crucial regulatory components in defense response. These findings provide comprehensive transcriptomic insight into understanding the interactions of resistance gene pyramids in single lines and can facilitate the breeding of improved resistance in *Brassica napus*.

References

Adhikari KC, McDonald MR, Gossen BD (2012) reaction to *Plasmidiophora brassicae* pathotype 6 in lines of *Brassica* vegetables, Wisconsin fast plants, and canola. Hortic Sci 47(3):374–377

Agarwal A, Kaul V, Faggian R, Cahill DM (2009) Development and use of a model system to monitor clubroot disease progression with an Australian field population of *Plasmodiophora brassicae*. Austr Plant Pathol 38(2):120–127

Agarwal A, Kaul V, Faggian R, Rookes JE, Cahill DM, Ludwig-Müller J (2011) Analysis of global host gene expression during the primary phase of the *Arabidopsis thaliana-Plasmodiophora brassicae* interaction. Funct Plant Biol 38:1–16

Ahmad P, Hakeem KUR, Kumar A, Ashraf M, Akram NA (2012) Salt-induced changes in photosynthetic activity and oxidative defense system of three cultivars of mustard (*Brassica juncea* L). Afr J Biotechnol 11(11):2694–2703

Akaba M, Kaneko Y, Hatakeyama K, Ishida M, Bang SW, Matsuzawa Y (2008) Evaluation of clubroot resistance in *Raphanus sativus* monosomic addition lines. Breed Res 10:205

Alix K, Lariagon C, Delourme R, Manzanares-Dauleux MJ (2007) Exploiting natural genetic diversity and mutant resources of *Arabidopsis thaliana* to study the *A. thaliana–Plasmodiophora brassicae* interaction. Plant Breed 126(2):218–221

Ayers GW (1972) Races of *Plasmodiphora brassicae*: infecting crucifer crops in Canada. Can Plant Dis Surv 52(3):77–81

Ayers GW, Lelacheur KE (1972) Genetics of resistance in rutabaga to races of *Plasmodiophora brassicae*. Can J Plant Sci 52:897–900

Baggett JR (1983) Clubroot-resistant cabbage breeding lines Oregon 100, 123, 140, and 142. Hortic Sci 18:112–114

Baggett JR, Kean D (1985) Clubroot-resistant broccoli breeding lines OSU CR-2 to OSU CR-8. Hortic Sci 20:784–785

Batley J, Dolatabadian A, Yang H, Severn-Ellis A, Alamery S, Tollenaere R, Bayer P, Hurgobin B, Golicz AA, Edwards D (2016) The more the merrier? Investigating copy number variation in *Brassica* disease resistance. In: Plant and animal genome conference (PAG) Asia (Singapore)

Bayer PE, Ruperao P, Mason AS, Stiller J, Chan CKK, Hayashi S, Long Y, Meng J, Sutton T, Visendi P, Varshney RK, Batley J, Edwards D (2015) High-resolution skim genotyping by sequencing reveals the distribution of crossovers and gene conversions in *Cicer arietinum* and *Brassica napus*. Theor Appl Genet 128:1039–1047

Bayer PE, Hurgobin B, Golicz AA, Chan CKK, Yuan Y, Lee H, Renton M, Meng J, Li R, Long Y, Zou J, Bancroft I, Chalhoub B, King GJ, Batley J, Edwards D (2017) Assembly and comparison of two closely related *Brassica napus* genomes. Plant Biotechnol J. https://doi.org/10.1111/pbi.12742

Bernoux M, Ellis JG, Dodds PN (2011) New insights in plant immunity signalling activation. Curr Opin Plant Biol 14:512–518

Bradshaw JE, Gemmell DJ, Wilson RN (1997) Transfer of resistance to clubroot (*Plasmodiophora brassicae*) to swedes (*Brassica napus* L. var. *napobrassica* Peterm) from *B. rapa*. Ann Appl Biol 130:337–348

Brodmann D, Schuller A, Ludwig-Muller J, Aeschbacher RA, Wiemken A, Boller T, Wingler A (2002) Induction of trehalase in *Arabidopsis* plants infected with the trehalose-producing pathogen *Plasmodiophora brassicae*. Mol Plant-Microbe Interact 15(7):693–700

Brun H, Chèvre AM, Fitt BDL, Powers S, Besnard AL, Ermel M et al (2010) Quantitative resistance increases the durability of qualitative resistance to *Leptosphaeria maculans* in *Brassica napus*. New Phytol 185:285–299

Buczacki ST, Toxopeus H, Johnston P, Dixon TD, Hobolt LA (1975) Study of physiological specialisation in *Plasmodiophora brassicae*: proposals for attempted rationalisation through an international approach. Trans Br Mycol Soc 65:295–303

Bus A, Hecht J, Huettel B, Reinhardt R, Stich B (2012) High throughput polymorphism detection and genotyping in *Brassica napus* using next-generation RAD sequencing. BMC Genomics 13:281

Cai C, Wang X, Liu B, Wu J, Liang J, Cui Y, Cheng F, Wang X (2017) *Brassica rapa* genome 2.0: a reference upgrade through sequence re-assembly and gene re-annotation. Mol Plant 10:649–651

Cao T, Srivastava S, Rahman MH, Kav NV, Hotte N, Deyholos M, Strelkov SE (2008) Proteome-level changes in the roots of *Brassica napus* as a result of *Plasmodiophora brassicae* infection. Plant Sci 174(1):97–115

Cao T, Manolii VP, Hwang SF, Howard RJ, Strelkov SE (2009) Virulence and spread of *Plasmodiophora brassicae* [clubroot] in Alberta, Canada. Can J Plant Pathol 31:321–329

Carlsson M, Von Bothmer R, Merker A (2004) Screening and evaluation of resistance to downy mildew (*Peronospora parasitica*) and clubroot (*Plasmodiophora brassicae*) in genetic resources of *Brassica oleracea*. Hereditas 141:293–300

Cavell A, Lydiate D, Parkin IAP, Dean C, Trick M (1998) Colinearity between 30-centimorgan segment of *Arabidopsis thaliana* chromosome 4 and duplicated regions within the *Brassica napus* genome. Genome 41:62–69

Chalhoub B, Denoeud F, Liu S, Parkin IA, Tang H, Wang X, Chiquet J, Belcram H, Tong C, Samans B, Corréa M, Silva CD, Just J, Falentin C, Koh CS, Clainche IL, Bernard M, Bento P, Noel B, Labadie K, Alberti A, Charles M, Arnaud D, Guo H, Daviaud C, Alamery S, Jabbari K, Zhao M, Edger PP, Chelaifa H, Tack D, Lassalle G, Mestiri I, Schnel N, Paslier M-CL, Fan G, Renault V, Bayer PE, Golicz AA, Manoli S, Lee T-H, Thi VHD, Chalabi S, Hu Q, Fan C, Tollenaere R, Lu Y, Battail C, Shen J, Sidebottom CHD, Wang X, Canaguier A, Chauveau A, Bérard A, Deniot G, Guan M, Liu Z, Sun F, Lim YP, Lyons E, Town CD, Bancroft I, Wang X, Meng J, Ma J, Pires JC, King GJ, Brunel D, Delourme R, Renard M, Aury J-M, Adams KL, Batley J, Snowdon RJ, Tost J, Edwards D, Zhou Y, Hua W, Sharpe AG, Paterson AH, Guan C, Wincker P (2014) Early allopolyploid evolution in the post-neolithic *Brassica napus* oilseed genome. Science 345:950–953

Chatterjee D, Banga S, Gupta M, Bharti S, Salisbury PA, Banga SS (2016) Resynthesis of *Brassica napus* through hybridization between *B. juncea* and *B. carinata*. Theor Appl Genet 129:977–990

Chattopadhyay AK, Moitra AK, Bhunia CK (2001) Evaluation of *Brassica* species for resistance to *Plasmodiophora brassicae* causing club root of rapeseed mustard. Indian Phytopathol 54 (1):131–132

Chen ZJ (2007) Genetic and epigenetic mechanisms for gene expression and phenotypic variation in plant polyploids. Annu Rev Plant Biol 58:377–406

Chen J, Jing J, Zhan Z, Zhang T, Zhang C, Piao Z (2013) Identification of novel QTLs for isolate-specific partial resistance to *Plasmodiophora brassicae* in *Brassica rapa*. PLoS ONE 8:e85307

Chen L, Zhang X, Xu H, Song B, Fan X (2016) Introgression of clubroot resistance into an elite pak choi inbred line through marker-assisted introgression breeding. Plant Breed 135:471–475

Chen J, Piaoa Y, Liua Y, Lia X, Piao Z (2018) Genome-wide identification and expression analysis of chitinase gene family in *Brassica rapa* reveals its role in clubroot resistance. Plant Sci 270:257–267

Cheng F, Liu S, Wu J, Fang L, Sun S, Liu B, Li P, Hua W, Wang X (2011) BRAD, the genetics and genomics database for *Brassica* plants. BMC Plant Biol 11(1):136

Cheng F, Wu J, Fang L, Sun S, Liu B, Lin K, Bonnema G, Wang X (2012) Biased gene fractionation and dominant gene expression among the subgenomes of *Brassica rapa*. PLoS ONE 7:e36442

Cheng F, Wu J, Cai C, Fu L, Liang J, Borm T, Zhuang M, Zhang Y, Zhang F, Bonnema G, Wang X (2016) Genome resequencing and comparative variome analysis in a *Brassica rapa* and *Brassica oleracea* collection. Sci Data 3:160119

Cheung F, Trick M, Drou N, Lim YP, Park JY, Kwon SJ, Kim J-A, Scott R, Pires JC, Paterson AH, Town C, Bancroft I (2009) Comparative analysis between homoeologous genome segments of *Brassica napus* and its progenitor species reveals extensive sequence-level divergence. Plant Cell 21:1912–1928

Chiang MS, Crete R (1970) Inheritance of clubroot resistance in cabbage (*Brassica oleracea* L. var *capitata* L.). Can J Genet Cytol 12:253–256

Chiang MS, Crete R (1976) Diallel analysis of the inheritance of resistance to race 6 of *Plasmodiophora brassicae* in cabbage. Can J Plant Sci 56:865–868

Chiang MS, Crete R (1983) Transfer of resistance to race 2 of *Plasmodiophora brassicae* from *Brassica napus* to cabbage (*B. oleracea* spp. *capitata*). V. The inheritance of resistance. Euphytica 32:479–483

Chiang MS, Crete R (1985) Acadie: a clubroot resistant cabbage cultivar. Can J Plant Sci 65:223–235

Chiang MS, Crete R (1989) Richelain: a clubroot resistant cabbage cultivar. Can J Plant Sci 69:337–340

Chiang MS, Chiang F, Grant GH (1977) Transfer of resistance to race 2 of *Plasmodiophora brassicae* from *Brassica napus* to cabbage (*B. oleracea* var. *capitata*). Euphytica 26:319–336

Chiang BY, Chiang MS, Grant WF, Crete R (1980) Transfer of resistance to race 2 of *Plasmodiophora brassicae* from *Brassica napus* to cabbage (*B. oleracea* spp. *capitata*). IV. A resistant 18- chromosome B1 plant and its B2 progenies. Euphytica 29:47–55

Cho KS, Han YH, Lee JT, Hur EJ, Yang TJ, Woo JG (2002) Pathogenic differentiation of *Plasmodiophora brassicae* and selection of Chinese cabbage cultivars resistant to clubroot disease in highland. Korean J Breed 34(3):168–173

Chu M, Yu F, Falk KC, Liu X, Zhang X, Chang A, Peng G (2013) Identification of the clubroot resistance gene *rpb1* and introgression of the resistance into canola breeding lines using a marker-assisted approach. Acta Hortic 1005:599–605

Chu M, Song T, Falk KC, Zhang X, Liu X, Chang A, Lahlali R, McGregor L, Gossen BD, Yu F, Peng G (2014) Fine mapping of *Rcr1* and analyses of its effect on transcriptome patterns during infection by *Plasmodiophora brassicae*. BMC Genomics 15(1):1166

Clarke WE, Higgins EE, Plieske J, Wieseke R, Sidebottom C, Khedikar Y, Batley J, Edwards D, Meng J, Li R, Lawley CT, Pauquet J, Laga B, Cheung W, Iniguez-Luy F, Dyrszka E, Rae S, Stich B, Snowdon RJ, Sharpe AG, Ganal MW, Parkin IAP (2016) A high-density SNP genotyping array for *Brassica napus* and its ancestral diploid species based on optimised selection of single-locus markers in the allotetraploid genome. Theor Appl Genet 129:1887–1899

Colhoun J (1958) Clubroot disease of crucifers caused by *Plasmodiophora brassicae* Woron. A monograph. Phytopathol Pap 3:1–109

Comai L (2005) The advantages and disadvantages of being polyploid. Nat Rev Genet 6:836–846

Crisp P, Crute IR, Sutherland RA, Angell SM, Bloor K, Burgess H, Gordon PL (1989) The exploitation of genetic resources of *Brassica oleracea* in breeding for resistance to clubroot (*Plasmodiophora brassicae*). Euphytica 42(3):215–226

Crute IR (1986) The relationship between *Plasmodiophora brassicae* and its hosts: the application of concepts relating to variation in inter-organismal associations. Adv Plant Pathol 5:1–52

Crute IR, Pink DAC (1989) The characteristics and inheritance of resistance to clubroot in *Brassica oleracea*. Asp Appl Biol 23:57–60

Crute IR, Gray AR, Crisp P, Buczacki ST (1980) Variation in *Plasmodiophora brassicae* and resistance to clubroot disease in Brassicas and allied crops - a critical review. Plant Breed Abst 50:91–104

Crute IR, Phelps K, Barnes A, Buczacki ST, Crips P (1983) The relationship between genotypes of three *Brassica* species and collections of *Plasmodiophora brassicae*. Plant Pathol 32:405–420

de Vallavieille-Pope C, Ali S, Leconte M, Enjalbert J, Delos M, Rouzet J (2012) Virulence dynamics and regional structuring of *Puccinia striiformis* f. sp. *tritici* in France between 1984 and 2009. Plant Dis 96:131–140

de Wit PJ (2007) How plants recognize pathogens and defend themselves. Cell Mol Life Sci 64:2726–2732

Delmas CEL, Fabre F, Jolivet J, Mazet ID, Richart Cervera S, Delière L, Delmotte F (2016) Adaptation of a plant pathogen to partial host resistance: selection for greater aggressiveness in grapevine downy mildew. Evol Appl 9:709–725

Delourme R, Foisset N, Horvais R, Barret P, Champagne G, Cheung WY, Landry BS, Renard M (1998) Characterisation of the radish introgression carrying the Rfo restorer gene for the Ogu-INRA cytoplasmic male sterility in rapeseed (*Brassica napus* L.). Theor Appl Genet 97:129–134

Deora A, Gossen BD, McDonald MR (2012) Infection and development of *Plasmodiophora brassicae* in resistant and susceptible canola cultivars. Can J Plant Pathol 34:239–247

Deora A, Gossen BD, McDonald MR (2013) Cytology of infection, development and expression of resistance to *Plasmodiophora brassicae* in canola. Ann Appl Biol 163:56–71

Devos S, Prinsen E (2006) Plant hormones: a key in clubroot development. Commun Agric Appl Biol Sci 71:869–872

Devos S, Laukens K, Deckers P, Van Der Straeten D, Beeckman T, Inzé D, Van Onckelen H, Witters E, Prinsen E (2006) A hormone and proteome approach to picturing the initial metabolic events during *Plasmodiophora brassicae* infection on *Arabidopsis*. Mol Plant-Microbe Interact 19:1431–1443

Dias JS, Ferreira ME, Williams PH (1993) Screening of Portuguese cole landraces (*Brassica oleracea* L.) with *Peronospora parasitica* and *Plasmodiophora brassicae*. Euphytica 67:135–141

Diederichsen E, Sacristan MD (1994) The use of ovule culture in reciprocal hybridization between *B. campestris* L. and *B. oleracea* L. Plant Breed 113:79–82

Diederichsen E, Sacristán MD (1996) Disease response of resynthesized *Brassica napus* L. lines carrying different combinations of resistance to *Plasmodiophora brassicae* Worn. Plant Breed 115:5–10

Diederichsen E, Wagenblatt B, Schallehn V, Deppe US, Sacristán MD (1996) Transfer of clubroot resistance from resynthesised *Brassica napus* into oilseed rape—identification of race-specific interactions with *Plasmodiophora brassicae*. Acta Hortic 407:423–430

Diederichsen E, Deppe U, Sacristan MD (2003) Characterization of clubroot resistance in recent winter oilseed rape material. In: Proc 11th Intl Rapeseed Cong, vol 1. Copenhagen, Denmark, pp 68–70

Diederichsen E, Beckmann J, Schondelmeier J, Dreyer F (2006) Genetics of clubroot resistance in *Brassica napus* 'Mendel'. Acta Hortic 706:307–311

Diederichsen E, Frauen M, Linders EGA, Hatakayema K, Hirai M (2009) Status and perspectives of clubroot resistance breeding in crucifer crops. J Plant Growth Regul 28:265–281

Diederichsen E, Frauen M, Ludwig-Muller J (2014) Clubroot disease management challenges from a German perspective. Can J Plant Pathol 36:85–98

Dinesh-Kumar SP, Baker BJ (2000) Alternatively spliced N resistance gene transcripts: their possible role in tobacco mosaic virus resistance. Proc Natl Acad Sci U S A 97:1908–1913

Dixon GR (1988) Shetland cabbage as a source of resistance to *Plasmodiophora brassicae*. In: Proc 5th Intl Cong Plant Pathol, Kyoto, Japan, p 178
Dixon GR (2009) The occurrence and impact of *Plasmodiophora brassicae* and clubroot disease. In: Dixon GR (eds) *Plasmodiophora brassicae* (clubroot) – a plant pathogen that alters host growth and productivity. J Plant Growth Regul 28:194–202
Dixon GR, Robinson DL (1986) The susceptibility of *Brassica oleracea* cultivars to clubroot. Plant Pathol 35:101–107
Dixon GR, Doodson JK, Beeney BW, Davies H, Moxon RH (1972) Studies of resistance in swede seed stocks to clubroot (*Plasmodiophora brassicae*). Plant Var Seeds 12:456–463
Dixon GR, Wilson F, Britt CP (1986) Calabrese lines resistant to *Plasmodiophora brassicae* (clubroot). Tests of agrochemicals and cultivars No 7. Ann Appl Biol 104:44–45
Dodds PN, Rathjen JP (2010) Plant immunity: towards an integrated view of plant–pathogen interactions. Nat Rev Genet 11:539–548
Donald EC (2005) The influence of abiotic factors and host plant physiology on the survival and pathology of *Plasmodiophora brassicae* of vegetable *Brassicas*. Ph.D Thesis University of Melbourne, Australia, p 215
Donald EC, Porter IJ (2009) Integrated control of clubroot. J Plant Growth Regul 28:289–303
Donald EC, Porter IJ (2014) Clubroot in Australia: the history and impact of *Plasmodiophora brassicae* in *Brassica* crops and research efforts directed towards its control. Can J Plant Pathol 36(S1):66–84
Donald EC, Cross SJ, Lawrence JM, Porter IJ (2006a) Pathotypes of *Plasmodiophora brassicae*, the cause of clubroot, in Australia. Ann Appl Biol 148(3):239–244
Donald EC, Porter IJ, Faggian R, Lancaster RA (2006b) An integrated approach to the control of clubroot in vegetable brassica crops. Acta Hortic 706:283–300
Donald EC, Jaudzems G, Porter IJ (2007) Pathology of cortical invasion by *Plasmodiophora brassicae* in clubroot resistant and susceptible Brassica oleracea hosts. Plant Pathol 57:201–209
Donald EC, Jaudzems G, Porter IJ (2008) Pathology of cortical invasion by *Plasmodiophora brassicae* in clubroot resistant and susceptible *Brassica oleracea* hosts. Plant Pathol 57:201–209
Dordas C (2008) Role of nutrients in controlling plant diseases in sustainable agriculture: a review. Agron Sustain Dev 28:33–46
Durrant WE, Dong X (2004) Systemic acquired resistance. Annu Rev Phytopathol 42:185–209
Edwards D, Batley J, Snowdon RJ (2013) Accessing complex crop genomes with next-generation sequencing. Theor Appl Genet 126:1–11
Fagard M, Launay A, Clément G, Courtial J, Dellagi A, Farjad M, Krapp A, Soulié MC, Masclaux-Daubresse C (2014) Nitrogen metabolism meets phytopathology. J Exp Bot 65:5643–5656
Fan J, Doerner P (2012) Genetic and molecular basis of nonhost disease resistance: complex, yes; silver bullet, no. Curr Opin Plant Biol 15(4):400–406
Feng J, Hwang SF, Strelkov SE (2012a) Analysis of expressed sequence tags derived from a compatible *Plasmodiophora brassicae*-canola interaction. Can J Plant Pathol 34:562–574
Feng J, Xiao Q, Hwang SF, Strelkov SE, Gossen BD (2012b) Infection of canola by secondary zoospores of *Plasmodiophora brassicae* produced on a nonhost. Eur J Plant Pathol 132:309–315
Feng J, Cao T, Rennie DC, Strelkov SE, Hwang SF (2014) Host-parasite interactions in clubroot of crucifers. Can J Plant Pathol 36:113–321
Feng J, Jiang J, Feindel D, Strelkov SE, Hwang SF (2016) The gene Cr811 is present exclusively in pathotype 5 and new emerged pathotypes of the clubroot pathogen *Plasmodiophora brassicae*. Eur J Plant Pathol 145:615–620
Figdore SS, Ferreira ME, Slocum MK, Williams PH (1993) Association of RFLP markers with trait loci affecting clubroot resistance and morphological characters in *Brassica oleracea* L. Euphytica 69(1/2):33–44
Fredua-Agyeman R, Rahman H (2016) Mapping of the clubroot disease resistance in spring *Brassica napus* canola introgressed from European winter canola cv. 'Mendel'. Euphytica 211:201–213

Fuchs H, Sacristan MD (1996) Identification of a gene in *Arabidopsis thaliana* controlling resistance to clubroot and characterization of the resistance response. Mol Plant-Microbe Interact 9:91–97

Gao S, Liu T, Li Y, Wu Q, Fu K, Chen J (2012) Understanding resistant germplasm-induced virulence variation through analysis of proteomics and suppression subtractive hybridization in a maize pathogen *Curvularia lunata*. Proteomics 12:3524–3535

Gao F, Hirani AH, Liu J, Liu Z, Fu G, Wu C, McVetty PBE, Li G (2014) Fine mapping a clubroot resistance locus in Chinese cabbage. J Am Soc Hortic Sci 139(3):247–252

Gibriel HA, Thomma BP, Seidl MF (2016) The age of effectors: genome-based discovery and applications. Phytopathology 106:1206–1212

Golicz AA, Bayer PE, Barker GC, Edger PP, Kim H, Martinez PA, Chan CK, Severn-Ellis A, McCombie WR, Parkin IA, Paterson AH, Pires JC, Sharpe AG, Tang H, Teakle GR, Town CD, Batley J, Edwards D (2016) The pangenome of an agronomically important crop plant *Brassica oleracea*. Nat Commun 7:13390

Gossen BD, Kasinathan H, Cao T, Manolii VP, Strelkov S, Hwang SF, Mcdonald MR (2013) Influence of pH and temperature on infection and symptom development of clubroot in canola. Can J Plant Pathol 35:294–303

Gossen BD, Deora A, Peng G, Hwang SF, McDonald MR (2014) Effect of environmental parameters on clubroot development and the risk of pathogen spread. Can J Plant Pathol 36:37–48

Gossen BD, Strelkov SE, Manolii VP, Rennie DC, Cao T, Hwang SF, Peng G, McDonald MR (2015) Spread of *Plasmodiophora brassicae* on canola in Canada, 2003–2014: old pathogen, new home. Can J Plant Pathol 37:403–413

Grandclement C, Thomas G (1996) Detection and analysis of QTL based on RAPD markers for polygenic resistance to *Plasmodiophora brassicae* Wor. in *Brassica oleracea* L. Theor Appl Genet 93:86–90

Grandclement C, Laurens F, Thomas G (1996) Genetic analysis of resistance to clubroot (*Plasmodiophora brassicae* Woron.) in two *Brassica oleracea* groups (ssp. *acephala* and ssp. *botrytis*) through diallel analysis. Plant Breed 115(3):152–156

Gravot A, Grillet L, Wagner G, Jubault M, Lariagon C, Baron C, Deleu C, Delourme R, Bouchereau A, Manzanares-Dauleux MJ (2011) Genetic and physiological analysis of the relationship between partial resistance to clubroot and tolerance to trehalose in *Arabidopsis thaliana*. New Phytol 191:1083–1094

Gravot A, Deleu C, Wagner G, Lariagon C, Lugan R, Todd C, Wendehenne D, Delourme R, Bouchereau A, Manzanares-Dauleux MJ (2012) Arginase induction represses gall development during clubroot infection in *Arabidopsis*. Plant Cell Physiol 53:901–911

Grsic S, Kirchheim B, Pieper K, Fritsch M, Hilgenberg W, Ludwig-Müller J (1999) Induction of auxin biosynthetic enzymes by jasmonic acid and in clubroot diseased Chinese cabbage plants. Physiol Plant 105:521–531

Gustafsson M, Falt AS (1986) Genetic studies on resistance to clubroot in *Brassica napus*. Ann Appl Biol 108:409–415

Hagimori M (1995) Clubroot disease resistance and other characteristics of the progeny of somatic hybrid between Japanese radish and cauliflower. Acta Hortic 392:77–80

Hagimori M, Nagaoka M, Kato N, Yoshikawa H (1992) Production and characterization of somatic hybrids between Japanese radish and cauliflower. Theor Appl Genet 84:819–824

Hansen M (1989) Genetic variation and inheritance of tolerance to clubroot (*Plasmodiophora brassicae* Wor.) and other quantitative characters in cabbage (*Brassica oleracea* L.). Hereditas 110:13–22

Hansen CE, Flengsrud R, Kopperud C (1994) Two-dimensional gel electrophoresis of proteins in healthy roots and clubroots of *Brassica oleracea*. Acta Agric Scand 44:123–128

Happstadius I, Ljungberg A, Kristiansson B, Dixelius C (2003) Identification of *Brassica oleracea* germplasm with improved resistance to Verticillium wilt. Plant Breed 122:30–34

Harbert DJ (1969) A simple effective embryo culture technique for *Brassica.* Euphytica 18:425–429

Hasan M, Strelkov SE, Howard RJ, Rahman H (2012) Screening of *Brassica* germplasm for resistance to *Plasmodiophora brassicae* pathotypes prevalent in Canada for broadening diversity in clubroot resistance. Can J Plant Sci 92:501–515

Hatakeyama K, Fujimura M, Ishida M, Suzuki T (2004) New classification method for *Plasmodiophora brassicae* field isolates in Japan based on resistance of F1 cultivars of Chinese cabbage (*Brassica rapa* L.) to clubroot. Breed Sci 54:197–201

Hatakeyama K, Fujimura M, Ishida M, Suzuki T, Sato T (2006) Classification of pathogenicity of *Plasmodiophora brassicae* field isolates in Japan based on resistance of F1 cultivars of Chinese cabbage (*Brassica rapa* L.) to clubroot. Acta Hortic 706:323–328

Hatakeyama K, Suwabe K, Tomita RN, Kato T, Nunome T, Fukuoka H, Matsumoto S (2013) Identification and characterization of *crr1a,* a gene for resistance to clubroot disease (*Plasmodiophora brassicae* Woronin) in *Brassica rapa* L. PLoS One 8(1):e54745

Hatakeyama K, Niwa T, Kato T, Ohara T, Kakizaki T, Matsumoto S (2017) The tandem repeated organization of NB-LRR genes in the clubroot resistant CRb locus in *Brassica rapa* L. Mol Gen Genomics 292:397–405

Hayashida N, Takabatake Y, Nakazawa N, Aruga D, Nakanishi H, Taguchi G, Sakamoto K, Matsumoto E (2008) Construction of a practical SCAR marker linked to clubroot resistance in Chinese cabbage, with intensive analysis of HC352b genes. J Jpn Soc Hortic Sci 77:150–154

Hirai M (2006) Genetic analysis of clubroot resistance in *Brassica* crops. Breed Sci 56:223–229

Hirai M, Harada T, Kubo N, Tsukada M, Suwabe K, Matsumoto S (2004) A novel locus for clubroot resistance in *Brassica rapa* and its linkage markers. Theor Appl Genet 108:639–643

Hirani AH, Li G (2015) Understanding the genetics of clubroot resistance for effectively controlling this disease in *Brassica* species, plants for the future, Hany El-Shemy, Intech Open. https://doi.org/10.5772/60936

Hirani AH, Gao F, Liu J, Fu G, Wu C, Yuan Y, Li W, Hou J, Duncan R, Li G (2016) Transferring clubroot resistance from Chinese cabbage (*Brassica rapa*) to canola (*B. napus*). Can J Plant Pathol 38:82–90

Hirani AH, Gao F, Liu J, Fu G, Wu C, McVetty PBE, Duncan RW, Li G (2018) Combinations of independent dominant loci conferring clubroot resistance in all four turnip accessions (*Brassica rapa*) from the European clubroot differential set. Front Plant Sci 9:1628

Horikoshi N, Tairako K (2002) Occurrence of clubroot in Japanese radish caused by *Plasmodiophora brassicae* Worn. in Fukushima (in Japanese with English summary). Ann Rep Prot Plant N Jpn 53:58–60

Huang Z, Peng G, Liu X, Deora A, Falk KC, Gossen BD, McDonald MR, Yu F (2017) Fine mapping of a clubroot resistance gene in Chinese cabbage using SNP markers identified from bulked segregant RNA sequencing. Front Plant Sci 8:1448

Hwang SF, Ahmed HU, Zhou Q, Strelkov SE, Gossen BD, Peng G, Turnbull GD (2011) Influence of cultivar response and inoculum density on root hair infection of canola (*Brassica napus*) by *Plasmodiophora brassicae*. Plant Pathol 60:820–829

Hwang SF, Cao T, Xiao Q, Ahmed H, Manolii VP, Turnbull G, Gossen BD, Peng G, Strelkov S (2012) Efficacy of fungicide, seeding date and seedling age on clubroot severity, seedling emergence and yield of canola. Can J Plant Sci 92(6):1175–1186

Irani S, Trost B, Waldner M, Nayidu N, Tu JY, Kusalik AJ, Todd CD, Wei YD, Bonham-Smith WC (2018) Transcriptome analysis of response to *Plasmodiophora brassicae* infection in the *Arabidopsis* shoot and root. BMC Genomics 19:23

Irani S, Todd CD, Wei Y, Bonham-Smith PC (2019) Changes in phenylpropanoid pathway gene expression in roots and leaves of susceptible and resistant *Brassica napus* lines in response to *Plasmodiophora brassicae* inoculation. Physiol Mol Plant Pathol 106:196–203

Ito S, Tanaka S, Miyanaga C, Takabayashi S, Yano S, Kameya-Iwaki M (1996) Constitutive and Inducible Proteins in the Root of Chinese Cabbage Infected with *Plasmodiophora brassicae*. J Phytopathol 144(2):89–95

James RV, Williams PH, Maxwell DP (1978) Inheritance and linkage studies related to resistance in *Brassica campestris* L. to *Plasmodiophora brassicae* race 6. Eucarpia Cruciferae Newslett 3:27
Johnston TD (1970) A new factor for resistance to club root in *Brassica napus*. Plant Pathol 19 (4):156–158
Jones JD, Dangl JL (2006) The plant immune system. Nature 444:323–329
Jubault M, Lariagon C, Simon M, Delourme R, Manzanares-Dauleux M (2008) Identification of quantitative trait loci controlling partial clubroot resistance in new mapping populations of *Arabidopsis thaliana*. Theor Appl Genet 117(2):191–202
Kamei A, Tsuro M, Kubo N, Hirai M (2007) QTL analysis of clubroot resistance in radish (*Raphanus sativus* L.). Breed Res 9:83
Karling JS (1968) The plasmodiophorales, 2nd edn. Hafner Publishing Company, New York, p 256
Kato T, Hatakeyama K, Fukino N, Matsumoto S (2012) Identification of a clubroot resistance locus conferring resistance to a *Plasmodiophora brassicae* classified into pathotype group 3 in Chinese cabbage (*Brassica rapa* L.). Breed Sci 62(3):282–287
Kazan K, Lyons R (2014) Intervention of phytohormone pathways by pathogen effectors. Plant Cell 26(6):2285–2309
Keen NT, Williams PH (1969) Translocation of sugars into infected cabbage tissues during clubroot development. Plant Physiol 44:748–754
Kitamura S (1958) Varieties and transitions of radish. In: Nishiyama I (ed) Japanese radish. Japan Sci Soc Press, Tokyo, pp 1–19
Kobayashi I, Murdoch LJ, Kunoh H, Hardham AR (1995) Cell biology of early events in the plant resistance response to infection by pathogenic fungi. Can J Bot 73:S418–S425
Kolmer JA (1996) Genetics of resistance to wheat leaf rust. Annu Rev Phytopathol 34:435–455
Kroll TK, Lacy GH, Moore LD (1983) A quantitative description of the colonization of susceptible and resistant radish plants by *Plasmodiophora brassicae*. Phytopathol Z 108:97–105
Kuginuki Y, Ajisaka H, Yui M, Yoshikawa H, Hida K, Hirai M (1997) RAPD markers linked to a clubroot-resistance locus in *Brassica rapa* L. Euphytica 98:149–154
Kuginuki Y, Yoshikawa H, Hirai M (1999) Variation in virulence of *Plasmodiophora brassicae* in Japan tested with clubroot-resistant cultivars of Chinese cabbage (*Brassica rapa* L. ssp. *pekinensis*). Eur J Plant Pathol 105(4):327–332
Lagercrantz U (1998) Comparative mapping between *Arabidopsis thaliana* and *Brassica nigra* indicates that *Brassica* genomes have evolved through extensive genome replication accompanied by chromosome fusion and frequent rearrangements. Genetics 150:1217–1228
Lagercrantz U, Lydiate DJ (1996) Comparative genome mapping in *Brassica*. Genetics 144:1903–1910
Lamb CJ, Lawton MA, Dron M, Dixon RA (1989) Signals and transduction mechanisms for activation of plant defenses against microbial attack. Cell 56:215–224
Lammerink J (1964) Pathologic specialisation of *Plasmodiophora brassicae* Worn. N Z J Agric Res 7:37–41
Lammerink J (1967) The inheritance of clubroot resistance in *Brassica napus* L. N Z J Agric Res 10 (1):109–115
Lammerink J (1969) Inter-specific transfer of clubroot resistance from *Brassica campestris* L. to *B. napus* L. N Z J Agric Res 13:105–110
Landry BS, Hubert N, Crete R, Chiang MS, Lincoln SE, Etoh T (1992) A genetic map for *Brassica oleracea* based on RFLP markers detected with expressed DNA sequences and mapping of resistance gene to race 2 of *Plasmodiophora brassicae*. Genome 35:409–420
Laperche A, Aigu Y, Jubault M, Ollier M, Guichard S, Glory P, Strelkov SE, Gravot A, Manzanares-Dauleux MJ (2017) Clubroot resistance QTL are modulated by nitrogen input in *Brassica napus*. Theor Appl Genet 130:669–684
Larkan NJ, Lydiate DJ, Parkin IAP, Nelson MN, Epp DJ, Cowling WA, Rimmer SR, Borhan MH (2013) The *Brassica napus* blackleg resistance gene LepR3 encodes a receptor-like protein triggered by the *Leptosphaeria maculans* effector AVRLM1. New Phytol 197:595–605

Laurens F, Thomas G (1993) Inheritance of resistance to clubroot (*Plasmodiophora brassicae* Worn.) in kale (*Brassica oleracea* ssp. *acephala*). Hereditas 119:253–262

LeBoldus JM, Manolii VP, Turkington TK, Strelkov SE (2012) Adaptation to *Brassica* host genotypes by a single-spore isolate and population of *Plasmodiophora brassicae* (clubroot). Plant Dis 96:833–838

Lee J, Izzah NK, Choi BS, Joh HJ, Lee SC, Perumal S, Seo J, Ahn K, Jo EJ, Choi GJ, Nou IS, Yu Y, Yang TJ (2016) Genotyping-by-sequencing map permits identification of clubroot resistance QTLs and revision of the reference genome assembly in cabbage (*Brassica oleracea* L.). DNA Res 23:29–41

Leister D (2004) Tandem and segmental gene duplication and recombination in the evolution of plant disease resistance gene. Trends Genet 20:116–122

Li L, Luo Y, Chen B, Xu K, Zhang F, Li H et al (2016) A genome-wide association study reveals new loci for resistance to clubroot disease in *Brassica napus*. Front Plant Sci 7:1483

Liang Z, Chen K, Li T, Zhang Y, Wang Y, Zhao Q, Liu J, Zhang H, Liu C, Ran Y, Gao C (2017) Efficient DNA-free genome editing of bread wheat using CRISPR/Cas9 ribonucleoprotein complexes. Nat Commun 8:14261

Liu SL, Adams KL (2010) Dramatic change in function and expression pattern of a gene duplicated by polyploidy created a paternal effect gene in the Brassicaceae. Genome Biol Evol 27:2817–2828

Liu S, Liu Y, Yang X, Tong C, Edwards D, Parkin I, Parkin IAP, Zhao M, Ma J, Yu J, Huang S, Wang X, Wang J, Lu K, Fang Z, Bancroft I, Yang T-J, Hu Q, Wang X, Yue Z, Li H, Yang L, Wu J, Zhou Q, Wang W, King GJ, Pires JC, Lu C, Wu Z, Sampath P, Wang Z, Guo H, Pan S, Yang L, Min J, Zhang D, Jin D, Li W, Belcram H, Tu J, Guan M, Qi C, Du D, Li J, Jiang L, Batley J, Sharpe AG, Park B-S, Ruperao P, Cheng F, Waminal NE, Huang Y, Dong C, Wang L, Li J, Hu Z, Zhuang M, Huang Y, Huang J, Shi J, Mei D, Liu J, Lee T-H, Wang J, Jin H, Li Z, Li X, Zhang J, Xiao L, Zhou Y, Liu Z, Liu X, Qin R, Tang X, Liu W, Wang Y, Zhang Y, Lee J, Kim HH, Denoeud F, Xu X, Liang X, Hua W, Wang X, Wang J, Chalhoub B, Paterson AH (2014) The *Brassica oleracea* genome reveals the asymmetrical evolution of polyploid genomes. Nat Commun 5:3930

Liu C, Yang Z, He P, Munir S, Wu Y, Ho H, He Y (2018) Deciphering the bacterial and fungal communities in clubroot-affected cabbage rhizosphere treated with *Bacillus subtilis* XF-1. Agric Ecosyst Environ 256:12–22

Lof ME, Van Der Werf W (2017) Modeling the effect of gene deployment strategies on durability of plant resistance under selection. Crop Prot 97:10–17

Loutre C, Wicker T, Travella S, Galli P, Scofield S, Fahima T, Feuillet C, Keller B (2009) Two different CC-NBS-LRR genes are required for Lr10-mediated leaf rust resistance in tetraploid and hexaploid wheat. Plant J 60:1043–1054

Lovelock DA (2013) Salicylic acid and its role in defence against *Plasmodiophora brassicae*. Deakin University, Geelong

Lovelock DA, Donald CE, Conlan XA, Cahill DM (2013) Salicylic acid suppression of clubroot in broccoli (*Brassicae oleracea* var. *italica*) caused by the obligate biotroph *Plasmodiophora brassicae*. Austr Plant Pathol 42:141–153

Ludwig-Müller J, Prinsen E, Rolfe S, Scholes J (2009) Metabolism and plant hormone action during clubroot disease. J Plant Growth Regul 28(3):229–244

Ludwig-Müller J, Jülke S, Geiß K, Richter F, Mithöfer A, Šola I, Rusak G, Keenan S, Bulman S (2015) A novel methyltransferase from the intracellular pathogen *Plasmodiophora brassicae* participates in methylation of salicylic acid. Mol Plant Pathol 16(4):349–364

Ma L, Borhan MH (2015) The receptor-like kinase SOBIR1 interacts with *Brassica napus LepR3* and is required for *Leptosphaeria maculans AvrLm1*- triggered immunity. Front Plant Sci 6:933

Manzanares-Dauleux MJ, Barret P, Thomas G (2000a) Development of a pathotype specific SCAR marker in Plasmodiophora brassicae. Eur J Plant Pathol 106(8):781–787

Manzanares-Dauleux MJ, Delourme R, Baron F, Thomas G (2000b) Mapping of one major gene and of QTLs involved in resistance to clubroot in *Brassica napus*. Theor Appl Genet 101 (5/6):885–891

Martin GB, Bogdanove AJ, Sessa G (2003) Understanding the functions of plant disease resistance proteins. Annu Rev Plant Biol 54:23–61

Matsumoto E, Yasui C, Ohi M, Tsukada M (1998) Linkage analysis of RFLP markers for clubroot resistance and pigmentation in Chinese cabbage. Euphytica 104:79–86

Matsumoto E, Hayashida N, Sakamoto K, Ohi M (2005) Behavior of DNA markers linked to a clubroot resistance gene in segregating populations of Chinese cabbage (*Brassica rapa* ssp. *pekinensis*). J Jpn Soc Hortic Sci 74:367–373

Matsumoto E, Ueno H, Aruga D, Sakamoto K, Hayashida N (2012) Accumulation of three clubroot resistance genes through marker-assisted selection in Chinese cabbage (*Brassica rapa* ssp. *pekinensis*). J Jpn Soc Hortic Sci 81:184–190

McDonald BA, Linde C (2002) Pathogen population genetics, evolutionary potential and durable resistance. Annu Rev Phytopathol 40:349–379

McHale LK, Haun WJ, Xu WW, Bhaskar PB, Anderson JE, Hyten DL, Gerhardt DJ, Jeddeloh JA, Stupar RM (2012) Structural variants in the soybean genome localize to clusters of biotic stress response genes. Plant Physiol 159:1295–1308

Mehta A, Brasileiro A, Souza DS, Romano E, Campos MA, Grossi-de-Sa MF et al (2008) Plant–pathogen interactions: what is proteomics telling us? FEBSJ 275:3731–3746

Morgner M (1995) Mikroskopische Untersuchungen der Reaktionen empfindlicher und resistenter Genotypen verschiedener Brassica-Arten gegenuber *Plasmodiophora brassicae* (Worn.), dissertation

Moriguchi K, Kimizuka-Takagi C, Ishii K, Nomura K (1999) A genetic map based on RAPD, RFLP, isozyme, morphological markers and QTL analysis for clubroot resistance in *Brassica oleracea*. Breed Sci 49:257–265

Mun J, Yu H, Park S, Park B (2009) Genome-wide identification of NBSencoding resistance genes in *Brassica rapa*. Mol Genet Genomics 282:617–631

Murakami H, Tsushima S, Akimoto T, Murakami K, Goto I, Shishido Y (2000) Effects of growing leafy daikon (*Raphanus sativus*) on populations of *Plasmodiophora brassicae* (clubroot). Plant Pathol 49:584–589

Mysore KS, Ryu CM (2004) Nonhost resistance: how much do we know? Trends Plant Sci 9:97–104

Nagaoka T, Doullah MA, Matsumoto S, Kawasaki S, Ishikawa T, Hori H, Okazaki K (2010) Identification of QTLs that control clubroot resistance in *Brassica oleracea* and comparative analysis of clubroot resistance genes between *B. rapa* and *B. oleracea*. Theor Appl Genet 120 (7):1335–1346

Neik TX, Barbetti MJ, Batley J (2017) Current status and challenges in identifying disease resistance genes in *Brassica napus*. Front Plant Sci 8:1788

Netea MG, Quintin J, Van der Meer JWM (2011) Trained immunity: a memory for innate host defense. Cell Host Microbe 9:355–361

Neuhaus K, Grsic-Rausch S, Sauerteig S, Ludwig-Muller J (2000) *Arabidopsis* plants transformed with nitrilase 1 or 2 in antisense direction are delayed in clubroot development. J Plant Physiol 156:756–761

Niks RE, Marcel TC (2009) Nonhost and basal resistance: how to explain specificity? New Phytol 182:817–828

Ning LY, Lin ZW, Gu JW, Gan L, Li YH, Wang H, Miao L, Zhang L, Wang B, Li M (2018) The initial deficiency of protein processing and flavonoids biosynthesis were the main mechanisms for the male sterility induced by SX-1 in *Brassica napus*. BMC Genomics 19:806

Nomura K, Minegishi Y, Kimizuka-Takagi C, Fujioka T, Moriguchi K, Shishido R, Ikehashi H (2005) Evaluation of F2 and F3 plants introgressed with QTL for clubroot resistance in cabbage developed by using SCAR markers. Plant Breed 124:371–375

Nordborg M, Hu TT, Ishino Y, Jhaveri J, Toomajian C, Zheng HG, Bakker E, Calabrese P, Gladstone J, Goyal R, Jakobsson M, Kim S, Morozov Y, Padhukasahasram B, Plagnol V, Rosenberg NA, Shah C, Wall JD, Wang J, Zhao KY, Kalbfleisch T, Schulz V, Kreitman M, Bergelson J (2005) The pattern of polymorphism in *Arabidopsis thaliana*. PLoS Biol 3:e196

Novo JVJ, Pascual J, Lucas RS, Romero-Rodriguez C, Ortega MR, Lenz C, Valledor L (2014) Fourteen years of plant proteomics reflected in "Proteomics": moving from model species and 2-DE based approaches to orphan species and gel-free platforms. Proteomics 15:5–6

O'Neill CM, Bancroft I (2000) Comparative physical mapping of segments of the genome of *Brassica oleracea* var. *alboglabra* that are homologous to sequenced regions of chromosomes 4 and 5 of *Arabidopsis thaliana*. Plant J 23:233–243

Osaki K, Tanaka S, Ito S (2004) A case of breakdown in clubroot-resistant cultivars of Chinese cabbage. J Phytopathol 70:66

Osaki K, Tanaka S, Ito S (2008a) Pathogenicity of *Plasmodiophora brassicae* populations from small, spheroid, resistant-type clubroot galls on roots of clubroot-resistant cultivars of Chinese cabbage (*Brassica rapa* L. subsp. *pekinensis*). J Gen Plant Pathol 74:242–245

Osaki K, Fujiyama S, Nakayama A, Shimizu Y, Ito S, Tanaka S (2008b) Relation between pathogenicity and genetic variation within *Plasmodiophora brassicae*. J Gen Plant Pathol 74:281–288

Osborn TC, Butrulle DV, Sharpe AG, Pickering KJ, Parkin IAP, Parker JS (2003) Detection and effects of a homeologous reciprocal transposition in *Brassica napus*. Genetics 165:1569–1577

Otani F, Maruyama S, Tsukada M, Nagase Y (1982) Studies on the breeding of clubroot resistance in Chinese cabbage and Chinese mustard II. Process of breeding the clubroot resistant line of Nozawana, a variety of Chinese mustard, and its characteristics (in Japanese with English summary). Bull Nagano Veg Ornam Crop Exp Sta Jpn 2:1–8

Oxley S (2007) Clubroot disease of oilseed rape and other brassica crops. Technical Note, TN 620

Pang W, Liang S, Li X, Li P, Yu S, Lim YP, Piao Z (2014) Genetic detection of clubroot resistance loci in a new population of *Brassica rapa*. Hortic Environ Biotechnol 55:540–547

Pang W, Fu P, Li X, Zhan Z, Yu S, Piao Z (2018) Identification and mapping of the clubroot resistance gene *crd* in Chinese cabbage (*Brassica rapa* ssp. *pekinensis*). Plant Sci 9:653

Parkin IAP, Gulden SM, Sharpe AG, Lukens L, Trick M, Osborn TC (2005) Segmental structure of the *Brassica napus* genome based on comparative analysis with *Arabidopsis thaliana*. Genetics 171:765–781

Parkin IAP, Koh C, Tang H, Robinson SJ, Kagale S, Clarke WE, Town CD, Nixon J, Krishnakumar V, Bidwell SL, Denoeud F, Belcram H, Links MG, Just J, Clarke C, Bender T, Huebert T, Mason AS, Pires JC, Barker G, Moore J, Walley PG, Manoli S, Batley J, Edwards D, Nelson MN, Wang X, Paterson AH, King G, Bancroft I, Chalhoub B, Sharpe AG (2014) Transcriptome and methylome profiling reveals relics of genome dominance in the mesopolyploid *Brassica oleracea*. Genome Biol 15:R77

Pasold S, Ludwig-Müller J (2013) Reduction of clubroot (*Plasmodiophora brassicae*) formation in Arabidopsis thaliana after treatment with prohexadione-calcium, an inhibitor of oxoglutaric acid-dependent dioxygenases. Plant Pathol 62:1357–1365

Pasold S, Siegel I, Seidel C, Ludwig-Müller J (2010) Flavonoid accumulation in *Arabidopsis thaliana* root galls caused by the obligate biotrophic pathogen *Plasmodiophora brassicae*. Mol Plant Pathol 11:545–562

Peng SS, Ren ZH, Huang XL, Liu MJ, Sun LF, Liu EM (2013) Physiological race identification on clubroot of cruciferous crops caused by *Plasmodiophora brassicae* in Hunan. J Changjiang Veg 6:46–49

Peng G, Lahlali R, Hwang SF, Pageau D, Hynes RK, McDonald MR, Gossen BD, Strelkov SE (2014) Crop rotation, cultivar resistance, and fungicides/biofungicides for managing clubroot (*Plasmodiophora brassicae*) on canola. Can J Plant Pathol 36:99–112

Peng Y, Huang Y, Ye Lei, Liu Zhenlin, Yang Hui, Shang Jing (2016) Occurrence and morphology of small spheroid galls on clubroot rsistant Chinese cabbage. Eur J Plant Pathol, 145:591–599.

Piao ZY, Deng YQ, Choi SR, Park YJ, Lim YP (2004) SCAR and CAPS mapping of CRb, a gene conferring resistance to *Plasmodiophora brassicae* in Chinese cabbage (*Brassica rapa* ssp. *pekinensis*). Theor Appl Genet 108:1458–1465

Piao Z, Ramchiary N, Lim YP (2009) *Plasmodiophora brassicae* (clubroot) – a plant pathogen that alters host growth and productivity. J Plant Growth Regul 28:252–264

Pires JC, Gaeta RT (2011) Structural and functional evolution of resynthesized polyploids. In: Schmidt R, Bancroft I (eds) Genetics and genomics of the brassicaceae. Springer, New York, pp 195–214

Pryor DE (1940) The effect of some mineral nutrients on the development of clubroot of brassicas. J Agric Res 61:149–160

Quirino BF, Candido ES, Campos PF, Franco OL, Krüger RH (2010) Proteomic approaches to study plant–pathogen interactions. Phytochemistry 71:351–362

Quiros CF, Grellet F, Sadowski J, Suzuki T, Li G, Wroblewski T (2001) *Arabidopsis* and *Brassica* comparative genomics: sequence, structure and gene content in the ABI1-Rps2-Ck1 chromosomal segment and related regions. Genetics 157:1321–1330

Rafiqi M, Bernoux M, Ellis JG, Dodds PN (2009) In the trenches of plant pathogen recognition: role of NB-LRR proteins. Semin Cell Dev Biol 20:1017–1024

Rahman H, Shakir A, Jakir Hasan M (2011) Breeding for clubroot resistant spring canola (*Brassica napus* L.) for the Canadian prairies: can the European winter canola cv. Mendel be used as a source of resistance? Can J Plant Sci 91:447–458

Rahman H, Peng G, Yu F, Falk KC, Kulkarni M, Selvaraj G (2014) Genetics and breeding for club root resistance in Canadian spring canola (*Brassica napus* L.). Can J Plant Pathol 36:122–134

Ren L, Xu L, Liu F, Chen K, Sun C, Li J, Fang X (2016) Host range of *Plasmodiophora brassicae* on cruciferous crops and weeds in China. Plant Dis 100:933–939

Rennie DC, Holtz MD, Turkington TK, Leboldus JM, Hwang SF, Howard RJ, Strelkov SE (2015) Movement of *Plasmodiophora brassicae* resting spores in windblown dust. Can J Plant Pathol 37:188–196

Ricarova V, Kaczmarek J, Strelkov SE, Kazda J, Lueders W, Rysanek P, Manolii V, Jedryczka M (2016) Pathotypes of *Plasmodiophora brassicae* causing damage to oilseed rape in the Czech Republic and Poland. Eur J Plant Pathol 145:559–572

Rocherieux J, Glory P, Giboulot A, Boury S, Barbeyron G, Thomas G, Manzanares-Dauleux MJ (2004) Isolate-specific and broad-spectrum QTLs are involved in the control of clubroot in *Brassica oleracea*. Theor Appl Genet 108:1555–1563

Rolfe SA, Strelkov SE, Links MG, Clarke WE, Robinson SJ, Djavaheri M, Malinowski R, Haddadi P, Kagale S, Parkin IAP, Taheri A, Borhan MH (2016) The compact genome of the plant pathogen *Plasmodiophora brassicae* is adapted to intracellular interactions with host *Brassica* spp. BMC Genomics 17:272

Rowe RC (1980) Evaluation of radish cultivars for resistance to clubroot (*Plasmodiophora brassicae*) race 6 for midwestern United States. Plant Dis 64:462–464

Saito M, Kubo N, Matsumoto S, Suwabe K, Tsukada M, Hirai M (2006) Fine mapping of the clubroot resistance gene, *Crr3*, in *Brassica rapa*. Theor Appl Genet 114:81–91

Sakai T, Liu HJ, Iwabuchi M, Kohno-Murase J (1996) Introduction of a gene from fertility restored radish (*Raphanus sativus*) into *Brassica napus* by fusion of X-irradiated protoplasts from a radish restorer line and iodacetoamide-treated protoplasts from a cytoplasmic male-sterile cybrid of *B. napus*. Theor Appl Genet 93:373–379

Sakamoto K, Saito A, Hayashida N, Taguchi G, Matsumoto E (2008) Mapping of isolate-specific QTL for clubroot resistance in Chinese cabbage (*Brassica rapa* L. ssp. *pekinensis*). Theor Appl Genet 117:759–767

Saude C, Alan MK, Gossen BD, Mc Donald MR (2012) Effect of host resistance and fungicide application on clubroot pathotype 6 in green cabbage and napa cabbage. Hortic Technol 22 (3):311–319

Saxena RK, Edwards D, Varshney RK (2014) Structural variations in plant genomes. Brief Funct Genomics 13:296–307

Scheben A, Batley J, Edwards D (2017a) Genotyping-by-sequencing approaches to characterize crop genomes: choosing the right tool for the right application. Plant Biotechnol J 15:149–161

Scheben A, Wolter F, Batley J, Puchta H, Edwards DC (2017b) Towards CRISPR/Cas crops – bringing together genomics and genome editing. New Phytol 216:682–698

Schranz ME, Lysak MA, Mitchell-Olds T (2006) The ABC's of comparative genomics in the Brassicaceae: building blocks of crucifer genomes. Trends Plant Sci 11:535–542

Seaman WL, Walker JC, Larson RH (1963) A new race of *Plasmodiophora brassicae* affecting Badger Shipper cabbage. Phytopathology 53:1426–1429

Shah N, Li Q, Qiang X, Liu J, Huang F, Zhan Z, Qin P, Zhou X, Yu W, Li Z, Zhang C (2020) *CRb* and *PbBa8*.1 synergically increases resistant genes expression upon infection of *Plasmodiophora brassicae* in *Brassica napus*. Gene 11:1–19

Sharma K, Gossen BD, Greenshields D, Selvaraj G, Strelkov SE, Mcdonald MR (2012) Reaction of lines of the rapid cycling Brassica collection and *Arabidopsis thaliana* to four pathotypes of *Plasmodiophora brassicae*. Plant Dis 97:720–727

Sharma K, Gossen BD, Greenshields D, Selvaraj G, Strelkov SE, McDonald MR (2013) Reaction of lines of the rapid cycling Brassica collection and *Arabidopsis thaliana* to four pathotypes of *Plasmodiophora brassicae*. Plant Dis 97:720–727

Siemens J, Nagel M, Ludwig-Muller J, Sacristan MD (2002) The interaction of *Plasmodiophora brassicae* and *Arabidopsis thaliana*: parameters for disease quantification and screening of mutant lines. J Phytopathol 150(11–12):592–605

Siemens J, Keller I, Sarx J, Kunz S, Schuller A, Nagel W, Schmülling T, Parniske M, Ludwig-Muller J (2006) Transcriptome analysis of *Arabidopsis* clubroots indicate a key role for cytokinins in disease development. Mol Plant-Microbe Interact 19:480–494

Siemens J, González MC, Wolf S, Hofmann C, Greiner S, Du Y, Rausch T, Roitsch T, Ludwig-Muller J (2011) Extracellular invertase is involved in the regulation of clubroot disease in *Arabidopsis thaliana*. Mol Plant Pathol 12:247–262

Somé A, Manzanares MJ, Laurens F, Baron F, Thomas G, Ouxel FR (1996) Variation for virulence of *Brassica napus* L., amongst *Plasmodiophora brassicae* collections from France and derived single-spore isolates. Plant Pathol 45(3):432–439

Song T, Chu M, Lahlali R, Yu F, Peng G (2016) Shotgun label-free proteomic analysis of clubroot (*Plasmodiophora brassicae*) resistance conferred by the gene *Rcr1* in *Brassica rapa*. Front Plant Sci 7:1013

Springer NM, Ying K, Fu Y, Ji T, Yeh CT, Jia Y, Wu W, Richmond T, Kitzman J, Rosenbaum H, Iniguez AL, Barbazuk WB, Jeddeloh JA, Nettleton D, Schnable PS (2009) Maize inbreds exhibit high levels of copy number variation (CNV) and presence/absence variation (PAV) in genome content. PLoS Genet 5:e1000734

Stotz HU, Mitrousia GK, DeWit PJGM, Fitt BDL (2014) Effector triggered defence against apoplastic fungal pathogens. Trends Plant Sci 19:491–500

Strandberg JO, Williams PH (1967) Inheritance of clubroot resistance in Chinese cabbage. Phytopathology 57:330

Strelkov SE, Tewari JP, Hartman M, Orchard D (2005) Clubroot on canola in Alberta in 2003 and 2004. Can Plant Dis Surv 85:72–73

Strelkov SE, Tewari JP, Smith-Degenhardt E (2006) Characterization of *Plasmodiophora brassicae* populations from Alberta, Canada. Can J Plant Pathol 28:467–474

Strelkov SE, Hwang SF, Howard RJ, Hartman M, Turkington TK (2011) Progress towards the sustainable management of clubroot [*Plasmodiophora brassicae*] of canola on the Canadian Prairies. Prairie Soils Crops J 4:114–121

Strelkov SE, Hwang SF, Manolii VP, Cao T, Feindel D (2016) Emergence of new virulence phenotypes of *Plasmodiophora brassicae* on canola (*Brassica napus*) in Alberta, Canada. Eur J Plant Pathol 145:517–529

Strelkov SE, Hwang SF, Manolii VP, Cao T, Fredua-Agyeman R, Harding MW, Peng G, Gossen BD, Mcdonald MR, Feindel D (2018) Virulence and pathotype classification of *Plasmodiophora brassicae* populations collected from clubroot resistant canola (*Brassica napus*) in Canada. Can J Plant Pathol 40(2):284–298

Stuttmann J, Hubberten HM, Rietz S, Kaur J, Muskett P, Guerois R, Bednarek P, Hoefgen R, Parker JE (2011) Perturbation of *Arabidopsis* amino acid metabolism causes incompatibility with the adapted biotrophic pathogen *Hyaloperonospora arabidopsidis*. Plant Cell 23:2788–2803

Subramaniam S, Wang X, Freeling M, Pires JC (2013) The fate of *Arabidopsis thaliana* homeologous CNSs and their motifs in the paleohexaploid *Brassica rapa*. Genome Biol Evol 5:646–660

Suwabe K, Tsukazaki H, Iketani H, Hatakeyama K, Fujimura M, Nunome T, Fukuoka H, Matsumoto S, Hirai M (2003) Identification of two loci for resistance to clubroot (*Plasmodiophora brassicae* Woronin) in *Brassica rapa* L. Theor Appl Genet 107:997–1002

Suwabe K, Tsukazaki H, Iketani H, Hatakeyama K, Kondo M, Fujimura M, Nunome T, Fukuoka H, Hirai M, Matsumoto S (2006) Simple sequence repeat-based comparative genomics between *Brassica rapa* and *Arabidopsis thaliana*: the genetic origin of clubroot resistance. Genetics 173:309–319

Suwabe K, Suzuki G, Nunome T, Hatakeyama K, Mukai Y, Fukuoka H, Matsumoto S (2012) Microstructure of a *Brassica rapa* genome segment homoeologous to the resistance gene cluster on *Arabidopsis* chromosome 4. Breed Sci 62:170–177

Svitashev S, Schwartz C, Lenderts B, Young JK, Mark Cigan A (2016) Genome editing in maize directed by CRISPR–Cas9 ribonucleo-protein complexes. Nat Commun 7:13274

Tamura K, Peterson D, Peterson N, Stecher G, Nei M, Kumar S (2011) MEGA5: molecular evolutionary genetics analysis using maximum likelihood, evolutionary distance, and maximum parsimony methods. Mol Biol Evol 28:2731–2739

Tanaka S, Ito S (2013) Pathogenic and genetic diversity in *Plasmodiophora brassicae* (clubroot) from Japan. J Gen Plant Pathol 79:297–306

Tanaka S, Mizui Y, Terasaki H, Ito S (2006) Distribution of clubroot disease of a cruciferous weed, *Cardamine flexuosa*, in major isolated islands, Hokkaido and Okinawa in Japan. Myco Sci 47 (2):72–77

Thomma BPHJ, Nürnberger T, Joosten MHAJ (2011) Of PAMPs and effectors: the blurred PTI-ETI dichotomy. Plant Cell 23:4–15

Thuma BA, Rowe RC, Madden LV (1983) Relationships of soil temperature and moisture to clubroot (*Plasmodiphra brassicae*) severity on radish in organic soil. Plant Dis 67:758–762

Tjallingii F (1965) Testing clubroot resistance in turnips in Netherlands and the physiologic specialization of *Plasmodiophora brassicae*. Euphytica 14:1–22

Tomita H, Shimizu M, Doullah AU, Fujimoto R, Okazaki K (2013) Accumulation of quantitative trait loci conferring broad-spectrum clubroot resistance in *Brassica oleracea*. Mol Breed 32:889–900

Toxopeus H, Janssen AMP (1975) Clubroot resistance in turnip II. The slurry screening method and clubroot races in the Netherlands. Euphytica 24:751–755

Toxopeus H, Dixon G, Mattusch P (1986) Physiological specialisation in *Plasmodiophora brassicae*: an analysis by international experimentation. Trans Br Mycol Soc 87:279–287

Truco MJ, Hu J, Sadowski J, Quiros CF (1996) Inter-and intra-genomic homology of the *Brassica* genomes: implications for their origin and evolution. Theor Appl Genet 93:1225–1233

Tsunoda S (1980) Eco-physiology of wild and cultivated forms in *Brassica* and allied genera. In: Tsunoda S, Hinata K, Gomez-Campo C (eds) *Brassica* crops and wild allies. Japan Sci Soc Press, Tokyo, pp 109–132

Ueno H, Matsumoto E, Aruga D, Kitagawa S, Matsumura H, Hayashida N (2012) Molecular characterization of the *CRa* gene conferring clubroot resistance in *Brassica rapa*. Plant Mol Biol 80(6):621–629

Van De Wouw AP, Cozijnsen AJ, Hane JK, Brunner PC, Mcdonald BA, Oliver RP, Howlett BJ (2010) Evolution of linked avirulence effectors in *Leptosphaeria maculans* is affected by genomic environment and exposure to resistance genes in host plants. PLoS Pathol 6:e1001180

Van Loon LC, Rep M, Pieterse CMJ (2006) Significance of inducible defense-related proteins in infected plants. Annu Rev Phytopathol 44:135–162

Vera Cruz CM, Bai J, Ona I, Leung H, Nelson RJ, Mew TW, Leach JE (2000) Predicting durability of a disease resistance gene based on an assessment of the fitness loss and epidemiological consequences of avirulence gene mutation. Proc Natl Acad Sci U S A 97:13500–13505

Voorips RE, Visser DL (1993) Examination of resistance to clubroot in accessions of *Brassica oleracea* using a glasshouse seedling test. Eur J Plant Pathol 99:269–276

Voorrips RE (1995) *Plasmodiophora brassicae* – aspects of pathogenesis and resistance in *Brassica*. Euphytica 83(2):139–146

Voorrips RE (1996) Production, characterization and interaction of single-spore isolates of *Plasmodiophora brassicae*. Eur J Plant Pathol 102:377–383

Voorrips RE, Kanne HJ (1997) Genetic analysis of resistance to clubroot (*Plasmodiophora brassicae*) in *Brassica oleracea*. II. Quantitative analysis of root symptom measurements. Euphytica 93(1):41–48

Voorrips RE, Jongerius MC, Kanne HJ (1997) Mapping of two genes for resistance to clubroot (*Plasmodiophora brassicae*) in a population of doubled haploid lines of *Brassica oleracea* by means of RFLP and AFLP markers. Theor Appl Genet 94(1):75–82

Vorrips RE, Kanne HJ (1997) Genetic analysis of resistance to clubroot (*Plasmodiophora brassicae*) in *Brassica oleracea*. I. Analysis of symptom grades. Euphytica 93:31–39

Walker JC, Larson RH (1951) Progress in the development of clubroot-resistant cabbage. Phytopathology 41:37

Wallenhammar AC (1996) Prevalence of *Plasmodiophora brassicae* in a spring oil seed rape growing area in central Sweden and factors influencing soil infestation levels. Plant Pathol 45:710–719

Wallenhammar AC, Johnsson L, Gerhardson B (1999) Clubroot resistance and yield loss in spring oilseed turnip rape and spring oilseed rape. In: New Horizons of an old crop: Proceedings of 10th. International Rapeseed Congress. (eds) Wratten N, Salisbury PA. Canberra, Australia

Walters DR, Bingham IJ (2007) Influence of nutrition on disease development caused by fungal pathogens: implications for plant disease control. Ann Appl Biol 151:307–324

Wang J, Huang Y, Li XL, Li HZ (2011) Research progress in clubroot of crucifers. Plant Prot 37:153–158

Wei L, Jian H, Lu K, Filardo F, Yin N, Liu L, Qu C, Li W, Du H, Li J (2016) Genome-wide association analysis and differential expression analysis of resistance to Sclerotinia stem rot in *Brassica napus*. Plant Biotechnol J 14:1368–1380

Weisaeth G (1977) Hodekalsortene 'Respla' og 'Resista' foredlinger med klumprotresistens. Inst, for gronnsakdyrking. Norges Landbr hogsk Meld 74:937–939

Werner S, Diederichsen E, Frauen M, Schondelmaier J, Jung C (2008) Genetic mapping of clubroot resistance genes in oilseed rape. Theor Appl Genet 116:363–372

Williams PH (1966) A system for the determination of races of *Plasmodiophora brassicae* that infect cabbage and rutabaga. Phytopathology 56(6):624–626

Williams PH, Keen NT, Strandberg JO, McNabola SS (1968) Metabolite synthesis and degradation during club root development in cabbage hypocotyls. Phytopathology 58:921

Williamson CJ, Bradshaw JE (1989) Sources of resistance to *Plasmodiophora brassicae* used in breeding programmes at SCRI. XIIth Eucarpia Congress. Vort Pflanzenz 15:17/12

Wit F (1964) Inheritance of reaction to clubroot in turnips. Hortic Res 5:47–49

Wit F, van de Weg M (1964) Clubroot resistance in turnips (*Brassica campestris* L.). I. Physiological races of the parasite and their identification in mixtures. Euphytica 13:9–18

Wolfe MS, McDermott JM (1994) Population genetics of plant pathogen interactions: the example of Erysiphe graminis-Hordeum vulgare pathosystem. Ann Rev Pytopathol 32:89–113

Wu J, Lee DY, Wang Y, Kim ST, Baek SB, Kim SG, Kang KY (2014a) Protein profiles secreted from phylloplane of rice leaves free from cytosolic proteins: application to study rice-*Magnaporthe oryzae* interactions. Physiol Mol Plant Pathol 88:28–35

Wu S, Shan L, He P (2014b) Microbial signature-triggered plant defense responses and early signaling mechanisms. Plant Sci 228C:118–126

Xing GM, Long MH, Tanaka S, Fujieda K (1989) Clubroot resistance in Brassicoraphanus (hybrid of cabbage radish). Plant Dis 33:189–194

Xu X, Liu X, Ge S, Jensen JD, Hu F, Li X, Dong Y, Gutenkunst RN, Fang L, Huang L, Li J, He W, Zhang G, Zheng X, Zhang F, Li Y, Yu C, Kristiansen K, Zhang X, Wang J, Wright M,

McCouch S, Nielsen R, Wang J, Wang W (2012) Resequencing 50 accessions of cultivated and wild rice yields markers for identifying agronomically important genes. Nat Biotechnol 30:105–111

Xue S, Cao T, Howard RJ, Hwang SF, Strelkov SE (2008) Isolation and variation in virulence of single-spore isolates of *Plasmodiophora brassicae* from Canada. Plant Dis 92(3):456–462

Yamagishi H, Yoshikawa H, Ashizawa KH, Yui S (1986) Effects of resistant plants as a catch crop on the reduction of resting spores of clubroot (*Plasmodiophora brassicae* Worn.) in soil. J Jpn Soc Hortic Sci 54:460–466

Yang TJ, Kim JS, Kwon SJ, Lim KB, Choi BS, Kim JA, Jin M, Park JY, Lim MH, Kim HI, Lim YP, Kang JJ, Hong JH, Kim CB, Bhak J, Bancroft I, Park BS (2006) Sequence-level analysis of the diploidization process in the triplicated Flowering Locus C region of *Brassica rapa*. Plant Cell 18:1339–1347

Yang J, Liu D, Wang X, Ji C, Cheng F, Liu B, Hu Z, Chen S, Pental D, Ju Y, Yao P, Li X, Xie K, Zhang J, Wang J, Liu F, Ma W, Shopan J, Zheng H, Mackenzie SA, Zhang M (2016) The genome sequence of allopolyploid *Brassica juncea* and analysis of differential homoeolog gene expression influencing selection. Nat Genet 48:1225–1232

Yoshikawa H (1980) Breeding for clubroot resistance in Chinese cabbage. In: First International Symposiumon Chinese Cabbage, Tsukuba Science City, Japan, (eds) Talekar NS, Selleck GW (Tainan: HongWen PrintingWorks)

Yoshikawa H (1981) Breeding for clubroot resistance in Chinese cabbage. In: Talekar NS, Griggs TD (eds), Chinese cabbage. In: Proceedings of the 1st international symposium, Tsukuba, Japan, pp 405–413

Yoshikawa H (1993) Studies on breeding of clubroot resistance in cole crops. Bull Natl Res Inst Veg Ornam Plants Tea Jpn Ser A 7:1–165

Yu S, Zhang F, Yu R, Zou Y, Qi J, Zhao X et al (2009) Genetic mapping and localization of a major QTL for seedling resistance to downy mildew in Chinese cabbage (Brassica rapa ssp. pekinensis). Mol Breed 23:573–590. https://doi.org/10.1007/s11032-009-9257-z

Yu F, Zhang X, Huang Z, Chu M, Song T, Falk KC, Deora A, Chen Q, Zhang Y, McGregor L, Gossen BD, McDonald MR, Peng G (2016) Identification of genome-wide variants and discovery of variants associated with *Brassica rapa* clubroot resistance gene Rcrl through bulked segregant RNA sequencing. PLoS ONE 11:e0153218

Yu F, Zhang X, Peng G, Falk KC, Strelkov SE, Gossen BD (2017) Genotyping-by-sequencing reveals three QTL for clubroot resistance to six pathotypes of *Plasmodiophora brassicae* in *Brassica rapa*. Sci Rep 7:4516

Zhang XC, Gassmann W (2003) RPS4-mediated disease resistance requires the combined presence of RPS4 transcripts with full-length and truncated open reading frames. Plant Cell 15:2333–2342

Zhang T, Zhao Z, Zhang C, Pang W, Choi SR, Lim YP, Zhongyun P (2014) Fine genetic and physical mapping of the CRb gene conferring resistance to clubroot disease in *Brassica rapa*. Mol Breed 34:1173–1183

Zhang H, Feng J, Zhang S, Zhang S, Li F, Strelkov SE, Sun R, Hwang SF (2015) Resistance to *Plasmodiophora brassicae* in *Brassica rapa* and *Brassica juncea* genotypes from China. Plant Dis 99:776–779

Zhang CW, Wei YP, Xiao D, Gao LW, Lyu SW, Hou XL, Bouuema G (2016) Transcriptomic and proteomic analyses provide new insights into the regulation mechanism of low-temperature-induced leafy head formation in Chinese cabbage. J Proteome 144:1–10

Genomics of Clubroot Pathogen and Pathogenesis

12

Abstract

The main characteristics of the *P. brassicae* genome are its small size of 24.2–25.41 Mb with high gene density and compactness. The genome contains genes potentially involved in host hormones manipulation. The enzymes (CAZymes) involved in synthesis, metabolism, and transport of carbohydrates are present. Mitochondrial genome is variable (34–180 kbp) in size because of variation in the intergenic regions and intron content. There is higher G+C content in mitochondrial sequence than in nuclear genome. Molecular mechanisms of pathogenesis revealed expression of *Y10* gene during plasmodial stages of *P. brassicae*. A trehalose-6-phosphate synthase gene and *PbSTK* L1 genes are expressed during resting spores formation. A serine protease (*PRO1*) gene is expressed during host infection as well as in dormant and germinating resting spores. More than 58 genes are up-regulated in the secondary zoospores relative to 55 genes down-regulated in primary zoospores. The genes *PbPP2A* and *PbHMG* are expressed at the early phases of infection. The gene *BjN1T1* expression in *B. juncea* promotes over production of auxins to form root galls. The main plant hormones pathways modified are cytokinin biosynthesis, auxin homeostasis and salicylic acid, and jasmonic acid metabolism. The methyltransferase gene of *P. brassicae* acts by methylation of salicylic acid, benzoic acid, and anthranilic acids to suppress salicylic acid-induced defense in the host plant. Bioinformatics have been used to identify putative effectors/genes of *P. brassicae* for pathogenesis. Transcriptomics analysis has revealed 162 candidates' enzymes/genes involved in lipid metabolism differentially expressed in the *P. brassicae* life cycle. The rate of infection and pathogen development is most rapid in susceptible host in comparison to resistant. The transcriptomic and proteomic approaches have elucidated mechanisms of *P. brassicae* infection and coordination in host reactions. Functional genomics approaches like expressed sequence tags (EST's), serial analysis of gene

G. S. Saharan et al., *Clubroot Disease of Crucifers*,
https://doi.org/10.1007/978-981-16-2133-8_12

expression (SAGE), massively parallel signature sequencing (MPSS), next generations sequencing (NGS), RNA sequencing (RNA-seq), microarray analysis, and construction of suppression subtractive hybridization (SSHcDNA) liberty have been used to achieve better insight into the molecular mechanisms of host–parasite interaction and expression of functional genes. Microarray chips, two-dimensional electrophoresis, and high-throughput sequencing techniques like metabotyping, laser microdissection, transcriptional profiling, and miRNA sequencing have been exploited to observe expression of genes regulating metabolic pathways, and signal transduction at early events during pathogenesis. Differential expression of genes in host shoot and root has been observed during *P. brassicae* infection. The members of TF families', viz., MYB, bHLH, and AP2/EREBP are highly up- or down-regulated in both shoot and root. However, both shoot and root also responded to pathogen in a number of common ways such as induction of JA, ABA, and shikimate pathways metabolism. The hypoxia genes *PDC1* and *PDC2* are induced during secondary infection in cortical cells of root galls. Several BrSWEET genes of clade 1 and clad III up-regulated on infection suggest close relationship between sugar translocation and *P. brassicae* growth. RNA's like arth-mi R160, arth-miR164, and arth-miR167 are involved in pathogenesis and target auxin-response-factor (ARF) in auxin homeostasis at clubroot development. Immunopropilin gene (*PbCYP3*) is differentially regulated in *P. brassicae* life stages at germinating, and maturing spores and plasmodial formation in infected *Brassica* hosts. Hydrolases and peptidases are the most common differentially regulated secreted proteins in *P. brassicae*-infected *B. napus*. The *Pb3-BSMT* gene (benzoic acid/salicylic methyltransferases) inactivates host defense by converting salicylic acid to methyl salicylate.

Keywords

Clubroot pathogen and pathogenesis · Early phase of pathogen genomics · Nuclear genome · Mitochondrial genome · Genomics of pathotypes · Pathogenesis · Expression and regulation of genes · Transcriptomics of genome · Functional characterization of gene · Effector genes · Cloning of protease gene · Differential expression of lipids genes · Omics data to elucidate mechanisms · Expression of genes for metabolic pathways · Expression of host genes · Host immunity by effectors genes · Host organ specific genes · Induction of hypoxia genes · SWEET genes · Host specific genes · Role of miRNAs · Induction of immunophilin genes

12.1 Introduction

Clubroot, caused by the obligate telluric biotroph protist *Plasmodiophora brassicae* Wor., is one of the economically most important diseases affecting *Brassica* crops in the world. The pathogen leads to severe yield losses worldwide. Clubroot development is characterized by the formation of galls on the roots of infected plants.

Above-ground symptoms include wilting, stunting, yellowing, and premature senescence. The pathogen has a complex life cycle comprising three stages: survival in soil as spores, root-hair infection, and cortical infection. First, resting spores can survive for a long period (more than 20 years) in the soil. Secondly, germination of haploid resting spores releases primary zoospores which infect root hairs, leading to intracellular haploid primary plasmodia. In the third stage, after nuclear division, secondary zoospores may be released into the soil and penetrate the cortical tissues. The step of released zoospores in the soil before invading the root cortex is not universally accepted among the research community. During this last stage, the pathogen develops into secondary multinucleate diploid plasmodia which cause the hypertrophy and hyperplasia of infected roots into characteristic clubs. Secondary plasmodia finally release a new set of haploid resting spores.

In the last decade of the twentieth century genomics research was initiated with the use of *Brassica* plant *Arabidopsis thaliana* as a model of host-pathosystem. During the first decade of present century, the genomics research has been strengthened with the application of omics approaches like genomics, transcriptomics, proteomics, biometabolomics, epigenomics, metagenomics, pathogenomics, phylogenomics, and bioinformatics. Now we are better equipped to identify, isolate clone, map, and locate the genes for pathogenomics, and host resistance to deploy and study their effects in augmenting host resistance through quantitative host resistance loci. Genes can be transferred from one species to another for better management of crucifer's biotic and abiotic stresses and having desirable attributes with high yield and quality.

The sequencing of pathogen (*P. brassicae*) and *Brassica* hosts (*A. thaliana, B. carinata, B. juncea, B. napus, B. nigra, B. oleracea, B. rapa*) genome has allowed molecular and genetical discoveries to understand the genetic relationship of different virulence genes for pathogenesis and R-genes of the Brassicaceae having interactions within and between members of host species and pathogen pathotypes. During the last decade, many virulence genes and R genes have been identified and functionally characterized to facilitate the discovery of host–pathogen interactions. The genetic and molecular mechanisms of pathogenesis and disease resistance have been determined. It has provided deeper insight into the molecular mechanisms and host–pathogen co-evolutionary arms races along with accurate molecular breeding using gene editing and gene silencing techniques for the betterment of the crucifers at the whole genome level. The genomics of host resistance to clubroot has been covered in Chap. 11 of this book.

12.2 Early Phase of *Plasmodiophora brassicae* Genomics

During the early phase of investigations, the number of chromosomes of *P. brassicae* has been a matter of debate for many years. Early light microscopy studies reported haploid numbers of $n = 4$ (Terby 1924), $n = 5$ (Tommerup and Ingram 1971), and $n = 8$ (Jones 1928). Therefore, karyotypic analysis based on serial sections of synaptonemal complexes in pachytene nuclei was carried out and revealed a haploid chromosomal number of 20 (Braselton 1982). Later studies using pulse-field gel electrophoresis reduced the number of chromosomes to 6 (Bryan et al. 1996) and 13 (Ito et al. 1994). Both groups reported chromosomal sizes of ca 0.5–2.2 Mb.

During investigations on the genomic DNA of *P. brassicae*, the presence of host DNA in the resting spores and zoospores of *P. brassicae* was found (Bryngelsson et al. 1988). It was speculated that the continuous uptake of host DNA can be of importance for the evolution of *P. brassicae*. It may be the existence of a sequence which hybridized with Brassica DNA in the *P. brassicae* genomic DNA (Buhariwalla and Mithen 1995). The cloned sequence showed homology to a tandem repeat element from *B. nigra* (Gupta et al. 1990). However, there is no evidence at present that host DNA is stably integrated into the genome of *P. brassicae*, although the continuous uptake of DNA does provide a route by which horizontal DNA transfer may occur. Also the mechanism of DNA uptake by the pathogen remains to be elucidated. Plasmodia of *P. brassicae* have been observed to engulf host organelles and cytoplasm (Buczacki 1983), which may provide a means of host DNA uptake.

To detect resting spores by polymerase chain reaction (PCR), specific primers were developed for *P. brassicae* DNA. The PCR is a sensitive tool, which is increasingly used for the detection and identification of plant pathogens (Henson and French 1993). Several investigations have shown that it is possible to isolate genomic DNA from resting spores of *P. brassicae* (Bischoff 1994; Bryngelsson et al. 1988; Buhariwalla and Mithen 1995; Buhariwalla et al. 1995; Ito et al. 1994). The cloning of several putative repetitive/high copy number DNA elements led to the development of specific PCR primers for the amplification of polymorphic DNA from *P. brassicae* (Buhariwalla et al. 1995) and was the first step towards detection of fungal resting spores by molecular methods. Ito et al. (1997) have attempted to isolate ipt (isopentenyl transferase)-like sequences from *P. brassicae*, because it is known that plasmodia can convert adenine to trans-zeatin (Muller and Hilgenberg 1986). Although a *P. brassicae*-specific DNA sequence was obtained by using PCR primers based on the ipt sequence of *Agrobacterium tumefaciens*, it did not show any homology to ipt. The sequence was conserved among all *P. brassicae* isolates tested and, thus, can be used for the detection of the fungus. This set of primers can specifically amplify a single-copy DNA sequence from *P. brassicae* DNA, whereas attempts to amplify DNA from other fungi including *Polymyxa betae*, *Polymyxa graminis,* and *Olpidium brassicae* gave negative results and can be, thus, used as a diagnostic tool.

An approach for the isolation of differentially induced proteins during pathogenesis can be the separation of proteins from infected and control roots on two-dimensional gel electrophoresis and subsequent comparison of the pattern. Hansen et al. (1994) examined protein patterns of *Brassica oleracea* roots infected with *P. brassicae* and found that several proteins present in healthy roots were reduced in clubroots. Ito et al. (1996) investigated the two-dimensional protein pattern of a susceptible and resistant response in roots of Chinese cabbage infected with two different *P. brassicae* isolates. They observed proteins which were enhanced in the susceptible response, repressed in susceptible response, enhanced in the resistant response, and repressed in resistant response. N-terminal sequencing of a protein increased in the susceptible response revealed homology to pathogenesis-related protein group 5 (Ito et al. 1996). A previous study on PR-proteins during pathogenesis showed no clear correlation between the activity of peroxidase and chitinase isoenzymes in roots of a susceptible and a resistant cultivar of Chinese cabbage and resistance/tolerance to clubroot (Ludwig-Muller et al. 1994). However, in the susceptible interaction, a novel chitinase isoenzyme with a pi of 7.3 was induced (Ludwig-Muller et al. 1994).

It has been shown in the interaction of plants with arbuscular mycorrhizal fungi that it was possible to isolate differentially expressed genes from the host plant and the AM fungus (Martin-Laurent et al. 1997; Murphy et al. 1997). Using a PCR-coupled subtractive hybridization technique (Buchanan-Wollaston and Ainsworth 1997) differentially expressed genes during clubroot development in Chinese cabbage were isolated. Three independent differentially expressed clones were isolated from a subtractive cDNA library constructed from root tissue 28 days after inoculation and sequenced. One was shown to have homology to small heat shock proteins (Ludwig-Muller 1999).

12.3 Main Features of the Nuclear Genome

For a better understanding of the pathogenesis and metabolic potential of *P. brassicae*, Daval et al. (2018) generated a de novo assembled genome of the eH isolate (belonging to pathotype 1). A paired-end shotgun library consisting of 73,082,688 reads (read length of 101 bp and inserts size of 430 bp) and approximately 225-fold genome sequence coverage was obtained. Two additional paired-end libraries were also generated. A 3 kb insert mate-pair Illumina library resulted in 97,629,456 reads (read length of 101 bp and insert size of 3200 bp) and a 5 kb insert mate-pair Illumina library resulted in 81,813,222 reads (read length of 101 bp and insert size of 5050 bp). Both libraries displayed also approximately 225-fold genome sequence coverage. The final assembly consisted of 136 gap-filled scaffolds with a N50 of 741,293 bp and the total sequence length was 24,559,637 bp with a 59.29% GC content. Basic information about the assembled genome and predicted genes is presented in Table 12.1, in comparison with the assembly of e3 and ZJ-1 genomes, while the ZJ-1 resources corresponded to draft data, the assemblies of eH and e3 shared the same levels of quality. The main features of the eH genome were indeed similar to the reference e3 genome, particularly concerning GC content (Fig. 12.1), but the eH genome displayed a less fragmented assembly (Table 12.1) and a higher number of predicted genes (12,811 in eH and 9730 in e3). Fig. 12.1 shows a visual representation of the eH genome's first 43 scaffolds (which represent 95% of the total genome sequence) compared to the other available *P. brassicae* sequences (Rolfe et al. 2016). The gene density, calculated as the number of genes every 20 kb, was nine on average. This gene density is relatively high as previously described, and contributes to the small genome size compared to the two known genomes of

Table 12.1 Genome statistics of *Plasmodiophora brassicae* assemblies. BUSCO analyses were done on genome sequences with the Ensembl Protists database (Daval et al. 2018)

Isolates	eH	e3 Schwelm et al. (2015)	ZJ-1 Bi et al. (2016)
Assembly accession	POCA00000000.1	GCA_001049375.1	GCA_002093825.1
Assembly assessment			
Genome assembly size (nt)	24,559,637	24,050,826	24,831,078
Number of assembled scaffolds	136	165	667
Mean scaffold size (nt)	180,585	145,762	37,228
Median scaffold size (nt)	11,351	36,592	2588
Minimal scaffold size (nt)	509	1868	2001
Maximal scaffold size (nt)	1,480,576	1,189,627	1,479,440
N50 (nt)	741,293	472,887	510,270
Genomic G + C content (%)	59.29	59.37	59.38
Busco 3.0.2			
Complete and single-copy	192	191	184
Complete and duplicated	3	3	10
Fragmented	3	3	3
Missing	17	18	18
Annotation statistics			
Number of protein-coding genes	12,590		
Average gene size	1398 nt		
Average exon size	335 nt		
Average exon number per gene	3.65		
Average intron size	66 nt		
Average intron number per gene	2.65		
Genes with 5′UTR	25%		
Genes with 3′UTR	35%		
Gene with transcript evidence	6736		
Gene without transcript evidence	6065		

free-living Rhizaria, *Bigelowiella natans* (~100 Mb) and *Reticulomyxa filosa* (~320 Mb).

Gene similarity between the eH and e3 isolates was studied by comparing the gene sequences of both genomes. Using a stringent query span of 80% (meaning that the percentage of alignment coverage of the query sequence should be at least 80%, making sure that genes with common regions), 60% of the genes found in the eH genome showed similarity with the e3 genes. Among the 6654 alignments displaying the query span of 80%, 4702 were with 100% gene identity and only

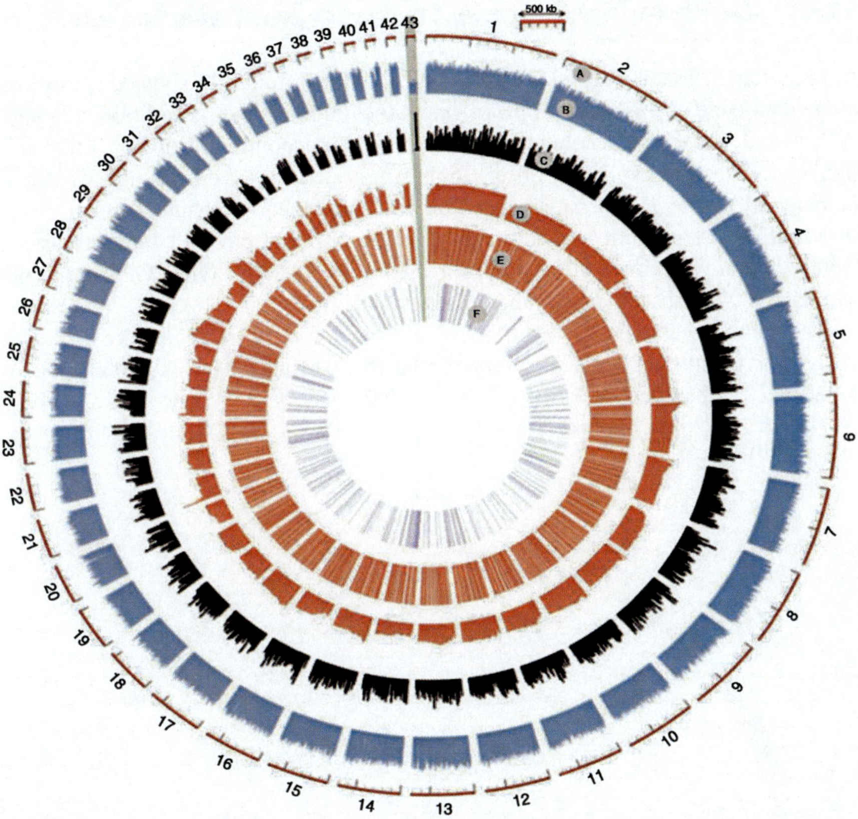

Fig. 12.1 Visual representation of the eH genome first 43 scaffolds. In the first outer circle (**a**) are represented the first 43 scaffolds of the assembled eH genome that correspond to 95% of the total genome sequence. In the second and third outer circles (**b, c**), the percentage in GC and the gene density (calculated as the number of genes every 20 kb) is indicated, respectively. In the two following inner circles, is represented the gene similarity between the isolates eH and e3: D circle pictures the relative rate every 10 kb of the eH genes number to the eH genes number that have similarity to e3 genes (80% threshold); (**e**) circle displays this similarity as lines when genes from eH and e3 have 80% similarity. The inner circle shows the gene similarity between the isolates eH and ZJ-1 ((**e**) purple lines; and eH and Led09, (**e**) green lines). The gray scaffold 43 represented the mitochondrial genome (Daval et al. 2018)

52 were with<60% gene identity (value often used to define genes as highly divergent). For the ZJ-1 isolate's genome, annotated regions are not available which made it unfeasible to carry out such accurate comparisons between genes and only contigs were compared which include non-coding regions. The scaffolds #44 to #242 (corresponding to 5% of the total genome) and in particular scaffolds #88 to #242 were more difficult to assemble and thus few genes were annotated in these regions.

12.4 Mitochondrial Genome, Structure, and Gene Content

In the circular genome representation (Fig. 12.1), scaffold #43 showed a low GC content and one match with a *P. brassicae* mitochondrial gene. In addition, as a high AT content was found in some Plasmodiophorid mitochondrial genomes, for example, in *S. subterranea*, these observations suggested that the scaffold #43 could correspond to the *P. brassicae* mitochondrial genome. To ensure that no errors occurred in its assembly, a new assembly of this scaffold #43 was performed using software specifically designed for organelle assembly (NOVOPlasty). Both assemblies were similar suggesting consistent results.

The newly assembled sequence of eH scaffold #43 is 102,962 bp, with a 24.67% GC content (Fig. 12.2) and was very similar (Max score = 91,831, Total score = 1.728×10^5, Query cover = 90%, Evalue = 0.0, Identity = 99%) to scaffold #49 of

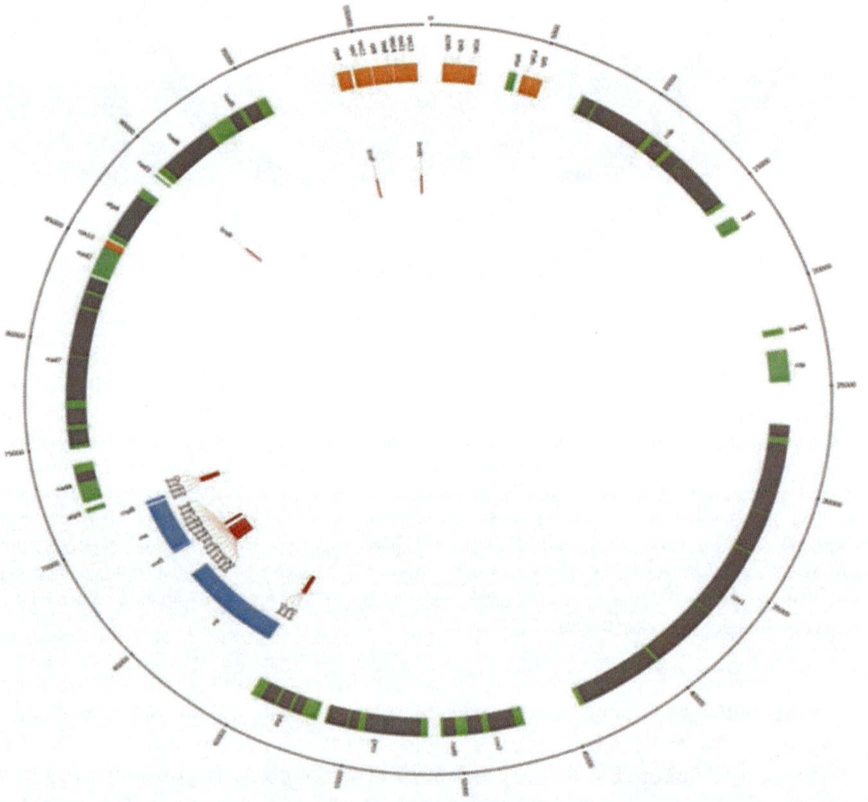

Fig. 12.2 Circular DNA map of the mitochondrial genome of *Plasmodiophora brassicae*. The outer circle shows the position of protein-coding genes with exons represented in green and introns in gray, and the position of ribosomal protein-coding genes in orange. The inner ring depicts the position of structural RNAs: rRNA and one ncRNA (rnpB) in blue, and tRNA in red (Daval et al. 2018)

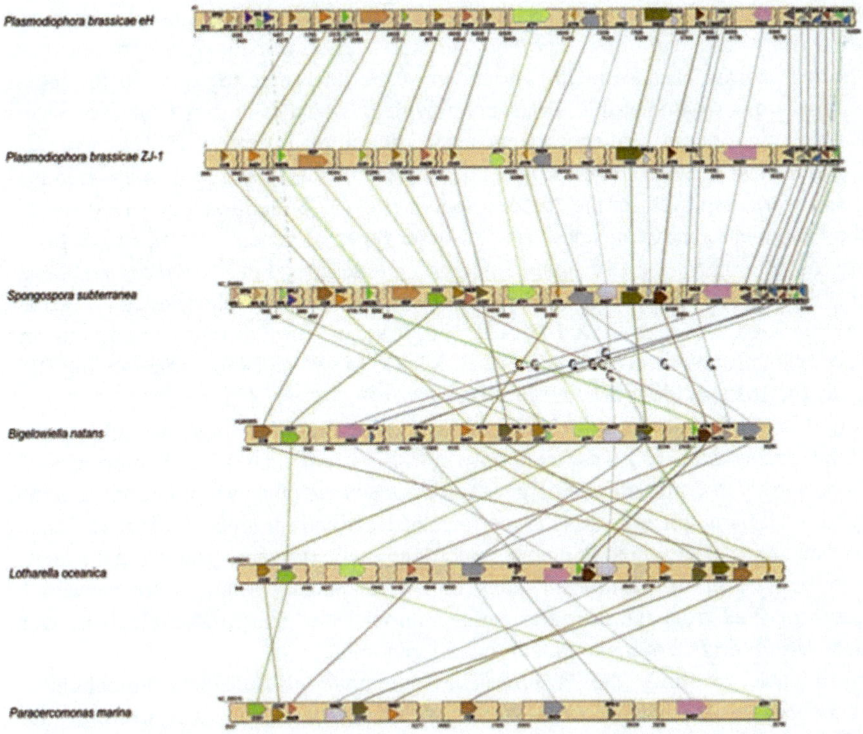

Fig. 12.3 Syntenic comparison between mitochondrial genomes. The names of the genes are indicated. Lines connecting regions of sequence similarity are drawn. Dotted lines show missing sequences (Daval et al. 2018)

the genome of the *P. brassicae* isolate ZJ-1 [10]. The size of the mitochondrial *P. brassicae* sequence (102,962 bp) was larger than that of *S. subterranea* (37,699 bp). Such diversity in genome size was previously described in Chlorarachniophytes, a small group of algae from the super group Rhizaria phylum Cercozoa, in which the mitochondrial genomes ranged from 34 to 180 kbp. Genome sizes vary considerably because of variation in the intergenic regions and intron content, which both contribute to mitochondrial genome plasticity. In *P. brassicae*, numerous long introns were found in mitochondrial genes in contrast to the *Lotharella oceanica* (Rhizaria group) and *B. natans* mtDNAs (Tanifuji et al. 2016), To assess the extent of mitochondrial genome rearrangements more closely, mitochondrial genomes of *P. brassicae* (eH and ZJ-1), *S. subterranea*, *B. natans*, *L. oceanica,* and *Paracercomonas marina* were aligned using the simple Synteny alignment tool and the protein sequences. A genome synteny map is shown in Fig. 12.3. Despite the significant difference in total size, mitochondrial genomes of *P. brassicae* eH and ZJ-1, and *S. subterranea* exhibited a similar architecture in which conservation of homologous genes and gene order were identified, but some genes are missing in ZJ-1. The organization of the three other mitochondrial sequences was also in part conserved in the content of genes but in a lesser extent concerning the gene order. The genome sizes varied considerably between these species, mainly because of the variations of the intergenic regions and the intron content, making impossible an accurate synteny's analysis using gene sequences. Both mitochondrial sequences were also quite similar between *P. brassicae* and *S. subterranea,* when the blast best-hit analyses of the *P. brassicae* mitochondrial genes were displayed for the "protein sequences" (32 protein-coding genes) and for the "complete gene sequences" (the 28 other types of genes, 60 genes were found among which 56 were also found among the 59 described in *S. subterranea*. Genes coding for proteins involved in respiration and oxidative phosphorylation (17 genes: ATP 1–7, 4 L, 9, COB, COX 1–3, ATP 1, 6, 8, 9), and for ribosomal subunit proteins (10 for small subunit: RPS 4, 7, 8, 10, 11, 12, 13, 14, 19, and 4 for large subunit: RPL 5, 6, 14, 16) was identified. As in *S. subterranea*, 24 genes coding for tRNA were found in the *P. brassicae*, but tRNH described in *S. subterranea* was not found in *P. brassicae* and tRNQ was found only in *P. brassicae* (and not in *S. subterranea*). Interestingly, the protein-coding genes in *P. brassicae* displayed a mosaic structure with small parts spread across the scaffold, as described for mitochondrial sequences of the chlorarachniophyte *B. natans* (belonging to the Rhizaria group) and the model cryptophytes protist *Guillardia theta*; both algal nuclear genomes are mosaics of genes derived from the host and endosymbionts. No transposable elements were detected in the *P. brassicae* mitochondrial genome.

In order to verify the *P. brassicae* taxonomic classification, mitochondrial sequences of the *COX1* gene were used in phylogenetic analyses. For this, 20 sequences of the *COX1* gene from diverse taxa were retrieved from NCBI (not found for ZJ-1 isolate), aligned, and used to create a maximum likelihood phylogenetic tree (Fig. 12.4), knowing that analyses with the Bayesian method gave the same results. The *COX1* gene was chosen because it was retrieved from more species than the other mitochondrial genes. Both *P. brassicae* and *S. subterranea* strains were clustered in a branch with very high bootstrap and a posterior probability of 0.99, confirming that both were closely related and belonged to the order of Plasmodiophoridae. The different known orders inside the Rhizaria were also relatively well related since families of Cercozoa (Plasmodiophoridae, Chlorarachniophyceae, Cercomonadidae, Euglyphidae, and Cyphoderiidae) were clustered in a clade supported by a posterior probability superior to 0.9.

Assessment of five insensitive lines under controlled conditions with application of B substantially reduced clubroot severity in insensitive lines. Comparison of two sensitive and two insensitive lines each of *B. napus* and *B. rapa* using Boron K-edge X-ray absorption near-edge structure (XANES) spectra from a synchrotron-based soft-X-ray beamline showed that some insensitive lines were able to extract B efficiently from a soil that contained very low levels of B. The spectra of sensitive and insensitive lines differed when additional B was applied, which demonstrated that the insensitive lines stored B (and other elements such as S and P) differently than the sensitive lines. Differences in the amount and type of storage likely affect the susceptibility to clubroot (Gossen et al. 2020).

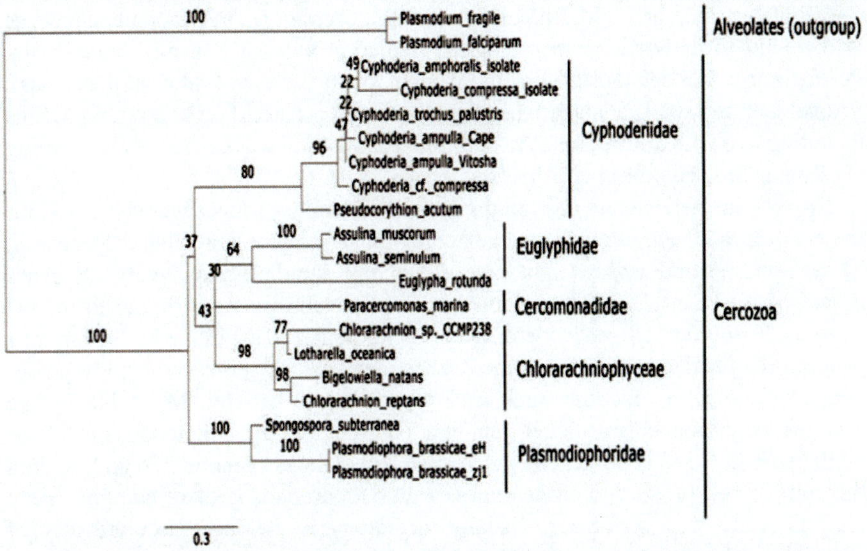

Fig. 12.4 Maximum likelihood phylogenetic tree of the COX1 mitochondrial sequences. The Alveolates cluster was used to root the tree. Bootstrap values are shown for each node. The scale bar corresponds to 0.3 substitutions per site (Daval et al. 2018)

12.5 Nuclear Genome Annotation

For annotation, the RNA-seq of resting spores was used and resulted in 23,540,954 reads. Other transcriptomes obtained *in planta* from other studies did not improve the annotation, showing that the transcriptional machinery was operational in resting spores. This allowed Schwelm et al. (2015) to well compare results to the e3 reference genome for which seven transcriptomes were used, including those of resting and maturing spores. The genome annotation was improved compared to the e3 reference genome by interrogating several external databases. The annotation data predicted 12,590 protein-coding genes in eH genome (Table 12.1). In total, 8989 genes were associated with GO terms, 4523 with EC numbers and 8231 with InterPro signatures. In rare cases of conflicting annotation, the results from manual annotation and Blast2GO were favored. Carbohydrate-active enzyme sequences of both the e3 and eH genomes were compared using dbCAN, a web server and DataBase for automated Carbohydrate-active enzyme Annotation. For the eH isolate, 287 enzymes were predicted. Eighty nine Glycoside Hydrolases (GH), 111 Glycosyltransferases, 47 Carbohydrate Esterases (CE), 7 Polysaccharide Lyases, 23 Carbohydrate-Binding Modules (CBM), and 10 Auxiliary Activities. GH18, CE4, and CBM18 were the most abundant enzymes (respectively 19, 21, 7 were predicted) within some of these groups. Among these 287 eH genes with predicted carbohydrate-active enzyme activity, 223 matched with e3 genes. The list of CAZymes was similar in both the eH and e3 genomes.

Using Signal P and TMHMM to predict the secretome, 741 protein sequences were found to harbor a secretion signal prediction without transmembrane helix. Furthermore, 543 sequences corresponded to small proteins (<450 amino acids). Schwelm et al. (2015) analyzed the e3 genome and predicted 553 secreted proteins including 416 with a secretome. An additional annotation was carried out by blasting eH data against clubroot published sequences.

Ninety-nine percent and one hundred percent identity matches were observed for the *P. brassicae* methyltransferase gene and the serine protease gene, respectively. Three hundred and eighty eight out of the four hundred and twenty-six genes described by Bi et al. (2016) involved in lipid metabolism were found in the eH genome. From the 118 genes identified by Feng et al. (2013) in the Led09 isolate's genome, 23 matched the eH genome. Genes identified as being involved in pathogenicity (i.e., with functions such as P-type ATPase activity, Mps1 binder-like protein) were mapped to the eH genome. To complete the annotation, out of the 9730 genes in the e3 genome, 6626 were matched to the eH genome. To go further in the comparison of eH and e3 genomes, a global comparison of orthologous gene clusters study was performed. Usually, orthology is the most accurate way of describing differences and similarities in the composition of genomes from different species, but it can also be used to compare strains intra species. The number of orthologous gene clusters was 8168, among which 7485 ortho groups were shared by e3 and eH isolates, and 683 clusters were specific to one of the two isolates, showing a certain relatedness between both stains.

The annotations were added to the corresponding eH genes using python scripts querying the Uniprot database. Using this updated annotation, the genes were classified into subsystems whose main terms are described in Fig. 1.1 Chap. 1. Most of the annotated genes were classed in the protein-binding (1433) GO term category. This class, not accurately defined for their function, still remained high because of the lack of functional data in Plasmodiophorid genomics resources. Ninety-six genes encoding potential RNA-binding proteins having at least one RNA-binding domain were identified. This protein family could play a role in *P. brassicae* pathogenicity, as described in *Ustilago maydis*, the basidiomycete causative agent of corn smut disease.

From the InterPro (IP) database, the ankyrin repeat (258 and 259) was found to be the major protein domain. The ankyrin repeat is the most common protein–protein interaction motif in nature. It is predominantly found in eukaryotic proteins, but also in many bacterial pathogens which employ various types of secretion systems to deliver ankyrin-containing proteins into eukaryotic cells where they mimic or manipulate various host functions.

In eukaryotes, proteins with ankyrin repeats are involved in many cellular functions, such as transcriptional regulation, signal transduction, cell cycle but also in virulence in Amoebae, suggesting an adaptation to their intracellular localization. The high number of ankyrin repeats in the *P. brassicae* genome could suggest a role in its capacity to invade host plant cortical cells.

To summarize the long list of GO terms by removing redundancy, a semantic similarity-based graph was constructed with REViGO. This allowed a representative subset of the terms to be found using a clustering algorithm that relies on semantic similarity measures. The Revigo Treemap showed a high representation of the GO terms linked to protein phosphorylation. Changes in the phosphorylation state have a broad impact on cells because they affect many processes such as the sub-cellular protein localization, protein–protein interactions, or signaling cascades. Moreover, interference of the host phosphorylation machinery is a common strategy used by pathogens to promote infection and development in host tissues, particularly in plant pathogenic fungi. A better understanding of the regulation of protein phosphorylation could help in developing targets for regulation of *P. brassicae* infection processes. Another GO term found frequently in the *P. brassicae* genome annotation was related to lipid metabolism. This is consistent with Bi et al. (2016) who reported proteins associated with lipid droplets in resting spores of *P. brassicae* and showed that lipid conversion is high in this protist during cortex infection.

Finally, the codon usage, a pattern that describes the degeneracy of the genetic code and illustrates differential usage of synonymous codons, was deciphered in nuclear genome versus mitochondrial genome. Using the genetic invertebrate mitrochondrial code from NCBI, the codon usage of the *P. brassicae* mitochondrial ORFs was determined. This showed that, as usually described in other species, the codon usage was not identical between nuclear and organelle in *P. brassicae*. As the major pattern in codon usage was a result of differences among the genes in the frequencies of G+C, which is coherent with the higher G+C content in mitochondrial sequence than in nuclear genome (Daval et al. 2018).

12.5.1 Genomics of *P. brassicae* Pathotypes

The details on the analysis of *P. brassicae* pathotypes genome have been given in Chap. 5 at 5.11. Readers are requested to see.

12.6 Genomics of *Plasmodiophora brassicae* Pathogenesis

The National Center for Biotechnology Information (NCBI) database for *P. brassicae*-originated accessions has yielded 180 nucleotide sequences, 138 expressed sequence tags, and 53 proteins. Of these sequences, only a small number have a potential importance in pathogenicity, largely based on sequence similarity with genes from other species. Most of these sequences are derived from cDNA libraries constructed by suppression subtractive hybridization (SSH). Among them, there are 76 *P. brassicae* genes isolated from a SSH library constructed between RNA from *P. brassicae*-infected and uninfected *Arabidopsis* tissue (Bulman et al. 2006). Many of the sequences are predicted to contain signal peptides for translocation of the encoded proteins. Using *Brassica rapa* L. clubroot galls, Sundelin et al. (2011) isolated about 140 clones. Half of these clones, which represented 24 unisequences (10 of which are newly characterized), originated from *P. brassicae*. Through the use of next generation sequencing (NGS) technologies, Burki et al. (2010) generated 3167 *P. brassicae*-specific contigs from infected *B. rapa*. These sequences, however, are very short, and their functions or relevance to clubroot development is not known. Feng et al. (2010) constructed a SSH library at the early stages of infection (7 days after inoculation) of canola roots by *P. brassicae*. The library consisted of 797 clones that represented 439 unigenes. Thirty-two of these genes are demonstrated to be of a *P. brassicae* origin. Differential expression of a subset of these genes in the resting spores, and at 7 days after inoculation was observed by quantitative PCR (qPCR), suggesting that some of these genes may play a role during pathogenesis. In addition to these libraries, a few other genes have been identified from *P. brassicae*, and postulated to be related to its pathogenicity. Ito et al. (1997) identified a gene (Y10) whose expression is exclusively correlated with the vegetative plasmodial stage of *P. brassicae*. Brodmann et al. (2002) found a trehalose-6-phosphate synthase gene with expression correlated with an accumulation of trehalose in resting spores. Ando et al. (2006b) demonstrated that the expression of PbSTKL1 increases strongly beginning 30 days after inoculation and is coincident with resting spore formation. Two cDNAs, PbBrip9, and PbCC249 are found to be strongly expressed at stages of disease development corresponding to the occurrence of sporulating plasmodia (Siemens et al. 2009b).

One *P. brassicae* gene, a serine protease (PRO1), has been functionally studied, and proven experimentally to be important for pathogenesis (Feng et al. 2010). Southern analysis and specific PCR amplification indicated that PRO1 is a single-copy gene present in a broad range of *P. brassicae* pathotypes. Northern analysis revealed expression of PRO1 during plant infection, as well as in dormant and germinating resting spores. The expression pattern assessed by Northern analysis was further confirmed by qPCR in a later study (Feng et al. 2013). The optimal temperature of Pro1 is 25 °C, and it exhibits maximum activity at pH 6.2–6.4, and minimal activity at pH 6.8–7.6. These values coincide with the temperature and pH conditions favorable for *P. brassicae* resting spore germination in the field. When Pro1 is used to treat canola root exudates, it enhances the stimulatory effect of the root exudates on *P. brassicae* resting spore germination, indicating that Pro1 plays a role during clubroot pathogenesis by stimulating resting spore germination through its proteolytic activity (Feng et al. 2014).) DNA sequences spanning or flanking 24 *P. brassicae* genes have been identified (Bulman et al. 2007) with sequencing close to 44 kb of genomic DNA. Some general characters of *P. brassicae* genes emerged from this preliminary genome survey: (1) *P. brassicae* genes are rich in spliceosomal introns with consensus splice sites, and branch-point sequences similar to those found in other eukaryotes, (2) as a consequence of being intron-rich, splicing is an important feature of gene expression, (3) *P. brassicae* transcription is likely to begin from initiator elements rather than TATA-box containing promoters, and (4) intergenic distances are short, ranging from 44 to 470 bp. Feng et al. (2013) reported a gene expression study in which all database-available *P. brassicae* genes (118) were investigated by dot-blot hybridization and qPCR. The expression of these genes in primary and secondary zoospores was compared. Both dot blot and qPCR identified up- and down-regulated genes, and the correlation between these two techniques was confirmed. Quantitative PCR indicated that 58 genes are up-regulated in the secondary zoospores relative to the primary zoospores, where 55 genes are down-regulated.

12.7 Expression and Regulation of Genes at Host–Pathogen Interaction

The functions of *P. brassicae* genes are hampered due to the small number of *P. brassicae* genes isolated, and the limited homology of many *Plasmodiophora* sequences with known genes. The approximately 100 partial cDNA fragments or full-length clones of *P. brassicae* have been isolated up to now. About half of these *P. brassicae* genes, has strongly indicated by DNA or amino acid similarities with characterized genes of other organisms, but until now limitedfunctions for *P. brassicae* genes have been proven experimentally. The remaining DNA fragments from *P. brassicae* reveal little or no similarity to sequence data in the databases, reflecting the paucity of genomic data available for protists from the Cercozoa (Keeling et al. 2005). Nucleotide or amino-acid-sequence similarity with known genes from other organisms can give a hint as to gene function, but the presence of a protein domain in a sequence might also be misleading.

The small roots of *A. thaliana* facilitate the observation of clubroot symptoms in this host plant, revealing a reasonably reproducible pattern of disease development under controlled environmental conditions (Kobelt 2000; Kobelt et al. 2000; Siemens et al. 2002, 2006a). Correlating transcription with distinct stages of symptom development caused by the pathogen will be an opportunity to define landmarks, and milestones of disease development with a possibility for gaining insights into the function of *P. brassicae* genes. A prominent actin cytoskeleton has been detected by immune-histochemistry over that part of the life cycle after secondary infection (Kobelt 2000). The PbActin and PbGST products appeared strongly expressed, and well detectable over the whole infection cycle within *A. thaliana* roots, and can therefore be used as indicators for an actively growing pathogen (Siemens et al. 2009a, b).

Two unidentified single-copy genes, *PbBrip9* and *PbCC249*, were strongly expressed at disease stages corresponding with the occurrence of sporulating plasmodia (Siemens et al. 2009b), whereas PbSTKL1 transcription coincided with resting spore formation (Ando et al. 2006b). The expression of gene fragment Y10 was exclusively correlated with the vegetative plasmodial stage (Ito et al. 1999), and the transcription of PbTPS correlated with an accumulation of trehalose (Brodmann et al. 2002). Finally, preliminary evidence indicates that the genes *PbPP2A* (protein phosphatase 2A) and *PbHMG* (DNA binding protein with high-mobility group box) are preferentially expressed in the early phases of infection in *B. rapa* roots (Sundelin 2008). The gene *BjN1T1* expression in *B. juncea* during infection promotes over production of auxins to form galls in infected roots (Liu et al. 2018). Altogether, the very first molecular landmarks of clubroot disease development in *Arabidopsis* are established. The whole genome transcription data from the host plant *A. thaliana* has revealed new insights into clubroot development (Siemens et al. 2006a, b; Jubault et al. 2008; Ludwig-Muller et al. 2009). Combining *Arabidopsis* microarray data with expression analyses of *P. brassicae* genes will provide a strong new approach to clubroot research in the future.

As with other biotrophic pathogens, it is likely that *P. brassicae* secretes proteins into the host cell to effect modifications of cell structure. Predicted proteins with signal peptides from *P. brassicae* are thus of much interest (Sundelin 2008). The *P. brassicae* sequences with signal peptides known to date, two matched with functionally characterized genes with roles in pathogenicity, namely, PbPPI1, which contains an FKBP family peptidyl-prolyl cis-trans isomerase domain, and PbPDA1, which contains a "Nod B-type" chitooligosaccharide deacetylase domain (Bulman et al. 2006). Surface-exposed or secreted PPIase domain-containing proteins have been found in several intracellular pathogens (Engleberg et al. 1989; Moro et al. 1995). The chitooligosaccharide deacetylase family includes members known to deacetylate substrates, including xylan and chitin, which might play roles in cell wall penetration during myxamoeboid movement of secondary plasmodia. The intracellular pathogen *P. brassicae* lives in close contact with the host, therefore, using a plant-derived RNAi strategy might work when an important parasitism gene of *P. brassicae* is identified (Siemens et al. 2009a, b).

12.8 Life Stages Specific Transcriptomics of *Plasmodiophora brassicae* Genome

Plasmodiophora brassicae genomes of one reference genome from a European isolate, and additional re-sequencing of five Canadian and one Chinese isolates have been characterized. The pathotype classification of these isolates is difficult to compare because these have been characterized using different differential hosts sets. The main characteristics of the *P. brassicae* genome is its small size (24.2–25.5 Mb), and high gene density. Thirteen chitin synthases were described, suggesting an important role for this gene family in resting spore formation. Secreted proteins, described to play a role in plant pathogenic organisms in suppressing the plant defense responses, and modifying the host metabolism, were previously predicted in silico in *P. brassicae*, but for most of them, no putative function could be assigned. The *P. brassicae* genome also contains genes potentially involved in host hormone manipulation, such as the auxin-responsive Gretchen Hagen 3, isopentenyl-transferases, methyl transferase, and cytokinin oxidase. The clubroot genome also appears to be lacking several metabolic pathways, a characteristic of the eukaryotic biotrophic plant pathogens. The missing genes encode proteins involved in sulfur, and nitrogen uptake, as well as the arginine, lysine, thiamine, and fatty acid biosynthesis pathways. In addition, only a few carbohydrate-active enzymes (CAZymes), involved in the synthesis, metabolism, and transport of carbohydrates, have been observed (Bi et al. 2016; Rolfe et al. 2016; Spanu et al. 2010).

The life stage-specific transcriptomics data related to the molecular events in the life cycle of *P. brassicae* was summarized by Schwelm et al. (2016a, b). During spore germination, and at the primary zoospore stage, enzymes involved in chitinous cell wall digestion are highly expressed. Several metabolism pathways such as the starch, citrate cycle, pentose phosphate pathway, as well as pyruvate, and trehalose metabolism were also very active. Although the molecular mechanisms involved in primary root-hair infection are unknown, the secondary phase has been better described. During re-infection by the secondary zoospores in root hairs or the root cortex, genes involved in basal and lipid metabolism were highly expressed. The plant clubroot symptoms (hyperplasia and hypertrophy of the infected tissues leading to the gall phenotype) were found to result from the deregulation of the host plant primary and secondary metabolism as well as from the modification of plant hormone homeostasis by the pathogen. The main plant hormone pathways modified were cytokinin biosynthesis, auxin homeostasis, and salicylic acid, and jasmonic acid metabolism. However, the mechanisms by which the pathogen may drive plant hormone homeostasis are still not completely worked out. At the end of the *P. brassicae* life cycle, secondary plasmodia transform into resting spores. In these structures, trehalose could play an important role for their long-term survival capacity. The description of the current transcriptomics data in *P. brassicae* shows that some important steps in *P. brassicae* developmental stages are still not well-characterized at the molecular level (Siemens et al. 2009a; Wagner et al. 2012; Ludwig-Muller et al. 2009, 2015). Daval et al. (2018) presented the genome of eH, a second European isolate. Refinement of the annotation of the eH genome led to the identification of the mitochondrial genome sequence, which was found to be bigger than that of *Spongospora subterranea*, another plant parasitic Plasmodiophorid phylogenetically related to *P. brassicae*. New pathways were also predicted, such as those for the synthesis of spermidine, a polyamine up-regulated in clubbed regions of roots. A *P. brassicae* pathway genome database was created to facilitate the functional study of metabolic pathways in transcriptomics approaches. These available tools can help in understanding the regulation of *P. brassicae* metabolism during infection, and in response to diverse constraints.

12.9 Functional Characterization of Gene During Pathogenesis

In order to manipulate plant defenses and enable pathogen colonization, many eukaryotic biotrophic plant pathogens have evolved advanced strategies to deliver effector proteins into the host cell during infection. Successful infection relies primarily on the success of the release of the effectors, which in many cases are responsible for the suppression of plant immunity. The initial recognition of conserved microbial features, known as pathogen-associated molecular patterns (PAMPs), leads to PAMP-triggered immunity (PTI) in the host. PAMP-triggered immunity is a first level of immune response that can be overcome by effector proteins produced by adapted pathogens. Resistance (R) to adapted pathogens is achieved through specific recognition of effectors, also known as avirulence proteins, by corresponding R proteins produced by the plant host. This effector-R protein recognition constitutes the second level of immune response, effector triggered immunity (ETI). *P. brassicae* is a well adapted pathogen of *Brassica* hosts; though indirect evidence suggests lack of either response, both PTI and ETI have not been well-characterized in the host-*P. brassicae* pathosystem (Dodds and Rathjen 2010; Jones and Dangl 2006; Siemens et al. 2011).

Understanding and identifying the proteins that comprise the secretome of *P. brassicae* is an important step towards identifying the complementary R proteins in the plant host. Putative effectors from biotrophic plant pathogens such as oomycetes and fungi are emerging from the sequencing and assembly of their genomes or transcriptomes, followed by comparative analysis with candidate effector genes. This strategy has been followed by Schwelm et al. (2015) and Rolfe et al. (2016) to identify putative effectors released by *P. brassicae* during the infection process. However, the obligate parasitic nature of *P. brassicae* has made it impossible to obtain a complete genome, with six draft genomes barely annotated for five Canadian pathotypes and one European pathotype. Transcriptomes of *P. brassicae* infecting *B. napus* and *Arabidopsis thaliana* in the Canadian study, and infecting *B. rapa*, *B. napus*, and *B. oleracea* var. *capitata* in the European study, have identified some candidate effectors. However, while the information obtained in both studies is a good start, the effectors specifically responsible for *P. brassicae* infection and subsequent disease progression have still to be identified (Pérez-López et al. 2018). Therefore, the overexpression of a benzoic acid/salicylic acid methyltransferase-encoding gene [PBRA_T000444 in *P. brassicae* European pathotype 3(Pbe3), PbPT3Sc00 026_A_1.308_1 in *P. brassicae* pathotype 3 (Pb3) draft genomes; GenBank accession number AFK13134] during the second week of infection, with expression peaking 3 and 4 weeks after infection has been recorded. Salicylic acid (SA) is essential for the activation of plant defense. In the plant cell, regulation of active SA is managed through the maintenance of different inactive forms of SA, such as methyl salicylate (MeSA). The methyltransferase identified in *P. brassicae* has been characterized in detail by Ludwig-Muller et al. (2015), identifying its role in the methylation of salicylic, benzoic, and anthranilic acids, thereby contributing to the suppression of the salicylic acid-induced defense in a plant host.

12.10 The Effector Proteins (Genes)

The basic criteria to identify candidate secreted effector proteins are: proteins with a signal peptide (within the initial 60 amino acids at the N-terminus), no transmembrane domains, small size between 300 to 450 amino acids, and mostly species-specific. These parameters have been used in the identification of putative *P. brassicae* effectors (Table 12.2). In addition, several other characteristics, motifs, and domains have been associated with effector proteins and have been used to improve identification and functional characterization of *P. brassicae* effectors/genes as follows:

Table 12.2 Steps to identify putative effectors within the secretome of *Plasmodiophora brassicae* in the European strain *Pbe3* and the Canadian strain *Pb3* (Pérez-López et al. 2018)

	Europe strain (*Pbe3*)	Canadian strain (*Pb3*)
Signal Peptide/No transmembrane domain[a]	533[b]	590[b]
D score >0.7	NA	431
Size (small-secreted proteins)[c]	416	221
Overexpression[d]	300	NA
Plant-free library[e]	92	NA

NA not applied
[a]In the Canadian strain, other sub-cellular localization signals were also used to remove putative proteins from further analysis
[b]Putative secreted proteins
[c]For the European strain the cutoff was <450 aa, while for the Canadian protein proteins <300 aa were selected
[d]At least 10 expected fragments per kilobase of transcript per million fragments (FPKM)
[e]FPKM log2 fold change >5 in plant-free library

12.10.1 Cysteine-Rich Proteins

Cysteine-rich small proteins have been identified as effectors in several plant pathogens, especially fungi, such as *Cladosporium fulvum* (syn. *Passalora fulva*), an asexual extracellular fungal pathogen of tomato. In *C. fulvum*, cys-rich effectors can inhibit and protect against plant hydrolytic enzymes, such as proteases, glucanases, and chitinases. Cys-rich small-secreted proteins have also been identified as major effectors in the obligate biotrophic pathogens *Melampsora laricipopulina*, and the Asian soybean rust fungus *Phakopsora pachyrhizi*, where one of the cys-rich small proteins identified as an effector has been shown to suppress plant immunity. However, its role in *P. brassicae* is not known.

12.10.2 Rxlr Motifs

The motif RxLR, arginine-any amino acid-leucine arginine, has been identified in the N-terminal of some oomycete and fungal effectors. Although the function of the RxLR motif in effector proteins remains unclear, it has been shown to be necessary for translocation into the host cell and to elicit immune responses in plant cells. Curiously, despite the divergence between *P. brassicae* and oomycetes, RxLR motifs have been reported in effectors in both the *P. brassicae* Canadian strain Pb3 and the European strain Pbe3, but the IDs of these RxLR protein-encoding genes have yet to be determined. Many putative effectors containing the RxLR motif also contain the second conserved motif, DEER (aspartate-glutamate-glutamatearginine), located toward the C-terminus (Yaeno et al. 2011).

12.10.3 Chitin-Binding Domains

Chitin, a recognized microbial PAMP, is a major structural component of fungal cell walls. Some fungal effectors have been shown to contain chitin-binding domains that are able to protect the pathogen against plant chitinases. These effectors can also act as scavengers of chitin fragments released by the pathogen during infection, thereby, avoiding a PAMP-triggered immunity response by the host plant. The resting spores of *P. brassicae*, formed at the end of the pathogen life cycle, contain chitin in their cell walls. The presence of the carbohydrate/chitin-binding (CBM18) domain, enriched in the plasmodiophorid secretome, suggests that these putative effectors might be involved in the formation and possibly the germination of resting spores. A blastp (https://blast.ncbi.nlm.nih.gov) search with a chitin-binding domain protein, identified in the genome of Canadian strain Pb3 (PbPT3Sc00048_S_5.266_1), showed identity with a *Fusarium fujikuroi* chitinase (GenBank accession number CCT72994). In the Pbe3 genome, chitin recognition proteins, like the protein encoded by the PBRA_002543 gene, have also been identified (Schwelm et al. 2015).

12.10.4 Protease/Protease Inhibitors

Sequence identity between plant pathogen effectors and other protein sequences is often low such that the assessment of functions based on putative orthology alone has been limited. In many cases, the three-dimensional structure of the protein, the disulfide bond pattern, and the cysteine spacing have been used to identify protease and/or protease inhibitors as putative effectors in obligate biotrophic soil-borne plant pathogens. These effectors target host proteins/proteases during infection, thereby, manipulating the host response to infection. During resting spore formation, Pbe3 over expresses releases Kazal-like serine protease (PBRA_001430), and papain protease inhibitors. The Kazal family of serine protease inhibitors, characterized by the presence of ten cysteine residues including the characteristic CX7CX6YX3CX2–3C signature, has been reported as effectors in fungi and oomycetes.

A putative serine protease (GenBank accession number AM411657) was identified among the *P. brassicae* genes that were expressed during infection. This protease carried a predicted signal peptide sequence but lacked homologs in other plant pathogens. Further, studies identified the serine protease as Pro1, a member of the S28 family of proteases that, due to its proteolytic activity, may play a role during clubroot pathogenesis by stimulating resting spore germination. Curiously, none of these studies suggested that this protein was an effector, probably because its suggested role occurs outside of the plant cell (Feng et al. 2010).

12.10.5 Nuclear Localization Domains

Differing from many fungi and oomycetes, *P. brassicae* is an intracellular pathogen. Another criterion used to identify putative effectors from obligate intracellular pathogens has been the presence of nuclear localization domains, which allows effectors to directly modulate plant gene expression. The effectors capable of migrating to the plant cell nucleus have only been described in bacteria for many years, but more recently these motifs together with nuclear localization of effectors have been described in fungi and nematodes. In bioinformatics pipelines designed to identify putative effectors, the inclusion of steps to remove proteins containing sub-cellular localization signals will remove these effectors, although researchers could analyze these amino acid sequences directly using the online tool Target P 1.1 Server (http://www.cbs.dtu.dk/services/TargetP). To date, none of the studies on the secretome and putative effectors of *P. brassicae* have detected or identified secreted proteins with such domains and none of the putative secreted proteins reported for *P. brassicae* have been identified as containing nuclear localization domains (Schwelm et al. 2015).

12.10.6 Pexel Motif

Plasmodiophora brassicae is evolutionarily closer to the malaria parasite, *Plasmodium* spp., than to oomycetes or fungal pathogens. While there are many differences between the immune system of animals and plants, they both share the common characteristic of being targeted by pathogen effectors. The discovery of the Pexel motif (Plasmodium export element) was a ground breaking finding that contributed to the understanding of the infection process of Plasmodium. Pexel is a pentameric motif present in the N-terminal portion of all the proteins translocated through the parasitophorous vacuole membrane. It is comprised of a positively charged, hydrophilic amino acid in position one (Arg or Lys), a hydrophobic amino acid in position three (Leu or Ile), and another less conserved amino acid in position 5 (predominantly Asp, Glu, or Gln), with non-charged amino acids in positions two and four (Ser, Thr, Cys, Met, Asn, or Gln). The N-terminal domain of an effector protein from the soybean cyst nematode *Heterodera glycines*, containing unique sequence similarity to domains of an effector of *Plasmodium* spp., indicates that the use of analogous effectors by highly diverse parasites of plants and animals occurs and is worth exploring (Marti et al. 2004).

12.10.7 Plant Pathogenic Plasmodiophorids

Plasmodiophora brassicae is not the only plant pathogenic protists, nor is it the only plasmodiophorid affecting economically important crops: the group includes *Spongospora subterranea*, the causal agent of powdery scab on potato tubers, and *Polymyxa* spp., *Polymyxa graminis,* and *Polymyxa betae*, which affect graminaceous plants and Chenopodiaceae plants, respectively. Genomic sequences, although limited, are available from *S. subterranea* and comprehensive *S. subterranea* transcriptomic datasets are available from root galls. These data suggest intron-rich genes and an enrichment of chitin-related enzymes in the *S. subterranean* transcriptome. Transposable elements are more expressed in *S. subterranea* than in *P. brassicae*, but evidence for recombination in *S. subterranea* is limited and there is little understanding of sexual recombination in phytomyxids. A study of secreted proteins in *S. subterranea* has been carried out, identifying the benzoic acid/salicylic acid methyltransferase-encoding gene overexpression (Schwelm et al. 2015, 2017).

12.10.8 Effectors (Genes) in Clubroot Life Cycle

Clubroot has three main stages to its life cycle, (1) survival in the soil and germination of resting spores, (2) root-hair or epidermal cell infection, and (3) cortical infection, although the pathogen has also been observed in phloem (Kageyama and Asano 2009). The serine protease, Pro1, is thought to be involved in step 1, the germination of resting spores (Feng et al. 2010). It is from step 2 to 5 of the life cycle, it is expected that other proteases and protease inhibitors, such as cys-rich proteases and methyltransferases will be produced to suppress plant immunity, thereby preventing the host plant from mounting responses such as programmed cell death and increasing the probability of success of the infection process (Siemens et al. 2011). During the formation of primary and secondary plasmodia, infected root tissues develop into swollen galls; it is expected that *P. brassicae* will secret an array of effector proteins triggering growth, expansion, and differentiations of infected host cells. During zoospore maturation, as expected, *P. brassicae* effectors with chitin-binding domains will express to remove chitin fragments that otherwise could trigger PAMP-associated immunity during late stages of the *P. brassicae* life cycle (Fig. 8.5, Chap. 8). Based on previous studies, might be also expected overexpression of effectors implies on manipulating plant defense such as PbBSMT (Malinowski et al. 2012; Ludwig-Muller et al. 2015) and effectors disturbing plant meristematic activity during the formation of secondary zoospore and secondary plasmodia. While there are still steps in the life cycle of *P. brassicae* that remain to be clarified, the study of effectors and their roles in life cycle completion and clubroot disease progression is an arena waiting to be explored (Pérez-López et al. 2018).

12.10.9 Coherent Pipeline to Identify Effectors in Future

Computational prediction is an excellent starting point to screen for putative effectors, to identify functional domains, and to help us understand the evolution, distribution, and characteristics of effectors. Utilizing the information available, Pérez-López et al. (2018) have designed a coherent pipeline aimed at identifying putative effectors involved in the infection process by *P. brassicae,* and subsequent clubroot disease progression in its plant host (Fig. 8.6, Chap. 8). The pipeline makes use of tools to identify the origin of reads, and identify motifs, and functions of the predicted secreted proteins. The bioinformatics tools referred to in the pipeline, together with their descriptions and websites, are listed in Table 8.1, Chap. 8. This pipeline is only a suggestion, although key parameters should be set in order to run it properly.

Bioinformatics analysis is the route to the identification of candidate effector proteins, but laboratory confirmation of function will always be required. Validation of the RNA-seq data through qPCR or ddPCR, identification of the sub-cellular localization of the candidate effector proteins through transient expression or stable transformation, and identification of the plant proteins interacting with the pathogen effectors are some of the logical next steps towards the identification of *P. brassicae* effector proteins (Bulman et al. 2018; Irani et al. 2018)

12.11 Cloning of Protease Gene in *Plasmodiophora brassicae* for Pathogenesis

A serine protease gene (*PRO1*) was cloned from *P. brassicae* and its molecular characteristics were investigated by Feng et al. (2010). Southern analysis and specific polymerase chain reaction (PCR) amplification indicated that *PRO1* is a single-copy gene present in a broad range of *P. brassicae* pathotypes. Northern analysis revealed that the expression of *PRO1* was induced during plant infection, and that the quantity of transcript fluctuated according to the stage of pathogenesis. Amino acid sequence analysis suggested that the encoded protein (Pro1) belongs to the S28 family of proteases, with a predicted signal peptide, and a theoretical molecular mass of 49.4 kDa. The open reading frame (ORF) of *PRO1* was transferred into *Pichia pastoris,* and Pro1 was heterologously produced. Pro1 showed proteolytic activity on skimmed milk and *N*-succinyl-Ala-Ala-Phe-7-amido-4-methylcoumarin, and the activity could be inhibited by serine protease inhibitors with the chelating agent ethylenediaminetetraacetic acid. The optimal temperature of Pro1 was 25 °C, and it exhibited high activity at pH 6.0–6.4. These values coincide with the temperature, and pH conditions favorable for *P. brassicae* resting spore germination in the field. When Pro1 was used to treat *Brassica* root exudates, it enhanced the stimulating effect of the root exudates on *P. brassicae* resting spore germination, indicating that Pro1 may play a role during clubroot pathogenesis by stimulating resting spore germination through its proteolytic activity (Feng et al. 2010).

12.12 Differential Expression of Lipids Genes

The lipid droplets are present in the resting spores, and secondary plasmodia of *P. brassicae* as observed by transmission electron microscopy (TEM). Nile red staining showed that different zoosporic stages of *P. brassicae* contained lipid droplets. Two hundred and ninety-five lipid droplet-associated proteins were identified by proteomic study, and seven types of fatty acid compositions were chemically characterized using GC-MS in lipid droplets. Genome integrated transcriptome, proteome of its lipid synthesis, storage, and metabolism are very essential aspect in *P. brassicae* life cycle (Bi et al. 2016). Lipid droplet-associated proteins were identified, and categorized into nine groups. Transcriptome analysis of these proteins during three different zoosporic stages revealed differences in gene expression pattern. GO enrichment analysis revealed that these proteins associated with lipid droplets were mainly linked to biosynthesis and metabolism. GC-MS analysis revealed that lipid droplets contain seven types of free fatty acids, saturated fatty acids C16:0 and C18:0, and unsaturated fatty acids C18:1Δ9, C18:1Δ11, C18:2, C20:4, and C20:5. *P. brassicae* accumulated a large amount of triacylglycerols (TAGs). Bi et al. (2016) systematically analyzed the putative proteins involved in TAG biosynthesis, and its metabolic pathway. KEGG pathway analysis defined 3390 genes, including 167 genes involved in lipid metabolism. Transcriptome analysis revealed that 162 candidate enzymes involved in lipid metabolism were differential expressed.

12.13 Effect of Host Resistance to *Plasmodiophora brassicae* Infection

Infection of root hairs by *P. brassicae* occurs in a broad range of plants, not only in susceptible host cultivars but also in resistant cultivars (Diederichsen et al. 2009) and even non-host plants (Feng et al. 2012b). Root hairs serve as a niche for pathogen increase prior to invasion of the root cortex, but also provide an opportunity for close interaction between host and pathogen, which facilitates the evolution of mechanisms by which *P. brassicae* can break down basal resistance to cortical infection (Feng et al. 2013). In susceptible plants, both root-hair infection and clubroot severity increase with higher inoculum density (Hwang et al. 2011b, c). Root-hair infection occurs more frequently with a compatible isolate (susceptible reaction) compared with an incompatible isolate (resistant reaction) and, subsequently, fewer cells within the root become infected when the host is resistant (Tanaka et al. 2006). The compatible isolate forms secondary plasmodia with many nuclei and eventually resting spores are formed in the host root tissue. In contrast, plasmodia formed by the incompatible isolate remain immature with only a small number of nuclei and do not produce resting spores (Hwang et al. 2011a, b; Deora et al. 2013). It suggest that resistance in host species is associated with a small amount of suppression of infection during primary infection, but a much greater suppression of secondary infection and subsequent plasmodial development in the root cortex. One important difference between the resistant and susceptible host reactions is that in a resistant host, the secondary thickening of the cell walls in the xylem are not degraded and there are fewer cell wall breakages (Donald et al. 2008). The influence of cultivar resistance on root-hair infection was studied across a range of inoculum densities in canola in one resistant and one susceptible cultivar (Hwang et al. 2011b). Subsequently it was observed that root-hair infection in two resistant and three susceptible canola cultivars (Hwang et al. 2012a). Infection and clubroot severity were higher and plant height was reduced in the susceptible cultivars relative to the resistant cultivars. Root-hair infection and the amount of *P. brassicae* DNA (assessed using qPCR analysis) increased in both resistant and susceptible cultivars over time, but the increases were greater in the susceptible cultivars. Also, there was a strong linear relationship between root infection and the amount of DNA of *P. brassicae* in the root hairs. The slow increase in the amount of pathogen DNA and root-hair infection observed in the resistant cultivars, and the sharp increases observed between 4 and 10 days after inoculation in the susceptible cultivars, indicate that secondary infection and pathogen development are occurring quickly in the susceptible cultivars. However, sharp declines in detectable levels of pathogen DNA in the seedling roots occurred during the transition from zoospore development in the root hairs to secondary infection (Hwang et al. 2012a). These levels quickly rebounded as the pathogen spreads into the root cortex after secondary infection (Gludovacz 2013).

The rate of infection and pathogen development is most rapid in susceptible canola lines (Deora et al. 2012, 2013). However, resistant plants showed reduced growth and delayed development when inoculated with resting spores of an avirulent pathotype (Hwang et al. 2012a, b; Deora et al. 2013), which indicates that resistance is an active process. In at least one vegetable *Brassica* line, microscopic analysis has revealed substantial levels of pathogen infection in inoculated roots of symptomless plants (Gludovacz 2013; Gludovacz et al. 2013). The mechanism for this response has not yet been determined, but it is possible that it involves genes for tolerance to *P. brassicae*. Successful management of *P. brassicae* through the deployment of genetically resistant cultivars will require very careful control of susceptible weeds and canola volunteers to maximize the impact of resistant cultivars on reducing resting spore levels in the soil and to prevent the build-up of virulent populations (Hwang et al. 2014).

12.14 Analysis of Omics Data to Elucidate Mechanisms of Pathogenesis

Suppression subtractive hybridization (SSH) has been conducted between RNA from *P. brassicae*-infected and uninfected *Arabidopsis* tissue, and a total of 76 new *P. brassicae* gene sequences have been identified (Bulman et al. 2006). Many of the sequences are predicted to contain signal peptides for translocation of the encoded proteins. Again using SSH, but starting with *B. rapa* rather than *Arabidopsis* clubroot galls, Sundelin et al. (2011) isolated about 140 clones. Half of these clones, which represented 24 unisequences, originated from *P. brassicae*. Out of these, 10 unisequences are newly characterized *P. brassicae* genes. By using next generation sequencing (NGS) technologies, Burki et al. (2010) generated 27,333 contigs assembled from 454 reads from *P. brassicae*-infected *B. rapa*. Among these contigs, 3167 are from *P. brassicae*. A few genes have been postulated to be related to the pathogenicity of *P. brassicae* (Ito et al. 1999; Brodmann et al. 2002; Ando et al. 2006a; Siemens et al. 2009a), mainly based on the observation that the expression of the gene is correlated with certain stage of disease development. One gene, a serine protease (*PRO1*), has been heterologously expressed and proven experimentally to be important for resting spore germination (Feng et al. 2010).

The transcriptomic and proteomic approaches have elucidated the mechanisms of *P. brassicae* infection and coordination in plant reactions. Using a full genome Affymetrix chip, Siemens et al. (2006a, b) found that more than 1000 *Arabidopsis* genes are differentially expressed in *P. brassicae*-infected roots at either 10 or 23 days after inoculation versus non-infected roots. Two proteomic studies, both using two-dimensional gel electrophoresis and quantitative image analysis in conjunction with mass spectrometry-based protein identification, illustrated changes in protein abundance at 4 days after inoculation (dai) in *Arabidopsis* (Devos et al. 2006), and at 0.5, 1, 2, and 3 dai in *B. napus* (Cao et al. 2008). The observed differential expression on mRNA or protein levels at the different stages of pathogenesis suggested that the corresponding genes might be involved in the plant response to the clubroot pathogen. Functional genomic approaches provide powerful tools for identifying expressed genes. Among these techniques, expressed sequence tags (ESTs, Adams et al. 1991), serial analysis of gene expression (SAGE, Velculescu et al. 1995), and massively parallel signature sequencing (MPSS, Brenner et al. 2000) have been successfully employed. Due to its relative simplicity and inexpensive nature, single pass EST sequencing has been the most widely used method to characterize genes associated with cellular development, with biotic and abiotic stress in plant research.

Although many EST clones lack the complete sequences of mRNAs, they can be very helpful to investigate the biological function of particular genes, especially in poorly annotated genomes. Additionally, it is possible to obtain full-length clones starting from ESTs. EST resources are also important for comparative transcriptomics because gene expression profiles can be obtained by analyzing ESTs generated under varying conditions. The whole transcriptome sequencing using next generation sequencing (NGS) technologies or RNA sequencing (RNA-seq) has been started to be utilized to reveal the complex landscape of the transcriptome from various organisms at an unprecedented level of sensitivity and accuracy (Ozsolak and Milos 2011). Compared with traditional EST sequencing and other high-throughput technologies such as microarrays, NGS and RNA-seq offer a near-complete snapshot of a transcriptome with a higher dynamic range of expression level. However, common NGS platforms including Illumina, SOLiD, and 454 suffer from the disadvantage that the sequence reads are often very short (35–500 bp). As a consequence, high technologies on transcriptome assembly are required after NGS.

Despite the advantages of NGS and RNA-seq, EST cloning incorporated with SSH (Diatchenko et al. 1996) has been used to identify genes involved in host–pathogen interactions. This method reduces, by subtractive hybridization, the cloning of abundantly expressed housekeeping genes or genes commonly expressed in both control and treated plants, thereby normalizing expressed cDNA profiles during library construction. Moreover, it significantly enhances, by suppression PCR, the chances of cloning differentially expressed genes. This is particularly important because many pathogenesis-related genes are expressed at low levels, and can be limited to a particular tissue or cell type. This technique has been used to isolate plant and pathogen genes that are expressed during infection (Huang et al. 2007). Both of the two cDNA libraries of *P. brassicae*-infected plants, one using *Arabidopsis* (Bulman et al. 2006) and the other *B. rapa* (Sundelin et al. 2011), were constructed by using the SSH technique.

The construction of a SSH cDNA library followed by EST analysis was used to achieve a better insight into the genes expressed during the compatible interaction between a susceptible *Brassica* cultivar and *P. brassicae*. The infection of plants by pathogenic and the subsequent establishment of disease involves substantial changes in the biochemistry and physiology of both partners. Analysis of genes that are expressed during these interactions represents a powerful strategy to obtain insights into the molecular events underlying these changes. Feng et al. (2012a) have used suppression subtractive hybridization (SSH) and expressed sequence tag (EST) analysis to investigate gene expression during the early stages of colonization of *Brassica* roots by *P. brassicae*. A cDNA library was constructed by SSH which consisted of 797 clones that represented 439 unigenes. Thirty-two of these genes were demonstrated to be of *P. brassicae* origin. The remaining 407 genes, which were of a *Brassica* origin, were subjected to gene ontology and *in silico* analyses. Real-time PCR analysis of ten *P. brassicae* and seven *Brassica* genes indicated that seven of the former and five of the latter were up-regulated at 7 days after infection, suggesting the importance of these genes in pathogenesis or host resistance.

Molecular mechanisms of *P. brassicae* infection during primary and secondary zoospores cannot be differentiated morphologically. When inoculated onto healthy *Brassica* roots, secondary zoospores cause primary infections similar to those caused by primary zoospores (Feng et al. 2012b). In contrast, when inoculated onto *Brassica* roots that are already under primary infection, primary zoospores can cause secondary infections without going through a primary infection phase (Feng et al. 2013). Thus, from a developmental biology standpoint, neither morphogenesis nor functional differentiation can clearly distinguish between these two kinds of zoospores. The primary and secondary zoospores may be essentially same or they undergo a divergence. No matter the state of divergence between the spore types, the "arms race" with the host would induce changes in the secondary zoospores. These changes are most likely at the gene expression level, and can be both genetic and epigenetic in nature. Knowledge of the molecular differences between primary and secondary zoospores may be useful for rational clubroot resistance breeding efforts as well as for the development of pesticides effective against *P. brassicae*. If these two kinds of zoospores are the same with respect to gene expression, a gene or quantitative trait loci (QTL) for resistance against either primary or secondary infection will be also effective against the other. Thus, the selection for resistance and subsequent genetic studies can reliably be conducted at the primary infection stage, which would save time and resources. This scenario is also applicable to fungicide development. Despite this importance, data on gene expression in these two kinds of zoospores is limited with information on the genetic makeup of *P. brassicae*, with only a small number of expressed sequence tags (ESTs) and protein sequences available in GenBank. Similarly, very few *P. brassicae* genes have been functionally characterized (Feng et al. 2010). Most of the database-available genes are derived from cDNA libraries that were constructed using the suppression subtractive hybridization techniques (Bulman et al. 2006; Feng et al. 2012b; Sundelin et al. 2011), which are enriched in genes up-regulated during pathogenesis. Thus, these genes are all good candidates for gene expression studies during the different stages of pathogenesis. All database-available of *P. brassicae* genes were analyzed in silico. To compare the gene regulation profiles of the primary and secondary zoospores, these genes were subjected to both dot-blot and real-time PCR analyses to understand clubroot pathogenesis and provide gene candidates for gene functional studies, which will in turn facilitate breeding for resistance and the identification of potential targets for the development of novel pesticides.

The molecular basis of pathogenicity of *P. brassicae* has been determined with specific primers designed and used to amplify genomic fragments of *P. brassicae* genes that represent all database-available proteins and ESTs. The PCR products were blotted on membranes and hybridized with digoxigenin-labeled double-stranded cDNA, derived from either primary or secondary zoospores of *P. brassicae*. The same primers were also used in real-time PCR against the single-stranded cDNA synthesized from the two types of zoospores. Both dot blot and real-time PCR identified up- and down-regulated genes provide the correlation between these two techniques. Real-time PCR indicated that 58 genes were up-regulated in the secondary zoospores relative to the primary zoospores, whereas 55 were down-regulated. These data suggest that different mechanisms are utilized by the pathogen in causing primary and secondary infections. The expression patterns of genes with known or putative functions suggest the relative importance of these genes during pathogenesis. In contrast, highly expressed or regulated genes with unknown function can be further studied in identifying pathogenicity factors (Feng et al. 2013).

12.15 Expression of Host Genes at Primary Stage of Infection

Wide-scale gene expression analysis offers an opportunity to explore the molecular basis of plant–pathogen interactions, particularly with respect to mechanisms of resistance and basal defense, as well as host versus non-host resistance, and biotrophy versus necrotrophy (Oliver and Ipcho 2004; Andreasson et al. 2005; Glazebrook 2005; Wise et al. 2007). Microarray technology has enabled the identification of defense-related genes for many plant–pathogen interactions as well as the mechanisms by which plants develop resistance.

The infection of *P. brassicae* occurs in two distinct phases: the primary phase, which occurs predominantly in root hairs, and the secondary phase, which occurs predominantly in the root cortex (Ingram and Tommerup 1972). It is the secondary phase that produces the characteristic disease symptoms of root tissue hypertrophy and hyperplasia, or "clubbing". Broadly speaking, in resistant hosts, the primary phase of the life cycle can still be observed, but not in the secondary phase (Dekhiujen 1979; Takahashi et al. 2006; Donald et al. 2008). In a compatible (susceptible) interaction with *Arabidopsis thaliana* (L.) Heynh., the pathogen can proceed through its entire life cycle from 3 to 28 days (Siemens et al. 2002). *P. brassicae* attachment and penetration of the root hair can occur from 4 days after inoculation onwards (Agarwal et al. 2009). This occurs after a resting spore releases a single motile zoospore that attaches to the wall of the root hair and encysts by means of a tubular structure (the rohr and the stachel), injecting its protoplast into the host cytoplasm (Aist and Williams 1971). Between 5 and 12 days after inoculation, plasmodia and then zoosporangia (producing secondary zoospores) develop within the root hairs with little visual sign of damage to the host cells. Between 10 and 13 days after inoculation, *P. brassicae* undergoes a transition from the primary phase to the secondary phase of its life cycle as observed by Devos et al. (2005) and Siemens et al. (2006a, b) depending on the isolate used (single spore or field population). Within 28 days after inoculation, the secondary plasmodia multiply into resting spores within the cortical cells proliferating into galls in *A. thaliana* (Agarwal et al. 2009). Siemens et al. (2006a, b) have reported the use of microarray analysis to examine gene expression following inoculation of a susceptible *A. thaliana* ecotype Col-0 with *P. brassicae* at 10 and 23 days after inoculation (DAI) using a single-spore isolate. These two time points were within the so-called secondary infection stage, which takes place in the root cortex, the 10-day time point being described as the early secondary plasmodial stage and the 23-day time point as the late secondary stage, by which time the disease has progressed substantially in the roots to develop galls. Their study found wide-scale changes in the expression of genes in several important functional groups including those involved in growth, sugar phosphate metabolism, defense, and plant hormone synthesis. The two time points analyzed, however, were relatively late in the interaction for studying the potentially important events that take place during initial pathogen invasion. Proteomic-based approaches have also been reported (Devos et al. 2006; Cao et al. 2008). Devos et al. (2006) examined the susceptible response in *A. thaliana* roots at 4 days after inoculation (DAI) with a field population of *P. brassicae*. Most of the differentially regulated proteins (12%) were involved in plant defense, cell differentiation, hormone metabolism, and detoxification. Cao et al. (2008) examined the susceptible response in canola (*Brassica napus* L.) roots 12, 24, 48, and 72 h after inoculation using the field population of *P. brassicae* collected from central Alberta, Canada. Only 20 protein spots in total were differentially regulated including proteins involved in lignin biosynthesis, cytokinin metabolism, glycolysis, intracellular calcium homeostasis, and the detoxification of reactive oxygen species (ROS). Further, a review of the interactions of *P. brassicae* with its hosts (Ludwig-Muller and Schuller 2008) emphasized that although the components of early signaling in the interaction between *A. thaliana* and leaf pathogens are well understood, there are relatively few studies with root pathogens and indeed none conducted with *P. brassicae*. A comparative study between resistance and susceptibility is not possible for the interaction of *P. brassicae* with *A. thaliana*, as almost all the

A. thaliana ecotypes tested so far are susceptible to *P. brassicae* (Siemens et al. 2002; Agarwal et al. 2009). Also, the enormous variation in the virulence of *P. brassicae* populations reported worldwide (Jones et al. 1982) often makes it difficult to determine the host's genomic and metabolomic responses.

Microarray analysis was used to investigate changes in host gene expression during the primary stages of the interaction between the susceptible plant *Arabidopsis thaliana* (L.) Heynh ecotype Col-0 and the biotrophic pathogen *Plasmodiophora brassicae* Woronin. Analyses were conducted at 4, 7, and 10 days after inoculation (DAI) and revealed significant induction or suppression of a relatively low number of genes in a range of functional categories. At 4 DAI, there was induced expression of several genes known to be critical for pathogen recognition and signal transduction in other resistant host–pathogen interactions. As the pathogen further colonized root tissue and progressed through the primary plasmodium stage for production of zoosporangia at 7 and 10 DAI, respectively, fewer genes showed changes in expression. The microarray results were validated by examining a subset of induced genes at 4 DAI by quantitative real-time reverse transcriptase PCR (RT-qPCR) analysis all of which correlated positively with the microarray data. The two *A. thaliana* mutants *jar1* and *coi1* tested were found to be susceptible to *P. brassicae*. The involvement of defense-related hormones in the interaction was further investigated (Agarwal et al. 2011) and the findings indicate that addition of salicylic acid can suppress clubroot disease in *A. thaliana* plants (Fig. 7.17 Chap. 7).

12.16 Expression of Genes for Metabolic Pathways During Infection of *Plasmodiophora brassicae*

Microarray chips, two-dimensional electrophoresis, and high-throughput sequencing technology have been used to study gene expression in the host plant in response to *P. brassicae* infection. Siemens et al. (2002, 2006a, b, 2009a, b) explored the gene expression of *A. thaliana* Col-0 inoculated with *P. brassicae* after 10 and 23 days using a microarray chip. They found that more than 1000 host genes were differentially expressed in the infected roots as compared to the control roots. Starch, sulfur, and secondary metabolism, auxin, and cytokinin synthesis, and the expression of transport-related genes changed significantly, whereas that of genes associated with defense and lignin synthesis did not. In addition, some flavonoid genes were also differentially expressed (Siemens et al. 2006a, b). Devos et al. (2006) and Cao et al. (2008) discovered protein changes in *Arabidopsis* following inoculation with *P. brassicae* using two-dimensional electrophoresis. Devos et al. (2006) observed 35 up-regulated, and 11 down-regulated proteins 4 days after *P. brassicae* inoculation, which were mainly associated with defense, cell metabolism, cell differentiation, and active oxygen activity. Cao et al. (2008) observed changes in the expression of 20 proteins (with 13 spots increasing and seven spots decreasing) 12, 24, and 48 h post-inoculation, including lignin synthesis, cytokinin synthesis, calcium steady-state, glycolysis, and oxygen activity. Agarwal et al. (2011) detected

147, 27, and 37 differentially expressed genes (DEGs) in *A. thaliana* after 4, 7, and 10 days, respectively, using a microarray chip. They further deduced that changes observed at 4 days post-infection (dpi) were mainly related to host and pathogen recognition and signal transduction. Using transcriptome analysis, Chen et al. (2015) observed major changes between sensitive and resistant varieties of *B. rapa* 0–96 h post-infection in metabolism, transport, and signal transduction. Nowadays, new high-throughput technologies are gradually being used to study the interaction between *P. brassicae* and the host, such as metabotyping, laser microdissection coupled to transcriptional profiling coupled and miRNA sequencing (Wagner et al. 2012; Schuller et al. 2014; Verma et al. 2014). The *P. brassicae* genome was sequenced, which will be very convenient for studies on the interaction between host plants and *P. brassicae* in the future (Schwelm et al. 2015; Rolfe et al. 2016). In order to clarify the early events taking place between the pathogen and the host, Zhao et al. (2017) analyzed differentially expressed genes and pathways in *A. thaliana* 24 and 48 h following infection with *P. brassicae*, using transcriptome analysis. In addition, the molecular mechanism of *A. thaliana's* response to *P. brassicae* in the early infection phase was described. Zhao et al. (2017) demonstrated changes on gene expression and pathways in *A. thaliana* during early infection with *P. brassicae* using transcriptome analysis, and identified 1903 and 1359 DEGs at 24 and 48 h post-inoculation (hpi), respectively. Flavonoids and the lignin synthesis pathways were enhanced, glucosinolates, terpenoids, and proanthocyanidins accumulated and many hormonal- and receptor-kinase related genes were expressed, caused by *P. brassicae* infection during its early phase (Fig. 7.18, Chap. 7). Therefore, the early interaction between *A. thaliana* and *P. brassicae* plays an important role in the entire infection process. In order to confirm the results of the RNA-seq analysis, 40 DEGs were selected for qRT-PCR, including 10 PR genes, three MYB transcription factor genes, one WRKY gene, 11 genes involved in the IAA, SA, JA, and cytokinin signaling pathways, 11 genes associated with secondary metabolism and four genes related to the cell wall.

12.17 Suppression of Host Immunity by Clubroot Effectors Genes

Crucifers host defenses are manipulated to enable pathogen colonization by biotrophic plant pathogens like *P. brassicae* to evolve advanced strategies to deliver effector proteins (genes) into the host cell during infection. Successful infection relies primarily on the success of the release of the effectors, which in many cases are responsible for the suppression of plant immunity. The initial recognition of conserved microbial features, known as pathogen-associated molecular patterns (PAMPs), leads to PAMP-triggered immunity (PTI) in the host. PAMP-triggered immunity is a first level of immune response that can be overcome by effector proteins produced by adapted pathogens. Resistance (R) to adapted pathogens is achieved through specific recognition of effectors, also known as avirulence proteins, by corresponding R proteins produced by the plant host. This effector-R protein recognition constitutes the second level of immune response, effector triggered immunity (ETI). *P. brassicae* is a well adapted pathogen of *Brassica* hosts; though indirect evidence suggests lack of either response, both PTI and ETI have not been well-characterized in the host-*P. brassicae* pathosystem. Identification of the proteins that comprise the secretome of *P. brassicae* is an important step towards understanding the complementary R proteins in the plant host. Putative effectors from biotrophic plant pathogens such as oomycetes and fungi are emerging from the sequencing and assembly of their genomes or transcriptomes, followed by comparative analysis with candidate effector genes. Schwelm et al. (2015) and Rolfe et al. (2016) identified putative effectors released by *P. brassicae* during the infection process. However, the obligate parasitic nature of *P. brassicae* has made it impossible to obtain a complete genome, with six draft genomes barely annotated for five Canadian pathotypes and one European pathotype. Transcriptomes of *P. brassicae* infecting *B. napus* and *Arabidopsis thaliana* in the Canadian study, and infecting *B. rapa*, *B. napus*, and *B. oleracea* var. *capitata* in the European study, have identified some candidate effectors. The overexpression of a benzoic acid/salicylic acid methyltransferase-encoding gene [PBRA_T000444 in *P. brassicae* European pathotype 3(Pbe3), PbPT3Sc00 026_A_1.308_1 in *P. brassicae* pathotype 3 (Pb3) draft genomes; GenBank accession number AFK13134] during the second week of infection, with expression peaking three and four weeks after infection has been observed (Schwelm et al. 2015; Rolfe et al. 2016; Ludwig-Muller et al. 2015). Salicylic acid (SA) is essential for the activation of plant defense. In the plant cell, regulation of active SA is managed through the maintenance of different inactive forms of SA, such as methyl salicylate (MeSA). The methyltransferase identified in *P. brassicae* has been characterized in detail by Ludwig-Muller et al. 2015, identifying its role in the methylation of salicylic, benzoic, and anthranilic acids, thereby contributing to the suppression of the salicylic acid-induced defense in a plant host (Pérez-López et al. 2018).

12.18 Host Organ Specific Expression of Genes During *Plasmodiophora brassicae* Infection

Plasmodiophora brassicae has evolved several strategies to avoid host recognition and subsequent cell death to establish an intracellular parasitism within living host cells, enabling it to extract/absorb nutrients from its host and to direct cell proliferation of host root tissues in order to produce a nutrient-sink gall habitat. There is limited resistant germplasm in the Brassicaceae, so exploring and understanding the genomics of disease progression in this host plant will contribute to the development of a clubroot resistance breeding strategy. Irani et al. (2018) provided a broad-spectrum view of the *Arabidopsis* transcriptomic response to *P. brassicae* inoculation, in above- and below-ground tissues of the host plant. They observed more DEGs at later stages of the disease establishment, with the largest transcriptomic changes occurring in the 24 dpi samples of both shoot and root. Although the root is the infected organ, the number of DEGs and the fold change in the DEGs across all three time points sampled was higher in the shoot (Fig. 7.19, Chap. 7), suggesting that metabolic and cellular processes of the shoot are in fact more affected by clubroot disease progression than those of the infected root. A higher number of DEGs in the shoot is probably a result, in some part, of a general stress response to the disease, as shown by the down-regulation of photosynthesis and light reaction genes, whereas, gene expression in diseased roots would predominantly be a specific response to the pathogen attack. A similar down-regulation of genes involved in photosynthesis, together with an up-regulation of pathogen protection and oxidative stress genes, has been reported in shoots in response to inoculation of roots with *Rhizobacterium*. Although spatially separated, the above- and below-ground tissues of plants do interact with each other and results suggest that future studies are needed to address this interaction in response to root pathogens (Rojas et al. 2014).

The identification and functional classification of highly DEGs showed that the expression of many of these genes was co-regulated or oppositely regulated in shoot and root. In the shoot, as expected, DEGs involved in photosynthesis were down-regulated after root inoculation with *P. brassicae* and subsequent clubroot progression. Such a down-regulation has previously been reported for genes involved in photosynthesis, biosynthesis of chlorophyll, and the small subunit of rubisco in response to pathogen attacks. Furthermore, several genes involved in the degradation of starch and triacylglycerols (lipid storage molecules) and the Calvin cycle were also down-regulated in the shoot. Following decreased water and nutrient uptake by the host plant, the general down-regulation of these genes in the shoot of infected plants is predictable. A decrease in starch and soluble carbohydrates in the leaves of *Arabidopsis*, five weeks after infection with *P. brassicae*, has previously been reported. These findings were compatible with a reduced rate of photosynthesis in *Arabidopsis* leaves. Contrary to these decreases, tryptophan (shikimate pathway metabolite) biosynthesis genes were highly induced in infected shoot, especially at 24 dpi. Increased levels of tryptophan-derived metabolites (e.g., camalexin and indol glucosinolates) have been reported in a number of plant species in response to various pathogen infections/disease progressions. Increased transcript levels for chalcone synthase (CHS) were also found in the shoot at 20 and 24 dpi. Under stress conditions, chalcone compounds produced through the shikimate pathway which, serve as antioxidants and antimicrobials (phytoalexins).

Infected *Arabidopsis* plants exhibited impaired growth and induced senescence, particularly at 20 and 24 dpi. Over this time period, the genes encoding cell wall structural proteins were mainly down-regulated in infected shoot. In addition, expression of many cell wall modification and degradation genes were turned off in the shoot over the 17, 20, and 24 dpi time points. ET signaling genes were induced in shoot following inoculation, peaking at 24 dpi. Depending on the pathogen and the host plant, ET signaling can play different roles during plant disease. One of the main suggested functions for ET in response to non-necrotrophic pathogens is induction of programmed cell death (Consonni et al. 2010; Sanchez-Vallet et al. 2010).

In infected root, expression of starch and sucrose biosynthesis and glycolysis pathway genes was mainly up-regulated. An increase in various sugars and starch has been previously reported in *P. brassicae*-infected *Arabidopsis* roots, whereas the suppression of invertase expression in transgenic plants resulted in enhanced tolerance to *P. brassicae*. An increased availability of carbon sources at the infection sites is a way to resource for the metabolic activities of the pathogen as well as for the increased cell division of the host plant, resulting in galls at these sites. Further, DEGs involved in lipid biosynthesis and elongation is up-regulated at 17 and 20 dpi in root. The higher abundance of these genes might be a response for the pathogen lipid demand. An up-regulation of lipid transfer genes was found in infected *Arabidopsis* roots when galls and resting spores are present, suggesting the formation of a lipid reserve. This is consistent with the observation that accumulation of abundant lipid droplets in the pathogen cytoplasm. Increased expression of phospholipases in infected root, mainly at 24 dpi, may be attributed to their roles in the synthesis of JA, oxylipins, and other important plant defense signaling molecules. However, since phospholipases hydrolyze phospholipids, an increased expression may indicate an increased level of catabolization of phospholipids by the host plant (Canonne et al. 2011).

A group of cell wall modification proteins, including expansins, xyloglucan endotrans-glucosylase/hydrolases (XTHs), and endo-β-1,4-glucanases (EGases) facilitate cell wall expansion and function in regulating the plasticity and rheology of the cell wall. The up-regulation of cell wall modification genes in the root is likely linked to cell wall loosening and cell expansion in infected root. However, several DEGs encoding cell wall degradation enzymes were strongly down-regulated in infected root at 24 dpi. *P. brassicae* induces regulation of cell wall modification proteins involved in an alteration of cell wall structure during infection. Moreover, in root, an up-regulation of *SAUR* genes that are an early auxin-responsive gene family is consistent with the suggestion that IAA induces cell division and elongation during gall formation in *P. brassicae*-infected *Arabidopsis*. An involvement of auxin in the host–*P. brassicae* interaction has been observed by Ludwig-Muller et al. (2015).

In both shoot and root at 24 dpi, the expression of several genes in major branches of the shikimate pathway is up-regulated. At 17 dpi, genes involved in the biosynthesis of sinapate esters through the phenylpropanoid branch of the shikimate pathway were up-regulated in both tissues. Both a higher level of transcription of shikimate pathway genes, together with an increased production of the pathway's metabolites in plants, has been reported in response to pathogenic attacks. In *P. brassicae*-infected roots, an accumulation of phenylpropanoid pathway compounds or a higher expression of related genes has been reported. The shikimate pathway metabolites have important roles in both root and shoot responses in infected host plants (Irani et al. 2018).

Transcriptome levels for genes in the JA biosynthesis pathway are elevated at all three time points in shoot and at 24 dpi in root. However, several genes involved in the JA response pathway are strongly down-regulated in root at 24 dpi, perhaps by the pathogen, indicating that the transcript regulation of JA response genes might be manipulated by the pathogen. *P. brassicae* is able to manipulate a number of hormone pathways and can alter hormone homeostasis in the host plant to its benefit. On the other hand, JA response genes may be down-regulated once the pathogen is established and having successfully disarmed the plant immune response by 24 dpi. An enhanced JA level was reported, in Chinese cabbage roots, 3–5 weeks after infection with *P. brassicae*. High levels of jasmonate-related compounds were also reported in *Arabidopsis* ecotype Col-0 roots, 21 days after inoculation with *P. brassicae*. In addition, during secondary infection by *P. brassicae,* the JA pathway is induced in susceptible *Arabidopsis* Col-0 but not in the partially resistant ecotype, Bur-0. The JA pathway is involved in clubroot disease during *P. brassicae* secondary infection (Rolfe et al. 2016; Lemarie et al. 2015).

A number of DEGs in the ABA response pathway are up-regulated in both shoot and root at 24 dpi. The level of ABA has been reported to increase in *P. brassicae*-infected *B. rapa* plants during gall formation. Moreover, up-regulation of ABA response and signaling genes during late time points after *P. brassicae* infection has been reported. Up-regulation of ABA response genes might be a response to abiotic stress conditions in infected tissues that are subsequently under water and nutrient stress. However, it has also been proposed that ABA plays a role in the induction of plant defense response genes and JA-induced defense genes (Adie et al. 2007).

Members of the TF families, MYB, bHLH, and AP2/EREBP, were highly up- or down-regulated in both shoot and root (Table 7.10, Chap. 7). In *Arabidopsis*, the functions of MYB TFs are very diverse, including anthocyanin, and flavonol biosynthesis, terminal differentiation and cell cycle regulation. Similarly, members of the bHLH family play various roles in the regulation of ABA and gibberellins signaling pathways, stomata development, and root development in *Arabidopsis*. The comparison of expression profiles of TFs, in shoot and root tissues, highlights the major differences in response of each tissue to the infection. As shown in Table 7.10 (Chap. 7), only a few TFs are commonly up- or down-regulated in both infected shoot and root (Feller et al. 2011). *Arabidopsis* has a complex transcriptomic response to *P. brassicae* infection, and in general, the plant response to pathogen infection of the root is different between shoot and root. Irani et al. (2018) proposed a simplified model of the major molecular changes in shoot and root in response to *P. brassicae* infection and clubroot disease progression (Fig. 7.20, Chap. 7). In the root, there is a general induction of genes related to cell wall modification, starch/sucrose biosynthesis and transport to provide energy. In contrast, in the shoot, and possibly as a result of the impairment of nutrient and water uptake, a down-regulation of photosynthesis, chlorophyll synthesis, starch/sucrose biosynthesis, and an inhibition of synthesis of proteins and enzymes involved in cell wall formation and its modification. As a result, the leaves show symptoms of stress such as reduced growth and necrosis. However, the shoot and root also responded to the pathogen in a number of common ways, such as an induction of JA biosynthesis, ABA response, and the production of shikimate pathway metabolites.

12.19 Induction of Hypoxia Genes in Roots of Infected Host by *Plasmodiophora brassicae*

The induction of alcohol fermentation in roots is a plant adaptive response to flooding stress and oxygen deprivation. Available transcriptomic data suggest that fermentation-related genes are also frequently induced in roots infected with gall forming pathogens. Hypoxia responses in *Arabidopsis* roots during infection by the clubroot pathogen *P. brassicae* were assessed by Gravot et al. (2016). The hypoxia-related gene markers Pyruvate decarboxylase 1 (PDC1), Pyruvate decarboxylase 2 (PDC2), and Alcohol dehydrogenase 1 (ADH1) were induced during secondary infection by two isolates of *P. brassicae*, eH and e2. The PDC2 was highly induced as soon as 7 days post-inoculation (dpi), i.e. before the development of gall symptoms, and GUS staining revealed that ADH1 induction was localized in infected cortical cells of root galls at 21 dpi. Clubroot symptoms were significantly milder in the pdc1 and pdc2 mutants compared with Col-0, but a null T-DNA insertional mutation of ADH1 did not affect clubroot susceptibility. The Arg/N-end rule pathway of ubiquitin-mediated proteolysis controls oxygen sensing in plants. Mutants of components of this pathway, *ate1ate2* and *prt6*, that both exhibit constitutive hypoxia responses, showed enhanced clubroot symptoms. In contrast, gall development was reduced in quintuple and sextuple mutants where the activity of all oxygen-sensing Group VII Ethylene Response Factor transcription factors (ERFVIIs) is absent (erfVII, and prt6 erfVII). The induction of PDC1 and PDC2 during the secondary infection of roots by *P. brassicae* contributes positively to clubroot development, and which is controlled by oxygen-sensing through ERFVIIs. The absence of any major role of ADH1 in symptom development may also suggest that PDC activity could contribute to the formation of galls through the activation of a PDH bypass.

12.20 SWEET Genes Trigger Gall Formation During *Plasmodiophora brassicae* Infection

After infection of the plant roots, *P. brassicae* colonization leads to swelling roots and gall formation, eventually inhibiting roots to uptake nutrients and water from the soil. Sugars are the main carbon source for the living organisms including pathogens and their host plants. In plants, sugars are synthesized in the photosynthetic organs and allocated into the sink organs via long-distance transport. Sugars produced in the living plants not only play key roles for plant growth and development, but also are transferred or secreted into the plant surface. Sugars secreted by plants could be used as carbon supply for pathogen growth (Patrick 1989; Chen et al. 2010). Especially, biotrophic microbes like *P. brassicae* obtain nutrients from host cells by extracting sugars (Manck-Götzenberger and Requena 2016; Roman and Rathjen 2017). In *P. brassicae*, it has been reported that 2% sucrose solution can promote germination of resting spores (Guo 2001). In addition, accumulation of soluble sugars (hexoses and sucrose) are found in the galls of *P. brassicae*-infected *Arabidopsis thaliana*

(Evans and Scholes 1995; Brodmann et al. 2002) and *Brassica* plants (Williams et al. 1968; Keen and Williams 1969), indicating that the galls might act as a newly established carbon sink during clubroot development. Sugar transporters, including well-known gene families such as MSTs (monosaccharide transporters) (Toyofuku et al. 2000), SUTs (sucrose transporters) (Reinders et al. 2008, 2011), and SWEETs (sugar will be eventually exported transporters) (Chen et al. 2012) are responsible to translocate monosaccharides or disaccharides.

Transcriptome studies revealed that several genes involved in sugar transport and metabolism strongly up-regulated during gall formation in clubroot-infected *Arabidopsis* (Siemens et al. 2006a, b; Malinowski et al. 2012). A transcriptome profiling performed on both shoot and root tissues of *A. thaliana* challenged by *P. brassicae* also pointed that genes involved in sucrose and starch biosynthesis are differently expressed in shoot and root (Irani et al. 2018). These transcriptomic and metabolic experiments provided insights that sugar accumulation is important for gall formation, probably depending on some sugar transport genes. Moreover, during the process of plant-pathogen co-evolution, the pathogens have evolved a mechanism to compete for sugars with infected host cells to sustain their life cycle (Doidy et al. 2012; Lemoine et al. 2013). Competition for sugars at the plant–pathogen interface is likely controlled by the sugar transporters (Toyofuku et al. 2000; Doidy et al. 2012), thus regulation of those transporters would give a new insight into disease control.

SWEETs are a class of proteins responsible for transporting sugar across cell membranes, involving in diverse physiological processes of various plant species, such as in reproductive development (Ge et al. 2000; Yang et al. 2006; Guan et al. 2008), leaf senescence (Quirino et al. 1999; Seo et al. 2011), and plant responses to abiotic and biotic stresses in *Arabidopsis* and rice (Chu et al. 2006; Yang et al. 2006; Antony et al. 2010; Chen et al. 2010; Liu et al. 2011). SWEETs genes are evolutionally conserved in both prokaryotes and eukaryotes (Gamas et al. 1996; Artero et al. 1998; Chong et al. 2014; Jian et al. 2016), but plant species possess more SWEETs than that in animals and prokaryotes (Yuan and Wang 2013). Phylogenetic analysis of plant SWEET genes revealed that they can be grouped into 4 clades as firstly defined in *A. thaliana* (Chen et al. 2010). SWEETs grouped into clades I and II have priority to transport hexoses, while clade III and IV proteins are preferentially transport sucrose and fructose, respectively (Chen et al. 2010, 2012; Lin et al. 2014). Characterization of SWEET proteins indicated that they contain 7 transmembrane (TM) helices carrying two MtN3/saliva domains in plants and animals, while only one MtN3/saliva domain with 3 TM helices exists in SWEET proteins of prokaryotes (Wang et al. 2014). SWEETs localized on the plasma membrane to import or export sugars (Chen et al. 2012). However, SWEET is a distinct transporter family which mediates energy-independent influx and efflux of sugars compared to MSTs and SUTs, both of which are energy-dependent (Reinders et al. 2008, 2011; Yuan and Wang 2013). Some SWEETs harboring two MtN3/saliva domains functioned as low-affinity glucose and sucrose transporters, such as glucose transporters AtSWEET4, AtSWEET5, AtSWEET7, and AtSWEET13, and sucrose transporters AtSWEET11 and AtSWEET12 in

Arabidopsis as well as OsSWEET11 and OsSWEET14 in rice (Chen et al. 2010, 2012).

SWEETs are also crucial to regulate carbon transport in parasitism and pathogens interaction. Some members of the SWEET genes have been demonstrated as the targets of extracellular pathogens. During the host–microbe interaction, the pathogen infection regulates expression patterns of the SWEET genes to facilitate the pathogen obtaining the sugar. Indeed, induction of SWEET genes upon pathogen infection has also been reported in *Arabidopsis*, rice, and other plant species. In *Arabidopsis*, at least 9 SWEET members (AtSWEET 2, 4, 7, 8, 10, 11, 12, 15, and 17) were differentially regulated following infection of different types of pathogens (Ferrari et al. 1994; Chen et al. 2010). In the *Arabidopsis–P. brassicae* interaction, *P. brassicae* infection strongly induced the expression of AtSWEET 15 during the gall formation (Siemens et al. 2006a, b).

The role of SWEETs during *P. brassicae*-induced gall formation in Chinese cabbage was examined for sugar contents in roots and leaves of the Chinese cabbage clubroot-resistant (CR) and clubroot-susceptible (CS) near-isogenic lines (NILs) after infection of *P. brassicae*. In addition, the response of *B. rapa* SWEET (hereafter BrSWEET) sugar transporters to *P. brassicae* infection as well as clubroot disease development in *Arabidopsis* sweet11 mutant was examined. The pathogen lives inside roots, and hijacks nutrients from the host plants. It is suggested that clubroot galls created an additional nutrient sink in infected roots. Sugar contents in leaves and roots before and after *P. brassicae* infection using a pair of Chinese cabbage near-isogenic lines (NILs), carrying either a clubroot-resistant (CR) or clubroot-susceptible (CS) allele at the CRb locus were assessed. *P. brassicae* infection caused significant increase of glucose and fructose contents in the root of CS-NIL compared to CR-NIL, suggesting that sugar translocation and *P. brassicae* growth are closely related. Among 32 *B. rapa* SWEET homologs, several BrSWEETs belonging to Clade I and III were significantly up-regulated, especially in CS-NIL upon *P. brassicae* infection. Moreover, *Arabidopsis* sweet11 mutant exhibited slower gall formation compared to the wild-type plants. *P. brassicae* infection probably triggers active sugar translocation between the sugar producing tissues and the clubbed tissues, and the SWEET family genes are involved in this process (Li et al. 2018).

12.21 Expression of Host Specific Genes

In the infection process of *P. brassicae*, the different *Brassica* species have differential responses in the expression of genes at early, middle, and late stages of infection. The stage-specific gene expression of *P. brassicae* can be better studied in *B. rapa* roots due to longer time taken for homogeneous development of plasmodial stages in comparison to *A. thaliana* which has short life span.

12.21.1 *Brassica rapa*

Orphan genes are often associated with biotic and abiotic stress, and they usually function in stress-response signal transduction pathways. Using RNA sequencing, transcriptome profiling was compared between clubroot-resistant and clubroot-susceptible *B. rapa* genotypes after infection with *P. brassicae* during early infection (Chen et al. 2015). Jiang et al. (2018) reassembled respective RNA sequencing data to analyze the BSGs expression patterns, and the results indicated that 180 unique BSGs were supported by RNA-Seq data (FPKM > 2 for least one sampling stage (Fig. 7.21, Chap. 7). Further, 24 BSGs exhibited high expression abundance at least in one sampling stage (FPKM > 50). Seventy-eight BSGs displayed constitutive expression at every point (FPKM > 2), and 32 BSGs showed period-specific expression (FPKM > 2). Deora et al. (2012) indicated that *P. brassicae* is blocked at middle or late stages of infection in resistant *Brassica* crops. However, BSGs may play more important roles in middle or late periods (4–20 days after inoculation (dai)). To confirm this hypothesis, roots from *B. rapa* plants were collected at 10 dai with *P. brassicae* (referred to CK and S10), and 52 successfully amplified BSGs in Chiifu were analyzed using semi-qPCR. As expected, 78.85% (41 genes) of analyzed genes responded to *P. brassicae* treatments, and 39 out of 41 BSGs were up-regulated (Fig. 7.21b, Chap. 7). However, only 8 out of these 52 BSGs have transcript support (FPKM > 2) (Fig. 7.21, Chap. 7). The results strongly suggest that BSGs play important roles in *P. brassicae* stress responses (Jiang et al. 2018).

12.21.2 *Arabidopsis thaliana*

The crucifer *A. thaliana* has been used to characterize the expression of several *P. brassicae* genes during pathogenesis (Siemens et al. 2009a, b). *A. thaliana* is a host of *P. brassicae* and the small size of the roots facilitates the observation of disease development. However, clubroot development in *Brassica* takes a long time and the roots show more homogeneous developmental stages of plasmodia in comparison to *A. thaliana* (Gustafsson and Falt 1986). Therefore, the short life span of this small weedy species might prevent researchers from identifying genes that are expressed in a stage-specific manner during the interaction. The expression of several *P. brassicae* genes over the time course of the host–pathogen interaction in *Brassica rapa* was analyzed by semi-quantitative RT-PCR in order to find molecular markers for the actively growing pathogen, and for distinct developmental stages of *P. brassicae*, early developmental stages in root hairs, or young vegetative plasmodia shortly after secondary infection. The gene expression was analyzed in three *B. rapa* host lines with a virulent isolate, to test whether semi-quantitative RT-PCR can be used to characterize host–pathogen interactions in crop plants. To find landmarks for specific clubroot development stages, expression of several *P. brassicae* genes was analyzed during the infection process in a universally susceptible *Brassica rapa* line (cv. Granaat, ECD 05). The results revealed that the genes *Pb-YPT*, *Pb-Brip9*, and *Pb-PSA* were regulated during infection of

P. brassicae, *Pb-Brip9* and *Pb-PSA* were expressed in early stages of infection. Wu et al. (2012) provide evidence that semi-quantitative RT-PCR could be used for discriminating developmental stages of *P. brassicae* in *B. rapa*. Stage-specific gene expression of *P. brassicae* should be studied in *B. rapa* roots due to longer disease development in comparison to *Arabidopsis thaliana* (Wu et al. 2012).

12.22 Role of miRNAs in Clubroot Development During Pathogenesis

MicroRNAs (miRNAs) are a highly conserved class of small non-coding RNAs that regulate gene expression by post-transcriptional repression. Emerging evidence indicates that hosts' endogenous small RNAs represent an important mechanism of control in plant immune responses and hormone signaling at the time of stress. RNAs like ath-miR160 and ath-miR167 are involved in pathogenesis and target the auxin-response-factor (ARF). Another microRNA, ath-miR164, has been implicated in auxin homeostasis and lateral root development, which may have a bearing on clubroot development. Therefore, it is conceivable that miRNAs may be involved in mediating important plant processes following infection with *P. brassicae*, which results in the development of root galls, leading to stunting and decreased productivity. Now, microarrays of miRNAs have become available for high-throughput global analysis of miRNAs expression patterns in response to pathogens. This valuable tool has been further exploited in a number of plants to study the miRNA: mRNA interactions in tomato, rice, and *Arabidopsis* which may lead to the identification of miRNAs that are involved in plant responses to specific stresses, including *P. brassicae* challenge (Li et al. 2010; Mallory et al. 2005; Guo et al. 2005; Liu et al. 2008). Clubroot disease of Brassicaceae family has potential to substantially limit *Brassica* production. A miRNA microarray-based approach was used to identify miRNA expression and regulation during the disease initiation and progression (Verma et al. 2014). Results revealed the useful basis towards understanding the dynamics of miRNAs related to *P. brassicae* infection of *Brassica*. The differential expression analysis of miRNA during pathogenesis of *P. brassicae* was carried out at two developmentally distinct time points. The expression of various miRNAs was observed to be modulated during the process of clubroot initiation and development. Many of these miRNAs were involved in the regulation of gene activity by targeting, among others, TFs, ARFs, MYB, MADS box, AP2 and ERFs. Many of these are known to be involved in stress-responses and developmental events related to pathogenesis. A diverse array of miRNAs and their target genes are modulated during the pathogenesis of *Brassica* by *P. brassicae*. It enhances the understanding related to the biological processes involved in *P. brassicae*–*Brassica* interactions and may lead to unique ways to generate disease resistant phenotypes (Fig. 7.22, Chap. 7)

12.23 Induction of Immunophilin Genes During *Plasmodiophora brassicae* Infection

Immunophilins are ubiquitous proteins with properties that allow them to regulate protein structure, activity, and stability (Wang and Heitman 2005; Hanes 2015). They operate either by peptidyl-prolyl isomerization of selected targets, as chaperons or by binding of small legends. The immunophilins comprise three structurally unrelated subfamilies: the cyclophilins (CYPs), the FK506-binding proteins (FKBPs), and the parvulin-like proteins (Hanes 2015). Besides modulating the formation of *cis–trans* isomers of proline (Galat 2003), immunophilins have two important amino acid properties: (a) prolyl-isomerase activity, which catalyzes the rotation of the X-Pro peptide bonds from the cis to trans configuration, a rate-limiting step in protein folding (Wang and Heitman 2005), and (b) affinity to bind to immunosuppressive drugs. The stramenopile human parasite *Blastocystis* sp. lives under anaerobic conditions partly like *P. brassicae* (Gravot et al. 2016), and secretes a range of immunophilins with potential effector functions, which could lead to cell death (apoptosis) in the host tissue (Denoeud et al. 2011).

Another animal parasite, the protozoan *Toxoplasma gondii* uses a cyclophilin (TgCYP18) to manipulate host cell responses (Ibrahim et al. 2009). Among plant pathogens, some cyclophilins are able to inhibit RNA replication of plant viruses (Lin et al. 2012; Kovalev and Nagy 2013), while some effector proteins interact with plant cyclophilins, and thereby induce plant defense responses (Domingues et al. 2010, 2012; Kong et al. 2015). To monitor the presence of immunophilin encoding genes in the genome of *P. brassicae*, and to analyze potential candidates for function, the molecular amenable plant pathogen *M. oryzae* as heterologous test system was used (Singh et al. 2018).

Immunophilins, a group of proteins known to have various cellular functions, including involvement in plant defense and pathogen virulence was analyzed in clubroot pathogen. Searches in the *P. brassicae* genome resulted in 20 putative immunophilins comprising of 11 cyclophilins (CYPs), 7 FK506-binding proteins (FKBPs), and 2 parvulin-like proteins. RNA-seq data showed that immunophilins were differentially regulated in enriched life stages such as germinating spores, maturing spores, and plasmodia, and infected *Brassica* hosts (*B. rapa, B. napus,* and *B. oleracea*). *PbCYP3* was highly induced in all studied life stages and during infection of all three *Brassica* hosts, and hence was selected for further analysis. *PbCYP3* was heterologously expressed in *Magnaporthe oryzae* gene-inactivated Δ*Cyp1* strain. The new strain Δ*Cyp1*+ overexpressing *PbCYP3* showed increased virulence on rice compared to the Δ*Cyp1* strain. These results suggest that the predicted immunophilins and particularly *PbCYP3* are activated during plant infection. *M. oryzae* is a well-studied fungal pathogen and could be a valuable tool for future functional studies of *P. brassicae* genes, particularly elucidating their role during various infection phases (Singh et al. 2018).

12.24 Analysis of Host Response Through Transcriptome Data

Two RNA-Seq experiments were performed to gain insight into the genes involved in host penetration, infection, and gall formation. RNA extracted from the roots of the clubroot-susceptible *B. napus* line DH12075 sampled at weekly intervals (from 1 to 6 weeks) after inoculation with Pb3 spores was sequenced. Sequences generated for RNA prepared from the roots of DH12075 plants mock inoculated with water at the same time points served as controls. In addition, Rolfe et al. (2016) sequenced RNA from control and infected roots or hypocotyls of *A. thaliana* plants at the onset (16 DPI) and at late stages (26 DPI) of gall formation with three independent biological replicates. This second experiment used a population of *P. brassicae* described by Malinowski et al. (2012). Analysis of differential expression in infected *B. napus* found significant changes in the expression of 617 Pb3 genes during the 6-week course of infection (genes in the *B. napus* data were considered to be differentially regulated if they showed a twofold or greater change in expression in at least 2 consecutive samples and an adjusted p value of ≤ 0.05). Within the differentially expressed genes there were genes involved in carbohydrate processing, amino-sugar and nucleotide sugar metabolism, transporters, cytoskeleton, motor proteins, and flagella-associated proteins. Of the differentially regulated transcripts, 41 were genes encoding for predicted secreted proteins with 14 of them being SSPs. Hydrolases and peptidases were the most common differentially regulated secreted proteins. The infected *A. thaliana* samples, ~25 million paired-end reads were obtained from each sample. For the 16 DPI samples, ~13 % (3.5 million) of the paired reads matched the *P. brassicae* genome in both hypocotyl and root samples. For the 26 DPI samples, 48 % of the paired reads (12 million) from hypocotyl tissue and somewhat less (34%, 9 million) from root tissue matched the *P. brassicae* genome. The root and hypocotyl samples at each time point were very similar to each other. Only 53 and 32 genes showed differential expression (2-fold or greater with a corrected probability of $p \leq 0.05$) between the 16 DPI and 26 DPI hypocotyl and root tissue, respectively. In contrast, gene expression differed markedly between the 16 and 26 DPI samples within each tissue (1661 genes for hypocotyl tissue and 957 genes for root tissue, of which 763 were in common). Genes that were differentially up- or down-regulated ($\geq$twofold, $p \leq 0.05$ after correction for false discovery) between 16 and 26 DPI samples in both tissue types were identified using DESeq2 and GO terms analyzed using BLAST2GO (Rolfe et al. 2016).

Several GO terms associated with cytoskeletal and membrane activities (cytoskeletal protein binding, plasma membrane, ion transport, cytoskeleton, cell differentiation, protein transport) were down-regulated at 26 DPI compared with 16 DPI, indicative of a shift in metabolic processes associated with the change from plasmodial growth form to spore formation. Many genes associated with general cellular metabolism (carboxylic, lipid, and organo-nitrogen metabolic processes, catalytic activity) showed the opposite pattern, with enrichment in the later stages of gall development and spore formation (Rolfe et al. 2016).

12.25 Manipulation of Host Resistance

A number of lines of evidence support the view that *P. brassicae*, in common with many other biotrophic pathogens, manipulates the defense responses of its hosts. Analysis of differentially expressed putative Pb3 effectors within the first six weeks post-inoculation provided evidence for the manipulation of plant defense by the induction of protease inhibitors and possibly suppression of salicylic acid (SA)-induced defense mechanism. Pair wise comparisons of *B. napus* genes differentially expressed in response to Pb3 infection using the application HORMONOMETER showed that salicylic acid mediated responses at the earliest stages of infection was negatively correlated with the *A. thaliana* gene expression profile in response to SA. In addition, the expression of a *P. brassicae* predicted secreted protein (PbPT3Sc00026_A_1.308_1) identified as a benzoic acid/salicylic acid methyltransferase (Pb3-BSMT) (NBCI accession number AFK13134) was extremely low in *B. napus* infected roots during the first week after inoculation, but increased in two week and peaked at weeks 3 and 4. Likewise, expression was higher in infected *A. thaliana* tissue at 16 DPI than at 26 DPI. A similar expression pattern for this gene was reported recently by Ludwig-Muller et al. (2015). Rolfe et al. (2016) confirmed the methyltransferase activity of this enzyme and suggested a possible role for this enzyme in inactivating plant defense by converting SA (Salicylic Acid) to MeSA (Methyl Salicylate), which acts as a mobile signal in systemic acquired resistance (SAR) in plants.

References

Adams MD, Kelley JM, Gocayne JD, Dubnick M, Polymeropoulos MH, Xiao H (1991) Links complementary DNA sequencing: expressed sequence tags and human genome project. Science 252:1651–1656

Adie BAT, Pérez-Pérez J, Pérez-Pérez MM, Godoy M, Sanchez-Serrano JJ, Schmelz EA, Solano R (2007) ABA is an essential signal for plant resistance to pathogens affecting JA biosynthesis and the activation of defenses in *Arabidopsis*. Plant Cell 19:1665–1681

Agarwal A, Kaul V, Faggian R, Cahill DM (2009) Development and use of a model system to monitor clubroot disease progression with an Australian field population of *Plasmodiophora brassicae*. Austr Plant Pathol 38:120–127

Agarwal A, Kaul V, Faggian R, Rookes JE, Ludwig-Muller J, Cahill DM (2011) Analysis of global host gene expression during the primary phase of the *Arabidopsis thaliana–Plasmodiophora brassicae* interaction. Funct Plant Biol 38(6):462–478

Aist JR, Williams PH (1971) The cytology and kinetics of cabbage root hair penetration by *Plasmodiophora brassicae* Worn. Can J Bot 49:2023–2034

Ando S, Tsushima S, Tagiri A, Kamachi S, Konagaya KI, Hagio T, Tabei Y (2006a) Increase in *BrAO1* gene expression and aldehyde oxidase activity during clubroot development in Chinese cabbage (*Brassica rapa* L.). Mol Plant Pathol 7:223–234

Ando S, Yamada T, Asano T, Kamachi S, Tsushima S, Hagio T, Tabei Y (2006b) Molecular cloning of PbSTKL1 gene from *Plasmodiophora brassicae* expressed during clubroot development. J Phytopathol 154:185–189

Andreasson E, Jenkins T, Brodersen P, Thorgrimsen S, Petersen NH, Zhu S, Qiu JL, Micheelsen P, Rocher A, Petersen M, Newman MA, Bjorn Nielsen H, Hirt H, Somssich I, Mattsson O, Mundy

J (2005) The MAP kinase substrate MKS1 is a regulator of plant defense responses. EMBO J 24:2579–2589

Antony G, Zhou JH, Huang S, Li T, Liu B, White F, Yang B (2010) Rice *xa13* recessive resistance to bacterial blight is defeated by induction of the disease susceptibility gene Os-*11N3*. Plant Cell 22:3864–3876

Artero RD, Terol-Alcayde J, Paricio N, Ring J, Bargues M, Torres A, Torres A, Perez-Alonso M (1998) Saliva, a new drosophila gene expressed in the embryonic salivary glands with homologues in plants and vertebrates. Mech Dev 75:159–162

Bi K, He Z, Gao Z, Zhao Y, Fu Y, Cheng J, Xie J, Jiang D, Chen T (2016) Integrated omics study of lipid droplets from *Plasmodiophora brassicae*. Sci Rep 6:36965

Bischoff M (1994) PCR- und RAPD-Analysen an *Plasmodiophora brassicae* Worn. und Vertretern der Gattung Brassica L.: identifikation einer Nitrilasesequenz in Brassicaceen.- Diploma Thesis, J. W. Goethe-Universität, Frankfurt am Main, Germany

Braselton JP (1982) Karyotypic analysis of *Plasmodiophora brassicae* based on serial thin sections of pachytene nuclei. Can J Bot 60:403–408

Brenner S, Johnson M, Bridgham J, Golda G, Lloyd DH, Johnson D (2000) Gene expression analysis by massively parallel signature sequencing (MPSS) on microbead arrays. Nat Biotechnol 18:630–634

Brodmann D, Schuller A, Ludwig-Muller J, Aeschbacher RA, Wiemken A, Boller T, Wingler A (2002) Induction of trehalase in *Arabidopsis* plants infected with the trehalose-producing pathogen *Plasmodiophora brassicae*. Mol Plant-Microbe Interact 15(7):693–700

Bryan RJ, Trese AT, Braselton JP (1996) Molecular karyotypes for the obligate, intracellular, plant pathogens, *Plasmodiophora brassicae* and *Spongospora subterranea*. Mycologia 88 (3):358–360

Bryngelsson T, Gustafsson M, Gréen B, Lind-Halldén C (1988) Uptake of host DNA by the parasitic fungus *Plasmodiophora brassicae*. Physiol Mol Plant Pathol 33(2):163–171

Buchanan-Wollaston V, Ainsworth C (1997) Leaf senescence in *Brassica napus*: cloning of senescence related genes by subtractive hybridization. Plant Mol Biol 33:821–834

Buczacki ST (1983) *Plasmodiophora*. In: Buczacki ST (ed) Zoosporic plant pathogens. Academic, London, pp 161–191

Buhariwalla H, Mithen R (1995) Cloning of a *Brassica* repetitive DNA element from resting spores of *Plasmodiophora brassicae*. Physiol Mol Plant Pathol 47:95–101

Buhariwalla S, Greaves S, Magrath R, Mithen R (1995) Development of specific PCR primers for the amplification of polymorphic DNA from the obligate root pathogen *Plasmodiophora brassicae*. Physiol Mol Plant Pathol 47:83–94

Bulman S, Siemens J, Ridgway HJ, Eady C, Conner AJ (2006) Identification of genes from the obligate intracellular plant pathogen, *Plasmodiophora brassicae*. FEMS Microbiol Lett 264 (2):198–204

Bulman S, Ridgway HJ, Eady C, Conner AJ (2007) Intron-rich gene structure in the intracellular plant parasite *Plasmodiophora brassicae*. Protist 158(4):423–433

Bulman S, Richter F, Marschollek S, Benade F, Jülke S, Ludwig-Müller J (2018) *Arabidopsis thaliana* expressing PbBSMT, a gene encoding a SABATH-type methyltransferase from the plant pathogenic protist *Plasmodiophora brassicae*, show leaf chlorosis and altered host susceptibility. Plant Biol 21(1):120–130. https://doi.org/10.1111/plb.12728

Burki F, Kudryavtsev A, Matz MV, Aglyamova GV, Bulman S, Fiers M, Keeling PJ, Pawlowski J (2010) Evolution of Rhizaria: new insights from phylogenomic analysis of uncultivated protists. BMC Evol Biol 10:377

Canonne J, Marino D, Jauneau A, Pouzet C, Briere C, Roby D, Rivas S (2011) The *Xanthomonas* type III effector *XopD* targets the *Arabidopsis* transcription factor *MYB30* to suppress plant defense. Plant Cell 23:3498–3511

Cao T, Srivastava S, Rahman MH, Kav NNV, Hotte N, Deyholos MK, Strelkov SE (2008) Proteome-level changes in the roots of *Brassica napus* as a result of *Plasmodiophora brassicae* infection. Plant Sci 174:97–115

Chen LQ, Hou BH, Lalonde S, Takanaga H, Hartung ML, Qu XQ, Guo WJ, Kim JG, Underwood W, Chaudhuri B, Chermak D, Antony G, White FF, Somerville SC, Mudgett MB, Frommer WB (2010) Sugar transporters for intercellular exchange and nutrition of pathogens. Nature 468:527–532

Chen LQ, Qu XQ, Hou BH, Sosso D, Osorio S, Fernie AR, Frommer WB (2012) Sucrose efflux mediated by SWEET proteins as a key step for phloem transport. Science 335:207–211

Chen J, Pang W, Chen B, Zhang C, Piao Z (2015) Transcriptome analysis of *Brassica rapa* near-isogenic lines carrying clubroot-resistant and-susceptible alleles in response to *Plasmodiophora brassicae* during early infection. Front Plant Sci 6:1183

Chong J, Piron MC, Meyer S, Merdinoglu D, Bertsch C, Mestre P (2014) The SWEET family of sugar transporters in grapevine: VvSWEET4 is involved in the interaction with *Botrytis cinerea*. J Exp Bot 65:6589–6601

Chu Z, Yuan M, Yao L, Ge X, Yuan B, Xu C, Li X, Fu B, Li Z, Bennetzen JL, Zhang Q, Wang S (2006) Promoter mutations of an essential gene for pollen development result in disease resistance in rice. Genes Dev 20:1250–1255. https://doi.org/10.1101/gad.1416306

Consonni C, Bednarek P, Humphry M, Francocci F, Ferrari S, Harzen A, Loren van Themaat EV, Panstruga R (2010) Tryptophan-derived metabolites are required for antifungal defense in the *Arabidopsis* mlo2 mutant. Plant Physiol 152:1544–1561

Daval S, Belcour A, Gazengel K, Legrand L, Gouzy J, Cottret L, Lebreton L, Aigu Y, Mougel C, Manzanares-Dauleux MJ (2018) Computational analysis of the *Plasmodiophora brassicae* genome: mitochondrial sequence description and metabolic pathway database design. Genomics 111(6):1629–1640. https://doi.org/10.1016/j.ygeno.2018.11.013

Dekhiujen HM (1979) Electron microscope studies on the root hairs and cortex of a susceptible and a resistant variety of *Brassica campestris* infected with *Plasmodiophora brassicae*. Eur J Plant Pathol 85:1–17

Denoeud F, Roussel M, Noel B, Wawrzyniak I, Da Silva C, Diogon M, Viscogliosi E, Brochier-Armanet C, Couloux A, Poulain J, Segurens B, Anthouard V, Texier C, Blot N, Poirier P, Ng GC, Tan KSW, Artiguenave F, Jaillon O, Aury J-M, Delbac F, Wincker P, Vivarès CP, Alaoui HE (2011) Genome sequence of the stramenopile *Blastocystis*, a human anaerobic parasite. Genome Biol 12:R29

Deora A, Gossen BD, McDonald MR (2012) Infection and development of *Plasmodiophora brassicae* in resistant and susceptible canola cultivars. Can J Plant Pathol 34:239–247

Deora A, Gossen BD, McDonald MR (2013) Cytology of infection, development, and expression of resistance to *Plasmodiophora brassicae* in canola. Ann Appl Biol 163:56–71

Devos S, Vissenberg K, Verbelen JP, Prinsen E (2005) Infection of Chinese cabbage by *Plasmodiophora brassicae* leads to a stimulation of plant growth: impacts in cell wall metabolism and hormone balance. New Phytol 166(1):241–250

Devos S, Laukens K, Deckers P, Van Der Straeten D, Beeckman T, Inzé D, Van Onckelen H, Witters E, Prinsen E (2006) A hormone and proteome approach to picturing the initial metabolic events during *Plasmodiophora brassicae* infection on *Arabidopsis*. MPMI 19(12):1431–1443

Diatchenko L, Lau YF, Campbell AP, Chenchik A, Moqadam F, Huang B, Lukyanov S, Lukyanov K, Gurskaya N, Sverdlov ED, Siebert PD (1996) Suppression subtractive hybridization: a method for generating differentially regulated or tissue-specific cDNA probes and libraries. Proc Natl Acad Sci U S A 93:6025–6030

Diederichsen E, Frauen M, Linders EGA, Hatakeyama K, Hirai M (2009) *Plasmodiophora brassicae* (clubroot) –a plant pathogen that alters host growth and productivity. J Plant Growth Regul 28:264–281

Dodds PN, Rathjen JP (2010) Plant immunity: towards an integrated view of plant– pathogen interactions. Nat Rev Genet 11(8):539–548

Doidy J, Grace E, Kühn C, Simonplas F, Casieri L, Wipf D (2012) Sugar transporters in plants and in their interactions with fungi. Trends Plant Sci 17:413–422. https://doi.org/10.1016/j.tplants.2012.03.009

Domingues MN, De Souza TA, Cernadas RA, de Oliveira MLP, Docena C, Farah CS, Benedetti CE (2010) The *Xanthomonas citri* effector protein PthA interacts with citrus proteins involved in nuclear transport, protein folding and ubiquitination associated with DNA repair. Mol Plant Pathol 11:663–675

Domingues MN, de Campos BM, de Oliveira ML, de Mello UQ, Benedetti CE (2012) TAL effectors target the C-terminal domain of RNA polymerase II (CTD) by inhibiting the prolyl-isomerase activity of a CTD-associated cyclophilin. PLoS One 7:e41553

Donald EC, Jaudzems G, Porter IJ (2008) Pathology of cortical invasion by *Plasmodiophora brassicae* in clubroot resistant and susceptible *Brassica oleracea* hosts. Plant Pathol 57:201–209

Engleberg NC, Carter C, Weber DR, Cianciotto NP, Eisenstein BI (1989) DNA sequence of mip, a *Legionella pneumophila* gene associated with macrophage infectivity. Infect Immun 57:1263–1270

Evans JL, Scholes JD (1995) How does clubroot alter the regulation of carbon metabolism in its host? Asp Appl Biol 42:125–132

Feller A, Machemer K, Braun EL, Grotewold E (2011) Evolutionary and comparative analysis of MYB and bHLH plant transcription factors. Plant J 66:94–116. https://doi.org/10.1111/j.1365-313X.2010.04459.x

Feng J, Hwang R, Hwang SF, Strelkov SE, Gossen BD, Zhou Q, Peng G (2010) Molecular characterization of a serine protease Pro1 from *Plasmodiophora brassicae* that stimulates resting spore germination. Mol Plant Pathol 11:503–512

Feng J, Hwang SF, Strelkov SE (2012a) Studies into primary and secondary infection processes by *Plasmodiophora brassicae* on canola. Plant Pathol. https://doi.org/10.1111/j.1365_3059.2012.02612.x

Feng J, Xiao Q, Hwang SF, Strelkov SE, Gossen BD (2012b) Infection of canola by secondary zoospores of *Plasmodiophora brassicae* produced on a nonhost. Eur J Plant Pathol 132:309–315

Feng J, Hwang SF, Strelkov SE (2013) Assessment of gene expression profiles in primary and secondary zoospores of *Plasmodiophora brassicae* by dot blot and real time PCR. Microbiol Res 168:518–524

Feng J, Cao T, Rennie DC, Strelkov SE, Hwang SF (2014) Host–parasite interactions in clubroot of crucifers. Can J Plant Pathol 36:113–121

Ferrari S, Galletti R, Denoux C, Lorenzo GD, Ausubel FM, Dewdney J (1994) Resistance to *Botrytis cinerea* induced in *Arabidopsis* by elicitors is independent of salicylic acid, ethylene, or jasmonate signaling but requires *Phytoalexin Deficent3*. Plant Physiol 144:367–379. https://doi.org/10.1104/pp.107.095596

Galat A (2003) Peptidylprolyl cis/trans isomerases (immunophilins): biological diversity, targets, functions. Curr Top Med Chem 3:1315–1347

Gamas P, Niebel FC, Lescure N, Cullimore J (1996) Use of a subtractive hybridization approach to identify new *Medicago truncatula* genes induced during root nodule development. Mol Plant-Microbe Interact 9:233–242. https://doi.org/10.1094/MPMI-9-0233

Ge YX, Angenent GC, Wittich PE, Peters J, Franken J, Busscher M, Zhang LM, Dahlhaus E, Kater MM, Wullems GJ, Creemers-Molenaar T (2000) *Nec1*, a novel gene, highly expressed in nectary tissue of *petunia hybrida*. Plant J 24:725–734. https://doi.org/10.1046/j.1365-313x.2000.00926.x

Glazebrook J (2005) Contrasting mechanisms of defense against biotrophic and necrotrophic pathogens. Annu Rev Phytopathol 43:205–227. https://doi.org/10.1146/annurevphyto.43.040204.135923

Gludovacz TV (2013) Clubroot development on Canola and Cabbage in relation to soil temperature and resistance. University of Guelph, Guelph

Gludovacz TV, Gossen BD, McDonald MR (2013) Evidence for tolerance to *Plasmodiophora brassicae* in a cabbage cultivar. Can J Plant Pathol 35:111

Gossen BD, Wang D, McLean A, Tu J, Lahlali R, Karunakaran C, Peng G, McDonald MR (2020) Boron sensitivity and storage in Brassica napus and its potential for clubroot management. Research Square. https://doi.org/10.21203/rs.3.rs-41078/v1

Gravot A, Richard G, Lime T, Lemarié S, Jubault M, Lariagon C, Lemoine J, Vicente J, Robert-Seilaniantz A, Holdsworth MJ, Manzanares-Dauleux MJ (2016) Hypoxia response in *Arabidopsis* roots infected by *Plasmodiophora brassicae* supports the development of clubroot. BMC Plant Biol 16(1):251. https://doi.org/10.1186/s12870-016-0941-y

Guan YF, Huang XY, Zhu J, Gao JF, Zhang HX, Yang ZN (2008) *Ruptured pollen grain1*, a member of the MtN3/saliva gene family, is crucial for exine pattern formation and cell integrity of microspore in *Arabidopsis*. Plant Physiol 147:852–863. https://doi.org/10.1104/pp.108.118026

Guo XH (2001) Biological characteristics and pathogenesis of *Plasmodiophora brassicae*. Southwest Agriculture University, Chongqing

Guo HS, Xie Q, Fei JF, Chua NH (2005) MicroRNA directs mRNA cleavage of the transcription factor NAC1 to downregulate auxin signals for *Arabidopsis* lateral root development. Plant Cell 17:1376–1386

Gupta V, Jagannathan V, Lakshmikumaran M (1990) A novel AT-rich tandem repeat of *Brassica nigra*. Plant Sci 68:223–229

Gustafsson M, Falt AS (1986) Genetic studies on resistance to clubroot in *Brassica napus*. Ann Appl Biol 108:409–415

Hanes SD (2015) Prolyl isomerases in gene transcription. Biochim Biophys Acta Gen Subj 1850:2017–2034

Hansen CE, Flengsrud R, Kopperud C (1994) Two-dimensional gel electrophoresis of proteins in healthy roots and clubroots of *Brassica oleracea*. Plant Sci 44(1):23–128

Henson JM, French R (1993) The polymerase chain reaction and plant disease diagnosis. Annu Rev Phytopathol 31:81–109

Huang X, Li Y, Niu Q, Zhang K (2007) Suppression subtractive hybridization (SSH) and its modifications on microbiological research. Appl Microbiol Biotechnol 76:753–760

Hwang SF, Ahmed HU, Zhou Q, Strelkov SE, Gossen BD, Peng G, Turnbull GD (2011a) Influence of cultivar resistance and inoculum density on root hair infection of canola (*Brassica napus*) by *Plasmodiophora brassicae*. Plant Pathol 60(5):820–829. https://doi.org/10.1111/j.1365-3059.2011.02457.x

Hwang SF, Ahmed HU, Zhou Q, Strelkov SE, Gossen BD, Peng G, Turnbull GD (2011b) Influence of cultivar resistance and inoculum concentration on root hair infection of canola (*Brassica napus*) by *Plasmodiophora brassicae*. Plant Pathol 60:820–829

Hwang SF, Strelkov SE, Gossen BD, Turnbull GD, Ahmed HU, Manolii V, Howard RJ (2011c) Soil treatments and amendments for amelioration of clubroot of canola. Can J Plant Sci 91:999–1010

Hwang SF, Ahmed HU, Zhou Q, Strelkov SE, Gossen BD, Peng G, Turnbull GD (2012a) Assessment of the impact of resistant and susceptible canola on *Plasmodiophora brassicae* inoculum potential. Plant Pathol 61:945–952

Hwang SF, Strelkov SE, Feng J, Gossen BD, Howard RJ (2012b) *Plasmodiophora brassicae*: a review of an emerging pathogen of the Canadian canola (*Brassica napus*) crop. Mol Plant Pathol 13:105–113

Hwang SF, Howard RJ, Strelkov SE, Gossen BD, Peng G (2014) Management of clubroot (*Plasmodiophora brassicae*) on canola (*Brassica napus*) in western Canada. Can J Plant Pathol 36:49–65. https://doi.org/10.1080/07060661.2013.863806

Ibrahim HM, Bannai H, Xuan X, Nishikawa Y (2009) *Toxoplasma gondii* cyclophilin 18-mediated production of nitric oxide induces bradyzoite conversion in a CCR5-dependent manner. Infect Immun 77:3686–3695

Ingram DS, Tommerup IC (1972) The life history of *Plasmodiophora brassicae* Woron. Proc R Soc Lond Ser B 180:103–112

Irani S, Trost B, Waldner M, Nayidu N, Tu J, Kusalik A, Todd CD, Wei Y, Bontham-Smith PC (2018) Transcriptome analysis of response to Plasmodiophora brassicae infection in the *Arabidopsis* shoot and root. BMC Genomics 19:23

Ito S, Yano S, Tanaka S, Kameya-Iwaki M (1994) The use of resting spore spheroplasts in the DNA analysis of *Plasmodiophora brassicae*. Ann Phytopathol Soc 60:491–495

Ito S, Tanaka S, Miyanaga C, Takabayashi S, Yano S, Kameya-Iwaki M (1996) Constitutive and inducible proteins in the root of Chinese cabbage infected with *Plasmodiophora brassicae*. J Phytopathol 144:89–95

Ito S, Maehara T, Tanaka S, Kameya-Iwaki M, Yano S, Kishi F (1997) Cloning of a single-copy DNA sequence unique to *Plasmodiophora brassicae*. Physiol Mol Plant Pathol 50:289–300

Ito S, Maehara T, Maruno E, Tanaka S, Kameya-Iwaki M, Kishi F (1999) Development of a PCR-based assay for the detection of Plasmodiophora brassicae in soil. J Phytopathol 147:83–88

Jian H, Lu K, Yang B, Wang T, Zhang L, Zhang A, Wang J, Liu L, Qu C, Li J (2016) Genome-wide analysis and expression profiling of the *SUC* and *SWEET* gene families of sucrose transporters in oilseed rape (*Brassica napus* l.). Front Plant Sci 7:1464. https://doi.org/10.3389/fpls.2016.01464

Jiang M, Dong X, Lang H, Pang W, Zhan Z, Li X, Piao Z (2018) Mining of *Brassica*-specific genes (BSGs) and their induction in different developmental stages and under *Plasmodiophora brassicae* stress in *Brassica rapa*. Int J Mol Sci 19(7):E2064. https://doi.org/10.3390/ijms19072064

Jones PM (1928) Morphology and cultural history of *Plasmodiophora brassicae*. Arch Protistenkd 62:313–327

Jones JDG, Dangl JL (2006) The plant immune system. Nature 444:323–329

Jones DR, Ingram DS, Dixon GR (1982) Factors affecting tests for differential pathogenicity in populations of *Plasmodiophora brassicae*. Plant Pathol 31:229–238

Jubault M, Lariagon C, Simon M, Delourme R, Manzanares-Dauleux M (2008) Identification of quantitative trait loci controlling partial clubroot resistance in new mapping populations of *Arabidopsis thaliana*. Theor Appl Genet 117(2):191–202

Kageyama K, Asano T (2009) Life cycle of *Plasmodiophora brassicae*. J Plant Growth Regul 28:203–211

Keeling PJ, Burger G, Durnford DG, Lang BF, Lee RW, Pearlman RE, Roger AJ, Gray MW (2005) The tree of eukaryotes. Trends Ecol Evol 20:670–676

Keen NT, Williams PH (1969) Translocation of sugars into infected cabbage tissues during clubroot development. Plant Physiol 44:748–754

Kobelt P (2000) Die Verbreitung von sekundaren Plasmodien von *Plasmodiophora brassicae* (Wor.) im Wurzelgewebe von *Arabidopsis thaliana* nach immunhistologischer Markierung des plasmodialen Zytoskeletts. PhD thesis, Free University Berlin, Berlin, Germany

Kobelt P, Siemens J, Sacristan M (2000) Histological characterization of the incompatible interaction between *Arabidopsis thaliana* and the obligate biotrophic pathogen *Plasmodiophora brassicae*. Mycol Res 104(2):220–225

Kong G, Zhao Y, Jing M, Huang J, Yang J, Xia Y, Kong L, Ye W, Xiong Q, Qiao Y, Dong S, Ma W, Wang Y (2015) The activation of *Phytophthora* effector Avr3b by plant cyclophilin is required for the nudix hydrolase activity of Avr3b. PLoS Pathol 11:e1005139

Kovalev N, Nagy PD (2013) Cyclophilin A binds to the viral RNA and replication proteins, resulting in inhibition of tombus viral replicase assembly. J Virol 87:13330–13342

Lemarie S, Robert-Seilaniantz A, Lariagon C, Lemoine J, Marnet N, Jubault M, Manzanares-Dauleux MJ, Gravot A (2015) Both the jasmonic acid and the salicylic acid pathways contribute to resistance to the biotrophic clubroot agent *Plasmodiophora brassicae* in *Arabidopsis*. Plant Cell Physiol 56:2158–2168. https://doi.org/10.1093/pcp/pcv127

Lemoine R, Camera SL, Atanassova R, Dédaldéchamp F, Allario T, Pourtau N, Bonnemain J-L, Laloi M, Coutos-Thévenot P, Maurousset L, Faucher M, Girousse C, Lemonnier P, Parrilla J,

Durand M (2013) Source-to-sink transport of sugar and regulation by environmental factors. Front Plant Sci 4:272. https://doi.org/10.3389/fpls.2013.00272

Li JP, Zhu YQ, Li BJ (2010) Transmission of clubroot of cruciferous vegetables. Chin Veg 5:21–22

Li B, Takahashi D, Kawamura Y, Uemura M (2018) Plasma membrane proteomics of *Arabidopsis* suspension-cultured cells associated with growth phase using nano-LC–MS/MS. Methods Mol Biol 1696:185–194. https://doi.org/10.1007/978-1-4939-7411-5_12

Lin JY, Mendu V, Pogany J, Qin J, Nagy PD (2012) The TPR domain in the host Cyp40-like cyclophilin binds to the viral replication protein and inhibits the assembly of the tombusviral replicase. PLoS Pathol 8:e1002491

Lin IW, Sosso D, Chen LQ, Gase K, Kim SG, Kessler D, Klinkenberg PM, Gorder MK, Hou B-H, Qu X-Q, Carter CJ, Baldwin IT, Frommer WB (2014) Nectar secretion requires sucrose phosphate synthases and the sugar transporter SWEET9. Nature 508:546–549. https://doi.org/10.1038/nature13082

Liu HH, Tian X, Li YJ, Wu CA, Zheng CC (2008) Microarray-based analysis of stress-regulated microRNAs in *Arabidopsis thaliana*. RNA 14:836–843

Liu Q, Yuan M, Zhou Y, Li X, Xiao J, Wang S (2011) A paralog of the MtN3/saliva family recessively confers race-specific resistance to *Xanthomonas oryzae* in rice. Plant Cell Environ 34:1958–1969. https://doi.org/10.1111/j.1365-3040.2011.02391.x

Liu C, Yang Z, He P, Munir S, Wu Y, Ho H, He Y (2018) Deciphering the bacterial and fungal communities in clubroot-affected cabbage rhizosphere treated with Bacillus Subtilis XF-1. Agric Ecosyst Environ 256:12–22

Ludwig-Muller J (1999) *Plasmodiophora brassicae*, the causal agent of clubroot disease: a review on molecular and biochemical events in pathogenesis/*Plasmodiophora brassicae*. J Plant Dis Protect 106(2):109–127

Ludwig-Muller J, Schuller A (2008) What can we learn from clubroots: alterations in host roots and hormone homeostasis caused by *Plasmodiophora brassicae*. Eur J Plant Pathol 121:291–302

Ludwig-Muller J, Thermann P, Pieper K, Hilgenberg W (1994) Peroxidase and chitinase isoenzyme activities during root infection of Chinese cabbage with *Plasmodiophora brassicae* Woron. Physiol Plant 90:661–670

Ludwig-Muller J, Prinsen E, Rolfe S, Scholes J (2009) Metabolism and plant hormone action during clubroot disease. J Plant Growth Regul 28(3):229–244

Ludwig-Muller J, Julke S, Geiss K, Richter F, Mithofer A, Sola I, Rusak G, Keenan S, Bulman S (2015) A novel methyltransferase from the intracellular pathogen *Plasmodiophora brassicae* methylates salicylic acid. Mol Plant Pathol 16:349–364

Malinowski R, Smith JA, Fleming AJ, Scholes JD, Rolfe SA (2012) Gall formation in clubroot-infected Arabidopsis results from an increase in existing meristematic activities of the host but is not essential for the completion of the pathogen life cycle. Plant J 71(2):226–238

Mallory AC, Bartel DP, Bartel B (2005) MicroRNA-directed regulation of *Arabidopsis* auxin response factor17 is essential for proper development and modulates expression of early auxin response genes. Plant Cell 17:1360–1375

Manck-Götzenberger J, Requena N (2016) *Arbuscular mycorrhiza* symbiosis induces a major transcriptional reprogramming of the potato *SWEET* sugar transporter family. Front Plant Sci 7:487. https://doi.org/10.3389/fpls.2016.00487

Marti M, Good RT, Rug M, Knuepfer E, Cowman AF (2004) Targeting malaria virulence and remodeling proteins to the host erythrocyte. Science 306:1930–1933

Martin-Laurent F, van Tuinen D, Dumas-Gaudot E, Gianinazzi-Pearson V, Gianinazzi S, Franken P (1997) Differential display analysis of RNA accumulation in arbuscular mycorrhiza of pea and isolation of a novel symbiosis-regulated plant gene. Mol Gen Genet 256:37–44

Moro A, Ruiz-Cabello F, Fernandez-Cano A, Stock RP, Gonzalez A (1995) Secretion by *Trypanosoma cruzi* of a peptidyl-prolyl cis-transisomerase involved in cell infection. EMBO J 14:2483–2490

Muller P, Hilgenberg W (1986) Isomers of zeatin and zeatin riboside in clubroot tissue: evidence for trans-zeatin biosynthesis by *Plasmodiophora brassicae*. Physiol Plant 66:245–250

Murphy PJ, Langridge P, Smith SE (1997) Cloning plant genes differentially expressed during colonization of roots of *Hordeum vulgare* by the vesicular arbuscular mycorrhizal fungus *Glomus intraradices*. New Phytol 135:291–301

Oliver RP, Ipcho SVS (2004) Arabidopsis pathology breathes new life into the necrotrophs-vs.-biotrophs classification of fungal pathogens. Mol Plant Pathol 5:347–352

Ozsolak F, Milos PM (2011) RNA sequencing: advances, challenges and opportunities. Nat Rev Genet 12:87–98

Patrick JW (1989) Solute efflux from the host at plant-microorganism interfaces. Funct Plant Biol 16:53–67. https://doi.org/10.1071/PP9890053

Pérez-López E, Waldner M, Hossain M, Kusalik AJ, Wei Y, Bonham-Smith PC, Todd CD (2018) Identification of *Plasmodiophora brassicae* effectors-a challenging goal. Virulence 9 (1):1344–1353

Quirino BF, Normanly J, Amasino RM (1999) Diverse range of gene activity during *Arabidopsis thaliana* leaf senescence includes pathogen-independent induction of defense-related genes. Plant Mol Biol 40:267–278. https://doi.org/10.1023/A:1006199932265

Reinders A, Sivitz AB, Starker CG, Gantt J, Stephen W, John M (2008) Functional analysis of LjSUT4, a vacuolar sucrose transporter from *Lotus japonicus*. Plant Mol Biol 68:289–299. https://doi.org/10.1007/s11103-008-9370-0

Reinders A, Sivitz AB, Ward JM (2011) Evolution of plant sucrose uptake transporters. Front Plant Sci 3:22. https://doi.org/10.3389/fpls.2012.00022

Rojas CM, Senthil-Kumar M, Tzin V, Mysore KS (2014) Regulation of primary plant metabolism during plant-pathogen interactions and its contribution to plant defense. Front Plant Sci 5:17

Rolfe SA, Strelkov SE, Links MG, Clarke WE, Robinson SJ, Djavaheri M, Malinowski R, Haddadi P, Kagale S, Parkin IAP, Taheri A, Borhan MH (2016) The compact genome of the plant pathogen *Plasmodiophora brassicae* is adapted to intracellular interactions with host *Brassica* spp. BMC Genomics 17:272. https://doi.org/10.1186/s12864-016-2597-2

Roman RV, Rathjen JP (2017) Apoplastic sugar extraction and quantification from wheat leaves infected with biotrophic fungi. Methods Mol Biol 1659:125–134. https://doi.org/10.1007/978-1-4939-7249-4-11

Sanchez-Vallet A, Ramos B, Bednarek P, López G, Piślewska-Bednarek M, Schulze-Lefert P, Molina A (2010) Tryptophan-derived secondary metabolites in *Arabidopsis thaliana* confer non-host resistance to necrotrophic *Plectosphaerella cucumerina* fungi. Plant J 63:115–127. https://doi.org/10.1111/j.1365-313x.2010.04224.x

Schuller A, Kehr J, Ludwig-Muller J (2014) Laser microdissection coupled to transcriptional profiling of Arabidopsis roots inoculated by *Plasmodiophora brassicae* indicates a role for brassinosteroids in clubroot formation. Plant Cell Physiol 55:392–411. https://doi.org/10.1093/pcp/pct174

Schwelm A, Fogelqvist J, Knaust A, Juelke S, Lilja T, Bonilla-Rosso G, Karlsson M, Shevchenko A, DhandapaniV CSR, Kim HG, Park JY, Lim YP, Ludwig-Mueller J, Dix-elius C (2015) The *Plasmodiophora brassicae* genome reveals insights in its life cycle and ancestry of chitin synthases. Sci Rep 5:11153

Schwelm A, Berney C, Dixelius C, Bass D, Neuhauser S (2016a) The large subunit rDNA sequence of *Plasmodiophora brassicae* does not contain intra-species polymorphism. Protist 167 (6):544–554

Schwelm A, Dixelius C, Ludwig-Muller J (2016b) New kid on the block – the clubroot pathogen genome moves the plasmodiophorids into the genomic era. Eur J Plant Pathol 145:531–542

Schwelm A, Badstober J, Bulman S, Desoignies N, Etemadi M, Falloon RE, Gachon CMM, Legreve A, Lukeš J, Merz U, Nenarokova A, Strittmatter M, Sullivan BK (2017) Not in your usual top 10: protists that infect plants and algae. Mol Plant Pathol 19(4):1029–1044. https://doi.org/10.1111/mpp.12580

Seo PJ, Park JM, Kang SK, Kim SG, Park CM (2011) An *Arabidopsis* senescence-associated SAG29 regulates cell viability under high salinity. Planta 233:189–200. https://doi.org/10.1007/s00425-010-1293-8

Siemens J, Nagel M, Ludwig-Muller J, Sacristan MD (2002) The interaction of *Plasmodiophora brassicae* and *Arabidopsis thaliana*: parameters for disease quantification and screening of mutant lines. J Phytopathol 150(11–12):592–605

Siemens J, Keller I, Sarx J, Kunz S, Schuller A, Nagel W, Schmulling T, Parniske M, Ludwig-Muller J (2006a) Transcriptome analysis of *Arabidopsis* clubroots indicate a key role for cytokinins in disease development. Mol Plant-Microbe Interact 19(5):480–494

Siemens J, Keller I, Sarx J, Kunz S, Schuller A, Nagel W, Schmulling T, Parniske M, Ludwig-Muller J (2006b) Transcriptome analysis of *Arabidopsis thaliana* clubroots. Acta Hortic 706:301–306

Siemens J, Bulman S, Rehn F, Sundelin T (2009a) Molecular Biology of *Plasmodiophora brassicae*. J Plant Growth Regul 28:245–251

Siemens J, Graf H, Bulman S, In O, Ludwig-Muller J (2009b) Monitoring expression of selected *Plasmodiophora brassicae* genes during clubroot development in *Arabidopsis thaliana*. Plant Pathol 58:130–136

Siemens J, González MC, Wolf S, Hofmann C, Greiner S, Du Y, Rausch T, Roitsch T, Ludwig-Muller J (2011) Extracellular invertase is involved in the regulation of clubroot disease in *Arabidopsis thaliana*. Mol Plant Pathol 12:247–262

Singh K, Tzelepis G, Zouhar M, Rysanek P, Dixelius C (2018) The immunophilin repertoire of *Plasmodiophora brassicae* and functional analysis of PbCYP3 cyclophilin. Mol Gen Genomics 293:381–390

Spanu PD, Abbot JC, Amselem J, Burgis TA, Soanes DM, Stüber K, van Themaat EVL, Brown JKM, Butcher SA, Gurr SJ, Lebrun MH, Ridout CJ, Schulze-Lefert P, Talbot NJ, Ahmadinejad N, Ametz C, Barton GR, Benjdia M, Bidzinski P, Bindschedler LV, Both M, Brewer MT, Cadle-Davidson L, Cadle-Davidson MM, Collemare J, Cramer R, Frenkel O, Godfrey D, Harriman J, Hoede C, King BC, Klages S, Kleemann J, Knoll D, Koti PS, Kreplak J, López-Ruiz FJ, Lu X, Maekawa T, Mahanil S, Micali C, Milgroom MG, Montana G, Noir S, O'Connel RJ, Oberhaensli S, Parlange F, Pendersen C, Quesneville H, Reinhardt R, Rott M, Sacristán S, Schmidt SM, Schön M, Skaminoti P, Sommer H, Stephens A, Takahra H, Thordal-Christensen H, Vigouroux M, Weßling R, Wicker T, Panstruga R (2010) Genome expansion and gene loss in powdery mildew fungi reveal tradeoffs in extreme parasitism. Science 330:1543–1546

Sundelin T (2008) Identification of expressed *Plasmodiophora brassicae* genes during plant infection and molecular diagnosis of plant pathogens. Ph D thesis, University of Copenhagen, Copenhagen

Sundelin T, Jensen DF, Lubeck M (2011) Identification of expressed genes during infection of Chinese cabbage (*Brassica rapa* subsp. *pekinensis*) by *Plasmodiophora brassicae*. J Eukaryot Microbiol 58:310–314

Takahashi H, Ishikawa T, Kaido M, Takita K, Hayakawa T, Okazaki K, Itoh K, Mitsui T, Hori H (2006) *Plasmodiophora brassicae*-induced cell death and medium alkalization in clubroot-resistant cultured roots of *Brassica rapa*. J Phytopathol 154(3):156–162

Tanaka S, Mido H, Ito S (2006) Colonisation by two isolate of *Plasmodiophora brassicae* on a clubroot-resistant cultivar of Chinese cabbage *Brassica rapa* L. Spp *pekinensis*. J Gen Plant Pathol 72(4):205–209

Tanifuji G, Archibald JM, Hashimoto T (2016) Comparative genomics of mitochondria in chlorarachniophyte algae: endosymbiotic gene transfer and organellar genome dynamics. Sci Rep 6:21016

Terby J (1924) Les divisions sporogoniques du *Plasmodiophora brassicae* Woron. Bull CI Sci Acad 5:519–537

Tommerup IC, Ingram DS (1971) The life cycle of *Plasmodiophora brassicae* Woron. in *Brassica* tissue cultures and intact roots. New Phytol 70:327–332

Toyofuku K, Kasahara M, Yamaguchi J (2000) Characterization and expression of monosaccharide transporters (*OsMSTs*) in rice. Plant Cell Physiol 41:940–947. https://doi.org/10.1093/pcp/pcd016

Velculescu VE, Zhang L, Vogelstein B, Kinzler KW (1995) Serial analysis of gene expression. Science 270(5235):484–487

Verma SS, Rahman MH, Deyholos MK, Basu U, Kav NNV (2014) Differential expression of miRNAs in *Brassica napus* root following infection with *Plasmodiophora brassicae*. PLoS ONE 9(1):e86648. https://doi.org/10.1371/journal.pone.0086648

Wagner G, Charton S, Lariagon C, Laperche A, Lugan R, Hopkins J, Frendo P, Bouchereau A, Delourme R, Gravot A, Manzanares-Dauleux MJ (2012) Metabotyping: a new approach to investigate rapeseed (*Brassica napus* L.) genetic diversity in the metabolic response to clubroot infection. Mol Plant-Microbe Interact 25:1478–1491

Wang P, Heitman J (2005) The cyclophilins. Genome Biol 6:226

Wang Y, Wu F, Bai J, He Y (2014) *BrpSPL9* (*Brassica rapa* ssp. *pekinensis* SPL9) controls the earliness of heading time in Chinese cabbage. Plant Biotechnol J 12:312–321

Williams PH, Keen NT, Strandberg JO, McNabola SS (1968) Metabolic synthesis and degradation during clubroot development in cabbage hypocotyls. Phytopathology 58:921–928

Wise RP, Moscou MJ, Bogdanove AJ, Whitham SA (2007) Transcript profiling in host–pathogen interactions. Annu Rev Phytopathol 45:329–369

Wu L, Siemens J, Li S, Ludwig-Müller J, Gong Y, Zhong L, He J (2012) Estimating *Plasmodiophora brassicae* gene expression in lines of *B. rapa* by RT-PCR. Sci Hortic 133:1–5

Yaeno T, Li H, Chaparro-Garcia A, Schornack S, Koshiba S, Watanabe S, Kigawa T, Kamoun S, Shirasu K (2011) Phosphatidyl inositol monophosphate-binding interface in the oomycete RXLR effector AVR3a is required for its stability in host cells to modulate plant immunity. Proc Natl Acad Sci U S A 108:14682–14687

Yang B, Sugio A, White FF (2006) Os8N3 is a host disease-susceptibility gene for bacterial blight of rice. Proc Natl Acad Sci U S A 103:10503–10508. https://doi.org/10.1073/pnas.0604088103

Yuan M, Wang S (2013) Rice *MtN3/saliva/SWEET* family genes and their homologues in cellular organisms. Mol Plant 6:665–674. https://doi.org/10.1093/mp/sst035

Zhao Y, Bi K, Gao Z, Chen T, Liu H, Xie J, Cheng J, Fu Y, Jiang D (2017) Transcriptome analysis of *Arabidopsis thaliana* in response to *Plasmodiophora brassicae* during early infection. Front Microbiol. https://doi.org/10.3389/fmicb.2017.00673

Disease Management 13

Abstract

Clubroot disease management has been aimed to target all the three stages of disease cycle. The first most important step is to bring inoculum load in the soil below minimum threshold level of resting spores. The second and third steps are to suppress primary root hairs infection and secondary cortical infections, respectively, which results in club formations to continue life cycle of pathogen, *P. brassicae*. Any activity that results in transport of soil or crop residues infested with *P. brassicae* from one point to another has the potential to disseminate the pathogen. The most effective cultural control is through removal of diseased plants, improved drainage, deep plowing, ridging, long crop rotation (more than 3 years), using clean and treated seed, avoiding application of raw manures, sanitization (cleaning and disinfecting of machinery, equipment, vehicle, tools, footwear, seed drill, plant materials, irrigation water, etc.), and manipulating date of sowing keeping in mind that sow the crop seed when soil temperature is too cold for clubroot pathogen development. Use of clean, treated seeds, crop rotation, bait crops with resistant, susceptible, and non-*Brassica* plants reduce clubroot inoculum in the infected fields. Soil solarization in combination with low rates of fumigation (Dazomet) can be a means of clubroot control. Post harvest weed management along with susceptible host plants is a very crucial factor to reduce inoculum potential in the fields. Several simple low-cost cultural strategies may be useful for effective management of disease in the endemic areas. Intercropping of *Brassica* and flaxseed reduces clubroot with increase in yield. Soil amendment with lime, wood ash, or calcium cyanamide increases soil pH and is an important clubroot management strategy. Judicious use of nutrients like N, P, K, Ca, B, molybdenum, and organic matter helps in management of clubroot through maintenance of pH with conducive effects on host growth and suppressive soil environment for pathogen infection and development. Several chemicals used as soil drench, surfactant, and soil fumigants control clubroot

G. S. Saharan et al., *Clubroot Disease of Crucifers*,
https://doi.org/10.1007/978-981-16-2133-8_13

through inhibition of resting spores germination and suppression of root hairs and cortical infection. Clubroot has been managed with the use of biological agents of bacterial, fungal, and plant products origin which produce antimicrobial metabolites against *P. brassicae* and colonize host roots to induce resistance. Commercial biofungicides like Serenade, Prestop, Mycostop, and granulate formulations Plasma soil, and Root Shield Plus with fungicides like fluazinam, and cyazofamid are very effective against clubroot. Disease management through host resistance is the most economical and ecofriendly strategy. Disease-resistant cultivars of *Brassica* crops have been developed and are being used to manage clubroot in countries like Canada, China, Australia, Japan, Germany, and Korea. Integrated approaches using cultural, chemical, biological, and host resistance have been extremely effective for clubroot management strategy along with high yield of cruciferous crops. In Latin American countries, clubroot has been managed with modification of cultural operations depending on soil environment, pH, and inoculum load in each country, in addition to the growing of resistance cvs. and using biological and chemical control strategies.

Keywords

Disease management · Cultural practices · Sanitization · Date of sowing · Use of bait crops · Soil solarization · Weed control · Crop scouting · Inter cropping · Soil aeration · Seed cleaning and fungicidal treatment · Soil amendments · Use of suppressive soils · Nutrient management · Use of chemicals · Biocontrol agents · *Bacillus subtilis* · Plant growth promoting formulations · Use of resistant cultivars · Integrated disease management

13.1 Introduction

Crop rotation has been the simplest and most effective means of managing this disease. However, this strategy may not be practical, because *P. brassicae* resting spores are capable of remaining viable in the soil for 20 years or more. The application of lime to increase soil pH and suppress disease development has also been an effective approach for the management of clubroot. However, this requires repeated application of large amounts of lime and is likely not a feasible option. Fungicide control has also been attempted but results are often inconsistent and the associated cost may be prohibitive. The application of calcium cyanamide is considered one of the most effective means of control but is also quite expensive. The efficacy of bait crops to reduce the inoculum loads of *P. brassicae* resting spores under field and greenhouse conditions has been advocated. The disease severities are slightly reduced following two cycles of cruciferous bait crops (*Brassica* or Chinese cabbage) compared with non-cruciferous crops (red clover, perennial ryegrass, orchard grass, or bent grass). However, bait crops had no effect on clubroot severity at locations where concentrations of resting spores are high. Seed treatment with fungicides with a long residual period is also a disease management tool. Two biological control agents, *Bacillus subtilis* and *Gliocladium catenulatum* have been found to be as effective as the fungicides fluazinam and cyazofamid under greenhouse conditions. The manipulation of seeding dates has been shown to reduce disease severity in cropping systems. The most desirable and cost-effective means of controlling this disease is clubroot-resistant cultivars.

In many other host–pathogen systems, the deployment of resistant cultivars has resulted in the selection of increased virulence in the pathogen population, potentially reducing the durability and stability of resistance. The best option for clubroot management is to integrate all available management strategies with multifaceted approaches. Integration of strategies including use of host resistance, crop rotation, nutrient supplements, and fungicides is often proposed for management of clubroot disease. Host resistance is a cornerstone of clubroot control, especially on crops such as canola or oilseed rape (*Brassica napus* L.). Calcium, boron, and pH levels in the soil or application of fungicides may also reduce the severity of clubroot disease. The naturally occurring microorganisms in the soil act as antagonists to clubroot pathogen. A wide range of fungicides have been evaluated for efficacy against this pathogen, primarily on *Brassica* vegetable crops. A soil-drench treatment with Terraclor (pentachloronitrobenzene) controlled *P. brassicae* on cole crops, while fluazinam and cyazofamid, the so-called reduced-risk products, have shown efficacy on several cruciferous vegetable crops. The biocontrol agent (BCA) can colonize the root and/or rhizosphere and durable root protection may be achieved. These BCAs include Serenade® (*Bacillus subtilis*), Prestop® [*Clonostachys rosea* f. *catenulate* (Gilman and Abbott) Schroers], Mycostop® (*Streptomyces griseoviridis*), and Root Shield® (*Trichoderma harzianum* Rifai).

Genetic resistance to clubroot has been found in all the major *Brassica* species providing an important management tool through resistance breeding. Most of the resistant genotypes identified so far appear to be race or pathotype-specific, although some have shown a broader resistance spectrum. Clubroot-resistant cultivars have been developed for cruciferous vegetables and oilseed rape. Multiple races populations or pathotypes of *P. brassicae* may exist in a given geographical region which serves as sources of variation that can lead to the erosion of host resistance. Almost all clubroot resistant cultivars have single gene-based resistance and are vulnerable to shifts in the virulence structure of the pathogen populations. Depending on the resistance gene involved, clubroot disease resistance can be eroded rapidly, as observed on oilseed rape and Chinese cabbage. It is estimated that lengthy break of 4–6 years substantially reduce the clubroot disease potential. The "half-life" of *P. brassicae* resting spores in field soils has been estimated at about 3.5–5 years and this poses a dilemma regarding the value and practicality of crop rotation for managing clubroot disease on cruciferous crops. A faster declining rate of *P. brassicae* resting spores has been observed when infested plots were planted with non-host crops or left fallow for 1–3 years.

13.2 Disease Management by Cultural Practices

It is primarily based on the principle of disease avoidance through removal of diseased plants, improved drainage, deep plowing, ridging, and long crop rotation of more than 3 years considering longevity of pathogens resting spores in the soil. Long crop rotation is required to reduce soil inoculum to levels below a disease causing threshold value. The most effective cultural control for management of clubroot is to ensure that fields remain free of the pathogen. Any activity that results in transport of soil or crop residues infested with *P. brassicae* from one point to another has the potential to disseminate the pathogen (Strelkov et al. 2011). A study on the distribution of infected plants within clubroot-infested fields showed that the incidence of infected plants was highest at the field entrances and lowest at distances of 150 m and 300 m from the entrance (Cao et al. 2009). This indicates that farm equipment laden with infested soil and crop debris was likely to be responsible for introducing the pathogen, so sanitization of infested equipment is a critically important component of disease avoidance strategies.

Another potential mechanism for the long-distance dissemination of *P. brassicae* is seed and tubers harvested from infested fields, which have been shown to carry detectable levels of resting spores (Rennie et al. 2011). Field trials which assessed the frequency of seed to seedling transmission of clubroot from resting spores inoculated onto canola seed have not resulted in clubroot symptoms in the susceptible canola crop or on highly susceptible hosts sown on the same site the following year. However, the success of transmission under greenhouse conditions (Rennie et al. 2011) still raises concerns that dissemination of the pathogen on seed may represent a viable mechanism for dispersal of clubroot disease. It is important to note that the movement of infested soil and crop residues on farm equipment, machinery, and vehicles is a much more common and consistent mechanism of spread than seed-borne transmission. The amount of soil, and therefore the number of resting spores, carried on equipment far exceeds that found on even the most heavily infested seed lots. Nevertheless, transmission of resting spores on seed and tubers could lead to the dissemination of *P. brassicae* over longer distances than might typically be associated with the movement of farm machinery. This could also result in the introduction of novel races or pathotypes to a region where previously these were absent. As such, the possibility of seed-borne transmission should not be ignored in the development of clubroot-containment strategies, and farmers should avoid planting self-saved/farm-saved common, untreated seeds harvested from clubroot-infested fields (Alberta Clubroot Management Committee 2010).

Another important potential mechanism of *P. brassicae* dispersal is irrigation water pumped from canals, creeks, and reservoirs contaminated with runoff from clubroot infested fields. Contaminated water can infest large areas of non-infested fields and should not be applied to plantings of cruciferous vegetables (Howard et al. 2010) or other susceptible crops such as canola. In fact, this practice should be discouraged for all agricultural fields given the longevity of clubroot resting spores in soil and the possibility that susceptible rotational crops could eventually be grown in these fields. Similarly, resting spores can survive passage through livestock, so application of raw manure from animals that have been fed or pastured on clubroot-infected fodder onto clean fields should be avoided (Howard et al. 2010). Temperature and moisture contents are known to be important for the successful eradication of *P. brassicae* resting spores from composted residues infested with clubroot (Noble and Roberts 2004; Fayolle et al. 2006; Wichuk et al. 2011), although research on the effects of the composting process is limited and questions have been raised as to how the temperatures achieved in composting could destroy all of *P. brassicae* resting spores. Plowing in roots of plants with a relatively low level of clubroot galls (disease index 20) has been shown to add 1×10^5 spores per gram of soil (Murakami et al. 2004). Considering that this concentration of resting spores in soil can cause severe clubroot under favorable conditions (Murakami et al. 2002a, b), the importance of rouging (the practice of removing diseased plants) should not be underestimated, particularly as a tool to manage new outbreaks.

Crop rotation with non-host species is commonly practiced to avoid clubroot, but very few long-term studies on the effect of crop rotation on the survival of resting spores of *P. brassicae* have been conducted. Wallenhammar (1996) demonstrated a half-life for *P. brassicae* spore inoculum of 3.6 years. Therefore, at high soil inoculum concentrations, extremely long rotations are needed to reduce soil inoculum to levels below a disease-causing threshold. Wallenhammar (1996) found that 17.3 years were required to reduce the level of infestation in soil from 100% infection to below that which could be detected using a plant bioassay. Over a 5-year period, continuous fallow, or cultivation with a clubroot-resistant Japanese radish caused the greatest decrease in resting-spore viability. Spinach, potato, strawberry, and welsh onion were the least effective at reducing *P. brassicae* populations of the 11 crops used (Ikegami 1985). Reduction of *P. brassicae* spores using crop rotation depends upon the removal of cruciferous weeds (Colhoun 1958), including shepherd's purse, wild radishes, and mustards, which are also hosts of *P. brassicae*. A number of non-cruciferous hosts are known in which at least the primary stages of the pathogen's life cycle can be completed (Colhoun 1958). The influence of these hosts on resting-spore populations is still largely unknown. A three year's rotation of canola with cereals and pulses is considered sustainable (Cathcart et al. 2006; Peng et al. 2014).

Root exudates from several non-host plant species such as leek (*Allium porrum* L.), winter rye (*Secale cereale* L.), and perennial ryegrass (*Lolium perenne* L.) stimulate germination of resting spores of *P. brassicae* (Friberg et al. 2005) in nutrient solution. Exudates from *L. perenne* stimulated spore germination more than those from the susceptible host species *B.* rapa var. *pekinensis*. The potential of these plants to be used to manage *P. brassicae* was studied in subsequent glasshouse and field trials, but there was no evidence that any of them would be useful in the sanitation of *P. brassicae*-infested soils within short periods (Friberg et al. 2006).

Biofumigant crops such as high-glucosinolate-containing cultivars of *B. rapa* and *B. napus* have been shown to reduce soil inoculum of *P. brassicae* (Cheah et al. 2001, 2006). Because these crops are hosts for *P. brassicae*, their use in rotation with vegetable Brassicas carries significant associated risk. Cheah et al. (2001) also found that addition of leaf and stem pieces from biofumigant crops could reduce inoculum of *P. brassicae*, and it is possible that development of biofumigant crop mulches or meals might provide the required glucosinolate mediated effect without the risk of growing a susceptible crop in rotation

13.2.1 Sanitization

Any activity that moves *P. brassicae* infested soil or infected crop residues from one field to another has the potential to spread clubroot (Strelkov et al. 2011). Sanitization (sanitation) is the process of cleaning and disinfecting or otherwise decontaminating hard surfaces (machinery, equipment, vehicles, tools, footwear), seeds, plant materials, water and/or soil infested with pathogens. In a plant agriculture setting, such practices are meant to eradicate pathogens, eliminate the risk of introducing them, or slow their spread from infested to non-infested fields or beyond localized areas within already infested fields. To prevent the transfer of clubroot-infested soil and infected crop residues to new sites, field equipment, tools, vehicles, and the like should be cleaned and, where possible, disinfect them prior to moving them from infested to clubroot-free fields. This process has been recommended for use on vegetable farms for many years (Miller et al. 1996; Tremblay et al. 1999; Donald et al. 2006), but its use in oil yielding *Brassica* crops production systems has been a relatively recent application. Sanitization mechanism protocols and procedures have been exploited at a large scale for canola production in Canada (Hwang et al. 2014). Sanitization involves three key steps: (1) rough cleaning using scraping, brushing, or blowing to remove bulk soil and crop debris from contaminated surfaces; (2) fine cleaning using pressure washing, scrubbing, or compressed air to remove any remaining residues; and (3) disinfection by applying an effective biocide to the cleaned surfaces and allowing at least 20 min of contact time to ensure that any remaining spores are killed (Canola Council of Canada 2011).

Unfortunately, many farmers find rigorous sanitization protocols to be excessively time consuming and labor intensive, since field equipment typically has many internal and external surfaces that may be exposed to soil or to spore-bearing dust and infected crop residues. Similar challenges are faced by petroleum, construction, and transportation companies, which often work in clubroot-infested fields in areas such as central Alberta. The Canadian Association of Petroleum Producers (CAPP) has recognized the challenges faced by its members and has produced a plan for mitigating the spread of clubroot through petroleum industry-related activities (Canadian Association of Petroleum Producers (CAPP) 2008). The most critical steps in sanitizing clubroot-infested farm machinery, equipment, and vehicles are the rough and fine cleaning, which should aim to remove up to 99% of the clubroot contamination (Canola Council of Canada 2011). At this point, the surfaces should be free of visible soil and plant material. The application of a disinfectant, such as 1–2% active ingredient sodium hypochlorite solution, on pre-cleaned surfaces will serve to kill or inactivate residual resting spores. A contact time of 20–30 min is required to achieve adequate spore mortality. In on-farm demonstration trials in southern Alberta, it took at least 2 hours to clean and disinfect a 12 m wide field cultivator and up to 4 hours for a large tractor (Hwang et al. 2015). This work was done using a mobile sanitation unit that contained a commercial pressure washer, air compressor, and other equipment that would typically be used by farmers for cleaning machinery, field equipment, and vehicles such as farm trucks. Tractors and tillage equipment generally carry the heaviest loads of infested soil (Canola Council of Canada 2011). For example, a 12 m wide cultivator that had worked in a moist, clubroot-infested field in southern Alberta had 50 kg of soil on its shovels, shanks, and tires, whereas the large tractor used to pull it had 150 kg of soil on the tires, and frame. Heavy soil loads were also observed on a double disc unit working in the same field. At harvest time in an infested canola field, swathers and combines were observed to carry large amounts of crop debris in the form of straw, chaff, and seed. Like soil, infested crop debris can be transported over long distances on machinery, vehicles, and equipment and may serve to spread clubroot into new areas. While chlorine has been shown to be useful for disinfesting water containing *P. brassicae* resting spores (Datnoff 1987; Datnoff et al. 1987), it is highly corrosive and therefore may not be desirable for repeated applications to bare metal on tools and field equipment. Likewise, sensitive electronic equipment, electrical systems and panels may also be damaged by water and/or corrosive disinfectants, thus, compromising equipment warranties and longevity. In such cases, it is recommended to use brushes or compressed air to remove loose dust and debris (Canola Council of Canada 2011).

A few studies have compared the relative efficacy of chemical disinfectants against the resting spores of *P. brassicae*. Donald et al. (2002) reported that none of the nine commercial disinfectants tested completely inactivated clubroot resting spores and, in fact, most were ineffective. Hypochlorite was the only treatment that resulted in a significant reduction in clubroot severity when treated spores were used to inoculate broccoli seedlings. They evaluated the efficacy of 10 commercial disinfectants against clubroot resting spores obtained from diseased canola roots. Infectivity declined sharply as disinfectant concentration increased over the range of dosages tested for the following products: General Storage Disinfectant, Hyperox®, KleenGrow™, EcoClear™, SaniDate®, Thymox™, Virkon®, and Industrial Bleach. It was noted that clubroot resting spores subjected to pressurized heat (autoclaved at 121 °C, 20 min) or dry heat (oven at 80 °C, 12 h) exhibited gross changes in their pathogenic activity and caused reduced disease in broccoli and Chinese cabbage seedlings. However, neither treatment resulted in 100% inactivation of the resting spores. In aqueous suspensions of resting spores to seven temperature regimes (40, 50, 60, 70, 80, 90, and 100 °C) for various lengths of time and then inoculated canola seedlings with the treated spores to determine if they were still infective. All treatments reduced infectivity in direct proportion to the exposure times, which ranged from 30 min to 72 h in increments of 0.5 or 1.0 h. The sharpest declines occurred at 80, 90, and 100 °C, after which most spores were rendered nonviable after as little as 30 min of treatment. At 40 and 50 °C, spores remained infective following 48 h of thermal treatment. Donald et al. (2002) investigated the use of ultraviolet light and calcium, boron, and potassium salts for their ability to inactivate clubroot spores in water solutions. Exposure to UV light resulted in reduced pathogenic activity. This was correlated ($r = 0.80$) with reductions in the severity of root galling on vegetable seedlings that had been inoculated with treated spores. By contrast, the ionic treatments resulted in relatively small reductions in pathogenic activity and there was no correlation with the degree of symptom expression.

13.2.2 Date of Sowing

Clubroot development is strongly favored by high soil moisture (Karling 1968) and temperatures near 25 °C (Feng et al. 2010; Sharma et al. 2011). Temperature has an important impact on each component of the disease cycle (Gossen et al. 2012). Hence, a key means of avoiding infection in canola and other long season *Brassicas* may be through the manipulation of the timing of seeding to minimize infection. Cool soil temperatures have been shown to inhibit pathogen development (Hwang et al. 2011a). Older plants have been shown to be less susceptible to infection, and the infection that does occur is less likely to affect yield (Hwang et al. 2011a). Early seeding dates reduced clubroot severity by 10–50% (only statistically significant at one of two sites) and increased yield by 30–58% (Gossen et al. 2012). Similar results were obtained by manipulating seeding date to minimize clubroot in Shanghai pak choy (*B. rapa* subsp. *chinensis* var. *communis* Tsen and Lee) and Chinese flowering cabbage (*B. rapa* subsp. *chinensis* var. *utilis* Tsen and Lee) over several years at sites in central Canada (McDonald and Westerveld 2008; Gossen et al. 2009, 2012; Adhikari 2010). Clubroot severity was highest in the July seeding (warm temp.) slightly lower in June and August and lowest in the May and September seeding dates (low temp.). Hwang et al. (2012a, b) reported that early seeding of canola reduced clubroot severity and increased yield, although it also reduced seedling emergence. Previous research has indicated that younger canola seedlings are more susceptible to infection than older seedlings (Hwang et al. 2011a). A resistant cultivar exhibited few or no symptoms of clubroot when inoculated 10–25 days after seeding, but some plants became infected when inoculated at 5 days after seeding. The susceptibility of canola roots to infection by *P. brassicae* declines with increasing age, perhaps as a result of the thickening of cell walls (Mellano et al. 1970) and formation of other barriers that limit pathogen colonization. It is likely that early seeding may be a useful tool for clubroot management on canola when used along with other crop management options, including host plant resistance, soil amendment, and fungicide application.

To assess the impact of temperature regime on clubroot incidence, and severity on two short-season Asian vegetable crops, Shanghai pak choy (*B. rapa* subsp. *chinensis* var. *communis*) and Chinese flowering cabbage (*B. rapa* subsp. *chinensis* var. *utilis*) under field conditions effect of seeding date were assessed by Gossen et al. (2012). Each crop was seeded at about monthly intervals from May to September on a naturally infested site near Bradford, Ontario in 2008 and 2009. A soil-drench treatment with cyazofamid fungicide (Ranman® 400 SC) was included at

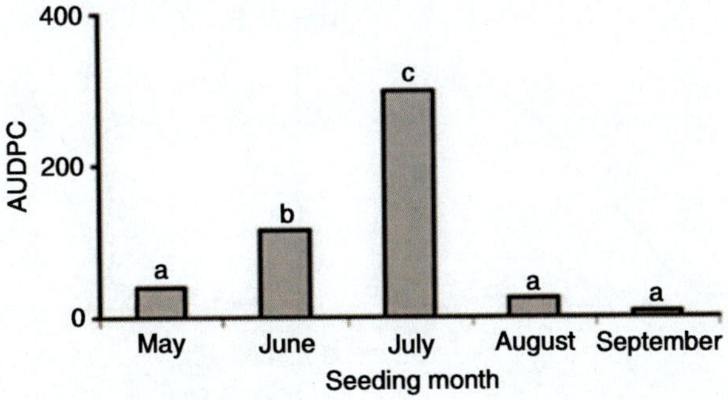

Fig. 13.1 The mean effect of seeding date on area under the disease progress curve (AUDPC) estimates of clubroot severity (disease severity index) at Holland Marsh, ON, 2008–2009, across host, fungicide treatment, and year. Bars with the same letter do not differ at $P = 0.05$ based on Tukey's Test (Gossen et al. 2012)

each seeding date to assess the interaction of fungicide and temperature. Clubroot severity was assessed at weekly intervals from 2–3 week after planting until about 6 week after planting using a 0–3 scale. Clubroot incidence and severity were consistently higher on Shanghai pak choy than Chinese flowering cabbage, but both host crops exhibited a similar pattern of symptom development. Clubroot severity was highest in the July seeding (warmest mean temperature), slightly lower in June and August and lowest in the May and September seeding dates (lowest temperatures) (Figs. 13.1 and 13.2). Clubroot levels were also positively correlated with rainfall. Cyazofamid fungicide consistently reduced clubroot levels. In case of short season crops, farmers can select seeding dates to avoid losses from clubroot, even in heavily infested fields.

13.2.3 Use of Bait Crops (Stimulates for Resting Spore's Germination)

Induction of resting spore germination in the absence of host plants has been proposed as a possible tool in the management of clubroot (Friberg et al. 2005). Resting spores of *P. brassicae* can survive for many years in the soil environment. However, following germination, they must quickly infect a host in order to ensure continued survival (Suzuki et al. 1992; Takahashi 1994). The concentration of viable resting spores in the soil is an important factor in the subsequent development of *P. brassicae* epidemics (Murakami et al. 2002a, b). The reduction of viable spore density to approximately 10^4–10^5 spores mL^{-1} of soil was required to reduce the clubroot disease index to 50% of that observed at 10^8 spores mL^{-1} as reported (Hwang et al. 2011b). However, extremely long rotations may be needed to reduce high inoculum concentrations to a level below a disease-causing threshold (Wallenhammar 1996; Donald and Porter 2009). Therefore, inducing resting spore germination in the absence of host plants to accelerate reduction in spore populations can be a useful component of an integrated clubroot management program (Friberg

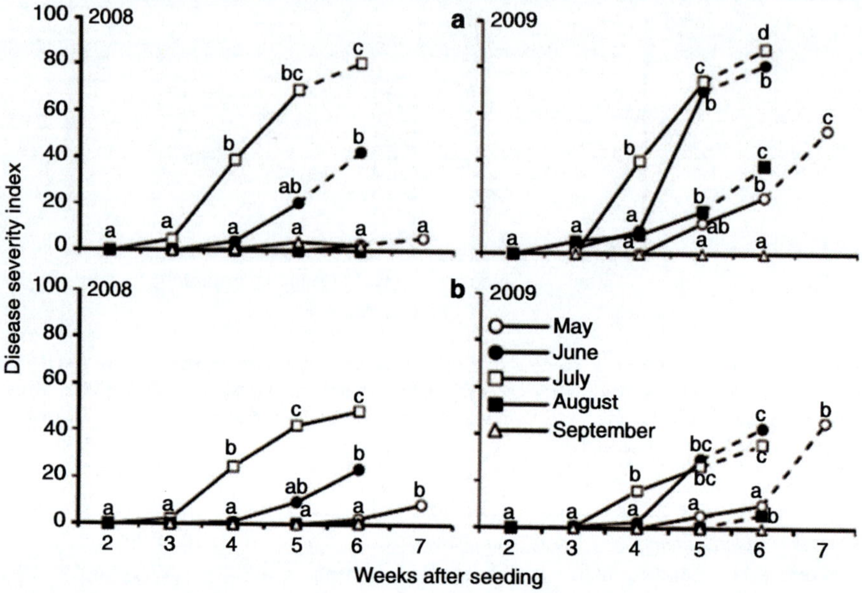

Fig. 13.2 Clubroot severity over time on Shanghai pak choy (**a**), and Chinese flowering cabbage (**b**) planted at monthly intervals at Holland Marsh, ON, 2008–2009. Solid line = clubroot development to optimum harvest maturity, dotted line = beyond optimum maturity. Values within a single seeding date followed by the same letter do not differ at $P = 0.05$ based on Tukey's test (Gossen et al. 2012)

et al. 2006). Many plants can induce germination of resting spores of *P. brassicae* without becoming infected. A crop that stimulates resting spore germination (i.e., a bait crop) could be planted to reduce pathogen populations in heavily infested fields. Host crops, if they are killed before the formation of clubs and resting spores, could also act as bait crops.

Root hair infection by *P. brassicae* has been reported in a wide range of non-brassica plant species, including common velvet-grass (*Holcus lanatus* L.), perennial ryegrass (*Lolium perenne* L.), Indian Cress (*Tropaeolum majus* L.) (Webb 1949; MacFarlane 1952), and strawberry (*Fragaria ananassa* Duch.) (Lugauskas et al. 2003). Subsequent cortical infection and development of resting spores have been reported in several non-cruciferous plant species, such as *T. majus* and beet (*Beta vulgaris* L.) (Ludwig-Muller et al. 1999). Root exudates of perennial ryegrass (Rod and Robak 1994), lettuce (*Lactuca sativa* L.) (Ikegami 1985; Robak 1996), leek (*Allium ampeloprasum* var. *porrum* (L.) Gay), rye (*Secale cereale* L.), and red clover (*Trifolium pratense* L.) induce resting spore germination (Friberg et al. 2006). Perennial ryegrass was effective in reducing the resting spore numbers in one study (Rod and Robak 1994), but was not effective in another (Robak 1996).

Studies in northern Europe using bait crops to stimulate germination of resting spores did not result in effective reductions in clubroot (Friberg et al. 2005, 2006). Moreover, the benefits of using non-host plants for baiting or in crop rotations came under question when fallow treatments without bait plants reduced resting spore levels to an extent similar to the most promising non-host crops (Ikegami 1985; Robak 1996). By contrast, a 5-year fallow period or the continuous cultivation of a clubroot-resistant Japanese radish (*Raphanus sativus* var. *longipinnatus* Bailey) resulted in a substantial decrease in resting spore populations (Ikegami 1985).

On the Canadian Prairies, the extensive scale of the clubroot epidemic on canola necessitates approaches that can be implemented across a large area, and bait crops were one of the few potential management options that met this requirement. However, environmental conditions and production systems on the Canadian Prairies are quite different from other countries, so information directly applicable to the Canadian situation was required to evaluate the potential for use of bait crops for clubroot management. In a heavily infested commercial field in Alberta, clubroot incidence and severity on canola were lower following a cruciferous bait crop versus a non-cruciferous crop or a cereal (Ahmed et al. 2011). Similarly, when sequences of bait crops (canola, ryegrass, and fallow treatments) were assessed, clubroot severity in the subsequent canola crop was lowest in the canola–fallow sequence. Clubroot severity was consistently high in the fallow–fallow sequence, but the results of the other sequences with canola were not as consistent across two repetitions of the experiment. The assessment of a bait crop at two other field sites in Alberta only had a small impact on resting spore populations and there was no effect on subsequent clubroot severity (Hwang et al. 2011b). Resting spore concentrations at these sites were about 1×10^6 spores per g soil, so a small reduction in the spore load was difficult to detect and would not be expected to affect clubroot severity. Resistant cultivars also stimulate the germination of resting spores, but few or no viable spores are produced, and this may result in a reduction in inoculum pressure (Hwang et al. 2011b). Long-term studies are needed to detect significant reductions in clubroot inoculum and severity in heavily infested field sites and to evaluate whether bait crops could be useful at locations where inoculum levels are low to moderate. The impact of bait crops on inoculum potential and clubroot severity is too small and inconsistent to be useful for managing this disease in commercial fields.

Bait cropping with resistant, susceptible, and non-brassicas plants reduces clubroot in the infected fields. Reducing resting spore populations or initial levels of inoculum in infested soil by prior cultivation of bait crops, including both resistant and susceptible Brassicas and non-brassicas, in order to stimulate resting spore germination before main cultivation of the *Brassica* crop, is one of the effective means of integrated management of clubroot disease (Friberg et al. 2006; Bhattacharya and Dixon 2010). A field experiment was conducted in highly infested plots (resting spore concentration of 1×10^7/g soil) in Kalimpong in the Darjeeling Hills to determine the effectiveness of different species of *Brassica* and non-brassica species as bait crops on disease severity and yield in cauliflower, cabbage, and local Indian leafy vegetables such as rayosak. The result of 2 years of field trials indicated that resistant hosts such as radish and toria sarson are the most effective bait crops when cultivated for 30–35 days before main crops like cabbage, cauliflower, and rayosak are sown (Bhattacharya and Dixon 2010).

13.2.4 Soil Solarization

Thermal inactivation of *P. brassicae* caused by the heating of air beneath a polyethylene sheet, a technique known as soil solarization, has been shown to control diseases caused by a number of fungal pathogens especially soil-borne inoculum, including *Verticillium* spp., *Pythium* spp., *Fusarium* spp., *Thielaviopsis basicola*, *Sclerotium rolfsii*, and *Rhizoctonia solani* (Katan 1980; Pullman et al. 1981). The efficacy of the treatment has been shown to depend upon achieving lethal time temperature combinations for a given pathogen (Pullman et al. 1981; Horiuchi et al. 1982). Soil solarization, therefore, is best suited to warm climates and has been effective in Israel (Katan et al. 1976; Grinstein et al. 1979; Jacobsohn et al. 1980) and California (Pullman et al. 1979). Soil solarization for the control of clubroot has been evaluated in southern Australia. Solarization for at least 4 weeks caused a significant reduction in the number of viable clubroot resting spores in the upper 10 cm of soil, as determined by bioassay, at two locations (Porter and Merriman 1983, 1985). Disease development was monitored at one of the two sites and clubroot severity was significantly reduced for the first 6 weeks of the crop. An increased yield from 0 to 14 t/ha of Chinese cabbage was associated with solarization, however, this figure is well below that which would be expected from a commercial crop. Porter et al. (1991) evaluated a range of fumigants in combination with soil solarization as a means of providing more consistent results. The extent of disease control was influenced by soil type and inoculum density. In heavy clay soils, soil solarization in combination with low rates of fumigants gave significantly better control than any of the treatments used alone. However, on sandy soils, which were less severely infested with *P. brassicae*, either dazomet or solarization used alone was as effective as the combined treatment. There have been a number of factors that have prevented Australian *Brassica* growers from adopting soil solarization as a means of clubroot control. The most important of these has been the cost and difficulties of using plastic and the timing of solarization. Dazomet has also not been adopted because it is extremely expensive to apply at dosage rates effective for clubroot control. Cheaper and more reliable treatments are used for clubroot control in cool climate regions (Donald and Porter 2009).

13.2.5 Weed Control

Cruciferous weeds may act as alternative hosts for the clubroot pathogen and should be controlled within the cropping season, as well as within the break between the crops used in rotations with susceptible vegetables. Some examples of potential weed hosts for *P. brassicae* include *Armoracia rusticana* (horseradish, red cole), *Brassica hirta* (white mustard), *Brassica kaber* (wild mustard), *Camelina sativa* (camelina, false flax), *Camelina microcarpa* (small-seeded false flax), *Capsella bursa-pastoris* (shepherd's purse), *Erysimum asperum* (western wallflower), *Lepidium campestre* (pepperwort), *Rorippa islandica* (marsh cress), *Rorippa sylvestris* (creeping yellow cress, yellow field cress), *Sisymbrium altissimum*

Table 13.1 Effect of method, and timing of plant termination (7, 21, and 35 days post-inoculation; dpi) on clubroot incidence (%) and disease severity index (%) in oilseed rape plants inoculated with *Plasmodiophora brassicae* resting spores under greenhouse conditions (Zamani-Noor and Rodemann 2018)

	Clubroot incidence (%)			Disease severity index (%)		
Treatment	7 dpi	21 dpi	35 dpi	7 dpi	21 dpi	35 dpi
Positive control	10.0 ± 12.0	100.0 ± 0.0b	100.0 ± 0.0 b	3.3 ± 3.8	80.0 ± 23.7 b	98.0 ± 10.6 b
Early mechanical treatment		25.0 ± 10.0a	22.0 ± 12.0 a		8.3 ± 12.6 a	7.6 ± 3.4 a
Early herbicide treatment		20.0 ± 10.0a	20.0 ± 16.0 a		6.6 ± 5.4 a	6.9 ± 6.3 a
Late mechanical treatment			100.0 ± 0.0 b			80.0 ± 14.4 b
Late herbicide treatment			100.0±0.0 b			68.2± 28.8 b

Data are pooled across two runs (i.e., repetitions). Means-standard deviations within the same column followed by the same letters do not differ at $P \leq 0.05$, calculated by Tukey's HSD test

(tumbling mustard), *Sisymbrium officinale* (hedge mustard), *Thlaspi arvense* L. (stinkweed), *Rumex* spp., *Barbarea vulgaris* (wintercress, yellow rocket), and *Hesperis matronalis* (dame's rocket) (Howard et al. 2010).

Timing of post-harvest weed management is a very crucial factor for clubroot management. Besides susceptible host plants, oilseed rape (OSR) volunteer seedlings as well as *Brassica* crops, and weeds can play a critical role in the maintenance of resting spore populations in infested fields (Friberg et al. 2005; Ahmed et al. 2011; Peng et al. 2011; Hwang et al. 2013). Therefore, post-harvest management of OSR volunteer plants by mechanical destruction or herbicide application can be used, which would otherwise allow *P. brassicae* populations to increase during the time between harvest of the previous winter OSR and sowing of new crop. The volunteers, *Brassica* crops, and weeds play a critical role in the predisposition of the disease. To determine the effect of timing of foliar application of the herbicide glyphosate or mechanical destruction of OSR volunteers in reduction of clubroot severity and resting spore production, a series of studies was conducted under controlled conditions with a susceptible OSR cultivar and an isolate of *P. brassicae* under condition. Plants were inoculated by injecting a spore suspension beside the root hairs at growth stage 11–12 (BBCH scale) and were terminated at 7 (early) or 21 (late) days post-inoculation (dpi). Under controlled conditions, the first symptoms on roots were observed as early as 7 dpi. The early application of glyphosate as well as early mechanical destruction resulted in significant reduction in the development of clubroot symptoms, root fresh weight, and the number of resting spores/g root. The effect of volunteer management on clubroot severity in the succeeding OSR was studied by inoculating plants with the resting spores obtained from treated clubbed roots. Inoculated OSR exhibited root clubs similar to the initial symptoms after 35 dpi. Plants that were inoculated with spore suspension from early treated roots resulted in significant reductions in clubroot incidence and severity

(Table 13.1). Conversely, plants inoculated with the spore suspension from the late treated roots displayed levels of clubroot similar to the plants inoculated with the spore solutions of positive controls (Zamani-Noor and Rodemann 2018).

13.2.6 Crop Scouting

Regular examination of cruciferous crops for clubroot symptoms may help to pinpoint new field infestations or to monitor disease spread within or between fields. Scouting should begin in transplant beds and continue through to harvest. Plants should be inspected for both foliar and root symptoms. Some root rot diseases may produce foliar symptoms, e.g. stunting, yellowing, wilting that mimic those of clubroot, so plants should be examined carefully for characteristic root galls containing resting spores. In addition, a genetic disorder known as "hybridization nodules" wherein roots develop hard, spherical swellings can also be mistaken for clubroot (Howard et al. 2010).

13.2.7 Inter-cropping

Yellow sarson and flaxseed (*Linum usitatissimum* L.) cultivated as an inter-crop in clubroot-infested plots reduced clubroot incidence and enhanced yield. Possibly, hydrogen cyanide present in root exudates of the flaxseed (Poulton 1990) acts as an inhibitor to resting spore germination and club formation on yellow sarson. Clubroot incidence was drastically reduced (30–50%) in this trial (Sen 1993).

13.2.8 Effect of Soil Aeration

The CO_2 concentrations in the soil air of the high-humic AndisoI are 2–4 times as high as those of the low-humic Andisol during the growing period of cabbage under Japan conditions. Influence of CO_2 concentration in the soil air on the occurrence of clubroot disease was examined in turnip in pot culture experiments under controlled soil water matric potential and CO_2 concentration in the soil air. It was found that the increase in the CO_2 concentration promoted the occurrence of clubroot disease in the high-humic and the humic Andisols at high soil moisture levels. In the field as well as in the pot culture experiments in which crucifers were severely damaged by the disease, the decrease in the soil water pH induced by CO_2 accumulation exceeded the value of 0.5 in the high-humic Andisol. The promotive effect of a high CO_2 concentration on the occurrence of the disease is ascribed to the fact that the decrease in the soil water pH by CO_2 accumulation in soil air may activate the germination of resting spores and impair the activity of crucifers' roots. On the other hand, the occurrence was negligible even at high soil moisture levels in the low-humic Andisol. This suppressive character of the low-humic Andisol is attributed to the physical and chemical properties of the soil surface other than the high drainage ability and high gas diffusivity. However, in the field, the high drainage ability, high gas diffusivity, and subsequently small decrease in soil water pH may also contribute to the suppression of the disease occurrence in the low-humic Andisol (Osozawa et al. 1994).

13.2.9 Seed Cleaning and Fungicidal Treatments

Seed contaminated with soil containing spores of *P. brassicae* may represent a mechanism for long-distance spread of this pathogen (Warne 1943). The potential for dissemination of *P. brassicae* resting spores as external contaminants on the seed of crucifers and other crops, including seed of spring wheat (*Triticum aestivum* L.) and tubers of potato (*Solanum tuberosum* L.) harvested from clubroot-infested fields has been reported. Seed cleaning, removal of lumps of soil, crop residue, and shriveled seed reduce the risk of transmission of clubroot with seed (Rennie et al. 2011).

The impact of fungicidal seed treatments on clubroot of canola was also investigated under greenhouse and field conditions in western Canada. Many of the fungicides assessed, including Dynasty® 100FS (azoxystrobin), Helix Xtra® (thiamethoxam + difenoconazole + metalaxyl + fludioxonil), Prosper™ FX (clothianidin + carbathiin + trifloxystrobin + metalaxyl), Vitavax® RS (carbathiin + thiram), and Nebijin® (flusulfamide), reduced infection on canola seedlings under greenhouse conditions, but had no effect on clubroot severity in heavily infested fields (Hwang et al. 2011c). It is possible that the amounts of these products on the seeds were insufficient to eliminate the large numbers of resting spores in the surrounding soil environment, which would still have been available to infect the seedlings after emergence. By contrast, Rod (1992) reported that fungicidal seed treatment may delay infection by *P. brassicae,* and thereby reduce the severity of clubroot symptoms on affected plants. The seed treatment with biological control agents (BCAs) has potential when used in combination with crop rotation that reduces the inoculum levels of the soil (Peng et al. 2014).

13.2.10 Low-Cost Benefit Cultural Practices

Very simple environmental benign, low cost with high benefit cultural operation can be very useful for clubroot management in the areas where pathogen has been endemic. Following key strategies may be useful for effective management of clubroot of crucifers without any additional cost of cultivation: (1) Use of long crop rotations between successive *Brassica* crops, e.g. $\geq$3 years for light infestations and $\geq$5 years for moderate to heavy infestations; (2) Control of volunteer canola, mustard, and cruciferous weeds in infested fields; (3) Following recommended sanitation measures, i.e. clean, wash and disinfection to reduce the risk of spreading clubroot on infested machinery, equipment, and vehicles; (4) Use of direct seeding and other soil conservation practices to reduce the number of tillage operations in fields; (5) Minimizing machinery and vehicle traffic to and from infested fields; (6) If fields are infested only at the entrance, creating a new access well away from the original point of entry; (7) Scouting canola and mustard fields regularly and carefully; (8) Restricting the use of straw, hay, green feed, silage, and manure from infested farms; (9) Avoiding the use of common seed (e.g., cereals, oilseed, and pulses) from infested fields which can introduce resting spores present in earth to non-infested fields, and (10) use of clean and treated seeds and growing of resistant varieties.

13.3 Soil Amendments

Various forms of agricultural soil liming influence clubroot development under field. Clubroot development is often favored by acidic soil conditions (Karling 1968; Myers and Campbell 1985; Rastas et al. 2012). Thus, maintaining a pH of 7.2 or higher with application of various soil amendments especially forms of lime has been used as a strategy aimed at reducing clubroot of cruciferous crops (Murakami et al. 2002a, b). Lime is available in various forms: agricultural lime (calcium carbonate and calcitic lime), dolomitic lime (calcium and magnesium carbonate), hydrated lime (calcium hydroxide), and quicklime (calcium oxide). Agricultural and dolomitic limes are relatively slow-acting, whereas hydrated lime and quicklime are more reactive and so act much more quickly to raise the soil pH. Application of slow acting forms of lime should be made in the autumn months (September to December in western Canada) to allow time for the amendments to disperse into the soil profile prior to spring planting. The substantial quantities have to be applied because a large proportion of the material applied will not yet have dispersed or dissolved into an active form. Fast acting limes are more suitable for spring application. Finely, ground amendments are potentially more reactive than coarse formulations and so they alter the pH more rapidly. Annual applications of soil amendments may be required to raise the pH and to maintain it at desirable levels during the growing season (Hwang et al. 2011c). The impact of soil amendment with lime, wood ash, and calcium cyanamide on clubroot severity, and seed yield was assessed in field trials on canola in central Alberta over several growing seasons. Applications of calcium carbonate at 5.0 or 7.5 t ha^{-1} or wood ash at 7.5 t ha^{-1} increased plant cover and height, even under heavy inoculum pressure (Hwang et al. 2008). Calcium carbonate reduced clubroot severity and increased yield compared with the non-treated control, but the disease reduction and yield increase were too small to justify the expense of lime application (Hwang et al. 2011c).

The form of lime affects the amount of calcium supplied per unit weight of lime, the neutralizing value of the lime, and the rate of release of calcium (Campbell and Grethead 1989). The neutralizing value (NV) of a lime is a measure of how effective that lime is at reducing soil acidity. Pure calcium carbonate has a neutralizing value of 100%. All other limes have neutralizing values expressed relative to that of pure calcium carbonate. A higher neutralizing value reflects a greater ability of that lime (relative to calcium carbonate) to neutralize soil acidity. Burnt lime (calcium oxide), with a higher neutralizing value and pH (NV = 160%, pH 12–12.5) than calcium carbonate-based limes (NV = 80–100%, pH 7.5), has effectively controlled clubroot in a large number of regions throughout Australia, especially when integrated with other methods (Porter et al. 2004).

The particle size of lime has a large influence on the reactivity of the lime. Finely ground limes have a greater surface area to contact with hydrogen ions than coarser limes (Leonard and Bolland 1996) and potentially can be better distributed throughout a soil profile. Dobson et al. (1983) and Tremblay et al. (2005) reported significantly reduced incidence of clubroot in soil limed with a fine lime stone than with a coarser fraction. Dobson et al. (1983) suggested that the source of the lime is less important than the surface area of the lime particles. There is a variation in soil pH (measured in 0.01M $CaCl_2$) up to two pH units from soil micro sites (5-g samples) in a moist silt loam roto tilled with lime to achieve an average soil pH of 6.7. In poorly mixed soils control failures can be explained in terms of the growing root passing through acidic micro zones where the pathogen can remain active. This finding supports the earlier work of Haenseler and Moyer (1937) who reported pH ranges of 5.73–8.45 in micro soil samples taken from a composite sample with a pH of 7.87 following 8 years of liming. Timing of the application of liming materials is critical so that soil pH is highest at transplanting. The number and maturation of root hair infections is minimized by exposure to alkaline pH within 3 days of inoculation. Prolonged exposure beyond 3 days gave no further reduction in root hair infection number or maturation but did suppress the development of symptoms of disease. Most significant reductions in disease index required alkaline pH applied over a 3–7 days period after inoculation (Webster 1986). Different soil types are known to vary considerably in their response to application of liming materials. The pH of soils with a high buffering capacity, such as those found in Santa Cruz and Monterey counties, California (Welch et al. 1976), is difficult to adjust. In other soils termed "lime non-responsive" by Myers et al. (1981), lime failed to control clubroot even where the pH of the soil was changed to 7.7. Campbell et al. (1985) correlated soil pH and extractable cations (calcium and magnesium) with control and proposed that cation concentration was a second important consequence of liming that affects disease control. They predicted that the actions of pH and cation concentration on the development of clubroot are not independent.

13.3.1 Mode of Action of Calcium in Pathogen Life Cycle and Pathogenesis

The association of calcium, in the form of lime, with the suppression of clubroot disease was well established even before the causal organism was identified (Anon 1853). Thereafter calcium as limes or to a lesser extent fertilizers such as calcium cyanamide were recommended as practical means for reducing the progress of clubroot disease (Colhoun 1958; Karling 1968). The mode of action of calcium and associated effects on soil pH has been elucidated. Fletcher et al. (1982) recognized that although pH influenced the progress of clubroot disease, this was affected by the concentrations of calcium ions in the soil solution. Campbell et al. (1985) concluded that there was an interaction between pH, extractable calcium, and magnesium, which must exceed approximately 14 meq/100 mg of soil from native minerals or with fertilizers, such as calcium cyanamide or lime treatments. These relationships were studied by Dixon and Webster (1988) and by Webster and Dixon (1991a, b) who demonstrated details of the stages in the life cycle of *P. brassicae* and the associated development of pathogenesis that are regulated by calcium ions, pH values, and host factors. Each of these components, while affected by the others, has independent effects. These operate throughout the pathogen life cycle from the germination of motile primary zoospores, penetration, and colonization of root hairs through the formation of primary plasmodia, sporangia, and then onto the infection of cortical cells and the secondary stages of pathogenesis which culminate in gall formation. Increasing soil calcium or alkalinity reduces the rate of maturation of *P. brassicae* in root hairs with retarding its speed and the quantities of plasmodia which form sporangia. MacFarlane (1958) had demonstrated that in controlled experiments, pH values of 5 to 6 had little influence on the level of successful infection by primary zoospores penetrating root hairs. Successful infection diminished greatly as pH increased and reached 8.0. This effect was further explained by Donald and Porter et al. (2004) who showed that at acidic pH values, the progress of infection into root hairs was only delayed by the presence of significant amounts of calcium (20 mM). As pH became increasingly alkaline, the amount of calcium required to reduce root hair infection lessened. Subsequently Takahashi et al. (2006) found that when *P. brassicae* encounters resistant cultivars, there is an alkalization of the environment in the rhizosphere which retards the capacity of the pathogen for invasion. The cell death in *Arabidopsis thaliana* resistant to *P. brassicae* was associated with increasing alkalization of the root environment and that did not occur in cultures of susceptible plants. Niwa et al. (2007) emphasized the role of pH in moderating the development of clubroot disease in field experiments. Soils which were suppressive of *P. brassicae* had elevated calcium contents and alkaline pH and considerable microbial activity. Webster (1986) found that nitrate-nitrogen could also be associated with reduced root hair invasion by *P. brassicae*. At least in part, this might be explained by findings that on root surfaces there is a localized increase in pH associated with nitrate-nitrogen and a decline in pH around root hairs when ammonium is the source of nitrogen (Marschner et al. 1986). Yano et al. (1991) investigated by quantitative analysis the relationships between calcium in spore suspension and effects of inoculation in culture solutions with resting spores. Results suggested that the presence of calcium (Ca_2+ ions) promoted the germination of *P. brassicae* resting spores. That exposes the resultant delicate primary zoospores to predation by soil-borne bacteria and may also reduce their motility. By analogy with the water molds (Oomyces), that also rely on flagella for motility, calcium ions in solution adversely affect the movement of zoospores (Donaldson and Deacon 1993).

Ground burnt agricultural (GBA) lime (76% calcium oxide, 12% magnesium oxide) application in the field is quite effective, because it is reactive with water and can effectively raise soil pH in a short period of time (generally at least 1.0 pH unit).

Table 13.2 The effect of lime and nutrient treatments on soil pH, clubroot severity, and marketable yield of broccoli in field trials conducted at Werribee, Victoria (Donald and Porter 2014)

Treatment/rate of application[a]	Soil pH[b]		Clubroot severity[c] (0–3)	Clubroot severity[c] (1–9)	Marketable yield (t ha^{-1})	
Year[d]	1996	1997	1996	1997	1996	1997
Untreated	7.3	6.9	1.5	5.2	5.4	5.1
Ground burnt	8.1	7.7	1.1	3.1	6.1	8.5
Agricultural lime (76% CaO, 12% MgO) 1 t ha^{-1}						
Ground burnt	8.4	8.0	0.7	2.5	6.3	9.5
Agricultural lime (76% CaO, 12% MgO) 2.5 t ha^{-1}						
Quicklime[b] (82% CaO, 4.4% MgO)1 t ha^{-1}	8.0	7.7	1.1	2.5	5.2	9.7
Calcium nitrate and boron (125, 225 and 225 kg ha^{-1} AT, 10 DAT and 20 DAT resp.)	7.0	7.1	0.9	4.6	7.4	6.0
Calcium nitrate and boron and lime (rates as above)	7.7	7.5	0.3	2.1	8.1	10.0
Calcium cyanamide (1 t ha^{-1}, 7–10 DBT)	7.9	7.1	0.9	3.3	10.1	9.9
Gypsum (4 t/ha, 7–10 DBT)	7.1	6.9	1.5	4.6	5.4	5.8
Dynamic lifter™	7.0	6.2	1.6	5.0	9.2	7.0
LSD (P = 0.05)	0.2	0.3	0.7	1.9	1.8	2.7

[a]AT refers to at transplanting, DAT days after transplanting, and DBT days before transplanting
[b]Samples collected 3 weeks after transplanting in 1996 and 6 weeks after transplanting in year 1997 pH was measured in water
[c]Visually assessed using 0–3 (1996) or 1–9 (1997) scale. As disease developed slowly in the 1996 trial, the data reported are from the 9-week post-transplanting assessment. In 1997, disease developed more quickly and the data reported are from the 6-week post-transplanting assessment
[d]In the first year of trial work, delivery problems led to this product being applied 3 days, rather than 7–14 days, before transplanting. This led to some phytotoxic effects; therefore, plant fresh weight and yield data for this product in 1996 are not indicative of its potential

It also can be applied at half the rate of calcium carbonate as it has a higher neutralizing value and pH (NV = 160%, pH 12–12.5) and finer particle size than calcium carbonate-based limes (NV = 80–100%, pH 7.5). GBA lime and similar products such as Quicklime™ (82% CaO, 4.4% MgO) must be applied a week before transplanting to avoid phytotoxicity (Table 13.2). Calcium and boron both have widely accepted roles in the inhibition of clubroot disease (Webster and Dixon 1991a, b; Donald and Porter 2009) and the interaction of pH with calcium is well known (Myers and Campbell 1985; Donald and Porter 2004). Products such as gypsum (calcium sulfate) and calcium nitrate that contain calcium but do not cause a significant increase in soil pH failed to significantly ($P = 0.05$) reduce symptoms of disease when applied alone or cause a consistent, significant increase in marketable yield. By contrast, when calcium nitrate was applied together with lime, the resulting reduction in root galling and increase in marketable yield were greater than the effect of either product used alone (Table 13.2). This supports laboratory results (Donald and Porter 2004) which suggest that pH is the dominant factor influencing the

Table 13.3 The effect of organic amendments on clubroot severity of broccoli in field trials conducted over 3 years on the same site at Boneo, Victoria, compared with fumigation using metham sodium and the standard fertilizer practice (Donald and Porter 2014)

	Clubroot severity[d]		
Main treatment	Year 1	Year 2	Year 3
Standard practice	0.050	0.20[b]	0.288[bc]
Metham sodium	0.050	0.13[b]	0.113[c]
Compost	0.075	0.16[b]	0.713[b]
Chicken manure	0.100	0.10[b]	1.613[a]
Silage	0.100	0.28[b]	0.150[c]
Lignite—High rate	–	0.58[a]	2.050[a]
Lignite—Low rate	–	–	0.875[b]
LSD (P = 0.05)	ns	0.287	0.486
Fertilizer effect	ns	–[e]	ns
Interaction	ns	–	ns

ns not significant (5% level)
[d]Clubroot severity visually assessed on a 0–3 scale where 0 = 0%, 1 = 1–10%, 2 = 11–50%, 3 = >50%
[e]Not measured on low fertilizer treatments

disease outcome. At low pH values, disease is severe regardless of calcium application and at high pH, disease severity is reduced irrespective of the amount of calcium applied. At intermediate pH values, there is a role for calcium in the control of clubroot. In the laboratory, there was a significant effect of added calcium at soil pH values between 6.5 and 7.5. At pH 8, there was no effect of added calcium (Donald and Porter 2004). In the complex soil environment, the relative effects of calcium and lime will depend upon a number of factors, including the buffering capacity of the soil, the soil organic matter content, and the calcium:magnesium ratio. In general terms, however, these trials demonstrate the potential benefits of the use of integrated strategies for disease control that combine several different effects such as pH, boron, and calcium. Three split applications of calcium nitrate formulations amended with boron in the first 3 weeks after transplanting were the most effective method of application of this treatment. While there may be scope to reduce the frequency of application, these trials suggest that early frequent applications of calcium are required to protect the young transplants from infection (Table 13.2). Organic products, although often increasing yields due to their fertilizer value and benefits to soil health, have not been shown to be effective for control of clubroot. This is because in most cases they acidify soils and increase the cation exchange capacity (CEC) in soils. The resultant increase in CEC and moisture levels also appears to contribute to a more conducive environment for disease. Early studies with Dynamic Lifter™ (pelletized chicken manure, Table 13.2) and more recent studies evaluating lignite, raw chicken manure, composted green waste and silage showed that none could reduce disease, and lignite in particular promoted disease (Porter et al. 2005, Table 13.3).

13.3.2 Effect of Soil Environment Through Suppressive Soils

In the rhizosphere, *P. brassicae* most probably encounters greatest numbers of antagonistic microorganisms. This is because the primary zoospores are less than 1.5 μm diameter, and hence will have limited supplies of energy and not be capable of moving far in soil moisture films from the resting spore to the host root surface (Dixon 2014) in response to host root exudates (Mattey and Dixon 2015). It is the suppressive or conducive environment in the rhizosphere which *P. brassicae* zoospores encounter as they attempt penetration of the root hairs which determines the success or failure of subsequent invasion and disease development. Rhizosphere is occupied by beneficial as well as pathogenic organisms and plant root systems. The rhizoplane (root surface) and surrounding rhizosphere act as "supporting complex microbial communities that can influence nutrient availability and the ability of pathogens to colonize roots." This complexity in the microbial community and resultant suppressiveness or conduciveness is influenced by several agricultural practices (Raaijmakers et al. 2009; Raaijmakers and Mazzola 2016). This encourages beneficial rhizosphere 499 microorganisms such as the bacteria *Pseudomonas*, *Burkholderia,* and *Bacillus* which adversely affect the population density, dynamics, and metabolic activities of soil-borne pathogens (Raaijmakers et al. 2009). Increasing the activities of microorganisms capable of suppressing pathogens is a route by which sustainable husbandry may be achieved as advocated by Mazzola (2004) particularly in relation to take-all disease of cereals (*Gaeumannomyces graminis*). Various soil amendments indicate that stimulates of soil microbial populations encourage chitinolytic bacteria which are capable of degrading the resting spores of *P. brassicae* because they contain chitin in the cell wall (Buczacki and Moxham 1983; Dixon and Walsh (1998). Horiuchi et al. (1983) observed that calcium cyanamide enhanced the suppression of clubroot in association with the use of soil solarization. Studies of soil-borne bacteria such as *Bacillus subtilis* established that they are antagonists of *P. brassicae* (Einhorn et al. 1991). Calcium and pH are important factors for the stimulation of populations of antagonistic microbes. In Taiwan, clubroot suppressive soils are associated with pH values more than 7.4 and with a calcium content of more than 1210 ppm. The local irrigation water is (Hseith and Wang 1986) highly calcareous but suppressivity is partially related to soil physicochemical factors such as texture, pH, organic matter content, and exchangeable potassium, sodium, magnesium, and calcium.

Applications of calcium cyanamide are associated with increased numbers of microbes in soil (Allison 1924; Mukerji 1932). Soil biological health is measured by the amount of amylase, and protease activity increases where calcium cyanamide is used, whereas the opposite effects are obtained with applications of ammonium sulfate (Bosch and Amberger 1983). Calcium hydroxide may be released from calcium cyanamide, which then raises alkaline soil pH, further encouraging microbial activity and benefiting soil fertility (Verona 1970). The subsequent release of nitrogen via DCD encourages populations of heterotrophic microbes (Nommik 1958). Overall, the breakdown of calcium cyanamide increases the heterogeneity of soil microbe populations and in turn this raises fertility. Crowther and Richardson

(1932) found that calcium cyanamide is most effective when applied as routine annual applications thereby, maintaining environments conducive for increased microbial activity.

In Brittany, Rouxel et al. (1985) assessed conducive and suppressive soils and established a relationship between pH values, soil moisture, and inoculum load, and clubroot disease. Page (2001) clearly demonstrated that applications of calcium in various forms of lime could be associated with the development of soil suppressiveness towards *P. brassicae*. In Scottish field studies, the development of soil suppressiveness towards *P. brassicae* resulting from the interactions between calcium and benign soil microbes was observed. This relationship was explored further by Murakami et al. (2002a, b) identifying the effects of reducing inoculum load as suppressiveness increased resulting in reduced disease expression associated with exchangeable soil calcium. A link between calcium and increased microbial activity was developed by Niwa et al. (2007, 2008) in relation to suppressive and conducive soils. They demonstrated that calcium is associated with increased microbial activity and reduced germination of *P. brassicae* resting spores and in turn this suppresses the development of clubroot disease. They also demonstrated that soil-borne microbes may change their metabolic activities when calcium cyanamide is present. The diversity and species richness of soil populations of beneficial bacteria was enhanced by calcium cyanamide as Niwa et al. (2008) observed by increasing amounts of rRNA fragments. Subsequently, Junwei et al. (2013) reported that applications of calcium cyanamide encourage soil bacterial community diversity and richness. Potentially that could increase soil suppressiveness and probably result in a decline in soil-borne plant pathogens. If suppressiveness is accepted as a component of soil health, then it also implies promoting a broader spectrum of soil inhabitants such as earthworms. It is also noted that beneficial bacteria such as nitrogen-fixing bradyrhizobia are increased by the presence of calcium cyanamide (Tewari et al. 2004).

Calcium cyanamide fertilizer was the first commercialized source of artificial nitrogen. The uses of calcium cyanamide have mainly centered on specialized applications for high-value field and glass house vegetable and salad crops. It is associated with the promotion of soil health, quality, and reducing the impact of soil-borne pathogens such as *P. brassicae* (Dixon et al. 1987). Additionally, the presence of the dimer DCD in the commercially formulated fertilizer and its formation in soils treated with calcium cyanamide reduces nitrification and slows nitrate pollution. Properties of this fertilizer for slowly releasing nitrogen into the rhizosphere linked with the ready availability of calcium encourage the development of microbial diversity and suppressiveness towards soil-borne pathogens. It is encouraging interest in calcium cyanamide for application in oilseed rape crops which are now increasingly damaged by clubroot disease (Dixon 2010; Diederichsen et al. 2014). Consequently, there is wider recognition of calcium cyanamide as a potentially valuable component of environmentally sustainable integrated pest management (IPM) strategies for this destructive disease worldwide (Tremblay et al. 2005; Donald and Porter 2014). This strategy involves encouraging suppressive soils containing microorganisms that are antagonistic towards soil-borne pathogens

(Dixon and Tilston 2010). Developing soil suppressiveness towards *P. brassicae* is an established technique following studies those of Arie et al. (1998), Einhorn et al. (1991), and Lahlali et al. (2014). Increasingly, accurate means for targeting such fertilizers into the root zone are becoming available following the development of automated robotic sensing of varying fertilizer requirements across crops through changes in foliar nitrogen status (Gianquinto et al. 2003) combined with increased efficiency in granule distribution into the canopy (Antille et al. 2015) with potential increases in efficacy and cost savings. As a result, the cost is likely to be reduced and efficacy increased. Artificial nitrogenous fertilizers have made incalculable contributions to human welfare for over a century. This role in crop husbandry is likely to continue, but the manner by which this is achieved will be modified as needs for sustaining environmental security, biodiversity, and human health increases. There are already requirements for protecting Nitrate Vulnerable Zones (NVZs) and the dangers of raised nitrate concentrations in crops (Pimpini et al. 1970) have been long recognized. Consideration should be given to using nitrogen fertilizer formulations that are not accompanied by these problems (Sieling and Kage 2010) and which encourage beneficial microorganisms (Yao et al. 2000). That improves soil health and quality which is internationally recognized as means for increasing agricultural productivity and conserving and sustaining biological diversity. Increasingly multidisciplinary research is providing knowledge which will enable improved exploitation of the benefits offered by benign soil inhabiting microbes in the repair and maintenance of soil health.

Agriculture is the sector of human activity which is most likely to be affected by climate change (Rosenzweig et al. 2014). The urgency of this situation is further emphasized by the large-scale meta-analysis by Challinor et al. (2014) showing the relationship between rising temperatures and declining crop yields in coming decades. The use of nitrogen fertilization will have particular relevance in mitigating some of these effects. In this process, compounds such as calcium cyanamide, which encourage natural soil microbial antagonisms towards plant pathogenic organisms and have limited environmental impact, can offer considerable benefits for crop husbandry (Dixon 2017).

13.3.3 Use of Calcium Cyanamide

In addition to the many forms of lime that have been used to amend clubroot-contaminated land, calcium in the form of calcium cyanamide has also been widely used (Karling 1942). As with lime, the mode of action of calcium cyanamide is not fully understood. Upon reaction with soil moisture, calcium cyanamide breaks down initially to hydrated lime and hydrogen cyanamide, which then forms urea, ammonia, and nitrate in sequence. The release of hydrated lime increases soil alkalinity and releases Ca^{2+} upon hydrolysis (liming effect). By comparing calcium cyanamide with corresponding amounts of its hydrolysis products, Walker (1935) found the fungicidal value of calcium cyanamide to be greater, indicating that its fungitoxicity is a result not only of its breakdown products but also of reactive intermediates.

Cornforth (1971) found that the intermediate anion CN^{2-} is probably fungitoxic. Phytotoxicity (Naiki and Dixon 1987; Williamson and Dyce 1989) is also reportedly due to the action of this intermediate anion during glasshouse tests using seedlings. Care must therefore be taken to use the appropriate recommended period between product application and transplantation to ensure complete decomposition of calcium cyanamide. This rate of decomposition depends upon soil type, temperature, application rate, humidity, crop type, and cultivation technique (Klasse 1996). Product efficacy also depends upon the particle size of the product, with the fine-particle components being more effective than larger granules (Donald et al. 2004). Humpherson-Jones et al. (1992) indicated that application of calcium cyanamide at 1500–1600 kg product/ha consistently provided good disease control and was as effective as the then standard Calomel™ treatments. Product efficacy was increased when the period between treatment application and planting was increased to between 14 and 21 days.

Depending on soil conditions, calcium cyanamide is broken down into urea and dicyandiamide within 5–20 days (Klasse 1996). Dicyandiamide slows the breakdown of ammonia into nitrate. This type of slow release nitrogenous fertilizer is less polluting than other types of nitrogenous fertilizer because groundwater contamination through leaching of excess nitrate is reduced. Williamson and Dyce (1989) and Naiki and Dixon et al. (1987) proved a direct effect of calcium cyanamide on the viability of resting spores. Incubation of resting spores in solutions containing 200 mg calcium cyanamide per liter for 1 day reduced the number of zoosporangial clusters from 228 in the untreated control to 18, and this was greater than that of four other chemicals tested (Naiki and Dixon 1987). Application of calcium cyanamide enhances elements of biocontrol of *P. brassicae*. In spite of the liming, fertilizer, fungicidal, and environmental benefits of this product, broadcast application of calcium cyanamide can be a relatively expensive treatment option in countries where the cost of transport from the production facility, together with prices, can have a significant impact on the cost of this product. This high cost of application of calcium cyanamide can be reduced, however, by approximately two thirds by incorporation of the product into bands along the transplant row (Donald et al. 2004; Tremblay et al. 2005; Donald and Porter 2009).

13.4 Nutrient Management

The treatment of soil with standard doses of fertilizers (N: P: K, 80: 40: 40 kg ha^{-1}) along with lime (CaO, calcium oxide, quicklime) applied at 3 t ha^{-1} increased the soil pH from 5.7 to 7.2 and reduced clubroot disease incidence significantly in India. In plots of yellow sarson, there was a reduction in disease incidence of 19.6% in treated versus untreated controls. The effect on disease severity was also pronounced, with a reduction in the size and number of clubs on plants in treated plots (Chattopdhyay 1991). The effects on clubroot disease incidence in Indian mustard of calcium borate, calcium carbonate, and ammonium molybdate along with a standard dose of fertilizer (N: P: K, 80:40:40 kg ha^{-1}) applied to highly infested plots with resting spore population of 2×10^7/g soil and pH 5.6 were analyzed (Sen 2005). The results of four consecutive years of trials (1996–2000) showed drastic reductions in the rate of club formation and root/shoot ratio with a combination of calcium, boron, and molybdenum and also with only boron + molybdenum treatments. The optimization of yield of mustard and highest yield of 916 kg ha^{-1} were achieved with a combination of the standard dose of N, P, K fertilizers, and calcium + boron + molybdenum. An effect of molybdenum in clubroot disease suppression in coarse sandy soil (pH 5.6) in acidic conditions has been demonstrated (Sen 2005). Pryor (1940) reported a 60% reduction of disease development in susceptible cultivars grown in potassium-deficient media. Walker and Hooker (1945) reported a similar effect of potassium, obtaining a consistently higher disease index at higher concentrations of potassium. Several studies suggest that calcium or magnesium in lime may affect disease development independent of pH (Myers and Campbell 1985; Murakami et al. 2002a, b). This interrelationship between mineral nutrients in the soil system is complex. Palm (1963) reported interrelationship between calcium and boron affecting the potential ability of lime to control clubroot. The response curve for calcium was found to vary according to the amount of available boron, leading to conclude that in limed soils, boron deficiency may limit the ability of the calcium ion to inhibit disease development. At pH 5.9, sharp peaks of infection occurred at 2 milli-equivalent calcium per liter and 7 milli-equivalent potassium per liter. The nature of the infection curve for calcium suggested that the primary action of calcium above 2 milli-equivalent/L was not against *P. brassicae* but that excess calcium might have changed the rate of differentiation of epidermal cell walls, making them more resistant to infection. Calcium response curves showed that when boron was omitted from the solution, the effect of increasing concentrations of calcium was suppressed. Large concentrations of potassium (8–16 milli equivalent/L) were required to inhibit infection. Palm (1963) indicated that abundant supplies of potassium following infection encouraged an expansion of gall tissue. Infection and establishment of the pathogen and subsequent gall development were affected by changes in available calcium, the ratio of calcium to potassium, pH, and the ratio of calcium to boron. In the complex soil environment, therefore, these or any number of other mineral interactions may over-ride the effect of pH.

A non-phosphate buffer system was developed by Myers and Campbell (1985). The results obtained using this sand solution system indicated that although pH is probably the most important factor influencing the development of disease prior to liming, high concentrations of calcium and magnesium may give control at pH below 7.2. Similarly, low concentrations may explain control failures using lime at pH above 7.2. Both calcium and magnesium exhibited pH-dependent inhibition of disease. These results support field evidence (Fletcher et al. 1982) using calcium and sodium salts, which suggests that the calcium ion (as gypsum, calcium sulfate) decreases the incidence and severity of clubroot, although not to the extent of changing pH by application of carbonates of calcium or sodium (Donald and Porter 2009).

13.4.1 Use of Boron

Boron (B) slows the development of *Plasmodiophora brassicae* (clubroot) during infection of root hairs (primary infection) and the root cortex (secondary infection) of several vegetable *Brassica* spp. Deora et al. (2011) assessed the impact of B application rates (0, 0.25, 0.5, 1, 2, 4, 8, 16, 32 kg ha^{-1}), application timing, and commercial formulations of B (Solubor, Boron Max, Boron) on primary and secondary development of clubroot in canola. Under controlled conditions, increasing rates of B application reduced root hair infection and subsequent development of primary and secondary infection. However, phytotoxicity to canola seedlings occurs at rates higher than 2 kg B ha^{-}1. Application of 2 kg B ha^{-1} reduces overall root hair infection only slightly, from 81 to 65%, but delays the development of each stage of primary infection. There are no substantial differences in reduction of incidence and severity of the disease by B whether it is applied before root-hair infection (pre-emergence) or before cortical infection (post-emergence) or split into two applications (pre-emergence + post-emergence). All three formulations of B exhibited similar responses. In field trials, 4 kg ha^{-1} was the most effective rate that produced no phytotoxic symptoms. Clubroot severity at 6 weeks after seeding was reduced by 64% compared with a non-treated control. This indicates that B application can be an important component of an integrated management program to manage *P. brassicae* in canola (Fig. 13.3).

Clubroot severity in canola (*Brassica napus*) can be suppressed by addition of boron (B), but excess B produces phytotoxicity. However, some lines of a closely related species, *B. rapa*, are relatively insensitive to high levels of B. Assessment of 150 accessions of B. napus and *B. rapa* treated with 0, 8, and 16 kg B ha^{-1} as disodium octaborate tetrahydrate (trade name Solubor) identified several lines that were relatively insensitive to high levels of B. Boron-insensitive lines generally had

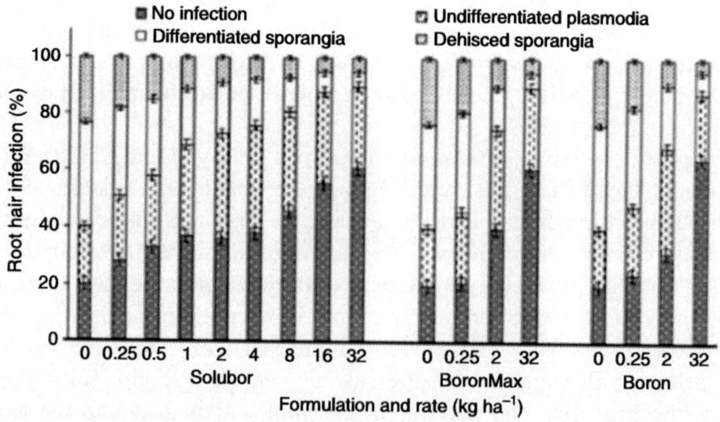

Fig. 13.3 Mean proportions of canola seedling root hairs infected with various developmental stages of *Plasmodiophora brassicae* for seedlings grown in sand treated with different rates of boron in three formulations under controlled conditions (Deora et al. 2011)

slightly lower clubroot severity than sensitive lines, even in the control with no added B. Application of B reduced clubroot slightly in the most sensitive lines, but produced a large and significant reduction in severity in the most insensitive lines.

13.4.2 Use of Calcium

A direct effect of calcium on the viability of resting spores of *P. brassicae* has been established for high (1 M) calcium concentrations (Myers and Campbell 1985). A similar direct effect of pH on the viability of resting spores has been reported (Lee and Hsieh 1992). Because pH levels in excess of 10 were required to observe significant decrease in viability and the effect of calcium was significant only at very low inoculum concentrations (102 spores/g), it can be assumed that both pH and calcium must act in some other way in most agricultural soils. Tissue analyses conducted during the controlled experiments of Myers and Campbell et al. (1985) suggest pH-dependent uptake of calcium by the host. The effect of calcium is more than merely structural. Calcium reduces root hair infection but also inhibits the production of differentiated and dehisced sporangia of *P. brassicae* within infected root hairs. The developmental stages of *P. brassicae* that occurs within root hairs, 4–7 days after inoculation are most vulnerable to high calcium. In solution culture, exposure to high calcium (30 or 55 milli-equivalent per liter) was most effective at reducing root hair infection over days 0–3, and exposure over days 0–7 provides the most effective reduction in disease index (Webster 1986). Root hair stages are also the most vulnerable to high pH. Both calcium and pH reduce the number of primary infections, and high calcium in combination with high pH provides more effective disease control than either acting alone (Webster 1986). Dixon and Webster (1988) proposed an intracellular effect of calcium and pH on the development of *P. brassicae* in host tissue and suggested that such intracellular processes might be more significant than any effect on the pathogen in the soil or during host invasion.

Calcium is commonly the major cation of the middle lamella of cell walls, where it is bound to polygalacturonic acids (pectins) as calcium pectate and is essential for the structural strength of cell walls. It has been proposed that calcium pectate increases tolerance to the action of cell wall-degrading enzymes (Punja et al. 1986). Deficient cells cease growing, become disorganized and discolored, and under severe deficiency die (Epstein 1972). Calcium provides this structural strength by cross-linking the pectic chains of the middle lamella (Marschner 1995). Calcium is also essential for the normal membrane structure (by cross-linking of phospholipids and proteins), transport, and retention of ions. When calcium is limiting, low-molecular weight compounds move from the cytoplasm into the apoplasm. These effects on cell wall and membrane stability are likely to inhibit the invasion of *P. brassicae*. Zoospore motility and chemotaxis play a critical role in the early stages of infection by *P. brassicae*. Potentially, a role exists for calcium regulation of both these processes. Although there is no direct evidence in *P. brassicae*, studies using *Phytophthora* indicate that zoospore motility is reduced in calcium-deficient media (Cho and Fuller 1989). Immunolabeling studies have shown high concentrations of the calcium-binding proteins, calmodulin (Gubler et al. 1990), and centrin (Hardham 1992) associated with *Phytophthora* flagella. Calmodulin is distributed differently in the two flagella, this may elicit a differential response to changes in external calcium, thereby governing directional movement (Gubler et al. 1990). High levels of calcium can impair flagellar motion and the movement of zoospores toward the host root.

In the field, calcium nitrate has been widely used to manage clubroot (Page 2001; Donald et al. 2006). Calcium nitrate directly affects the viability of *P. brassicae* resting spores and can reduce the extent of root galling even when applied after infection (Page 2001). Dixon and Page (1998) explained a more complex relationship between calcium, nitrogen, and boron in the primary stages of the life cycle of *P. brassicae* from the germination of resting spores through to the development of primary sporangia in the root hairs. The growth of primary plasmodia into sporangia in root hairs was delayed by nitrate–nitrogen, calcium, and boron.

13.4.3 Use of Organic Matter

Field assessments across a wide range of soil types and environments indicated that reduction in clubroot severity was more consistent in soils with very high organic matter content than in the typical mineral soils found on the Canadian Prairies (Gossen et al. 2014). The incorporation of FYM and FSC at 2.5% (w/w) to the conducive soil suppresses the clubroot infection and the finer particles (p5 mm) of FSC inhibit the infection and increase soil pH more effectively. Neutralization of the conducive soil by Ca $(OH)_2$, $CaCO_3$, and KOH suppressed the infection, but the effectiveness of KOH was less than those of $Ca(OH)_2$ and $CaCO_3$. Acidification of the suppressive soils by H_2SO_4 promotes the infection. The involvement of soil biota in the disease suppression was investigated using the sterilized (G-ray irradiation) suppressive soils with respect to soil pH. The G-ray irradiation promotes the infection at pH 5.5, but no infection was observed at pH 7.4 irrespective of the sterilization status. It suggests that soil pH is a major factor in disease suppression by organic matter application and that Ca and soil biota play certain roles in the suppression under the influence of soil pH (Niwa et al. 2007).

13.4.4 Clubroot Control Through Soil Management

The severity of clubroot is directly related to the density and distribution of the resting spores of *P. brassicae* in the soil and to the soil environment. Soil physical factors, such as texture, moisture, and temperature, and chemical factors, such as pH, calcium, nitrogen, and boron levels, can affect the entire disease cycle, from the germination of *P. brassicae* resting spores to the infection and colonization of host cells (Donald and Porter 2009; Dixon 2009; Gossen et al. 2014). These same factors can also affect the soil microbial community structure and diversity (Strelkov et al. 2016), as well as the development of cruciferous plants (May et al. 2007).

The physical, chemical, and biological soil attributes, in turn, are directly affected by soil type, organic matter level and composition, climatic conditions, the management strategy adopted, and the type and amount of inputs applied (Tkacz et al. 2015; Bonanomi et al. 2018). Some inputs normally used in the production of vegetables, such as fungicides, mineral fertilizers, and organic waste, may contain copper in their composition as an active principal agent, micronutrient or as an impurity (Dalolio et al. 2017). Copper is a potentially toxic metal with fungicidal and bactericidal action (Fones and Preston 2013). Cupric fungicides, which have been used since the nineteenth century to control plant diseases, can lead to copper accumulation in the soil and affect its microbiota (Wightwick et al. 2013). Studies concerning the interaction of soil physical and chemical factors, soil microbial diversity, and clubroot development are scarce due to the complexity and dynamism of the interactions involved. This information is even rarer in tropical regions such as the mountain agro ecosystems in Brazil, where soils are weathered and acidic. Soil microorganisms play an important role in nutrient cycling and, in many cases, in plant nutrition and development (Rousk et al. 2015), as well as disease suppression (Bonanomi et al. 2018). Similarly, soil characteristics and plant types affect the soil microbiota. At the rhizosphere level, plants may select distinct microbiota, but their composition and stability will depend on the soil characteristics (Tkacz et al. 2015). Plants of the family Brassicaceae can produce bioactive compounds with strong bactericidal action, such as glucosinolates. These compounds, as well as the products of their hydrolysis, can affect the microbiota (Ishida et al. 2014). To suppress soil phytopathogens and to reduce the severity of their corresponding disease, practices favoring the establishment and diversification of soil microbiota can be employed, such as addition of organic matter, acidity correction, and mineral fertilization (Bonanomi et al. 2018). Some studies have reported the potential of using microorganisms that naturally occur in the soil to reduce clubroot (Gossen et al. 2015).

In previous studies with family farmers in Nova Friburgo, Brazil (Bhering et al. 2017), a close relationship between soil attributes was found, particularly between acidity and exchangeable aluminium content and clubroot severity associated with losses in cauliflower. That is, increasing the availability of exchangeable aluminium under high acidity conditions restricts the development of roots and increases the losses caused by the disease. Not performing soil acidity amelioration with liming and application of high doses of non-composted poultry litter are important components in the management adopted in the region that contribute to accentuate soil acidity. Bhering et al. (2020) investigated the relationships between soil attributes, bacterial population composition, and clubroot intensity in a tropical mountain agro-ecosystem of Brazil. Soil and plant samples were collected from 17 cauliflower-producing localities in Nova Friburgo, Brazil. The physical and chemical attributes, the bacterial population composition, the number of pathogen resting spores in the soil samples, and the clubroot severity were evaluated. Based on bacterial population composition, two distinct groups of soils were identified, which varied in attributes such as organic matter, potential acidity, pH, and disease severity. The organic matter level was associated with differences in soil acidity, bacterial population composition, and disease intensity, but there was not a cause and effect relationship between bacterial population composition and clubroot severity. Under the conditions of this agro-ecosystem, soils with increased organic matter presented higher acidity potential (H + Al), which was strongly correlated with clubroot severity. Thus, high acidity potential can be a key factor leading to clubroot development in mountain regions of Brazil.

13.5 Disease Management by Chemicals

Consistent control of clubroot has been reported for only a small number of chemical actives. Among the most effective and widely used has been mercurous chloride (Calomel™). The high mammalian toxicity of mercury and the persistence of this chemical in the environment have led to the now almost complete withdrawal of Calomel from the world market. Among the most widely evaluated group of chemicals for the control of clubroot are the benzimidazoles and their precursors. In a comprehensive glasshouse evaluation of this group of chemicals, Buczacki (1973) identified four—benomyl, thiophanate, thiophanate methyl, and NF 48—with the potential to control clubroot by soil incorporation. All of these are precursors of the actives methyl benzimidazole-2-yl-carbamate (MBC) and ethyl benzimidazol-2-ylcarbamate (EBC). The efficacy of thiophanate methyl was subsequently confirmed by Doyle and Clancy (1986) who reported that control could be improved by the addition of sulfur at 15 mg/kg. Incorporation of either benomyl or thiophanate methyl at 100 mg/kg significantly reduced the severity of disease and increased the average head weight of cabbage by 451 and 462 g, respectively (Buczaki et al. 1976). Results using dipping solutions have been inconsistent (Buczaki et al. 1976; Maric et al. 1990). Tate and Eales (1982) recommended a mechanized transplanting drench, delivering 120 ml of chemical solution per plant, as the most effective method of application of benomyl. Combining root dipping and drenching resulted in the development of symptoms of phytotoxicity that led to depressed yield. A large number of related chemicals, the alkylene bisdithiocarbamates, are also active against clubroot. Among these, maneb (manganese ethylene bisdithiocarbamate), mancozeb (manganese ethylene bisdithiocarbamate complex with zinc), and zineb (zinc ethylene bisdithiocarbamate) are among the most effective (Buczacki and Cadd 1976). Zineb has since been used as an alternative to benomyl transplant drenches, the latter chemical proving phytotoxic to cauliflower (Tate and Eales 1982). Colhoun (1958) observed that the chlorinated nitrobenzenes can provide substantial control only under conditions where heavy attacks of clubroot do not occur. Environmental concern over the accumulation of PCNB and hexachlorobenzene, a technical impurity, in soils (Nishimura et al. 1980) has led to the restriction of PCNB in most countries. The phasing out of PCNB, once widely used in Japan, together with the intensity of Japanese cropping systems, a result of only limited land being suitable for *Brassica* production in that country, has led to a number of new chemicals being developed for clubroot control in Japan by joint private–publically funded development. Among these, trichlamide, flusulfamide, fluazinam, and cyazofamid have been extensively evaluated. N-(1-alkoxy-2, 2, 2-trichloroethyl)-2-hydroxybenzamides were evaluated by Ohmori et al. (1986) as an alternative to PCNB in a greenhouse trial. The C_4–C_6 derivatives of this compound were found to give superior control of clubroot compared with PCNB. Of these, the C_4 derivative [trichlamide, WL 105 305 (NK 483)] was most effective. In a subsequent field evaluation, broadcast incorporation of trichlamide at 30 kg a.i./ha provided control of clubroot equal to that obtained using PCNB (Ohmori et al. 1986). Dixon and Wilson (1984) also reported a reduction in the severity of disease symptoms using trichlamide. This effect was, however, significant only at 1.5 times the rate recommended by the manufacturer (45 kg a.i./ha). Naiki and Dixon et al. (1987) reported that trichlamide has a significant effect on *P. brassicae* 7 days after plants become infected and therefore, effect occurs during the secondary stage of development of the pathogen within the cortical tissue of hosts.

Flusulfamide provided consistently better control of clubroot than thiophanate methyl. The suspension concentrate was the most effective of the two formulations tested, however, there was no significant difference between the three rates used (0.6, 0.9, and 1.2 kg a.i./ha) for either formulation (Dixon et al. 1994). In contrast, Cheah et al. (1998), using a site more severely infested with clubroot than the three sites used by Dixon et al. (1994), report that 2.4 kg a.i./ha of flusulfamide was required to cause a significant reduction in clubroot symptoms. Flusulfamide is not directly toxic to *P. brassicae* resting spores, rather it suppresses the development of clubroot by inhibition of germination of *P. brassicae* by adsorption onto the cell walls (Tanaka et al. 1999). A number of related compounds, the benzenesulfonanilides, have been evaluated for the control of clubroot by Shimotori et al. (1996). None were more active than flusulfamide. Flusulfamide is widely used in Japan and is also registered in New Zealand for control of clubroot in vegetable Brassicas. Fluazinam (Shirlan™ or Omega™) [3-chloro-N-(3-chloro-5-trifluoro-methyl-2 pyridyl)-α, α, α-trifluoro-2,6-dinitro-ptoluidine], a pyridinamine fungicide acts to interrupt the production of energy in fungal pathogens by an uncoupling effect on oxidative phosphorylation (Guo et al. 1991). The activity of fluazinam against resting spores of *P. brassicae* and inhibition of subsequent root hair and cortical stages of infection of Chinese cabbage has been demonstrated by Suzuki et al. (1995). Field control of clubroot using fluazinam has been reported in Australia (Ransom et al. 1991; Porter et al. 1998; Donald et al. 2001) and New Zealand (Cheah et al. 1998). Donald (2005) found that plant dips were the cheapest and simplest method of applying fluazinam to *Brassica* crops, however, this method of application was not pursued because even at the lowest rate used (0.25 g fluazinam a.i./L), some of the seedlings developed blindness (loss of a functional growing tip). Rates lower than 0.5 ml of product/L (0.25 g a.i./L) were considered too small to be delivered accurately in a commercial situation. A strategic method of application (Donald et al. 2001) was developed for fluazinam to overcome problems with existing drench methods, such as poor infiltration and large volumes of water required for plant drenches and phytotoxicity of in-furrow applications (Porter et al. 1998). Using this method fluazinam could be incorporated at 1.5 L a.i. per treated ha into the transplant row using small rotary hoes mounted between the tractor and the transplanter. Because the treated area is reduced (to approximately one third the total area), the effective rate (L/treated area) of the fungicide can be increased and the amount of chemical used per hectare (L/visual ha) and hence treatment cost to the grower is reduced. This method of application consistently increased the marketable yield of broccoli and cauliflower. In one trial, banded soil incorporation increased the marketable yield of both crops by at least 80% compared with the other commercial methods of application currently in use (Donald et al. 2001). The volume of water applied was also reduced from 2500 to 500 L/ha.

Cyazofamid applied at 0.3 mg/L inhibits the germination of resting spores of *P. brassicae* by approximately 80% (Mitani et al. 2003). When applied to soil (1–3 mg/kg dry soil), root hair infections and gall formation caused by *P. brassicae* were strongly inhibited. At the highest rate (3 mg/kg dry soil) complete control of clubroot was achieved (Mitani et al. 2003). In UK glasshouse trials, a post-transplanting drench (200 ml suspension containing 0.08 g a.i. per plant) completely suppressed clubroot disease. The plug-plant dip (10 ml suspension containing 0.04 g a.i. per plant) also reduced disease severity by 78% (Townley and Fox 2003). The control of clubroot using cyazofamid is higher than fluazinam (Townley 2005; Miller et al. 2007). In contrast to fluazinam, phytotoxic effects on *Brassica* plants were not observed for cyazofamid drench or plug-plant dip applications (Townley and Fox 2003). Consistent control of clubroot in Chinese cabbage and cabbage has also been obtained using AG3 phosphonate as a post planting drench in greenhouse, microplot, and field trials (Abbasi and Lazarovits 2006a, b). In field trials with bok choy (*Brassica rapa* var. *chinensis*), one (0.07, 0.14, and 0.21% a.i.) or two (0.07% a.i) post planting drench applications reduced the incidence of clubroot by between 52 and 87%, but this effect was not consistent in a subsequent field trial.

The fungicides that received the most attention in Canada have been fluazinam (Allegro®, Omega®) and cyazofamid (Ranman™); fluazinam is registered against *P. brassicae* on vegetable crops, while cyazofamid is registered for the control of other diseases of vegetables excluding clubroot in Canada. Cyazofamid when applied as a soil drench to Shanghai pak choy (*B. rapa* L. subsp. *chinensis*) and Chinese flowering cabbage (*B. rapa* subsp. *chinensis*) grown on a high organic matter (om) soil [~70% om, pH 6.7] reduced clubroot incidence and disease severity substantially (7 of 10 trials) where inoculum pressure was high (Gossen et al. 2012). The reduction of disease severity ranged from 44 to 89% based on the area under a disease progress curve. In a trial with directly seeded Chinese cabbage grown in a similar high organic matter soil, fluazinam reduced the disease severity index (DSI) from 44 to 20% while cyazofamid reduced DSI to 8% (Peng et al. 2011). In controlled environment trials, these two fungicides provided 100% clubroot control on Shanghai pak choy (Adhikari 2010). Fluazinam and cyazofamid appear to be more effective on short-season crops such as pak choy and Chinese flowering cabbage. If used on longer-season crops, that are grown from transplants, a high water volume may be required to improve efficacy. Indeed, research in Australia has shown that fluazinam is most effective when incorporated into the soil in a band prior to transplanting (Donald et al. 2001). The modes of action for most of the fungicides against infection and symptom development caused by *P. brassicae* largely remain unknown. Flusulfamide (NebijinTM SSC) has been reported to inhibit germination of resting spores of *P. brassicae* (Tanaka et al. 1999). Incubating resting spores with NebijinTM SSC for 2 days, however, did not reduce resting spore viability or their ability to cause clubroot disease on Chinese cabbage "Granaat." It was also believed that cyazofamid would be able to inhibit resting spore germination, root-hair infection, and development of clubroot symptoms (Mitani et al. 2003). Considering the vulnerability of zoospores in the pathogen life cycle (Dixon 2009), it is likely that most fungicides would act against primary and/or secondary zoospores.

Cyazofamid was most effective on canola when incorporated into the soil prior to seeding, but was only effective at low to moderate inoculum pressures (Hwang et al. 2008). Similarly, fluazinam was effective under low inoculum pressure (Ahmed et al. 2011). PCNB (Terraclor 75% WP), when incorporated into the soil before seeding, reduced clubroot severity more consistently than other fungicides in heavily infested fields (Hwang et al. 2008). Fungicide seed treatments have also been examined. Dynasty 100 FS (azoxystrobin), NebijinTM SSC, Vitavax RS (carbathiin + thiram), Prosper FX (carbathiin + trifloxystrobin + metalaxyl), and Helix Xtra (difencazole + metalaxyl + fludioxonil) reduced infection under controlled greenhouse conditions, but none of the treatments was effective in field trials (Hwang et al. 2011a). Seed treatments may have a role only in reducing the risk of seed-borne dissemination of *P. brassicae* carried on seed produced in infested fields (Rennie et al. 2011). These results indicate that fungicide seed treatments are ineffective in heavily infested fields. In general, limitations in product delivery and the lack of an option for soil-drench applications can contribute to the low efficacy of fungicides observed on canola relative to vegetable crops.

13.5.1 Use of Surfactants

A number of surfactant products have been evaluated for their ability to control clubroot (Humpherson-Jones 1989, 1993). In a preliminary glasshouse study in which seven surfactant compounds were evaluated, only the dioctyl sulfosuccinates (Monawet MO-70 and Manoxol OT) and alkyl phenol ethylene oxide (Agral™) provided effective disease control at 500 and 2000 lg/g, respectively. The use of Agral resulted in the largest consistent yield increase and is the only product that is not phytotoxic to young plants (Humpherson-Jones 1993). In field trials, liquid formulations of several nonionic surfactants are effective as split (root ball soak followed by 0.2% transplant hole application or 0.2% transplant hole application followed by 0.2% surface drench 10 days later) or single applications (0.5% transplant hole application), however, when applied in the absence of disease pressure, all of these treatments are phytotoxic compared with lower rates of application (Hildebrand and McRae 1998). Dixon et al. (1997) reported that 2% Agral and 15–20 ppm boron applied with the starter fertilizer (75 ml per plant) resulted in a period of approximately 14 days in which the transplant could establish its root system free of infection. In an alternative approach, Cheah et al. (1999) used the adjuvant Silwet™ at 0.1% combined with a chemical soil drench (50 ml/plant) to achieve almost complete control (97%) of clubroot in a heavily infested field site. Addition of the adjuvant significantly reduced symptoms of disease beyond that which was obtained using the chemical (flusulfamide) alone. Although this treatment resulted in significant leaf burn, with some modification of rate and possibly method of application there may be potential for surfactants to be used in this way.

13.5.2 Use of Soil Fumigants

Soil fumigants have been evaluated for efficacy against clubroot on canola under field conditions in central Alberta. Application of Vapam, a commercial aqueous formulation of metham-sodium (dithiocarbamate; sodium N-methyldithiocarbamate), and several other fumigants could reduce clubroot severity on cabbage (*B. oleracea* L. var. *capitata*) (White and Buczacki 1977). Dazomet applied at 200–400 kg/ha controlled disease with increase in head weight of cabbage and Chinese cabbage (Buczacki and White 1979; Ahmad 1994). Complete control of clubroot on Chinese cabbage has been obtained by Cetas (1958) using a drench application of metham sodium at 0.12 L/m^2 sealed with 2.5 cm water. Band application of metham sodium is the most promising method under field conditions. When in contact with moist soil, Vapam is largely converted to methyl isothiocyanate, a volatile compound that diffuses as a gaseous fumigant through the soil and has considerable nematicidal, fungicidal, and phytocidal activity (Smelt and Leistra 1974). Preliminary tests conducted in Alberta showed that the application of Vapam HL® (metham sodium 42%) one week prior to planting reduced clubroot severity on canola. In trials under controlled conditions, application of Vapam HL® at 0.4–1.6 ml L^{-1} soil resulted in a 12- to 16-fold reduction in primary and secondary infection and clubroot severity in canola seedlings (Hwang et al. 2014). Moreover, the residual effect of Vapam HL® at the higher rates (0.8 and 1.6 ml L^{-1} soil) reduced clubroot development and improved plant health in a subsequent crop compared with the non-treated control. Soil moisture influences the efficacy of Vapam. Application of Vapam HL® at soil moisture levels in the range of 10–30% had a larger effect on both clubroot severity and plant growth than at lower soil moisture levels. Therefore, it appears that Vapam HL® at a higher rate has potential utility for the eradication of isolated clubroot infestations in commercial canola fields. A downside of the application of metham-sodium, however, is that it may cause localized air pollution (Saeed et al. 2000). Thus, a standard method of application that limits the evaporation of the vapor (methyl-dithiocarbamate) will be required before any recommendation can be made to farmers.

The fungicides fluazinam and flusulfamide and the surfactant alkyl phenyl ethylene oxide (Agral™) were selected for evaluation in broad-scale studies in Victoria based on initial reports from Ransom et al. (1991), Dixon et al. (1994), and Humpherson-Jones (1993), respectively. These were compared with ground burnt agricultural lime (76% CaO, 12% MgO), an existing fungicide treatment of PCNB and two existing fumigant treatments, metham sodium and dazomet (Table 13.4).

Table 13.4 Treatments and application details for chemicals used in clubroot control trials conducted during 1994–1996 (Donald and Porter 2014)

Treatment	Rate of application[a]	Method of application	Time of application
Untreated control	–	–	–
Metham sodium (42.3% a. i.)	212 or 423 L	Injected into preformed beds, rolled and irrigated	1–2 weeks before transplanting
Dazomet (98% a.i.)	196 kg	Band-incorporation into top 10 cm	1–2 weeks before transplanting
Fluazinam (50% a.i.) Seedling dip	0.1[b] or 0.3 ml a.i. $^{-1}$	20-min seedling dip	Before transplanting
In-furrow soil drench	0.38[b] or 0.75 L[c]	Drench open furrow by hand	At transplanting
Spot plant drench Pentachloro-nitrobenzene (75% a.i.)	0.75 L[c]	Drench plants, 100 mL plant^{-1}	Immediately after transplanting
Seedling dip	6 g a.i. L^{-1}	20-min seedling dip	Before transplanting
In-furrow (IF) drench	13.5 or 15 kg[c]	Drench open furrow by hand	At transplanting
Spot plant drench	13.5 or 15 kg[c]	Drench plants, 100 mL plant^{-1}	Immediately after transplanting
Flusulfamide (5% a.i.)	1.5 L	Incorporated into top 15 cm of preformed beds and irrigated	1–2 weeks before transplanting
Agral™ (60% a.i.)	5 L[c]	Drench plants, 100 mL plant^{-1}	Immediately after transplanting
Ground burnt agricultural lime (76% CaO, 12 % MgO)	0.5, 1 or 5 t Product ha^{-1}	Incorporated into top 15 cm of preformed beds and irrigated	1–2 weeks before transplanting

[a]Expressed as rate of active ingredient per hectare (a.i. ha^{-1}) unless otherwise stated
[b]Used as a combination dip and in-furrow treatment at Mornington in 1996 only
[c]Applied in 5000 L water ha^{-1}

Dip or in-furrow soil drench applications of the fungicides PCNB and fluazinam suppressed clubroot as effectively as high rate application (423L a.i. ha^{-1}) of the soil fumigant metham sodium. These methods of fungicide application, however, were associated with phytotoxicity that resulted in lower than anticipated yields (Table 13.5). Spot drench application (100 ml plant^{-1}) of either fungicide, while not as effective at suppressing symptoms of clubroot as the dip or in-furrow drenches, consistently increased the marketable yield of broccoli (Table 13.5).

The methyl isothiocyanate (MITC) generator fumigants, metham sodium and dazomet, have been evaluated extensively and shown to be very effective for

Table 13.5 The effect of chemical treatments on clubroot severity and marketable yield of broccoli in field trials conducted at Werribee, Victoria (Donald and Porter 2014)

Treatment	Rate of application[a]	Clubroot rating[b]			Marketable yield (t ha^{-1})
		1994	1995	1994	1995
Untreated	–	2.4	2.1	0.3	3.1
Metham sodium	212 L	2.4	1.3	2.5	7.8
Metham sodium	423 L	–	0.7	–	11.4
Dazomet	196 g	2.4	1.5	0.3	4.7
Fluazinam (dip)	0.3 mL a.i. L^{-1}	–	0.2	–	5.1
Fluazinam (IF drench)[c]	0.75 L	–	0.7	–	4.9
Fluazinam (spot drench)[d]	0.75 L	0.3	1.8	6.2	8.4
PCNB (dip)	6 g a.i. L^{-1}	–	0.7	–	5.6
PCNB (IF drench)[c]	15 kg	–	1.0	–	5.6
PCNB (spot drench)[d]	15 kg	0.6	1.5	4.5	7.1
Agral™ (spot drench)[d]	5 L	–	0.7	–	6.4
GBA lime	0.5 (1994) and 5 (1995) t product	2.5	1.2	1.8	6.7
LSD ($P = 0.05$)		0.7	0.84	1.6	2.9

[a]Expressed as rate of active ingredient per hectare (a.i. ha^{-1}) unless otherwise stated
[b]Visually assessed using 0–3 scale, 4 weeks after transplanting
[c]Drench applied into open furrow (IF) by hand, seedlings planted into the furrow
[d]100 mL poured over the transplants

clubroot control. The efficacy of the fumigant treatments against clubroot was shown to be influenced by soil type. Very low rates of the fumigant dazomet were shown to be very effective treatments in the sandy soils of Victoria (Boneo soil, Table 13.6), but ineffective in heavy clay loams soil of Werribee (Table 13.5). In contrast, metham sodium was more effective in the heavy clay soils of Werribee, although the lowest rate used (212 L a. i. ha^{-1}) was probably marginal for this soil type and produced some inconsistent results (Tables 13.5 and 13.6). A prototype machine consisting of a small tool bar to which two spray jets and two small rotary hoes were fitted was developed to distribute the fungicide evenly within the plant root zone along the planting rows. Each rotary hoe was mounted in a metal box 23 cm wide to retain treated soil in the planting row following tillage. Incorporation of fluazinam (Shirlan; 50% a.i. SC; Crop Care Australasia, at a rate of 1.5 L a.i. in 2500 L ha^{-1} water) using this purpose-built machine was consistently more effective than either a spot drench (1.5 L a.i. ha^{-1} in 5000 L ha^{-1} water applied at 100 mL plant^{-1}) or the "grower preferred" continuous spray (1.5 L a.i. ha^{-1} in 2500 L ha^{-1} water) over the plants immediately after transplanting (Donald et al. 2001). This method of application consistently increased the marketable yield of broccoli and cauliflower. In one instance, yields of both crops were increased by at least 80% compared with other

Table 13.6 The effect of treatments on clubroot rating and marketable yield of broccoli at Boneo, Victoria in 1996 (Donald and Porter 2014)

Treatment	Rate of application[a]	Clubroot rating[b]		Plant fresh weight (g)		Marketable heads cut (%)	Mean head weight (g)	Marketable yield (t/ha)
		3 WAT	6 WAT	3 WAT	6 WAT			
Untreated	–	0	1.1	15.1	345	95.6	231.2	7.55
GBA lime	1.0 t product ha^{-1}	0	1.1	17.9	367	92.5	231.1	7.54
Metham sodium	212 L	0	0.1	17.2	415	94.1	264.0	8.83
Metham sodium and GBA lime[c]		0	0.5	18.9	323	91.7	240.6	7.79
Fluazinam continuous spray	0.75 L	0	0.4	14.8	294	94.1	231.7	7.61
Fluazinam dip and in-furrow drench	0.1 mL a.i. L^{-1} and 0.38 L	0	0.5	6.9	205	94.9	283.1	9.45
PCNB	13.5 kg	0	0.3	11.1	277	96.6	242.8	8.73
Flusulfamide	1.5 L	0	1.2	16.2	343	94.2	216.4	7.37
Dazomet	196 kg	0	0	18.2	458	93.7	272.6	10.42
LSD ($P = 0.05$)			0.6	4.2	75	5.13	39.41	1.84

[a]Expressed as rate of active ingredient per hectare (a.i. ha^{-1}) unless otherwise stated
[b]Visually assessed using 0–3 scale; WAT-weeks after treatment
[c]Rate and method of application as described for single product treatments at top of table

commercial methods of application in use (Donald et al. 2001). The method was effective in a range of soil types and remained effective when the volume of water used was reduced to 500 L ha^{-1}.

13.6 Disease Management by Biological Control Agents

Several studies have illustrated the potential for reducing clubroot using naturally occurring microorganisms (Narisawa et al. 1998; Arie et al. 1999; Cheah et al. 2001; Joo et al. 2004; Lee et al. 2008). Some of these organisms produce antimicrobial metabolites against *P. brassicae* (Arie et al. 1998; Kim et al. 2004), while others colonize roots (Hashiba et al. 2003; Usuki and Narisawa 2007) and induce resistance to the disease (Morita et al. 2003). Microbial control of clubroot is attractive because certain soil microbes colonize root and/or the rhizosphere and provide durable protection. The potential biocontrol agents reported are *Bacillus subtilis* (Serenade), *Gliocladium catenulatum* (syn. *Clonostachys rosea* f. *catenulate*) (Prestop), *Streptomyces griseoviridis* (Mycostop), *S. lydicus* (Actinovate), and *Trichoderma harzianum* (Root Shield), which have been registered in Canada for control of other soil-borne diseases. These biocontrol agents may be integrated for clubroot control (Peng et al. 2011). Soils that are suppressive to clubroot exist widely throughout the world. Biological agents associated with the activity of microorganisms are an important factor determining the suppressiveness of some soils (Takahashi 1994; Murakami et al. 2000a). Young et al. (1991) proposed that a phenolic compound, gentisic acid, is involved in disease suppression. Niwa et al. (2007) studied disease suppression in soils that had a 15-year history of amendment with high rates of organic matter. Although calcium and soil biota influence suppression of clubroot, their effect is moderated by soil pH, which is the major factor associated with disease suppression by organic matter. Page (2001) also found pH, calcium, and soil biota to be responsible for suppression of clubroot in Scottish soils. Several attempts have been made to control clubroot using soil-borne fungi (Djatnika 1991; Cheah and Page 1997; Cheah et al. 2000) or bacteria (Einhorn et al. 1991; Cheah et al. 2000). Reports of field control using these biological organisms are limited (Cheah and Page 1997) and there are no commercial biocontrol products available. In an alternative approach, Arie et al. (1998) demonstrated complete control of clubroot in greenhouse tests following application of culture broth of an isolate of *Phoma glomerata* (Corda) Wollenw and Hochapfel to Chinese cabbage and turnip (*B. rapa*), cabbage and broccoli (*B. oleracea*) hosts. This effect is caused by epoxydon (5-hydroxy-3-(hydroxymethyl)-7-oxabicyclo [4.1.0] hept-3-en-2-one) and may be due to antiauxin properties of this chemical.

The bait crops are effective only if infection is not too severe and they are used in conjunction with liming (Harling and Kennedy 1991). Murakami et al. (2004) caution that any reduction in soil resting-spore numbers resulting from the use of bait crops can be lost immediately upon plowing of clubbed roots in the subsequent crop. The high cost of seed, additional cultivation, and the time that the land is out of production (most bait crops are hoed in 6 weeks after planting), together with an inability to accurately predict soil inoculum load pre- or post-bait crop have limited the use of this method of disease management. Several radish bait crops are recommended to reduce the soil inoculum load to an acceptable level, further increasing the treatment cost. Another compound, a dry powder extract of *Posidonia australis* Hook. F., a species of seagrass, has been shown to stimulate germination of *P. brassicae* and reduce disease severity (Hata et al. 2002), but there are no commercially available germination stimulants with proven efficacy against *P. brassicae.*

Mycorrhizal fungi, known to inhibit disease development in a number of plant species (Linderman 1994), are rare or absent in the Brassicaceae (Brundrett 1991). A root endophytic fungus, *Heteroconium chaetospira* (Grove) M.B. Ellis, which was isolated from Chinese cabbage was shown to inhibit clubroot in both sterile and non-sterile soil (Narisawa et al. 1998). This was the first successful report of suppression of clubroot using a root endophyte. Teruyoshi et al. (2001) reported that 18 of 666 rhizoplane fungi caused a significant reduction in symptoms of clubroot in *B. rapa*. Out of these, two were isolates of the root endophytic fungus *H. chaetospira*. Further greenhouse and field experiments with this endophyte indicated that it is an effective biocontrol agent against *P. brassicae* in Chinese cabbage at low to moderate soil moisture and at pathogen resting-spore densities at or below 105 spores per gram of soil (Narisawa et al. 2005). Bait or decoy crops (radish, leafy daikon) are commercially available. These crops induce spore germination. The bait crop becomes infected; however, the crop is plowed under before the pathogen completes its life cycle. In this way, the number of clubroot spores in the soil is reduced further than can be achieved with other rotation crops (Yamagishi et al. 1986; Murakami et al. 2000b). In addition to radish, canola and mustard are also suitable bait crops.

Seven biological control agents and two fungicides were evaluated as soil drenches or seed treatments for control of clubroot by Peng et al. (2011). Under growth cabinet conditions, a soil-drench application of formulated biocontrol agents *Bacillus subtilis* and *Gliocladium catenulatum* reduced clubroot severity by more than 80% relative to pathogen-inoculated controls on a highly susceptible canola cultivar (Table 13.7). This efficacy was similar to that of the fungicides fluazinam and cyazofamid. Under high disease pressure in greenhouse conditions, the biocontrol agents were less effective than the fungicides. Additionally, all of the treatments delivered as a seed coating were less effective than the soil drench. In field trials conducted in 2009, different treatments consisting of a commercial formulation of *B. subtilis, G. catenulatum*, fluazinam or cyazofamid were applied as an in-furrow drench at 500 L ha^{-1} water volume to one susceptible and one resistant cultivar at two sites seeded to canola in Alberta and one site of Chinese cabbage in Ontario. There was no substantial impact on the susceptible canola cultivar, but all of the treatments reduced clubroot on the susceptible cultivar of Chinese cabbage, lowering disease severity by 54–84% (Fig. 13.4). There was a period of 4 weeks without rain after the canola was seeded, which likely contributed to the low treatment efficacy on canola. Under growth cabinet conditions, fluazinam and *B. subtilis* products became substantially less effective after 2 weeks in a dry soil, but cyazofamid retained its

Table 13.7 Effect of biocontrol agents and fungicides applied as a soil drench on the disease severity index (DSI) of clubroot, caused by *Plasmodiophora brassicae*, on canola grown under growth cabinet conditions ($n = 3$) (Peng et al. 2011)

Treatment	DSI (%)	Clubroot suppression (%)
Inoculated control	37 abc[a]	0
Biocontrol agent		
Trichoderma harzianum	47 a	−27
Streptomyces lydicus	40 ab	−8
Bacillus subtilis var. amyloliquefaciens	29 bcd	22
Gliocladium virens	24 bcde	35
Streptomyces griseoviridis	14 defg	61
Gliocladium catenulatum	7 efg	81
Bacillus subtilis	3 fg	91
Fungicide		
Fluazinam	3 fg	91
Cyazofamid	3 fg	91

[a]Means followed by the same letter(s) do not differ based on a protected LSD test at $P \leq 0.05$

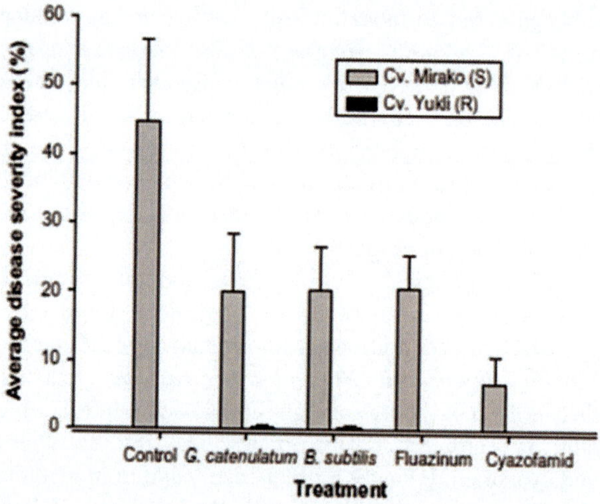

Fig. 13.4 Mean clubroot severity, caused by *Plasmodiophora brassicae*, on Chinese cabbage as affected by biofungicide/fungicide treatments and cultivar resistance near Bradford, Ontario in 2009 ($n = 4$) (Peng et al. 2011)

efficacy for at least 4 weeks (Fig. 13.5). *Bacillus subtilis* XF-1 isolated from cabbage rhizosphere displays an extremely efficient biocontrol activity on clubroot (Guoru et al. 2009; Liu et al. 2018). The production of fengycin-type cyclopeptides (Li et al. 2012, 2013) and PBT1, a novel antimicrobial protein from the biocontrol agent *B. subtilis* XF-1, have strong inhibitory effects against *P. brassicae* (Zhao et al. 2016). *B. subtilis* XF-1 can be used to manage clubroot and have potential to control other fungal diseases. Recombinant plasmid pQE 81L-PBT1 has been constructed using the coding code gene of Pbt1 and plasmid PQE 81-L which has been expressed in *Escherichia coli* BL-21. The fusion protein of BL-21 exhibits significant detrimental effects on *P. brassicae* spores. A highly potent strain of *Bacillus subtilis*,

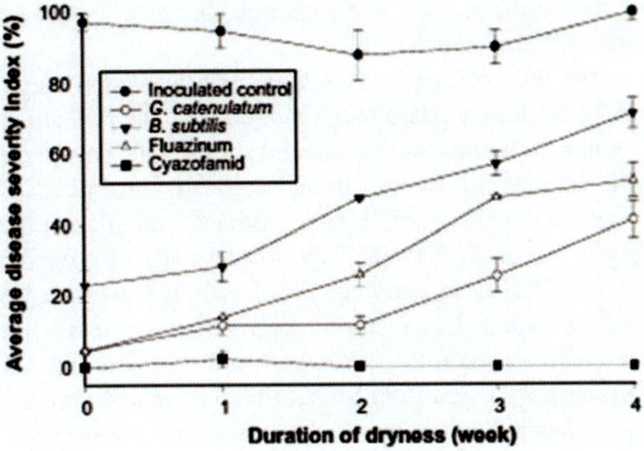

Fig. 13.5 Effect of post-seeding dryness duration on efficacy of biofungicides/fungicides against clubroot, caused by *Plasmodiophora brassicae*, on canola under growth cabinet conditions ($n = 2$) (Peng et al. 2011)

XF-1, which was isolated from the rhizosphere of Chinese cabbages with severe clubroot in Guandu District of Kunming, Yunnan Province, China, had strong inhibitory effects on both the survival and germination of resting spores of *P. brassicae*.

Several strategies were evaluated to efficiently manage clubroot in Chinese cabbage, and different mechanisms were investigated. The ability of Chinese cabbage root hairs to reduce infection by *P. brassicae* in the growth and efficiency of colonization of XF-1 on seedling roots was also studied. Responses to the suppression of clubroot increased with increased seedling age in pot and field conditions, and the rate of non-infectious root hair was enhanced by the inoculation of pathogens, suggesting that seedling age impacts the infection by *P. brassicae*. The application of *B. subtilis* XF-1 liquid culture for treatment during the developmental stages of infected Chinese cabbage seedlings gradually decreased the control effect. Moreover, soaking seeds with *B. subtilis* XF-1 culture reduced clubroot disease incidence rate and disease incidence by 40% and 69%, respectively. These results indicate that prevention and treatment by *B. subtilis* XF-1 at an early stage is the best way to reduce the damage caused by clubroot disease (He et al. 2019).

Many of the common microorganisms found in the rhizosphere and root cortex of most vegetable and field crops do not flourish in association with roots of *Brassica* spp. (Usuki and Narisawa 2007). This characteristic of *Brassica* crops may limit the potential for using traditional rhizosphere colonizers that have biofungicidal properties for management of clubroot. However, previous studies have demonstrated the potential for using naturally occurring microorganisms to reduce clubroot severity on *Brassica* vegetables (Arie et al. 1998; Narisawa et al. 1998; Cheah et al. 2000; Joo et al. 2004). Some of these potential biological control agents (BCA) produce antimicrobial metabolites (Arie et al. 1998; Kim et al. 2004), while others colonize roots and induce resistance to clubroot (Hashiba et al. 2003; Usuki and Narisawa 2007). An important impact on the efficacy of biofungicides is the method of application. Application as a soil drench, or to seedling roots prior to transplanting, can be used to saturate the root zone of vegetable crops with the BCA and optimize its efficacy.

The life cycle of *P. brassicae* consists of a primary infection phase in root hairs and a secondary phase in the root cortex. The primary and secondary zoospores that initiate each phase are fragile and short-lived. Zoospores appear to be a weak link in the pathogen life cycle (Dixon 2006), so, clubroot management strategies generally target the zoospores. Under optimal conditions, the production of secondary zoospores peaks between 7 and 14 days after first exposure to the pathogen (Sharma et al. 2011b), so early application of biofungicides has been selected for most biocontrol agents. Several commercial biofungicides, including Serenade® (*Bacillus subtilis*), Prestop® (*Gliocladium catenulatum*), Mycostop® (*Streptomyces griseoviridis*), and Root Shield Plus® (*Trichoderma harzianum*), and the promising new fungicides fluazinam and cyazofamid are very useful in addition to the endophytic fungus *Heteroconium chaetospira* (Grove) M.B. Ellis strain B2HB1 isolated from forest soils in British Columbia, Canada. Three endophytic fungal isolates and four rhizobacterial isolates reduce clubroot severity by more than 75% relative to the positive control and an additional 10 isolates reduce clubroot severity by 50–75%. A granular formulation of *H. chaetospira* also reduces clubroot severity by more than 80% relative to the control (Lahlali et al. 2012). However, applying *H. chaetospira* at seeding is much less effective. It appears that root colonization is required for *H. chaetospira* to effectively reduce clubroot, but colonization takes time because *H. chaetospira* grows very slowly. More extensive root colonization by *H. chaetospira* resulted in more effective suppression of *P. brassicae* infection and subsequent clubroot development (Lahlali et al. 2012). A microarray assay of gene expression in canola roots inoculated with *H. chaetospira* demonstrated that several defense-related genes are up-regulated, especially those that control the jasmonic-acid (JA)/ethylene (Et) pathways. Also, the transcriptional expression of JA and Et genes was measured with quantitative PCR (qPCR) and the activity of JA and Et genes was enhanced substantially relative to non-treated plants (Lahlali et al. 2012). These results indicate that endophytic development of *H. chaetospira* stimulated gene expression that regulates host defense-related pathways in canola roots. The requirement for applying *H. chaetospira* a week prior to exposure to *P. brassicae* makes the BCA impractical for management of clubroot in canola. However, this approach might be feasible for cruciferous vegetable transplants because the treatment can be applied to seedlings in the greenhouse before they are transplanted into infested fields.

The bio-fungicides Serenade and Prestop provide 85–100% efficacy when applied as a soil drench at seeding in controlled environment studies (Peng et al. 2011), which is similar to the efficacy of fluazinam and cyazofamid. However, Mycostop and Actinovate are more effective than Serenade and Prestop in a trial on Shanghai pak choy (*B. rapa* L. ssp. *chinensis*) (Adhikari 2010). In general, Serenade and Prestop are less effective than the synthetic fungicides when disease pressure was high (Peng et al. 2011). Both Serenade and Prestop contain secondary metabolites that have a role in the efficacy of these biofungicides, but antibiosis is not the only mode of action for biocontrol exhibited by these two biofungicides

(Lahlali et al. 2011). The efficacy of selected BCAs was compared with that of several commercial biofungicides, including Serenade (Bayer Crop Science, Germany), Prestop (Verdera Oy, Finland), Mycostop (Verdera Oy), Actinovate® (*Streptomyces lydicus*, Natural Industries, USA), and Root Shield (Bio Works Inc. USA) that are registered in Canada, as well as against the synthetic fungicides fluazinam and cyazofamid. Serenade and Prestop were generally more effective than the indigenous BCAs and as effective as the synthetic fungicides, especially under low *P. brassicae* inoculum pressure (Peng et al. 2011). Serenade and Prestop, with 85–100% reduction in clubroot disease severity when applied as a soil drench after seeding, are also more effective than the other biofungicides in controlled environments (Peng et al. 2011). In contrast, Mycostop and Actinovate are more effective on Shanghai pak choy in a greenhouse trial (Adhikari 2010). Frequently, Serenade or Prestop is slightly less effective than the fungicides fluazinam and cyazofamid when disease pressure is high (Peng et al. 2011). Both biofungicides contain secondary metabolites that have a role in their activity against clubroot. Additionally, the microbes in these products also colonize the rhizosphere of canola and induce plant resistance through up-regulation of the genes that are involved in phenylpropanoid (BnOPCL, BnCCR), jasmonic acid (BnOPR2), and ethylene (BnACO) pathways (Lahlali et al. 2011, 2013; Lahlali and Peng 2014). Development of *P. brassicae* and consequently clubroot symptoms may be partially suppressed by these biofungicides via induced systemic resistance, which is reflected to some extent by the reduced amount of *P. brassicae* DNA in the root detected using quantitative PCR (qPCR) at 2 weeks after inoculation. This level of DNA also is positively correlated with reduced clubroot severity. These biofungicides do not reduce the germination or viability of *P. brassicae* resting spores (Lahlali et al. 2011; Lahlali and Peng 2014), so suppression of clubroot disease probably results from the reduction of root-hair and/or cortical infection by primary and secondary zoospores (Lahlali et al. 2013), as well as the colonization of cortex by plasmodia.

Some *Brassica* vegetables are grown typically from transplants, and therefore a biofungicide can be applied to the seedlings in plug trays, allowing time for BCA to colonize the roots before the plants are transplanted into the infested soil. Serenade and Prestop reduced clubroot severity by about 50% on Chinese cabbage grown on a high organic matter soil, when applied in the seed furrow. This efficacy is equivalent to that of fluazinam (Peng et al. 2011). The biofungicides Root Shield and Serenade are ineffective when applied as a soil drench treatment around cabbage transplants (Adhikari 2010; Kasinathan 2012). In a controlled environment, Serenade and fluazinam appeared to be more sensitive than Prestop and cyazofamid to dry soil conditions (<30% water content and 23/18 °C with a 16-h photoperiod), especially when the situation persisted for more than 2 weeks (Peng et al. 2011).

On Shanghai pak choy, however, Mycostop was more consistently effective while Actinovate was effective only at low inoculum levels (Adhikari 2010). Soil types have potential influence on the efficacy of Serenade and Prestop against clubroot severity on canola and Shanghai pak choy. The biofungicides were applied to four types of growth media: muck soil (pH 6.2, 68% organic matter), mineral soil, (pH 6.8, 3.5% organic matter), non-calcareous coarse sand (pH 6.5, no organic matter), and soil-less mix (pH 6.0 and an unknown amount of organic matter) (Sunshine LB-2 Mix Basic, Sun Gro Horticulture Canada Ltd, Seba Beach, AB). These media were selected to represent the soil types on which field trials had been conducted previously (mineral and muck soils), or the growing media used for controlled-environment studies (sand and soilless mix). The efficacy of both biofungicides was influenced by the growing medium. Prestop reduced clubroot on canola in muck and mineral soils by 40%, while Serenade was only effective on sand (Kasinathan 2012). In a further trial where the growing media were watered with only light sprays that resulted in noticeably less compaction, the biofungicides reduced clubroot severity, with Serenade being slightly more effective than Prestop. This indicates that the excessive amount of water used in earlier trials to stimulate infection by *P. brassicae* might have limited the activity of the biofungicides, perhaps by leaching some of the product components away from the root zone. It is possible that the efficacy of biofungicides would not vary substantially when used in mineral and muck soils, but adding the sand and soil-less mix in controlled-environment tests may produce additional variations in levels of organic matter, moisture, and possibly compaction, which appear to increase the complexity of microbial interactions and variation in the outcome of biocontrol.

A formulation that balances the need of BCA for survival and antagonism against the target pathogen would enhance the performance of the biofungicide. Formulation ingredients are important for preserving the viability and efficacy of biopesticides and also need to be economical and compatible with standard application equipment (Burges 1998; Hynes and Boyetchko 2011). The managing clubroot disease, formulations that deliver the BCA to the rhizosphere as a seed dressing or granular inoculant can provide the most efficient root protection. *Bacillus subtilis*, which produces spores resistant to adverse conditions, has a distinct advantage with regard to the selection of formulation ingredient and processes (Schisler et al. 2004). A seed-treatment formulation was developed using the complex coacervate reaction to micro encapsulate *B. subtilis* (Serenade) (Mayya et al. 2003; Hynes et al. 2010) and was applied to the seed coat at a rate of 1×10^5 cfu seed^{-1}. A soil applied granular formulation (0.7 mm dia.) made of corn starch and peat mixed with a *B. subtilis* suspension (1×10^{11} cfu mL^{-1}) via extrusion and spheronization (Hynes and Boyetchko 2011) was developed. There was no change in the *B. subtilis* titer with these formulations after 8–14 week of storage at room temperature, and their efficacy was equivalent to that of Serenade in aqueous form under controlled conditions (Peng et al. 2014). Heat sterilized soil and application of a combination of *B. megaterium*, *Clostridium tyrobutyricum*, and *Saccharomyces cerevisiae* have the effective control for clubroot with improvement in the growth and yield of Chinese cabbage (Gao Yu 2014).

13.6.1 Disease Management with Plant Products and Biocontrol Agents

Under Indian conditions, Bhattacharya and Pramanik (1998) assessed the effect of plant products and biocontrol agents on two phases of the life cycle of *P. brassicae*. Indian neem tree (*Azadirachta indica* A. Juss.) oil emulsion in water (2% v/v) emulsified with 0.1% Tween 80 (Polyoxyethylene (20) sorbitan monooleate) when applied as a soil drench resulted in maximum inhibition (80%) of root hair infection of yellow sarson during the primary phase of colonization in a laboratory study. Only a very few small clubs were produced following this treatment. Field trials conducted with different formulations of the neem product (granular and suspension) revealed significant reductions in clubroot disease severity relative to the control (Bhattacharya and Pramanik 1998). There was little difference found between the effects of different doses and formulations of neem products when applied under field conditions. Inhibition of root hair infection by 10–15% following treatment with the antagonistic rhizobacteria *Streptomyces graminifaciens* G11 and *Pseudomonas fluorescens* H237 was observed in an in vitro test. These two antagonists also reduced club formation and disease severity values by 15–20% relative to the control in glasshouse experiments. The effects of antagonist rhizobacteria were less pronounced relative to neem products (Bhattacharya et al. 2014).

13.6.2 Mechanism of Action of *Bacillus subtilis*

The mechanism of action of biofungicide was investigated how the timing of application of the biofungicide Serenade (*Bacillus subtilis* QST713) or its components (product filtrate and bacterial cell suspension) influence infection of canola by *Plasmodiophora brassicae* under controlled conditions. The biofungicide and its components were applied as a soil drench at 5% concentration (vol/vol or equivalent CFU) to a planting mix infested with *P. brassicae* at seeding or at transplanting 7 or 14 days after seeding (DAS) to target primary and secondary zoospores of *P. brassicae*. Quantitative polymerase chain reaction (qPCR) was used to assess root colonization by *B. subtilis* as well as *P. brassicae*. The biofungicide was consistently more effective than the individual components in reducing infection by *P. brassicae*. Two applications were more effective than one, with the biofungicide suppressing infection completely and the individual components reducing clubroot severity by 62–83% (Table 13.8). The biofungicide also reduced genomic DNA of *P. brassicae* in canola roots by 26–99% at 7 and 14 DAS, and the qPCR results were strongly correlated with root hair infection (%) assessed at the same time ($r = 0.84$–0.95) (Fig. 13.6). qPCR was also used to quantify the transcript activity of nine host-defense-related genes in inoculated plants treated with Serenade at 14 DAS for potential induced resistance (Fig. 13.7). Genes encoding the jasmonic acid (BnOPR2), ethylene (BnACO), and phenylpropanoid (BnOPCL and BnCCR) pathways were up-regulated by 2.2- to 23-fold in plants treated with the biofungicide relative to control plants. This induced defense response was translocated to the

Table 13.8 Efficacy of Serenade and its components applied at seeding or transplanting of canola against clubroot[a] (Lahlali et al. 2013)

Zoospores targeted	Inoculation/treatment timing		Mean disease severity index (%)[b]	
	At seeding[c]	At transplanting	Transplanted at 7 DAS	Transplanted at 14 DAS
Control	*Plasmodiophora brassicae* + water	Water	83.7 e	100.0 e
A–primary	*P. brassicae* + product filtrate	Water	66.7 de	80.9 e
	P. brassicae + *Bacillus subtilis*	Water	40.5 cde	47.6 cd
	P. brassicae + Serenade	Water	19.1 bc	0.0 a
B–secondary	*P. brassicae* + water	Product filtrate	57.1 de	54.8 cd
	P. brassicae + water	*B. subtilis*	85.7 e	52.4 cd
	P. brassicae+ water	Serenade	23.8 bcd	16.6 b
C–primary and secondary	*P. brassicae* + Product filtrate	Product filtrate	21.4 bcd	23.8 bc
	P. brassicae + *B. subtilis*	*B. subtilis*	14.3 b	61.9 cd
	P. brassicae + Serenade	Serenade	0.0 a	0.0 a

[a]Means in a column with the same letters do not differ (least significant difference, $P = 0.05$)
[b]Measured on plants 4 weeks after transplanting. DAS = days after seeding. Some infection might have occurred prior to transplanting, especially in root hairs, but clubbing symptoms were not visible during transplanting
[c]*P. brassicae* was added with or without a Serenade component

foliage (determined based on the inhibition of infection by *Leptosphaeria maculans*). It is possible that antibiosis and induced resistance are involved in clubroot suppression by Serenade. Activity against the infection from both primary and secondary zoospores of *P. brassicae* may be required for maximum efficacy against clubroot (Lahlali et al. 2013).

13.6.3 Clubroot Control by Plant Growth-Promoting Formulations

The ability of two organic plant growth stimulants, mainly based on algal extracts, amino acids, and phosphonate, to reduce clubroot formation in the model plant *Arabidopsis thaliana* and two economically important *Brassica* species, *Brassica rapa* (Chinese cabbage) and *Brassica napus* (oilseed rape) was investigated by Kammerich et al. (2014). A commercial liquid (Frutogard®) and a granulate (Plasma Soil®) formulation were used to find optimum conditions for both control of the pathogen and plant growth. Both formulations are able to significantly reduce gall formation after *Plasmodiophora brassicae* infection on Chinese cabbage, but Plasma Soil® gave better effects, possibly as a result of the continuous supply of

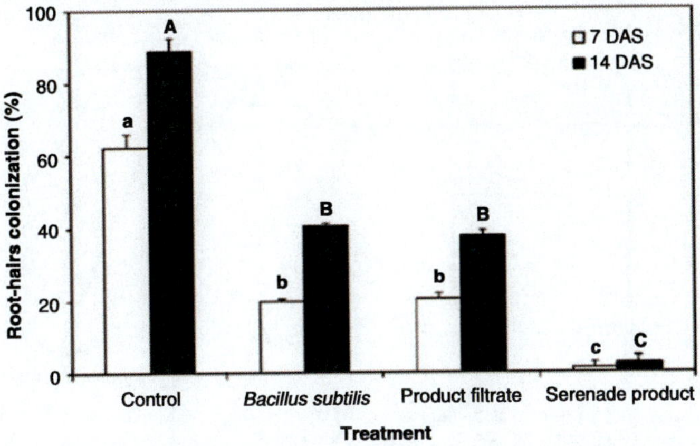

Fig. 13.6 Influence of selected Serenade components on infection of canola root hairs by *Plasmodiophora brassicae* at 7 and 14 days after seeding (DAS). Bars represent the average of six replicates for each treatment. Mean values of root-hair infection at 7 DAS (lower case letters) or 14 DAS (uppercase letters) marked with different letters are significantly different (least significant difference, $P < 0.05$). Capped lines represent standard errors (Lahlali et al. 2013)

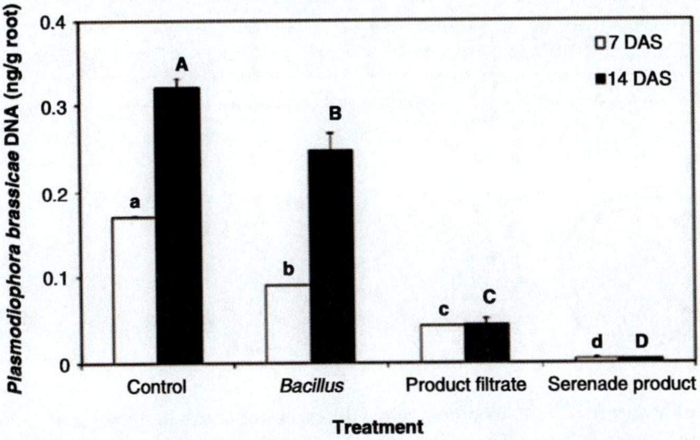

Fig. 13.7 Quantitative polymerase chain reaction quantification of *Plasmodiophora brassicae* DNA in canola roots, as affected by Serenade and its components at 7 and 14 days after seeding (DAS). Roots of each plant were washed and rinsed to remove most of the microbes on root surfaces prior to testing. Bars represent the average of six replicates of each treatment. Mean amounts of *Plasmodiophora brassicae* DNA detected in canola roots at 7 DAS (lowercase letters) or 14 DAS (uppercase letters) marked with different letters are significantly different (least significant difference, $P < 0.05$). Capped lines represent standard errors (Lahlali et al. 2013)

the components to the soil. Individual components did not have the same effect. Clubroots on oilseed rape can also be reduced. In contrast, club formation is not reduced in the model plant *Arabidopsis thaliana*. The interesting differences in the induction of resistance in the different species have been recorded. Cross-sections of

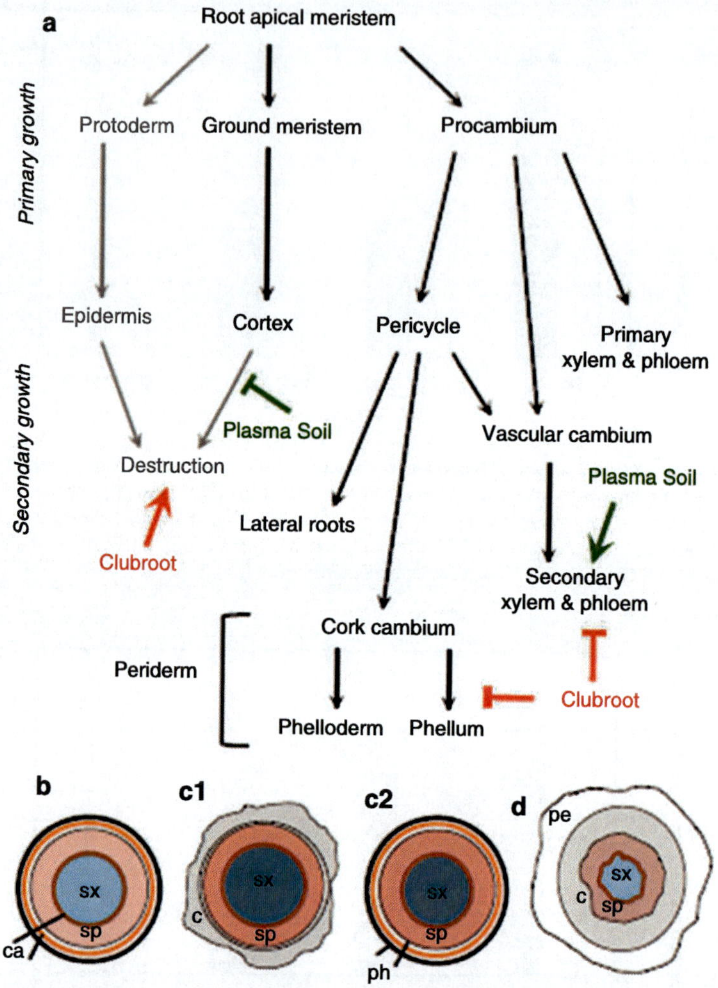

Fig. 13.8 (a) Model for the development of the different cell layers during primary and secondary root growth of *Brassica rapa*. The effect of clubroot and of additional plasma soil treatment is shown (online, in red and green, respectively). (**b–d**) Drawing to illustrate the different cell layers in four different situations. (**b**) Control roots (uninfected); (**c**) roots inoculated with *Plasmodiophora brassicae* and treated with Plasma Soil; (**d**) *P. brassicae*-infected roots. (**c1**) shows a transition stage between primary and secondary growth, (**c2**) the mature secondary growth stage. The dense cell layers of the secondary phloem appear as darker color in (**b**). Secondary xylem is depicted in blue (sx), secondary phloem in red (sp), cambium in orange/brown (ca), cortex in gray (c), and periderm in white (pe). A thick black line for the phellem (ph) symbolizes several cell files, a thin line only one cell file (Kammerich et al. 2014)

infected and treated roots stained for different macromolecules (callose, lignin, suberin) indicated differences in anatomy as a result of the two formulations. An application of the granulate formulation Plasma Soil^R in clubroot control is useful (Fig. 13.8).

13.6.4 Disease Management with Fungal Bioagents

Piriformospora indica, plant-root-colonizing fungus of Basidiomycetes, is capable to reduce the formation of clubroot in *B. campestris* sp. chinensis which is extremely susceptible to the root pathogen *Plasmodiophora brassicae*. Roots of the *P. brassicae*-infected plant were colonized with *P. indica*, as demonstrated by staining technique and microscopy. Like that, enhanced biomass was observed in *P. indica* co-inoculated plants by comparing with plants grown without *P. indica* as well as the plants infected with the pathogen only. Co-inoculation of the plant with both *P. indica* and *P. brassicae* reduced gall formation up to 61.60% by comparing plants infected with only *P. brassicae* as shown by the disease index. Higher level of flavonoids was also observed in *P. indica* inoculated plants which contribute to enhancing the plant immune system. In *B. campestris* sp. *chinensis* clubroot development can be delayed with the inoculation of *P. indica*. Expression of *P. brassicae* genes which were differentially expressed during disease development was monitored using quantitative polymerase chain reaction and was corroborated with delayed pathogen development. The results suggest that *P. indica* and perhaps other beneficial endosymbionts can be an effective tool for the management of clubroot infection (Khalid et al. 2020).

13.7 Disease Management Through Resistant Cultivars

Sources of resistance against *P. brassicae* have been identified at different locations and have been used to develop clubroot resistant cvs. Please see Chap. 11 at 11.15 for details. Breeding of resistant cultivars is one of the most effective approaches for minimizing crop loss caused by *P. brassicae* infection (Diederichsen et al. 2009). Genetic resistance to clubroot can vary from broad-spectrum resistance, effective against several races or pathotypes of *P. brassicae*, to highly specific resistance effective only against one particular strain of the pathogen (Somé et al. 1996; Diederichsen et al. 2006). The durability of resistance is influenced by the composition of pathotypes within the region intended for deployment of that resistance source (Strelkov et al. 2011). The analysis of *P. brassicae* populations from China has revealed some pathogenic diversity in the virulence characteristics of the pathogen, with at least eight different pathotypes (1, 2, 4, 7, 9, 10, 11, and 13) identified using the Williams (1966) differential set (Shen et al. 2009; Ding et al. 2013; Ji et al. 2013; Peng et al. 2013; Wang et al. 2013; Zhao et al. 2013).

Breeding for clubroot resistance (CR) in China today focuses on Chinese cabbage, European cabbage, and canola. There are more than 10 private companies and

20 public institutions breeding resistant cultivars of these crops. A concerted effort to produce clubroot resistant hybrids has resulted in the release of several cultivars into the Chinese market ("CR589," Degao "CR117," "Kanggen 51," "Kanggen 25" and "Kanggen 39," "CR Huimin," "CR Weimin" and "CR Aimin," "Kangda No. 3," "Shennong 106" of Chinese cabbage, "Xiyuan 6" of European cabbage, and "A 35," "Huayou 7," "Yunyoushuang 1" of canola) (Li et al. 2006; Piao et al. 2010; Geng et al. 2012; Zhang et al. 2012, 2013). While the availability of these cultivars represents important progress in the management of clubroot in China, the cropping of resistant crops will have to be carefully organized to prevent erosion of resistance. The virulence composition of *P. brassicae* has been shown to change quickly in response to the cropping of resistant cultivars, and previous experience in other countries has shown that genetic resistance can be eroded (Manzanares-Dauleux et al. 2000; Diederichsen et al. 2006; Hirai 2006; Yang et al. 2011; LeBoldus et al. 2012). In Canada, for instance, a break of 3 years is recommended between resistant canola cultivars grown on clubroot-infested fields, but resistance stewardship is complicated by a lack of knowledge on the sources of resistance used in commercial hybrids (Strelkov et al. 2011).

13.8 Integrated Control of Clubroot

The most effective clubroot management strategy is to integrate all the methods in best combination depending on soil environment and inoculum load in a specific area. The integration of all cultural, chemical, biological, and host resistance methods results in higher production of crucifers crops with best control of clubroot disease.

13.8.1 Prerequisites of Disease Management Strategies

The ability to predict disease and expected yield loss is the foundation upon which an integrated clubroot control strategy can be implemented. In its simplest form, this involves a sound knowledge of site history, including the severity of disease in the most recent *Brassica* crop, the rotational history, site preparation, and treatments applied in previous crops. This information is readily available to most growers, but often not recorded. A range of more sophisticated serologic (Wakeham and White 1996) and molecular methods have been developed in the last 15 years to detect *P. brassicae* (Faggian et al. 1999a; Ito et al. 1999; Wallenhammar and Arwidsson 2001; Cao et al. 2007). These techniques offer rapid and reliable means for detecting and/or quantifying *P. brassicae*. When combined with sound local knowledge of factors such as soil type and climatic conditions and basic site history, these techniques can be powerful decision support tools for farmers. *Plasmodiophora brassicae* has been detected in irrigation pond sediment (Datnoff et al. 1984), dams, bores, and *Brassica* seedling nurseries (Faggian et al. 1999b). In these cases, the source of contamination of water sources is likely to have been runoff from fields infested with the pathogen. Limited studies have been conducted to try and reduce the viability of *P. brassicae* and its movement in irrigation water. Datnoff et al. (1987) reported that 2 mg chlorine per liter water reduced clubroot in cabbage. Treatment of irrigation water with 200 mg chlorine per liter also reduced the incidence of clubroot in the field but reduced plant growth and yield. In addition to the effects on plant growth, treatment of large volumes of water for irrigation purposes appears presently impractical. In a controlled study, resting spores of *P. brassicae* remained viable in water for 34 months, and repeated irrigation with water containing as few as 10 spores per ml resulted in root galling (Donald 2005). The resting spores settled in undisturbed columns of water at a rate of 25 cm/day. In a farm dam, most of the resting spore population could be expected to be concentrated in the sediment at the bottom of the dam and that the risk of spreading disease through irrigation with contaminated farm dam water could be reduced by locating the irrigation intake pipe in the stillest part of the dam, mounted on a float to source water from near the surface.

Plastic seedling trays being returned for reuse have been identified as a high risk of *P. brassicae* contamination in nurseries (Faggian et al. 1999b). Often these trays have been placed on the ground and are contaminated with farm soil on their return. There are no commercial disinfectants designed specifically for control of *P. brassicae*. Evaluation of a range of commercial disinfectant products available in Australia found that with the exception of hypochlorite (1000 ppm), all were ineffective against *P. brassicae* at the rates used (Donald et al. 2002). Hypochlorite is cheap and readily available; however, its efficacy decreases rapidly in the presence of high organic matter loads. For this reason, a high-pressure prewash is recommended to remove soil and organic matter before any disinfection treatment. In Australia, guidelines have been developed for growers, nurserymen, and contractors, detail strategies for the prevention and quarantine of new infections, disinfection, and the production and distribution of *Brassica* transplants (Porter and Cross 1995). These guidelines have slowed the spread of clubroot within isolated growing regions.

Plasmodiophora brassicae has been the most important soil-borne pathogen of vegetable *Brassica* crops, including broccoli, cauliflower, cabbage, Brussels sprouts (*Brassica oleracea* L.), Chinese cabbage, turnip (*Brassica rapa* L.), and a number of oil yielding *Brassica* spp. Apart from these, several weed 330 genera and 3700 species of the family Brassicaceae are potential hosts of *P. brassicae* (Dixon 2009; Hwang et al. 2014). Resting spores of *P. brassicae* are readily transmitted on or in anything that carries contaminated soil. This includes farm machinery, boots, and hooves of grazing animals, infected transplants, and surface flood water. The number of resting spores in the soil can increase rapidly under continuous cropping systems (Murakami et al. 2004). These spores can remain viable for more than 20 years, even in the absence of a suitable host (Wallenhammar 1996; Donald and Porter 2009). Visible symptoms of clubroot, such as root galling and wilting, become obvious only when soil concentrations of *P. brassicae* exceed the threshold value of inoculum. This threshold depends upon soil type and is also influenced by the plant species or cultivar, but values lower than 10 resting spores per g soil have been reported

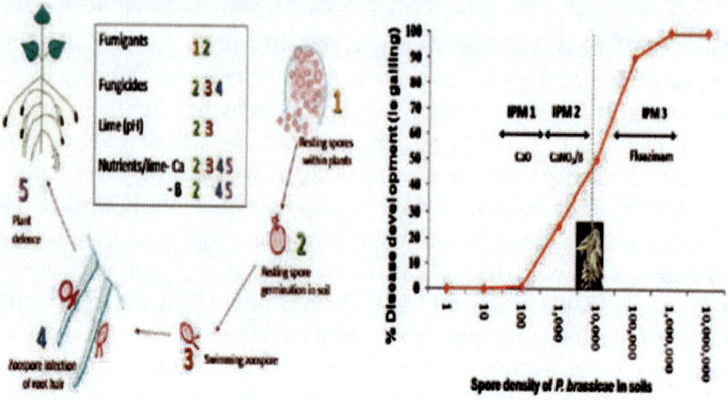

Fig. 13.10 (Left panel) Potential site of action of chemical and nutrient treatments on the *Plasmodiophora brassicae* life cycle and (right panel) a schematic relationship of the impact of three integrated treatments (GBA lime (76% Ca0, 12% MgO), calcium nitrate with boron (Nitrabor™) and fluazinam) on inoculum (log10 spores/g soil) and clubroot expression (Donald and Porter 2014)

(Murakami et al. 2002a, b). Generally, however, when soil inoculum loads are less than 102 resting spores per g soil, clubroot infection is low (disease indices per 30; Murakami et al. 2002a, b) and there is little impact on crop yield or aboveground symptoms. As it is not yet possible to predict and avoid infested land until symptoms of disease become obvious, spores can be spread inadvertently in soil and water from one field to another even though the farmer may consider the field to be clean. For these reasons, adequate control has been difficult to achieve. Traditionally, clubroot has been managed in vegetable Brassicas by rotation with non-brassica crops and application of agricultural lime (calcium carbonate) to change pH (Colhoun 1958). The increased use of cell-grown transplants, continuous cropping, and the trend toward shorter rotations has increased both the area of land affected and the inoculum load in the soil resulting in an increase in crop losses due to clubroot (Donald and Porter 2009; Hwang et al. 2014). Several factors, together with an increased awareness of the need for better soil health management and sustainable production practices, have been conducive to the development of an integrated approach to the management of clubroot of crucifers. Such an approach, including detection of *P. brassicae* and prediction of yield loss due to clubroot, identification of hygiene risks in nurseries, and the development of methods to minimize these risks together with in-field cultural methods, manipulation of soil pH, calcium and boron amendment, strategic use of pesticides, and the integration of these methods, has been extremely effective in crucifer crops production.

To develop integrated control strategies approach, one has to rely on knowledge of the pathogen life cycle, inoculum densities, the mode of action and site of infection, and strategic application of effective methods of control (Porter et al. 2005; Donald et al. 2006, Fig. 13.10). On low pH sites with soil pH (water) between 6.4 and 6.9, the most effective control strategies were treatments that consisted of combinations of two or three treatments including lime (Tables 13.9 and 13.10;

Table 13.9 Soil pH, disease severity (visually assessed 1–9 scale 6 weeks after transplanting) and marketable yield following application of various treatments. Trials were conducted during 1997 at Werribee and Lindenow (L), Victoria using two host species (Donald et al. 2006)

	Soil pH (water)			Root score (1–9)			Marketable yield (t ha^{-1})		
Treatment[a]	Broc	Cauli	Broc (L)	Broc	Cauli	Broc (L)	Broc	Cauli	Broc (L)
Untreated	6.9	6.9	6.4	5.2	7.9	4.9	5.1	2.6	3.7
GBA lime	7.7	7.2	7.2	3.1	5.6	4.1	8.5	9.5	5.9
$CaNO_3$/B	7.1	6.7	6.6	4.6	6.5	5.6	6.0	2.9	3.4
$CaNO_3$/B and GBA lime	7.5	7.1	7.1	2.1	3.8	3.9	10.0	13.5	6.5
Metham	7.0			3.8			8.6		
Metham and GBA lime	7.6			1.6			11.0		
Fluazinam	7.1	6.9	6.7	5.2	7.0	4.8	6.0	3.6	4.3
Fluazinam and GBA lime	7.5	7.6	7.1	3.7	4.8	4.3	8.4	16.6	7.0
Metham and fluazinam	7.0			3.4			8.0		
Fluazinam and $CaNO_3$/B		6.6	6.4		5.6	4.6		7.0	4.7
Metham and fluazinam and GBA lime	7.6			1.7			10.0		
Fluazinam and $CaNO_3$/B and GBA lime		7.2	7.4		4.2	2.3		17.7	8.5
LSD ($P = 0.05$)	0.3	0.5	0.5	1.9	1.9	2.1	2.7	6.8	2.7

[a]Fluazinam was sprayed over the plants in a continuous band immediately after planting. It was applied at 0.75 kg a.i. ha^{-1} in 2500 L ha^{-1} in the Werribee broccoli trial. In the other trials (Werribee cauliflower and Lindenow broccoli), a higher rate 1.5 kg a.i. ha^{-1} in 2500 L ha^{-1} was used. Metham sodium and GBA lime were applied 7–10 days before transplanting. Metham sodium was injected into preformed beds at 500 L ha^{-1} and GBA lime was broadcast at 1 t ha^{-1} and immediately incorporated. Calcium nitrate and boron (CaNO3/B, Nitrabor™) was sprinkled along the transplant rows at 125, 225, and 225 kg ha^{-1} at transplanting 10 and 20 days after transplanting, respectively

Fig. 13.9). As previously mentioned, the preferred treatments adopted were based on fertilizer treatments using calcium oxide (CaO) based lime, calcium nitrate, and boron and if necessary, fluazinam as shown in Fig. 13.10. With the exception of the fluazinam/lime combination used on broccoli at Werribee where the rate of fluazinam was marginal (0.75 kg a.i. ha^{-1}), the two-way treatment combinations (either boronated calcium nitrate ($CaNO_3$/B) blend, fluazinam, or fumigants) that included lime were consistently more effective at reducing root galling due to clubroot and increasing marketable yield than the lime-alone treatment. While the root galling observed as a result of the combination treatments was generally not significantly less than could be achieved by application of GBA lime alone, this

Table 13.10 Additional profit (AU \$ ha^{-1} above the untreated control) resulting from treatment application for clubroot (Donald and Porter 2014)

Treatment[a]	Additional profit[b] ($ AU returns above untreated)		
	Werribee (broccoli)	Werribee (cauliflower)	Lindenow (broccoli)
Untreated	0	0	0
CaO, GBA lime	2928	6702	1950
$CaNO_3$/B	447	−183	−836
$CaNO_3$/B and CaO	4212	10,274	2153
Metham	2768	–	–
Metham and CaO	4899	–	–
Fluazinam	652	465	26
Fluazinam and CaO	2824	13,281	2548
Metham and fluazinam	1862	–	–
Fluazinam and $CaNO_3$/B		3408	−5
Metham and fluazinam and CaO	3768	–	–
Fluazinam and $CaNO_3$/B and CaO	–	13,954	3527

[a]Fluazinam was sprayed over the plants in a continuous band immediately after planting. It was applied at 0.75 kg a.i. ha^{-1} in 2500 L ha^{-1} in the Werribee broccoli trial. In other trials (Werribee cauliflower and Lindenow broccoli), a higher rate 1.5 kg a.i. ha^{-1} in 2500 L ha^{-1} was used. Metham sodium and GBA lime were applied 7–10 days before transplanting. Metham sodium was injected into preformed beds at 500 L ha^{-1} and GBA lime was broadcast at 1 t ha^{-1} and immediately incorporated. Calcium nitrate and boron ($CaNO_3$/B, Nitrabor™) was sprinkled along the transplant rows at 125, 225, and 225 kg ha^{-1} at transplanting 10 and 20 days after transplanting, respectively
[b]Based on \$1 kg^{-1} produce and the following product costs: CaO (GBA lime®) AU \$190 ha^{-1}, $CaNO_3$/B (Nitrabor™) AU \$469 ha^{-1}, Metham sodium (Metham™) AU \$750 ha^{-1}, fluazinam AU \$270 ha^{-1} (at 0.75 kg a.i. ha^{-1}—Werribee broccoli trial) and AU \$540 ha^{-1} (at 1.5 kg a.i. ha^{-1} other trials) (Donald et al. 2006)

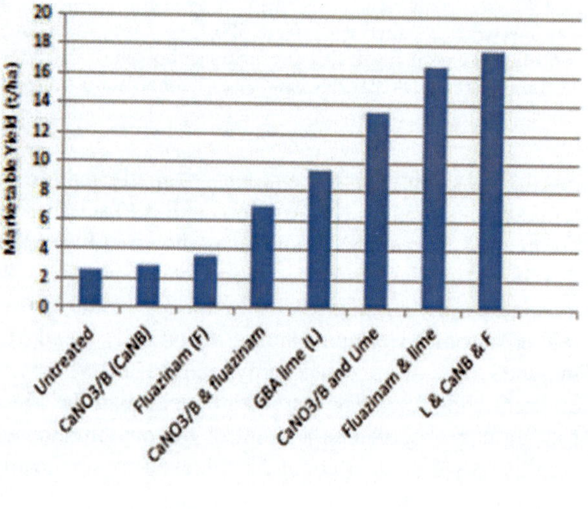

Fig. 13.9 Effect of three nutrient and chemical treatments either alone or integrated together on the marketable yield (t ha^{-1}) of cauliflower affected by clubroot at Werribee, Victoria in 1997. LSD ($P = 0.05$) 6.8 (Donald and Porter 2014)

translated to between 0.6 and 2.5 t ha^{-1} increased yield of broccoli or between 4 and 7.1 t ha^{-1} more cauliflower than could be achieved by application of lime alone (Table 13.9). This produced an increase in crop profitability of between AU $203 and AU$ 6579 than could be obtained using lime alone (Table 13.10). The three-way treatment combinations were consistently among the most effective treatments used and most profitable (Tables 13.9 and 13.10, Fig. 13.9). At some sites in Victoria (the Lindenow site), which have high disease pressure, the only treatment to significantly reduce root galling due to clubroot was the three-way treatment combination of fluazinam, $CaNO_3$/B blend (Nitrabor™), and lime (Table 13.9). At other sites, however, the additional benefit in terms of increased yield from the addition of the fungicide treatment did not always translate directly to increased profitability, due to the additional treatment costs and growth reduction occurring with some methods of application of fluazinam or low disease pressure. On some sites where disease pressure was extremely high, either fumigation or fluazinam applied using the optimum method incorporated around the root zone were the only treatments which gave added economic benefits as the nutrient treatments did not reduce the disease effectively alone. Owing to difficulties in the application of fluazinam, growers preferentially manage clubroot using calcium and boron fertilizers, application of calcium oxide-based lime to increase soil pH, rotation with non-brassica crops and sowing of badly infested fields in the cooler autumn and winter months when soil temperatures are less conducive to development of disease.

13.9 Clubroot Management in Latin America

Clubroot management in Latin American countries has been assessed depending on cultural conditions of cruciferous crops and soil conditions in each country. The disease management has been very effective through cultural, chemical, biological, and using resistant cultivars.

13.9.1 Assessment of Management Practices

The pros and cons of the main strategies used to manage clubroot worldwide, and the strategies that have been incorporated in Latin America, are summarized in Table 13.11. Clubroot management has been based on two main strategies: avoiding pathogen introduction into clean fields and reducing clubroot incidence and severity, mainly via resistant cultivars, when a field is already infested (Gossen et al. 2015). Pathogen introduction into fields can be minimized through farm/nursery hygiene and biovigilance to minimize movement of resting spores. Resting spores *of P. brassicae* can adhere to soil particles, which are spread to new fields mainly on machinery or with eroded soil (Rennie et al. 2015). Sanitization of machinery hasbeen identified as an important strategy to slow the spread of the pathogen to new fields in many parts of the world (Donald and Porter 2009; Diederichsen et al. 2014; Hwang et al. 2014). In most regions, thorough sanitization of all field

Table 13.11 Main management practices for clubroot disease in temperate regions and their current use status in Latin America (Boteroa et al. 2019)

Practice	Pros	Cons	Latin America
Hygiene	Avoid entry of the pathogen Low cost	Time consuming	Not used
Soil tests	Avoid infested fields Easy to perform (bioassay)	Time consuming	Not used
Resistant cultivars	Most efficient strategy	Only monogenic resistance available Resistance easily overcome	Only resistant cabbage and cauliflower available
Liming	Easily incorporated into production system	Expensive (large amounts of lime required) Inconsistent	Most soils are strongly acidic, so raising pH is expensive
Crop rotation	Easily incorporated into production system Reduces inoculum Inexpensive	Must control cruciferous weeds	Pathogen survival is not known
		Long rotations may be required	Crops of aromatic species may be effective
Synthetic pesticides	Can be effective	Impact on environment Difficult registration	Only flusulfamide effective
Biological control	Environmentally friendly	Inconsistent	Focused on *Trichoderma* spp. Results inconsistent

equipment has not been widely adopted because of the time and expense of the sanitation protocols (Hwang et al. 2014). In Latin America, unlike Canada or European countries with more experience managing clubroot, there are no described or commonly used sanitation protocols. Clubroot inoculum can also be moved to new fields in water (Datnoff et al. 1984; Yu et al. 2013). Treatment of irrigation water with chlorine (200 mg L^{-1}) can reduce clubroot incidence, but can also reduce plant height, fresh weight, and stand count (Datnoff et al. 1987). A study from Colombia reported that irrigation with water that included sediments collected from an open irrigation system resulted in 20–47% clubroot symptoms in susceptible broccoli (Navas et al. 2017). However, no effective strategy for managing resting spore numbers in open irrigation systems is available.

Once the pathogen has been detected in a field, environmental conditions and factors such as soil compaction and pH can have a substantial effect on clubroot infection and symptom development (Gossen et al. 2014, 2016). These factors have been studied in considerable detail in temperate regions, but this work is only just beginning in Latin America. In one recent paper, soil factors such as pH had a substantial impact on clubroot levels in cauliflower production in high mountain areas in Brazil (Bhering et al. 2017), but much more research is required in this area. Cultural practices can be used to change conditions to reduce clubroot infection and development. One of the most common strategies employed in temperate areas is liming, where the objective is to raise the pH and Ca^{2+} concentration in the surface layer of soil (Donald and Porter 2009). Liming has been used for many years to reduce clubroot (Myers and Campbell 1981), but is not consistently effective (McDonald and Westerveld 2008; Gossen et al. 2013). Although liming has been widely used worldwide, it may be impractical in Latin America (except possibly Argentina) because of the highly acidic nature and buffering capacity of the soils in the region (IGBP-DSI 1998). Liming may be prohibitively expensive because such large amounts of lime are required to increase soil pH. The large volumes of lime required may also contribute to making the efficacy of liming inconsistent. The limestone applied at up to 4 t ha^{-1} in Brazil increased soil pH from 6.2 to basic pH values and reduced clubroot levels in cauliflower (Santos et al. 2017). In contrast, the application of 2.1 t ha^{-1} of dolomite limestone to a cabbage field in Colombia raised pH from 6.2 to 6.5 but did not reduce clubroot, probably because the treatment did not sufficiently change the soil pH or Ca^{2+} concentration (Botero et al. 2015). In two other studies in Brazil, a lack of success in reducing clubroot with liming was attributed to high levels (unspecified) of pathogen inoculum in the soil (Ruaro et al. 2010; Kowata-Dresch and May-De Mio 2012).

In temperate soils, cropping rotations with more than 2 years between cruciferous crops has been proposed as another strategy to reduce resting spore concentration in infested fields (Peng et al. 2015; Ernst et al. 2019; Gossen et al. 2019). This practice has promise but has not been evaluated in tropical regions, where the survival of resting spores over time is not known. The weeds of the Brassicaceae family can be present over the entire year in tropical regions and would need to be effectively managed throughout the cropping rotation for this approach to be effective. One cultural practice developed specifically for tropical regions is to grow aromatic species such as *Petroselinum hortense* and *Arctium minus* just before a cruciferous crop to reduce clubroot incidence and severity (Hasse et al. 2007). Both species substantially reduced soil inoculum levels, although the mechanism underlying this strategy is not well understood (Hasse et al. 2007).

13.9.2 Disease Control Through Host Resistance

The clubroot-resistant hybrid green cabbage cultivar Tekila and clubroot-resistant cauliflower cultivar "Clarify" both carry a single dominant clubroot resistance gene and are widely grown in Colombia, Ecuador, and Mexico (Diederichsen et al. 2009; Syngenta 2016; Dakouri et al. 2018). Similarly, clubroot-resistant cultivars of cabbage (cv. Kilaton) and cauliflower (cv. Clapton) from Thompson and Morgan have been marketed in Brazil, along with some hybrids with moderate resistance to clubroot, such as the Chinese cabbage cultivars Akiko (AF 70), Harume (AF 69), and Matsuri (AF 72) from Sakata (Carvalho 2012). It is important that these resistant hybrids not be over used. Frequent cropping of a resistant cultivar in fields with a high spore load produces a strong selection pressure towards virulence in the pathogen population (Holtz et al. 2018), which can quickly result in resistance breakdown (Kuginuki et al. 1999; Tanaka and Ito 2013; Strelkov et al. 2016). Therefore, it will be necessary to monitor the response of resistant cultivars to ensure continued resistance to the pathotypes prevalent in Latin America.

13.9.3 Disease Management by Chemical Fungicides and Fumigants

Very few fungicides are effective against clubroot (Hwang et al. 2017) and several effective fungicides are not available in many countries because of their potentially negative impact on the environment (Dixon 2014). In Japan and New Zealand, flusulfamide has been used to reduce clubroot incidence (Donald and Porter 2009). In Latin America, flusulfamide has been evaluated only in Brazil, together with quintozene and chlorothalonil in heavily infested cauliflower, cabbage, and Chinese cabbage fields (Kowata-Dresch and May-De Mio 2012). A reduction in clubroot severity was achieved with 20 L ha^{-1} of flusulfamide applied at the beginning of the cropping season. However, combining liming, soil solarization, and flusulfamide might be required to provide a reduction of clubroot incidence in highly infested fields (Kowata-Dresch and May-De Mio 2012).

Soil fumigants have been shown to be highly effective against clubroot (Hwang et al. 2017), but fumigants have been banned in some countries because of their potential negative impact on the environment (Donald and Porter 2009). Tarping treated fields with specialized plastic sheets after application of the fumigant increases the efficacy of the treatment and reduces the potential for unintended exposure because the active ingredient is retained in the soil (McDonald et al. 2019), but effective use of fumigants requires specialized training and equipment that are not always available to farmers in Latin America (Lathrop and Bushur-Hallam 1995).

13.9.4 Disease Management by Biological Control

Biological control, with the potential to be an efficient, environmentally friendly tool to manage clubroot, has captured the attention of clubroot researchers in Latin America (Alabouvette et al. 2006; Dixon 2014). Unfortunately, clubroot incidence is not consistently reduced under field conditions in temperate regions (Peng et al. 2011), making biological control potentially problematic for commercial producers (Botero et al. 2015). Biocontrol research in Latin America has focused on inundation applications of *Trichoderma* species, with the specific biocontrol agent and application frequency being important factors in effective clubroot management (Botero 2016). A single application of *T. koningiopsis* isolate Th003 or *T. brevicompactum* is effective in reducing clubroot incidence by up to 40% in cabbage, while *T. asperellum* shows a greater reduction in clubroot incidence as the number of applications increases (Botero et al. 2015). *Trichoderma harzianum* reduced clubroot incidence by 33% on broccoli in Venezuela (Garcia et al. 2006). In contrast, application of a commercial biocontrol product based on *T. harzianum* has no effect on clubroot incidence in cauliflower in Brazil (Santos et al. 2017) or on broccoli in Colombia (Castillo and Guerrero 2008) and Peru (Carpio 2014). The application of *T. harzianum* together with *Bacillus subtilis* did not reduce clubroot severity on cabbage in Peru (Pachari 2011). The application of organic material, known as "bioles" in some Latin American countries, increased the size of broccoli plants affected by *P. brassicae* in Costa Rica, but there was no information regarding its effect on clubroot incidence or severity (Xiu 2018).

Biofumigation is the application of massive amounts of organic matter to create an anaerobic soil environment resulting from dramatically increased microbial respiration, which also results in the release of volatile compounds. Biofumigation has limited use in Latin America. In Guatemala, the use of cattle manures, chicken manure, and residues of broccoli reduced the incidence of *P. brassicae* on cabbage (Hernandez 2002). Novel approaches to clubroot management include a nutritional supplement, Agrisafe AG-250 (Syngenta), which increased the root mass and reduced damage caused by *P. brassicae* on broccoli in Ecuador (Almachi 2013), and a combination of a fungicide Agroamonio (EcoAgro), and a root-stimulating product, Forti-raiz (EcoAgro), that reduced the incidence of clubroot on broccoli in Ecuador (Telenchana 2015).

References

Abbasi PA, Lazarovits G (2006a) Reduction in the incidence and severity of clubroot caused by *Plasmodiophora brassicae* on pokchoy and cabbage with soil applications of AG3 phosphonate. Can J Plant Pathol 28:342–371

Abbasi PA, Lazarovits G (2006b) Effect of soil application of AG3 phosphonate on the severity of clubroot of pokchoy and cabbage caused by *Plasmodiophora brassicae*. Plant Dis 90 (12):1517–1522

Adhikari KKC (2010) Effect of temperature, biofungicides and fungicides on clubroot of selected *Brassica* crops. M.Sc. thesis, University of Guelph, Guelph, Ontario, Canada

Ahmad A (1994) Clubroot disease of Chinese cabbage in the Cameron highlands. Teknol Sayur-sayuran Jil 10:17–20

Ahmed HU, Hwang SF, Strelkov SE, Gossen BD, Peng G, Howard RJ, Turnbull GD (2011) Assessment of bait crops to reduce inoculum of clubroot (*Plasmodiophora brassicae*) of canola. Can J Plant Sci 91:545–551

Alabouvette C, Olivain C, Steinberg C (2006) Biological control of plant diseases: the European situation. Eur J Plant Pathol 114:329–341

Alberta Clubroot Management Committee (2010) Alberta clubroot management plan. AGDEX 140/638-2. Alberta Agriculture and Rural Development

Allison FE (1924) The effect of cyanamid and related compounds on the numbers of microorganisms in soil. J Agric Res 28:1159–1166

Almachi VEP (2013) Evaluacion de Dos Hibridos de Brocoli (*Brassica oleracea* var. *italica*) a Tres Metodos Alternativos, Aplicados a Dos Dosis, para el Control de Hernia de la Col (*Plasmodiophora brassicae* Worn.), en la Hacienda Limache–Mulalo–Cotopaxi. Cotopaxi, Ecuador: Universidad Tecnica de Cotopaxi, Agronomist thesis.

Anon (1853) The disease of finger and toe in turnips. Trans Highland Agric Soc Scotland 3:118–140

Antille DL, Gallar L, Miller PCH, Godwin RJ (2015) An investigation into the fertilizer particle dynamics off the disc. Appl Eng Agric 31:49–60

Arie T, Kobayashi G, Okada G, Kono Y, Yamaguchi I (1998) Control of soil borne clubroot disease of cruciferous plants by epoxydon from *Phoma glomerata*. Plant Pathol 47:741–748

Arie T, Kobayashi Y, Kono Y, Gen O, Yamaguchi I (1999) Control of clubroot of crucifers by *Phoma glomerata* and its product epoxydon. Pestic Sci 55:602–604

Bhattacharya I, Dixon GR (2010) Management of clubroot disease (*Plasmodiophora brassicae*) of Brassicas using trap cropping techniques. Acta Hortic 867:157–164

Bhattacharya I, Pramanik M (1998) Effect of different antagonist rhizobacteria and neem products on clubroot of crucifers. Indian Phytopathol 51:87–90

Bhattacharya I, Dutta S, Mondal S, Mandal B (2014) Clubroot disease on *Brassica* crops in India. Can J Plant Pathol 36(S1):154–160

Bhering AS, do Carmo MG, Matos TM, Lima ES, Sabrihno NM (2017) Soil factors related to the severity of clubroot in Rio de Janeiro, Brazil. Plant Dis 101:1345–1353

Bhering AS, Carmo MGF, Coelho IS, Lima ESA, de Carvalho CF, Saraiva ALRF, Passos SR, Amaral Sobrinho NMB (2020) Soil management in a mountain agro ecosystem and clubroot disease. Plant Pathol 69:302–309

Bonanomi G, Lorito M, Vinale F, Woo SL (2018) Organic amendments, beneficial microbes, and soil microbiota: toward a unified framework for disease suppression. Annu Rev Phytopathol 56:1–20

Bosch M, Amberger A (1983) Influence of long-term fertilizing with different forms of nitrogen fertilizer on pH, humic fractions, biological activity and dynamics of nitrogen of an arable brown earth (Einflu⍰ langjahriger Düngung mit verschiedenen N-formen auf pH-Wert, Humusfraktionen, bilogische Aktivität und Stickstoffdynamik einer Acker Braunerde). Zeitschrift. Z Pflanzenernähr Bodenkd 146:714–724

Botero RA (2016) Effect of three *Trichoderma* species on Clubroot disease in cabbage. Bogota, Colombia: Universidad Nacional de Colombia, MSc thesis

Botero RA, Gomez I, Benitez E, Garcia C (2015) Liming with dolomite reduces the efficacy of the biocontrol fungus *Trichoderma koningiopsis* against cabbage clubroot. Agron Colomb 33:49–57

Boteroa A, Garcia C, Gossenc BD, Strelkova SE, Todd CD, Bonham-Smith PC, Perez-Lopez E (2019) Clubroot disease in Latin America: distribution and management strategies. Plant Pathol 68:827–833

Brundrett M (1991) Mycorrhizas in natural ecosystems. Adv Ecol Res 21:171–313

Buczacki ST (1973) Glasshouse evaluation of some systemic fungicides for control of clubroot of brassicae. Ann Appl Biol 74:85–90

Buczacki ST, Cadd SE (1976) Size distribution of resting spores of *Plasmodiophora brassicae*. Trans Br Mycol Soc 67:133–136

Buczacki ST, Moxham SE (1983) Structure of the resting spore wall of *Plasmodiophora brassicae* revealed by electron microscopy and chemical digestion. Mycol Res 81:221–231

Buczacki ST, White JG (1979) The value of soil sterilants for the control of clubroot on a field scale. Plant Pathol 28:36–39

Buczaki ST, Cadd SE, Ockendon JG, White JG (1976) Field evaluations of systemic fungicides and derivatives of dithiocarbamic acid for the control of clubroot. Ann Appl Biol 84:51–56

Burges HD (1998) Formulation of microbial biopesticides: beneficial microorganisms, nematodes and seed treatments. Kluwer Academic Publishers, Dordrecht, p 412

Campbell RN, Grethead AS (1989) Control of clubroot of brassicas by liming. In: Engelhard AW (ed) Soilborne pathogens: management of disease with macro- and microelements. APS Press, St. Paul, pp 90–101

Campbell RN, Greathead AS, Myers DF, de Boer GJ (1985) Factors related to control of clubroot of crucifers in the Salinas Valley of California. Phytopathology 75:665–670

Canadian Association of Petroleum Producers (CAPP) (2008) Best management practices: clubroot disease management; [cited 2013 Apr 23]. Available from http://www.capp.ca/getdoc.aspx?DocId=139848andDT=PDF

Canola Council of Canada (2011) The History of Canola [online] Available from http://www.canolacouncil.org/oil-and-meal/what-is-canola/the-history-of-canola/ [accessed12 December 2013]

Cao T, Tewari J, Strelkov SE (2007) Molecular detection of *Plasmodiophora brassicae*, causal agent of clubroot of crucifers, in plant and soil. Plant Dis 91:80–87

Cao T, Manolii VP, Hwang SF, Howard RJ, Strelkov SE (2009) Virulence and spread of *Plasmodiophora brassicae* [clubroot] in Alberta, Canada. Can J Plant Pathol 31:321–329

Carpio CCL (2014) Compost Mas Dos Cepas De Trichoderma harzianum y T. viride en la Produccion de Brocoli (Brassica oleracea L. var. italica) cv. Legacy en el Valle de Chilina. Arequipa, Peru: Universidad Nacional de San Agustin, Agronomist thesis

Carvalho MV (2012) Efecto de la Asociacion de Plantas Aromaticas con *Brassica* spp. en el Control de la Hernia de las Cruciferas (*Plasmodiophora brassicae* Woron.). Cordoba, Brazil: Universidad de Cordoba, Instituto de Sociologia y Estudios Campesinos ISEC, PhD thesis

Castillo J, Guerrero O (2008) Efecto de controladores biologicos sobre la hernia de las cruciferas en Tabio, Cundinamarca. Inventum 5:30–40

Cathcart RJ, Topinka AK, Kharbanda P, Lange R, Yang RC, Hall LM (2006) Rotation length, canola variety and herbicide resistance system affect weed populations and yield. Weed Sci 54 (4):726–734

Cetas RC (1958) The use of sodium methyl dithiocarbamate for the control of clubroot of crucifers. Plant Dis Rep 42:324–328

Challinor AJ, Watson J, Lobell DB, Howden SM, Smith DR, Chetri N (2014) A meta-analysis of crop yields under climate change and adaptation. Nat Clim Chang 4:287–291

Chattopdhyay AK (1991) Studies on the control of clubroot disease of rapeseed mustard in West Bengal. Indian Phytopathol 44:397–398

Cheah L-H, Page BBC (1997) *Trichoderma* spp. for potential biocontrol of clubroot of vegetable brassicas. Proc N Z Plant Prot Conf 50:150–153

Cheah LH, Page BBC, Koolaard JP (1998) Soil-incorporation of fungicides for control of clubroot of vegetable brassicas. Plant Prot Soc 130:133

Cheah L-H, Page BBC, Falloon RE (1999) Integrated control of clubroot of vegetable Brassicas. In: Australasian plant pathology society 12th biennial conference, Canberra, p 135

Cheah LH, Veerakone S, Kent G (2000) Biological control of clubroot on cauliflower with *Trichoderma* and *Mumeijun* spp. N Z Plant Protect 53:18–21

Cheah LH, Kent G, Gowers S (2001) *Brassica* crops and a *Streptomyces* sp. as potential biocontrol for clubroot of Brassicas. N Z Plant Prot Soc 54:80–83

Cheah LH, Gowers S, Marsh AT (2006) Clubroot control using Brassica break crops. Acta Hortic 706:329–332

Cho CW, Fuller MS (1989) Observations of the water expulsion vacoule of *Phytophthora palmivora*. Protoplasma 149:47–55

Colhoun J (1958) Club root disease of crucifers caused by *Plasmodiophora brassicae* Woron. – a monograph. The Commonwealth Mycological Institute, Kew, Surrey, Phytopathological Paper No 3

Cornforth IS (1971) Calcium cyanamide in agriculture. Soils Fert 34:463–468

Crowther EM, Richardson HL (1932) Studies on calcium cyanamide. 1. The decomposition of calcium cyanamide in the soil and its effects on germination, nitrification and soil reaction. J Agric Sci 22:300–334

Dakouri A, Zhang X, Peng G, Falk KC, Gossen BD, Strelkov SE, Yu F (2018) Analysis of genome-wide variants through bulked segregant RNA sequencing reveals a major gene for resistance to *Plasmodiophora brassicae* in *Brassica oleracea*. Sci Rep 8:17657

Dalolio FS, Silva JN, Oliveira ACC et al (2017) Poultry litter as biomass energy: a review and future perspectives. Renew Sust Energ Rev 76:941–949

Datnoff LE (1987) Efficacy of chlorine for decontaminating water infested with resting spores of *Plasmodiophora brassicae*. Plant Dis 71:734

Datnoff LE, Lacy GH, Fox JA (1984) Occurrence and populations of *Plasmodiophora brassicae* in sediments of irrigation water sources. Plant Dis 68:200–203

Datnoff LE, Kroll TK, Lacy GH (1987) Efficacy of chlorine for decontaminating water infested with resting spores of *Plasmodiophora brassicae*. Plant Dis 71:734–736

Deora A, Gossen BD, Walley F, McDonald MR (2011) Boron reduces development of clubroot in canola. Can J Plant Pathol 33(4):475–484

Diederichsen E, Beckmann J, Schondelmeier J, Dreyer F (2006) Genetics of clubroot resistance in *Brassica napus* 'Mendel'. Acta Hortic 706:307–311

Diederichsen E, Frauen M, Linders EGA, Hatakayema K, Hirai M (2009) Status and perspectives of clubroot resistance breeding in crucifer crops. J Plant Growth Regul 28:265–281

Diederichsen E, Frauen M, Ludwig-Muller J (2014) Clubroot disease management challenges from a German perspective. Can J Plant Pathol 36(Suppl. 1):85–98

Ding YH, Jian YC, Yu YJ, Wang WH, Geng LH, Kang JG (2013) Identification of pathotype of *Plasmodiophora brassicae* on crucifer vegetables in eight provinces of China. Chin Veg 16:85–88

Dixon GR (2006) The biology of *Plasmodiophora brassicae* Worn.—a review of recent advances. Acta Hortic 706:271–282

Dixon GR (2009) *Plasmodiophora brassicae* in its environment. J Plant Growth Regul 28 (3):212–228

Dixon GR (2010) Calcium and pH as parts of a coherent control strategy for clubroot disease (*Plasmodiophora brassicae*). Acta Hortic 867:151–156

Dixon GR (2014) Clubroot (*Plasmodiophora brassicae* Woron.) – an agricultural and 788 biological challenge worldwide. Can J Plant Pathol 36(S1):5–18

Dixon GR (2017) Managing clubroot disease (caused by *Plasmodiophora brassicae* Worn.) by exploiting the interactions between calcium cyanamide fertilizer and soil microorganisms. J Agric Sci 155(4):527543

Dixon GR, Page LV (1998) Calcium and nitrogen eliciting alterations to growth and reproduction of *Plasmodiophora brassicae* (clubroot). Acta Hortic 459:343–349

Dixon GR, Tilston EL (2010) Soil-borne pathogens and their interaction with the soil environment. In: Dixon GR, Tilston EL (eds) Soil microbiology and sustainable crop production. Springer, Dordrecht, pp 197–272

Dixon GR, Walsh UF (1998) Suppression of plant pathogens by organic extracts - a Review. Acta Hortic 469:383–390

Dixon GR, Webster MA (1988) Antagonistic effects of boron, calcium and pH on pathogenesis caused by *Plasmodiophora brassicae* (clubroot) - a review of recent work. Crop Res 28:83–95

Dixon GR, Wilson F (1984) Field evaluation of chemicals for control of clubroot (*Plasmodiophora brassicae*). Proc. Crop Protection in Northern Britain Conference, Dundee, pp 400–405

Dixon GR, Naiki T, Webster A, Wilson F (1987) Integrated use of boron, calcium cyanamide and nitrogen for control of clubroot (*Plasmodiophora brassicae*). In: Crop protection in Northern Britain. The British Crop Prot Council, Farnham, pp 399–404

Dixon GR, Craig MA, Burgess PJ, Thomas J (1994) MTF 651: a new soil-applied fungicide for the control of plasmodial fungi. Proc Brighton Crop Prot Conf Pests Dis 2:541–548

Dixon GR, Kennedy R, Baird L, Kenyon DM (1997) Report of field experimentation with Boron and Agral for the control of clubroot (*Plasmodiophora brassicae*) using field experimentation in Scotland and England. Final report for FV 177 project to the Horticultural Development Council, West Malling Kent and to Borax Consolidated Ltd., Guildford, Surrey, p 25

Djatnika I (1991) Prospects of biological control of clubroot. Cruciferae Newslett 14–15:142

Dobson RL, Gabrielson RL, Baker AS, Bennett L (1983) Effects of lime particle size and distribution and fertilizer formulation on clubroot disease caused by *Plasmodiophora brassicae*. Plant Dis 67:50–52

Donald EC (2005) The influence of abiotic factors and host plant physiology on the survival and pathology of *Plasmodiophora brassicae* of vegetable Brassicas. PhD Thesis University of Melbourne, Australia, p 215

Donald C, Porter IJ (2004) A sand solution culture technique used to observe the effect of calcium and pH on root hair and cortical stages of infection by *Plasmodiophora brassicae*. Aus Plant Pathol 33(4):585–589

Donald EC, Porter I (2009) Integrated control of clubroot. J Plant Growth Regul 28:289–303

Donald EC, Porter IJ (2014) Clubroot in Australia: the history and impact of *Plasmodiophora brassicae* in *Brassica* crops and research efforts directed towards its control. Can J Plant Pathol 36(S1):66–84

Donald EC, Porter IJ, Lancaster RA (2001) Band incorporation of fluazinam (Shirlan) into soil to control clubroot of vegetable *Brassica* crops. Aus J Exp Agric 41:1223–1226

Donald EC, Porter IJ, Lancaster RA (2002) Strategic application of lime, fertilizers and fungicides for improved control of clubroot. Cruciferae Newslett 24:81–82

Donald EC, Lawrence JM, Porter IJ (2004) Influence of particle size and application method on the efficacy of calcium cyanamide for control of clubroot of vegetable Brassicas. Crop Prot 23 (4):297–303

Donald EC, Porter IJ, Faggian R, Lancaster RA (2006) An integrated approach to the control of clubroot in vegetable *Brassica* crops. Acta Hortic 706:283–300

Donaldson SP, Deacon JW (1993) Changes in motility of *Pythium* zoospores induced by calcium and calcium-modulating drugs. Mycol Res 97:877–883

Doyle OPE, Clancy KJ (1986) Integrated chemical applications for the control of clubroot in Brassicas. Proc Br Crop Prot Conf 8:923–930

Einhorn G, Bochow H, Huber J, Krebs B (1991) Methodological studies to detect antagonists of the clubroot pathogen *Plasmodiophora brassicae* Worn. Archiv. Arch Phytopathol Plant Protect 27 (3):205–208

Epstein E (1972) Mineral nutrition of plants: principles and perspectives. John Wiley and Sons, New York, p 412

Ernst TW, Kher S, Stanton D, Rennie DC, Hwang SF, Strelkov SE (2019) *Plasmodiophora brassicae* resting spore dynamics in clubroot resistant canola (*Brassica napus*) cropping systems. Plant Pathol 68:399–408

Faggian R, Bulman SR, Lawrie AC, Porter IJ (1999a) Specific PCR primers for the detection of *Plasmodiophora brassicae* in soil and water. Phytopathology 89:392–397

Faggian R, Donald C, Porter IJ, Lawrie AC (1999b) Epidemiology of recent clubroot outbreaks using PCR to trace sources of inoculum. In: Aus Plant Pathol Soc 12th biennial Conf, Canberra, p 342

Fayolle L, Noble R, Coventry E, Aimé S, Alabouvette C (2006) Eradication of *Plasmodiophora brassicae* during composting of wastes. Plant Pathol 55(4):553–558

Feng J, Hwang RU, Hwang SF (2010) Molecular characterization of a serine protease Pro1 from *Plasmodiophora brassicae* that stimulates resting spore germination. Mol Plant Pathol 11:503–512

Fletcher JT, Hims MJ, Archer FC, Brown A (1982) Effects of adding calcium and sodium salts to field soils on the incidence of clubroot. Ann Appl Biol 100:245–251

Fones H, Preston GM (2013) The impact of transition metals on bacterial plant disease. FEMS Microbiol Rev 37:495–519

Friberg H, Lagerlof J, Ramert B (2005) Germination of *Plasmodiophora brassicae* resting spores stimulated by a non – host. Eur J Plant Pathol 113(3):275–281

Friberg H, Lagerlof J, Ramert B (2006) Usefulness of non host plants in managing *Plasmodiophora brassicae*. Plant Pathol 55(5):690–695

Gao Yu XG (2014) Development of an effective nonchemical method against *Plasmodiophora brassicae* on Chinese cabbage. Int J Agron 2014:307367

Garcia R, Riera R, Zambrano C, Gutierrez L (2006) Desarrollo de unfungicida biologico a base de una cepa del hongo *Trichoderma harzianum* proveniente de la region andina venezolana. Fitosanidad 10:115–121

Geng XC, Shen XQ, Guo YF, Ma SF, Hong YT (2012) Breeding of anti-clubroot AB lines type I of orange head Chinese cabbage. Acta Agric Boreali 21:155–161

Gianquinto G, Sambo P, Bona S (2003) The use of SPAD-502 chlorophyll meter for dynamically optimising the nitrogen supply in potato crop: a methodological approach. Acta Hortic 627:217–224

Gossen BD, McDonald MR, Hwang SF, Kalpana KC (2009) Manipulating seeding date to minimize clubroot (*Plasmodiophora brassicae*) damage in canola and vegetable *Brassicas*. Phytopathology 99:S45

Gossen BD, Adhikari KKC, McDonald MR (2012) Effect of seeding date on development of clubroot in short-season *Brassica* crops. Can J Plant Pathol 34(4):516–523

Gossen BD, Kasinathan H, Cao T, Manolii VP, Strelkov S, Hwang SF, McDonald MR (2013) Influence of pH and temperature on infection and symptom development of clubroot in canola. Can J Plant Pathol 35:294–303

Gossen BD, Deora A, Peng G, Hwang SF, McDonald MR (2014) Effect of environmental parameters on clubroot development and the risk of pathogen spread. Can J Plant Pathol 36 (Suppl. 1):37–48

Gossen BD, Strelkov SE, Manolii VP, Rennie DC, Cao T, Hwang SF, Peng G, McDonald MR (2015) Spread of *Plasmodiophora brassicae* on canola in Canada, 2003–2014: old pathogen, new home. Can J Plant Pathol 37:403–413

Gossen BD, Kasinathan H, Deora A, Peng G, McDonald MR (2016) Soil type, soil compaction and biofungicides affect clubroot (*Plasmodiophora brassicae*). Plant Pathol 65:1238–1245

Gossen BD, Al-Daoud F, Dumonceaux T, Dalton JA, Peng G, Pageau D, McDonald MR (2019) Comparison of techniques for estimation of resting spores of *Plasmodiophora brassicae* in soil. Plant Pathol 68(5):954–961

Grinstein A, Katan J, Abdul-Razik A, Zeydan O, Elad Y (1979) Control of *Sclerotium rolfsii* and weeds in peanuts by solar heating of soil. Plant Dis Rep 63:1056–1059

Gubler F, Jablonsky PP, Duniec J, Hardham AR (1990) Localization of calmodulin in flagella of zoospores of *Phytophthora cinnamomi*. Protoplasma 155:233–238

Guo Z, Miyoshi H, Komyoji T, Haga T, Fujita T (1991) Uncoupling activity of a newly developed fungicide, fluazinam [3-chloro-N- (3-chloro-2,6-dinitro-4-trifluoromethylphenyl)-5-trifluorormethyl-2-pryidinamine]. Biochim Biophys Acta 1056:89–92

Guoru X, Cuiying Z, Qingfeng L (2009) Fungistatic activity of crude proteins of *Bacillus subtilis* XF-1. Plant Prot 35:92–95

Haenseler CM, Moyer TR (1937) Effect of calcium cyanamide on the soil 863 microflora with special reference to certain plant parasites. Soil Sci 43:133–152

Hardham AR (1992) Cell biology of pathogenesis. Annu Rev Plant Physiol Plant Mol Biol 43:491–526

Harling R, Kennedy SH (1991) Biological control of *Plasmodiophora brassicae* using a bait crop. Med Forum 56:159–170

Hashiba T, Morita S, Narisawa K, Usuki K (2003) Mechanisms of symbiosis and disease suppression of root endophytic fungus. Jpn Soc Biosci 77:130–133

Hasse I, Mio M-D, Larissa L, Neto L, Costa VDA (2007) The effect of pre-plantation with medicinal plants in the *Plasmodiophora brassicae* control. Summa Phytopathol 33:74–79

Hata S, Sumi Y, Ohi M (2002) Dry powder and extract of *Posidonia australis* Hook. F., a species of sea grass, stimulate the germination of the pathogen *Plasmodiophora brassicae* and control clubroot of Chinese cabbage. J Jpn Soc Hortic Sci 71(2):197–202

He P, Cui W, Munir S, He P, Li X, Wu Y, Yang X, Tang P, He Y (2019) *Plasmodiophora brassicae* root hair interaction and control by *Bacillus subtilis* XF-1 in Chinese cabbage. Biol Control 128:56–63

Hernandez REC (2002) Efecto Biofumigante de Diversas Fuentes deMateria Organica en el Cultivo de Brocoli (*Brassica oleracea* var. *italica*). Chimaltenango, Guatemala: Universidad Rafael Landivar, Guatemala, Agronomist thesis

Hildebrand PD, McRae KB (1998) Control of clubroot caused by *Plasmodiophora brassicae* with nonionic surfactants. Can J Plant Pathol 20:1–11

Hirai M (2006) Genetic analysis of clubroot resistance in *Brassica* crops. Breed Sci 56:223–229

Holtz MD, Hwang SF, Strelkov SE (2018) Genotyping of *Plasmodiophora brassicae* reveals the presence of distinct populations. BMC Genomics 19:254

Horiuchi S, Hori M, Takashi S, Shimizu K (1982) Factors responsible for the development of clubroot suppressing effect in soil solarization. Bull Chugoku Natl Agric Exp Stn E 20:25–48

Horiuchi S, Hori M, Takashi S, Shimizu K (1983) Factors responsible for the development of clubroot-suppressing effect in soil solarization. Bull Chugoku Natl Agric Expt Stn 20:25–48

Howard RJ, Strelkov SE, Harding MW (2010) Clubroot of cruciferous crops–new perspectives on an old disease. Can J Plant Sci 32:43–57

Hseith WH, Wang J (1986) Investigation on suppressive soils of clubroot of crucifers in Taiwan. Plant Prot Bull 28:353–362

Humpherson-Jones FM (1989) Glasshouse evaluation of fungicides, biocides and surfactants for control of clubroot. Ann Appl Biol 114(Supplement):36–37

Humpherson-Jones FM (1993) Effect of surfactants and fungicides on clubroot (*Plasmodiophora brassicae*) of Brassicas. Ann Appl Biol 122:457–465

Humpherson-Jones FM, Dixon GR, Craig MA, Ann DM (1992) Control of clubroot using calcium cyanamide - a review. Proc Brighton Crop Protect Conf Pests Dis 98(4):1147–1154

Hwang SF, Strelkov SE, Turnbull GD, Manolii V, Howard RJ, Hartman M (2008) Soil treatments and amendments for management of clubroot on canola in Alberta. Can J Plant Pathol 90:410

Hwang SF, Ahmed HU, Strelkov SE, Gossen BD, Peng G, Turnbull GD (2011a) Influence of cultivar resistance and inoculum density on root hair infection of canola (*Brassica napus*) by *Plasmodiophora brassicae*. Plant Pathol 60:820–829

Hwang SF, Ahmed HU, Strelkov SE, Gossen BD, Turnbull GD, Peng G, Howard RJ (2011b) Seedling age and inoculum density affect clubroot severity and seed yield in canola. Can J Plant Sci 91:183–190

Hwang SF, Strelkov SE, Gossen BD, Turnbull GD, Ahmed HU, Manolii VP (2011c) Soil treatments and amendments for amelioration of clubroot on canola. Can J Plant Pathol 91:999–1010

Hwang SF, Cao T, Xiao Q, Ahmed H, Manolii VP, Turnbull G, Gossen BD, Peng G, Strelkov S (2012a) Efficacy of fungicide, seeding date and seedling age on clubroot severity, seedling emergence and yield of canola. Can J Plant Sci 92(6):1175–1186

Hwang S-F, Strelkov SE, Feng J, Gossen BD, Howard RJ (2012b) *Plasmodiophora brassicae*: a review of an emerging pathogen of the Canadian canola (*Brassica napus*) crop. Mol Plant Pathol 13:105–113

Hwang SF, Ahmed HU, Zhou Q, Rashid A, Strelkov SE, Gossen BD, Peng G, Turnbull GD (2013) Effect of susceptible and resistant canola plants on *Plasmodiophora brassicae* resting spore populations in the soil. Plant Pathol 62:404–412

Hwang SF, Howard RJ, Strelkov SE, Gossen BD, Peng G (2014) Management of clubroot (*Plasmodiophora brassicae*) on canola (*Brassica napus*) in western Canada. Can J Plant Pathol 36(Suppl. 1):49–65

Hwang SF, Ahmed HU, Zhou Q, Turnbull GD, Strelkov SE, Gossen BD, Peng G (2015) Effect of host and non-host crops on *Plasmodiophora brassicae* resting spore concentrations and clubroot of canola. Plant Pathol 64:1198–1206

Hwang SF, Ahmed HU, Strelkov SE, Zhou Q, Gossen BD, McDonald MR, Peng G, Turnbull GD (2017) Suppression of clubroot by dazomet fumigant. Can J Plant Sci 98:172–182

Hynes RK, Boyetchko SM (2011) Improvements to the pesta formulation to promote survival and dispersal of *Pseudomonas fluorescens* BRG100, green foxtail bioherbicide. Pest Technol 5:80–87

Hynes RK, Chumala PB, Hupka D, Peng G (2010) A complex coacervate formulation for delivery of *Colletotrichum truncatum* 00-003B1. Weed Technol 24:185–192

IGBP-DSI (1998) SOILDATA (v. 0). A program for creating global soil-property data bases. https://nelson.wisc.edu/sage/data-and-models/atlas/maps.php? Datas etid=20 and include related links=1 and dataset=20. Accessed 8 June 2018

Ikegami H (1985) Decrease of clubroot fungus by cultivation of different crops in heavily infested soil. Res Bull Fac Agric 50:19–32

Ishida M, Hara M, Fukino N, Kakizaki T, Morimitsu Y (2014) Glucosinolate metabolism, functionality and breeding for the improvement of Brassicaceae vegetables. Breed Sci 64:8–59

Ito S, Maehara T, Maruno E, Tanaka S, Kameya-Iwaki M, Kishi F (1999) Development of a PCR-based assay for the detection of *Plasmodiophora brassicae* in soil. J Phytopathol 147:83–88

Jacobsohn R, Greenberger A, Katan J, Levi M, Alon H (1980) Control of Egyptian broomrape (*Orobanche aegyptiaca*) and other weeds by means of solar heating of the soil by polyethylene mulching. Weed Sci 28:312–316

Ji HW, Ren L, Chen KR, Xu L, Liu F, Sun CC, Li J, Liu SY, Fang XP (2013) Identification of physiological races of club root and resistance of rape cultivars to *Plasmodiophora brassicae*. Chin J Oil Crop Sci 35:301–206

Joo GJ, Kim YM, Kim JW, Kim WC, Rhee IK, Choi YH, Kim JH (2004) Biocontrol of cabbage clubroot by the organic fertiliser using *Streptomyces* sp. AC-3. Korean J Microbiol Biotechnol 32(2):172–178

Junwei MA, Wanchun SUN, Qiaogang YU, Qiang FU, Jianrong FU (2013) Effects of cyanamide fertiliser on microbial community structure of continuous cropping soil. J Zhejiang Univ 39:281–290

Kammerich J, Beckmann S, Scharafatb I, Ludwig-Muller J (2014) Suppression of the clubroot pathogen *Plasmodiophora brassicae* by plant growth promoting formulations in roots of two *Brassica* species. Plant Pathol 63:846–857

Karling JS (1942) The Plasmodiophorales, 1st edn. Karling JS, New York

Karling JS (1968) The Plasmodiophorales. Hafner Publishing Company, Inc., New York, NY

Kasinathan H (2012) Influence of pH, temperature, and biofungicides on clubroot of canola. M.Sc. thesis, University of Guelph, Guelph (ON). Available from http://atrium.lib.uoguelph.ca/xmlui/bitstream/handle/10214/3547/HemaKasinathanThesisMay12012.pdf

Katan J (1980) Solar pasteurization of soils for disease control: status and prospects. Plant Dis 64:450–454

Katan J, Greenberger A, Alon H, Grinstein A (1976) Solar heating by polyethylene mulching for the control of diseases caused by soil borne pathogens. Phytopathology 66:683–688

Khalid M, Hui N, Hayat K, Huang D (2020) Suppression of clubroot (*Plasmodiophora brassicae*) development in *Brassica campestris* sp. *chinensis* L. via exogenous inoculation of *Piriformospora indica.* J Radiation Res. Appl Sci 13(1):180–190

Kim S, Shin C, Moon S, Yi YS, Choi GJ, Cho KY, Song J, Lim Y (2004) Isolation and characterization of *Streptomyces* sp. KACC 91027 against *Plasmodiophora brassicae*. J Microbiol Biotechnol 14:220–223

Klasse HJ (1996) Calcium cyanamide-an effective tool to control clubroot-a review. Acta Hortic 407:403–409

Kowata-Dresch LS, May-De Mio LL (2012) Clubroot management of highly infested soils. Crop Prot 35:47–52

Kuginuki Y, Yoshikawa H, Hirai M (1999) Variation in virulence of *Plasmodiophora brassicae* in Japan tested with clubroot-resistant cultivars of Chinese cabbage (*Brassica rapa* L. ssp. *pekinensis*). Eur J Plant Pathol 105:327–332

Lahlali R, Peng G (2014) Suppression of clubroot by *Clonostachys rosea* via antibiosis and induced host resistance. Plant Pathol 63:447–455

Lahlali R, Peng G, McGregor L, Gossen BD, Hwang SF, McDonald M (2011) Mechanisms of the biofungicide Serenade (*Bacillus subtilis* QST713) in suppressing clubroot. Biocontrol Sci Tech 21:1351–1362

Lahlali R, Peng G, McGregor L, Geissler L, McDonald MR, Gossen BD, Hwang SF (2012) Suppression of clubroot by *Heteroconium chaetospira* may be mediated by the modulation of jasmonic acid pathway signalling. Can J Plant Pathol 34(2):320

Lahlali RL, Peng G, Gossen BD, McGregor L, Yu FQ, Hynes RK, Hwang SF, McDonald MR, Boyetchko SM (2013) Evidence that the biofungicide Serenade (*Bacillus subtilis*) suppresses clubroot on canola via antibiosis and induced host resistance. Phytopathology 103(3):245–254

Lahlali RL, Mcgregor L, Song T, Gossen BD, Narisawa K, Peng G (2014) *Heteroconium chaetospira* suppresses clubroot by inducing host resistance via ethylene, jasmonic acid and auxin pathway signalling. PLoS ONE 4:e94144

Lathrop KW, Bushur-Hallam CK (1995) U.S. and Mexican regulation of methyl bromide: comparing pesticide regulations after NAFTA. Oklahoma Law Rev 48:289

LeBoldus JM, Manolii VP, Turkington TK, Strelkov SE (2012) Adaptation to Brassica host genotypes by a single-spore isolate and population of *Plasmodiophora brassicae* (clubroot). Plant Dis 96:833–838

Lee ML, Hsieh WH (1992) Influence of pH, sodium and potassium salts on viability of resting spores of *Plasmodiophora brassicae*. Plant Pathol Bull 1:31–36

Lee SO, Choi GJ, Choi YH, Jang KS, Park DJ, Kim CJ, Kim JC (2008) Isolation and characterization of endophytic actinomycetes from Chinese cabbage roots as antagonists to *Plasmodiophora brassicae*. J Microbiol Biotechnol 18(11):1741–1746

Leonard L, Bolland M (1996) Soil acidity: a reference manual. Agriculture Western Australia, miscellaneous publication 1/96

Li WF, He JM, Li SK (2006) Studied progress on the root swells sickness of mustard family plant. Southwest Chin J Agric Sci 19:560–563

Li XY, Mao ZC, Wang YH, Wu YX, He YQ, Long CL (2012) ESI LC-MS and MS/MS characterization of antifungal cyclic lipopeptides produced by *Bacillus subtilis* XF-1. J Mol Microbiol Biotechnol 22:83–93

Li XY, Mao ZC, Wang YH, Wu YX, He YQ, Long CL (2013) Diversity and active mechanism of fengycin-type cyclopeptides from *Bacillus subtilis* XF-1 against *Plasmodiophora brassicae*. J Microbiol Biotechnol 23(3):313–321

Linderman RG (1994) Role in VAM fungi in biocontrol. In: Pfleger FL, Linderman RG (eds) Mycorrhiza and plant health. APS Press, St. Paul, pp 1–26

Liu C, Yang Z, He P, Munir S, Wu Y, Ho H, He Y (2018) Deciphering the bacterial and fungal communities in clubroot-affected cabbage rhizosphere treated with *Bacillus subtilis* XF-1. Agric Ecosyst Environ 256:12–22

Ludwig-Muller J, Bennett RN, Kiddle G, Ihmig S, Ruppell M, Hilgenberg W (1999) The host range of *Plasmodiophora brassicae* and its relationship to endogenous glucosinolate content. New Phytol 141(3):4443–4458

Lugauskas A, Repeckienė J, Uselis N, Rasinskienė AB (2003) Problems of a longtime strawberry growing in one plot. Hortum Cultis 2:59–68

MacFarlane I (1952) Factors affecting the survival of *Plasmodiophora brassicae* Worn. in the soil and its assessment by a host test. Ann Appl Biol 39:239–256

MacFarlane I (1958) A solution – culture technique for obtaining root – hair or primary infection by *Plasmodiophora brassicae*. J Gen Microbiol 18:720–732

Manzanares-Dauleux MJ, Delourme R, Baron F, Thomas G (2000) Mapping of one major gene and of QTLs involved in resistance to clubroot in *Brassica napus*. Theor Appl Genet 101 (5/6):885–891

Maric A, Stojin V, Mitrovic P (1990) Cabbage clubroot (*Plasmodiophora brassicae*) in Semberia and possibilities of its control by fungicides. Zaštita Bilja 41:13–20

Marschner H (1995) Mineral nutrition of higher plants, 2nd edn. Academic, London, p 889

Marschner J, Römheld V, Horst WJ, Martin P (1986) Root-induced changes in the rhizosphere: importance for mineral nutrition of plants. Z Pflanzenernahr 149:441–456

Mattey M, Dixon GR (2015) Premature germination of resting spores as a means of protecting *Brassica* crops from *Plasmodiophora brassicae* Worn. (clubroot). Crop Prot 77:27–30

May A, Tivelli SW, Vargas PF, Samra AG, Sacconi LV, Pinheiro MQ (2007) A Cultura da Couve-flor. Boletim Tecnico IAC, no. 200. Campinas, Brazil: Instituto Agronomico

Mayya KS, Bhattacharyya A, Argillier JF (2003) Micro-encapsulation by complex coacervation: influence of surfactant. Polym Int 52:644–647

Mazzola M (2004) Assessment and management of soil microbial community structure for disease suppression. Annu Rev Phytopathol 42:35–59

McDonald MR, Westerveld SM (2008) Temperature prior to harvest influences the incidence and severity of clubroot on two Asian Brassica vegetables. Hortic Sci 43:1509–1513

McDonald MR, Gossen BD, Robson J, Al-Daoud F (2019) Interaction of solarization, fumigation and totally impermeable film for the management of clubroot (*Plasmodiophora brassicae*) on *Brassica* crops. Acta Hortic 1270:17

Mellano HM, Munnecke DE, Endo RM (1970) Relationship of seedling age to development of *Pythium ultimum* on roots of *Antirrhinum majus*. Phytopathology 60:935–942

Miller SA, Rowe RC, Riedel RM (1996) Clubroot of crucifers extension factsheet HYG-3118-96. Ohio State University, Columbus, p 2

Miller SA, Mera JR, Baysal F (2007) Evaluation of fungicides for the control of clubroot of broccoli, 2006. Plant Disease Management Reports (online). Report 1, V004. The Am Phytopathol Soc, St. Paul, MN

Mitani S, Sugimoto K, Hayashi H, Takii Y, Ohshima T, Matsuo N (2003) Effects of cyazofamid against *Plasmodiophora brassicae* Woronin on Chinese cabbage. Pest Mgt Sci 59:287–293

Morita S, Azuma M, Aoba T, Satou H, Narisawa K, Hashiba T (2003) Induced systemic resistance of Chinese cabbage to bacterial leaf spot and Alternaria leaf spot by the root endophytic fungus, *Heteroconium chaetospira*. J Gen Plant Pathol 69:71–75

Mukerji BK (1932) Studies on calcium cyanamide. II. Microbiological aspects of nitrification in soils under varied environmental conditions. J Agric Sci 22:335–347

Murakami H, Tsushima S, Shishido Y (2000a) Soil suppressiveness to clubroot disease of Chinese cabbage caused by *Plasmodiophora brassicae*. Soil Biol Biochem 32:1637–1642

Murakami H, Tsushima S, Akimoto T, Murakami K, Goto I, Shishido Y (2000b) Effects of growing leafy daikon (*Raphanus sativus*) on populations of *Plasmodiophora brassicae* (clubroot). Plant Pathol 49:584–589

Murakami H, Tsushima S, Kuroyanagi Y, Shishido Y (2002a) Reduction of resting spore density of *Plasmodiophora brassicae* and clubroot disease severity by liming. Soil Sci Plant Nutr 48 (5):685–691

Murakami H, Tsushima S, Shishido Y (2002b) Factors affecting the pattern of dose response curve of clubroot disease caused by *Plasmodiophora brassicae*. Soil Sci Plant Nutr 48(3):421–427

Murakami H, Tsushima S, Akimoto T, Kuoryanagi Y, Shishido Y (2004) Quantitative studies on the relationship between plowing into soil of clubbed roots of preceding crops caused by *Plasmodiophora brassicae* and disease severity in succeeding crops. Soil Sci Plant Nutr 50 (8):1307–1311

Myers DF, Campbell RN (1981) Clubroot of Brassicas in California: soils respond differently to lime for clubroot control. Phytopathology 71:1005–1006

Myers DF, Campbell RN (1985) Lime and the control of clubroot of crucifers: effects of pH, calcium, magnesium, and their interactions. Phytopathology 75:670–673

Myers DF, Campbell RN, Greathead AS (1981) Clubroot of Brassicas in California: soils respond differently to lime for clubroot control. Phytopathology 71:1005–1006

Naiki T, Dixon GR (1987) The effects of chemicals on developmental stages of *Plasmodiophora brassicae* (clubroot). Plant Pathol 36:316–327

Narisawa K, Tokumasu S, Hashiba T (1998) Suppression of clubroot formation in Chinese cabbage by the root endophytic fungus, *Heteroconium chaetospira*. Plant Pathol 47:206–210

Narisawa K, Shimura M, Usuki F, Fukuhara S, Hashiba T (2005) Effects of pathogen density, soil moisture, and soil pH on biological control of clubroot in Chinese cabbage by *Heteroconium chaetospira*. Plant Dis 89(3):285–290

Navas S, Botero-Ramirez A, Urbano D, Urquijo J, Garcia-Dominguez C (2017) Assessment of the presence of *Plasmodiophora brassicae* viable inoculum in the irrigation sediments in the

Savannah of Bogota. https://apsnet.confex.com/apsnet/2017/meetingapp.cgi/Paper/5576. Accessed 7 June 2018

Nishimura T, Tsuge S, Sakamote T, Suzuki S, Kawahara T (1980) Residues of pentachloronitrobenzene and its impurity hexachlorobenzene in soil and various crops. Bull Agric Chem Insp Stn 20:38–45

Niwa R, Kumei T, Nomura Y, Yoshida S, Osaki M, Ezawa T (2007) Increase in soil pH due to Ca – rich organic matter application causes suppression of clubroot disease of crucifers. Soil Biol Biochem 39:778–785

Niwa R, Nomura Y, Osaki M, Ezawa T (2008) Suppression of clubroot disease under neutral pH caused by inhibition of spore germination of *Plasmodiophora brassicae* in the rhizosphere. Plant Pathol 57:445–452

Noble R, Roberts SJ (2004) Eradication of plant pathogens and nematodes during composting: a review. Plant Pathol 53:548–568

Nommik H (1958) On decomposition of calcium cyanamide in the soil. Acta Agri Scand 8:404–440

Ohmori K, Nakagawa T, Koike K (1986) Evaluation of N-(1-alkoxy2, 2, 2-trichloroethyl)-2-hydroxybenzamides for clubroot control in Chinese cabbage. Plant Dis 70:51–53

Osozawa S, Iwama H, Kubota T (1994) Effect of soil aeration on the occurrence of clubroot disease of crucifers. Soil Sci Plant Nutr 40(3):445–455

Pachari L (2011) *Trichoderma harzianum* y *Bacillus subtilis* en el manejo integrado de la 'potra de la col' (*Plasmodiophora brassicae*) en broccoli (*Brassica oleracea* L. var. *italica*) cv. Legacy en Chilina-Arequipa. Arequipa, Ecuador: Universidad Nacional de San Agustin, Agronomist thesis

Page LV (2001b) Studies of components for a potential integrated control system for *Plasmodiophora brassicae*, Ph.D thesis. University of Strathclyde, Glasgow

Palm ET (1963) Effect of mineral nutrition on the invasion and response of turnip tissue to *Plasmodiophora brassicae* Worn. Contrib Boyce Thompson Inst 22:91–112

Peng G, McGregor L, Lahlali R, Gossen BD, Hwang SF, Adhikari KK, Strelkov SE, McDonald MR (2011) Potential biological control of clubroot on canola and crucifer vegetable crops. Plant Pathol 60:566–574

Peng SS, Ren ZH, Huang XL, Liu MJ, Sun LF, Liu EM (2013) Physiological race identification on clubroot of cruciferous crops caused by *Plasmodiophora brassicae* in Hunan. J Changjiang Veg 6:46–49

Peng G, Lahlali R, Hwang SF, Pageau D, Hynes RK, McDonald MR, Gossen BD, Strelkov SE (2014) Crop rotation, cultivar resistance, and fungicides/biofungicides for managing clubroot (*Plasmodiophora brassicae*) on canola. Can J Plant Pathol 36(sup1):99–112

Peng G, Pageau D, Strelkov SE, Gossen BD, Hwang S-F, Lahlali R (2015) A >2-year crop rotation reduces resting spores of *Plasmodiophora brassicae* in soil and the impact of clubroot on canola. Eur J Agron 70:78–84

Piao ZY, Wu D, Wang M, Zhang T (2010) Marker-assisted selection of near isogenic lines for clubroot resistant gene in Chinese cabbage. Acta Hort Sin 8:1264–1272

Pimpini F, Venter F, Wünsch A (1970) The influence of different nitrogen forms and increasing nitrogen doses on the content of total nitrogen and of nitrate in cauliflower plants. Landwirtsch Forsch 23:363–370

Porter I, Cross S (1995) Clubroot hygiene strategy. Institute for Horticultural Development, Agriculture Victoria, Victoria, p 4

Porter IJ, Merriman PR (1983) Effects of solarization of soil on nematode and fungal pathogens at two sites in Victoria. Soil Biol Biochem 15:39–44

Porter IJ, Merriman PR (1985) Evaluation of soil solarization for control of root diseases of row crops in Victoria. Plant Pathol 34:108–118

Porter IJ, Merriman PR, Keane PJ (1991) Soil solarization combined with low rates of soil fumigants controls clubroot of cauliflowers, caused by *Plasmodiophora brassicae* Woron. Aust J Exp Agric 31:843–851

Porter IJ, Donald EC, Cross SJ (1998) Field evaluation of fluazinam against clubroot (*Plasmodiophora brassicae*) of cruciferous vegetable crops. Ann Appl Biol 132(Supl):12–13

Porter I, Donald C, Lancaster R, Faggian R (2004) Integration of classical and molecular approaches to successfully control clubroot of crucifers. In: British Society for plant pathology presidential meeting 2004, University of Aberdeen, 5–10 September, 2004, p 25

Porter IJ, Cross SJ, Donald EC. 2005. Integrated control of clubroot, *Plasmodiophora brassicae*, of crucifers – a model system for soil borne pathogens. 15th Australasian Plant Pathology Society Conference Handbook; 2005 Sept 26–29; Geelong (Australia)

Poulton JE (1990) Cyanogenesis in plants. Plant Physiol 94:401–405

Pryor DE (1940) The effect of some mineral nutrients on the development of clubroot of Brassicas. J Agric Res 61:149–160

Pullman GS, De Vay JE, Garber RH, Weinhold AR (1979) Control of soil-borne fungal pathogens by plastic tarping of soil. In: Schippers B, Gams W (eds) Soil-borne plant pathogens. Academic, London, pp 439–446

Pullman GS, DeVay JE, Garber RH (1981) Soil solarization and thermal death: a logarithmic relationship between time and temperature for four soil borne plant pathogens. Phytopathology 71:959–964

Punja ZK, Carter JD, Campbell GM, Rossell EL (1986) Effects of calcium and nitrogen fertilizers, fungicides, and tillage practices on incidence of *Sclerotium rolfsii* on processing carrots. Plant Dis 70:819–824

Raaijmakers JM, Mazzola M (2016) Soil immune responses. Science 352:1392–1393

Raaijmakers JM, Paulitz TC, Steinberg C, Alabouvette C, Moemme-Loccoz Y (2009) The rhizosphere: a playground and battlefield for soil borne pathogens and beneficial organisms. Plant Soil 321:535

Ransom LM, Collins D, Taylor PA, Baker A (1991) A new weapon to beat clubroot. In: Australasian Plant Pathology Society 8th biennial Conference, Sydney, Australia, p 8

Rastas M, Latvala S, Hannukkala A (2012) Occurrence of *Plasmodiophora brassicae* in Finnish turnip rape and oilseed rape fields. Agric Food Sci 21:141–158

Rennie D, Manolii VP, Cao T, Hwang SF, Howard RJ, Strelkov SE (2011) Direct evidence of surface infestation of seeds and tubers by *Plasmodiophora brassicae* and quantification of spore loads. Plant Pathol 60:811–819

Rennie DC, Holtz MD, Turkington TK, Leboldus JM, Hwang SF, Howard RJ, Strelkov SE (2015) Movement of *Plasmodiophora brassicae* resting spores in windblown dust. Can J Plant Pathol 37:188–196

Robak J (1996) The effect of some crop rotations on decrease of clubroot, *Plasmodiophora brassicae*, in soils. Brighton Crop Protection Conference: Pests and Diseases 1996 Farnham, UK British Crop Prot Council pp 647–651

Rod J (1992) The effect of seed treatment on clubroot (*Plasmodiophora brassicae* Worn.) incidence. Ochr Rostl 28:17

Rod J, Robak J (1994) The effect of some crops on inactivation of clubroot (*Plasmodiophora brassicae* Worn.) in soils. Ochrana Rostlin 30:267–272

Rosenzweig C, Elliott J, Deryng D, Ruane AC, Muller C, Arneth A, Boote KJ, Folbeth C, Glotter M, Khabarov N, Neumann K, Piontek F, Pugh TAM, Schmid E, Stehfest E, Yang H, Jones JW (2014) Assessing agricultural risks of climate change in the 21st century in a global gridded crop model intercomparison. Proc Natl Acad Sci 111:3268–3273

Rousk J, Hill PW, Jones DL (2015) Priming of the decomposition of ageing soil organic matter: concentration dependence and microbial control. Funct Ecol 29:285–296

Rouxel F, Briard M, Lejeune B (1985) Studies of soil receptiveness to clubroot caused by *Plasmodiophora brassicae*: experiments on responses of a series of vegetable soils in Brittany. Progress on pest management of field vegetables. Proc. CEC/IOBC Experts' Group, Rennes, France 20-22 November 1985. (eds) Cavallorao R, Pelerents CJ). Rotterdam, The Netherlands: AA Balkema, pp 145–152

Ruaro L, Neto L, Vda C, Motta ACV (2010) Efeito do pH do solo em diferentes niveis de concentracao de inoculo no controle de *Plasmodiophora brassicae*. Summa Phytopathol 36:16–20

Saeed IAM, Rouse DI, Harkin JM (2000) Methyl isothiocyanate volatilization from fields treated with metam-sodium. Pest Magt Sci 56:813–817

Santos CA, Sobrinho NMBA, Costa ESP, Diniz CS, Carmo MG (2017) Liming and biofungicide for the control of clubroot in cauliflower. Pesquisa Agropecuaria Trop 47:303–311

Schisler DA, Slininger PJ, Behle RW, Jackson MA (2004) Formulation of *Bacillus* spp. for biological control of plant diseases. Phytopathology 94:1267–1271

Sen P (1993) Control of clubroot disease of rape and mustard attained by intercropping. Indian Agric 37:127–128

Sen P (2005) Antagonistic effect of Ca, B and Mo on club-root disease of rape-mustard. Indian Agric 49(1/2):13–16

Sharma K, Gossen BD, McDonald MR (2011) Effect of temperature on primary infection by *Plasmodiophora brassicae* and initiation of clubroot symptoms. Plant Pathol 60:830–838

Sharma K, Gossen BD, McDonald MR (2011b) Effect of temperature on cortical infection by *Plasmodiophora brassicae* and clubroot severity. Phytopathology 101:1424–1432

Shen XQ, Nie K, Wu Q, Zhang YG, Meng XH (2009) Initial research report on differentiation identification of Chinese cabbage clubroot main physiological races. Chin Veg 8:59–62

Shimotori H, Yanagida H, Enomoto Y, Igarashi K, Yoshinari M, Umemoto M (1996) Evaluation of benzenesulfonanilide derivatives for the control of crucifers clubroot. J Pest Sci 21:31–35

Sieling K, Kage H (2010) Efficient N management using winter oilseed rape: a review. Agron Sustain Dev 30:271–279

Smelt JH, Leistra M (1974) Conversion of metham-sodium to methyl isothiocyanate and basic data on the behaviour of methyl isothiocyanate in soil. Pestic Sci 5:401–407

Somé A, Manzanares MJ, Laurens F, Baron F, Thomas G, Rouxel F (1996) Variation for virulence on *Brassica napus* L. amongst *Plasmodiophora brassicae* collections from France derived from single spore isolates. Plant Pathol 45:432–439

Strelkov SE, Manolii VP, Rennie DC, Xiao Q, Cui D, Hwang SF (2011) The occurrence of clubroot on canola in Alberta in 2010. Can Plant Dis Surv 91:109–111

Strelkov SE, Hwang SF, Manolii VP, Cao T, Feindel D (2016) Emergence of new virulence phenotypes of *Plasmodiophora brassicae* on canola (*Brassica napus*) in Alberta, Canada. Eur J Plant Pathol 145:517–529

Suzuki K, Matsumiya E, Ueno Y, Mizutani J (1992) Some properties of germination-stimulating factor from plants for resting spores of *Plasmodiophora brassicae*. Ann Phytopathol Soc Jpn 60:699–705

Suzuki K, Sugimoto K, Hayashi H, Komyoji T (1995) Biological mode of action of fluazinam, a new fungicide for Chinese cabbage clubroot. Phytopathol Soc 61:395–398

Syngenta (2016) Tekila. https://www.syngenta.com.co/tekila. Accessed 8June 2018

Takahashi K (1994) Biological agents affecting the viability of resting spores of *Plasmodiophora brassicae* Worn. in soil without host roots. Ann Phytopathol Soc 60:667–674

Takahashi H, Ishikawa T, Kaido M, Takita K, Hayakawa T, Okazaki K, Itoh K, Mitsui T, Hori H (2006) *Plasmodiophora brassicae* induced cell death and medium alkalization in clubroot-resistant cultured roots of *Brassica rapa*. J Phytopathol 154(3):156–162

Tanaka S, Ito S (2013) Pathogenic and genetic diversity in *Plasmodiophora brassicae* (clubroot) from Japan. J Gen Plant Pathol 79:297–306

Tanaka S, Kochi SI, Kunita H, Ito SI, KameyaIwaki M (1999) Biological mode of action of the fungicide, flusulfamide, against *Plasmodiophora brassicae* (clubroot). Eur J Plant Pathol 105:577584

Tate KG, Eales RF (1982) Clubroot control in cabbage and cauliflower using mechanized transplanting drenches. N Z J Exp Agric 10:409–412

Telenchana PMY (2015) Aplicacion de Productos Sello Verde en el Manejo de la Hernia de las Crucifer as (*Plasmodiophora brassicae*) en el Cultivo de Brocoli (*Brassica oleracea* var. *avenger*), en las Condiciones Agroecologicas de Izamba. Cevallos, Ecuador: Universidad Tecnica de Ambato, Agronomist thesis

Teruyoshi H, Kazuhiko N, Toshiko A, Takehiro O, Miho Y (2001) Biological control of clubroot disease in Chinese cabbage using endophytic fungus, *Heteroconium chaetospora*. Res Inst Environ Sci 19:147–150

Tewari K, Suganuma T, Fuijikake H, Ohtake N, Sueyoshi K, Takahashi Y, Ohyama T (2004) Effect of deep placement of N fertilisers and different inoculation methods on bradyrhizobia on growth, N-fixation activity and N-absorption rate of field grown soybean plants. J Agron Crop Sci 190:46–58

Tkacz A, Cheema J, Chandra G, Grant A, Poole PS (2015) Stability and succession of the rhizosphere microbiota depends upon plant type and soil composition. ISME J 9:2349–2359

Townley D (2005) Control of clubroot disease using modern antiprotozoal fungicides. PhD thesis R9541, University of Reading

Townley D, Fox TV (2003) Control of clubroot disease using cyazofamid and fluazinam fungicides. Plant Pathol 2:72

Tremblay N, Bêlec C, Lawrence H, Carisse O (1999) Clubroot of crucifers-control strategies. Agriculture and Agri-Food Canada, Saint-Jean-Sur-Richelieu, p 3

Tremblay N, Belec C, Coulombe J, Godin C (2005) Evaluation of calcium cyanamide and liming for control of clubroot disease in cauliflower. Crop Prot 24(9):798–803

Usuki F, Narisawa K (2007) A mutualistic symbiosis between a dark septate endophytic fungus, *Heteroconium chaetospira*, and a non-mycorrhizal plant, Chinese cabbage. Mycologia 99:175–184

Verona O (1970) The effect of calcium cyanamide on some groups of lower fungi [Der Einfluss des Kalkstickstoffs auf einige Gruppen von niederen Pilzen]. Landwirtsch Forsch 23:36–50

Wakeham AJ, White JG (1996) Serological detection in soil of *Plasmodiophora brassicae* resting spores. Physiol Mol Plant Pathol 48:289–303

Walker JC (1935) Calcium cyanamide in relation to control of clubroot in cabbage. J Agric Res 51:183–189

Walker JC, Hooker WJ (1945) Plant nutrition in relation to disease development: II. Cabbage clubroot. Am J Bot 32:487–490

Wallenhammar AC (1996) Prevalence of *Plasmodiophora brassicae* in a spring oilseed rape growing area in central Sweden and factors influencing soil infestation levels. Plant Pathol 45:710–719

Wallenhammar AC, Arwidsson O (2001) Detection of *Plasmodiophora brassicae* by PCR in naturally infested soils. Eur J Plant Pathol 107:313–321

Wang WH, Yu YJ, Ding YH, Zhang FL, Yu SC, Zhang DS et al (2013) Physiological race identification of clubroot (*Plasmodiophora brassicae* Worn.) from Changyang county, Hubei province and resistant resource screening for Chinese cabbage breeding. Chin Veg 12:55–60

Warne LGG (1943) A case of club-root of Swedes due to a seed-borne infection. Nature 3861:509

Webb PCR (1949) Zoospores, believed to be those of *Plasmodiophora brassicae*, in the root hairs of non-cruciferous plant. Nature 163:608

Webster MA (1986) pH and nutritional effects on infection by *Plasmodiophora brassicae* Worn. and on clubroot symptoms. PhD thesis, University of Aberdeen, p 273

Webster MA, Dixon GR (1991a) Calcium, pH and inoculum concentration influencing colonization by *Plasmodiophora brassicae*. Mycol Res 95:64–73

Webster MA, Dixon GR (1991b) Boron, pH and inoculum concentration influencing colonization by *Plasmodiophora brassicae*. Mycol Res 95:74–79

Welch N, Greathead AS, Inman J, Quick J (1976) Clubroot control in Brussels sprouts using lime for pH adjustment. Calif Agric:10–11

White JG, Buczacki ST (1977) The control of clubroot by soil partial sterilization: a review. Ann Appl Biol 85:287–300

Wichuk KM, Tewari JP, McCartney D (2011) Plant pathogen eradication during composting: a literature review. Composting Sci Utilizat 19(3):244–266

Wightwick AM, Salzman SS, Reichman SSM, Allinson G, Menzies NW (2013) Effects of copper fungicide residues on the microbial function of vineyard soils. Environ Sci Pollut Res 20:1574–1585

Williams PH (1966) A system for the determination of races of *Plasmodiophora brassicae* that infect cabbage and rutabaga. Phytopathology 56:624–626

Williamson CJ, Dyce PE (1989) The effect of calcium cyanamide on the reaction of swede cultivars to populations of *Plasmodiophora brassicae*. Plant Pathol 38:230–238

Xiu PA (2018) Efectos de Bioles en Brocoli (*Brassica oleracea*) y lechuga (*Lactuca sativa*) en la Zona Horticola de Cartago, Costa Rica. Turrialba, Costa Rica: Centro Agronomico Tropical de Investigacion y Ensenanza (CATIE), Masters thesis

Yamagishi H, Yoshikawa H, Ashizawa KH, Yui S (1986) Effects of resistant plants as a catch crop on the reduction of resting spores of clubroot (*Plasmodiophora brassicae* Worn.) in soil. J Jpn Soc Hortic Sci 54:460–466

Yang YZ, Zhang XL, Yang WG, Tian WW, Wang Q (2011) Studies on the resistance of rape variety to *Plasmodiophora brassicae* and screening. Chin Agric Sci Bull 27:278–282

Yano S, Tanaka S, Kameya-Iwaki M, Katumoto K (1991) Relation of Ca2+ efflux to germination of resting spores of clubroot fungus. Bull Fac Agric 39:105–112

Yao H, He Z, Wilson MJ, Campbell CD (2000) Microbial biomass and community structure in a sequence of soils with increasing fertility and changing land use. Microb Ecol 40:223–237

Young CC, Cheng KT, Waller GR (1991) Phenolic compounds in conducive and suppressive soils on clubroot disease of crucifers. Soil Biol Biochem 23:1183–1189

Yu X, Wu Y, Mao Z, He Y (2013) Water-mediated dissemination and chemical control of cabbage clubroot disease. J Huazhong Agric Univ 1:48–53

Zamani-Noor N, Rodemann B (2018) Reducing the build-up of *Plasmodiophora brassicae* inoculum by early management of oilseed rape volunteers. Plant Pathol 67(2):253–504

Zhang T, Wu D, Zhao Z, Wang Z, Piao ZY (2012) Development of near isogenic lines for clubroot resistance in Chinese cabbage and their assessment. Mol Plant Breed 10:722–730

Zhang JW, Zhang JW, Sun MY (2013) Breeding of CR589, a new Chinese cabbage cultivar resistant to clubroot. J Changjiang Veg 10:16–17

Zhao Y, Bai YJ, Miao ZY, Li Y, Zhao KH (2013) Identification of physiological races of *Plasmodiophora brassicae* causing clubroot in Chinese cabbage from Northeast China. J Hunan Agric Univ 39:176–178

Zhao J, Wu YX, Ho HH, Chen ZJ, Li XY, He YQ (2016) PBT1, a novel antimicrobial protein from the biocontrol agent Bacillus subtilis XF-1 against *Plasmodiophora brassicae*. Eur J Plant Pathol 145:583–590

Techniques

14

Abstract

Techniques are sound base and backbone of getting reliable, reproducible, and well analyzed research results in any research project, and also very useful methods to demonstrate research methodology to graduate students. Several standardized techniques from basic to advanced molecular level have been included in this chapter for the researchers of cruciferous crop diseases. The clubroot pathogen identification and detection from soil and host tissues at different stages of host–pathogen interaction can be revealed by biological, pathological, microscopical, serological, and molecular techniques. Procedures to determine inoculum load as well as pathogens resting spores, number in soil and viability percent have been elucidated. A protocol is available on callus culture to target weak link in the life cycle of pathogen. Histopathological methods to study infection process in response to host resistance can reveal developmental stages of the pathogens, morphological, and cellular structures. Development of dual culture protocol can be helpful to obtain axenic culture of *P. brassicae* and study infection cycle in different cruciferous species. Pathogenicity assays in Petri dishes, pots, and field conditions have been demonstrated. A very precise and simple technique to assess the responses of lime-based fertilizers, and nutrients are elaborated. A procedure to develop clubroot sick plot and screening of *Brassica* germplasm under field and green house conditions has been described. Germplasm evaluation techniques are given for field and controlled conditions through artificial inoculation of the pathogen. The protocol of modern molecular techniques like PCR and qPCR has been described to detect pathogen in soil, and host tissues with very high precision level and sensitivity. Single spore isolation technique is very useful for determination of pathogenic variability. Techniques are described for isolation of resting spores; inoculum preparation; testing of resting spores germination; detection of seed infestation; detection of host pathogen at different stages of disease development; detection of

G. S. Saharan et al., *Clubroot Disease of Crucifers*,
https://doi.org/10.1007/978-981-16-2133-8_14

pathogen using loop mediated isothermal DNA amplification; quantification of auxins and cytokinin; DNA extraction from infected roots; soil, plant, fungal, and bacterial material; root hairs infection; staining of resting spores; transcriptomic analysis; *in planta* quantification of pathogen; identification of pathotypes, and designation of clubroot-resistant genes, and *P. brassicae* pathotypes.

Keywords

Techniques · Pathogen inoculum preparation · Host pathogen interaction · Callus culture · Germplasm evaluation technique · Artificial pathogen inoculation · Single spore isolation · PCR detection · Isolation of resting spores · Seed infestation · Loop-mediated isothermal DNA amplification · Real time PCR assay · Gene expression analysis · Quantification of auxins and cytokinins · Genetic transformation · Pathogenic variability · Production of pure genotypes isolates · Molecular detection and quantification · Disease assessment

14.1 Introduction

Reliable and repeatable research results can be achieved through the use of standardized techniques developed by experts, and refined subsequently with the advancement in science and technology. Several techniques, bioassays, methods, procedures, tools, protocols, etc. have been developed, used, and standardized by clubroot scientists to accomplish better management of the disease and have in depth insight in the life cycle of *P. brassicae*. The following techniques are very useful for *Brassica* researchers and graduate students to get reliable results. Those who are working on the development of research techniques can use these as basis to validate, refine, and update their research pursuits.

14.2 Pathogen Inoculum Preparation for Artificial Inoculation

Plasmodiophora brassicae cannot be cultured axenically. Galls containing resting spores may be stored at −20 °C for several years with minimal loss of viability. When required, the galls are brought to room temperature and macerated either mechanically or with a pestle and mortar. The resultant suspensions are filtered through nylon or muslin sheets removing host debris and then centrifuged at approximately 3000 rpm for 10 min. This yields pellets of resting spores that are free from macro-material. The supernatant liquid is decanted and discarded. Resting spore pellets are resuspended in sterile water and re-centrifuged until clean preparations, free from starch granules, are obtained. Inoculum may be applied directly to seedling roots by dipping them, it may be mixed with compost into which seed or seedlings are placed or aliquots of standardized concentration of spores (normally 1×10^4–10^6 resting spores ml^{-1}) may be applied to the surface of compost in which seedlings will be grown (Dixon 1976). Inoculated test plants should be maintained at a basal temperature of 20–25 °C up to 6 weeks at which time galling symptoms will be apparent and may be assessed by the method of Dixon (1976). Worldwide physiologic specialization was analyzed by Toxopeus et al. (1986). On the micro geographic scale there is substantial physiologic variation, whereas on the macro scale, it is more limited and reflects the dominant species of *Brassica* most commonly cultivated in a particular geographic region.

In another method to prepare resting spore's inoculum of *P. brassicae*, 5 g of air dried clubroot galls are soaked in 200 ml distilled water for 2 h to soften the tissue before grinding in a mortar and pestle. After filtering through seven layers of cheesecloth, the spore concentration is estimated using an hemocytometer and adjusted to 5.0×10^7 spores/ml. *Brassica* seeds are sown in a soilless potting mix and plants are grown under 22 °C, a 16 h/8 h light/dark photoperiod in a growth chamber. Five-days-old seedlings of *Brassica* are inoculated by pipetting 5 ml of resting spore suspension (5.0×10^7 spores/ml) around the base of the seedlings shoots, where it meets the soil (Tu et al. 2018).

14.3 Host–Pathogen Interaction

Many studies on host plant–*P. brassicae* interactions have been performed under artificial inoculation conditions in growth chambers and greenhouses, that is, flooding potting mix with resting spore suspensions (Sharma et al. 2011), growing a host plant in slurries of resting spores (Hwang et al. 2011a, b), and using sand-liquid culture (Agarwal et al. 2009; Deora et al. 2013). However, observations of the infection process, especially during the early stages of infection, have proven difficult due to adherent soil and sand particles not easily removed by washing, subsequent root hair or root surface damage during root sample preparation and the presence of soil-borne bacteria, fungi, and oomycetes on and/or inside host tissues, increasing the complexity of this system for elucidating the *P. brassicae* life cycle. In addition to the inoculation in soil and sand, several dual culture systems such as callus culture (Takahashi et al. 2001), hairy root culture induced by *Agrobacterium* (Asano et al. 2000), cell suspension culture (Asano and Kageyama 2006), and adventitious root culture (Takahashi et al. 2006) have been developed for maintaining an axenic culture of *P. brassicae* within infected host tissues. Many of these studies have used *Brassica napus* or *Brassica rapa* as host species and provided adequate systems for studying *P. brassicae*–host cell interactions avoiding the influence of other microbe contaminants.

Various histological methods have been used to study *P. brassicae*–host plant interactions (Schuller and Ludwig-Müller 2016). Due to its obligate biotrophic nature, *P. brassicae* is not amenable for transgenic research, thus impeding the availability of fluorescent protein-tagged strains for pathogen visualization during the infection process. Abundant electron-dense lipid droplets have been identified in *P. brassicae* plasmodia and resting spores (Deora et al. 2013; Williams and McNabola 1967). The presence of enriched lipid droplets within pathogen structures allows for the pathogen to be distinguished from surrounding host cells of infected root tissues and may, therefore, serve as targets for histological staining. Lipid droplets have a core of neutral lipids, mainly triacylglycerides and sterol esters. In many studies, 9-diethylamino-5H-benzo [a] phenoxazine-5-one (Nile red), a hydrophobic and metachromatic dye with poor solubility and fluorescence in water (Rumin et al. 2015), has been reported as a general lipid dye for both polar and non-polar lipids using different excitation and emission wavelength ranges (Diaz et al. 2008) in a wide range of biological systems (Guzman et al. 2010; Kimura et al. 2004; Kopischke et al. 2013; Miquel et al. 2014; Rumin et al. 2015). Detection of non-polar, neutral lipids by Nile red staining is coupled with a short excitation wavelength such as EX/EM wavelengths of 488/550–650 nm, whereas detection of polar lipids can be achieved by Nile red staining using a long excitation wavelength of 530–560 nm (Guzman et al. 2010; Li et al. 2015; Rumin et al. 2015). Nevertheless, Nile red staining for intracellular lipid droplets is being used for identifying *P. brassicae* structures within clubroot tissues warrants further investigation and potential procedure optimization.

Tu et al. (2018) have developed a two-step dual culture protocol to obtain axenic cultures of *P. brassicae* from *Brassica* galls and to conduct pathogenicity assays on *Brassica* and *Arabidopsis* seedlings grown on medium plates in Petri dishes using pure inoculum. In addition, they have established and optimized live cell imaging techniques, by specific labeling of intracellular lipid droplets, the nucleus, and cell walls of the pathogen, that enable the identification of initial stages of the *P. brassicae* infection cycle. The live cell imaging based on an axenic culture system has proved to be a valuable tool for investigating the cellular and molecular mechanisms underlying *P. brassicae*–host interactions.

To establish a new protocol, based on a two-step axenic culture of *P. brassicae* with its host tissues, for easy and *in planta* observation of cellular interactions between *P. brassicae* and host plants, Tu et al. (2018) have suggested two steps axenic culture system: first, co-culture of *P. brassicae* with infected *Brassica* root tissues, on growth-medium plates, enables the propagation of *P. brassicae* that serves as pure inoculum for pathogenicity assays, and second, the pure inoculum is subsequently used for pathogenicity tests on both *Brassica* and *Arabidopsis* seedlings grown on medium plates in Petri dishes. During the first axenic culture, a staining protocol has been established by which the pathogen is fluorescently labeled with Nile red and calcofluor white, thus allowing *in planta* observation of pathogen development. In the pathogenicity assays, axenic cultures of *P. brassicae*, in calli, remain fully virulent and complete its life cycle in both canola and *Arabidopsis* roots grown in Petri dishes. Thus, established protocols for *in planta* detection of *P. brassicae* structures and the live cell imaging of *P. brassicae—Arabidopsis* interactions provide a novel strategy for improving our detailed knowledge of *P. brassicae* infection in host tissues.

14.4 Establishment of Callus Culture

The callus culture system can be established according to Bulman et al. (2011) with minor modifications. *Brassica* clubroot galls are harvested 5 weeks after inoculation. After washing thoroughly in tap water, galls are surface-sterilized with 70% ethanol for 1 min, then bleach (30%) with 0.05% tween (1.4% sodium hypochlorite) for 20 min, followed by three washings in sterile distilled water. After this second sterilization, the sections are cut into 0.5 cm segments and are placed on a medium consisting of Gamborg B5 basal medium plus 3% sucrose, 300 mg/l timentin, and 1.5% agar. Lateral root segments of non-inoculated plants are used as control, because mature main roots are inflexible and too difficult to cut into small segments.

After 2 week incubation in the dark at 23 °C, gall segments showing callus growth on the cut edges are transferred to fresh medium minus timentin. At the same time, a thin section is removed from each cultured segment, divided into two pieces, with one piece placed into Luria–Bertani (LB) liquid medium and one onto Potato Dextrose Agar (PDA) medium to confirm that the gall/callus segment is free of any endophytic bacteria and fungi. Each medium is observed for 3 days to confirm no bacterial or fungal growth. Endophytic microbe-free calli are used to inoculate host seedlings (Tu et al. 2018).

14.5 Pathogenicity Assay of Canola and *Arabidopsis* Seedlings Grown in Petri Dishes

Pathogenicity testing of *Brassica,* and *Arabidopsis* seedlings grown on medium plates in Petri dishes, *Brassica* cultivar Westar, and *Arabidopsis thaliana* Col-0 ecotype can be used. Westar seeds, surface sterilized for 1 min in 70% ethanol, followed by 20 min in 30% bleach and three washes in sterile water, are grown in square Petri dishes on half-strength Murashige and Skoog (MS) medium in a growth chamber at 23 °C with a 16 h/8 h, light/dark photoperiod (light intensity 120 μmol/ms). Approximately 10 *Brassica* seeds or 15–20 *Arabidopsis* seeds are germinated and grown on MS plate in Petri dish. After 4 days, each seedling is inoculated with 400 μl of 5.0×10^7 spores/ml resting spore suspension, freshly prepared from callus, by pipetting onto the roots. Petri dishes are sealed thoroughly with Parafilm to confine the internal environment. The inoculation procedure in *Arabidopsis* is similar with one difference: The time for bleach sterilization is reduced to 5 min. The infection process is observed under transmission light microscopy. The pathogenicity tests are repeated at least three times on both *Arabidopsis* and *Brassica* seedlings. *Arabidopsis* transgenic lines, expressing PMA4-GFP (Nottingham *Arabidopsis* Stock Centre, stock number N799380 (Teh and Moore 2007)) and AT4G19150-GFP (*Arabidopsis* Biological Resource Centre, stock number CS84731 (Cutler et al. 2000)) are infected using the same inoculation procedure (Tu et al. 2018).

14.6 Host Inoculation for Pathogenicity Test

To extract *P. brassicae* resting spores from dried rot tissue, approximately 1 g of dry root material is placed in a mortar and ground to a powder using a pestle. Twenty milli-liters of sterile distilled water is added, and the homogenate is then mixed and filtered through six layers of cheesecloth. To extract spores from frozen tissue, a modification of the procedure described by Williams (1966) can be employed (Tewari et al. 2005). Collections of *P. brassicae* from soil samples are recovered by planting 5-days-old Chinese cabbage (*B. rapa* var. *pekinensis*) "Granaat" seedlings (the universal susceptible) in the infested soil; plants are grown for 6 weeks, the galled roots are harvested, and the spores are extracted as described above. Resting spores are quantified with a hemocytometer, and used immediately or stored at 4 °C for a maximum of 2 days. Suspensions of *P. brassicae* resting spores, adjusted to a concentration of 1×10^7 spores/ml using sterile distilled water, are used to inoculate the host plants to be tested. Five-days-old seedlings, germinated on filter paper in Petri dishes, are inoculated by dipping the roots in the spore suspension. Controls are dipped in sterile distilled water. Seedlings are immediately planted in 4×4 cm plastic pots filled with Metro-Mix 290 soil at a density of one seedling per pot. The pots are thoroughly watered and transferred to a growth cabinet kept at 21:18 °C (day: night) with a 16 h photoperiod. The soil is kept saturated with water for the first week after inoculation, and then watered and fertilized as required (Strelkov et al. 2006)

Extraction of *P. brassicae* resting spores and host inoculation were conducted as described by Strelkov et al. (2006) for dried root tissues. The spore concentration was adjusted to 1×10^7 to 1×10^8 resting spores per milliliter with sterile distilled water and used for inoculation of the host roots. The roots of 7-days-old seedlings (pre-germinated on moistened filter paper in Petri dishes) were inoculated by dipping them in the resting spore suspension for 10 s. The inoculated seedlings were then planted immediately in $6 \times 6 \times 6$ cm plastic pots filled with Metro-Mix 290 Growing Medium at a density of one seedling per pot. The pots were transferred to a greenhouse kept at 20 ± 2 °C with a 16 h photoperiod (light intensity of 300 μmol $m^{-2}\ s^{-1}$), with the soil kept saturated with water for the first week after inoculation. Subsequently, the seedlings were watered and fertilized as required. Twelve seedlings of each host genotype were inoculated with each population of *P. brassicae* and inoculation were repeated total of four times (Cao et al. 2009).

14.7 In Vitro Culture of *Plasmodiophora brassicae*

The Plasmodiophorales are generally considered to be obligate parasites (Webster 1980), although there have been several reports that plasmodial plant pathogens can be cultured in the laboratory. Diriwachter et al. (1979) reported that spores of the plasmodia *Spongospora subterranea* f. sp. *subterranea*, obtained from surface sterilized diseased potato tubers, germinated readily on a simple agar medium to release amoebae which formed colonies, mixed with colonies of bacteria. After a few days, the amoebae encysted, which in turn released amoebae when transferred to fresh medium. These cultures can be maintained indefinitely when transferred to fresh medium in the presence of contaminating bacteria. They also described the development of disease symptoms on tomato plants inoculated with a mixture of amoebae and resting spores obtained in vitro. Jones (1928) described the isolation of eight cultures of *P. brassicae* which were maintained under laboratory conditions for 2 months. The cultures were obtained from surface sterilized diseased roots which were placed in sterile distilled water. The organisms isolated were described as amoebae and were associated with bacteria. If the cultures were left without sub-culturing, the amoebae encysted. It was also shown that disease symptoms could be produced on turnip plants inoculated with the amoebae and cysts.

P. brassicae is capable of at least a short-lived period of saprotrophic amoebal growth. Spores extracted from surface sterilized blocks of clubroot tissue are incubated at 25 °C on strength Ringer's agar, spread previously with a suspension of *Escherichia coli.* After a period of 5–10 days, outgrowth of amoebae commonly appears around the inoculum. Amoebae are subcultured on to fresh Ringer's- *E. coli* plates, where they continue to grow, and eventually encyst. An amoebal cyst suspension is prepared from these plates and used to inoculate Chinese cabbage *(Brassica campestris* var. *chinensis)*, and Savoy cabbage (*B. oleracea*) seedlings. Out of 27 amoebal isolates tested, nine were found induced typical clubroot disease symptoms. However, when these putative positive *P. brassicae* amoebal isolates are subcultured, the subsequent generation of amoebal cysts failed to induce disease symptoms. A random genomic library is constructed in pUCl8 using *BamH* 1-digested DNA extracted from *P. brassicae* resting spores. The inserts from nine individual clones are then used to probe DNA extracted from 12 amoebal cultures, three showed positive hybridization to the *P. brassicae* probes but this hybridization signal was lost after subculture of the amoebae (Arnold et al. 1996).

14.8 Germplasm Evaluation Technique

The various techniques to evaluate the *Brassica*/cruciferous germplasm, varieties, cultivars, lines/ genotypes are explained as follows:

14.8.1 Clubroot Assessment Scale

The assessment of host resistance, a nine-grade scale has been used by Wesolowska (2014) at Poland to categories resistance against *P. brassicae*. The nine-grade scale based on root symptoms established by Williams (1985), and modified by Reby and Michalik (2005) (Fig. 14.1) was used. The extent and severity of the clubroot symptoms were assessed visually according to a nine-grade scale: I—without visible disease symptoms; III—singular galls on lateral secondary roots; V—larger galls on lateral roots; VII—enlarging gall on the main and lateral roots; IX—large singular gall on root neck without root.

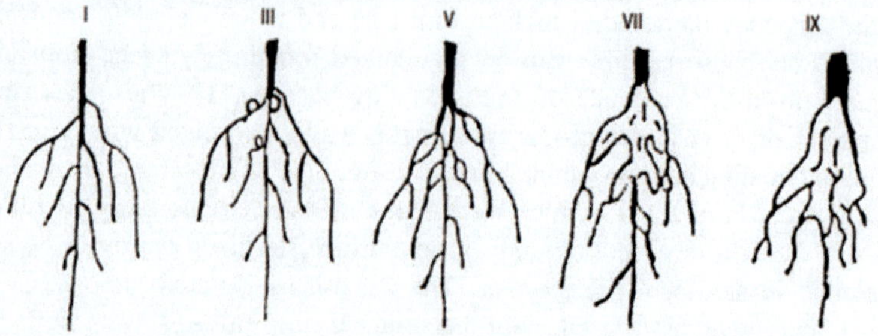

Fig. 14.1 Nine-grade scale based on clubroot symptoms (Wesolowska 2011)

The disease index (DI) was calculated using the nine-grade scale according to the formula:

$$\mathrm{DI} = \frac{\sum(n\, x\, p)}{\sum n}$$

where *n* is the number of plants in the indicated class, *p* class I–IX.

The disease indexes of the tested objects were calculated according to plant susceptibility at the eight-week seedling stage and evaluated once more during harvest, on the clubroot-infected field.

14.8.2 Development of Clubroot Sick Plot

The clubroot field was established at Poland by adding harvest remains and the soil from the field where heavily clubbed plants of white cabbage were observed by scattering on the surface of the area. The oilseed rape, which is susceptible to clubroot, was sown and plants of different cabbage cultivars artificially inoculated by *P. brassicae* were cultivated. Each year, clubbed plant roots were mixed with the soil for extra infection with the pathogen in the field, which was carried out after the autumn disease evaluation and harvests of tested plants. As a result, the field was contaminated by a mixture of different *P. brassicae* pathotypes. The procedures concerning the seed inoculation method and screening tests were the same as described in detail by Wesolowska (2011).

14.8.3 Germplasm Evaluation

A wide range of different methods to evaluate clubroot susceptibility screening have been developed. The soil inoculation method and dipping method seem to be the most popular. In such methods, plant materials are inoculated either by sowing seeds in an infected substrate or by dipping the roots of approximately 10-day-old seedlings in a resting spore suspension (Robak 1991). A double-stage evaluation system for the clubroot resistance was used by Wesolowska (2014). This procedure was established by Reby and Michalik (2005). Firstly, cabbage seed samples were inoculated by submerging them into the water pathogen spore suspension. This method is a modification of the one described by Lewis and Brokenshire (1978). The seed inoculation technique was found to be simple and not time-consuming. Additionally, after the plantlet stage screening, the resistant plants were transplanted into the clubroot-infested field for a second evaluation. This way of testing on adult plants in field conditions was described by Manzanares-Dauleux et al. (2000).

An evaluation of clubroot development in 8-week-old cabbage plants from the standard objects showed an influence of the pathogenicity level of the *P. brassicae* isolates used for seed inoculation on plant reaction. All of them were virulent against the susceptible HTM line plants. Heavily infested root systems were observed in plants from seeds inoculated by the *K* isolate from Krojczyn (DI-8.7), and the *P* isolate from Pobiednik (DI-8.0). All of the plants from the HTM line and "Badger Shipper" derived from the P isolate combinations were completely infested. The isolate from Pobiednik was found to be particularly aggressive not only to the susceptible HTM and "Badger Shipper" but also to the clubroot-resistant "Kilaxy" F1. The cultivar disease index calculated for that was 6.5, and only 9% of the plants were classified as healthy. Plants derived from Oregon123 were less susceptible to the P isolate (DI - 2.3) and characterized by 65% of the plants without clubroot galls.

14.8.4 The Germplasm Evaluation Technique Under Indian Conditions

The screening program can be started from the area where soil pH is at the level of 5.7 and severity of clubroot is very high. In every year 50–60 genotypes can be tested in the sick plot and the same is repeated for two seasons. The genotypes which show moderately susceptible or susceptible reactions are discarded and the entries which show resistance are retained and further tested for 3–4 years to avoid any chance of escape. Entries of different *Brassica* species viz, *B. campestris* var. *toria*, *B. campestris* var. *yellow sarson*, *B. campestris* var. *brown sarson*, *B. juncea*, *B. chinensis*, *B. napus*, *B. carinata*, *B. nigra*, *B. oleracea* var. *capitata*, *B. oleracea* var. *botrytis*, and *B. oleracea* var. *gongylodes* are tested in the sick plot. Each entry is sown in a single row of 4m in length with 50 × 10 cm spacing. A susceptible check (YSB-9) is planted after every two test entries. Irrigation is given at 30, and 60 days after sowing. Fertilizer dose of N.P.K at 80:40:0 kg/ha is applied or as per requirement.

After 70 days, plants are uprooted. Soil is carefully removed from the roots and washed thoroughly. The degree of severity is rated as described by Seaman et al. (1963). No clubs is 0; few small clubs on secondary roots is 1; considerable clubbing on the lateral roots is 2; and severe clubs on primary and secondary roots are 3. The number of plants in class 0 is multiplied by 0; in class 1 by 10; in class 2 by 60; in class 3 by 100, and the sum of the products is divided by the total number of plants to give the disease index. Initiation of formation of clubs in roots takes place between

25 and 32 days after sowing in the susceptible cultivar YSB-9. More than ninety per cent plants of the susceptible check are affected, bearing large size clubs in primary and secondary roots throughout the plot confirming the uniform distribution of pathogen in the sick plot (Chattopadhyay et al. 2001).

14.8.5 Greenhouse Screening for Clubroot Resistance

Screening of *Brassica* germplasm with *P. brassicae* pathotypes has been described by Hasan et al. (2012). Details of how to classify pathotypes, methods of inoculum preparation, and procedures of inoculation, data collection, and statistical analysis have been described.

(i) Isolates

Five single spore derived isolates of *P. brassicae* (pathotypes 2, 3, 5, 6, and 8), collected from Canadian *Brassica* crops and infested soil (Xue et al. 2008), were used to evaluate the *Brassica* germplasm for resistance to clubroot. These pathotypes were classified based on Williams (1966) system. These single spore derived isolates were reproduced on the universally susceptible Chinese cabbage cultivar "Graanat" (*B. rapa* L. var. *pekinensis*) and galls were preserved in −20 °C temperature.

(ii) Preparation of Inoculum

Resting spore suspensions (inocula) were prepared on the day of inoculation from frozen galls following the protocol of Williams (1966) as described by Strelkov et al. (2007). Approximately 2.0–2.5 g of clubbed roots were ground in a mortar with a pestle by adding 50 ml of sterile de-ionized water (sdH_2O). The homogenate was filtered through eight layers of cheesecloth to remove the plant debris. The resting spores in suspension were quantified with a hemocytometer and adjusted to minimum concentration of 1.0×10^7 resting spores ml^{-1}, to be the optimum concentration for successful inoculation (Voorips and Visser 1993).

(iii) Inoculation

Seeds were germinated under room temperature on moistened Whatman No. 1 filter paper in 90 × 15 mm size Petri dishes, and the plates were kept moist with tap water to provide adequate moisture for germination of the seeds. Seven-days-old seedlings were inoculated by dipping the entire root in the resting spore suspension for 10 s (Nieuwhof and Wiering 1961). The inoculated seedlings were immediately planted in 9 or 18 cell trays filled with sunshine professional growing mix. The seedlings were grown in a greenhouse at 21 ± 2 °C with a 16-h photoperiod. To ensure the optimum conditions for resting spore germination and root-hair infection by the zoospores, the trays were kept saturated with water for the first week. From the second week, the trays were watered once a day. Quick release mixed fertilizer, viz., 15-30-15 (N-P-K) was applied once a week to provide nutrients for the seedlings.

(iv) Data Collection

Seedlings were evaluated for clubroot resistance 42–45 days after inoculation. Roots were dug out, washed in tap water and examined for gall formation. The severity of gall development was rated on a 0 to 3 scale as described by Kuginuki et al. (1999), where 0—no galling; 1—one or few small galls on the lateral roots; 2—moderate galling on the lateral roots; and 3—severe galling on the lateral roots or on the main root.

(v) Evaluation of Germplasm

Evaluation of the *Brassica* germplasm was conducted in two phases. In the first phase, all 275 *Brassica* lines were evaluated for resistance to pathotypes 2, 3, 5, 6, and 8. For this, nine seedlings of each line/cultivar were evaluated against each pathotype in one replication. A total of 59 cultivars/lines exhibiting resistance to multiple pathotypes in the first screening were selected for evaluation in a second, replicated screening against the pathotype 3, and 5. Eighteen seedlings of each line were tested against each pathotype, with the experiment repeated three times. Chinese cabbage (*B. rapa* var. *pekinensis*) cultivar Granaat (ECD 05) was used as a susceptible control.

(vi) Statistical Analysis

Percent resistance for each cultivar/line against each pathotype was calculated using the following formula:

$$\text{Percent resistance }(\%) = \frac{\text{Number of resistant seedlings}}{\text{Total no.of seedlings inoculated}} \times 100$$

where seedlings with no visible gall at 42–45 days after inoculation were considered as resistant.

The disease score was calculated for each cultivar/line using the following formula (Strelkov et al. 2007):

$$\text{Average score} = \sum (n \times 0 + n \times 1 + n \times 2 + n \times 3)/N$$

where N is the total number of plants, n is the number of plants in each class and 0, 1, 2, and 3 are the symptom severity classes, as described above. An index of disease (ID) was also calculated for each cultivar/line, using the method of Horiuchi and Hori (1980) as modified by Strelkov et al. (2006):

$$\text{ID }(\%) = \frac{\sum (n \times 0 + n \times 1 + n \times 2 + n \times 3)}{N \times 3} \times 100$$

where n is the number of plants in each class, N is the total number of plants, and 0, 1, 2, and 3 are the symptom severity classes. The ID value was calculated for each cultivar/line against each treatment from each replication.

The reaction of different cultivar/linc against different pathotypes was analyzed with SAS software version 9.2 (SAS 2008). Experiments were conducted in a split-plot design where cultivars/lines were main plots and pathotypes were the sub-plots.

Residuals for ID were calculated using SAS PROC GLM. These residuals were used to estimate skewness and kurtosis with SAS PROC UNIVARIATE, and homogeneity of variance by PROC GLM. However, skewness and kurtosis of ID residual indicated that the data were not normally distributed and residual variances were not homogeneous. Therefore, data from replicated experiments were analyzed according to a generalized linear mixed model of split-plot design by using the SAS PROC GLIMMIX, where pathotypes and cultivars/lines were considered as fixed effects and replication as the random effect.

14.9 Artificial Pathogen Inoculation Techniques

Different species of *Brassica* germplasm have been used for inoculum extraction and conducive growth conditions after inoculation as per methods suggested by Buczacki et al. (1975). The different inoculation techniques followed are:

1. *The slurry method*: In which 400 ml of the spore suspension is mixed with 6000 g soil and the seeds are sown in the slurry (Toxopeus and Janssen 1975).
2. *The modified slurry methods*: Where after mixing of spore suspension and soil, 7-days-old seedling are planted in slurry.
3. *A root dip method*: Proposed by Johnston (1968) in which the roots of 7-days-old seedlings are dipped in the spore suspension for 2 h.
4. *Seed Dipping*: Dipping of seeds in spore suspension for 6 h before sowing is also recommended.
5. *An Injection method*: Injection method was also suggested by Schoeller and Grunewald 1987. In this method, 7-days-old seedlings are planted in the non-infected soil. Close to each plantlet, a hole is made in which 2 ml of the spore suspension is added so that roots of the plantlets are contaminated with the suspension.

After making a comparison of these inoculation techniques, Schoeller and Grunewald (1987) reported that only with the modified slurry method, the root dip method, and the injection method could infect 100% of tested plants. Rod (1988) described partly modified inoculation method which is very easy and rapid to use under glasshouse or phytotron conditions.

14.10 Single Spore Isolation

A wild population of *P. brassicae* consists of a mixture of pathotypes even in a field plot or within a gall tissue. Accordingly, single spore isolation is essential for purification of pathotypes. It is difficult since *P. brassicae* cannot be artificially cultured, except for callus culture of infected host tissue with the pathogen. Although isolates have been obtained by inoculating seedlings with a single resting spore, the techniques are laborious and often fail to give a high success rate (Kageyama et al. 1995).

14.10.1 Isolation of Single Resting Spores

Clubroot galls were collected from Prefectures of Aichi and Ibaraki in Japan, and the pathogens were designated as Toyohashi and Ibaraki populations, respectively. They were maintained and multiplied in highly susceptible Chinese cabbage, *Brassica campestris* L. var. *chinensis* (L.) Makino, cvs. Granaat (ECD line 05), and Muso, or turnip, *B. campestris* var. *rapifera* Metz, cv. Taibyohikari by using common method of Williams (1966). The clubroot tissues were recovered from the inoculated Chinese cabbage or turnip plants, followed by crushing in sterilized distilled water in a Petri dish with tweezers. Resting spores were easily suspended when clubbed tissues were severely rotted. Spore suspension was filtered through eight layers of gauze, and adjusted to the concentration of 2×10^3 spores/ml, so as to contain one single spore in 0.5μl of the suspension. A 0.5 μl drop of the suspension was placed on a piece of sterilized cover glass with a micropipette. The drop was checked at least three times under an inverted microscope to ensure that it carried exactly one spore. The drop was added with one micro-liter of sterilized distilled water in order to prevent the spore from desiccation. The cover glasses with drops were transferred to Petri dishes with moistened filter paper until employed.

14.10.2 Inoculation to Plant with Single Spore

Seeds of Chinese cabbage (cvs. Muso or Granaat) and turnip (cv. Taibyohikari) were surface sterilized by immersing in 70% ethanol for ca. 1min, and subsequent 0.5% sodium hypochlorite for 30 min, and rinsed three times with sterilized distilled water. The seeds were placed in Petri dishes with moistened filter paper and incubated at 25 °C for 1 or 2 days. The cover glass carrying the single spore was placed on the surface of autoclaved soil with which a polycarbonate bottle ($6 \times 6 \times 10$ cm) was filled. The young seedling was placed over the drop on the cover glass, and covered with a small amount of potting soil named as the cover glass method. The bottles with the inoculated seedlings were covered with polypropylene lids for a week to keep humidity and to enhance infection with the pathogen. The seedlings were kept in a growth chamber at 25 °C with a 16-h photoperiod for 3 weeks. The plants were then transferred to another growth chamber maintained at 20 °C at night and 25 °C in the daytime, and grown for a further 5-weeks period. The gall formation on roots was graded with the scale from 0 (no clubs) to 3 (severe swelling on lateral and tap roots) according to Naiki et al. (1978).

14.10.3 Pathogenicity Test

The host lines of European Clubroot Differential (ECD) set (Buczacki et al. 1975) were used to compare pathogenicity between the single spore isolates and their original populations. ECD lines 01 or 04 were not used because all these hosts had never been attacked by Toyohashi and Ibaraki populations in the preliminary experiments. Inoculum level obtained from the galls which had been formed with the single spore inoculation was not enough to use to test pathogenicity because a large amount of inoculum is required to inoculate 11 host lines of ECD. Therefore, multiplication of inoculum was conducted by using cover glass method in which Inoculum was obtained by crushing the galls with a mortar and a pestle. Host plants were inoculated by modifying the technique known as slurry method (Toxopeus and Janssen 1975). Inoculum for pathogenicity test was obtained by macerating the multiplied galls, filtering the suspension through eight layers of gauze and centrifuging the filtrate. The spore suspension was mixed with autoclaved soil (1:2, w/w) to make slurry. The spore concentration in the slurry was adjusted to 0.3–3.0 × 10^6/ml. On each pot (5 cm in diam. and 10 cm high) containing autoclaved soil, a hole of 3 cm diam. and 2 cm deep was made with a centrifuge tube. Approximately 20 ml of the slurry was poured into the hole in each pot, and six seedlings of one-day-old were immediately transplanted on the slurry. The pots were placed on trays filled with water up to 1–2 cm high and incubated on greenhouse benches. The number of plants per pot was thinned afterward to five plants. The test for each ECD line was replicated five times. The galls were graded 6–8 weeks after inoculation, as described. The disease severity was calculated with the following method. The number of plants in index 0 was multiplied by 0; in index 1 × 10; in index 2 × 80; and in index 3 × 100, and the sum was divided by the total number of plants (Buczacki et al. 1975).

14.10.4 Inoculation of Differential Host

Resting spores were collected from the galled roots of 8-week-old Granaat plants inoculated with single spore isolates of *P. brassicae*, and the spore concentration was diluted to approximately 1 × 10^7 spores/ml with sdH_2O. One-week-old seedlings, which were pre-germinated on moistened sterile filter paper, were inoculated by dipping the roots in the spore suspension for 10 s. The inoculated seedlings then were planted immediately in 4 × 4-cm plastic pots filled with Metro-Mix 290 soil, placed in flat containers without drainage holes at the bottom, and transferred to a greenhouse maintained at 21 and 18 °C (in light and darkness, respectively) with a 16-h photoperiod (natural light supplemented by high-pressure sodium light). The soil was kept saturated using tap water (pH 6.0) for the first week after inoculation, and then watered with regular tap water as required. Fertilizer (15N-30P-15K) was applied once a week (Xue et al. 2008).

14.11 PCR Detection of Clubroot Pathogen

A simple, one-step polymerase chain reaction (PCR) protocol was developed to detect the pathogen in plant, and soil samples at Canada (Cao et al. 2007). The primers TC1F and TC1R, based on a *P. brassicae* partial 18S ribosomal RNA (rRNA) gene sequence from GenBank, yielded a 548-bp product in the optimized

PCR. A second pair of primers, TC2F, and TC2R, which amplified a fragment of the 18S and internal transcribed spacer (ITS) 1 regions of the rDNA repeat, also was tested and produced a 519-bp product. Neither set of primers amplified any DNA fragment from non-infected host plant, nor non-infested soil, or common soil fungi and bacteria tested in this study. Quantities of 100 fg or less of total *P. brassicae* DNA, or 1×10^3 resting spores per gram of soil, could be detected consistently using these primers and PCR protocol, corresponding to an index of disease of 11% or lower when the soil was bioassayed. The protocol also enabled detection of *P. brassicae* in symptomless root tissue 3 days after inoculation with the pathogen. Therefore, the PCR assay described by Cao et al. (2007) provides a reliable diagnosis for routine detection of *P. brassicae* in plant and soil materials in a specific and rapid manner.

14.11.1 PCR Technique to Identify Pathogen

PCR amplification was conducted to confirm the pathogen. DNA was extracted from root galls and healthy roots (control) of Shepherd's purse (*Capsella bursa-pastoris*) (SP). Two primers, TC2F (5′-AAACAACGAGTCAGCTTGAATGCTAGTGTG-3′) and TC2R (5′-CTTTAGTTGTGTTTCGGCTAGGATGGTTCG-3′) were used to detect *P. brassicae* (Cao et al. 2007). No PCR amplifications were observed with the control DNA as template. A fragment of the expected size (approximately 520 bp) was obtained when DNA was amplified from diseased roots of SP. These results suggest that the pathogen in the galled roots of SP is *P. brassicae*. Pathogenicity of *P. brassicae* in SP was tested on plants of both SP and Chinese cabbage (CC) (*B. campestris* ssp. *pekinensis*). A resting spore suspension prepared from naturally infected SP roots was mixed with a sterilized soil in two plastic pots, resulting in a final concentration of 5×10^6 spores/g soil. Soil treated with the same volume of sterile water was used as a control. Seeds of SP and CC were pre-germinated on moist filter paper for 2 days (20°C), and seeded into the infested and control pots. One seed per pot was planted for CC and four seeds per pot for SP. The pots were placed in a chamber at 15–25 °C under 12 h light and 12 h dark. Plants in each pot were uprooted after 4 weeks and the roots of each plant were washed under tap water and rated for clubroot disease. No disease symptoms were observed in the control treatments of SP or CC. Plants of both species showed symptoms of clubroot, with the disease incidence of 62.5% and 100% on SP and CC, respectively. The pathogen was isolated from diseased roots of each plant and confirmed as *P. brassicae* based on morphological characteristics and PCR detection (Ren et al. 2014).

14.12 Isolation of Resting Spores

Plasmodiophora brassicae resting spores can be isolated according to the procedure of Castlebury et al. (1994). Approximately 2 g of fresh galled roots of canola cv. Westar or Chinese cabbage (*B.* *rapa* var. *pekinensis* or cv. Granaat were homogenized in a blender (Waring, Torrington, CT) in 50 ml of sterile distilled water (sdH_2O) for 2 min at high speed, and the resultant homogenate was filtered through eight layers of cheesecloth. The filtrate was centrifuged at 4 °C in 50-ml centrifuge tubes at 4000 g for 10 min in a swinging-bucket rotor. The supernatant was discarded and the pellet was resuspended in 5 ml of 50% sucrose and centrifuged again at 2000 g for 10 min. The resulting supernatant was transferred to a new 50-ml tube and diluted with 30 ml of sdH_2O before centrifuging at 4000 g for 10 min. The pellet obtained was resuspended in 5 ml of sdH_2O and centrifuged as above to remove any remaining sucrose. The supernatant was removed and the resting spore pellet resuspended in 5 ml of sdH_2O. The resting spores then were purified using a continuous gradient of LUDOX (HS-40 colloidal silica, 40 wt% suspensions in water, as described by Castlebury et al. (1994). The purified *P. brassicae* resting spore suspension normally was stored at 4 °C for less than 48 h before DNA extraction. Spore concentrations were quantified using a hemocytometer (Cao et al. 2007).

14.12.1 Inoculum Preparation of Spores

The resting spores for plant inoculation were prepared as described by Feng et al. (2013a). Galls were homogenized in 10% (w/v) sucrose in a blender. The slurry was passed through eight layers of cheese cloth and suspension was centrifuged at 50 g for 5 min. After transfer into a new tube, the supernatant was centrifuged at 2000 g for 5 min. The upper layer of the resultant pellet was suspended in 5 ml of water by gentle pipetting and transferred into new tube containing 40 ml of water. After centrifugation at 2000 g for 5 min the supernatant was removed. The resting spores in the pellet were resuspended in 2 ml of water and spore concentration was adjusted to 1×10^7 spores/ml. The resting spores were surface disinfected with colistin sulfate at 1 μg/ml and vancomycin hydrochloride at 1 μg/ml at 4 °C in the dark for 24 h. The spores were used as inoculum after removal of the antibiotics by washing twice with 40 ml of distilled water. For DNA extraction, the resting spores were prepared with the same protocol except that after overnight treatment with antibiotics at 4 °C, the spore suspension were further treated at 37 °C with 10 units/ml of DNase I for 2 h to remove genomic DNA from the host plant or other organisms released into the suspension during homogenization of galls (Zhang et al. 2015).

14.12.2 Resting Spore Purification and Single Spore Isolation

Currently, there is not a standard resting spore purification protocol that satisfies the requirement for molecular studies. The routinely used protocols as described by Bryngelsson et al. (1988) and Asano et al. (1999) are sufficient in quality for inoculation purposes, but not for DNA or RNA extraction when contamination from other organisms needs to be excluded. Similarly, although there are several reports on the isolation of single-spores of *P. brassicae* (Xue et al. 2008), the techniques described are not capable of producing sufficient volumes of material. Therefore, while these techniques may be suitable for many purposes, further improvements are needed to develop protocols that would enable the purification and recovery of much larger numbers of single resting spores, which would increase efficiency and facilitate molecular studies (Feng et al. 2014).

(i) Isolation of Zoospores

A protocol to isolate secondary zoospores was reported by Feng et al. (2012). By washing the roots of infected canola seedlings 7 days after inoculation, the secondary zoospores were isolated without contamination of resting spores and demonstrated that the collected secondary zoospores could initiate infection. This technique allowed investigation of the infection processes of primary and secondary zoospores separately, and will be useful in genetic studies in which resistance can be partitioned into different types based on the infection stage at which it is expressed. This study has stimulated several additional studies to investigate *P. brassicae* gene expression (Feng et al. 2013a) and the plant responses to primary and secondary infection (Deora et al. 2012; Feng et al. 2013c).

(ii) *In Vitro* Dual Culture

Bulman et al. (2011) standardized an in vitro dual culture system to study the genomics of *P. brassicae* and related protist species. In the absence of exogenous plant growth regulators, stable, long-term *B. rapa* cell callus cultures infected by *P. brassicae* were produced and demonstrated to be an excellent starting material for gene discovery. This technique can also be used on a related plasmodiophorid species, *Spongospora subterranean* (Wallr.) Lagerh., Feng et al. (2014) constructed a pilot scale DNA library and found that almost all of the DNA clones were from the protist rather than the plant host. This technique provides an avenue to remove plant contaminants from the pathogen materials, which is critical in genomics studies.

14.12.3 Effect of Root Exudates and pH on Resting Spore Germination

The various steps used to test the germination of resting spores are as follows:

(i) Extraction and Surface Sterilization of Resting Spores

Resting spores were extracted from fully-grown and decaying canola root galls collected from clubroot-infested fields near Edmonton and Leduc, Alberta, Canada and stored at 20 °C. Within 2–4 weeks, the frozen galls were washed under running tap water to remove soil particles. The outer layer of the gall was removed with a knife, and the remaining part of the gall was surface sterilized in 70% ethanol for 1 min, and then with 1% sodium hypochlorite (NaOCl) for 20 min. Five grams of the surface sterilized gall tissue was rinsed three times with sdH_2O, aseptically cut into small pieces and homogenized in 50 ml sdH_2O with a blender. The homogenate was filtered through eight layers of sterile cheese cloth, and the filtrate was centrifuged at

4000 g for 10 min. The resulting spore pellets were washed five times with sdH_2O and the spores were then surface sterilized according to the procedure of Asano et al. (2000). The spore suspension was adjusted to a final concentration of 6×10^7 spores/ml with sdH_2O.

(ii) Preparation of Root Exudates

Six hundred seeds from each of canola, Chinese cabbage and perennial ryegrass were surface sterilized with 1% sodium hypochlorite for 3 min and then rinsed five times with sdH_2O. The seeds were incubated for 1 week on moistened, sterilized filter paper in 9-cm diameter Petri dishes (at a density of 20 seeds per dish) to allow them to germinate. The roots of the seedlings (2–3 cm) were surface sterilized in a streptomycin sulfate solution (50 ppm) for 5 min, and rinsed in sdH_2O three times. Sixty seedlings of each plant species were transferred to 9-cm diameter Petri dishes containing 20 ml of 1:5 (v/v) dilution of modified Hoagland's solution (MacFarlane 1970) herein referred to as "nutrient solution, NS" and arranged so that the seedlings remained in an upright position. The Petri dishes were carefully transferred into 11.4 L clear plastic containers (41.9 × 29.2 × 15.9 cm), which contained several layers of water-soaked sterilized paper towel at the bottom to maintain a humid environment. The root exudates prepared from each plant species was collected after two weeks with a syringe, subjected to filter-sterilization (0.2 mm) along with the nutrient solution. The concentration of the root exudate preparation was standardized on an ml per root fresh weight (g) basis. The standardized exudate preparation was stored in a 50 ml centrifuge tube at 4 °C and used the next day.

(iii) Root Exudate Treatment

Germination of resting spores after treatment with root exudates from canola, Chinese cabbage, and perennial ryegrass was evaluated by incubating 0.2 ml of spore suspension with 1.5 ml of each root exudate preparation in 2 ml centrifuge tubes at 25 °C in the darkness. The germination medium also included 0.1 mg/ml of cefotaxime sodium, a broad-spectrum bacterial antibiotic to prevent bacterial growth (Asano et al. 2000). Resting spore germination was assessed every 24 h over a 6-day period. Control treatments consisted of 0.2 ml of spore suspension incubated in 1.5 ml of 1:5 dilution of nutrient solution or 1.5 ml sdH_2O. At each time point, 60 ml of each spore/root exudates suspension was fixed in Aceto-Orcein stain, and three slides were prepared from each replicate of each treatment. On each slide, 100 resting spores were examined along a transect under the microscope. Resting spores were considered to have germinated if they appeared empty and remained unstained by Aceto-Orcein, and could be distinguished from red-stained non-germinated spores (Naiki and Dixon 1987). After 6 days of incubation, aggregation of resting spores made spore counting difficult, so the experiment ended. The experiment was performed using a randomized complete block design with eight-replicate tubes prepared for each treatment (exudates and controls) for each time point, and was repeated once. The data from the two trials performed were combined for statistical analysis, since preliminary analysis of variance did not show any interactions between trials and treatments.

(iv) Chemical Analysis of Root Exudates

Root exudate preparations from each plant species, the nutrient solution control, and the solvent control (sdH_2O) were subjected to chemical analysis following Friberg et al. (2005). The concentrations of nitrogen (N), potassium (P), sulfur (S), calcium (Ca), magnesium (Mg), and boron (B) as well as pH values were analyzed using the Inductively Coupled Plasma (ICP) Method.

(v) pH Treatment

For pH treatment, 25 ml aliquots of either canola root exudates prepared in nutrient solution or nutrient solution alone were separately dispensed into each of nine sterilized tubes for adjustment of pH levels using a combination of 1M Trizma acid (Tris hydroxymethyl aminomethane HCl) and 1M Trizma-base with a final Tris concentration of 25 mM. The pH levels were adjusted to 5.0, 5.5, 6.0, 6.5, 7.0, 7.5, 8.0, 8.5, and 9.0. A 250 μl aliquot of freshly prepared resting spore stock suspension (6×10^7 spores/ml^{-1}) was added to each of the nine tubes and thoroughly mixed. This final resting spore suspension was distributed into 15 replicate Eppendorf tubes with 0.7 ml tube^{-1} for each pH treatment and incubated at 25 °C in darkness. Three samples were collected at each sampling time (2, 3, 4, 5, and 6 days after treatment) from each pH treatment. Each sample was treated with 20 ml of Aceto-Orcein and stored at 4 °C until used for microscopic determination of germination (Rashid et al. 2013).

14.13 Pathogen Detection at Different Stages of Disease Development

To assess the ability to detect *P. brassicae* infection prior to the development of visible symptoms, 7-days-old canola cv. Westar seedlings were germinated on filter paper in Petri dishes and inoculated by dipping the roots in a resting spore suspension (1×10^7 spores/ml) as previously described (Strelkov et al. 2007). Control plants were dipped in sdH_2O. Seedlings then were planted in 4 × 4-cm plastic pots filled with Metro-Mix soil at a rate of one seedling per pot and placed in a greenhouse at 22 + 2 °C with a 16-h photoperiod (natural light supplemented with artificial lighting). To ensure infection, the soil was kept saturated with water for the first week after inoculation, and then fertilized and watered as required. Seedlings were harvested at 3, 10, 24, and 36 days after inoculation and the roots were washed in H_2O and examined for symptom development. Total DNA then was extracted from the tissue according to the protocols of Rogers and Bendich (1985, 1995) and used as a template in the PCR protocol described by Cao et al. (2007) (see Sect. 14.11).

14.14 Detection of Seed Infestation

Using quantitative PCR, DNA of *Plasmodiophora brassicae*, the causal agent of clubroot, was detected and quantified on canola, pea, and wheat seeds, as well as on potato tubers, all harvested from clubroot-infested fields in Alberta, Canada. Quantifiable levels of infestation were found on seven of the 46 samples analyzed, and ranged from 1.0×10^3 to 3.4×10^4 resting spores per 10 g seeds, the vast majority (80–100%) of resting spores on these samples were viable, as determined by Evan's blue vital staining. However, the levels of infestation found were generally lower than that required to cause consistent clubroot symptoms in greenhouse plant bioassays. While the occurrence of *P. brassicae* resting spores on seeds and tubers harvested from clubroot-infested fields suggests that seed borne dissemination of this pathogen is possible, practices such as commercial seed cleaning may be sufficient to effectively mitigate this risk (Rennie et al. 2011).

14.15 Detection of *Plasmodiophora brassicae* Using Loop Mediated Isothermal DNA Amplification

Loop mediated isothermal DNA amplification (LAMP) technique has been developed to detect *P. brassicae*. The set of three primer pairs was designed using LAMP software. The detection was performed with the GspSSD polymerase, isolated from bacteria *Geobacillus* sp., with strand displacement activity. DNA extraction from clubbed roots obtained from farmers' fields of oilseed rape infected by *P. brassicae* was done using a modified CTAB method. The reaction was performed for 60 min at 62 °C. The visual detection was done using CF X 96 Real-Time PCR Detection System or Gerie II. The detection with LAMP proved its usefulness, it was easy, fast, and accurate and independent of plant age. The detection limit was 5 spores per 1 μl of the spore suspension, so LAMP was less sensitive than quantitative PCR tests. However, the method is cheap and simple, so, it is a good alternative, when it comes to practical use and the assessment of numerous samples (Kaczmarek et al. 2014).

14.16 A Real-Time PCR Assay to Detect *Plasmodiophora brassicae* Under Field Conditions

A protocol using real-time polymerase chain reaction (PCR) for the direct detection and quantification of *Plasmodiophora brassicae* in soil samples has been developed and used on naturally and artificially infested soil samples containing different concentrations of *P. brassicae* under Swedish soil samples collected from seven fields across three locations. Species-specific primers and a TaqMan fluorogenic probe were designed to amplify a small region of *P. brassicae* ribosomal DNA. Total genomic DNA was extracted and purified from soil samples using commercial kits. The amount of pathogen DNA was quantified using a standard curve generated by including reactions containing different amounts of a plasmid carrying the

P. brassicae target sequence. The PCR assay was optimized to give high amplification efficiency and three to four copies of the target DNA sequence were detected. Regression analysis showed that the standard curve was linear over at least six orders of magnitude ($R^2 > 0.99$) and that the amplification efficiency was >92%. The detection limit in soil samples corresponded to 500 resting spores g^{-1} soil. The inter sample reproducibility was similar to, or higher than, that of assays for other pathogens quantified in soil samples. Bait plants were used to validate the real-time PCR assay. The protocol developed was used to investigate the spatial distribution of *P. brassicae* DNA in different fields and a significant difference was found between in-field sampling points. The reproducibility of soil sampling was evaluated and showed no significant differences for samples with low levels of inoculum, whereas at higher levels differences occurred. Indicator kriging was used for mapping the probability of detecting *P. brassicae* within a 2-ha area of a field. A threshold level of 5 fg plasmid DNA g^{-1} soil, corresponding to approximately 3×10^3 *P. brassicae* resting spores g^{-1} soil has been suggested for growing resistant cultivars. These results provide a robust and reliable technique for predicting the risk of disease development and for assessing the distribution of disease within fields (Wallenhammar et al. 2012).

In China: SYBR Green I real-time PCR assay has been developed to detect and quantify *Plasmodiophora brassicae* ribosomal DNA (rDNA) and internal transcribed spacer (ITS). A pair of primers PBF1/PBR1 was designed based on the conservative region of rDNA-ITS of *P. brassicae*. The positive plasmid pB12 was obtained and used as the template to create standard curve. The specificity, sensitivity, and reproducibility of real-time PCR were evaluated, respectively. Naturally and artificially infested soil samples containing different concentrations of *P. brassicae* were detected. The results demonstrated that standard curve established by recombinant plasmid had shown a fine linear relationship between threshold cycle and template concentration. The melting curve was specific with the correlation coefficient of 0.995, and the amplification efficiency was 93.8%. The detection limit of *P. brassicae* genomic DNA was approximately 40 copies per 25 μl. The sensitivity of the assay was at least 100-fold higher than conventional PCR. Only DNA from *P. brassicae* could be amplified and detected using this assay, suggesting the high specificity of this assay. The coefficient of variation was less than 3%, indicating that the PCR method revealed high reproducibility. The detection limit in soil samples corresponded to 1 000 resting spores g^{-1} soil. Bait plants were used to validate the real-time PCR assay. This developed real-time PCR assay allows for fast and sensitive detection of *P. brassicae* in soil and should be useful in disease management and pathogen interception so as to prevent further spread of *P. brassicae* (Jin-ping et al. 2013).

14.16.1 Bioassay Technique of Wallenhammar et al. (2012)

Bait plants were used to validate the real-time PCR assay developed. The level of *P. brassicae* in naturally infested soil samples was assessed using a bioassay method previously developed by Wallenhammar (1996) and modified by Wallenhammar et al. (2000). The pots used were 90 × 90 × 95 mm^3 (vefi). Infected plants were scored according to the following classes: 0 = no galls; 1 = enlarged lateral roots; 2 = enlarged tap root; 3 = enlarged napiform tap root; 4 = enlarged napiform tap root, lateral roots healthy; and 5 = enlarged napiform tap root, lateral roots infected. A disease severity index (DSI) was calculated according to the equation:

$$\mathrm{DSI} = \frac{\sum(N1 \times 1)}{(N \times 4)} \times 100\%$$

where *N*, class no.1, no. of plants in each class; *N* total no. of plants

Samples of approximately 500 g were withdrawn for further DNA detection and dried stored at room temperature until extraction.

(i) DNA Extraction from Infected Roots, Fungal Material, and Plant Material

A modified E.Z.N.A. SP Plant DNA Miniprep Kit extraction method was used to extract DNA from plant and fungal material. A total of 30–130 mg of fresh mycelia, apothecia or infected root tissue was incubated with two to six stainless steel beads (2–3 mm in diameter) in 1.5-ml cetyltrimethyl ammonium bromide (CTAB) solution (20 g CTAB, 12.11 g Tris, 81.8 g NaCl, and 7.4 g Na_2-ethylene diamine tetraacetic acid (EDTA) L^{-1}; pH = 8.0) and 20 μl proteinase K (20 mg ml^{-1}, GeneScan) at 60 °C for 2 h after 30 s shaking at speed setting 4.5–5.5 using a Hybaid Ribolyser (Hybaid). The suspension was then centrifuged at >10,000 g for 5 min and 1 ml of the supernatant was mixed with 5μl RNase A in a 1.5-ml tube and incubated at 60 °C for 15 min. After centrifugation at 20,000 g for 1 min, 600 μl supernatant were mixed with 210 μl SP2 buffer and incubated in a refrigerator for 5 min. The suspension was centrifuged at >10,000 g for 10 min, and a 400-μl portion of the supernatant was transferred to a 1.5-ml tube and mixed with 600 μl SP3 buffer. The supernatant was then transferred to a HiBind DNA Mini Column, washed and eluted in 100μl elution buffer according to the manufacturer's protocol. All DNA solutions were stored at −20 °C.

(ii) DNA Extraction from Soil

Soil samples were air dried at room temperature (20 °C), plant material was removed by hand and the soil was pulverized using a mallet. DNA was extracted from 350-mg soil samples (two replicates per sample) using the Fast DNA SPIN Kit for Soil according to the manufacturer's instructions with the following modifications. After cell lysis for 30 s at speed setting 5.5, samples were centrifuged at 14,000 g for 5 min. At the 10th step, 600 μl of the supernatant were discarded and 600 μl of the resuspended mixture transferred to a SPIN Filter. DNA was eluted in 100μl DES. Further purification of the eluate was performed using the Wizard DNA Clean-Up System and Illustra Micro Spin S-300 HR Columns according to the manufacturers' instructions with the following modifications. In the Promega kit, 900 μl of the resin was used and incubated for 5 min at 60 °C together with the DNA solution. DNA

Table 14.1 Primer and probe sequences used for the amplification and detection of the *Plasmodiophora brassicae* target sequence (Wallenhammar et al. 2012)

Primer/probe	Sequence[a]
Primer forward (PbF)	5Φ-AAA CAA CGA GTC AGC TTG AAT GC-3Φ
Primer reverse (PbR)	5Φ-TTC GCG CAC AAG CAC TTG-3Φ
Probe (PbP)	5Φ-CGC GCC ATG CGA CAC TGT TAA ATT G-3Φ

[a]Bases 4–23 of PbF are the same bases as in primer PbITS6 (Faggian et al. 1999a, b)

was eluted in 100 µl. The Micro Spin S-300 HR Columns were centrifuged for 2 min before adding the DNA eluate.

(iii) Primers and Probes

Primers and a probe with putative specificity to *P. brassicae* as mentioned below were designed using a ribosomal RNA gene sequence available in GenBank (accession number AF231027) and amplified a 103-bp fragment. The forward primer was based on the sequence of the forward primer used by Faggian et al. (1999a, b). The 5′ terminal reporter dye of the probe was VIC and the 3′ quencher was TAMRA. Primer and probe sequences were checked for identity against other known sequences using the BLASTN program (http://www. ncbi.nlm. nih. gov/blast/ Blast.cgi). Primers were purchased from MWG-Biotech.

(iv) Real-Time PCR

Real-time quantitative PCR was performed using the 7300 Real-Time PCR System in a total volume of 25 µl. The reaction mixture included 1 × Taq Man Universal PCR Master Mix No AmpErase UNG, 0.12 µM PbF, 0.6 µM PbR, 0.2 µM PbP and 5 µl template DNA. Soil DNA extracts was diluted fivefold before analysis and were analyzed in duplicate. The thermal cycling conditions consisted of an initial denaturation for 10 min at 95 °C followed by 45 cycles at 95 °C for 15 s and finally 59 °C for 60 s (Table 14.1).

(v) Quantification

Since *P. brassicae* is an obligate biotroph, it is difficult and laborious to extract pure genomic DNA from spores to use as a standard. Therefore, the amount of pathogen DNA was quantified using a standard curve generated by including reactions containing different amounts of a plasmid carrying the *P. brassicae* target sequence. The target DNA used to generate the calibration curves was a 607-bp sequence, amplified using a universal forward primer [NS7: 5′- to fg (10^{-15} g)] plasmid DNA g^{-1} soil by taking into account that 350 mg soil were used for the DNA extraction and that 1 µl of the original DNA extract was analysed. By analyzing the series of artificially infested soil samples containing different amounts of *P. brassicae* resting spores, it was possible to convert fg plasmid DNAg^{-1} soil to number of *P. brassicae* resting spores g^{-1} soil. Amplification efficiency (E) was calculated as a percentage using the equation:

$$E = 100 \times (10 - 1/k - 1)$$

where k is the slope of the equation describing the plot of Ct versus the logarithm of the DNA amount. GAGGCAATAACAGGTCTGTGATGC-3′) (White et al. 1990) and a *P. brassicae*-specific reverse primer (Pb3R: 5′-ACCATACCCAGGGCGATTG-3′), which was cloned into pGEM-T vectors according to the manufacturer's instructions. Plasmid DNA was prepared using a QI Aprep Spin Mini prep Kit, and the DNA concentration was determined spectrophotometrically. A tenfold dilution series of the plasmid standard was analyzed in triplicate using the real-time PCR assay and Ct values were then plotted against the amount of plasmid DNA to create a standard curve. Soil samples with low levels of *P. brassicae* DNA outside the range of the standard curve were quantified by extrapolating the standard curve to facilitate analysis of the results. The starting quantity of DNA in soil samples was transformed from ag (10^{-18} g) plasmid DNA.

(vi) Evaluation of Inhibition and Repeatability

Inhibition of the PCR reaction by contaminants in the DNA extracts was assessed for the soil samples analyzed in this study by analyzing a separate PCR reaction containing the soil DNA extract and 2 μl of the *P. brassicae* plasmid standard containing approximately 150 gene copies. The same amount of plasmid standard was added to a water control. The percentage inhibition effect (I) was calculated using the equation:

$$\mathrm{I} = 100 \times \frac{\{(Q\mathrm{s} + \mathrm{sp}) - (Q\mathrm{s} + Q\mathrm{sp})\}}{Q\mathrm{s} + Q\mathrm{sp}}$$

where Q_s is the quantity of the unknown sample, Q_{sp} is the quantity of the standard spiked in water, and $Q_{\mathrm{s+sp}}$ is the quantity of the unknown sample spiked with the standard. The inter sample repeatability was evaluated on separate extractions of 10 soil subsamples quantified in a single real-time PCR assay.

(vii) Distribution of *P. brassicae* Inoculum and Risk Assessment

Indicator kriging was used for mapping the probability of detecting *P. brassicae* within a 2-ha area of field D at farm 2 (Gotene). This method is a nonlinear form of ordinary kriging in which the original data are transformed to a binary scale (0–1) (Burrough and McDonnell 1998). This approach, which has been applied frequently in soil mapping (Van Meirvenne and Goovaerts 2001; Lark and Ferguson 2004), is appropriate when the goal is to produce an estimate of the probability that a given threshold is exceeded. In this case, observations with a data value (z) of less than or equal to a specified threshold (T) (here T was set to 5.0 fg plasmid DNA g^{-1} soil) were assigned a value of 0. Other observations were given a value of 1. The transformed data were then used in variogram analysis and ordinary kriging (Webster and Oliver 1992; Burrough and McDonnell 1998). The map displays the probability of T being exceeded. The software GS+ (Gamma Design) was used for the geostatistical analysis and ARCGIS (ESRI) was used for mapping (Wallenhammar et al. 2012).

14.16.2 Bioassay Technique of Jin-ping et al. (2013)

Bait plants *Brassica pekinensis* (Lour.) Rupr. variety Zhongbai were used to validate the developed real-time PCR assay method. The level of *P. brassicae* in naturally infested soil samples was assessed using a bioassay method previously developed by Wallenhammar (1996) and modified by Wallenhammar et al. (2000). The pots used were 40 × 40 × 40 mm (vefi). Infected plants were scored according to the following classes: 0 = no galls; 1 = enlarged lateral roots; 2 = enlarged tap root; 3 = enlarged napiform tap root; 4 = enlarged napiform tap root, lateral roots healthy; and 5 = enlarged napiform tap root, lateral roots infected. A disease severity index (DSI) was calculated according to the equation:

$$\mathrm{DSI} = \frac{\sum(N1 \times 1)}{(N \times 4)} \times 100\%$$

where *N* class no.1, no. of plants in each class; *N* total no. of plants

(i) DNA Extraction from Infected Roots, Fungal Material, Bacterial Material, and Soil

All the mycelia, apothecia were incubated in solid media, and then transferred into liquid media for 7 days at 28 °C. A modified CTAB method was used to extract DNA from infected root tissue, fungal, and bacterial material. Frozen mycelium was placed in 1.5 ml micro tubes, 600 μl of CTAB extraction buffer (2% CTAB, 0.3 mol L^{-1} NaCl, 1 mol L^{-1} Tris-HCl, 0.5 mol L^{-1} EDTA (pH 8.0)) was added, and the tubes were vortexed and then incubated at 65 °C for 30 min. Chloroform-isoamyl alcohol (24:1 mixture of chloroform and isoamyl alcohol) (1 volume) was added to each tube, and the tubes were mixed and centrifuged for 10 min at 13,000 g. The aqueous phase was transferred to a new tube, and the chloroform extraction was repeated. DNA was precipitated overnight at −20 °C after the addition of 0.1 volume of 3 mol L^{-1} sodium acetate (pH 8.0) and 2 volumes of cold 100% ethanol. The supernatant was discarded, and the pellets were washed with 70% ethanol and then dried by vacuum centrifugation. DNA was resuspended in 80 μl of TE (10 mmol L^{-1} Tris-HCl, 0.1 mmol L^{-1} EDTA (pH 8.0)) and then diluted 1:100 for use in PCRs in TE. The internal transcribed spacer (ITS) region of *P. brassicae* was based on GenBank (Y12831) and used for designing the primer PBF1 (5′-GCTCCTGCGTGTCGCTGTTATT-3′) and PBR1 (5′-CTGCGTTCTTCATCGTTGTGA-3′). Soil samples were air dried at room temperature (20 ± 2) °C, plant material was removed by hand and the soil was pulverized using a mallet. DNA was extracted twice independently from 200 mg soil samples using the DNA SPIN Kit according to the manufacturer's instructions.

(ii) Real-time PCR Assay

DNA concentration was determined spectrophotometrically. A tenfold dilution series of the plasmid standard was analyzed in triplicate using the real-time PCR assay. Each reaction of 25 μl contained 10 ng of plasmid pB12, 0.5 μl of each 15 μmol L^{-1} primer (PBF1 and PBR1), 12.5 μl of 2× SYBR Green master mix, and 0.4 μl of a 1000× diluted reference dye. The PCR reaction was set up in triplicate for each DNA sample. Prior to amplification, an initial denaturation step was performed (95 °C for 3 min) ensuring complete denaturation of the DNA and activation of the *Taq* polymerase. The reactions were performed in a iCyeler iQ5 Sequence Detection System. This was followed by 40 cycles of 30 s at 95 °C, 30 s at 60 °C, and 35 s at 72 °C. Fluorescence was detected after each cycle. Ct values were then plotted against the amount of plasmid DNA to create a standard curve.

(iii) Sensitivity and Reproducibility of Real-Time PCR

The tenfold serial dilutions of the *P. brassicae* genomic DNA were made to measure the sensitivity of the assay. Every dilution repeated five times. Real-time PCR was performed using the optimal reaction system. Conventional PCR amplification was performed as control. The inter sample repeatability was evaluated on 10 separate soil DNA extracts of *P. brassicae* from one soil samples analyzed in the same real-time PCR. Standard deviation (SD) of the Ct values was determined based on the 10 replicates for each amount of target DNA.

(iv) The Detection of Artificially and Naturally Infested Soils

The artificially and naturally infested soil samples were detected using the optimal real-time PCR reaction system. The artificially infested soils sample that contained 1.0×10^8, 1.0×10^7, 1.0×10^6, 1.0×10^5, 1.0×10^4, 1.0×10^3, 5.0×10^2, 10, and 0 resting spores g^{-1} soil (DW) were measured to validate the sensitivity of the real-time PCR and determine the threshold level of clubroot. The total of 160 soil samples of wide type and the sterilized soil samples from lab were detected using real-time PCR and conventional PCR. The thermal cycling parameters of conventional PCR were initial denaturation at 94 °C for 3 min followed by 40 cycles consisting of denaturation at 94 °C for 1 min, annealing at 56 °C for 1 min, and extension at 72 °C for 1.5 min. A final extension at 72 °C for 7 min was done at the end of the amplification. Negative controls (no DNA template) were used in every experiment to test for the presence of contamination in reagents (Jin-ping et al. 2013).

14.16.3 PCR and Quantitative PCR

Primers for the identification of the pathotypes –specific gene(S) were derived from Feng et al. (2013b) with other primers used. The genomic locations of the primers

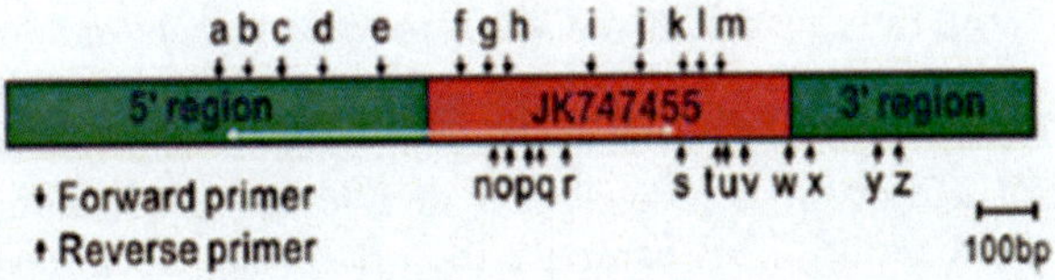

Fig. 14.2 Diagram of the genomic region of the *Cr 811* gene, JK 747455 GenBank accessions number of the expressed sequence tag that is a part of *Cr811* "5" region and "3" region. Sequences obtained by genome walking a to z primers whose sequences are linked. The white line denotes the sequenced cDNA (Zhang et al. 2015)

are illustrated in Fig. 14.2. PCR analyzes were conducted in an Eppendorf Mater cycler thermal cycler. Except in the genome walking analysis, all PCR analyzes were run in a final volume of 20 μl in the Promega master mix with 0.25 μM of each primer and 100 ng in the template; for all primers, the annealing temperature was 52 °C and /or 56 °C, all elongation time was 45s, and the cycle number was 35. Quantitative PCR (qPCR) assays were conducted in a Step One Real-Time PCR System using the GoTaq 2- step RT qPCR System (Promega) for RNA analysis or the power SYBR Green PCR Master Mix (Life Technologies) for DNA analysis. The *P. brassicae Actin 1* gene (GenBank accession number AAR88383) was used for normalization. Primer efficiency was verified using standard curves generated from the purified PCR product of the targeted fragments. The quantity of *Cr811* relative to *Actin1*, at the DNA or RNA level was calculated via the delta-delta method, with relative expression or amount $= 2^{-(\Delta CT\ sample - \Delta CT\ control)}$. All qPCR amplification was performed in triplicate for each of three template preparations derived from different biological samples. Data from qPCR were analyzed for statistical significance using the general linear model (GLM) procedure of the Statistical Analysis System. All data were subjected to one-way analysis of variance and when appropriate, a multiple range test was conducted by Fisher's least significant difference (LSD) at $P \leq 0.01$ (Zhang et al. 2015).

14.16.4 Quantification, and Control Assessment of Clubroot with qPCR

Conventionally the detection of *P. brassicae* is carried out by gall observations and bait tests. With the advent of molecular techniques, PCR amplification of DNA has been used to aid detection and study the diversity of *P. brassicae* (Buhariwalla et al. 1995; Buhariwalla and Mithen 1995; Cao et al. 2007; Faggian et al. 1999a, b; Faggian and Strelkov 2009; Ito et al. 1997, 1999; Wallenhammar and Arwidsson 2001). Methods of *P. brassicae* PCR detection are often based on ribosomal RNA (rRNA) gene internal transcribed spacer regions (ITS) (Ito et al. 1999). Detection and quantification is central to investigating the control of *P. brassicae*. Wakeham et al. (2017) quantified *P. brassicae* spores using real-time PCR (qPCR) and investigated the effect of Limex (LimeX70) and conventional lime-based products on clubroot control. Limex is a form of $CaCO_3$ produced as a by-product of the sugar beet refining process (Harling 2007). Limex contains P_2O_5, MgO, and SO_3. The addition of calcium carbonate to clubroot contaminated land has been shown to improve control. However, large quantities of calcium carbonate are required to produce consistent results. This can raise the pH of the soil to levels which are inconsistent with the other crop production in the rotation (potatoes). The number of *P. brassicae* resting spores in soil was monitored in different control treatments and the subsequent effect on crop yield was ascertained.

14.16.5 Assessment of the Response of lime-Based Fertilizers with qPCR

Determining *P. brassicae* infection within a field prior to crop planting has long posed a problem for choosing appropriate control treatments. The options for control of this pathogen are limited, and are based on adjusting pH with soil amendments. The quantitative polymerase chain reaction (qPCR) was investigated for measurement of this pathogen in different control treatments. The qPCR was capable of reliably quantifying *P. brassicae* at levels greater than and including 10^3 resting spores/g soil. The assay was used to study the effect of lime-based products (LimeX) on the incidence of the clubroot pathogen in field trials with broccoli crops grown on contaminated land. The results showed that variation occurred in clubroot resting spore levels in treated and untreated plots during the crop growing period. In year one, there was a 96% decrease in spore load during the growth of the crop. Treatment with LimeX resulted in a greater marketable head weight of broccoli in 2 years of the field trials, and significantly reduced gall numbers on the roots in 1 year. The rate of lime (calcium carbonate) application was not found to have a significant effect, however, a greater reduction in clubroot was observed at higher LimeX concentrations (Wakeham et al. 2017).

14.16.6 PCR and Quantitative PCR Analysis of Infested Soil Samples

Total genomic DNA's extracted from all soil samples collected from infested fields with a Fast DNA Spin kit for soil as per the manufacturer's instructions. An aliquot of 250 mg of soil is used per sample, with three replicates. Infestation by *P. brassicae* is detected by conventional PCR analysis (Cao et al. 2007). The amplification products are run on 1% agarose gels stained with SYBR Safe (Invitrogen), and visualized under UV light for quantification.

Quantitative PCR analysis is performed on diluted DNA (1/10 v/v) with sdH_2O to avoid inhibition of the reaction. The quantitative PCR amplification is conducted (Rennie et al. 2011). The qPCR amplifications are carried out in a StepOnePlus Real Time PCR System in a 10 μl volume reaction (2.5 μl template DNA, 2.5 μl of 3.2 mM DR1F and DR1R primer mix), and 5 μl Dynamite qPCR Master mix. The reaction conditions consist of 95 °C for 2 min (initial denaturation step), followed by

35 cycles of 95 °C for 15s (denaturation step) and 60 °C for 60 s (annealing, extension, and fluorescence measurement step). A standard curve for the quantification of *P. brassicae* spores is generated with DNA isolated from known quantities of spores, which can be purified as per Castlebury et al. (1994) with modifications made by Cao et al. (2007). The 100 g of galled tissue and 500 ml of water are blended, and the homogenate passed through a 450 μm metallic sieve. The resting spores are isolated by centrifugation in a 50% sucrose gradient, quantified with a hemocytometer (VWR), and adjusted to 1×10^8 resting spores/ml. Total DNA for the standard curve is extracted, and serially diluted with sdH_2O at tenfold intervals down to a dilution of 1×10^2 resting spores/ml. DNA from each dilution in the series is used as a template in the qPCR analysis according to Rennie et al. (2011). The correlation between the amount of pathogen DNA detected, and the soil pH is determined for the used samples (Ricarova et al. 2016).

14.16.7 Detection of Seed Infestation Using qPCR

The details are given at Sect. 14.14 in this chapter.

14.17 Clubroot Identification and Detection Methods

The clubroot identification and detection methods can generally be divided into: biological/ pathological, symptomatic, microscopic, serologic, and use of modern molecular techniques.

14.17.1 Biological and Pathological Bioassay Method

One of the simplest and easiest methods is using a susceptible species to detect the *P. brassicae* presence in a soil sample. Susceptible plants are grown in the soil suspected of the *P. brassicae* presence in controlled conditions for 5–6 weeks. After this time plants can be examined for the presence of disease symptoms on roots in the form of galls. Chinese cabbage cv. Granaat is often used as susceptible plant, as it is a universally susceptible host of all *P. brassicae* pathotypes. This method is time-consuming as well as laborious and can be used only if the inoculum reaches a concentration of at least 1000 spores per 1 g of soil (Faggian and Strelkov 2009). The limitation can also be a lack of space in greenhouses. The bioassay results bring only information about the pathogen presence in the soil, but not about its quantity (Hwang et al. 2012).

14.17.2 Microscopy

Another diagnostic tool includes the microscopic examination of root hairs and root cortex for the presence of *P. brassicae* infection sites (MacFarlane 1952; Hwang et al. 2011b). The soil can also be examined microscopically with a fluorescent dye being used to aid the spore detection (Takahashi and Yamaguchi 1988). Ricarova et al. (2016) evolved the protocol for microscopic detection with two fluorophores, which can even detect the viability of spores. The first fluorophore–Calcoflor White M2R–binds to chitin in the cell walls of undamaged spores. The second fluorophore—ethidium bromide—penetrates damaged and nonviable spores. Then viable spores are fluorescing with blue color and nonviable ones with red. A correlation between the blue light intensity and index of disease (ID) was also confirmed (Takahashi and Yamaguchi 1988). This method is time-consuming and laborious again and also requires well-trained personnel. Staining can also be performed in clubs or on the whole tissue suspected of infection by *P. brassicae*. Different chemicals which bind to chitin are used. For viability determination, staining with Evans blue (Rennie et al. 2011) or orcein (Rod 1996) may be used. Nonviable spores absorb dye inside and viable spores have stained just cell walls (Naiki and Dixon 1987).

14.17.3 Serology

Serological tests are based on polyclonal antisera (Lange et al. 1989; Wakeham and White 1996). Tests based on dip-stick method, indirect ELISA, and immune fluorescences have achieved the sensitivity of 100 spores per 1 g of soil (Wakeham and White 1996). The methods based on polyclonal antisera bring results which can be affected by variability in antiserum specificity and sensitivity. Therefore the test based on monoclonal antiserum should be used, because it would be more specific and also more accessible (Faggian and Strelkov 2009)

14.17.4 Molecular Methods

Primers and PCR (polymerase chain reaction) techniques are well developed to amplify specific sections of the *P. brassicae* genome (Hwang et al. 2012). PCR brings quick and reliable results about pathogen presence or absence in different kinds of samples (Hwang et al. 2012). Designed primers amplify the ITS regions (section on ribosomes) (Chee et al. 1998; Faggian et al. 1999a, b; Wallenhammar and Arwidsson 2001; Cao et al. 2007), one primer pair is based on isopentyl transferase amplification (Ito et al. 1999). PCR protocol developed by Cao et al. (2007) is commonly used for commercial purposes for *P. brassicae* detection from plant tissue or from soil (Hwang et al. 2012). This is the first "one step" PCR protocol which can identify as little as 100 fg of pathogen DNA or even less in a sample, which corresponds to 1000 spores per 1 g of soil or 11% of disease index. Common PCR assay provides information about the presence/absence of pathogen in a sample (Hwang et al. 2012). On the other

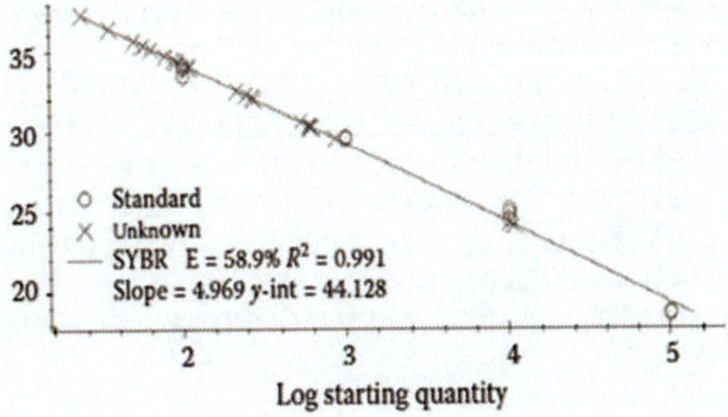

Fig. 14.3 Quantitative PCR standard curve of *Plasmodiophora brassicae* detection *x*-axis: amount of DNA, *y*-axis: Cq—quantification cycle; standards (circles): known amount of spores, daggers: unknown samples (Ricarova et al. 2016)

hand, the qPCR method (Q-RT PCR; quantitative polymerase chain reaction) can detect only the presence but also the amount of *P. brassicae* in a sample (Hwang et al. 2012). Several protocols have been developed (Sundelin et al. 2010; Rennie et al. 2011; Wallenhammar et al. 2012). The qPCR method is based on detection of fluorescent signal during the amplification of pathogen DNA. The protocol of Rennie et al. (2011) was developed for SYBR green signal. The number of spores is counted as the amount of DNA reported by machine in proportion to the amount of sample entering the reaction (Fig. 14.3). This method can detect 1000 spores and more per 1 g of soil with the level of precision in thousands (Rennie et al. 2011). For more details, see techniques given at 14.16–14.16.6.

14.18 Gene Expression (Transcriptomic) Analysis

RNASeq analysis was conducted for gene expression, RNA was prepared from hypocotyls or the upper 1 cm of control and infected plants (*A. thaliana* Cal 0) 16 and 26 DPI. Three independent biological replicates were used for the transcriptomic analysis and a second set of three biological replicates, grown at a different time, used for qRTPCR analysis. RNASeq was performed at the Applied Genomics and Analytical Technologies Dept., National Research Council of Canada using an Illumina Hi-Seq 2500 platform. Approximately 25 million reads were obtained from each sample. Paired-end reads were mapped using Stampy (Lunter and Goodson 2011) to the *Arabidopsis* TAIR10 genome (https://www.arabidopsis.org/) and *P. brassicae* genome. Over 99% of the reads matched one genome or the other. For the 16 DPI samples ~13% (3.5 million) of the reads matched the *P. brassicae* genome in both hypocotyl and root samples. For the 26 DPI samples, 48% of the reads (12 million) matched the *P. brassicae* genome from hypocotyl tissue and 34% (9 million) reads from root tissue. Reads matching predicted transcripts were counted using HTSeq Count (Anders et al. 2015) using the default "union" method and differential gene expression analyzed using DESeq2 (Love et al. 2014). Phylogenetic trees were constructed using MEGA6 (Tamura et al. 2013). Fold-change data were overlaid on metabolic pathways using Mapman (Usadel et al. 2009). Quantitative RT-PCR was performed as described in

Malinowski et al. (2012) using the following primers ARR4 (At1g10470) forward 5′-GCCTGGAATGACTGGTTATGAG-3′, reverse 5-′-AGAAATCTTGAGCACCTTCCTC-3′; ARR5, (At3g48100), forward 5-′-CTTCATTAGCATCACCGAAACTTC-3′, reverse 5′-AAGCTCTTGTCGCACTATCG-3′; ARR6, (At5g62920), forward 5′--GTCGGTTTTGAGGATTTGAAGG-3′, and reverse 5′-AGGCAAAATG-TTCTCGGAGG-3′, ARR7, (At1g19050), forward 5′--CGTAAAGTCATCGAGAGGTTGC-3′, reverse 5-′-TCCGTCACTATCAAATTCACCTTC-3′. All results were normalized against 18S rRNA expression.

A microarray analysis of control and infected ipt 1; 3; 5; 7 tissue was performed using control and infected root and hypocotyl tissue 16 DPI. RNA was prepared from duplicate samples and hybridized to Affymetrix Aragene-1-1-ST microarrays using the NASC International Affymetrix Service (http://www.affymetrix.arabidopsis.info/). Expression data contained in the CEL files provided from this service were normalized using the R package "oligo" with the RMA option (Carvalho and Irizarry 2010) and differential expression calculated using "limma" (Smyth 2005). Data are available from Array Express (www.ebi.ac.uk/arrayexpress) under the accessions E-MTAB-4201 (microarray) and E-MTAB-4176 (RNA-Seq) (Malinowski et al. 2016).

14.19 Quantification of Auxins and Cytokinins

The quantification of auxins and cytokinins protocols have been developed. For analysis of cytokinins, ~50 mg of fresh material was collected (hypocotyl +1 cm root) and frozen immediately in liquid nitrogen. Samples were extracted in modified Bielesky buffer (methanol/water/formic acid, 15/4/1, v/v/v; Hoyerova et al. 2006) using an internal standard of stable isotopically labelled cytokinins (0.5 pmol of CK bases, ribosides, N-glucosides, 1 pmol of O-glucosides and nucleotides). Material was extracted by bead beating in a Precellys tube containing 2 × 3 mm diameter tungsten carbide beads. Samples were beaten for 2 × 5 min at 27 Hz with cooling in between. All other extraction procedures followed Svacinova et al. (2012), including modifications described by Antoniadia et al. (2015). Auxin measurements were made from ~20 mg of material extracted in 50 mM Na-phosphate buffer pH 7 including 1% diethyl-dithiocarbamic acid with the same bead beating procedure as above according to Novak et al. (2012). All samples were analyzed by ultra-high performance liquid chromatography (UHPLC) coupled to a Xevo® TQ-S triple quadrupole mass spectrometer equipped with an electro spray interface (Malinowski et al. 2016).

14.20 Genetic Transformation

A protocol for the genetic transformation of *P. brassicae* has been recently reported (Feng et al. 2013b), in which a protoplast preparation was superseded with lithium acetate treatment and the selection step was omitted. With this protocol, the transformed germinating/germinated resting spores were used with two fungal expression vectors and then inoculated them on canola plants. Approximately 50% of the galls obtained contained resting spores from which transforming DNA could be detected by PCR analysis. Quantitative PCR and genome walking indicated that the transforming DNA was integrated into the *P. brassicae* genome. Transcript of the transforming gene was also detected by reverse transcription qPCR from selected transformants. Following another round of inoculation, the transforming DNA could still be identified from the resultant galls, indicating the stable inheritance of the transforming DNA. This protocol makes strain labeling and gene knockout studies possible in *P. brassicae* (Feng et al. 2014).

14.21 In Planta Quantification of *Plasmodiophora brassicae*

In planta quantification of *P. brassicae* by whole-cell fatty acid (WCFA) measurements and real-time polymerase chain reaction (PCR) has been developed by Sundelin et al. (2010). Arachidonic acid (ARA, 20:4) was the most abundant fatty acid in resting spores and was only found in infected roots, which indicates a potential of ARA as a biomarker for *P. brassicae*. A real-time PCR assay was developed using primers designed from the internal transcribed spacer region of the ribosomal DNA. Using these primers, it was possible to detect *P. brassicae* in infected roots 10 days after germination of plants sown in infested soil. A bioassay showed that the amounts of ARA found by WCFA analysis and the DNA found by real-time PCR in infected plants were well correlated. These measurements also correlated with the soil spore content and the assessed disease incidence and disease severity scores. WCFA analysis and real-time PCR are good tools for *P. brassicae* quantification that can be applied to basic studies of the pathogen and in resistance screens.

Neither biochemical nor molecular methods have been developed for quantification of *Plasmodiophora brassicae* in roots or soil during the entire life cycle of the pathogen. Such methods would make it possible to give precise and objective measures of *P. brassicae* root infection level and biomass estimates, which can be used in basic studies of the biology of the pathogen and when screening for plant resistance or during development of other control measures.

14.21.1 Preparation of Resting Spore Inoculum

Infected Chinese cabbage (*B. rapa* subsp. *pekinensis*) roots showing severe clubroot symptoms were collected from a Danish cabbage field heavily infested with *P. brassicae*, and the clubbed roots were subsequently washed and stored at −18 °C. The frozen roots were crushed, mixed with sterile water, blended for approximately 3 min, and filtered through gauze followed by filtration through 30-μm nylon mesh. The suspension was centrifuged at 2500 g for 5 min and the supernatant was discarded. Pellets were washed with sterile water and centrifuged three times as described above. Finally, pellets were resuspended in sterile water and the concentration of *P. brassicae* spores was determined using a hemocytometer. The spore suspensions were kept at 5 °C until use.

14.21.2 Soil Bioassay

Field soil was sterilized with dry heat at 110 °C for 24 h and mixed with quartz sand 1:1 (wt/wt). The following minerals were mixed into the soil mixture: NH_4NO_3 (86 mg kg^{-1} of soil), KH_2PO_4 (44 mg kg^{-1} of soil), K_2SO_4 (70 mg kg^{-1} of soil), $CaCl_2$ (70 mg kg^{-1} of soil), $CuSO_4 \times 5H_2O$ (2.2 mg kg^{-1} of soil), $ZnSO_4 \times 7H_2O$ (5 mg kg^{-1} soil), $MnSO_4 \times 7H_2O$ (10 mg kg^{-1} of soil), $CoSO_4 \times 7H_2O$ (0.33 mg kg^{-1} of soil), $NaMoO_4$ (0.2 mg kg^{-1} of soil), and $MgSO_4 \times 7H_2O$ (20 mg kg^{-1} of soil). Spore suspensions (100 ml) were mixed into 1-kg portions of the soil mixture and filled in 1600-ml pots. The following densities of *P. brassicae* resting spores per gram of soil were obtained: 0, 1.0×10^2, 2.0×10^2, 2.6×10^2, 7.8×10^2, 9.4×10^2, 1.6×10^3, 2.0×10^3, 4.7×10^3, 7.8×10^3, 1.0×10^4, 4.7×10^4, 1.4×10^5, and 1.0×10^6. Ten seeds of oilseed rape cv. Caracas were sown in each pot and, subsequently, the pots were watered to 75% of field capacity. Four replicates per treatment were randomized in four blocks and placed on a greenhouse bench. Natural daylight was supplemented with 16 h of light per day from Osram daylight lamps providing a minimum of 150 μmol of photo synthetically active radiation (PAR) m^{-2} s^{-1}. Minimum day and night temperatures were 20 and 16 °C, respectively. One week after germination, the number of plants was reduced to six plants per pot. In order to determine how early infection could be detected using real-time PCR, another bioassay was carried out. Quartz sand (1 kg) was inoculated with 200 ml of spore suspension (0, 5.0×10^3, and 1.0×10^6 spores/g of soil) and filled in 800-ml pots. Each spore concentration was made in four replications. Ten seeds of Chinese cabbage cv. Granaat were sown in each pot. The pots were placed in a growth chambers with 16 h of light (36W/950, 200 μmol PAR m^{-2} s^{-1}) and 8 h of darkness. Day and night temperatures were maintained at 20 and 15 °C, respectively. Roots were harvested in liquid N_2 at 10, 15, 25, and 35 days after sowing and stored at −80 °C. This experiment was carried out twice.

14.21.3 Disease Assessment and Fresh Weight Measurements

Oilseed rape plants were harvested and the roots were washed 35 days after sowing. The number of plants with clubroot symptoms was determined and DI was assessed for each replication. DSSs were assessed, using a scale with four classes modified after Buczacki et al. (1975) and Voorrips and Kanne (1997), where 0 = no symptoms, 1 = swollen main root or small sporadic attack with clubs <2 mm, 2 = clubs ≤6 mm, and 3 = clubs >6 mm or clubs account for more than half of the root system. Following disease assessment, roots were cut from shoots with a scalpel and their fresh weights were determined. Shoots were dried at 80 °C for 24 h before recording the shoot dry weight. Subsequently, roots were lyophilized for 24 h (0.5 h Pa, −85 °C) and pulverized in a steel mill. The plant material was stored at −20 °C.

14.21.4 WCFA Analysis of Resting Spores, and Roots

WCFA were extracted from the pulverized root samples (25 mg) from the soil bioassay described above ($n = 4$) and from resting spores used as *P. brassicae* inoculum ($n = 2$) employing a four-step fatty acid extraction procedure according to a method by Sasser (1990): (1) saponification, (2) methylation, (3) extraction, and (4) base wash. To enable quantification of the extracted fatty acid methyl esters, an internal standard, nonadecanoate fatty acid methyl ester, was added to each sample. Analyses of fatty acid methyl esters were performed using the software package Sherlock (version 6.0; MIDI Inc., Newark, DE) with an HP Chem station, and an HP6890 GC fitted with a 25-m fused silica capillary column (HP part no. 19091B-102) and hydrogen as carrier gas. The injector temperature was 250 °C, and the detector temperature was 300 °C. The temperature program was as follows: initial temperature of 170 °C increasing to 270 °C at 5 °C/min. In all, 2 μl per sample was injected. The MIDI software automatically controlled all gas chromatography operations, including calibration, subsequent sample sequencing, peak integration, and naming. Calibration standards contained a mixture of straight chain saturated and hydroxy fatty acid methyl esters with a length of 10–20 carbon (MIDI part mo. 1200A).

14.21.5 DNA Extraction

Genomic DNA was extracted from 100 mg of the lyophilized or frozen plant material mentioned above, using the DNeasy Plant Mini Kit following the manufacturer's instructions. Pure *P. brassicae* DNA used to prepare standard series in real-time PCR was extracted from resting spores prepared from clubbed roots as described above. The extracted spores were washed at least three times with sterile water. Streptomycin was added to a concentration of 100 μg/ml in order to limit bacterial growth. The isolated spores were treated with DNase I (100 U/ml) in 1× DNase I reaction buffer (10 mM Tris-HCl, 2.5 mM $MgCl_2$, and 0.5 mM $CaCl_2$) for 6 h at 37 °C to remove plant DNA contamination. The DNase I was heat inactivated at 75 °C for 10 min. The spores were washed with sterile water and lyophilized before DNA extraction using the DNeasy Plant Mini Kit (Qiagen). All DNA concentrations were measured using the NanoDrop ND-1000.

14.21.6 Real-Time PCR Analysis of Roots

The internal transcribed spacer (ITS) region of *P. brassicae* was assessed from GenBank (Y12831) and used for designing the primer Pb4-1

(TACCATACCCAGGGC GATT) using Primer3 (http://www.frodo.wi.mit.edu/primer3/). After design of the primer, it was checked by alignment that the primer sequence was found in all *P. brassicae* ITS sequences available in GenBank. This primer was used in combination with the primer PbITS6 (CAACGAGTCAGCTTGAATGC) (Faggian et al. 1999a, b). In combination, the two primers amplify a 139-bp fragment of the ITS region. PCR was performed using the real-time PCR system Mx3000P. Each reaction of 25 μl contained 10 ng of genomic DNA, 10 pmol of each primer, 12.5 μl of 2× SYBR Green master mix (Stratagene), and 0.4 μl of a 1000× diluted reference dye (Stratagene). The PCR reaction was set up in triplicate for each DNA sample. A standard series made with serial diluted pure *P. brassicae* DNA of known concentration was included on each plate. In addition, a similar standard series was used to determine the minimum amount of pure *P. brassicae* DNA that could be detected using real-time PCR. Prior to amplification, an initial denaturation step was performed (95 °C for 10 min) ensuring complete denaturation of the DNA and activation of the Taq polymerase. This was followed by 40 cycles of 15 s at 95 °C, 30s at 58 °C, and 30 s at 72 °C. Fluorescence was detected after each cycle. After amplification, a melting curve analysis was performed in order to ensure that only one PCR product was amplified and the PCR product was sequenced by MWG Biotech.

14.21.7 Data Analysis

Comparison of the WCFA composition of Chinese cabbage roots with and without *P. brassicae* infection was done by one-way analysis of variance of the individual fatty acids (arc sine transformed). Data on plant growth parameters (root and top weight), disease development (DI and DSS), and *P. brassicae* quantification (WCFA analysis and real-time PCR) were subjected to analysis of variance assuming a normal distribution. Treatment variances were stabilized by appropriate transformation of data, if necessary, and treatment means were separated by least significant difference values. Correlations between methods of disease assessment, quantification of *P. brassicae* DNA and ARA, and the plant growth parameters were tested using Kendall tau correlation. Hypotheses were rejected at $P \leq 0.05$. All data were analyzed by PC-SAS (Sundelin et al. 2010).

14.22 Root-Hair Infection Test

14.22.1 Sand Culture Method

Experiments were conducted using coarse sand. The sand was washed five times with water, sterilized by autoclaving, and dried at room temperature on a lab bench. Plastic cups of 50 ml volume were filled with sterilized sand and then each cup was supplemented with 20 ml of nutrient solution that had been adjusted to different pH values: 5.0, 5.5, 6.0, 6.5, 7.0, 7.5, 8.0, 8.5, and 9.0. These nutrient solutions with different pH values were either mixed with resting spores (1×10^8 spores ml^{-1}) prepared as described earlier or used without resting spores as a control for each pH treatment. The cups were supplemented with equal volumes of nutrient solution at the respective pH levels during the course of the experiment to maintain a high level of moisture for each treatment. Each replicate cup was transplanted with five seedlings of a clubroot susceptible canola (cv. 45H26) and left to grow under greenhouse conditions (16 h photoperiod at ~22°/16 °C day/night temperature, and 125μmol m^{-2} s^{-1} light intensity) using a split-plot design with five replicates. After twelve days, plants were uprooted following measurement of height, roots, cleaned in running water, fixed in FAA (formalin aceto-alcohol; 5 ml formaldehyde, 5 ml acetic acid, and 90 ml 50% ethanol) solution, and stored at room temperature.

14.22.2 Soil Solution Method

Experiments were conducted using soil collected from a naturally infested clubroot field near Edmonton, Alberta, Canada. The population ("CDCN04-1") of *P. brassicae* in this soil was previously described as pathotype 5 or ECD (European Clubroot Differential) 16/15/0 based on the differentials of Williams (1966) or Buczacki et al. (1975), respectively (Strelkov et al. 2006). Soil was thoroughly mixed with Sunshine Mix SS in ratios of 1:3 and 1:5 (v/v). These two soil mixtures were separately used to fill 450-ml white plastic cups containing four perforations at the bottom. The pH solutions were prepared by proportionately mixing a 50 mM Tris-acid and a 50 mM Tris-base solution following the Sigma buffer mixing table to create solutions with pH levels of 5.0, 5.5, 6.0, 6.5, 7.0, 7.5, 8.0, 8.5, and 9.0. The experiments were conducted under greenhouse conditions following a split-plot design with inoculum concentration as the main plot and pH treatments as the sub-plots, and repeated once. Twelve cups (replicates) were used for each pH treatment. Each cup was seeded with five susceptible canola seeds (cv. 45H26). The cups were placed in trays containing respective pH solutions and maintained water-soaked throughout the experimental period. Seedling emergence, root-hair infection, plant height, and disease severity were recorded, respectively, at 5, 10, 6.20 (measured at two different dates), and 30 days after seeding (DAS). To examine root-hair infection, six cups from each pH treatment were randomly collected, plants from each cup were carefully separated and cleaned under running tap water, and their roots were fixed in FAA solution. A visual assessment of disease severity on canola plants was conducted on the remaining six cups using a 0–3 scale (0 ¼ no gall formation, 1 ¼ a few small galls, 2 ¼ moderate galling, 3 ¼ severe gall formation) following Kuginuki et al. (1999). The experiment was repeated once.

14.22.3 Counting of Root-Hair Infection

Root-hair infection was determined following the procedure described previously (Hwang et al. 2011a). Briefly, roots were cut into small pieces (1 cm), mounted on a

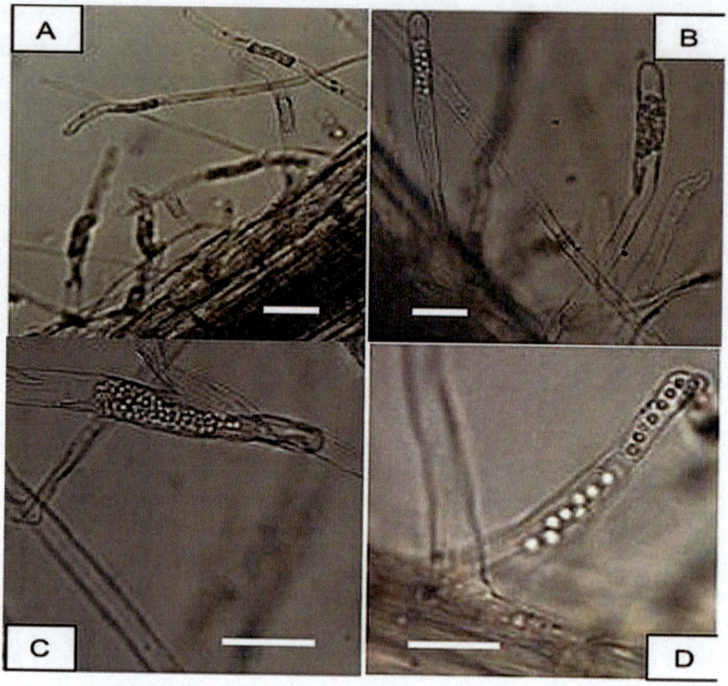

Plate 14.1 Different stages of canola root-hair infection by *Plasmodiophora brassicae*. (**a**) Primary plasmodial stage; (**b**, **c**) Early zoosporangial stages; (**d**) Late zoosporangial stage. Root-hair infection was examined using 10–12-day old seedlings. Horizontal bars ¼ 10 μm (Rashid et al. 2013)

microscope slide with 5% (v/v) glycerol, and examined using a Zeiss AX10 microscope. Three slides were prepared per replicate for each pH treatment, and one hundred root hairs were assessed from each slide. A root hair was considered infected when a primary plasmodium or a zoosporangium was visible inside the root hair (Plate 14.1).

14.22.4 Data Analysis

All data were subjected to analysis of variance (ANOVA) using PROC GLM of the SAS statistical package (version 9.2, SAS Institute, and differences among treatments were compared using Duncan's multiple range test ($P \leq 0.05$).

14.23 Evans Blue Staining of *Plasmodiophora brassicae* Resting Spores

14.23.1 Isolation of *P. brassicae* Resting Spores

Symptomatic canola roots were obtained from 12-week-old plants growing in a naturally infested field near Brooks, Alberta. The roots were washed and intact clubroot galls were collected and stored at −20 °C until further processing. Resting spores were obtained by thawing frozen galls overnight at room temperature and then placing them in an air dryer set at 20 °C for 7 days. Dried galls were ground into a coarse powder with a model 4 Thomas- Wiley laboratory mill using a mil coarse screen. Three grams of ground gall material were suspended in 25 ml of 5% glycerol solution prepared in distilled water. The suspension was agitated for 5–10 min at room temperature and then filtered through eight layers of cheese cloth. The resulting spore suspension was mixed thoroughly for 5 min at room temperature with magnetic stirrer. Resting spore suspension more than 4 h old were discarded and not used. Final concentration of resting spores suspension were between 1.0×10^8 and 2.2×10^8 spores per ml^{-1}.

14.23.2 Heat Treatment

Five milliliter aliquots of spore suspension were placed into 8 ml (13 × 100 mm) glass tubes, capped, and incubated in an AccBlock model D1100 digital dry heat block that was pre heated and held at the predefined experimental temperature. After exposure, tubes were removed from the heat block and kept at room temperature. Five replicate tubes were exposed at each temperature and time, including a non-heated (room temperature) control. To determine the effect of staining time on uptake of Evans blue stain by *P. brassicae* resting spores a sample of resting spores was treated at 80 °C for 72 h. resting spores were evaluated every 30 min for the first 2 h, then hourly up to 12 h, and then at 24, 48, and 72 h. A non-heated (0 h) sample was also evaluated.

14.23.3 Evans Blue Staining

Immediately after heat treatment, spore suspension were gently mixed and 50 μl transferred to a 1.5 ml micro-centrifuge tube containing an equal vol. of freshly prepared Evans blue stain solution (Sigma–Aldrich) prepared at 20 mg ml^{-1}. The staining solution was incubated uncovered at room temperature for the prescribed exposure time. Resting spore staining was visualized using bright field microscopy with a Zeiss Axio imager M2 microscope at 1000× magnification. A total of 100 spores per sample were evaluated for each of five replicates and number of spores having cytoplasm stained dark blue were recorded. The present viable spores were calculated for each. All experiment described herein employed an overnight stain exposure time of 18–20 h with the exception of the time course staining experiment that compared multiple staining exposure times.

14.23.4 Clubroot Bioassays

There were 14 canola seedlings per treatment, replicated five times using the clubroot susceptible cultivar 45H31. After the heat treatment, 5 ml of treated resting spore suspension were poured into a disposal 35 mm Petri dish and the roots of each canola seedlings were swished in the spore suspension for ca. 10 s, and then transplanted into a prepared Cone tainer™ root trainer containing ~55 g of pre-moistened sunshine Mix # 4 medium. After transplanting the seedling base was inoculated by pipetting 200 μl of spore suspension from the second tube, into a small depression on the media surface. The *P. brassicae* resting spore concentration in each sample was approximately 1.0×10^8 spores per ml^{-1}. Based on volume of spore suspension used in the bioassay, each Cone tainer™ would have contained approximately 3.6×10^5 spores per g^{-1} of growth chamber at ambient humidity and a light/ dark cycle of 16 h at 23 °C with maximum light intensity and 8 h at 18 °C with no light. Each treatment was placed in a separate tray to prevent any cross contamination. Plants were gently hand watered five times each week to saturation with tap water and fertigated two days each week to saturation with prepared nutrient solution (N-225 ppm, p-61 ppm, K-302 ppm, Ca-190 ppm and pH 6.0). Canola plants were grown under these conditions for 6 weeks after which their roots were gently removed from the Cone– tainer™, carefully rinsed with tap water, and then visually rated for clubroot symptoms. Disease incidence was calculated as the percent plants with symptoms of clubroot. Disease severity was estimated using a visual scale ranging from 0–3 as follows: 0 (no disease), 1 (small galls on less than one third of roots), 2 (small to medium galls on one third to two thirds of roots), and 3 (medium to large galls on more than two thirds of roots) Kuginuki et al. 1999). The index of disease was calculated as described by Strelkov et al. (2018).

$$\mathrm{ID}\ (\%) = \frac{\sum(n \times 0 + n \times 1 + n \times 2 + n \times 3)}{N \times 3} \times 100$$

where n is the number of plants in class. N is the total number of plants, 0, 1, 2, and 3 are the symptom severity classes (Harding et al. 2018)

14.24 Propidium Monoazide (PMA) Treatment and qPCR Detection of Viable *Plasmodiophora brassicae* resting Spores

14.24.1 PMA Treatment

Resting spore suspension containing 2.2×10^8 spores per ml^{-1} were treated at 70, 80, and 90 °C for 0–5 h. Samples were collected at each exposure time and paired samples evaluated using Evans blue staining or PMA qPCR. One hundred microliter of heat treated resting spore suspensions were treated with PMA by adding 5 μl PMA at a concentration of 1 mM (final concentration = 47.62 μM). The tubes were shaken at 300 rpm in the dark at room temperature for 30 min and then set on ice and treated with a 500 W halogen light for 15 min at 20 cm above the sample. DNA was immediately extracted from each sample and quantitative (q) PCR was performed.

14.24.2 DNA Extraction and qPCR

All DNA extraction was performed using the Qiagen DNeasy Power soil kit. All extracted DNA was dissolved in 50 µl water. A qPCR primer set specific to the rDNA region of *P. brassicae*, CRqF2 (CTAGCGCTGCATCCCATATC), and CRqR2 (TGTTTCGGCTAGGATGGTTC) and the probe PB1 (56-FAM/CCATGTGAA/ZEN/CCGGTGAC/31ABkFQ) were manually designed according to the Gen Bank accession number KX011135. Using this primer set, qPCR analysis was conducted in a BioRad Sso Advanced Univ probe qPCR master mix with a CFX connect cycler. The qPCR program and cycling parameters were one (1) cycle at 95 °C for 2 min followed by 40 cycles at 95 °C for 5 s and 60 °C for 30 s. All DNA samples in all qPCR analysis were prepared as three technical repeats where all components of qPCR reactions mixtures were identical and contained 2 µl of DNA template. A primer efficiency standard curve analysis was done using resting spores of a field populations of *P. brassicae* prepared as three 100 µl suspension each containing 4×10^7 spores. DNA was extracted from these three samples and each of the resultant DNAs was serially diluted six times by tenfold each. Quantitative PCR analysis was performed on these samples and a standard curve was generated. Based on the equation of the primer efficiency, spore number in the original DNA samples could be calculated from the qPCR CT value. To determine the efficiency of the PMA qPCR, a resting spore suspension was prepared from the field population at a concentration of 1.5×10^8 spores per ml^{-1}. An aliquots of this suspension was treated at 95 °C for 10 min. After cooling to room temperature, aliquots of this suspension were mixed with untreated viable spores. These suspensions were prepared in 100 µl final volumes and thus consisted of 1.5×10^7 spores/tube. Three sets of such serial suspension were prepared to comprise three biological repeats (Harding et al. 2018).

14.25 Identification of Pathogenic Variability

The pathogenic variability has been determined with the differential sets to designate races/pathotypes with procedures developed by Williams (1966), Buczacki et al. (1975) with ECD-European Clubroot Differential set, Some et al. (1996), and Strelkov et al. (2018) with Canadian clubroot differential sets (CCD). In some countries, scientists have used modified differential sets using their national resistant cvs. as additional host to get better picture of identification, and distribution of virulence on cruciferous species. The *Brassica* species and varieties/genotypes used are given in Chap. 11 in Table 11.11. There are several limits of differentials sets used so far in the absence of isogenic lines, and use of isolates as populations of pathogen. At present time, there is a need to develop isogenic lines of each *Brassica* species and use single spore isolate for the pathogenic variability in *P. brassicae* at global level. The procedure of getting single spore isolates, preparation of inoculum, inoculation method, evaluation of host differentials, and analysis of data has been described in the techniques included here.

14.26 Novel System of Designation, and Nomenclature of Clubroot-Resistant Genes, and Pathotypes of *Plasmodiophora brassicae*

The designation and name of the host resistance genes governing clubroot resistance has not been standardized at International level. Researchers have followed different systems and patterns to designate a resistance gene in *Brassica* crops against *P. brassicae* (Chap. 11, Table 11.11). It becomes difficult to compare, confirm, and validate each other's results. Following uniform pattern at International level will avoid ambiguity among scientific community. The designation of *R* gene should indicate host species from which the gene has been identified along with the pathogen against which it is effective. A binomical nomenclature system is proposed to designate *R* genes in *Brassica* species, viz., BrCR1, BrCR2, . . .n for *Brassica rapa*; BoCR1, BoCR2, . . .*n* for *B. oleracea*; BnCR1, BnCr2, . . .*n* for *B. napus, and* so on. The first two letters in this system indicate *Brassica* species and second two letters denote the number of clubroot *R* genes. Similar pattern can be followed to designate pathotypes of *P. brassicae* viz., BrPb1, BrPb2, . . .*n* for *B rapa*; BoPb1, BoPb2, . . .*n* for *B. oleracea*; BnPb1, BnPb2, . . .*n* for *B. napus,* and so on indicating host species (first two letters) which have recognized a pathotypes (second two letters) after interaction in the form of a avirulence gene (AVR gene) of the pathogen along with R gene of the host. This scheme can be extended up to genotype level once isogenic lines in each *Brassica* species are developed and available. Like International Code of Botanical Nomenclature (ICBN) for naming an organism, a pattern or uniform system of designation and nomenclature of *R* genes and AVR genes should be followed with proper code of conduct.

14.27 Production of Pure Genotypes Isolates of *Plasmodiophora brassicae*

Plant material *P. brassicae* infected root hairs were collected from *Brassica rapa* cv. Chorus. Clubroot susceptible *B. napus* cv. Loras plants were used for most inoculations with single resting spores or infected root hairs. Single resting spores were also transferred on *B. rapa* seedlings (*B. rapa* ssp. *chinensis* cv. Pak Choy, *B. rapa* ssp. *pekinensis* cv. Granaat verb. Cantoner Witkrop, and *B. rapa* ssp. *trilocularis* cv. Yellow Seeded Sarson).

14.27.1 Inoculum Preparation

Infected roots were macerated in distilled water using a blender and an "Ultra Turrax T25" homogenizer at 8000 revolutions/min. The homogenate was filtered through gauze (25–50 μm) and washed three times by centrifugation in distilled water (2830 g, 5 min). The suspension was adjusted to the required density (see below) after estimation of the spore density with a hemocytometer.

14.27.2 Plant Cultivation

Single spore inoculation experiments were made in a greenhouse compartment that hitherto had not been used for experiments with *P. brassicae*. Temperatures ranged in general from 18 to 25 °C except for the experiment with "e-4.01" (20–45 °C). Supplementary light was provided to maintain a 16 h day length. For this purpose, 2000 W and 400 W HQI-type high-pressure mercury discharge lamps were used. For some experiments, a growth chamber was used where the temperature was maintained at about 26 °C, day length was 16 h (Philips TLD 50 W, 83/84HF), as mentioned below.

14.27.3 Inoculations with Single or Low Amounts of Resting Spores

Seeds were surface sterilized for 15 min in detergent containing 5% (v/v) sodium hypochlorite and rinsed three times in sterile distilled water. Subsequently they were germinated on moist filter paper in Petri dishes in the culture room for 4 days. For the identification of single resting spores 1 cm^2 blocks were cut out from 0.8%water agar in Petri dishes. Droplets of 5 μl of a spore suspension containing 200 resting spores/ml of isolate "a" were placed in the holes on the moist surface of the plastic Petri dish and examined under an inverted microscope at 320-fold magnification. A seedling was placed with the root into the droplet when it could be confirmed that it contained only a single resting spore, the closed Petri dishes were then kept overnight in the culture room. The seedlings were transplanted in clay pots (5 cm diameter) that had never been used for clubroot tests and that were filled with commercial soil ("Einheitserde P," pH 5–6, supplemented with sand (10%, v/v)). All pots and substrates had been autoclaved for 60 min at 121 °C and 1 bar. High soil humidity was maintained by placing the pots in glass vessels with a reservoir of sterile tap water. The water had previously been adjusted to pH 5.5 with phosphate buffer. After 20 days, each plant was carefully transplanted into a larger plastic pot (9 cm diameter). Supplementary substrate had previously been steam sterilized (5 h, 90 °C). Special care was taken to avoid any mixing of the soil from different pots or cross contaminations by splashing water. Disease symptoms were assessed 8 weeks after inoculation. Clubroot severity was rated using the 0–3 scale suggested by Buczacki et al. (1975). Controls consisted of noninoculated plants that were cultivated next to the inoculated plants. Inoculations with low spore numbers were treated in the same way except for the fact that inoculation was made by pipetting different volumes of the same spore suspension directly on the root hairs of the susceptible seedling.

14.27.4 Inoculations with Single Infected Root Hairs

Seeds of Chinese cabbage cv. Chorus was germinated for 4–5 days on moist filter paper in Petri dishes at room temperature. Germinated plants were placed in sterile clay pots (15 cm diameter) with a 1:1 blend (v/v) of commercial soil ("Einheitserde P") and perlite that was under layered by 1.5 cm peat. Seedlings were inoculated with 2 ml of a suspension containing 2×10^8 resting spores/ml which was injected into the soil along the root. The soil humidity was kept at field capacity by a moist fleece sub-layer. Six to eleven days after inoculation, single plantlets were carefully cleaned in water before shaking them for 30 min in distilled water containing Tween 20. To remove the detergent, the roots were rinsed three times in distilled water. They were examined for root hairs containing a single sporangiosorus with an inverted microscope at ×40 to ×200 magnifications. Root hairs containing mature sporangiosori were detached with a lancet and examined for the presence of a single sporangiosorus and for adherent resting spores or debris. Selected root hairs were seized individually with a pair of fine tweezers. Lancet and tweezers were sterilized in ethanol between all inoculations.

Each inoculated or control seedling was individually placed in a sterile small clay pot (5 cm diameter) containing commercial soil supplemented with sand (10 %, v/v). All pots and soil had been autoclaved for 60 min at 121 °C. A small, chemically sterilized, plastic basin constantly filled with 0.5–1 cm fresh tap water was placed under each pot. Control plants were distributed at random among every set of plants inoculated with infected root hairs. As described earlier, each plant was carefully transplanted in a larger plastic pot 3–4 weeks later. Ten weeks after inoculation the substrate was carefully removed from the plants which were examined for disease symptoms as described above (Diederichsen et al. 2016).

14.28 *Plasmodiophora brassicae* Root-Hair Interaction

The sterilized seeds were germinated on wet filter paper for 48 h (25 °C/20 °C day/night temperatures with a photoperiod of 14 h light and 10 h dark). A black bag was wrapped approximately 50 ml centrifuge tubes to exclude light. Half strength Hoagland nutrient solution as culture solution was added to the tubes and adjusted to a pH of 6.0. The tubes were covered with plastic wraps with 0.5 cm diameter holes. The seedlings were fixed in round holes, and placed in top of the tubes, with single seedling per tube. To inoculate the culture solution, 1 ml spore suspension containing 5×10^7 spores ml^{-1} was added to the solution on different days (1, 5, 10, 15, 20, 25, and 30 days), to give a final concentration of approximately 10^6 spores ml^{-1}. All tubes were watered daily and cleaned weekly with culture solution.

The nutrition solution was replaced after tubes were cleaned. A total of 10 plants per treatment were used. All plants were drawn out 9 days after inoculation, rinsed with clean tap water, and cut into 1 cm sections (Luo et al. 2014). The segment was stained with FAA Phloxine B and observed with an optical microscope in 3 microscopic fields of the same root using a ×20 magnification (Donald et al. 2008). Root hairs with primary infection (reported as percent of total) were counted in each field of view. Each experiment included ten roots from different plants (a total of 30 microscopic fields examined per experiment). The experiment was repeated three times (He et al. 2019).

14.29 Molecular Detection and Quantification

As an obligate biotroph, detection of *P. brassicae* is challenging as the pathogen cannot be cultured. A range of detection methods including plant bioassays, fluorescence microscopy, serology, and DNA-based methods have been used to detect *P. brassicae* in soil, water, and plant samples. These methods have been reviewed by Faggian and Strelkov (2009). In Australia, visual or microscopic examination of bait plants grown in a sample of soil under controlled conditions has been used historically to estimate soil inoculum load. Visual assessment of gall development is time-consuming and requires considerable amounts of glasshouse space. Microscopic observation of root-hair infection is faster but several workers have reported that no linear relationship exists between root-hair infection and subsequent gall development (MacFarlane 1952; Naiki et al. 1978). A fluorescent staining technique utilizing two fluorochromes, calcofluor white M2R and ethidium bromide, to differentiate between viable and nonviable resting spores was developed by Takahashi and Yamaguchi (1988, 1989), but the presence of fluorescing soil particles and other artifacts made it difficult to identify and enumerate *P. brassicae* resting spores in soil samples. Even when soil was excluded from the test system, the fluorescent stain did not correlate well with disease severity measured using bait plants in two out of three experiments conducted by Donald et al. (2002). Faggian et al. (1999a) developed a nested PCR detection protocol directed at the ribosomal genes (rDNA) and internal transcribed spacer (ITS) regions of *P. brassicae*. The assay enabled detection down to 10 spores per gram and was developed as a commercial test in 2001 by the Victorian Department of Primary Industries. Its use was predominantly to resolve disputes between growers, determining sources of spread, and within research programs to predict soil inoculum loads (Faggian and Strelkov 2009). The method used to extract DNA from soil was further refined by Faggian and Parsons (2002) utilizing a bead beating mill and polyvinyl polypyrrolidone (PVPP) spin columns to remove unknown inhibitory compounds which had previously resulted in poor amplification efficiency from some soil types. The development of a quantitative real-time PCR assay can be used it to predict disease severity in field plots. Large-scale field evaluation of the quantitative assay was conducted by Donald et al. (2004) who reported good correlation between actual and predicted disease severity at 32 of the 55 sites sampled. However, DNA of *P. brassicae* was not detected in 23 samples from a range of soil types in which disease developed. The fastest and simplest of these methods, the use of Calgon™ (100% sodium hexametaphosphate), together with pH adjustment, was recommended as a pre-extraction treatment for all soil samples before testing for clubroot. Subsequently, Faggian et al. (2007) used a PCR reaction additive (T4 Gene 32 Protein) to reduce the inhibitory effects of humic/fulvic acids and reported 10-fold greater sensitivity using this method compared with Calgon™ treatment. Treatment with T4 Gene 32 Protein is a post-extraction procedure that involves the addition of protein to the PCR reaction mix. It is much quicker and simpler than Calgon™ treatment, and more effective. It also has a completely different mode of action, since it acts directly on the DNA polymerase and enables the enzyme to function in the presence of inhibitors. Faggian et al. (2007) concluded that their results using the T4 Gene 32 Protein and the finding that updated soil extraction kits were able to extract DNA from soil more reliably than previous iterations, suggested that the mechanism of action of Calgon™ may not involve the soil–spore interaction at all. Instead, the more likely scenario is that the pH adjustments inherent in Calgon™ treatment somehow reduce the co-purification of humic/fulvic acids during downstream DNA extraction procedures. Faggian et al. (2007) were also able to demonstrate improved yields of DNA extracted from soil using Fast DNA Spin Kits with the addition of skim milk powder as an adsorption competitor to reduce the number of binding sites on soil. In spite of the substantial amount of work undertaken to develop and optimize PCR-based assays in Australia and to overcome the issues of poor yield of extracted DNA and inhibition of the PCR reaction in some soil types, commercial use of the final diagnostic assay in Australia was limited due largely to its high cost. A low-cost alternative monoclonal antibody technique has been used with success to detect and quantify fungal plant pathogens (Thornton et al. 2004). Unlike polyclonal antisera, monoclonal antisera can be produced indefinitely. These tests are also very inexpensive, can be conducted on site by the farmer and, potentially, a single lateral flow device can be used to detect or quantify multiple diseases. The PCR-based molecular detection test (Faggian et al. 1999a) was used to detect *P. brassicae* in dams, boreholes, and nurseries in a *Brassica* production region in Victoria, Australia (Faggian et al. 1999b). Of particular concern in this study was the identification of contamination by *P. brassicae* on plastic seedling trays being returned to the nursery for reuse and in dust. Frequently, these trays were dipped in a disinfectant bath before reuse; however, laboratory screening indicated that many of the commercial disinfectant compounds in use were ineffective at either reducing or eradicating *P. brassicae* spore viability (Donald et al. 2002).

The potential for irrigation water contaminated with *P. brassicae* to cause clubroot in seedlings or crops was investigated in a series of experiments (Donald 2005). Spore suspensions containing 108 spores mL^{-1} were stored at 2 °C for periods of 1–3.5 years. No relationship was found between the ability of the stored suspensions to cause root galling in bait plants and the duration of storage. Some suspensions that were approximately 1 year old failed to cause root galling in bait plants while others that were almost 3 years old remained infective. Subsequent daily irrigation of bait plants with water containing 0, 10^1, 10^2, 10^3, 10^4 or 10^6 spores ml^{-1}

indicated that even at the lowest concentration of inoculum, half of the bait plants became infected. In spite of this, irrigation with water collected from contaminated farm dams in Victoria has failed to induce symptoms of clubroot in pot trials. In farm dams, resting spores might be expected to settle to the bottom of the dam over time. This rate of settling of *P. brassicae* resting spores in an undisturbed body of water was estimated at 25 cm day^{-1} using settling columns (Donald and Porter 2004).

14.30 Disease Assessment

Several disease assessment scales have been developed to assess disease intensity under field and controlled conditions. The details of the scales used for recording disease index to get information on disease severity under different situations, and to test host resistance are described in Chap. 2 at Sect. 2.7.

References

Agarwal A, Kaul V, Faggian R, Cahill DM (2009) Development and use of a model system to monitor clubroot disease progression with an Australian field population of *Plasmodiophora brassicae*. Austr Plant Pathol 38:120–127

Anders S, Pyl PT, Huber W (2015) HT Seq—a Python frame work to work with high-throughput sequencing data. Bioinformatics 31(2):166–169

Antoniadia I, Plackovác L, Simonovika B, Dolezalc K, Turnbull C, Ljunga K (2015) Cell-type specific cytokinin distribution within the *Arabidopsis* primary root apex. Plant Cell. https://doi.org/10.1105/tpc.15.00176

Arnold DL, Blakesley D, Clarkson JM (1996) Evidence for the growth of *Plasmodiophora brassicae in vitro*. Mycol Res 100:535–540

Asano T, Kageyama K (2006) Growth and movement of secondary plasmodia of *Plasmodiophora brassicae* in turnip suspension-culture cells. Plant Pathol 55(1):145–151

Asano T, Kageyama K, Hyakumachi M (1999) Surface disinfestations of resting spores of *Plasmodiophora brassicae* used to infect hairy roots of *Brassica* spp. Phytopathology 89:314–319

Asano T, Kageyama K, Hyakumachi M (2000) Germination of surface disinfected resting spores of *Plasmodiophora brassicae* and their root hair infection in turnip hairy roots. Mycol Sci 41 (11):49–54

Bryngelsson T, Gustafsson M, Gréen B, Lind-Halldén C (1988) Uptake of host DNA by the parasitic fungus *Plasmodiophora brassicae*. Physiol Mol Plant Pathol 33(2):163–171

Buczacki ST, Toxopeus H, Johnston P, Dixon TD, Hobolt LA (1975) Study of physiological specialisation in *Plasmodiophora brassicae*: proposals for attempted rationalization through an inter- national approach. Trans Br Mycol Soc 65:295–303

Buhariwalla H, Mithen R (1995) Cloning of a *Brassica* repetitive DNA element from resting spores of *Plasmodiophora brassicae*. Physiol Mol Plant Pathol 47:95–101

Buhariwalla S, Greaves S, Magrath R, Mithen R (1995) Development of specific PCR primers for the amplification of polymorphic DNA from the obligate root pathogen *Plasmodiophora brassicae*. Physiol Mol Plant Pathol 47:83–94

Bulman S, Candy JM, Fiers M, Lister R, Conner AJ, Eady CC (2011) Genomics of biotrophic, plant-infecting plasmodiophorids using in vitro dual culture. Protist 162(3):449–461

Burrough PA, McDonnell RA (1998) Principles of geographical information systems. Oxford University Press, Oxford

Cao T, Tewari J, Strelkov SE (2007) Molecular detection of *Plasmodiophora brassicae*, causal agent of clubroot of crucifers, in plant and soil. Plant Dis 91:80–87

Cao T, Manolii VP, Hwang S-F, Howard RJ, Strelkov SE (2009) Virulence and spread of *Plasmodiophora brassicae* (clubroot) in Alberta, Canada. Can J Plant Pathol 31:321–329

Carvalho BS, Irizarry RA (2010) A framework for oligo-nucleotide microarray preprocessing. Bioinformatics 26(19):2363–2367

Castlebury LA, Maddox JV, Glawe DA (1994) A technique for the extraction and purification of viable *Plasmodiophora brassicae* resting spores from host tissue. Mycologia 86:458–460

Chattopadhyay AK, Moitra AK, Bhunia CK (2001) Evaluation of *Brassica* species for resistance to *Plasmodiophora brassicae* causing club root of rapeseed mustard. Indian Phytopathol 54 (1):131–132

Chee HY, Kim WG, Cho WD, Jee HJ, Choi YC (1998) Detection of *Plasmodiophora brassicae* by using polymerase chain reaction. Korean J Plant Pathol 14:589–593

Cutler SR, Ehrhardt DW, Griffitts JS, Somerville CR (2000) Random GFP:cDNA fusions enable visualization of subcellular structures in cells of *Arabidopsis* at a high frequency. Proc Natl Acad Sci U S A 97(7):3718–3723

Deora A, Gossen BD, McDonald MR (2012) Infection and development of *Plasmodiophora brassicae* in resistant and susceptible canola cultivars. Can J Plant Pathol 34:239–247

Deora A, Gossen BD, McDonald MR (2013) Cytology of infection, development and expression of resistance to *Plasmodiophora brassicae* in canola. Ann Appl Biol 163:56–71

Diaz G, Melis M, Batetta B, Angius F, Falchi AM (2008) Hydrophobic characterization of intracellular lipids in situ by Nile Red red/yellow emission ratio. Micron 39(7):819–824

Diederichsen E, Wagenblatt V, Schallehn V (2016) Production of pure genotype isolates of *Plasmodiophora brassicae* Wor.- comparison of inoculations with root hairs containing single sporangiosori or with single resting spores. Eur J Plant Pathol 145:621–627. https://doi.org/10.1007/s10658-016-0876-z

Diriwachter G, Herter G, Gindrat D (1979) Observations sur la culture de *Spongospora subterranea* (Plasmodiophoromycetes) en milieu gelose. Berichte Schweizerische Botanische Gesellschaft 89:105–113

Dixon GR (1976) Methods used in Western Europe and USA for testing *Brassica* seedlings resistance to clubroot (*Plasmodiophora brassicae*). Plant Pathol 25(3):129–134

Donald EC (2005) The influence of abiotic factors and host plant physiology on the survival and pathology of *Plasmodiophora brassicae* of vegetable Brassicas. Melbourne University, Melbourne

Donald EC, Porter IJ (2004) A sand solution culture technique used to observe the effect of calcium and pH on root hair and cortical stages of *Plasmodiophora brassicae*. Austr Plant Pathol 33:585–589

Donald EC, Lawrence JM, Porter IJ (2002) Evaluation of a fluorescent staining technique as an indicator of pathogenicity of resting spores of *Plasmodiophora brassicae*. Austr Plant Pathol 31:373–379

Donald EC, Czerniakowski B, Porter IJ. (2004) Optimizing a diagnostic test for clubroot. Final report for project VG03022. Sydney (Australia): Horticulture Australia Limited

Donald EC, Jaudzems G, Porter IJ (2008) Pathology of cortical invasion by *Plasmodiophora brassicae* in clubroot resistant and susceptible *Brassica oleracea* hosts. Plant Pathol 57 (2):201–209

Faggian R, Parsons S (2002) A Rapid diagnostic test for clubroot, final report for project VG99008. Horticultural Research and Development Corporation, Sydney

Faggian R, Strelkov SE (2009) *Plasmodiophora brassicae* (clubroot)–a plant pathogen that alters host growth and productivity. J Plant Growth Regul 28:282–287

Faggian R, Bulman SR, Lawrie AC, Porter IJ (1999a) Specific PCR primers for the detection of *Plasmodiophora brassicae* in soil and water. Phytopathology 89:392–397

Faggian R, Donald C, Porter IJ, Lawrie AC (1999b) Epidemiology of recent clubroot outbreaks using PCR to trace sources of inoculums. In: Proceedings of the 12th Australasian Plant Pathology Society Conference; 1999 Sept 27–30; Canberra (Australia)

Faggian R, Kaul V, Donald C, Porter IJ, Kennedy R, Wakeham A (2007) On farm diagnostic kits for *Brassica* diseases, final report for project VG04059. Horticulture Australia Limited, Sydney

Feng J, Xiao Q, Hwang SF, Strelkov SE, Gossen BD (2012) Infection of canola by secondary zoospores of *Plasmodiophora brassicae* produced on a non host. Eur J Plant Pathol 132:309–315

Feng J, Hwang SF, Strelkov SE (2013a) Genetic transformation of the obligate parasite *Plasmodiophora brassicae*. Phytopathology 103:1052–1057

Feng J, Hwang SF, Strelkov SE (2013b) Assessment of gene expression profiles in primary and secondary zoospores of *Plasmodiophora brassicae* by dot-blot and real-time PCR. Microbiol Res 168:518–524

Feng J, Hwang SF, Strelkov SE (2013c) Studies into primary and secondary infection processes by *Plasmodiophora brassicae* on canola. Plant Pathol 62:177–183

Feng J, Cao T, Rennie DC, Strelkov SE, Hwang SF (2014) Host-parasite interactions in clubroot of crucifers. Can J Plant Pathol 36(Suppl.1):113–321

Friberg H, Lagerlof J, Ramert B (2005) Germination of *Plasmodiophora brassicae* resting spores stimulated by a non – host. Eur J Plant Pathol 113(3):275–281

Guzman HM, De la Jara A, Duarte LC, Presmanes KF (2010) Estimate by means of flow cytometry of variation in composition of fatty acids from *Tetraselmis suecica* in response to culture conditions. Aquac Int 18(2):189–199

Harding MW, Hill TB, Yang Y, Daniels GC, Hwang SF, Strelkov SE, Howard RJ, Feng J (2018) An Improved Evans blue staining method for consistent, accurate assessment of *Plasmodiophora brassicae* resting spore viability. Plant Dis 103(9):2330–2336

Harling R (2007) Clubroot control using novel and sustainable methods. Final report DEFRA project HH3227TFV

Hasan M, Strelkov SE, Howard RJ, Rahman H (2012) Screening of *Brassica* germplasm for resistance to *Plasmodiophora brassicae* pathotypes prevalent in Canada for broadening diversity in clubroot resistance. Can J Plant Sci 92:501–515

He P, Cui W, Munir S, He P, Li X, Wu Y, Yang X, Tang P, He Y (2019) *Plasmodiophora brassicae* root hair interaction and control by *Bacillus subtilis* XF-1 in Chinse cabbage. Biol Control 128:56–63

Horiuchi S, Hori M (1980) A simple greenhouse technique for obtaining high levels of clubroot incidence. Bull Chugoku Natl Agric Exp 17:33–55

Hoyerova K, Gaudinova A, Malbeck J, Dobrev PI, Kocabek T, Solcova B, Travnickova A, Kaminek M (2006) Efficiency of different methods of extraction and purification of cytokinins. Phytochemistry 67(11):1151–1159

Hwang SF, Ahmed HU, Zhou Q, Strelkov SE, Gossen BD, Peng G, Turnbull GD (2011a) Influence of cultivar resistance and inoculums density on root hair infection of canola (*Brassica napus*) by *Plasmodiophora brassicae*. Plant Pathol 60:820–829

Hwang SF, Ahmed HU, Zhou Q, Strelkov SE, Gossen BD, Turnbull GD (2011b) Influence of cultivar response and inoculum density on root hair infection by *Plasmodiophora brassicae*. Plant Pathol 60:820–829

Hwang SF, Strelkov SE, Feng J, Gossen BD, Howard RJ (2012) *Plasmodiophora brassicae*: a review of an emerging pathogen of the Canadian canola (*Brassica napus*) crop. Mol Plant Pathol 13:105–113

Ito S, Maechara T, Tanaka S, Kameya-Iwaki M, Yano S, Kishi F (1997) Cloning of single-copy DNA sequence unique to *Plasmodiophora brassicae*. Physiol Mol Plant Pathol 50:289–300

Ito S, Maehara T, Maruno E, Tanaka S, Kameya-Iwaki M, Kishi F (1999) Development of a PCR-based assay for the detection of *Plasmodiophora brassicae* in soil. J Phytopathol 147:83–88

Jin-ping L, Yan L, Yan-xia S, Xue-wen X, Ali C, Baoju L (2013) Development of a real-time PCR assay for *Plasmodiophora brassicae* and its detection in soil samples. J Integr Agric 12 (10):1799–1806

Johnston TD (1968) Clubroot in *Brassica*: a standard inoculation technique and the specification of races. Plant Pathol 17:184–187

Jones PM (1928) Morphology and cultural history of *Plasmodiophora brassicae*. Arch Protistenkunde 62:313–327

Kaczmarek J, Latunde-Dada AO, Irzykowski W, Cools HJ, Stonard JF, Brachaczek A, Jedryczka M (2014) Molecular screening for avirulence alleles *AvrLm1* and *AvrLm6* in airborne inoculum of *Leptosphaeria maculans* and winter oilseed rape (*Brassica napus*) plants from Poland and the UK. J Appl Genet 5:529–539

Kageyama K, Kamimura Y, Hyakumachi M (1995) A simple inoculation method with a single resting spore of *Plasmodiophora brassicae*. Ann Phytopathol Soc 61:415–418

Kimura K, Yamaoka M, Kamisaka Y (2004) Rapid estimation of lipids in oleaginous fungi and yeasts using Nile red fluorescence. J Microbiol Methods 56(3):331–338

Kopischke M, Westphal L, Schneeberger K, Clark R, Ossowski S, Wewer V, Fuchs R, Landtag J, Hause G, Dörmann P, Lipka V, Weigel D, Schulze-Lefert P, Scheel D, Rosahl S (2013) Impaired sterol ester synthesis alters the response of *Arabidopsis thaliana* to *Phytophthora infestans*. Plant J 73(3):456–468

Kuginuki Y, Hiroaki Y, Hirai M (1999) Variation in virulence of *Plasmodiophora brassicae* in Japan tested with clubroot-resistant cultivars of Chinese cabbage (*Brassica rapa* L. ssp. *pekinensis*). Eur J Plant Pathol 105:327–332

Lange L, Heide M, Hobolth L (1989) Serological detection of *Plasmodiophora brassicae* by dot immunobinding and visualization of the serological reaction by scanning electron microscopy. Phytopathology 79:1066–1071

Lark RM, Ferguson RB (2004) Mapping risk of soil nutrient deficiency or excess by disjunctive and indicator kriging. Geoderma 118:39–53

Lewis SJ, Brokenshire T (1978) A comparison of plant inoculation techniques for the determination of *Plasmodiophora brassicae* populations. Proc. Woronin +100 Conference, Madison, pp 54–67

Li R, Sun R, Hicks GR, Raikhel NV (2015) *Arabidopsis* ribosomal proteins control vacuole trafficking and developmental programs through the regulation of lipid metabolism. Proc Natl Acad Sci U S A 112(1):E89–E98

Love MI, Huber W, Anders S (2014) Moderated estimation of fold change and dispersion for RNA-seq data with DESeq2. Genome Biol 15(12):550

Lunter G, Goodson M (2011) Stampy: a statistical algorithm for sensitive and fast mapping of Illumina sequence reads. Genome Res 21(6):936–939

Luo HC, Chen GK, Liu CP, Huang Y, Xiao CG (2014) An improved culture solution technique for *Plasmodiophora brassicae* infection and the dynamic infection in the root hair. Australas Plant Pathol 43(1):53–60

MacFarlane I (1952) Factors affecting the survival of *Plasmodiophora brassicae* Worin. the soil and its assessment by a host test. Ann Appl Biol 39:239–256

MacFarlane I (1970) Germination of resting spores of *Plasmodiophora brassicae*. Trans Br Mycol Soc 55:97–112

Malinowski R, Smith JA, Fleming AJ, Scholes JD, Rolfe SA (2012) Gall formation in clubroot-infected *Arabidopsis* results from an increase in existing meristematic activities of the host but is not essential for the completion of the pathogen life cycle. Plant J 71(2):226–238

Malinowski R, Novak O, Borhan MH, Spíchal L, Strnad M, Rolfe SA (2016) The role of cytokinins in clubroot disease. Eur J Plant Pathol 145:543–557

Manzanares-Dauleux MJ, Barret P, Thomas G (2000) Development of a pathotype specific SCAR marker in *Plasmodiophora brassicae*. Eur J Plant Pathol 106(8):781–787

Miquel M, Trigui G, d'Andrea S, Kelemen Z, Baud S, Berger A, Deruyffelaere C, Trubuil A, Lepiniec L, Dubreucq B (2014) Specialization of oleosins in oil body dynamics during seed development in *Arabidopsis* seeds. Plant Physiol 164(4):1866–1878

Naiki T, Dixon GR (1987) The effects of chemicals on developmental stages of *Plasmodiophora brassicae* (clubroot). Plant Pathol 36:316–327

Naiki T, Kageyama K, Ikegami H (1978) The relation of spore density of *Plasmodiophora brassicae* Worn. to the root hair infection and club formation in Chinese cabbage (Studies on the clubroot of cruciferous plant II). Ann Phytopathol Soc 44:432–439

Nieuwhof M, Wiering D (1961) Testing cabbage plants for clubroot resistance. Euphytica 10:191–200

Novak O, Henykova E, Sairanen I, Kowalczyk M, Pospísil T, Ljung K (2012) Tissue-specific profiling of the *Arabidopsis thaliana* auxin metabolome. Plant J 72(3):523–536

Rashid A, Ahmed HU, Xiao Q, Hwang SF, Strelkov SE (2013) Effects of root exudates and pH on *Plasmodiophora brassicae* resting spore germination and infection of canola (*Brassica napus* L.) root hairs. Crop Prot 48:16–23

Reby E, Michalik B (2005) Ocena podatności odmian wzorcowych kapusty z genami odporności na krajowe izolaty kiły (*Plasmodiophora brassicae*). In: Michalik B, Żurawicz E (eds) Zmienność genetyczna i jej wykorzystanie w hodowli roślin ogrodniczych. ISiK, Skierniewice, pp 37–42

Ren L, Fang XP, Sun CC, Chen KR, Liu F, Li M, Xu L (2014) First report of clubroot on *Capsella bursa-pastoris* caused by *Plasmodiophora brassicae* in Sichuan Province of China. Plant Dis 98 (5):687

Rennie DC, Manolii VP, Cao T, Hwang SF, Howard RJ, Strelkov SE (2011) Direct evidence of surface infestation of seeds and tubers by *Plasmodiophora brassicae* and quantification of spore loads. Plant Pathol 60:811–819

Ricarova V, Kaczmarek J, Strelkov SE, Kazda J, Lueders W, Rysanek P, Manolii V, Jedryczka M (2016) Pathotypes of *Plasmodiophora brassicae* causing damage to oilseed rape in the Czech Republic and Poland. Eur J Plant Pathol 145:559–572

Robak J (1991) Variability of the pathotypes of *Plasmodiophora brassicae* Wor. present in Poland and their pathogenicity to the cultivars and breeding lines of *Brassica oleracea* [in Polish]. Habilitation monograph. Skierniewice: Research Institute of Vegetable Crops

Rod J (1988) New inoculation method for the testing of *Brassica* for resistance to *Plasmodiophora brassicae* Worn. Crucferae Newslett 13:102–103

Rod J (1996) Reports – agent of clubroot of crucifer. Brno 37(Special Issue):1–45

Rogers SO, Bendich AJ (1985) Extraction of DNA from milligram amounts of fresh, herbarium and mummified plant tissues. Plant Mol Biol 5:69–76

Rogers SO, Bendich AJ (1995) Extraction of total cellular DNA from plants, algae and fungi. In: Gelvin SB, Schilperoort RA (eds) Plant molecular biology manual. Kluwer Academic Publishers, Dordrecht, pp 1–8

Rumin J, Bonnefond H, Saint-Jean B, Rouxel C, Sciandra A, Bernard O, Cadoret JP, Bougaran G (2015) The use of fluorescent Nile red and BODIPY for lipid measurement in microalgae. Biotechnol Biofuels 8:42

SAS (2008) Analytics Software and Solution Institute, Version 9.2

Sasser M (1990) Identification of bacteria through fatty acid analysis. In: Clement Z, Rudolph K, Sands DC (eds) Methods in phytobacteriology. Akadémiai Kiadó, Budapest, pp 199–204

Schoeller M, Grunewald J (1987) Comparison of inoculum techniques for testing Brassica seedling resistance against *Plasmodiophora brassicae*. Worn - Cruciferae Newslett 12:86–87

Schuller A, Ludwig-Müller J (2016) Histological methods to detect the clubroot pathogen *Plasmodiophora brassicae* during its complex life cycle. Plant Pathol 65:1223–1237

Seaman WL, Walker JC, Larson RH (1963) A new race of *Plasmodiophora brassicae* affecting Badger Shipper cabbage. Phytopathology 53:1426–1429

Sharma K, Gossen BD, McDonald MR (2011) Effect of temperature on cortical infection by *Plasmodiophora brassicae* and clubroot severity. Phytopathology 60:830–838

Smyth GK (2005) Limma: linear models for microarray data. In: Gentleman R, Carey V, Huber W, Irizarry R, Dudoit S (eds) Bioinformatics and computational biology solution susing R and bioconductor. Springer, New York, pp 397–420

Some A, Manzanares MJ, Laurens F, Baron F, Thomas G, Ouxel FR (1996) Variation for virulence of *Brassica napus* L., amongst *Plasmodiophora brassicae* collections from France and derived single-spore isolates. Plant Pathol 45:432–439

Strelkov SE, Tewari JP, Smith-Degenhardt E (2006) Characterization of *Plasmodiophora brassicae* populations from Alberta, Canada. Can J Plant Pathol 28:467–474

Strelkov SE, Manolii VP, Cao T, Xue S, Hwang SF (2007) Pathotype classification of *Plasmodiophora brassicae* and its occurrence in *Brassica napus* in Alberta. J Phytopathol 105:706–712

Strelkov SE, Hwang SF, Manolii VP, Cao T, Fredua-Agyeman R, Harding MW, Peng G, Gossen BD, Mcdonald MR, Feindel D (2018) Virulence and pathotype classification of *Plasmodiophora brassicae* populations collected from clubroot resistant canola (*Brassica napus*) in Canada. Can J Plant Pathol 40(2):284–298

Sundelin T, Christensen CB, Larsen J, Moller K, Lübeck M, Bodker L, Jensen B (2010) *In planta* quantification of *Plasmodiophora brassicae* using signature fatty acids and real-time PCR. Plant Dis 94:432–438

Svacinova J, Novák O, Plackova L, Lenobel R, Holík J, Strnad M, Dolezal K (2012) A new approach forcytokinin isolation from *Arabidopsis* tissues using miniaturized purification: pipette tip solid-phase extraction. Plant Methods 8(1):1–14

Takahashi K, Yamaguchi T (1988) A method for assessing the pathogenic activity of resting spores of *Plasmodiophora brassicae* by fluorescence microscopy. Ann Phytopathol Soc 54:466–475

Takahashi K, Yamaguchi T (1989) Assessment of pathogenicity of resting spores of *Plasmodiophora brassicae* in soil by fluorescence microscopy. Ann Phytopathol Soc 55:621–628

Takahashi H, Muraoka S, Ito K, Mitsui T, Hori H, Kiso A (2001) Resting spores of *Plasmodiophora brassicae* proliferation only in the callus of clubroot disease-susceptible turnip but increases PAL activity in the callus of clubroot disease-resistant turnip. Plant Biotechnol 18 (4):267–274

Takahashi H, Ishikawa T, Kaido M, Takita K, Hayakawa T, Okazaki K, Itoh K, Mitsui T, Hori H (2006) *Plasmodiophora brassicae*-induced cell death and medium alkalization in clubroot-resistant cultured roots of *Brassica rapa*. J Phytopathol 154:156–162

Tamura K, Stecher G, Peterson D, Filipski A, Kumar S (2013) MEGA6: Molecular evolutionary genetics analysis version 6.0. Mol Biol Evol 30(12):2725–2729

Teh OK, Moore I (2007) An ARF-GEF acting at the Golgi and in selective endocytosis in polarized plant cells. Nature 448:493–496

Tewari JP, Strelkov SE, Orchard D, Hartman M, Lange RM, Turkington T (2005) Identification of clubroot of crucifers on canola (*Brassica napus*) in Alberta. Can J Plant Pathol 27:143–144

Thornton CR, Groenhof AC, Forrest R, Larnotte R (2004) A one-step, immunochromatographic lateral flow device specific to *Rhizoctonia solani* and certain related species, and its use to detect and quantify *R. solani* in soil. Phytopathology 94:280–288

Toxopeus H, Janssen AMP (1975) Clubroot resistance in turnip -II. the 'slurry' screening method and clubroot races in the Netherlands. Euphytica 24:751–755

Toxopeus H, Dixon GR, Mattusch P (1986) Physiological specialization in *Plasmodiophora brassicae*: an analysis by international experimentation. Trans Br Mycol Soc 87:279–287

Tu J, Bush J, Bonham-Smith P, Wei Y (2018) Live cell imaging of *Plasmodiophora brassicae*—host plant interactions based on a two step axenic culture system. Microbiol Open 8:e765

Usadel B, Poree F, Nagel A, Lohse M, Czedik-Eysenberg A, Stitt M (2009) A guide to using MapMan to visualize and compare omics data in plants: a case study in the crop species, Maize. Plant Cell Environ 32(9):1211–1229

Van Meirvenne M, Goovaerts P (2001) Evaluating the probability of exceeding a site-specific soil cadmium contamination threshold. Geoderma 102:75–100

Voorips RE, Visser DL (1993) Examination of resistance to clubroot in accessions of *Brassica oleracea* using a glasshouse seedling test. Eur J Plant Pathol 99:269–276

Voorrips RE, Kanne J (1997) Genetic analysis of resistance to clubroot (*Plasmodiophora brassicae*) in *Brassica oleracea*. II. Quantitative analysis of root symptom measurements. Euphytica 93:41–48

Wakeham AJ, White JG (1996) Serological detection in soil of *Plasmodiophora brassicae* resting spores. Physiol Mol Plant Pathol 48:289–303

Wakeham AJ, Faggian R, Kennedy R (2017) Assessment of the response of *Plasmodiophora brassicae* in contaminated horticultural land, using lime based fertilizer concentrations. Food Energy Secur 6:122

Wallenhammar AC (1996) Prevalence of *Plasmodiophora brassicae* in a spring oil seed rape growing area in central Sweden and factors influencing soil infestation levels. Plant Pathol 45:710–719

Wallenhammar AC, Arwidsson O (2001) Detection of *Plasmodiophora brassicae* by PCR in naturally infested soils. Eur J Plant Pathol 107(3):313–321

Wallenhammar AC, Johnsson L, Gerhardson B (2000) Agronomic performance of partly clubroot-resistant spring oilseed turnip rape lines. J Phytopathol 148:495–499

Wallenhammar AC, Almquist C, Soderstrom M, Jonsson A (2012) In field distribution of *Plasmodiophora brassicae* measured using quantitative real-time PCR. Plant Pathol 61:16–28

Webster J (1980) Introduction to fungi, 2nd ed., Plasmodiophoromycetes, pp 42–53

Webster R, Oliver MA (1992) Sample adequately to estimate variograms of soil properties. J Soil Sci 43:177–192

Wesolowska M (2011) Isolate pathogenicity recognition of *Plasmodiophora brassicae* Worn. in different areas of Poland. Folia Hortic 23(2):165–174

Wesolowska M (2014) Identification of *Plasmodiophora brassicae* Worn. isolate suppressing clubroot resistance in 'Kilaxy' F_1 white cabbage. Folia Hortic 26:57–62

White TJ, Bruns T, Lee S, Taylor J (1990) Amplification and direct sequencing of fungal ribosomal RNA genes for phylogenetics. In: Innis MA, Gelfand DH, Sninsky JJ, White TJ (eds) PCR protocols: a guide to methods and applications. Academic, San Diego, pp 315–322

Williams PH (1966) A system for the determination of races of *Plasmodiophora brassicae* that infect cabbage and rutabaga. Phytopathology 56(6):624–626

Williams PH (1985) CrGC Resource Book. Dept. of Plant Pathology, University of Wisconsin, Madison

Williams PH, McNabola SS (1967) Fine structure of *Plasmodiophora brassicae* in sporogenesis. Can J Bot 45(9):1665–1669

Xue S, Cao T, Howard RJ, Hwang SF, Strelkov SE (2008) Isolation and variation in virulence of single-spore isolates of *Plasmodiophora brassicae* from Canada. Plant Dis 92(3):456–462

Zhang H, Feng J, Zhang S, Zhang S, Li F, Strelkov SE, Sun R, Hwang SF (2015) Resistance to *Plasmodiophora brassicae* in *Brassica rapa* and *Brassica juncea* genotypes from China. Plant Dis 99:776–779

Clubroot of Crucifers: An Introspection Epilog

15

Abstract

The clubroot of crucifers caused by *P. brassicae* is a very serious devastating disease in more than 88 countries of the world with average loss of 10–15% yield and 2–6% in oil content. The pathogen is obligate, non-axenic, microscopic, single celled, soil borne with dormant, well protected resting spores which can survive for more than 20 years. Its genome is small in size (24.2–25.5 mb) with high gene density. It has two stages of infections as primary root hair and secondary cortical cell infection. The pathogen is highly variable and exists as mixture of pathotypes in the soil and even in single clubbed roots. It produces morphological structures like resting spores, zoospores, and plasmodia. Its life cycle is very complex consisting of different zoosporic stages, formation of primary and secondary plasmodia and resting spores inside the host cells. Its ultra-structures have been revealed through electron and scanning electron microscopy of infected host galls. The occurrence, distribution, and pattern of *P. brassicae* pathotypes/ virulence have been characterized from more than 20 countries of the world. The pathotypes have been designated as serial numbers, triplicates, and letters. Molecular markers have been developed for rapid identification and detection of pathotypes infecting Brassica species. CR811 gene has been identified in pathotypes 3 and 5 which plays role in host resistance and pathogenicity. Various physical, chemical and biological components of the soil environment have effect on the survival, growth, and reproduction of this pathogen being solely soil borne in nature. Infection process by the pathogen is initiated after germination of resting spores triggered by host exudates at 20–25 °C temperature. The clubroot development and severity of pathogenesis is regulated by metabolites, growth regulators, sugars, hormones, enzymes, proteins, lipids, and glucosinolates. Auxin and cytokinin causes increased cell division and cell elongation. Genes Pb-Bripa and Pb PGA are expressed at early stages of infection. The pathogen successfully completes its disease cycle through

G. S. Saharan et al., *Clubroot Disease of Crucifers*,
https://doi.org/10.1007/978-981-16-2133-8_15

three stages of life cycle. The disease development is affected by temperature, soil pH, soil type, rainfall, drainage, boron, calcium, nitrogen, inoculum potential, host resistance, weeds, cultural practices and interactions of soil environ. The change in status and metabolism of hormones and compounds like auxins, cytokinins, glucosinolates, camalexin, brassinosteroids, gibberellin, JA, ethylene, flavonoids, and sugars have effects on club root phenotypes. Host resistance is government by qualitative and quantitative genes. Several QTL's have been identified and mapped on chromosomes of Brassica species conferring resistance. Club root resistance in Brassica species is inherited by many dominant, recessive, additive and complementary genes. Club root resistance have been transferred in Brassica species through marker assisted selection, interspecific hybridization, distant hybridization and mutational approaches. Histological and systemic resistance has been identified in Brassica species. Molecular mechanisms of pathogenesis, plasmodial formation, and resting spores germination have been revealed through expression of genes Y10, trehalose-6- phosphate synthase and PbSTKL1. The genes PbPP2A and PbHMG are expressed at the early phases of infection. Multiomics approaches have revealed expression of several genes for infection, pathogenesis, metabolic pathways and regulation of molecular mechanisms of host resistance. Clubroot control has been aimed to target all the three stages of disease cycle. Clubroot management has been advocated through integration of approaches like cultural, chemical, biological, use of judicious nutrition and host resistance. Several standardized procedures, protocols, method and techniques have been developed to get better insight in Plasmodiophora-crucifer's host pathosystem. However, there is always scope of improvement in our technology to get desired results for breeding high yielding disease resistant Brassica varieties for wider cultivation.

Keywords

Clubroot epilogue · Disease · The pathogen · Ultra structure · Electron microscopy · Pathogenic variability · Perpetuations and survival · Infection and pathogenesis · Disease cycle · Epidemiology and forecasting · Biochemistry of host pathogen interaction · Host resistance · Disease management

15.1 Introduction

The clubroot of crucifers caused by *Plasmodiophora brassicae* Woronin was observed for the first time in 1736 on turnip. During the last 284 years, it has become a very serious devastating disease in more than 88 countries of the world on economically and nutritionally very important precious crops of oilseeds and vegetables like rapeseed (Canola), mustard, cauliflower, cabbage, Chinese cabbage, Brussels sprouts, Kale, Kohlrabi, radish, rutabaga, and turnip. It is causing very heavy losses in these crops ranging from 10 to 100% with an average loss of 10–15% at global level with decrease in 2–6% in oil content, and 50% increase in chlorophyll content in oil. Its host range is very wide with infection on more than 330 genera and 3700 species of crucifers besides several weeds and non-host plants where it survives in their rhizosphere. It is a consistent and potential threat to cruciferous crops covering 2.69% of total cropped area of the world with an area of 1384.8 million ha and 147.69 million metric tonnes production. Its causal organism was identified and established as *P. brassicae* Woronin in 1875 after 140 years of first report in 1736. Its biology, ecology, *Brassica*–host–pathogen interaction, and management strategies were unveiled only in the last few decades. The pathogen is obligate, biotrophic, protists, non-axenic, microscopic, single-celled, intracellular, soil borne, highly variable in virulence, produces well-protected and robust resting spores, resilient to harsh soil environment, and can survive in soil for more than 20 years. It has two stages of infections after resting spores germination by primary zoospores as primary root-hairs infection, and secondary zoospores as secondary cortical cells infection resulting in clubroots as characteristics symptoms of the disease. *Plasmodiophora brassicae* genome is compact, small in size (25 Mb), with high gene density, and has 13 chitin synthesis genes. Potential pathogenicity genes have been identified. In the soil environment, surviving resting spores germinate to release mobile biflagellate primary zoospores which exist in the soil moisture film for very short time to cause primary root-hair infection following multiplication and causing secondary stage infection of root cortical cells through secondary zoospores to induce massive host cells hypertrophy by growth regulators production exhibiting clubroot symptoms. Such clubroots are full of resting spores (inoculum load) which act as source of primary inoculum and liberate back into the soil to continue life cycle of the pathogen. The different components of pathogens life cycle, viz., zoospores, sporangia, and sporangiospore develop and grow well protected within the host cells. Clubroot has a very complex disease cycle, and secretive life style of *P. brassicae* where even the occurrence, and position of meiosis is still unclear. Being in the protected host cells, and complex soil environment, it is a big challenge for the researchers to target a weak link in the disease cycle to manage this disease. Once the disease in introduced in an area, its inoculum (resting spores) multiplies with an exponential rate through cultivation of cruciferous crops in the same field. Scientists have generated technology to manage the disease by cultural, chemical, physical, biological, host resistance, and with integrated approaches but with limited success. Host resistance governed by few major genes breaks down within few years due to evolution of new virulent pathotypes through host selection. It may be effective and durable with CR genes pyramiding approach to develop multigene resistant cvs. Other control measures are not fulgent due to complex soil environment, and multifaceted interactions of biotic and abiotic factors in soil. During the last few decades, substantial progress has been made with advancement in modern sciences and technology to dissect *P. brassicae* life cycle in dual culture, and scanning fine structures by electron microscopy, rapid identification, and detection of *P. brassicae* from seed, soil, host tissues, and water, identification, isolation, cloning, and mapping of CR genes; detection of metabolic pathways, and modification of plant hormone homeostasis by the pathogen; identification of pathotypes during host–pathogen interaction using molecular approaches of omics, genomics, proteomics, and transcriptomics. However, there are still several gaps (chapter 16) to fill in for fully comprehension and effective management of clubroot of crucifers.

15.2 Clubroot Perspective

Clubroot of crucifers was recorded as a disease of turnip in 1736. Several causes were suspected till Woronin in 1875 established its causal organism which he names as *Plasmodiophora brassicae* in 1878. The characteristics distinguishing symptoms are visible only after uprooting the infected plants in the form of clubroot. The disease is distributed on all continents of the world on cruciferous economic vegetables and oil yielding crops causing heavy quantitative and qualitative yield losses. The pathogen is obligate, non-axenic, microscopic, single-celled, soil borne with dormant, well-protected resting spores and has two stages of infections as primary root-hair and secondary cortical cells infection. For the majority of its life cycle, the pathogen exists in a highly protected environment and has immediate access to all the nutrition that is required for its growth and reproduction. Being protected in this manner is a big challenge to control this pathogen. The pathogens survival source is resting spores which are very robust and well protected by five spore walls composed of fungal chitin, and carbohydrates which render the spore's resilient to degradation by extracellular enzymes produced by predatory soil organisms. *P. brassicae* genome is small in size (24.2–25.5 Mb) and has high gene density. It contains genes potentially involved in host hormone metabolism and metabolic pathways. The *P. brassicae* is a pathogen of cultivation and there has been continuous co-evolution of cruciferous hosts with *P. brassicae*–pathogen interactions. The characteristics symptoms of the disease, the root galls ultimately function as strong metabolic sinks in infected plants drawing nutrients and other resources away from the shoots and leaves contributing to above-ground symptoms development. Several genes governing quantitative and qualitative resistance to *P. brassicae* have been identified and characterized in different *Brassica* species. However, only a few genes have been postulated or demonstrated to be related to the pathogenicity of *P. brassicae*. The isolates of the pathogen display great variations in the form of races or pathotypes which show tendencies for overcoming resistance sources from *Brassica* species. Biologically *P. brassicae* is an intrinsically interesting pathogen which has a novel life cycle well adapted to the hosts which provide food and energy for use by the pathogen for its future generations. Several traditional and modern molecular techniques have been developed and are being used to detect and quantify pathogen from soil, water, and host tissues with fair amount of precision and sensitivity. Disease management through cultural, chemical, biological, nutrition, soil regulation along with integrated approach has been advocated but still there are number of challenges before researchers to get satisfactory results. The interactions of physical, chemical, biological, and interactions between *P. brassicae* and *Brassica* hosts and non-hosts in the soil environment are very complex, and challenging to determine conduciveness and suppressiveness conditions for the disease and pathogen.

15.3 The Disease Clubroot

The disease is known by the name clubroot due to its characteristics symptoms of infected hypertrophied root cortex cells leading to the formation and development of root galls. The importance of the disease can be judged by its distribution in more than 88 countries of the world on more than 330 genera and 3700 species of plants, with more than 60 vernacular names causing 10–100 damage to vegetables and oil yielding crops. The pathogen can survive in soil for more than 20 years in the form of resting spores in the soil, and difficult to eradicate. The disease which was discovered about 284 years ago in 1736 (1742–1744) on turnip in England as a serious and contagious form has now distributed on all continents of the world wherever cruciferous vegetations are available. Clubroot is a potential threat to cruciferous crops covering 1384.8 m ha area with 147.69 mT production at global level which is 2.69% of total cropped area of the world. There is a direct and strong relationship between disease incidences, degree of soil infestation, and crop yield loss. Host resistance helps to reduce yield losses. All cultivated *Brassica* vegetables and oil yielding crops including several cruciferous weeds are susceptible to this disease. The pathogen can survive in the root cortex of non-crucifers plants also. Symptoms may vary depending on the growth stage of the crop when infection occurs. The characteristics symptoms of the disease appear on the roots as spindle like shredded, knobby, or club shaped swellings. The clubroot has been assessed in the form of incidence and intensity using different rating scales, descriptive as well as pictorial to assess yield losses, host resistance, and efficacy of technology. The disease is usually monocyclic, and zoospore movement in the soil is limited. However, any activity that results in transport of infested soil from one point to another has the potential to disseminate clubroot.

15.4 The Pathogen *Plasmodiophora brassicae*

The pathogen *Plasmodiophora brassicae* Woronin is a very serious, soil borne, intracellular, obligate, non-axenic, endoparasites, wide spread, and destructive to cruciferous crops all over the world causing clubroot disease which is a nutrient sink to the host plant. The pathogen is highly variable, and exists as mixture of pathotypes in the soil, and even in single-clubbed roots. The isolates differ in preferences and aggressiveness towards different cruciferous hosts. The survival of the pathogen in soil leads to infection, development, and multiplication in host tissues, and in soil. It produces morphological structures like resting spores, zoospores, and plasmodia. The pathogen has a very complex life cycle consisting of different zoosporic stages (primary and secondary), formation of primary and secondary plasmodia, and resting spores inside the host cells. Plasmodia provoke hypertrophy and hyperplasia leading to the development of galls that obstruct nutrient and water transport to the infected plants. The only host-free stages are the short-lived biflagillated zoospores, and resting spores. Resting spores are extremely resilient to harsh environmental conditions, and survive in the soil for more than 20 years. Although, traditionally, *Plasmodiophora* was classified as fungus but now molecular phylogenetic analysis of the small subunit (SSU) ribosomal RNA, actin, and polyubiquitin genes suggests that it belongs to the protists phylum Cercozoa in the kingdom Rhizaria, and Plasmodiophorida order. During the sexual reproduction, meiosis appears to occur in the plasmodium before resting spore formation but how and when karyogamy takes place is not exactly known. The *P. brassicae* genome is compact with small size (24.2–25.5 Mb), high gene density, and with scarcity of repeats. In its genome, there is reduction and overlapping of intergenic distances. High degree of variation in field populations indicates sexual recombination. The molecular biology techniques have revealed *P. brassicae* genome size, recombination pattern, chromosome number, number of genes, gene structure, gene functions during pathogenesis, disease developmental stages, pattern of gene expression, and signaling for metabolic pathways.

15.5 Electron Microscopy and Ultra-Structures

During infection process, small galls emerge from epidermal cells in the elongation zone of host roots, and abundant secondary plasmodia, and resting spores can be seen within the parenchymal cells of such galls. At early infection stages in host cells, primary plasmodia contain complex cellular components including nuclei, lipid droplets, an endomembrane system and visible in epidermal cells of the root elongation zone. Cortical invasion by *P. brassicae* shows disruption of cell wall, presence of vesicles or inclusion bodies inside the cell wall, cell wall thickness in combination with plasmodesmata, and elongation and/or disorganization of host nuclei. Multinucleate, intracellular plasmodia are visible within the clubs in the cortex and pericycle cells. Gall development amplifies merismatic activity within the vascular cambium and phloem parenchyma cells in the region of the hypocotyl. It also perturbs vascular development with reduction in xylem, and increase in phloem parenchyma cells in infected plants. Scanning electron microscopy of infected host galls revealed initial plasmodia of the pathogen as spherical, smooth surfaces with average diameter of 3.7 μm increasing to average size of 4.2 μm in diameter. Young plasmodia are round with partial rough surfaces and average 8 μm in diameter. Vegetative plasmodia are smooth surface, spherical, and 45 μm in diameters which almost fill the host cells. Young resting spores are connected by fine filamentous materials, and mature resting spores have spiny surfaces. In resting sporangia during primary zoosporogenesis, single zoospores are differentiated within each sporangium being initiated by the emergence of two flagella in the tight space formed by invagination of plasma membrane within the sporangium. A deep grove is formed on the zoospore cell body by further invagination of the plasma membrane. Two flagella appear to coil around the zoospore cell body in parallel along this groove.

A nucleolus like region is visible in the mature zoospores. A partially opened germ pore is seen in the sporangium containing the mature zoospores. Zoospores have electron dense mitochondria and well-developed endoplasmic reticulum. Mitotic nuclear division is characterized by a persistent nucleolus, bipolar centrioles paired end-to-end, densely stained chromatin, and a complex array of membranes that surrounded and ramify through the spindle. Chromatin begins to condense in prophase, and is aligned at metaphase in a reticulate plate on the nuclear equator. Large amounts of membrane cisternae within the spindle are visible at metaphase. Several concentric sheets of perinuclear endoplasmic reticulum surround the spindle, and give the mitotic nucleus the superficial appearance of having an intact nuclear envelop.

15.6 Pathogenic Variability

Pathogenic variability in *P. brassicae* was detected long back during the 1930s on the basis of differential interactions with cruciferous species. First generation widely used host differentials all over the world included genotypes from *B. napus* var. *napobrassica, B. napus* var. *rapifera, B. oleracea* var. *capitata, B. rapa* var. *pekinensis, B. rapa* ssp. *rapifera,* and *B. napus* var. *napus*. The occurrence, distribution, and pattern of *P. brassicae* virulence/pathotypes have been characterized from Australia, Canada, China, Czech Republic, Finland, France, GDR, Iceland, India, Japan, Korea, New Zealand, Norway, Poland, Sweden, Turkey, UK, USA, and USSR. Molecular techniques like PCR, RAPD, and RFLP/AFLP have also been used to characterize *P. brassicae* isolates. Although these studies have revealed high degree of polymorphisms among as well as within *P. brassicae* populations but no correlations have been established with the pathogenicity. Modified differentials sets have been used at Canada, China, Japan, Korea, and Turkey to get more clear picture of pathotypes in their countries by including genotypes of *Brassica* having local/national importance. Numerous pathotypes of *P. brassicae* have been identified from more than 20 countries of the world with designations as serial numbers (1, 2. . .n), triplicates (16/14/15, 16/2/14. . .n), and letters (A, B, C,. . .X1, P1, P2. . .n). However, the characterization of pathotypes with respect to virulence spectrum, and erosion of clubroot cultivars is from very limited countries. For rapid identification, and detection of pathotypes infecting *Brassica* species, molecular markers have been developed. A *CR 811* gene has been identified in pathotypes 3 and 5 which plays an important role in host resistance, and *P. brassicae* pathogenicity. Isolate/pathotype specific molecular markers are very useful tools to identify genes to test, and differentiate unknown pathotypes. Differential sets development, and used so far have several limitations to detect, and identify new pathotypes as well as distribution of virulence all over the world. Since the sets consist of genetically heterogeneous *Brassica* species which is against to the International standard differentials host with near-isogenic lines. Similarly, isolates used so far are composed of genetically heterogeneous populations instead of genetically pure single-spore isolates. Similarly, designation and nomenclature of pathotypes have limitations since different systems have been followed (serial number), triples (binary notation) which makes difficult to compare, analyze, and detection of new pathotypes. Following this system, similar pathotypes have been reported from different countries which are not consistent with the biology of the pathogen. The intermediate and fluctuating results of certain deferential hosts following inoculation with the population of *P. brassicae* indicate importance of single-spore isolates for virulence analysis prevalent world over.

15.7 Perpetuation and Survival of the Pathogen

Plasmodiophora brassicae causing clubroot disease of crucifers is highly evolved and very successful to perpetuate and survive to complete its life cycle in all kinds of ecological niche it occupies. It produces robust, well-protected soil-borne resting spores which can withstand adverse conditions to survive for a very long period of more than 20 years to respond speedily once a compatible host comes in contact to cause infection through primary zoospores emerging from the perennating spores. Various physical, chemical, and biological components of the soil environment have effect on the survival, growth, and reproduction of this pathogen being solely soil borne in nature. Resting spores germination is triggered as a result of signals initiating by the root exudates of compatible hosts in the soil environ. The resultant motile zoospores move rapidly to the root-hairs surface to penetrate and colonize. The short period between germination and penetration is one of the greatest vulnerable stages in the life cycle of the pathogen. At this stage, perpetuation, and survival of the pathogen is affected by numerous factors in the soil environment. Various factors influence the inoculum potential, viability, and invasion capacity of the pathogen. The different factors interacting with resting spores and zoospores of the pathogen are soil moisture, temperature, light intensity, soil texture and structure, spore load, pathotypes, host resistance, soil pH, acidity and alkalinity, amount of various chemicals like calcium, boron, nitrogen, magnesium, molybdenum, calcium cyanamide, calcareous seaweed, soil fauna and flora, plant growth regulators, extracts of compost, rhizosphere components, residue of the pesticides applied, host and non-host plants, and cultural activity to manage the pathogen. Interactions with the molecular and biochemical cellular environment are also a factor of recent origin to develop resistant cvs. of crucifers.

15.8 Infection and Pathogenesis

The infection process by *P. brassicae* is initiated soon after the germination of resting spores triggered by host exudates under favorable temperature conditions of 20 to 25 °C. Resting spores in the soil germinate, and release primary zoospores that penetrate root hairs to form primary plasmodia. The primary plasmodia form multinucleate zoosporangia that produce uninucleate secondary zoospores. These secondary zoospores are released into the soil and infect root cortical cells and develop into motile secondary plasmodia. Cortical infection results in hyperplasia and hypertrophy of the roots cells which produces swollen, rounded clubroot symptoms. Clubroot development and severity of the pathogenesis is induced, and regulated by several metabolites, growth regulators, sugars, hormones, enzymes, proteins, lipids, glucosinolates, etc., during host–pathogen interactions. The involvement of auxin and cytokinin as pathogenicity factors causes increased cell division and cell elongation during late infection stages. Molecular mechanism of pathogenesis revealed a serine protease (Pro1) gene important for resting spore germination. During infection and pathogenesis *P. brassicae* members of the TF families, MYB, bHLH, and APZ/EREBP are highly up- or down-regulated in both shoot and root of host plants. *P. brassicae* infection triggers active sugars translocation between the sugars producing tissues, and the clubbed tissues, and SWEET family genes are involved in this process. *Brassica* specific genes (BSGs) play important roles in *P. brassicae* stress responses. Genes Pb-Bripa and PbPGA are expressed in early stages of infection, miRNAs are expressed during the process of clubroot interaction and development. G protein-coupled receptors (GPCR) signal transduction pathways play important roles in the growth, development, and pathogenesis of *P. brassicae* to their cruciferous host species extensively.

15.9 Disease Cycle

Plasmodiophora brassicae causing clubroot disease of cruciferous crops very successfully completes its disease cycle through three stages of life cycle; first, long time survival of resting spores in the soil, second root-hairs infection by the primary zoospores, and third cortical infection by secondary zoospores. In the infected root cells, the pathogen develop and root exudates into secondary plasmodia which cause hypertrophy followed by gall formation and generation of new resting spores. Calcium ions trigger the germination of the resting sores. The germination of mature resting spores is affected by pH, humidity, temperature, and other inorganic ions as well as biological factors. The zoospores released from resting spores survive for short periods of time. In dual culture, the development of *P. brassicae* in hairy roots is similar to that found in intact plants. The secondary plasmodia in the inner cortical cells are distributed by cytoplasmic streaming and host cell division. The role of effectors/gene has been determined in the disease cycle. The serine protease Pro1 is involved in the germination of resting spores. During root-hair infection by primary zoospores, primary plasmodia, secondary zoospore, secondary plasmodia, proteases, and methyltransferases are produced to suppress plant immunity for the success of infection process.

15.10 Epidemiology and Disease Forecasting

Clubroot of crucifers is strictly a disease of soil environment; therefore, its development is influenced by environmental parameters of soil, and their interactions with all abiotic and biotic soil inhabitants. The disease development is affected by factors such as temperature, soil pH, soil type, rainfall, and drainage; micronutrients like boron, calcium, and nitrogen; inoculum potential, host resistance, weeds, cultural practices, and interactions of all these factors in a soil complex environment. Temperature below 17 °C inhibits the development of clubroot at all stages of its disease cycle. Alkaline pH also reduces infection and symptoms development. However, alkaline pH does not eliminate clubroot when other conditions are conducive for infection. Nutrients cannot limit the clubroot development when soil moisture in the rhizosphere during primary and secondary infection favor disease development. Both calcium and nitrate form of nitrogen, and calcium cyanamide inhibit the development of plasmodium into sporangia in root hairs, and reduce the variability of resting spores, and the ability of primary zoospores to invade the host. Root-hairs infection, cortical infection, and disease severity are high at 25 °C temperature and pH of 5.0–6.0 but decrease as pH increases, and temperature decreases. Clubroot severity is positively correlated with inoculum density in the field. Host resistance inhibits cortical infection by the pathogen. Disease forecasting is based on congenial factors for disease development.

15.11 Biochemistry of Host–Pathogen Interaction

The life cycle of *P. brassicae* is entirely different from other biotrophic pathogens. The intimate relationship between host and pathogen leads to change in hormonal status, cellular development of pathogen, and sources sink relations of clubbed roots and shoot of host plant. In the infected plants growth-promoting hormones such as auxins, cytokinin, and changes in secondary metabolites are associated with club development which exhibits their effect either directly on growth hormones or indirectly via their bioactive properties. The change in status, and metabolism of hormones, and compounds like auxins, cytokinin, glucosinolates, camalexin, brassinosteroids, gibberellin, jasmonic acid, ethylene, flavinoids, and sugars have effects on clubroot phenotypes. The metabolic sink results in the accumulation of starch grains in clubbed tissues. The different pathways and hormonal signals result in a homeostasis between the pathogen and the host plant resulting in galls formation and/or suppression of gall growth. A protein (PbGH3) can modify auxins and jasmonic acid produced in infected root galls. The resilient resting spores of the pathogen contain chitin in their cell walls. A trehalase encoding gene is primarily expressed in spores to release glucose as an energy source in the germination process. The *PLeff* genes are expressed in plasmodia, the *Heff* genes are expressed in the transcriptomes of the host, and the *Leff* genes are expressed late in the life cycle during spore formation. A complete pathway for the spermidine biosynthesis has been predicted in the *P. brassicae.*

15.12 Host Resistance

Genetical and molecular mechanisms of host resistance in the form of qualitative and quantitative genes have been expressed against *P. brassicae* in the economically important and widely grown cruciferous species, viz., *B. rapa*, *B. oleracea*, *B. napus*, *R. sativus,* and *Arabidopsis thaliana.* Mapping of quantitative and qualitative traits loci (QTL's) have revealed several (12) dominant loci in *B. rapa* more than 22 QTL's in *B. oleracea,* and more than 19 QTL's in *B. napus* for clubroot resistance to different pathotypes of *P. brassicae*. Clubroot resistance in *Brassica* species is inherited by many dominant, recessive, additive, and complementary genes. Substantial progress has been made in breeding clubroot resistance genotypes/cvs. of *B. rapa*, *B. oleracea*, *B. napus*, *R. sativus,* and *A. thaliana*. Molecular mechanisms to identify host resistance have been extensively exploited through genetic mapping of CR loci, genomics, transcriptomics, and proteomics approaches. Multiple pathotypes specific QTL's conferring resistance to *P. brassicae* have been identified and functionally characterized by RNA sequencing and genotyping-by-sequencing to facilitate cloning of R-genes. To counteract the role of mitogen in modulation of CR QTL's, new sources of resistance have been found among the oilseed rape accessions under low and high N conditions. Marker-assisted selection of CR genes has ensured accumulation of more than one R gene into a single CR cv. for more durability under field conditions against different pathotypes of *P. brassicae.* To trigger the pathogen-associated molecular pattern (PAMP's) immunity thirty three chitinase genes have been identified in *B. rapa* with their structure, evolution, and chromosome location. Map-based cloning and characterization of *Crr1A* genes which confers resistance to *B. rapa* encodes a toll-interlinked-1 receptor/nucleotide binding set/Leucine-rich repeat (TIR-NB-LRR) protein expressed in the stele and cortex of hypocotyl and root, where secondary infection of pathogen occurs but not in root hairs. Differential R-genes expression and their functional mechanisms during life cycle of pathogen to confer resistance at different stages of infection and pathogen development have been elucidated. Phenylpropanoid pathway genes (lignin biosynthesis) are up-regulated in CR *B. rapa.* Clubroot resistance has been transferred in *Brassica* through molecular marker-assisted selection, interspecific hybridization, distant hybridization, and mutational approaches. Effective management of R cvs. of *Brassica* can enhance durability and cost-effective control of clubroot. The nomenclature of CR genes has not been standardized in *Brassica* species. A basal defense mechanism is activated by a virulent pathogen through induction of trehalase gene. Cellular resistance is expressed through accumulation of ROS or lignification of epidermal or cortical cells in the resistant cvs. Small spheroid galls of clubroot act as an indicator of host resistance. Histological resistance is expressed through deposition of callose in response to pathogen invasion. Systemic acquired resistance is associated with up-regulation of pathogenesis-related genes PR-1 and PR-2.

15.13 The Genomics of Clubroot Pathogen and Pathogenesis

The main characteristics of the *P. brassicae* genome are its small size of 24.2–25.41 Mb with high gene density and compactness. The genome contains genes potentially involved in host hormones manipulation. The enzymes (CAZymes) involved in synthesis, metabolism, and transport of carbohydrates are present. Mitochondrial genome is variable (34–180 kbp) in size because of variation in the intergenic regions and intron content. There is higher G+C content in mitochondrial sequence than in nuclear genome. Molecular mechanisms of pathogenesis revealed expression of *Y10* gene during plasmodial stages of *P. brassicae*. A trehalose-6-phosphate synthase gene and *PbSTK* L1 genes are expressed during resting spores formation. A serine protease (*PRO1*) gene is expressed during host infection as well as in dormant and germinating resting spores. More than 58 genes are up-regulated in the secondary zoospores relative to 55 genes down-regulated in primary zoospores. The genes *PbPP2A* and *PbHMG* are expressed at the early phases of infection. The gene *BjN1T1* expression in *B. juncea* promotes over production of auxins to form root galls. The main plant hormones pathways modified are cytokinin biosynthesis, auxin homeostasis and salicylic acid, and jasmonic acid metabolism. The methyltransferase gene of *P. brassicae* acts by methylation of salicylic acid, benzoic acid, and anthranilic acids to suppress salicylic acid-induced defense in the host plant. Bioinformatics have been used to identify putative effectors/genes of *P. brassicae* for pathogenesis. Transcriptomics analysis has revealed 162 candidates' enzymes/genes involved in lipid metabolism differentially expressed in the *P. brassicae* life cycle. The rate of infection and pathogen development is most rapid in susceptible host in comparison to resistant. The transcriptomic and proteomic approaches have elucidated mechanisms of *P. brassicae* infection and coordination in host reactions. Functional genomics approaches like expressed sequence tags (EST's), serial analysis of gene expression (SAGE), massively parallel signature sequencing (MPSS), next generations sequencing (NGS), RNA sequencing (RNA-seq), microarray analysis, and construction of suppression subtractive hybridization (SSHcDNA) liberty have been used to achieve better insight into the molecular mechanisms of host–parasite interaction and expression of functional genes. Microarray chips, two-dimensional electrophoresis, and high-throughput sequencing techniques like metabotyping, laser microdissection, transcriptional profiling, and miRNA sequencing have been exploited to observe expression of genes regulating metabolic pathways and signal transduction at early events during pathogenesis. Differential expression of genes in host shoot and root has been observed during *P. brassicae* infection. The members of TF families', viz., MYB, bHLH, and AP2/EREBP are highly up- or down-regulated in both shoot and root. However, both shoot and root also responded to pathogen in a number of common ways such as induction of JA, ABA, and shikimate pathways metabolism. The hypoxia genes *PDC1* and *PDC2* are induced during secondary infection in cortical cells of root galls. Several BrSWEET genes of clade 1 and clad III up-regulated on infection suggest close relationship between sugar translocation and *P. brassicae* growth. RNA's like arth-mi R160, arth-miR164, and arth-miR167 are involved in pathogenesis and target auxin-response-factor (ARF) in auxin homeostasis at clubroot development. Immunopropilin gene (*PbCYP3*) is differentially regulated in *P. brassicae* life stages at germinating, and maturing spores and plasmodial formation in infected *Brassica* hosts. Hydrolases and peptidases are the most common differentially regulated secreted proteins in *P. brassicae*-infected *B. napus*. The *Pb3-BSMT* gene (benzoic acid/salicylic methyltransferases) inactivates host defense by converting salicylic acid to methyl salicylate.

15.14 Disease Management

Clubroot disease management has been aimed to target all the three stages of disease cycle. The first most important step is to bring inoculum load in the soil below minimum threshold level of resting spores. The second and third steps are to suppress primary root-hairs infection and secondary cortical infections, respectively, which results in club formations to continue life cycle of pathogen, *P. brassicae*. Any activity that results in transport of soil or crop residues infested with *P. brassicae* from one point to another has the potential to disseminate the pathogen. The most effective cultural control is through removal of diseased plants, improved drainage, deep plowing, ridging, long crop rotation (>3 years), using clean and treated seed, avoiding application of raw manures, sanitization (cleaning and disinfecting of machinery, equipment, vehicle, tools, footwear, seed drill, plant materials, irrigation water, etc.), and manipulating date of sowing to sow the crop seed when soil temperature is too cold for clubroot pathogen development. Use of clean, treated seeds, crop rotation, bait crops with resistant, susceptible, and non-*Brassica* plants reduce clubroot inoculum in the infected fields. Soil solarization in combination with low rates of fumigation (Dazomet) can be a means of clubroot control. Post-harvest weed management along with susceptible host plants is very crucial factor to reduce inoculum potential in the fields. Several simple low-cost cultural strategies may be useful for effective management of disease in the endemic areas. Intercropping of *Brassica* and flaxseed reduces clubroot with increase in yield. Soil amendment with lime, wood ash, or calcium cyanamide increases soil pH, and is an important clubroot management strategy. Judicious use of nutrients like N, P, K, Ca, B, Molybdenum, and organic matter helps in management of clubroot through maintenance of pH with conducive effects on host growth, and suppressive soil environment for pathogen infection and development. Several chemicals used as soil drench, surfactant, and soil fumigants control clubroot through inhibition of resting spores germination, and suppression of root hairs, and cortical infection. Clubroot has been managed with the use of biological agents of bacterial, fungal, and plant products origin which produce antimicrobial metabolites against *P. brassicae*, and colonize host roots to induce resistance. Commercial biofungicides like Serenade, Prestop, Mycostop, granulate formulation Plasma soil, and Root Shield Plus with fungicides like fluazinam and cyazofamid are very effective against clubroot. Disease management through host resistance is the most economical and ecofriendly strategy. Disease resistant cultivars of *Brassica* crops have been developed, and are being used to manage clubroot in the countries like Canada, China, Australia, Japan, Germany, and Korea. Integrated approaches using cultural, chemical, biological, and host resistance have been an extremely effective clubroot management strategy along with high yield of cruciferous crops. In Latin American countries, clubroot has been managed with modification of cultural operations depending on soil environment, pH, and inoculum load in each country, in addition to the growing of resistance cvs. and using biological and chemical control strategies.

15.15 Techniques

Techniques are sound base and backbone of getting reliable, reproducible, and well analyzed research results in any research project, and also very useful methods to demonstrate research methodology to graduate students. Several standardized techniques from basic to advanced molecular level have been included in this chapter for the researchers of cruciferous crop diseases. The clubroot pathogen identification and detection from soil and host tissues at different stages of host–pathogen interaction can be revealed by biological, pathological, microscopical, serological, and molecular techniques. Procedures to determine inoculum load as well as pathogens resting spores, number in soil, and viability percent have been elucidated. A protocol is available on callus culture to target weak link in the life cycle of pathogen. Histopathological methods to study infection process in response to host resistance can reveal developmental stages of the pathogens, morphological, and cellular structures. Development of dual culture protocol can be helpful to obtain axenic culture of *P. brassicae* and study infection cycle in different cruciferous species. Pathogenicity assays in Petri dishes, pots, and field conditions have been demonstrated. A very precise and simple technique to assess the responses of lime-based fertilizers and nutrients is elaborated. A procedure to develop clubroot sick plot, and screening of *Brassica* germplasm under field and green house conditions has been described. Germplasm evaluation techniques are given for field and controlled conditions through artificial inoculation of the pathogen. The protocol of modern molecular techniques like PCR and qPCR has been described to detect pathogen in soil and host tissues with very high precision level and sensitivity. Single spore isolation technique is very useful for determination of pathogenic variability. Techniques are described for isolation of resting spores, inoculum preparation, testing of resting spores germination, detection of seed infestation, detection of host pathogen at different stages of disease development, detection of pathogen using loop mediated isothermal DNA amplification, quantification of auxins and cytokinin, DNA extraction from infected roots, soil, plant, fungal, and bacterial material, root-hairs infection, staining of resting spores, transcriptomic analysis; *in planta* quantification of pathogen; identification of pathotypes, and designation of clubroot-resistant genes, and *P. brassicae* pathotypes.

Future Priority Areas of Clubroot Research for Its Better Management 16

Abstract

Clubroot of crucifers is very widely distributed and destructive at global stage. The disease is soil borne with abundant amount of resting spores of the pathogen which are extremely resilient to harsh environmental conditions and cannot be eradicated once introduced in an area. These resting spores contain high amount of chitin in their cell walls which helps them to survive for longer period. The information generated on clubroot from all over the world reviewed in this book indicates enough scope of research to fill up the existing gaps for better management of disease at international levels. The following areas of research will strengthen our understanding of clubroot management more effectively at large. The secretive life style of pathogen needs more research on nuclear fusion and meiosis, and growth of pathogen in axenic culture to get deeper insight in weak links of pathogens life cycle. Multiomics approaches for complete sequencing of P. *brassicae* transcriptome, molecular characterization of effector genes for pathogenesis and disease progression are high priority areas of research. The co-relationship between root hair and cortex infection and infection process by primary and secondary zoospores is very poorly known. Genetic networks involved in P. *brassicae* host–pathogen interaction require detailed analysis. There is a need to standardize differential hosts as isogenic lines along with standardization of pathotypes and R-genes nomenclature. Generation of data on molecular and cellular biology, metabolic pathways, and signal molecules for expression of genes to suppress gall formation is very crucial. Deeper insight into disease cycle at different development stages can reveal weak point to target for disease management. The disease epidemiology and forecasting require more attention. The impact of climate change and evolution of new virulence, in the pathogen influencing durability of R-genes should be assessed. The pyramiding of QTLs and R-genes against pathotypes is a key lever to increase host resistance. The development of transgenic crop varieties for disease management and high

G. S. Saharan et al., *Clubroot Disease of Crucifers*,
https://doi.org/10.1007/978-981-16-2133-8_16

yield is required. An integrated crop management module is crucial for this soil borne, highly variable and long period survival pathogen.

Keywords

Future priorities · Effective management · Research gaps

16.1 Introduction

Clubroot of crucifers is very widely distributed and destructive at global stage. The disease is soil borne with abundant amount of resting spores of the pathogen which are extremely resilient to harsh environmental conditions and cannot be eradicated once introduced in an area. These resting spore contain high amount of chitin in their cell walls which helps them to survive for longer period. The information generated on clubroot from all over the world reviewed in this book indicates enough scope of research to fill up the existing gaps for better management of disease at international levels. The following areas of research will strengthen our understanding of clubroot management more effectively at large.

16.2 Clubroot Research Priorities

16.2.1 Pathogen

a. The stages and positions of nuclear fusion and meiosis in the life cycle of *P. brassicae* need more attention of researchers to reveal secretive life style of the pathogen.
b. Some of the important steps in *P. brassicae* development stages are still not well characterized at the molecular level. There is a need to grow the pathogen in axenic culture to get more insight on weak links in pathogen life cycle.
c. The complete sequencing of *P. brassicae* transcriptome will speed up to reveal many molecular aspects of the host–pathogen interaction.
d. To detect, locate, and identify the time and period of pathogens vulnerable stage i.e., primary zoospores production in the natural soil environment under different interacting factors can help to manage the disease through inoculum suppression strategies. This is the singularly most vulnerable part of the entire life cycle of *P. brassicae*. Flagella, the motile organs of the zoospores provide movement in the soil but little is known about its extent, time, and potential of spread.
e. Molecular characterization of *P. brassicae* genes during pathogenesis at different stages of disease progression.

16.2.2 Infection

a. The co-relationship between root-hair infection (primary) and root cortex infection (secondary) is very poorly known. There is a need to improve our understanding of the mechanisms of infection, colonization process, the developmental stages of the host, and host–pathogen interactions at molecular level to device management strategies.
b. Time limitations to the viability of gall section may be overcome by the dual culture system of secondary plasmodia in suspension cells, where movement and division effects can be observed more easily compared with the normal development of disease in roots.
c. The role of effectors/genes in the completion of pathogen life cycle and disease progression needs more in depth study.
d. The infection process by primary and secondary zoospores and their control by molecular mechanisms in not clear.
e. Detailed analysis of the transcriptomic and proteomic levels can strengthen our understanding of genes and genetic networks involved in the mechanisms of *P. brassicae* host–pathogen interaction.

16.2.3 Pathogenic variability

a. There is a need to standardized host differentials in each *Brassica* species in the form of isogenic lines, and use them at international level to know status, and distribution of pathogenic variability in *P. brassicae*. Similarly designation and nomenclature of races/ pathotypes and host resistance gene should be standardized at world level with uniform pattern for validation and use by all *P. brassicae* research groups. It will ease an international collaborative network to create a world database on host–pathogen interaction status at molecular level. It will also reveal dynamics of R-genes breakdown and virulence pattern of *P. brassicae* races/pathotypes.
b. Molecular techniques which can rapidly identify the virulence on different *Brassica* species to facilitate breeding R cultivars of cruciferous crops may be strengthened.
c. It is essential to determine how differences in vigor, aggressiveness, infection capacity, virulence spectrum are derived and come into action by different pathotypes to plan durable resistant cvs. of *Brassica*.

16.2.4 Biochemistry

a. The understanding of the regulation of *P. brassicae* metabolites during infection, and in response to various biotic and abiotic constraints needs more research.
b. Our understanding of the molecular, and cellular biology, and biochemistry of the pathogen remains limited. More precise and repeatable technology is required to

get insight to biochemical mechanisms of host–pathogen interaction and resistance.

c. Generation of data is required on metabolic pathways leading to signal molecules and expression of genes to suppress gall formation in different cruciferous species.
d. The role of enzymes in digestion of cell wall components during pathogenesis of cortical tissues is not characterized.

16.2.5 Disease cycle

a. Being strictly biotrophic pathogen, its axenic culture is difficult therefore, so far how and when karyogamy takes place in its life cycle is not clearly known but meiosis occurs in the plasmodia before resting spore formation.
b. There is a need to detect exact period of host-free stages of short lived bi-flagellated zoospores in the life cycle of the pathogen to plan disease management strategy.
c. The cortical infection process needs more clear picture about zoospores whether they are released from the root hair into the soil or move inside the root hair or both.
d. Determination of structure and pattern of *P. brassicae* field population will benefit to plan novel strategies of disease management.

16.2.6 Epidemiology

a. Epidemiology of the disease has not been worked out properly because of strictly soil borne nature of the pathogen. Defining the above ground environmental conditions conducive or suppressive to clubroot forms a major part of research interest linked with effects of soil type, structure, texture, and nutrients composition along with other soil fauna and flora interactions.
b. Since innate resistance to *P. brassicae* infection in Brassicas is limited, an alternative strategy is to explore how anatomical and histological features of the host can significantly decrease the infection rate or progress of the disease.
c. The information on the interaction of abiotic and biotic factors influencing clubroot initiation, development, and suppression needs validation at different locations because of controversial results obtained so far.

16.2.7 Resistance and genomics

a. Identification, mapping, cloning, and characterization of R genes in each *Brassica* species and presence and absence of R genes in the *Brassica* A and C genomes need more significant insight.

b. The impact of climate change on durability of R genes and evolution of new virulence is also a priority area of research.
c. There is a need to identify R genes which can prevent early infection and establishment of the pathogen and also curtail subsequent pathogen inoculum production, and consequent secondary spread of the disease.
d. Sequencing of host and pathogen genomes is essential to understand genetic relationships of different R genes with virulence within and between divergent groups of crucifers.
e. Understanding the mechanisms of race-specific and/ or non-race specific R genes will be useful to get insight into combinations of R genes for durability and deployment of cultivars effectiveness. R genes pyramiding through marker-assisted selection can be a better strategy.
f. There is a need to broaden the genetic base for clubroot resistance by both natural variation and transgenic strategies.
g. To work with genetically homogeneous material, we require the generation of single spore lines for effective and stable sources of R genes from different cruciferous gene pool.
h. Transient gene-expression has been used to analyze gene of obligate pathogens this techniques can be achieved in *P. brassicae* infected plant cells and in secondary plasmodia of *P. brassicae* by particle bombardment.
i. There is need to combine molecular markers of both host and pathogen to understand the progress of disease in relation to resistance and virulence mechanisms operations.
j. The information on co-evolution of cruciferous species with P. brassicae interaction at different geographical areas is very essential to plan breeding disease resistant cultivars.
k. Identification of more QTL's or genome wide association study (GWAS) in *Brassica* species can strengthen breeding disease resistant program.
l. The potential of all A, B, and C genomes of the *Brassica* species should be exploited to harbor R genes against *P. brassicae.*
m. The pyramiding of qualitative and quantitative resistance is considered today as a key level to increase the durability of host resistance against rapidly evolving pathogen population of *P. brassicae*.
n. The role of chitinase gene in host resistance may be determined in depth since chitin is the major component of resting spores cell walls providing protection against harsh climate and soil environment.

16.2.8 Disease management

a. Management of pathogen infection at different developmental stages of the host (seedling v/s adult) and in different host organs is essential aspect for increasing *Brassica* production.

b. An integrated crop management approach including all strategies of crop husbandry practices and use of resistant cultivars in appropriate rotation considering distribution of pathotypes needs attention of researchers.
c. Identification and role of chitinolytic enzymes producing bacteria as a tool to degrade resting spores of the pathogen in the soil may be useful to suppress primary source of inoculum.
d. Identification and characterization of suppressive soils and interacting factors can be helpful to plan disease management strategies in better way.
e. Rhizosphere biology of resistant and susceptible hosts can reveal role of interacting biotic and abiotic factors in the soil environ influencing survival and germination of resting spores.
f. There is a need to resolve controversial results on disease management by application of calcium, nitrogen, boron, management of soil pH, soil suppressiveness, soils physical and chemical factors and their interactions.

Index

G. S. Saharan et al., *Clubroot Disease of Crucifers*,
https://doi.org/10.1007/978-981-16-2133-8

C

D

E

H